BALL REDBOOK

18th Edition

VOLUME 2
CROP PRODUCTION

Edited by Jim Nau

BALL PUBLISHING

Ball Publishing
622 Town Road
West Chicago, Illinois 60185
www.ballpublishing.com

Cover designed by Christine Truesdale.
Cover photography by Mark Widhalm, copyright © 2011 Ball Publishing. All rights reserved.
Interior designed by Bay Graphics, Walworth, Wisconsin.
Edited by Jayne VanderVelde, A+ Editorial Services, Tinley Park, Illinois.
Printed by Walsworth Print Group, Marceline, Missouri.

Library of Congress Cataloging-in-Publication Data

Ball redbook. — 18th ed.
v.
Includes index.
Contents: v. 1. Greenhouses and equipment / edited by Chris Beytes — v. 2. Crop production / edited by Jim Nau.
ISBN-13: 978-1-883052-67-6 (v. 1)
ISBN-10: 1-883052-67-X
ISBN-13: 978-1-883052-68-3 (v. 2)
1. Floriculture. 2. Plants, Ornamental. 3. Ornamental plant industry. I. Beytes, Chris, 1960– II. Nau, Jim, 1958–
SB405.B254 2011
635.9—dc23
2011035401

ISBN: 978-1-883052-68-3
Printed in the United States of America.
1 2011

Dedication

This eighteenth edition, published on the eve of the *RedBook's* eightieth anniversary, is dedicated to its founder, George J. Ball, and to his son Vic. It was George's vision that created what are today the Ball Horticultural Company, *GrowerTalks* magazine, and the *Ball RedBook.* For his entire career until his death in 1997, Vic Ball was the industry's champion of education, communication, and sharing. We at Ball Publishing are proud to continue the journey along the path that these two pioneers blazed for us all. As George wrote, circa 1930, "Our books and trade papers are the most powerful source of information and exchange of ideas that we have." We believe that still holds true in 2011, even as "books and trade papers" are augmented by the Internet and all that it offers.

Contents

Preface vii
Acknowledgments ix

Part 1

Introduction 3

Water, Media and Nutrition

1. Water Quality 9
2. Media 21
3. Plant Nutrition 33
4. Managing pH for Container Media 43
5. Water, Media and Nutrition Testing 49
6. Temperature 67
7. Light 83
8. Growth Regulators 95

Pest Control

9. Managing Insects and Mites 107
10. Managing Diseases 121
11. Controlling Soilborne Pests 131
12. Managing Resistance 137

Propagation

13. Propagating Seed Crops 143
14. Propagating Vegetative Crops 155
15. Indexing for Disease 173
16. Postharvest Care and Handling of Flowering Potted Plants . 177
17. Sustainability 193

Part 2: Crop Culture A–Z 199

Appendix: USDA Hardiness Zone Map 737
Bibliography 739
Subject Index 751
Plant Index 765

Preface

Most growers worldwide would agree that if they could only have one book on their office shelf, it would be the *Ball RedBook.*

The first edition of the *Ball RedBook,* published in 1932, sold for twenty cents, and was titled *Ball Red Book, Miscellaneous Flower Seed Crops.* George J. Ball penned the manuscript in longhand. At that time, almost all growers produced cut flowers, and most were also florists. That first edition featured cutting-edge crop culture on cut flowers such as asters (*Callistephus*), stock (*Matthiola*), snapdragons (*Antirrhinum*), larkspur (*Consolida*), calendula, sweet peas (*Lathyrus*), mignonette (*Reseda odorata*), zinnias, *Clarkia* (*Godetia*), centaurea, gerbera, *Didiscus,* and *Scabiosa.* The only bedding plants were petunias, candytuft (*Iberis*), marigolds (*Tagetes*), and lupine.

In today's floriculture industry, the bread-and-butter commodity cut flower production has moved offshore to Colombia and Ecuador, and most greenhouse producers focus on producing high-value, quick-turning bedding plants, perennials, foliage, and flowering pot plants. There are niche growers producing cut flowers, and, interestingly, many of the crops written about in the first *Ball RedBook* are viable, profitable cut flowers today. Incidentally, updated crop culture for most of these crops appears in this eighteenth edition.

As the industry has changed, so has the *Ball RedBook.* Vic Ball took over the *RedBook* editing duties from his father and improved it with each subsequent edition. Over the years, the size of the book has increased to accommodate an expanding list of crops grown from seed and cuttings.

The *Ball RedBook* also has always touched on the technology side of the industry, with Vic sharing the contents of his notebooks filled with comments from growers all over the United States, Canada, and Europe on new ideas such as hydroponics, Dutch trays, roll-out bedding plants, open-roof greenhouses, round-robin production systems, transplanters, and more. There was no innovation that excited growers about which Vic Ball wasn't interested. His passion for encouraging innovation among growers and sharing information about growers was boundless. Vic was an inspiration to those of us who were fortunate to work with him and to every grower he encountered. Vic served as editor of the sixteenth edition of the *RedBook* and passed away in 1997, shortly after it was published.

When we published the seventeenth edition in 2003, we expanded the book into two volumes in order to devote even more pages to the topics of greenhouse production and crop culture: *Volume 1: Greenhouses and Equipment* and *Volume 2: Crop Production.* Each volume is complete in its own right as a stand-alone book. Together, however, the volumes include enough practical information to set anyone interested in becoming a greenhouse grower on the road to success. Existing growers who have relied on the *Ball RedBook* as their "first consulted" reference text will find the volumes to be an invaluable resource.

This eighteenth edition continues the tradition of excellence set by the previous seventeen editions. Long-time *GrowerTalks* editor Chris Beytes again handled the editing duties on Volume 1, while Jim Nau, Greenhouses and Gardens Manager for Ball Horticultural Company, tackled Volume 2. An experienced horticulturist with several books to his credit, including the *Ball Culture Guide* and *Ball Perennial Manual*, Jim worked tirelessly for over a year to update the volume with many new crops.

Volume 1: Greenhouses and Equipment covers the basics of greenhouse structures and the tools needed to produce and market quality crops. This latest edition includes updated sections on open-roof greenhouse technology, water sanitation, "lean flow," sustainability, robotics, and alternative energy sources such as biomass boilers and solar and wind power.

Volume 2: Crop Production covers the basics of floricultural production in the greenhouse. Written in laymen's terms, the book is divided into two parts. Part 1 presents the basics of growing—including broad topics such as water, media, nutrition, temperature, light, and postharvest, as well as applied subjects such as insect and disease control and growth regulators—all in grower-friendly text with many tables and graphics. Part 2 is a cultural encyclopedia of every important greenhouse crop, from annuals to cut flowers, perennials, and potted plants. Dozens of contributors lent their expertise to its pages. There, you'll find propagation, growing on, pest control, troubleshooting, and postharvest information presented in an easy-to-use format.

We hope you find the *Ball RedBook* to be helpful, useful, and inspirational. Now, as Vic would have said, *"Let's go!"*

Chris Beytes

Jim Nau

Acknowledgments

No work of this size is possible without the support and direct input of many people. The chapters in Part 1 of this volume are credited directly to their authors, and those authors are acknowledged here. Authors of various sections in Part 2 are also credited here. The information in Crop Culture A–Z was compiled using three primary sources: past editions of the *Ball RedBook*, *GrowerTalks* magazine, and the Ball Seed Company Cultural Database. These sources were supplemented with hundreds of magazine articles, company catalogs, research reports, text and reference books, websites, and personal observations. A bibliography appears at the end of the book.

The following authors were instrumental in the creation of this, and I thank each of them for their contribution:

Ron Adams, Adams Application Services Inc., Greer, South Carolina
Robert Anderson, University of Kentucky, Lexington, Kentucky
Matthew Blanchard, Michigan State University, East Lansing, Michigan
Todd Cavins, Sun Gro Horticulture, Stillwater, Oklahoma
Raymond Cloyd, Kansas State University, Manhatten, Kansas
Simon Crawford, Gourmet Genetics, Whichford, England
Margery Daughtrey, Cornell University, Riverhead, New York
August De Hertogh, North Carolina State University, Raleigh, North Carolina
Ron Derrig, Ron Derrig Horticultural Consulting, Apopka, Florida
Bill DeVor, Greenheart Farms, Arroyo Grande, California
John Dole, North Carolina State University, Raleigh, North Carolina
James Faust, Clemson University, Clemson, South Carolina
Paul Fisher, University of Florida, Gainesville, Florida
William Fonteno, North Carolina State University, Raleigh, North Carolina
Bob Frye, The Plantation, Lincoln, Nebraska
James Gibson, Fafard Inc., Anderson, South Carolina
Jerry Gorchels, PanAmerican Seed Company, West Chicago, Illinois
P. Allen Hammer, Dummen USA, West Lafayette, Indiana
Paul D. Hammer, Dummen USA, Westerville, Ohio
Will Healy, Ball Horticultural Company, West Chicago, Illinois
Gary Hennen, Ogelsby Plants International, Alva, Florida
Kerry Herndon, Kerry's Bromeliads, Homestead, Florida
Ed Higgins, Ball Horticultural Company, West Chicago, Illinois
Jim Kennedy, Ball FloraPlant, West Chicago, Illinois
Gerald Kinro, Department of Agriculture, Honolulu, Hawaii
Mike Klopmeyer, Darwin Plants, West Chicago, Illinois
Peter Konjoian, Konjoian's Educational Services, Andover, Massachusetts
Harvey Lang, Syngenta Flowers, Gilroy, California
Joyce Latimer, Virginia Tech, Blacksburg, Virginia
Dan Lehman, Syngenta Flowers, Boulder, Colorado
Ria T. Leonard, University of Florida, Gainesville, Florida

Ingram McCall, North Carolina State University, Raleigh, North Carolina
Chad T. Miller, Kansas State University, Manhattan, Kansas
Marvin Miller, Ball Horticultural Company, West Chicago, Illinois
Robert O. Miller, Dahlstrom & Watt, Smith River, California
William Miller, Cornell University, Ithaca, New York
Terril A. Nell, University of Florida, Gainesville, Florida
Paul Nelson, North Carolina State University, Raleigh, North Carolina
Kerstin Ouellet, Pen & Petal, Fallbrook, California
Wayne Poole, Ogelsby Plants International, Fort Lauderdale, Florida
Grace Romero, W. Atlee Burpee Company, Warminster, Pennsylvania
Erik Runkle, Michigan State University, East Lansing, Michigan
Don Snow, Syngenta Flowers, Gilroy, California
Roger Styer, Fides Oro, Santa Paula, California
Jan Van der Meij, Ball FloraPlant, West Chicago, Illinois
Colleen Warfield, Ball Horticultural Company, West Chicago, Illinois
Brian Whipker, North Carolina State University, Raleigh, North Carolina
Anne Whealy, Proprietary Rights International, Roanoke, Texas
Jennifer Duffield White, Editor-at-Large, *GrowerTalks* magazine, Missoula, Montana
Gary Wilfret, Manatee Floral Inc., Palmetto, Florida
Jack Williams, Paul Ecke Ranch, Encinitas, California (deceased)

Special thanks to Chuck Otto, Ball Horticultural Company, West Chicago, Illinois, who reviewed a wide range of crops and provided input on cultural details; Lloyd Traven of Peace Tree Farm (Kintnersville, Pennsylvania), who took the time to review many of the herb and vegetable cultures; and Debbie Hamrick, who edited the seventeenth edition of the *Ball RedBook.* Her help and guidance on how to manage editing the eighteenth edition was most insightful. While I had written books, I had not edited one. With this many authors, Debbie was invaluable in helping me through the process.

Many thanks also to the Ball Publishing staff, including Adriana Heikkila, who kept me organized; Production Manager Kathy Wootton, who coordinated the making of this book and kept us on schedule in stellar fashion; and Christine Truesdale, our talented Creative Director, who thought she'd signed on to design magazines.

Finally, I want to thank Ball Publishing for asking me to edit the crops in the *Ball RedBook.* Plants, including crop production, are near and dear to my heart, something I probably learned while using the twelfth edition of *Ball Redbook* in my horticultural studies at Iowa State University many years ago. The *Ball RedBook* was a required reference. I hope you find that it's an essential part of your business as well.

Jim Nau
West Chicago, Illinois

Part 1

Introduction

Jim Nau

Horticulture is in the throes of change. While we once considered our industry recession-proof, the global economic recession of 2008 hit all sectors of business, including gardening. I thought it pertinent to include a *GrowerTalks* magazine article by Marvin Miller from September 2010, on the most recently available United Stated Department of Agriculture (USDA) floriculture statistics. (For a complete version of this article, go to http://tinyurl.com/notjusttheweather.)

Not Just the Weather

At first glance, the recently released USDA Floriculture Crops Summary suggests 2009 was a dismal year for the industry compared to 2008.

Sales were down 6.9% overall to $3.829 billion and were off for each of the industry's major segments. Some might blame the overall economy. Others might point to the weather. But a closer look at the long-term trends quickly suggests that more is at play here than either of these issues alone.

For growers with at least $100,000 in wholesale sales, the data are quite revealing. First, total finished floriculture sales for these growers were $3.328 billion in 2009, down 6% from 2008. Sales of all industry segments were down, including those involving indoor product uses; this challenges the notion that only the weather is involved. Indeed, sales of domestically produced cut flowers were off 13.9% to $359.2 million, and sales of cut greens declined 19.4% from 2008 to 2009's $74 million. Potted flowering plant sales were off 5.4% to $632.4 million in 2009, and foliage plant sales declined 11% to $454.3 million during the same period.

When it comes to bedding/garden plants, the 2008-to-2009 period saw a decline in sales of 2.4%. Sales of annuals rose 0.6% to $1.317 billion, but this increase was more than offset by a steeper 9.8% decline in perennial sales, which dropped to $491.7 million. Hence, total segment sales were $1.808 billion in 2009. Should the overall decline be sustained after the 2009 numbers are revised next year, this will be the first decrease in bedding/garden plant sales ever recorded since these data have been collected, going all the way back to 1976.

A "Major Shift" Occurring

While the varying declines of the different industry segments have shifted the relative shares of finished floricultural sales, the more telling comparison is from 2005 to 2009, in which definite market shifts can be seen. Indeed, over this period, cut flowers, cut greens, and foliage plants have yielded share to bedding/garden plants and, to a lesser extent, potted flowering plants. Except for bedding/garden plants, all of these segments were smaller in 2009 than in 2005 in the fifteen-state survey.

The 6% decline in sales for the larger growers from 2008 to 2009, for the fifteen states from which data are collected, represents the largest sales decline ever recorded, but it's the third drop registered in the last four years. And while there's a chance the 2009 revised data to be released next year will alter the story somewhat, the reality is there's a major shift occurring in sales and in the industry.

Grower numbers were down 12.3% in 2009. However, numbers of the largest 45.2%, those reporting wholesale sales of at least $100,000, were off just 3.3%. These larger growers accounted for 96.3% of all estimated sales in the fifteen states covered. Both for the total grower numbers, as well as for those "larger grower" numbers, the 2009 tally was the second consecutive year of declining grower numbers.

NC moves to fourth place

Another shift seen in the data is in the state rankings. While California, Florida, and Michigan remain the top three states for wholesale sales, with 24.4%, 18.2%, and 10.4% of the survey's value, respectively, North Carolina has edged out Texas for fourth place, with 6.6% versus the Texas share of 6.4% of the value. These five states now account for 66% of the survey's total value. Ohio, New York, New Jersey, Pennsylvania, Washington, Illinois, and Oregon are the next states. In total, these twelve states represent nearly 95% of the survey total and are the only states in the survey to report $100 million or more in sales.

Bedding/garden

Among the sub-segments of the bedding/garden plant market, several longer-term shifts have been revealed in recent years. For example, from 2005 to 2009, the dollar sales of hanging baskets and vegetables have increased both overall and in share for the sub-segments. While the growth has not been continuous over the period, the sales of vegetables increased over 53% from 2005 to 2009, while dollar sales of hanging baskets increased over 17% during this period. In contrast, the sales of flats of annuals have dropped almost 11% over this time frame, while the share for this sub-segment is also down. Dollar sales of pots of annuals and dollar sales of herbaceous perennials are relatively unchanged over the 2005-to-2009 period.

Trends in unit sales are unfortunately not quite as robust. From 2005 to 2009, unit sales of flowering hanging baskets increased 3.5%, unit sales of flats of vegetables were down 2.5%, but unit sales of pots of vegetables were up nearly 64%. Unit sales of flats of annuals were down 19%, and unit sales of pots of annuals were down nearly 16%. Unit sales of pots of herbaceous perennials were down nearly 14%.

Given the relatively lackluster growth of dollar sales for annuals over the 2005-to-2009 period (a total of 1.5% increase), it would not be surprising if there were no dramatic trends on a species-by-species basis; however, this is not the case. In fact, other than the 53% increase in vegetable dollar sales noted above, there were only two genera exhibiting increased sales. Petunia dollar sales increased 12.2% and vegetative geranium sales increased 4.3% in dollar volumes over the four-year period. Dollar sales of New Guinea impatiens were off 13.6% from 2005 to 2009, while dollar sales of *Impatiens walleriana* were off 11.3% over this period. Begonia dollar sales were also off 10.1%. Though also down, the dollar sales of seed geraniums and marigolds were both off less than 2%, while dollar sales of pansies and violas were off 3.7%. Dollar sales of all other flowering annuals were off 1.1% over the period.

Among the thirty-two different genus and container combinations reported in the bedding/garden plant data, there were particular standouts for both positive and negative reasons (see charts on page 7). For example, dollar sales of potted vegetables increased 124% over the 2005-to-2009 period in the fifteen-state survey, while flats of vegetables were up 6% in dollar volumes. Among hanging basket sales, dollar volumes were down considerably for begonias (off 18%), impatiens (off 17.5%), and New Guinea impatiens (off 19%) and off less dramatically for seed geraniums (down 5%). Yet, hanging basket sales were up nearly 21% for vegetative geraniums, up 398% for marigolds (on a very low base), up 86% for pansy/viola baskets, up 43% for petunia baskets, and up 29% for "all other flowering hanging baskets."

Potted flowering plants

Sales of potted flowering plants were off 5.4% in dollars in 2009 from the 2008 levels to $632.36 million. Even though total dollar sales declined, the share of the total finished plant market increased fractionally from 18.9% to 19.0% (since several other segments experienced greater declines in sales). Unit sales dropped 11.5% for the 2009 year from 2008 levels. In comparison, the 2005-to-2009 period has shown an over 18% decline in the total potted flowering plant units in the fifteen states but less than a 1% decline in dollar volume for the category.

There's a new leader in potted flowering plants. Though the industry still sells more poinsettia units than any other pot crop in a year's time, orchids have overtaken the poinsettia as the top pot crop in revenues. The fifteen-state survey reported $159.6 million in orchid sales for 2009 compared with $145.1 million in poinsettia revenues.

Foliage plants

Total foliage plant sales in 2009 were $454.3 million, off 11% from 2008 levels. Foliage plant sales really lost market share, dropping from 14.4% of the finished product value in 2008 to 13.7% of the 2009 market. Over the 2005-to-2009 period, foliage plant sales have dropped three out of four years, and total sales dollars are off by one-third.

Cut flowers

Cut flower sales dropped 13.9% from 2008 to 2009's $359.2 million. Though gladioli unit sales increased nearly 24% and dollar sales were up almost 16% from 2008, and cut orchid blooms were off 3% but dollar sales were up almost 87%, the sales of all other separately enumerated cut flower types dropped significantly from 2008 to 2009 in both units and dollars (with the exception of gerbera, which experienced a nearly 12% decline in units but only an 8% decline in dollars, all other declines were of a double-digit percentage reduction in both units and dollars).

Cut cultivated greens

Sales of cut cultivated greens dropped 19% from 2008 to about $74 million in 2009. Leatherleaf fern sales were off 9% in bunches and 12% in dollars in 2009 from the 2008 levels. In total, there were 31.6 million bunches of leatherleaf ferns reported valued at $29.8 million in 2009. Sales of "all other cut cultivated greens" were off nearly 24% to $44.2 million.

Propagative materials

Though not included in the discussion above of finished floricultural product sales, the USDA does report on sales of propagative materials and unfinished plants. Overall sales in this category were off 6.8% in 2009 from 2008 levels, to $357.7 million. Sales of this category dropped for annuals (10.5%), perennials (2.2%), foliage plants (13.3%), and cut flowers (30.7%), but increased for potted flowering plants (0.8%) and cut cultivated greens (378.4%).

The Horizon

There's no doubt the current economic conditions are impacting the industry, but there are many arguments that other factors are also at play. Certainly, poor weather during the prime selling seasons doesn't help, but there have been numerous reports of satisfactory weather during the prime selling weekends of spring, and of stable customer counts, but with a dramatic decline in sales due to significantly lower average sales per customer. The only conclusions we can draw are that either the economy is having a dramatic impact on our industry or the industry has struggled to remain relevant in the eyes of the consumer, or both.

One suggested solution is to offer customers some lower price points as an added selection option, especially in locales hard hit by high unemployment.

Another observation notes generational consumption patterns. Some consumption, such as food, is largely generationally neutral, while other consumption is related to demographic considerations, such as marriage, first homes, college tuitions, and so on; and there's concern that Gen X is about 12 million consumers smaller than either the Baby Boomers or Gen Y. However, on the surface this concern is more an issue for nursery products like trees and shrubs for first-time homebuyers.

Perhaps a more salient observation suggests that our industry has relied for too long on traditional and impulse sales and has ignored the new buying habits of consumers. Today's consumers are looking for additional reasons to justify their spending. Trends include a greater desire to save, the recognition that previous spending habits only garnered a lot of "meaningless stuff," and more introspective consumers who are spending more time evaluating competing products to discern the greater values. The Internet and social media are two tools that allow consumers easy access to information supporting their new decision-making processes.

Perhaps our industry needs to learn to market its products using "more than pretty" as the reason the consumer should be buying. The industry certainly has much research at the ready that can support claims of everything from economic development, environmental improvements, and lifestyle enhancements. These arguments are aimed at helping consumers recognize the industry's products as necessities for their well-being rather than a luxury. Many of the studies supporting such claims can be found on the America in Bloom website (www.americainbloom.org) under "Community Resources."

In any case, nationwide prosperity isn't likely to be just around the corner. Difficult issues, including immigration reform, health care, labor issues, overseas conflicts, international trade, and the value of the dollar all may impact our industry in the next few years. Competition to woo consumers is also an issue, as is quality, pricing, service, and value. In the end, our industry will need to court consumers in a smarter manner than ever before.

Dollar Distribution - Sales of Annuals - 2005

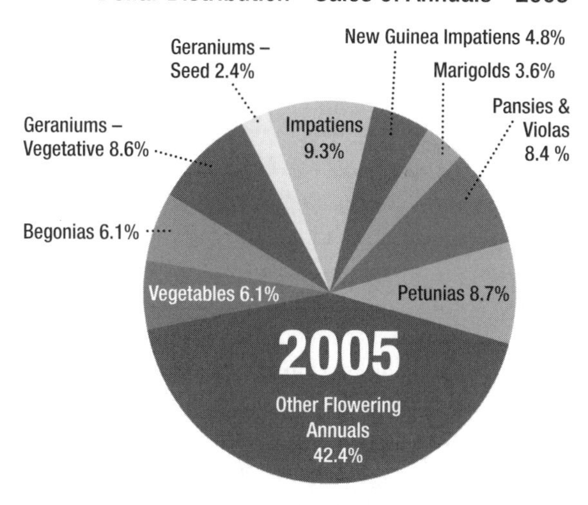

Dollar Distribution - Sales of Annuals - 2009

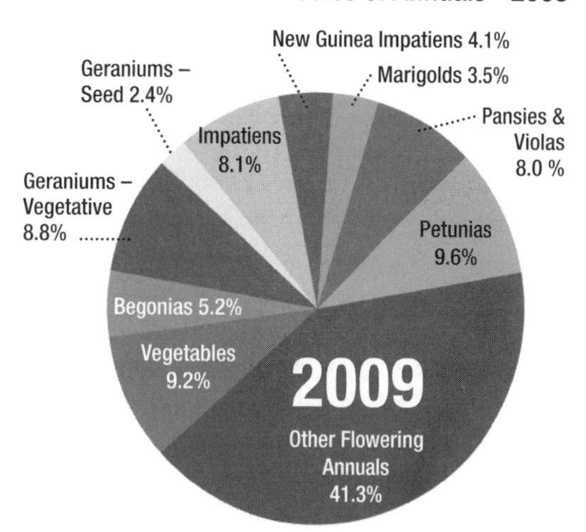

The 2005 U.S. Floriculture Production Pie $3.604 Billion in Sales*

(Finished Production Only - Wholesale Value)

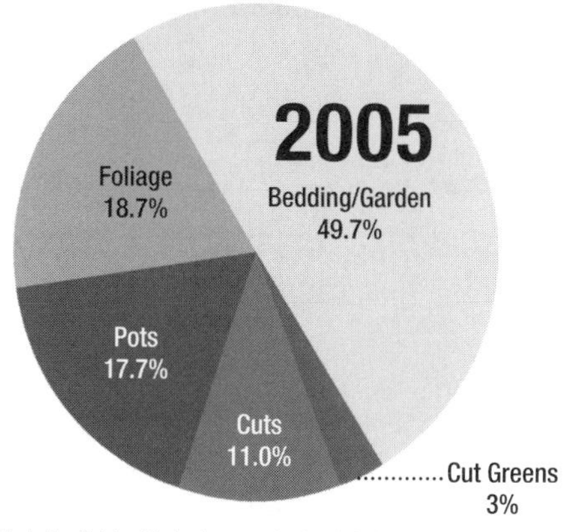

The 2009 U.S. Floriculture Production Pie $3.328 Billion in Sales*

(Finished Production Only - Wholesale Value)

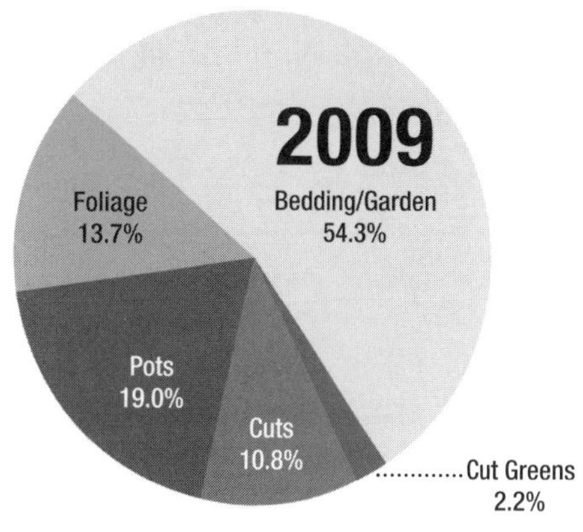

*Includes finished floriculture production (wholesale value) only for firms with $100,000 or more in wholesale ("farm gate") sales in 15 states.

USDA/NASS, Agricultural Statistics Board; Floriculture Crops-2005 Summary and 2009 Summary

Sales of Bedding/Garden Plants - 2005 Dollars

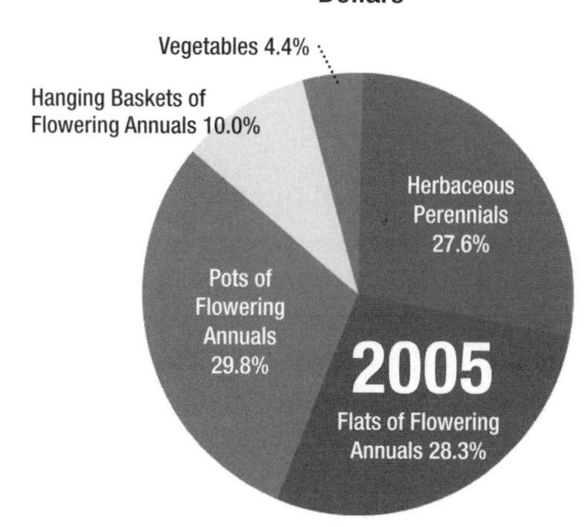

Sales of Bedding/Garden Plants - 2009 Dollars

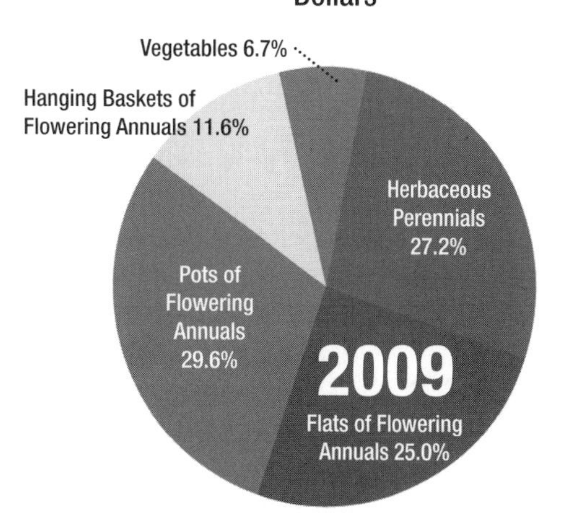

A Note About Volume 2

In the material that follows, you'll find much of the information you need to grow greenhouse crops.

Part 1 consists of chapters written by industry experts on the critical points of growing crops: light, temperatures, water, media, nutrition, etc.

In Part 2, you'll find crop culture for more than two hundred specific crops—including over seventy species that are new to the *Ball RedBook.* In addition, existing crop cultures have been updated to include additional methods of propagation. In previous editions, the culture information presented was general. Plug and production crop times were not specific; this has been enhanced in the eighteenth edition to provide ranges of times to make it easier to plan a crop. Granted, you will have to adjust these crop times to your location as well as the time of the year. This part of the book has been put together using several main sources. First and foremost has been the Cultural Database from the Ball Seed Company. Individual academic, grower, and industry contributors and experts have supplied several of the sections and reviewed others.

Within the bibliography, you'll discover a useful list of books and publications that you may want to add to your own bookshelf and of websites to bookmark and visit again and again.

As a grower, the *Ball RedBook* is *the* book to buy if you can only have one book on your shelf.

1

John M. Dole, Paul Fisher, and Brian E. Whipker

Water Quality

Poor quality water—whether it is high in electrical conductivity (EC), alkalinity, or waterborne pathogens—can make a grower's life difficult. Water with a high EC can reduce seed germination, inhibit rooting of cuttings, and increase the likelihood of root and crown diseases. Water with an excessively high pH and alkalinity increases the likelihood of nutrient deficiency problems. The presence of waterborne pathogens in the water can eventually result in diseased plants. These problems are greatly magnified if you grow plugs and cuttings, because young plants are particularly sensitive to water quality.

Before we address water quality considerations, however, we need to discuss the practice of watering. For most crops the best plant growth is obtained when plants are irrigated just prior to wilting. Water stress reduces photosynthesis and slows growth. In addition, the cells of drought-stressed plants do not expand to their full potential, resulting in shorter, more compact plants. While limited drought stress can be beneficial for bedding plants such as bedding impatiens or tomatoes, plant quality is reduced on a majority of crops. Excessive drought stress can cause lower-leaf yellowing and damage roots, leading to disease. Excess water can lead to anaerobic conditions in substrate, also encouraging root diseases, nutrient deficiencies resulting from poor root function, and reduced root growth.

Generally the substrate should be watered thoroughly to the point that at least a small amount of water can be seen coming out of the bottom of the containers or beds. Insufficient watering can mean that part of the root-ball remains dry, or that salts such as sodium and chloride accumulate in the substrate. Irrigating with an excessively high volume of water will leach fertilizers from the growing substrate and increase the potential for runoff into the environment.

When hand watering, be sure to use a nozzle that breaks up the flow of the water and reduces the force of the water hitting the plants and the substrate. Using hoses without breakers at high volume will flush substrate out of the containers or cause compaction, resulting in reduced soil aeration. Automated irrigation systems are covered in volume 1 of the *RedBook.*

What Is High-Quality Water?

A variety of measures are used to determine chemical water quality (table 1-1).

One of the most important factors is the electrical conductivity (EC), a measure of total soluble salts. Water with a low EC (0.0 to 0.5 mS/cm) is advantageous because high soluble-salt levels are less likely to accumulate in the root substrate, and leaching is not necessary. Plant species vary in their tolerance to high EC, which can stunt plant growth, induce wilting even though the substrate is moist, and cause marginal leaf burn especially in sensitive crops such as New Guinea impatiens, pentas, and ferns. High substrate EC levels can be managed by periodic leaching. See chapter 3 for more information on EC management. Reverse osmosis (discussed in High Soluble Salts) can be used when water EC is too high.

Other equally important measures are pH and alkalinity. The ideal water pH is in the range of 5.4 to 7.0, and alkalinity between 0.8 to 1.3 meq (40 to 65 ppm HCO_3^-), although a broader range in levels can be successfully managed. Water pH is important for dissolution and efficacy of chemicals and pesticides. Check the ideal solution pH range of agrichemicals from the pesticide label or manufacturer. Water that is high in alkalinity (which can be considered as dissolved liming content) results in higher pH of the growing substrate (the "substrate pH") over time, resulting in deficiencies in micronutrients such as iron. Conversely, water with low alkalinity is often also low in dissolved calcium and magnesium (which therefore need to be incorporated in fertilizer) and increases the likelihood that iron and manganese toxicity will result from excessive uptake of those elements when the substrate pH is low. Treatment options for alkalinity are discussed in the Alkalinity section.

Table 1-1. Characteristics of High-Quality Irrigation Water

CHARACTERISTIC	DESIRED LEVEL	UPPER LIMIT
Soluble salts (EC)	0.2–0.5 mS/cm	0.75 mS/cm for plugs 1.5 mS/cm for general production
Soluble salts (total dissolved solids)	128–320 ppm	480 ppm for plugs 960 ppm for general production
pH	5.4–6.8	7.0
Alkalinity ($CaCO_3$ equivalent) Bicarbonates	 40–65 ppm (0.8–1.3 meq/L) 40–65 ppm (0.70–1.1 meq/L)	 150 ppm (3 meq/L) 122 ppm (2 meq/L)
Hardness ($CaCO_3$ equivalent)	<100 ppm (2 meq/L)	150 ppm (3 meq/L)
Sodium (Na)	<50 ppm (2 meq/L)	69 ppm (3 meq/L)
Chloride (Cl)	<71 ppm (2 meq/L)	108 ppm (3 meq/L)
SAR[1]	<4	8
Nitrogen Nitrate (NO_3) Ammonium (NH_4)	<5 ppm (0.36 meq/L) <5 ppm (0.08 meq/L) <5 ppm (0.28 meq/L)	10 ppm (0.72 meq/L) 10 ppm (0.16 meq/L) 10 ppm (0.56 meq/L)
Phosphorus (P) Phosphate (H_2PO_4)	<1 ppm (0.3 meq/L) <1 ppm (0.01 meq/L)	5 ppm (1.5 meq/L) 5 ppm (0.05 meq/L)
Potassium (K)	<10 ppm (0.26 meq/L)	20 ppm (0.52 meq/L)
Calcium (Ca)	<60 ppm (3 meq/L)	120 ppm (6 meq/L)
Sulfates (SO_4)	<30 ppm (0.63 meq/L)	45 ppm (0.94 meq/L)
Magnesium (Mg)	<5 ppm (0.42 meq/L)	24 ppm (2 meq/L)
Manganese (Mn)	<1 ppm	2 ppm
Iron (Fe)	<1 ppm	5 ppm
Boron (B)	<0.3 ppm	0.5 ppm
Copper (Cu)	<0.1 ppm	0.2 ppm
Zinc (Zn)	<2 ppm	5 ppm
Aluminum (Al)	<2 ppm	5 ppm
Fluoride (F)	<1 ppm	1 ppm

[1] SAR, sodium absorption ratio, relates sodium to calcium and magnesium levels.

The content of individual nutrients (i.e., N, P, K, Ca, Mg, S, B, Cu, Fe, Mn, Mo, and Zn) of the water should also be checked, along with non-nutrient ions such as Na, Cl, Al, and F. When fertigating with water-soluble fertilizers, the nutrient solution is a blend of ions in the water plus the ions added in the injected fertilizers and/or acid. Therefore, while low levels of some nutrients can be beneficial, high levels of one or more nutrients may require adjustments in the fertilizer program. If the water has high levels of nitrogen, calcium, or magnesium, less of those nutrients can be added as fertilizers. High nitrogen levels can be especially prevalent in areas with sandy soil, shallow wells, or intensive agriculture. Unfortunately, high levels of calcium, magnesium, and iron can be antagonistic to other nutrients, such as manganese or boron, and can reduce their uptake. Water that is specifically high in calcium and magnesium is known as hard water. Most plant species are tolerant of high calcium and magnesium levels, but overhead irrigation with hard water can leave unsightly white salt deposits on the foliage, especially with mist propagation. High micronutrient levels (particularly boron) can be phytotoxic. Sodium and chloride increase the water EC and can disrupt nutrient uptake, without contributing to plant growth. Ions such as aluminum and fluoride are not nutrients but can be toxic to plant growth above threshold concentrations. Treatments for preventing problems with specific ions are listed below.

Testing your water: Commercial labs

A number of commercial labs conduct water analysis for alkalinity and micronutrients. A partial listing is in table 1-2.

Testing your water: In-house

In-house testing of your water sample is an economical and easy way to monitor your pH, EC, and alkalinity levels. To determine the water pH, a quality pH meter is required. Combination pH and EC meters are an excellent choice. Expect to pay $150 or more for a combination meter, and be sure to follow instructions on storage, calibration, and use. To determine the water alkalinity level, a colormetric alkalinity test kit is required. Expect to pay $30 or more for a kit.

Remember that EC is a measure of the combination of all ions including nutrients such as N, P, K, Ca, and Mg, plus other ions, such as sodium and chloride, which are not needed for growth. Nutrient-specific tests are available for ions such as potassium. However, generally pH, EC, and alkalinity are the main onsite tests, and for detailed analysis of other elements it is more cost effective and accurate to send samples to a commercial laboratory.

Alkalinity

A major factor increasing substrate pH over time is the amount of alkalinity in the irrigation water. High substrate pH (discussed in chapter 4) results in micro-

Table 1-2. Water-Testing Labs

The following labs specialize in irrigation water, fertilizer solution, greenhouse substrate, and plant tissue samples. Check with the individual labs about the procedures for submitting samples and for current pricing, which typically range from $24 to $65 per sample. Standard tests usually include pH, EC, alkalinity, and a number of individual ions. Most testing laboratories also offer plant tissue and substrate tests.

A & L Laboratories 2790 Whitten Road, Memphis, TN 38133 (800) 264-4522, (901) 213-2400 www.allabs.com	North Carolina Department of Agriculture & Consumer Services, Agronomic Services Division 4300 Reedy Creek Road, Raleigh, NC 27607-6465 (919) 773-2655 www.ncagr.gov/agronomi/uyrsoln.htm
Cornell University Nutrient Analysis Laboratory 804 Bradfield Hall, Ithaca, NY 14853 (607) 255-4540 soiltest@cornell.edu cnal.cals.cornell.edu/	JR Peters Laboratory 6656 Grant Way, Allentown, PA 18106 (866) 522-5752 info@jrpeterslab.com www.jrpeterslab.com
Fafard Horticultural Services 1471 Amity Road, Anderson, SC 29621 (800) 722-7645 fafardlab@fafard.com http://www.fafard.com/?p=113	Quality Analytical Laboratories 403 East 11th Street, Panama City, FL 32401 (850) 872-9595 www.qal.us
MMI Laboratories/SunGro Analytical Services 183 Paradise Boulevard Suite 108, Athens, GA 30607 (706) 548-4557 pamaaron@mmilabs.com www.mmilabs.com, www.sungroanalytical.com	Scotts Testing Lab 300 Speedway Circle, Lincoln, NE 68502 (877) 467-8522 fred.hulme@scotts.com, keith.santner@scotts.com www.scottstestlab.com

nutrient deficiencies in crop plants. The relative concentration of carbonate types—carbonates (CO_3^{2-}), bicarbonates (HCO_3^-), and carbonic acid (H_2CO_3) — is the main buffering system controlling irrigation water pH and substrate solution pH. If the irrigation water contains a high concentration of carbonates and bicarbonates, the substrate pH can rise to undesirable levels during plant production.

Alkalinity levels

High alkalinity levels in irrigation water can limit plant growth and cause losses for producers of container-grown nursery and greenhouse crops. High alkalinity typically occurs in coastal areas or in locations with limestone bedrock. Much of the well water in the Midwest and Great Plains of the United States, Florida, southern Ontario, and the prairie provinces of Canada contain high levels of alkalinity.

The level of alkalinity in irrigation water can vary with well location, well depth, and time of year. A standard water analysis usually includes pH, EC, and alkalinity. Growers may also want to test for macro- and micronutrients in their water. Water tests are recommended for each well and should be done annually. Because plugs contain a small volume of substrate with little buffering capacity, plug producers should consider monthly water tests.

Evaluating and neutralizing your alkalinity level

As alkalinity increases, the appropriate management options change in terms of the desirability of acid injection to neutralize alkalinity, nitrate versus ammonium fertilizer selection, and blending with more pure water sources (table 1-3). Regardless of the method you select to manage alkalinity, neutralizing alkalinity is required for operations that have water alkalinity levels above 2 meq. Select a method that best suits your operation. Conduct a routine analysis of root substrate to monitor pH and nutrient levels and to ensure that fertility and alkalinity neutralization programs are on target.

A number of methods can be used to overcome high alkalinity in irrigation water: acid injection, fertilizer modification, or blending with another water source low in alkalinity (such as captured rainfall or reverse osmosis–treated water). Every greenhouse varies in water quality, root substrate type, fertilizer type (i.e., acidic or basic), watering practices, container size, and length of time a crop is grown. Therefore, because alkalinity is the main component influencing the production system, neutralize alkalinity first and then determine a fertilizer strategy. Excessive alkalinity levels will need to be neutralized by adding acid, which is discussed below, or fertilizer, as discussed in chapter 4.

Acid

Acid is injected into the irrigation water to neutralize the alkalinity. The amount of acid to use depends on the starting pH, the alkalinity level of the irrigation water, and the target endpoint alkalinity level desired. In general, a target endpoint alkalinity of around 2 meq is recommended for most crops. This should result in an endpoint water pH of 6.0–6.2. This target endpoint allows for seasonal variations of alkalinity that naturally occurs in wells, limits the potential

Table 1-3. Classification of Irrigation Water Quality Based on Alkalinity[1]

ALKALINITY RANGE (MEQ)[2]	CLASS	COMMENTS
0–1.0	Low	Pure water with little buffering capacity. Growers will need to monitor substrate pH to ensure the pH remains within the acceptable range. A high nitrate (basic reaction) fertilizer is generally required to avoid a drop in substrate pH.
1.5–4.0	Marginal	Consider acid injection if substrate pH tends to rise. Neutral-to-acidic reaction fertilizers (at least 25% of the nitrogen in ammonium form) may be sufficient to balance alkalinity without acidification.
4.0–6.0	High	Acid injection required.
>6.0	Very high	Acid injection required. May require blending with another water source or use of reverse osmosis.

[1] The recommended target endpoint alkalinity level (after a corrective measure has been incorporated) is around 2 meq of alkalinity for most pot and bedding plants. Levels for plugs are around 1 meq of alkalinity.

[2] Conversion factors: 1 meq of alkalinity = 50 ppm calcium carbonate equivalent or 61 ppm bicarbonate

problem of plants that naturally acidify the root substrate (for example, geraniums, dianthus, *Celosia*, begonias, and others), and allows for errors in measuring acids. Operations that produce plugs and are willing to monitor their alkalinity level weekly may desire to neutralize to 1 meq of alkalinity (resulting in a water pH near 5.7) to have greater control of their substrate pH.

Deciding which acid to use

The common acids used for alkalinity control are phosphoric (H_3PO_4) (75 and 85%), sulfuric (H_2SO_4) (35 and 93%), and nitric (HNO_3) (61.4 and 67%). Each acid supplies nutrients to the plants. For instance, one ounce of each acid added per 1,000 gal. (7.4 mg/l) of water would supply: 2.92 ppm phosphorus (P) with 75% phosphoric acid, 1.14 ppm sulfur (S) with 35% sulfuric acid, or 1.47 ppm nitrogen (N) with 61.4% nitric acid. The fertility regime may need to be altered to accommodate the added nutrients. All acids are dangerous because of their caustic characteristics, with the most relatively safe being phosphoric, then sulfuric, then nitric. Citric acid can also be used but is the most costly.

The most commonly used chemical is battery acid, which is 35% sulfuric acid. It is the least expensive, is moderately safe, and provides sulfur. Phosphoric acid is suited for operations needing to neutralize up to 1 meq of alkalinity. When higher amounts of alkalinity must be neutralized, the amount of P provided far exceeds the requirements of the plants. High levels of P can result in excessive plant stretch, especially for plugs. Alternatively, some growers use phosphoric acid to provide their plants with sufficient levels of P and neutralize the remaining alkalinity with sulfuric or nitric acid. Some growers select nitric acid because it supplies N and allows them to decrease the amount of N fertilizer applied. After adding acid, retest the water after one day and again after two to three weeks to double-check the water pH and alkalinity levels.

Determining the amount of acid to add

An online tool that calculates the amount of acid to add to your irrigation water is available from the University of New Hampshire website: http://extension.unh.edu/Agric/AGGHFL/Alkcalc.cfm.

Water Treatments for Ions

A number of options are available to treat your water if water quality is poor. The first option is to locate a high-quality water source (such as municipal water, well water, or surface water from a river or other source) to blend with poor-quality water. If using poor-quality well water, check with a hydrologist to see if another well could be drilled, as water quality can vary with the depth of the well. Often it is difficult and expensive to find another water source, and water treatments will need to be considered.

High soluble salts

Reverse osmosis (RO) is the most common method to remove ions from water and reduce EC. Reverse osmosis forces water through a semipermeable membrane, leaving 90 to 99% of the soluble salts behind. One drawback to RO is the large quantity of wastewater produced (30 to 60% of original volume), which contains high amounts of salts. Disposal of this brine should be handled carefully due to environmental and regulatory concerns. Proper filtration and maintenance are essential in order to keep an RO unit operating smoothly.

Other water treatment systems (i.e., deionization, distillation, and electrodialysis) are available for treating water with a high ion content but are currently more expensive than RO or do not produce the volume of water needed during production. With deionization, water flows over ion-exchange resins that remove the ions. The resins are usually solid beads with either positive or negative charges. Deionization is most feasible when highly pure water is required and the water has a low initial EC. In distillation, the water is boiled and the resulting steam is condensed into pure water, which leaves behind the salts, particulates, and nonvolatile compounds. In electrodialysis, water is passed between cation- and anion-permeable membranes. When an electrical current is applied, ions migrate through the membranes, leaving pure water. Neither distillation nor electrodialysis is used on a large commercial scale in the greenhouse, but advances in technology may make them more feasible in the future.

If water treatment is not an option for handling water with a high EC, cultural practices can be used to reduce the problem. Increased leaching rates will prevent soluble salts from building up in the substrate and will prevent plant damage. High EC water is particularly challenging for the production of seedlings and cuttings. Buying in plugs and rooted cuttings instead of propagating your own and using high-quality water for propagation are options. For growers using recirculating irrigation water, controlled-release fertilizers can also be used to reduce the nutrient content of water. Generally, proper use of controlled-release fertilizers will allow the plants to take up a greater percentage of the nutrients applied. Consequently, less fertilizer is leached out of the pots with controlled-release fertilizers, which helps to keep the EC of the recirculated water low.

Specific ions

Water can occasionally be high in individual ions without having a high overall salt content. With some of these ions, specific treatments or cultural practices are required.

Iron and manganese

Subsurface water can be high in a reduced, soluble form of iron and manganese, which oxidizes upon contact with air into a less soluble, rust-colored form. The oxidized form is responsible for the brown- to rust-colored stains on plants and equipment. The iron and manganese can be removed by instigating the oxidation process prior to using the water. The water is sprayed into a tank or pond, which rapidly oxidizes the iron and manganese into insoluble forms that are allowed to settle to the bottom of the tank or pond. The tanks or ponds must be large enough to treat sufficient water to allow the iron and manganese to settle before the water is used.

Calcium and magnesium

The calcium and magnesium in hard water can be replaced by potassium in a process known as water softening. The total salt content is not reduced, but the potassium can act as a fertilizer and the amount of potassium can be reduced or eliminated in the fertilizer solution. *Note:* Water softening should not be confused with home water softening, which replaces the calcium and magnesium with sodium, which can be damaging to plants.

Carbonates and bicarbonates

Carbonates and bicarbonates can be eliminated by acid injection, as described earlier.

Fluoride

While avoiding fluoride is generally the best strategy, sometimes there is no other choice but to use water high in fluoride. Activated alumina or activated carbon can absorb fluoride from water. Water must be pH 5.5 for alumina to be effective, and an ion-exchange system can be used to obtain that pH. In addition, maintaining a substrate pH of 6.0 to 6.5 will make the fluoride relatively insoluble and prevent fluoride toxicity.

Boron

Anion-exchange systems can be used to remove boron, which occurs in water as negatively charged borate. Anion-exchange systems act on all anions.

Control of Waterborne Pathogens, Algae, and Biofilm

As recirculating irrigation systems become more common, many growers are concerned about the possibility of spreading plant pathogens such as *Pythium* or *Phytophthora* through irrigation water. Waterborne microorganisms also include algae. Algae can be an aesthetic problem on substrate surfaces, can form a surface layer that reduces water absorption into growing substrate, can encourage pests such as shore flies, and can be a worker hazard by causing slippery floors. Biofilm is a complex of microorganisms that forms in irrigation lines and can cause emitters to clog and result in increased labor costs in order to clean lines and in a decreased working life of equipment such as drip lines.

Water treatment for microorganisms is especially desirable if any of the following situations apply to your operation:

• Recycling irrigation water has high potential for contamination with pathogens, microbes, and algae, which can then be distributed back onto other crops. If regulations require you to recycle water or reduce runoff, then water treatment is an important risk-management step to avoid crop losses.

• Your water source is a surface pond, river, or lake. In a survey of twenty-four grower locations around the United States, University of Florida research found much higher microbial load in catchment basins and tanks than in municipal- or well-water sources.

• Your operation propagates cuttings or plugs. The combination of warmth and high humidity encourages algae growth, and young plugs and liners are the most disease-sensitive crop stage.

• An existing or potential disease problem that could be spread by water is present. Many pathogens can be distributed in irrigation systems, and organisms such as the water molds *Pythium* and *Phytophthora* are commonly found in collection tanks and basins. Crops susceptible to quarantine or other potentially devastating pathogens such as *Phytophthora ramorum* (sudden oak death) or *Ralstonia solanacearum* require a non-recycled source or a high level of water treatment.

• Excess algae growth or clogged irrigation lines from biofilm are present.

• Your operation has a food-safety risk. As ornamental plant growers diversify into production of herbs, vegetables, and fruit crops, especially for the retail market, it is essential that plants are free of human pathogens such as *E. coli*.

• Persistent pesticides, growth regulators, and herbicides need to be removed. This may be achieved through some treatments that also control microorganisms, such as oxidation, activated carbon filtration, or membrane filtration.

Sanitizing water technology options

A challenge for growers is to choose which options are best suited to their situations. In most cases, a combination of approaches is best. For example, a grower might choose to shock his irrigation lines with chlorine dioxide to remove biofilm, improve filtration to reduce organic load and remove larger particles of peat and plant material that harbor pathogens, and use copper ionization as a continuous treatment.

Each technology will treat certain problems in specific situations, but no single technology is always the "best" solution. Consider the financial tradeoff between higher initial capital costs to purchase and install the system versus operating cost to treat each 1,000 gal. (3,785 l) used (table 1-4). Smaller-scale growers or those aiming to only treat a few greenhouse zones may prefer a low capital-cost/higher operating-cost solution. For large water volumes, a low operating cost is essential. When budgeting for installation, the cost to change water flow, treatment tanks, etc. may exceed the purchase cost of the treatment unit itself.

Residual control indicates that the treatment is effective throughout the entire irrigation system. For example, with a well-designed system you can measure residual-free chlorine at the water outlet furthest from the point of injection. Reverse osmosis and UV do not have residual control and are designed to sanitize water at the point of treatment only.

Calcium hypochlorite (solid granules or tablets), sodium hypochlorite (liquid bleach), or chlorine gas all form hypochlorous acid (a strong sanitizer favored below pH 7.5) or hypochlorite (a weak sanitizer favored above pH 7.5) in solution with water. Injection of 1 to 2 ppm of free chlorine on a continuous basis provides a low-cost sanitizing option. Many systems dose at a constant injection ratio with flow rate. However, the safest and most effective way to chlorinate is to have in-line monitoring and control of water pH (through acidification) and oxidation-reduction potential (ORP) to increase the sanitizing power. Inline control increases capital cost but can decrease the applied ppm and improve results. Manual sensors and ORP meters can also be used to monitor chlorine.

Chlorine dioxide is a very effective material for removing biofilm in pipes as a shock treatment, is less sensitive to pH than chlorine, and can also be applied on a continual basis. Products are available that can be mixed in a single stock tank and dosed with an injector. This single-tank system has low installation costs, but the operating cost is higher than other options in table 1-4. Many growers, therefore, use this system in areas where sanitation is a high priority and there is low total-water demand, such as misting young plants.

Another option to inject chlorine dioxide is onsite mixing from two or three tanks. The tank dosing and monitoring equipment can either be purchased or leased to reduce the installation cost. A multi-tank option also reduces the operating cost of chlorine dioxide when applied water volumes are high (at least 10,000 gal. [37, 854 l] per day). ORP inline sensors or manual meters can be used to monitor either single- or multi-tank levels of chlorine dioxide, in addition to manual test-strip measurements of residual chemical at the furthest outlet.

Copper ionization is widely used by plug and liner growers because, although the installation cost is relatively high, the operating cost (electricity and the occasional replacement of copper rods or plates) is low. Copper is applied at 1 to 2 ppm for pathogen and algae control. Improvements have been made in copper ionization equipment to provide consistent dosage even when water EC or flow rate varies, which is a common situation when using water-soluble fertilizers.

EPA-registered products based on hydrogen dioxide/activated peroxygen chemistry are more stable and active than hydrogen dioxide (also called hydrogen peroxide). Activated peroxygens include other ingredients, notably peroxyacetic acid (PAA) or octanoic acid, in addition to hydrogen dioxide. Activated peroxygen products are less sensitive to organic matter and pH than chlorine, which enhances their residual effect. Products are injected with a proportioner, and test kits are available to manually measure active ingredient concentration. Inline monitoring and control is currently in development.

Hypochlorous acid (trademarked as Oxcide) is applied as a shock or continuous treatment to reduce biofilm and scale in irrigation lines. This product may be a good choice where clogging of irrigation equipment is a major issue.

Ozone systems are now available with a high level of control with inline monitoring of ORP, ozone dosage, and injection. Like copper, this is a system with high initial costs but low operating costs (mainly electricity). Some growers have had poor results using ozone in the past because of undersized units, limited control, and low dissolution of ozone gas into the irrigation water. A few growers are now using ozone in flood-floor greenhouses, and research is underway to evaluate the effects on the nutrient solution and microbes.

Reverse osmosis (RO) is a membrane-filtration effect that removes the most troublesome ions from water other than boron, reducing EC (high salt) problems. Reportedly, RO membranes, which have extremely fine pores (below 0.001 microns), also remove almost all pathogens, including viruses. Some vegetable growers are starting to combine coarse primary filtration with a finer membrane filter for algae and pathogen removal.

Slow sand or constructed wetlands are more ecological options that use the natural soil, microbial, and plant growth processes to remove contaminants in water. Treatment cost varies widely depending on design, and a specialized engineering firm should be consulted.

Ultraviolet light is another example of a high capital cost, low operating cost technology. Check that the turbidity of your water source is low enough to allow for successful penetration of UV light. Other treatments such as ozone and filtration increase water clarity and, therefore, UV efficacy.

In summary, water treatment is one part of an overall sanitation and nutrient management plan. Not all growers need to treat water if they have low risk for waterborne pathogens, algae, biofilm, or chemical contaminants. Overall system design, installation and operating costs, and the technology's residual effect are among factors to consider in your choice of technology.

Filtration

Many automated irrigation systems require water filtration for optimum performance to prevent plugging. Filtration is easier and less expensive than repeatedly cleaning or replacing nozzles and emitters. The efficacy of sanitizing chemicals for controlling pathogens, biofilm, and algae is also greatly increased with the removal of organic load from the water, including peat particles, plant parts, and sediment.

You should obtain advice from an irrigation expert before installing a greenhouse-wide filtration system because one size does not fit all. A greenhouse operation using municipal water and one-directional irrigation may need only 200 mesh (75 micron) filtration to operate efficiently. A second location, using pond

Table 1-4. Comparison of Installation and Operating Cost and Residual Activity of Several Water Treatment Technologies

ACTIVE INGREDIENT	EXAMPLE BRAND NAMES	INSTALLATION COST	OPERATING COST	RESIDUAL
calcium hypochlorite (solid)	Accu-Tab (PPG)	● to ●●*	●	■
chlorine gas	Regal Gas (Chlorinators Incorporated)	● to ●●*	●	■
sodium hypochlorite (liquid)	Hanna Instruments	● to ●●*	●	■
chlorine dioxide one tank injection	Ultra-Shield (BASF), Selectrocide (Selective Micro Technologies)	●	●●●	■
chlorine dioxide 2 or 3 tank system	AquaPulse Systems, CH_2O	●	●	■
copper ionization	Aqua-Hort, Superior Aqua	●●●	●	■
hydrogen dioxide/ activated peroxygen	SaniDate 12.0, ZeroTol (BioSafe Systems), Xeroton-3 (Phyton Corporation)	●	●● to ●●●	■
hypochlorous acid	Oxcide (Chem Fresh)	●	●●	■
ozone	Dramm, Pure-O-Tech, TrueLeaf Technologies	●●●	●	■
reverse osmosis	Various	●●●	●	
slow-sand filtration or constructed wetlands	Consult an environmental engineer	●●●	●	
UV light	Priva, Pure-O-Tech	●●●	●	
Notes	Other brand names are available in some categories. No endorsement or criticism by the authors is implied.	Based on a 5-acre greenhouse: ● < $2000; ●● $2001 to $10,000; ●●● >$10,000; *In-line monitoring and control increases installation cost but can reduce operating cost.	Cost to treat 1,000 gal. at continuous label rate: ● < $0.25; ●● $0.26 to $1; ●●● >$1	■ Residual of active ingredient, byproducts, or ORP

Table 1-5 Filtration Options for Greenhouses and Nurseries*

WHAT TO FILTER OUT		SCREEN/MESH FILTRATION		MEDIA FILTRATION			MEMBRANE FILTRATION			
		COARSE 4–50 mesh (5,000–300 micron)	FINE 50+ mesh (<300 micron)	SAND	SLOW SAND/ BIO-FILTER	PAPER/ FABRIC 5–50 micron	MICRO 1–0.1 micron	ULTRA 0.1–0.01 micron	NANO 0.01–0.001 micron	REVERSE OSMOSIS (RO) <0.001 micron
INORGANIC PARTICLE	Debris	++	++	++ Small load only		++				
	Sand	+	++	++ Small load only	Not intended for large amounts of solids.	++	Particles other than intended for a specific membrane will shorten its life span or destroy it.			
	Silt	–	++	+ Small load only	Excess solids will clog bio-active zone	++	Proper pretreatment of water is essential.			
ORGANIC PARTICLE	Debris	++	++	++ Small load only		++				
	Soil particles	+	++ Small load only	++ Small load only		++				
	Algae, biofilm	–	++ Small load only	++ Small load only	+ In small amounts	++	Will clog membranes			
	Pathogens	–	–	Minor effect	++	Minor effect	+ Except viruses	++	++	++
DISSOLVED INORGANICS	Salts, iron	–	–	–	–	–	–	–	–	++
	$CaCO_3$ (hard water)	–	–	–	–	–	–	–	++	++
DISSOLVED ORGANICS	Humic acids	–	–	–	–	–	–	–	++	++
	Pesticides herbicides	–	–	–	–	–	–	–	++	++
NOTES		Mainly prefiltration; drippers, nozzles need 120+ mesh.	Substantial dirt loads require backflush systems.	Backflush is standard; do not use for heavy dirt loads.	Use for low flow only. Prefiltration is needed for heavy dirt loads.	These can handle heavy dirt loads in one step.	These require lower pressure than RO. Membranes are tailored to specific applications. Rejection rates (discharged portion of the feed water carrying concentrated waste) are generally smaller than RO.			RO removes almost everything. It is typically back-blended with supply water.

Dimensions: 1 micron = 1,000 nm = 1/1,000 mm = 0.00004"
Legend for efficacy: ++ indicates good, + fair, and – not effective filtration.

*Permission for use granted by Ratus Fischer, Ph.D., fischerecoworks.com water that receives runoff from the operation and also uses flood floors, may need several stages of filtration to achieve better than 600 mesh (25 micron) filtration. Even at this level of filtration, *Pythium* zoospores will not be trapped. Table 1-5 summarizes some filtration options.

Testing for Plant Pathogens and Sanitizing Agents in Water

Water testing is the only way to know whether a plant disease outbreak is being caused by irrigation water, and it helps pinpoint where in your irrigation system the contamination is occurring. Testing also ensures that your water treatment system is working correctly.

Onsite tests of active ingredients

If you have a water treatment system in place, when was the last time you checked if it is working correctly? All water treatment systems require maintenance and checking—they are not equipment you can install and forget. For example, copper systems rely on controlled electrolysis to form soluble copper ions, often resulting in corroded connections and plates or rods that require maintenance. Problems with copper and other systems are sometimes not identified until a rise in algae level is observed—not the best monitoring method!

Onsite tests allow you to check the concentration of control agents such as copper, chlorine, hydrogen dioxide, and oxidation-reduction potential (ORP). These tests include manual colorimetric tests, handheld meters, or in-line controls for continual dosage systems. Onsite tests have the advantage of being low cost and rapid, allowing repeat measurements and the tracking of trends over time just as you can track pH or EC.

Powerful oxidizers, such as chlorine, are hazardous to worker safety if the injection system malfunctions. If the manufacturer provides or recommends a test kit, make sure the kit is used regularly (at least once a month) and has not exceeded its shelf life. Because colorimetric tests (and especially test strips) are inherently subjective, train one person to do the tests. With in-line control systems, train staff to check that sensors are calibrated and ensure they have a technical and commonsense understanding of the system.

Microbes and pathogens

Laboratory tests of biological water quality are available from water treatment companies, university plant pathology laboratories, and private microbiology laboratories. Onsite detection methods for biological water quality (i.e., aerobic bacteria, algae, and pathogen detection) are being developed but are currently in the research stage. Information is available at www.watereducationalliance.org on new monitoring techniques.

Certain water tests are non-specific and measure, for example, the microbial density (i.e., colony-forming units per milliliter or cfu/mL) of total aerobic bacteria. Most bacteria and fungi in water samples are likely to be beneficial or benign, rather than pathogenic. Greenhouses are not sterile environments, and attempting to completely sterilize all water and surfaces, as well as the growing substrate, would require high doses of chemicals that are likely to be phytotoxic to crops. Removing beneficial organisms can also increase the likelihood that pathogenic organisms will cause disease. High colony counts of non-specific microorganisms present in a sample do not indicate a pathogen problem but rather represent total microbial density. There is currently little standardization across private horticulture laboratories on how samples are processed, and control thresholds have not yet been established. A threshold of 10,000 cfu/mL of aerobic bacteria is generally recommended to reduce clogging of drip lines and micro-emitters. Total colony counts of bacteria sampled both before and after the point of treatment in the irrigation line can be an indicator of the general efficacy (or failure) of a treatment system. Sampling at points along an irrigation system can also identify where conditions favor the development of biofilm.

Plant pathology laboratories can use selective agar media to identify the presence or absence of pathogens in irrigation water, typically to the genus level (for example, the University of Massachusetts Diagnostic Lab will test for *Pythium*, *Phytophthora*, *Rhizoctonia*, and *Fusarium*, tel. 413-545-3209, http://www.umassextension.org/agriculture/index.php/services/plant-problem-diagnostics). Identification of pathogens to the species level requires mycologists who specialize in particular genera. Some labs also have access to more specific DNA analysis techniques (for example, the University of Guelph Laboratory Services, tel: 877-863-4235 or 519-767-6299, http://www.guelphlabservices.com/). Irrigation water may be contaminated with many species of *Pythium* and *Phytophthora*, although not all these species are pathogenic to floriculture crops. If a laboratory reports a positive presence of *Pythium*, *Phytophthora*, *Fusarium*, or *Rhizoctonia* at the genus level, it is prudent to assume that a pathogen is present and treat accordingly.

Sampling water for microbial analysis

For consistent results it is best to have the same person collect and ship water samples. Typical sample sites include:

• Water source: Sample from the point closest to the source.

• Pre- and post-treatment: Sample from a short distance before and after water treatment.

• Furthest outlet: A sample from the farthest point from the source, such as a watering hose or mist-head typically used for irrigation.

The laboratory will provide guidelines on the water sample volume required and may provide water and shipping containers. Generally 16 oz. (500 ml) is enough. It should be collected in a sterile container. An unopened, noncarbonated distilled water bottle from a supermarket is generally adequate, although spring water can sometimes contain organisms. Run the irrigation line for three minutes or longer, if needed, to completely flush the line. Remove the bottle cap at the site, and do not contaminate the sample with your hands. Empty and refill the bottle three times with water from the source being tested, and then fill sample bottles no more than three-quarters full to allow some gas exchange.

Send water samples within twelve hours. Refrigerate if not shipping immediately and insulate the container, if possible. With microbial analysis, it is important to inform the laboratory before sending samples. The integrity of the sample is more likely to be maintained if the sample is shipped overnight and early in the week, so the sample does not sit over a weekend.

2

Ron Adams and William Fonteno

Media

Dr. William Fonteno and I have combined more than seventy-five years of commercial growing media experience into this chapter. Our experiences include studying, researching, testing, educating, and manufacturing professional growing mixes. It is from these experiences that we have drawn to discuss various aspects of commercial growing mixes. Everyday we are asked questions regarding horticultural media. Basic questions often involve physical and chemical properties. We present a few of these questions in this chapter. In the following discussion, issues range from the physical and chemical properties of media to how media interact with plant pathogens and economic issues. As you read, always remember that you have control over changing a growing medium's physical properties before filling the container but seldom can you make any beneficial changes once pots are filled and the plants are growing. Most often it is the grower's experience and application of that experience that make for an economical crop, not the type or brand of medium chosen. We will address the following questions:

General
Which mix is best suited for me?
If growing mix is not dirt (mineral soil), then what is it?
Is mineral soil still used in growing media?
What are the most commonly used components in mixes?
What about using wood in substrates? Are they reliable?
How do I make an organic mix?

Physical properties
How does texture and particle size affect the media?
Does the initial media moisture level affect porosity in a container?
How can I test the initial moisture level in my mix? How much should I have?
How can I add moisture to my bagged mix?
Does the manufacturer add wetting agents to the mix?
How does moisture level affect shrinkage in the container after filling?
How much should I compact the medium when filling the containers?
What happens when a mix is too compacted?
Is porosity one of the biggest factors of a mix?

Chemical properties
What are some other critical aspects of media?
If I use constant liquid feed, do I need nutrients in my mix?
What is buffering capacity?

Media biology
What should I do if I see mold growing in my mix?
Do bark mixes suppress disease?
Are mixes sterile?

Handling issues
How should I store my mix?
How long should I store my mix?
Can I use mix left over from last season?
Can I prefill containers prior to the start of the season?
What other things should I be concerned about when storing mix?

Postproduction
Are there mixes that improve the postharvest keeping quality of a plant?

General

Which mix is best suited for me?

There are many well-formulated commercial mixes on the market. Choose a supplier that will provide a consistent quality for the intended purpose. Use the right mix for the right crop and container. For example, there are mixes specifically developed for plug production. They are designed for small cells and not for use in large containers. If you are growing geraniums, you might need to use a mix with a higher pH. If you are mixing your own medium, then you will need to choose the component mixture suited to your location and growing needs.

Should you mix your own or buy a commercial mix? In order to make this decision, you should consider the cost of components, equipment, labor, and quality control. In general, operations with less than 50,000 ft.^{2} (4,800 m^2) of growing space benefit from purchasing commercial mixes. Operations over 100,000 ft.^{2} (9,600 m^2) can afford the cost of mixing their own. However, under most circumstances, it is beneficial to use a commercially prepared mix.

Generally, it is better to select a company rather than a particular mix. Most reputable companies will have a mix that will suit your needs. Over the last several years, many companies have improved their equipment and now have the ability to easily customize mixes for individual growers. The three things to look for when picking a company are: service, quality control, and technical support. Can the company deliver on time? Can it help you when you need additional mix in a hurry? Are its terms reasonable? Does the company have a good quality-control program? Does it test its mixes often? Lastly, if you have a problem, is it willing to help? Can it help you with fertility or watering problems that are not a result of the mix? Good companies want you to be successful.

Technically, a good greenhouse mix can be used anywhere to grow just about any crop. However, there do tend to be some regional differences. Northern growers tend to use more peat-based mixes, while growers in the South tend to use mixes that contain pine bark. This is based more on the geographic proximity to sources—peat bogs and pine saw mills—than on mix performance. There are also mixes made specifically for certain crops, such as geranium mix, pansy/vinca mix, and plug mix. Growers can choose which mix best suits their needs. Some companies also offer the option of buying in prefilled pots or flats.

If growing mix is not dirt (mineral soil), then what is it?

Most growing media in North America are composed of lightweight organic and mineral components that are typically 10–20% solids by volume. That means that each container has 80–90% pore space. (See figure 2-1.) This is the main difference between field soil and growing media. Experienced growers learn to "manage" these spaces by supplying proper amounts of water and nutrients.

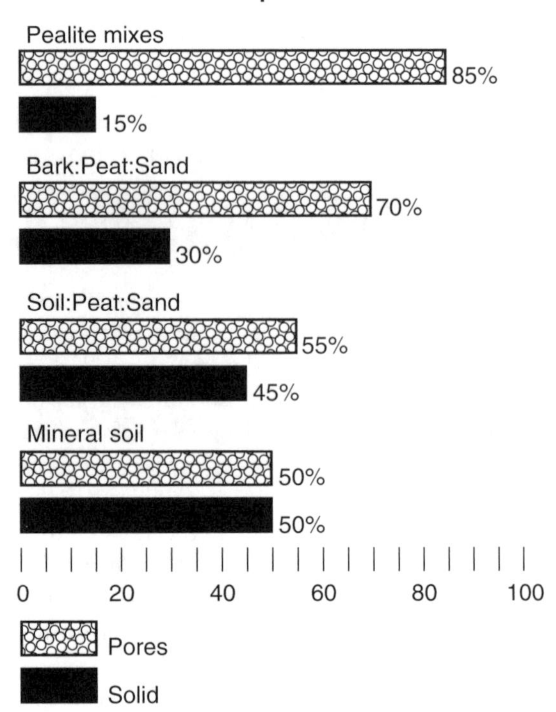

Figure 2-1. Total pore space and solid percentages for various media

Blends of organic and mineral substances provide plant support, air, and a water/nutrient reservoir to make available nutrients and gases to the plant as it develops a root system for exchanging the solutes. We know that certain plants can extract specific nutrients out of the medium solution, which causes growers to supply very specific nutrients. A good example is the zonal geranium (*Pelargonium*), which will readily take calcium out of the medium solution and replace it with hydrogen ions driving the growing mix pH down. Once calcium reaches a critical level in the medium, there is an increased uptake of iron and manganese in the lower leaves, and the new growth is limited in expansion. Severe calcium deficiency can show as a leaf edge burn or distortion in the new growth. Roots are often underdeveloped with a lack of calcium.

Is mineral soil still used in growing media?

Yes, it can be used but is not commonly available from commercial blenders because of the uniformity of minerals from different locations. Mineral soil needs to be sterilized to remove potential plant pathogens; it is also harder to find mineral soil that does not contain herbicides used on agricultural crops; and its bulk density is quite high, making it heavy to ship any distance. There are other components that are used with greater chemical consistency, are more economical, and are essentially sterile.

Mineral soil is ten to fifty times heavier than other components used in growing media. This excess weight can significantly increase shipping and handling costs because the pots are much heavier. After consistency, the next biggest problem is drainage. The fine texture of mineral soil makes it drain very poorly in containers that are less than 12" (30 cm) deep. This causes the growing medium to drain more slowly and have less air space than other components.

What are the most commonly used components in mixes?

Peat moss, bark, coir, perlite, vermiculite, rock wool, expanded polystyrene, formaldehyde foams, mineral soil, and muck peat soils are the most commonly used components in mixes. These components can be blended together or used alone to provide a medium acceptable for plant growth and development. Oftentimes, it's the grower's practices that affect how these components can be blended for an acceptable growing medium. Most mixes are either peat or pine bark based. Pine bark–based mixes decompose more slowly and, therefore, shrink less over time, making them well suited for large planters or for crops lasting longer than five months.

Most media have 30–60% organic matter, such as peat moss, pine bark, or coir, and the balance in vermiculite, perlite, or polystyrene beads (figure 2-2). (Polystyrene is inexpensive, but it will float and work its way out of a medium, where wind can carry it all over the greenhouse and even to adjacent properties.) The blend depends on availability, the cost of components, and the needs of the grower. The type or amount of a particular component is not as important to the plant as what the final mixture of components does for the root system. More important is the quality of the components and the attention to detail in blending them.

Growers who mix their own media need to be especially aware of changes in raw ingredients from lot to lot. Careful inspection is a must to prevent variability in mix performance from batch to batch. For peat, check pH, soluble salts, and wettability. The presence of weed seeds varies from one harvest year to another. For pine bark, check pH, soluble salts, and temperature. Hot bark means that it is not mature compost. Composted pine bark should not be over 14°F (8°C) higher than ambient temperatures. If the pH is low, the salts are up, or the temperature is high, leach the bark, pile it no higher than 12' (3.6 m) high, wait for about two weeks, and test it again.

What about using wood in substrates? Are they reliable?

In recent years, new processing technologies have made ground wood available for mixes. Whole trees (WT), with or without limbs and leaves, are chipped or shredded and used in nursery and, more recently, greenhouse mixes. Wood sawdust, a byproduct of the lumber industry, has been around for many years and generally shunned by growers because it broke down easily and tied up nitrogen in the process. However, current wood sources and processing technologies have minimized many of the previous concerns attributed to sawdust. The most widely tested tree species is loblolly pine (*Pinus taeda*), although other pines, firs, etc. are currently under examination by researchers.

WT wood today has much larger particle size than sawdust and is remarkably stable. Breakdown is generally no more than with fresh bark. There can be some nitrogen tie up, but it can be easily compensated in the fertility plan. The economics are about the same as pine bark. Growers can substitute for bark at about the same cost. However, with a finer grind, the wood can be used as a peat substitute. This should provide a cost savings for the grower.

However, the real benefit may be as an alternative source to traditional components. Bark supplies have declined over 50% in the last ten years. Now with the uncertainty of supply created by the national program to increase biofuel production, more of the existing supplies may be diverted to fuel. Growers need a stable, sustainable component like peat and bark.

The biggest fear of growers is nitrogen tie up and reduced plant growth. Research at Virginia Tech shows a need for three to four pounds of slow-release fertilizer for woody plants and a 100 ppm–nitrogen increase in liquid feed for herbaceous plants over traditional peat or bark. But these values are necessary when using the ground wood at 100% of the mix. At

10–30%, there is very little increased nitrogen needed during crop production.

When ground and passed through a ¼- to ⅝-in. screen, the wood can be used as a bark substitute. For supplementing peat, the screen size must be about ¼- to $^{3}/_{16}$-in. Commonly, growers can use volumes of 10–30% wood with almost no noticeable change in growth or fertility. Initially, growers will substitute WT for a portion of their bark or peat fraction to be on the safe side. They will then adjust the amount depending on growth results. Large national mix manufacturers are now providing WT wood in some of their blends with good success.

At the time of this printing, ground wood is being used cautiously in the greenhouse industry. Within three to five years, we expect this to be a standard choice for growers.

How do I make an organic mix?

There is an increased demand for organic mixes for vegetable and herb production. Several commercial companies are currently supplying organic mixes that can be used for organic production. If you want to make your own, then you can use some of the same base ingredients as have been used for years in peatlite mixes. They are Canadian sphagnum peat moss, vermiculite, perlite, and processed bark fines. What needs to be changed are the fertilizers and wetting agent. Chemical wetting agents are not allowed in organic mixes. Fertilizer sources have to be organically derived or mineral derived. Bone meal and blood meal can be incorporated instead of a soluble fertilizer. Neil Mattson of Cornell University and Stephanie Burnett from the University of Maine have made the following suggestion for a basic organic mix.

Figure 2-2. Cornell Mix (Organic)	
Peat moss	0.6 yd.^{3} (0.46 m^3)
Course vermiculite	0.2 yd.^{3} (0.15 m^3)
Perlite	0.2 yd.^{3} (0.15 m^3)
Bone meal	10 lbs. (4.6 kg)
Dolomitic limestone	5 lbs. (2.3 kg)
Blood meal	5 lbs. (2.3 kg)

There are variations of this basic mix where perlite can be substituted for more of the vermiculite. Over the years bark, coir, rice hulls, leaf compost, and vermicompost have also been substituted for some of the peat moss, vermiculite, and perlite. It is more difficult to add an organic product, because it would have to be separately certified as organic prior to incorporation.

Since there is no wetting agent incorporated, adding additional water into the mix at blending and prior to potting is a must for uniform watering.

Physical Properties

Does texture affect the media?

Yes, texture is important. Texture often refers to the particle size and the air porosity level. Commonly its relation to container size is what determines what the texture should be. Container size varies from small plug cells to large nursery containers. Therefore, you want to vary the particle size. You could not uniformly pack course particles in a plug tray, nor would you want to fill a large container with finely textured particles. Texture should vary according to the container to maximize the air-porosity level.

Texture is important for the individual components and will affect the mix. For example, if your current supply of peat moss or perlite is more dusty and finer in texture than last year's supply, the current mix will drain more slowly and have less air space than last year's mix. If growers buy their components to mix their own media, they must be wary of bargains. Just as in most things, you get what you pay for. Good-quality components are more consistent from batch to batch. They are also more expensive (but not much).

Texture can affect drainage, as shown in table 2-2. However, the size of the container can have a greater affect than texture. Most of our containers are very short: 3–6" (8–15 cm) in pots, 1–3" (2.5–8 cm) in flats, and less than 1" (2.5 cm) for plugs. All mixes drain by the force of gravity. However, gravity is "sensed" internally in the mix. In other words, the shorter the container, the less the mix senses gravity. Consequently, the same mix will drain differently depending on the container. The less it drains, the less air space is in the mix. For example, table 2-2 shows that a common peat vermiculite mix will have air space of 20% by volume after drainage. The same mix will have only 13% in a 4" (10 cm) pot. In a 48-cell bedding plant flat, the drainage is even less and the air space is only 8%. However, a 288-plug cell, which is about 1" (2.5 cm) tall, has only 3% air space. The less air space, the slower roots grow and the more susceptible the roots are to root rot diseases.

Table 2-1. Component Specifications

	TOTAL POROSITY (% VOLUME)		TOTAL POROSITY (% VOLUME)		
Component	**Mid-range**	**Maximum/ Minimum**	**Mid-range**	**Maximum/ Minimum**	**Bulk density**
Compost (yard waste)	83–87	88/80	7–18	44/3	0.23–0.28
Peat moss (Canadian sphagnum)	89–94	97/87	20–12	27/6	0.06–0.10
Perlite	68	82/65	28–32	46/14	0.15–0.17
Pine bark (0.5"/1 cm)	75–80	83/73	19–24	41/6	0.20
Pink bark (fines)	75	78/73	16	20/10	0.25
Rock wool (hydrophilic)	90	94/87	20	25/19	0.20
Sand	38	38/36	3	4/2	1.40
Soil	50		3		1.2
Vermiculite	78–80	85/74	6–10	26/4	0.16–0.18
4 bark:1 sand	76		10		0.46
3 bark:1 sand:1 peat	71		5		0.50
3 soil:1 bark	54		4		1.08
3 bark:1 peat	81		16		0.17
Peat:vermiculite	88	90/82	9	13/6	0.14
Peat:perlite	78	82/76	16	27/14	0.12
Peat:rock wool	88	92/74	13–17	28/9	0.16
Commercial greenhouse mixes	87	90/81	9	19/9	0.14

Table 2-2. Container Size Effect on Air Porosity

	VALUES FOR A MIX AFTER IT IS WATERED AND ALLOWED TO DRAIN				
	6" (15 cm)	**4" (10 cm)**	**48 cell**	**288-plug tray**	**648-plug tray**
Air	20%	13%	8%	3%	0.5%
Water	67%	74%	79%	84%	86.5%
Solid	13%	13%	13%	13%	13%

Does the initial media moisture level affect porosity in a container?

Moisture is critical. Most manufacturers like to have about 50% moisture in a medium at the time of shipping. Baled peat moss mixtures will normally have less than 50% moisture at shipping. The moisture level at the time of shipping is lower than what the mix needs to be when filling the trays because shipping water is expensive and unnecessary. Growers can add the needed moisture prior to filling the containers.

Dry media (less than 50%) will shrink when initially watered in a pot. However, if a grower increases the moisture content in the mix before it is put in the pot, it will not shrink and will actually have higher aeration. The "trick" is to add the moisture to the mix the night before potting, let the mix absorb the water under cover so it cannot dry out, and then fluff it right before use. This effect of less shrinkage and more aeration is even greater in smaller containers, where drainage is less. This method is not always possible with the super bales and bulk mixes that are used today; but if you are experiencing inconsistencies of water in either cells or pots, then the lack of initial adequate moisture in the mix might be the reason for the inconsistency.

How can I test the initial moisture level in my mix? How much should I have?

There are various methods that can be used to test the initial moisture level, but a practical one is the squeeze test. The squeeze test consists of taking a handful of mix and squeezing it hard. If water runs out between your fingers and the mix makes a hard ball, then there is too much moisture. If the mix ball easily falls apart after squeezing, then there is too little moisture. At the right moisture level, the mix ball should remain intact but easily separate when poked with a finger.

At 50% moisture, there is about 20 gal. of water/yd.^{3} of mix, or about 2.2 gal. in a 3 ft.^{3} (104 l/m^3) bag (table 2-3.) As a rule of thumb, growers can add 1 gal. of water/3 ft.^{3} (47 l/m^3) bag without worry. Peat-based mixes can get another gallon (two total). It is best if the mix is left to absorb the water overnight before use. Even more can be added if the mix starts out with less than 50% moisture.

Table 2-3. Water Content Found in a Cubic Yard of Mix at Specific Moisture Contents*

% MOISTURE	WATER/YD.^{3} (GAL.)
0	0
10	2
20	5
33	10
50	20 (pots)
60	30
67	40 (cells)
72	50
75	60

* Medium is a peat:vermiculite mix.

How can I add moisture to my bagged mix?

It is best to add water into the bag a day before using. Very dry mixes can absorb as much as 3–5 gal. (11–19 l) of water and uniformly moisten all particles through capillary action overnight. If water is added just prior to filling the trays, then there will be dry pockets in the mix. Also mix that has just been wet has not had time to fully absorb the water and can appear "tacky" and stick to the equipment. However, fluffing the mix left to equilibrate overnight will result in a mix that is fluffier, softer, and actually will fill more pots than a drier mix.

Does the manufacturer add wetting agents to the mix?

Yes. The mode of action of wetting agents is the same as the soap you use to wash dishes. It makes water "wetter" by decreasing its surface tension. Unfortunately, this same action can be detrimental to plants. In fact, out of the thousands of surfactants used in other industries, only a handful are safe enough to be used in mixes. Once placed in mixes, these surfactants will lose their effectiveness over time. The older the mix, the harder it will be to wet. After about eight months, check the mix for easy wetting. If it is hard to wet, you can add a liquid wetting agent drench to revive the wetting properties of the mix.

Wetting agents are typically added to aid the medium's ability to absorb water by reducing surface tension of the water/particles. The wetting agent does not change the physical state of dry particles, as moisture.

Dry mix can shrink when hydrated, decreasing the air porosity, so it is better to have the right amount of moisture in the medium when the containers are filled.

How does moisture level affect shrinkage in the container after filling?

Wetting and drying out can decrease the pore space between the particles in production, causing shrinkage. Watering practices also affect the amount of compaction. Overhead watering with a breaker or sprinkler can compact the medium, decreasing porosity. Also, peat particles are elastic, and the swelling and contracting with moisture over time lowers porosity. Vermiculite particles can compact with watering, reducing pore space.

The greatest amount of shrinkage occurs when a grower waters in pots that contain a dry mix. The mix can shrink 20–30% in dry peatlite mixes. Some growers try to compensate for this shrinkage by pressing more dry mix into the container when filling the container. This can reduce shrinkage, but it will also increase compaction. To reduce shrinkage, put more water into the mix before potting.

How much should I compact the medium when filling the containers?

Most often, if the initial watering completely wets the mix, then the mix will settle to an acceptable level. Compacting the mix will decrease the porosity. Once the medium is compacted in the container, it cannot be changed without repotting the plant.

Compaction at the time of filling the containers is highly probable and can cause the greatest pot-to-pot variation. For example, a peatlite mix in a 1020 bedding plant flat (48-cell pack) can have 9% aeration when filled and brushed (table 2-4). If that same flat is tapped two or three times on the bench to "settle" the mix, the air space drops to 4% (over half). Just as important, the amount of unavailable water increases from 21–26%. This means that compaction decreases both air space and available water. If the flats are stacked on top of one another so that the one above nests into the cell below, the lower tray can have 2% air space and 30% unavailable water.

If the mix has at least 50% moisture in it, you should fill the container to overflowing and brush or strike the excess so it is level with the rim of the container. Do not stack filled containers so that they nest into one another. When planting, you can firm the mix around the young plant for support. No other compression or compaction is required. Once the pot is watered in, the pore size distribution is set and the air- and water-holding capacities are fixed in place. The closer you adhere to the filling, brushing, and watering-in plan, the more aeration you will have in that container.

What happens when a mix is too compacted?

A medium that is too compact will be slow to dry out, decreasing the ability to feed the plant correctly and increasing the potential for root rots such as *Pythium*. A too-compact medium also increases the potential for algae to develop on the medium surface. Algae growth attracts fungus gnats and shore flies that can infect the plant with *Pythium* or can feed directly on the plant.

If the mix is too compact, there will be less air and more water in the mix. This reduces young roots' ability to penetrate deep into the mix. Generally, if a mix has the proper aeration, you should see roots at the bottom of the container rather quickly after transplanting. This would be three to seven days for flats and seven to ten days for 6.5" (17 cm) pots.

Is porosity one of the biggest factors of a mix?

Yes. Having and maintaining porosity is one if the most critical aspects of plant production. You can

Table 2-4. How Compaction of Medium into Bedding Plant Flats (48-Cell) Affects Air Space, Unavailable Water, and Available Water*

LEVEL OF COMPACTION	AIR SPACE (PERCENTAGE)	UNAVAILABLE WATER (PERCENTAGE)	AVAILABLE WATER (PERCENTAGE)
Light	9	21	58
Medium	4	26	56
Heavy	2	30	52

* Medium is a peat:vermiculite mix.

always adjust your irrigation to increase the amount of water needed, but you cannot increase the amount of air present in the mix. Air porosity is fixed at the time of potting and only decreases over time. Fortunately, good root growth can maintain or even slightly increase aeration. A healthy root system maintains porosity and aids in water penetration and wettability. At the time of transplant, the physical properties of the mix "dominate" the young roots, dictating how fast and far they can grow. In three to five weeks, the root system should be sufficiently robust to dominate the container. High porosity causes rapid root growth and allows the root system to take over the container quickly. Poor aeration produces slow root growth and can delay or prevent the roots from complete exploration of the container.

Chemical Properties

What are some other critical aspects of media?

Starter charge

Most manufacturers try to balance the initial nutrition of the mix with a starter charge so the plants have what they need to start. With no starter charge, a mix may have too low of a nutrition level that may be hard to balance and maintain, because water quality can dramatically impact the nutritional balance soon after planting. Most commercial growing mixes have a nutrient charge that is balanced for starting plant growth. Cornell Mix A, which became the basis of soilless media, had a nutrient charge comprised of the following ingredients per cubic yard: 1 lb. 8 oz. calcium or potassium nitrate; 1 lb. super phosphate; 5 lbs. limestone; 2 oz. FTE 503 (fritted trace elements); and wetting agent.

Today's commercial mixes are adaptations from the basic Cornell Mix A. Each manufacturer has its own adaptation of Cornell's peatlite nutrient incorporation. It can be as simple as a commercial soluble-fertilizer formulation such as 20-10-20 peatlite, Daniels Liquid fertilizer, or other formulations developed by the manufacturer or soluble-fertilizer producers.

Whatever the starter charge origins, the concentration (EC) varies with the crop. For seedlings, an EC reading of 0.75–1.2 mS/cm (saturated media extract) should be used. For most crops, a standard charge is 1.5–2.0. For a few special crops, a high starter charge, 2.0–3.49, may be needed.

Note that the nature of a starter charge is generally to provide a flush of nutrients quickly and then return to a more normal level. Typically nutrient charges last three to four irrigations and are then gone, so it is best to test the medium two weeks after planting to see the initial impact of the water-nutrient balance.

Alkalinity and pH

Both are critical. Low-alkalinity water will dissolve calcium and magnesium, lowering the medium solution pH, while a high calcium-carbonate alkalinity water will increase media solution pH, potentially limiting minor element uptake.

Light level, mix temperature

These two environmental factors also play a role in the plant's ability to establish in a medium. Start the fertilizer program after the plant begins to root in after planting. Certain plants, such as rooted cuttings, can be fed at planting since many cuttings (e.g., chrysanthemums) have not had a lot of feeding since the cuttings were taken from the stock plants.

If I use constant liquid feed, do I need nutrients in my mix?

That is often a personal choice. Remember, though, that when you start a feeding program without nutrients in the medium, you will have to provide a complete nutrient package, including minor elements. It is best to have a balanced mix to begin with and then supplement major elements once the plants have started to grow.

Almost all crops that you start by transplanting young plants from seed or cuttings will benefit from a starter charge. Heavy feeders, such as poinsettias and pot mums, do better when a starter charge is used. Light feeders, such as New Guinea impatiens, also do well with a starter charge in the mix. Plug growers may or may not use starter charges, depending on the specific crop. In general, limestone, a wetting agent, and a starter charge are commonly used in commercial mixes.

What is buffering capacity?

This is the medium's ability to minimize chemical changes from applied solutions or the plant's effect on the medium. You can grow in a medium that only provides support and air exchange, such as Oasis foam. When using such a medium, you have to satisfy all of the other requirements for the plant. If a high clay-mineral soil is used, the plant can have difficulty exchanging nutrients from the medium solution.

Growing media with a high buffering capacity require less fertilizer and seldom have micronutrient problems. Media with a low buffering capacity, such as sand, cannot hold these nutrients and require frequent fertilization. Also, lower buffering capacity means the mix is more susceptible to changes in pH and other nutrient problems.

Media Biology

What should I do if I see a mold growing in my mix?

Often a nonpathogenic mold will develop in the mix. That is associated with being stored in a building without temperature control, where the temperatures will vary widely during the day. It can also develop in the area between the bag and the mix when direct sunlight hits the bag. This mold often goes away after planting. Sometimes mushrooms will develop as a fruiting body of the mold.

The mold growth is almost always nonpathogenic to your crop. However, one mold can harm your crop by preventing water from penetrating into the mix. This fungus is found in pine-bark storage piles and has a gray threadlike structure (mycelium) that repels water. It grows quickly and causes the media to remain dry, even after repeated irrigations. There are several mold genera that are found in bark. The most problematic is an *Ostracoderma* slime mold. Experts believe that several species may be involved.

The problem is more common in outdoor nursery containers and much less common in greenhouses. It is rarely found in commercial mixes. Growers who mix their own media with a majority of pine bark as a base may see this problem. While there is no scientific evidence to prove it, pine-bark processors have reported that a drench of ZeroTol (a hydrogen dioxide product) has proven effective to reduce the effects of the mold. Often the mold can be physically broken up in a pot-filling machine before potting.

Do bark mixes suppress disease?

A bark mix that has been adequately composted will suppress some root and crown rots. During composting, beneficial microorganisms colonize on the bark. The beneficial microorganisms compete with the disease organisms. Once the beneficials are in place, the pathogens have no place to live. There are several factors involved in this process, and only the manufacturer can tell you if its bark processing generates the suppression.

Aging and processing does not always generate the beneficial microorganisms. Just because bark goes through a heating process does not mean that it will suppress disease in the container. The composting process causes rapid heating, which destroys most fungal and bacterial organisms. As the compost pile cools, new organisms (the vast majority being beneficial to plants) rush to repopulate the bark. These organisms need air, water, and favorable temperatures to thrive, just like other plants. If the bark dries out, the organisms will die. If the bark pile composts (heats up) unevenly, the beneficial repopulation will be spotty. The bark you use may well have disease-suppressive properties, but you cannot tell by looking at or feeling it. Most bark processors do not test for biological activity, so suppression can be variable. Remember, if you use a fungicide drench, you can affect the beneficial organisms as well.

Are mixes sterile?

Yes and no. Most commercial mixes are not sterilized but are comprised of components that are essentially sterile by their processing or nature. For instance, perlite and vermiculite are heated during their expansion at temperatures of 1,000°F (538°C) or higher. Sphagnum peat moss develops under very acidic conditions, where plant pathogens normally do not develop. Composted bark will attain temperatures of 140°F (60°C), minimizing diseases and weed seeds.

Most mixes are not pasteurized. However, because of the properties listed above, mixes do not usually come with pathogens. If you mix your own media, you have a greater chance of introducing pathogens. For example, any mineral soil must be pasteurized before use. Storing bulk mix on the ground can also introduce pathogens. You can reduce disease possibilities by disinfecting mixing and filling equipment prior to use.

Handling Issues

How should I store the mix?

It is best to store the mix inside, protected from the elements (i.e., heat, cold, rain, and direct sun). If the mix is stored outside, it should be covered with black plastic and the bags should not be in direct contact with the ground.

Most commercial mixes are available stacked and shrink-wrapped on pallets. Many have black plastic "caps" that reduce light penetration. This provides the best protection from the elements. Indoor storage with temperature control is always best. Mix exposed to high temperatures can release nutrients and dry out quickly. Mix exposed to heavy rains can gain excess moisture. High light can cause the plastic bags to deteriorate and split when handled. Mix exposed to cold temperatures can freeze. Frozen bags should be brought into a warm area and allowed to thaw before use, which could take several days, depending on the severity of the freeze. A cold mix combined with cold water (less than 55°F [13°C]) can retard root growth. This combination causes young plants to start slowly, lengthens the time for proper root penetration, and can increase cropping time. If in doubt, measure the mix temperature with a soil thermometer from your horticultural supply dealer.

How long should I store the mix?

Newly mixed medium is designed to be used right away. The longer you keep it, the more it changes. Keep up with the age of your unused mix. Be sure to use the oldest mix first. Many times a lone pallet of mix stays in the back of the holding area, and a new supply of mix gets placed in front of it. Rotate your old mix to the front.

It is best to use the mix within five to six months. Physical and chemical changes can occur with long-term storage. Physical changes include a loss or gain of moisture, depending on storage conditions. Chemical changes that can occur are an increase in pH as the moisture allows the lime to dissolve; a decrease in nitrogen levels, depending on biological activity; and a loss of activity in the wetting agent due to biological activity.

Can I use mix left over from last season?

If you use mix from last season, test it first. Does it wet up adequately? If not, you might need to add a wetting agent. Is the pH high or low? You may need to add additional calcium. Where is the nitrogen level? If it is too low, then crops planted into it may require additional feeding.

Most likely, last season's mix will be drier and harder to wet than a new season's mix. Check the moisture content and wettability. Corrections include adding water before potting and/or an initial irrigation with a liquid wetting-agent drench. Another option is blending with fresh mix. You should be able to blend 10% of the old mix with 90% of fresh mix with little problem. Depending on the condition of the old mix, you may increase it up to 25%. Make sure the blending is thorough to prevent pot-to-pot variations.

Can I prefill containers prior to the start of the season?

You can prefill containers; however, try to keep the time between prefilling and use to a minimum. If you use flats with solid bottoms, you can stack them directly on top of one another, as long as you have "plastic on top of plastic." Generally, a month or less is best. Remember, the mix has physical, chemical, and *biological* properties. The longer the mix sits in the containers, the more difficult it is to regenerate the biology when used.

This does not mean that you should not prefill your containers, only that you need to be aware of the pros and cons of prefilling. Make sure that the mix does not dry out too much before planting. Palletize and wrap the prefilled containers to minimize drying. To avoid compaction in the individual cells or containers, a structured sheet should also separate the layers of pots. Nesting can cause compaction of the mix, decreasing air porosity.

What other things should I be concerned about when storing mix?

Again, keep the mix from gaining or losing moisture. If additional moisture is added to the mix during storage, you might incur fungus gnat and shore fly populations. If algae are allowed to develop in the mix during storage, the algae will attract fungus gnats and shore flies. You could pick up diseases in storage depending on the basic sanitation of the area.

If you are mixing your own bark-based medium and storing it in bulk, remember to monitor temperatures in the pile to prevent overheating. Storing mix in bulk will cause internal temperatures to rise above 110°F (42°C) and excessive drying can occur. This heating and drying can stimulate the growth of the water-repellant fungi described above. Monitor and turn these piles regularly.

Postproduction

Are there mixes that improve the postharvest keeping quality of a plant?

Some commercial growing-mix producers have a labeled postharvest mix. It is comprised of a standard mix with water-absorbing (hydrophilic) polymers added into the mix. These polymers have been around since the early 1970s, getting their start in baby diapers and later being modified for horticultural use. Results from polymers are inconsistent. Generally, they are incorporated into the mix. They can be effective in postproduction situations but can also alter water and fertility practices in production.

Another proven way to improve the postharvest handling of plants is to drench the medium with a wetting agent seven to ten days prior to shipping the plants to retail. (This is in addition to the wetting agent that may come in the mix.) Adding the additional wetting agent allows the medium to fully hydrate, restoring the medium to its full water-holding potential. This can be of great benefit in rehydrating plants that become water stressed in a busy retail situation.

If you have been having problems at retail with the plants drying out inconsistently, you might want to consider these options. Make sure you are adding the polymer prior to planting. The drench with the wetting agent should be applied far enough ahead of shipping to allow the mix to be fully hydrated at the time of shipping.

Editors' note: For more information, check out the following North Carolina State University websites:

Horticultural Substrates Laboratory: www.substrateslab.com

Wood Substrates: www.ncsu.edu/project/woodsubstrates/index.html

Commercial Floriculture: www.floricultureinfo.com.

3

Brian E. Whipker, Todd J. Cavins, James L. Gibson, John M. Dole, Paul V. Nelson, and William Fonteno

Plant Nutrition

Nutrition is an area with abundant information and technology available today. Better fertilizers, refined analytical techniques, inexpensive diagnostic equipment, quick and easy access to information, and well-trained technical support personnel are available. Even so, many growers experience nutritional problems because so much of the focus of the information and the approach to nutrition has been on solving problems rather than preventing them. In today's marketplace, this is no longer acceptable. Nutritional problems can delay crops, reduce quality, and lower cost efficiency. The good news is that we can prevent the majority of the nutritional problems by monitoring two simple parameters—pH and electrical conductivity (EC). By reading and studying this chapter, you will learn how to prevent nutritional problems by doing just that. This chapter discusses the principles of pH and EC, how to select a fertilizer, common nutritional problems, and how to conduct a monitoring program.

Electrical Conductivity (EC)

Soluble salts are the total dissolved salts in the medium and are measured by EC. A conductivity meter measures the passage of electrical current through a solution. The higher the EC value, the greater the electric current moving through the solution. The EC of the medium provides insight to the nutrient status of the crop. Keep in mind that not all of the salts measured by an EC meter are fertilizer salts. An EC meter measures the sum of *all* salts in a solution but does not provide details about the type or amount of each salt present.

Be aware of substrate EC levels, because excess salts can accumulate when leaching during irrigation is insufficient, the amount of fertilizer applied is greater than the plant requires, and/or the irrigation water contains a high amount of soluble salts.

Excessively high EC values are associated with poor shoot and root growth. Symptoms often begin on the lower leaves as chlorosis and progress to necrotic leaf tips and margins. If the medium is allowed to dry, plants may wilt because of root-tip dieback, which further inhibits water and nutrient uptake. High EC has also been linked with the increased incidence of *Pythium* root rot.

In contrast, when EC values are too low, plant growth can be stunted or leaf discoloration can result from the lack of nutrients. Usually nitrogen is the most typical nutrient deficiency (lower leaf yellowing). However, lower leaf purpling (phosphorus deficiency), interveinal chlorosis of the lower leaves (magnesium deficiency), or lower leaf interveinal chlorosis and marginal necrosis (potassium deficiency) can also occur.

Factors affecting EC

Fertilizers

Potting mix EC values can be used to monitor the nutrient status of the crop. EC levels are more consistent with constant liquid feed (CLF) irrigation than with periodic fertilization. Mix EC levels with a periodic fertilization regime can vary due to the length of time after fertilization that the sample was taken and the number of clear irrigations applied between the fertilizer applications. The nutrient contribution of slow-release fertilizers can also be monitored with regular medium testing.

Fertilizers contribute to the EC of the medium. The most common ones are nitrate (NO_3), ammonium (NH_4), phosphate (PO_4), potassium (K), calcium (Ca), magnesium (Mg), sulfate (SO_4), and chloride (Cl). Organic materials also contribute to the EC when a portion changes from an insoluble to a soluble form. EC can also be used to monitor the accuracy of a fertilizer injector.

Irrigation water

The EC of the irrigation water contributes to the medium's EC. Elevated EC levels can be caused by high concentrations of calcium (Ca), magnesium (Mg), sulfate (SO_4), sodium (Na), chloride (Cl), or bicarbonate (HCO_3) in the irrigation water. Excess bicarbonates not only increase EC but can also increase the medium pH to detrimental levels (see chapter 2).

Electrical conductivity levels above 0.5 mS/cm can be detrimental during seed germination or when rooting cuttings. Established plants are usually more tolerant of elevated EC levels than younger plants.

Irrigation method

Different irrigation methods provide different amounts of leaching. Leaching prevents excess soluble salts from building up in the medium. However, if leaching is required on a routine basis, you may want to lower your fertility levels if the irrigation water is not the source of EC. Excessive leaching or excessive rainfall with outdoor production can lead to low EC problems.

Medium type

Some media have a fertilizer starter charge that increases the media EC. These mixes are generally used for growing established plants. The starter charge lasts for a short time, and its effectiveness and contribution to EC will decrease after several irrigations. The components in the mix can also influence the EC. Sphagnum peat has a low chemical content, thus, a relatively low EC. Composted pine bark usually has a slight EC charge and can provide manganese, copper, and zinc to the plant. If pine bark is not composted well, it can immobilize nitrogen. Coir usually has an elevated EC level (more than 0.5 mS/cm) compared to peat and may not be suitable for seed germination or rooting cuttings. EC levels of compost can be extremely variable; these are not recommended for floriculture crops. Test any new mix prior to use to ensure that it meets specifications.

Crop factors

When establishing an EC monitoring program, match the fertilizer rate with the nutrient demands of the crop. Two main parameters need to be considered: the crop's nutrient demands and the stage of development.

Nutrient demands

Crops vary in their fertility requirements for optimal growth. Most greenhouse crops grow best when the PourThru EC levels are 2.0–3.5 mS/cm, but some crops are sensitive to high EC and grow optimally at 1.0–2.6 mS/cm. Other crops require a high level of fertility, and EC levels of 2.6–4.6 mS/cm are optimal. The guidelines in table 3-1 can be used to establish EC target ranges for a crop.

Crop development stage

The nutrient demands of a crop vary by developmental stage. Plugs and rooted cuttings require lower fertility levels; nutritional needs increase as they become established. Actively growing plants have the highest nutrient demands, with demands decreasing as plants produce buds and flowers. In fact, flower longevity is increased when fertilizer rates are decreased or terminated at visible bud or just prior to flowering for most crops.

An EC monitoring plan should be adapted for each crop's nutrient demands and development stage. Use table 3-1 to establish target EC levels for actively growing plants using the PourThru extraction method. Table 3-2 contains examples of how to adjust target EC levels to account for the establishment and flowering stages of growth. Table 3-3 contains interpretative EC values for the 1:5, 1:2, SME, and PourThru extraction procedures, which allows comparisons among the methods and will help in establishing target values.

Selecting a Fertilizer

A number of fertilizers are available for floriculture production, including a few that have crop-specific names (e.g., Poinsettia Special or Geranium Special). No one fertilizer is appropriate for all situations. How do you select a fertilizer to use? A number of fertilization strategies will work. The primary goals are to provide adequate, but not excessive, levels of all the essential nutrients to the crop and to ensure that the fertility program maintains substrate pH and electrical conductivity (EC) levels within the acceptable range. Five factors need to be considered to achieve a successful fertility program: (1) Should you use an acidic or basic fertilizer? (2) How much phosphorous should you apply? (3) Does the fertilizer supply all the essential elements? (4) What is the potassium-calcium-magnesium ratio? (5) Are micronutrients supplied?

Acidic versus basic fertilizers

Knowing the optimal pH range for each crop will help determine if an acidic or basic fertilizer should be used. In general, a high proportion of nitrate nitrogen will cause the pH to increase, while a high percentage of ammoniacal nitrogen will lower the pH. Also, keep

Table 3-1. The Relative Nutrient Requirements of Actively Growing Greenhouse Crops, with EC Ranges for Both the SME and PourThru Methods

Use this classification system and the examples provided in table 3-3 for the PourThru method to determine the suggested target EC ranges for the entire crop production cycle. Assumes less than 30% of the nitrogen is from urea.

No additional fertilizer required		***Medium*** SME EC of 1.5–3.0 mS/cm PourThru EC of 2.0–3.5 mS/cm	
Crocus *Hippeastrum* *Narcissus*		*Allium* *Alstroemeria* *Bougainvillea* *Brassica* *Calendula* *Campanula* *Capsicum* *Centaurea* *Cleome* *Clerodendrum* *Consolida* *Crossandra* *Dahlia* *Dianthus* *Exacum* *Helianthus* *Hibiscus* *Hydrangea* *Ipomoea*	*Kalanchoe* *Lilium hybrida* *Lilium longiflorum* *Lobelia* *Lobularia* *Lycopersicon* (tomato) *Oxalis* *Pelargonium* (zonal) *Petunia* *Phlox* *Platycodon* *Portulaca* *Ranunculus* *Rosa* *Schlumbergera* *Senecio* *Verbena*
Light SME EC of 0.76–2.0 mS/cm PourThru EC of 1.0–2.6 mS/cm		***Heavy*** SME EC of 2.0–3.5 mS/cm PourThru EC of 2.6–4.6 mS/cm	
Aconitum *Ageratum* *Anemone* *Anigozanthos* *Antirrhinum* *Asclepias* *Aster* *Astilbe* *Begonia* *Caladium* *Calceolaria* *Celosia* *Cineraria* *Coleus* *Cosmos* Cuttings (during rooting) *Cyclamen*	*Freesia* *Gerbera* *Gloxinia* *Impatiens* *Orchids* *Pelargonium* (seed) *Plugs* *Primula* *Rhododendron* *Saintpaulia* *Salvia* *Sinningia* *Streptocarpus* *Tagetes* *Viola* *Zantedeschia* *Zinnia*	*Chrysanthemum* *Euphorbia* (poinsettia) *Petunia* (vegetative)	

Adapted from: Bunt, A. C. 1988. *Media and Mixes for Container-Grown Plants.* London: Unwin Hyman. pp. 309.
Devitt, D. A., and R. L. Morris. 1987. "Morphological Response of Flowering Annuals to Salinity." *Journal of the American Society of Horticultural Science.* 112:951–955.
Dole, J., and H. Wilkins. 1999. *Floriculture: Principles and Species.* New York: Prentice Hall.
Hofstra, G., and R. Wukasch. 1987. "Are You Pickling Your Pansies?" *Greenhouse Grower.* Sept: 14–17.
Nelson, P. V. 1996. "Macronutrient fertilizer programs," pp. 141–170. In: D. W. Reed. *Water, Media, and Nutrition for Greenhouse Crops.* Batavia, Ill.: Ball Publishing.
Wilkerson, D. C. "Soilless Growing Media and pH." *Texas Greenhouse Management Handbook.* pp. 30–34, 45–47.

in mind the alkalinity content of your irrigation water. Do you need an acidic fertilizer to neutralize the alkalinity, or is your water pure? For pure water, is a basic fertilizer required to prevent pH drop and offer some buffering capacity?

Amount of phosphorus to apply

Most crops require only 5–10 ppm phosphorus (P) for adequate growth. In fact, P applications should be limited if compact growth is required. Remember, compact growth is not beneficial for cut flowers

Table 3-2. Suggested Media PourThru EC Ranges for Floricultural Crops Grown in Soilless Media

These values are guidelines; make adjustments based on your growing practices. Assumes less than 30% of the nitrogen is supplied from urea.

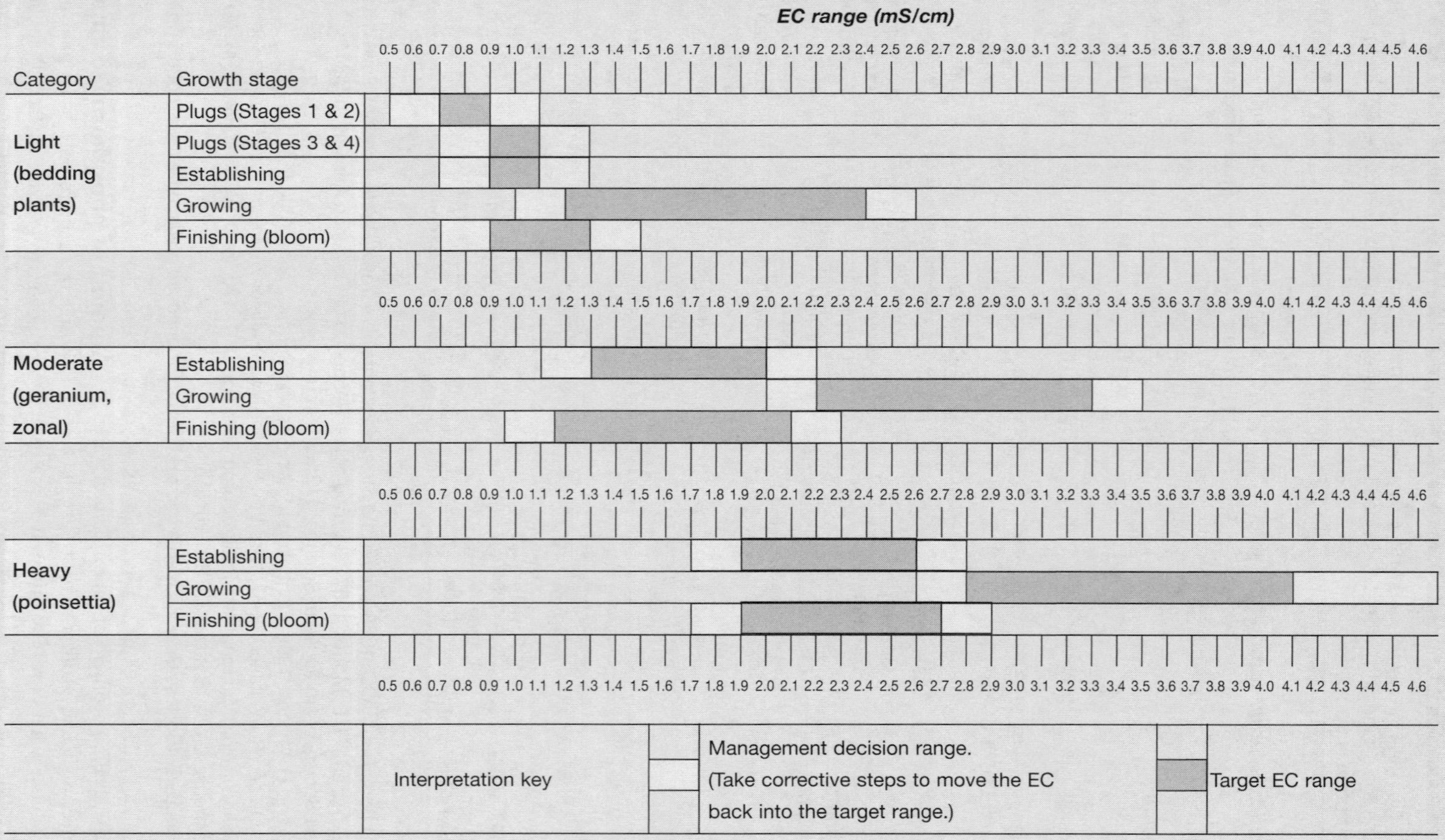

EC range (mS/cm)

0.5 0.6 0.7 0.8 0.9 1.0 1.1 1.2 1.3 1.4 1.5 1.6 1.7 1.8 1.9 2.0 2.1 2.2 2.3 2.4 2.5 2.6 2.7 2.8 2.9 3.0 3.1 3.2 3.3 3.4 3.5 3.6 3.7 3.8 3.9 4.0 4.1 4.2 4.3 4.4 4.5 4.6

Category	Growth stage
Light (bedding plants)	Plugs (Stages 1 & 2)
	Plugs (Stages 3 & 4)
	Establishing
	Growing
	Finishing (bloom)
Moderate (geranium, zonal)	Establishing
	Growing
	Finishing (bloom)
Heavy (poinsettia)	Establishing
	Growing
	Finishing (bloom)

0.5 0.6 0.7 0.8 0.9 1.0 1.1 1.2 1.3 1.4 1.5 1.6 1.7 1.8 1.9 2.0 2.1 2.2 2.3 2.4 2.5 2.6 2.7 2.8 2.9 3.0 3.1 3.2 3.3 3.4 3.5 3.6 3.7 3.8 3.9 4.0 4.1 4.2 4.3 4.4 4.5 4.6

Interpretation key	
(unshaded)	Management decision range. (Take corrective steps to move the EC back into the target range.)
(shaded)	Target EC range

Table 3-3. EC Interpretation Values (mS/cm) for Various Extraction Methods[1]

Values are based on actively growing plants, which have medium nutrient requirements.

1:5	1:2	SME	PourThru[2]	Indication
0–0.11	0–0.25	0–0.75	0–1.0	**Very low.** Nutrient levels may not be sufficient to sustain rapid growth.
0.12–0.35	0.26–0.75	0.76–2.0	1.0–2.6	**Low.** Suitable for seedlings, bedding plants, and salt-sensitive plants.
0.36–0.65	0.76–1.25	2.0–3.5	2.6–4.6	**Normal.** Standard root zone range for most established plants. Upper range for salt-sensitive plants.
0.66–0.89	1.26–1.75	3.5–5.0	4.6–6.5	**High.** Reduced vigor and growth may result, particularly during hot weather.
0.9–1.10	1.76–2.25	5.0–6.0	6.6–7.8	**Very high.** May result in salt injury due to reduced water uptake. Reduced growth rates likely. Symptoms include marginal leaf burn and wilting.
>1.1	>2.25	>6.0	>7.8	**Extreme.** Most crops will suffer salt injury at these levels. Immediate leaching is required.

[1] Adapted from: *On-Site Testing of Growing Media and Irrigation Water.* 1996. British Columbia Ministry of Agriculture.

[2] Due to the variability of PourThru-technique results, growers should always compare their results to the SME method to establish acceptable ranges.

and usually not a consideration for many perennials. One of the most common fertilizers used, 20-10-20, applied at the rate of 200 ppm nitrogen (N) will supply 44 ppm of P—more than required! Phosphorus levels can be managed by using a fertilizer such as 15-0-15, which does not contain P, or by using a low-P fertilizer such as 13-2-13 or 15-5-15 calcium magnesium. (Keep in mind that these are all basic fertilizers and will influence the pH management strategy.)

Supplying all of the essential elements

Are adequate calcium (Ca), magnesium (Mg), and sulfur (S) supplied by your irrigation water or media, or do they need to come from the fertilizer? A fertilization rate of 50–100 ppm Ca and 25–50 ppm Mg through a constant liquid fertilization will provide ample Ca and Mg. For many locations, sufficient Ca and Mg already occur in the irrigation water. For many areas of the Southeast and Northeast, Ca and Mg additions are required, which is especially important if you are using a fertilizer such as 20-10-20 that does not contain Ca or Mg.

If your fertilizer does not contain adequate Mg, another option is magnesium sulfate (Epsom salts/ $MgSO_4{\cdot}7H_2O$) applications. Apply at a rate of 1–2 lbs./100 gal. (1.2–2.4 g/l) water each month. Do not mix with other fertilizers to avoid possible precipitations with calcium and phosphate. This rate of Epsom salts also supplies adequate levels of sulfur.

While Ca is required, avoid adding too much, especially for pansies (*Viola*). Boron deficiency can be easily induced by excess Ca applications, specifically with the use of calcium nitrate [$Ca(NO_3)_2{\cdot}4H_2O$].

The potassium-calcium-magnesium ratio

Potassium (K) fertilization rates above 200 ppm can have an antagonistic effect on calcium (Ca) or magnesium (Mg) uptake by the plant. Supplying the plants with a K-Ca-Mg ratio (ppm) of 4:2:1 will limit any antagonisms. See above for recommended Ca and Mg fertilization rates.

Micronutrients

Do you need to add micronutrients (i.e., micros)? Most fertilizers include micronutrients, but if you are formulating your own fertilizers such as mixing calcium nitrate ($Ca[NO_3]_2 \cdot 4H_2O$) and potassium nitrate (KNO_3), micronutrients may be lacking. Also, if you are growing poinsettias, additional molybdenum (Mo) is highly recommended.

Based on these criteria, it is likely that no single fertilizer will fulfill all the requirements for every crop you grow. What to do? Rotate fertilizers. For example, you could use a basic fertilizer such as 15-2-20, 15-3-30, or 13-2-13 to provide adequate levels of Ca, Mg, and micros but have low amounts of P and ammoniacal nitrogen to increase the medium's pH. Rotate with an acidic fertilizer such as 21-5-20, which also has a low amount of P, to lower the medium's pH. These are just a few examples of what will work. Evaluate what other fertilizers are available to you, because many will fulfill your requirements. Keep in mind that if your fertility program is working for you, there may be no need to change it.

Whichever fertilizer you select, conduct medium tests either in-house or through a commercial lab to ensure that your fertility program is on target. Ideally, testing should be conducted every two weeks, and the results should be plotted to detect trends before any deficiency or toxicity symptoms appear.

Nutritional Problems

Why do nutrient problems occur?

Deficiencies and toxicities occur in greenhouse crops for many reasons other than simply a shortage or overapplication of nutrients. Unless you know why a nutrient deficiency occurs, you will be unable to formulate a plan for preventing its reoccurrence in future crops. The following are some of the common reasons for deficiencies and toxicities.

pH

One factor that can introduce nutrient problems in greenhouse crops is the substrate pH. The general pH range for greenhouse crops in a soilless medium is 5.4–6.8, but maintaining the pH at 5.6–6.2 is recommended. Low uptake of nutrients—particularly boron (B), copper (Cu), iron (Fe), manganese (Mn), and zinc (Zn)—can occur if the medium's pH is above 6.5. Certain macronutrients such as calcium (Ca) and magnesium (Mg) can become less available at pH values below 5.4. Toxicities (Mn and Fe) can also occur if pH values drop too low. (See chapter 4 for more information on managing pH.)

Improperly working equipment

An improperly working fertilizer proportioner can cause nutrients to be less than optimal in the medium. Calibrate your injector weekly. Equipment failure can cause multiple nutrient deficiencies.

Water stress (overwatering)

Constant saturation of the medium can lead to macro- and micronutrient deficiencies. When overwatering reduces oxygen levels, root growth can be limited and water uptake slowed. Elements such as calcium (Ca) are transported via water flow, and deficiency symptoms can develop rapidly on new growth. Also, the inactivity of root systems due to saturated conditions can lead to insufficient uptake of all nutrients, particularly iron (Fe) or phosphorus (P).

Soluble salts levels

Soluble salts are the total dissolved salts in the medium at any given time and are measured by electrical conductivity (EC). When the EC of the medium is too low, plant growth is stunted and mineral deficiencies are observed. Low salts can be due to excessive leaching, too many clear-water irrigations between fertilizations, or an improper injector ratio. Deficiency symptoms such as lower leaf yellowing (N), lower leaf purpling (P), and lower leaf interveinal chlorosis (Mg) are common when EC values are below 0.25 mS/cm (1:2 extraction), 0.75 mS/cm (SME extraction), or 1.0 mS/cm (PourThru extraction).

When the EC of the medium is too high, plant growth is stunted and flowering can be reduced or delayed. High salts can be due to inadequate leaching, too infrequent clear-water irrigations, or an improper injector ratio. Problems can occur when EC values are above 1.26 mS/cm (1:2 extraction), 3.5 mS/cm (SME extraction), or 4.6 mS/cm (PourThru extraction).

Conduct an in-house EC monitoring program using the PourThru method to avoid most EC problems.

Mineral antagonisms

When certain elements are provided in excess to plants, uptake of other nutrients may be hindered. One example of a mineral antagonism is the nitrogen-potassium (N-P) interaction; a 1:1 nitrogen-potassium ratio is recommended for most floriculture crops. Another antagonism is the potassium-calcium-magnesium (K-Ca-Mg) interaction. Any one of these elements in excess can cause a decrease in the uptake of the other; therefore, a ratio of 4:2:1 should be used. Excess phosphorus can also decrease zinc, iron, manganese, and copper uptake. Excess calcium reduces boron uptake.

Low temperature

Temperature can also play a role in the introduction of nutrient deficiencies. One classic example is the effect of low temperature (less than 55°F [13°C]) on the uptake of phosphorus in tomato (*Lycopersicon*). Purpling of the lower foliage is the common symptom. Geraniums (*Pelargonium*) also can express phosphorus deficiency when they are grown too cool in the spring.

Disease

Organisms such as *Pythium* impair roots, causing inefficient uptake of nutrients. Iron deficiency (upper-leaf interveinal chlorosis) is often the first deficiency to occur if root rot pathogens infect the root system. Foliar diseases, particularly fungal diseases, can cause chlorosis of leaf tissue, which is a direct reflection of harvesting nutrients from plant cells.

Typical symptoms of nutrient disorders

Knowing why a deficiency has occurred is a crucial tool in the identification of a nutrient disorder. Another important aspect in diagnosis is the location on the plant where the symptom is expressed. Understanding these principles will enable you to diagnose more correctly and will, in most cases, pinpoint the macro- or micronutrient disorder.

Nitrogen deficiency

The older leaves become uniformly chlorotic. After considerable time, older leaves become necrotic and may drop off. Purple to red discoloration may develop in older leaves, as they turn chlorotic in species such as *Begonia, Tagetes* (marigold), and *Viola* (pansy).

Ammonium toxicity

On older plants with flower buds, margins of older leaves curl upward or downward, depending on the plant species. Older leaves develop chlorosis. Necrosis follows chlorosis on the older leaves. Fewer roots form and, in advanced toxicities, root tips become necrotic, often with an orange-brown color.

On seedlings and bedding plants, young leaves develop chlorosis (most often interveinal) and margins curl up or down, depending on species.

Phosphorus deficiency

The plant becomes severely stunted, and the foliage becomes deeper green than normal. In some species the older leaves develop purple coloration. In garden chrysanthemums the veins of the lower leaves turn purple. Older leaves then develop chlorosis, followed by necrosis. Roots become longer than normal when the deficiency is moderate.

Older leaves of foliage plants may lose their sheen and become dull green; this is followed by red, yellow, and blue pigments showing through the green, particularly on the undersides of the leaves along the veins. These symptoms spread across the leaf. Older leaves may abscise; otherwise, necrosis develops from the tip toward the base.

Potassium deficiency

The margins of older leaves become chlorotic and then necroctic immediately after. Similar necrotic spots may form across the blades of older leaves but are more numerous toward the margin. Soon the older leaves become totally necrotic. Seedlings and young bedding plants are more compact and deeper green than normal prior to the formation of chlorosis and necrosis on older foliage. Some foliage plants will develop oily spots on the undersides of older leaves, which then become necrotic.

Calcium deficiency

Symptoms are expressed at the top of the plant. Young leaves may develop variable patterns of chlorosis and distortion, such as dwarfing, strap-like shape, or crinkling. The edges of leaves may become necrotic. Shoots stop growing. Petals or flower stems may collapse. Roots are shortened, thickened, and branched.

In foliage plants, in addition to the above symptoms, the older leaves may become thick and brittle. In *Philodendron scandens oxycardium* and *Epipremnum aureum,* calcium deficiency has symptoms similar to that of a mobile nutrient. Yellow spots occur in the basal half of older leaves. These spots enlarge into irregular yellow areas containing numerous scattered, oil-soaked spots.

Magnesium deficiency

Older leaves develop interveinal chlorosis. In several species, pink, red, or purple pigmentation will develop in the older leaves following the onset of chlorosis.

For foliage plants with pinnately (netted) veined leaves, bronze-yellow chlorosis begins at the upper margin of older leaves, progressing downward along the veins, leaving a green, V-shaped pattern at the top of the leaf. As chlorosis progresses down the leaf, a green, V shape of tissue remains at the bottom. Eventually, the tip and base become chlorotic. Necrosis follows chlorosis in the same pattern.

Sulfur deficiency

Foliage over the entire plant becomes uniformly chlorotic. Sometimes the symptoms tend to be more pronounced toward the top of the plant. While symptoms on the individual leaf look like those of nitrogen deficiency, sulfur deficiency can be easily distinguished from nitrogen deficiency, because nitrogen deficiency begins in the lowest leaves.

Iron deficiency

An iron deficiency shows up as interveinal chlorosis of young leaves. Young leaves of seedlings sometimes develop general chlorosis rather than interveinal chlorosis. In late stages, the leaf blade may lose nearly all pigment, appearing white.

Iron toxicity

Iron toxicity mainly affects African marigolds (*Tagetes*), geraniums (both seed and vegetative), basil (*Ocimum*), cosmos, dahlia, nasturtium, pepper (*Capsicum*), strawflower (*Bracteantha*), tomato, and zinnia. Marigolds develop bronzing on recently fully expanded leaves. The bronzing consists of numerous pinpoint-sized spots that begin yellow and quickly turn bronze. Affected leaves become necrotic. On other crops, the older leaves develop numerous, pinpoint-sized, black necrotic spots across the blade. The entire leaf may die as the spots enlarge.

Manganese deficiency

This deficiency appears as black spots on vegetative strawflower, *Bracteantha bracteata* 'Florabella Pink'. Symptoms begin as interveinal chlorosis of young leaves, which is sometimes followed by tan spots in the chlorotic areas between the veins.

Manganese toxicity

Toxicity often begins with interveinal chlorosis of young leaves due to iron deficiency caused by high manganese antagonism of iron uptake. Manganese toxicity takes the form of burning of the tips and margins of older leaves or formation of reddish-brown spots on older leaves. The spots are initially about 0.0625" (1.6 mm) in diameter and are scattered over the leaf. Spots become more numerous and eventually coalesce into patches.

Zinc deficiency

Young leaves are very small and internodes are short, giving the stem a rosette appearance. These leaves are also chlorotic in varying patterns but tend toward interveinal. In kalanchoe, zinc deficiency can express itself as a fasciation (or a flattened, highly branched stem).

Copper deficiency

Young leaves develop interveinal chlorosis; however, the tips and lobes of these leaves may remain green. Next, the youngest fully expanded leaves rapidly become necrotic. The sudden death of these leaves resembles desiccation.

Boron deficiency

Symptoms can begin as an incomplete formation of flower parts, such as fewer petals, small petals, sudden wilting, or collapse of petals, and notches of tissue missing in flower stems, leaf petioles, or stems. Death of the bud gives rise to branching, followed by death of the new buds, which eventually leads to a proliferation of shoots termed a witch's broom. Other symptoms include short internodes; crinkling of young leaves; corking of young leaves, stems, and buds; and thickening of young leaves. Chlorosis of young leaves may occur but not in any definite pattern. Roots become short and thick with eventual death of root tips.

Additional symptoms in foliage plants can include brittle stems and leaves, as well as necrotic spots (black and sunken) on stems just below the nodes.

Nodal roots on vine plants may become thick and short and abscise, and vines may become highly curled at the nodes.

Boron toxicity

The margins of older leaves become necrotic with a characteristic reddish-brown color. Necrotic spots may also develop across the leaf blade but tend to be concentrated at the margins.

Molybdenum deficiency

These symptoms apply to poinsettias, the only greenhouse floral crop molybdenum deficiency is known to affect. The margins of leaves halfway up the stem become chlorotic, presenting a silhouette appearance, and then quickly become necrotic. Symptoms spread up and down the plant. Leaves may also become misshapen, resembling a half-moon pattern with some crinkling.

Paul Fisher

Managing pH for Container Media

The most common nutritional problems occur in greenhouse crops when the pH of the growing medium (the "substrate pH") is outside the optimum range. pH is a measure of the acidity (low pH = acid) or basicity (high pH = basic, also called alkaline) of the growing medium. The substrate pH is important because it affects a chain of events:

Substrate pH determines micronutrient solubility in the growing medium.

low pH = very soluble
high pH = less soluble

↓

Plant takes up soluble nutrients through roots.

↓

Nutrients are transported into leaves and growing points.

excess = toxicity
adequate = healthy
insufficient = deficiency

↓

Toxicity or deficiency results in stunted growth and poor plant appearance.

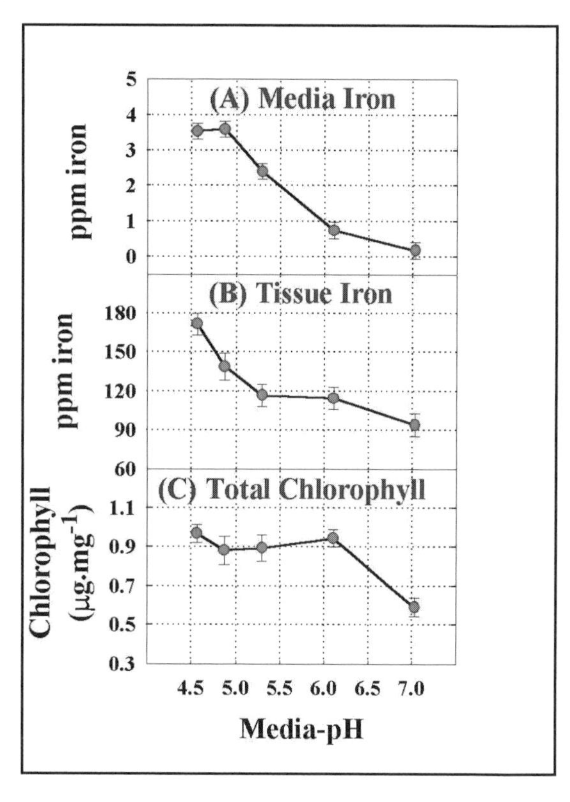

Figure 4-1. The effect of growing petunias at different substrate levels on (A) the medium's iron content (from a saturated media extract using deionized water as the extractant), (B) iron content in the tissue, and (C) chlorophyll content. As pH increased, there was a decrease in the available iron in the medium and less uptake of iron into the leaves. At the highest pH, plants showed chlorosis (lack of chlorophyll) because of iron deficiency. Based on research by Brandon Smith and Paul Fisher, University of New Hampshire, and William Argo, Blackmore Company.

Plants take up dissolved nutrients through their roots. The substrate pH drives the chemical reactions that determine whether nutrients are either available for root uptake (soluble) or unavailable for uptake (insoluble). Several nutrients are affected by pH, but the most important are phosphorus and most micronutrients, especially iron, manganese, copper, zinc, and boron (which decrease in solubility at high pH levels), and molybdenum (which increases in solubility at high pH levels). At high substrate pH levels, calcium and magnesium are often at high levels in the growing medium, but this is because of a high application rate of limestone, rather than solubility of those nutrients.

The optimum pH range for most crops growing in a soilless medium is 5.6–6.4, because in this range micronutrients are soluble enough to satisfy plant needs without becoming so soluble as to be toxic.

Recognizing the Problem

At high substrate pH, micronutrients (especially iron) become less soluble in the medium and deficient in plant tissue. Iron is required to produce chlorophyll (the green pigment in leaves). If iron is insoluble in the medium because of high pH, then new leaf tissue grows faster than roots can supply the necessary iron. Symptoms of the resulting iron deficiency are chlorosis (yellowing from lack of chlorophyll) in the younger foliage, which is sometimes interveinal (between leaf veins). As deficiency becomes more severe, foliage becomes almost completely white and necrotic (dead) areas form at the growing points.

At low substrate pH, iron and manganese are highly soluble. Excess micronutrients can accumulate in plant tissue and cause chlorosis and necrosis (dead tissue) on leaf margins and as leaf spots. The damage tends to occur in older leaves, because the longer a leaf grows on the plant, the more time it has to accumulate excess micronutrients.

Why Do pH Problems Arise?

Some of the reasons substrate pH can be too high or low include poor buffering of soilless media, lime, a wide range in crops, the fertilizer type, and water alkalinity.

Poor buffering of soilless media

Since the 1980s, the move away from using soil in greenhouse container media has resulted in less buffering (i.e., chemical resistance to pH change). In peat- and bark-based media, a change of up to 1 pH unit in a week can sometimes occur in commercial crops. Although use of soilless media has many benefits (i.e., uniformity, consistency, aeration, and sterility), one downside is that pH is very likely to change over time, even if the media start out at the optimum pH range at the time of planting. pH can drift up or down depending on the balance of factors, including water alkalinity, lime activity, acidification of the media by plant roots, and use of an acidic- or basic-reaction fertilizer. As a result, it is important to not just blame problems on the media but to understand how grower management can cause pH to change over time.

Lime

Lime is mixed into media to raise pH to around 6.0, because both peat and bark are acidic. Limes differ in their source, chemistry, and particle size, which causes them to vary in how reactive they are and in how long they continue to react during crop growth. Fast-reacting limes that rapidly neutralize the acidity of peat or bark include hydrated lime (calcium hydroxide) or fine calcitic or dolomitic carbonate limes with the majority of particles passing through a 200-mesh screen. Larger limestone particles (20 to 100 mesh) react more slowly and provide some buffering (resistance) to a pH drop during crop production. Highly reactive limes are ideal for short-term crops such as plugs and flats, whereas a mix of reactive (200 mesh or finer) and slower-reacting (20 to 100 mesh) lime particles are suitable for longer-term container crops. If the incorrect type or quantity of lime is used during substrate mixing, pH can either be out of range at the start of the crop or drift over time. If you mix your own media, consult a fertilizer or media company to obtain a suitable type of lime and run small batch tests to check how much lime is needed to bring pH up to the target level. If you change your source of lime, peat, bark, or vermiculite, you will need to retest your recipe. If you consistently run into problems with high or low pH and you have correctly matched the fertilizer type with water alkalinity, consider changing the lime type or rate.

Wide range in crops

Species differ in their nutritional needs and can be separated into three nutritional groups based on their efficiency at taking up micronutrients.

Petunia group: This group is iron inefficient and prone to iron deficiency at high pH, especially combined with low fertilizer concentration. Examples include snapdragon (*Antirrhinum*), *Calibrachoa*, vinca (*Catharanthus*), *Diascia*, *Nemesia*, petunia, and pansy (*Viola*). Grow at a lower pH range of 5.4–6.2 to increase solubility of micronutrients. This group is often misdiagnosed as a "high-feed" group—they do not necessarily require more nitrogen (N), phosphorus (P), and potassium (K) than other crops, but they are especially sensitive to high pH and the need for adequate iron.

General group: Included here are chrysanthemum, impatiens, ivy geranium (*Pelargonium peltatum*), and poinsettia (*Euphorbia pulcherrima*). Grow at a moderate pH range of 5.6–6.4.

Geranium (*Pelargonium*) group: These plants are iron efficient and prone to iron/manganese toxicity at low pH, especially when combined with high fertilizer concentration. Included are marigold (*Tagetes*), seed and zonal geraniums (*Pelargonium*), New Guinea impatiens, and *Eustoma*. Grow at a higher pH range, 6.0–6.6, to limit the solubility of micronutrients.

Some of the strategies growers use to manage multiple crops include:

1. Train staff to recognize the likely problems with each group of plants (i.e., micronutrient deficiency for petunia and toxicity for geranium).

2. Select fertilizers or limestone rates for plant groups. For example, for the petunia group, use a more acidic fertilizer (more ammonium) or less preplant lime than used with the geranium group.

3. Organize greenhouse zones so that plants within the same nutritional group can be managed with one injector or fertilizer tank.

4. If you are using one injector for everything, you may need to apply flowable lime to the geranium group if the pH is below 6.0 or supplement iron for the petunia group if pH is above 6.3.

5. Monitor and manage EC and pH together. Managing EC and pH go hand in hand. High fertilizer concentration and high EC increase the risk of toxicity for the geranium group grown at low pH. In contrast, deficiency symptoms in the petunia group are more likely with low fertilizer concentration and low EC in combination with high pH.

6. Maintain a pH of 6.0–6.2 when growing a range of species (e.g., mixed baskets).

Fertilizer type

You cannot measure the acidic or basic reaction of a water-soluble fertilizer using a pH meter in the stock tank. Rather, the tendency of a water-soluble fertilizer to change the substrate pH depends on the form of nitrogen used (i.e., ammonium, nitrate, or urea). The nitrogen form has a consistent effect on pH: Ammonium and urea are both acidic (drop pH), whereas nitrate is somewhat basic (raises pH).

The acidic or basic reaction of a water-soluble fertilizer is written on the bag as an acidic or basic "calcium carbonate equivalency" (CCE), which is a relative measure of the tendency of the fertilizer to raise or lower the substrate pH. The CCE of acidic and basic fertilizers is affected by how much of the total nitrogen in the fertilizer is contributed by ammonium and urea. "Acidic" fertilizers such as 20-10-20 tend to contain more than 24% of all nitrogen in the ammonium or urea form, and "basic" fertilizers such as 13-2-13 generally have less than 24% of all nitrogen as ammonium or urea.

When you are selecting a fertilizer to raise or lower pH, keep in mind the following points:

• Nitrate (basic) increases the substrate pH only when plant roots take up the fertilizer. Therefore, if plants are small or stressed and not growing, nitrate has little influence on pH.

• Ammonium (acidic) can cause pH to go down even if the plant is small or is not growing, because soil bacteria acidify the medium through a process called nitrification.

• Ammonium is less effective at lowering pH in cool, saturated media, because nitrification is inhibited. In addition, ammonium toxicity in plants can occur in cool, wet conditions, because plants are more likely to take up excess ammonium.

• Sometimes ammonium will not drop pH at all, because other factors (especially high lime or water alkalinity) can have a stronger effect on pH than the fertilizer.

• Be aware that although you can use the ammonium level in the fertilizer to raise or lower pH (table 4-1) or balance water alkalinity (table 4-2), high ammonium may cause ammonium toxicity.

• Many high-ammonium fertilizers also have a high proportion of phosphorus, which can lead to excess growth and the need for growth regulator applications.

• Urea fertilizers tend to be acidic, but urea is less acidic than ammonium.

• A "neutral" fertilizer in practice is a fertilizer that leads to a stable substrate pH over time in your situation, given other factors such as the water alkalinity, growing medium, plant species, watering practices, and fertilizer concentration applied. The CCE on a fertilizer bag is, therefore, only a guideline relative to other fertilizers.

Water alkalinity

Water alkalinity is a measure of the basic ions, mainly bicarbonates and carbonates, dissolved in the water. Alkalinity can be thought of as the "liming content" of the water. Irrigating with a highly alkaline water (above 150 ppm $CaCO_3$ of alkalinity) can cause substrate pH to increase over time, because you are effectively adding dissolved lime with each irrigation. An important concept often misunderstood by growers is that water alkalinity directly affects substrate pH over time, whereas pH of the water source has little effect on the substrate pH. Irrigating with a water source

Table 4-1. Examples of the Acidity or Basicity of Three Water-Soluble Fertilizers						
FERTILIZER	% AMMONIUM	% UREA	% NITRATE	% TOTAL NITROGEN	% ACIDIC NITROGEN = (AMMONIUM + UREA)/ TOTAL N)	CALCIUM CARBONATE EQUIVALENCY (CCE) (LBS./ TON)
20-10-20	8	0	12	20	40	429 acidic
17-5-17	4	0	13	17	24	0 (neutral)
13-2-13	0.6	0	12.8	13.4	4	330 basic

with 200 ppm of alkalinity and water a pH of 7.5 will tend to raise substrate pH much more than water with 50 ppm of alkalinity and water pH of 8.5.

High alkalinity can be managed by increasing the ammonium (acid) level in the fertilizer (table 4-2), blending with a more pure water source, reverse osmosis, or neutralizing the alkalinity through injection of an acid.

Alkalinity can be tested onsite or in a commercial laboratory with a colorimetric test kit. (Note: The paper test strips that are available are too imprecise for greenhouse needs.) Laboratories report alkalinity in a number of different ways, including ppm or mg/l calcium carbonate ($CaCO_3$) equivalents of alkalinity (which is used in this section), milliequivalents (meq) of $CaCO_3$ alkalinity, and ppm or mg/l bicarbonate. To convert between units, 50 ppm $CaCO_3$ alkalinity = 1 meq $CaCO_3$ alkalinity = 61 ppm or mg/l bicarbonate. For a complete discussion of alkalinity, see chapter 1.

Regular Testing

Basing fertilizer decisions on regular tests (every one to two weeks) of your substrate pH, substrate EC, and the EC of the fertilizer solution solves 90% of nutritional problems by alerting growers to problem trends before plants are stressed. A soil test of pH is also an easy way to confirm a suspected pH problem. Monitoring other factors (e.g., root diseases, greenhouse temperatures, pest problems, and high or low EC) help rule out these problems, because many factors other than the substrate pH can cause similar symptoms.

Correcting pH Problems

The following recommendations for raising or lowering media pH are intended for crops already under severe stress. Preventing pH problems is better than relying on a cure, and these actions are intended for crops that would be unsalable without intervention. Phytotoxicity or staining is very likely with these chemicals, and applications should be tested on a small number of plants before applying to the entire crop. Necrotic tissue will not recover, and the goal is to produce new, healthy foliage that will cover damage.

Correcting low pH

When pH falls below the optimum range, stop acidifying water (if acid is being injected) and shift to a nitrate-based fertilizer (e.g., 13-2-13 or 15-5-15). Further action is needed if pH has not risen within a week and plants are becoming stressed, especially for a species in the geranium group when pH is below 6.0 or other crops with a pH below 5.4. Consider soil drenches with either flowable lime or potassium bicarbonate. Other options (i.e., hydrated lime or potassium hydroxide) have special uses but are less reliable as a corrective liming material.

Several factors affect the choice between flowable lime and potassium bicarbonate. Flowable lime has a more predictable and stable effect on substrate pH without increasing EC. Potassium bicarbonate is easier to apply, however, and should be used on flood floors or when applied through low-volume drippers. Both liming materials are fast acting and show most of their effect on pH within one day. Following a drench, you can reapply after five days if pH is not up to the optimum range.

To minimize phytotoxicity from flowable lime or potassium bicarbonate, apply in cool weather so the material does not dry quickly on foliage, avoid splashing foliage during application, immediately rinse foliage with a fine spray, and apply with generous leaching to maximize the effect at low concentration.

Table 4-2. Fertilizers and Alkalinity

Approximate guidelines to matching fertilizers with water alkalinity in order to achieve a stable pH over time

CALCIUM CARBONATE EQUIVALENCY (LBS./TON)	% ACIDIC NITROGEN = (AMMONIUM + UREA)/ TOTAL N	EXAMPLES	ALKALINITY CONCENTRATION (PPM $CACO_3$) THAT PROVIDES A STABLE SUBSTRATE pH
>500 acidic	>50%	20-20-20, 21-7-7	250–300
150–500 acidic	40%	20-10-20, 21-5-20	150–250
150 acidic–150 basic	20-30%	20-0-20, 7-5-17	60–150
>150 basic	<15%	13-2-13, 14-0-14	30–60

Other tips for applying flowable lime:

• Apply at 4 qts./100 gal. (10 ml/l or a 1:100 ratio).

• You can use an injector to dilute the solution, but lime particles can be very abrasive. Immediately clean equipment after application.

•Do not apply through drippers or on flood floors, because it will clog equipment and leave residue.

Other tips for applying potassium bicarbonate:

• Apply at 2 lbs./100 gal. (2.4 g/l).

• Can be delivered through emitters or on flood floors.

• One day after application, apply a basic fertilizer (e.g., 13-2-13) with moderate leaching to wash out salts and to reestablish nutrient balance.

• It is likely that you will need to repeat applications.

Correcting high pH

Several actions may be necessary when the substrate pH is too high.

Use a high-ammonium fertilizer combined with low alkalinity

Select a high-ammonium (very acidic) fertilizer such as 21-7-7. The effect on pH can sometimes be slow (more than one to two weeks) especially in cool, wet conditions or with small plants growing in large containers. Repeated applications of ammonium in cool, dark conditions may also cause toxic levels of ammonium to accumulate in leaf tissue.

If you have the necessary equipment and the alkalinity is greater than 80 ppm, acidify water to drop the irrigation water pH to around 5 (which gives near-zero alkalinity). Continue until substrate pH is in the target range. For the appropriate acid rate for your water source, see the alkalinity section of chapter 1.

Correct micronutrient deficiencies

Masking the symptoms of high pH with micronutrient applications can be very effective for keeping plants alive and healthy when grown under high media-pH conditions. However, unless your customers continue the iron sprays or drenches or transplant the plants soon after receiving them, plant quality will suffer. Use tissue analysis to find out which nutrient is deficient. Iron deficiency is most common; however, if a different nutrient such as manganese is deficient, then applying iron may worsen the problem because of antagonistic effects.

Iron comes in different forms that vary in solubility at high pH levels. The best iron forms as a drench at high pH are Iron-EDDHA (sold as Sprint 138 or Dissolvine QFe6) and Iron-DTPA (Sprint 330 or Dissolvine DFe11). These materials are more effective than Iron-EDTA and iron sulfate, which have low solubility at high pH.

The recommended application rate for an iron drench is 5 oz./100 gal. (39 ml/100 l) of either Iron-EDDHA (provides 22.5 ppm iron) or Iron-DTPA (37.5 ppm iron). The solutions should be applied with generous leaching, followed immediately by a washing of foliage to avoid leaf spotting. All options are economical. Iron-DTPA (Sprint 330) can be purchased from greenhouse and nursery suppliers. Ask for Iron-EDDHA (Sprint 138) from a fertilizer representative.

Foliar sprays are also somewhat effective, especially if iron chlorosis is mild. Suggested iron forms and rates for iron sprays are Iron-EDTA (60 ppm iron, equals 6.1 oz./100 gal. [48 ml/100 l]) or Iron-DTPA at 60 ppm iron (8 oz./100 gal. [63 ml/100 l]). Repeat applications are likely to be needed every five days because the iron is not transported to new leaves and the plant can grow out of a foliar spray. Phytotoxicity is likely. After applying foliar sprays to a test group, wait three days to check for damage before applying to the entire crop.

The spray application method is very important. Tips for maximum effectiveness of foliar sprays:

- Include an organosilicone surfactant (e.g., Capsil).
- Apply in the early morning on cool, cloudy days for a gradual drying of leaves in order to increase uptake and reduce spotting.
- Spray both sides of leaves because penetration may be better on the underside of leaves, where the cuticle is thinner.

Consider acid drenches in extreme cases

An iron-sulfate drench can rapidly reduce substrate pH, but phytotoxicity is very likely. A trial is needed on a small number of plants before applying to the entire crop. To minimize phytotoxicity, apply during a cool morning. Avoid contact with foliage, and immediately rinse foliage after application. Apply with generous leaching.

Aluminum sulfate should only be used to drop pH for hydrangeas, because it may cause nutrient imbalances in other crops. Flowable or elemental sulfur is sometimes used to drop pH in the nursery trade, but it tends to cause a gradual reduction in pH over time that is difficult to control, because microbial action is needed for the sulfur to be effective. Sulfuric acid is also sometimes used as a soil drench at pH levels from 1.5 to 2.0, but this approach is highly caustic to people, foliage, and equipment.

Iron sulfate provides iron (which is usually deficient in plants at high pH levels) in addition to causing a temporary drop in substrate pH. However, this material increases EC (1.2 mS/cm at 2 lbs./100 gal. [2.4 g/l]). The excess iron (2 lbs./100 gal. [2.4 g/l] provides 500 ppm iron) may cause imbalances if pH falls below 6.0. Iron sulfate should never be used with iron-efficient species (i.e., geranium group) or long-term crops. The maximum recommended rate is 2 lbs./100 gal. (2.4 g/l). Higher rates up to 6 lbs./100 gal. (14.4 g/l) will cause a greater drop in pH but also increase the risk of phytotoxicity.

Other tips to applying iron sulfate:

1. Store dark and dry. Iron sulfate oxidizes over time and has a 6.5-month shelf life. Mix in water with a pH below 7.0 and only use if the final solution is not cloudy.

2. Iron sulfate can stain medium and plastic subirrigation benches black.

3. Leach heavily with a complete fertilizer after one week to try to remove excess iron and restore nutrient balance.

Prevention Better Than a Cure

As a final note, remember that substrate pH problems should never cause crop losses if you set up a sensible nutrient-management program that is suited to your crop types and water source, establish a regular monitoring regime, and develop strategies that will keep pH and EC on track.

The author thanks the American Floral Endowment, the University of New Hampshire and University of Florida Agricultural Experiment Stations, and the Anna and Raymond Tuttle Environmental Horticulture Fund for supporting this research. Dr. William Argo, Linda Bilodeau, Jeremy Bishko, Jinsheng Huang, Connie Johnson, Brandon Smith, and Ron Wik contributed to the research presented in this section.

Brian E. Whipker, Todd J. Cavins, James L. Gibson, John M. Dole, Paul V. Nelson, and William Fonteno

Water, Media, and Nutrition Testing

Providing proper crop nutrition is important for growth and development as well as plant health. Sampling the plant tissue or medium is a simple check of the nutritional status of the crop and can provide clues about a crop's performance before deficiency or toxicity symptoms appear. Foliar analysis is especially useful when you need to determine micronutrient levels in the plant or need confirmation for an apparent nutritional symptom. Testing the irrigation water provides information on pH, electrical conductivity (EC), and alkalinity levels. High alkalinity can significantly increase substrate pH over time. The pH of the substrate dramatically influences the availability of various nutrients to the plants. In addition, you can use a fertilizer solution analysis as a way to double-check mixing procedures or to check the calibration of the fertilizer injectors. Below are instructions for taking representative samples for laboratory analysis. A selected list of labs specializing in greenhouse media, irrigation water, fertilizer solution, and plant tissue samples is provided in table 1-2 in chapter 1.

Laboratory Testing

Greenhouse media testing

Conducting routine analysis is essential in determining the nutrient status of the crop and facilitates the possibility of preventing future issues. Samples should be sent to a lab every three to four weeks to check nutrient levels. A standard analysis usually includes pH, EC, nitrate nitrogen (NO_3-N), ammonium nitrogen (NH_4-N), phosphorus (P), potassium (K), calcium (Ca), and magnesium (Mg). In some cases you may also want to test for sulfur and certain micronutrients (boron [B], copper [Cu], iron [Fe], manganese [Mn], molybdenum [Mo], and zinc [Zn]) in the medium. Generally, testing for micronutrient concentrations in the medium is recommended only under special circumstances and not for routine analysis. Plant tissue analysis is a better indicator of micronutrient concentrations.

The medium sample should be representative of the crop or problem you wish to analyze. The procedure to take a medium sample is:

1. For routine analysis, collect samples from five to ten containers and combine them into a single sample. Collect the sample by either removing a wedge-shaped piece from the top to the bottom of the pot, excluding the top 0.5" (13 mm) of the medium, or by pinching a handful of medium from the center third of the pot.

2. Thoroughly mix the samples together to make a single, homogeneous sample. Remove any large roots or debris.

3. Sample problem plants separately from the rest of the crop.

4. Put samples in a plastic bag that is labeled with your name, your address, the crop, and the sample's location.

Follow the same procedure every time you sample, so you can compare results and detect trends over time. You will need to collect 1–2 cups of medium for the analysis. Smaller volumes can be submitted, but the results may not be as accurate. Wet new medium to container capacity by placing the sample in a growing container and watering it until drainage occurs. After drainage stops, the sample can be bagged and mailed. With mail delivery time, this allows two days for the lime to react, allowing the lab to obtain correct pH readings. Keep in mind that it can take a week or more for the limestone to react, so pH values will likely increase over the first few weeks.

Irrigation water testing

Alkalinity can be a major factor affecting pH changes in the medium. The higher the alkalinity level, the quicker the pH of a soilless medium will increase. Water tests are recommended for each source and should be conducted one to four times per year. A standard analysis usually includes pH, EC, and alkalinity.

You may want to test the water for the following macro- and micronutrients: nitrogen (N), phosphorus (P), potassium (K), calcium (Ca), magnesium (Mg), sulfur (S), boron (B), chloride (Cl), copper (Cu), iron (Fe), manganese (Mn), molybdenum (Mo), and zinc (Zn). Testing for nutrients should be performed for each source at least once per year. Excessive alkalinity levels will need to be neutralized by acid addition. Check to see if the lab provides an acid addition rate.

To obtain a water sample, do the following:

1. Allow the water to run for five minutes to clear the line.
2. Rinse a clean, plastic, 2-qt. (2 l) container two to three times with the water to be tested.
3. Fill the container completely and cap it tightly.
4. Label the bottle with your name, your address, and the type of analysis requested.
5. Mail the sample within twenty-four hours.

Fertilizer solution testing

Fertilizer solution tests are recommended for checking the accuracy of an injector or to check mixing procedures. Testing by commercial labs should be performed two to four times per year. In addition, you should conduct weekly in-house EC testing of an injector's accuracy. A standard analysis usually includes: pH, EC, NO_3-N (nitrate nitrogen), NH_4-N (ammonium nitrogen), P, K, Ca, and Mg. Before submitting a sample, accurately weigh out the fertilizer and thoroughly mix it to have a representative sample.

Follow these steps to get an accurate sample:

1. Allow the fertilizer solution to run for five minutes to obtain a representative sample.
2. Rinse a clean, plastic, 2-qt. (2 l) container two to three times with the water to be tested.
3. Fill the container completely and cap it tightly.
4. Label the bottle with your name, your address, and the type of analysis requested.
5. Mail the sample within twenty-four hours.

Plant tissue testing

Foliar analysis can be conducted to determine the nutrient status of the crop or for problem solving. Plant tissue analysis is especially useful when you need to determine micronutrient levels in the plant. A standard analysis usually includes macronutrients (N, P, K, Ca, and Mg) and micronutrients (B, Cu, Fe, Mn, Mo, and Zn).

The leaf tissue samples should be representative of the crop or problem you wish to analyze.

1. For routine analysis, collect the most recently matured, fully expanded leaves from twenty to thirty plants (more leaves are required for plants with small leaves) and combine into a single sample. In general, to have an adequate sample, you will need around fifty leaves.
2. Sample problem plants or benches individually.
3. If the crop has been hand-watered or foliar nutrient sprays have been used, gently wash the leaves in distilled water for ten to twenty seconds to remove surface contaminants.
4. Samples should be sent in a *paper bag* that is labeled with your name, your address, the crop, and the sample's location. To prevent rotting, avoid using plastic bags.
5. Mail the sample within twenty-four hours. Try to collect the sample at the beginning of the week so delivery will not be delayed over the weekend.

Plant analysis standards for many of the minor floriculture crops have not been established. Submit a problematic and nonproblematic sample for a comparison. Follow the same procedure every time you sample, so you can compare results and detect trends over time.

Table 5-1. Guidelines for Greenhouse Media Analyzed by the Saturated Media Extraction (SME) Method

NUTRIENT	MEDIA NUTRIENT LEVELS (IN PPM)				
	LOW	ACCEPTABLE	OPTIMUM	HIGH	VERY HIGH
NO_3-N	0–39	40–99	100–199	200–299	>300
P	0–2	3–5	6–10	11–18	>19
K	0–59	60–149	150–249	250–349	>350
Ca	0–79	80–199	>200		
Mg	0–29	30–69	>70		

Adapted from Michigan State University

Interpretation

Interpretation tables for media samples tested by the saturated media extraction (SME) method (table 5-1) and for plant tissue analysis samples (table 5-2) are provided.

In-House Testing

Sampling for in-house testing

Sampling the medium for pH and electrical conductivity (EC) with PourThru, 1:2, or press extraction method is a quick and simple check of the nutritional status of a crop. The PourThru, 1:2, and press extractions allow a rapid, onsite determination of pH and EC values. The values provide clues about a crop's performance before deficiency or toxicity symptoms appear.

Sampling procedures

Sampling results are only valid if they represent the whole crop. How you should set up a sampling program, which crops you should sample, and how many samples you should collect are all details to consider.

For routine analysis, collect and analyze a minimum of five individual containers or five cell packs (i.e., bedding plant flats), or press one row of plants from five different plug trays. Results from each of the five samples can then be averaged for a single "interpretation value." (Do not combine the five samples for a single test.) If the five values vary widely in pH or EC, you may need to increase the sample size.

Testing frequency

Ideally, sample all crops weekly, but this may not be practical. Since time availability is a concern, consider selecting the "top ten crops of concern," based on economic value or nutritional problems. Divide the ten crops into two groups, and test each group every other week.

The recommended frequency of testing depends on the container size, as smaller substrates are more susceptible to rapid changes in pH than larger, more buffered volumes. For plugs, test pH and EC two to three times per week. Weekly testing should be sufficient for crops such as flats of bedding plants or 4" (10 cm) pots. For crops grown in large containers (such as 6" [15 cm] pots), monitoring every two weeks should be sufficient.

Crop factors

Consider taking separate samples within a crop if there are variations in medium type (i.e., different manufacturers or ingredients), fertilizer type or rates (i.e., acidic or basic types), or planting dates.

If steps to correct pH or EC problems have been taken, then resample the plants weekly. If needed, sample as frequently as every two days. If results seem atypical, consider resampling before making drastic changes, especially to the substrate pH.

PourThru steps

The PourThru method is an easy, quick way to test medium samples in-house. Following are instructions on how to perform this method.

1. Irrigate the crop one hour before testing. Make sure the medium is saturated. If the water supplied by the automatic irrigation system varies, then water the containers or flats by hand. If you use constant liquid feed, irrigate with fertilizer solution as usual. If using periodic feeding, irrigate with clear water and test a day or two before fertilizing or test on the same day in the fertilizing cycle each time.

2. Place a saucer under the container. After the container has drained for thirty to sixty minutes, place plastic saucers under the containers to be sampled. If you are testing seedlings in bedding plant flats, place cell packs in saucers.

3. Pour enough distilled water on the surface of the media. Make sure you pour enough water on the media to get 1.7 oz. (50 ml) of leachate in the saucer. The amount of water needed will vary with container size, crop, and environmental conditions. Use the values in table 5-3 as guides.

4. Collect leachate for pH and EC. Make sure you get about 1.7 oz. (50 ml) of leachate each time you test. Leachate volumes over 2.5 oz. (70 ml, based on 6" [15 cm] pot extractions) will begin to dilute the sample and give you lower EC readings.

5. Calibrate the pH and EC meters prior to testing. The test results are only as good as the last meter calibration. Calibrate the instruments every day they are used. Always use fresh, standard solutions, and never pour used solution back into the original bottle.

6. Test the samples for pH and EC. Test the leachate as soon as possible. Electrical conductivity will not vary much over time if there is no evaporation of the sample. However, the pH can change within two hours. Record the values on the chart specific to each crop.

Table 5-2. Tissue Nutrient Levels of High-Quality Greenhouse Plants

Note that the indicated values are only general recommendations. The optimum values will vary among firms because of different cultural practices, climate, and cultivars.*

CROP	%					PPM				
	N	P	K	Ca	Mg	Fe	Mn	Zn	Cu	B
Adiantum	1.5–2.5	0.4–0.8	2.0–3.0	0.2–0.3	0.2–0.4	—	—	—	—	—
Aechmea	1.5–2.0	0.4–0.7	1.5–2.5	0.5–1.0	0.4–0.8	—	—	—	—	—
Aglaonema	2.5–3.5	0.2–0.4	2.5–3.5	1.0–1.5	0.3–0.6	—	—	—	—	—
Alstroemeria	3.8–7.6	0.3–0.7	3.7–4.8	0.6–1.5	0.2–0.6	175–275	60–200	35–110	5–15	5–50
Antirrhinum	1.0–5.3	0.2–0.6	2.2–4.1	0.5–1.4	0.5–1.0	70–135	60–185	30–55	5–15	20–40
Asparagus	1.5–2.5	0.3–0.5	2.0–3.0	0.1–0.3	0.1–0.3	—	—	—	—	—
Aster	2.2–3.1	0.2–0.7	3.3–3.7	1.0–1.7	0.2–0.4	162–180	65–273	26–121	—	37–46
Begonia x *hiemalis*	3.4–4.6	0.4–0.8	2.0–3.5	0.7–2.4	0.3–0.8	80–390	35–190	20–30	5–10	35–130
Begonia x *semperflorens-cultorum*	4.4–5.2	0.3–0.6	3.4–4.2	0.7–4.2	0.4–1.0	100–260	90–355	50–65	10–15	30–40
Caladium	3.6–4.9	0.4–0.7	2.3–4.1	1.1–1.6	0.1–0.3	65–90	110–135	125–135	5–10	95–145
Catharanthus	4.9–5.4	0.4–0.5	2.9–3.6	1.4–1.6	0.4–0.5	95–150	165–300	40–45	5–10	25–40
Chrysanthemum	4.0–6.5	0.3–1.0	4.5–6.5	1.0–2.0	0.4–0.7	30–350	60–500	15–50	25–75	50–100
Cyclamen	2.9–5.0	0.4–1.0	1.2–4.5	0.3–1.3	0.4–1.3	150–550	100–500	30–100	5–20	70–350
Dianthus	3.2–5.2	0.2–0.3	2.5–6.0	1.0–2.0	0.2–0.5	100–300	50–150	25–75	10–30	30–100
Dieffenbachia	2.5–3.5	0.2–0.4	3.0–4.5	1.0–1.5	0.3–0.8	—	—	—	—	—
Dracaena deremensis 'Janet Craig'	2.0–3.0	0.2–0.3	3.0–4.0	1.5–2.0	0.3–0.6	—	—	—	—	—
Dracaena deremensis 'Warneckii'	2.5–3.5	0.1–0.3	3.0–4.5	1.0–2.0	0.5–1.0	—	—	—	—	—
Dracaena fragrans 'Massangeana'	2.0–3.0	0.1–0.3	1.0–2.0	1.0–2.0	0.5–1.0	—	—	—	—	—
Euphorbia (poinsettia)	4.0–6.0	0.3–0.6	1.5–3.5	0.7–1.8	0.3–1.0	100–300	60–300	25–60	2–10	25–75
Exacum	3.8–5.3	0.3–0.7	2.3–3.4	0.5–0.8	0.4–0.7	55–155	70–165	25–85	5–75	25–60
Ficus benjamina	1.8–2.5	0.1–0.2	1.0–1.5	2.0–3.0	0.4–0.8	—	—	—	—	—
Foliage plants (general)	1.5–3.5	0.2–0.4	1.0–4.0	0.5–2.0	0.3–0.8	31–300	50–150	16–50	6–20	25–100
Freesia	2.7–5.6	0.4–1.2	3.1–5.9	0.4–1.0	0.3–1.8	80–115	30–540	40–110	5–130	30–100
Fuchsia	2.8–4.6	0.4–0.6	2.2–2.5	1.6–2.4	0.4–0.7	95–335	75–220	30–45	5–10	25–35
Gerbera	2.7–4.1	0.3–0.7	3.1–3.9	0.4–4.2	0.3–2.8	60–130	30–260	19–80	2–10	19–50
Helianthus	5.0–6.0	0.7–0.8	5.4–6.3	2.2–2.5	0.6–0.8	—	67–99	77–115	5–8	43–53
Hibiscus	3.5–4.5	0.2–0.6	2.0–2.9	1.9–2.3	0.5–0.7	60–75	135–180	35–50	5–10	35–50
Hydrangea	2.0–3.8	0.3–2.5	2.5–6.3	0.8–1.5	0.2–0.4	85–115	100–345	50–105	5–10	20–25
Impatiens, bedding	3.9–5.3	0.6–0.8	1.8–3.5	2.8–3.3	0.6–0.8	405–885	50–490	65–70	10–15	45–105
Impatiens, New Guinea	3.3–4.6	0.3–0.8	1.2–2.7	0.7–2.7	0.3–0.8	75–300	100–250	40–85	5–14	40–80
Kalanchoe	2.5–5.0	0.2–0.5	2.0–4.8	1.1–4.5	0.4–1.0	75–200	60–250	25–80	5–20	30–60

(Continued)

Table 5-2. Tissue Nutrient Levels of High-Quality Greenhouse Plants *(Continued)*

Note that the indicated values are only general recommendations. The optimum values will vary among firms because of different cultural practices, climate, and cultivars.*

CROP	%					PPM				
	N	P	K	Ca	Mg	Fe	Mn	Zn	Cu	B
Liatris	2.7–3.3	0.2–0.3	1.2–2.3	1.1–1.5	0.4–0.5	200–230	163–178	86–94	—	24–31
Lilium	2.4–4.0	0.1–0.7	2.0–5.0	0.2–4.0	0.3–2.0	100–250	50–250	30–70	5–25	20–25
Pelargonium, zonal	3.8–4.4	0.3–0.5	2.6–3.5	1.4–2.0	0.2–0.4	110–580	270–325	50–55	5–15	40–50
Pelargonium, ivy	3.4–4.4	0.4–0.7	2.8–4.7	0.9–1.4	0.2–0.6	115–270	40–175	10–45	5–15	30–100
Pelargonium, regal	3.0–3.2	0.3–0.6	1.1–3.1	1.2–2.6	0.3–0.9	120–225	115–475	35–50	5–10	15–45
Pelargonium, seed	3.7–4.8	0.4–0.7	2.5–3.9	0.8–2.1	0.2–0.5	120–200	110–285	35–60	5–15	35–60
Petunia	2.8–5.8	0.5–1.2	3.5–5.5	0.6–4.8	0.3–1.4	40–700	90–185	30–90	5–15	20–50
Philodendron	2.0–3.0	0.2–0.3	3.0–4.5	0.5–1.5	0.3–0.6	—	—	—	—	—
Primula	2.5–3.3	0.4–0.8	2.1–4.2	0.6–1.0	0.2–0.4	78–155	50–90	40–45	5–10	30–35
Rhododendron	2.0–3.0	0.2–0.5	1.0–1.6	0.5–1.6	0.2–0.5	50–300	60–150	26–60	5–15	31–100
Rosa	3.5–4.5	0.2–0.3	2.0–2.5	1.0–1.5	0.28–0.32	70–120	80–120	20–40	7–15	40–60
Saintpaulia	2.2–2.7	0.2–0.9	1.5–6.0	0.6–1.7	0.7–1.1	70–320	35–490	20–80	5–30	30–200
Salvia	3.0–4.5	0.3–0.7	3.5–5.0	1.5–2.5	0.3–0.6	—	—	—	—	—
Schefflera (Brassaia)	2.5–3.5	0.2–0.4	2.5–3.5	1.0–1.5	0.3–0.6	—	—	—	—	—
Schefflera (Dizygotheca)	2.0–2.5	0.3–0.4	1.5–2.5	0.5–1.0	0.2–0.3	—	—	—	—	—
Schlumbergera	2.7–3.7	0.5–0.9	6.2–7.0	0.7–0.9	1.6–2.2	105–110	35–130	50–65	10–15	65–70
Sinningia	3.3–3.8	0.3–0.5	4.5–5.0	1.5–2.2	0.4–0.5	70–150	95–170	20–35	5–20	30–35
Streptocarpus	2.0–3.5	0.1–0.7	4.8–5.5	1.2–1.9	0.3–0.5	90–260	130–300	85–130	15–20	55–65
Syngonium	2.5–3.5	0.2–0.3	3.0–4.5	0.4–1.0	0.3–0.6	—	—	—	—	—
Viola x *wittrockiana*	2.5–4.5	0.25–1.00	2.5–5.0	0.6–3.0	0.25–0.75	30–300	25–300	20–100	5–40	20–80

* Adapted from Dole and Wilkins. *Floriculture: Principles and Species* (Upper Saddle River, N.J.: Prentice Hall, 1999), pp. 74–75.

1:2 steps

1. Collect medium. Remove the plant from the container and remove the medium from the root ball. Continue gathering the medium from at least five plants to obtain a representative sample. Mix the media together to obtain a homogeneous sample and place 1 cup (225 ml) of medium into a clean container. If a controlled-release fertilizer is present, removing the capsules is optional. Regardless, ensure no capsules are damaged. (Damaged capsules will release fertilizer and provide nonrepresentative EC values.) It is best to use the PourThru method when slow-release fertilizer is present.

2. Add water. Add a volume of distilled water equal to two times the volume of the medium.

3. Mix. Thoroughly mix medium and water. Allow the mixture to set for a minimum of twenty minutes. (This allows for equilibration.)

4. Calibrate the pH and EC meters prior to testing. The test results are only as good as the last calibration. Calibrate the instruments every day they are used. Always use fresh, standard solutions, and never pour used solution back into the original bottle.

5. Test the samples for pH and EC. Test the slurry twenty minutes to one hour after mixing. Take care not to damage the pH or EC probe on medium components when inserting into the slurry. Record the values on the chart specific to each crop.

Press steps

1. Irrigate the crop one hour before testing. Make sure the medium is saturated. Testing should be per-

Table 5-3. Amount of Distilled or Deionized Water to Apply to Various Containers to Obtain 1.7 oz. (50 ml) of Extract*

CONTAINER SIZE	WATER TO ADD**	
	MILLILITERS	OUNCES
4" (10 cm)	50	1.7
6" (15 cm)	75	2.5
6.5" (16 cm) azalea	100	3.3
1 qt. (1.1 l)	75	2.5
4 qt. (4 l)	150	5.0
12 qt. (12 l)	350	11.7
CELL PACKS 606 (36 plants) 1203 (36 plants) 1204 (48 plants)	50	1.7

* Containers should be brought to container capacity thirty to sixty minutes before applying these amounts.
** These amounts are estimates. Actual amounts will vary depending on crop, medium type, and environmental conditions. Water should be distilled or deionized.

formed after fertilization to ensure that the results are representative of the nutrition being provided to the plugs. Testing can be done after a clear water irrigation, but keep in mind that very different results will likely be obtained. The bottom line is to be consistent by being aware of when testing takes place in relation to the fertilizer application.

2. Choose plugs to sample. Select one row (or equivalent) of plants from five different plug trays. This will ensure representative pH and EC values of the entire crop. Collect and test the leachate from each of the five flats as a sample. Do not combine multiple flats for one sample.

3. Place a container under the plug flat. Select a container that is wide enough to ensure catching the leachate without tipping over.

4. Collect leachate. To collect the leachate, simply press on the top of the plug medium. (The leachate will be pressed out of the drainage hole.) Take care not to damage the plant when squeezing. (The plant will be pushed over but should recover in one to two days.) Two to three drops of leachate from each plug should be sufficient.

5. Calibrate the pH and EC meters prior to testing. The test results are only as good as the last calibration. Calibrate the instruments every day they are used. Always use fresh, standard solutions, and never pour used solution back into the original bottle. Use pH and EC meters that only require a few drops of solution for accurate measures. Although they are generally more expensive than economy meters, the capability to test smaller amounts of leachate will reduce the number of plugs that need to be pressed.

6. Test the samples for pH and EC. Test the leachate as soon as possible. Electrical conductivity will not vary much over time if there is no evaporation of the sample. However, the pH can change within two hours.

Testing, interpreting, and managing medium EC

EC charts have been developed for recording values obtained with the PourThru, 1:2, and press extraction methods (table 5-4). During the first two weeks after transplanting, gradually increase the medium EC to the target level (table 3-2 in chapter 3). Also, as the crop flowers, gradually decrease the EC levels.

If the medium EC rises into or above the upper EC decision range, take action to reduce EC. If the medium EC drops into or below the lower EC decision range, take action to increase EC.

Adjusting medium EC

The medium EC changes over time due to many factors. The five main factors are: (1) initial components and amendments in the media, including the media type and initial nutrient charge; (2) fertility regime; (3) the crop's nutrient demands; (4) the crop's development stage; and (5) the EC of the irrigation water. During production the crop being grown or the medium composition cannot change, but adjustments to the fertilization program can.

Steps to lower EC

The following (in preferred order) are immediate steps to lower the medium's EC.

1. Lower fertilizer rates. Decrease the fertilization rate or decrease the frequency of fertilization (i.e., irrigate with clear water).

2. Leach with clear water to reduce the salts level. The medium should be irrigated to allow for 20% leaching and then irrigated again immediately. The medium should then be allowed to dry to the usual stage. If further leaching is required, the double-irrigation treatment can be repeated. Recheck the EC values to make sure they are within the acceptable range.

Steps to increase EC

Listed below (in preferred order) are immediate steps to raise media EC.

1. Increase fertilizer rates. A corrective nitrogen fertilization will return the lower leaves to the normal green color within one to two weeks. Do not overapply. It is important to correct nutrient deficiencies when symptoms first appear because lower leaf drop or necrosis cannot be reversed.

2. Increase the fertilization frequency. Use constant fertilization and discontinue any clear water irrigation until the EC levels are within the acceptable range.

If you are using calcium nitrate [$Ca(NO_3)_2$] plus potassium nitrate (KNO_3), remember to supply phosphorus (P), magnesium (Mg), and micronutrients to the plants. If you are using 20-10-20 or 20-20-20, remember to supply calcium (Ca) and magnesium (Mg) to the plants. If you are using a urea-nitrogen source, EC values will be lower than a conventional fertilizer using a nitrate or ammoniacal-nitrogen source.

Table 5-4. EC Interpretation Values for Various Extraction Methods[1]

Values are based on actively growing plants, which have moderate nutrient requirements.

1:5	1:2	SME	PourThru[2]	INDICATION
0–0.11	0–0.25	0–0.75	0–1.0	**Very low.** Nutrient levels may not be sufficient to sustain rapid growth.
0.12–0.35	0.26–0.75	0.76–2.0	1.0–2.6	**Low.** Suitable for seedlings, bedding plants, and salt-sensitive plants.
0.36–0.65	0.76–1.25	2.0–3.5	2.6–4.6	**Normal.** Standard root zone range for most established plants. Upper range for salt-sensitive plants.
0.66–0.89	1.26–1.75	3.5–5.0	4.6–6.5	**High.** Reduced vigor and growth may result, particularly during hot weather.
0.9–1.10	1.76–2.25	5.0–6.0	6.6–7.8	**Very high.** May result in salt injury due to reduced water uptake. Reduced growth rates likely. Symptoms include marginal leaf burn and wilting.
>1.1	>2.25	>6.0	>7.8	**Extreme.** Most crops will suffer salt injury at these levels. Immediate leaching is required.

1 Adapted from *On-site Testing of Growing Media and Irrigation Water.* 1996. British Columbia Ministry of Agriculture.

2 Due to the variability of PourThru-method results, growers should always compare their results to the SME method to establish ranges.

Keeping records

Keeping accurate records of routine analyses, such as media pH, EC, and water quality, can provide great insight when making decisions about corrective procedures. Knowing the pH and EC history of a crop can save time, labor, and money by aiding in selecting the proper corrective procedures or production regime.

For example, if a pH value of 5.9 is obtained on a four-week-old geranium grown in a soilless medium, the value may or may not be an indication of a problem. If the pH has been 5.9–6.0 for several weeks, then a pH adjustment may not be necessary. If an alteration is desired, then a simple change in fertilizer choice may increase or maintain the pH in the desired range.

However, if the previous pH values have been 6.5–6.7, indicating that the current value of 5.9 is a dramatic drop in pH, then more extreme corrective procedures, such as applying flowable limestone, may be warranted. Knowing values from earlier in the crop cycle may influence pH management decisions, regardless of the current values.

What to record

Ideally, pH and EC records should be kept on all crops; however, this is not practical. pH and EC records should be collected on crops with a significant amount of sales or on crops with reoccurring production problems.

Initial values of the medium should also be checked and recorded. Each time a new shipment of medium is received from the manufacturer or mixed onsite, check the pH and EC values. Gradual differentiation in the calibration of mixing equipment may go undetected. However, if records have been kept, the trends will be visible.

Also record water quality values. Since water quality analysis is generally not performed as frequently as other analyses, changes in water values, such as alkalinity, EC, or micronutrient concentrations, may go undetected. Even if values are within acceptable ranges, observing trends of increased or decreased concentration may allow you to anticipate a problem and give you ample time to adjust fertilizer regimes, to secure a new water source, or to install a filtration system.

Charts

Charts are great tools to visually depict trends. Be sure to record the factor being tested, the crop, and dates. For crop medium pH and EC analysis, it may be a good idea to list the dates according to crop schedule (i.e., sow date, transplant date, and sell date) in addition to calendar dates.

Calibrating Meters

Every greenhouse grower should own a pH and EC meter. These meters are important analysis tools. Why? Because immediate decisions can be made about a crop's nutritional status based on the medium pH and EC values.

pH and EC meter principles

By definition, pH is the negative log of the hydrogen ion activity of a medium solution or a measure of acidity or basicity of a solution. The more hydrogen ions within the solution, the lower (more acidic) the pH value will be, and vice versa. How does a pH meter accurately provide you with the pH value? Modern meters have a single probe that contains two types of electrodes. One electrode (measurement cell) determines the hydrogen ion activity, while the reference electrode offers constant voltage output. As the probe is submersed into an aqueous solution, electrical current passes from the reference electrode into the measurement cell, then along a silver/silver chloride wire to the meter. The pH value (in millivolts) displayed is the difference between the measurement cell and the electrical current produced by the reference electrode.

An EC meter measures the concentration of *all* soluble salts dissolved in a solution, but it does not determine which salts are present at specific concentrations. The meter measures an electric current that passes through a solution via a probe with two metal prongs. Electrical current flows between the two prongs; the higher the salt concentration, the easier it is for electrical current to move through the solution, thus a higher EC value.

Electrical conductivity is measured in millisiemens per centimeter (mS/cm) or millimhos per centimeter (mmhos/cm). Some EC meters or combination pH/EC meters express salt concentration in total dissolved solids (TDS). This is expressed as parts per million (ppm), which is mS/cm multiplied by 640 or 700. These and other conversion values are listed in table 5-5.

Calibration solutions

Some pH meters require two pH calibration solutions of 7.0 and 4.0 (the range of most soilless media), but many of the portable units have only a single-point calibration. Some instrument companies also carry pH 6.0 standards that work well for meters requiring only a single-point calibration. EC meters generally require one standard. If possible, use a calibration solution reported in mS/cm (or dS/cm, which is equivalent), as most meters and analytical labs report values in mS/cm. However, converting values from ppm or µmho/cm (table 5-6) is relatively easy. Always store the meters and solutions at room temperature. Avoid fluctuating temperatures and high humidity; *do not store the meters and solutions in the greenhouse.* Throw away decanted calibration solutions after use. The solutions should never be recycled back into the original solution container.

Electrode care

Meters and probes are sensitive tools that need to be maintained properly. Store the pH electrode according to manufacturer instructions. Probes should never be stored in distilled or deionized water. With proper maintenance and care, electrodes should last for two or more years. However, all pH electrodes eventually fail because of constant ion exchange between the two electrodes on the probe. Avoid scratching the probe because the electrode will not function properly.

Temperature

Why is temperature so important when taking measurements of the medium? At different temperatures, hydrogen ions move at different rates in solution. Thus, the pH probe will measure greater or fewer hydrogen ions, depending on the solution temperature. Changing the solution temperature takes time; therefore, most meters have an automatic temperature compensation (ATC) function. This eliminates the need for mathematical conversions to adjust the solution pH for temperature.

Power source

Along with ATC functions, many meters have automatic shut-off functions. Some meters do not have this characteristic, which can result in shortened battery life. Always keep fresh backup batteries for the meter. Most pen-type meters run on watch batteries, which can be quite costly and frustrating to replace if there is not an automatic shut-off function. Fresh batteries expedite calibration and acquisition of pH and EC values.

Calibration

Outlined below are steps for calibration and use of a combined pH and EC meter (i.e., Hanna Model HI 9811). Meters must be calibrated prior to use. (Each

Table 5-5. Units for Expressing Electrical Conductivity (EC)

METHOD	ABBREVIATION	UNITS	EXAMPLE
Millisiemens	mS/cm	EC x 10^{-3}/cm	2.25 mS/cm
Millimhos	mmhos/cm	EC x 10^{-3}/cm	2.25 mmhos/cm
A new term for millimhos is millisiemens, which is the metric (SI) unit of expressing electrical conductance. There is no change in value, just terminology.			
Decisiemens	dS/m	EC x 10^{-1}/m	2.25 dS/m
Decisiemens per meter is the common term used in scientific literature to express electrical conductance. The term *deci-* means one-tenth and the term *milli-* means one-thousandth, so a deci- is one hundred times greater than a milli-. While expressing dS/m, the denominator is given in terms per meter (m) and for mS/cm the denominator is given in terms per centimeter (cm). One meter contains 100 cm, therefore when comparing values in dS/m and mS/cm, the zeros cancel out mathematically, and the decimal point appears at the same place for both units (i.e., 2.25 dS/m = 2.25 mS/cm).			
	mho x 10^{-5} /cm	EC x 10^{-5}/cm	225 mho x 10^{-5}/cm
Some labs prefer to express EC as a whole number (i.e., 225); therefore, the decimal point is shifted two places to the right.			
Micromhos	µmhos/cm	EC x 10^{-6}/cm	2250 µmhos x 10^{-6}/cm
The term *micro-* means one-millionth and is one thousand times smaller than a milli-.			

day it is used should be sufficient.) Please note the steps for proper use and storage of the meter.

pH electrode calibration and use

1. Pour enough pH 7.0, 6.0, or 4.0 solution into a small cup (3 oz. [85 ml] or similar) to submerge the tip of the probe in solution.
2. With the meter on, press the pH button.
3. Place the probe into the room-temperature pH 7.0, 6.0, or 4.0 solution. Make sure not to immerse the probe too deeply (not more than 1.5" [4 cm]). Stir gently and wait approximately ten seconds. Adjust the pH calibration dial until pH 7.0, 6.0, or 4.0 appears on the display, depending upon which calibration solution is used.
4. The meter is now ready to use. Rinse the electrode with water, and then immerse the electrode in the sample. Stir gently and wait until the display stabilizes. Read and record the pH of the sample.
5. After use, rinse the electrode with water and store the electrode with a few drops of pH 7.01 standard or a recommended storage solution in the protective cap. Always replace the protective cap after use.

EC electrode calibration and use

1. Pour enough of the calibration solution into a small cup to submerge the tip of the probe in solution.
2. Turn on the meter and place the probe into the calibration solution. Stir gently and wait approximately twenty seconds for the reading to stabilize. Turn the calibration dial to match the solution value. For example, if using 1.41 mS/cm solution, adjust the dial until 1.41 mS/cm appears on the meter. If using 12.88 mS/cm solution, adjust the dial until the meter reads 12.88 mS/cm.
3. The meter is now ready to use. Rinse the electrode with water, and then immerse the electrode (don't let it touch the bottom of the container) in the sample. Stir gently and wait until the display stabilizes. Read and record the EC for the sample.
4. After use, rinse the electrode with water and store the electrode in the protective cap or as recommended by the manufacturer. Always replace the protective cap after use.

Measuring injector accuracy via EC

An EC meter can be used to measure the accuracy of a fertilizer injector. Table 5-7 provides EC values for numerous fertilizers marketed by several fertilizer companies. The expected EC value is given for various concentrations of each fertilizer based on nitrogen concentrations from 50–1,000 parts per million (ppm). Many companies print these EC values on the fertilizer bag or make them available in the technical literature. Presented in table 5-8 is a similar pattern of EC values for six single salts used for formulating fertilizers. Table 5-9 presents EC values for three "homemade" fertilizers that are commonly formulated by individual greenhouse firms using individual salts (recipes for these are given in table 5-10).

The electrical conductivity units used in tables 5-7 through 5-10 is millisiemens per centimeter (mS/cm), which is identical to millimhos per centimeter (mmhos/cm) or decisiemens per meter (dS/cm). If you are more familiar with EC being reported in mmhos/cm (mho x 10^{-3}/cm), µmho/cm (mho x 10^{-6}/cm), or mho x 10^{-5}/cm, table 5-6 provides conversion factors.

To test the accuracy of the injector, follow these steps:
1. Calibrate the electrical conductivity (EC) meter using a standard solution.
2. Take an EC reading of the clear irrigation water. Flush the line completely (for about five minutes) prior to taking the sample.
3. Take an EC reading of final fertilizer solution. Make sure the fertilizer solution has been running through the line for about five minutes prior to taking the sample.
4. Subtract the EC reading of the clear water from the fertilizer solution.
5. Compare the value determined in step 4 with values on the fertilizer bag (if available) or to values listed in tables 5-7, 5-8, and 5-9.
6. Adjust the injector or the fertilizer concentration if the value measured is not within 5% of the target EC value.

The following is an example of the testing steps to check an injector for accuracy using 250 ppm nitrogen from Sun Gro's Technigro Plus 20-9-20 applied with a 1:100 injector.
1. Calibrate the EC meter.
2. The EC reading of the clear irrigation water was 0.34 mS/cm.
3. The EC reading of the final fertilizer solution coming out of the hose was 2.01 mS/cm.
4. The EC contributed by the fertilizer salts was 1.67 mS/cm (2.01 - 0.34 = 1.67).

Table 5-6. Conversion Factors among Electrical Conductivity (EC) Units		
FROM	TO	MULTIPLY BY:
mmhos/cm or mS/cm or dS/m	mho x 10^{-5}/cm	100
mho x 10^{-5}/cm	mmhos/cm or mS/cm or dS/m	0.01
mmhos/cm or mS/cm or dS/m	µmho or mho x 10^{-6}/cm	1,000
µmho or mho x 10^{-6}/cm	mmhos/cm or mS/cm or dS/m	0.001
mmhos/cm or mS/cm or dS/m	ppm	670[1]
ppm	mmhos/cm or mS/cm or dS/m	0.00149251
mho x 10^{-5}/cm	ppm	6.701
ppm	mho x 10^{-5}/cm	0.149251
µmho or mho x 10^{-6}/cm	ppm	0.6701
ppm	µmho or mho x 10^{-6}/cm	1.49251

[1] Some labs report EC in the terms of ppm or convert EC to ppm. Although 670 is the basis used in this example, the conversion factor can vary from 640 to 700. This conversion factor is an average because of the variability in the type of fertilizer salts contributing to the EC of the medium in each sample. This conversion should be considered a broad approximation. Expressing EC in terms of mS/cm or mmhos/cm is the preferred method.

5. According to table 5-7, the EC of a 250-ppm solution of Sun Gro's Technigro Plus 20-9-20 should be 1.73.

6. The test reading (1.67 mS/cm) was within 5% of the target value, so no adjustment was needed.

Author's note: Copies of the PourThru Manual *(a forty-four-page guide to the PourThru monitoring method),* Plant Root Zone Management Manual *(an eighty-eight-page guide to managing, monitoring, and preventing root zone problems,) and the four-part series of color photos of the most common plant nutritional disorders and corrective procedures can be ordered from the North Carolina Commercial Flower Growers' Association, (919) 334-0093, or at www.nccfga.org.*

Table 5-7. EC Values for Dilution Levels for Manufacturers' Fertilizers

Electrical conductivity (EC in mS/cm) values for fourteen dilution levels ranging from 50–1,000 ppm nitrogen for each of numerous greenhouse fertilizers produced by these manufacturers.

FERTILIZER	NITROGEN CONCENTRATION (PPM)													
	50	100	150	200	250	300	350	400	500	600	700	800	900	1,000
Daniels														
10-4-3	0.19	0.38	0.56	0.80	0.94	1.13	1.32	1.50	1.85	—	—	—	—	—
Sun Gro Technigro														
24-7-15 Plus	0.23	0.45	0.67	0.89	1.11	1.33	1.55	1.77	2.21	2.65	3.09	3.53	3.97	4.41
20-18-20	0.24	0.47	0.71	0.94	1.18	1.41	1.64	1.88	2.35	2.81	3.28	3.75	4.22	4.69
20-18-18 Plus	0.24	0.48	0.71	0.95	1.19	1.42	1.66	1.89	2.37	2.84	3.31	3.78	4.25	4.73
20-9-20 Plus	0.34	0.69	1.04	1.38	1.73	2.08	2.42	2.77	3.47	4.16	4.85	5.55	6.24	6.94
17-5-24 Plus	0.38	0.76	1.14	1.52	1.90	2.28	2.66	3.04	3.80	4.56	5.32	6.08	6.84	7.60
16-17-17 Plus	0.37	0.73	1.10	1.46	1.83	2.19	2.55	2.92	3.65	4.37	5.10	5.83	6.56	7.29
15-0-15 Plus	0.37	0.75	1.13	1.50	1.88	2.26	2.63	3.01	3.77	4.52	5.27	6.03	6.78	7.54
13-2-13 Plus	0.42	0.83	1.25	1.66	2.08	2.49	2.91	3.32	4.15	4.98	5.81	6.64	7.47	8.30
Greencare														
17-5-17	0.32	0.64	0.96	1.28	1.60	1.92	2.24	2.56	—	—	—	—	—	—
13-2-13	0.34	0.68	1.02	1.36	1.70	2.04	2.38	2.72	—	—	—	—	—	—
Masterblend														
30-10-10	0.07	0.14	0.21	0.28	0.35	0.42	0.49	0.56	0.70	0.84	0.98	1.12	1.26	1.40
27-15-12	0.11	0.21	0.32	0.42	0.53	0.63	0.74	0.84	1.05	1.26	1.47	1.68	1.89	2.10
25-10-10	0.08	0.17	0.25	0.34	0.42	0.50	0.58	0.67	0.84	1.00	1.18	1.34	1.51	1.70
25-5-30	0.13	0.27	0.40	0.54	0.67	0.81	0.94	1.08	1.35	1.62	1.89	2.16	2.43	2.70
25-5-20	0.13	0.27	0.40	0.54	0.67	0.81	0.94	1.08	1.35	1.62	1.89	2.16	2.43	2.70
25-0-25	0.15	0.3	0.45	0.60	0.75	0.90	1.05	1.20	1.50	1.80	2.10	2.40	2.70	3.00
21-7-7 Acid	0.28	0.56	0.84	1.12	1.40	1.68	1.96	2.24	2.80	3.36	3.92	4.48	5.04	5.60
21-7-7 Neutral	0.21	0.42	0.63	0.84	1.05	1.26	1.47	1.68	2.10	2.52	2.94	3.36	3.78	4.20
20-20-20	0.20	0.4	0.60	0.80	1.00	1.20	1.40	1.60	2.00	2.40	2.80	3.20	3.60	4.00
20-19-18	0.20	0.4	0.60	0.80	1.00	1.20	1.40	1.60	2.00	2.40	2.80	3.20	3.60	4.00
20-15-25	0.21	0.42	0.63	0.84	1.05	1.26	1.47	1.68	2.10	2.52	2.94	3.36	3.78	4.20
20-10-20	0.33	0.65	0.98	1.30	1.63	1.95	2.28	2.60	3.25	3.90	4.55	5.20	5.85	6.50
20-5-30	0.22	0.45	0.67	0.90	1.12	1.35	1.57	1.80	2.25	2.70	3.15	3.60	4.05	4.50
20-2-20	0.31	0.62	0.93	1.24	1.55	1.86	2.17	2.48	3.10	3.72	4.34	4.96	5.58	6.20
20-0-20	0.34	0.67	1.01	1.34	1.67	2.01	2.34	2.68	3.35	4.02	4.69	5.36	6.03	6.70
17-0-17	0.35	0.7	1.05	1.40	1.75	2.10	2.45	2.80	3.50	4.20	4.90	5.60	6.30	7.00
16-4-12	0.34	0.68	1.02	1.36	1.70	2.04	2.38	2.72	3.40	4.08	4.76	5.44	6.12	6.80
15-30-15	0.31	0.62	0.93	1.24	1.55	1.86	2.17	2.48	3.10	3.72	4.34	4.96	5.58	6.20

(Continued)

Table 5-7. EC Values for Dilution Levels for Manufacturers' Fertilizers *(Continued)*

Electrical conductivity (EC in mS/cm) values for fourteen dilution levels ranging from 50–1,000 ppm nitrogen for each of numerous greenhouse fertilizers produced by these manufacturers.

FERTILIZER	NITROGEN CONCENTRATION (PPM)													
15-17-17	0.37	0.74	1.11	1.48	1.85	2.22	2.59	2.96	3.70	4.44	—	—	—	—
15-16-17	0.34	0.69	1.03	1.38	1.72	2.07	2.41	2.76	3.45	4.14	4.83	5.52	6.21	6.90
15-15-15	0.32	0.65	0.97	1.30	1.62	1.95	2.27	2.60	3.25	3.90	4.55	5.20	5.85	6.50
15-11-29	0.34	0.69	1.03	1.38	1.72	2.07	2.41	2.76	3.45	4.14	4.83	5.52	6.21	6.90
15-10-30	0.35	0.7	1.05	1.40	1.75	2.10	2.45	2.80	3.50	4.20	4.90	5.60	6.30	7.00
15-5-25	0.44	0.87	1.30	1.74	2.18	2.61	3.05	3.48	4.35	5.22	6.09	6.96	7.83	8.70
15-3-20	0.35	0.7	1.05	1.40	1.75	2.10	2.45	2.80	3.50	4.20	4.90	5.60	6.30	7.00
15-0-15	0.37	0.75	1.12	1.50	1.87	2.25	2.62	3.00	3.75	4.50	5.25	6.00	6.75	7.50
14-0-14	0.37	0.75	1.12	1.50	1.87	2.25	2.62	3.00	3.75	4.50	5.25	6.00	6.75	7.50
13-2-13	0.37	0.75	1.12	1.50	1.87	2.25	2.62	3.00	3.75	4.50	5.25	6.00	6.75	7.50
12-36-14	0.53	1.05	1.58	2.10	2.63	3.15	3.68	4.20	5.25	6.30	7.35	8.40	9.45	10.50
12-0-43	0.48	0.95	1.42	1.90	2.37	2.85	3.32	3.80	4.75	5.70	6.65	7.60	8.55	9.50
10-30-20	0.48	0.95	1.43	1.90	2.38	2.85	3.33	3.80	4.75	5.70	6.65	7.60	8.55	9.50
10-15-20	0.58	1.15	1.73	2.30	2.88	3.45	4.03	4.60	5.75	6.90	8.05	9.20	10.35	11.50
9-45-15	0.60	1.2	1.80	2.40	3.00	3.60	4.20	4.80	6.00	7.20	8.40	9.60	10.85	12.00
Scotts Peters Excel														
21-5-20	0.32	0.63	0.95	1.26	1.58	1.89	2.21	2.52	3.15	3.78	4.41	5.04	5.67	6.30
15-5-15	0.35	0.69	1.04	1.38	1.73	2.07	2.42	2.76	3.45	4.14	4.83	5.52	6.21	6.90
15-2-20	0.39	0.77	1.16	1.54	1.93	2.31	2.70	3.08	3.85	4.62	5.39	6.16	6.93	7.70
13-2-13	0.38	0.75	1.13	1.50	1.88	2.25	2.63	3.00	3.75	4.50	5.25	6.00	6.75	7.50
10-0-0	0.35	0.70	1.05	1.40	1.75	2.10	2.45	2.80	3.50	4.20	4.90	5.60	6.30	7.00
Scotts Peters														
24-8-16	0.23	0.45	0.68	0.90	1.13	1.35	1.58	1.80	2.25	2.70	3.15	3.60	4.05	4.50
21-7-7	0.26	0.52	0.78	1.04	1.30	1.56	1.82	2.08	2.60	3.12	3.64	4.16	4.68	5.20
20-20-20	0.21	0.41	0.62	0.82	1.03	1.23	1.44	1.64	2.05	2.46	2.87	3.28	3.69	4.10
20-10-20	0.31	0.62	0.93	1.24	1.55	1.86	2.17	2.48	3.10	3.72	4.34	4.96	5.58	6.20
13-2-13 No Minors	0.37	0.74	1.11	1.48	1.85	2.22	2.59	2.96	3.70	4.44	5.18	5.92	6.66	7.40
5-11-26	0.73	1.45	2.18	2.90	3.63	4.35	5.08	5.80	7.25	8.70	10.15	11.60	13.05	14.50
Scotts Peters Peatlite														
20-10-20	0.30	0.59	0.89	1.18	1.48	1.77	2.07	2.36	2.95	3.54	4.13	4.72	5.31	5.90
20-10-20 No Boron	0.29	0.57	0.86	1.14	1.43	1.71	2.00	2.28	2.85	3.42	3.99	4.56	5.13	5.70
20-2-20	0.30	0.60	0.90	1.20	1.50	1.80	2.10	2.40	3.00	3.60	4.20	4.80	5.40	6.00
18-8-17	0.33	0.65	0.98	1.30	1.63	1.95	2.28	2.60	3.25	3.90	4.55	5.20	5.85	6.50
17-3-17	0.33	0.65	0.98	1.30	1.63	1.95	2.28	2.60	3.25	3.90	4.55	5.20	5.85	6.50

(Continued)

Table 5-7. EC Values for Dilution Levels for Manufacturers' Fertilizers *(Continued)*

Electrical conductivity (EC in mS/cm) values for fourteen dilution levels ranging from 50–1,000 ppm nitrogen for each of numerous greenhouse fertilizers produced by these manufacturers.

FERTILIZER	NITROGEN CONCENTRATION (PPM)													
15-16-17	0.31	0.62	0.93	1.24	1.55	1.86	2.17	2.48	3.10	3.72	4.34	4.96	5.58	6.20
15-5-25	0.39	0.77	1.16	1.54	1.93	2.31	2.70	3.08	3.85	4.62	5.39	6.16	6.93	7.70
15-3-25 Plus Iron	0.40	0.79	1.19	1.58	1.98	2.37	2.77	3.16	3.95	4.74	5.53	6.32	7.11	7.90
15-0-15	0.36	0.71	1.07	1.42	1.78	2.13	2.49	2.84	3.55	4.26	4.97	5.68	6.39	7.10
10-30-20	0.43	0.86	1.29	1.72	2.15	2.58	3.01	3.44	4.30	5.16	6.02	6.88	7.74	8.60
Peters Specialty Water-Soluble Fertilizer														
30-10-10	0.09	0.18	0.27	0.36	0.45	0.54	0.63	0.72	0.90	1.08	1.26	1.44	1.62	1.80
20-20-20	0.26	0.51	0.77	1.02	1.28	1.53	1.79	2.04	2.55	3.06	3.57	4.08	4.59	5.10
Champion Water-Soluble Fertilizer														
21-18-18	0.19	0.37	0.56	0.74	0.93	1.11	1.30	1.48	1.85	2.22	2.59	2.96	3.33	3.70
21-8-18	0.30	0.59	0.89	1.18	1.48	1.77	2.07	2.36	2.95	3.54	4.13	4.72	5.31	5.90
Jack's Professional														
30-10-10	0.34	0.68	1.02	1.36	1.70	2.04	2.38	2.72	3.40	4.08	4.76	5.44	6.12	6.80
27-15-12	0.11	0.22	0.33	0.44	0.55	0.66	0.77	0.88	1.10	1.32	1.54	1.76	1.98	2.20
25-9-17	0.28	0.55	0.83	1.10	1.38	1.65	1.93	2.20	2.75	3.30	3.85	4.40	4.95	5.50
25-5-15 HFP	0.29	0.57	0.86	1.14	1.43	1.71	2.00	2.28	2.85	3.42	3.99	4.56	5.13	5.70
24-8-16	0.20	0.40	0.60	0.80	1.00	1.20	1.40	1.60	2.00	2.40	2.80	3.20	3.60	4.00
22-5-16 FeED	0.34	0.67	1.01	1.34	1.68	2.01	2.35	2.68	3.35	4.02	4.69	5.36	6.03	6.70
21-8-18	0.32	0.64	0.96	1.28	1.60	1.92	2.24	2.56	3.20	3.84	4.48	5.12	5.76	6.40
21-7-7 Neutral	0.31	0.62	0.93	1.24	1.55	1.86	2.17	2.48	3.10	3.72	4.34	4.96	5.58	6.20
21-7-7 Acid	0.30	0.60	0.90	1.20	1.50	1.80	2.10	2.40	3.00	3.60	4.20	4.80	5.40	6.00
21-5-20 LX	0.31	0.62	0.93	1.24	1.55	1.86	2.17	2.48	3.10	3.72	4.34	4.96	5.58	6.20
21-2-18	0.36	0.72	1.08	1.44	1.80	2.16	2.52	2.88	3.60	4.32	5.04	5.76	6.48	7.20
20-8-20	0.40	0.80	1.20	1.60	2.00	2.40	2.80	3.20	4.00	4.80	5.60	6.40	7.20	8.00
20-5-19 LX	0.32	0.64	0.96	1.28	1.60	1.92	2.24	2.56	3.20	3.84	4.48	5.12	5.76	6.40
20-5-30	0.25	0.50	0.75	1.00	1.25	1.50	1.75	2.00	2.50	3.00	3.50	4.00	4.50	5.00
20-3-19 FeED	0.32	0.64	0.96	1.28	1.60	1.92	2.24	2.56	3.20	3.84	4.48	5.12	5.76	6.40
20-20-20 CP	0.21	0.42	0.63	0.84	1.05	1.26	1.47	1.68	2.10	2.52	2.94	3.36	3.78	4.20
20-19-18	0.20	0.40	0.60	0.80	1.00	1.20	1.40	1.60	2.00	2.40	2.80	3.20	3.60	4.00
20-10-20 GP	0.32	0.64	0.96	1.28	1.60	1.92	2.24	2.56	3.20	3.84	4.48	5.12	5.76	6.40
20-10-20 PL	0.32	0.64	0.96	1.28	1.60	1.92	2.24	2.56	3.20	3.84	4.48	5.12	5.76	6.40
20-0-20	0.24	0.48	0.72	0.96	1.20	1.44	1.68	1.92	2.40	2.88	3.36	3.84	4.32	4.80
17-5-26	0.40	0.80	1.20	1.60	2.00	2.40	2.80	3.20	4.00	4.80	5.60	6.40	7.20	8.00
17-5-19 FeED	0.39	0.78	1.17	1.56	1.95	2.34	2.73	3.12	3.90	4.68	5.46	6.24	7.02	7.80
17-15-17	0.29	0.58	0.87	1.16	1.45	1.74	2.03	2.32	2.90	3.48	4.06	4.64	5.22	5.80
17-4-17 LX	0.36	0.72	1.08	1.44	1.80	2.16	2.52	2.88	3.60	4.32	5.04	5.76	6.48	7.20
17-3-19 FeED	0.37	0.74	1.11	1.48	1.85	2.22	2.59	2.96	3.70	4.44	5.18	5.92	6.66	7.40
17-0-17	0.35	0.70	1.05	1.40	1.75	2.10	2.45	2.80	3.50	4.20	4.90	5.60	6.30	7.00

(Continued)

Table 5-7. EC Values for Dilution Levels for Manufacturers' Fertilizers *(Continued)*

Electrical conductivity (EC in mS/cm) values for fourteen dilution levels ranging from 50–1,000 ppm nitrogen for each of numerous greenhouse fertilizers produced by these manufacturers.

FERTILIZER	NITROGEN CONCENTRATION (PPM)													
16-4-20	0.35	0.70	1.05	1.40	1.75	2.10	2.45	2.80	3.50	4.20	4.90	5.60	6.30	7.00
16-4-17 FeED	0.35	0.69	1.04	1.38	1.73	2.07	2.42	2.76	3.45	4.14	4.83	5.52	6.21	6.90
16-4-12	0.35	0.70	1.05	1.40	1.75	2.10	2.45	2.80	3.50	4.20	4.90	5.60	6.30	7.00
16-2-15 LX	0.39	0.78	1.17	1.56	1.95	2.34	2.73	3.12	3.90	4.68	5.46	6.24	7.02	7.80
15-5-25	0.41	0.82	1.23	1.64	2.05	2.46	2.87	3.28	4.10	4.92	5.74	6.56	7.38	8.20
15-5-15 LX	0.38	0.76	1.14	1.52	1.90	2.28	2.66	3.04	3.80	4.56	5.32	6.08	6.84	7.60
15-4-15 FeED	0.38	0.76	1.14	1.52	1.90	2.28	2.66	3.04	3.80	4.56	5.32	6.08	6.84	7.60
15-30-15	0.29	0.58	0.87	1.16	1.45	1.74	2.03	2.32	2.90	3.48	4.06	4.64	5.22	5.80
15-2-20 FeED	0.37	0.74	1.11	1.48	1.85	2.22	2.59	2.96	3.70	4.44	5.18	5.92	6.66	7.40
15-16-17	0.34	0.68	1.02	1.36	1.70	2.04	2.38	2.72	3.40	4.08	4.76	5.44	6.12	6.80
15-15-15	0.35	0.7	1.05	1.40	1.75	2.10	2.45	2.80	3.50	4.20	4.90	5.60	6.30	7.00
15-11-29	0.48	0.96	1.44	1.92	2.40	2.88	3.36	3.84	4.80	5.76	6.72	7.68	8.64	9.60
15-10-30	0.35	0.7	1.05	1.40	1.75	2.10	2.45	2.80	3.50	4.20	4.90	5.60	6.30	7.00
15-0-15	0.37	0.74	1.11	1.48	1.85	2.22	2.59	2.96	3.70	4.44	5.18	5.92	6.66	7.40
15-0-14 LX	0.37	0.74	1.11	1.48	1.85	2.22	2.59	2.96	3.70	4.44	5.18	5.92	6.66	7.40
15-0-0 LX	0.37	0.74	1.11	1.48	1.85	2.22	2.59	2.96	3.70	4.44	5.18	5.92	6.66	7.40
14-5-38 LX	0.39	0.78	1.17	1.56	1.95	2.34	2.73	3.12	3.90	4.68	5.46	6.24	7.02	7.80
13-2-13 LX	0.41	0.82	1.23	1.64	2.05	2.46	2.87	3.28	4.10	4.92	5.74	6.56	7.38	8.20
13-0-44	0.48	0.95	1.43	1.90	2.38	2.85	3.33	3.80	4.75	5.70	6.65	7.60	8.55	9.50
12-36-14	0.53	1.05	1.58	2.10	2.63	3.15	3.68	4.20	5.25	6.30	7.35	8.40	9.45	10.50
12-3-15	0.50	1	1.50	2.00	2.50	3.00	3.50	4.00	5.00	6.00	7.00	8.00	9.00	10.00
12-0-43	0.48	0.95	1.43	1.90	2.38	2.85	3.33	3.80	4.75	5.70	6.65	7.60	8.55	9.50
10-30-20	0.49	0.98	1.47	1.96	2.45	2.94	3.43	3.92	4.90	5.88	6.86	7.84	8.82	9.80
10-0-0 LX	0.40	0.8	1.20	1.60	2.00	2.40	2.80	3.20	4.00	4.80	5.60	6.40	7.20	8.00
9-45-15	0.60	1.2	1.80	2.40	3.00	3.60	4.20	4.80	6.00	7.20	8.40	9.60	10.80	12.00
5-50-18	0.49	0.97	1.46	1.94	2.43	2.91	3.40	3.88	4.85	5.82	6.79	7.76	8.73	9.70
Plant Marvel's Nutriculture														
30-10-10	0.11	0.22	0.33	0.43	—	0.66	—	0.85	1.10	—	—	—	—	—
28-18-8	0.10	0.20	0.30	0.40	—	0.60	—	0.80	1.00	—	—	—	—	—
25-15-10	0.15	0.31	0.46	0.62	—	0.92	—	1.23	1.55	—	—	—	—	—
25-10-20	0.16	0.32	0.49	0.65	—	0.98	—	1.30	1.63	—	—	—	—	—
25-5-20	0.14	0.30	0.42	0.61	—	0.90	—	1.20	1.50	—	—	—	—	—
25-0-25	0.15	0.30	0.45	0.61	—	0.92	—	1.22	1.52	—	—	—	—	—
24-8-16	0.21	0.42	0.63	0.85	—	1.27	—	1.70	2.12	—	—	—	—	—
21-8-18	0.32	0.64	0.96	1.28	—	1.92	—	2.56	3.20	—	—	—	—	—
21-7-7 N	0.18	0.36	0.54	0.72	—	1.07	—	1.43	1.80	—	—	—	—	—
21-7-7 A	0.31	0.61	0.92	1.22	—	1.83	—	2.44	3.05	—	—	—	—	—
20-20-20	0.21	0.41	0.62	0.82	—	1.23	—	1.64	2.05	—	—	—	—	—

(Continued)

Table 5-7. EC Values for Dilution Levels for Manufacturers' Fertilizers *(Continued)*

Electrical conductivity (EC in mS/cm) values for fourteen dilution levels ranging from 50–1,000 ppm nitrogen for each of numerous greenhouse fertilizers produced by these manufacturers.

FERTILIZER	NITROGEN CONCENTRATION (PPM)													
20-10-20	0.31	0.62	0.94	1.25	—	1.88	—	2.50	3.13	—	—	—	—	—
20-7-20	0.33	0.65	0.99	1.30	—	1.95	—	2.60	3.25	—	—	—	—	—
20-7-19	0.30	0.60	0.90	1.20	—	1.80	—	2.40	3.00	—	—	—	—	—
20-5-30	0.23	0.47	0.70	0.93	—	1.39	—	1.86	2.33	—	—	—	—	—
20-5-20	0.33	0.65	0.98	1.30	—	1.96	—	2.62	3.25	—	—	—	—	—
20-0-20	0.21	0.41	0.62	0.82	—	1.23	—	1.64	2.05	—	—	—	—	—
19-26-14	0.21	0.42	0.63	0.84	—	1.25	—	1.67	2.09	—	—	—	—	—
18-6-18	0.34	0.68	1.01	1.37	—	2.04	—	2.74	3.40	—	—	—	—	—
18-3-18	0.34	0.68	1.01	1.37	—	2.04	—	2.74	3.40	—	—	—	—	—
17-17-17	0.27	0.54	0.80	1.07	—	1.61	—	2.14	2.68	—	—	—	—	—
17-5-17	0.34	0.68	1.01	1.37	—	2.04	—	2.70	3.40	—	—	—	—	—
17-0-17	0.35	0.70	1.05	1.40	—	2.10	—	2.80	3.50	—	—	—	—	—
16-4-12	0.33	0.68	1.01	1.35	—	2.04	—	2.70	3.37	—	—	—	—	—
15-30-15	0.32	0.64	0.96	1.28	—	1.93	—	2.57	3.21	—	—	—	—	—
15-20-25	0.33	0.66	1.00	1.37	—	2.05	—	2.74	3.42	—	—	—	—	—
15-10-30	0.35	0.71	1.06	1.42	—	2.12	—	2.83	3.54	—	—	—	—	—
15-5-30	0.37	0.74	1.11	1.47	—	2.21	—	2.95	3.68	—	—	—	—	—
15-5-25	0.38	0.76	1.14	1.52	—	2.28	—	3.04	3.80	—	—	—	—	—
15-5-15	0.36	0.73	1.09	1.45	—	2.18	—	2.90	3.63	—	—	—	—	—
15-3-20	0.35	0.70	1.05	1.40	—	2.10	—	2.80	3.50	—	—	—	—	—
15-3-18	0.35	0.71	1.06	1.42	—	2.12	—	2.83	3.54	—	—	—	—	—
15-0-30	0.36	0.71	1.07	1.49	—	2.14	—	2.85	3.55	—	—	—	—	—
15-0-15	0.34	0.69	1.03	1.38	—	2.06	—	2.75	3.44	—	—	—	—	—
14-3-20	0.35	0.71	1.06	1.42	—	2.12	—	2.83	3.55	—	—	—	—	—
14-0-14	0.37	0.75	1.12	1.50	—	2.25	—	3.00	3.75	—	—	—	—	—
13-2-13	0.37	0.75	1.12	1.50	—	2.25	—	3.00	3.75	—	—	—	—	—
13-0-44	0.48	0.95	1.41	1.88	—	2.83	—	3.77	4.73	—	—	—	—	—
12-45-10	0.36	0.71	1.07	1.42	—	2.13	—	2.84	3.55	—	—	—	—	—
12-31-14	0.42	0.84	1.25	1.67	—	2.51	—	3.34	4.18	—	—	—	—	—
12-4-12	0.17	0.35	1.07	1.43	—	2.14	—	2.86	3.58	—	—	—	—	—
10-30-20	0.48	0.96	1.44	1.96	—	2.88	—	3.84	4.95	—	—	—	—	—
10-20-30	0.50	0.99	1.50	1.99	—	2.99	—	3.98	4.79	—	—	—	—	—
7-40-17	0.60	1.2	1.80	2.40	—	3.60	—	4.80	6.00	—	—	—	—	—
5-40-17	1.43	2.85	4.28	5.70	—	8.55	—	11.40	14.25	—	—	—	—	—
4-25-35	1.30	2.6	3.90	5.20	—	7.80	—	10.40	13.00	—	—	—	—	—

Table 5-8. EC Values for Dilution Levels of Fertilizer Salts

Electrical conductivity (EC in mS/cm) values for fourteen dilution levels ranging from 50–1,000 ppm nitrogen or magnesium for each of six salts commonly used for formulating greenhouse fertilizers

FERTILIZER SALTS	NITROGEN CONCENTRATION (PPM)													
	50	100	150	200	250	300	350	400	500	600	700	800	900	1,000
Ammonium nitrate NH_4NO_3 (34% N)	0.23	0.46	0.69	0.92	1.15	1.38	1.61	1.84	2.30	2.76	3.22	3.68	4.14	4.60
Ammonium sulfate $(NH_4)_2SO_4$ (21% N)	0.45	0.90	1.35	1.80	2.25	2.70	3.15	3.60	4.50	5.40	6.30	7.20	8.10	9.00
Sodium nitrate $NaNO_3$ (16% N)	0.43	0.86	1.29	1.72	2.15	2.58	3.01	3.44	4.30	5.16	6.02	6.88	7.74	8.60
Potassium nitrate KNO_3 (14% N)	0.48	0.95	1.42	1.90	2.37	2.85	3.32	3.80	4.75	5.70	6.65	7.60	8.55	9.50
Calcium nitrate $Ca(NO_3)_2$ (15.5% N)	0.37	0.74	1.11	1.48	1.85	2.22	2.59	2.96	3.70	4.44	5.18	5.92	6.66	7.40
FERTILIZER SALTS	**MAGNESIUM CONCENTRATION (PPM)**													
	50	100	150	200	250	300	350	400	500	600	700	800	900	1,000
Epsom salts $MgSO_4 \cdot 7H_2O$ (10% Mg)	0.38	0.75	1.13	1.50	1.88	2.25	2.63	3.00	3.75	4.50	5.25	6.30	6.75	7.50

Table 5-9. EC Values for Dilution Levels for Fertilizers

Electrical conductivity (EC in mS/cm) values for fourteen dilution levels ranging from 50–1,000 ppm nitrogen for each of three fertilizers commonly formulated by greenhouse firms (recipes for these three fertilizers are given in table 5-10).

FERTILIZER	NITROGEN CONCENTRATION (PPM)															
	50	100	150	200	250	300	350	400	450	500	550	600	700	800	900	1,000
15-0-15	0.49	0.85	1.22	1.58	1.95	2.31	2.68	3.04	3.41	3.77	4.14	4.50	5.23	5.96	6.69	7.42
23-0-23	0.35	0.66	0.97	1.28	1.59	1.90	2.21	2.52	2.83	3.14	3.45	3.76	4.38	5.00	5.62	6.24
20-10-20	0.38	0.69	1.00	1.32	1.63	1.94	2.25	2.56	2.87	3.18	3.49	3.80	4.42	5.04	5.67	6.29

Table 5-10. Fertilizer Formulations

Quantity (oz.) of salts to dissolve in 100 gal. of water to make 15-0-15, 23-0-23, or 21-10-20 fertilizer formulations at concentrations of 50–700 ppm each of nitrogen (N) and potassium (K).

		CONCENTRATION OF N AND K_2O[Z]							
FERTILIZER	%NH_4^+[Y]	50	100	200	300	400	500	600	700
15-0-15	0 to 4.5[X]								
potassium nitrate (13.75-0-44.5) +		1.50	3.0	6.0	9.0	12.0	15.0	18.0	21.0
calcium nitrate (15.5-0-0)		2.98	6.0	11.9	17.9	23.8	29.8	35.7	41.7
23-0-23	35								
potassium nitrate (13.75-0-44.5) +		1.50	3.0	6.0	9.0	12.0	15.0	18.0	21.0
ammonium nitrate (34-0-0)		1.36	2.7	5.4	8.1	10.9	13.6	16.3	19.0
20-10-20	40								
potassium nitrate (13.75-0-44.5) +		1.50	3.0	6.0	9.0	12.0	15.0	18.0	21.0
ammonium nitrate (34-0-0) +		1.16	2.3	4.7	7.0	9.3	11.6	14.0	16.3
monoammonium phosphate (12-62-0)		0.55	1.1	2.2	3.3	4.4	5.5	6.6	7.7

[Z] The first two formulations do not contain phosphorus, while in the 20-10-20 formulation phosphorus (P_2O_5) is present at half the concentration of nitrogen.

[Y] Percent of the total nitrogen that is in the ammoniacal (NH_4^+) form. The remainder is nitrate (NO_3^-) nitrogen.

[X] If calcium nitrate contains 1% ammoniacal nitrogen (most available sources do), the 15-0-15 fertilizer formulated will contain 4.5% of its total nitrogen in the ammoniacal form.

6

Matthew Blanchard and Erik Runkle

Temperature

Temperature influences many plant processes and attributes including seed germination, rooting, flowering, production time, plant architecture, and plant quality. Temperature primarily controls the rate of plant development (e.g., time to develop a leaf or time to flower). In comparison, light primarily influences the accumulation of plant biomass (e.g., shoot growth, branching, and stem thickness). Temperature interacts with light throughout crop production and, thus, both environmental factors should be considered when determining appropriate temperature set points.

Temperature is most commonly measured in the United States using degrees Fahrenheit (°F), although scientists and people in many other countries use degrees Celsius (°C). To convert from °F to °C, subtract 32 from the °F value, and then divide by 1.8. For example, to convert 70°F to °C: (70°F – 32) ÷ 1.8 = 21.1°C. To convert from °C to °F, multiply the °C value by 1.8 and then add 32. For example, to convert 15°C to °F: (15°C × 1.8) + 32 = 59°F.

A major advantage of growing crops in a greenhouse is the ability to control temperature. Without adequate temperature controls, a grower is more often at the mercy of the weather. Temperature control is a compelling reason to move outdoor crops into a greenhouse structure, especially when it can be managed with a computerized environmental control system. The ability to manage temperature more easily enables a grower to produce a floriculture crop that is marketable on a predetermined date.

Proper temperature management requires an understanding of how temperature influences plant growth and development. Knowledge of how crops respond to temperature makes it possible to optimize the greenhouse environment so that production time is minimized and crops are grown as efficiently as possible. This chapter discusses the fundamentals of temperature, how it can be measured, and how temperature influences the production of plants and their products. Much of the information known about floriculture crop responses to temperature has been generated through decades of university and industry research.

Air, Medium, and Plant Temperature

During crop production, the development of a plant is influenced by the factors that influence temperature at the growing points of plants (i.e., shoot and root tips). Air temperature generally has the largest effect on the development of leaves and flowers, whereas media temperature primarily influences root growth. Other environmental factors, such as humidity and light intensity, can also influence plant temperature.

Air temperature

Temperature recommendations for crop production are often based on air temperature, because it is very easily measured. Most greenhouse climate control computers measure air temperature, and a specific, user-defined set point is achieved with heating and venting. Common instruments used to measure air temperature include thermometers, thermocouples, thermistors, and resistant temperature detectors (RTD). All of these instruments differ in cost, accuracy, and installation requirements.

Instruments to measure temperature can be connected to data loggers or computers to continuously measure and store data. Data storage allows for temperature to be graphically monitored and the average temperature to be calculated over time. This data retrieval can help make production decisions on temperature settings. Handheld meters are also available to measure air temperature, but they are primarily used to spot-check greenhouse temperatures.

Air temperature sensors should be placed in a location near the crop that is representative of the growing area, such as the center of the greenhouse, and should be positioned near the canopy of plants. Sensors should also be shielded from the sun and supplemental lighting and aspirated with a fan to provide a minimum airflow of approximately 10 ft. per second (3 m/second). Otherwise, the temperature readings may not be accurate.

Medium temperature

The temperature of the growing medium is important during seed germination and the rooting of cuttings. A general recommendation is to maintain a medium temperature during propagation of 72–77°F (22–25°C). However, the recommended medium temperature for propagation differs among species. For example, a recommended medium temperature for the germination of seed impatiens (*Impatiens walle-rana*) is 72–76°F (22–24°C), while it is 65°F (18°C) for nemesia (*Nemesia strumosa*). Recommended media temperatures can also vary depending on the stage of seed germination or rooting of cuttings.

Since misting is often used during propagation, the evaporation of water can lower the temperature of the medium, leaves, and air. Misting or irrigating with cold water can also directly decrease medium temperature. Medium temperature should be closely monitored with a temperature probe stuck into the center of the rooting media. A thermocouple can be inserted into the medium and connected to a data logger for continuous measurement and recording. To increase medium temperature, under-bench heating (i.e., root-zone heating) is often used to accelerate rooting.

During finish production, medium temperature is sometimes important to monitor, especially when growing on a heated floor. A root-zone heating system can reduce greenhouse-heating costs, because the heat is delivered close to the plants. As the medium is heated, warm air rises and heats the stems and leaves of the plant. However, since the actual plant shoot-tip temperature controls the crop time during finish production, plant temperature should still be monitored when using root-zone heating.

For some crops, the medium temperature is especially important to control during early stages of production, when shoot growth occurs below or near the media surface. For example, a recommended media temperature for Easter lilies (*Lilium longiflorum*) is 63–65°F (17–18°C) during the precooling phase and 60–65°F (16–18°C) from the end of cold treatment to shoot emergence.

Plant temperature

The actual temperature of a plant affects how slow or fast a crop develops. Thus, it is important to consider the actual plant temperature during production and not just the medium or air temperature. Actual plant temperature is determined by the transfer of energy between the plant and the surrounding environment and is influenced by many factors including the amount of solar (shortwave) radiation absorbed by the plant, emitted infrared (long-wave) radiation, conduction, convection, and transpiration. Depending on the environmental conditions, plant temperature can be several degrees warmer or cooler than air temperature.

Heat can move from the air to the plant, or vice versa, through a process called convection. Convection is influenced by the temperature gradient between the air and the plant and increases as air movement around the leaf increases. This is one of several reasons to operate horizontal airflow (HAF) fans in the greenhouse, especially at night. HAF fans not only increase air temperature uniformity throughout the greenhouse, but they also increase convective heat transfer between the air and the plant. Air movement in the greenhouse can help raise plant temperature at night and, as a result, reduce crop time.

During the day, air temperature, shortwave radiation, and transpiration have the largest effect on plant temperature. Shortwave radiation that is not absorbed by leaves for photosynthesis is either reflected or transmitted through the leaf and can increase plant temperature. As radiation from the sun increases, plant temperature increases unless leaves can dissipate heat through transpiration or convection. When light intensity is high, leaf temperature can be several degrees warmer than air temperature. A study with poinsettia (*Euphorbia pulcherrima*) showed that as total shortwave radiation increased from 500–3,000 f.c. (5.4–32 klux, 100–600 $\mu mol \cdot m^{-2} \cdot s^{-1}$, or 50–305 $W \cdot m^{-2}$), plant temperature increased by 3°F (1.8°C).

Plant temperature can also increase above the air temperature if light is delivered from high-intensity lighting, such as high-pressure sodium or metal halide lamps. These light sources emit high amounts of infrared radiation, which increase the thermal load on the plant. For example, vinca (*Catharanthus roseus*) exposed to supplemental light from high-pressure sodium lamps at 760 f.c. (8.1 klux or 100 $\mu mol \cdot m^{-2} \cdot s^{-1}$) were 3.1°F (1.7°C) warmer than plants grown in darkness. The higher plant temperature under these lamps can actually be beneficial in accelerating crop development and reducing production time.

Plants are able to dissipate heat through the evaporation of water (transpiration). Transpiration primarily occurs during the day, but some species transpire at night. In some environments, transpiration can lower the leaf temperature by as much as 5°F (3°C). Plant transpiration generally increases at higher temperatures and decreases at lower temperatures. If humidity levels are low and temperatures are warm, leaf temperatures can drop considerably below air temperature because of increased transpiration (assuming water is available for root uptake). Conversely, leaf temperatures can be much higher than the surrounding air temperature when transpiration is reduced, such as under high humidity and high temperatures. During warm periods, some greenhouse growers provide cooling with mist or fog. As water drops evaporate from the leaf's surface, temperatures can drop 8°F (4°C) or more.

During the night, convection and the transfer of long-wave radiation between the plant and the greenhouse structure often have the greatest effect on plant temperature. Long-wave radiation is invisible energy that all objects emit. The amount of long-wave radiation that is emitted is influenced by an object's temperature and physical properties. The lower an object's temperature, the less long-wave radiation is emitted.

During cold winter nights, the glazing temperature of a greenhouse can be lower than the inside air and plant temperature. As the glazing temperature decreases, the amount of long-wave radiation emitted by the plant can exceed that emitted by the glazing. This net loss of long-wave radiation can cause plant temperature to decrease below air temperature. One solution to reduce the loss of long-wave radiation to cold greenhouse glazing is to use a retractable thermal screen. A research study at Michigan State University showed that when outside temperatures were below freezing, New Guinea impatiens (*Impatiens hawkeri*) grown in a glass-glazed greenhouse under a thermal screen extended at night had a shoot-tip temperature up to 4°F (2.3°C) higher than plants without a screen.

Since air and plant temperatures are often different, it is important to know the actual plant temperature of a crop. The best tool to measure plant temperature is an infrared thermometer. Infrared thermometers measure the temperature of an area that increases as the distance from the crop increases. Thus, these should be positioned so that they measure only plant canopy temperature, and not the temperature of media, walls, floors, etc. Some sophisticated greenhouse climate control systems incorporate an infrared thermometer, and the greenhouse environment is controlled based on the measured leaf temperature of a crop.

Average Daily Temperature

The average daily temperature (ADT) is the primary influence on the rate of plant development (i.e., time to flower or to unfold a leaf). The ADT is the mathematical average temperature over a series of twenty-four-hour periods and can be calculated as:

ADT = [(day temperature × hours) + (night temperature × hours)] ÷ 24

If more than two temperature set points are delivered over a twenty-four-hour period, such as 62°F (17°C) from midnight to 8 A.M., 72°F (22°C) from 8 A.M. to 6 P.M., and 66°F (19°C) from 6 P.M. to midnight, then the equation needs to be expanded. To calculate ADT in this scenario, for each temperature period, multiply the temperature by the number of hours. Then add all of these values together, and divide by twenty-four hours. The ADT in this example would be 67.2°F (19.5°C).

Generally, the warmer the ADT, the faster a plant develops. This is analogous to how fast you drive your automobile to get to work. The faster you drive, the earlier you arrive at work. Similarly, the warmer a crop is grown, the quicker it will develop and become ready for market. Therefore, if you lower the ADT in the greenhouse, the plants will take longer to become marketable. This concept applies to plugs, flats, potted crops, hanging baskets, and any other size of plant or container. There are also other factors that influence crop timing, including photoperiod and the average daily light integral (DLI), which are discussed in other chapters.

Plants are able to integrate temperature day by day and week by week from the time they enter a greenhouse until they leave. So, if crops are grown cool one week and warm another week, then their developmental rate is a function of the average temperature for both weeks. Some growers may grow cooler early in the spring and increasingly warmer later in the spring. Crop timing will depend on the cumulative average temperature delivered, beginning when the crop first enters the greenhouse. When using temperature integration to control developmental rate, crop scheduling will only be correct if temperatures delivered are between the base and optimum temperature (see below).

The effect of changing ADT on production time depends on the species, the magnitude of the change, and the original temperature set point. For example, crop timing of petunia and vinca at different ADTs is illustrated in figure 6-1. Lowering the temperature by 5°F (2–3°C) has a somewhat small effect at warm temperatures and has a larger effect at cooler temperatures. For example, lowering the ADT by 5°F (2–3°C) from 65 to 60°F (18 to 16°C) delays a petunia crop (from seed) by about thirteen days, while lowering the temperature from 60 to 55°F (16 to 13°C) delays petunia by twenty-two days.

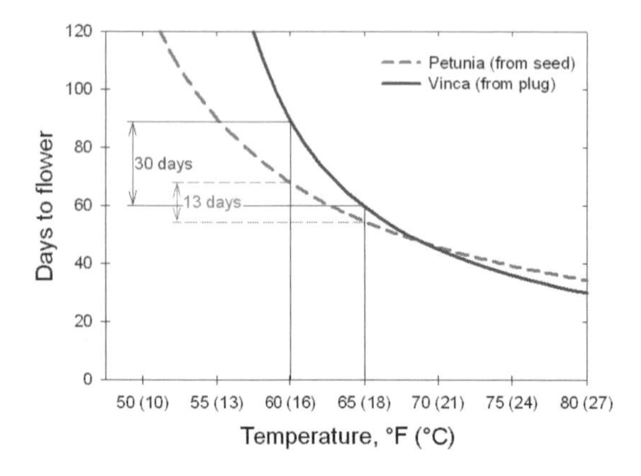

Figure 6-1. The effect of temperature on time to flower of petunia (*Petunia × hybrida*) from seed and vinca (*Catharanthus roseus*) from a small plug. When temperature is decreased, there is a larger delay in flowering for plants that are cold sensitive (vinca) compared to plants that are cold tolerant (petunia).

The effect of lowering the temperature can have a more dramatic effect on cold-sensitive crops such as vinca compared to cold-tolerant crops such as petunia. For example, lowering the temperature from 65 to 60°F (18 to 16°C) increases time to flower of vinca (from a plug) by about thirty days—much longer than the delay in petunia with the same temperature decrease.

Angelonia (*Angelonia angustifolia*) and French marigold (*Tagetes patula*) are other examples of crops that respond differently to temperature (figure 6-2). At an ADT below 60°F (16°C), flower development of angelonia is delayed considerably, while French marigold continues to develop. For example, lowering ADT from 63 to 58°F (17 to 14°C) lengthens production time of angelonia by 63% and of French marigold by only 19%. Therefore, crops such as angelonia should not be grown cool.

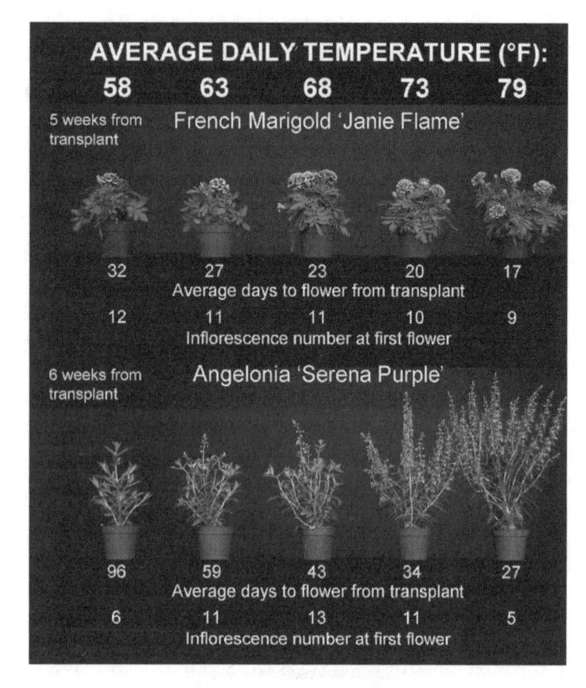

Figure 6-2. The effects of ADT on time to flower and number of inflorescences (at first flowering) in French marigold 'Janie Flame' and angelonia 'Serena Purple'. Plants were grown under a sixteen-hour photoperiod and an average DLI of 10 $mol \cdot m^{-2} \cdot d^{-1}$. Photographs were taken five or six weeks after transplant from a 288-cell plug tray that was grown under long days.

Many greenhouse crops produce a certain number of leaves before flower initiation, which allows the rate of progress towards flowering to be tracked by counting the number of leaves that unfold each day. The rate of unfolding is controlled by the ADT, and the leaf-unfolding rate can be accelerated or slowed down by raising or lowering the ADT. Easter lily growers are familiar with this leaf-counting technique to track plant development and ensure that their crops are on schedule.

Base Temperature

The base temperature is a cool temperature below which a plant stops growing (figure 6-3). The base temperature is species specific and typically ranges from 30–50°F (-1–10°C). For example, cold-sensitive vinca has a base temperature of 50°F (10°C). Angelonia, browallia (*Browallia speciosa*), New Guinea impatiens (*Impatiens hawkeri*), poinsettia, and zinnia (*Zinnia elegans*) also have high base temperatures.

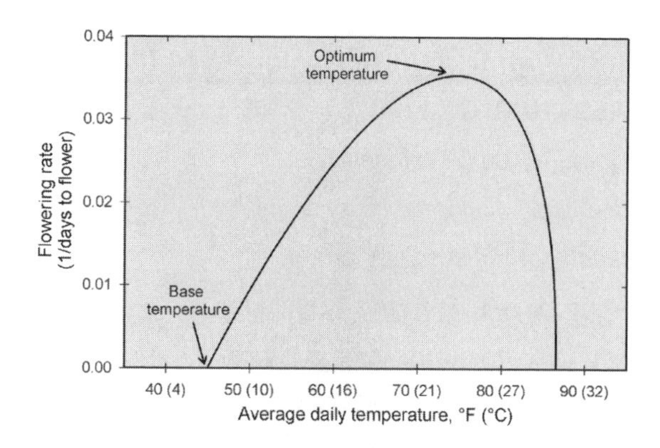

Figure 6-3. The effect of ADT on flower developmental rate (one divided by the number of days to flower) in cosmos (*Cosmos sulphureus*) grown under a sixteen-hour photoperiod and an average DLI of 10 mol·m^{-2}·d^{-1}. As ADT increases, flowering rate increases between the base temperature (45°F [7°C]) and the optimum temperature (75°F [24°C]).

We can put plants into temperature-response categories based on their estimated base temperature. We consider "cold-tolerant" crops to be those with a base temperature of less than 40°F (4°C). Examples of cold-tolerant crops include snapdragon (*Antirrhinum majus*), French marigold, pansy (*Viola* × *wittrockiana*), and many herbaceous perennials. Experienced growers can often predict which crops have a low base temperature, because they are usually grown cooler than others.

We categorize plants because they differ in how they respond to changing greenhouse temperatures; cold-sensitive plants are more responsive to lowering the greenhouse temperature than cold-tolerant species. So, decreasing the greenhouse temperature set point will delay crop timing more with cold-sensitive crops. (Table 6-1 contains a list of plants categorized by their base temperatures.) Ideally, crops with large differences in base temperatures should be grown in separate greenhouses with different temperature set points.

Table 6-1. Plants Categorized by Base Temperature

Base temperature is the temperature at or below which plant development ceases. "Cold-tolerant crops" are those with a base temperature lower than 40°F (4°C), "intermediate crops" are those with a base temperature of 40–45°F (4–7°C), and "cold-sensitive crops" are those with a base temperature greater than 45°F (7°C).*

COLD-TOLERANT CROPS (BASE TEMPERATURE < 40°F [4°C])	
Antirrhinum majus (snapdragon)	*Lilium longiflorum* (Easter lily)
Aquilegia canadensis (red columbine)	*Lilium* sp. (Asiatic and Oriental lily)
Aquilegia × *hybrida* (hybrid columbine)	*Lobularia maritima* (alyssum)
Astilbe chinensis (astilbe)	*Matthiola incana* (stock)
Calendula officinalis (pot marigold)	*Nemesia strumosa* (nemesia)
Campanula carpatica (Carpathian bellflower)	*Osteospermum* sp. (osteospermum)
Cineraria × *hybrida* (cineraria)	*Pericallis* × *hybrida* (cineraria)
Coreopsis grandiflora (coreopsis)	*Petunia* sp. (petunia; some cultivars)
Coreopsis verticillata (thread-leaved tickseed)	*Phlox subulata* (moss phlox)
Delphinium grandiflorum (Siberian larkspur)	*Pisum sativum* (pea)
Dianthus chinensis (dianthus)	*Platycodon grandiflorus* (balloon flower)
Diascia sp. (twinspur)	*Rudbeckia fulgida* (black-eyed Susan)
Gaillardia × *grandiflora* (blanket flower)	*Scabiosa caucasia* (pincushion flower)
Gaura lindheimeri (gaura)	*Schlumbergera truncata* (Thanksgiving cactus)
Gerbera jamesonii (gerbera)	*Sedum* sp. (stonecrop)
Hemerocallis sp. (daylily)	*Tagetes erecta* (African marigold, some cultivars)
Hosta hybrida (hosta)	*Tagetes patula* (French marigold)
Leucanthemum × *superbum* (Shasta daisy)	*Viola* × *wittrockiana* (pansy)

(Continued)

Table 6-1. Plants Categorized by Base Temperature *(Continued)*	
INTERMEDIATE CROPS (BASE TEMPERATURE OF 40–45°F [4–7°C])	
Aquilegia flabellata (fan columbine)	*Lobelia erinus* (blue lobelia)
Calibrachoa × *hybrida* (calibrachoa)	*Nicotiana* × *sanderae* (nicotiana)
Capsicum annuum (pepper)	*Oenothera fruticosa* (sundrops)
Chrysanthemum × *grandiflorum* (chrysanthemum)	*Pelargonium* × *hortorum* (geranium)
Cosmos sulphureus (cosmos)	*Petunia* sp. (petunia, some cultivars)
Cyclamen persicum (cyclamen)	*Phlox paniculata* (garden phlox)
Dahlia sp. (dahlia)	*Rudbeckia hirta* (black-eyed Susan)
Gazania rigens (gazania)	*Salvia splendens* (red salvia)
Heliotropium arborescens (heliotrope)	*Tagetes erecta* (African marigold, some cultivars)
Impatiens wallerana (seed impatiens)	*Verbena* sp. (verbena)
COLD-SENSITIVE CROPS (BASE TEMPERATURE > 45°F [7°C])	
Ageratum houstonianum (floss flower)	*Hibiscus* sp. (hibiscus)
Angelonia angustifolia (angelonia)	*Impatiens hawkeri* (New Guinea impatiens)
Begonia hiemalis (Rieger begonia)	*Lycopersicon esculentum* (tomato)
Begonia × *semperflorens-cultorum* (fibrous begonia)	*Musa ornata* (banana)
Browallia speciosa (browallia)	*Pennisetum setaceum* 'Rubrum' (purple fountain grass)
Caladium bicolor (caladium)	*Pentas lanceolata* (pentas)
Capsicum annuum (pepper)	*Phalaenopsis* sp. (phalaenopsis orchid)
Catharanthus roseus (vinca)	*Portulaca grandiflora* (moss rose)
Celosia argentea (celosia)	*Rosa* × *hybrida* (Rose)
Cleome hassleriana (cleome)	*Saintpaulia ionantha* (African violet)
Colocasia sp. (elephant ears)	*Salvia farinacea* (blue salvia)
Euphorbia pulcherrima (poinsettia)	*Torenia fournieri* (wishbone plant)
Gomphrena globosa (globe amaranth)	*Zinnia elegans* (zinnia)

* Information is based primarily on research at Michigan State University and supplemented with published, research-based articles.

Floriculture crops are often grown about 20–30°F (11–17°C) higher than their base temperatures. At temperatures closer to the base temperature, plant development is often very slow. There are a few times when growing near the base temperature is desirable. One example is when plants need to be held because the markets are not available to receive plants. Another example is when perennials or bulbs are provided with cool temperature treatments to satisfy a vernalization response to induce flowering.

Optimum Temperature

As ADT increases above the base temperature, the plant developmental rate increases and becomes fastest at the optimum temperature (figure 6-3). The optimum temperature for a crop is not based on plant-quality attributes, and thus the optimum temperature is not necessarily the most desirable growing temperature. As temperature increases beyond the optimum value, development slows as plants experience heat stress.

Therefore, in most instances, crops are grown above the base temperature but not above the optimum temperature. The optimum temperature varies among species and can be around 70°F (21°C) for cool-season crops, such as pansy and alyssum (*Lobularia maritima*), or as high as 90°F (32°C) for warm-season crops, such as vinca and hibiscus (*Hibiscus* sp.). The optimum temperature can decrease as the plant matures. For example, temperatures are higher during propagation than they are during finished crop production and are sometimes even lower as plants become ready for market. Unfortunately, the optimum temperature has only been determined on a handful of crops.

Other factors interact with temperature to either limit or accelerate plant growth and development. The optimum temperature for a crop can change based on the availability of light and CO_2. For example, under higher light levels, the photosynthetic rate can increase and the optimum temperature for development may be higher. If CO_2 is injected into the air, plants can be grown at a higher temperature and increase plant growth. Any one of these three factors—light, temperature, and CO_2—can limit the others. So even at the optimum temperature, if light levels are not high enough, photosynthesis and plant growth will be reduced. In an airtight greenhouse during winter, CO_2 may be limiting plant growth, even though you are growing at the optimum temperature and providing sufficient light. Growers should be aware of how these factors interact to influence plant growth and development.

Cropping Time

The production of ornamental crops can be influenced by several environmental and cultural factors, including ADT, starting material size, finished plant size, photoperiod, DLI, and use of plant growth retardants. Generally, the larger and more mature the starting plant and the smaller the size of the finish container, the shorter the finish crop production time. For example, finish time of calibrachoa 'Superbells Red' (*Calibrachoa* × *hybrida*) grown in 4.5" (11 cm) pots and at an ADT of 70°F (21°F) can be reduced by over two weeks by starting with a 50-cell plug (40 mm liner) versus a 144-cell plug (20 mm liner).

Studies have been performed with many floriculture crops under different temperature and light conditions to generate crop-timing data. This information allows growers to efficiently schedule crops to be finished for predetermined dates. Table 6.2 compares the effects of ADT on time to flower from transplant of 288-cell bedding plant plugs that were grown under a sixteen-hour-long day and a DLI of 9–11 $mol \cdot m^{-2} \cdot d^{-1}$. Finish conditions assume a sixteen-hour-long day and

Table 6-2. Crop Comparison of Effects of ADT on Crop Maturation

The effects of average daily temperature (ADT) on the time to flower from transplant of 288-cell bedding plant plugs vary based on the crop. Plugs were grown under a sixteen-hour-long day and a DLI of 9–11 $mol \cdot m^{-2} \cdot d^{-1}$. Finish conditions assume long days and an average DLI of 14 $mol \cdot m^{-2} \cdot d^{-1}$.

BEDDING PLANT CROP	SERIES	DAYS FROM TRANSPLANT TO FLOWER AT ADT			
		58°F (14°C)	65°F (18°C)	72°F (22°C)	79°F (26°C)
Angelonia angustifolia (angelonia)	'Serena'	95	51	35	27
Antirrhinum majus (snapdragon)	'Montego'	39	30	24	20
Dianthus chinensis (dianthus)	'Super Parfait'	71	52	41	34
Pelargonium × *hortorum* (seed geranium)	'Florever'	89	63	49	40
Pentas lanceolata (pentas)	'Graffiti'	123	70	49	38
Petunia × *hybrida* (petunia grandiflora)	'Dreams'	34	26	21	17
Tagetes patula (French marigold)	'Janie'	32	25	20	17
Verbena × *hybrida* (verbena)	'Obsession'	50	33	25	20
Zinnia elegans (zinnia)	'Dreamland'	77	48	35	28

an average DLI of 14 mol·m^{-2}·d^{-1}. Crop time varies considerably among species. For example, at an ADT of 72°F (22°C), pentas 'Graffiti' (*Pentas lanceolata*) takes four weeks longer to flower than French marigold 'Janie Flame'.

Research at the University of Florida and Michigan State University has shown that a similar production time can be achieved by substituting time in the liner stage for time in the finished container. For example, when starting with calibrachoa 'Superbells Red' in 105-cell liners (1" [25 mm]) that are four weeks old, plants required eight weeks to finish in a 12" (30 cm) hanging basket (four liners per basket). In comparison, only four weeks were needed to finish the hanging basket when starting with 18-cell liners (2.8" [70 mm]) that were eight weeks old. In both scenarios, the total production time was similar, twelve weeks. A financial analysis could be performed to determine if the cost of growing or purchasing larger liners outweighs the savings from reduced production time in the finished container.

High Temperature

High temperature stress can occur in crops when the day temperature, night temperature, or both are above some crop-specific temperature. Many crops become heat stressed at night temperatures above 77°F (25°C) and day temperatures above 90°F (32°C). High temperature stress occurs at temperatures above the optimum temperature for the fastest development rate. At these warm temperatures, flower initiation and development can be delayed, which is commonly known as "heat delay." For example, chrysanthemum and poinsettia flower initiation is delayed when the night temperature is above 77–79°F (25–26°C). Chrysanthemums are most likely to experience heat delay in the summer months, when blackout cloth is used to artificially create short days. Heat builds up underneath the blackout cloth, raising temperatures. In some crops, the delay in flowering at a high temperature is associated with the development of more leaves before flower initiation.

High temperatures can also reduce plant photosynthesis. The rate of photosynthesis generally increases proportionally with temperature until some maximum value [95°F (35°C) for most species]. Above this temperature, any further increase in temperature decreases photosynthesis and reduces crop growth. Plants also respire more as temperature increases. If the amount of carbohydrates lost through respiration exceeds the amount gained through photosynthesis, then growth decreases. Therefore, high-night and cool-day temperatures are not often used together for this and other reasons.

Plant quality can be negatively affected by high temperature, especially under low light conditions. As temperature increases, flower size, flower number, and branching can decrease. For example, flower bud number in pansy can be up to 77% less if grown at 86°F (30°C) compared with 68°F (20°C). One reason for the decrease in flower and branch number at high temperatures is because plants flower so quickly that they have little time to harvest light before flowering. Figure 6-4 illustrates the effect of forcing temperature on flower size in campanula 'Birch Hybrid' (*Campanula*) when grown under the same light conditions.

Figure 6-4. Influence of forcing temperature on flower size of campanula 'Birch Hybrid'

Photo courtesy of Alison Frane, Michigan State University.

Some crops can tolerate short-term exposure to high temperatures. However, prolonged exposure can cause flower buds to abort or necrosis to develop on leaves. The thermal tolerance of crops can vary among species and be influenced by plant maturity and recent growing conditions. A temperature that causes heat stress on one species may actually be favored by another species. For example, at constant temperatures above 86°F (30°C), most cultivars of dahlia will die, but many petunias will continue to flower. Other examples of heat-loving crops include globe amaranth (*Gomphrena globosa*), pentas, black-eyed Susan (*Rudbeckia hirta*), and zinnia (*Zinnia elegans*). Tropical species such as ficus (*Ficus* sp.) and philodendron (*Philodendron* sp.) also grow best at these high temperatures. Some new cultivars of bedding and potted plants have been selected for improved tolerance of and performance at high temperatures.

Low Temperature

As with high temperature, the concept of "low temperature" is relative and situational. Low temperature is sometimes provided intentionally, such as to "hold" plants to slow down crop growth or to vernalize perennial and bulb crops to induce flowering. However, low temperatures can damage plants in different ways and can create conditions that are more favorable for the spread of pathogens such as botrytis.

At cold temperatures, plants develop slowly and, thus, do not take up water or nutrients very quickly. Temperature can play a big role in nutrient deficiencies. For example, at temperatures below 55°F (13°C), phosphorus uptake is inhibited in tomato (*Lycopersicon*); purpling of lower foliage is the typical symptom. Geranium (*Pelargonium*) can also show phosphorus deficiency when grown too cool in the spring.

Cold water damage on leaves may appear as spots. It's most notable on highly cold-sensitive crops such as *Aglaonema*, dieffenbachia, philodendron, and African violet (*Saintpaulia*). To prevent problems, use water that has been conditioned to a temperature similar to the air temperature for sensitive crops.

Chilling and freezing injury

Depending on the crop, temperature, and duration, low temperature can cause unintended damage to plants. Chilling injury can occur in actively growing, cold-sensitive crops that are exposed to low, above-freezing temperatures. This cool temperature, especially when it occurs rapidly, can damage cell membranes and cause cells to collapse, resulting in death of tissue. Examples of crops sensitive to chilling injury are angelonia, tomato, coleus (*Solenostemon*), and gloxinia (*Sinningia*). Freezing injury occurs when plants are exposed to temperatures below 32°F (0°C). Plant damage then appears once the temperature begins to increase. Symptoms of minor chilling and freezing injury include lesions, discolorations, and a water-soaked appearance on leaves. An extended period of low-temperature exposure can lead to plant death. Generally, actively growing cold-tolerant crops should not be exposed to temperatures below 40°F (5°C), and cold-sensitive crops should not be exposed to temperatures below 50°F (10°C).

Vernalization

The flowering of many cold-hardy herbaceous perennials is initiated by an extended period of cold, which is known as a vernalization treatment. Some perennials do not flower without a cold treatment, while others flower more rapidly and have a stronger flowering response (such as more flowers) when provided with a cold treatment. Table 6-3 lists some of the perennials that require a vernalization treatment for flowering.

Most perennials propagated by cuttings can be satisfactorily vernalized in liner trays. However, some perennials propagated by seed must develop a minimum number of leaves before they become receptive to a cold treatment. For example, several older varieties of columbine (*Aquilegia*) should have at least eight to ten leaves per plant (not per cell) before starting a cold treatment. Less mature plants may be juvenile and not responsive to the cooling treatment. In many cases, subsequent flowering is improved when plants are cooled in larger plug sizes rather than smaller ones. The highest plant quality, regardless of propagation technique, is usually achieved when plants are planted in their final containers, allowed to grow for at least three weeks at warm temperatures, and then vernalized.

Researchers at Michigan State University have quantified how many herbaceous perennials respond to a range of cooling temperatures and durations. (Visit http://www.flor.hrt.msu.edu/perennials for more information.) The most effective cooling temperature range for a large number of species is 40–45°F (4–7°C), especially when short durations (four to six weeks) of cooling are provided. With longer cooling durations, the effective temperature range broadens. In general, six to eight weeks of cooling at 40–45°F (4–7°C) saturates the vernalization response of most (but not all) species.

Many perennials continue to actively grow—albeit slowly—when held at low temperatures above freezing. Therefore, plants should be provided with light and water during above-freezing cold treatments. In refrigerated coolers, a nine-hour photoperiod that delivers at least 25–50 f.c. (269–538 lux or 3–7 $\mu mol \cdot m^{-2} \cdot s^{-1}$) of light has been adequate to maintain plants at 41°F (5°C) for up to fifteen weeks. The need for light decreases as the cooling temperature decreases, but for highest plant quality and survival, providing light is recommended at above-freezing temperatures. In general, the photoperiod during cooling does not influence the vernalization response. Fluorescent lamps are desirable if cooling is provided in refrigerated chambers because of their potential for uniform light distribution, although their light output declines with temperature.

From late fall through early spring, perennials can also be successfully vernalized in cold climates in their final containers, either in a minimally heated greenhouse or outdoors under a thermal blanket. Outdoor cooling is typically not a problem when snow covers the blanket, which provides an additional layer of insulation. Extreme temperature fluctuations (without snow cover) can cause plant stress, and insufficient cool temperatures can prevent a complete vernalization response. In addition, when the temperature rises

Table 6-3. Herbaceous Perennials that Require Vernalization
Some herbaceous perennials require vernalization for complete and uniform flowering. For best results, young plants should be transplanted into their finish container, grown for at least several weeks, and then cooled.*
Ajuga 'Bronze Beauty', 'Chocolate Chip', 'Golden Beauty' (bugleweed)
Alchemilla mollis (lady's mantle)
Aquilegia (columbine)
Armeria maritima 'Cottontail' and related species (sea thrift)
Astilbe × *hybrida* (astilbe)
Amsonia (bluestar)
Baptisia australis (false indigo)
Cimicifuga racemosa (bugbane)
Dianthus gratianopolitanus 'Bath's Pink' (cheddar pink)
Dianthus deltoides 'Shrimp', 'Zing Rose', others (maiden pink)
Dicentra eximia 'Luxuriant' and related (fringed bleeding heart)
Euphorbia amygdaloides 'Purpurea' (wood spurge)
Euphorbia polychroma (cushion spurge)
Filipendula purpurea 'Kakome' (meadowsweet)
Geranium dalmaticum (Damaltian cranesbill)
Geranium 'Johnson's Blue' (cranesbill)
Geranium sanguineum 'New Hampshire Purple' (bloody cranesbill)
Geum chiloense 'Mrs. Bradshaw' (Grecian rose)
Heuchera (coral bells)
Heucherella 'Cranberry Ice', 'Dayglow Pink', 'Viking Ship' (heucherella)
Iberis sempervirens 'Alexander's White', 'Snowflake', others (candytuft)
Lychnis coronaria 'Angel Blush' (rose campion)
Phlox divaricata (woodland phlox)
Phlox subulata (moss phlox)
Potentilla atrosanguinea 'Mrs. Willmott' (cinquefoil)
Pulmonaria saccharata 'Peirres Perfect Pink' (lungwort)
Saxifraga *'Aureopunctata'*, 'London Pride', 'Triumph' (London pride)
Thalictrum aquilegifolium (meadow rue)

* Adapted from *Greenhouse Grower* articles published by Michigan State University researchers.

in the spring, covers need to be promptly removed so that plants can receive light and water to sustain their growth. Of the various cooling techniques, outdoor vernalization is the most unpredictable and uncontrollable, yet it is usually the least expensive.

Regardless of the vernalization technique, some plants go dormant and lose their leaves during cooling. Therefore, foliage should be kept dry to help prevent problems with diseases. In addition, maintain a slightly moist media, if possible, to avoid dehydration of plants (when the media is too dry) and root rot (when the media is too wet).

DIF

Greenhouses are traditionally operated with warm days and cooler nights, for example 70°F (21°C) during the day and 62°F (17°C) at night. Providing higher daytime temperatures allows growers to take advantage of solar radiation and the natural heat gain inside the greenhouse structure during the day. Cool nights conserve fuel, since heaters do not have to operate as much to maintain a lower temperature.

Through research conducted at Michigan State University, Royal Heins and his graduate students learned that the way temperature is delivered during the day and night can influence plant height. By providing warm days and cool nights, growers were inadvertently causing plants to grow tall, and, in many cases, increasing the need for growth-regulating chemicals. They developed the term "DIF" and helped growers incorporate this concept into their production schedule.

DIF refers to the mathematical DIFference between the day and night temperature. A 74°F (23°C) day and a 62°F (17°C) night equates to a +12°F (+7°C) DIF. If the temperatures are reversed, 62°F days and 74°F nights result in a –12°F (–7°C) DIF. If the day and night temperature are the same, then there is no DIF (0 DIF). Most plants grown at a positive DIF will be taller than plants grown at a zero or negative DIF, assuming that the ADT is the same (figure 6-5). Also, the greater the DIF is, the stronger the response. Thus, plants grown at a –15°F (–8°C) DIF are typically shorter than those grown at a –5°F (–3°C) DIF.

Figure 6-5. The influence of day and night temperature difference (DIF) on stem elongation in Easter lily. The shorter plant on the left was grown at 57°F (14°C) day and 86°F (30°C) night [–29°F (–16°C) DIF], and the taller plant on the right was grown at 86°F (30°C) day and 57°F (14°C) night [+29°F (+16°C) DIF]. *Photo courtesy of John Erwin and Royal Heins, Michigan State University.*

DIF typically has no effect on crop timing; rather, it influences the distance from one leaf to the next. As discussed previously, plants develop in response to the twenty-four-hour ADT. For example, crop timing is similar if plants are grown at a constant temperature of 65°F (18°C) or a day/night of 70/60°F (21/16°C) if the day and night are each twelve hours long. If growers lower the night temperature without increasing the day temperature by the same amount, crops will be delayed.

The DIF concept can be applied in a number of ways. Using greenhouses equipped with automated temperature sensors and an environmental control computer, a desired DIF program can be easily programmed. Without automated temperature control, using DIF is more difficult, but it can still be accomplished (see the DIP section).

Many crops respond to DIF, and there are hundreds of growers who use the technique every day with desirable results (table 6-4). However, crops such as asters, foliage plants, kalanchoes, zonal geraniums (*Pelargonium*), marigolds (*Tagetes*), curcubits (i.e., cucumbers, squash, and watermelon), and most Dutch bulbs respond very little or not at all to DIF. In general, crops originating from temperate climates are more responsive to DIF than crops from tropical climates.

Most flowering pot plant growers use graphical tracking to plot the height of their crops each day and use DIF to adjust height control. Graphical tracking is discussed in detail in the Graphical Tracking table within the *Euphorbia* section. Graphical tracking is a great tool to use for holiday pot plant production where crops must meet minimum and maximum size requirements for buyers as well as be marketable on predetermined dates.

There are some downfalls to using DIF. First, DIF may not be practical when outdoor temperatures or light intensities are high; attaining a cool day temperature may not be possible. Second, fuel cost for heating a greenhouse is usually less expensive with a positive

Table 6-4. Effect of DIF and DIP on Stem Elongation

DIF AND TEMPERATURE DROP (DIP) AFFECT THE STEM ELONGATION OF CROP SPECIES IN DIFFERENT WAYS.[z]

No entry = Research has not been performed.
* = small or no effect
** = medium or strong effect

	DIF	DIP
Antirrhinum majus (snapdragon)	*	
Aster novi-belgii (aster)	*	
Begonia × hiemalis (Rieger begonia)	**	
Begonia × tuberhybrida (tuberous begonia)	**	*
Calceolaria × herbeohybrida (pocketbook plant)	**	
Campanula carpatica (Carpathian bellflower)	**	
Campanula isophylla (Italian bellflower)	**	
Capsicum annuum (pepper)	**	
Catharanthus roseus (vinca)	**	
Celosia argentea (celosia)	**	
Chrysanthemum × grandiflorum (chrysanthemum)	**	*
Citrullus lanatus (watermelon)	*	
Cosmos sulphureus (cosmos)	**	
Cucurbita sp. (cucumber, melon, squash)	*	*
Cyclamen persicum (cyclamen)	**	*
Dahlia sp. (dahlia)	**	
Dianthus carthusianorum (Carthusian Pink)	**	
Dianthus chinensis (dianthus)	**	
Euphorbia pulcherrima (poinsettia)	**	**
Foliage crops (many species)	**	*
Fuchsia × hybrida (fuchsia)	**	*
Gerbera jamesonnii (gerbera)	**	
Hyacinthus sp. (hyacinth)	*	
Impatiens wallerana (seed impatiens)	**	*

(Continued)

DIF compared with a zero or negative DIF. However, fuel cost savings depends on the time of year, greenhouse and heating characteristics, location, and temperature set points. For example, a grower in Grand Rapids, Michigan, delivering an ADT of 65°F (18°C) could lower his March fuel bill by an estimated 18% by switching from a –10°F (–6°C) DIF to a +10°F (+6°C) DIF. A producer in Charlotte, North Carolina, growing crops at an ADT of 72°F (22°C) could lower her April fuel bill by 27% by switching from a 0 DIF to a +10°F (+6°C) DIF. Producers may want to use the Virtual Grower computer program (available at no charge at http://www.virtualgrower.net) to estimate fuel costs at different DIF regimens.

A high negative DIF can also cause downward curving of leaves on some crops such as Easter lilies. Leaves reorient to a more normal position when temperatures approach a zero or positive DIF. Also, running cold daytime temperatures with some high negative DIF regimens can cause foliar chlorosis on some crops. Again, the situation is temporary and remedied by returning temperatures to a zero or positive DIF.

DIP

An alternative way to manipulate temperature to control plant elongation is to drop the temperature during the early morning, which is referred to as a temperature drop, or DIP. In many crops, plant elongation is

Table 6-4. Effect of DIF and DIP on Stem Elongation *(Continued)*

DIF AND TEMPERATURE DROP (DIP) AFFECT THE STEM ELONGATION OF CROP SPECIES IN DIFFERENT WAYS.[z]

No entry = Research has not been performed.
* = small or no effect
** = medium or strong effect

	DIF	DIP
Kalanachoe blossfeldiana (kalanchoe)	*	*
Lilium sp. (Asiatic, Easter, and Oriental lily)	**	**
Lycopersicon esculentum (tomato)	**	**
Narcissus sp. (daffodil)	*	
Pelargonium (cutting geranium)	*	
Pelargonium (seed geranium)	**	*
Petunia sp. (petunia)	**	*
Phaseolus sp. (bean)	*	
Pisum sp. (pea)	**	
Rosa sp. (rose)	*	*
Saintpaulia ionantha (African violet)	*	
Salvia splendens (red salvia)	**	*
Senecio cineraria (dusty miller)	**	
Solanum tuberosum (potato)	**	*
Streptocarpus nobilis (streptocarpus)	**	
Tagetes patula (French marigold)	*	
Tulipa sp. (tulip)	*	
Verbena × hybrida (verbena)	*	*
Verbena bonariensis (tall verbena)	**	
Viola × wittrockiana (pansy)	**	
Zinnia elegans (zinnia)	**	

[z] *Information is based on research at Michigan State University and published research-based articles.*

greatest during the first daylight hours. Plants are most responsive to low temperature treatments as darkness becomes day, from approximately thirty minutes before sunrise until two or three hours after sunrise. Using a temperature drop, growers can lower temperatures about an hour before sunrise and run those temperatures for the first two to three hours of the day. Generally, the greater the temperature drop, the larger the effect on inhibiting stem extension. Remember that lowering the temperature in the morning reduces the ADT. To keep crops on time, increase the afternoon and/or night temperature slightly to offset the low temperature drop.

Results with a temperature drop are often almost as good as if temperatures were lowered for the entire day (negative DIF). For example, North Carolina State University recommends 64°F (18°C) nights as a minimum for poinsettia production. To use a morning DIP, they recommend dropping temperatures to 58–60°F (14–16°C) about one hour prior to sunrise and to keep daytime temperatures as low as possible throughout the day. The goal for many growers in warm climates is not to run a negative DIF, but to keep the DIF as low as possible.

Energy-Efficient Production

For growers in cold climates, the cost to heat a greenhouse can represent 10–15% of the total production costs. There has been increasing interest in lowering fuel costs given volatile and generally increasing energy costs, as well as the desire to grow crops more sustainably. Some technological improvements to greenhouses can reduce heating costs, such as the use of retractable energy curtains at night and purchasing new, energy-efficient heaters or boilers. Energy costs can also be cut by managing the greenhouse temperature and through precise scheduling, so that crops are in the greenhouse for a minimum period of time.

Growers can gain a better understanding of how infrastructure and growing environment influence energy consumption using the Virtual Grower software program, developed by Jonathan Frantz and colleagues in the Application Technology Research Unit of the USDA-ARS in Toledo, Ohio. This program, available in English and Spanish at http://www.virtualgrower.net, enables growers to simulate their own greenhouses and predict how changes or investments could impact growing environments, heating costs, and crop responses. The program is simple to install and can be used as a tool to assist growers in making decisions.

Once the program is downloaded and installed, you build a virtual greenhouse with the same characteristics as your existing greenhouse, or one that you hope to build. You'll enter such information as location, greenhouse dimensions and design, greenhouse construction materials, heating system characteristics, fuel type and price, heating set points, and presence and type of energy curtain. It only takes about ten to fifteen minutes to enter the data for a greenhouse range, and you can save the information so you do not need to reenter it the next time you use Virtual Grower. The program uses your information to estimate your heating costs per month and per square foot per month. By changing one or more variables, such as raising the day and lowering the night temperatures, you can predict how your fuel costs would change.

There are a few limitations to the program, including the assumption of "normal" temperatures during the winter. (What is "normal," and when was the last time we had a "normal" year of weather?) There are also some important assumptions in the plant growth section, including the photoperiod and starting plant size. Nevertheless, the program provides unbiased, research-based information that can be used to help you to help make greenhouse and crop culture decisions.

Grow warm or grow cool?

The Virtual Grower program enables growers to predict the amount of energy needed to maintain a desired temperature at different times of the year. When combined with information on the temperature's effect on crop timing, growers can identify the most energy-efficient growing temperatures at different times of the year. In the past few years, researchers at Michigan State University and the University of Minnesota have been quantifying how temperature controls crop timing on a wide range of ornamental herbaceous crops. For example, ageratum (*Ageratum houstonianum*) grown in 288-cell plug trays under a sixteen-hour-long day and then transplanted and grown under long days take approximately sixty-one days to flower at 58°F (14°C), forty-three days at 63°F (17°C), thirty-three days at 68°F (20°C), and twenty-seven days at 73°F (23°C). With this information, growers can determine transplant dates so that plants are in flower for predetermined market dates when grown at different temperatures. For example, for first flowering of ageratum on April 5, 288-cell plugs need to be transplanted on February 3 if grown at an ADT of 58°F (14°C) and long days or on March 3 if grown at an ADT of 68°F (20°C). Because of the substantial delay in flowering when grown cool, more energy may be consumed when growing at cooler temperatures than if grown warmer with a shorter finish time.

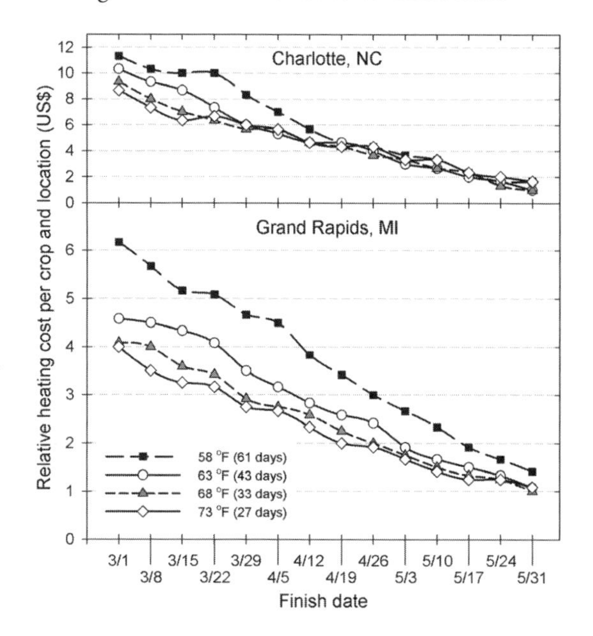

Figure 6-6. Example of how temperature set points, finish date, and location can influence greenhouse heating costs for a crop of ageratum using estimates from the Virtual Grower software program. Estimated costs are relative to each location.

Figure 6-6 presents estimated relative greenhouse-heating costs for growing ageratum using Virtual Grower. The costs are based on a per-crop basis in two locations for finish dates in March, April, and May. In Grand Rapids, Michigan, more energy for heating was used to grow ageratum cool throughout the spring season. In the winter, when daily heating costs are highest, it was cheaper to produce the crop warm to finish it in a shorter period of time. As the finish date ran into May, there were fewer differences in heating costs at different finish temperatures, but heating costs were always greatest when ageratum was grown cool at 58°F (14°C). Lower temperatures increasingly delayed crop development and increased total crop-heating costs.

Using the same approach for a greenhouse in Charlotte, North Carolina, growing ageratum at 58°F (14°C) was more expensive on a per-crop basis until mid-April. For most dates, the total energy consumed to grow the crop at 63–73°F (17–23°C) was similar. However, growing warmer allows growers to turn crops more quickly, which opens space for other crops. So why doesn't every grower turn up the heat and produce crops more quickly? First, the information generated from the Virtual Grower program depends on the location, greenhouse characteristics, and crop. That's why using Virtual Grower for your own greenhouse is so important: Results vary from one greenhouse to the next, so the program's utility hinges on spending a little time to generate the most meaningful results. Second, under light-limiting conditions in the early spring, growing some species warm can produce low-quality plants. Therefore, plants that are typically grown cool, such as pansy, petunia, and snapdragon, should generally not be grown warmer than the low 70s (°F, or low 20s [°C]), at least until light conditions are higher.

7

James E. Faust

Light

Light is the driving force for photosynthesis, thus it is fundamental to greenhouse production. There are many facets of light that growers need to consider, since each facet can influence plants differently. In this chapter, we will examine the effects of light intensity, daily light integral, photoperiod, and spectral light quality on plant growth and development. We will also discuss how to manage and measure light in a commercial greenhouse.

What is light? The three primary ways to think about light are:

1. Light is energy that plants can capture to create food (carbohydrates) in a process called photosynthesis, which fuels growth.

2. Light provides a means for the plant to determine the time of year by measuring the length of day and night.

3. Light contains all the colors of the rainbow, and the specific colors can affect plant growth and flowering.

The first section of this chapter will discuss these three aspects of light and their effects on plant growth and flowering. The second section will deal with issues related to the greenhouse light environment. These issues include the use of shadecloth, the effect of hanging basket production on the amount of light delivered to the greenhouse bench crops, and the use of artificial light sources in commercial greenhouses and their effects on plant growth and flowering. Finally, we will cover the challenging topic of light units and measuring light. Light is a very difficult parameter to discuss due to (1) the myriad number of units used to make measurements and (2) the difficulty in making useful light measurements inside greenhouses where the light distribution patterns are spatially variable and constantly changing.

Plant Responses to Light

Light intensity

During photosynthesis, a plant takes water from the soil, carbon dioxide (CO_2) from the air, and energy contained in sunlight to create sugars and starches that can be moved within the plant to provide fuel for growth. Photosynthesis and thus plant growth are greatly affected by the amount of light absorbed by the plant. The amount of light delivered to a plant at any given second is called the *light intensity*. Figure 7-1 shows the effect of light intensity on whole-plant photosynthesis of zinnias. The important point to notice is that most plants are nearly saturated with light at light intensities considerably less than full sunlight at noon (10,000 f.c. [108 klux]). Thus, the biggest impact of light on photosynthesis occurs as the light intensity increases from low intensities to moderate light intensities. For example, figure 7-1 demonstrates that increasing the light intensity from 0 to 2,000 f.c. (22 klux) has a large effect on photosynthesis (75% of the maximum photosynthesis rate), increasing the light intensity from 2,000 to 4,000 f.c. (22 to 43 klux) has an additional 20% increase in photosynthesis, while increasing the light intensity from 4,000 to 6,000 f.c. (43 to 65 klux) increases photosynthesis by only an additional 5%.

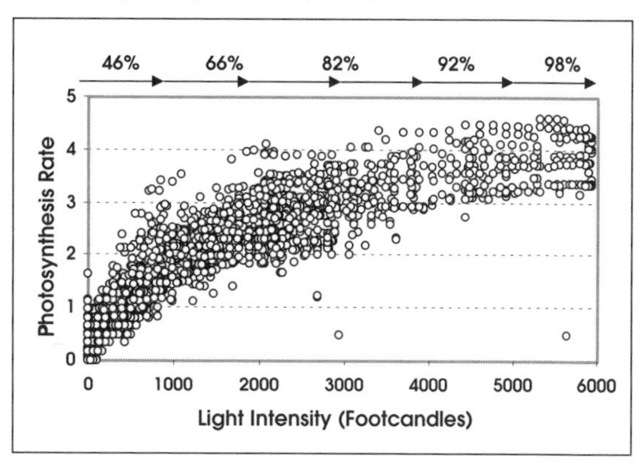

Figure 7-1. The effect of light intensity (measured in f.c.) on whole-plant photosynthesis of zinnia. The percentages reflect the portion of the total possible photosynthesis achieved.

The maximum light intensity measured outside varies throughout the year. During the summer, the maximum light intensity outside is about 10,000 f.c. (108 klux, 2,000 μmol·m^{-2}·s^{-1}, or 1000 W·m^{-2}) during the midday under perfectly clear skies. During the winter, the maximum light intensity measured outside is only approximately 5,000 f.c. (54 klux, 1,000 μmol.m^{-2}·s^{-1}, or 500 W·m^{-2}). These maximum values may only occur rarely in parts of the country that regularly experience overcast winter and hazy summer sky conditions.

Daily light integral

The term *daily light integral* is used to describe the total quantity of light delivered over the course of an entire day. The daily light integral is reported as the number of moles (particles of light) per day. The advantage of an integrated measurement over an instantaneous measurement can be best demonstrated with an analogy. If you want to know how much rain fell during the course of a day, you would place a bucket outdoors and record the volume of water collected. Whereas, recording the intensity of rainfall at one instant (i.e., the raindrops per second) would be of little value. Similarly, knowing the quantity of light delivered throughout the day is much more useful than making an instantaneous foot-candle (f.c.) measurement in the middle of the day. The daily light integral is a relatively new concept for the greenhouse industry. However, it has some advantages, because plant growth is often closely correlated to the daily light integral.

Figure 7-2 demonstrates the effect of the time of year and sky conditions on the light intensity and daily light integral delivered to a greenhouse crop. The maximum light intensity delivered to a greenhouse crop on a cloudy day in December was 900 f.c. (9.7 klux), while the daily light integral was 3 moles/day. The maximum light intensity was about 2,000 f.c. (22 klux) during both a sunny day in December and a cloudy day in June; however, the daily light integral in December was 9 moles/day compared to 12 moles/day in June. The cloudy summer day has a 25% higher daily light integral than the sunny winter day because the day length is longer in the summer, so there is more time to accumulate, or absorb, sunlight. An instantaneous foot-candle measurement cannot take day length into account. For this reason, daily light integral is the preferred method of quantifying the amount of light delivered to greenhouse crops. A sunny day in June resulted in a maximum light intensity of 4,600 f.c. (50 klux) and a daily light integral of 26 moles/day, more than eight times the amount of light delivered during the cloudy December day. Although the outside daily light integral during the summer may exceed 40 moles/day, the daily light integral measured inside greenhouses is rarely higher than 20 to 25 moles/day due to interception of light by the greenhouse infrastructure and shadecloth.

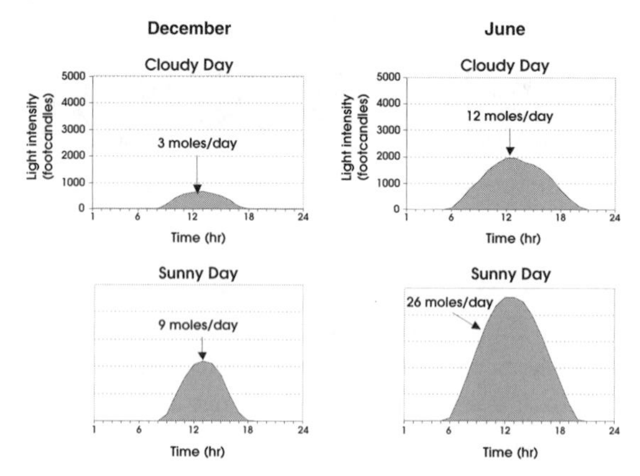

Figure 7-2. The daily light integral (moles/day) delivered to a greenhouse crop on a cloudy day and a sunny day in December and June. The gray area indicates the quantity of light delivered to the crop.

Daily light integral maps have been developed to allow growers to estimate the amount of light being delivered to different locations in the United States for each month of the year (figure 7-3). These maps are discussed in detail in the section on outdoor light levels later in this chapter.

Plant Growth Responses to Daily Light Integrals

The amount of light that a plant receives influences root growth, shoot growth (branching, stem diameter, and leaf size), and flowering (flower initiation, flower number, and time to flower). Table 7-1 provides a generalization of plant responses to different light levels.

Very low light conditions (below 5 moles/day or 500–1,000 f.c. [5.4–11 klux]) typically result in poor-quality plant growth and flowering. Under low light conditions, the plant lacks sufficient energy to produce a high-quality plant. The plants often have just one thin primary stem with very little lateral branching. There may be insufficient light to support flowers, so flowering may be delayed, flowers may be very small, or few, if any, flowers may be produced. Only a few crops, such as African violets, can produce acceptable plants under very low light conditions.

The quality of the growth that occurs under low light conditions (5–10 moles/day or 1,000–2,000

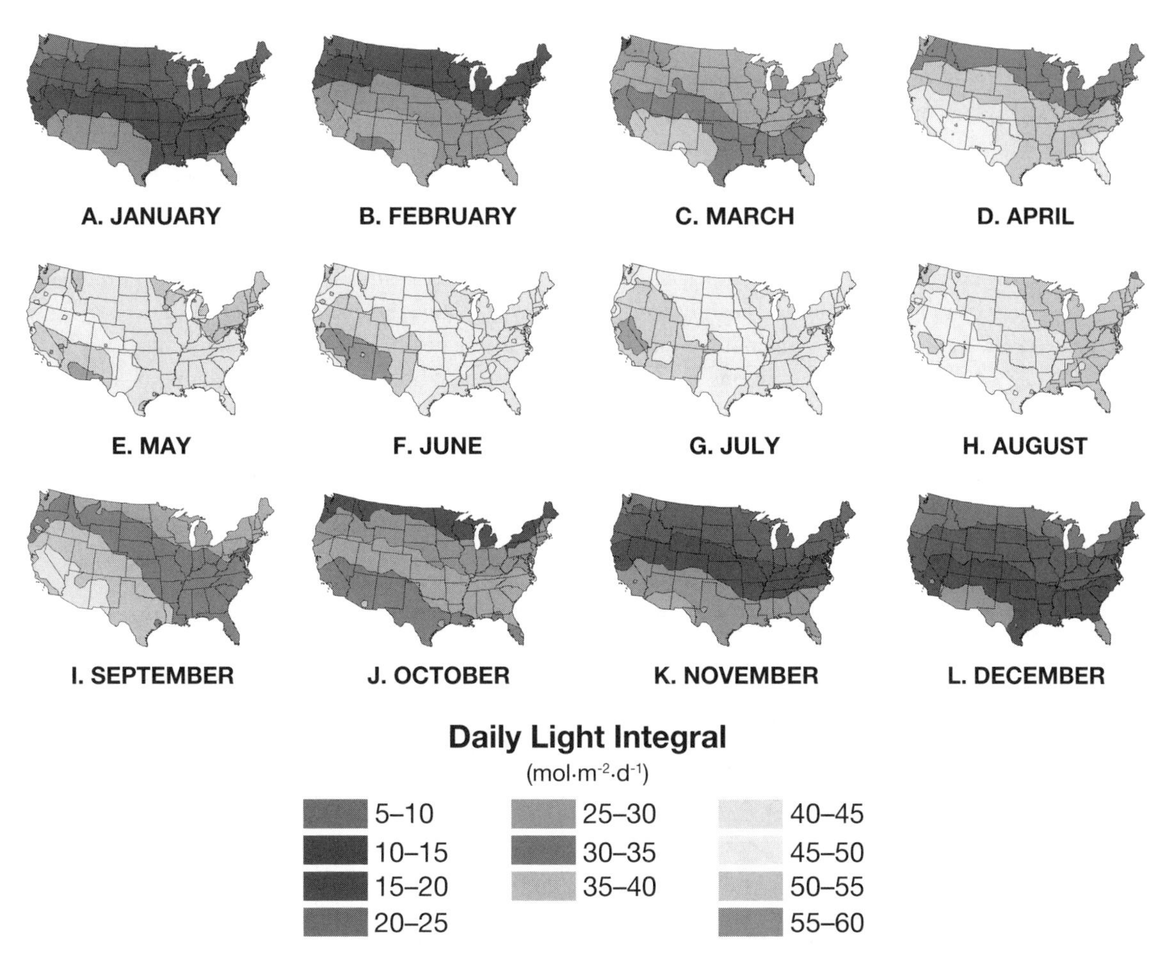

Figure 7-3. The monthly mean outdoor daily light integrals (moles/day) delivered across the contiguous United States. Typical greenhouses allow only 40–60% of the outdoor light to be transmitted to the bench or floor.

Table 7-1. Generalized Plant Responses to Different Light Levels

RELATIVE LIGHT LEVEL INSIDE A GREENHOUSE	DAILY LIGHT INTEGRAL[1] (MOLES/DAY)	FOOT-CANDLES (F.C.)[1] MEASURED AT NOON	GENERALIZED PLANT GROWTH RESPONSE
Very Low	<5	500–1,000	Poor quality[2]
Low	5–10	1,000–2,000	Minimum acceptable quality[3]
Medium	10–20	2,000–4,000	Good quality[4]
High	>20	>4,000	Excellent quality[5]

[1] It is not possible to make a direct conversion between an instantaneous light measurement (f.c.) and the daily light integral (moles/day), so this table does not suggest a direct mathematical conversion.
[2] "Shade crops" may produce acceptable plants, but quality is still not as good as possible.
[3] High-quality propagation can be achieved at this light level.
[4] Most species grow very well at this light level.
[5] Greenhouse temperatures often become excessive at this light level, but plants grow excellently with sufficient cooling.

f.c. [11–22 klux]), largely depends on the greenhouse temperatures. Under cool growing conditions (below 65°F [18°C]), plant quality can be quite good. For example, northern European growers compensate for low light conditions by growing their crops cool. Crop time is increased, but the quality is good. Cool temperatures allow the leaves and flowers to develop slowly, which allows the plant more time to accumulate energy from sunlight to produce healthy leaves and flowers. In contrast, high temperatures (above 75°F [24°C]) during low light conditions result in poor-quality growth. Under warm temperatures, the plant is developing new leaves and flowers very quickly, but there is insufficient energy from sunlight to produce substantive leaves and flowers.

Contrary to popular belief, low light does not always cause an increase in plant height or "stretch" (figure 7-4). In nature, no competitive advantage exists for a plant to grow taller after several cloudy days, since increased height will not result in increased light interception. In contrast, tightly spaced plants will elongate rapidly due to competition for light with neighboring plants; however, that is a spectral light–quality response, not a light-intensity or light-quantity response. (Light quality will be discussed later in this chapter.) Under low light conditions, individual leaf size may increase, leaves will be thinner, lateral branching is reduced, stems are thinner, and flower number is reduced. Thus, the quality of growth under low light is relatively poor, and the perception is that the plants have stretched because the canopy appears loose and relatively open.

Figure 7-4. The effect of daily light integral on the growth and flowering of zinnias. Flower diameter (top), branching and plant height (middle), and root growth (bottom) increase as the daily light integral increased. From left to right, plants were grown at 90, 70, 50, and 0% outside shade, and the daily light integrals delivered to these plants were 3, 7, 15, and 38 moles/day.

Plant growth is usually commercially acceptable for most greenhouse crops grown under moderate light conditions (10–20 moles/day or 2,000–4,000 f.c. [22–43 klux]). Plants flower normally with acceptable branching and flower number. It is relatively easy to manage watering under moderate light conditions compared to higher light levels.

The highest-quality, greenhouse-grown bedding plants, stock plants, and herbaceous perennials are usually produced under high light conditions (20–30 moles/day or 4,000–6,000 f.c. [43–65 klux]). These crops often produce commercially acceptable crops at moderate light conditions; however, the quality will improve further under high light conditions. High light conditions provide extremely well-branched plants, high flower numbers, large flowers, and excellent root growth (figure 7-4). The highest yields of greenhouse-grown, cut flower crops and greenhouse vegetables are typically grown under high light levels.

Many species produce superior-quality plants outdoors compared to inside greenhouses. This is due to higher light levels (30–60 moles/day or 6,000–10,000 f.c. [65–108 klux]) and cooler plant temperatures. Plant temperatures can be cooler outdoors during the summer due to increased air movement, lower relative humidity, and thermal cooling due to the exposure to the open sky.

Excessively high light may result in a change in leaf orientation and shape. Leaves grown under excessively high light produce a more vertical and curled leaf blade in order to avoid light interception. For example, the leaves of young pansy plants may curl when grown above 25 moles/day (outdoor summer light levels), even if temperatures are cool. Excessive light can also result in heat stress or "sunburn" of some species. Sunburn is most likely to occur when a plant has not been acclimated to high light conditions. Plant water use and evaporation increases as sunlight increases, so water management can be more difficult under high light conditions. Root death can also occur on the south side of dark-colored containers due to excessively high soil temperatures resulting from direct outdoor sunlight.

Daily light integral requirements for different species

The daily light integrals delivered inside a greenhouse are nearly always less than 30 moles/day. So, table 7-2 separates greenhouse (2–30 moles/day) and outdoor daily light integrals (30–50 moles/day). The table lists the different light requirements for many different species. The bars next to each line differentiate the light levels required for "minimum acceptable quality," "good quality," and "high quality" crops. These are subjective categories, but the intent is to demonstrate that plant quality typically improves as the daily light integral increases from low to moderate levels, then there is a relatively wide range of light levels that produces high-quality plants.

Environmental conditions can alter plant responses to light. In particular, temperature and water availability affect plant responses to light. The species that are particularly sensitive to temperature or water are noted in the table. Cool temperatures often allow plants to be grown at lower light levels, while the minimum light levels are higher at warm temperatures. Similarly, high temperatures can limit the amount of light that "cool" crops can absorb without suffering from heat stress. For example, lobelia and fuchsia tolerate full sunlight if the temperatures are moderate, but require partial shade when temperatures are warm.

Water can also modify the light requirement. For example, caladiums are typically considered to be shade plants, while they are grown outdoors in full sunlight in Florida for tuber production. However, ample water is always provided. Drought stress under full sunlight conditions is damaging to many species, since drought forces the stomata on the leaves to close. As a result, transpiration ceases, leaf temperatures increase, and sunburn can occur.

Table 7-2. Daily Light Integrals for Various Greenhouse Crops

SPECIES	DAILY LIGHT INTEGRAL									
	GREENHOUSE									OUTDOOR
	1	2	4	6	8	10	12	15	20	25–50
Cuttings in propagation										
African violet *(Saintpaulia ionantha)*										
Ferns *(Pteris, Adiantum)*										
Maranta										
Phalaenopsis (orchid)										
Spathiphyllum										
Forced spring bulbs										
Aglaonema										
Boston fern *(Nephrolepis)*										
Bromeliads										
Dieffenbachia										
Dracaena										
Streptocarpus										
Caladium					x					
Coleus (shade)										
Hosta					x					
Christmas cactus						y				

(Continued)

Table 7-2. Daily Light Integrals for Various Greenhouse Crops *(Continued)*

SPECIES	DAILY LIGHT INTEGRAL									
	GREENHOUSE									OUTDOOR
	1	2	4	6	8	10	12	15	20	25–50
Cyclamen										
Exacum										
Gloxinia *(Sinningia speciosa)*										
Heuchera										
Begonia × *hiemalis*										
Impatiens spp.										
New Guinea impatiens *(I. hawkeri)*						xy				
Poinsettia *(Euphorbia pulcherrima)*										
Dusty miller *(Senecio cineraria)*										
Dutch iris, cutflowers *(I.* × *hollandica)*										
Fibrous begonia *(B.* × *semperflorens-cultorum)*										
English ivy *(Hedera)*										
Ivy geranium *(Pelargonium peltatum)*						y				
Kalanchoe										
Lobelia						y				
Primula						y				
Schefflera										
Asiatic and Oriental lily *(Lilium* × *hybrida)*										
Easter lily *(Lilium longiflorum)*										
Fuchsia							y			
Hydrangea										
Miniature rose										
Chrysanthemum (potted)										
Ageratum										
Alyssum *(Lobularia)*										
Dianthus										
Gazania										
Gerbera										
Herbs										
Hibiscus rosa-sinensis										
Red salvia *(S. splendens)*										
Snapdragon *(Antirrhinum)*										
Succulents & cacti										
Zonal geranium *(Pelargonium* × *hortorum)*										

(Continued)

Table 7-2. Daily Light Integrals for Various Greenhouse Crops *(Continued)*

SPECIES	DAILY LIGHT INTEGRAL									
	GREENHOUSE									OUTDOOR
	1	2	4	6	8	10	12	15	20	25–50
Angelonia										
Aster										
Blue salvia *(S. farinacea)*										
Celosia										
Coleus (sun)										
Cosmos										
Croton										
Dahlia										
Ficus benjamina										
Chrysanthemum (garden)										
Gomphrena										
Lantana										
Marigold *(Tagetes)*										
Pansy *(Viola)*									y	
Perennials (full sun)										
Petunia										
Scaevola										
Verbena										
Vinca *(Catharanthus roseus)*										
Zinnia										
Cut flowers										
Greenhouse vegetables										

x = Requires ample water to perform well at high light levels.
y = Requires cool or moderate temperatures to perform well at high light levels.

Minimum acceptable quality
Good quality
High quality

Photoperiod

Day length or photoperiod are somewhat misleading terms, since plants actually measure the length of the night in order to determine the time of year. In nature, photoperiod is a more reliable indicator of changing weather patterns than temperature, since each day is the same from year to year, while temperatures and light levels vary from day-to-day and week-to-week. So, photoperiod is an effective means to signal to plants the proper times to perform certain tasks in order to avoid temperature stress (cold or hot) in temperate climates or drought associated with dry seasons in tropical climates.

Flowering is the most common factor affected by photoperiod. Plants can be categorized as short-day plants, long-day plants, or day-neutral plants. Short-day and long-day plants can be further subdivided into facultative or obligate photoperiod requirements.

Obligate short-day plants require short days for flowering—flowering does not occur under long days (e.g., poinsettias). Facultative short-day plants benefit from short days—the plants will flower under long days or short days, but flowering is fastest under short days (e.g., African marigolds). Day-neutral plants are not affected by photoperiod (e.g., impatiens).

Photoperiod responses can vary between varieties with a particular species. For example, garden chrysanthemums are short-day plants. Early-season varieties are facultative short-day plants, so they flower under long days (i.e., sixteen-hour days), but their time to flower is slower than under short days. Late-season varieties are obligate short-day plants, so they require short days before flowers will develop. Mid-season varieties are facultative short-day plants if the temperatures are cool (65°F [18°C]), but they are obligate short-day plants if temperatures are warm (75°F [24°C]).

Other plant processes can also be affected by photoperiod. For example, plants that store starch in their roots or stems prior to winter begin to develop storage organs, such as tubers, under short-day conditions. Similarly, short days can induce dormancy of certain herbaceous perennials.

Spectral Light Quality

Light can be separated into the different colors of the rainbow, or the light spectrum. Each color represents different wavelengths of light. For example, blue light occurs at 400–500 nanometer wavelengths, while red light occurs at 600–700 nanometer wavelengths. When plants respond to specific wavelengths in the light spectrum, they are said to respond to the spectral light quality. The most familiar light-quality issue of importance to greenhouse growers is the ratio of red light to far-red light.

Sunlight has a balanced ratio, almost 1:1, of red and far-red light. Leaves use red light for photosynthesis, thus leaves readily absorb red light but do not absorb very much far-red light. As a result, far-red light reflects off the foliage or passes straight through the leaves. Consequently, the light measured near plants is often low in red light relative to far-red light yielding a low red-to-far-red-light ratio. Plants use this information to determine if there are neighboring plants competing for light. If a neighboring plant is shading another plant, then the competing plants will potentially benefit from growing taller. A taller plant will intercept more light than its neighbor, allowing for increased plant growth. Thus, plants in the greenhouse that are spaced close together will elongate very rapidly due to the low red light, high far-red light environment (figure 7-5). Shade provided from shadecloth or clouds does not have the same affect on stem elongation, because these shade sources remove equal amounts of red and far-red light. Thus, shadecloth and clouds do not have a significant impact on spectral light quality.

Figure 7-5. Red salvia plants grown at pot-to-pot spacing (left) and at staggered spacing in a shuttle tray (right). The two plants received the same light quantity (daily light integral) and are flowering at the same time, but the tightly spaced plants have longer internodes in response to the altered spectral light quality (i.e., the low red-to-far-red-light ratio caused by the close proximity of neighboring plants).

Canopy closure

Canopy closure is a term used to describe the proximity of neighboring plants. When the leaves of neighboring plants begin to overlap, the red-to-far-red ratio begins to be impacted, resulting in an increase in stem elongation. Thus, the degree of canopy closure, or leaf overlap, is an indicator of the potential stem elongation to be expected from a crop in the coming weeks. Canopy closure should be factored into the plant growth regulator (PGR) decision-making process, since a "closed" canopy is more likely to elongate rapidly than an "open" canopy in which the neighboring plants have not yet begun to overlap.

Artificial lighting

Artificial lighting can alter the spectral light quality, since electrical lamps can have a different ratio of red and far-red light than sunlight. Incandescent lamps provide a large quantity of far-red light relative to red light. Consequently, plants grown under incandescent lamps often stretch more than plants grown in sunlight. High-pressure sodium lamps produce more red light than far-red light, so stretching does not typically occur on plants grown under these lamps.

Natural light

Sunlight does not always provide the same red-to-far-red-light ratio throughout the day. Twilight, or the light at sunrise and sunset, tends to be higher in far-red light than red light compared to the sunlight delivered throughout the day. When black cloth is closed over a greenhouse crop in the late afternoon and reopened after sunrise to provide short-day conditions, the plants may actually elongate more slowly than plants grown without black cloth since the plants are not "seeing" the low red light and high far-red light provided at twilight. The duration of twilight increases as latitude increases, so twilight is a more important issue for northern growers.

Outdoor Light Levels

The average monthly outdoor daily light integrals delivered across the contiguous United States are shown in figure 7-3. The mean outdoor daily light integral ranges from 5–10 moles/day across the northern states in December to 55–60 moles/day in the southwestern states in May through July. During the winter months, the differences in daily light integral primarily occur between the northern and southern regions, while during the summer months the differences in daily light integral primarily occur between the eastern and western states.

The daily light integral changes rapidly during the spring and fall. For example, the mean outdoor daily light integrals for Columbia, Missouri, are 28 moles/day in March, 37 moles/day in April, and 42 moles/day in May. Similarly, the daily light integrals change from 42 to 33 to 25 to 16 moles/day in August, September, October, and November, respectively. Thus, the light available for plant growth increases dramatically during spring production and decreases dramatically during fall production. These changes have a tremendous impact on plant growth and quality.

The daily light integral maps only describe the amount of light delivered from the sun to the outside of the greenhouse. The amount of light actually reaching the crop on the greenhouse bench is affected by the greenhouse light transmission.

Managing Light in Greenhouses

Light transmission

Light must be transmitted through the greenhouse structure to be delivered to the plants. Surprisingly, only 35–70% of light measured outside the greenhouse typically reaches the greenhouse crop. Don't be fooled by the high transmission percentage of most greenhouse-glazing materials when considering greenhouse light transmission. Polyethylene and glass may transmit more than 90% of the light that hits the material in a perpendicular orientation; however, the greenhouse infrastructure (i.e., posts, gutters, trusses, etc.) allows 0% transmission, and sunlight that hits the glazing material at a low angle, such as in the winter or early or late in the day, is often reflected outward, away from the greenhouse. Thus, the actual amount of light transmitted to the greenhouse crop may be much less than expected. Factors such as dust, condensation, hanging baskets, and shadecloth will reduce greenhouse light transmission even further. The only way to know the greenhouse light transmission for a particular greenhouse is to make some measurements.

Growers can measure greenhouse transmission by measuring the light intensity outside the greenhouse and then quickly measuring the light intensity inside the greenhouse. These measurements must be made when the light intensity is not changing rapidly, so either clear-sky or uniformly overcast conditions are preferred. Overcast conditions have the benefit that the greenhouse light environment is relatively uniform, although the transmission percentage may be slightly higher on cloudy days compared to sunny days. On sunny days, the light-intensity measurements recorded in the greenhouse will vary tremendously depending on the position of the sensor. Also, if the light intensity is changing from moment to moment, it is difficult to make comparable measurements inside and outside the greenhouse, unless you have two sensors and radio communication. Greenhouse light transmission is calculated with the following equation:

$$\text{Greenhouse light transmission (\%)} = \left(1 - \frac{LI_{out} - LI_{in}}{LI_{out}}\right) \times 100$$

where: LI_{out} = Light intensity measurement made outdoors

LI_{in} = Light intensity measurement inside the greenhouse at canopy level

Daily light integrals inside the greenhouse can be estimated by multiplying the greenhouse light transmission percentage by the daily light integrals indicated on the contour maps in figure 7-3. For example, 25–30 moles/day are typically delivered to South Carolina in October. If the greenhouse light transmission is 50%, then we could expect that greenhouse to receive 12.5 to 15 moles/day during October; this is plenty of light to grow a "high-quality" finished poinsettia crop but is in the "good-quality" range for pansies (table 7-2). For this reason, outdoor fall pansy crops are typically superior to greenhouse-grown pansies, assuming outdoor temperatures are not excessively high.

Retractable-roof greenhouses provide the benefit of providing higher light levels than greenhouses permanently covered with a glazing material. Crops requiring higher light levels will benefit from this environment and will produce higher-quality crops. However, one should not assume that the amount of light delivered to a greenhouse crop inside a retractable-roof greenhouse is equivalent to outdoor light levels. The infrastructure and folded roof will still absorb a significant amount of light. For example, 20% or more of the outdoor light may be intercepted by the infrastructure resulting a greenhouse light transmission of 80% or less. Thus, the daily light integrals measured inside a retractable-roof structure may be 30–40 moles/day during the summer, compared to the 50 moles/day measured outdoors. (See the discussion on retractable shade curtains later in this chapter.)

Light distribution patterns

Gutters, shade curtains, and black cloth can cast relatively large and dark shadows on crops growing on the benches below. The shadow pattern cast by these objects is determined by their orientation. Gutters or curtains placed in a north-to-south orientation will cast a shadow that is constantly changing while the sun moves east to west throughout the day. In contrast, gutters or curtains placed in an east-to-west orientation cast a relatively fixed shadow as the sun moves through the sky. Fixed shadows create a poorer light distribution pattern that results in poorer uniformity for flowering and for water usage. As a result, plants growing within the shadow pattern may flower a few days to a week slower than other plants in the same greenhouse. Also, uniform watering can be challenging, since the shaded plants will dry out more slowly than the neighboring plants receiving more light. Thus, a north-to-south orientation of gutters and curtains is preferable in most production situations.

Shading

Shading is often used from late spring to early fall for the primary purpose of reducing the heat stress that results from high light levels entering the greenhouse. Most bedding plants and herbaceous perennials grow very well in full sun, so the primary benefit of shade is reduced greenhouse temperatures. Evaporative cooling pads help to offset the heat load caused by high light, thus effective greenhouse cooling allows the grower to provide higher light levels before heat stress is problematic. Cuttings on propagation benches often require very low light levels (5 moles/day or 1,000 f.c. [11 klux]) to prevent water stress during the first days or weeks after being stuck. Excessively low light (below 2.5 moles/day or 500 f.c. [5.4 klux]) can delay rooting. The light levels in propagation should continue to increase as the cuttings begin to root and are weaned to normal greenhouse light levels.

Shade can be provided by placing shadecloth on the inside or outside of the greenhouse or by painting a shading compound (whitewash) on the outside of the greenhouse glazing material. Shadecloth can be fixed into position for the entire season, or it can be opened and closed with a retractable shade-curtain system. Retractable shade systems allow growers to provide higher light intensities in the morning and afternoons when the temperatures are moderate and full shade when heat stress is a concern. The daily light integral delivered to a crop with a retractable curtain system can be considerably higher than a fixed curtain system. Figure 7-6 demonstrates the effect of a permanent shade curtain versus a retractable shade curtain on the daily light integral delivered to the bench crop. In this example, the retractable curtain system allows the greenhouse crops to receive an additional 25% more light compared to the fixed shade system. Note that the shade opens at a lower light intensity than it closes. This is done for two reasons: (1) so that the curtain is not opening and closing with each passing cloud and (2) this provides the crops with more shade in the hotter afternoon compared to the cooler morning. Also, note that the shade drops the light intensity down to about 3,000 f.c. (32 klux), which is sufficient to continue to drive photosynthesis. This example utilizes a 50% shade curtain. Shade curtains typically intercept 40–80% of sunlight. The proper percentage of shade will depend on the crops being grown and the effectiveness of the greenhouse cooling system. However, in most cases, the 40–60% shade curtains are most desirable. Higher shade percentages can reduce the light intensity in the greenhouse to below the desired levels for adequate growth. During the summer, greenhouses are frequently hot regardless of the amount of shading. In these cases, heavy shade often serves to excessively reduce plant quality by creating a low light environment while the temperatures continue to be very warm.

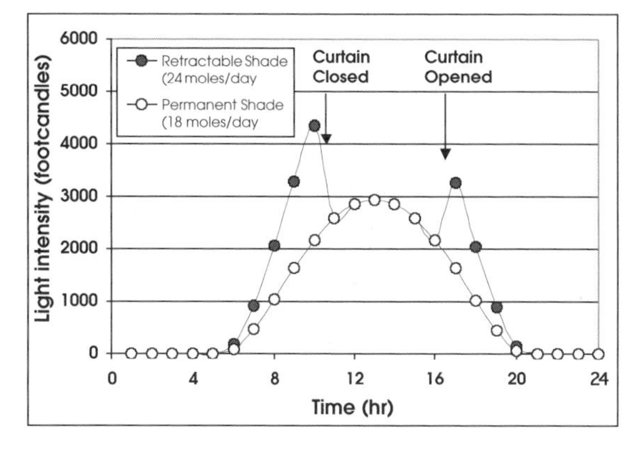

Figure 7-6. Comparison of the light intensity and daily light integral provided to a greenhouse crop under a permanent shade curtain and a retractable shade curtain (50% shade)

Retractable shade curtains can be controlled by a time clock, temperature set points, outdoor light set points or a combination of these methods. There is no perfect technique, but, in general, it makes sense to provide as much light as the plant requires for proper growth unless a temperature stress point is reached. Leaf and flower development are severely reduced in many species when the average daily temperature (twenty-four-hour temperature) exceeds 80–85°F (27–29°C). Thus, a common procedure is to operate the curtains based on an outdoor light set point that is overridden if the temperature becomes excessively high. Since greenhouse temperatures tend to be much hotter in the afternoons, it is desirable to provide less shade in the morning than in the afternoon. Finally, it is easiest to manage the greenhouse environment when the shadecloth is set to be fully open or fully closed. Partial closing or opening creates hotspots on the bench during sunny days that can create water-management challenges, especially in propagation houses or with young crops on the bench.

Shade curtains are also effective thermal blankets. In retractable systems, the curtains are often closed at night to reduce heat loss from the greenhouse. Aluminized shade curtains are more effective as thermal blankets than black-colored shade curtains. Curtains with transparent plastic strips are also more effective as thermal blankets than shade curtains with open spaces between the strands of fabric.

Hanging baskets

Overhead hanging baskets intercept light that would otherwise be delivered to the bench crop. For example, figure 7-7 displays a "plant's eye" view of hanging baskets placed overhead. The hanging baskets without plants intercepted 12% of the light that would have otherwise reached the bench crop, while the hanging baskets with plants intercepted 28% of the light. Too many hanging baskets overhead can obviously reduce plant quality on the bench crops by reducing the daily light integral delivered to those crops. The color of the hanging basket container also impacts the light environment. White containers reflect sunlight and thus intercept less light than green containers.

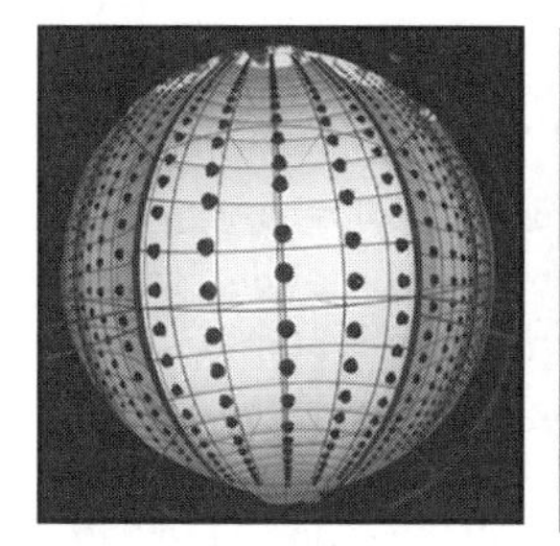
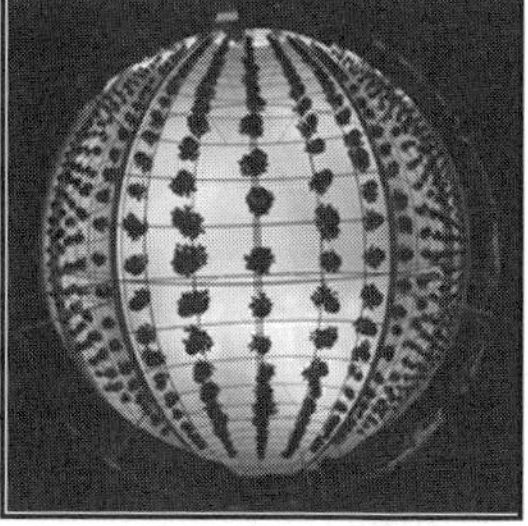

Figure 7-7. Plant's eye view of overhead hanging baskets. Green baskets *without* plants (left) arranged at this density (1.3 baskets/yd.2) intercepted about 12% of the sunlight, while the same hanging basket density *with* plants (right) intercepted 28% of the sunlight.

How many hanging baskets can be grown overhead? This question is usually answered through trial and error based on observations of the bench crop; however, the maps in figure 7-3 combined with a greenhouse light transmission measurement can provide a reasonable estimate for planning purposes. For example, in Columbus, Ohio, the average outdoor daily light integral for March is 20–25 moles/day (we will assume 20 moles/day in this example to account for a below-average year). A greenhouse transmission measurement indicates that this greenhouse transmits 50% of the outdoor light without any hanging baskets, resulting in 10 moles/day being delivered to the bench crop during March. If we make the assumption that 8 moles/day is adequate to grow a good-quality bench crop (table 7-2), then in Columbus during

March, we can place hanging baskets overhead to intercept 2 moles/day or 20% of the light that would have otherwise reached the bench crop. Greenhouse light transmission measurements can be taken again after baskets are hung to estimate the percentage of light intercepted by the hanging baskets.

Artificial lighting

There are two different purposes for using artificial lighting. First, lamps can be used to provide additional light to increase photosynthesis and plant growth. This is termed *supplemental lighting*. Secondly, lamps can also be used to alter the photoperiod perceived by the plants. This is termed *photoperiodic lighting*. Supplemental lighting requires much higher light intensities to be effective compared to photoperiodic lighting. For example, lamps used for supplemental lighting deliver 250–750 f.c. (2.7–8.1 klux) for eight to sixteen hours, while photoperiodic lighting only requires 10 f.c. (108 lux) for four hours in the middle of the night, which is termed *night interruption lighting*. Thus, photoperiodic lighting does not provide sufficient light for any significant photosynthesis, or plant growth, to occur. This issue is often confused in crop culture guides where a crop is said to "benefit from lighting in the winter." In this statement, it is unclear as to whether the author is recommending supplemental or photoperiodic lighting. Providing supplemental versus photoperiodic lighting can have dramatically different effects on plant growth and flowering.

Supplemental lighting

Supplemental lighting increases the quantity of light, or daily light integral, delivered to a plant resulting in increased plant growth (i.e., more roots, shoots, and flowers). Supplemental lighting is usually provided with high-pressure sodium or metal-halide lamps. The typical intensities provided by these lamps are 250–750 f.c. (2.7–8.1 klux). Lamps that deliver 300 f.c. (3.2 klux) for twelve hours from high-pressure sodium lamps provide a daily light integral of 2 moles/day. This is a relatively small amount compared to the daily light integrals delivered by the sun during much of the year (figure 7-3). However, during overcast winter conditions, a greenhouse crop may receive less than 2 moles/day. Thus, supplemental lighting increases the daily light integral by 100% or more and will have a substantial impact on plant growth. If a crop receives 12 moles/day, then supplemental lighting would only increase the daily light integral to 14 moles/day (a 17% increase) and would have a relatively small impact on the growth and quality of most species (table 7-2). Consequently, supplemental lighting is most commonly used on high-value crops in northern greenhouses during the winter months.

The time of day that lights are on is also important. Light delivered by supplemental lighting is most effective when the ambient light intensities are low. As shown in figure 7-1, increasing the light intensity from 1,000–1,500 f.c. (11–16 klux) has a much larger impact on photosynthesis than increasing the light intensity from 5,000–5,500 f.c. (54–59 klux).

Photoperiodic lighting

Photoperiodic lighting is provided to manipulate plant perception of the length of the night. Flowering is the most common response to photoperiod. So, photoperiod manipulation provides growers with a switch that can be used to force a plant to grow vegetatively or to flower. Photoperiodic lighting has traditionally been provided with incandescent lamps; however, high-pressure sodium, metal-halide, and fluorescent lamps work equally well. Lamps can be mounted on irrigation booms that can be sent up and down the greenhouse bay throughout the middle of the night. Any given plant in the greenhouse should receive at least 10 f.c. (108 lux) of light for five minutes out of every thirty minutes. Incandescent lamps produce more far-red than red light, so plants receiving a photoperiodic lighting treatment with incandescent lamps will elongate more rapidly than under normal conditions. Compact fluorescent lamps are beginning to replace incandescent lamps for use in night interruption lighting.

LED lighting

Much research is currently exploring the use of light-emitting diodes (LEDs) for greenhouse applications. Once the price of these lamps becomes more competitive in the marketplace, we can expect to see many opportunities to use them in greenhouse production for both supplemental and photoperiodic lighting purposes.

8

Brian E. Whipker, Ingram McCall, and Joyce Latimer

Growth Regulators

Regulating plant growth is an integral part of the production program of many greenhouse crops. Plant growth regulators (PGRs) are chemicals that are designed to affect plant growth and/or development. They are applied for specific purposes to elicit specific plant responses. Although there is much scientific information on using PGRs in the greenhouse, it is not an exact science. Achieving the best results with PGRs is a combination of art and science—science tempered with a lot of trial and error and a good understanding of plant growth and development.

For best results, PGRs should be handled as production tools, like water and fertilizer. PGRs should be an integrated part of your crop-production program. They should not be used as crutches for poor management of other cultural practices. However, PGRs should be used in conjunction with a number of nonchemical—both biological and physical—control options to manipulate plant growth so that well-proportioned, compact plants are produced. We will first examine the nonchemical control options and then detail how to maximize including chemical control options in your production program.

Biological Control

Growing varieties that are less vigorous is often the first step available to growers. While this may work in theory, it may not be commercially practical. Many customers demand specific color or characteristics, and shorter-growing varieties with these attributes may not be available. Thus, physical or chemical controls must also be incorporated into a production plan.

Physical Control

Knowing how the growing environment and cultural practices can affect plant growth will help in managing a crop's growth. There are a number of physical control options available: container size, timing, water stress, nutrient stress, mechanical conditioning, light quality and quantity, pinching, and temperature.

Container size

Restricting roots can be used to control plant growth. This is done by utilizing a small container or by increasing the number of plants per pot. This method works especially well when other production parameters such as ample light and proper nutrition are provided.

Timing

One of the most-effective methods of controlling excessive plant growth is crop timing. Simply staggering the finishing times of a crop at two- to three-week intervals is very effective with many crops, such as bedding plants. This ensures that a new supply of plants will always be available, avoiding the need to hold a crop, which can become leggy.

Water stress

A traditional method of controlling plant growth is to withhold water. Water stress can be used on a number of crops, including impatiens and tomatoes. Allow the plants to wilt slightly between waterings, but do not allow them to reach the permanent wilting point. This will lead to shorter plants but may have the undesirable effect of reduced plant quality, delayed flowering, or premature bolting (as with brassicas).

Nutrient stress

Reducing or withholding fertilizer tends to slow overall plant growth. Limiting the amount of nitrogen (N) to 50–100 ppm will help control growth of many crops, such as bedding plants. The type of nitrogen supplied can also impact plant growth. Relying on nitrate nitrogen instead of ammoniacal-nitrogen or urea-nitrogen forms (which encourage lush growth) will also help limit leaf expansion.

Phosphorus (P) also promotes plant growth by encouraging stem elongation. Plug producers commonly use low-phosphorus fertilizers, such as 13-2-13 (calcium magnesium) or 15-0-15 (which does not contain phosphorus), to help limit stem elongation.

Mechanical conditioning

Brushing plants has been found to be a very effective way of controlling plant height (30–50%) of many vegetable transplants or herbs. This is especially important for these crops since limited chemical control measures are available other than Sumagic (uniconazole). Brushing involves the movement of a PVC pipe or wooden dowel rod over the top third of the plant. Researchers at the University of Georgia suggest that the plants should be brushed daily for about forty strokes to obtain the greatest effect. The foliage should be dry to avoid damaging the leaves. This method is not effective on brassicas such as cabbage or broccoli and should not be used if foliar diseases are present.

Light quality and quantity

Higher light quality tends to limit plant elongation, thus resulting in shorter plants. Low light conditions caused by late spacing, crowding, or too many hanging baskets overhead can lead to leggy plants and should be avoided. Photoperiod also can be used to control plant growth. This practice is widely used with pot chrysanthemums by providing taller varieties with only one week of long days (LD) to limit vegetative growth when compared to shorter varieties that receive three weeks of LD to promote growth.

Pinching

Pinching (also pruning or sheering) can be used to increase branching, improve the shape of the plant, and decrease the height of the plant. However, labor costs of pinching and the potential delay in flowering that may occur may make this method of height control an economically unfeasible option.

Temperature

Temperature manipulation can be used very effectively in controlling plant growth. Details about differential temperatures (DIF) are covered in chapter 6.

Chemical Growth Control

The most common plant growth regulators (PGRs) used in greenhouse crop production are the plant growth retardants. Quality standards dictate that most container-grown greenhouse crops be compact, have short internodes, have a height consistent with the container they are grown in, and have strong stems. Although short or dwarf cultivars exist for many greenhouse crop species, chemicals that further reduce plant height and increase the compactness and strength of the plant are often required. PGRs may also be used to slow growth or "hold" plant material in the greenhouse.

The class of plant hormones called gibberellins influences plant growth by affecting cell elongation. If synthetically produced gibberellins are applied to a plant, it will become tall and spindly. In contrast, if gibberellin production in the plant is reduced, it will be shorter and stronger with thicker stems and smaller leaves. Therefore, most of the commercially available growth retardants function by inhibiting gibberellin synthesis. There are a number of commercial growth retardants used in greenhouse crop production. Each label has specific recommended dose ranges, recommendations, and precautions (table 8-1).

Commonly used PGRs

Some of the most commonly used greenhouse PGRs include daminozide, chlormequat chloride, ancymidol, flurprimidol, paclobutrazol, and uniconazole. *Note:* Ancymidol, flurprimidol, paclobutrazol, and uniconazole are persistent on plastic surfaces and in soil. Do not reuse flats, pots, or soil from treated plants, especially for plug production of sensitive crops.

Daminozide

Sold under the commercial names B-Nine, Compress WSG, and Dazide, this material is applied only as a foliar spray because it is rapidly broken down when applied to the substrate. It is highly mobile in the plant and will rapidly move from the point of application to all parts of the plant. It is most commonly applied at rates between 1,250–5,000 ppm. Daminozide is effective on most crops except lilies. It is highly effective in controlling growth of seedlings in plug flats, and it is most effective in cooler climates.

Chlormequat chloride

Sold under the commercial names Chlormequat E-Pro, Citadel, and Cycocel, this material is one of the most widely used PGRs in agriculture because it is also used to prevent lodging in grain crops. In greenhouse crops, it is most commonly used on poinsettias, geraniums, osteospermum (figure 8-1), and hibiscus. It is usually applied as a foliar spray at 1,000–3000 ppm. Foliar chlormequat-chloride applications often result in a phytotoxic response (chlorosis), but the symptoms are acceptable because they are usually covered up with new leaf growth. Substrate drenches are also effective at controlling excessive growth. However, because rates are similar to what are used with foliar sprays, the practice is usually not a cost-effective option in the United States. In certain crops (i.e., poinsettias, geraniums, and herbaceous perennials), a mixture of daminozide and chlormequat chloride (both at reduced rates) may be used. This usually provides for greater height control and reduces the potential for phytotoxicity.

Figure 8-1. Two foliar sprays of Cycocel at 1,500 ppm provided excellent control of osteospermum plant stretch.

Ancymidol

Sold under the commercial names Abide and A-Rest, this chemical is effective at much lower rates than either daminozide or chlormequat chloride. The concentrations applied are usually in the range of 10–200 ppm for foliar sprays and 0.15–0.5 mg per 6" (15 cm) container for substrate drenches. Ancymidol readily moves through the plant and is usually used on crops where other chemicals are not effective (most notably in bulb crops) or on very high-value crops (i.e., plugs). Growers often prefer the use of ancymidol on plugs because of its lack of phytotoxicity and because its limited residual allows the plugs to grow out of the growth control effects after being transplanted, thus making it a "safer" PGR to apply. Phytotoxicity may occur from applications of high rates of ancymidol (especially under high temperatures) and usually appears as necrotic spots.

Flurprimidol

Sold under the commercial name Topflor, this chemical is a relatively recent introduction into the US market, although it has been available in Europe since the 1990s. Flurprimidol is closely related to ancymidol chemically but has a greater degree of activity. Most commercial spray application rates are between 0.5–50 ppm (figure 8-2). Flurprimidol is also one of the most cost-effective growth retardants to use as a drench, with recommended-usage rates in a range similar to uniconazole on most plants.

Figure 8-2. Topflor foliar sprays of 40–50 ppm were effective in controlling salvia 'Red Hot Sally' growth when grown in 1203-cell packs. Left to right: untreated control, 40 ppm, and 50 ppm.

Paclobutrazol

Sold under the commercial names Bonzi, Downsize (labeled for drench applications only), Paczol, Florazol, and Piccolo, paclobutrazol is a member of the family of PGRs known as triazoles. (Uniconazole is in this family as well.) These chemicals do not readily move within the plant since they are transported in the xylem and not in the phloem. Therefore, triazoles are absorbed by the leaves but cannot be transported out of the leaves to other parts of the plant. Because of this fact, it is important that when applied as a foliar spray, the triazole solution should contact the plant stems. The triazole plant growth retardants are the most persistent (longest lasting) of the plant growth retardants. These materials are active at very low rates, and thus the potential for error and crop overdose is greater than with other plant growth retardants.

Paclobutrazol is the most widely used growth retardant for greenhouse-grown floriculture crops in the United States. It is commonly applied as a foliar spray, and the trial rates of 5–90 ppm are listed for experimental use, but most commercial spray application rates are between 1–50 ppm. It is also effective as a substrate drench. It can be applied as a single, high-dose drench of 2–200 ppm to provide season-long control of excess growth (rates vary among plant species and cultivars). Additionally, the application of low-dose drenches of 0.1–1 ppm can be used to provide temporary control of plant growth, allowing greenhouse managers to apply additional drenches as needed.

Uniconazole

Sold under the commercial names Concise and Sumagic, uniconazole is applied as both a foliar spray and substrate drench (figure 8-3). Experimental use rates of 1–50 ppm are listed on the label, but most commercial spray application rates are between 0.5–25 ppm. Uniconazole can also be used as a drench, at rates 50% lower than recommended for paclobutrazol. This chemical is commonly used on perennials because it is highly effective on a very broad range of plant species.

Figure 8-3. Sumagic substrate drenches are effective at controlling excessive stretch of calla lily 'Crystal Blush'. Left to right: untreated control and 1, 2, and 4 mg of active ingredient per pot.

Other PGRs Used in Greenhouse Production

Not all PGRs are used to control plant height. Others are used to cause flower bud abscission, increase branching, promote flowering, and stimulate shoot elongation.

Abscisic acid (ABA)

Sold under the commercial name ConTego, abscisic acid (ABA) is one of the newest materials being introduced into greenhouse production. ABA is used for enhancing postharvest performance by extending the shelf life of plants. An application of ABA as a foliar spray at concentrations of 250–1,000 ppm induces stomatal closure, thus reducing water loss. Plants simply use less water and dry out more slowly, thus avoiding wilt.

Dikegulac sodium

Augeo is the commercial dikegulac-sodium product currently registered for greenhouse use. Predecessors to this product were Atrimmec and Atrinal. This formulation temporarily stops shoot elongation, thereby promoting lateral branching. It is thus a pinching agent for greenhouse crops including azalea, *Elatior begonia*, bougainvillea, clerodendron, fuchsia, gardenia, grape ivy, geranium, kalanchoe, lantana, lipstick vine, shrimp plant, *Schefflera arboricola*, verbena, and some herbaceous perennials (figure 8-4). Some phytotoxicity and distorted growth can occur with dikegulac sodium, so sufficient time is required to allow new plant growth to cover any damaged leave

Figure 8-4. Augeo at 400 ppm (right) nearly doubled the number of lateral branches on *Gaillardia aristata* 'Gallo Yellow' while controlling vegetative growth and improving the upright habit of the plant.

Ethephon phosphonic acid

Sold under the commercial name Florel, this material is absorbed by the plant tissue and, due to a change in pH once absorbed into the plant cells, releases ethylene. This acid is used to promote flower bud abortion and vegetative branching in crops. Although it is used in many situations, it is most commonly used where vegetative cuttings are being produced and in hanging basket production. The formulation is applied as a foliar spray at concentrations of 250–500 ppm. It may also be used to promote flowering in bromeliads.

Benzyladenine (BA)

Sold under the commercial name Configure, benzyladenine (BA) is used to promote branching and increase flower set. Configure has specific label recommendations for Christmas cactus (100–200 ppm foliar spray) (figure 8-5), *Echinacea* (300–900 ppm foliar spray), and hosta (500–1,000 ppm foliar spray) and a supplemental-use label allowing for experimental applications on any annual, perennial, foliage, or tropical plant grown in a greenhouse. Optimal results occur when the plant is actively growing and is physiologically receptive for growth or flower promotion. BA has been very effective in improving branching in both plugs and finished plants of many herbaceous perennial crops (figure 8-6). BA does not readily move within the plant, therefore complete coverage is required.

Figure 8-5. Configure applied as a foliar spray at 100 ppm to vegetatively growing Christmas cacti increases the number of lateral shoots.

Figure 8-6. *Gaura lindheimeri* 'Siskyou Pink' treated with 600 ppm Configure (right) had more shoots per pot, more lateral breaks, more inflorescences, and more flowers than untreated control plants (left).

Gibberellins (GA)

This PGR is sold under the commercial names Florgib and ProGibb T&O and can be applied when additional stem elongation is desired in such plants such as tree forms of azaleas, poinsettias (figure 8-7), and geraniums. For the production of tree forms of these species, the general recommendation is to apply a 50 ppm gibberellin (GA) foliar spray after plants are approximately 6" (15 cm) tall. Depending on the desired height and size, another one or two GA applications may be applied. Gibberellins can also be applied to promote growth and overcome over-application of gibberellin-inhibiting PGRs. For this use, the general recommendation is to apply 1–3 ppm GA as a foliar spray and check for growth stimulation after five days. The application can be repeated if additional shoot elongation is desired.

Figure 8-7. Extra stem growth of tree forms of plants can be achieved with the application of gibberellins to encourage cell elongation.

Benzyladenine + gibberellin combinations

Sold under the commercial names Fascination and Fresco, these combination products are used on potted lilies as foliar sprays to avoid lower-leaf yellowing and leaf drop. A typical recommendation is to apply a foliar spray at 25–100 ppm. The actual concentration depends on timing (early- versus late-production applications) and species. Sprays are also beneficial at prolonging the flower life of potted lilies.

Application timing

There are a number of factors that influence the efficacy of PGRs. Remember, growth retardants do not reduce plant size. They reduce the plant's growth rate. You must apply the growth retardant prior to the "stretch." Look for recommendations on the PGR label for the appropriate application time. These recommendations will be given in terms of plant development or plant size as opposed to production time.

Generally, PGRs should be applied just prior to rapid shoot growth. This is generally one to two weeks after transplanting a plug, after the roots are established and as the plant resumes active growth; on pinched plants, it is after the new shoots are visible and starting to elongate. Softer PGRs, like daminozide, chlorme-

Table 8.1. Comparing Attributes of PGRs

ATTRIBUTES		PLANT GROWTH REGULATOR (PGR)							
Chemical		Ancymidol	Chlormequat chloride	Daminozide	Daminozide + Chlormequat chloride	Ethephon phosphonic acid	Flurprimidol	Paclobutralzol	Uniconazole
Trade name(s)		Abide, A-Rest	Chlormequat E-Pro, Citadel, Cycocel	B-Nine, Compress WSG, Dazide	–	Florel	Topflor	Bonzi, Downsize, Paczol, Florazol, Piccolo	Concise, Sumagic
Active ingredient		0.0264%	11.8%	85%	–	3.9&	0.38%	0.4%	0.055%
Activity level		++	+	+	++	+	+++	+++	+++
Multiple applications needed		++	+++	+++	++	++	+	+	+
Application type[1]	Foliar spray	yes	yes	yes	yes	yes	yes	yes	yes
	Substrate drench	yes	yes	no	no	no	yes	yes1	yes
	Dips	cuttings	cuttings	cuttings	–	cuttings	bulbs, cuttings	bulbs, cuttings	bulbs, cuttings
Chemical absorption	Ease of absorption	+++	+	+	+	++	+++	+++	+++
	Time (hours)	0.5–1.0	4	18–24	18–24	12–16	0.5–1.0	0.5–1.0	0.5–1.0
	Factors that improve absorption	high humidity, limited air movement, cloudy days, early morning or late afternoon applications							
	Translocation within the plant	+++	+++	+++	+++	–	+	+	+
Absorption sites	Leaves	+++	+++	+++	+++	+++	++	++	++
	Stems	+	+	–	+	–	++	++	++
	Roots	++	+	–	–	–	+++	+++	+++
Typical concentrations	Foliar sprays (ppm or mg/l)	15–50	1,000–3,000	1,250–3,000	Daminozide 750–2,500 + Chlormequat 750–1,500	250–1,000	1–50	1–50	0.5–25
	Drench (mg active ingredient per pot)	0.15–0.5	–	300–3,000	–	–	0.01–4.0	0.01–8.0	0.01–4.0
Other factors	Do pink bark substrates affect drenches?	++	–	+	–	–	++	++	++
	Phytotoxicity potential	+	+	+++	+	++	+ (Do not apply to stressed plants)	+	+
	Overdose potential	+	+	++	++	++	+++	+++	+++
	Influence of water pH	–	–	–	–	pH 4.0 optimal	–	–	–
Shelf life	In the bottle (years)	<3	<2	<2	–	indefinite	<4	<4	<2
	Mixed solution	within 24 hours	within 24 hours	within 24 hours	within 24 hours	within 4 hours	within 24 hours	within 1week	within 24 hours

– = Not applicable Degree of activity: (+) least to (+++) greatest [1] Check label for legal uses quat chloride, or ancymidol, can be applied earlier in the crop cycle than the more potent PGRs, like uniconazole, paclobutrazol, or flurprimidol, wherein applications are made when the plant has developed sufficient canopy growth. For example, with fall pansies (*Viola*), paclobutrazol labels recommend an application when the plant is 2" (5 cm) in diameter, while uniconazole labels recommend applying it when the plant is near final height, 3" (8 cm) tall.

This is where the art of plant growth regulation is most important. You must learn how your crop grows and when to intervene to obtain the desired results. Remember to note details of crop development in your records of PGR treatments. For example, due to the weather conditions, next year you may need to treat at seven days after transplanting instead of at the ten days after transplanting that you used this year. You must gauge when rapid elongation will likely occur and treat to counter it.

Many growers use multiple applications of growth retardants to better control plant growth. A single application at a high rate early in the plant production cycle may be excessive if growing conditions are not as good as expected. An early application at a lower rate provides more flexibility, but the tradeoff is the additional labor involved with a second application if it becomes necessary. Some growers improve crop uniformity by using multiple applications at lower rates to affect small corrections in plant growth.

How will you know how well the PGR treatments really work? The only way to confirm the efficacy of a PGR is to leave a few representative plants untreated. These "check plants" offer valuable insight into ways to adjust future PGR applications.

Be aware that excessive rates of many of these PGRs can cause persistent growth reductions in the flat or even in the landscape. It is always a good idea to evaluate the long-term effects of your treatments by growing some out for yourself and talking with your customers. Be careful to avoid late applications, especially of paclobutrazol or uniconazole, as they may delay flower opening on bedding plants. Labels for all of the PGRs have recommendations for how late you can make an application in order to avoid a delay in flowering or smaller flowers. However, drench applications of paclobutrazol have provided excellent control of poinsettia height very late in the production cycle without causing the reduction in bract size accompanying late spray applications. Learn the art of using PGRs for plant growth regulation.

Application guidelines

Spray applications

Plants to be treated with PGRs should be healthy, turgid, and unstressed—never wilted. The label will identify the target tissue for that PGR. For example, daminozide is only effective as a foliar spray, whereas paclobutrazol and uniconazole sprays must reach the stems. Uptake and effectiveness of a PGR depend on selecting the application technique that will ensure proper coverage of the target tissue. Leaf surfaces should be dry for foliar applications, and the best uptake of PGRs from spray applications will occur under low-stress, low-drying conditions. This is more critical for daminozide and ethephon than for some of the newer chemistries like the triazoles. Overhead irrigation after treatment with daminozide or ethephon should be delayed for eighteen to twenty-four hours to avoid washing the material off the leaves.

The triazoles—paclobutrazol and uniconazole—are absorbed primarily by stem tissue and then translocated upwards in the plant. Therefore, consistent and complete coverage of the stems is necessary for uniform effect. In other words, if the stem of one lateral receives an inadequate amount of spray, it will grow faster than the others, resulting in a poorly shaped plant, most noticeable in potted crops like poinsettia or chrysanthemum. Both the foliage and stems take up ancymidol and flurprimidol. In addition, all four of these compounds are very "soil active," meaning they may be adsorbed to particles in the media and become available to the plant through root uptake.

The label will provide a recommended application volume for sprays, especially for chemicals that are soil active. All foliar applications of PGRs should be applied on an area basis (i.e., uniformly spray the area where the plants are located with the recommended volume of solution). Do *not* spray individual plants or spray to reach a subjective target like "spray to glisten-ing." Since every applicator will have a slightly different definition of these goals, there will be no way of recommending appropriate rates or obtaining predictable results. For soil-active PGRs, dosage is dependant on both the concentration of the solution and the volume of that solution applied in the treated area. Therefore, to improve predictability, the label-recommended spray application rates are generally set at 2-qt. (1.9 l) finished spray per 100 ft.^{2} (9.3 m^2), which is sufficient to cover the plant and permit a small amount of runoff onto the medium. It is also considered to be a comfortable walking pace for applicators with handheld sprayers.

With the soil-active PGRs, precautions should be taken to avoid overapplication with sprays. Spray applications require more attention to detail because overspray material lands or drips onto the medium. The overspray from a 2 qt. per 100 ft.^{2} (1.9 l/9.3 m^2) application is a part of the recommended dosage. However, if your application volume exceeds that recommendation, then your application dosage also exceeds the recommendation.

Recognizing that stem coverage is necessary for the triazoles, you may need to apply a higher volume to large or dense plants to obtain adequate coverage. In fact, the paclobutrazol label recommends a spray volume of 3 qt. per 100 ft.^{2} (2.8 l/9.3 m^2), for "larger plants with a well-developed canopy." Adjust the concentration you apply accordingly. This demonstrates the importance of recordkeeping (see the Recordkeeping section). Always consider the rates presented on PGR product labels, or from any other resource, to be a guideline to assist you in developing your own rates based on your growing conditions and application methods.

The relationship of rate and volume can be exploited when treating multiple crops with different PGR needs. With a single solution of PGR in the spray tank, you can apply the label-recommended volume to attain your basic application dosage, or you can apply additional volume to crops that need additional growth regulation to attain a higher dosage. Application volume is another tool that you can use to maximize your efforts and reduce time mixing or reloading higher concentrations of PGR solutions.

Spray equipment

To assure proper spray volumes, your compressed air sprayer should be equipped with a pressure gauge and regulator, and you should consistently use the same nozzle for all PGR applications. Your sprayer should be calibrated by determining the output of the chemical with the selected nozzle at the selected pressure within a specified time period. Using this information, you can apply a known amount of material to a known area. Spray droplet size also affects response with smaller droplet sizes providing better coverage but only up to a point. Mist- or fog-type applicators do *not* provide adequate volume for coverage of plant stems and the medium and, therefore, have not been effective when used with compounds like paclobutrazol and uniconazole. PGR applicators should be trained to uniformly apply a given amount of clear water in the greenhouse before they make PGR applications. Uniformity of the application is critical to the uniformity of the crop response.

Applying drenches

Applying PGRs to the surface of the substrate has several advantages over foliar sprays. Drenches generally have less effect on flowering or flower size and tend to provide longer-lasting growth regulation than sprays. Drenches are easier to apply uniformly than sprays, because the drench volume is easily measured and, when applied to moist media, it is easy to obtain good distribution of the PGR in the media. Therefore, the resulting growth regulation is frequently more uniform. The product label specifies the recommended volumes for drench applications to different size pots or types of media. In general, 4 fl. oz. (118 ml) of drench solution is applied to a 6" (15 cm) "azalea" pot, and that volume is adjusted up or down with pot size to obtain a volume where about 10% of the solution runs out the bottom of the pot when the media is moist. Remember that the amount of active ingredient applied to the plants depends on both the concentration (ppm) of the solution and the volume applied. ***Read the label.***

Alternative application methods

Alternative methods of applying PGRs directly to the media have been developed and are described on the label. For example, ancymidol, flurprimidol, and paclobutrazol are labeled for application through the irrigation system ("chemigation"). These are generally labeled for flood (subirrigation), drip irrigation, and overhead sprinkler systems. Again, rates vary with the volumes used and the method of application. Paclobutrazol applied once by subirrigation requires 50–75% of the amount of paclobutrazol that is applied in a typical drench application. Pressure-compensated drippers are recommended for use with PGRs to more accurately regulate the volume of solution applied to each pot. Read and exactly follow the label for chemigation applications, especially with regard to the safety of municipal water supplies.

Growers are using three other methods of providing a drench-type application of soil-active PGRs on a more economical scale: media surface application sprays, "sprenches," and "watering in."

Media surface application

Media surface application sprays are made to the surface of the media of filled flats or pots. The treatment is applied at normal-to-high spray volumes. However, since it is applied to the media surface, it is activated by irrigation and is available to the plant in the root zone. Both paclobutrazol and uniconazole are labeled for this method of application. Rates are lower than used for sprays, but higher than used for drench applications.

Sprenches

"Sprenches" are high-volume foliar sprays that result in additional runoff into the media, providing a drench effect. Rates are lower than those recommended for spray rates.

Watering in

The third technique, "watering in," is a type of chemigation in which the PGR is injected into the irrigation water and applied at each irrigation at very low rates of active ingredient. Only PGRs labeled for chemigation can be used for watering in.

All of these application methods use the relationship between rate and volume to provide the desired control. Again, you must develop techniques that fit your production methods and your growth management preferences.

Liner dips

Liner dips, or drenches, are another specialized way to use soil-active growth retardants. Although many of the soil-active PGRs have been tested, only Paczol (paclobutrazol) is labeled for this application. The root system of rooted liners or plugs is dipped into a solution of the PGR (or they may be thoroughly drenched in the plug tray). Extensive work has been conducted at the University of Florida on this application method.

Liners should be "dry," which is defined as the root-ball being ready for irrigation, but plants are not under drought stress. Time in the solution is not critical; thirty seconds to two minutes is sufficient for saturation of the root-ball. Liners may be planted immediately or held up to ten days without loss of PGR effect. There is no loss of effectiveness of the dip solution during treatment, so the remaining solution can be disposed of by using it as a drench on appropriate crops.

Advantages of the liner dip include early control of very vigorous crops and flexibility of the treatment with respect to not having to handle plants during the restricted entry interval (REI). This application method has been very effective on herbaceous perennials like *Coreopsis* (figure 8-8). The liner dip is especially useful in combination plantings where the more vigorous plants can be treated prior to planting without reducing the growth of the slower plants in the group. The liner dip rates should be selected to provide early control of plant growth. Additional applications can be made as necessary for longer-term crops.

Figure 8-8. *Coreopsis* 'Sweet Dreams' treated with Paczol liner dips at the time of planting showed good to excessive control of plant height at six weeks after treatment with 0, 0.7, 1.3, 2.0, 2.7, or 3.4 ppm Paczol (left to right).

Be aware of bark

For many years, the adage in PGR drenches has been "Bark ties up soil-active PGRs." However, new research shows that this is not necessarily true. As long as the bark is properly aged before the media is mixed, it has little effect on the availability of these soil-active PGRs to the plant roots. Again, you must identify PGRs and rates that work with your production system.

Environmental conditions

Environmental conditions can have a significant impact on the efficacy of a PGR. Generally, a healthy, unstressed plant growing under low-evaporative conditions (i.e., early in the morning or late in the afternoon) is most responsive to treatment. To maximize uptake, the chemical must remain in contact with the leaf long enough to be absorbed. This time varies for the different PGRs, but generally foliar uptake is enhanced with slower drying conditions, which in turn increases the effectiveness of the treatment. This is especially important with foliar uptake of PGRs like daminozide, chlromequat chloride, benzyladenine, or ethephon phosphonic acid. Plants treated with daminozide or ethephon phosphonic acid should not be overhead irrigated for at least eighteen to twenty-four hours after treatment, but plants treated with flurprimidol, paclobutrazol, or uniconazole may be irrigated one hour after treatment. Read the label for any warnings on how irrigation or environmental conditions will affect plant response to the PGR treatment.

Recordkeeping

Recommended PGR rates to apply to most of the commonly grown floricultural crops are listed on the product label. These should be used as guidelines, and adjustments should be made for your particular location, cultivar, stage of development, and weather conditions. Keeping notes on your application methods and the results of your PGR treatments will allow you to improve the consistency of your own application methods and establish rates and volumes appropriate for your production system. Note the concentration and the volume applied, the stage of development of the crop (i.e., the number of leaves, approximate height, and presence of flowers), fertilization program, weather conditions, and the environmental conditions under which the PGR was applied. Always keep a few plants untreated for comparison, especially if you are new to using PGRs.

Preparing PGR solutions

The dose to apply to a crop is based on two factors: the solution concentration and the volume of solution applied per area. Foliar sprays and drenches require an even application to obtain consistent results. To accomplish this, base the dose on: (1) measuring out a known amount of chemical, (2) adding it to a known volume of water, and (3) applying the growth regulator to a known bench area or apply a known volume of drench to each pot or plant. The volume of drench applied increases with the pot size. (Specifics are listed on each product label.)

When mixing PGRs, take care to accurately measure and apply the chemical. Table 8-2 provides mixing guidelines for foliar sprays. Drench applications vary by pot size and desired dose, so refer to the product label for exact mixing instructions. As always, the label contains the legal mixing information. The University of New Hampshire Cooperative Extension website (http://extension.unh.edu/Agric/AGGHFL/AGGHFL.htm) contains a web-based tool, called PGRCALC, to help you calculate PGR rates.

Summary

The degree of growth regulation caused by PGRs is impacted by all other phases of plant culture. Remember that you have to fit PGRs into your own production program. Plan ahead to achieve the best results from PGRs; do not use them as an afterthought when the plants are out of control. You cannot "shrink" an overgrown plant!

The multitude of variations possible in application methods, cultivars and species grown, and growing conditions make it impossible to recommend specific rates for all operations. Use the product labels as a resource for the use of PGRs on a variety of crops, using the lower of suggested effective rates for starting your own trials. These, or any other rate recommendation, should be used as guidelines, and adjustments should be made for your particular location, cultivar, stage of development, and weather conditions. Whenever you treat your crop, hold back a few untreated plants so that you can judge the effectiveness of your treatment. Remember that methods of application have significant effects on results. Develop your own program, recording your procedures and results. Then test and refine your program.

Table 8-2. Mixing Rates for Plant Growth Regulators Used as Foliar Sprays				
Chemical	Spray solution ppm (mg/l)	fl. oz./gal. of final solution	ml/gal. final solution	ml/l final solution
Ancymidol	1	0.48	14.34	3.79
	5	2.43	71.70	18.94
	10	4.85	143.39	37.88
	50	24.24	716.93	189.39
Chlormequat chloride	500	0.54	16.04	4.24
	1,000	1.08	32.08	8.47
	1,500	1.63	48.12	12.71
	3,000	3.25	96.24	25.42
Paclobutrazol	1	0.032	0.95	0.25
	5	0.160	4.73	1.25
	10	0.320	9.46	2.50
	50	1.600	47.32	12.50
Ethephon phosphonic acid	250	0.81	23.93	6.32
	500	1.62	47.86	12.64
	1,000	3.24	95.73	25.29
Flurprimidol	5	0.160	4.78	1.26
	10	0.320	9.55	2.52
	50	1.61	47.75	12.61
Uniconazole	1.0	0.26	7.57	2
	2.5	0.65	18.93	5
	5.0	1.28	37.85	10
	10.0	2.56	75.71	20
	Spray solution ppm (mg/l)	Dry oz./gal. of final solution	g/gal. final solution	g/l final solution
Daminozide	1,000	0.16	4.45	1.18
	2,500	0.39	11.13	2.94
	5,000	0.79	22.26	5.88

Raymond A. Cloyd

Managing Insect and Mite Pests

A wide diversity of insect and mite pests feed on greenhouse-grown floriculture crops. This chapter will discuss the management strategies that greenhouse growers can use to deal with insect and mite pests, including cultural, physical, pesticidal, and biological. In addition, various scouting techniques will be addressed. Finally, we will detail the biology of, the damage done by, and the specific management strategies for the major greenhouse pests.

Cultural Management

Watering

Overwatering plants may increase the numbers of certain insects, such as fungus gnats and shore flies. Excessive moisture levels keep the surface of the growing medium moist, resulting in algae accumulation that provides breeding sites for fungus gnats and shore flies. In addition, overwatering stresses plants, thus increasing their susceptibility to other plant-feeding insects and mites.

Fertility

High or low fertility levels can increase crop susceptibility to certain insects and mites. Many plants grown under optimal light and nutrient conditions produce natural chemical defenses that protect them from plant-feeding insects and mites. Any changes in light and nutrition can compromise these defenses, thus increasing the potential for insect and mite problems.

High fertility levels can increase the concentration of soluble salts in the growing medium, which stresses plants and increases their susceptibility to insects and mites. The soft, succulent tissue resulting from excess fertilization is often easier for insects and mites to penetrate with their mouthparts.

In addition, plants respond to high fertility levels, especially from excess levels of nitrogen, by allocating more nutrients to new growth. This provides pests with easier access to nutrients they need to develop and reproduce. Consequently, insects and mites can develop faster and cause more injury to the crop. Higher levels of nitrogen may also increase the female reproductive ability of some insects including whiteflies and aphids.

Sanitation

Weeds and plant debris within the greenhouse provide sites that insects and mites can use to survive and then spread to the main crop. Furthermore, weeds harbor plant pathogens, most notably viruses that can be obtained by insects and then transmitted to crops when they feed. Weeds that serve as reservoirs for viruses, specifically the tospoviruses—impatiens necrotic spot virus (INSV) and tomato spotted wilt virus (TSWV)—include chickweed, lambsquarters, nightshade, oxalis, shepherd's purse, pigweed, and bindweed.

Weeds also provide refuge for whiteflies, aphids, leafminers, thrips, and spider mites, which can move from desiccating weeds onto the main crop.

Plant debris such as leaves, flowers, and growing media may also provide a refuge for insects and mites. Insects and mites can migrate to fresh plant material as plant debris dries out (desiccates). Also, leftover growing media provide sites for fungus gnat adults to lay eggs and western flower thrips to pupate.

Physical/Mechanical Management

Screening

Screening greenhouse openings such as vents and sidewalls can prevent or restrict the entry of insects into greenhouses. This may reduce the number of insecticide applications required during the growing season. The appropriate screen size depends on the insect species, whether it be whiteflies (462 microns [0.018"]), aphids (340 microns [0.013"]), leafminers (640 microns [0.025"]), or thrips (192 microns [0.0075"]). It is best to select insect screening based on the smallest insect to be excluded. Screen-construction types include woven, knitted, and film. The use of screening will reduce airflow through the greenhouse, and screens with smaller holes are more resistant to airflow. However, increasing the surface area through which air passes can compensate for the reduced airflow. This can be accomplished by building a large, wooden box frame, which is then attached to the greenhouse and retrofitted with insect screening.

The National Greenhouse Manufacturers Association (NGMA) has information on how much screening is needed in order to avoid restricting airflow and burning out fan motors. However, not all greenhouses are amenable to screening. If it is not possible to screen the entire greenhouse(s), then concentrate on screening the windward sides. Screening greenhouses is only effective when used with other management techniques such as scouting and plant inspection. Screens must be cleaned regularly to remove debris, which may reduce airflow. However, do not use a high-pressure spray or brush to clean the screens. In addition, be sure to turn off fans before cleaning screens.

Scouting

Scouting, or monitoring, is an important component of every insect and mite management program in greenhouse production systems. However, before any sticky card is placed inside the greenhouse or plants are visually inspected, greenhouse producers need to determine the goals of their scouting program.

Scouting goals

The goals of any scouting program may consist of one or a combination of the following:

1. Reduce insecticide/miticide use

A reduction in the amount of insecticides/miticides you use may decrease the selection pressure placed on insect and mite populations, which leads to a lower likelihood of populations developing resistance to insecticides and miticides. This may also lead to fewer problems with phytotoxicity, resulting in the production of healthier plants.

2. Determine efficacy of management strategies

The effectiveness of cultural, physical, chemical, and biological pest management strategies needs to be evaluated to ascertain if current strategies are working. This will help determine if any adjustments to a pest management program are required.

2. Evaluate population dynamics and trends

This will determine seasonal abundance and fluctuations in pest populations during the crop-growing season, which will lead to properly timing applications of insecticides and miticides before pest populations build up to damaging levels. In addition, this will determine the times of the year when insecticides or miticides are not needed because populations are low.

3. Improve use of insecticides and miticides

Timing applications of insecticides and miticides to kill the most vulnerable life stage of a pest, such as the larvae, nymphs, and adults, will improve their effectiveness and result in higher pest mortality.

Scouting tools

Certain tools or materials are needed in order to initiate a proper scouting program based on the goals outlined above. Scouting and recordkeeping materials include colored sticky cards (blue or yellow), potato wedges, potato sticks, a 10x hand lens, a clipboard, flags, data sheets, and a map of each greenhouse. Data gathered from the scouting program may be computerized and placed in a spreadsheet or database.

Scouting techniques

Once the goal(s) and materials of the scouting program have been determined, it is then important to understand the scouting techniques used to ascertain the relative number of insects and mites. Scouting techniques may be categorized as either passive or active.

Passive scouting

Passive scouting techniques involve the use of devices (traps) such as colored sticky cards, potato wedges, or potato sticks that attract or lure insects. Sticky cards are effective in capturing the adult stages of thrips, fungus gnats, shore flies, whiteflies, leafminers, and winged aphids. Blue or yellow sticky cards may be used. Blue sticky cards are generally used when thrips is the dominant insect pest attacking the crop. Yellow sticky cards attract a wide variety of flying insects, including adult thrips. With yellow sticky cards, it is also easier to see insects and to differentiate pests from growing medium that may be present on the card.

For most insects, place colored sticky cards just above the crop canopy and set them (attached to a bamboo stake with a clothespin) so that they can be adjusted as the crop increases in size. For fungus gnat adults, yellow sticky cards should be placed horizontally near growing medium and below the crop canopy, because this is where fungus gnat adults are most active. In all cases, only one side of a yellow sticky card may be used at one time (leave the protective wax paper on the unused side of the sticky card for the first week). This allows for one sticky card to be used for two weeks. Scout at least once per week, and record the number of insects captured on the sticky cards on data sheets. The number of sticky cards to use within a crop varies with the greenhouse operation, but the general rule is one or two sticky cards per 1,000 ft.^{2} (95 m^2). However, additional sticky cards may be needed if the crop is susceptible to the viruses transmitted by thrips.

In addition to placing sticky cards within the crop, it is important to locate them near greenhouse openings such as doors, vents, and sidewalls. This will detect the migration of insects from outside the greenhouse. Place sticky cards among the crop near openings, because this is where an infestation will most likely begin. Weeds located outside the greenhouse are a source of insects migrating into the greenhouse, thus sticky cards placed near openings will detect insects moving off weeds. Sticky cards may also be positioned underneath benches in greenhouses with gravel or soil floors. This will detect the presence of thrips, fungus gnats, and shore flies and determine if they are pupating in these areas.

Potato wedges or sticks are used to scout for fungus gnat larvae. For potato wedges, cut potatoes into 0.25" (6 mm) pieces and then firmly place them into the growing medium. Allow the wedges to sit for forty-eight hours, and then remove them and look for fungus gnat larvae feeding on the wedges. Fungus gnat larvae are approximately 0.125–0.25" (3–6 mm) long and translucent with a black head. Count the number of larvae on the wedges and record this number on a data sheet. For potato sticks, cut potatoes into sticks that are shaped like french fries, 3–5" (8–13 cm) long, and 0.25" (6 mm) wide. Insert them into the growing medium, leaving approximately 0.25" (6 mm) sticking out. Potato sticks are effective in scouting for fungus gnat larvae that reside deep within the growing medium and when plants are large and have an extensive root system. Furthermore, they are effective in determining larval populations in bulb crops such as Easter lilies (*Lilium longiflorum*). Remove sticks after forty-eight hours, count the number of larvae, and record the number on a data sheet.

Active scouting

Active scouting techniques involve visual inspection—looking for insect and mite pests on the underside of leaves or inspecting entire plants. Randomly select twenty plants within a greenhouse section or a crop-growing area. Use these plants as indicators, mark or flag them to determine the pest infestation levels and record the number from each plant. Also, note the life stage (eggs, larvae, pupae, and/or adults) to help time the applications of insecticides and miticides.

To save time and labor costs, concentrate scouting efforts on susceptible crops. In addition to scouting different susceptible crop types, different cultivars (e.g., chrysanthemum) should be scouted, as certain cultivars may be more susceptible to insect or mite pests than others.

The active scouting technique works well for non-flying pests such as immature thrips, whitefly nymphs, young and wingless aphids, mealybugs, and spider mites. Placing a white sheet of paper (8.5" x 11" [A4]) and gently striking the plant foliage can also demonstrate the presence of spider mites and thrips. This will dislodge spider mites and thrips and force them to land on the paper, where they can be seen moving around.

Conventional and Alternative Pesticides (Pest Control Materials)

Greenhouse producers tend to rely on pesticides (or pest control materials) such as insecticides and/or miticides as the major method of dealing with insect and mite pests. The insecticides and miticides used in greenhouses may be classified as either conventional or alternative.

Conventional

Conventional insecticides and miticides include those in the major chemical classes such as organophosphates (acephate and chlorpyrifos), carbamates (methiocarb), and pyrethroids (bifenthrin, cyfluthrin, fenpropathrin, fluvalinate, lambda-cyhalothrin, and permethrin). These materials have a broad spectrum of pest activity and kill a wide range of insect and/or mite pests. However, they are specific in the way they kill a target pest (mode of action), which makes them susceptible to insect or mite populations developing resistance.

In order to prolong the longevity of conventional insecticides and miticides, it is important to rotate modes of action to avoid insect and mite populations becoming resistant. The key is to rotate modes of activity, not chemical classes, because some chemical classes have very similar modes of activity. For example, the organophosphates and carbamates, despite being different chemical classes, both have identical modes of activity (acetylcholinesterase inhibitors). Therefore, using acephate (Orthene) and then switching to methiocarb (Mesurol) is not a correct rotation scheme. Similarly, although acequinocyl (Shuttle), pyridaben (Sanmite), and fenpyroximate (Akari) are in different chemical classes (napththoquinone, pyridazinone, and phenoxypyrazole), all three are active on the energy production system and thus should not be used in succession. The modes of activity of the major chemical classes are presented in table 9-1. In general, rotate different modes of activity every two to three weeks (during one generation cycle); however, this depends on the time of year.

Alternative

Alternative insecticides and miticides include insect growth regulators (azadirachtin, buprofezin, cyromazine, diflubenzuron, fenoxycarb, kinoprene, novaluron, and pyriproxyfen), insecticidal soap (potassium salts of fatty acids), horticultural oils (petroleum and paraffinic-based oils), neem oils (clarified hydrophobic extract of neem oil), entomopathogenic fungi (*Beauveria bassiana*), bacteria (*Bacillus thuringiensis*), and microorganisms (*Saccharopolyspora spinosa*). In general, these materials have a broad mode of action in the way they kill target pests. These materials are also generally less directly harmful to natural enemies or biological control agents than conventional pest control materials. However, many alternative insecticides and miticides have a narrow spectrum of pest activity or are selective in the range of pests they control. Some only control one group of pests (e.g., mites), whereas others may kill two to three different types of insects or mites.

Contact and systemic

Many insecticides and miticides registered for use in greenhouses kill pests either by contact or ingestion. However, several materials have either systemic or translaminar properties. Systemic insecticides are those in which the active ingredient is taken up into the plant tissues and is transported (translocated) to other locations in the plant, where it will affect insect pests. These may be applied to the foliage or as a drench or granule to the growing medium. Most systemics do not have activity on spider mites. An example of a systemic insecticide is imidacloprid (Marathon). Some insecticides and miticides have translaminar, or local systemic activity. These

Table 9-1. Major Chemical Classes and Modes of Activity

CHEMICAL CLASS	MODE OF ACTIVITY
1. Organophosphates and carbamates	Inhibit the enzyme cholinesterase. This prevents the termination of nerve-impulse transmission.
2. Pyrethroids	Destabilize nerve cell membranes.
3. Macrocyclic lactone	Affects gamma-amino butyric acid (GABA)–dependent chloride ion channels, inhibiting nerve transmission.
4. Neonicotinoids	Work on central nervous system, causing overstimulation and blockage of the postsynaptic nicotine acetylcholine receptors.
5. Insect growth regulators	Chitin synthesis inhibitors, juvenile hormone mimics, and ecdysone antagonists. Chitin synthesis inhibitors prevent the formation of chitin, which is an essential component of an insect's exoskeleton. Juvenile hormone mimics cause insects to remain in a young stage. Ecdysone antagonists block the molting hormone.
6. Soaps and oils	Damage the waxy layer of the exoskeleton of soft-bodied insects, which results in desiccation, or the smothering of insects by covering the breathing pores (spiracles).
7. Mitochondria electron transport inhibitors	Inhibit energy production by preventing the synthesis of adenosine triphosphate (ATP).

materials penetrate leaf tissues and form a reservoir of active ingredient within the leaf. This provides residual activity against foliar-feeding insects and mites. Examples of insecticides and miticides with translaminar activity include abamectin (Avid), pyriproxyfen (Distance), chlorfenapyr (Pylon), spinosad (Conserve), etoxazole (TetraSan), spiromesifen (Judo), and acephate (Orthene). Systemic insecticides should be applied when plants have an extensive, well-established root system and are actively growing. This will lead to greater uptake of the active ingredient through water-conducting tissues. Applying systemic insecticides during sunny days will also lead to increased uptake of the active ingredient through the transpiration stream. In contrast, uptake is inhibited when plants do not have well-established root systems. In addition, high humidity and low light conditions can lead to reduced uptake of systemic insecticides. Any delayed uptake of the active ingredient may result in the material taking longer to kill insect pests. Systemics are more effective when plants are herbaceous rather than woody, particularly on stem-feeding insects, such as aphids.

Placement and timing

Thorough, uniform spray coverage is absolutely essential for controlling insect and mite pests. Understanding the biology of the major greenhouse plant-feeding pests will help greenhouse producers determine the pest-feeding location so that spray applications can be directed to specific plant parts. For example, sprays of contact insecticides and miticides must get to leaf undersides, where spider mites and whiteflies are normally located.

Insecticides and miticides are less effective if the vulnerable life stage of an insect or mite pest is not present. In general, contact and systemic insecticides do not affect the egg and pupal stages of many insects. For example, control will be minimal if western flower thrips eggs and pupae are the dominant life stages present. Young nymphs that emerge from eggs and adults that emerge from pupae will have escaped exposure from a previous application. This is especially true with short-residual materials. As a result, repeat applications are usually necessary. In general, the young (crawlers or nymphs) are most susceptible to applications of insecticides and miticides. Proper scouting can help detect the presence of vulnerable pest life stages so that an insecticide or miticide can be applied accordingly.

Insecticides should be applied during the early morning or late afternoon because this is normally when most insects such as aphids, whiteflies, and thrips are active. Short-residual, contact insecticides and miticides are minimally effective when insects and mites are less active. Applying pest control materials during hot, sunny days may result in rapid drying and a reduction in control. In contrast, applying horticultural oils (petroleum or paraffinic-based oils) during cloudy weather or under high relative humidity (70%) may result in phytotoxicity because the material takes longer to dry. Another concern when applying sprays in the evening is the promotion of foliar diseases, because leaves may remain moist for an extended period of time. Enhance drying through heating, venting, and the use of horizontal airflow (HAF) fans to alleviate potential problems with foliar diseases.

Biological Control

Biological control involves using living organisms or natural enemies, such as parasitic wasps (parasitoids), predators, and pathogens, to manage or regulate insect and mite pest populations in greenhouses. A female parasitoid inserts her eggs into an insect host, the eggs hatch into young larvae that consume the insect's internal contents, and then mature into adults that eat a hole in the dead insect and fly away. Parasitoids don't kill insects immediately, but they do reduce reproduction and fitness. Parasitoids are generally specific to the pest species and life stage preferred for attack. Predators consume portions of or eat the entire insect. They generally feed on all insect life stages including eggs, young, and adults. Pathogens such as bacteria, fungi, and nematodes work in a very similar manner to parasitoids in that they use the insect or mite pest as a food source by consuming the internal contents. Both parasitoids and pathogens are slower acting in terms of killing insect pests than predators.

In order to obtain sufficient control or regulation, it is important to release natural enemies early, before insect or mite populations reach outbreak populations. Release them immediately upon arrival, as biological control agents, in general, have a very short shelf life. In addition, check biological control agents prior to release to make sure they are alive.

For questions on the use of biological control, it is best to consult a biological control supplier beforehand. Commercially available biological control agents for the major greenhouse insect and mite pests are listed in table 9-2.

Major Greenhouse Insect and Mite Pests

Thrips

Western flower thrips (*Frankliniella occidentalis*) is the predominant thrips species that feeds on a wide range of horticulture and floriculture crops grown in greenhouses and has a distribution throughout the United States and worldwide.

Biology

The life cycle of western flower thrips consists of egg, nymph, pupa, and adult. Western flower thrips are small [less than 0.08" (2 mm) long] insects with piercing-sucking mouthparts. Adult females lay eggs into leaves. Females may live up to forty-five days and can lay between 150–250 eggs during their lifetime. Eggs hatch into nymphs, which feed on leaves and flowers. Thrips may pupate in flowers, leaf litter, or growing media. Later, adults emerge and feed primarily on flowers. The life cycle is temperature dependent, generally requiring two to three weeks to complete development from egg to adult. During cooler weather it takes thrips longer to complete their development. However, as the temperature increases, the thrips life cycle is shortened, which can cause populations to rise dramatically. For example, the life cycle may be completed in seven to ten days at 85°F (29°C). As a result, thrips populations may fluctuate over the course of the growing season.

Damage

Western flower thrips cause direct damage by feeding on flowers and leaves. Thrips feed on developing flowers prior to opening, causing premature bud abortion or distortion of flowers. Thrips feeding on leaf buds before they open can result in leaf scarring. Flowers and leaves fed upon by thrips have a silvery appearance. Feeding injury to geranium (*Pelargonium*) leaves often resembles edema (small blisters). Western flower thrips also cause indirect damage by vectoring the tospoviruses tomato spotted wilt virus (TSWV) and impatiens necrotic spot virus (INSV).

Management

Implementing sanitation practices, such as removing weeds, old plants, and old growing medium debris, are important in avoiding thrips problems. Removing weeds from within the greenhouse will eliminate hiding places (refugia) and sources of the tospoviruses. Thrips can acquire viruses from many weed types and then transmit them during feeding to the main crop. Plant debris should be removed because this provides hiding places for thrips, especially during spray applications. Growing media left on benches or on greenhouse floors provide ideal sites for thrips to pupate. In addition, remove old stock or mother plants, as these can harbor thrips and serve as sources of viruses. Another potentially effective strategy is to detach old or extra blooms (especially yellow and blue) from plants in order to reduce the number of thrips eggs and adults.

If feasible, screening greenhouse openings including vents and sidewalls will restrict the movement of thrips migrating into the greenhouse, especially during spring and summer, when weeds and other outdoor crops (corn, soybeans, and vegetables) are present.

There are a number of insecticides that are effective in managing thrips including spinosad (Conserve), methiocarb (Mesurol), abamectin (Avid), chlorfenapyr (Pylon), pyridalyl (Overture), and acephate (Orthene). However, relying exclusively on insecticides to manage thrips will most likely lead to the development of resistant thrips populations. Pest control materials with contact and/or translaminar activity are primarily used to control thrips, as systemics are generally unable to move into the flower parts (petals and sepals) where thrips normally feed. Once thrips enter terminal or flower buds, it is very difficult to obtain adequate control. Thus, insecticides must be applied before this.

Most insecticides with thrips activity only kill the nymph and adult stages; having no effect on the eggs and pupae. As a result, repeat applications will be required to kill those life stages that escaped exposure from previous applications, including nymphs that were in the egg stage and adults that were previously in the pupal stage. This is especially important when different life stages are present simultaneously or there are overlapping generations.

During cooler temperatures, the life cycle is extended compared to warmer temperatures; thus, the frequency of applying an insecticide may depend on the season. This can influence the number of applications needed per week and also reduce the potential for resistance developing in thrips populations

Biological control of thrips is difficult. It is best to consult a supplier beforehand. Biological control agents that are available are listed in table 9-2.

Aphids

Aphids are one of the most difficult insect pests to control or suppress in greenhouse production systems and may be a problem on many horticulture and floriculture crops, including bedding plants, vegetables, and chrysanthemums.

There are a number of different aphid species that attack plants, including green peach aphid (*Myzus persicae*), melon/cotton aphid (*Aphis gossypii*), foxglove aphid (*Aulacorthum solani*), chrysanthemum aphid (*Macrosiphoniella sanborni*), rose aphid (*Macrosiphum rosae*), and potato aphid (*Macrosiphum euphorbiae*). Aphid color will vary with the particular host plant fed upon and should not be used for identification.

Biology

Aphids are small; approximately 0.04–0.10" (1–2 mm) long, soft-bodied insects that possess tubes (cornicles) on the end of their abdomens. Male aphids are usually absent from the greenhouse. Females do not need to mate to reproduce; this process is called parthenogenesis. Females give birth to live female offspring that can start producing their own young (nymphs) in seven to ten days. Each of these females in turn can give birth to sixty to one hundred live young per day for a period of twenty to thirty days. This rapid reproductive ability can create tremendous population explosions within a short period of time. Aphid reproduction depends on plant quality and nutrition. Aphids in greenhouses are normally wingless; however, winged forms will develop when the host plant is crowded with aphids or when plant quality declines, increasing the aphid distribution throughout a greenhouse.

In general, aphids feed on new terminal growth and on the underside of leaves. However, aphid distribution on a plant varies with the species. For example, green peach aphid is commonly located on the upper leaves and stems, whereas melon aphid is distributed throughout the plant canopy.

Damage

Aphids cause direct damage or injury by removing plant fluids with their piercing-sucking mouthparts.

Table 9-2. Biological Control Agents for the Major Insect and Mite Pests in Greenhouses

PEST	PARASITOIDS	PREDATORS	BENEFICIAL NEMATODES
Fungus gnats		*Hypoaspis miles, Atheta coriaria*	*Steinernema feltiae*
Whiteflies	*Encarsia formosa, Eretmocerus eremicus, Eretmocerus mundus*	*Delphastus cataliniae*	
Spider mites		*Phytoseiulus persimilis, Mesoseiulus (=Phytoseiulus) longipes, Neoseiulus (=Amblyseius) californicus, Galendromus occidentalis, Amblyseius (=Neoseiulus) fallacis, Feltiella acarisuga*	
Thrips		*Amblyseius (=Neoseiulus) cucumeris, Amblyseius (=Iphiseius) degenerans, Amblyseius swirkskii, Hypoaspis miles, Orius insidiosus*	
Aphids	*Aphidius colemanii, Aphidius matricariae, Aphidius ervi*	*Chrysoperla* spp. (green lacewing), *Hippodamia convergens* (ladybird beetle), *Aphidoletes aphidimyza*	

Feeding on new growth results in young leaves appearing crinkled, curled, or distorted. They may also cause plant stunting. In addition, aphids produce a clear, sticky liquid material called honeydew, which acts as a growing medium for black sooty mold fungi. This can detract from or reduce the aesthetic quality of plants. The presence of white cast skins (due to molting) can also detract from the beauty of plants. In addition, aphids are capable of transmitting many destructive viruses.

Management

Weed removal is important in reducing aphid problems. Aphids feed on many broadleaf and grassy weed species commonly found in and around greenhouses. Weeds in greenhouses, which are not normally sprayed, serve as a reservoir for aphids and can support large populations. In addition, greenhouse producers should avoid overfertilizing plants because aphids are highly attracted to and feed on plants containing excess nitrogen. Aphid reproductive capacity increases when plants are overfertilized with nitrogen-based fertilizers.

Chrysanthemum, sweet potato vine (*Ipomoea*), and other plants and plant cultivars vary in their susceptibility to different aphid species. To prevent large aphid populations from establishing, monitor more frequently those plants that are highly susceptible to aphids.

To manage an aphid problem, it is critical to identify the particular aphid species, because there is variation in sensitivity to insecticides and acceptance by biological control agents, especially parasitoids. Insecticides that may be effective in managing or suppressing aphid populations include imidacloprid (Marathon), acephate (Orthene), pymetrozine (Endeavor), potassium salts of fatty acids (insecticidal soap), petroleum oil (PureSpray Green/SuffOil-X), paraffinic oil (Ultra-Fine Oil), dinotefuran (Safari), bifenthrin (Talstar), thiamethoxam (Flagship), and clarified hydrophobic extract of neem oil (Triact). Most insecticides have contact activity, so thorough plant coverage is essential. When using systemic insecticides (granules or drenches), be sure to make applications to every plant container because those that are missed may serve as a reservoir for aphids. In addition, rotate insecticides with different modes of action in order to avoid aphid populations developing resistance. Another method that may be effective in preventing resistance is to spot-spray localized areas where aphids are most likely to enter the greenhouse such as vents, sidewalls, and doors.

Before using a biological control, identify the aphid species because the various parasitoids are specific in regards to the aphid species they will attack (see table 9-2). In addition, releases of either parasitoids or predators must be initiated before aphid populations are extensive and causing plant injury.

Whiteflies

The major whitefly species encountered in greenhouses throughout the growing season include the greenhouse whitefly (*Trialeurodes vaporariorum*) and the sweet potato whitefly B-biotype (*Bemisia tabaci*). Another whitefly species, the bandedwing whitefly (*Trialeurodes abutiloneus*), may be present in greenhouses from late summer through early fall.

Biology

Whiteflies are generally located on leaf undersides, where they use their piercing-sucking mouthparts to feed within the vascular tissues (phloem) of plants. The whitefly life cycle consists of an egg, three nymphal stages, pupa, and adult. Adult whiteflies are white, narrow, and approximately 0.08–0.1" (2–3 mm) long. Adult females deposit eggs on the underside of mature leaves, often in a crescent-shaped pattern. The spindle-shaped eggs are white when first laid and turn dark gray (greenhouse whitefly) or amber-brown (sweet potato whitefly B-biotype) with time. Eggs hatch in approximately ten to twelve days at temperatures of 65–75°F (18–24°C). The tiny, first nymphal stage, or crawler, emerges from the egg, crawls a short distance [0.25–0.33" (6–8 mm)] and eventually settles down to feed. It does not move from this spot until emerging (after undergoing two additional nymphal stages) from the pupa as an adult. Approximately four to five days before adult emergence, the nymph enters a pupal stage, or more correctly a fourth nymphal stage. At this time, the red eyespots of the developing adult are visible through the insect skin or pupal case.

It takes both greenhouse whitefly and sweet potato whitefly B-biotype an average of thirty-two to thirty-nine days to develop from egg to adult at temperatures of 65–75°F (18–24°C). The life cycle takes less time as temperatures increase, especially during spring and summer. Approximately sixteen to twenty days of the developmental time are spent in life stages (egg and pupa) that are tolerant of many foliar and systemic insecticides. A single female whitefly can lay eggs one to three days after emerging as an adult. A female whitefly can lay up to two hundred eggs and live about twenty-five to thirty days, depending on environmental conditions.

Greenhouse whitefly and sweet potato whitefly B-biotype identification can be confirmed using the pupal and adult stages. Greenhouse whitefly pupae have parallel sides that are perpendicular to the leaf surface, giving a disk- or cake-shaped appearance. The pupae also have a fringe of hairs (setae) around the edges. Greenhouse whitefly adults have a white body and white wings, with the wings held nearly horizontal over the body. Sweet potato whitefly B-biotype pupae appear more rounded, dome-shaped, and without parallel sides. The pupae have no hairs around the edges. Sweet potato whitefly B-biotype adults have a light yellowish body with white wings; the wings are held roof-like over the body at a 45° angle.

Damage

The immature or nymphal stages of whitefly cause direct plant damage by feeding on plant fluids with their piercing-sucking mouthparts, which may result in leaf yellowing, leaf distortion, plant stunting, plant wilting, and possibly plant death. In addition, the immature stages produce a clear, sticky liquid material called honeydew, which acts as a growing medium for black sooty mold fungi. This can detract from or reduce the aesthetic quality of plants. The presence of large numbers of whitefly adults may be a visual nuisance, which may reduce crop salability. Heavily infested plants, if disturbed, may produce a cloud of flying whitefly adults. These adults may also be very obvious when at rest on leaf undersides. Whiteflies are capable of transmitting various diseases, including bacteria and viruses.

Management

Proper irrigation and fertility can decrease plant susceptibility to whiteflies. Maintain plant health and avoid overfertilizing, especially with nitrogen. The source of nitrogen may influence whitefly reproduction. For example, it has been shown that higher numbers of whitefly eggs are found on poinsettia (*Euphorbia*) fertilized with ammonium nitrate than those fertilized with calcium nitrate.

Sanitation is another important management strategy that can reduce problems with whiteflies. Eliminate plant debris and old stock plants. Whiteflies in the pupal stage may emerge as adults from leaves that have withered or dried. Place discarded plant material as far away from the greenhouse as possible or bag it, so whiteflies from discarded plant material cannot reenter the greenhouse. In addition, avoid leaving dried-up plant material in open garbage containers. Always place plant material debris in garbage containers with tight-sealing lids. Furthermore, remove weeds from within the greenhouse (i.e., pots, underneath benches, and on floors) and around the greenhouse exterior. Certain weeds such as sow thistle, oxalis, chickweed, velvet leaf, and dandelion are very attractive to and may serve as a reservoir for whiteflies.

Inspect whitefly-sensitive plants including hibiscus, fuchsia, salvia, lantana, gerbera, and poinsettia before bringing them into your greenhouse. Detaching the lower leaves of plants may remove whitefly eggs and pupae.

Insecticides that may be effective in managing or suppressing whitefly populations include imidacloprid (Marathon), pyriproxyfen (Distance), kinoprene (Enstar II/AQ), acetamiprid (TriStar), potassium salts of fatty acids (insecticidal soap), spiromesifen (Judo), novaluron (Pedestal), Beauveria bassiana (BotaniGard/Naturalis), pyridaben (Sanmite), thiamethoxam (Flagship), pymetrozine (Endeavor), petroleum oil (PureSpray Green/SuffOil-X), and paraffinic oil (Ultra-Fine Oil). Systemic insecticides are primarily used to manage whiteflies, especially on poinsettia. Systemic insecticides may be applied as a drench or granule to the growing medium, or as a foliar spray. Applications to the growing medium may provide up to eight weeks of whitefly control or suppression, if the application is timed appropriately. In order to avoid dealing with whiteflies before shipping or sale of the crop, it is best to use another insecticide early on, such as an insect growth regulator, insecticidal soap, insect-killing fungus (*Beauveria bassiana*), or horticultural oil. These insecticides are more effective early in the crop cycle, when whitefly populations are typically lower. In addition, the smaller plant size makes it easier for spray applications to penetrate the foliage and obtain adequate coverage on leaf undersides.

Biological control agents for whitefly (both greenhouse whitefly and sweet potato whitefly B-biotype) are listed in table 9-2.

Mites

The two-spotted spider mite (*Tetranychus urticae*) is the primary mite pest encountered in greenhouse production systems, especially in late spring through summer. Two-spotted spider mites are a problem for several reasons, including their small size, which makes them difficult to see; primarily feeding on leaf undersides, which makes them difficult to detect; rapid life cycle; high female reproductive capacity; wide host range; and their ability to develop resistance to miticides. There are other mite pests that attack greenhouse-grown crops, including cyclamen mite (*Stenotarsonemus pallidus*), broad mite (*Polyphagotarsonemus latus*), and bulb mite (*Rhizoglyphus echinopus*).

Biology

Two-spotted spider mites prefer warm, dry conditions with low relative humidity. They are oval shaped and may be yellow-orange, green, or red in color. Two-spotted spider mite adults have two dark spots on both sides of the abdomen. Males are smaller than females. Females live about thirty days and lay tiny, spherical, transparent eggs on leaf undersides. They can lay between fifty to two hundred eggs. The eggs hatch into six-legged larvae that undergo two eight-legged nymphal stages before reaching adulthood. The life cycle from egg to adult takes one to two weeks to complete, depending on temperature. For example, the life cycle from egg to adult takes fourteen days to complete at 70°F (21°C) and seven days at 85°F (29°C).

Cyclamen and broad mites are very small, approximately 0.05" (1 mm) long, about the size of a pinhead. They are both too small to be seen with the naked eye. Cyclamen mites are white or green to pale brown in color. Broad mites vary in color from yellow to dark green. Female broad mites are elongated and twice as long as males. Cyclamen mites are also elongated with the head and legs located on one end of the body. Female broad mites lay eggs on the underside of leaves in dark, moist places on plants, whereas female cyclamen mites lay eggs on the upper leaf surface. Broad mite eggs hatch into six-legged larvae that feed on plant tissues. They undergo an inactive, eight-legged nymph stage, followed by an active, eight-legged adult. Female cyclamen mites lay two to three eggs each day. These eggs hatch in two to eleven days into active larvae that enter a resting stage and then molt into an adult. The life cycle from egg to adult takes approximately ten days to complete for broad mites and eighteen days for cyclamen mites. However, development is temperature dependent. Both broad and cyclamen mites require high relative humidity for survival. All life stages may be present during the growing season.

Bulb mite adults are slow moving, tiny (approximately 0.03" [0.8 mm] in length), and pearly white in color. They often have two brown spots on their bodies. Female bulb mites lay between one hundred to four hundred white, elliptical eggs in groups or individually on the surface of a bulb or near decaying tissue. The eggs hatch into oval, white, six-legged larvae. Bulb mites then undergo three eight-legged, nymphal stages before reaching adulthood. The life cycle from egg to adult may take approximately forty days to complete. However, this is dependent on temperature. For example, at 77°F (25°C), the life cycle may be completed in twelve days. Bulb mites generally feed in groups.

Damage

Two-spotted spider mites tend to be located on older leaves. They primarily feed on leaf undersides and remove chlorophyll (green pigment) from within plant cells with their stylet-like mouthparts. Two-spotted spider mites generally feed near the midrib and veins of plants. Damaged leaves appear stippled with small, silvery-gray to yellowish speckles. Webbing may be present if populations are extensive.

Cyclamen and broad mites primarily feed in the buds of plants. Their feeding causes disfiguration of shoots, such as leaves and flowers. Distorted leaves are curled, cupped downward, glossy or purplish in color, and very hard or thickened. Damage is most noticeable when buds and leaves mature. Although plants do not normally die, they are not very attractive or salable. The presence of broad or cyclamen mites typically occurs after plant injury is noticeable as opposed to actually detecting the mites themselves.

Bulb mites feed on a variety of bulb crops, including freesia, gladiola, hyacinth, narcissus, lily, tulip, and iris. They feed primarily on rotting bulbs or decaying plant material, but they will burrow into and feed on healthy bulbs. Bulb mites are commonly associated with decaying plant matter as a result of damage caused by fungus gnat larvae or soilborne plant pathogens. They may also feed on the leaves and stems of lilies. Bulb mite feeding, especially when high numbers are present, can cause stunting, leaf yellowing, and possibly distorted growth. In addition, bulb mite feeding may create wounds that allow entry of soilborne pathogens such as *Pythium*, rhizoctonia, and *Fusarium*. These wounds may also provide entry sites for fungus gnat larvae.

Management

Implementing proper cultural practices will greatly reduce problems with two-spotted spider mites. First, avoid overfertilizing plants with excess nitrogen, which results in the production of soft, succulent tissue that is attractive to these mites. This also increases their ability to produce more eggs. Second, discard old plant material between cropping cycles, because it can serve as a source of two-spotted spider mites when the next crop is started. Third, avoid water-stressing plants, as this increases susceptibility to mites. Finally, remove weeds from within the greenhouse, because many weeds serve as a refuge for mites.

Pest control materials with miticidal properties may be used to manage or suppress spider mite populations including abamectin (Avid), bifenazate (Floramite), pyridaben (Sanmite), cyfluthrin (Decathlon), bifenthrin (Talstar), acequinocyl (Shuttle), fenbutatin-oxide (ProMite), and fenpyroximate (Akari). However, they need to be used properly. These miticides have contact activity, so thorough coverage of plant parts, especially the underside of leaves, is essential. Some miticides such as abamectin (Avid), chlorfenapyr (Pylon), spiromesifen (Judo), and etoxazole (TetraSan) have translaminar activity, which means that the material resides within the leaf tissue, providing a reservoir of active ingredient even after spray residues have dried. In this case, coverage is not as critical, as mites will take up the material when they feed within the leaf tissues. Biological control agents for two-spotted spider mites are listed in table 9-2.

Cyclamen, broad, and bulb mites are more difficult to manage or suppress, because once feeding damage is noticed, it is normally too late. In addition, they are very hard to reach with miticides due to their cryptic habit. First and foremost, it is important to discard any plants exhibiting symptoms as soon as possible to prevent further spread. Scouting the crop regularly and being aware of specific plants that are most susceptible to these mites will alleviate problems.

Pest control materials with miticidal properties that may have activity on cyclamen and broad mites include abamectin (Avid), spiromesifen (Judo), pyridaben (Sanmite), fenpyroximate (Akari), and chlorfenapyr (Pylon). However, these materials need to be applied preventatively in order to protect the crop from mite feeding.

Controlling fungus gnats, disposing of all plants exhibiting symptoms of bulb mite feeding, avoid damaging bulbs, and using soilless growing media will go a long way in alleviating problems with bulb mites. There are no miticides commercially available for controlling bulb mites.

Fungus gnats and shore flies

Fungus gnats (*Bradysia* spp.) and shore flies (*Scatella* spp.) are considered major greenhouse insect pests throughout the United States. They are both generally a problem under moist conditions, especially during propagation and plug production, particularly before plants develop well-established root systems. However, they can be a problem year-round.

Biology

Fungus gnats have a life cycle consisting of egg, four larval, pupal, and adult stages. A generation can be completed in twenty to twenty-eight days, depending on temperature. Fungus gnat adults are winged, 0.125" (3 mm) in length with long legs and antennae. They tend to fly around the surface of the growing medium and live approximately seven to ten days. Females deposit between one hundred and two hundred eggs into the cracks and crevices of the growing medium. They are highly attracted to growing media containing peat moss and pine bark. Eggs hatch into white, transparent or slightly translucent, legless larvae that are approximately 0.125" (3 mm) long. A characteristic diagnostic feature of fungus gnat larvae is the presence of a black head capsule that is absent in shore fly larvae. Larvae are generally located within the top 1–2" (3–5 cm) of the growing medium. However, they may also be found in the bottom of containers near drainage holes. They are highly attracted to cuttings before callus formation.

Shore flies have a life cycle consisting of egg, three larval, pupal, and adult stages. A generation can be completed in fifteen to twenty days, depending on temperature. Adult shore flies resemble houseflies. They are 0.125" (3 mm) long with black bodies. Each wing usually has at least five light-colored spots. Antennae and legs are short, and the head is small.

Larvae are opaque yellowish-brown with no head capsule and 0.25" (6 mm) in length. Shore fly adults are stronger fliers than fungus gnats.

Damage

Fungus gnats are a problem in crop production systems for several reasons. First, extensive populations of adults flying around may affect crop salability. Second, both the adult and larval stages are capable of disseminating and transmitting diseases. Third, larvae cause direct plant injury to roots, causing plant stunting and wilting. In addition, they create wounds that may allow secondary soilborne plant pathogens such as *Pythium* and *Fusarium* to enter. Finally, larvae will tunnel into cuttings and stems, which may result in plant death and makes management with either insecticides or biological controls very difficult.

Shore flies are less likely than fungus gnats to cause direct plant damage. Shore flies are a concern because they are more noticeable flying around plants and are easily seen, especially when captured on yellow sticky cards. Shipping plants with large numbers of shore fly adults flying around may reduce crop marketability. Although adult shore flies are generally considered a nuisance pest, they may leave black fecal deposits on plant leaves that may affect the plant's aesthetic quality. Shore fly larvae primarily feed on algae located on the surface of the growing medium. They may also be found within the growing medium, but they do not normally feed on plant roots.

Management

Proper sanitation is important in alleviating fungus gnat and shore fly problems; this includes removing weeds, old plant material, and old growing medium. Weeds growing underneath benches may create a moist environment that is conducive for fungus gnat and shore fly development.

Avoid overwatering and overfertilizing plants, as this leads to conditions that promote algae growth. It is essential to eliminate the buildup of algae, as both fungus gnats and shore flies breed in algae. Keep floors, benches, and cooling pads free of algae by using a disinfectant such as hydrogen dioxide (ZeroTol) or a material containing quaternary ammonium salts (Green Shield, Triathlon, or Physan 20).

Insecticides may be used to manage fungus gnats and shore flies; however, they must be used in conjunction with algae control. As such, do not rely solely on insecticides to manage fungus gnats and shore flies. Many of the insecticides registered for fungus gnats and shore flies are insect growth regulators (IGRs). These include azadirachtin (Azatin/Ornazin/Molt-X), cyromazine (Citation), diflubenzuron (Adept), fenoxycarb (Precision), kinoprene (Enstar II/AQ), and pyriproxyfen (Distance). IGRs are only effective on larval stages; they have no direct effect on adult activity. The microbial (bacteria) insecticide *Bacillus thuringiensis* subsp. *israelensis* (Gnatrol) has activity on fungus gnat larvae but no activity on shore fly larvae.

Conventional insecticides are also used to manage fungus gnats and shore flies; these include chlorpyrifos (DuraGuard) and chlorfenapyr (Pylon) for the larvae and bifenthrin (Talstar) and chlorpyrifos + cyfluthrin (Duraplex) for the adults.

Insecticides for larval control or suppression are generally applied as drenches or "sprenches" into containers, or they are applied directly to gravel or soil floors. Both adult fungus gnats and shore flies may be controlled or suppressed with conventional sprays or aerosols. However, because shore flies may be located throughout the greenhouse, uniform distribution of the material is essential.

Biological control agents that may be effective in controlling or suppressing fungus gnat larval populations are the beneficial nematode *Steinernema feltiae*, the soil-predatory mite *Hypoaspis miles*, and the rove beetle, *Atheta coriaria*. All three biological control agents attack the fungus gnat larvae. They can be applied to the growing medium or the floor. They need to be applied early before the fungus gnat population builds up.

Biological control of shore flies is generally not effective because shore flies live and develop under wet conditions, which are not conducive for survival of most biological control agents.

Snails and slugs

Snails and slugs are not insects but belong to a group of organisms referred to as mollusks, which also includes oysters and clams. Snails possess a hard outer shell, whereas slugs have a soft, shiny body without the shell (they are often referred to as "naked snails"). They feed on a wide variety of plant types. Some of their favorite ornamental plants include petunia, marigold, hosta, and orchid.

Biology

Snails range in size from 1.0–1.5" (3–4 cm) long. Slugs range anywhere in size from 0.125–6.0" (0.05–15 cm) in length. Most snails and slugs are dark or light gray, tan, yellow, green, or black; however, they can be lavender to purple in color, depending on the species. Some snails and slugs have dark spots with patterns. Snails and slugs lay clusters of translucent, pearl-shaped eggs under debris or buried beneath the soil surface. They lay jellylike clusters of approximately twenty-five eggs apiece. Snails and slugs, depending on the species, can lay between twenty to one hundred eggs several times per year. It takes approximately two years for snails to mature, whereas slugs reach maturity in about one year.

Damage

Snails and slugs have chewing/rasping mouthparts and cause plant damage by creating irregularly shaped holes in leaves with either smooth or tattered edges. They feed using a structure called a radula, which is in the mouth and is covered with small teeth. This allows them to scrape away the surface of plant tissue. Leaves may be sheered off entirely. Snails and slugs tend to feed on succulent foliage, such as seedlings and herbaceous plants including hosta, impatiens, petunia, chrysanthemum, *Leucanthemum,* lily, narcissus, marigold, primrose, begonia, and lobelia. They will also feed on gerbera daisy and poinsettia.

Management

The primary way to avoid snail and slug problems is through habitat modification, which involves eliminating all places where snails and slugs can hide. Remove weeds, leaves, groundcovers (under benches), and undesirable plants in and around the greenhouse. In addition, remove all wood debris and rocks. Pesticides (molluscides) with the active ingredient metaldehyde (Deadline), methiocarb (Mesurol), or iron phosphate (Sluggo) are available for use in greenhouses. Metaldehyde kills by both contact and ingestion, whereas iron phosphate must be ingested. Metaldehyde does not kill snails and slugs directly. It works by paralyzing the pests and causing them to secrete excess amounts of mucous. Methiocarb is a true nerve poison, as it interferes with nerve-impulse transmission. Both metaldehyde and methiocarb are less effective in areas of the greenhouse that are dry. As a result, it is recommended to irrigate before applying these materials in order to promote snail and slug activity. Iron phosphate can be used around pets and wildlife, and it remains potent for a longer period of time than metaldehyde. The material contains a bait that attracts and kills slugs. Iron, which is a heavy metal, is toxic to snails and slugs. It works within three to six days.

All of these materials should be applied in the evening for maximum effectiveness. These materials may need to be reapplied regularly, depending on the amount of watering performed in the greenhouse.

10

Margery Daughtrey

Managing Diseases

Diseases are an inevitable part of growing crops in the greenhouse. In years past, many growers routinely relied on pesticides to prevent diseases, without first learning other control measures or understanding the disease cycle. However, relying only on pesticides results in very poor disease control because the severity of disease is so greatly dependent upon environmental conditions. Skillful disease management requires knowing the disease, its symptoms, and what specific cultural and environmental conditions lead to its development. We will begin this chapter with a discussion of how you can manage disease problems by focusing on the greenhouse environment and using crop culture as your first line of defense. Then, we'll present the major root rot and foliar diseases along with their symptoms, the environmental conditions that foster them, and their control measures, including chemicals. By first managing your crop to prevent disease, you will find that the pesticides you do apply will be more effective and your crops will be healthier.

Focus First on Managing Water

For foliar diseases, water management is key: The number of hours that the plant surface sits wet will determine whether fungus spores can successfully germinate and penetrate the plant. Bacteria, also, are better able to infect through leaf stomata and hydathodes when there is protracted leaf wetness. Wet foliage is your enemy.

The goal should be to supply enough water for the plants' needs, but to keep the surfaces dry between waterings. Stock plants must be kept free from disease organisms (known as "pathogens") because the high-moisture conditions during propagation favor disease development. Any spores that might be present on the cutting will find a ready environment to strike once the cutting is on the warm, moist rooting bench.

If you irrigate from overhead, do so with a minimum of splashing and irrigate early enough in the day that the foliage can dry before nightfall. At sunset, ventilating and heating for a short time will help to drive the moisture-laden air out of the greenhouse and prevent the condensation on plant surfaces that would otherwise occur. Additionally, the greenhouse should be arranged to optimize air movement between plants: Open wire-mesh benches are desirable for permitting air movement up between plants. Metal or plastic benching is also easier to disinfest than are wooden benches. Air movement, which is critical for foliar disease control, is accomplished through the strategic use of fans and vents.

For root disease management, again water is the key, and again the idea is to avoid excessive amounts of it that will favor the pathogens and stress the crops. Choose a growing medium that allows excellent drainage: High air-pore space will provide good oxygen supply to the roots and discourage root rot pathogens. If you are subirrigating, be careful to avoid excessively long flooding periods. Redesign any growing surfaces that allow puddling. Provide the appropriate fertility level to each crop, avoiding overfertilization that may trigger a *Pythium* species attack. Also, avoid letting the crop dry down excessively, because an injured root system is also an invitation to *Pythium* root rot.

Scouting & Symptoms

We know that a plant is diseased by the symptoms that appear on foliage, stems, roots, or flower parts. Growers who learn to "read" their crops and pick up on the first indications of disease will be the most successful at minimizing their impact. The formal way to watch your crop is called scouting, or monitoring. The process works best if one employee has this duty assigned and regularly (at least once per week) walks through the operation looking for indications of crop health problems.

Regular scouting and detailed records are important for disease management over time—note which varieties were affected and when an outbreak occurred (both calendar date and environmental conditions), as well as what pest management actions were undertaken and their results. Another part of the scout's job should be to watch for and record the appearance of phytotoxicity symptoms following pesticide applications. Familiarity with symptoms of contagious diseases (caused by microorganisms) and physiological diseases (caused by cultural stresses such as nutrient imbalances, water stress, air pollution, phytotoxicity, etc.) will be an advantage to the scout, but a keen eye and a willingness to learn are also key attributes.

The basic symptom types are spots, cankers, soft rots, root rots, wilts, powdery mildews, and rusts. Leaf spots can be described by their color, shape (round, vein-bounded, or irregular in outline), size, rim color, presence or absence of yellow halos, and their distribution on the leaf. More aggressive leaf spots that may extend into stems are sometimes referred to as blights. Cankers are dead areas in stem tissue. Soft rots, often caused by bacteria, are used to describe mushy tissue degradation in plant storage tissues (bulbs, corms, or rhizomes). Root rots, caused primarily by fungi or oomycetes, may vary somewhat in the coloring of the infected tissue. *Thielaviopsis basicola*, for example, causes a very black root discoloration in infected pansies.

Plants with a vascular wilt disease have the same external appearance as plants suffering from drought stress. However, because many vascular wilts result from xylem infection, there may also be internal clues: brown, black, or purple discoloration of the vascular system in stems, corms, or bulbs.

With powdery mildews, the pathogen itself is usually visible as white fungal colonies on leaves, stems, or flower parts, so the disease can be recognized directly rather than indirectly through symptoms. Rust diseases are similar in that the fungus produces spores in easily visible pustules on the surface of the plant, usually on the underside of the leaves. The rust spores are often highly colorful, spilling out in yellow, orange, brown, or white masses from pustules.

Viruses produce their own distinctive range of symptoms, including mosaics, line patterns, and concentric ring spots. These symptoms are not produced by any of the other pathogens. Viroids are similar to viruses in the symptoms that they cause, but they are composed purely of genetic material, without the protein coat that is characteristic of virus particles.

Phytoplasmas are single-celled organisms, similar to bacteria but without cell walls. The symptoms that they cause look similar to virus symptoms. They are vectored by leafhoppers and live in the phloem of plants, causing curious symptoms of stunting, yellowing of foliage, and greening of flower parts. Although rarely seen in greenhouses because it is not seed-borne, the phytoplasma disease aster yellows is commonly seen in cut flower production of China asters (*Callistephus*), *Cosmos*, coreopsis, and other composites.

Observe symptoms on plants closely and make note of the pattern of their occurrence in the greenhouse. A check of recent spray records will help to determine whether phytotoxicity is a likely explanation for symptoms. If a contagious disease is suspected, forward all pertinent information to a university or private diagnostic lab, along with a carefully packed group of plants showing the range of symptoms. Specialized labs are available for virus detection, but a university or private lab with a broad range of diagnostic techniques can often best handle general problem diagnoses. Guidebooks such as the *Ball Field Guide to Diseases of Greenhouse Ornamentals* or the *Compendium of Flowering Potted Plant Diseases* (available from APS Press) are useful for making an onsite, educated guess of what may be causing the symptoms.

Root Diseases

Pythium

Pythium species are a group of pathogens that growers feel quite familiar with: They know them as root rotters causing crop losses in poorly drained mixes. Recently, studies of DNA have shown that *Pythium* (along with *Phytophthora* species and downy mildews) is more closely related to algae than to fungi. *Pythium* should now be called an oomycete rather than a fungus.

At the same time that their taxonomic classification has changed radically, pathogenic *Pythium* spp. have also increased in importance on greenhouse crops. Widespread use of soilless media has reduced the impact of these oomycetes for a generation, but recently the introduction of recirculating irrigation systems has provided a mechanism to foster the spread and development of *Pythium* spp. within greenhouse-grown crops. We are also learning that we need to be more precise in identifying a *Pythium* disease problem—the particular species of *Pythium* causing the problem and the fungicide sensitivity of the strain should be determined in order to make informed management decisions.

Four species of *Pythium* are currently problematic in greenhouse crops: *P. ultimum*, *P. irregulare*, *P. cryptoirregulare*, and *P. aphanidermatum*. *P. ultimum* usually does not form a swimming spore stage, whereas the other three form zoospores and thus are particularly favored by subirrigation systems. Recirculation of the irrigation water allows buildup and distribution of oomycete pathogens. Studies in Pennsylvania, New York, and Massachusetts have shown that *P. irregulare* and *P. cryptoirregulare* are found on a wide range of greenhouse crops (including geranium and impatiens), whereas *P. aphanidermatum* is most often seen in poinsettia and chrysanthemum. *P. ultimum*, once the most common greenhouse species, is now relatively rare. A high level of resistance to the fungicides metalaxyl and mefenoxam (Subdue and Subdue MAXX) is being noted frequently in strains of *Pythium* spp. isolated from greenhouse crops with root rot.

Symptoms

Pythium species commonly cause damping-off, attacking germinating seeds (preemergence) or the young roots (postemergence). On older plants, these pathogens are primarily root rotters, killing roots back from their tips. The outer cortex of the root is darkened and softened and easily pulls off the hard, vascular core of the root. In some cases, the *Pythium* spp. will grow up into the stem base causing "black leg" (as on geranium) or black streaks (as on New Guinea impatiens). In other instances, the impact on the roots is not directly visible to the naked eye, but attacks on the fine feeder roots result in stunting of the plant. The aboveground effects of *Pythium* spp. include yellowing and browning of lower leaves, wilting, and death.

Environmental influences

The different species of *Pythium* have different optimum temperatures: Cool temperatures favor *P. ultimum*; intermediate temperatures favor *P. irregulare* and *P. cryptoirregulare*; and high temperatures favor *P. aphanidermatum*. Poorly drained media or anything causing root injury favor all *Pythium* spp.

Management

Sanitation is critical. Remove dying plants promptly, and clean benches and floors with approved greenhouse disinfectants between crops. Inspect the root systems of rooted cuttings before transplant to learn whether *Pythium* root rot symptoms are present. Use mixes with high air-pore space, which offer good drainage. Maintain a fungus gnat control program, as these insects are known to vector *Pythium* spp. Avoid excessive nitrogen fertilization because overfertilization will trigger attack by these organisms. Grow plants prone to *Pythium* root rot (e.g., snapdragon, calibrachoa, geranium, and poinsettia) with particular care because cultural errors in water or nutrient management can result in root rot disease losses. Use protectant fungicide drenches on crops that are highly susceptible. (For some crops, such as poinsettia, protection against *Rhizoctonia* stem rot should be supplied along with the *Pythium* root rot control.) Do not use Subdue or Subdue MAXX exclusively: These products must be rotated with other materials less vulnerable to resistance development. If Subdue or Subdue MAXX resistance has been identified in crops grown in a recirculating irrigation system, do not return to the use of that fungicide until the system has been thoroughly disinfested. Options for chemical control include: cyazofamid (Segway); etridiazole (Truban, Terrazole, and Banrot); fluopicolide (Adorn); fenamidone (Fenstop); metalaxyl/mefenoxam (Subdue, Subdue MAXX, and Hurricane); phosphonates (such as Aliette, Alude, Fosphite, and K-Phite); and propamocarb hydrochloride (Banol). Biological controls, including Actinovate, RootShield, SoilGard, Cease, and Companion, should be used on a preventive basis only in concert with careful sanitation practices.

Phytophthora

Phytophthora is closely related to *Pythium* (it is also an oomycete), but its species usually attack plants more aggressively than those of *Pythium*. Several *Phytophthora* species currently cause problems on greenhouse crops: *P. cryptogea* is often found on gerbera; *P. nicotianae* on pansy, gloxinia, fuchsia, azalea, and poinsettia; and *P. drechsleri* on poinsettia. Like *Pythium*, *Phytophthora* is favored by wet growing conditions. *Phytophthora* spp., however, are more likely to attack the root crown or stem rather than the roots.

Symptoms

Phytophthora diseases are more likely to kill plants than to stunt them. Symptoms include root rot, stem base cankers, and brown lesions at the base of leaves attached to cankered stems.

Environmental influences

The diseases caused by *Phytophthora* species thrive during propagation under mist, which is ideal for the formation of spore structures (sporangia) that will produce zoospores for spread. Water splash, worker handling, or subirrigation easily move zoospores or sporangia from plant to plant. Generally, high humidity and warm to hot greenhouse temperatures favor *Phytophthora* root and crown rots.

Management

Careful control of *Phytophthora* root and crown rots on stock plants (for fuchsia, poinsettia, etc.) is important so that cuttings will be free from inoculum as they go into propagation. Remove diseased plants promptly—this will reduce disease spread by eliminating an inoculum source. Careful cultural management is not enough for control of these diseases: If this oomycete is present, fungicides are needed to prevent disease expression and spread of the pathogen. The same materials used for control of *Pythium* root rot will work against *Phytophthora* diseases, and in some cases will be more effective against *Phytophthora* spp. In addition, strobilurin fungicides (CompassO, Cygnus, Disarm, Heritage, Insignia, and Pageant) work well against *Phytophthora* spp., as does dimethomorph (Stature SC). Resistance to Subdue and Subdue MAXX has been found in *P. nicotianae* and is likely to occur in other species, so take care to use a rotational scheme rather than using these exclusively. Biological controls generally appear to be less effective against *Phytophthora* than against *Pythium* spp.

Rhizoctonia

Rhizoctonia solani is a common soil fungus that frequently becomes a contaminant, causing plant losses in greenhouse culture. It does not produce spores, but rather is moved from place to place along with particles of soil moved by water or wind. It can also be passed from greenhouse to greenhouse on diseased cuttings or plants. The hyphae of the fungus are very sticky, so it may easily be spread on workers' hands during transplanting. *Rhizoctonia solani* has a wide host range, including virtually all herbaceous and woody greenhouse crops. Younger, more succulent plants are more susceptible than mature ones. Rogue out the diseased plants or trays during plug production to avoid later losses in packs or pots.

Symptoms

Rhizoctonia solani is the most common cause of damping-off of young seedlings, toppling them after attack at the soil line. The fungus may have a similar effect on older plants, causing a brown to black canker at the stem base that often girdles and kills the plant. Plants with dense canopies grown under warm, high humidity conditions may also be subject to a rhizoctonia web blight, in which the fungus grows up the stem and onto the leaves.

Environmental influences

Rhizoctonia solani does not need cultural errors in order to attack greenhouse crops: It thrives under typical greenhouse humidity levels and temperatures, particularly during spring and summer months. High humidity at the surface of the growing medium allows it to grow rapidly; overwatering is not necessary for *R. solani* to attack plants.

Management

Sanitation is critical. The dust blown in from outside the greenhouse or soil from the greenhouse floor can supply inoculum. There are no cultural controls effective against this disease. Frequent light waterings are more conducive to *Rhizoctonia* canker than thorough waterings made less frequently. Fungicides may be used preventively for crops that have repeatedly been found to be vulnerable to *R. solani*, or drenches may be made on an as-needed basis when scouted crops develop typical symptoms. Diseased plants should be discarded prior to making a protective drench application to the rest of the crop. Fungicides for suppression of *R. solani* include thiophanate-methyl (3336, 6672, Fungo, Allban, 26/36, and Banrot); fludioxonil (Medallion and Hurricane); PCNB (Terraclor); iprodione (Chipco 26019, 26 GT, and Sextant); triflumizole (Terraguard); and strobilurins (CompassO and Heritage). When *Pythium* or *Phytophthora* spp. are also of concern, combination drenches with fungicides controlling oomycetes are desirable. Biocontrol materials, including those with *Trichoderma* spp., *Bacillus subtilis*, or *Streptomyces* spp. as active ingredients, will help protect plants against *R. solani* if used preventively.

Fusarium wilt

Some *Fusarium* species, such as *F. avenaceum* on lisianthus and *F. solani* on many different crops, cause diseases of plant stems or roots. However, host-specialized forms of *F. oxysporum* are usually the cause of vascular wilt diseases. Thus, even though diseases caused by *F. oxysporum* are quite serious on their hosts, they do not present a threat to other crops (those in other plant families) in the greenhouse. The fungi that can cause *Fusarium* wilt may also cause root rot; however, their major impact comes from their ability to colonize the vascular system of their hosts and cause a systemic infection.

Fusarium wilt diseases are frequently seen in carnation, chrysanthemum, coreopsis, cyclamen, and basil. Inoculum is seed-borne in basil and probably in cyclamen as well. Symptomless infected cuttings (with latent infections) are the primary means for introduction of the disease in pot or garden mum crops. Bed-grown cut flower crops of carnation or chrysanthemum may pick up inoculum from imperfectly pasteurized soil in the ground beds. Susceptibility to the *Fusarium* wilt diseases varies from variety to variety.

Symptoms

Because the xylem of the plant is infected, the effects of *Fusarium* wilt are usually indistinguishable from water stress symptoms. Infected plants may be stunted, and lower leaves will turn yellow and then brown and die. Leaves of chrysanthemums may appear scorched. The vascular system of a cross-sectioned cyclamen corm has a reddish-brown to purple discoloration. The xylem may also be darkly discolored in chrysanthemum or basil stems. Wilting and death eventually result from *Fusarium* wilt. The disease cannot be distinguished visually from *Verticillium* wilt; a lab diagnosis is needed to separate the diseases caused by these two vascular wilt fungi.

Environmental influences

High temperatures and low pH generally favor *Fusarium* wilt diseases. High levels of ammonium nitrogen are also conducive to disease.

Management

Clean stock systems (that employ culture indexing) are essential to successfully combat *Fusarium* wilt in vegetatively propagated plants such as chrysanthemum and carnation. Clean seed sources are essential for *Fusarium*-free production of basil and cyclamen. Although the fungal pathogens causing these wilt diseases cannot parasitize hosts outside of their narrow host ranges, these fungi do survive in greenhouses as saprophytes indefinitely. This means that greenhouse sanitation is very important: Benches and floors where *Fusarium* wilt has occurred on a crop should be carefully disinfested, and fungus gnats and shore flies should be kept under control.

Scout crops susceptible to *Fusarium* wilt carefully for signs of the disease, and rogue out plants with symptoms immediately. Splashing spread is particularly important in the case of cyclamen: Small numbers of diseased plants may be the source of infections for many others early in production, when plants are grown pot-to-pot and watered from overhead. Subirrigation systems may spread *Fusarium* inoculum from plant to plant, so these are not a good choice for crops prone to *Fusarium* wilt. Disease spread is not much of a problem for growers finishing garden chrysanthemum, but losses may be high within a few varieties that presumably arrived harboring latent infections.

Fungicides are of limited use in the control of *Fusarium* wilt diseases. Sprays or drenches of materials such as thiophanate-methyl (3336, 6672, Allban, Fungo, and Banrot), fludioxonil (Medallion and Hurricane), azoxystrobin (Heritage) or triflumizole (Terraguard) are among those which may slow the plant-to-plant spread or slow the course of disease, but complete protection against this disease has not been obtained with fungicides. Combining helpful fungicide treatments with careful sanitation practices, scouting, using a growing medium pH of 6.2 or above, and calcium-nitrate fertilization will reduce disease. Clean seed and stock sources are imperative for disease-free crops. Use less-susceptible cultivars whenever possible.

Foliar Diseases

Powdery mildew

All powdery mildew diseases look fairly similar, but there are actually many different species of powdery mildew fungi that will affect different greenhouse crops. These fungi are unusual pathogens in that they are largely external to the plant. The powdery mildew grows across the leaf surface, penetrating individual epidermal cells with structures called haustoria in order to absorb nutrients from the plant host. Crops prone to powdery mildew include begonia, African violet, verbena, poinsettia, rose, petunia, New Guinea impatiens, gerbera, hydrangea, rosemary, and tomato.

Symptoms

The white powdery appearance of typical powdery mildew colonies is created by the mass of spores (conidia) produced on the surface of the leaf. The symptoms are usually easy to identify, especially when they appear on the upper leaf surfaces. On highly susceptible plants such as begonia and verbena, scout the undersurfaces of the lower leaves in order to find the first symptoms of the disease. Sometimes, as with powdery mildew on poinsettia, there will be a chlorotic spot visible on the upper leaf surface to indicate the area where the powdery mildew fungus has colonized the undersurface. Plants in the family Crassulaceae (jade plant, sedum, and kalanchoe) react to powdery mildew by producing brown, scabby spots, and only close examination with a hand lens will allow you to see the powdery mildew growth on the lesions.

Environmental influences

The powdery mildews are, as a group, very well suited to prosper in greenhouses. The diseases are usually favored by high humidity. The different powdery mildews have different optimum temperatures: powdery mildew of rose, *Podosphaera* (sect. *Sphaerotheca*) *pannosa* f. sp. *rosae*, is most severe under spring and summer conditions in the Northeast, while powdery mildew on poinsettia (*Oidium* sp.) will only start to develop obvious colonies and spread in the fall, after greenhouse temperatures cease to reach 86°F (30°C) during the day. Daily variations in humidity regulate the powdery mildew life cycle. The release of conidial inocula typically occurs in late morning, following a sharp dip in humidity. These conidia are spread throughout the greenhouse by air currents.

Management

Learn the time of year that powdery mildew is a threat to the crops that you grow, and scout carefully for the earliest signs of the disease. Once scouting has determined the start of a powdery mildew outbreak, use a combination of environmental deterrents and chemical/biological sprays to keep it in check. Use fans to circulate the air, and space plants adequately. Heat and ventilate at sunset to reduce nighttime humidity in the greenhouse. Grow the varieties that are the least susceptible to powdery mildew.

There are extensive fungicide options for powdery mildew management. Experiment until you find an effective rotation between two systemic materials with different modes of action or between a systemic and a protectant material. The rotation is important for resistance management—powdery mildew fungi are very likely to develop resistance to materials that target a single site. (For more information, see the discussion on managing resistance in chapter 12). Some of the materials available for powdery mildew control are azoxystrobin (Heritage); chlorothalonil (Daconil and PathGuard); chlorothalonil and thiophanate-methyl (Spectro); copper (Camelot, Kocide, and Phyton-27); fenarimol (Rubigan); horticultural spray oil (Ultra-Fine Oil); hydrogen dioxide (ZeroTol); kresoxim-methyl (Cygnus); hydrogen peroxide, peroxyacetic acid, and octanoic acid (Xeroton 3); myclobutanil (Eagle and Hoist); clarified extract of neem oil (Triact 70); piperalin (Pipron); potassium bicarbonate (MilStop); pyraclostrobin (Insignia); pyraclostrobin + boscalid (Pageant); thiophanate-methyl (3336, 6672, Allban, and Fungo); triadimefon (Strike); Trichoderma harzianum (RootShield); trifloxystrobin (CompassO); and triflumizole (Terraguard).

Alternaria leaf spot

There are many miscellaneous fungal leaf spots that will affect greenhouse crops from time to time. One of the more common of these is *Alternaria* leaf spot. Because the *Alternaria* species are seed-borne fungi, the disease can affect even young seedlings during plug production. Usually the species of *Alternaria* are specific for a particular crop, so it would not be expected to see the disease move from one plant species to another. In seed-grown crops, the problem may develop in a single variety but spread to others grown nearby.

Symptoms

Alternaria leaf spots are usually brown in color and either circular or irregular in outline. There is often a yellow halo around the spot. On impatiens, the spots are round with a pigmented border, scattered across the leaf surface; sporulation may be seen (with a hand lens) on the pale centers of older lesions. On poinsettia and zinnia, the circular or vein-bounded brown spots develop yellow halos and look very similar to the *Xanthomonas* (bacterial) leaf spots that occur on those crops. On dusty miller, dark brown, irregularly shaped lesions develop in the leaves attacked by an *Alternaria* sp.

Environmental influences

Alternaria leaf spots are generally favored by warm, moist conditions. Splashing water is the primary source for spreading spores.

Management

A scouting program is a good weapon against *Alternaria* leaf spot in seed-grown crops. Prompt detection will mean that a few infested flats can be discarded prior to using protectant treatments on the remainder of the crop. Since these diseases are not commonplace, using pesticides responsively rather than preventively is an option for growers who monitor their crop closely. After roguing out visibly diseased flats, cultural practices should be adjusted to minimize water splashing and the duration of leaf wetness. Fungicides may be used as protectants, but be careful not to use a thiophanate-methyl material (such as Fungo, Allban, 3336, or 6672), as this chemistry will be ineffective against *Alternaria* species. Materials containing mancozeb (Protect T/O and Dithane); iprodione (Chipco 26019 and 26/36); chlorothalonil (Daconil); or copper (Camelot and Phyton 27) are the most useful for *Alternaria* leaf spot management.

Bacterial diseases caused by *Xanthomonas*

There are many bacterial diseases of plants caused by different *Xanthomonas* species. Each of the pathogens is very host-specialized, so the diseases are not exchanged except within the same plant family. Some of the most common and damaging *Xanthomonas* leaf spots and blights include those on nephthytis, zinnia, begonia, poinsettia, and geranium. Because the geranium disease called "bacterial blight of geranium" causes the greatest dollar losses, its avoidance will be discussed in more detail.

Symptoms

Leaf spot symptoms caused by *Xanthomonas campestris* pv. *pelargonii*, a pathogen of *Pelargonium* and *Geranium* species, are small, round, brown lesions, 0.06–0.125" (2–3 mm) in diameter. Brown wedges may also form in the leaves. A yellow halo or large areas of chlorosis may accompany either of these symptoms. Wilting of individual leaves or the whole plant results from systemic infection. The roots are not affected and remain white, which helps to distinguish bacterial blight from wilting caused by *Pythium* root rot. Wilting caused by bacterial blight looks very similar to wilting caused by another bacterium, *Ralstonia solanacearum*, which causes the systemic disease Southern wilt. *Ralstonia solanacearum*, however, will not cause leaf spots.

Environmental influences

Bacterial diseases cause symptoms most rapidly and spread most easily in warm temperatures. Bacterial blight of geraniums caused by *Xanthomonas campestris* pv. *pelargonii* may escape detection on stock plants during the winter, whereas cuttings in propagation in the spring will show symptoms quickly. The bacteria will spread by splashing, so new leaf spots will appear on plants adjacent to those that originally showed symptoms. The bacteria can be picked up by the root systems and can travel through subirrigation to infect large numbers of plants.

Management

Culture indexing to provide clean stock is at the heart of the process for avoiding bacterial blight of geraniums. Unfortunately, during the increase of stock from the indexed plants, sometimes recontamination occurs. Growers should be on the alert for leaf spot or wilt symptoms on their geranium even if they have purchased cuttings from a reliable propagator: This disease typically appears in at least a few geranium crops each year, for one reason or another.

Scout for symptoms regularly, and either use in-house immunostrip test kits or send plants off to a university or private lab for testing. Contact your supplier as soon as suspicious symptoms are noted, and be sure that a professional diagnostician confirms any in-house guesses or tests before you rogue out large quantities of plants—many other diseases have symptoms similar to bacterial blight. Early detection significantly reduces losses; throwing out a section of diseased plants promptly allows the rest of the crop to be finished for sale. Do not mingle plants from multiple suppliers because then a problem from one source will be spread into plants from other clean sources.

Do not keep plants over from one season to the next. Growing hardy *Geranium* species in the same operation as greenhouse *Pelargonium* species or hybrids is very dangerous, as these herbaceous perennials quite often are infected with *X. campestris* pv. *pelargonii*. The bacteria cause only a leaf spot disease on *Geranium* species, but when the disease is transferred to a *Pelargonium* sp., both leaf spots and wilt will occur.

All kinds of *Pelargonium* crops are susceptible, although symptoms will be seen most often in ivy and zonal types grown from cuttings. Never grow hanging baskets of ivy geraniums over a zonal crop (either cutting or seedling), as bacteria could drip down onto the lower crop. Seedling geraniums may contract the disease if they are brought into contact with a cutting crop, but they have never been implicated as a source of inoculum for an epidemic.

Water with a minimum of splashing; use individual-tube watering for stock plants. If possible, do not subirrigate geraniums, as this practice allows the entire crop to be exposed to inoculum of the bacterial blight disease, increasing the chance of extensive losses. Sprays with a copper material may help to slow the course of spread but should not be relied upon to provide total disease control. Roguing out diseased plants and those adjacent to them is critically important once the disease has been diagnosed.

The tospoviruses: INSV and TSWV

Impatiens necrotic spot virus (INSV) is the virus most likely to cause crop losses in bedding plants or flowering potted plants. This tospovirus, along with its close relative tomato spotted wilt virus (TSWV), is vectored by western flower thrips (WFT). Both INSV and TSWV have a very wide host range, potentially infecting nearly all non-woody crops. However, INSV is often seen in greenhouses on impatiens, cineraria, gloxinia, cyclamen, begonia, browallia, and various other species and is rarely seen outdoors. TSWV, on the other hand, is rare in greenhouses but is often found outdoors on herbaceous perennials in areas where the virus is endemic in the local weeds.

Symptoms

The symptoms of INSV and TSWV are indistinguishable from one another. Symptoms will vary with plant variety and with the age of the plant. Symptoms include stunting, spots, zonate spots, concentric rings, "oak leaf" line patterns, browning at the petiole end of the leaf, stem cankers, and wilt. The color of symptoms varies from white to yellow to brown to black, depending on the host. Unless the grower is very familiar with the symptoms on a particular crop, virus-suspect samples should be sent out to a lab to verify the presence of TSWV or INSV. Serological tests, usually enzyme-linked immunosorbent assay (ELISA) or PCR tests, are used to confirm the presence of INSV or TSWV. Test kits to identify INSV and TSWV are also available for in-house use.

Environmental influences

The severity of INSV and TSWV symptoms will vary somewhat according to the time of year, as temperature and day length affect the susceptibility of the host plants. The thrips vector population will rise in the spring as temperatures rise, and this increases the frequency of virus transmission. There are no direct environmental or cultural controls for virus management, but indirect measures such as keeping the greenhouse free of weeds may be critical for gaining control of a tospovirus problem.

Management

Management of these virus diseases is primarily a matter of managing the thrips vector. The only way to directly reduce the disease is to bag and remove symptomatic plants from the greenhouse. Plants with INSV or TSWV that are kept in the greenhouse can continually provide a virus-laden food source for larval thrips. When these larvae become adults, they will transmit the virus as they fly from plant to plant and feed. Strategically plan which crops will be grown together in your operation. Grow plants from seed separately from plants from cuttings. INSV or TSWV (and thrips) can be introduced on plugs, cuttings, or prefinished plants. Inspect new plants on arrival for virus symptoms and thrips. If growing a very sensitive crop (e.g., gloxinia), keep it isolated from new plant material. See the discussion on thrips management in the chapter 9 for details on how to manage the vector population. Use sticky cards to monitor for thrips. Indicator plants (certain cultivars of petunia or fava beans) can also be used to monitor for virus symptoms. (This is a valuable technique for crops such as cyclamen that may take two months to show symptoms).

Botrytis

The fungus *Botrytis cinerea* causes gray mold disease, named after the fuzzy growth of spores that appears on killed plant parts. This is probably the most common disease in greenhouse culture worldwide because *B. cinerea* thrives on the high humidity that greenhouses provide. All crops are susceptible to some degree, and flowers especially so. Botrytis management is critical for gerbera, rose, lisianthus, snapdragon, and many other species grown as cut flowers. The bedding and potted plant crops that are particularly vulnerable include geranium, lisianthus, exacum, poinsettia, petunia, bacopa, and fuchsia. Lilies are susceptible to the omnipresent *B. cinerea* but are also host to *Botrytis elliptica*, which causes oval leaf spots and flower blight on plants in the lily family.

Symptoms

Botrytis cinerea causes leaf spots that tend to be roughly circular or wedge shaped. These continue to enlarge whenever humidity is favorable. The spreading lesions may encompass entire leaves and then grow down the petiole into the stem. The resulting cankers may girdle the stem and kill the area above the canker—wilted branches or wilted plants are the result. Spotting on flowers may begin as tiny, round spots (white, red, or tan in color) and then expand to blight the entire flower. Symptoms include damping-off of young seedlings and dieback from cutting stubs.

Environmental influence

Botrytis blight can affect crops over a wide temperature range, at any temperature below 90°F (32°C), but the fungus is most favored by cool temperatures (55°F [13°C]). High humidity and poor air circulation increase gray mold disease, because the fungus needs free water on plant surfaces for establishing new infections and high humidity for growth and spore production. Even four hours of continuous leaf wetness is sufficient for infection when other factors are optimum.

Management

Humidity should be kept below 85% at all times. Unpack boxed plants promptly, do not overcrowd plants on the bench, and use fans and vents to move air. Heat and ventilate at sunset to drive out moisture-laden air and to prevent condensation on plants during the night. One group of fungicides used for botrytis blight control in the past—the benzimidazoles (Benlate, 3336, 6672, Allban, and Fungo)—is no longer effective because of the development of resistance in the *B. cinerea* population. Dicarboximides (Chipco 26019) are now often found to be less effective than they were when they were first introduced to the greenhouse industry, but they still are partly effective. Some of the older fungicides with multi-site action (e.g., coppers and chlorothalonil) are still effective against *B. cinerea.* Newer materials such as azoxystrobin (Heritage), fenhexamid (Decree), fludioxonil (Medallion), pyraclostrobin + boscalid (Pageant), and trifloxystrobin (CompassO) are effective, but users are encouraged to rotate between fungicides with different modes of action to slow the development of resistance. (See the discussion on managing resistance in chapter 12.)

Rust

Rust diseases are only rarely a problem in the greenhouse, but they can build up quickly if undetected. Learn which of your crops are susceptible to a rust disease so that you can be alert for the symptoms. Fuchsia, chrysanthemum, geranium, aster, and snapdragon are examples of crops with fairly common rust diseases. Chrysanthemum is susceptible to both brown rust (*Puccinia chrysanthemi*) and white rust (*Puccinia horiana*), with the latter subject to quarantine regulations in the United States. The rust diseases on snapdragon and aster are most often seen outdoors during cut flower production rather than in the greenhouse. Rust diseases are host-specific and thus tend to infect only a single crop species, but they can build up to serious epidemics on their host plant under favorable environmental conditions.

Symptoms

Rust fungi often cause spots on the upper surfaces of leaves. Chrysanthemum and geranium may show yellow spots; fuchsia may show necrotic patches with purple borders. The way to conclusively identify a rust, however, is by examining the undersurface of the leaves. This is where the fungus will produce its spores, within pustules. In the case of geranium rust, the pustules are arranged in concentric rings and the spores are a rich chocolate brown; on fuchsia, the pustules cover the undersurface of the lesion and are a yellow-orange color. White rust of chrysanthemum produces beige to white, raised pustules on the leaf undersurface: *This disease must be reported to your state horticultural inspector.*

Environmental influences

Rust diseases are favored by conditions similar to those that favor botrytis blight. The spores of rust fungi need a wet leaf surface continuously for a number of hours in order to germinate and penetrate. For geranium rust, *Puccinia pelargonii-zonalis*, only three hours of continuous free water are necessary for infection when the temperature is 61–70°F (16–21°C).

Management

Avoiding rust diseases requires attention to the duration of leaf wetness. Continuous cycles of infection may occur under wet growing conditions; splashing water easily spreads spores. Water early in the day, especially if irrigating from overhead, and avoid condensation by heating and ventilating at sunset. Fungicides should be used to protect against new infections from the first sign of a rust disease. Many of the same materials that are effective against powdery mildews are also effective against rusts.

Protectant fungicides include chlorothalonil (Daconil) and mancozeb (Dithane and Protect T/O); systemic fungicides include triadimefon (Strike) and myclobutanil (Hoist and Eagle). Strobilurins (such as CompassO, Heritage, and Insignia/Pageant), which are somewhat systemic, may also give good rust control.

Rotating between systemic and contact materials is often a good policy for rust management. It is important to remember that the pigmented spores of rust fungi are well designed for survival: Removing visibly infected plants from the greenhouse does not remove all of the inoculum. In the case of geranium rust, the spores stay viable in the greenhouse for at least three months. This means that even after the plants with symptoms have all been removed, rust fungicide treatments must be continued until the end of the cropping cycle.

11

Gerald Kinro

Controlling Soilborne Pests

Editor's note: The following chapter is an edited reprint from the seventeenth edition of the Ball RedBook. *Although the content has not been updated (except to remove out-of-date resources), the information contained within was too important to leave out of this edition.*

Soilborne plant pests cover a wide spectrum. They include insects, fungi, bacteria, nematodes, and weeds. Some are benign and easy to manage, while others, if left alone, can hinder or even destroy a crop. Some are easily managed while others are persistent and very difficult to control.

Traditional cultural practices such as crop rotation, although useful, have limitations and are not fully effective. Over time, certain pathogens can increase their numbers to levels that require intervention if crops are to be grown successfully.

Over the years, a number of pre- and postplant methods have evolved for controlling soilborne pathogens. Most use chemical controls. Concerns over short- and long-term human and environmental effects, however, have led to the United States Environmental Protection Agency (EPA) canceling use of some these agrochemicals. Examples are heptachlor, ethylene dibromide, dibromochloropropane, and the fumigant methyl bromide.

A number of nonchemical strategies have been researched, validated, and employed commercially. Using heat ranks high among these methods. With heat, there are fewer long-term effects to the environment, no phytotoxicity, a shorter aeration time, and no worker exposure to toxic chemicals. Most of all, heat can be effective. Figure 11-1 shows the temperatures at which various pathogens are killed.

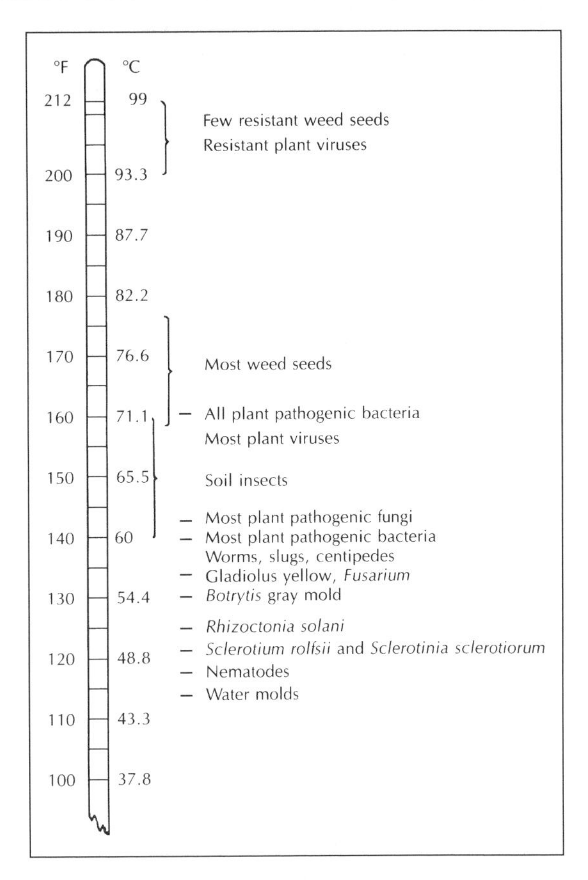

Figure 11-1. This thermometer graph shows the temperatures necessary to kill pathogens and other soilborne organisms harmful to plants. Temperatures shown are for thirty-minute exposures under moist conditions.

Heating

The basic principle behind heating field areas is to place a tarp over the treated area and introduce heat under the covering. With the tarp retaining the heat for a prescribed time, pathogens exposed to above-ambient temperatures die. In order for this procedure to work, a number of steps are required.

Till the soil to be heated until it is friable and free of large clods. Then rake the soil smooth, as if you are preparing a seed bed. Large clods or other debris create air pockets that inhibit efficient heating. As in fumigation, heat is effective only up to the depth of till. A clean, flat surface will also prevent accidental puncturing of the tarp.

Moist soil works best, for soilborne pests are more susceptible to moist, hot soil than to dry soil. If the soil is dry, water the areas to be heated before laying the tarp. Many growers lay irrigation systems such as drip systems or soaker hoses to be kept under the tarp to ensure adequate moisture. Moisture conducts the heat, which is another reason soils must be moist. However, overly wet soils will take longer, for it takes a tremendous amount of heat to increase the temperature of the excess water. Saturated soil takes a tremendous amount of heat to increase the temperature of all the excess water. Conversely, dry soil does not conduct heat because of the large amount of dead air space, which acts as an excellent insulator.

Place a tarp or other covering over the treated area. Using continuous sheets is best for disease and nematode control because the entire area is disinfested. Smaller strips, 2–3" (61–91 cm) wide, are often more convenient and economical for many bed-grown vegetable crops.

Seal the edges of the tarp to prevent loss of heat. This is usually done by covering the edge with soil or by other means such as "snakes" used by commercial structural fumigators. "Snakes" are tubes about twenty feet (6 m) long filled with water or sand. They are placed on the covering edges and provide weight to hold it in tight contact with the ground surface. Heavy link chain also works well.

Finally, apply heat either mechanically or through solar radiation. These two methods are discussed in more detail below.

Mechanical heat

This method, known as "steam pasteurization," simply uses a steam generator or motor to supply heat. The object here is to raise the temperature under the sheet to desired levels and to maintain it for at least thirty minutes.

Many growers with boilers can convert them to generate steam. Contact your boiler dealer to find out if your boiler is capable of generating steam that can be used to pasteurize beds.

While stationary generators can be used indoors, field heating has fewer options, and portable units are the only practical means of delivery. There are commercially produced soil steaming wagons and steam generators available on a custom-made basis. These generate heat by burning petroleum fuel. They can be easily moved into place and can provide enough heat in a short period of time. If you choose to use steam heat, the tarp must be able to withstand the temperature of the steam. There must also be some provision for venting in case internal pressure builds up under the covering.

Apply moisture to beds and fields that will be steam pasteurized several days in advance if weeds are a problem. That way seed will soften and start to sprout, which makes them easier to kill during steaming. Apply steam at 180°F (83°C) for a minimum of thirty minutes to kill fungi, bacteria, and weed seeds. Use a thermometer designed for this purpose.

Avoid steaming in very hot weather, as buildup of ammonia or nitrite nitrogen could occur. Why does this happen? Peat, manure, and other forms of organic matter in soils contain nitrogen. Before plants can use the nitrogen, it has to be converted to the nitrate form. Several types of bacteria do this. Some bacteria turn organic nitrogen to ammonia. Some bacteria convert it to nitrate nitrogen. When the soil is heated to 180°F (83°C), most of the bacteria that convert organic matter to nitrate are killed. However, the bacteria that convert organic matter to ammonia are hardier and often survive. Therefore, several weeks after steaming it is not unusual that high ammonia nitrogen is found in the soil. Often levels are high enough to burn roots.

The problem more typically occurs in hot weather. While just about all soils have some organic matter that is decomposing throughout the year, during the summer organic matter is decomposing most rapidly, and it uses much of the available oxygen in the soil. The bacteria that convert the excess ammonia to nitrate nitrogen also need oxygen. Not only that, but it seems that the bacteria that produce ammonia seem to be stimulated at higher temperatures.

You can solve this problem in several ways.

1. Use high-quality, long-lasting peat or other types of organic matter that break down slowly. Sphagnum peat will not normally release enough ammonia after steaming to cause damage. Avoid manure and compost since they break down rapidly after steaming, releasing toxic quantities of ammonia.

2. For cut flower growers who steam after every crop, allow fertility to run down as the previous crop finished. Do not fertilize after the crop shows color, and leach soils during the last irrigations.

3. Keep soil cultivated to encourage air to enter the lower soil.
4. Adding 40% superphosphate (4 lbs./100 ft.^{2} [195 g/m^2) before steaming or gypsum (4 lbs./100 ft.^{2} [195 g/m^2) immediately after steaming can help tie up free ammonia.
5. Leach heavily after steaming (4 gal./ft.^{2} [162 l/m^2] applied in multiple applications), or let the soil stand for two or more weeks after treating.
6. Consider amending soils with organic matter after steaming, as soils lower in organic matter do not experience as many problems.

Natural heat

Using the sun's heat is an option for those living in areas where there are long periods of uninterrupted sunlight. This process is called soil solarization. The degree of success with solarization depends on the intensity of heat, its penetration depth into the soil, and its duration. Therefore, it works best with long, hot, sunny days, July or August being ideal. It can be done, however, during autumn and spring in areas with adequate sunlight if the heat treatment is kept longer. Table 11-1 summarizes points of comparison between steam pasteurization and solarization.

Six to eight weeks during the summer should be adequate for most soilborne pests that can be controlled by solarization, with periods of up to three months during cooler temperatures. As in mechanical heating, the soil must be wet.

You will need sheets of clear plastic as your sheeting to allow penetration of sunlight as well as to contain the heat. White or black materials do not transmit enough solar energy to raise soil temperatures. Also, thick plastic, despite being "clear," is more reflective than thin plastic, losing much of the transparency. Thinner sheets (0.5–1 mil) work best. They are less costly and transmit more solar energy. However, they tear or puncture more easily. As a compromise, thicker plastic (2–3 mils) should be used only when damage from high winds is likely.

Figure 11-2. Hot sun, clear plastic, and prepared moist soil are all essential for successful solarization.

Since the tarps will be out for several weeks, you may wish to purchase those that are treated against ultraviolet (UV) light. UV can weaken or damage plastic and make it difficult to remove. Using two layers of thin plastic sheeting separated by a thin insulating layer of air increases soil temperatures and the overall effectiveness of a solarization treatment.

Table 11-1. Comparison of Steam Pasteurization and Solarization

	Steam pasteurization	*Solarization*
Efficacy	Very effective if hot enough	Less effective than steaming, but still effective
Treating large areas	Not practical	Possible, but treated areas are out of production for the duration of treatment
Exposure time	Half-hour	Several weeks, minimum
Machinery required	Steamer, soil tiller	Soil tiller
Tarp	Must be durable to high temperatures	Must be clear plastic and withstand weeks of exposure
Weather dependency	Doesn't require sun	Requires long periods of sun
Energy requirements	Petroleum to power steam	Uses the sun

For effective solarization, the edges of tarps laid over raised beds must be buried in the adjoining furrows. In addition, the tarp must fit tightly over the soil surface. Patch any holes immediately with duct tape to prevent heat loss. Crops may be seeded or planted directly into slits or holes in UV-stabilized strip mulches after soil solarization is completed.

Table 11-2 lists many of the pests that have been successfully controlled using solarization.

Disadvantages of heat

There are disadvantages to the use of heat for in-ground situations.

1. Due to depth limitations, heat is not a suitable method for many deep-rooted crops.

2. Limited efficacy of heat along the tarp's borders may allow increases of pest populations along the edges of treated areas. These could spill into the rest of the treated area. This disadvantage is the same as if a fumigant were used under the covering.

3. Cultivation after heating could bring healthy weed seeds to the soil surface, where they could germinate.

4. When using generators, there are limits to the amount of heat a unit can emit, thus limiting the size of the area treated at one time. Also, high energy and labor costs make using heat expensive. With this in mind, using steam generators does not appear to be practical for large acreage.

5. Solarization will keep land out of production for a relatively long time. Practicality is governed by how much land the grower is willing to keep unplanted and for how long.

6. Both solarization and steam pasteurization require time and labor to do the job properly.

The Integrated Approach

The recommended approach to countering pest problems is through integrated pest management (IPM). Recognizing the imperfections of every means of pest control, IPM utilizes all methods available including resistant varieties, physical control, biological control, cultural practices, and, if the grower chooses, chemical control. It requires frequent scouting for pests. It mandates sanitation to prevent the introduction of pests into a "clean" area. Heat, a form of physical control, can be part of an effective IPM program. It merits consideration, and, if feasible, can be a suitable replacement for some lost tools.

Table 11-2. Some Pests Managed by Solarization

Fungi
Fusarium
Pythium
Phytophthora
Rhizoctonia
Sclerotinia
Verticillium
Bacteria
Several root rots and blights
Nematodes
Heterodora spp.
Meloidogyne spp.
Pratylenchus spp.
Weeds
Abutilon spp.
Brassica spp.
Eleusine spp.
Many species of annual grasses and broad leafs. It will kill seeds of many perennials as well.
Insects
Insects and their eggs are very susceptible to heat treatment.

Pesticides

Pesticides may be fumigant or non-fumigant. Many are classified as restricted-use pesticides by federal and local agencies and require applicators to be certified. When using a pesticide, the applicator must follow all pesticide label directions.

Fumigant pesticides

These kill a wide range of soilborne organisms by emitting toxic chemicals to voids, where they contact and kill pests. As in heating, the soil must be tilled for best results. The fumigant must be sealed into the soil with a tarp or with water. After fumigation, aerate the soil to allow the fumigant to escape. There are fumigants still available for agricultural use such as chloropicrin; 1,3-dichloropropene; and metam-sodium. Methods of application vary. For example, metam-sodium is incorporated into the soil as a liquid or granule. The treated area is then watered heavily to seal the fumigant. Other fumigants may require specialized equipment such as injectors and shanks.

Non-fumigant pesticides

These include insecticides, nematicides, fungicides, and herbicides. They are less effective against a broad range of pests than are fumigants. Some nematicides include carbofuran and fenamiphos. Insecticides may kill on contact or are translocated throughout the host plant and kill insects that feed on it. Herbicides and fungicides are incorporated into the soil or are surface-applied.

Many chemical tools are being reviewed as mandated by the Federal Insecticide Fungicide and Rodenticide Act (FIFRA) and the Food Quality Protection Act (FQPA). As a result, growers may yet lose more pesticides.

Biofumigation

In recent years, the chemicals of plants themselves have been used to control pests. Crucifers, especially canola, when used as a green manure cover crop, produce isothiocyanate and glucosinate compounds that help control a wide range of soilborne pests. Research has shown that this method works well in broad acre cropping as well as for horticultural applications. Biofumigation reduces disease incidence by reducing the quantity of the inocula. Some growers have taken this farther by green manuring the cover crop, tilling the soil well, and then applying heat through solarization.

Living mulches

A living mulch is nothing more than a groundcover growing with the main crop. It evolved from some familiar ideas, such as no-till farming and IPM. Erosion losses prompted no-till farming as a solution. As the name implies, there is no tilling and no soil disruption except for the small area where seeds are planted. In this system, weeds are generally killed with herbicides and left in the field. The crop is then planted into the areas treated with herbicide. Concerns over the environment revived IPM. IPM uses all means of pest control and aims to reduce the dependency on pesticides. Old methods such as cultural practices, resistant varieties, and biological control come into play to keep pests in check.

The fusion of these two leads to a living mulch. The one difference between a no-till planting and the living mulch is that in the living mulch system, the ground may be tilled once when the cover crop is

Figure 11-3. The ground must be tilled for heat to be successful. Here, growers are using crucifers as a green manure (biofumigation). They hope to enhance the heat treatment by using the pest control properties of the green manure.

planted. Traditional mulches are valuable in controlling weeds. Living mulches do more. We know the benefits of crop rotation in pest control. A common practice is to introduce a crop not eaten by the pest into the rotation to deter or even repel the previous crop's pests. The living mulch goes a step further by extending the benefits of a non-host, rotational cover crop into the production phase of the desired crop.

Many have experienced natural control of insects and diseases through the use of groundcovers. Growers have found that nematodes and whiteflies that attack plants in the Solanaceae family (eggplant, tomato) are dramatically reduced when using Rhodes grass (*Chloris gayana*) as a cover. In this case, the mulch was not a reservoir for the pest, but for its natural enemies. The cover also has a masking effect, hiding the crop from its insect pests. In California, Miguel Altierri of the University of California found that legume groundcovers help control aphids in broccoli. His living mulch reflected sunlight in a manner that repelled the insect.

Author's note: For more information, the reader should consult Pests of the Garden and Small Farm: A Grower's Guide to Using Less Pesticide *by Mary Louise Flint (University of California Press, 1998).*

12

Raymond A. Cloyd and Margery L. Daughtrey

Resistance Mitigation

Resistance mitigation is a strategy emphasized by entomologists and plant pathologists to preserve the effectiveness of currently available pesticides. Certain insects such as the western flower thrips (*Frankliniella occidentalis*) are notorious for developing resistance to some of the pesticides that were effective previously. Some fungal pathogens have also developed resistance to certain fungicides. *Botrytis cinerea* strains in greenhouses are commonly found to show some level of resistance to fungicides in two classes: dicarboximides (e.g., Chipco 26GT) and benzimidazoles (e.g., 3336 and Fungo). Bacterial plant pathogens have also, in some cases, developed resistance to copper or antibiotics. Once resistance to a pesticide has developed and the resistant population has been distributed throughout the industry, it is unlikely that greenhouse producers can ever again use the pesticide effectively. This is because resistance genes often continue to reside in the pest population long after selection pressure has been reduced.

Every greenhouse should implement resistance-mitigating strategies when dealing with insects, mites, and plant pathogens so as to avoid problems. Many pesticide label recommendations state to rotate pesticides by indicating that a particular material should only be used once or twice before switching to another pesticide. This information is for your benefit: The usefulness and longevity of pesticides will be conserved by the judicious use of pesticides as described on the label.

Resistance Basics

The concepts of resistance mitigation are somewhat different for insects, mites, and pathogens. However, whichever pest you are trying to control, it is important to first understand the concept of resistance and how it affects your ability to manage insects, mites, and diseases. Resistance is manifested as the ability of some individuals in a population to survive a pesticide application because of genetic changes, primarily caused by previous exposure to a pesticide. Resistance may be enhanced in a pest population due to selection and interbreeding of the survivors from pesticide applications, which then increases the frequency or proportion of resistant individuals in the pest population. From the grower's perspective, resistance is the ability of pest organisms to survive applications of one or more pesticides that were previously effective.

Resistance is an inherited trait. The gene(s) for resistance may already exist in the insect, mite, or plant pathogen population. Pesticide use or the frequency of application creates selection pressure that favors the survival of resistant individuals. This means that surviving insects, mites, and plant pathogens can transfer resistant genes to their offspring. Both the occurrence of mutations in genes and the amount of selection pressure may affect the capacity of insect, mite, or pathogenic fungal or bacterial populations to become resistant to pesticides. Organisms, such as insects, mites, fungi, and bacteria, that have high reproductive capacity tend to have an unstable arrangement of genes and thus are more likely to have mutations that may help them to quickly adapt to changes in their environment. This level of instability leads to an increased probability of developing resistance to pesticides. This potential for resistance development is manifested similarly in human diseases caused by bacteria. Because human bacterial pathogens are highly vulnerable to mutation and other processes, resulting in genetic change, this leads to a high probability for them to develop resistance to the antibiotics prescribed by medical doctors. The indiscriminate use of antibiotics places undue selection pressure on these organisms, which then encourages the promotion of resistant strains.

Every time an insect, mite, fungal, or bacterial population is exposed to a pest control material, there is the potential for resistance selection, which then increases the frequency or proportion of resistant genes within that population.

Factors that may influence the rate of resistance development are general operational procedures; insect, mite, and pathogen biological characteristics; and greenhouse conditions. One general operational procedure that will impact the rate of resistance development is applying pesticides on a routine basis, regardless of the population dynamics of the pest. Such a strategy unnecessarily increases selection pressure, thus increasing the frequency or proportion of resistant individuals in a pest population. Also, the continuous application of the same pesticide or the use of pesticides with similar modes of action for an extended period of time will increase selection pressure.

Biological characteristics of insects, mites, and plant pathogens that may increase the rate of resistance include rapid development time (short generation time), high reproductive rate (large number of offspring or spores produced per generation), high mobility, and wide host range. All of these characteristics may lead to increased exposure to the pesticides applied and thus higher selection pressure.

Greenhouse conditions that may lead to the development of resistance include environmental conditions (temperature and relative humidity) that are conducive for insect, mite, and plant pathogen development. The greenhouse structure encloses pests and greatly restricts the movement of susceptible individuals into the population. Resistant insects, for example, remain in the greenhouse and breed, whereas susceptible individuals from areas not treated with pesticides are less likely to enter the greenhouse to breed with resistant insects. Furthermore, intensive year-round production in greenhouses provides a continuous food supply for insects, mites, and pathogens. As a result, greenhouse producers tend to make more pesticide applications per year than would ordinarily occur in a seasonally grown outdoor crop, thus increasing selection pressure.

There are three ways that the movement of pests may result in resistance problems. First, insects or pathogens migrating from one crop to another within the greenhouse or from one greenhouse to another on infested or infected plants may have already been exposed to repeated applications of a particular pesticide. Second, receiving shipments of pest-infested plants may speed the development of resistance in your greenhouse because a large percentage of these pests may already possess genes for resistance because of treatments made in the source greenhouse. Finally, insects or fungi that enter the greenhouse from nearby field or vegetable crops may have been exposed to agricultural pesticides that are similar to the ones used in greenhouses. All three of these scenarios will increase the chances that pests in a greenhouse may already possess resistant genes.

The rate of resistance development may depend on the season. The number of insecticide or fungicide applications, based on the population dynamics of the insect population or disease pressure, may vary throughout the year. The development of insecticide resistance, for example, may be enhanced during spring and summer, because this is when insect populations are more abundant and multiplying faster, thus requiring more frequent insecticide applications.

Now that you have some background information on resistance and understand the conditions that may enhance the ability of insects, mites, and plant pathogens to develop resistance to pesticides, we will provide some strategies that greenhouse producers can implement to avoid or minimize problems with resistance.

How to Avoid Problems Associated with Resistance

Vary pest management strategies

Implement a pest management program that includes cultural, physical, pesticidal (pest control material), and biological methods to minimize pest problems. It is important to conduct sound cultural practices, such as proper watering and fertilizing, and sanitation practices, including the removal of weeds and old plant material debris. These efforts will reduce the number of pesticide applications required and thus the amount of selection pressure placed on pest populations. Also, scouting crops and monitoring for pests regularly throughout the growing season will determine the population dynamics of pest populations and help greenhouse growers to time pesticide applications accordingly (or determine if an application is necessary).

Rotate pesticides or pest control materials with different modes of action

Relying exclusively on one type of pesticide or pest control material to manage or suppress pests will result in high selection pressure on the pest population, which will increase the speed of resistance development. It is important to rotate pesticides with different modes of activity.

Use a pesticide or pest control material for at least the duration of one generation of an insect or mite population before rotating to a pesticide or pest control material with a different mode of activity. Insecticides, acaricides (miticides), and fungicides in new chemical classes are still being registered for use on greenhouse-grown crops. These new materials will help with resistance mitigation only if they are used properly. Many pesticides now contain label information that will orient greenhouse growers towards resistance-mitigation strategies—they include statements warning that only one or two applications can be made before a pesticide with a different mode of action must be used.

Use pesticides with nonspecific modes of action

Do not rely exclusively on systemic insecticides and fungicides that have specific (single-site) modes of action. Pesticides with specific modes of action, which target a single gene, mechanism, or biological pathway, are the most vulnerable to the development of resistance. Many systemic fungicides have specific modes of action that make them vulnerable to resistance development, whereas protectant fungicides are often far more stable in their effectiveness over many years of use. For insecticides, miticides, and fungicides, products with multi-site action are less susceptible to resistance developing in the target pest population. Multi-site or broad mode of action products include insecticidal soaps, horticultural oils, bicarbonates, neem compounds, selective feeding blockers, insect growth regulators, and biological control organisms, as well as many of the older fungicides, insecticides, and miticides. See table 12-1 for a listing of some common fungicides, their classes, and whether they possess single-site or multi-site modes of action. The FRAC Codes provided by the Fungicide Resistance Action Committee are also given: Fungicides with the same FRAC code number have the same modes of action. See table 12-2 for a listing of insecticides and miticides, their chemical classes, and whether they have specific or broad modes of action. Also, the IRAC (Insecticide Resistance Action Committee) designations or code numbers are provided. Insecticides and miticides with the same IRAC code numbers have the same mode of action.

Minimize the use of pesticide mixtures

Greenhouse producers sometimes prefer to mix together pesticides in order to manage a wide variety of pests. However, tank-mixing pest control materials with similar modes of action may lead to resistance development in insect or mite pest populations. Insect or mite pest populations that have acquired resistance to two different materials with different modes of action (multiple resistance) are very difficult to manage and may limit management options. Use tank mixes of pesticides with specific modes of action only when absolutely necessary to obtain control. Tank-mixing pesticides with specific modes of action with pesticides that have multi-site action may mitigate the development of resistance: Some fungicide labels recommend this practice. It is important to mix together only pesticides that have different modes of action. Some insecticide and fungicide products contain a combination of more than one active ingredient. In some cases, combination products are designed purely to extend the spectrum of pest activity, while in others the mixture is intended to slow the development of resistance to a single-site or specific mode of action material or to combine two products with different modes of action so as to enhance management of target pests.

Use minimum label rates of insecticides and miticides

Constant use of the highest label rate of an insecticide or miticide may result in limited options when this rate fails to provide control. The lower label rate may be just as effective as the highest label rate. If the lower label rate fails, it is then possible to use the middle or higher label rate. Most importantly, this extends the longevity of insecticides and miticides. In the case of fungicide application, however, it is thought that the use of less-effective rates contributes to faster resistance development. For diseases such as botrytis blight, powdery mildews, rusts, and downy mildews (which all exhibit extremely rapid inoculum buildup), use a labeled rate that provides excellent control, not a moderate level of control.

Table 12-1. Chemical Class and Mode of Action of Fungicides for Greenhouse Use			
FUNGICIDE (ACTIVE INGREDIENT)	**CHEMICAL CLASS**	**MODE OF ACTION**	**FRAC CODE***
3336, 6672, Allban, Fungo (thiophanate-methyl)	benzimidazole	single-site	1
Adorn (fluopicolide)	acylpicolide	unknown	U9 (not known)
Aliette, Avalon, Flanker (fosetyl-Al)	phosphonate	unknown	33
Alude, Fosphite, K-Phite (phos acid salts)	phosphonate	unknown	33
Banrot (thiphanate-methyl + etridiazole)	benzimidazole + 1, 2, 4-thiodiazole	single-site + unknown (different purposes)	1 + 14
Camelot (copper salts)	copper	multi-site	M1
Chipco 26GT, 26019, Sextant (iprodione)	dicarboximide	single-site	2
Compass O (trifloxystrobin)	strobilurin	single-site	11
Cygnus (kresoxim-methyl)	strobilurin	single-site	11
Daconil, PathGuard (chlorothalonil)	chloronitrile	multi-site	M5
Decree (fenhexamid)	hydroxyanilide	single-site	17
Disarm O (fluoxastrobin)	dihydro-dioxazine	single-site	11
Endorse (polyoxin D zinc salt)	polyoxin	single-site	19
Fenstop (fenamidone)	imidazolinone	single-site	11
Heritage (azoxystrobin)	strobilurin	single-site	11
Hoist, Eagle (myclobutanil)	triazole	single-site	3
Hurricane (mefenoxam + fludioxonil)	acylalanine + phenylpyrrole	single-site + single-site (different purposes)	4 + 12
Insignia (pyraclostrobin)	strobilurin	single-site	11
Junction (copper + mancozeb)	copper + EBDC	multi-site + multi-site	M1 + M3
Medallion (fludioxonil)	phenylpyrrole	single-site	12
Milstop (potassium bicarbonate)	bicarbonate	unknown	Not classified
Pageant (pyraclostrobin + boscalid)	strobilurin + carboxamide	single-site + single-site	11 + 7
Phyton-27 (copper sulphate)	copper	multi-site	M1
Pipron (piperalin)	piperidine	single-site	5
Protect T/O, Dithane (mancozeb)	EBDC	multi-site	M3
Segway (cyazofamid)	cyanoimidazole	single-site	21
Spectro (chlorothalonil + thiophanate-methyl)	chloronitrile + benzimidazole	both	M5 + 1
Stature SC (dodemorph)	morpholine	single-site	40
Strike, Bayleton (triadimefon)	triazole	single-site	3
Subdue MAXX (mefenoxam)	acylalanine	single-site	4
Terraclor (PCNB)	aromatic hydrocarbon	not clear	14
Terraguard (triflumizole)	imidazole	single-site	3
Triact 70 (extract of neem oil)	neem oil	multi-site	Not classified
Truban, Terrazole (etridiazole)	1, 2 ,4-thiadiazole	unknown	14
Ultra-Fine Oil (horticultural spray oil)	paraffinic oil	multi-site	Not classified
Xeroton 3 (hydrogen peroxide, peroxyacetic acid and octanoic acid)	peroxide	multi-site	Not classified
ZeroTol (hydrogen dioxide)	peroxide	multi-site	Not classified
Zyban (mancozeb + thiophanate-methyl)	EBDC + benzimidazole	multi-site and single-site	M3 + 1

* The FRAC Codes indicate different modes of action for fungicides. Rotations for mitigation of resistance development should be made between/among materials with different FRAC Codes.

Table 12-2. Chemical Class and Mode of Action of Insecticides and Miticides for Greenhouse Use

ACTIVE INGREDIENT (TRADE NAME)	CHEMICAL CLASS	MODE OF ACTION	IRAC CODE*
Acephate (Orthene)	Organophosphate	Specific	1B
Chlorpyrifos (DuraGuard)	Organophosphate	Specific	1B
Dichlorvos (Vapona)	Organophosphate	Specific	1B
Methiocarb (Mesurol)	Carbamate	Specific	1B
Bifenthrin (Talstar/Attain)	Pyrethroid	Specific	3
Cyfluthrin (Decathlon)	Pyrethroid	Specific	3
Fenpropathrin (Tame)	Pyrethroid	Specific	3
Fluvalinate (Mavrik)	Pyrethroid	Specific	3
Lambda-cyhalothrin (Scimitar)	Pyrethroid	Specific	3
Permethrin (Astro)	Pyrethroid	Specific	3
Chlorpyrifos + cyfluthrin (Duraplex)	Organophosphate + pyrethroid	Specific	1B + 3
Abamectin (Avid)	Macrocyclic lactone	Specific	6
Imidacloprid (Marathon)	Neonicotinoid	Specific	4A
Acetamiprid (TriStar)	Neonicotinoid	Specific	4A
Dinotefuran (Safari)	Neonicotinoid	Specific	4A
Thiamethoxam (Flagship)	Neonicotinoid	Specific	4A
Azadirachtin (Azatin, Ornazin, Molt-X)	Insect growth regulator	Broad	18B
Cyromazine (Citation)	Insect growth regulator	Broad	17
Diflubenzuron (Adept)	Insect growth regulator	Broad	15
Etoxazole (TetraSan)	Insect growth regulator	Broad	10B
Fenoxycarb (Precision, Preclude)	Insect growth regulator	Broad	7B
Kinoprene (Enstar II/AQ)	Insect growth regulator	Broad	7A
Novaluron (Pedestal)	Insect growth regulator	Broad	15
Pyriproxyfen (Distance)	Insect growth regulator	Broad	7C
Buprofezin (Talus)	Insect growth regulator	Broad	16
Potassium salts of fatty acids (M-Pede, insecticidal soap)	Insecticidal soap	Broad	Unclassified mode of action
Clarified hydrophobic extract of neem oil (Triact)	Horticultural oil	Broad	Unclassified mode of action
Paraffinic-based oil (Ultra-Fine Oil)	Horticultural oil	Broad	Unclassified mode of action
Petroleum-based oil (PureSpray Green, SuffOil-X)	Horticultural oil	Broad	Unclassified mode of action
Beauveria bassiana (BotaniGard, Naturalis)	Biological	Broad	Unclassified mode of action

(Continued)

Table 12-2. Chemical Class and Mode of Action of Insecticides and Miticides for Greenhouse Use *(Continued)*

Spinosad (Conserve)	Microbial	Broad	5
Bacillus thuringiensis subsp. *kurstaki* (DiPel)	Microbial	Broad	11B2
Bacillus thuringiensis subsp. *israelensis* (Gnatrol)	Microbial	Broad	11A1
Nicotine (Fulex/Plant Products)	Botanical	Specific	4B
Pyrethrin (Pyrethrum)	Botanical	Specific	3
Fenpyroximate (Akari)	Phenyl pyrazole	Specific	21
Pymetrozine (Endeavor)	Pyridine azomethine	Broad	9B
Flonicamid (Aria)	Pyridincarboxamid	Broad	9C
Bifenazate (Floramite)	Carbazate	Specific	25
Hexythiazox (Hexygon)	Carboximide	Specific	10A
Clofentezine (Ovation)	Tetrazine	Specific	10A
Chlorfenapyr (Pylon)	Pyrrole	Specific	13
Pyridaben (Sanmite)	Pyridazinone	Specific	21
Fenbutatin-oxide (ProMite)	Organo-tin compound	Specific	12B
Acequinocyl (Shuttle)	Napththoquinone	Specific	20B
Pyridalyl (Overture)	Unclassified	Specific	Unclassified mode of action
Spiromesifen (Judo)	Tetronic acid	Specific	23
Spirotetramat (Kontos)	Tetronic acid	Specific	23

* The IRAC (Insecticide Resistance Action Committee) codes indicate the different modes of action for insecticides and miticides registered for use in greenhouses. Rotation programs designed to mitigate resistance should include as many products (e.g., insecticides and/or miticides) with different modes of action as possible.

Spot spray for insects and mites

Targeting spray applications where insect and mite pests are most concentrated or localized avoids exposing the entire pest population to a pesticide application. Spraying only those areas or plants with extensive numbers of pests avoids placing undue selection pressure on the entire pest population, thus leaving some susceptible individuals. A fungal control strategy, however, is different, as thorough spraying of the entire greenhouse area is most desirable when trying to control diseases such as powdery mildew, downy mildew, botrytis blight, and rusts.

Time applications to target the most vulnerable life stage

This is the "weakest link in the chain" concept in which insects and mites have a developmental stage or stages that are more susceptible to pesticides than are other life stages. Generally, the immature or young stages are more susceptible than older life stages (i.e., adults) to conventional or alternative pesticides. Targeting early developmental stages may reduce the number of insecticide or miticide applications required, and thus lessen the amount of selection pressure placed on a pest population.

Take a "pesticide break"

Some greenhouse producers use biological controls for a portion of the year or on some crops to suppress insect, mite, or disease problems. Anything that can be done to reduce selection pressure or the frequency of applying pesticides should help with resistance mitigation. For diseases, biological controls may help in protecting crops when disease pressure is low. Under highly conducive environmental conditions or when inoculum is abundant, biological controls will not provide the level of control that may be attained with appropriate, effective fungicides.

13

Roger C. Styer

Propagating Seed Crops

Growers of all sizes can produce their own plugs. Equipment investments can be as little as buying a wand seeder, filling flats by hand, and germinating and growing on existing benches or ground. However, if numbers increase, more investment is needed in seeders, flat fillers, top coaters, watering tunnels, germination chambers, under-bench heating, cooling and shading, boom irrigators, and robotic transplanters. Trained personnel familiar with growing plugs also becomes a necessity. Anything that can go wrong with a plant can happen quickly in a plug tray. A good plug grower is the most technically trained grower with the best eye for growing and watering. For these reasons, many growers will buy in plugs from brokers and other growers rather than committing to growing their own plugs. Other growers will grow most of their own plugs but buy in the most difficult crops. Regardless, each grower needs to make his own economic decisions about growing plugs or buying them.

Stages of Plug Growth

To make it easier to understand and more specific for crop recommendations, plug production is divided into four different stages, from the beginning of germination to the time of transplanting. In Stage 1, the primary root (or radicle) emerges from the seed. The seedling enters Stage 2 after the radicle emerges, during which the radicle penetrates the soil and the stem (hypocotyls) and seed leaves (cotyledons) emerge. Stages 1 and 2 comprise the germination stages, meaning that all germination should be finished by the end of Stage 2. During Stage 3, the true leaves grow and develop. In Stage 4, the seedling is ready for shipping, transplanting, or holding. Stages 3 and 4 make up the growth stages, as most of the growth of the seedlings occurs during this time. The environmental and cultural conditions needed for each stage is tied into the seedling development. Table 13-1 summarizes the conditions needed for most plug crops.

Table 13-1. Conditions for Plugs

CONDITION	STAGE 1 ——►	STAGE 4
Temperature	High	Low
Moisture	High	Low
Light	Low	High
Nutrition	Low	High

Stage 0: Equipment and techniques

All activities associated with plug production begin with Stage 0. This includes how you buy your seed and plug trays and the specific equipment you employ to fill trays, seed, cover, and water trays. Your germination percentage, seedling uniformity, and optimum root growth will be greatly influenced by the actions taken in Stage 0. The basics of Stage 0 are: (1) selecting the best seed to use, (2) selecting the appropriate plug tray size for your needs, (3) properly filling the trays with growing medium, 4) placing a seed into the center of each plug cell, (5) covering the seed uniformly if needed, and (6) suitably watering in the trays.

Plug trays

There are a wide variety of plug tray sizes, depths, and shapes from which to choose. Determine your needs based on which containers you will use for finishing, your schedule, the type of crop, and how crops will be transplanted. Generally, plugs are grown in trays containing 288 or 512 cells. The smaller the finished container, the smaller the plug size can be. For faster finished crop times, use a larger plug. The depth of the plug cell will influence the rooting. Deeper cells drain better but take a little longer to root in. Some crops need a deep root system in the plug to survive in the field or container. Square-shaped cells provide better roots than round cells, but hexagonal or octagonal cells are the best. Ventilated plug trays have holes between cells for better drying in the middle of the tray and less disease.

Plug trays can be recycled if they are properly washed and disinfected. Be sure to keep them in disinfectant long enough to kill disease spores. Cover trays with plastic when stacked for future use. Many growers do not reuse plug trays due to disease or algae problems or damage due to robotic transplanters or handling. However, thick reusable plug trays are now available that can withstand robotic transplanters and handling, but the cost per tray is much higher.

Flat filling

Plug trays can be filled by hand or machine. If you are producing less than a thousand plug trays, it may be more economical to fill by hand. Many growers are now linking a flat filler to their seeder, top coater, and watering tunnel to have an in-line system. Make sure you have enough room for all of the Stage 0 equipment you need.

If you are using a flat filler, make sure it is adjusted for filling the type of plug trays you are using. Follow the manufacturer's recommendations for making adjustments and check brushes, belts, and paddles to ensure they are not worn out. Always keep enough medium in the hopper for proper filling. Never store medium in the hopper for longer than a day, as you risk blowing dust and disease spores into the medium.

The plug medium should have the right moisture content to properly fill the trays, whether by hand or machine. There is not enough moisture in the medium right out of the bag. Use the squeeze method for determining moisture. Add enough moisture to the medium and mix it up. Take a handful, squeeze it, and release the pressure. The medium should retain its shape. When you press on it with your finger, it should start to fall apart. This is the appropriate amount of moisture to properly fill the trays. If there is too little moisture, then there will be more dust and settling of the medium after handling the trays, resulting in less air porosity in the mix and less fill in the trays. If there is too much moisture, then air pockets will develop in the cells during filling or the equipment will not function properly. If you fill trays ahead of time, cross-stack filled trays before running them through the seeder in order to avoid compacting the mix.

After filling the trays and before seeding, a dibble roller or bar is needed. Medium to large seed needs more room in the plug cell to sit and be covered up enough with medium or vermiculite. Dibbling the plug trays will accomplish this task. The bigger the seed, the deeper the necessary dibble.

Seeders

A wide variety of seeders is available, both new and used. Prices range from under $1,000 to more than $30,000. Before purchasing a seeder, evaluate your needs for the number of trays per week, type of plug trays, types of seed to sow, desired accuracy, and seeder cost and reliability. Gather as much information about the different seeders from suppliers, look them over at tradeshows, visit other growers who are using them, and ask all of your questions before buying. Many operations will need more than one seeder or seeder type to sow all of the plugs needed.

Equipment commonly used with a seeder includes dibblers (discussed under Flat filling), top coaters, and watering tunnels. Top coaters are used to cover seed evenly with medium, vermiculite, or perlite for better germination. Watering tunnels provide enough moisture to transport trays to the chamber or bench for germination.

Perhaps the most important aspects of seeders are the seeder operator and maintenance. Today's seeders are extremely accurate and easy to use if the equipment is maintained properly, so be sure to follow the supplier's maintenance schedule. It is imperative that you have trained and proficient seeder operators. You cannot produce a quality plug if the seed never gets into the plug cell!

Seed

There are numerous choices in seed products. Some seeds are modified to make them easier to sow. Marigolds (*Tagetes*) are detailed, tomatoes (*Lycopersicon*) are defuzzed, and ageratum is dewinged. Small seed such as begonia and petunia are pelleted. Odd-shaped seed such as marigold, dahlia, and zinnia are coated to make them easier to pick up and place into the plug cell.

Other seeds are offered as primed or pregerminated to speed up the germination process and to obtain higher plant stands. Primed pansy (*Viola*) seed will germinate under warmer temperatures, while pregerminated impatiens seed will germinate faster and with higher stands. Since all enhanced seed products have an increased cost and a shorter storage life, you need to carefully evaluate each one for suitability in your production system. Order only the amount of seed to be used within the season, and store all seed at 42°F (5°C) and 25–35% relative humidity.

Stages 1 and 2: Germination

Successfully germinating seed is a direct result of a grower's ability to provide the microenvironment needed around that seed for that particular crop, including temperature, moisture, light, and oxygen. The required levels of these factors vary with each crop.

Covering seed

Seed of some crops needs to be covered to get the best germination or rooting in. Generally, the larger the seed, the more covering it needs. A covering will provide more humidity or moisture around the seed for germination, may provide dark conditions to improve germination, and help the initial root to go into the soil. Improper covering will result in reduced or erratic plant stands. Crops that should be covered include aster, brassica, calendula, capsicum, *Catharanthus, Cosmos,* dahlia, dianthus, gazania, *Lycopersicon, Pelargonium,* phlox, primula, ranunculus, salvia, *Tagetes,* verbena, *Viola,* and zinnia.

You can cover the seed in a number of ways. Most growers cover with the soilless medium they are using in the plug tray, with vermiculite, or with perlite. Other growers use plastic film or frost cover. When using a soilless medium, do not place too much of the mix on top of the seed, as too much moisture could build up around the seed and suffocate it. Medium- to coarse-grade vermiculite is more effective, as it provides more air but still maintains humidity around the seed. Make sure vermiculite is uniformly popped or expanded. Perlite is impossible to overwater, but it can get too dry very quickly. Plastic film maintains humidity very well but provides no help with darkness or rooting in. Frost cover functions the same way, yet allows air and water in. These latter two coverings are used more for crops with small seed, such as begonia, to help with maintaining moisture during the long germination period.

Temperature

Every crop grown from seed has an optimal temperature for germination. However, you will need to be able to group crops to provide reasonable germination temperatures in your own facility. Table 13-2 shows temperature groups for many common seed-grown annuals. Failing to provide a reasonable substrate temperature during germination will result in lower stands, more erratic and slower germination, or possibly more diseases.

Table 13-2. Germination Substrate Temperatures for Stage 1 of Plug Production

77°F (25°C)	72°F (22°C)	67°F (19°C)
Ageratum	*Antirrhinum*	Phlox
Begonia	Aster	*Primula*
Browallia	Brassica	*Ranunculus* (60°F [16°C])
Capsicum	Calendula	*Viola*
Catharanthus	Coreopsis	
Celosia	Cosmos	
Coleus	Dahlia	
Gomphrena	Dianthus	
Hypoestes	*Eustoma*	
Impatiens (less than 78°F [26°C])	Gazania	
Lobelia	*Leucanthemum*	
Lobularia	*Lycopersicon*	
Nicotiana	*Matthiola*	
Petunia	*Papaver*	
Portulaca	*Pelargonium*	
Salvia	*Senecio*	
Verbena	*Tagetes*	
	Zinnia	

Many plug growers use germination chambers, which provide the proper temperature and moisture for the best germination. They also germinate crops requiring cooler conditions on the bench during the winter and early spring, while germinating warmer crops in the chamber. During the summer and early fall, the reverse is true: Cool crops are germinated in the chambers and warm crops on the bench.

Growers who germinate only on benches need to monitor soil temperatures closely and provide different temperature zones for germination. Under-bench heating is the best method for bench germination. Remember, it is substrate temperature, not air temperature, that you need to control. Above-bench heating may create substrate temperatures 3–5°F (1–3°C) lower than air temperatures. Use a small thermometer with a bulb or a digital thermometer to measure soil temperatures. Remember that the temperature of the water during the winter may be below 50°F (10°C), so this will reduce soil temperature. If you are germinating seed on a bench, be sure to use tempered water (heated to 70°F [21°C]).

Moisture

Providing the right amount of moisture during germination is the hardest part of the process. Again, you can group crops based on different moisture levels. Table 13-3 lists the moisture levels of a wide range of bedding plant annuals. To help understand the moisture levels on the chart, wet means the substrate is glistening on the surface, water is readily apparent to touch and is easy to squeeze from the media, and trays are heavy. The medium level means no glistening, the substrate is moist to the touch, and trays are not as heavy. Dry means that the substrate feels dry to the touch but still has moisture and trays are lighter. Crops requiring dry conditions are always covered with medium, vermiculite, or perlite. The moisture should stay in the covering, not the substrate.

Too much moisture may bury small seed or suffocate covered seed and lower germination. Too little moisture will cause erratic and lower stands. Maintaining proper moisture levels in seedling flats is easier in many ways than in plug trays, which have individual cells that dry out faster. Avoid overwatering seedling flats, especially if they are covered. Use a fine mist or fog (10–80 micron particle size) for moisture management, especially on the bench. Some crops should be grown drier in Stage 2 to reduce stretch and improve rooting. Begonia needs to be kept uniformly moist through early Stage 3, when its first true leaf is half expanded. At that time, the shallow roots will start going down further into the plug cell and the surface can be dried off. Improper moisture management previously will reduce stands and create nonuniform seedling growth in this crop.

Make sure plug trays are evenly watered, particularly around the edges. Germination chambers can provide fog to help hold moisture added after seeding. Stationary mist nozzles, boom irrigation, or fog can provide manually bench germination moisture. If watering germinating trays by hand, use a fine-mist nozzle, such as a Foggit nozzle, to avoid compacting the medium or burying small seed. Applying a fine mist rather than water droplets will also help to prevent algae from growing on trays.

Light

For many crops, light is not needed to begin the germination process but is important to finish it. This fact is most significant for growers using germination chambers without lights. Keep the trays in the chamber for a minimum time and then move them to the bench to satisfy the light requirement for the latter part of germination. Leaving crops in the chamber longer may actually reduce the germination. Crops such as antirrhinum, begonia, impatiens, petunia, and primula fit into this category. Consult table 13-4 for a more complete list of crops with special dark/light requirements. Growers using bench germination do not need to provide any further light.

Covering a crop with a soilless medium, vermiculite, or perlite does not interfere with the light requirement unless the covering is too heavy. Gerbera seed can safely be covered with coarse vermiculite if trays are kept on the drier side. Some crops like total darkness for best germination, such as *Catharanthus* and cyclamen. This can be achieved with proper covering (no seed or soil visible) or a dark germination chamber.

Breaking dormancy

Even if growers provide the proper germination conditions as indicated previously, there are a number of crops that may not germinate properly or at all. This seed may be dormant (alive but unable to germinate) and may need special treatments to break out of its dormancy. Many perennial varieties have dormancy problems, for example.

Table 13-3. Media Moisture Levels for Stages 1 and 2						
CROP	STAGE 1			STAGE 2		
	WET	MEDIUM	DRY	WET	MEDIUM	DRY
Ageratum		x			x	
Antirrhinum		x				x
Aster			x			x
Begonia	x			x		
Brassica		x				x
Browallia		x			x	
Calendula			x			x
Capsicum		x				x
Catharanthus	x				x	
Celosia		x			x	
Coleus		x			x	
Coreopsis			x			x
Cosmos			x			x
Dahlia			x			x
Dianthus		x			x	
Eustoma		x			x	
Gazania			x			x
Gomphrena		x				x
Hypoestes		x			x	
Impatiens	x				x	
Lobelia		x			x	
Lobularia		x			x	
Lycopersicon		x				x
Matthiola		x			x	
Nicotiana		x			x	
Papaver		x			x	
Pelargonium		x			x	
Petunia		x			x	
Phlox			x			x
Portulaca		x				x
Primula		x			x	
Ranunculus	x			x		
Salvia		x				x
Senecio		x			x	
Tagetes		x				x
Verbena			x			x
Viola	x				x	
Zinnia			x			x

Table 13-4. Crops Requiring Either Light or Dark for Germination

LIGHT	DARK
Antirrhinum	*Catharanthus*
Begonia	Cyclamen
Eustoma	Phlox
Gerbera	
Impatiens	
Nicotiana	
Petunia	
Primula	

Seed pretreatments are designed to overcome impermeable seed coats, chemical inhibitors, immature embryos, or a combination of these. Hard seed can be soaked overnight in hot (not boiling) water or scarified (breaking the seed coat) by rubbing between two pieces of sandpaper. Seed with chemical inhibitors can be soaked in gibberellic acid (GA) solution for up to twelve hours or soaked in hot water. Other seed will respond to moist cold stratification. Place seed in moist peat moss in resealable plastic bags, put in a refrigerator (42°F [5°C]) for two to twelve weeks, and then remove, sow, and germinate normally. Still other seed is dormant after harvesting but will overcome dormancy during storage at room temperature or cooler temperatures up to a year.

Germination facilities

Growers can germinate seed in chambers or on the bench. The germination chamber is an environmentally controlled room that growers can build themselves. The chamber is designed so that trays may be stacked vertically on movable carts or racks and then rolled out or taken out of the room for observation or moving to the greenhouse for growing on. Temperature is closely controlled with heating and cooling systems. Watering the plug trays after seeding provides moisture. A fog system maintains this moisture in the chamber. Light may or not be provided, but be aware that some crops require light for germination. Not oversaturating the medium or burying the seed improves the oxygen level of the seed. You can control the chamber's environment with thermostats, timers, or computers. The timing of each crop placed into the germination chamber is critical. If trays are left too long in the chamber, the seedlings will stretch rapidly and be ruined. If taken out too soon, germination and uniformity may be reduced.

Germinating plug trays in the greenhouse allows easy access to trays for visual inspection. You can devote a portion of a greenhouse to plug germination if you maintain the proper conditions. Provide root-zone heating for best soil temperature control. An overhead fog or fine-mist system can supply the necessary moisture and can be controlled with timers or by visual inspection. You can also water trays by hand but use a fine-mist nozzle (Foggit nozzle).

Some growers cover plug trays with porous materials (frost covers) that permit water to pass through but at a much-reduced rate compared to uncovered trays. (For more on this, see the section on covering seed.) This method helps maintain a uniform moisture environment. The major drawback is the difficulty for the grower to check the trays. These porous coverings can be cleaned and reused. When removing such coverings from the plug trays, take care not to pull up seedlings.

Water, Media, and Nutrition

While water quality, media, and nutrition are covered in their own chapters, these three production elements are critical to growing a quality plug crop. In the section that follows, you will find broad comments as they pertain to growing plugs. Be sure to also consult the appropriate chapters prior to attempting to grow plugs.

Water quality

Use a qualified laboratory to test your water supply at least twice a year. Look at three important measurements: alkalinity, soluble salts (EC), and nutrients in the water. Alkalinity in the water influences your substrate pH and subsequent nutrient availability of fertilizer solutions. Keep alkalinity levels at 60–80 ppm by injecting acid if necessary. If alkalinity is naturally lower than 60 ppm, you may need to buffer the water with low levels of calcium/magnesium fertilizer, such as 14-0-14. The soluble salts content of the water should be less than 0.75 mS/cm. Know what salts are involved if levels are higher than recommended. The amounts of nutrients in the water should also be considered. Sodium (Na) competes with calcium (Ca) and magnesium (Mg) and is detrimental to good root growth. Keep sodium at less than 40 ppm and the sodium adsorption ratio (SAR) at less than 2.0.

Boron (B) should be less than 0.5 ppm to avoid toxicity. Water pH should be 5.5–7.0 for best solubility and efficacy of chemicals added to the water, such as pesticides and growth regulators.

Media quality

Buy a quality plug mix from a reputable manufacturer, and have it tested at regular intervals by a qualified lab. A good plug medium must be able to hold water but still have enough air porosity (2–10%). How you handle the plug medium in filling the plug trays will also determine if you lose air porosity. Most commercial plug mixes now contain peat moss and perlite, not vermiculite, for better air porosity.

The pH of the medium should start at 5.5–5.8 for most crops. For *Eustoma,* impatiens, *Pelargonium,* and *Tagetes,* increase the mix pH starting at 6.0–6.2 within two weeks of seeding. Keep pH at less than 6.5 at all times to prevent micronutrient deficiencies.

Know the amount and concentration of soluble salts in your plug mix to start. Keep the starter charge at less than 1.0 mS/cm (SME). Germination will be increased and early seedling stretch will be more controllable if the starter charge is less than 1.0 mS/cm, with low ammonium (NH_4) and phosphorus (P). If your plug mix has little or no starter charge, you need to start feeding early with 50 ppm nitorgen of a balanced fertilizer low in ammonium.

You should be concerned about the medium's pH because it affects the availability of nutrients to the plants (table 13-5). In a peat-based medium, nutrients are readily available when the mix pH is 5.5–6.5. At a high pH, micronutrients such as iron and boron are tied up, whereas calcium becomes more available. At low pH, micronutrients become too available, whereas phosphorus is more readily leached and calcium is tied up. With cool soil temperatures, ammonium can become toxic at low mix pH.

Too much of one nutrient can tie up another nutrient. Calcium competes with magnesium; iron competes with manganese; and sodium competes with calcium, magnesium, and potassium. Try to maintain a 2:1 ratio between calcium and magnesium and between iron and manganese. When sodium presents a problem in water (greater than 40 ppm), you need to increase levels of calcium, magnesium, and potassium.

Now that you know why substrate pH is important, you also need to know what controls substrate pH. The four factors are: (1) alkalinity in the water, (2) lime in the medium, (3) the type of fertilizer used, and (4) the crop being grown. First, think of alkalinity as lime in the water. The higher the alkalinity level, the more lime you are adding each time you are watering. That tends to raise the substrate pH quickly, especially in small plug cells. Typically, you can control high alkalinity by injecting acid with a target range for alkalinity of 60–80 ppm.

Second, the lime in the medium is added to offset the acidic nature of the peat moss. How much lime, what type of lime, the particle size, and how often the medium is watered determine how fast the lime will react. To determine your starting point, test your medium before using it. Then test the medium in plug trays for the first three weeks without fertilizing and, using good-quality water, with normal moisture levels for germinating and growing. If your pH climbs rapidly within the first three weeks, you may have too much lime in your mix. Refer to the substrate pH ranges above for best growth.

Table 13-5. Nutrient Availability Changes with pH in a Peat-Based Medium

As pH changes, so does nutrient availability. At less than 5.5, many nutrients are either in excess or deficient.

PH RANGE →	
5.5 EXCESS	**6.5 EXCESS**
Manganese (Mn)	Calcium (Ca)
Iron (Fe)	Nitrogen (N)
Boron (B)	
Copper (Cu)	
Zinc (Zn)	
Sodium (Na)	
Ammonium (NH_4)	
READILY AVAILABLE	
DEFICIENT	**DEFICIENT**
Calcium (Ca)	Iron (Fe)
Magnesium (Mg)	Manganese (Mn)
Phosphorus (P)	Boron (B)
Potassium (K)	Copper (Cu)
Sulfur (S)	Zinc (Zn)
Molybdenum (Mo)	Phosphorus (P)
Magnesium (Mg)	

Third, fertilizers can be acidic or basic. Potential acidity or basicity can be expressed for common fertilizers (table 13-6). Note that fertilizers high in ammonium tend to be acidic, while fertilizers low in ammonium but high in calcium tend to be basic.

Finally, the crop itself can influence pH. Geranium (*Pelargonium*) roots will slowly make the medium more acidic, whereas vinca (*Catharanthus*) roots tend to make the medium more basic (raises the media pH).

To control substrate pH, a plug grower needs to know not only these factors but also how they relate to each other. When you change one of these variables (such as a different media or fertilizer), you need to know how it will affect the other variables and ultimately media pH.

Selecting a fertilizer to control growth

Knowing what a particular fertilizer will do for substrate pH is important, but a plug grower also needs to know what it will do for plant growth. The three factors that you need to know are: (1) the percentage of total nitrogen expressed as ammonium (NH_4) or urea, (2) how much phosphorus (P), and (3) how much calcium (Ca) and magnesium (Mg) it contains. The percent of total nitrogen (N) expressed as ammonium or urea for common fertilizers can be seen in table 13-6. The greater the amount of ammonium or urea, the more shoot and leaf growth you will have. These fertilizers promote big, green leaves and rapid shoot growth but will not promote root growth or flowering. Fertilizers high in nitrate nitrogen (NO_3) (bottom half of table 13-6) will give more toned growth, better roots, and faster flowering.

Fertilizers high in ammonium typically are high in phosphorus (the middle number of the formulation). Phosphorus has been shown to play a major role in stretching of plug growth for many crops. Some phosphorus is needed to keep lower leaves green, but typically plug growers will use low-phosphorus fertil-

Table 13-6. Common Commercial Fertilizers[a]

FERTILIZER	NH_4 (%)[b]	POTENTIAL ACIDITY[c]	POTENTIAL BASICITY[d]	CA (%)[e]	MG (%)[e]
21-7-7	100	1,560		—	—
9-45-15	100	940		—	—
20-20-20	69	583		—	—
20-10-20	40	422		—	—
21-5-20 (Excel)	40	418		—	—
15-15-15[f]	52	261		—	—
15-16-17[f]	30	165		—	—
20-0-20	25	40		5	—
17-5-17	24	0	0	3	1
17-0-17	20		75	4	2
15-5-15 (Excel)	22		141	5	2
13-2-13	11		200	6	3
14-0-14	8		220	6	3
15-0-15	13		420	11	—

[a] List of some commercially available fertilizers used for plugs and bedding plants. Not all formulations are the same from every company. Check the label.

[b] NH_4 (%) is the total nitrogen percentage that is in the ammonium plus urea forms; the remaining nitrogen is nitrate.

[c] Pounds of calcium carbonate limestone required to neutralize the acidity caused by using one ton of the specified fertilizer.

[d] Application of one ton of the specified fertilizer is equivalent to applying this many pounds of calcium carbonate limestone.

[e] Only where % Ca or % Mg were 1% or greater.

[f] Contains sodium nitrate (nitrate of soda), which adds unwanted sodium to plugs.

izers to control height. These types of fertilizers also have more calcium and magnesium in them. Calcium is important in promoting root growth and toned plants, while magnesium is needed for green leaves and photosynthesis.

Typically, plug growers have alternated between 20-10-20 and a high-calcium fertilizer. Now, more growers are starting to use an intermediate fertilizer such as 15-5-15 or 17-5-17 to control the growth more than with 20-10-20. These growers will then use a high-calcium fertilizer such as 13-2-13 or 14-0-14 to supply more calcium and magnesium, especially during low-light conditions. Adjustments in fertilizer selection need to be made based on the season, weather conditions, crop, and stage of development. Remember, changes in fertilizer will affect substrate pH as well.

The quantity of fertilizer applied and the frequency of application will vary with the grower, time of year, irrigation system used, type of crop, desired growth control, and weather conditions. Generally, feed plugs after the first true leaves are emerging, unless it is taking two to four weeks to finish germination. Early feedings should be 50–75 ppm nitrogen (N) about every other watering. Once active growth occurs in Stage 3, step up the fertilizer to 100–150 ppm N at every other watering. You can constant-feed plugs, but use 50–80 ppm N each time and watch your substrate EC levels to avoid over- or underfeeding.

Environment

The basics of plug growth are greatly influenced by the environment. Each of several key factors—temperature, light, carbon dioxide (CO_2), and humidity—has individual effects on processes from germination to photosynthesis to flowering. These environmental factors also interact with each other and with nutrition to control plant growth and flowering; the effects vary with the crop and growth stage involved.

During the active growth phases of Stage 3, shoot and root growth increases proportionally as temperatures increase from 50–85°F (10–30°C). There is an optimal temperature for each crop for maximum growth, which may not be the same for flowering. Vinca (*Catharanthus*) needs a warmer temperature to grow than does pansy (*Viola*). For best results, try to group crops needing a warmer Stage 3 temperature together and provide proper soil temperatures for best growth (see table 13-2 for crop groupings). During Stage 4, soil temperature should be reduced to slow both root and shoot growth. Water plug trays in the morning to reduce diseases caused by wet foliage and cool temperatures.

Growing plugs during the winter and early spring in the northern United States can be a problem. Growers can control the temperature fairly well, but providing enough light is the main problem. Light levels are very limiting during the winter, getting as low as 150 f.c. (1.6 klux) inside the greenhouse. Leaves will use

Table 13-7. Crops That Benefit Most from Supplemental Lighting in Plug Trays

LONG DAYS	SHORT DAYS	GROWTH/EARLIER FLOWERING
Ageratum	*Salvia splendens**	*Begonia semperflorens*
Antirrhinum	African marigold (*Tagetes*)	*Catharanthus*
Begonia (tuberous)	*Zinnia elegans*	Dianthus
Dahlia	Cosmos	Gerbera
Eustoma	Celosia	Impatiens
Lobelia		*Pelargonium*
Nierembergia		*Pentas*
Petunia		
Salvia farinacea		
*Salvia splendens**		
Verbena		
Pansy (*Viola*)		

* Not all varieties

up to 3,000 f.c. (32 klux) for photosynthesis, above which the extra light turns into heat on the leaves. The result of low light levels is stretched growth and poor roots.

Using supplemental HID lighting will provide enough light for better plug growth. Plan on HID lights providing a uniform 400–500 f.c. (4.3–5.4 klux). Contact the manufacturer for calculations on the number of lights needed and their arrangement in the greenhouse. Turn on the lights at the end of the day and leave them on until midnight. This gives you a day length of sixteen to eighteen hours. Plugs are most receptive to this extra light once cotyledons have expanded (end of Stage 2). You can satisfy photoperiod requirements for key crops in the plug stage by using supplemental lighting (table 13-7). Other crops will respond with better growth due to greater light quantity. Be aware that substrate temperature is raised about 2–3°F (1–1.5°C) with lighting. Lights can be discontinued once natural day length is twelve hours or longer.

High humidity in the greenhouse causes plugs to stretch and be soft due to a lack of calcium uptake. Dehumidification cycles combined with morning watering will help reduce humidity levels. Ventilated plug trays will also help dry off the plugs quicker than solid trays will. Double-poly greenhouses are more humid with more drips than glass houses. Trays grown on benches with under-bench heating will dry down faster than trays grown on the ground.

Controlling temperature, light, humidity, watering, and nutrition will control plug growth. However, roots and shoots may benefit differently from the same environmental factors (table 13-8). Be aware of whether the shoot growth is ahead of the root growth or not. Too often, growers must use chemical growth regulators to overcome poor control of the environment and nutrition.

Chemical Growth Regulators

Well-balanced plugs with roots and shoots in proportion are the direct result of proper environment, moisture management, type of fertilizer used, and chemical plant growth regulators (PGRs). Most growth regulators have wide labels for usage and are effective on a range of bedding plants. A-Rest, B-Nine, and Cycocel are commonly used in the North, whereas Bonzi and Sumagic are used more in the South. Learn to apply the most forgiving growth regulators first. If you are starting out, it is easiest to learn to use B-Nine first, move to A-Rest and the tank mix of B-Nine + Cycocel next, and then apply Bonzi and Sumagic as you need stronger control. More growers are learning how to use A-Rest effectively and are getting away from overdoses resulting from Bonzi and cool weather. A-Rest is more effective than B-Nine or Cycocel alone and easier to use than Bonzi or Sumagic. Remember, B-Nine and Cycocel take a long time to get into the plant (at least four hours wet on leaves). A-Rest, Bonzi, and Sumagic all get into the plant quickly and are active in the substrate through the roots.

Under warm growing temperatures, more growers are using a B-Nine + Cycocel tank mix. Starting rates for the tank mix are 800 ppm of each up to 1,500 ppm Cycocel + 2,500 ppm B-Nine.

Starting rates for the other chemicals are: B-Nine (2,500–5,000 ppm), Cycocel (300–1,000 ppm), A-Rest (3–10 ppm), Bonzi (1–30 ppm), and Sumagic (0.5–15.0 ppm). When using Bonzi or Sumagic, understand that no one rate is good for all crops. Some crops need more chemical than other crops (table 13-9). When using Sumagic, start at about 20–50% of the ppm used for Bonzi. Remember, both Bonzi and Sumagic will have greater reaction if the weather gets cool around or after the spray application. Work with each of your growers to standardize how to apply Bonzi or Sumagic, as application rates are critical.

Table 13-8. Environmental and Cultural Factors Promoting Shoot or Root Growth

FACTOR	PROMOTES SHOOTS	PROMOTES ROOTS
Temperature	Increasing (50–80°F [10–27°C]) +DIF	Increasing (50–80°F [10–27°C]) –DIF
Light Intensity	Low (<1,500 f.c. [16 klux])	High (>1,500 f.c. [16 klux])
Moisture	High	Low
Nutrition	High NH_4 and P	High NO_3 and Ca
CO_2	High (1,000 ppm)	High (1,000 ppm)
Humidity	High	Low

Conduct your own PGR trials on ppm needed, when to apply, and how to apply any of these chemicals. Keep track of your environmental conditions and the varieties grown, as different chemical programs may be needed. Go faster over slow-growing varieties and slower over fast-growing varieties to get different amounts of chemical onto and into the plants.

Pests and Diseases

In plug production, overwatering, poor air movement, humid greenhouses, poor sanitation, and poor weed control cause most disease problems. Poor monitoring procedures, improper pesticide applications, a lack of insect screening, and poor weed control cause most insect and mite problems.

The biggest disease problems in plugs are root and crown rots, along with damping-off (i.e., *Pythium,* rhizoctonia, *Thielaviopsis,* and botrytis). You will see damping-off when you have drips, with crops that are multiple-seeded (e.g., *Lobularia* and portulaca), and with crops that are covered and overwatered (e.g., *Celosia* and portulaca). Vinca (*Catharanthus*) is one crop that is susceptible to many root rots and does not like to be overwatered or grown too cool. Manage root rots by disinfecting reused plug trays properly, practicing good sanitation in the greenhouse, avoiding high substrate pH and EC, and practicing good moisture management. You can drench susceptible crops monthly with Banrot, Subdue MAXX (low end of rate), Truban, or Terrazole + Cleary's 3336, OHP 6672, or Medallion, or use a biocontrol agent such as Rootshield or Actinovate.

Botrytis can become a problem when plugs are older and the plant canopy is tight. Look for dead and dying lower leaves and grayish spores on them. Avoid watering late in the day, dry down plants properly, transplant on time, have good air movement, and spray as needed under the canopy with a rotation of Decree, Medallion, Daconil, and Chipco 26019.

Fungal and bacterial leaf spots may appear as the plugs get older. Avoid water staying on leaves too long, especially with cooler temperatures. Spray as needed with a rotation of Phyton 27, Camelot, Medallion, Heritage, Compass, Pageant, Daconil, or Chipco 26019.

Table 13-9. Classification of Plug Crops by Optimum Rate Range for Bonzi and Sumagic*

	OPTIMUM RATE RANGE (PPM)		
GROUP	BONZI	SUMAGIC	CROP
High	15–30	7.5–15.0	Snapdragon (*Antirrhinum*)**
			Pentas
			Petunia
			Salvia**
			African marigold (*Tagetes*)**
Medium	7.5–15.0	2.5–7.5	*Ageratum*
			Celosia
			Coleus
			Dahlia
			French marigold (*Tagetes*)**
			Verbena
			Most other crops
Low	2.5–7.5	0.5–1.5	*Vinca (Catharanthus)*
			Impatiens
			Pelargonium
			Pansy (*Viola*)

* *Note:* Growers in the North start with the low-end rate or lower, especially during cooler weather. Also, growers can use low-end rates under retractable-roof (high light) conditions, even if hot.

** Sumagic works better than Bonzi.

The biggest insect problems in plugs are fungus gnats and shore flies. These insects are more than a nuisance—they can spread root rots, and fungus gnat larvae damage roots. Learn to tell the difference between these two insects. Fungus gnat adults have long, thin bodies and long antennae and are lazy fliers like mosquitoes. The larvae have black head capsules. Shore fly adults have short, thick bodies and short antennae and are strong flyers like flies.

Use sticky cards placed 2" (5 cm) above the crop to monitor for both insects. Control these insects by managing moisture and controlling algae on plug trays and under benches. Chemical control can be used to target larvae in the plug trays and under benches. Rotate between Adept, Distance, Azatin or Orzanin, and Citation. Knock down adults with aerosols or autofogs of Decathlon, Tame, Talstar, Conserve, or pyrethrum. For fungus gnat larvae, you can also use biological control agents, such as Gnatrol, Nemasys or Nemashield (nematodes), *Hypoaspis* mites, and Atheta beetles. Control algae by managing moisture, not compacting the soil surface, feeding properly, and using ZeroTol or X3.

Transplanting

Transplanting plugs is probably one of the most neglected procedures in the production and finishing of bedding plants. Plugs should be transplanted as soon as possible after they have reached the appropriate size. Moistened plugs should be pullable, with an intact root system. Plugs held too long become root-bound and growth stalls. The goal for the initial transplant is for roots to rapidly establish into the growing medium, which should have a soluble salts level less than 1.5 m^S/cm (SME). The growing medium should be a high-quality finishing mix that drains well, not a plug mix.

The transplant medium should be moist, but avoid saturated conditions for the first week after transplant for the same reason you do not want saturated plug medium in Stage 2 (poor root growth). Soil temperatures after transplanting should be 65–70°F (18–21°C) to encourage rooting within the first three to five days, after which temperatures can be reduced to normal growing-on levels. Fertilization is not recommended until one week after transplanting, or when the roots reach the side of the container. In general, use moderate light levels for newly transplanted flats and pots. You can increase light levels after the first week or so. Chemical growth regulators can be applied on crops that need them after the roots get to the side of the container. Fungicide drenches should be applied as needed after the roots start to come out of the plug root-ball (about three to five days). If needed, plug trays can be fertilized or drenched with fungicides right before transplanting to provide better growth and protection and to avoid saturation of media for the first ten to fourteen days after transplanting.

14

Mike Klopmeyer, Dan Lehman and C. Anne Whealy*

Propagating Vegetative Crops

Calibrachoa, Argyranthemum, Lantana, Diascia, Nemesia, Scaevola, and the list goes on and on. All of these plants have several things in common. First, these crops, from the grower's perspective, are sold with a higher profit margin than are many seed varieties. Most importantly, all of these, as well as thousands of other varieties, are propagated exclusively via cuttings. No longer is vegetative propagation confined to a small number of crops such as coleus, fuchsia, and *Pentas.* As the public's appetite for an ever-expanding palette of plant material has grown, so has the importance of vegetative propagation.

Many of these cutting-edge varieties offer unique characteristics such as flower color and form, plant habit, leaf shape, and overall plant texture. Many of these same varieties currently cannot be produced reliably or cost effectively from seed. In addition, the amount of time it takes for a new vegetative variety to be introduced is shorter than what is required for a seed introduction.

The Decision to Deal with Stock Plants

The bedding plant industry worldwide is experiencing a dramatic increase in the finishing and sale of floricultural crops propagated by cuttings. Crops such as geraniums (*Pelargonium*), New Guinea impatiens, verbena, and fuchsia, to name only a few, are gaining even more importance as an integral part of a grower's finished program. Many years ago most growers grew their own mother stock so they could control the quality and timing of their starter material. With today's economics of running a successful greenhouse business, decisions such as growing your own stock or buying in cuttings may play a significant part of the overall cost of production. Should you take a look at maintaining stock plants? This section will discuss the latest market trends in vegetative production and analyze the pros and cons of growing your own stock.

Market trends

It is only in the past twelve to fifteen years that the vegetative market has changed considerably as it relates to the supply of cuttings. Prior to the late 1980s, most vegetatively propagated crops such as *Pelargonium* were self-propagated by many growers. "Self" propagation can be defined as those growers who bought in stock plants each summer/fall and propagated cuttings from this stock for their own use. This market flourished, since growers wanted control of their production by having a reliable, disease-free supply of unrooted cuttings when they needed them. They could also finish these mother plants for spring sales at a higher price point.

Since the 1980s, new and more varieties have entered the market. Coupled with this influx of new varieties, more companies are providing growers with the opportunity to bring in unrooted cuttings from offshore locations in place of growing stock. For example, in the 1980s more than 70% of geraniums sold in the market were self-propagated. Now, less than 50% are self-propagated due to the proliferation of the offshore supply of unrooted cuttings.

Are unrooted cuttings for me?

The advantages of buying in unrooted cuttings are multifold:

1. Cuttings may be of higher quality since they were grown in the high light climates of Central America rather than the dark, winter months typical of stock growers in North America.
2. Growers do not have to take up valuable greenhouse space for stock and instead can use this space for more profitable propagation or finishing.
3. Buying in unrooted cuttings offers a more extensive variety list to choose from than from their own stock plants.

* Mike Klopmeyer and Dan Lehman wrote Propagating Vegetative Crops, while C. Anne Whealy provided information on the US Plant Patents and Legal Issues section.

4. Growers can bring in cuttings when they need them and not worry about scheduling their own stock and the associated hand labor.

Of course, there are disadvantages of *not* growing your own stock. They include the lack of control and your dependence on someone else growing the stock, the potential threat of disease if the outsourced stock is not managed properly, and the potentially higher cost of purchasing versus producing your own cuttings.

Many growers, however, do not fully understand their internal production costs to make a decision based strictly on the numbers for buying in compared with producing their own cuttings.

Crops to consider for your own stock

Many of the new vegetatively propagated varieties in the market today were bred and developed by specialist breeder/producers. Nearly all new varieties are protected from illegal propagation by a US Plant Patent (Plant Breeder's Rights in the rest of the world). In turn, the breeder/producers control the distribution and sale of these varieties through licensing. Some license agreements would allow for self-propagation or propagation with distribution to third parties, but there are many agreements where no propagation is allowed of the variety. The legal side of these agreements is explained in the US Plant Patents and Legal Issues section, beginning on page 167.

There are a number of varieties to consider for in your own stock program. These include geranium (*Pelargonium*), New Guinea impatiens, poinsettia (*Euphorbia*), and all of the non-patented varieties (i.e., accent plants) in the market today. Outside of the legal allowances under plant patent law, some important factors to consider when determining which varieties to grow as your own stock include: the growing difficulty, cost-effective yields/season, threat of major crop-destroying diseases, and appropriate greenhouse space and patience to grow your own stock.

Below are some general environmental and cultural factors to consider when growing your own stock plants.

Stock source

It is important to begin your stock season on the right foot. This includes starting with culture- and virus-indexed, disease- and insect-free mother stock. With any mother stock program where many cuttings are propagated from a single plant, the chances of building up and spreading disease are high. Be sure to understand the clean-stock program of the grower/supplier that is supplying your new mother stock. Further information on this vital element of successful stock production can be found in chapter 15.

Greenhouse location

Ideally, your stock program should be maintained in a physically separated greenhouse range or zone. This separation provides "isolation" of the clean mother stock from other noncertified plants in the greenhouse and crossover propagation and general service areas. Access to this stock area should be limited to specifically trained employees. Consider installing a simple wash station (running water with soap) and providing lab coats or aprons to protect the stock plants from direct contact with workers' clothing.

Light and temperature

The light and temperature requirements for stock plants depend on the species produced. Typically the conditions required for optimal finishing of the crop will also work for stock. For example, *Pelargonium* responds to high light with warm day temperatures and cool night temperatures. In North America during the winter months, it is difficult to attain consistent high light levels, so consideration should be made to provide supplemental lighting in the stock area using high intensity discharge (HID) lighting. High light levels can also initiate flower development that diverts the plant's energy into flowers rather than cuttings. Manual or chemical flower bud removal may be required to keep the stock vegetative.

You cannot cheat on providing adequate temperatures during the cold winter months. If the crop is not grown at temperatures optimal for growth, then cutting yields and quality will suffer. Bottom heat is more effective in stock production by providing a warmer medium for the root system and helping to maintain low relative humidity in the canopy for less problems with botrytis.

Media

Stock programs have typically longer crop times than your finished crop. Larger container sizes (i.e., 1-gal. [4 l] containers) will allow for adequate root development and sustain the plant's top growth for many months. Start your stock program using a clean, soilless medium from a reliable source. Your medium should contain some peat moss to help in water retention, yet have good drainage. The ability to leach the medium is important to maintain salt levels (fertilizer) in the appropriate range. For low pH–sensitive varieties such as *Pelargonium*, the medium should be supplemented with high-quality dolomitic limestone (8–10 lbs./yd.^{3} [4.7–5.9 kg/m^3]).

Fertilizer and growth regulators

There are two phases in stock plant growth and development. The first phase is the period immediately after planting and developing the scaffold from where the cuttings originate. Depending on the crop produced, this development phase may be from six to twelve weeks after planting. During this phase, the crop should be fed with a fertilizer such as 20-10-20 that will promote branching, leaf expansion, and "softer" stock growth. The higher percentage of ammonium nitrate in this type of fertilizer is the key factor. The use of a higher phosphorus feed will also provide some stem stretch to create a more open scaffold. If a tight scaffold is desired, alternate with 14-0-14 or 15-0-15. During this phase, the stock is pinched on a regular basis to promote basal branching and the development of a scaffold.

Using Florel as a foliar spray biweekly at concentrations ranging from 100–500 ppm can improve lateral branching. Florel has the advantage of stimulating lateral branching plus removing flower buds, thus saving the grower from the labor-intensive activities of manual pinching and flower bud removal.

The second phase of stock plant development is when the stock is ready to come into production and supply cuttings on a weekly basis. As this phase begins, it is important to lower the ammonium concentration in the fertilizer and raise the levels of potassium or calcium nitrate. Nitrate-based fertilizers will help tone up the stock and promote better rooting in the cuttings that are harvested. At all phases of stock development, it is important to maintain a steady concentration of soluble salts in the medium. For example, EC ranges of 0.5–1.5 mS/cm are adequate for impatiens, while high-feed crops such as *Pelargonium* should range from 1.5–3.0 mS/cm. Minimize clear water leaching, as that will cause wild fluctuations in the EC levels in the medium. Instead, if the soil EC levels are too high, then consider leaching with a feed solution that has a lower EC.

Diseases and pests

The major disease problem in stock production is botrytis. The stock canopy, where little air circulation occurs, is a prime location for botrytis to infect and spread. Proper environmental conditions that reduce the relative humidity in the canopy are important. Consider growing the stock using bottom heat and use adequate ventilation and heating at nightfall to reduce the relative humidity in the greenhouse. An active botrytis fungicide spray program may be required to maintain consistent quality.

Scouting for major insect pests through the growing season is a must. Many insects, including whiteflies, thrips, and fungus gnats, can go undetected in the canopy of the plants. If these pests are controlled early in the scaffold development, it is easier to maintain a lower pest incidence as the crop matures.

Stock plant maintenance

Growing your own stock plants will require manual hand labor at the appropriate time. If the stock is grown properly, it is important to keep it in good shape to sustain yields during the harvest period. In the stock plant development phase, manual pinching is needed. As pinches are made, it may also be necessary to remove larger leaves and flowers that may block light penetration into the canopy. Do not remove too many leaves and/or pinches on a given week, as this may harden up and stress the plant. As the stock gets older and fewer cuttings are needed for production, consider a hard cutback of the canopy and letting the stock plant flush out for flowering and sale.

Harvesting

The stock has been fed and the proper scaffold has been developed. Now there are cuttings that need to be harvested. The following are some key points to consider for harvesting.

1. Make sure your employees are adequately trained in taking cuttings. If they are not trained properly, they could easily destroy the stock plant or set it back so far that consistent yields are unattainable. Determine specifications for the unrooted cutting. Key specifications include stem length, node and leaf count, and taking the cutting straight across the stem base. Angled cuts will delay wound healing and rooting.

2. Use a sharp, high-quality knife or scalpel blade that has been adequately disinfected. An effective routine would be for each harvester to use two knives. While one is being used for harvesting (from five to ten stock plants), the other is soaking in disinfectant. Suggested disinfectants include quaternary ammonium compounds (at recommended rates for surface disinfection), trisodium phosphate (TSP), or 70% ethanol with flaming. Discourage harvesters from removing cuttings by breaking; this technique usually does not create a clean, cut end and also may excessively wound the stock plant.

3. Upon harvesting, remove excess leaves attached to the base of the cuttings, place in a clean plastic bag, and remove from direct sunlight immediately. Store cuttings in sealed plastic bags overnight in a cooler—*Pelargonium* at 40°F (4°C) and all other crops at 45–50°F (7–10°C)—and stick them the following day. This overnight cool treatment is effective for the cutting to regain lost turgidity and to initiate the wound healing process to enhance root formation.

Rooting Cuttings

After the cuttings arrive, or you have taken them yourself, you are ready for the next step: rooting the cuttings. Rooting cuttings is not an easy endeavor because the environment has to be just right and attention to detail is required. For success, it is vital that you apply the best management practices to this area of production. The following are guidelines for how to do just that.

Stages of vegetative propagation

There are five critical stages growers need to be familiar with to be successful with unrooted cuttings. These five stages are:

Stage 0: Prior to arrival
Stage 1: Arrival and sticking
Stage 2: Callusing
Stage 3: Root development
Stage 4: Toning the rooted cutting

Stage 0 is the foundation upon which each crop is built. Stage 1 is the first chance you have to assess the

Table 14-1. General Guidelines for Vegetative Propagation (

CROP	AVERAGE ROOT TIME (WEEKS)	ROOTING HORMONE	MIST (DAYS)	PINCHING	COMMENTS
Angelonia	4–5	No	7–12	Yes	Watch for iron deficiency in propagation.
Snapdragon (*Antirrhinum*)	3–4	No	7–10	Yes	Use as little mist as possible.
Argyranthemum	3–4	No	5–10	No	Watch for iron deficiency in propagation.
Bacopa (*Sutera*)	3–4	No	5–10	Yes	Pinch before transplanting.
Bidens	3–4	No	5–10	Yes	Watch for cutting stretch.
Bracteantha	3–4	Yes	5–10	No	Use as little mist as possible.
Calibrachoa	5	Optional	7–12	Yes	Use as little mist as possible.
Coleus	3–4	No	7–10	Optional	Watch for cutting stretch.
Cuphea	4–6	No	7–11	Yes	Use as little mist as possible.
Dahlia	4–5	Optional	7–12	No	Hormone will help rooting.
Diascia	4	Optional	7–12	Yes	Use as little mist as possible.
Evolvulus	3–4	No	7–12	Optional	Use as little mist as possible.
Geranium, ivy	4–5	Optional	5–10	No	Hormone may help rooting.
Geranium, zonal	3–4	Optional	5–10	No	Avoid delays in sticking.
Helichrysum	4	Optional	5–10	Optional	Hormone may help rooting.
Heliotrope	3–4	No	5–10	Optional	Use as little mist as possible.
Impatiens, mini	3–4	No	5–10	No	
Impatiens, New Guinea	3–4	No	5–10	No	Watch for botrytis.

(Continued)

quality of the raw product. Stage 2 is the most important opportunity to impact the overall quality of a crop. Stages 3 and 4 are different for each crop in terms of duration and visible results but determine how the final product, a rooted cutting, performs. Stage 4 serves as a transition from propagation to finishing and allows growers to mold the rooted cutting depending on their needs. As each stage is covered, we will discuss general benchmarks that you can use to measure your progress and fine-tune the system you have in place for handling unrooted cuttings of specialty garden plants. Table 14-1 provides some general guidelines for rooting cuttings of a wide range of spring plants.

Stage 0: Prior to arrival

Regardless of the species or variety, the production of high-quality rooted cuttings requires the same focus and attention to detail as does growing a finished crop. A cutting is only in a rooting area for three to four weeks, as compared to a two- to three-month-long finished crop, so decisions, such as how much mist to apply, may have to be adjusted sometimes several times a day as conditions change. The old axiom "No one plans to fail, they only fail to plan" has never been more true. The building blocks for producing high-quality rooted cuttings that will yield an excellent finished crop are laid in place during Stage 0. Above all, Stage 0 should focus on the requirements for strict sanitation in the propagation area. Whether your space is 200 or 200,000 $ft.^2$ (18.6 or 18,581 m^2), the propagation area must be held to very high standards for sanitation practices. These standards need to be established before the first cutting arrives and must be continuously monitored for compliance when unrooted cuttings are on the premises.

Stick order

One of the most critical steps in planning how unrooted cuttings will be handled are the decisions as

Table 14-1. General Guidelines for Vegetative Propagation (*Continued*)

CROP	AVERAGE ROOT TIME (WEEKS)	ROOTING HORMONE	MIST (DAYS)	PINCHING	COMMENTS
Impatiens, double	3–4	No	5–10	Optional	Watch for cutting stretch.
Impatiens, trailing	3–4	No	5–10	No	Watch for cutting stretch.
Ipomoea	3-4	No	5-10	No	Watch for stretch.
Lamium	3–4	No	10–13	No	
Lantana	4	Optional	5–10	Optional	Use as little mist as possible.
Lobelia	4–6	Optional	7–10	Optional	Hormone will help rooting.
Mimulus	4	Yes	5–10	Optional	Use as little mist as possible.
Monopsis	4–5	Yes	5–10	Yes	Use as little mist as possible.
Nemesia	4	Optional	5–10	Yes	Use as little mist as possible.
Nolana	4	No	5–10	No	Use as little mist as possible.
Osteospermum	4–5	Optional	7–12	Optional	Avoid wilting in prop.
Petunia	3–4	No	5–10	No	Use as little mist as possible.
Plectranthus	3–4	No	5–8	Optional	
Salvia	4–5	No	7–10	Yes	Use as little mist as possible.
Sanvitalia	3–4	No	8–10	Yes	Do not use etridiazole on it.
Scaevola	4-5	Optional	7– 12	Yes	Avoid wilting in prop.
Thunbergia	4–5	Yes	5–12	Yes	
Verbena	3–4	No	5–8	Yes	Use as little mist as possible.

to which plants need to be stuck first and which plants can be stored for sticking within the twenty-four-hour window that is appropriate for most unrooted cuttings. This stick order is critical due to the wide range of responses vegetative material will show after the stresses of shipping and storage. Table 14-2 shows the suggested order of sticking for most vegetative crops. Plants listed as group 1 are the most urgent and least advisable to store for more than one or two hours. Groups 2–4 are increasingly less likely to be damaged by storage at the recommended temperature and humidity for up to twenty-four hours and therefore can be more safely stored. It should always be the goal to stick unrooted cuttings as soon as possible, whether the plant material is shipped thousands of miles or is harvested from your own stock maintained.

Storage cooler

If cuttings are to be stored, the cooler should be given a thorough mechanical checkup to make sure the necessary temperature and humidity can be maintained. Storage temperatures can vary for different species. Geraniums can handle storage temperatures of 38–40°F (4–5°C), while *Ipomoea*, lantana, and impatiens are best stored above 48°F (9°C). If only one cooler is available, a good compromise would be to run a constant 50°F (10°C), with a relative humidity of 60–70%. The cooler must be free of any plant material or debris and disinfected prior to the first cuttings' arrival and weekly as new unrooted cuttings are expected. Likewise, any carts or racks used to store cuttings must be disinfected weekly.

Table 14-2. Sticking Priorities for Crops

CROP	STICKING PRIORITY 1 = MOST URGENT 4 = LEAST URGENT	CROP	STICKING PRIORITY 1 = MOST URGENT 4 = LEAST URGENT
Geranium, zonal	1	Dahlia	2
Poinsettia (*Euphorbia*)	1	Lobelia	2
Lantana	1	*Strobilanthes*	3
Geranium, ivy	1	*Angelonia*	3
Jamesbrittenia	1	*Argyranthemum*	3
Purslane	1	Bacopa (*Sutera*)	3
Heliotropium	1	*Bidens*	3
Thunbergia	1	*Bracteantha*	3
Calibrachoa	2	*Cuphea*	3
Coleus	2	*Helichrysum*	3
Diascia	2	Impatiens, New Guinea	3
Evolvulus	2	Impatiens, trailing	3
Impatiens, double	2	*Lamium*	3
Impatiens, mini	2	*Monopsis*	3
Verbena	2	*Sanvitalia*	3
Nemesia	2	*Scaevola*	3
Petunia	2	Portulaca	3
Salvia	2	*Gaura*	3
Snapdragon (*Antirrhinum*)	2	*Plectranthus*	3
Osteospermum	2	*Lysimachia*	4

Disinfectants

An important decision for you to make is which compound will be used throughout the area as a disinfectant. Currently, quaternary ammonium compounds (QAC) are the most reliable choice. These compounds are produced under several brand names and should be readily available from most chemical distributors. Quaternary ammonium compounds are commonly used for disinfecting almost everything within the greenhouse. They are superior to household bleach for many reasons. Foremost, bleach is not labeled for agricultural use, while QACs are. Bleach tends to have a very short useable life in the greenhouse, especially when it becomes contaminated with organic matter such as soil. QACs are stable much longer under all conditions in the greenhouse. Also, most QACs are less irritating to greenhouse employees' hands than other disinfectants.

There may be instances in which a different type of disinfectant is warranted. While QACs are good for bacteria and make good all around disinfectants, some spring plants are more susceptible to viruses that QACs may not be as good at preventing. When handling certain spring crops, such as calibrachoa, petunia, lobelia, and verbena, a disinfectant that prevents viruses from spreading may be a better alternative than a QAC. Some of these include nonfat dry milk, trisodium phosphate (TSP), or disinfectants containing potassium peroxymonosulfate.

Regardless of the disinfectant chosen, you should plan to treat all areas where cuttings will be handled prior to the first crop, as well as after each crop passes through the greenhouse during the season. Areas to be treated should include the cooler, the area for initial handling and processing, and the evaporative pad system. If possible, entrances into the immediate propagation area should be equipped with a footbath (2" [5 cm] depth) filled daily with a QAC. The traffic flow within the propagation area should be set up so that anyone entering this area would automatically pass through the footbath. This practice can serve as an excellent barrier to the introduction of pathogens into the area. All equipment should also be disinfected before and between all crops. Benches, trays, walkways, carts, and any container used to hold or transport the cuttings must be treated. There should be an area for employees to wash their hands prior to handling the cuttings. This area should include a scrub brush for nails and be equipped with hot water.

All propagation areas should be free of weeds to reduce the opportunity for insects to spread potentially devastating viruses. Breeding/production companies spend countless hours and dollars protecting stock plants from pathogens such as impatiens necrotic spot virus (INSV) and tobacco mosaic virus by eliminating insect populations. You should address these and other pathogens with the same sense of urgency.

Pest and disease control

You must ensure there is sufficient drainage so that no standing water will be present in the propagation area. Standing water and the algae that will quickly follow are breeding grounds for fungus gnats. Given the conditions that will exist in a typical mist area (i.e., high humidity, warm temperatures, and a plentiful supply of organic matter), fungus gnat populations will overwhelm even the most skillful growers once they are allowed to swell. Stop the problem before it ever begins, and make sure no freestanding surface water is present.

You should make one person responsible for the regularly scouting the propagation area for pests and disease. This person should be trained to spot the most common insects and pathogens for all crops in the propagation area. This person should be able to make appropriate spray recommendations as the need arises.

Environment

Environmental monitoring systems are critical in the propagation area. Heating, cooling, and shading systems must all be checked prior to the arrival of the first cuttings. Likewise, the mist and lighting systems must be functioning correctly. Unrooted cuttings are far less forgiving than finished material when environmental conditions stray from the ideal. Cuttings in a propagation area can be severely delayed or lost completely with a relatively short lapse in environmental system function. If a computerized system is used, this component must be regularly checked.

Benches

The benching system in the greenhouse is equally important to the success of rooted cutting production. Maintaining a clean, flat bench surface provides the best opportunity for uniform water and heat distribution to the tray.

Media

There are many growing media choices available. Technical support is available from all the breeder/

producers as well as distribution companies when trying to make these decisions. Rely on the expertise of these resources for advice regarding tray size and media. Synthetic and soilless media can all work, given the right cultural practices. Likewise, tray size and configuration can vary as much as media composition. The most common cell densities range from 105 cuttings per tray down to 84 cuttings per tray. Any configuration within this range will provide adequate space for growth and air movement while maximizing each square foot of propagation space. You may wish to explore different tray densities for different uses during the season. Some growers are using smaller cell sizes (105s) during the early season when space is at a premium. As the season winds down and propagation space frees up, a "quick crop" can be scheduled using multiple cuttings in a larger cell size (38s or 50s). Cuttings in the larger tray sizes may spend an additional two to three weeks in the tray as compared to those in a 105 tray. This additional time allows at least one pinch and reduces the amount of time required in the finished container.

Stage 1: Arrival to sticking

Stage 1 allows growers to put the planning done in Stage 0 into action. Regardless of whether the cuttings

Acclimatizing Tissue-Cultured Plantlets

Gary Hennen

Many crops today are offered as tissue-cultured plantlets in Stages II or III*. This form of vegetative propagation offers many advantages, primarily in that plants grown from tissue culture are disease free and quick growing. Tissue-cultured plantlets require special attention during the acclimation process from the laboratory to the greenhouse in order to avoid losses and ensure crop uniformity. As the laboratory is a relatively stress-free environment, the plants need to be slowly phased into greenhouse conditions.

Incoming Stage II microcuttings need relatively immediate attention, while Stage III plants can be held in a controlled environment for a few days without diminishing their quality. As plants are coming from a relatively sterile environment, worker and workplace sanitation is essential. Disinfect all tools, containers, and work areas routinely.

Most growers prefer to plant their tissue-cultured microcuttings into a peat-based medium. Use 70–80% quality the milled peat amended with perlite, vermiculite, or bark. Avoid coarse, large particles if using bark. Adjust the pH to 5.8–6.1 with incorporated dolomitic limestone. Incorporate a low concentration of slow-release fertilizer including trace elements. Moisten medium before filling trays.

Keep in mind that plants in the laboratory are grown under low light, only 200–600 f.c. (2.2–6.5 klux), and are kept very close to a uniform 78°F (25°C). Thus, while planting and during acclimation, protect the plants from the brighter greenhouse environment and extreme temperatures. Stage III plants are rooted in a lab medium containing high concentrations of sugar. Gently wash or shake off medium to prevent the growth of undesirable microorganisms once planted. Use a dibble when planting. Forceps with felt padding can be used to aid in planting very small microcuttings. Place plantlets just deep enough to anchor them into the medium. Use a spray bottle of water to mist the plantlets during handling. When a tray is completely planted, water them in thoroughly but gently with a mild solution of 20-20-20 (or similar analysis) at 100–150 ppm.

Plantlets directly from the lab have no cuticle and will require high humidity to survive. Use a mist or fog system to prevent desiccation. Increase the mist interval gradually during acclimation. An alternative is to cover the trays with clear lids or a layer of polyethylene film or cheesecloth. Be sure to provide ventilation, as necessary, to prevent heat buildup. Additional shading should be provided. Double the light levels every one to two weeks until plants are acclimated to their normal light requirements.

Do not allow the medium to become waterlogged. A regular fungicide program should prevent losses from root and stem rot diseases. The initial environment should be kept at 80–85°F (26–30°C) days and 70–75°F (21–24°C) nights. Bottom heat will help induce rooting.

* Tissue culture, like plug production and vegetative cutting production, refers to various stages in plant development during the propagation process. The stages for tissue culture do not correlate to the various stages for plug production or rooted cutting production. When a grower purchases plants from tissue culture, the supplier will offer the plants as Stage II or Stage III microcuttings.

are to be stuck immediately or placed in a cooler for sticking later, all boxes should be opened upon receipt and the cuttings examined closely. The cuttings should be turgid with a minimal amount of wilting. The cuttings should be a healthy shade of green, with an appropriate stem length and caliper as well as leaf number and leaf size. Cuttings should also be clearly labeled by variety tags and the number of cuttings per bag or shipping container. After the initial quality assessment is completed, cuttings should be checked against the packing slip and the original order for proper order fulfillment. If there are problems with any aspect of the order, notify the broker immediately. If notification is timely, replacement cuttings may be ordered for the following week with minimum disruption to scheduling.

If the cuttings are not going to be stuck immediately, the open boxes should be placed in the disinfected cooler for no more than twenty-four hours at 50°F (10°C) and 60–70% (minimum) relative humidity until sticking. If the cuttings are not fully turgid when the boxes are opened, they can be placed in the cooler overnight as a means to rehydrate and reestablish the integrity of the cuttings. Remember, the less time cuttings are stored, the greater the chances are for success of the finished crop.

If the cuttings can be stuck immediately and no storage is required, take the boxes to the propagation area once the quality and count check-in is complete. Use the arrived-upon "stick order" from Stage 0 to help determine when to take material from the staging/sorting area to the greenhouse. Once the cuttings are taken to the greenhouse, temperatures inside the boxes and bags will rise rapidly—even with moderate light levels—so be prudent with how much is taken at one time. At this point the chosen medium should be on the bench, watered in, and ready to receive the cuttings.

Turn on the mist system once sticking begins to reduce water stress and start to reestablish turgor in the cuttings. The frequency of mist applications will vary widely depending on the crop, local weather conditions, and the environment within the greenhouse. Remember that the goal during Stage 1 is to reduce the stress experienced by the cuttings and allow them to become firm and full of water as quickly as possible. Cuttings should be completely turgid by the following day. This should be accomplished with the least amount of watering to minimize disease problems in the coming days. Any wilting during the first three to five days will delay rooting or cause the rooting to be uneven. Inspect cuttings frequently and adjust the mist as needed. The cuttings should be turgid but not soaking wet. The same goes for the media. It should be moist, but not a puddle of water. If a rooting hormone is being used (see the Should I Use a Rooting Hormone? sidebar), apply it just prior to sticking. Place the cutting in the medium at a depth where the lowest set of leaves is not substantially covered but the cutting will be held upright after sticking is complete.

Any tags that were shipped with the cuttings from the breeder/producer must be stuck with the correct cuttings. Should a problem arise with the cuttings, the information on these tags will be extremely helpful in diagnosing the situation and arriving at the best course of action. It is strongly recommended that when a variety is completed, employees disinfect their hands with a quaternary ammonium compound, or one of the other disinfectants listed in Stage 0, before starting the next variety. This is an excellent way to reduce the spread of many pathogens, should they be present. A small bowl filled with the disinfectant or a spray bottle will work fine. The small amount of time required for this procedure is an inexpensive insurance policy. Disposable latex or vinyl gloves are another excellent way to limit the spread of many pathogens that might be present on the unrooted cuttings when they arrive. Employees should disinfect disposable gloves as outlined above.

Like the mist, the bottom heating system should already be operating and the substrate temperature should be within the range of 70–74°F (21–23°C) for most crops. Make use of technical support from breeders, producers, and field marketers to determine if a mix temperature outside this range is needed for a specific crop. The mix temperature should be monitored constantly for uniform distribution of heat across the bench and consistency throughout the day. In many areas of the country, growers should use tempered water during winter months to reduce the significant cooling effect cold water (mist) can have on the medium. If very cold water (mist) is applied, the fluctuations in substrate temperature can easily add days and weeks to rooting times. Water for the mist system should be heated to 85–90°F (29–32°C) before application.

During Stage 1, the air temperature in the propagation area should be kept no lower than what is

Should I Use a Rooting Hormone?

Any growers who have rooted even the smallest quantities of specialty garden plants have asked themselves this question as they stand in front of the trays with the cuttings in hand. In general, improvements in stock plant maintenance and cutting production have eliminated the routine use of rooting hormones. The added cost of application is usually not offset by the degree of improved rooting. However, there are several situations in the average greenhouse when a grower could justify the use of rooting hormones.

1. Substrate or air temperatures are routinely below optimum.
2. Mist coverage is slightly uneven.
3. Delivery of the cuttings to the greenhouse has been moderately delayed.
4. Light levels are greatly reduced during the rooting process (i.e., cloudy, short days).
5. Varieties are prone to slower, less even rooting even under ideal conditions.

The last item is perhaps the most important. With the public's insatiable appetite for "new and innovative" plant material, the likelihood of introducing varieties that perform differently than standard varieties is high. Using a rooting hormone may provide a bit of an insurance policy in dealing with unknown production factors for just such crops. Again, use the technical support available from the sales representative or the breeder/producer.

There are several rooting hormones available, all of which can be used successfully in the greenhouse. There are also a number of ways in which the hormones can be applied, with advantages and disadvantages to each. Some rooting hormones are used in the powder form, in which the base of the cutting is dipped before sticking. This method might be more time consuming, but the chance of spreading a disease is lessened. There are also water-soluble formulations and concentrated liquids that can be used as a liquid in which to dip the basal end of a cutting. The water-soluble forms can also be used as an overhead spray once the cuttings have been stuck. The parts per million (ppm) of IBA vary with this method, with rates starting at 50 ppm and going as high as 400 ppm. As with all chemicals, the label should be read carefully and followed closely as to rate and application method.

PRODUCT	ACTIVE INGREDIENT(S)	FORMULATION
Hormex No. 1	IBA (0.1%)	Powder
Hormex No. 3	IBA (0.3%)	Powder
Hormex No. 8	IBA (0.8%)	Powder
Hormex No. 16	IBA (1.6%)	Powder
Hormex No. 30	IBA (3.0%)	Powder
Hormex No. 45	IBA (4.5%)	Powder
Hormodin No. 1	IBA (0.1%)	Powder
Hormodin No. 2	IBA (0.3%)	Powder
Hormodin No. 3	IBA (0.8%)	Powder
Rhizopon AA #1	IBA (0.1%)	Powder
Rhizopon AA #2	IBA (0.3%)	Powder
Rhizopon AA #3	IBA (0.8%)	Powder
Rhizopon Water-Soluble Tablets	IBA (20%)	Tablets
Hortus IBA Water-Soluble Salts	IBA (20%)	Salts
Stim-Root #1	IBA (0.1%)	Powder
Stim-Root #2	IBA (0.4%)	Powder
Dip 'N Grow	IBA (1.0%), NAA (0.5%)	Liquid concentrate

required to keep substrate temperatures optimum and light levels at approximately 1,400–1,800 f.c. (15–19 klux). Again, this will vary somewhat with each crop, but these are starting points. As with water management, the goal is to eliminate stress on the cutting and regain turgor by reducing light levels and moderating the temperature. Do not apply fertilizer during this period. A small starter charge in the medium is acceptable, but the EC of the medium should be below 0.75–0.80 mS/cm (2:1 method). Some growers choose to apply a foliar feed during Stages 1 and 2 using primarily calcium nitrate. The usefulness of this practice for a broad range of plant types has not been proven. Timely handling/processing after arrival, applying the correct amount of water, and moving rapidly from Stage 1 to 2 will reduce the opportunity for cuttings to develop yellow foliage from leaching.

The final step in Stage 1 is to survey all aspects of propagation environment to make sure the conditions are in place to meet the needs of the cuttings and begin the callusing phase immediately. Mist coverage and frequency, soil temperature, light levels, and correct depth of sticking are all aspects that should be reviewed.

Stage 2: Callusing

Once cuttings are stuck and the environment has been set to match the needs of the unrooted cuttings, Stage 2 begins. Stage 2 is, in many ways, more passive and involves primarily monitoring all the environmental factors (i.e., light, temperature, and moisture), the health of the cuttings, and the possibility of pathogens (primarily botrytis). Two to three days after sticking is a good time to get a preventative spray for botrytis on the cuttings. The use of a nonionic surfactant, or wetting agent, in conjunction with the botrytis spray will help distribute the water more evenly on the leaf and allow for better absorption. (Because nonionic surfactants help plants better absorb water, they can also be used alone to help cuttings regain turgor and may lessen the misting requirements.) If needed, another application spray can be used seven to ten days after the first. Be sure to alternate chemicals and read labels before using.

Moisture management, more than any other factor, should be the No. 1 focus. In Stage 2, the object is to maintain near 100% relative humidity in the medium at the cut end of the stem without having constantly waterlogged medium. You should be able to squeeze water from the medium during Stage 2. The mist applied at nighttime will be drastically reduced or eliminated during this stage, while the daytime mist is slowly being reduced. Cuttings should be turgid first thing in the morning, with just enough mist applied during the day to prevent them from wilting. These conditions, coupled with the warm soil temperatures recommended in Stage 1, are ideal for rapid, uniform callus development and root initiation.

Callusing will actually begin immediately after the cutting is harvested, when a "plate" of dead cells forms over the cut end to reduce water loss and somewhat protect against pathogens. Most cuttings will stop the process at that point until they are placed under favorable environmental conditions. Once those conditions are present in the propagation area, callus formation and root initiation will rapidly progress. Depending on the species, the next sign of callusing will be a swelling of the tissue immediately up the stem and behind the plate formed at harvest. In some plants, such as zonal geranium, this swelling will be accompanied by a change in color to light green or white of this same small ring of tissue. Finally, the tissue will begin to crack and split open, and a rough, light tan to white area of tissue will develop. This is the callus tissue.

The appearance, color, and size of the callus will vary widely; in some species it will be difficult to see without a hand lens. Likewise, the time required for callus development will vary greatly. The point is that callus formation and root initiation are two different processes. In most cases, roots do not form from callus tissue. However, callus formation is an excellent indicator that all the correct environmental conditions are present for root formation. Callusing and root initiation are dependent on the same environmental conditions (i.e., warm temperatures and high humidity) and are a very positive sign of forward progress for the cuttings.

After callusing, root initials will be visible growing through the tissue. In some plants, such as coleus, root initials will also form along the stem well above the cut end. These are called preformed root initials, and they are present when the cutting is harvested. They only become "activated" after the cutting is harvested and put into favorable environmental conditions. Again, this stage is primarily about making sure all the needed factors are accounted for in the propagation environment.

Stage 2 light levels should be kept below 1,800–2,000 f.c. (19–22 klux) to reduce the stress on the cuttings. Substrate temperatures should be maintained in the range of 70–74°F (21–23°C), again with near 100% relative humidity in the medium. No fertilizer should be applied at any time during Stage 2, due to the absence of roots for nutrient uptake. Once roots are visible and the cutting begins to uptake water, cuttings will experience dramatic and daily changes. When this occurs, return to more active management of the plants and environment.

Stage 3: Root development

As cuttings leave Stage 2 and enter Stage 3, roots are visible, the cuttings are starting to show signs of approaching new top growth, and wilting is less frequent, even in the warmer parts of the day. Begin to aggressively manage all aspects of the environment to maximize root development and minimize excessive top growth.

The primary tool to manage this growth will be water management. Once roots are visible, begin to look for ways to apply less water and dry the medium down to promote root growth. As the cuttings begin to uptake water via their forming root systems, less mist will be needed. Adjust the starting and stopping time for mist, the duration of the "on" cycle, and increase the time between "on" cycles as the cuttings are slowly removed entirely from mist. Free water should only be present in the medium immediately after watering or fertilizer application. At all other times, the medium should be significantly drier than Stage 2, with only small amounts of water being removed from the medium when squeezed in the hand. This push toward drying the medium also allows for more frequent fertilizer applications.

At the first sign of rooting, start applying fertilizer at 100 ppm from a source low in phosphorous and ammoniacal nitrogen. Formulations such as 15-0-15, 15-5-15, or 13-2-13 are ideal, with nearly all of the nitrogen in the nitrate form. As rapidly as possible, levels should be increased to 150 ppm nitrogen with fresh water applied often enough to maintain an EC of 0.8–1.1 mS/cm.

Properly controlling light levels is another critical management tool during this period. Intensity can be increased to a range of 1,800–2,500 f.c. (19–27 klux), with shade applied in the hottest part of the day only when absolutely necessary. If possible, soil temperatures can be dropped to reduce unwanted stretch. An ideal soil temperature is 66–70°F (19–21°C) for most of Stage 3. Air temperatures can be dropped to reduce unwanted stem growth and perhaps improve stem caliper. The range for air temperature during Stage 3 should be 60–65°F (15–18°C), if this degree of environmental control is available. Apply a protective fungicide application as the cuttings come off mist to prevent the development of botrytis.

This is the first opportunity for growth regulators to be applied to the cuttings. Only the most vigorous species or varieties require growth regulators this early. In many cases, because of the nature of the variety and the skillful management of the environment, little if any growth regulators may be needed at any time while in propagation. Managing growth without growth regulators should be the goal of all growers.

Stage 4: Toning the rooted cuttings

The transition from Stage 3 to Stage 4 is as much a change in focus as a noticeable change in the rooted cuttings. For the remainder of the time the crop is in propagation, the primary focus is preparing the plant for the next step and accompanying environment. Getting the cuttings ready for transplanting and placing them into the general production area of the greenhouse is the goal. At the same time, root growth must continue so the root system fills out the cell and creates a tight root-ball that will survive the rigors of transplanting and sustain the plant immediately afterwards. As has been the case throughout the previous stages, use all environmental factors available to tone the cuttings.

Water management will continue to be the best method for managing growth and toning. The medium should be allowed to dry more between watering as you push the developing root system and toughen the plant with more noticeable water stress. However, extremes are not the goal, even during Stage 4. The cuttings should not experience even short periods of dramatic wilt. Rather, the medium should be allowed to dry to the point of being light brown with no free water present. Soil at this stage will crumble apart when squeezed. The medium should not be allowed to dry to the point of becoming tan to white and shrinking away from the walls of the cell.

The fertilization program should continue as another means of toning. The same formulations described earlier for Stage 3 should continue. Apply fresh water after every other fertilization, although this will vary depending on the frequency of fertilization. In areas where winter days are short and dark, rooted cuttings may only be fertilized once per week, in which case there will be little opportunity to apply fresh water. In situations like this, always try to apply enough fertilizer solution to ensure 10–15% leaching from the medium; this will lessen the likelihood of salts buildup. Substrate EC should be targeted for a range of 0.9–1.2 mS/cm. Supplemental iron, in the form of iron chelate, may be beneficial on certain crops at this time. Those crops may include calibrachoa, petunia, or verbena. A rate of 2 oz./100 gal. (1 ml/6.4 l) of water is a good starting point. Be sure to wash off foliage to prevent burning.

The light level in the propagation area should be raised, but not above a maximum of 2,500–4,000 f.c. (27–43 klux), depending on the crop. This will serve three purposes for the developing plants. First, branching will be maximized, which will lead to a greater number of flowers on the finished plant. Second, the higher light levels will shorten internodes and lessen stem stretch. Third, the leaves will become much thicker with a more protective cuticle layer. This will be of great benefit as the rooted cuttings are removed from the protected environment of propagation and into production greenhouses.

With no mist being applied, the relative humidity will fluctuate during the day and closely mirror that of the production greenhouses. This will further improve the tone of the plants.

The substrate temperature during Stage 4 should be kept at 64–66°F (18–19°C) to allow the root system to continue to develop, although at a slightly slower pace than in Stage 3. If the greenhouse is set up in such a way as to allow for each crop to be compartmentalized, rooted cuttings in Stage 4 can be kept at an air temperature of 58–62°F (14–16°C). This may seem excessively low, but it will serve as an ideal plant growth regulator. Stems will be thickened, leaf expansion will be kept in check, and stem stretch will be eliminated. Rooted cuttings grown under this regime will be a lighter green than those grown warmer. This condition is temporary, as plants will regain the appropriate darker green color once transplanted. If the propagation area does not allow for zones to be kept at such different temperatures, aim for air temperatures of 60–65°F (16–18°C), as in Stage 3.

If all the environmental factors are managed throughout the crop in such a way as to minimize unwanted top growth and eliminate stem stretch, many crops will require no chemical plant growth regulators while in propagation. Should growth regulators be needed, use them late in Stage 3 or anytime in Stage 4. As with all chemicals in the greenhouse, read and follow all label recommendations. Applying growth regulators to rooted cuttings presents many of the same challenges, as does application of these chemicals to plugs. Set up your own trials of not only the growth regulators themselves but also the rates and volumes applied, before making applications to large areas of your crops. Table 14-3 provides a basis from which you can develop your own trials.

Aside from chemical growth regulators, pinching is another method used to control height and help fill out plants. Table 14-1 highlights the crops that generally receive a pinch to remove the apical dominance. Manual pinching can be done with scissors, or machines, but care must be used to prevent transmitting any disease. Disinfecting tools and hands between trays or varieties is important. As mentioned earlier, there are a number of disinfectants available, so choose which one works best for the particular situation. Ethephon (Florel) is also an option in some instances to promote branching. Calibrachoa, petunia, and verbena are a few crops that may benefit from a ethephon application, especially when using larger plug sizes. Be sure to trial different rates and flowering responses, as some delay in flowering may occur. In general, 150–200 ppm is a good starting rate in propagation.

US Plant Patents and Legal Issues

Plant breeders protect their new vegetatively (asexually) propagated varieties in the United States with US Plant Patents and in other countries with Plant Breeder's Rights.

This section outlines the laws and issues surrounding intellectual property protection in the United States. The plant patent allows the plant breeder or patent holder (assignee) to exclude others from using their new varieties in any way. Therefore, the plant breeder, assignee, or their legal representative must license growers to allow them to propagate a protected variety by making stock plants and taking cuttings from those stock plants. In other words, it is illegal to propagate patented varieties without a license agreement issued by the plant breeder, assignee, or their legal representative. Increasingly, plant breeders and assignees are pursuing violators. If you are propagating plants at your greenhouse, it is your responsibility to know which varieties are protected by plant patents and to be sure that you have permission from the plant breeder, assignee, or their legal representative before you make stock plants and take cuttings.

Intellectual property in the form of new varieties of plants, plant phenotypes, and plant propagation methods can be protected in the United States by one or more of the following:

- US Plant Patents for new varieties of vegetatively (or asexually) reproduced plants, that is, propagated by cuttings, tissue culture, grafting, divisions, etc.;
- US Utility Patents for new varieties of vegetatively or seed-reproduced plants, phenotypes, or propagation processes;
- US Plant Variety Protection for new varieties of seed-reproduced plants; and

Table 14-3. Plant Growth Regulators for Vegetatively Propagated Crops

CROP	GROWTH REGULATOR	RATE(S) PPM	APPLICATION METHOD	TIMING (DAYS AFTER STICK)
Angelonia	B-Nine/Cycocel tank mix	1,500–2,500 B-Nine/750–1,000 Cycocel	Spray	14–28
Snapdragon (*Antirrhinum*)	B-Nine	1,500–2,500	Spray	14–21
	Bonzi	10–30	Spray	14–21
Argyranthemum	B-Nine	1,500–2,500	Spray	14–24
	Bonzi	5–10	Spray	14–24
Bidens	B-Nine	2,500–3,000	Spray	21–30
Bracteantha	Bonzi	5–10	Spray	21–30
Calibrachoa	B-Nine	2,500–3,000	Spray	21–35
	Bonzi	5–30	Spray	21–30
Coleus	B-Nine/Cycocel tank mix	1,500–2,500 B-Nine/750 Cycocel	Spray	10–21
Dahlia	B-Nine	2,500–3,500	Spray	10–14
Diascia	B-Nine	1,500–2,500	Spray	14–24
Geranium, ivy	B-Nine/Cycocel tank mix	1,500–2,500 B-Nine/750 Cycocel	Spray	18–24
Geranium, zonal	B-Nine/Cycocel tank mix	1,500–2,500 B-Nine/750–1,000 Cycocel	Spray	18–24
Impatiens, New Guinea	Bonzi	1–5	Spray	14–21
Impatiens, double	Bonzi	3–10	Spray	10–15
Impatiens, trailing	Bonzi	3–15	Spray	7–21
Lobelia	B-Nine	1,500–2,500	Spray	14–25
Nemesia	B-Nine	1,500–2,500	Spray	14–24
Osteospermum	B-Nine	1,500–2,500	Spray	10–14
	Sumagic	3–5	Spray	10–14
Petunia	Bonzi	2–4	Drench	15–24
	B-Nine	2,500	Spray	15–24
Sanvitalia	B-Nine	2,500–3,000	Spray	20–30
Verbena	B-Nine	1,500–2,500	Spray	15–20

- Trade secrets for new varieties or propagation processes.

The most popular type of protection for asexually reproduced plants is the US Plant Patent. US Plant Patents have been available since the enactment of the Plant Patent Act of 1930. According to the US Congressional Committee's report, "The purpose of the bill is to afford agriculture the same opportunity to participate in the benefits of the patent system as has been given industry and thus assist in placing agriculture on a basis of economic equality with industry. . . . It is hoped that the bill will afford a sound basis for investing capital in plant breeding and consequently stimulate plant development through private funds" (Plant Patent Act of 1930 enacted on June 17, 1920, as Title III of the Hawley-Smoot Tariff, chapter 497, 46 Stat. 703 [54H.R. Rep. No. 71-1129 (1930); S. Rep. No. 71-315 (1930)]. The bill was strongly supported by Luther Burbank and Thomas A. Edison. Edison testified at the congressional hearing, "Nothing that Congress could do to help farming would be of greater value and permanence than to give the plant breeder the same status as the mechanical and chemical inventors now have through the patent law. There are but few plant breeders. This [bill] will, I feel sure, give us many Burbanks" (US House of Representatives Committee on Patents, Report to Accompany H.R. 11372. Report No. 1129, April 30, 1930).

And indeed, it has done just that! Providing patent protection for plant breeders' new varieties has greatly stimulated plant breeding and development efforts, and the plethora of new superior varieties of plants introduced into the US market in recent years has confirmed Edison's prediction.

Plant patents protect the vast majority of all new vegetatively reproduced varieties of flowering plants, annuals, perennials, shrubs, trees, and foliage plants grown and sold in the United States. Plant patent applications are applied for in the name of an individual or individuals, and the applicant is referred to as an inventor. For new plant varieties, the inventor may be a plant breeder or a discoverer of a newly found seedling from a cross-, self- or open-pollination, a naturally occurring sport, or an induced mutation. The inventor or inventors may hold the rights to the patent in his, her, or their names, or the inventor or inventors may transfer, or assign, a portion of or the entire rights of the patent to another individual, entity, or a company, which is then referred to as the assignee. The rights can be further transferred or assigned provided that the original assignee agrees to the reassignment.

What is a US Plant Patent application?

A US Plant Patent application is essentially a "legal case" that is brought before the US Patent Office and prosecuted on its merits during the examination of the "case." To receive a granted plant patent, the inventor must demonstrate that the new variety is new and distinctly different from its antecedents and known related varieties in at least one distinguishing characteristic. A distinguishing characteristic might be plant size, flower color, flower form, or leaf color, for example. The inventor must declare that she is the original and sole inventor of the new variety. In addition, the inventor must declare that the new variety has been discovered in a cultivated area, such as a greenhouse or nursery. The requirement of discovery of the new variety in a cultivated area excludes the possibility of patenting the discovery of a wild species as it is not possible to patent plants that freely occur "in nature." Further, the inventor must declare that the new variety has been reproduced asexually and that the unique characteristics of the new variety are uniform and stable over successive generations of asexual reproduction.

In addition to the aforementioned declarations, the plant patent application must also include a mostly visual (phenotypic) detailed botanical description of the new variety, which clearly describes the new variety and differentiates it from its antecedents and known related varieties; the location and method by which the new variety has been asexually reproduced; and colored photographs of the new variety that accurately depict its unique characteristics. "Genetic fingerprinting" techniques may also be used to further describe the plant variety's unique characteristics; however, these techniques are usually more useful in the event of infringement and are typically not included in the plant patent application.

A plant patent application may be barred from becoming a granted US Plant Patent if the new variety has been sold, "offered for sale," shown at a tradeshow, appeared in a printed publication, or appeared on a web-page for more than one year in the United States prior to filing the application.

How long does the patent last?

For US Plant Patents filed prior to June 8, 1995, "the term of (the) patent is the longer of 17 years from the date of the grant of (the) patent or 20 years from the earliest effective filing date of the application subject to any statutory extension." For US Plant Patents filed on or after June 8, 1995, "the term of (the) patent is 20 years from the U.S. filing date, subject to any statutory extension" [General Agreement of Tariffs and Trade (GATT) Uruguay Round (PL 103-465), June 8, 1995]. However, applications that refer to previously filed applications or claiming priority will have a term of twenty years from the earliest filing date. After the plant patent has expired, the subject matter of the patent becomes public domain, which means that the plant becomes a "free variety," and it can be asexually reproduced, used, and/or sold without permission from the plant breeder or assignee.

Inventors typically procure the assistance of a federally registered patent agent or attorney to assist them with the preparation and prosecution of their patent applications.

What rights does the patent grant?

The granting of a US Plant Patent provides the inventor and/or assignee with "the right to exclude others from asexually reproducing the plant, and from using, offering for sale, or selling the plant so reproduced, or any of its parts, throughout the United States of America or from importing the asexually reproduced plant, or any parts thereof, into the United States of America" without the inventor's and/or assignee's permission [Plant Protection Act of 1930, 35 U.S.C. 163 Grant (amended 1998)]. The inventor, assignee, or their legal representative grants permission to others to propagate, use, offer for sale, and/or sell the new plant via license agreements. License agreements are legal contracts and are governed under standard contract law. Professional growers may enter into license agreements directly with the plant breeder, with the assignee, and/or with the inventor's or assignee's agent or legal representative. License agreements allow the licensee "usage" of the new varieties, so license agreements can be viewed as rental agreements. License agreements permit the licensee to rent the rights to propagate, use, and/or sell the new variety during the term and under the conditions of the agreement; and the "rent" paid by the licensee is the royalty.

The terms and conditions of license agreements vary with the actual variety, inventor or assignee, type of propagation method, and other considerations. However, all license agreements should specify the location where the variety can be reproduced, used, and sold; the types, sizes, and forms of the products to be produced; the term of the contract; the royalty rate; provisions for royalty enforcement, administration, and collection; labeling requirements; and conditions under which the contract can be terminated.

Plant breeding is similar to other types of businesses with expenses such as employees, equipment, and facilities. However, unlike other types of businesses, plant breeding requires a much longer investment of time and effort to produce its product: a new variety. For breeding of ornamental crops, it may take three to five years from the initial crossing of the parent varieties to commercialization of a new variety. During this time, plant breeders are not being compensated for their expenses and efforts. When the new variety is eventually introduced and commercialized, plant breeders finally begin to get paid for their years of work in the form of royalties. Compared to other types of businesses, plant breeding is considered a high risk because of the long-term commitment required to produce new varieties and their potentially short life, as a competing plant breeder's new variety has the potential to make the plant breeder's brand new variety "yesterday's news."

It is unlikely that plant breeders would risk this investment of resources and time if there were no provision to collect royalties. Royalties are essential for the continuation of successful plant breeding programs and the development of new plant breeding programs; royalties could be considered the lifeblood of our industry.

Who owns sports or mutations?

Commercially viable spontaneous or induced mutations of patented plant varieties may occur. Ownership of such mutations, or sports, is usually considered in license agreements. Licensees who discover such sports should promptly contact the owner of the parent variety regarding ownership considerations.

Market identity

It is important for the inventor and/or assignee to physically identify, that is, to provide proper marking, of the new plant with its variety name and patent number to afford protection under the US Patent Laws, and it is important for license agreements to require proper marking by the licensees. The variety name and patent number should be clearly displayed on all tags, containers, catalogs, invoices, websites, and other literature pertaining to the new variety. While there is no penalty for the patent owner to not provide or require proper identification, not properly identifying the plants could adversely affect the right to damages in the event of third-party infringement without knowledge of the patent. Properly identifying the new plant with the variety name and patent number presumes such knowledge.

Patent infringement

Propagation of a patented plant for any purpose without permission from the patent owner is illegal and an infringement of the plant patent owner's rights. In addition, offering for sale, selling, or using plants or plant parts derived from illegal propagation is also considered infringement. Proving infringement is presently the plant patent owner's burden; however, it is the professional grower's obligation to respect a plant patent owner's property. Nonpayment or underpayment of royalties by the licensees is also illegal and is theft.

Professional growers should be familiar with what may constitute infringement. Infringement may be defined as follows (adapted from the Plant Variety Protection Act of 1970, described in Public Law 91-577):

- asexual reproduction of the new variety;
- selling, offering for sale, exposing for sale, delivering, shipping, consigning, exchanging, soliciting an offer to buy, or any other transfer of title or possession of plants of the new variety;
- importation or exportation of the new variety into or from the United States;
- dispensing the new variety to another, in a form which can be asexually reproduced, without notice as to being a protected variety; or
- performing, instigating, or actively inducing performance of any of the aforementioned acts.

Granted US Plant Patents are listed on the US Patent and Trademark Office's website at www.uspto.gov. Information regarding pending plant patent applications is not publicly available. Therefore, to avoid potential infringement, it is highly advisable for professional growers to directly contact various plant breeding companies, marketing companies, sales agents, brokers, and patent owner's agents and representatives to determine if any new varieties are in the process of being protected and if license agreements are available before they contemplate propagating, using, or selling new varieties.

15

Jan VanDerMeij and Colleen Warfield with additional information provided by Mike Klopmeyer

Indexing for Disease

Vegetatively propagated floricultural crops are under constant threat of attack and carry the risk of spreading diseases caused by fungal, bacterial, and viral pathogens. The only effective control measure to prevent the introduction and/or spread of these diseases is to begin the production process with clean, certified, disease-indexed stock. What is clean stock? This section is devoted to explaining the processes required for producing disease-free plants.

What Is Disease Indexing?

Disease indexing, simply stated, is testing various plant parts (i.e., leaves, roots, and stems) for the presence of plant pathogens. If plant pathogenic fungi, bacteria, viruses, or viroids are present in the plant part tested, that plant is either destroyed or subjected to various procedures designed to eliminate the pathogen. Only plants in which pathogens are not detected are considered to be disease free, and these plants are subsequently increased in large numbers under strict sanitary conditions.

It is important to understand that although these plants index free of all targeted plant pathogens, they are still susceptible to them and can become infected if exposed. Thus, pathogen-free production procedures—growing mother stock in isolated greenhouses with clean substrate, pots, and benches and replacing this mother stock on at least an annual basis—are essential to successful production.

How Is Disease Indexing Done?

Indexing for bacterial and fungal pathogens

The first step in disease indexing is to test the variety or clone for the presence of bacterial and fungal pathogens. A very reliable method is to use culture indexing to select cuttings free of culturable, systemic bacteria and fungi. Systemic pathogens can exist in the water- and food-conducting tissues of a plant. Culture indexing is carefully conducted in the laboratory using cuttings removed from mother stock plants grown in an isolated "incubation" greenhouse zone. This incubation greenhouse is maintained at temperatures considered optimal for disease development. Surface-sterilized stem sections from each cutting are placed into a nutrient solution to encourage fungal and bacterial growth. If bacterial or fungal growth occurs in these solutions, the corresponding cutting is destroyed. Only cuttings that exhibit no bacterial or fungal growth in the nutrient solutions are kept and become the new mother stock. Stock plants are renewed three to four times during a twelve-month period, whereby cuttings are removed and culture-indexed at each sampling date. The advantage of this process is that culturable, vascular-inhabiting bacteria and fungi (both pathogenic and nonpathogenic) can be detected early and the infected stock plant can be eliminated. This culture-indexing process is very effective in detecting and eliminating plants that are infected with *Xanthomonas campestris* pv. *pelargonii*, the causal agent of bacterial blight of geranium, and *Ralstonia solanacearum*, the causal agent of southern bacterial wilt. This same technique can also be used in indexing plants maintained in tissue culture to ensure they are free of culturable fungal and bacterial contaminants.

Virus indexing

Once plants are certified free of culturable fungal and bacterial pathogens, they are ready for virus indexing Virus indexing is a term used for thoroughly testing the plant material for both known and unknown plant viruses. The three most common virus-indexing methods are ELISA, PCR (or RT-PCR), and biological indicator-plant inoculations. The ELISA technique is a low-cost, sensitive method that allows for relatively quick testing of a large number of plant samples using a combination of antibodies that specifically react with certain viruses. There are several suppliers of commercially available ELISA kits that target the major viral pathogens of floricultural crops. These ELISA kits are designed for laboratory use, but there are also easy-to-use, onsite rapid assay kits that growers can use to detect viruses such as tomato spotted wilt virus (TSWV), impatiens necrotic spot virus (INSV), calibrachoa mottle virus (CbMV), cucumber mosaic virus (CMV), potato virus Y (PVY), and tobacco mosaic virus (TMV). It should be noted that onsite-detection kits are also available for several bacterial and fungal pathogens, such as *Xanthomonas campestris* pv. *pelargonii, Ralstonia solanacearum,* and *Phytophthora* spp.

The polymerase chain reaction (PCR) is a highly sensitive technique that allows for the rapid and specific detection of plant pathogens including viruses, viroids, fungi, and bacteria in infected plant tissue. PCR selectively and exponentially amplifies a small portion of the pathogen's genetic material (DNA). Very low amounts of the pathogen can be detected because of the amplification steps in PCR. Even RNA-based viruses can be detected by PCR by first copying the viral RNA into DNA.

PCR-based detection assays are typically designed to detect a specific pathogen species. However, the assay may also be designed to detect pathogens above the species level. This is helpful when there is uncertainty as to which pathogen species may be present in an infected plant. For example, a single PCR assay designed to detect viruses belonging to the tobamovirus group would be able to detect tobacco mosaic virus (TMV), tomato mosaic virus (ToMV), ribgrass mosaic virus (RMV), and others. The plant tissue selected for testing is extremely important because only a very small amount of plant tissue is actually used in the PCR assay, posing a risk that a virus could be present in the plant but not present in the tissue sampled.

An older, but sensitive, virus-detection technique is to use biological indicator plants. Indicator plants are plant species that are naturally susceptible to many different plant pathogenic viruses and express relatively rapid, distinct disease symptoms when infected. Leaves are removed from the plant to be tested, ground in an inoculation buffer to release the plant sap, and then rubbed onto the cotyledons or leaf surface of various indicator plants. If a virus is present in the leaf sample, symptoms may appear on one or more indicator plants seven to twenty-one days after inoculation. This method of detection is slower and more laborious than ELISA and PCR techniques but provides a means to detect unknown viruses. Reliable indicator plant species have been found in the genera *Nicotiana* (tobacco), *Solanum* (potato), *Chenopodium* (lamb's quarter), *Cucumis, Phaseolus, Vicia,* and *Brassica.*

Each virus indexing technique has its strengths and weaknesses. The best technique is one that is sensitive, independent of plant sampling location, and affordable. Unfortunately, there is no single technique that fits all of these criteria. Even if the testing method is sensitive, it cannot be effective if the wrong plant parts are tested or the test is done at the wrong time of the year.

How to Eliminate Viruses?

If virus indexing reveals the presence of viruses, these plants need to be cleaned up. Eliminating viruses from infected plants is accomplished mainly with heat therapy and special tissue-culture techniques in the laboratory. These methods have been used through the years for vegetatively propagated crops including chrysanthemum, geranium (*Pelargonium*), carnation (*Dianthus*), and potato (*Solanum*). The method used is dependent on the type of virus that needs to be eliminated from the plant. For example, for viruses that are unevenly distributed in the plant, such as TSWV and INSV, the virus can be eliminated by removing the tiniest growing point, the meristem tip, and raising the plant artificially in a special tissue-culture growth medium in the laboratory. Most viruses are incapable of advancing into the plant's youngest growing portion.

Viruses that are evenly distributed in the plant, such as TMV) and CMV, are very stable and easily spread. These viruses require additional treatments, such as heat therapy (90–100°F [32–38°C]) for up to one month, that will lower virus concentration. The combination of virus-concentration reduction and meristem-tip culture is very effective in eliminating the major viruses of floricultural crops.

After four to six weeks in culture, the tiny meristem tip (0.04–0.08" [1–2 mm] in size) develops into a small, rooted plant that is taken into an insect-free greenhouse. Successful virus elimination from these plants needs to be verified by virus indexing. These clean plants can also be maintained in tissue culture in the laboratory to protect them from pathogens, insects, and other pests. They are only brought to the greenhouse for flower checks and buildup for production.

What Is Clean-Stock Production?

A considerable amount of time, money, and effort has been put into successfully eliminating fungal, bacterial, viral, and viriod pathogens. After the laboratory portion of this program, it is of utmost importance that certain "rules" are followed in the buildup and commercial production of these vegetatively propagated crops.

Successful clean-stock production programs adhere to the following rules.

Repeated testing

Careful repeated testing of mother stock plants assures that plants remain free of detectable pathogens.

Unidirectional flow

Commercial production of culture- and virus-indexed stock always occurs in a sequential series of steps through cleanup (i.e., elite nucleus and nucleus stock), increase (i.e., buildup), and production (i.e., unrooted and rooted cuttings to grower customers). No plants are ever sent backwards in the scheme (i.e., production to increase or increase to nucleus) without first repeating the culture- and virus-indexing processes.

Annual renewal

All stock plants at all levels are "flushed" out or discarded each crop cycle and replaced with newly indexed plants. Since this stock is still susceptible to infection after indexing, the probability of a breakdown in the sanitation system increases with the length of time the plants are grown in the greenhouse.

Isolation

The indexed plants must be grown in isolation or separated from non-indexed plants to prevent the introduction and spread of pathogens. Isolated zones should have a sanitation program including a wash station for all workers to utilize prior to entry. Many pathogens, such as bacteria and viruses, are easily transmitted on workers' hands or clothing. Lavatories with soap and running water as well as clean aprons should be used.

What Can Go Wrong?

The introduction of a crop-threatening pathogen into a clean-stock production program can occur through the use of non-certified stock; unsanitized greenhouses, growing materials, and tools; insects; and people. If one or more of these factors are overlooked prior to or during the production cycle, serious problems will ensue for the producer and, ultimately, the grower. Additionally, as the quantity of stock plants increases, sanitation controls naturally decrease. For example, only a few people are required for plant maintenance at the nuclear stock level, while at production levels potentially hundreds of people are involved in maintaining, harvesting, sticking, and finishing cuttings. How then can successful, disease-free production occur? Controlled clean-stock development, tight sanitation measures, intensive worker training, and enforcement of sanitation procedures are critical for successful disease-free production and finishing.

Top Ten Ways to Stay Clean

You have purchased plants that are disease-free from indexed stock. Now, how do you keep them that way in the greenhouse?

10. Ban algae. Keep algae at bay by avoiding standing water and scrubbing floors regularly, or treating them with an algaecide. Eliminating algae means fewer problems with algae-eating fungus gnats and shore flies that can spread plant pathogens and whose larvae may feed on tender young roots.

9. Isolate and separate disease-sensitive crops. Keep workflow under control with single entrances to greenhouses and by assigning employees to specific greenhouses and crops. This helps keep the spread of disease to a minimum.

8. Scout and isolate. Identify and train pest management scouts to inspect plants on a weekly basis and report any suspect symptoms or unusual findings. Isolate symptomatic plants. Send suspect plants to a professional diagnostic laboratory for an accurate diagnosis.

7. Set the tone. Emphasize the importance of sanitation from the top down in your company's organization. Appoint sanitation "champions" within your staff and hold sanitation offenders accountable.

6. Train and monitor. Train all workers on sanitation protocols and follow up with refresher sessions. Monitor procedures and processes regularly and adjust them as needed.

5. Make it user friendly. Make it easy to follow proper sanitation practices by providing plenty of faucets and easy-to-use footbaths, avoiding unnecessarily harsh disinfestants, supplying clean aprons or lab coats outside each growing zone, and making the greenhouses a comfortable place to work.

4. Provide wash stations. Equip simple wash stations outside of major growing zones with running water, soap, and fingernail brushes.

3. Emphasize hand washing. Hands transmit most diseases, and regular and thorough hand washing is the simplest and most effective preventive measure. Most disinfestants require the complete removal of all organic material to be effective.

2. Disinfest. All greenhouse surfaces should be disinfested prior to the start of each growing season or crop. Quaternary ammonium or hydrogen peroxide compounds used at the manufacturer -suggested rates are highly effective. Use caution with bleach-containing disinfestants, which release harmful chlorine gas that can affect greenhouse workers. In addition, the compound can corrode galvanized benches.

1. Start clean! With each new crop cycle, remove all plant material, pots, and soil from the production area; sweep benches and floors; and bring in clean medium, new or adequately sterilized pots, and clean plants. Make sure your suppliers are providing cuttings from certified stock.

16

Terril A. Nell and Ria T. Leonard

Postharvest Care & Handling of Potted Plants

The Importance of Quality

The postharvest performance of potted plants is determined by numerous factors that are dependent upon one another to ensure a high-quality, long-lasting plant. One weak link in the "chain" during production, shipping, or retail display and quality can be greatly compromised. Postharvest quality begins with the selection of each variety grown and is followed by the production, environmental, and cultural conditions plants are subjected to during production. The next critical steps in maintaining quality are during the postharvest handling stages during storage, transport, and retail and consumer display.

For a grower, quality might refer to a poinsettia crop that is of a uniform height and in a consistent stage of bloom. It may mean potted chrysanthemums with a maximum number of breaks. Or it might relate to foliage plants that meet size requirements as outlined by a mass-market buyer. All of these attributes are very important and are a part of the quality equation. However, it is key to understand that the value of products must extend well beyond the production bench. In a competitive, global marketplace, the necessity of a commitment to quality has never been greater.

As a grower, it is vital to tailor production practices to prolong postharvest life. The factors that affect plant longevity, and therefore plant quality, are easy to influence. In the production range, variety selection and production practices—including fertilizer type and amount, temperature, light, and root development—have the greatest impact on a plant's postharvest life. Beyond the greenhouse, shipping temperature and duration also wield great impact upon a plant's postharvest performance. But the full burden of prolonging a plant's shelf life does not fall solely to the grower and shipper. The retailer must also be committed to maintain proper handling and environmental conditions while plants are on display to keep the "quality chain" intact. It is also essential to educate buyers, retailers, and consumers in regard to proper plant handling. With the majority of potted plants currently being sold at mass retail outlets, this need is greater than ever. It makes no sense to deliver high-quality plants to a retail outlet only to have them be neglected and become unsalable.

So how can we, as an industry, produce quality plants from greenhouse to consumer? It is easiest to approach this if we realize that quality is a conformance to standards—standards to produce, ship, and display plants according to specifications designed to produce a plant of a desirable size and that will last for the consumer. These specifications include things such as uniform physical dimensions and appearance, adequate fertilizer and light levels during production, proper temperature control during transport, and sufficient light and watering regimes during display. Production, shipping, and retail practices that are developed to create high-quality plants will not only yield plants that look good and last longer, but will also increase sales and profits.

Factors That Affect Postharvest Quality

Variety selection

Variety selection is, without a doubt, the most important decision a grower can make to enhance postharvest performance. This is true of every type of potted plant available. Some varieties inherently last longer than others, even if they have been grown under ideal production conditions and handled properly during postharvest. Some varieties do not tolerate transport and storage conditions or are more sensitive to botrytis and ethylene than other varieties. Choosing varieties depends on many factors from the ease of production to market demand. When you add a long-lasting plant to this equation, you have a winner.

Often, varieties are selected that grow with relative ease and adequate flower production, that exhibit specific insect or disease tolerance, and that yield the most desirable characteristics. Typically, variety selection hinges on one of two things: visible characteristics of a plant (i.e., flower or leaf color, plant height, etc.) or production performance (i.e., timing or number of lateral breaks). The decision about which varieties to grow should also focus on plant longevity. If you want to improve your plant performance for the consumer, the first step is to choose varieties that last.

Choosing the right variety makes an enormous difference in influencing postharvest plant life. For example, leaf drop in poinsettia and premature leaf yellowing in chrysanthemum are no longer issues in the industry—when the correct varieties are grown and handled properly. Tests have shown that flower longevity can double or even triple when the right variety is selected. For example, our research has found a range in pot life among varieties ranging from twelve to thirty-eight days for chrysanthemum, twenty-two to sixty-six days for kalanchoe, sixteen to twenty-one days for gerbera, fourteen to thirty-eight days for hydrangea, and eleven to twenty-four days for potted rose. Therefore, identifying long-lasting, high-quality varieties is a critical step in enhancing postharvest performance.

There is relatively little published information comparing the postharvest life of different varieties within the numerous types of potted flowering and foliage plants. With the continual introduction of new varieties into the marketplace each year, a list like that would quickly become outdated. The constant flux of variety introductions can present challenges when choosing varieties. Some varieties are developed and introduced for specific attributes, such as color, form, size, and disease resistance, with little regard for shelf life. The more progressive breeders, however, do consider shelf life and incorporate this characteristic into their breeding programs.

Your best bet is to find out as much as you can about a variety. Contact the breeding company, a local extension agent, or a local university, attend allied tradeshows and trial gardens, or discuss with fellow growers who may have experience with particular varieties of interest. Another worthwhile strategy is to conduct your own tests to determine how well a variety holds up to transport conditions and how long it will last in simulated consumer conditions. We recommend conducting a simulated ship for three to five days in the same way the plant would be handled for shipping, whether in carts, trays, or sleeved in boxes. Maintain the temperature during this time as close to what the plant would experience on a truck during transit. You could either simulate a retail setting for the next few days or place the plant directly inside an office or home environment to evaluate its performance.

If there are disease problems at unboxing or premature flower or bud drop, these are signs that you have a poor-performing variety. Continually testing varieties will prove invaluable in determining which varieties to grow. In addition, you should always sell plants with the variety name on the tag or pot so you can receive feedback from your buyers and customers. That way, you can identify the very best, longest-lasting varieties and learn the varieties that do not ship or withstand display conditions very well.

A lot of attention has been given to poinsettia varieties over the last several years that have proved invaluable in promoting high-quality plants. The market has witnessed an unprecedented increase of new types and colors of poinsettias over the last decade. Poinsettia-breeding companies and university researchers have worked together in evaluating poinsettia varieties with an emphasis on postharvest quality. Over one hundred varieties have been grown yearly at university research institutions in various parts of the United States for production and postharvest performance evaluation. The information gathered has been used to identify the top-performing varieties that withstand shipping, handling, and retail conditions without leaf drop, epinasty ("drooped" or bent leaves and/or bracts, due to the petiole bending during sleeving), or bract edge burn. This has made poinsettia production easier for growers, has provided consumers with high-quality plants, and has supplied breeders with improved breeding stock.

We know poinsettia varieties differ dramatically in their responses to shipping and handling procedures and the home environment. For years, the greatest postproduction obstacles plaguing poinsettias were leaf and bract drop, leaf yellowing, and epinasty. Breeding efforts tackled these challenges, essentially conquering these issues. But some of the newer varieties came with their own share of issues, such as lateral stem breakage and bract edge burn. Sometimes these problems surface at the end of production, but most often they are revealed when retailers open shipping cartons and unload plants. Production techniques can address and curtail some postharvest issues, but again the surest way to limit problems like bract edge burn, leaf yellowing, bract fading, and lateral stem breakage is to start with varieties that are not susceptible to these problems.

Production practices

The most common target that growers and floral buyers seek when evaluating a crop is plant size and flower number. The floral buyer specifies plant height, pot size, and number of flowers and buds, and that is the mark that the growers strive to reach. Now it is time to add increased longevity to that plant production goal.

For years, the number one reason that growers shied away from the thought of changing production techniques to increase postharvest longevity of plants was because they thought it would increase production costs. That is an incorrect assumption. The production practices and environmental condition modifications required to produce long-lasting plants hinge around things like reducing and terminating fertilizer, lowering production temperatures at the end of a crop, and eliminating overwatering. These actions reduce production costs, not drive them up.

It is not uncommon for growers to manipulate production temperature and light levels to encourage root development, to increase branching, and to control height (using DIF). In the same way, growers can finesse fertilizer, temperatures, light, and watering practices to create a higher-quality plant.

Fertilizer

Of all the production factors, the nutritional program and fertilization practices have one of the greatest effects on pot life and quality. Application method (liquid and controlled release) does not affect postharvest life, but fertilizer rate, nutrient ratios, nitrogen source, and duration do have an impact on longevity. Calcium also has a major effect on a plant's postproduction life, especially for poinsettia. More recently, our research has shown that silicon can reduce powdery mildew on potted rose.

As growers, it is easy to list the benefits of plants raised with high fertilizer levels: larger, darker leaves and more intense flower colors. These things sound like the makings of a quality plant. But, when postharvest plant life is added into the quality equation, high fertilizer levels become a detriment. In general, the more fertilizer that a plant receives, the softer it becomes, the greater sensitivity it will have to adverse shipping and retail display conditions, and the less time it will last in an indoor environment.

Many growers use fertilizer rates that are unnecessarily high. For chrysanthemum, lowering fertilizer concentrations from 450 to 150 ppm at every watering can increase postharvest life by up to fifteen days, without having any impact on production quality and appearance. Reducing fertilizer rates have also been shown to dramatically reduce leaf drop and improve interior performance of potted foliage plants. It is hard to believe that reduced fertilizer concentrations will not detract from a crop's marketability, but it is true. One way to begin to decrease fertilizer that is applied with the irrigation is simply to reduce the amount of water leaching through pots. As an added bonus, reducing leaching and runoff are desirable environmentally as well.

Nutrient ratios

For most flowering potted plants, nitrate should comprise 60–70% of the nitrogen applied to crops. The remaining 30% of nitrogen should come from ammonium and/or urea. If you use fertilizers that are higher in ammonium-nitrogen sources, the effects show up in the leaves: You'll see leaf margins turn brown and premature leaf yellowing.

The next item to monitor is the nitrogen-potassium ratio. For the majority of the crop cycle, this ratio should be roughly 1:1. During the final two to three weeks of the crop, change the ratio to 0.5:1.0, to increase the potassium relative to the nitrogen. If you blend your own fertilizer, changing the ratio is simply a matter of modifying the components. If you use commercially blended fertilizer, look for products that offer the revised N-K ratio and reduced ammonium.

Terminating fertilizer

Terminating fertilizer about two to three weeks from the end of a crop (usually at bud color for many potted flowering plants) can add postharvest life. Growing medium, fertilizer level, crop, and variety all have an impact on the effectiveness of fertilizer termination. For most potted plants, fertilizer termination does not cause plants to look any less appealing.

For some plants, such as cyclamen, chrysanthemum, and poinsettia, fertilizer termination has a beneficial effect on postharvest life. Terminating fertilizer two weeks before flowering in cyclamen can increase pot life and increase the percentage of flowers that open in postharvest conditions. With certain chrysanthemum varieties, stopping fertilizer at the point of disbud can increase plant longevity from seven to eleven days. Reducing fertilizer rates have also been shown to dramatically reduce leaf drop and interior performance of potted foliage plants. However, for other crops such as lisianthus, terminating fertilizer yields yellow leaves and plants unacceptable for sale. Fertilizer termination also does not improve longevity for Easter lilies (*Lilium longiflorum*); fertilize these at standard rates until the point of shipping. For most spring-flowering bulb crops, no fertilizer is needed. Although in some cases, such as tulips, growers apply calcium nitrate to strengthen stems and therefore minimize stem topple. Our research found for potted carnation, gerbera, hiemalis begonia, hydrangea, and kalanchoe, fertilizer termination had no effect on quality or pot life. Therefore, terminating fertilizer at the end of production for these crops will reduce fertilizer costs and still provide high-quality plants.

If there is one secret to success for fertilizer termination it is this: You have to grow plants using optimal fertilizer rates. Fertilizer termination may have limited value if lower-than-recommended fertilizer rates are used throughout crop production. For most crops we tested, terminating fertilizer did not cause the plants to look any less appealing, but we used ideal concentrations during growth. It is best to follow closely each species' specific fertilizer requirements. If you are unsure about the effects of reducing or terminating fertilizer at the end of production, try it on a few plants before doing the entire crop. Each variety should also be tested as results may vary depending on variety.

Calcium sprays

Calcium levels in plants affect two major areas that are pivotal to quality: appearance and longevity. For poinsettias, bract edge burn is linked to calcium deficiency in expanding bracts.

As the postharvest problems of epinasty and leaf drop were conquered, bract edge burn arrived on the scene. Its unsightly necrotic spots on leaf margins detract from poinsettia appearance and quality and increase a bract's susceptibility to botrytis. Because bracts expand so rapidly during bract coloring, the bract tissue simply cannot take up calcium in large enough quantities, and necrotic spots result. Any environmental conditions that reduce water uptake, such as high humidity, low light, and low substrate and air temperatures, can lead to calcium shortages in bract tissue.

High ammoniacal nitrogen fertilizers (more than 30% ammonia and urea nitrogen) in the final three weeks of production have also been tied to bract edge burn in poinsettia. Thus, many poinsettias show high levels of bract edge burn with 14-14-14 Osmocote, which is approximately 50% ammoniacal nitrogen. If you use slow-release fertilizer, choose one that has a rate that limits nutrient release late in the production season. Otherwise, the high amounts of nitrogen accelerate bract enlargement and lead to calcium deficiency, bract edge burn, and possibly botrytis.

The problem intensifies when you discover that simply increasing the calcium in your fertilizer solution does not solve the problem. The best way to overcome bract edge burn is to spray calcium on foliage and bracts at a rate of 400 ppm. Begin sprays at first bract color and continue weekly until shipping. Spray with calcium chloride or chelated calcium. It is wise to avoid calcium nitrate since plants should not have additional nitrogen toward the end of the cycle. Add a wetting agent to the calcium solution before application.

Temperature and light

Two of the most basic parameters influencing plant growth and quality in general also have a great impact upon postharvest longevity of potted plants. Typically, growers manipulate temperature to control plant height, particularly finessing the relationship between day and night temperatures. This technique is called DIF (read more about DIF in chapter 6 and under *Euphorbia* [poinsettia] in the Crop Culture section). Air temperature also influences rate of flower development.

The key time to change your temperatures to maximize longevity is near the end of the crop (during the last two to four weeks of production). At this point, reducing night temperatures by 5°F (2–3°C) will not only increase plant longevity, but will also reduce your energy costs. Under lower night temperatures, flower color intensifies and overall longevity increases.

Growers who practice DIF height control on their crops should switch to a positive or zero DIF during the final two to four weeks of production. Negative DIF (low day/high night temperatures) decreases carbohydrate concentrations in plants, which, in the final weeks of a crop, will lead to reduced longevity. Using positive DIF at this point in a crop will not have a great impact on plant height, because stem elongation is slowed at this time. It is important to note that bract edge burn in poinsettia intensifies under any conditions that enhance bract enlargement—including high night temperatures. Avoiding a negative DIF after bracts begin to color can help to prevent bract edge burn. These temperature manipulation techniques hold true for poinsettia, chrysanthemum, lily, elatior begonia, azalea, and other crops.

When it comes to light, the effect upon plant quality is clear. In general, higher light levels increase longevity, reducing bud drop and premature flower drop. The impact of higher light levels upon the longevity of poinsettia, chrysanthemum, potted rose, or any other high-light crop is so great that, in northern climates where natural light levels in winter are low, supplemental lighting is supplied with high intensity discharge (HID) lighting until flowering.

Some growers shade their flowering potted plants as they begin to color. This is not a good practice and will decrease postharvest longevity, unless shading is needed to avoid flower burn. The combination of high light and cool night temperatures builds a good carbohydrate reserve the plant can use after leaving the greenhouse (when the plant enters low-light, interior conditions).

For many potted foliage plants an acclimatization period from high light to low light prior to sale is needed. It is commonly known that plants such as *Ficus* will drop all their leaves if moved directly from high light production conditions to an interior environment. Sensitive foliage plants can adapt to reduced light if they are produced or acclimatized under specific light levels. This process may take six weeks, six months, or even longer, depending on the species and size of the plant. To optimize interior performance, many foliage plants are grown at low light levels for the entire production period.

Watering

If there is one "unseen" aspect of interior longevity, it is root development. Healthy roots will continue to function long after the consumer carries the plant home. Unfortunately, because the root system is unseen, it is easily overlooked in the plant-quality equation until problems surface in the parts of a plant that are seen and showy: foliage and flowers. By that point, it is too late.

If growers want to grow quality plants, they need to produce pots that have a balanced amount of aboveground (leaves and flowers) and belowground (roots) parts. Irrigation practices and growing medium play the greatest role in determining a healthy root system. Some growers contend that they have discovered that the key to a long postharvest life is keeping salts and irrigation in balance to produce healthy roots.

Both overwatering and underwatering kill roots quickly. A soggy or dry medium is not conducive to healthy root development. Healthy roots are white and plump and grow in medium that smells almost sweet. In an unhealthy root system, it will be hard to distinguish roots from the medium (they will be darkly colored), and the medium will often have an unpleasant smell. It is easy to overwater plants in an effort to wash excess salts from the medium. The need to leach can be eliminated by cutting the fertilizer and reducing the water applied. Do not reduce irrigation without concomitantly reducing fertilizer, or the roots will be damaged. Do not forget that low light, low temperatures, high humidity, and reduced air movement all contribute to plants having lower water needs.

On the other end of the continuum, plants that are grown too dry—where the medium is allowed to dry out too much between waterings—will also lack healthy roots. Achieving an irrigation balance is key to growing high-quality, long-lasting plants.

High salts in the medium can also quickly erode healthy root systems. We used to preach leaching; but when you use the right fertilizer level, leaching is not an issue. The trick with salts is making sure you know three things: your water quality, the proper fertilizer level for your crop, and the pH of your growing medium. Plants grown under conditions of high salts have root systems similar in appearance and dysfunction to plants that are overwatered.

With most flowering potted plants, it is a good idea to reduce irrigation in the last two to four weeks prior to shipping. Limit watering at the same time you stop fertilizing plants and switch to a positive or zero DIF. Cutting back on irrigation at this time with poinsettia will also slow bract expansion and help to prevent bract edge burn.

Shipping and Transportation

Three major rules must be considered when preparing plants for shipping. Flowers must be at the proper stage of development, plants must be well watered, and ethylene-sensitive species must be treated. How and when the plants are shipped weighs heavily in determining the resulting postharvest life. If flowers are shipped before they are at the right stage of development, buds may never open. If plants are sleeved and boxed just after watering, retailers may open boxes to discover a severe botrytis infection. Understanding the basics of when and how to ship potted plants and following some basic steps will increase the plants' life beyond the greenhouse bench.

Stage of development

No matter what crop you grow and ship, there is a right time—and a wrong time—to ship it (table 16-1). There is not an easy, one-size-fits-all formula that suits all crops. You have to approach shipping on a crop-by-crop basis. The goal is to ship at a stage that ensures continued flower development and yields maximum quality at point of sale. Remember that the stage of marketability also depends on the time required to get the plant to market. Plants being shipped long distances may be able to be packed at a slightly less mature stage than those for local markets. Plants with many open flowers are more sensitive to mechanical damage and ethylene during shipping and handling. Shipping plants with open flowers can be a problem with hibiscus, Easter lily, Christmas cactus, and regal geranium, but is not a problem for poinsettia (bracts should be fully colored and pollen showing) or chrysanthemum.

Choosing the right moment in crop development to ship can add days or weeks to individual plant longevity. If you ship too early, blooms might not reach proper size or color—or worse, may never open at all. Buds may remain tightly closed or yellow and drop off. For some plants that are shipped too soon, flower buds may go on to open indoors, but petal color will be faded and washed out. This is especially true with dark-petaled flowers or those that have multiple flowers like cineraria, calceolaria, and lisianthus.

For plants shipped too late, generally with many open flowers, you rob valuable pot life from the consumer. Consumers may end up purchasing plants that linger a mere week or less. For instance, chrysanthemums lose a day or more of life for every day they remain in the greenhouse past the optimum shipping stage. The balance you want to achieve is to ship plants that have enough color appeal at time of sale to attract customers into a purchase, but not so much color that longevity is cut in half.

Most spring-flowering bulbs, such as tulip, lily, and narcissus are marketed at the very first sign of bud color. This stage is called "market tight" where buds are showing color and are "puffy" but not open. Other bulb species, like astilbe and dahlia, are marketed with some open flowers. To learn more about the proper stage of marketing for potted bulb species consult the *Holland Bulb Forcer's Guide, 5th ed.* by Gus De Hertogh (Ball Publishing, Batavia, Illinois).

Our work has shown that for some long-lasting crops, like hiemalis begonia and kalanchoe, you can

Principles of Postharvest Care and Handling
Grow long-lasting varieties.
Provide proper production, environmental, and cultural conditions during production.
Maintain a scouting program for disease and pest control and always recheck before shipping.
Consider terminating or reducing fertilizer levels during final phase of production on certain crops.
Acclimatize sensitive foliage plants to low light conditions prior to marketing.
Maintain sanitary conditions in the greenhouse, work areas, packinghouse, and coolers, and always remove plant debris.
Ship at the proper stage of development.
Make sure plants are well watered prior to shipping, and add a wetting agent, if needed.
Use specialized treatments that will enhance postharvest performance by protecting against ethylene, disease, and water loss.
Transport plants at the proper temperature for no longer than three days.
Keep plants away from ethylene sources and do not ship, store, or display near ripening fruit.
Choose reputable brokers and shippers that have temperature-controlled facilities.
Provide care instructions to retailers and consumers.

Table 16-1. Ethylene Sensitivity and Recommended Shipping Stage and Temperature for Potted Flowering and Foliage Plants

Crop	Ethylene symptom(s) and sensitivity rating[1]	Stage of marketability	Shipping temperature
POTTED FLOWERING			
Achimene	Flower/bud drop***	50% open flowers	55–65°F (13–18°C)
African violet	Flower wilt**	5 or more open flowers	55–65°F (13–18°C)
Ageratum	Leaf yellow/drop*	50% open flowers	45–50°F (4–10°C)
Anemone	Sepal shatter*	50% open flowers	35°F (2°C)
Aster	Flower/bud drop***	50% open flowers	35°F (2°C)
Azalea	Leaf drop*	25% open flowers	35–40°F (2–4°C)
Begonia, Rieger, Hiemalis, and Elatior	Flower/bud drop***	20% open flowers	50–60°F (10–16°C)
Bougainvillea	Flower/bract drop***	25–50% bracts colored	55–60°F (13–16°C)
Cactus, Christmas and Easter	Flower/bud drop***	Large bud stage	50–60°F (10–16°C)
Calceolaria	Flower/bud drop***	4–8 open flowers	40–55°F (4–13°C)
Calendiva	Flower wilt***	10–30% open	40–45°F (4–7°C)
Calendula	Not sensitive	50% open flowers	35–40°F (2–4°C)
Campanula	Flower wilt/bud drop**	1 open flower with buds	35–40°F (2–4°C)
Chrysanthemum	Stunted growth/leaf drop*	50% open flowers	35–40°F (2–4°C)
Cineraria	Wilt	5–6 open flowers	40–45°F (4–7°C)
Coreopsis	Flower drop*	Green or in flower	35–40°F (2–4°C)
Crossandra	Flower drop**	50% open flowers	50–55°F (10–13°C)
Cyclamen	Flower wilt/drop*	2–3 open flowers	40–45°F (4–7°C)
Cymbidium	Flower wilt**	Few open flowers	55–60°F (13–16°C)
Exacum	Flower wilt/drop; bud drop*	10–20% open flowers	55–60°F (13–16°C)
Fuchsia	Flower/bud drop; leaf yellowing***	At first flower	40–45°F (4–7°C)
Gardenia	Flower/bud drop*	Buds swollen/puffy	55–60°F (13–16°C)
Geranium	Buds not open; petal shatter; leaf yellowing***	0–20% flowers open	40–45°F (4–7°C)
Gerbera	Reduced flower life*	50% open flowers	40°F (4°C)
Gloxinia	Flower drop*	4–5 open flowers	60–65°F (16–18°C)
Hibiscus	Flower/bud drop***	2-3 open flowers	55–60°F (13–16°C)
Hydrangea	Flower/bud drop**	Florets fully colored	35–40°F (2–4°C)
Impatiens	Flower/bud/leaf drop**	50% open flowers	55–65°F (13–18°C)
Kalanchoe	Flower wilt/fading; buds not open***	10–30% open	40–45°F (4–7°C)
Lantana	Flower/bud drop***	50% open	45–50°F (7–10°C)
Lisianthus	Flower wilt/bud drop*	50–75% open flowers	35–40°F (2–4°C)
Lupine	Flower/bud drop/wilt***	50% open	35–40°F (2–4°C)
Marigold	Leaf/stem wilt*	1–2 open flowers	50–55°F (10–13°C)
Osteospermum	Wilt/epinasty**	1–2 flowers and bud color	40–45°F (4–7°C)
Pansy	Not sensitive	2–3 open flowers	40–45°F (4–7°C)
Pentas	Petal shatter***	First flower open	50–55°F (10–13°C)
Petunia	Flower/leaf wilting***	Start of flowering	35–40°F (2–4°C)
Periwinkle (*Catheranthus*)	Leaf/flower/bud drop***	Start of flowering	60–65°F (16–18°C)
Phalaenopsis	Flower drop/wilt***	2–5 open flowers	50–60°F (10–15°C)
Poinsettia	Leaf drop/epinasty*	Pollen showing	55–60°F (13–15°C)
Portulaca	Reduced flower life**	Several open flowers	45–50°F (7–10°C)

(Continued)

[1] Ethylene sensitivity rating: * slight; ** moderate; ***high. Ethylene sensitivity depends upon the cultivar within a species, the stage of plant development, the concentration of ethylene, the temperature during exposure, and the duration of exposure. Symptoms can appear from twenty-four hours to several days after exposure.

Table 16-1. Ethylene Sensitivity and Recommended Shipping Stage and Temperature for Potted Flowering and Foliage Plants *(Continued)*

Crop	Ethylene symptom(s) and sensitivity rating[1]	Stage of marketability	Shipping temperature
Rose	Flower/bud/leaf drop**	2–4 flowers/buds colored	35–40°F (2–4°C)
Salvia	Flower/bud drop***	25–50% open	50–55°F (10–13°C)
Snapdragon	Flower wilt/drop***	25–50% open flowers	35°F (2°C)
Streptocarpus	Flower wilt/drop**	25–50% open flowers	50–60°F (10–15°C)
Zinnia	Not sensitive	Varies with variety	35–40°F (2–4°C)
POTTED BULBS			
Allium	Unknown	Flower head expanding out of sheath	33–35°F (0.5–2°C)
Amaryllis	Flower wilt***	12" (30 cm) flower stalk	45–50°F (7–10°C)
Astilbe	Flower wilt/drop***	Primary inflorescence 30–50% colored	35°F (2°C)
Caladium	Bending of petiole***	Leaves fully expanded	55–60°F (13–15°C)
Crocus	Varies with cultivar*	"Sprout" stage	33-35°F (0.5–2°C)
Dahlia	Reduced flower life*	First flower fully open	45-50°F (7–10°C)
Freesia	Flower drop**	First flower open	35–40°F (2–4°C)
Hyacinth	Flower drop*	"Green bud" stage	33–35°F (0.5–2°C)
Iris, dwarf	Flower/bud/leaf drop*	"Sprout" stage	33–35°F (0.5–2°C)
Lily, Easter and Asiatic and Oriental hybrids	Bud drop**	First bud "puffy" and showing color	33–35°F (0.5–2°C)
Narcissus, daffodil	Reduced flower life**	"Pencil" stage	33–35°F (0.5–2°C)
Narcissus, paperwhites	Reduced flower life**	8–12" (20–30 cm) flower stalk	33–35°F (0.5–2°C)
Tulip	Flower/bud drop*	"Green bud" stage	33–35°F (0.5–2°C)
POTTED FOLIAGE			
Aglaonema spp.	Leaf yellowing*		55–60°F (13–16°C)
Anthurium scherzerianum	Reduced flower life*		55–65°F (13–18°C)
Aphelandra squarrosa	Leaf drop***		55–60°F (13–16°C)
Asplenium nidas	Not sensitive		55–60°F (13–16°C)
Capsicum spp.	Leaf/fruit drop***		55–65°F (13–18°C)
Chamaedorea elegans	Not sensitive		55–60°F (13–16°C)
Chlorophytum comosum	Leaf wilt**		55–60°F (13–16°C)
Codiaeum variegatum pictum (Garden croton)	Leaf yellow/drop**		60–65°F (16–18°C)
Dieffenbachia spp.	Leaf yellowing**		55–65°F (13–18°C)
Dracaena spp.	Leaf necrosis/yellowing**		60–65°F (16–18°C)
Epipremnum aureum	Leaf yellowing**		55–60°F (13–16°C)
Euphorbia splendens	Leaf yellowing**		55-60°F (13–16°C)
Ficus spp.	Leaf drop**		55–60°F (13–16°C)
Hedera helix	Not sensitive		50–55°F (10–13°C)
Nephrolepis exaltata	Leaflet drop*		55–60°F (13–16°C)
Philodendron scandens oxycardium	Leaf yellowing/drop**		55–60°F (13–16°C)
Plectranthus australis	Epinasty, flower/leaf drop*		55–60°F (13–16°C)
Polyscias fruticosa	Leaf drop**		60–65°F (16–18°C)
Radermachera sinica	Leaf drop***		50–55°F (10–13°C)
Schefflera arboricola	Leaf drop***		50–55°F (10–13°C)
Schefflera elegantissima	Leaf drop**		55–60°F (13–16°C)
Spathiphyllum spp.	Reduced flower life*		55–60°F (13–16°C)
Syngonium podophyllum	Not sensitive		55–60°F (13–16°C)

[1] Ethylene sensitivity rating: * slight; ** moderate; ***high. Ethylene sensitivity depends upon the cultivar within a species, the stage of plant development, the concentration of ethylene, the temperature during exposure, and the duration of exposure. Symptoms can appear from twenty-four hours to several days after exposure.

ship with only 10% of flowers open, and the remaining buds will continue to open and fully develop, provided they are not exposed to ethylene and are displayed in proper conditions. However, that is the exception. For chrysanthemum, cyclamen, and gerbera, ship with 25–50% open flowers. Hydrangea need to be fully colored but not overly mature. If you ship hydrangea too early, they will never color unless placed in high light areas. Pull lisianthus to ship when 75% of the blooms have opened; otherwise buds will not open. With poinsettia, do not ship until bracts are fully colored and cyathia are starting to show pollen. Otherwise, the secondary bracts will fail to expand and color properly.

The maturity of bedding plants is a vexing issue, because although such plants do best if they are sold when still growing rapidly, consumers show a preference for plants with color, which are often root-bound and will not perform as well in the garden. The target is to ship plants at the stage that yields maximum quality at point of sale and ensures continued flowering and longest flower life for the consumer.

It is interesting to note that market demand can and will influence stage of development for shipping. Kalanchoe, for example, last longer when marketed with only three to five flowers open. This is the European standard and is accepted among European consumers, who purchase the plants knowing that subsequent buds will open. In American markets, consumers favor plants with more color, so kalanchoe typically leave the greenhouse with 30–50% of flowers open.

Water plants the day before they are ready to leave the greenhouse so they do not dry out before they reach retail outlets. If you have a crop extremely sensitive to drying out, water the same day as shipping as long as you can be assured the water will no longer be dripping from the pot and has dried off the plant and pots. High moisture content during shipping, especially if plants are sleeved, can create conditions for diseases such as botrytis and mildew to take hold. The use of wetting agents, discussed later in the Specialized Treatment section, can aid in maintaining moisture in the potting medium during the stress of transport and retail conditions.

Do not forget to make a final check of each plant for insects or diseases at time of shipping. One infected or diseased plant can rapidly spread disease to other healthy plants, quickly rendering them unmarketable. Aphids or spider mites can cause a rapid decline in chrysanthemum, as can powdery mildew on potted rose. Also, remove any damaged or fallen leaves and flowers at the time of shipping to avoid these easily targeted disease substrates.

Transport systems

Whether you sleeve and box plants or use shipping carts and trays, the idea is to keep plants protected during transit. How you prepare plants for transit may depend on the crop, how long plants will be in transit, or what your customer prefers. In northern regions, growers must also make arrangements to protect plants from exposure to cold temperatures.

The core problem in transporting plants—no matter what method is used, carts or boxes—is providing an environment throughout the shipping process that maintains plant quality, reduces respiration and ethylene susceptibility, and prevents disease spread. As you determine your shipping conditions, make it your goal to provide a shipping environment that yields a plant that will be in the same quality, healthy condition upon arrival as when it left your shipping area.

Sleeves protect leaves and flowers from mechanical damage during transport. In general, choose sleeves that extend beyond the tops of plants another 2–3" (5–8 cm) for the best protection. Sleeve composition, whether made of paper, fiber, plastic, or mesh, will protect plants during transit. However, differences occur when it comes to moisture retention within the sleeve.

Sleeve material is a matter of concern when the medium is moist at the time of sleeving or when condensation occurs. If the potting mix is wet when the plants are sleeved and boxed, paper, fiber, and mesh sleeves will allow more moisture to escape from the pot and plant, while solid plastic sleeves will not. Once plants are sleeved and placed into boxes, the environment in that sealed box is nearly 100% relative humidity. This is the ideal condition for diseases to form and multiply rapidly. Another problem when using sleeves is condensation. Care must be taken when plants are sleeved in warm conditions and then cooled down in preparation for boxing and subsequent transport or when plants are cooled prior to shipping and are then allowed to warm up. At that point, the shift in plant and potting mix temperatures from warm to cool or cool to warm causes condensation to collect inside the sleeves, especially plastic ones that do not have holes, increasing the likelihood of disease development. It is best to water plants a minimum of six hours to a maximum of twenty-four hours before sleeving to prevent disease.

The cart-and-tray system eliminates the problem of humidity inside sleeves and shipping boxes altogether. When possible, transport plants on open-cart systems. You forgo the expense of sleeves and boxes, and the increased air movement around the plants reduces disease outbreaks. Also, many growers are now using permanent plant rings like those commonly used for poinsettias that support plants and prevent stems from breaking.

Transport conditions

Three major factors during transport can make or break a potted plant's postharvest life: (1) temperature, (2) shipping duration, and (3) ethylene. Temperatures that are too high or too low can speed a plant toward its death and physically damage leaves and flowers. Shipping duration beyond three days will, in many cases, dramatically reduce pot life. Ethylene exposure can cause flowers, buds, and leaves to drop or wilt. Understanding each of these factors and the way they affect plant postharvest life will improve the quality of your potted crops.

Temperature

Temperature during transit has a major impact on subsequent postharvest life and quality. Keeping plants at proper temperatures during transit and storage reduces respiration rates, conserves carbohydrates, reduces the incidence of diseases, and increases quality and flower life. Too high a temperature during transit causes plants to respire more quickly and consume their carbohydrate reserves. It is the carbohydrate reserve that plants rely on to keep leaves green, promote bud development, and further flower opening in retail and home environments. Therefore, reducing temperatures during transit is important to conserve these carbohydrate reserves and the key to postharvest longevity.

Specific shipping temperatures are dependent on plant species (table 16-1). Some flowering potted plant species like azalea, chrysanthemum, rose, and lily, can withstand cold temperatures during transit. Holding these species at 35°F (2°C) will improve the pot life for the consumer. Conversely, most tropical plant species, such as African violet, orchid, hibiscus, poinsettia, and many tropical foliage plant species need to be shipped at higher temperatures ranging between 50 and 60°F (10 and 15°C); otherwise severe damage or immediate death can occur. Many of these plants, including poinsettia, should not be exposed to temperatures below 50°F (10°C). Symptoms of chilling injury include bluing on bract edges of poinsettia, bract drop on bougainvillea, and bud drop on hibiscus. For African violet and gloxinia, exposure to temperatures below 50°F (10°C) will actually kill the entire plant. Most foliage plants are also highly susceptible to low temperatures during transportation and need to be shipped at temperatures between 55 and 60°F (13 and 15°C).

Keep in mind that most crops show increased sensitivity to ethylene as temperatures climb. So, it is vital to ship plants at the lowest possible temperature they can tolerate. For plants that are prone to both chilling injury and ethylene damage (e.g., begonia, cymbidium, exacum, hibiscus, bougainvillea, and schefflera), it is best to ship for short durations. Do all that you can to avoid exposing sensitive plants to ethylene and consider treating them with anti-ethylene compounds (see the Specialized Treatments section in this chapter).

Crops that cannot be shipped at low temperatures and are ethylene sensitive make perfect choices for local markets or for markets that can be reached within twenty-four hours. Use these crops to establish a niche market locally with retail garden centers and mass merchandisers. Promote these plants on their long-lasting natures and the fact that they can be obtained successfully locally.

Use reputable transport companies that specialize in ornamental plant transportation (never transport with fruits or vegetables). This way, you can be somewhat assured that there will be temperature controls in place. Unless the trucking company uses temperature recorders in their trucks, there is no way of knowing what temperatures your plants are subjected to during transit. Ask if records are available to confirm proper temperature control during transit. If not, consider placing temperature recorders in your boxes or carts and have them sent back once the plants arrive at their destination. A growing trend in the industry is the use of radio-frequency identification (RFID) technology to track location, temperature, humidity, and other vital environmental measurements during transit. This tracking technology is currently being used in boat shipments of cut flowers. It is also commonly employed throughout the food industry when transporting highly perishable products. Consider tracking your shipments to ensure proper temperature control.

Shipping duration

The ideal situation for any plant is to avoid shipping. Unfortunately, that's not the world in which we live. Many plants are grown thousands of miles from their point of sale. As you plan how you will get your crops to market, choose the shortest possible shipping duration. Three days is the maximum recommended shipping window for many flowering plants. After that, each additional hour begins to take postharvest life away from plants.

It is common in the industry to store and ship potted plants for more than three days. Most people judge this practice to be acceptable because the plants still look fine upon arrival. However, looking good at the time of unboxing does not mean the plant and flowers will last for the consumer. Long storage and transit conditions can cause major postharvest problems. Due to a long period of darkness, buds and flowers can drop and leaves can start to turn yellow. Dehydration can also occur as storage and transit time increases. If improper shipping temperatures are encountered during extended transport, the result could be a shipment of unmarketable plants.

Any time a plant spends in storage counts as part of that three-day shipping window. Storage times and shipping times are cumulative. For example, if you store a crop for twenty-four hours (holding it to make a holiday market window), then you had better only be subjecting that same crop to transport conditions for an additional two days, if you plan to deliver high-quality and long-lasting plants.

Ethylene

One of the trickiest channels of the shipping process that you must navigate is reducing the plants' exposure to ethylene. The entire transport cycle offers opportunity after opportunity for plants to encounter ethylene. The results of ethylene exposure are dramatic and can completely undo all contributions toward plant quality that you have made. If the air is free of ethylene, there are still other factors to consider. For example, you must consider the plants' internal ethylene production as a viable cause of this damaging gas. Regardless of the source, ethylene can make plants unmarketable, reduce quality, and shorten pot life.

Ethylene is a colorless and odorless gas, making it impossible for our human senses to detect. Ethylene enters the floral transport systems in a variety of ways. It is generated from combustion heaters in the greenhouse, forklift and transport vehicle exhausts, smoke (including cigarettes), banana ripening rooms in supermarket distribution warehouses, and ripening produce near floral retail display areas in supermarkets. Decaying plant material also generates ethylene.

Ethylene is also a vital plant hormone produced in plants. It regulates many plant functions and is considered a "stress" hormone that plants typically produce in response to stressful conditions, such as injury, extended darkness, vibration, and high temperatures. It is also produced during the natural plant-aging process. As a result, some plants can exhibit ethylene symptoms even if they have not been exposed to ethylene externally. New Guinea impatiens, for instance, can drop flowers just from the vibration of shipping. The vibration triggers internal ethylene production, causing adverse symptoms to appear.

When plants encounter ethylene, the results vary depending on species, cultivar, stage of plant development, temperature during exposure, duration of exposure, and ethylene concentration. The effect of ethylene can range from no response to a severe response with one or more parts of the plant affected. The sensitivity and effects of ethylene has been determined for many potted plant species (table 16-1). Some potted plants can respond negatively to ethylene levels as low as 0.01 to 0.1 ppm, while other plants do not respond until levels reach 1 ppm or higher. In some instances, ethylene sensitivity may vary from variety to variety within the same species. Therefore, it can be difficult to generalize and classify entire species.

Once a sensitive plant is exposed to ethylene, response times vary. Some crops react immediately; others do not show any symptoms until several days later. Typically, sensitive plants tend to be less sensitive at lower temperatures and more sensitive at higher temperatures. Ethylene can cause numerous symptoms on plants. For instance, flower petals may shatter, buds and leaves may drop or turn yellow, flowers may fail to open, and plants may die prematurely (figure 16-1). In some cases, flowers that are already fully open are more sensitive to ethylene than those that are closed. Susceptible foliage plant species can experience leaf drop or leaf yellowing in response to ethylene exposure.

Figure 16-1. Ethylene can be extremely damaging to plants. Exposing periwinkle (*Catheranthus*) to 1 ppm ethylene for two days at 70°F (21°C) induced flower and bud drop and leaf wilting (right) compared to plants not exposed to ethylene (left).

Treatment with anti-ethylene compounds provides protection against ethylene injury, but it does not last forever. Usually protection lasts seven to twelve days. Care should be taken not to ignore temperature management, variety selection, and the elimination of ethylene pollution just because plants have been treated with an anti-ethylene compound. While protection may be obtained in some situations and on some crops, anti-ethylene compounds will not solve all ethylene issues faced in the distribution of potted plants. Further discussion of treating plants with anti-ethylene compounds is discussed in the next section.

Specialized Treatments

Specialized treatments are available to reduce or prevent many postharvest problems of potted plants. Growers are now able to protect plants from ethylene injury, conserve water during transit and in retail, and eliminate many troubling diseases. While most disease and pest controls are implemented during production using chemical, biological, or environmental controls, ethylene protection, in contrast, is done right before or during transport.

Ethylene protection

EthylBloc is currently the only commercial product designed to protect plants from ethylene injury that is approved for use on some potted ornamentals. This product was approved by the US Environmental Protection Agency in 1999. You should check with your local Department of Agriculture to confirm legal use of this product.

The active ingredient in EthylBloc is 1-methylcyclopropene (1-MCP), a powder that when wetted becomes a gas. The gas diffuses into the plant and works by inhibiting ethylene action. Growers should review the label so proper application procedures are followed. Many researchers have shown this compound to be very effective in protecting numerous potted flowering, bedding, nursery, and foliage plants from ethylene injury (figure 16-2). It has also been shown to improve postharvest quality even when plants are not exposed to external ethylene, since it can protect plants from the ethylene it makes internally, usually in response to stress induced during transport.

Figure 16-2. *Aphelandra squarrosa* plants were protected from ethylene exposure (1 ppm ethylene for four days at 70°F [21°C]) when treated with EthylBloc gas (middle) prior to shipping or with an EthylBloc Sachet (right) placed inside the shipping box as compared to untreated plants (left).

Since the EthylBloc delivery system is a gas, plants need to be treated in air tight, sealed areas such as coolers, sealed tents, sealed rooms, or trucks. Plants need to be treated right before shipping by the grower or can be treated during shipping in trucks. Without a sealed area, however, many growers find it difficult to apply EthylBloc at their facilities. Truck kits are available for use inside transport trucks. Another more recent mode of delivery is the use of EthylBloc Sachets, which resemble tea bags containing the 1-MCP powder. These sachets are designed for use in smaller treatment areas, such as inside transport boxes. The number of sachets needed per box depends on the size and volume of the box. Any openings or precooling flaps in the box should be closed or taped shut. The sachets are dipped in water for a few seconds and placed inside the shipping box. The box lid is closed immediately, and the gas is released over time during shipping.

Treating plants with EthylBloc is a very easy, efficient way to protect plants from ethylene. Most studies have found that the protection lasts from seven to twelve days. Growers are strongly urged to utilize this technology. Plants are not only protected from the devastating effects of ethylene, but the use of EthylBloc can promote longer-lasting plants even when not exposed to ethylene. Product and purchasing information can be obtained from Floralife at www.Floralife.com or AgroFresh at www.agrofresh.com/ethlybloc.html.

Delay wilting

Many potted plants are grown in soil mixes that contain difficult to wet particles such as sphagnum peat moss and pine bark. Wetting agents are commonly incorporated into many commercial or grower-made mixes to ensure rapid and uniform wetting of the media during the initial wetting and the production cycle. They can also be added in liquid form for use in constant-feed fertilization injection programs. Wetting agents act by reducing the surface tension of the water/particles, thus maintaining a better balance of water distribution and drainage.

The additional water held in the media can improve plant growth, reduce irrigation frequency, or delay the onset of wilting in container-grown plants. Wetting agents, however, can lose their effectiveness over time during the crop cycle if not in a constant-feed program. Drenching the media with a wetting agent (such as AquaGrow L with PsiMatric Technology or Oasis Soax), seven to ten days prior to shipping can assist in restoring the medium to its full water-holding capacity and make it easier to rewet. This is a benefit that is especially useful during postharvest transit or retail display, during which time plants have a tendency to become water stressed or dry out from neglect. Plants that are in mixes containing wetting agents at the time of shipping are easier to maintain and often do not wilt as easily.

Not to be confused as a wetting agent, synthetic commercial gels referred to as hydrogels, such as polyacrylamide gels (PAM), have a different mode of operation in soilless media. They have tremendous water-absorbing capabilities, but the claim that they provide more water during marketing has been questionable. Some studies have shown a delay in watering needs early in production with little or no benefit later in production or postharvest. A study in 2005 by researchers at the US Department of Agriculture (USDA) and the University of Toledo found that New Guinea impatiens treated with PAM were smaller and the number of flowers and flower longevity decreased with increasing amounts of PAM. To confound the matter, the water-holding capacity of PAM is altered, and thereby reduced, in the presence of dilute salts containing divalent cations such as Ca^{2+} and Mg^{2+} (found in fertilizers for many potted flowering plants). Even tap water can reduce the water-holding capacity of PAMs by more than 70%. Therefore, gels are not generally recommended as a means to delay wilt in the postharvest setting for potted plants. Careful consideration and testing should be done on a crop-by-crop basis using your production practices and water supply if you are considering the use of gels.

Retail Handling and Display

By far, one of the largest challenges in maintaining the "chain of quality" lies at the final stage of distribution—the retailer. Over the last decade, the majority of sales of potted plants have occurred at supermarkets and mass-market retailers. Our industry benefits economically from this distribution mode that makes plants easily available for busy, time-stressed consumers through large mass-market chains or food warehouses set up for one-stop shopping. The problem, however, at any retailer, large or small, is ensuring that proper care and attention is given to plants upon arrival and during display.

Unfortunately, many retailers and mass outlets make a detrimental mistake when plants are left in sleeves or on carts until point of sale. Quality will quickly diminish in this situation, making plants unsalable or sending valuable customers home with a plant that will not last. For instance, the act of sleeving poinsettias can make bracts droop, a condition known as epinasty. The sleeve causes the upward bending of the leaves and bracts, which causes the plant to produce ethylene. While plants will bounce back once out of the sleeve, the longer they are sleeved the worse the condition becomes and the longer it takes them to recover. If poinsettias are left in sleeves too long, bracts, leaves, and cyathia can prematurely drop, leaves can yellow, and the incidence of bract edge burn can increase (figure 16-3).

Figure 16-3. Prolonged sleeving of poinsettia (right) causes premature bract, leaf, and cyathia drop and increases the incidence of bract edge burn.

Regardless of packaging, plants need to be set-up and checked immediately upon arrival at the retailer. Sleeves and pot covers need to be removed carefully and without delay, pots need to be removed from carts, and plants need to be properly spaced in flats or on shelves. This allows free air movement around the plants, reduces humidity buildup, prevents disease, and allows needed light in. Regrettably, many mass merchandisers leave sleeves on plants in retail displays. This is a major mistake. The sleeves make watering plants extremely difficult and checking the soil moisture level impossible. This neglect has a very negative impact on plant quality and pot life. Sleeves are only designed to protect plants during shipping. Plants can easily be re-sleeved at the time of consumer purchase to protect them during the journey home with the consumer.

Retail set-up is another time to scout for any diseases or pests that may have developed during the stress of transport. Plants should be carefully inspected for any infestations. Common diseases that can show up after shipping include botrytis and powdery mildew. Left unchecked, these diseases can rapidly spread and infect nearby plants. Retailers should be taught to identify signs and symptoms of diseases and pests. Any infected plants should promptly be discarded.

Upon arrival, plants need to be checked for watering as well. For most crops, the soil should stay moist to the touch during display, not too wet nor dry. Plants should never reach a wilting state. As mentioned earlier, use of a wetting agent will allow for more uniform watering and delay the time to wilt at retail. Some plants may never recover from wilt. Others may appear to regain turgidity, but the damage caused by wilting can be long lasting. Premature leaf, bud, and flower drop or yellowing can occur as a result of excessive water loss.

When watering, allow excess water to drain out the bottom of the pot. Discard any excess water, as pots left sitting in water may suffer from permanent root rot damage. This is a concern when pots are foil-wrapped or placed in preformed covers. In this case, discard the water collected in the pot cover. For this reason, it is best to display plants without covers. Avoid getting leaves and flowers wet when watering.

Each crop will have specific display light and temperatures regimes that are needed while in the display environment. Many plants, with the exception of most bedding plants, will not tolerate direct sunlight. Too high a light will scorch many foliage plant species and too low a light can induce premature leaf, flower, or bud drop. For most species, the indoor light level should be at least 70–100 f.c. (0.8–1.1 klux) or higher and temperatures maintained at 65–75°F (18–24°C). For cooler-loving species like lily and other potted bulbs, displaying at 40–45°F (4–7°C) will slow bud and flower development until purchase, allowing a longer flower life for the consumer to enjoy.

Take time to provide your retail customers with this vital care information. This is a necessary "link" in the "chain of quality" that is often lacking. Our industry needs to do everything possible to educate the retailer about proper handling practices. Consider onsite training performed by a grower or other trained personnel and provided to individuals responsible for plant care, or hold plant-care seminars. This will help to maintain quality until point of sale, and in turn, provide lasting plants for the consumer. Realize that each time you do one thing to increase the postharvest life of a plant, you are influencing future sales. Some of the recent mandates by retail outlets for pay-by-scan is forcing growers to not only provide quality plants but to maintain the plants until they are sold. We cannot emphasize this enough: Plants that last longer in the home environment spur repeat sales.

What kind of information should you give to your retail customers? Provide a step-by-step check-in process that retailers can follow when plants arrive. Indicate any daily upkeep tasks that will keep plants looking good, such as checking for watering needs and removing spent leaves or flowers. Choose simple, direct language.

Consumer Care

Just as in retail display, special care is needed to promote quality once the consumer gets his purchase home. The principles of proper watering and display temperature and light at the consumer level are the crucial last steps in the "chain of quality." One of the most successful and easiest ways to educate the consumer is by adding specialized-care tags to each type of plant. Ensure that tags are specific for each crop—a "one tag fits all" approach does not often work. The home care for African violets is different from the home care for poinsettias; each plant has its own individual watering and light condition requirements.

Tags should outline the specific light and temperature requirements. Instructions on how and when to water are critical, as are any special fertilizer requirements. Special instructions such as keeping away from drafts or direct sunlight should also be addressed. This information truly makes a difference in a plant lasting as long as possible in the consumer setting.

Additional information regarding the suitability of a plant to be repotted or planted in the landscape after blooms are spent inside should be noted. Potted miniature rose, chrysanthemum, gerbera, and hydrangea are some crops that fit into this category. When the consumer knows they are able to reuse a plant in this way, it can be a huge selling point.

Summary

Postharvest quality and pot life is influenced by numerous genetic, production, cultural, and environmental factors from the moment a plant is propagated and planted until it reaches the consumer's hands. As a grower, everything you do will influence the subsequent quality of your crop from your variety selection to fertilizer and irrigation practices to disease and pest control regimes to choice of shipper and buyer.

Just as important is following each crop's necessary and specific handling and care requirements in order to promote optimal postharvest performance. Treat plants to protect against disease, water stress, and ethylene exposure. Maintain proper temperatures during storage and transport in order to prolong pot life and prevent diseases. Make sure your retail customers and end consumers know how to display and care for the plants to complete the final step in the "chain of quality." Follow postharvest principles to ensure plants will perform for the consumer. Preventing problems before they arise is the key to maximizing postharvest performance of potted flowering and foliage plants.

References

Blessington, Thomas M., and Pamela C. Collins. *Foliage Plants: Prolonging Quality.* Batavia, Illinois: Ball Publishing, 1993.

Blodgett, Allyson M., David J. Beattie, and John W. White. 1995. "Growth and shelf life of *Impatiens* in media amended with hydrophilic polymer and wetting agent." *HortTechnology* 5(1): 38–40.

De Hertogh, August. *Holland Bulb Forcer's Guide, 5th ed.* Batavia, Illinois: Ball Publishing, 1996.

Dole, John M., and Harold F. Wilkins. *Floriculture Principles and Practices.* Upper Saddle River, New Jersey: Prentice Hall, 1999. p. 613.

Frantz, Jonathan M., James C. Locke, Dharmalingam S. Pitchay, and Charles R. Krause. 2005. "Actual performance versus theoretical advantages of polyacrylamide hydrogel throughout bedding plant production." *HortScience* 40(7): 2040–2046.

Gibson, James L., Brian E. Whipker, Sylvia Blankenship, Mike Boyette, Tom Creswell, Janet Miles, and Mary Peet. *Horticulture Information Leaflet 530: Ethylene: Sources, Symptoms, and Prevention for Greenhouse Crops.* July 2000. North Carolina State University. http://www.ces.ncsu.edu/depts/hort/floriculture/hils/HIL530.pdf (accessed March 7, 2011)

Gross, Kenneth C., Chien Y. Wang, and Mikal Saltveit. *USDA Agriculture Handbook No. 66: The Commercial Storage of Fruits, Vegetables, and Florist and Nursery Stocks.* 2004. United States Department of Agriculture. http://www.ba.ars.usda.gov/hb66/title.html (accessed February 25, 2011).

Nell, Terril A. *Flowering Potted Plants: Prolonging Shelf Performance.* Batavia, Illinois: Ball Publishing, 1993.

Nell, Terril A., and Michael S. Reid. *Flower & Plant Care: The 21st Century Approach.* Alexandria, Virginia: Society of American Florists, 2000.

Nowak, Joanna, and Ryszard M. Rudnicki. *Postharvest Handling and Storage of Cut Flowers, Florist Greens, and Potted Plants.* Portland, Oregon: Timber Press, 1990.

"Chain of Life Network." Perishables Research Organization. 2011. www.chainoflife.org (accessed February 25, 2011).

Reid, Michael S., Linda L. Dodge, Ann I. King, and Richard Y. Evans. "Postharvest Care and Handling of Potted Plants." *Flower and Nursery Report for Commercial Growers* (Summer 1991): 1–5. http://ohric.ucdavis.edu/Newsltr/fn_report/FNReportSu91.pdf (accessed on February 25, 2011).

Woltering, Ernst J. "Effects of ethylene on ornamental pot plants: A classification." *Scientia Horticulturae* 31, no. 3–4 (May 1987): 283–294.

The authors would like to acknowledge the supporters of their research program: American Floral Endowment; AgroFresh, Inc.; Floralife, Inc.; National Foliage Foundation; Paul Ecke Poinsettias; Poulsen Roser A/S; Smithers-Oasis; and the many growers, nursery suppliers, and shipping companies that provide plant material and transportation services.

17

Jennifer Duffield White

Sustainability

sustainable floriculture: producing and selling greenhouse or field crops in a manner that provides a profit for the business, minimizes the impact upon the environment, maximizes employee well-being, and benefits the community.

Before sustainability became a common term in the boardroom and on the sales floor, before it was "trendy" to talk about, it was still an issue, whether or not we used the word *sustainable.* Any grower who has worried about how to reduce fuel costs; improve employee conditions; manage water, chemicals, fertilizers, and other inputs; and better the operation's energy efficiency has worried about sustainability in one way or another.

In the broad sense, sustainability is about the wise use of the resources we have: people, the environment, and finances. This is sometimes referred to as *the three pillars of sustainability* or *the three P's*: people, planet, profit. It's about looking at the big picture in order to make both big and small decisions.

There are dozens of interpretations when it comes to the definition of sustainability. However, the one most cited is attributed to a United Nations committee, the Brundtland Commission, assigned to address the problem of the degradation of the human environment and natural resources:

Sustainable development is development that meets the needs of the present without compromising the ability of future generations to meet their own needs.

The national Sustainable Agriculture Service breaks sustainable agriculture into three points: (1) environmental stewardship, (2) farm profitability, and (3) a prosperous farming community. While these principles have been on the minds of many for decades, the most recent surge in sustainable practices was brought on by a confluence of factors. On the international level, we are dealing with an industrialized economy, environmental degradation, increasing environmental regulations, limited resources, an expanding population, climate change, and a resulting rise in production costs. Gone are the days of cheap fertilizer, cheap energy, and cheap plastics. On top of that, a rapid growth in consumer awareness and demand for green products created a huge surge in "green" marketing efforts in the United States and an increase in the number of discussions regarding sustainability at the national and international levels. With other industries already in quick pursuit of a "green" profile, the North American horticulture industry began to really rally around the concept of sustainability in 2007.

The horticulture industry uses three major resources: water, energy, and plastics (with petroleum origins) to produce its products. The high demand and limited supply of all of these resources created a strong and compelling argument for sustainability. It quickly became obvious that using these resources efficiently and wisely (or seeking out new technologies or practices) would be necessary to manage costs. It isn't hard to imagine a future where the current or next generation may have to deal with scarcity issues in regards to oil, traditional fertilizer and chemical inputs, plastics, and so on. In fact, many growers have already experienced the debilitating effects of droughts, high oil prices, and escalating input costs. In other words, natural resources and resulting products can make or break the bottom line, and turning to more sustainable practices—be it finding new input products or improving efficiency—is a natural way of survival. Not surprisingly, managing the first two Ps (people and planet) can boost your third P (profit).

Thus, businesses around the globe are clamoring to be more efficient—to improve both their current bottom lines and those in the future. They are also searching for alternative technologies. Sustainability is becoming essential for companies wishing to remain resilient to future changes.

Marketing

It is impossible to discuss the sustainability movement without addressing the issue of marketing. After all, marketing is the most visible proponent of sustainability, and more than a few folks have gotten on the "green" bandwagon just so they can advertise their product as "green." However, it's important to remember that marketing is merely public talk created to sell a product or a company's reputation. That is all fair and well, but marketing does not represent nor necessarily communicate all that sustainability represents. Marketing is not the sole driving force behind horticulture and other businesses pursuing sustainability.

In addition, as consumers have steadily demonstrated a desire for "greener" products, there has been an exponential rise in so-called "green," "environmentally friendly," and "natural" products on the shelves of every store. That, in turn has led to what is termed "greenwashing," defined as a misrepresentation of a product's or company's environmental attributes. In short, there's a lot of stuff out there that's packaged to look like a smart choice for the eco-conscience consumer, but it's not living up to its claims.

You can avoid the greenwashing controversy by providing transparency in regards to your company and product. Use specific terms and explain how and why the product is produced in a responsible manner. Don't rely on the term "environmentally friendly" to do the work for you. After all, what does that even mean? The more specific and honest you are, the more the consumer will trust you.

While there are certainly some dangers to marketing green, there's also a large group of consumers who want to make the right purchasing decision. In a university study on sustainable horticulture options by Chengyan Yue, Jennifer Dennis, Bridget Behe, Charles Hall, Benjamin Campbell, and Roberto Lopez, survey results showed that consumers were most interested in purchasing plants in more sustainable containers and plants that were produced locally.[1] Those with children under the age of twelve also showed an increased interest in certified organic plants. *Note:* The survey only included the states of Minnesota, Indiana, Michigan, and Texas. There are hot pockets of green-minded consumers in the United States, as well as areas where "eco" is *not* as desired. Know your demographics. Some audiences may respond most enthusiastically to locally grown, while others may want certified products, plants with health benefits, or packaging with compostable or recyclable contents.

Can you charge more for a sustainable product? There are many examples of company's charging a premium for their products—be it certified organic, offered in eco-friendly packaging, locally grown, or having any of the myriad of "green" attributes. But there are also plenty of anecdotes from growers who have not been able to charge a premium for their biodegradable pots or their certified sustainable flowers.

The same group of researchers who studied consumer interest in sustainable products (Yue, Dennis, Behe, Hall, Campbell, and Lopez) examined what consumers are willing to pay for sustainable plant packaging (i.e., compostable containers and/or those made from waste material). Their results, published in the November 2010 *Journal of Agricultural and Applied Economics*, found that consumers were willing to pay more for sustainable plant packaging, though the amount varied. They found that the higher the percentage of waste material composition in a pot, the higher the premium. In the study, participants were also willing to pay more for a container that was carbon saving, and they discounted containers that were labeled as carbon-intensive.[2]

In the last few years, in response to greenwashing, corporations in numerous industries have turned toward improving their overall corporate image, rather than just relying on specific products. Surveys have shown, as well, that consumers positively respond to the perceived reputation of a company. If your business is pursuing sustainability, tell your story. Tell your customers about your philosophy, your actions, and your history. How can your customers be part of your story?

Horticulture often claims to be the original green industry. And while, overall, it certainly has a number

1. Yue, Chengyan, Jennifer Dennis, Bridget Behe, Charles Hall, Benjamin Campbell, and Roberto Lopez. 2011. "Investigating consumer preference for organic, local, and sustainable plants." *HortScience* 46(4):610–615.

2. Yue, Chengyan, Charles R. Hall, Bridget K. Behe, Benjamin L. Campbell, Jennifer H. Dennis, and Roberto G. Lopez. 2010. "Are consumers willing to pay more for biodegradable containers than for plastic ones?" *Journal of Agricultural and Applied Economics* 42(4):757–772.

of things to improve on in this arena, there is a lot of truth to the idea that it has always been green. Let's not forget that plants have numerous environmental, psychological, and community benefits. In addition to the sustainable actions of a grower or retailer, don't ignore all the benefits that add up once the products leave the store.

Certifications

With greenwashing occurring in every consumer segment, manufacturers across the globe have turned to certifications and standards in order to support their marketing claims. While certification schemes are common in the international horticulture market, it is a rather new effort for North American horticulture, except for the certified organic movement, which until recently was more for harvested food products than it was for ornamentals or even vegetable starts.

Today, however, horticulture has a growing number of certifications—brought on, in part, by retailers—and presumably consumers—who want the assurance of certified products. You can find certified sustainable or certified organic cut flowers, container-grown ornamental plants, and edible plants. You can also purchase a number of input products (from media to pest control materials, containers, fertilizer, seed, and plant tags) that have various certifications.

Right now, North American growers are working with three main certification schemes for plants: USDA organic, VeriFlora, and MPS.

USDA organic is probably the most widely recognized certification in the typical American grocery store. Growers who are certified organic must be certified by an independent third party and adhere to a strict list of materials that can and cannot be used in production. (There are a number of third-party certifiers to choose from, and they range in both reputation and cost.) While greenhouse growers have found a solid niche for certified organic vegetables and herbs, few have found a matching demand for organic ornamental plants. Most notably, certified organic growers cannot use synthetic fertilizers or pest control materials. While organic has stringent regulations regarding chemicals and production practices, it does not address some of the larger issues of sustainability, such as water conservation, employees, energy usage, waste management, and so on.

VeriFlora is a sustainable certification for ornamental cut flower and container-grown plants. Growers are certified against the standard by SCS Certified, a large third-party certifier that certifies for a number of environmental labels. In this certification, growers are audited annually by SCS, which looks at everything from chemical usage to irrigation systems, employee relations, energy use, soil systems, waste management, habitat management, product quality, and community relations. In this program, the standard is one developed by SCS.

MPS, based in the Netherlands, offers a number of different certifications for growers. They have been certifying the international growing community since 1995, but they only entered the North American market in the last two years, when growers began to ask for another certification option. The MPS-ABC certificate certifies against a sustainable standard, but also gives growers a grade and monitors them on a monthly basis—the goal being to encourage improvement over time. Growers submit numbers on energy usage, chemical and fertilizer usage, water, employee safety, waste, and more each month, and they receive the certification after a year of submitting data.

On the input products side, manufacturers are making it easier for growers to source more sustainable materials. You can find plant containers that are certified to contain a particular percentage of post-consumer recycled material. There are fertilizer and pest-control materials that are approved for organic growing (by the Organic Materials Review Institute—OMRI). Peat moss suppliers have begun to certify their product as sustainably harvested. VeriFlora can certify "materials manufacturers" as sustainable (including those who make containers and biological pest control materials), and a number of other certifications exist, depending on the industry involved.

How to Be Sustainable

Some argue that it is impossible to calculate or ensure sustainability. After all, how do we know for sure how long certain resources will last and what a sustainable rate of use might be? We make educated guesses. We make improvements. And sometimes, what's good for one "P" might not be good for another "P," and you have to make tough decisions.

However, that's no reason to give up. The sustainability revolution is not about becoming 100% sustainable by a strict definition; it's about becoming "more sustainable"—a process of constant improvement.

Making decisions that are smart for the environment, smart for your business, and smart for your employees and the people who live in your community are not mutually exclusive.

The good news is that you're probably already doing a lot of things that have put you on the path to sustainability—practical and money-saving practices that are easy to implement or are just plain common sense. The flip side is that no one is all the way there; it's a series of innovations and steps in the right direction.

The tricky part of sustainability is that the answers aren't always black and white; what will work for one grower won't necessarily fit into another grower's business model or facility. While burning woodchips might be considered local, renewable, cheap fuel for one grower, it could be a risky investment and have more negative environmental effects for a grower in a different location. Likewise, certain compostable pots may be a no-brainer for some crops, and problematic for plants that grow in the container for longer periods. Pest management decisions always have to be tailored to the crop and facility. And when it comes to water, some locations may face troubles with nutrient runoff, while others may be looking at serious water supply issues. Thus, part of sustainability is really just about critical thinking. Evaluate the tradeoffs. Think beyond just the immediate numbers and ask yourself if it pays off in the long run. Know what environmental issues and regulations might affect your business. Know where you might encounter price volatility.

While there's no perfect sustainability equation for all greenhouses, there are a lot of great options to consider and technology to keep an eye on for the future. Some of the suggestions below may be just right for some operations and not as suitable for others.

Energy

Keeping your energy use down is a no-brainer. It also saves money. There are some easy investments, such as energy curtains, that pay off and start saving you money quickly. There are other investments, such as a new heating system, that require more time to pay off as well as lots of research about fuel availability, pros and cons, and future developments. The good news is that greenhouses are solar collectors. The bad news is that we still have to heat and cool them, and some days they gain too much heat while other days they lose too much.

Some things growers are doing to be more sustainable about their energy usage include

- Renewable energy: solar, wind, and geothermal;
- Biomass boilers (to burn corn, wood, and other waste materials),
- High-efficiency heaters,
- Energy curtains,
- In-floor heating,
- Environmental control systems,
- Energy-efficient lighting,
- Equipment, growing methods, and structures that reduce energy consumption,
- Energy audits,
- Choosing crops with low-energy needs,
- Altering crop production processes to save energy, and
- Choosing efficient structures to reduce heating/cooling needs.

Water & Fertilizer

When it comes to water, there are multiple issues at hand. For some regions, water availability is—or is about to become—a chronic issue. Others areas may fear only periodic droughts. However, regulations at local and state levels have also made water an important issue. Greenhouse operations have faced regulations on how much water they can use, as well as regulations about containing the runoff on their property and protecting nearby waterways. When it comes to water and fertilizer, think about both conservation/efficiency and containing any possible pollutants.

Some examples of what growers have done to mitigate these issues include

- Irrigation systems that reuse water (flood floors, troughs, etc.),
- Drip irrigation,
- Irrigation ponds,
- Rainwater collection systems,
- Fertilizer applications that reduce nutrient runoff,
- Compost teas,
- Sourcing fertilizer with more sustainable components,
- Reclamations systems that capture runoff in outdoor growing areas, and
- Filtering runoff water before it leaves the property.

Pesticides

Pest control continues to be a controversial topic, and the industry is faced with several opposing arguments, ranging from public rhetoric to anti-chemical lobbying groups to corporate rhetoric from the manufacturers themselves. Growers are taking a number of different approaches to "sustainable" pest management, such as cutting out conventional chemicals altogether or using integrated pest management (IPM), where they're using a combination of scouting, preventative culture, biologicals, and traditional chemicals. At the same time, what is "traditional" may be changing as chemical companies look to offer safer and reduced-risk chemicals. In short, pest management remains one of those case-by-case situations, where there are dozens of factors to consider when choosing a pest control solution. Growers must evaluate everything from the efficacy of the product to the age of the crop, the severity of infestation, the size of the area to be treated, the cost, and their own priorities in terms of what kinds of pest control materials they want to use.

In short, growers are making their pest management practices more sustainable in a number of ways, such as:

- Choosing their chemicals wisely and applying only when necessary,
- Using IPM techniques,
- Using biologicals and beneficial insects, and
- Improving conditions/safety for workers.

Crop Culture

When the price of input products, such as fertilizers, pesticides, and growing media, starts to rise, everyone always scrambles to figure out how to reduce costs. There has been much research on growing cooler crops, reducing fertilizer and pesticide use, finding alternative media supplements, and so on. Sometimes it is a matter of reducing. Other times, it may pay to look for an alternative. In this case, cost savings and environmental benefits often go hand in hand.

Growers have adjusted their growing to be more sustainable in a number of ways, including:

- Growing cultivars with lower pest and disease pressures,
- Choosing cultivars suited to the climate (i.e., drought tolerant, heat tolerant, cold tolerant, etc.),
- Choosing cultivars with shorter growing times,
- Adjusting crop schedules,
- Using IPM techniques,
- Focusing on soil/media health,
- Adjusting nutrition programs, and
- Sourcing input supplies with sustainability attributes in mind.

Offices & Transportation

Every manufacturer has offices and transportation to deal with as well as marketing. There are numerous sustainable options for these ubiquitous functions. Here's a sampling:

- Online ordering and e-mail to reduce paper use,
- Power usage management for computers and office equipment,
- Efficient lighting,
- Recycling and waste management,
- Fuel-efficient vehicles (biodiesel, hybrids, etc.),
- Modified shipping routes and methods,
- Efficient use of space in shipping trucks and racks,
- Transparent communication about sustainability,
- Promotion of the sustainable aspects of a product, and
- Promotion of locally grown products.

The Life Cycle

Manufacturers often speak of their products and sustainability in terms of life cycle. In other words: *What goes into producing and disposing of this product during its lifetime?*

When asking a supplier about the life cycle of their product, here is what you should look for:

- Raw materials used (and if they are renewable or recyclable),
- Energy used to produce the product,
- Waste generated in production,
- Transportation required to ship the product or the raw materials,
- Waste created at the end of the life cycle, and
- Whether or not the product is recycled, recyclable, compostable, or biodegradable.

Employees & Community

Don't forget the ever-important third pillar of sustainability: people. After all, a hard-working, loyal workforce does wonders for any company. Good relations with your neighbors also helps keep the peace (and it may help sell some plants along the way). Growers who give back to their communities will tell you that it's not only the right thing to do, it's also a wise business strategy.

Here are some ideas for sustaining good relations with your employees and community:

- Provide employee benefits (both monetary and otherwise),
- Schedule teambuilding activities,
- Ensure a safe, healthy and enjoyable working environment,
- Find ways to thank your employees (i.e., incentives, activities, gifts, food, time off, etc.),
- Involve employees in sustainability discussions and brainstorming,
- Get involved with community philanthropy,
- Join local environmental efforts, and
- Promote the benefits of plants and gardening.

Sustainability is an ongoing journey. Give yourself credit for what your business is already doing and create a vision for the future. While there is much change ahead, sustainability can help make each business more versatile and resilient in the face of those changes. Most importantly, though, don't forget that plants can be part of the solution.

Part 2:
Crop Culture A–Z

Abelmoschus

Abelmoschus esculentus
Common names: Okra; gumbo; lady's fingers
Annual vegetable

Okra is a mostly upright, heat-loving annual vegetable that excels in the southern United States but can be grown in more northern locations as well. Plants produce elongated, finger-like pods that are used in cooking. Plants are related to hibiscus and grow from 3–4' (0.9–1.2 m) tall, preferring full sun and warm to hot temperatures in the garden for best production. Flowers are single and white in color with a black throat in most cases.

Propagation

Sow seed to a 512-plug tray using one seed per cell and cover lightly with vermiculite. Germinated at 70–72°F (21–22°C), seedlings emerge in five to seven days and plugs will be salable three to five weeks after sowing.

Another method is to sow to a 288 or larger plug. Treat as you would a 512-plug tray, and plugs will be ready to transplant in five to six weeks.

While some growers use two or three seeds per cell, this is more beneficial when sowing to a 128 or larger plug rather than smaller plug trays.

Growing On

Transplant plugs to cell packs or 4" (10 cm) pots and allow another four to six weeks for a 32-cell pack or six to seven weeks for 4" (10 cm) pots to be salable green when grown at 70–75°F (21–24°C) days and 60–65°F (16–18°C) nights. Plant growth accelerates under warm and high light conditions. Equally, plants can begin to stretch. If the growth starts to become elongated, drop the night temperatures down to 58–60°F (14–16°C). Temperatures below 54°F (12°C) slow and restrict growth; these lower temperatures can also curl the leaves.

Toward the end of production, the lower foliage may begin to yellow and drop as the plants become root-bound in the cell pack. They can still be planted to the garden with great success. However, if the sales period is still several weeks away, transplant from a cell pack to 4" (10 cm) pots and allow another two to three weeks to fully root, growing at the temperatures noted above.

Abutilon

Abutilon × *hybridum*
Common names: Flowering maple; Chinese lantern
Tropical or semi-tropical; treated as an annual in cold-winter climates (or as a houseplant)

This old-time plant is known for its large, single flowers (2–3" [5–8 cm] across) in bright colors of apricot, red, white, salmon, cream, yellow, and coral. Plants grow from 18–36" (46–91 cm), depending on variety. The tropical-looking flowers on these heat-loving plants make abutilon a great product for bedding plant growers' spring and summer sales in pots, especially the seed variety 'Bella'. Plants perform best in morning sun and afternoon shade.

Propagation

Sow seed into a well-drained, disease-free medium with a pH of 5.5–6.3. Cover seed lightly with coarse vermiculite. Provide 100–200 f.c. (1.1–2.3 klux) of light during germination. Radicles will emerge in three to six days at 65–72°F (18–22°C). After germination, raise light levels to 1,000–2,500 f.c. (11–27 klux) and

reduce the temperature to 65–70°F (18–21°C). Begin fertilizing with 50–75 ppm 15-0-15 immediately and increase to 100–150 ppm once cotyledons have fully expanded. If growth slows, then use 20-10-20 for every third irrigation. Maintain EC at 1.0–1.5 mS/cm. After cotyledons have expanded, reduce humidity by allowing trays to dry slightly between irrigations. Do not allow seedlings to wilt.

As seedlings mature, increase light up to 5,000 f.c. (54 klux), if temperatures can be maintained. If growth regulators are required during the plug stage, then a one-time application of B-Nine at 2,500 ppm may be helpful.

Plug crop time is approximately four to five weeks in a 288-plug tray.

Abutilon may also be propagated vegetatively by stem cuttings.

Growing On

Pot plants use one plug per 4" (10 cm) pot, two to three per 6" (15 cm) pot, and four to six per 10" (25 cm) hanging basket. Select a medium with a pH of 5.5–6.5 and a moderate initial nutrient charge.

Grow on at 65–75°F (18–24°C) days and 60–70°F (15–21°C) nights. Temperatures below 60°F (15°C) may cause excessively compact, reduced growth. Temperatures above 85°F (30°C) may cause flower bud abortion.

Maintain light levels at 4,000–5,000 f.c. (43–54 klux). Supplemental light during low-light times of the year will shorten time to flower. Under low light conditions, plants will stretch. High light conditions promote a shorter plant overall.

Abutilon's growth is checked if plants dry out: Strive for uniform moisture at all times. Do not allow plants to severely wilt. Nutritional requirements are light to moderate. Use 150 ppm from a nitrate-based fertilizer at every other irrigation. Ammonium fertilizers promote large leaves. Too much nitrogen fertilization will result in excess growth and suppressed flowering.

Crop time in the spring is eight to ten weeks for 4 and 6" (10 and 15 cm) pots and ten to twelve weeks for 10" (25 cm) baskets. In the summer, crop times are two weeks faster.

If grown pot-tight, apply a Bonzi spray at 5 ppm two to three weeks after transplant. For separate 'Bella' colors, only one application should be needed. For 'Bella Mix', a second application at the same rate can be given two weeks later, if necessary.

To increase branching on 'Bella Mix', apply Florel at 500 ppm two weeks after transplant, combined with a pinch one week later. This treatment will delay flowering one to two weeks; however, it will make a nicer plant with three to four uniform branches. Space when the foliage touches the sides of the pot. This stimulates the basal branching for a symmetrical plant. Pinching is not needed for the separate 'Bella' colors.

Pests & Diseases

Aphids, mites, thrips, and whiteflies can attack abutilon. Botrytis, *Pythium*, and rhizoctonia can become disease problems. *Warning:* Abutilon is sensitive to pesticides containing hydrocarbon solvents such as emulsifiable concentrate formulations.

Varieties

Few seed-propagated varieties are still available on the market at this writing. They were often irregular to germinate and plants were uneven to finish. The main seed variety for the past few years has been 'Bella', which is a grower-friendly F_1 abutilon with flowers in apricot, coral, red, peach, rose, pink, ivory, and lemon yellow. Plants bloom continuously, although during the winter (in the home) the flowering is sparser.

Variegated abutilon, which is vegetatively propagated, is available from a wide range of liner suppliers and is often sold in the marketplace.

Achillea

Achillea sp.
Common name: Yarrow
Perennial, cut flower (Hardy to USDA Zones 3–8)

The many different types of *Achillea* are mainstays for perennial growers and specialty cut flower producers. Various species are highlighted at the end of this culture. They are great in a mixed border, adding a fine leaf texture, and they make excellent fillers in mixed bouquets.

Propagation

Achillea can be started from seed, stem-tip cuttings, or bare-root transplants.

Seed

Sow two to three seeds onto a disease-free medium in larger plug trays (288 or larger). Do not cover or cover sparingly with coarse vermiculite to help maintain humidity at the seed coat. Keep medium uniformly moist. Germinate at 65–70°F (18–21°C) in 100–400 f.c. (1.1–4.3 klux) of light. Maintain a pH of 5.5–5.8 and an EC less than 0.75 mS/cm. Radicles will emerge in three to five days.

Move trays to Stage 2 by reducing substrate temperature to 62–65°F (17–18°C). Once cotyledons have fully expanded, increase light to 500–1,000 f.c. (5.4–11 klux) and begin feeding weekly with 50–75 ppm 14-0-14. Stem and cotyledon emergence will take seven to ten days.

Developing true leaves in Stage 3 will take from twenty-eight to thirty-five days. Reduce substrate temperature to 60–65°F (16–18°C) and begin to allow trays to dry thoroughly between irrigations. Increase light to 1,000–1,500 f.c. (11–16 klux). Feed with 100–150 ppm 14-0-14 weekly. Allow the EC to increase to less than 1.0 mS/cm. If growth regulators are needed, use A-Rest, B-Nine, or Cycocel at this time.

Harden plants in Stage 4 over seven days prior to transplant or sale by lowering substrate temperature to 58–62°F (14–17°C) and increasing light to 1,500–2,500 f.c. (16–22 klux).

From sowing to a salable 288-plug tray will take from five to seven weeks for most varieties. For flowering plants during summer, sow seed from June to August the previous year for overwintering plants dormant in a cold frame. Plants are sold green in the spring in their container and will flower during summer. Mid- to late winter sowings will produce salable plants for the spring. However, they will not flower profusely during the summer.

Plants sown the year before will flower under natural conditions the following summer.

Cuttings

Cuttings can be stuck to various liner trays from 105 to 32 cells per flat, can be dipped in a rooting hormone, given 68–74°F (20–23°C) bottom heat, and then can be rooted under mist. Mist can be discontinued after seven to nine days, depending on the development of roots. Cuttings do not need to be pinched, and liner trays are well rooted after four to five weeks (depending on tray size). An application of a preventative fungicide spray is recommended after sticking the cuttings.

Bare-root transplants

Bare-root transplants are widely available from a number of commercial propagators. Pot up roots in January or February to 1-qt. (1.1 l) or 1-gal. (4 l) pots, with the crown just below the soil line. Water in thoroughly and allow the plants to root in the greenhouse at 65–72°F (18–22°C) days. Night temperatures should be 5–8°F (3–4°C) cooler.

Growing On

Seed

Seed sown to plug trays in summer is transplanted to 32-cell flats and grown on for four to six weeks prior to transplanting to 1-qt. (1.1 l) or 1-gal. (4 l) pots. These plugs are allowed to fully root and then are overwintered dormant for sales the following spring.

If buying in plugs after the first of the year, be sure to receive trays by February. This allows time for the plugs to bulk up prior to their sale as green plants in the spring. If the plugs are vernalized (given a cold treatment), then they will flower profusely during the summer months. If not, plants will flower but not as profusely.

Seed can also be sown from October to November for transplanting to containers in winter. These are grown cold during the winter (35–40°F [2–4°C] nights), once the roots have fully established in the final container.

Achillea flowers best when provided with a cold period (vernalization) to even up plant development and uniform flowering. To force plants into flower for spring sale, bulk seedlings up before putting them into cool treatment. Roots should be fully developed throughout the pot prior to cool treatment. Bulk plants up to at least a 70-liner (or a 32-liner as previously noted) size (i.e., 2.5" [6 cm] container) at 58–62°F (14–17°C) days and 55–60°F (13–15°C) nights. Maintain 2,000–3,000 f.c. (22–32 klux) of light and fertilize with 100–200 ppm 14-0-14 at every other irrigation. Maintain substrate pH at 5.5–6.2 and EC at 1.0. mS/cm.

Bulked-up plants can then be moved into cool storage. Vernalize at 41°F (5°C) for six weeks under 25–50 f.c. (269–538 lux) of light in a cooler or unheated greenhouse. Pots can be stored pot tight; fertilizer is unnecessary.

Provide long days to trigger flowering on vernalized plants through the forcing period (four weeks). Provide nine-hour days with four hours of night interruption, or use supplemental light for more than fourteen hours daily.

Force plants at 68°F (20°C) days and 65°F (18°C) nights. Allow plants to dry thoroughly between irrigations. Provide high light (3,000–5,000 f.c. [32–54 klux]). *Achillea* are not heavy feeders—do not overfertilize. Too much nutrition may cause plants to focus on leaf development, not flowering. Feed with 75–100 ppm from 15-0-15 at every other irrigation. Maintain pH at 5.5–6.2.

If required, height can be controlled with A-Rest, B-Nine, Bonzi, or Cycocel.

Cuttings

For cuttings taken in spring to early summer, transplant rooted liners to 1-gal. (4 l) containers during early to midsummer. Grow plants throughout the summer outdoors or in cold frames or similar environment.

For cuttings taken later (i.e., November to midwinter), allow four to six weeks to root. Transplant to 1-gal. (4 l) containers. Once fully rooted in final containers, expose the containers to 35–41°F (2–5°C) for six to nine weeks. This will increase overall flowering. Cold temperatures (vernalization) are not required for flowering, but they help to increase overall flowering as well as encourage earlier blooming.

From transplanting, quarts (1.1 l) are ready in six to nine weeks and gallons (4 l) from eleven to thirteen weeks.

Bare-root transplants

Transplants are available from a number of sources. Potted up in early February to 1-gal. (4 l) containers, plants root quickly upon potting (within six to eight weeks) and can be moved to a cold frame or similar environment once the roots reach the side of the container. Grow on with nights at 48–50°F (9–10°C) and days at ambient (natural daytime temperatures as long as they are above 55°F [13°C]). Plants are fully rooted and salable green by late April or May.

Cut Flowers

Work beds so they are loose and well drained. *Achillea* prefers a medium low in salts. For field production, plant roots or divisions at 1/ft.^{2} (9/m^2) using 12" (30 cm) centers. A spring or summer planting will begin yielding stems about four months later. An early fall planting in temperate or tropical areas can give production in winter/early spring. In very high light areas, the crop may benefit from being grown under 30–50% shade or Saran plastic wrap to maximize stem length. *Achillea* will benefit from dividing plants after the second year of production, as crowded plants have shorter stems and lower yield. Use one layer of support.

In the greenhouse, plantings from September to December will give a good flush from January to March. Plant 2.5 plants/ft.^{2} (27/m^2) and provide one layer of support.

Pests & Diseases

Aphids, scale, and spittlebugs can attack. Diseases such as downy mildew, powdery mildew, root rot, and yarrow leaf spot (*Entyloma achilleae*) may also become problems.

Varieties

As *Achillea millefolium* is the most common species of *Achillea*, there are a number of both seed and vegetatively propagated varieties.

'Cerise Queen' is a cerise-red, flowering variety more aptly described as rose red in its flower color. Plants can be 3' (90 cm) tall once they are established in the garden, but plants may also range from 2–2.5' (60–75 cm) tall. Both seed and vegetative forms exist. Seed-propagated plants are more vigorous, and the flower colors change in shade more readily than on plants that are vegetatively propagated.

'Summer Pastels' is a seed-propagated mix of colors that won an All-America Selections Award in 1990. Plants grow to 16" (41 cm) and spread to 16–18" (41–46 cm) across. Flower colors include salmon, cream, soft blush pink, and primrose yellow. Other colors will appear as more of the mixture is grown, but these are the predominant colors. This is an excellent variety especially recommended for landscaping.

As for true, vegetatively propagated varieties, the list of varieties is numerous. Refer to each propagator's catalog for detail.

Related Species

Achillea filipendulina can be propagated from either seed or cuttings. Highlighted by the popular variety 'Cloth of Gold', this species will not flower readily during the first year if sown from seed in December or later. While some flowering is noted from December or January sowings, it is slight at best. Follow the guidelines above under seed sowing for mid- to late summer sowings for flowering plants the following summer.

A. × 'Coronation Gold' (vegetatively propagated) is a well-known *Achillea* and is suitable as a perennial or cut flower. Landscapers love it because plants are maintenance free. It does require deadheading, since the flowers do not fall away once they die. Flowers are golden yellow. Follow the culture guidelines above for cuttings for propagation and growing on.

Achillea hybrida 'Moonshine' is an interspecific cross of *A. clypeolata* and *A. taygetea* and is vegetatively propagated. Flowers are sulfur-yellow atop fernlike foliage. 'Moonshine' is one of the perennial industry's lynchpins. Follow the culture guidelines above for cuttings for propagation and growing on.

A. ptarmica (seed and vegetatively propagated) is a white-flowering variety. 'The Pearl' is a well-known seed-propagated variety in this species. Seed sown in January or February will flower the same year as sown. However, there is a greater profusion of blooms if sown the previous year using the cultures highlighted above.

'Gypsy' is a more recent development with fully double-blooming, white flowers that is propagated by cuttings. This excellent selection can be grown using the steps noted above for cutting production.

Postharvest

Achillea flowers naturally in the summer. Perennial consumers will like the fact that most *Achillea* will reflower in about one month if plants are cut back after flowering.

Harvest cut flower stems when flower heads are fully open (pollen visible). Premature harvest reduces vase life and can cause flowers to drop or stems to wilt in the vase. Immediately after cutting, place stems in clean water. Use a bactericide or biocide at all times during postharvest.

Achillea may also be dried by hanging stems upside down. Do not harvest flowers too early as the results will be disappointing.

Achimenes

Achimenes sp.
Common names: Star of India; monkey-faced pansy; orchid pansy; hot-water plant
Pot plant, annual, tender perennial

Achimenes are members of the Gesneriad family, the same family as African violets and gloxinias. The popularity of achimenes has waned and surged since they were first introduced to England. Recently there has been a renewed interest. Achimenes are mainly grown as summer-flowering houseplants but are also well suited for use as pot plants, in mixed containers, and for hanging baskets. The single (or double), 1–3" (3–8 cm) trumpet-like flowers come in an array of colors including white, scarlet, salmon, pink, blue, lavender, purple, and even yellow. Plants bloom profusely year-round indoors and from early summer to fall outdoors. Their pubescent foliage adds interest and ranges from bright to dark green; some varieties possess bronze or burgundy undertones. Achimenes are more widely grown and distributed in Europe as compared to the United States, where rhizome or cutting availability is limited.

Propagation

Achimenes are typically propagated from pinecone-looking rhizomes and will flower in two to four months at 68–72°F (20–22°C) days and 60°F (16°C) nights. Stem tip and leaf cuttings are also used for clonal propagation and root readily, producing flowering plants in three months. Achimenes can be grown from seed, but crop variability increases. The tiny seeds (3 million/oz. [85 million/g]) are best sown in December to March and germinate in fourteen to twenty days at 75–81°F (24–27°C). Seed-grown plants will flower in five to six months.

Growing On

Three to five rhizomes can be planted less than 1" (2 cm) deep in a well-drained media. Plants grow best at high indirect (diffuse) light. Higher (direct) light can cause leaf and flower burn. Achimenes perform well with a constant feed of 200 to 300 ppm nitrogen (N), and potting media must be kept evenly moist, as premature dormancy is reported to occur when pots are allowed to dry out. Achimenes are a day-neutral plant, and flowers are initiated after the third or fourth leaf node. Plants are marketable in eight to sixteen weeks, depending on whether plants are started from rhizomes or stem cuttings.

After flowering ceases in autumn, the aerial vegetation begins to senesce and the rhizomes enter a dormant phase. Rhizomes can be harvested and stored in dry conditions, where they remain dormant for a substantial period of time—from two to five months, depending on the cultivar. Dormancy can be broken by storing rhizomes at 50°F (10°C) for forty days and then transferring to 72°F (22°C). After dormancy release, rhizomes can be directly planted into pots. Rhizomes that do not receive the minimum dormancy period have the potential to pupate, or develop new rhizomes at the apical meristem, further delaying shoot growth.

Pests & Diseases

A few pest problems are associated with achimenes. The long, tubular flower shape provides perfect conditions for thrips (*Echinothrips americanus*). Aphids, whiteflies, and cyclamen mites are also potential pests. *Botrytis cinerea* can be problematic, particularly when ventilation is poor and plants are not given enough space during production.

Troubleshooting

As with other gesneriads, it is important to irrigate with water at ambient room temperature or warmer during production. Achimenes are susceptible to leaf spotting due to cold water.

Some achimenes cultivars may benefit from PGR applications, but flower delay may occur. Sprays of Bonzi (25–100 ppm) or A-Rest (25 ppm) have been reported to reduce plant height and also suppress flowering. Other research has shown Florel applications (250 ppm) can reduce plant height and delay flowering. However, small trials should be conducted with your cultivars in your greenhouse.

Varieties

Commercial production of achimenes rhizomes is limited. Often rhizomes are sold as mixes. Smaller, specialty plant growers produce many different achimenes cultivars. Plant habits for different cultivars vary, ranging from spreading or pendulous flowering stems to compact, multi-branching flowering stems. Some cultivars include 'Prima Donna', 'Jewel Pink', 'Ambrose Verschaffelt', 'Blue Sparks', 'Snow Princess', 'Purple Prince', 'Charm', 'Cascade Violet Night', 'Schneewittchen', 'Linda', 'Tetraelfe', 'Rosenelfe', and 'Flamenco'.

Postharvest

Little research has been done regarding the postharvest care of achimenes. It has been reported that achimenes are highly sensitive to ethylene. Research has shown that flowers and flower buds begin to abscise within twenty-four hours of ethylene exposure. Sprays of silver thiosulfate decrease the negative effects of ethylene and increased plant longevity. EthylBloc treatments may also protect against ethylene. To avoid chilling injury, plants and rhizomes of achimenes should not be exposed to temperatures below 41°F (5°C).

Achmella

Achmella oleracea (also known as *Spilanthes*)
Common name: Eyeball plant; toothache plant
Annual

Also referred to as *Spilanthes* in some sources, most people in the commercial bedding plant world never even knew *Achmella oleracea* existed until Pan American Seed introduced 'Peek-A-Boo' with its olive-shaped, yellow and burgundy flowers. Take one look at 'Peek-A-Boo', and the word *fun* automatically springs forth from your lips. For growers, it is an easy crop and makes a striking addition to combination pots and planters or a novel, stand-alone pot. Landscapers love its whimsical appearance. It even has a medicinal use as a toothache pain reliever.

Plants grow from 12–15" (30–38 cm) tall and can spread as much as 24" (60 cm) across. Plants prefer full-sun locations.

Propagation

Sow seed onto a well-drained, disease-free, germination medium with a pH of 5.5–6.3 and a moderate initial starter charge. Cover seed lightly with coarse vermiculite. Germinate at 72–76°F (22–24°C). If germinating in a chamber, provide 10 f.c. (108 lux) of light during germination, or germinate on a bench. Keep medium uniformly moist. Radicles will emerge in four days.

Move trays to Stage 2 when half of the seedlings are germinated for stem and cotyledon emergence; increase light to 1,000–2,500 f.c. (11–27 klux) and reduce moisture levels once radicles are fully emerged. Provide soil temperatures of 65–72°F (18–22°C). Once cotyledons emerge, begin to fertilize with 50–75 ppm from 15-0-15, increasing the rate rapidly to 100–150 ppm as seedlings develop. Do not allow seedlings to wilt.

Move trays to Stage 3 to develop true leaves by providing soil temperatures of 65–70°F (18–21°C). Increase light levels to 3,000–4,000 f.c. (32–43 klux).

Harden plugs in Stage 4 for seven days prior to sale or transplanting by increasing light to 5,000 f.c. (54 klux) and lowering temperatures to 62–65°F (16–18°C).

Allow about five to six weeks from sowing for 288-cell plugs.

Growing On

Pot 1 plug/4" (10 cm) pot and 3 plugs/6" (15 cm) or 1-gal. (4 l) pot. Select a well-drained, disease-free medium with a moderate initial starter charge and a pH of 5.5–6.5. Grow on at 65–75°F (18–24°C) days and 62–65°F (16–18°C) nights. Provide light levels from 5,000–7,000 f.c. (54–75 klux).

Keep plants uniformly moist; do not allow wilting. Apply 200 ppm from 20-10-20 or another complete fertilizer weekly. Bonzi applied as a spray at 15 ppm about two weeks after transplant provides effective height control. Repeat one week later.

Pinch plants about three weeks after planting (one week after applying Bonzi) when the fourth set of true leaves has developed. Pinch plants back so two sets of true leaves remain. Apply the second Bonzi spray one week after the pinch.

'Peek-A-Boo' has a one-sided plant habit until flower buds are set. The main stem grows up, arches, and maintains dominance until the first flower buds are set. Once plants are budded, branching occurs naturally. Pinching results in a more uniform plant, although flowering is delayed by about one week and the plant will take on a more prostrate habit.

Crop time for a 4" (10 cm) pot is seven to eight weeks from potting up a 288 liner using one liner per pot.

Pests & Diseases

Spider mites can attack plants when flowering begins.

Two Bonzi applications of 15 ppm sprayed two and three weeks after transplanting make the plants more compact (center pot). Pinching between Bonzi applications creates a more symmetrical plant (left).

African Violet (see *Saintpaulia*)

Agastache

Agastache foeniculum
Common names: Anise hyssop; giant hyssop; blue giant hyssop
Herb, perennial (Hardy to USDA Zones 5–9)

This mostly upright-growing, perennial herb is noted mostly for its scented foliage. Plants range in height from 20–30" (51–76 cm) tall with flowers in either blue/purple or white. Plants are primarily perennial, although their best performance is Zones 6–9. While they are often perennial in the Midwest, some varieties are short lived—lasting from two to three years. See Varieties for more information.

Propagation & Growing On

Agastache is commonly propagated from both seed and cuttings.

Seed

Agastache is a fast crop from seed to finished, non-flowering plant. Seed sown during winter to March will flower the same year as sown. In fact, even though it is a perennial (perennials from seed usually require a longer crop time than annuals) green cell packs (804s with 32 cells per flat) are ready in eight to nine weeks from sowing, green 4" (10 cm) pots in ten to eleven weeks. Once planted to the garden in May, plants will flower in June or July.

For a 512-plug tray, use one seed per cell and lightly cover with coarse vermiculite and allow three to five weeks from sowing to a finished plug tray. Potted to 4" (10 cm) pots, the plants are salable (fully rooted, not yet in flower) in five to seven weeks for spring sales.

Note: Pinching may not be necessary for seed varieties when transplanting from plug trays to pots. This species branches pretty uniformly.

Cuttings

The following guidelines are based on the variety 'Blue Fortune'.

Unrooted cuttings (rooting hormone is not needed) can be stuck to a 105 or larger liner/plug tray and placed under mist for an average of six to ten days before moving to a greenhouse bench. Liners are ready to transplant to 1-qt. (1.1 l) or 1-gal. (4 l) pots in four to six weeks from sticking. Give one pinch upon transplanting to encourage branching.

If liners are transplanted to 1-gal. (4 l) pots in early February, the crop should finish green by the end of April to mid-May. Plants should be well rooted and, once planted to the garden in either late May or early June, will flower in mid- to late June.

Note: Take care during propagation to avoid saturated media under mist. Agastache can rot in this environment, which is why it may be beneficial to place plants in propagation chambers with low-volume mist.

While the production for a 105 tray is noted, plants are equally easy in a 72, 50, or larger cell sizes.

Varieties

The 'Golden Jubilee' seed-propagated, All-America Selections Award–winning variety bears golden-yellow to lime-green foliage topped with lavender-blue flowers. Plants grow from 24–30" (61–76 cm) tall and bloom all summer long. This is one of the hardier varieties from seed and is more tolerant of a Midwestern winter than some of the other seed-propagated varieties.

'Blue Fortune' is probably the most well-known variety of agastache on the perennial market. It bears lavender-blue flowers on plants that reach 24–36" (61–91 cm) tall and flowers all season. It is a hybrid of *A. rugosa* and *A. foeniculum.*

Ageratum

Ageratum houstonianum
Common name: Floss flower
Annual

Ageratum is one of those "must-grow" bedding plants. Most varieties are low growing and are ideal for borders or edging plants, although taller selections are available as well. Most selections have heights ranging from 6–16" (15–40 cm) and soft blue, white, or purple flowers. However, there are some landscape/cut flower varieties that grow tall as 30" (75 cm), and these are mostly purple or white. Plants prefer full sun and a well-drained location in the garden. Ageratum is a great attractor of bees and butterflies in the garden.

During the past decade vegetatively propagated selections have become more commonly available. They are noticeably more heat tolerant than the seed varieties in the South—the Dallas Arboretum has used them for years with great success. For the grower, ageratum from cuttings is easy to grow and works great in many different container sizes.

Propagation

Seed

Single sow pelleted seed onto disease-free, germination medium with a pH of 5.5–5.8. Do not cover. Germinate at 78–80°F (26–27°C) substrate temperatures. Radicles will emerge in two to three days.

Move trays to Stage 2 for stem and cotyledon emergence by lowering soil temperatures to 72–75°F (22–24°C). Begin fertilizing weekly with 14-0-14 as soon as cotyledons have expanded. Allow trays to dry slightly between irrigations. Stage 2 will take from five to seven days. Maintain light levels at 450–700 f.c. (4.8–7.5 klux).

Stage 3, developing true leaves, will last from fourteen to twenty-one days. Lower soil temperatures to 65–68°F (18–20°C) and allow trays to dry thoroughly between waterings. Fertilize weekly with 100–150 ppm, alternating between 20-10-20 and 14-0-14. A-Rest, B-Nine, or Bonzi can be added as required. Increase light levels to 1,000–2,500 f.c. (11–27 klux)

Harden plugs in Stage 4 for seven days prior to transplant or sale. Lower soil temperatures to 60–62°F (16–17°C) and continue to allow trays to dry between irrigations. Fertilize weekly with 14-0-14 at 100–150 ppm. Raise light levels to 2,500–4,000 f.c. (27–43 klux). Total plug time is four to five weeks for a 288 plug.

Cuttings

Root tip cuttings in Oasis or a disease-free medium at 70–73°F (21–23°C) substrate temperatures and 68–75°F (20–24°C) air temperatures. Lower air temperatures will create a stockier rooted cutting. Rooting hormones are not necessary. Apply mist to keep cuttings fully turgid. Once roots are visible, begin reducing mist frequency and duration. A broad-spectrum fungicide spray one to two days after sticking will help prevent disease during propagation. A wetting-type spray adjuvant such as Capsil may be included in this spray and will help rehydrate the cuttings.

Media pH should be between 5.8 and 6.2. When roots start developing, begin applying a calcium-magnesium (Ca-Mg, or cal-mag) fertilizer such as 15-5-15 at 100 ppm. This can be increased to 200 ppm under high light conditions once the liner is well rooted. Pinching is not generally required for vegetative ageratum, but if plants appear to be stretching under low light conditions a soft pinch will help increase branching. If rooted under high light conditions, PGRs are not generally necessary during the propagation phase. If necessary, use B-Nine at 1,500–2,500 ppm. A 105-liner tray with one cutting per cell will take from five to six weeks to be ready to transplant to larger containers.

Growing On

Transplant plugs or liners into a well-drained, disease-free mix with a pH of 5.5–6.2 and a moderate initial starter charge. Plant 1.1 liner/4" or 6" (10 or 15 cm) pot and 4–5 liners/10" or 12" (25 or 30 cm) patio planter. Be careful to avoid overwatering the crop immediately after planting.

Start growing the crop at 65–70°F (18–21°C) days and 60–65°F (15–18°C) nights until the plants are well established. It is important to provide adequate heating to promote healthy root growth, since ageratum are susceptible to root rots. Allow plants to dry moderately between irrigations to help avoid root rots.

Ageratum are moderate feeders. Alternate a cal-mag fertilizer such as 15-5-15 or 14-4-14 with an ammoniacal-based fertilizer such as 20-10-20 or 15-15-15 at 200 ppm. Monitor EC to maintain levels at 1.8–2.2. Maintain light levels as high as possible (3,000–4,000 f.c. [32–43 klux]), while maintaining moderate production temperatures.

In general, plants do not require pinching, as they have been bred to be very self-branching. Under high light conditions, PGRs are usually not necessary either. If environmental conditions that favor soft growth are being experienced, a spray application of B-Nine at 2,550 ppm or Sumagic at 2–5 ppm will help control growth. Since ageratum is a facultative, long-day plant, early crops grown under high light will flower without supplemental lighting; but in darker climates supplemental lighting will speed up flowering.

A 288-plug tray is most commonly used for seed varieties and will take from seven to nine weeks; although flowering plants in a 4" (10 cm) pot will take from eight to nine weeks. A 105-liner tray of the many vegetative varieties will be a little earlier to finish. Plants are salable in six weeks but not in flower until seven to eight weeks.

Pests & Diseases

Aphids and whiteflies are the most common pests on ageratum. Check the undersides of the leaves frequently. Whiteflies can become a serious problem. Both aphids and whiteflies can be controlled relatively easily with biologicals or systemic pesticides. The most serious diseases that trouble ageratum are root rots, especially near finish. Allow plants to dry between irrigations, especially during periods of cool, wet weather.

Troubleshooting

Allowing plants to remain wet for extended periods of time can lead to plant collapse from root rots. Single branches can also wilt and die from botrytis stem infections. Excessive nitrogen in the fertilizer, overfertilization under low light, and/or low light and overwatering can cause lack of flowering and too much vegetative growth.

Varieties

Historically our industry has offered seed-propagated varieties of ageratum in bedding flats. However, do not overlook the opportunity to sell this as a component plant for mixed containers in 4" (10 cm) pots. From seed, the more dwarf series such as 'Hawaii' are great for flats, while larger series, like 'High Tide', are best suited for the 1801- and 306-cell size. One plant per 4" (10 cm) pot or three plants per 6" (15 cm) pot are great for that instant color

For an unusual offering, try 'Blue Horizon' in 4" or 6" (10 or 15 cm) pots. This tall-growing ageratum is a great addition to the back of mixed borders and can be cut. 'Blue Horizon' is also a great landscape bedding plant item. Plants will flower from June through frost. Stems of 'Blue Horizon' may also be dried. 'Leilani' is a tall-growing ageratum that can be substituted for 'Blue Horizon'.

The 'Artist' series is vegetatively propagated and comes in 'Blue', 'Purple', 'Blue Violet', 'Rose', and 'Alto Blue'. Plants reach a garden height is 8–12"(20–30 cm) and are best in partial- to full-sun plantings. The 'Patina' series (also vegetatively propagated) comes in 'White', 'Blue', 'Purple', and 'Delft', a unique blue-and-white bicolor. Both series are excellent in the garden, although the 'Artist' series (especially 'Blue' and 'Purple') has excelled in more southern locations.

Postharvest

Plants are ready to sell when the first six to eight blooms open and visible buds for subsequent blooms are present. Display plants for sale under cool conditions, 65–70°F (18–21°C), and bright light. Applying light shade may be beneficial to moderate temperatures.

Aglaonema

Aglaonema spp.
Common name: Chinese evergreen
Tropical foliage plant

No mall or office building in America would be complete without the graceful and soothing leaves of *Aglaonema* planted in the interiorscape.

Propagation

Commercial propagation companies located in the southern United States perform most of the *Aglaonema* propagation. Cuttings, whether unrooted, rooted, or callused, are most often used and have between four and five leaves and root in four to five weeks; longer in northern greenhouses, especially under cloudy or colder portions of the year.

Growing On

Most northern growers buy in finished *Aglaonema* from foliage growers to resell in their local markets, as crop time is long—twenty-six to thirty-four weeks from a liner—and plants require very high production temperatures that are just not economical outside of Florida, California, or southern Texas, where plants can be produced in unheated or minimally heated structures. Foliage growers generally receive cuttings from farms in Central America.

Aglaonema prefers a well-drained, disease-free medium with a slightly acid to neutral pH. Use black or green pots instead of white ones. Plants are moderate feeders; use 100–200 ppm from a 2-1-2-analysis fertilizer at every irrigation. Moisture is critical to success. Do not allow *Aglaonema* to completely dry out; run them on the dry side. Watering at the right time is essential. Production light levels are 1,500–2,500 f.c. (16–27 klux). Grow plants warm, 80–90°F (27–32°C) days and 65–75°F (18–24°C) nights. Do not allow temperatures to dip below 58°F (14°C).

Pests & Diseases

Erwinia, a bacterial disease, can be a serious problem. Since there is no cure for bacterial disease, avoiding the problem through rigid sanitation procedures is the best control. (See the section on disease control under *Pelargonium hortorum,* p. 594, for a detailed list of cultural controls for bacterial diseases.) Plants infected with bacterial disease and those immediately surrounding the infected plants should be removed from the premises. Nematodes, *Fusarium,* and leaf spot can also be problems. Insect problems are rare.

Troubleshooting

Tip burn can be caused when soluble salts levels are too high. Fluoride toxicity is rare; maintain soil pH and calcium levels to avoid problems. Cold injury shows up when the upper sides of leaves look transparent and darkened. Spots will eventually become necrotic.

Varieties

'Silver Queen' is the most popular variety for its bright silvery-green leaves.

Postharvest

Plants may be sleeved and boxed for transport. Ship at 60–65°F (16–18°C). Plants may be shipped for seven to fourteen days.

Retailers and wholesalers should provide a minimum of 50 f.c. (538 lux) of light, if plants must be held prior to sale. Provide 70–75°F (21–24°C) temperatures.

In the interiorscape or home, maintain minimum light levels from 150–250 f.c. (1.6–2.7 klux) for best results. Do not allow *Aglaonema* to dry out too much, as lower leaves will yellow and die. Maintain interior temperatures at 60°F (15°C) or above for best results.

Ajuga

Ajuga reptans
Common name: Bugle weed (also written as bugleweed)
Perennial (Hardy from USDA Zones 3–9)

This stoloniferous perennial is frequently used as a groundcover or in combination containers. Its colorful foliage often persists throughout the winter in many areas across the United States. Plants range from 4–6" (10–15 cm) tall and up to 12" (30 cm) tall when in flower. Blooms are purple, pink, or white. Plants flower in May or June, but the foliage garners the most attention.

Propagation

Stem tip cuttings can be stuck to various liner trays including 32, 50, 72, or 105 cells per standard flat. Cuttings do not need rooting hormone to root, although during darker days of the year it is recommended for a more uniform cutting stand in the liner tray. Provide a media temperature of 68–74°F (20–23°C) and give mist. Mist can be discontinued after five to seven days.

Cuttings are well rooted in three to five weeks, depending on the time of the year. Maintain a soil pH of 6.0–6.5 and light levels of 3,000–5,000 f.c. (32–54 klux) to finish *Ajuga* liners. Pinching is generally unnecessary.

Growing On

Transplant rooted liners to 4.5" (11 cm) pots and allow five to seven weeks before sales. If using 1-gal. (4 l) pots, allow eleven to thirteen weeks. Grow on at 64–68°F (17–20°C) days (nights 5–7°F [3–4°C] cooler), until roots establish and then drop night temperatures to 55–60°F (13–16°C). Vernalization is not required for flowering.

Pests and Diseases

Aphids and whiteflies can attack *Ajuga*. Root and stem rots can also be a problem.

Alcea

Alcea rosea
Common name: Hollyhock
Annual, perennial (Hardy to USDA Zones 3–7)

Hollyhock's tall, stately spires are the epitome of the cottage garden. This old-fashioned garden favorite should be regarded as an annual or short-lived perennial in most locations. Some, if allowed to reseed, will flower in the garden the following year.

Treated as hardy annuals in the North, the plants will often come back at least one time. Experts suggest that the plants overwinter better in containers (gallon pots stored for spring sales) than they do in the garden. It is the diversity of their enemies—and their persistency—that often makes these plants short lived. Hollyhock can live for several years as a true perennial if plants are protected or mulched. In unprotected locations in areas of the country where frost penetrates deeply, hollyhocks often die out. They perform best when planted against a foundation or up against a barrier from hard winter winds.

Propagation

Sow seed into a disease-free, germination medium and cover lightly. Provide substrate temperatures of 70–72°F (21–22°C). Seed should germinate in five to ten days. Choose larger plug trays, such as 288s or 128s, as opposed to smaller trays to accommodate hollyhock's root system.

Perennial hollyhocks grown from seed sown in the winter do not flower well the following summer. To ensure that perennial hollyhock will flower its first year in the garden, sow seed in summer to early fall and overwinter in 6" (15 cm) or 1-gal. (4 l) pots inside a cold frame. These larger pots will be ready to sell green beginning in April and should flower for the consumer in the garden. Annual hollyhock varieties may be sown in January or February for green plant sales in the spring and flowering plant sales in June or July.

Transplant plugs as soon as they are well rooted in, about six to seven weeks after sowing. Hollyhocks have a taproot; to ensure plant growth is not inhibited, transplant them on time.

Growing On

Place hollyhocks in deep pots that will accommodate their taproots. They will not perform well in packs. Select a well-drained, disease-free medium with a moderate initial starter charge and a pH of 5.5–6.2. Use 1 plant/pot. Fertilize with 150–200 ppm from 20-10-20 at every other irrigation.

Allow seven to eight weeks for a 6" (15 cm) pot from a 288 plug. If using a 1-gal. (4 l) pot, allow ten to eleven weeks. Plants will be rooted but may not be to the bottom of the containers. Keep light levels at 4,000–5,000 f.c. (43–54 klux) and temperatures at 70–72°F (21–22°C) and nights at 55–60°F (13–15°C).

Pests & Diseases

Puccinia malvacearum, hollyhock rust, attacks plants in the summer after they have flowered. This disease can overwinter on debris left from the previous growing season. Therefore, it is vital to clean up dead leaves and plant parts in the fall. Newer varieties are more tolerant. Slugs can also be a problem.

Varieties

'Chater's Double' hybrid is a mix in copper, maroon, pink, red, rose, violet, white, and yellow. Flowers are semi-double and double, and plants will grow tall, up to 8' (2.4 m) in the garden. 'Chater's Double' will become a short-lived perennial in most locations.

Annual hollyhocks are shorter growing, from 2' (61 cm) to about 6' (2 m) in the garden. Among the varieties to try are 'Indian Summer Mix' in primarily white and yellow shades and 'Summer Carnival Mix' with double flowers in rose, pink, white, yellow, and carmine. A more recent selection, 'Spring Celebrities', is available in a number of separate colors and grows 18–24" (46–61 cm) tall. Annual hollyhocks frequently reseed themselves each year, if they are not pulled up from the garden too early in the fall. Annual varieties will flower in summer from a January or February sowing and potted to 1-gal. (4 l) containers.

Postharvest

Advise consumers to plant hollyhock in a well-drained, high-light spot and to stake plants once they reach 3' (1 m) tall. Hollyhock naturally flowers in the garden from late spring to early summer.

Alchemilla

Alchemilla mollis
Common name: Lady's mantle
Perennial (Hardy to USDA Zones 3–7)

This low-growing perennial displays pubescent, lobed leaves on plants from 12–18" (30–46 cm) tall in bloom and from 18–24" (46–61 cm) across. Flowers are yellow-green in color and dull in appearance. The unusual foliage is the primary focus of this crop and has no equal for form, height, and show in the perennial garden. The leaves are covered with a dense mass of fine hairs, giving the plant a velvety texture.

The flowers are actually apetalous (without petals) and measure 0.125–0.25" (0.3–0.6 cm) across. Due to their small size, the flowers, from a distance, appear to hang like a cloud or fog wavering above the plants in a breeze. As they age, the flowers turn a dull brown, which adds appeal to the midsummer garden.

Propagation & Production

Alchemilla can be sown from seed and is available as plugs or liners and as bare-root transplants.

Seed

Alchemilla is a relatively easy crop to germinate. Sow seed from May to July and lightly cover seeds with peat or vermiculite at this time of the year. This keeps the seeds from drying out when germinated on a greenhouse bench. If you are using a plug chamber, seed covering is not necessary.

If sowing to a 288-plug tray, sow three to four seeds per cell and allow seven to nine weeks from sowing before transplanting. You can either transplant to a 1-qt. (1.1 l) container to size up and then grow dormant/cold overwinter or transplant to a 32-cell pack (804 tray). The latter will be ready to pot up to 1-gal. (4 l) containers in five to seven weeks. Once transplanted to the gallon pots, these plants can be overwintered cold/dormant for spring sales once the roots have fully developed.

Another method is sowing in late August to a 72 liner using four to five seeds per cell and then transplanting to 1-qt. (1.1 l) containers in late October/November and growing in a cool greenhouse during the winter for flowering plants in late April to mid-May.

In general, the total crop time, including a cold period, for a flowering quart is twenty-six to thirty weeks for spring sales.

Bare-root transplants

Commercial propagation firms carry *Alchemilla* as bare-root transplants. These can be received during winter and potted to 2- or 3-qt. (1.9 or 3 l) or 1-gal. (4 l) containers. Place the crown just below soil level, shade during high light days until the roots establish, and keep the soil moist (but not wet). During the cloudy days of late winter and early spring, avoid keeping the roots saturated, because saturated roots can lead to root rots.

Potted up in mid-February, grown at 65–70°F (18–21°C) during the day and 58–60°F (14–16°C) nights until roots are established. Plants are salable green in May. Once rooted, move the containers to a cold frame or similar environment with nighttime temperatures from 48–55°F (9–13°C). Visually the plants will be salable in mid-May, although 1-gal. (4 l) pots may not be fully rooted but are salable nonetheless. Two-quart (1.9 l) containers are more thoroughly rooted by this time.

Related Species

Alchemilla erythropoda 'Alma' is a compact-growing selection with excellent basal branching. Highlighted by blue-green leaves, plants grow to 12" (30 cm) tall in flower.

'Alma' is seed propagated and is best grown following the notes provided above for *Alchemilla mollis*. However, in pots in the spring, the plants are more compact with a tight leaf structure, while *A. mollis* has a slightly fuller habit. 'Alma', however, will flower earlier than *A. mollis*; this is especially true of sowings made the previous year. Sowings after February 1 are not suggested for spring sales in quarts or larger. Equally, neither species will flower very well (if at all) with a late winter sowing.

Allium

Allium cepa
Common name: Onion
Vegetable

Propagation & Production

Onions can be sown directly into their final growing container. Sow several seeds (three to five per 48-, 52-, or 72-cell pack/liner tray) onto a well-drained medium; seeds can be left exposed to light or lightly covered with media or vermiculite. Germinate under mist at 70–75°F (21–24°C) soil temperatures. Seedlings emerge in four to ten days.

As seedlings develop, it is better to thin to one to three seedlings per cell in order to allow the crop to grow uniformly and to increase the size of garden-transplanted bulbs. Grow on at 65–70°F (18–21°C) and begin a constant liquid feed program at 100 ppm nitrogen from a complete fertilizer. Grow plants at 65–75°F (18–24°C) days and nights at 50–58°F (10–14°C) for best development. Crop time from seeding a 48 tray to a salable tray is nine to eleven weeks.

Onions are daylength sensitive. However, how sensitive they are depends on the variety, so carefully select the right variety for your location. Some bulbs initiate under long days, some under short days, and a few are day neutral. Long-day varieties are most commonly grown in northern locations of the United States and throughout Canada, while short-day varieties are grown in the southern United States. Most daylength-neutral varieties can be grown throughout North America—but it is best to carefully read a seed company's variety descriptions in its catalog or discuss varieties with your seed representative before purchasing.

Allium ampeloprasum var. *porrum*
Common name: Leeks
Vegetable

A relative of onions, leeks may be grown following the onion culture above. The critical factor in their production is that leeks take longer to produce in a cell pack than onions do. Using the culture noted above for onions, add another three to four weeks for salable transplants out of a 32 tray. Another major difference here: It is necessary to thin the plants to one plant per cell pack.

Plants have larger leaves, both in width and length, and the plants fill out a cell pack that visually look appealing to home gardeners while they are shopping for the plants in a garden center.

Allium schoenoprasum

Common name: Chive

Perennial herb (Hardy to USDA Zones 3–9)

Chive is a cool-season, perennial herb that is easily propagated by seed. While some references suggest taking division (which is very easy), this is better for the home gardener than the professional grower. Plants flower in June or earlier.

Chive is one of those crops that you can sow seed directly into a cell pack as opposed to a plug tray; although both can be done easily. Seed can be sown from December to February direct to cell packs; covered lightly with peat or vermiculite and germinated at 70–72°F (21–22°C). Place six to ten seeds/cell pack (using an 804 or a 32-cell flat) or ten to twelve seeds/4" (10 cm) pot. Seed germinates in seven to fourteen days. Move the cell packs to a greenhouse bench and grow on at 55–60°F (13–16°C) night (days 8–10°F [4–6°C] warmer), until the plants are established and then grow on at 55–60°F (13–15°C) days and 40–55°F (4–13°C) nights in a cold frame or similar environment from late March through late May. From sowing to a nonflowering, finished flat is thirteen to fifteen weeks, while a 4" (10 cm) pot is fifteen to seventeen weeks.

You do not need to use this many seeds, but these sowings provide for a quick finish. However, this many seeds and a warm night temperature (70°F [21°C]) can cause plants to stretch. If plants become weedy, shear them back to 2" (5 cm); plants will regenerate new shoots. However, if the plants are kept cool, then they do not require a shearing (e.g., 55–60°F [13–15°C] days and 40–55°F [4–13°C] nights).

Plug growers should place four to eight seeds/cell of a 288 tray in summer to early fall for transplanting to 1-qt. (1.1 l) containers. Once plants are established, provide constant liquid feed at 125–150 ppm and grow on at 60°F (15°C). Plug trays are ready to transplant in four to six weeks.

For larger, more robust, plants sold in spring in 1-qt. (1.1 l) or 6" (13 cm) pots, sow from summer to October or November and transplant to larger containers when ready. These can be sold in the spring and in many cases, depending on how cool they were finished once fully rooted, may begin flowering. Sowings made in January or later can be sold in cell packs or 4" (10 cm) pots.

Chives can experience problems with downy mildew and rust.

Allium giganteum

Cut flower bulb crop

The majestic, lilac-purple flowers of *Allium giganteum* sit high on stalks that can grow 40–50" (1.0–1.3 m) in length.

Although this ornamental allium can be grown from seed, flowering will not occur for three to five years. That is why commercial crops are started from bought-in bulbs.

As one would imagine, such a large, flowering plant would have big beginnings: Bulbs are large, starting at 7" (18 cm) and going up. Most growers purchase 8" (20 cm) bulbs for field planting. Beware of bulb rot caused by *Sclerotium cepivorum*. Make sure to buy bulbs from a reputable supplier to prevent introducing this disease into the crop.

Purchase precooled bulbs, or cool bulbs at 40°F (4°C) for eight to ten weeks prior to spring planting. Bulbs planted in the fall do not require cooling. Cooled bulbs require sixteen to eighteen weeks to flower once planted, when grown at 55–60°F (13–15°C) days and 40–55°F (4–13°C) nights.

Plant bulbs in a well-drained medium, deep enough to cover with 4" (10 cm) of medium. Plant at a density of 6 bulbs/ft.^{2} (64 bulbs/m^2). Plants will flower beginning in mid-May in the South to early July in northern regions and cool Pacific climates. In northern regions and consistently cool climates, flower harvest is spread over three to four weeks. In warmer regions, harvest can last from one to two weeks. Generally one bulb yields one flower. Bulbs can be left in the ground; you will be able to harvest flowers a second year.

Harvest flowers when at least one-third, but no more than half, of the flowers on the umbel have opened. Flowers will last about two weeks provided they have not been stored. Place in a postharvest solution adjusted to a pH of 4.0. Dutch growers commonly recommend using chlorinated water. Storage actually reduces vase life, so plan to sell stems as soon as they are harvested. Retailers should keep stems refrigerated at 40–45°F (5–7°C).

Other Cut Flower Alliums

Many other allium species are useful for cut flower production—most often as outdoor, in-bed crops. The main strategy is to plant bulbs in the fall and harvest from late fall through early summer. Table 1 lists many species and basic data on minimum bulb size for flowering, color, and height. Be careful in using this table as size varies a great deal between species (not only in bulb size and cost but also stem length at flowering). Some are lovely smaller flowers, while others are large, majestic specimens.

Table 1. Allium Species for Cut Flower Use

SPECIES	HEIGHT (IN.)	COLOR	MINIMUM SIZE (IN.)	APPROXIMATE FLOWERING TIME (ZONE 5)
A. aflatuense	32–36	Pink/violet	4+	Late May/early June
A. atropurpureum	16–28	Red-purple/purple	2+	June
A. cernum	14–16	Pink/red-purple	Plants	May/June
A. christophii	16–20	Metallic violet	4+	Mid-June
A. cowanii	14	White	2+	Late May
A. giganteum	50–60	Mauve/lilac	7+	July
A. macleanii	40–50	Pink/violet	6+	End of May/early July
A. moly	10	Yellow	2+	June
A. neapolitanum	8–12	White	2+	May/June
A. nigrum	24–36	Cream white	3+	June
A. roseum	14	Pink to white	1+	June
A. schubertii	16–20	Pink	5+	June/July
A. sphaerocephalon	20–28	Dark purple	2+	July
A. stipiatum	50–60	Violet	6+	May/June
A. unifolium	14	Bright pink	2+	June

Alstroemeria

Alstroemeria hybrids
Common names: Inca lily; princess lily
Cut flower, pot plant

Consumers encounter *Alstroemeria* frequently in the retail marketplace, as it is one of the world's top-ten cut flowers. Its exotic flowers look like a cross between an orchid and a lily. As a cut, when handled correctly, *Alstroemeria* has a tremendous postharvest life, making it an excellent consumer value.

Consistent exposure of *Alstroemeria* as a cut flower has helped to pave the way for *Alstroemeria* as a flowering pot plant as well. In milder climates, *Alstroemeria* is even perennial. Do not be surprised if it becomes more embraced by retailers and home gardeners more and more. Breeders are developing all-new pot *Alstroemeria* types, colors, and series that are easier to produce.

Growing *Alstroemeria* commercially is not for beginners. Young plants are expensive, and plants can throw blinds (shoots that do not flower) if they are not properly handled. Plants can even go dormant if they do not receive proper temperatures and light levels. Seasoned growers will easily control these production issues and find *Alstroemeria* to be a rewarding crop that is highly programmable.

Cut Flower Crops

Propagation

Alstroemeria can become affected by a large number of viruses. For that reason, and since the same plants can be cultivated for three to five years in the greenhouse, growers purchase disease-free plants from reputable breeders. Most often these plants get their start in a tissue culture lab and are produced in a clean-stock propagation system. Growers receive well-rooted young plants with actively growing shoots established in medium in deep growing pots.

Plants can also be multiplied by division. Prior to dividing, trim plants back to 6–8" (15–20 cm). Dig deep to uncover the active-growing point on the rhizome, which may be 12–14" (30–36 cm) down. Each rhizome division should consist of a single rhizome with an undamaged, blunt growing point, some aerial shoots, and fleshy storage roots.

Because the best varieties are patented and/or protected by breeder's rights, propagation is illegal unless you have a license agreement with the breeder. For that reason—and to ensure they are planting up clean stock—growers buy in plants.

Growing On

Under normal greenhouse conditions in most parts of the United States, plants flower from January until August. As a result, transplants are preferably planted in the summer and early fall, although they may be planted year-round.

Alstroemeria is typically grown in raised ground beds. Prior to the arrival of plants, be sure to pasteurize the soil and work the bed 16–18" (30–40 cm) deep in preparation for plants. Soil should be loose and well drained and have a high organic content. Incorporate superphosphate at 5 lbs./100 ft.^{2} (244 g/m^2) If soil tests show that calcium is low, use limestone if pH is also low or gypsum if pH is acceptable. Both may be incorporated at 5 lbs./100 ft.^{2} (244 g/m^2). Space plants 18–24" (45–60 cm) on center. Plant at the same depth as liners are growing in their pots. Many growers go ahead and set up wire mesh supports that can be raised as the plants grow. Support openings of 8" × 7" (20 × 17 cm) or 8" × 8" (20 × 20 cm) can be used. The crop will require three to four layers of support.

Some growers apply a fungicide drench with potassium phosphate at planting and again a month later if they do not observe vigorous root growth.

Water in with clear water after planting, but be careful not to keep beds too wet, as rhizomes at this stage can easily rot. Spot watering is preferred. Grow plants at 65–70°F (18–21°C) for four to eight weeks until they are established, prior to lowering temperatures for flower initiation.

As rhizomes become established, they will produce shoots. Remove weak shoots to encourage lateral breaks to form. You can remove weak shoots by pulling them out using a quick upward pull: Do not cut them. Be careful not to dislodge newly planted rhizomes. As plants become more established, this will not be an issue.

Irrigation is critical: Water plants with perimeter watering or trickle drip tubes running in the center of beds. Once plants are established, do not let them remain dry for very long. *Alstroemeria* is a heavy feeder and requires high nutrient levels once established. Feed weekly with 400 ppm nitrogen (N). Avoid ammonia-based fertilizers because ammonia is not readily converted to nitrate under cool growing conditions. Keep the EC below 1.3 mS/cm.

Provide growing temperatures of 65–70°F (18–21°C) days and 50–60°F (10–15°C) nights. Prolonged temperatures above 75°F (24°C) can decrease or stop flowering. Air temperatures are secondary to temperatures in the ground. Maintain soil temperatures of 55–61°F (12–16°C) to induce flowering. In warm months, flowering can be extended with soil cooling, where cold water is circulated in small pipes either buried in the ground or laid on top of beds.

Provide high light; *Alstroemeria* may even be grown outdoors in full sun, provided plants remain moist. Supplemental light hastens flowering and increases flower production. Apply 600 f.c. (6.5 klux) if you have access to supplemental lights at canopy height and increase photoperiod to sixteen hours of day length. Plants provided with supplemental light in the fall flower up to twelve weeks earlier and produce as much as 30% more flowers than plants grown under natural days. However, this response will not occur if plants are not also induced with cool temperatures.

Even if you are unable to use supplemental lights, plants will still benefit from long days (night interruption lighting with incandescent bulbs, 5 f.c. [54 lux]). Plants that receive sixteen hours of light flower faster than plants receiving only thirteen hours. Do not provide long days for the first forty-five to sixty days after a new planting. Lighting can be used on established plants from about September 1 to April 1 in northern latitudes.

Established *Alstroemeria* plants require thinning to increase light levels reaching the interior of plants to stimulate lateral branches and stagger flowering. Remove weak vegetative stems by pulling them out. During winter months, as much as 15–25% of vegetative shoots are pulled out until flowering begins.

Flowering

Alstroemeria produces two types of shoots: flowering and vegetative. Shoots emerge from underground rhizomes. Normally shoots that have unfolded more than thirty leaves are vegetative and will not flower. Flowering shoots produce a cyme with one to five flower buds.

Inducing flowering in *Alstroemeria* is a process involving cold temperatures and long-day photoperiods. Plants must receive cold treatment before they receive long days. When plants begin to flower, they will continue to flower until soil temperatures rise above 65–70°F (18–21°C) for extended periods.

Cold response is variety dependent with temperature requirements varying from 50–63°F (10–17°C) and for varying lengths of time. Follow the breeders' recommendations when providing the cold treatment.

Pests & Diseases

Snails or slugs can be a problem in field plantings. Aphids, caterpillars, and whiteflies can also attack plants. Thrips are the biggest insect problem and are especially difficult because they can transmit viruses. Any virus-infected plants should be removed from beds and destroyed immediately. Botrytis and root rot can plague plants during winter months.

Troubleshooting

Mottled leaves can be caused by nematode virus transmission to plants in ground beds. Be sure to sterilize beds before planting. Iron deficiency can result in yellow new growth or poor vigor. Add iron chelate to soil as a drench or spray.

Postharvest

Harvest *Alstroemeria* by gently pulling stems. Most postharvest issues with *Alstroemeria* are the direct result of flowers being harvested before they are ready. Flowers are ready for harvest when one to two flowers are open. Flowers harvested at the proper stage open fully and show bright colors. Immature flowers may not open properly and generally have faded colors. However, growers shipping stems long distances tend to harvest at an earlier stage so stems can be more tightly packed, which is why so many poor-quality *Alstroemeria* are in the market. If you must harvest stems for shipping, make sure to wait until at least two buds are showing color and are puffed, ready to open.

Remove foliage on the lower one-third of stems, grade, and bunch. For immediate shipping, put flowers in warm water with a sucrose-based floral preservative and bactericide. Hold stems at 45–50°F (7–10°C) for at least six to eight hours to cool them. To store stems long term, pulse with a silver-thiosulfate and sucrose solution for twenty-four hours, and then place them in the cooler in fresh preservative solution at 39°F (4°C).

Alstroemeria is sensitive to ethylene. Avoid placing stems near ripening fruit or vegetables, and be sure to ventilate display areas.

Flower color is enhanced when stems are held or displayed under at least 200 f.c. (2.2 klux) of light. Retailers will ideally recut stems upon receipt and pulse with a silver-thiosulfate solution. Be sure to remove all foliage from the stem segments in the water. Later place stems in a hydration solution with a pH of 3.5. Do not use water high in fluoride, as this can damage *Alstroemeria*.

Pot Plant Crops

Propagation

Many *Alstroemeria* pot plants are started from tissue-cultured plantlets and sold as young plants to growers. This is the most common method. However, *Alstroemeria* may also be started from seed. If using fresh, collected seed, scarify it first by pouring hot (water hot enough that an arm can stand) water over them and then allowing them to imbibe water for eight hours or overnight. Ideally, replace water once during imbibition. Purchased seed from a commercial seed house does not require this technique.

Sow seed into 200-plug trays and cover lightly with coarse vermiculite. Temperature requirements for optimal germination (Stage 1) are as follows: Provide one week at 72°F (22°C), followed by two weeks at 44°F (7°C), and then followed by one week at 65°F (18°C). Keep substrate uniformly moist and provide less than 2,000 f.c. (22 klux) of light. Maintain a substrate pH of 5.5–5.8 and an EC less than 0.75 mS/cm. Keep ammonium levels below 10 ppm.

Stage 2, stem and cotyledon emergence, lasts fourteen days. Provide soil temperatures of 65°F (18°C) and reduce moisture levels somewhat. Begin fertilizing with 75–100 ppm from 14-0-14 or a calcium/potassium-nitrate feed at every other irrigation as soon as cotyledons have expanded.

Stage 3, development of true leaves, lasts fourteen days. Continue to maintain 65°F (18°C) soil temperatures. Increase feed to 100–150 ppm and fertilize twice a week. To harden plugs prior to sale or transplant (Stage 4), provide seven days at 65°F (18°C) soil temperatures. Allow medium to thoroughly dry between waterings and fertilize at 200 ppm at every other watering. A 200-plug tray requires eight to ten weeks from sowing to be ready to transplant to larger containers.

Growing On

Liners

Plant one 2.5" (5 cm) liner or plug/6" (15 cm) pot. Because plants develop large root structures, use deep pots as opposed to shallow ones. Select a well-drained, disease-free, peat-based medium with a moderate starter charge and a pH of 5.8–6.5. Plant liners shallow so that rhizomes are 1" (3 cm) below the surface of the medium. If plants also have large, fleshy storage roots, they can be planted deeper. Do not cut back any foliage on plants, which will only delay flowering. However, remove any dead or damaged shoots.

Plant liners a minimum of twelve to sixteen weeks prior to sale. Or plant pots in the fall and hold in a cold frame: Fall plantings require little care other than occasional watering. After plants are established, lower temperatures to as low as 33–38°F (1–3°C). Even at these temperatures, roots will continue to grow. Then, thirteen to fourteen weeks prior to sale, cut back foliage completely and move pots to a greenhouse at 55–63°F (13–17°C). By cutting back foliage, plants will stay short. If you plant liners thirteen to fourteen weeks prior to sale, do not cut foliage back.

If you are growing potted *Alstroemeria* from bought in liners and are potting them twelve to sixteen weeks prior to sale, provide 64–68°F (18–20°C) days and 60–64°F (18–20°C) nights for the first four to six weeks to establish plants. For the remainder of the crop, grow on at 58–60°F (14–16°C) days and 54–58°F (12–14°C) nights and provide long days (greater than thirteen hours) through night interruption lighting or supplemental light (600 f.c. [6.5 klux]). Note that when seed-grown plants have reached the proper size for bud formation, they must receive twenty-one days at 50–60°F (10–16°C) to initiate buds. Before they can initiate buds, plants must have ten or more leaves. High soil temperatures will increase blind shoots and promote plant dormancy. Temperatures above 65°F (18°C) in the winter for more than six hours per day will prevent plants from forming flower buds.

Height control is best accomplished by managing the production environment and by choosing dwarf-growing varieties. Grow at cool temperatures, do not keep plants wet, and increase light intensity by providing adequate spacing (15" × 15" [38 × 38 cm]). Run moisture levels medium to dry and provide 4,000–6,000 f.c. (43–65 klux) of light. If you are using supplemental lights during winter months, do not light for more than fourteen hours daily. Fertilize at every watering with 200–250 ppm nitrogen (N) from 20-10-20, alternating with 14-0-14. Maintain the EC below 1.3 mS/cm and the pH at 5.8–6.5.

Alstroemeria also responds to pruning for height control: The more it is pruned, the more compact it grows. Every two to three weeks, "shape" plants by pulling out dead and weak stems. This encourages new shoots to grow shorter and encourages lateral breaks from the rhizome. Be careful not to pull plants from the pot! It may be necessary to treat some seed-grown varieties with growth regulators (Sumagic). *Alstroemeria* are also responsive to negative DIF.

Seed

Plants produced from 200s and transplanted to 6" (15 cm) pots will finish in twelve to sixteen weeks in the spring. Maintain long days during production and finishing.

Pests & Diseases

See the Pests & Diseases section under Cut Flower Crops.

Troubleshooting

Slow-developing plants or plants that do not flower may be signs of soil temperatures above 60°F (16°C), short days, low light, or poor rhizome development. Low fertility and wide temperature extremes can cause poor plant vigor. Low light, calcium deficiency, high temperatures, or overwatering can result in flower bud abortion. Fluoride contamination from superphosphate or perlite can lead to leaf scorch. Lobed leaves can be due to production temperatures above 65°F (18°C). To return plants to production of smooth margined leaves, provide correct growing temperatures.

Varieties

Princess lilies are the first genetic dwarf varieties from tissue culture. Their main characteristics include growing short without growth regulators, compact plants, and rich and continuous early flowering. Flowers are large and colorful.

'Jazze' is the first compact F_1 hybrid *Alstroemeria* from seed. Flowers are large and vibrant. The series is available in 'Deep Rose', 'Purple Rose', and 'Rose Frost'. 'Jazze' blooms all season when planted outdoors in cool climates.

Postharvest

Plants are ready for sale when one to two flowers are open. Pots can be boxed and shipped cool (39°F [4°C]) for up to six days. While flowers are sensitive to ethylene, silver thiosulfate treatment has not been effective on pot plants.

Retailers should display *Alstroemeria* under high light and cool temperatures.

Consumers can enjoy *Alstroemeria* as a flowering potted plant they discard, as an annual in the garden, or as a perennial in southern and West Coast locations. Remove flowering stems after they have bloomed to keep plants shorter. Mulch is helpful in keeping plant roots cool during warm months.

Hydroponic Cut *Alstroemeria* at Len Busch

In 1996, Al Reinarz of Len Busch Roses in Plymouth, Minnesota, began experimenting with hydroponic *Alstroemeria.* What began as a bed of rock wool, expanded clay pellets, and sand has grown into a large hydroponic operation.

Seeing no real production difference between the three media, Al decided to switch over to sand. While flower quality was improved in this new bed, there were few production differences between the 1996 and 1997 crops. After testing both beds, the difference was found to be air content: The first bed's sand had 18% airspace, while the second had only 12%. And while the initial solution had a pH of 5.8, the leachate in the first bed stayed at a consistent 7.1; the pH of the leachate in the second bed varied between 7.2 and 7.3. Calcium carbonate was another differing factor: 5% for the first bed and 8% for the second; the increase in the latter is attributed to the higher level of limestone in the sand.

Deciding to push forward, a year later 15" (38 cm) of sand was laid over a poly barrier in a brand-new, 0.74-acre (3,000 m^2) hydroponic *Alstroemeria* greenhouse. The poly was sloped for leachate collection, and four black poly pipes/bed were positioned 4" (10 cm) below the sand to facilitate soil cooling. Each of the thirty-six 3' × 157' (1 × 48 m) beds is watered six times per day by irrigation at a rate of 1 gal./100' (1 1/8 m) per minute. Their "recipe" was developed at the Aalsmeer Research Station in the Netherlands (see inset).

When they began recirculating irrigation water, they discovered that if they kept the leachate around 70%, it more closely resembled the analysis of the initial solution. A 6,075-gal. (23,000 l) make-up solution is now added every five to eight days. When the EC reaches about 2.5, the recirculation pump is shut off and the tank is filled with a new mix. Because the pH of the leachate is always around 7.1, phosphoric acid is injected as it returns to the main tank to bring the pH back to about 5.8.

Biological controls are employed to counter pests: *Aphidius ervi* for aphids, *Encarsia formosa* for whiteflies, and *Amblyseius cucumeris* for thrips. These biological solutions have been so successful that Al has not had to spray his crop for years.

With a 10% higher yield than in soil, Al soon had *Alstroemeria* "coming out of his ears." With this in mind, the team decided to rotate the plants every five years—and not just for the usual reasons, either: "When we dug up the original bed, it was still the highest-producing bed in the greenhouse," says Al. "But we just got tired of looking at the variety."

Watering Solution Developed by the Aalsmeer Research Station

CHEMICAL	CONCENTRATION (IN MILLI MOLS)
NH_{4+}	1.25
NO_{3-}	11.25
P-	1.25
K	6
Ca	2.875
Mg	1
SO_4-S	1.25

Reprinted from the seventeenth edition of the Ball RedBook.

Alternanthera

Alternanthera ficoidea
Alternanthera dentata
Common name: Joseph's coat
Annual

Alternanthera is well known by its common name as Joseph's coat and is enjoying a resurgence in popularity. Its popular name comes from its bright, multicolored foliage seen on some select varieties. However, the foliage color ranges from pure, dark brown or green selections to various colors per leaf to two-toned yellow, white, and other colors with green highlights. There are large-growing varieties that make excellent landscape additions, while others are more intermediate in height, and then there are dwarf selections that make great border or edging plants. In addition they make an excellent addition to any combination pot and brighten up any lightly shaded corner in the garden.

Propagation

The most common method of propagation is from stem cuttings. However, seed can be done as well, although seed is only available commercially for the variety 'Purple Knight'.

Seed

The information for seed propagation is based on the variety 'Purple Knight'. Sow seed into a well-drained, disease-free medium with a pH of 5.5–6.3. Cover seed lightly with coarse vermiculite. Radicles will emerge in three to four days at 72–76°F (22–24°C). After germination, raise light levels to 1,000–2,500 f.c. (11–27 klux) and reduce temperatures to 65–72°F (18–22°C). Begin fertilizing with 50–75 ppm 15-0-15 immediately and increase to 100–150 ppm once cotyledons have fully expanded. Every third irrigation use 20-10-20, if growth slows. Maintain EC at 1.0–1.5 mS/cm. After cotyledons have expanded, reduce humidity by allowing trays to dry slightly between irrigations. Do not allow seedlings to wilt.

As seedlings mature, increase light up to 5,000 f.c. (54 klux), if temperatures can be maintained. Plug crop time is approximately five to seven weeks in a 288 plug.

Cuttings

Alternanthera roots easily in three to four weeks in a 105- or 84-liner tray on most varieties. Dwarf or slowing-growing selections will require a week or two longer.

Maintain day temperatures of 68–75°F (20–24°C) and night temperatures of 65–70°F (18–21°C). In the first week, keep plants misted and apply shade (500–700 f.c. [5.4–7.5 klux] of light is all that is needed). Keep pH at 5.5–6.0 and maintain EC of less than 0.75 mS/cm, increasing to less than 1.0 mS/cm as cuttings finish. Apply foliar feed once a week during rooting with a 20-10-20 starting at 50–75 ppm and gradually increasing to 150–200 ppm as cuttings are finishing. Increase light levels as cuttings finish to 1,000–2,000 f.c. (11–22 klux).

Growing On

Whether using plugs or liners, use a well-drained medium with good water-holding capacity, a pH of 5.4–5.8, and a high starter charge. Use 1 plant/4" (10 cm) pot, 2–3 plants/6" (15 cm), and 4–6 plants/10" (25 cm) basket. Do not plant liners too deep—make sure the crown is even with the top of the medium.

Begin pinching vigorous-growing plants by hedging them back once roots reach the container wall. Most of the dwarf-growing varieties do not require pinching. Since plants are well branching, growth regulators are not required. Avoid overwatering, which can result in root loss.

Alternanthera has low to moderate light requirements: Light shade is ideal outdoors. Provide 1,000–2,000 f.c. (11–22 klux) of light to crops in the greenhouse; northern growers can provide higher light levels, provided temperatures are moderate. When light levels are too high, leaves will roll; however, too-low light levels cause plants to stretch. Plants are daylength neutral in their flowering, but flowering is profuse as light intensity improves. Flowers are inconspicuous—the foliage is the ornamental value of the plant.

Fertilization is an important production aspect. *Alternanthera* is a moderate feeder. Alternate irrigations with 20-10-20 and 15-0-15 at 200–250 ppm N. Do not incorporate slow-release fertilizers. Maintain EC levels of less than 1.0 mS/cm and pH of 5.4–5.8.

Crop times are four to six weeks when using 1 plant/4" (10 cm) pot from a liner (105 plants per tray) and six to eight weeks when using 2–3 plants/6" (15 cm) pot.

As for 'Purple Knight': If seeded to a 288-plug tray, plants will be ready to sell in seven to eight weeks from a 4" (10 cm) pot. One pinch will help to fill out the crown, although it is not required.

Pests & Diseases

Whiteflies, aphids, fungus gnats, and leaf miners can be problems. Root rots such as *Pythium* or rhizoctonia can cause plant collapse.

Troubleshooting

Too-low temperatures, high nitrate nitrogen, and low fertilizer levels combined with low light and/or overwatering can cause poor vegetative growth. Stretching occurs when plants are not receiving enough light. Insufficient light, overwatering, and a lack of nitrogen can result in poor foliage color.

Postharvest

Because cold temperatures adversely affect plant growth, do not offer plants for sale to consumers when ground temperatures are still cold. Display plants and encourage consumers to plant *Alternanthera* in lightly shaded areas with 1,000–2,000 f.c. (11–22 klux). Do not overwater plants.

Alyssum

Alyssum montanum
Common names: Madwort; yellow tuft; mountain gold
Perennial (Hardy to USDA Zones 3–7)

For *Alyssum saxatilis* (i.e., basket of gold), see *Aurinia saxatilis.*
For sweet alyssum, see *Lobularia.*

Closely related to the popular garden plant *Aurinia saxatilis, Alyssum montanum* differs in having smaller flowers (0.0625" [2 mm]) on more mounding plants with smaller leaves than *Aurinia saxatilis.* The flowers are a buff or light yellow color on plants that reach 18"(46 cm) tall when blooming. The plants are more erect than most other species in the genus but possess the same gray-green foliage.

Propagation

Cuttings and seed are the two most-common methods to propagate alyssums. Seed is the most common and easiest form of propagation, but plants may vary in performance. The habit and vigor of seed-propagated plants are usually uniform, though bud set and flowering may differ.

Seed

Sow two to four seeds per cell of a 288-plug tray, leaving seed exposed to light or lightly covering with vermiculite, and then germinate at 68–72°F (20–22°C). Seedlings will emerge in three to eight days and plug trays are salable in five to seven weeks after sowing. After germination, drop the night temperatures to 58–60°F (14–16°C) to reduce stretching. If sowing to open flats, seedlings can be transplanted within two weeks after sowing.

Cuttings

Take cuttings of double-flowering or variegated foliage varieties from early spring (before bud set) or in summer after flowering. Plants can be sheared back and fed to encourage shoot growth. A 72-liner tray will take from eight to ten weeks from sticking until plants are ready to transplant to larger containers.

Growing On

Seed

For spring-flowering plants, sow seed anytime from June until autumn. June and July sowings can be transplanted to 1-qt. (1.1 l) or 1-gal. (4 l) pots for overwintering dormant; although the foliage often persists even at these cold temperatures.

Sowings made in October or November are transplanted to 1-qt. (1.1 l), 1-gal. (4 l), or 6" (15 cm) pots and grown cold (38–45°F [3–7°C]) once the plants are well rooted in their final container. These will flower in the spring.

Seed sown from January onwards will produce salable plants in the spring, but they will not flower until the following year.

Cuttings

A 72-cell, or liner, tray can be potted to 1-qt. (1.1 l) or 1-gal. (4 l) containers during late July or August. Plants are overwintered in a cold frame or similar environment for flowering plant sales the following spring.

Amaryllis (see *Hippeastrum*)

Anemone

Anemone coronaria
Common name: Windflower
Cut flower, pot plant

With origins in southern Europe around the Mediterranean, *Anemone coronaria* is the most widely grown anemone species. Anemones are popular in Europe and Japan as greenhouse cut flowers and garden plants. Until the 1990s, all anemone cultivars were grown from corms produced by specialist growers. More recently, the F_1 hybrid 'Mona Lisa' was introduced.

Grown in cool houses, anemones are energy-efficient. Compared to many other cool cut flower crops, such as carnations, they are also less labor intensive, as they do not require staking, stringing, or disbudding.

Propagation

Sow seed in mid-March to mid-April in a well-drained, disease-free soilless medium with a pH of 6.8–7.0. Cover the seed lightly with medium or coarse vermiculite. Some growers use a fungicide drench to water in the seed. Keep trays moist but not fully saturated. Germinate at 60°F (15°C). Higher temperatures reduce germination percentage. Seed will germinate in five to seven days.

Move trays to Stage 2 for stem and cotyledon development. Maintain temperatures at 60°F (15°C). Increase light levels to 500–1,000 f.c. (5.4–11 klux) and apply 50–75 ppm of 14-0-14 as soon as cotyledons are fully expanded. When radicles begin to penetrate the medium, allow trays to dry down somewhat between irrigations, but never allow seedlings to wilt. Stage 2 takes from seven to fourteen days.

Move trays to Stage 3 to develop true leaves. Reduce temperatures to 50–55°F (10–13°C) and increase light levels to 1,000–2,000 f.c. (11–22 klux). Fertilize with 100–150 ppm once or twice a week alternating between 13-2-13 and 14-0-14. Allow trays to dry somewhat between irrigations, but do not let seedlings wilt. Stage 3 takes about thirty to thirty-five days.

Harden plugs prior to sale or transplanting by maintaining temperatures at 50–55°F (10–13°C) and light levels at 1,000–2,000 f.c. (11–22 klux). Allow trays to dry thoroughly between irrigations.

Allow eight to nine weeks for a 288-plug size. However, 'Mona Lisa' has a deep root system that does better in a deeper plug. Most commercial producers offer these plugs in larger cell sizes such as 72s, which will take a little longer to finish (ten to twelve weeks).

Growing On

Cut flower production

Growing your crop in a raised bed will have the most benefit with a heavy substrate, while flat beds will be more adequate for a well-drained substrate. If a soil-based medium is used, it should be sterilized. Mulch to keep the soil temperature down.

Before transplanting, make a soil test to determine specific fertilization needs. It is necessary that the pH of a soil-based medium remain neutral (pH 6.8–7.0) throughout the growing season. The pH of soil-less medium should be approximately 6.2–6.5. Add 2.0–2.5 lbs. of superphosphate/yd.^{3} of medium (about 1.2–1.5 kg/m^3).

The aim is to obtain 1–2 plants/ft.^{2} (11–22/m^2). Several bench-spacing plans are used in Europe. Most European crops are grown in ground beds. Often the walk is dug out, and the growing beds are raised 4 or 5" (10 or 13 cm) above the walk level. The goal is to ensure good drainage.

Scheme 1: Two-row beds. Two rows 12" (30 cm) apart, a 30" (76 cm) space, and two more rows 12" (30 cm) apart. Plants are spaced 3" (8 cm) apart in the rows. Harvesting is easy, and there is good air circulation around the plants.

Scheme 2: Four-row beds. Four rows spaced 10" (25 cm) apart with plants spaced 6–7" (15–18 cm) apart. A narrow, 16–18" (41–46 cm) walk separates the beds. This plan is used on ground beds with, again, soil mounded above the walk level. This spacing is also used on raised benches.

When transplanting, take care to prevent injury to the delicate root systems. You can stimulate initial root growth by adding 150 ppm of a starter solution such as 9-45-15, or 150 ppm of calcium nitrate if superphosphate has been added to the growing medium. After about two weeks, drench with a broad-spectrum fungicide to prevent root rot.

Water before noon to allow the foliage to dry completely before sundown. Take precautions to keep water off the foliage. If possible, it is better to use a ground-level watering system.

Base fertilizer use on substrate and water tests. Use fertilizers that are low in ammonium or urea nitrogen, as acid-forming fertilizers may decrease substrate pH when irrigation water has a low buffering capacity. Try 20-10-20 at 150–200 ppm. If no superphosphate has been added to the medium, it may be best to use 15-16-17 or 15-17-17 (no sodium) at 150–200 ppm. Leach with clear water occasionally to decrease the medium's soluble salts. Do not shock 'Mona Lisa' with high fertilizer rates or irregular water schedules, as they can cause cracking of the flower stems. For stronger stems, fertilize occasionally with calcium nitrate at 200 ppm.

Anemones grow best in cool, shady conditions. In areas of extremely high temperatures or high light intensity, heavy shading may be required. Shading should be removed in areas where cooler temperatures and cloudy weather persist throughout the growing season. During warm weather when cooler night temperatures are not possible, plants do better when grown in houses equipped with fan-and-pad cooling. The ideal production temperature is 55°F (13°C) to develop plants. As plants mature, lower temperatures to 42°F (5°C).

If using a 72 plug, it will take eight to ten weeks to flower upon transplanting.

Pot plant production

Much of the culture mentioned above pertains to pot culture as well. Transplant 1 plant/4" (10 cm) or 6" (15 cm) pot. Some growers have tried 3 plants/8" (20 cm) pot with very favorable results. The average crop time for 'Mona Lisa' in 4" (10 cm) pots is approximately twenty weeks from seed and ten to twelve weeks from 72-cell liners. The average temperatures after transplant should be 60–65°F (15–18°C) days and 55°F (13°C) nights.

Anemones require relatively high nutrient levels. Once the plugs have rooted-out, begin fertilizing with 200 ppm nitrogen (N) from 15-5-15 calcium magnesium once a week. Alternate with 200 ppm from a calcium-nitrate fertilizer. Since anemones grow best at cool temperatures, avoid ammonia-based fertilizers. Maintain an EC of 1.5–2.0 and a pH of 5.6–6.2 from transplant until finish.

To produce 'Mona Lisa' anemones as a pot crop, plant growth regulator treatments are needed. Apply Bonzi as a drench at a 2 ppm concentration about six weeks after transplant into a 4" (10 cm) pot. Plants should be full in the pots with no visible soil. The first visible buds should be showing. One application of Bonzi should be enough. Drench rates up to 4 ppm of Bonzi can be used with good results. The timing for the Bonzi treatment may vary depending on the size of the plug cell, the final container size, and the time of the year. During the warm summer season, the Bonzi treatment may be applied one week earlier if needed. Make sure that the crop has well-developed root mass before the drench application; the roots should fill the pot.

Pests & Diseases

Anemones are highly susceptible to a number of problems. Among the diseases to watch for are botrytis, *Colletotrichum,* downy mildew, *Pythium,* and rhizoctonia. Insect pests to scout for include aphids, thrips, and whiteflies.

Troubleshooting

Colletotrichum can cause stunted and gnarled leaves, irregular margins, and flower deformity or discoloration (in severe cases). High temperature and humidity promote the disease. Bud deformation may be caused by thrips.

Varieties

'Mona Lisa' produces strong 17" (43 cm) stems with 4–5" (10–13 cm) blooms. Plants are highly productive, and vase life is ten to fourteen days. Additionally, 'Mona Lisa' has fewer disease problems than plants grown from corms. With a good growing regime and a planting density of 12–15 plants/yd.^{2} (14–18/m^2), at least 125 stems/yd.^{2} (150/m^2) can be harvested. 'Mona Lisa' also makes a great large-pot, bedding plant crop in cool/coastal climates, such as California.

Among the varieties that can be grown from corms are the single-flowered 'De Caen' varieties and the double and semi-double 'St. Brigid' varieties. Although still widely grown as cut flowers, they suffer from having a poor color range, small flowers, and short stems. Another corm-raised cultivar is 'St. Piran', which has larger flowers and longer stems but a limited color range.

Postharvest

'Mona Lisa' flowers close naturally at night. Flowers are best cut as early as possible—before greenhouse temperatures rise and flowers begin to open. Cut the flower close to the crown, but use a clean, sharp knife, as this can become a site of disease infection. Sterilize the knife frequently.

Flowers can be cut at the closed-bud stage and stored for several days in water containing an antibacterial agent. This will produce a medium-sized flower. For maximum flower size, allow the flower to open and close once before cutting, bearing in mind that this will reduce the vase life.

Flowers can be kept in a cooler prior to shipping but should not be held for longer than fourteen days at 34°F (1°C). Stored blooms can be cooled to 40°F (4°C) in the retail environment.

Stems need not be shipped in water if markets are close by; however, do not ship stems lying down in a box, since "crooking" of the necks can occur. Always demand that boxes be shipped with flowers standing upright.

Anethum

Anethum graveolens
Common name: Dill
Herb, annual

Highlighted with edible seeds and foliage, dill flavors many foods. Highly prized for its use in making pickles, it can also be used as a filler for mixed cut flower arrangements.

Propagation & Production

Dill seed is best sown to plug trays and transplanted to either packs or pots or sown direct to cell packs (e.g., 32, 48, or 50 cells per flat) or 4" (10 cm) pots. Do not allow the plants to stretch or become overgrown as this will spoil garden performance. Growers often sow every two or three weeks to keep the crop fresh.

While dill can also be sown to an open flat and then individual seedlings can be transplanted to cell packs or pots, this is not recommended. Seedlings grow quickly and can become overgrown prior to transplanting. If transplanted in this condition, the resulting crop will be weak and the stems will flop, ruining garden performance.

For a 288-plug tray, sow two to three seeds per cell and cover lightly with vermiculite to maintain moisture around the seed or do not cover at all. Light aids germination, so do not cover with peat moss. Germinated at 68–70°F (20–21°C), the root radicle shows in four to six days and cotyledon expansion is complete in eight to thirteen days. Plug trays are ready to transplant in four to six weeks. Once transplanted to 4" (10 cm) containers, grow on at 55–60°F (13–16°C) nights. Plants are salable in six to seven weeks.

Seed can also be sown directly into the final cell pack or 4" (10 cm) pot (2–4 seeds/pot). Following the plug germination guidelines, cell packs will be ready to sell seven to nine weeks after sowing and nine to eleven weeks in 4" (10 cm) pots. To avoid stretching, grow on at cooler temperatures once plants are established in their final container.

Plants grow quickly, and the trick is to sell them while they are actively growing and not root-bound. Once roots fill out their growing container, plants will bolt and flower in the pack or pot, which diminishes garden performance. Sell plants when they are 6–7" (15–18 cm) tall or bump them into larger pots to keep roots from becoming bound in the final container.

Varieties

'Fern Leaf', an All-America Selections Award winner, is a dwarf selection, growing around 24" (61 cm) tall.

'Vierling' is a variety that reaches 40–50" (102–127 cm) tall and is suitable for the ornamental and cut flower markets as well as for culinary use.

Angelonia

Angelonia angustifolia
Common names: Summer snapdragon; summer orchid
Annual, cut flower

Angelonia has won the hearts of many growers, landscapers, and consumers. Its tolerance of weather extremes combined with the versatility of the genetics in the market make this one of the great new bedding plants of the twenty-first century. Used primarily for mixed patio containers for late spring and summer sales, *Angelonia* is used frequently in commercial landscapes as well. *Angelonia* flowers appear on spikes in purple, purple bicolor, pink, plum, raspberry, and white. Each orchid-like flower measures about 0.5–1" (1.5–3 cm) in diameter.

The common name of summer snapdragon refers to the flower form and not the cooler temperatures or garden treatment in which snapdragons excel. That being said, some question the common name, since there is little similarity between *Angelonia* and snapdragon flowers.

Angelonia is a tropical perennial native to Brazil. It loves hot, sunny conditions and plenty of water but is surprisingly drought tolerant.

Plants grow from 10–24" (25–61 cm) tall, depending on variety. In southern locations they are considerably taller with the persistence of heat and humidity.

Propagation

Angelonia are primarily propagated through cuttings; although one seed-propagated series, 'Serena', is also available.

Cuttings

Most suppliers offer unrooted cuttings. Stick cuttings immediately on arrival. If that is not possible, store them for no longer than twenty-four hours at 50°F (10°C). Select a rooting medium with a high starter charge. Rooting hormone is not required. Place newly stuck cuttings under intermittent mist for seven to nine days with substrate temperatures of 70–75°F (21–24°C) and air temperatures of 75–80°F (24–26°C) days and 68–70°F (20–21°C) nights. The medium should be sufficiently moist so that water is easily squeezed out. The mist frequency should vary as light levels and temperatures change during the day. VPD (Vapor Pressure Deficit) technology is very effective in controlling mist intervals. Tempered water in mist lines helps maintain soil temperature, a key factor in fat-root initiation. Provide 500–1,000 f.c. (5.4–11 klux) of light; retractable shade is ideal as it allows you to adjust light levels as cuttings root and mature. Apply 50–75 ppm from 15-0-15 as a foliar feed, if there is loss of foliage color. Callus should form within five days.

The first emerging roots signal that the cuttings are ready for the next stage in propagation, root development. Develop roots by lowering soil temperatures to 68–72°F (20–22°C) and maintaining air temperatures of 75–80°F (24–26°C) days and 68–70°F (20–21°C) nights. Increase light levels to 1,000–2,000 f.c. (11–22 klux) as cuttings begin to root. Gradually reduce soil moisture and mist frequency. Allow medium to begin drying slightly. Apply feed of 100 ppm from 15-0-15 alternating with 20-10-20. Increase the rate and frequency rapidly to 150 ppm as cuttings root out under higher light and lower humidity conditions. Roots should reach the bottom and sides of the propagule in ten to fourteen days.

During the root development stage, it is critical to control the top growth of the liners. Plants can tend to stretch. Pinch eighteen to twenty-one days after sticking. If you prefer a plant growth regulator, a tank-mix spray application of B-Nine (1,500 ppm) and Cycocel (750 ppm) applied ten to fourteen days after sticking will control internode stretch in propagation. Pinch at fourteen to eighteen days leaving four to five nodes below the pinch.

Harden (tone) cuttings for sale or transplant at 70–75°F (21–24°C) days and 62–68°F (16–20°C) nights for one week. Provide 2,000–3,000 f.c. (22–32 klux) of light. Fertilize with 150–200 ppm from 15-0-15 or 20-10-20. A 105 tray will take three to four weeks to root depending on the time of the year.

Seed

Sow into a well-drained, disease-free, soilless medium with a pH of 5.5–6.0 and a medium initial nutrient charge (EC 0.75 mS/cm with a 1:2 extraction). Plug tray size can vary from 406 to 128. Do not cover or bury the seed.

Stage 1 germination takes four to five days. Soil temperature should be 72–76°F (22–24°C). Light is required for germination, because the seeds will not germinate in the dark. Keep soil moist but not saturated during Stage 1 for optimal germination. Maintain 95% relative humidity (RH) until radicle emergence.

After radicle emergence during Stage 2, lower the temperature to 68–72°F (20–23°C), and move trays to a high light, seedling area. Keep the soil moist but not saturated and allow it to dry down a bit in order to stimulate the root system. Apply fertilizer at rate of less than 100 ppm nitrogen (N) from nitrate-form fertilizers with low phosphorous.

In Stage 3, lower the temperature to 65–70°F (18–21°C) with high light levels. Allow the media to further dry until the surface becomes light brown (i.e., level 2) before watering. Keep the moisture level on a wet-dry cycle (i.e., moisture levels from 4 to 2). Do not allow the seedlings to wilt, as they do not recover very well. Increase fertilizer to 100–175 ppm N. Growth regulators are generally not needed in plug stage. If necessary, a B-Nine/Alar (daminozide) 5,000 ppm spray can be used.

To finish the plug for transplant, the temperature should be lowered to 65–67°F (18–19°C) with as high light levels as possible. Repeat the moisture and fertility regime recommended for Stage 3. Plug time is five to six weeks for a 288-cell size.

Growing On

Rooted liners

Plant 1 cutting/4" (10 cm) pot, 1–2 cuttings/6" (15 cm) pot, and 3 cuttings/1-gal. (4 l) pot. Some growers prefer to establish plants first in 3" or 4" (8 or 10 cm) pots before they place *Angelonia* in larger containers such as combination planters. Cuttings are rooted and should be in flower in six to eight weeks in 4" (10 cm) pots if grown at warm temperatures and full sun in the greenhouse.

For larger pots (e.g., 1-gal. [4 l], 8" [20 cm], and patio pots), plants should be pinched ten to fourteen days after planting, when new growth is apparent. Pinch vegetation back to two to three nodes above first pinch (applied in liner stage). Pinch 4" (10 cm) pots once (in liner stage), and 6" (15 cm) or 1-gal. (4 l) pots two to three times. Growers have trialed Florel to encourage full, bushy plants. Apply the final application of Florel eight weeks prior to sale to ensure heavy flowering. Florel may cause leaf-tip burn. Florel has not demonstrated an ability to replace a manual pinch for the top-quality finish of *Angelonia* plants.

Grow in a warm greenhouse at 75–90°F (24–32°C) days and 65–70°F (18–21°C) nights in full sun for best performance. Northern growers should wait to bring cuttings in until their later spring plant shipments, as crops received too early will not grow until temperatures and light levels are higher. Winter crops benefit from additional supplemental light. Use moderate to low rates of fertilizer to keep flowering stems from becoming weak and floppy. Apply 150–200 ppm from 20-10-20 alternating with 15-0-15 with each irrigation. Leach with clear water at every third irrigation to prevent soluble salts problems. Keep medium uniformly moist; plants allowed to completely dry out may develop tip burn.

Height can be controlled partially by maintaining moderate fertilizer and moisture levels. For vigorous varieties, B-Nine (3,000 ppm) and Cycocel (1,500 ppm) are effective. Apply from one to three times starting ten to fourteen days after the first pinch.

Seed

Transplant into a well-drained, disease-free, soilless medium with a pH of 5.4–6.2 and a medium initial nutrient charge. Growing on temperatures should be 65–67°F (18–19°C) nights and 65–76°F (18–24°C) days. Daily average temperatures below 65°F (18°C) will slow crop growth rate dramatically. Keep light as high as possible, while maintaining the recommended temperatures. Avoid both excessive watering and drought.

Feed plants weekly at 175–225 ppm nitrogen (N), using predominantly nitrate-form fertilizer with low phosphorus and high potassium. Maintain the media EC at 1.5–2.0 mS/cm and pH at 5.8–6.2.

Normally plant growth regulators are not needed; however, later in the season or in the South where high temperatures can cause excessive growth, a tank mix of B-Nine/Alar (daminizide) 2,500 ppm mixed with Cycocel (chlormequat) 750–1,000 ppm is the most effective growth regulator for *Angelonia.* Cycocel rates can be adjusted depending on environmental conditions. Use lower rates under cooler and shorter daylength conditions, and higher rates under warmer and longer daylength conditions. Growth regulators can be started two weeks after transplanting and repeated as needed.

In warmer climates, a Bonzi (paclobutrazol) drench at 5–10 ppm (1.3–2.5 ml/l, 0.4% formulation) can be used two weeks after transplant instead of the B-Nine/Cycocel tank mix.

Do not pinch plants! Seed *Angelonia* has excellent, natural basal branching. Pinching will delay flowering and make the plant habit unattractive. The crop time from sowing to transplanting for 406- to 128-cell plug trays is five to six weeks. Table 2 below provides the scheduling for specific container sizes.

Cut Flowers

Angelonia is a prolific cut flower. Research at the University of Kentucky has demonstrated average yields of nine varieties of 78 stems/ft.^{2} (841 stems/m^2) for approximately twenty weeks of greenhouse production. 'AngelMist Purple' had the highest yield with 119 stems/ft.^{2} (1,285 stems/m^2), while 'AngelMist Deep Plum' had the lowest yield, with 58 stems/ft.^{2} (629 stems/m^2). Cut stems were harvested from 14–36" (36–91 cm) long and over 60% of the cut stems were over 18" (46 cm) long.

Hot summer greenhouse conditions, even without evaporative cooling, are fine as long as the plants get sufficient water. The plants tolerate average daily greenhouse temperatures of 76–85°F (24–29°C) and even many days with maximum daily temperatures of 100–105°F (38–40°C). Plants should be fertilized at 150–200 ppm for two to three weeks after harvest to get new shoots growing rapidly, and then the fertilizer should be reduced to zero until harvest. This allows strong, straight stems to form with few lateral branches.

A plant density of 7–8 plants/ft.^{2} (75–85 plants/m^2) is appropriate, and plants can be grown in 4" or 6" (10 or 15 cm) pots. One layer of cut flower mesh will support the plants properly. It seems best to cut stems as low as possible at harvest and allow new shoots to arise from the base. Additionally, remove all stems over the two- to three-week harvest period to allow new shoots to develop together. Expect two to three harvests, spaced six to seven weeks apart, during the summer.

Pests & Diseases

Angelonia is quite free of insects and disease. Whiteflies can attack if the insects are around in high numbers. Aphids and thrips can also become problems. Disease problems include botrytis, *Pythium,* and rhizoctonia.

Troubleshooting

Plants may become chlorotic from iron deficiency. Plants can collapse from the medium remaining wet

Table 2. Crop Time from Transplant from 406 to 288 Trays to Saleable Finished Container*

CONTAINER SIZE	PLANTS PER POT	WEEKS FROM TRANSPLANT	TOTAL WEEKS
306 pack	1	8–9	13–15
4–4.5" (10–11 cm) pot	1	9–10	13–15
6–6.5" (15–16 cm) pot	3	9–10	14–16
1-gal. (4 l) pot	3	9–10	14–16

* *Note:* When transplanted from a 128-tray, finish crop time for 'Serena' can be reduced by one to two weeks.

for extended periods of time or root rot caused by planting liners too deeply. High ammonia levels in the medium; too much fertilizer under low light conditions; low light; excess or late Florel applications; and/or high temperatures and high light can cause excessive vegetative growth. Low fertilizer levels and/or low light can lead to poor branching.

Varieties

If you choose the non-patented cultivars, pink ('Pandiana' and 'Pink Princess') and purple ('Purple', 'Hilo Princess', and 'Mandiana') are shorter and more appropriate for pots, while purple/white ('Blue Pacific' and 'Tiger Princess') and white ('Alba') are taller, more vigorous, and better suited for cut flowers. Be aware that these varieties may be infected with cucumber mosaic virus, which may affect other greenhouse crops.

The 'AngelMist' series has been disease indexed, cleaned up, and maintained in a clean-stock program. As a result, cuttings offer superior performance. 'Pink' and 'Deep Plum' are shorter, while 'Purple Stripe', 'Purple', 'White', and 'Lavender Improved' are taller and perform well as cut flowers. The 'Angelface' and 'Carita' series are also available in a range of colors.

From seed, the 'Serena' series is available in a number of separate colors and several mixes. It is an excellent selection and performs well in home gardens and professional landscapes.

Postharvest

Angelonia leaves and stems are covered with glandular hairs. These hairs exude sticky, aromatic compounds, and some workers might find this objectionable. Cut flower workers have to wash their hands after handling the stems and removing the lower leaves.

Garden center retailers should display finished pots under partial shade to reduce watering requirements. Keep plants moist, as pots allowed to dry out completely may abort flowers. Do not crowd plants on the retail bench, as air and light need to reach bottom leaves to maintain plant quality.

Instruct consumers to shear plants back by about 50% to improve plant habit, especially in warmer climates. These plants will reflower heavily in two to three weeks. *Angelonia* should be planted in full sun in the garden or landscape for the most color. In partial-sun beds the plants will not color as much as in full sun. Plants should be spaced at 8" (20 cm) centers in the landscape and will grow to about 12" (30 cm) tall.

Anthurium

Anthurium spp.
Pot plant, cut flower

Anthurium has been cultivated for many decades as a cut flower. Since the mid-1980s, anthurium's popularity as a flowering pot plant has increased dramatically, and it has become a popular addition to many foliage growers' product lines.

Anthurium is relatively easy to grow and has attractive foliage. When grown in the proper environment, plants produce long-lasting flowers year-round. A wide range of cultivars offer different flower sizes, shapes, and colors; some even offer delicate fragrances. Commercially, anthurium is grown throughout the world with the heaviest concentrations in the United States (in Florida) and the Netherlands.

Propagation

Young plants are primarily propagated by tissue culture and available commercially either as microcuttings or as 72- or 98-cell liner trays. Depending on the cultivar's inherent branching and flowering habit, young plant producers use 1–3 plants (microcuttings)/liner cell. Cultural conditions, especially light intensity, are very important for young plant production. If you are buying in liners, avoid using young plants grown under low light conditions.

Growing On

Select a well-drained, disease-free medium; anthurium will not tolerate saturated, poorly drained soil mixes. Best results are achieved with a 1-1-1 ratio of Canadian peat, composted pine bark (watch for particle size—not too much dust), and perlite or airlite. Avoid vermiculite except in 4" (10 cm) containers. In long-term crops, such as 6" (15 cm) pots and up, vermiculite compacts and will waterlog. Maintain the substrate pH at 5.5–6.5.

Most pot anthurium is sold in 6" and 8" (15 and 20 cm) containers, with a smaller percentage in 4" and

10" (10 and 25 cm) containers. Crop finish times will vary depending on variety, size, and cultural environment. Except in the case of *A. scherzerianum*, growers should consider anthurium a long-term floral crop. In Florida's subtropical climate, most 6" (15 cm) crops are finished in eight to ten months using 72- or 98-cell tray young plants. *A. scherzerianum* is usually grown in 3.5–6" (9–15 cm) containers and will finish in four to seven months. A young plant supplier will be able to give recommendations on the optimum container size and finish times for each individual variety.

Anthurium grows best with day temperatures of 78–90°F (25–32°C) and night temperatures of 70–75°F (21–24°C). Temperatures above 90°F (32°C) may cause foliar burning, faded flower color, and reduced flower life. Night temperatures of 40–50°F (4–10°C) can result in slow growth and yellowing of lower leaves. *A. scherzerianum* cultivars require lower temperatures in the range of 68–80°F (20–27°C) days and 60–70°F (15–21°C) nights. Anthurium will not tolerate frost or freezing conditions.

Irrigation is critical to a successful anthurium crop. Most anthurium species are native to tropical rain forests and are primarily epiphytic in nature, meaning that they get nutrients from the air and rain, not through their root systems. Thus, in their natural habitat, they receive ample, frequent water with good drainage. In cultivation, anthurium prefers an evenly moist medium, especially when actively growing. Overall, it is better to slightly underwater than overwater. Drying out may cause tip burn, root damage, and reduced growth rates, while overwatering can cause root damage and sudden yellowing of older leaves.

Moderate but consistent levels of a complete fertilizer are important. Magnesium requirements in anthurium plant tissue are higher than for most foliage crops, especially in warmer climates. Because of the long-term nature of anthurium crops, pay special attention to ensure continued magnesium availability. Incorporate 10 lbs. (4.5 kg) of dolomite and 3.5 lbs. (1.6 kg) Hi-Cal lime/yd.^{3} (0.8 m^3) of medium to balance the calcium and magnesium ratio. Regular foliar applications of magnesium sources (i.e., Epsom salts, magnesium nitrate, etc.) will help prevent magnesium deficiencies. After twenty-four to twenty-six weeks, top-dress with dolomite (3 tbsp./10" [44 ml/25 cm] pot) or another magnesium source to help ensure continued magnesium availability. Top-dressing Epsom salts is beneficial but short lived.

Avoid high nutrient levels, especially after planting young plants. Liquid fertilizer on a constant-feed program should not exceed 250 ppm nitrogen (N). On mature plants, occasional rates as high as 400 ppm N are acceptable but must be alternated with clear water. Tests have shown that plants given frequent doses of 300–400 ppm N grow slower, have lighter flower colors, and produce thick, deformed leaves. When using an overhead irrigation system to dispense liquid fertilizer, a quick rinse with pure water is beneficial since liquid fertilizer left on foliage can cause grayish, corky scars on leaves. With dry fertilizer applications, it is very important to water frequently to reduce salt buildup. When using time-release fertilizers, carefully consider crop times and, if necessary, reapply to avoid deficiencies.

Anthuriums grow under a wide range of light intensities, but their actual performance depends on the variety, elevation, temperature, and nutrition. Generally, most anthuriums grow well at light intensities ranging from 1,500–2,500 f.c. (16–27 klux). Light intensities higher than 2,500 f.c. (27 klux) can result in faded flower and leaf color. *A. scherzerianum* varieties are best grown at light intensities between 1,000–1,500 f.c. (11–16 klux).

Pests & Diseases

Implement preventive maintenance programs for mites, snails, slugs, worms, thrips, and whiteflies. Whiteflies are especially attracted to new growth and, once established, are difficult to eradicate. Many growers have experienced phytotoxicity on numerous anthurium varieties from using certain pesticides. Never apply pesticides while plants are under any form of stress (i.e., moisture or hot temperatures). Keep plants weed free to reduce the incidence of insects.

Many anthurium cultivars are susceptible to *Phytophthora,* rhizoctonia, *Pythium, Colletotrichum,* and pseudomonas. Prevent infection by keeping plants off the ground, providing good ventilation, and avoiding overhead irrigation during late afternoon or evening hours. As a matter of caution, all new pesticides should be used in a controlled test on a small percentage of each variety grown. Always allow four weeks for phytotoxic symptoms to appear. Most often, symptoms occur as distortion and/or discoloration of new growth.

Anthurium blight, caused by *Xanthomonas campestris* pv. *dieffenbachiae,* is by far the greatest challenge to the anthurium grower. While many of the pot varieties (specifically the 'Andreacola' varieties) are resistant or partially resistant, many of the hybrids with larger, showier flowers have no resistance. *Xanthomonas* is a bacterium. The pathogen *Xanthomonas campestris* pv. *dieffenbachiae* is specific to plants in the Araceae family and is most pathogenic on dieffenbachia, *Aglaonema,* and anthurium. The disease can easily spread from one genus to another in this group. *Syngonium* and *Pothos* also have the potential to host strains of *Xanthomonas* that may be pathogenic to anthurium.

There is no available chemical cure for *Xanthomonas* blight. While some chemicals are effective as a preventive measure, none of the fungicides/bactericides available today will actually cure an infection. Thus, the only effective way of controlling blight is via sanitation and prevention.

Sanitation

Keeping your anthurium crop clean is the most important aspect of production. Virtually all of the pot-type anthurium on the market today are produced from tissue culture. As plants will not survive in vitro infected with *Xanthomonas,* plants directly harvested from tissue culture can be considered free of infection. However, young plants weaned in the greenhouse are susceptible to infection. The same sanitation practices should be in place in young plant production as those practices effective in finished production.

Since *Xanthomonas* can exist in plants without any visible symptoms, it is wise to isolate incoming plant material for observation before introducing it into the production facility. Proper cultural practices and sanitation can be effective in the prevention and elimination of blight.

Keep foliage dry, if possible. Using drip irrigation and growing under a hard cover are essential when growing susceptible cultivars. Lower humidity will decrease guttation and can help dry the foliage faster if using overhead irrigation. Give plants ample spacing to allow for good air circulation. Avoid condensation by using fans, and do not hang plants above your anthurium crop. High nitrogen fertilizer makes plants more susceptible to infection. Preventive maintenance programs of copper-based fungicides alternated with bactericides of streptomycin or oxytetracycline can help prevent infection. However, copper can be phytotoxic to many cultivars!

Disinfect all benches, pots, and tools coming in contact with plant material. Prevent and eliminate standing water under benches. Routinely (preferably daily) rogue out any infected plants. This is most effective at the end of the day. Do this when foliage is dry, and remove all infected plants and plant parts from the greenhouse premises. Avoid any unnecessary movement through aisles. Avoid cross-contamination by isolating susceptible crops. *Aglaonema* imported from the tropics are frequently carriers. In-house vegetative propagation of dieffenbachia and *Aglaonema* should be kept far from tissue-cultured plants.

Varieties

Anthurium can be divided into four basic groups: *A. andreanum* cultivars; interspecific hybrids between *A. andreanum* cultivars and dwarf species currently referred to as 'Andreacola' types; *A. scherzerianum* hybrids; and foliage anthurium. *Anthurium andreanum,* a generally large, somewhat open-structured plant with large flowers, is commonly grown for cut flower production and sometimes adaptable to pot culture. New cultivars, selected specifically for pot culture, are more compact. *A. andreanum's* primary flower colors are white, pink, red, red-orange, and green. 'Andreacola' cultivars are small to intermediate in overall size, fuller, and more compact; they generally produce smaller but more numerous flowers than *A. andreanum* cultivars. 'Andreacola' cultivars tend to have thicker, dark green leaves and often show resistance to the more aggressive anthurium diseases. Primary flower colors are white, pink, red, and lavender. *A. scherzerianum*, the first widely cultivated anthurium pot plant, is small and compact with white, pink, and red flower colors. Foliage anthuriums come in numerous shapes and sizes and represent a minor portion of the total anthurium pot market. However, it should be noted that most foliage anthuriums are durable plants and offer distinct forms.

Antirrhinum

Antirrhinum majus
Common name: Snapdragon
Annual, cut flower, bedding plant

For vertical color, whether it is in the garden or a cut flower bouquet or arrangement, you cannot beat snapdragons. Their tall spires are increasingly becoming a year-round staple in the cut flower market. Bedding plant snapdragons are on the rise as well. Southern growers can sell snaps green or flowering in the fall for spectacular displays the following spring. New breeding in bedding plant snapdragons is introducing new versions of butterfly, or open snaps, onto the market. Vegetative upright and trailing snapdragons are also available. These add a new form (i.e., trailing) as well as additional flower colors.

For cut flower growers, snapdragons are not the easiest crop to grow: Careful scheduling of the proper varieties for flowering throughout the year is required, and postharvest can be a challenge since stems are geotropic, meaning they bend when they are stored flat!

Bedding plant growers will find snapdragons a breeze and an excellent crop for cooler greenhouse production.

Propagation

Snapdragons are commonly seed or vegetatively propagated.

Seed

Growing snapdragon plugs is not especially difficult, but it is easier to order plugs from commercial propagators when you need them than to schedule your own plug production, especially if you are growing cut flower snapdragons. Cut flower snapdragon plugs are available year-round from a number of suppliers. Bedding plant snapdragon plugs are generally planted in the spring; but if you are interested in a fall planting in the South, be sure to select those varieties that flower under short days.

Sow seed onto a sterile, germination medium with a pH of 5.5–5.8 and an EC less than 0.75 mS/cm. Since light is necessary for germination, cover with a light coating of vermiculite. This will help maintain better moisture around the seed as it germinates. Maintain substrate temperatures of 65–75°F (18–24°C); radicles will emerge in six to eight days. Maintain uniform substrate moisture but not saturated conditions. Snapdragons are *very sensitive* to a high starter charge in the mix, so keep ammonium levels less than 5 ppm.

Once radicles have emerged, move trays to Stage 2 (stem and cotyledon emergence), which will take about seven days. Increase light levels to 450–1,500 f.c. (4.8–16 klux) and apply 50–75 ppm from 14-0-14 once a week. Liquid fertilization may not be necessary if sufficient nutrition is incorporated in the growing mix before planting. Keep the mix evenly moist but not saturated. A light cover of vermiculite at this stage will help maintain moisture and prevent algae growth. Continue to maintain substrate temperatures. Make sure foliage is dry by nightfall to avoid disease problems.

Stage 3 (growth and development of true leaves) takes from twenty-one to twenty-eight days for 384s. Lower the soil temperatures to 62–65°F (17–18°C) and raise light levels to 1,000–2,500 f.c. (11–27 klux). Allow trays to dry between irrigations but avoid wilting. Increase fertilizer to 100–150 ppm weekly from 20-10-20 alternating with 14-0-14. Avoid ammonium-based fertilizers if growing below 65°F (18°C). Supplement with magnesium once or twice during Stage 3 using magnesium sulfate (16 oz./100 gal. [1.2 g/l]) or magnesium nitrate. Occasionally leach with clear water to reduce salts. Apply fungicides at the lowest recommended rate to control *Pythium,* rhizoctonia, and *Thielaviopsis.*

Harden plugs prior to transplant or sale by lowering soil temperatures to 60–62°F (16–17°C). Allow medium to dry thoroughly between watering. Continue to maintain light levels at 1,000–2,500 f.c. (11–27 klux). Feed with 100–150 ppm from 14-0-14 weekly.

A 288-plug tray will take from four to six weeks from sowing to be ready to transplant to larger containers.

Cuttings

Stick unrooted cuttings in Oasis wedges or a sterile medium with a pH of 5.6–5.8. Rooting hormone is not necessary unless the cuttings are not supple. Maintain substrate temperatures of 68–72°F (20–22°C) and air temperatures of 75–80°F (24–26°C) days and 68–70°F (20–22°C) nights. Provide 500–1,000 f.c. (5.4–11 klux) of light, and apply a foliar feed of 50–75 ppm from 15-0-15 as needed. Retractable shade is ideal since you will be able to adjust light levels as cuttings mature. Use mist for seven to ten days to maintain humidity around the cuttings; adjust the frequency based on light levels and temperature. Calluses will form in five to seven days.

Once calluses have formed and 50% of the cuttings are differentiating root initials, cuttings are ready for the next stage—root development. Continue maintaining soil and air temperatures. Increase light levels to 1,000–2,000 f.c. (11–22 klux) and increase fertilizer to 100–150 ppm weekly from 20-10-20, alternating with 15-0-15. Reduce mist as roots form, but avoid drying out air, as this will increase evapotranspiration, reducing the root zone temperature. Roots will develop over seven to fourteen days.

Harden cuttings for seven days prior to transplant or sale by reducing air temperatures to 75–80°F (24–26°C) days and 62–68°F (16–20°C) nights. Increase fertilizer to two times a week from 150–200 ppm from 20-10-20, alternating with 15-0-15.

If sticking cuttings into a 105 tray, allow four to five weeks for plants to be ready to transplant to larger containers.

Bedding Plant Crops

Growing On

Seed

Transplant plugs when they are ready, otherwise growth will be checked. Bedding plant snapdragon plugs can be transplanted into just about any size cell pack, although larger 18 or 36 packs will offer a longer-lasting product on the retail shelf since plants will be able to attain some size prior to flowering. You can also offer snaps in 4" (10 cm) pots or even put from three to five plants in 1-gal. (4 l) containers. Some retail growers are running with the concept of "knee-high" gardening and planting either cut snapdragon varieties or taller-growing bedding plant snaps into 6" (15 cm) or 1-gal. (4 l) containers for customers to plant in their own cutting garden. If you are looking for something that your customers will not find at the mass-market store, tall snapdragons are a great item. Large bedding plant growers cannot put enough of them on a truck to make them pay. If you are growing larger pots of taller-growing bedding plant snaps and you would like to create a spectacular flower show, you can pinch these plants once roots have reached the edge of the container.

If height control is an issue, A-Rest, B-Nine, and Cycocel are effective. Some growers use a B-Nine/Cycocel tank mix of 800–1,000 ppm of each.

If transplanting one plug per 36-cell flat, allow five to six weeks for green plant sales. For 4" (10 cm) pots, allow six to seven weeks for budded color for most varieties. This is true for all dwarf-growing selections. For taller-growing varieties, these crop times are appropriate for green plant sales. Varieties like 'Rocket', 'Sonnet', or 'Liberty' are best planted green to the garden. They will flower in several weeks; 'Rocket' takes the longest.

Cuttings

Transplant 1 snapdragon liner cutting/4" (10 cm) pot, 2 cuttings/6" (15 cm), and 4–5 cuttings/10" (20 cm) hanging basket.

Select a disease-free, sterile medium with a moderate initial starter charge and a pH of 5.5–6.2. Provide high light, 5,000–6,000 f.c. (54–65 klux). Low light levels promote plant stretch, reduce flowering, and increase crop time. Snapdragons are not daylength sensitive, but flowering is faster under long days.

Grow plants on at 72–75°F (22–24°C) days and 62–65°F (16–18°C) nights. Once plants are established, reduce temperatures to 45–50°F (7–10°C) nights to promote flowering. When flowering begins, keep night temperatures below 60°F (16°C) to maintain flowering.

Fertilize two times a week with 150–200 ppm alternating between 20-10-20 and 15-0-15. Increase the rate to 200–300 ppm as plants become established. If high salts become a problem, irrigate with clear water every third watering. Be sure to maintain fertilizer levels for trailing snapdragons, as low nutrition levels inhibit plant branching. When spikes begin to elongate, reduce fertilizer applications to increase postharvest life.

To help control diseases, avoid wetting foliage: If you must water overhead, make sure all foliage is dry well before nightfall.

Pinch vegetative snapdragons once roots have reached the edge of the container. No additional pinching should be necessary.

The crop time from transplanting a liner to a flowering 4" (10 cm) pot is six to eight weeks on all varieties.

Pests & Diseases

Snapdragons are relatively pest free in comparison to many crops. Aphids, mites, and thrips are the most common pests on mature plants, while fungus gnats and shore fly larvae damage seedlings. Some pesticides are known to damage snapdragon spikes: Read and follow the instructions on the label to avoid problems.

Fungal diseases that affect snapdragons include downy mildew, powdery mildew, *Pythium*, botrytis, rust, *Phyllosticta* blight, and anthracnose. TSWV and INSV can devastate plants.

Troubleshooting

High substrate pH may cause stunted or uneven seedling growth or poor root development. Very high pH may result in iron or boron deficiency. Yellow interveinal chlorosis of upper leaves may be caused by iron deficiency resulting from high pH or a cool, wet medium, which reduces iron uptake. Identify the cause of the deficiency. If it is high pH, lower the pH with one application of iron sulfate at 33 oz./100 gal. (2.5 g/l) applied to saturate the medium. If the deficiency is the result of cool, wet conditions, drying out the soil and warming the air temperature will improve growth in three to four days.

Plant collapse can be due to botrytis or keeping the medium wet for extended periods of time. Excessive foliage growth can be caused by high nitrogen concentration in the medium, delivering too much fertilizer under low-light conditions, or low light and overwatering. Drying plants out between irrigations or high soluble salts in the medium may cause foliage necrosis. Hanging-basket cutting snapdragons that do not branch may be growing in too little light or may have received insufficient fertilization in early stages of growth.

Varieties

Most bedding plant growers and retailers list two or three types of garden snaps on their sales sheet. The well-known F_1 'Floral Shower' series is the dwarf strain widely used for edging and mass plantings. Growing 6–8" (15–20 cm) tall, they flower well in packs and 4" (10 cm) pots. 'Chimes' and 'Montego' are also great snapdragons in this height class. Another dwarf-flowering selection is the 'Snapshot' series—a uniform selection that flowers at 6–10" (15–25 cm) tall in the garden. Also in the dwarf snapdragon class is the butterfly, or open-flowered, 'Bells' series.

In medium-tall snapdragons (14–18" [36–46 cm] tall), try the 'Crown' for regular flowers and 'La Bella' for butterfly flowers.

Growers looking for taller snaps—to grow in pots for the landscape trade—will like 'Liberty Classic' or 'Sonnet'. Both are grow to 24" (61 cm) tall or slightly less and can be used as field-grown cuts.

Some of the latest breeding advancements have been varieties like 'Speedy Sonnet', which flowers up to three to four weeks earlier under shorter days for spring flowering and still grows up to 24" (61 cm) tall. This is on the heels of the variety 'Solstice', which also flowers faster under shorter days. 'Solstice' can be used in southern climates for splashes of winter color and can be produced and marketed in color sooner in the spring than other varieties. 'Chimes' is a dwarf snapdragon that flowers early under short days.

On the tall end of the scale is the F_1 'Rocket', usually grown for garden-market cut flower purposes and reaching heights of 30–36" (76–91 cm). In some areas, 'Rocket' is grown for commercial cut flower purposes. Planted in the North by June 1, 'Rocket' can yield two harvests as a cut. Many bedding plant growers routinely offer 'Rocket' for consumers planting cut flower gardens.

From the vegetative market there are a multitude of varieties. These are mostly used for mixed combination containers and hanging baskets. Selections include the 'Luminaire' series growing form 8–10" (20–25 cm) tall and spreading up to 20" (50 cm). They are excellent in hanging baskets.

Postharvest

Do not store snapdragons in the dark. Store ready-to-sell plants in the greenhouse at 50–55°F (10–13°C). Snapdragons are sensitive to ethylene, so keep plants away from ripening fruits and vegetables. Display plants under high light but cool temperatures (65–70°F [18–21°C]).

Snapdragons make an excellent cut flower in the home garden. Consumers should cut flowers several times to encourage continuous flowering. When hardened off, snapdragons can survive temperatures of 25–28°F (–4––2°C). Instruct consumers in the South to mulch plants well when planting snapdragons in the fall.

Cut Flower Crops

The demand for forcing snapdragons is being met by a variety of different types of growers and through several different distribution channels. Snapdragons are grown throughout North, Central, and South America; Japan; and Europe. Producers in highland tropical and subtropical areas are also meeting demand. A grower's choice of distribution channels strongly affects the quantity produced, the maturity harvested, and the price received. Typically, a grower experiences the best prices when supplying local markets or shipping directly to the end user.

With increased worldwide competition for traditional mainline cut flowers, many growers have discovered forcing snapdragons to be a welcome way to diversify. Year-round product availability and ever-increasing color range are two key factors that have increased their use as a cut flower.

Propagation

All cut flower selections are propagated from seed. Many cut flower growers purchase plug trays from commercial propagators instead of sowing their own; although they can do this as well. Follow the bedding plant culture guidelines for sowing cut flower seed.

Growing On

Snapdragons are commonly grown in field soil or soilless medium directly in the ground or in raised beds or benches. Snapdragons grow best in a growing medium that allows adequate aeration to the roots yet holds a steady supply of moisture. The greater the aeration, the more forgiving the medium is to overwatering, but a medium with high aeration will require frequent irrigation. Growing mix in benches must be better aerated than a mix used to grow snapdragons directly in the ground, because the bench bottom creates a "perched water table" that limits water drainage.

The growing medium for raised benches should consist of less than 50% field soil, with the remaining percentage consisting of a mixture of several of the following: vermiculite, perlite, peat, composted bark, rice hulls, and so on.

Ground beds in locations with sandy loam soils may be suitable for growing snapdragons without any amendments. Heavy soils should be improved prior to planting by tilling in organic material, such as peat moss, rice hulls, compost, and decomposed manure. The growing medium for raised benches should consist of less than 50% field soil, with the remaining percentage consisting of a mixture of more than one of the following: vermiculite, perlite, peat moss, composted bark, rice hulls, and so on.

No matter what kind of medium you use, it must be free of disease-causing organisms. Most artificial media are naturally free of disease organisms and may be used without treatment. Disinfect the medium prior to planting—pasteurizing the medium with high temperatures (160°F [71°C]) from aerated steam for thirty minutes is the most common method. Chemical treatments are also possible.

Test the growing medium prior to planting. Fertility should be moderate, with an EC of 1.0–1.75 mS/cm, with less than 10 ppm ammonium. Substrate should have a pH of 5.5–6.5. The more mineral soil (field soil) included in the mix, the higher the optimum pH. An organic medium should have a pH at the lower end of this range. Amend the soil to adjust the pH several weeks prior to planting. Water the medium and retest before planting to determine if the desired changes have occurred.

Snapdragons are often described as "light feeders." Phosphorus and calcium are usually incorporated into growing mix prior to planting, and the other nutrients are supplied with a soluble fertilizer during growth. Superphosphate incorporated at 5 lbs./100 ft.^{2} (250 g/m^2) should supply sufficient phosphorous for the entire crop, except in a very porous medium. If tests show calcium is low, incorporate limestone (if the pH is too low) or gypsum (if pH is acceptable), either at 5 lbs./100 ft.^{2} (250 g/m^2). If phosphoric acid is used to modify water alkalinity, superphosphate may not be needed.

Snapdragon plugs are generally ready to transplant four to five weeks after sowing, when plugs pull easily from the plug tray and the second set of true leaves unfold. When buying in seedlings or plugs, acclimate them to greenhouse conditions for twenty-four hours prior to planting, then transplant quickly. Keep snapdragons actively growing by transplanting on time; delays in planting will subsequently cause delays in flowering and lower stem quality. If holding plugs is unavoidable, store plugs at 36–39°F (2–4°C) under fluorescent lights at 250 f.c. (2.7 klux) for fourteen hours per day. Treat with fungicide prior to storage to prevent botrytis.

Transplant plugs for greenhouse production on a spacing of 10–12 plants/net ft.^{2} (100–130/net m^2), decreasing to 8 plants/net ft.^{2} (85–90/net m^2) in seasons with low light. Treat with a fungicide to prevent botrytis prior to storage. Plant unpinched field crops on 9–12" (23–30 cm) centers; if you pinch plants, allow more space.

Irrigate seedlings with clear water after transplanting. Begin fertilizing at the next watering, using a well-balanced, low-ammonium (less than 40%) fertilizer at a rate of 150–200 ppm. Constant liquid fertilization with occasional clear water for leaching can be used until flower buds begin to swell. Once flower buds swell and show color, use clear water only.

Excessive side shoots are an indication of high moisture or fertility levels or just improper variety selection. Maintain a moderate to low substrate EC (less than 2.5 mS/cm) to avoid excessive side shoots. Irrigate with clear water, if necessary, to lower EC. Light, porous media are less prone to excessive nutrient and moisture levels, resulting in fewer side shoots. It is also important to choose snapdragon varieties from the correct response groups. When subjected to long days and high temperatures, varieties in Groups 1 and 2 (see table 3) tend to increase their side branching. If side shoots persist on edge rows, trim them off to increase light and air circulation reaching in the center of the bed.

Snapdragons need support during production. Two layers of support are the minimum, but three are preferred. Mesh sizes of 4" × 4" to 6" × 6" (10 × 10 cm to 15 × 15 cm) are the most commonly used and provide adequate support for the stems. Place the first level at 4–6" (10–15 cm) above the soil and the second level 6" (15 cm) above the first. Raise the upper level of the support as the stems lengthen. Keep support below the first flower.

Snapdragon growth and flowering response depend on the interaction of light quality, light quantity, light duration, temperature, CO_2 levels, humidity, and soil type, as well as other environmental factors. Snapdragons may be grown under various light intensities, provided appropriate varieties are used.

Snapdragons do not flower in response to day length; however, plants flower faster and with fewer leaves under long days. Under short days, plants tend to produce vegetation. Some varieties are not affected by day length, while others are. Make sure to select the right variety for the right time of year. Variety selection is more important for greenhouse-grown crops than for field crops.

The best quality is usually achieved with the highest light levels. Shading may be necessary in some climates for temperature control. While temperature affects overall growth rate, day length and light quality are the most important factors influencing flower initiation. Initiation in young plants occurs when they have five to ten pairs of leaves, depending on the response group and individual variety. Unusual environmental conditions during this critical stage (e.g., a long stretch of overcast weather) can greatly affect crop time.

Once flower initiation has occurred, night temperature has the greatest influence on flowering time and final quality. The ideal night growing temperature depends on the response group. For the highest-quality snapdragons, optimum night temperatures by variety are as follows:

Group 1: 45–50°F (7–10°C)
Group 2: 50–55°F (10–13°C)
Group 3: 55–60°F (13–16°C)
Group 4: above 60°F (16°C)

Generally, the lower temperature in each range gives the best quality at the expense of a longer crop time. The lower temperature is advisable during extended periods of low light.

The highest-quality snapdragons can be grown with supplemental HID lights. In this production method, snaps in Groups 3 and 4 can be grown year-round by lighting the plants when natural day lengths are less than twelve hours. Groups 1 and 2 are not recommended for HID culture because they initiate flowers too quickly, which causes short, weak stems. For northern states, begin supplemental HID lighting at week 36, started with two hours of additional light and increasing by one hour each week to ten hours by week 44. Maintain ten hours of supplemental lighting through week 2. Reduce supplemental lighting by one hour each week starting in week 3. Use 400–800 f.c. (4.3–8.6 klux). Optimize conditions by increasing fertilizer to 300–500 ppm, growing at night temperatures of 60–62°F (16–17°C) and by adding CO_2 at 800–1,200 ppm.

Pests & Diseases

See the pest and disease comments for snapdragon bedding plant crops.

Troubleshooting

See the troubleshooting comments for snapdragon bedding plant crops.

Varieties

Cut snapdragons can be produced year-round in most climates. The varieties are separated into four groups based on their optimal growing conditions (table 3).

Knowing the relationship between the flowering times of different varieties allows you to fine-tune crop scheduling. This is especially important in two situations: targeting a key holiday and scheduling a smooth transition between groups. If a white variety scheduled for Christmas harvest consistently blooms too short or too early, try a later-blooming white variety or sow the early white variety slightly later than recommended for that group, assuming all other factors are constant. The sowing and harvest dates presented in table 4 are purposely given in a range to account for variety differences and regional and environmental differences.

Normal weather variations from year to year can still complicate the most well-planned schedule. The fewer environmental controls available (e.g., heat or fans), the more buffers that must be added to guarantee a successful crop, such as using more than one variety or multiple sowing dates of a favorite variety.

Growers often cite the fall transition from Group 3 to Group 2 as the most difficult time to schedule a continuous succession of quality snapdragons. Excessively warm temperatures and high light at the young plant stage (late summer) can make Group 2 snapdragons bloom too early and too short. On the other hand, unusually cool nights, even after flowers

Table 3. Variety Groupings

CATEGORY	DAY LENGTH	LIGHT LEVELS	NIGHT TEMPS	
			°F	°C
Group 1	Short	Low	45–50	7–10
Group 2	Short (but not as short as Group 1)	Moderate	50–55	10–13
Group 3	Medium to long	Moderate to high	55–60	13–16
Group 4	Long	High	60+	16+

have initiated, can drastically lengthen the crop time of Group 3 varieties. Intermediate varieties (called late Group 2 and early Group 3 in the descriptions to follow) are excellent choices for harvest during this period. Alternatively, use the descriptions to choose varieties that help connect Group 2 to Group 3. The logical progression as daylight decreases is Group 3, early Group 3, late Group 2, and then Group 2.

Forced snapdragons are classified into the four groups according to their flowering response to a combination of environmental factors. There is some overlapping of groups and varieties, and some varieties do well in more than one situation. For instance, a variety included in Groups 1 and 2 will perform well throughout fall Group 2, winter Group 1, and spring Group 2 harvest periods (see table 5 for a listing of varieties).

Postharvest

Harvest flowers when at least five to seven florets are open. Premature harvesting leads to poor color development and reduced flower size, as flowers continue to open once stems are cut. This is especially critical on dark colors, such as rose and royal purple.

For maximum vase life, place snapdragon stems in water as soon as possible after cutting. Remove the foliage from the lower third of the stems, then grade and bunch. To condition for immediate use or shipping, place the flowers in warm water (70–75°F [21–25°C]) containing floral preservatives and hold at 45–50°F (7–10°C) for at least six to eight hours or overnight. Select a preservative that contains sucrose as well as 8-HQC (8-hydroxyquinoline citrate) or another bactericide to facilitate water uptake and inhibit stem plugging. Holding the stems in light (200 f.c. [2.2 klux]) enhances color development.

Shattering in response to ethylene can be a problem with some snapdragons. Many shatter-tolerant varieties exist, so the problem can be avoided with careful variety selection. Use a floral preservative containing an ethylene inhibitor such as silver thiosulfate (STS). Pulse stems for one hour or add STS to your overnight holding solution. Avoid natural sources of ethylene, such as ripening fruit. Ventilate and reduce temperatures to slow ethylene buildup.

Snapdragons should be stored and shipped upright at all times to prevent spike curvature. Place cut stems vertically as soon as possible after harvest; stems placed horizontally may begin to bend upward in as little as thirty minutes. To maintain flower quality, it is important to sleeve the upper portion of the snapdragon bunches and use tall, upright hampers for shipping.

Snapdragons can be stored for three to four days, dry or in water, at 40°F (4°C). If stored dry, rehydrate and condition in the same manner as for freshly cut snapdragons. For longer-term storage (five to ten days), select only the highest-quality stems, wrap each spike in plastic to prevent desiccation, and hold the stem in a preservative at air temperatures of 32–40°F (0–4°C).

Table 4. Snapdragon Scheduling

	NORTH*			SOUTH*		
GROUP	SOW (WEEK)	TRANSPLANT PLUGS (WEEK)	FLOWER (WEEK)	SOW (WEEK)	TRANSPLANT PLUGS (WEEK)	FLOWER (WEEK)
1	33–35	37–39	50–7	N/A	N/A	N/A
2	37–49	40–1	8–19	34–51	38–4	49–17
	30–32	34–38	44–49			
3	50–11	2–15	20–26	28–33	32–37	40–48
	25–28	28–33	37–43	2–10	5–14	18–24
4	13–23	16–27	27–36	11–26	15–31	25–39

N/A= Not applicable

Note: Times given are general guidelines. Conditions in certain areas may warrant deviations from these ranges.

* The North and South are separated by the 38th parallel, running from about San Francisco on the West Coast through Colorado Springs, Kansas City, St. Louis, and Louisville to Washington, D.C.

Table 5. Flowering Response for Snapdragon Varieties

VARIETY	COLOR	RESPONSE*
GROUP 1		
VARIETY	**COLOR**	**RESPONSE***
'Maryland Dark Orange'	Deep bronze	L
'Maryland Flame'	Orange yellow	M
'Maryland Light Bronze'	Light bronze	E
'Winter Euro Pink'	Pink	E
'Maryland True Pink'	Pink	E
'Maryland Yosemite Pink'	Pink	M
'Winter Pink'	Light pink	L
'Maryland Royal'	Purple	M
'Maryland Red'	Red	E
'Winter Euro Rose'	Deep rose	M
'Winter White'	White	E
'Winter Euro White'	White	L
'Maryland Ivory'	Ivory white	E
'Winter Euro Yellow'	Deep yellow	M
'Winter Yellow'	Deep yellow	M
GROUP 2		
VARIETY	**COLOR**	**RESPONSE***
'Maryland Appleblossom'	Pink and white bicolor	L
'Maryland Plumblossom'	White and purple bicolor	M
'Maryland Flame'	Orange yellow	E
'Maryland Dark Orange'	Deep bronze	M
'Maryland Light Bronze'	Light bronze	E
'Maryland Lavender'	Lavender	M
'Winter Pink'	Light pink	ML
'Maryland True Pink'	Pink	E
'Maryland Yosemite Pink'	Pink	M
'Maryland Shell Pink'	Light pink	M
'Maryland Royal'	Purple	E
'Maryland Red'	Red	E
'Monaco Red'	Deep wine red	L
'Maryland Flamingo'	Salmon rose	M
'Monaco Baltimore Rose'	Deep rose	L
'Monaco Violet'	Purple	L
'Monaco Rose'	Deep rose	L
'Winter Euro Rose'	Deep rose	M
'Maryland Ivory'	Ivory white	E
'Monaco White'	White	L
'Winter White'	White	E
'Maryland White'	White	M
'Winter Euro White'	White	M
'Winter Euro Yellow'	Yellow	M
'Monaco Yellow'	Yellow	ML
'Maryland Bright Yellow'	Yellow	L
'Winter Yellow'	Deep yellow	ME

(Continued)

Table 5. Flowering Response for Snapdragon Varieties *(Continued)*

VARIETY	COLOR	RESPONSE*
GROUP 3		
'Potomac Appleblossom'	White and pink bicolor	M
'Potomac Plumblossom'	White and purple bicolor	ME
'Potomac Early Orange'	Light bronze	E
'Potomac Dark Orange'	Deep bronze	M
'Potomac Orange'	Orange	M
'Potomac Early Pink'	Pink	ME
'Potomac Light Rose'	Pink	M
'Potomac Pink'	Pink	M
'Potomac Royal'	Purple	L
'Potomac Red'	Red	L
'Monaco Red'	Deep wine red	E
'Monaco Baltimore Rose'	Deep rose	E
'Monaco Rose'	Deep rose	E
'Monaco Violet'	Purple	E
'Potomac Rose'	Rose	ML
'Potomac Ivory White'	Ivory white	M
'Monaco White'	White	E
'Potomac Early White'	White	ME
'Monaco Yellow'	Yellow	E
'Potomac Soft Yellow'	Light yellow	ME
'Potomac Yellow'	Yellow	M
GROUP 4		
VARIETY	**COLOR**	**RESPONSE***
'Potomac Appleblossom'	White and rose bicolor	M
'Potomac Plumblossom'	White and purple bicolor	E
'Potomac Orange'	Orange	ME
'Potomac Light Rose'	Pink	ME
'Potomac Pink'	Pink	M
'Potomac Royal'	Purple	ML
'Potomac Rose'	Rose	ML
'Potomac Ivory White'	Ivory white	M
'Potomac Soft Yellow'	Light yellow	E
'Potomac Yellow'	Yellow	M

* E: Early; ME: Medium-Early; M: Medium; ML: Medium-Late; L: Late

Aquilegia

Aquilegia × *hybrida*
Common name: Columbine
Perennial (Hardy to USDA Zones 3–8)

Aquilegia is one of the most well-known and beautiful garden perennials. Many popular varieties today resulted from crosses between species within this genus; and many others will soon follow. Due to the intercrossing between species and varieties, there are a number of flower colors available, including carmine, blue, lavender, yellow, pink, rose, and white. Flower colors are often two tones or bicolors—petals are one color, while the sepals (calyx) are another. From seed, pure yellow, white, and violet blue are also available.

Most varieties on the market are single flowering, but double-flowering forms are also available. The scentless flowers can measure up to 3" (7 cm) across. Plants grow from 18–30" (46–76 cm) tall, depending on the variety, and will spread from 10–24" (25–61 cm) across. Plants flower in May or June.

Propagation

Seed is the most common method of propagation, and there is a wealth of varieties available. Many varieties are commonly offered as plugs or liners, and there are a select number of species types grown as bare-root transplants as well.

Sow seed onto a well-drained, disease-free, germination medium with pH of 5.5–5.8, and cover lightly with coarse vermiculite. Germinate at soil temperatures of 70–75°F (21–24°C). Maintain moderate moisture levels; medium should not be saturated. Keep ammonium levels below 10 ppm and sodium levels below 40 ppm. Radicles will emerge in two to three days.

After radicles have emerged, move trays to Stage 2 for stem and cotyledon emergence. Maintain substrate temperatures and increase light to 500–1,000 f.c. (5.4–11 klux). As soon as cotyledons have expanded, begin fertilizing with 50–75 ppm from 14-0-14, alternating with clear water. Maintain EC at less than 1.0 mS/cm. Stage 2 will take from fifteen to twenty-five days.

Move trays to Stage 3 to develop true leaves. Lower soil temperatures to 62–68°F (17–20°C) and increase light levels to 1,000–1,500 f.c. (11–16 klux). Increase fertilizer to 100–150 ppm from 14-0-14, alternating with 20-10-20. Fertilize during every second or third irrigation. Allow trays to dry slightly between irrigations. If growth regulators are required, A-Rest and B-Nine (3,000–5,000 ppm) may be used. Stage 3 will take from forty-four to forty-nine days for 288s.

Harden plugs in Stage 4 for seven days prior to sale or transplant by maintaining feed and temperature regimes. Increase light levels to 1,500–2,000 f.c. (16–22 klux).

As for crop time, many varieties germinate readily. In general, depending on variety or species, a 288-plug tray takes from five to seven weeks from sow to a salable plug while a 128-plug tray six to ten weeks. *Aquilegia* species take longer, while the newer hybrids are quicker to finish.

Because of long plug times and for convenience, many growers buy in plugs from commercial propagators. Aquilegia plugs can be bought in during late summer or early fall, potted, and overwintered for flowering plants in April or May or bought in from January to February for green sales in April and May.

Some plug suppliers offer vernalized large plugs, such as 50 cells, which can be delivered in January, forced into flower, and sold in color in the spring.

Aquilegia is also available as a bare-root transplant. While some are available in the late summer or early fall, some are available in the winter months as well.

Growing On

Plugs or bare-root plants purchased in the fall will make the largest, highest-quality pots. Pot up plugs or liners to 1-qt. (1.1 l) or 1-gal. (4 l) pots and overwinter dormant in a cold greenhouse or cold frame. You can also use multiple plugs per pot. No. 1 transplants can go into quarts or gallons. Select a well-drained, disease-free medium with a moderate initial starter charge and a pH of 5.5–6.2. A well-drained bark mix is excellent. Aquilegia requires eight to ten weeks of vernalization at 41°F (5°C) for flowering. However, before the cold treatment can begin, you have to bulk-up plants by getting them to fully root in the pot and become established. Bulk aquilegia under 1,500–2,000 f.c. (16–22 klux) of light. Allow pots to dry between irrigations. Fertilize at every other irrigation with 150–200 ppm from 15-0-15. Grow at 65–70°F (19–21°C) days and 55–60°F (13–16°C) nights. This process will take from six to eight weeks. The goal is to develop fifteen to twenty leaf nodes on plugs prior to the start of the cold treatment.

Provide the cold treatment in a cooler (under 50 f.c. [538 lux] of light) or cold frame at 41°F (5°C) for ten to twelve weeks. ('Cameo' will need only six to eight weeks; while at the twelve- to fifteen-leaf stage, 'Songbird' requires four to six weeks at 41°F [5°C]). During this time, do not fertilize plants.

Bring pots out of the cold treatment and force them in a cool greenhouse at 68°F (20°C) days and 60–65°F (16–18°C) nights. Provide 3,000–4,000 f.c. (32–43 klux) of light. Aquilegia responds well to supplemental light during the winter in dark areas of the country. Run plants relatively dry, allowing them to fully dry out between irrigations. Feed with 150–200 ppm from 15-0-15 or 20-10-20 at every other irrigation. Maintain a pH of 5.5–6.2 and an EC of 1.0 mS/cm. Flowering will take about four weeks and will continue for several weeks, provided you deadhead.

A-Rest and B-Nine (3,000–5,000 ppm) can be used to control height.

Pests & Diseases

Leaf miners can be a problem for aquilegia, rendering plants unsalable. If leaf miners attack leftover plants that have already flowered, cut back all the foliage and remove it. Plants will regenerate fresh foliage, and you will be able to sell plants out of flower in the summer. Watch for signs of crown and root rots; both are exacerbated when plants are overwatered.

Forcing Aquilegia 'Songbird'

JANUARY–MARCH CROPS

(Crop time sow to sale: 26–30 weeks)
Sow July–August in 392s.
Transplant to 50 cells at eight weeks.
Provide mum lighting (14 hours) to 50-cells for eight weeks.
Transplant to 5.5–6.5" (14–17 cm) pots.
Grow on under natural days for five weeks.
Begin mum lighting at the start of week 6 of production.
Continue lighting until flowering begins (for seven to eight weeks).

JULY–NOVEMBER CROPS

(Crop time sow to sale: 22–26 weeks)
Sow February–May in 392s.
Transplant to 50 cells at eight weeks.
Put trays in lighted cooler (100 f.c. [1.1 klux]/14 hours) at 41°F for four weeks.
Remove from cooler and transplant into 5.5–6.5" (14–17 cm) pots.
Plants will flower in four to six weeks.
For flowering after October 1, provide fourteen-hour days until flowering begins.

Varieties

If you are looking for a series of columbine to force for early perennial sales or as a flowering pot plant, be aware that each series responds differently. Many times responses among colors within the same series also differ.

In *A. hybrida*, the 'Songbird' series—'Blue Bird', 'Blue Jay', 'Bunting', 'Cardinal', 'Dove', 'Goldfinch', 'Nightingale', 'Robin', and 'Mix'—blooms its first year from seed, if started in November of the year before. 'McKana's Giant' is available as plugs or transplants but requires a longer vernalization to get it to flower. This tall-growing columbine will reach 30" (76 cm) and has extra-large flowers. Plant habit is open and casual. 'Biedermeier' is a compact mix growing to 8–10" (20–25 cm) tall. It should be treated like 'McKana's Giant' to get it to flower in the spring.

The 'Origami' series is drawing much attention for flowering pot and perennial production. Flowers are oversized and fabulous! If there is only one color you try, make it 'Origami White'—pots will walk out the door when they are in flower. As for flowering, most sowings done the year before up to February will flower readily in the garden the upcoming summer. However, the plants are not usually long lived in the garden—lasting for only a season or two.

In *A. flabellata* types, try the 'Cameo' series in 'Blue White', 'Blush', 'Pink White', 'Rose White', 'White', and a 'Mix'. This series is an excellent contender for early flowering pot plant sales. Plants are dwarf, growing to about 5–6" (13–15 cm) tall. 'Cameo' is two weeks earlier than 'Ministar'.

A new *A. vulgaris* series is 'Winky', which produces clusters of single flowers that sit just above foliage. Plants grow to 14" (36 cm) and come in four colors and a mix.

Postharvest

Aquilegia is sensitive to ethylene gas. Keep plants groomed by removing dead flowers and foliage to remove these sources of ethylene production.

Arabis

Arabis caucasica
Common names: Rock cress; wall cress
Perennial (Hardy to USDA Zones 3–7)

Arabis caucasica is a low-growing, mostly gray-green, rosetting perennial that grows to 6" (15 cm) tall in bloom. Flowers are white, purple, or rose pink. The plants spread from 6–8" (15–20 cm) across, slightly larger in cooler, coastal climates. The flowers are four-lobed, fragrant, and about 0.5" (1 cm) across. Double- and single-flowering varieties are available as well as variegated-leaf types. The plants do not become invasive.

Propagation

Arabis is commonly seed propagated, although basal or stem cuttings are often done on the double-flowering and variegated-leaf varieties.

Seed

Sow seed in late spring or summer, cover lightly with vermiculite or peat, and in two to three weeks transplant 1–3 seedlings/cell to an 804 tray (32 cells/flat). These are potted up into 1-qt. (1.1 l) or 1-gal. (4 l) pots during mid- to late summer, and plants are overwintered dormant or cold for spring sales. Plants will flower in April or May of the following year.

For 288-plug trays, sow 3–4 seeds/cell, lightly cover with coarse vermiculite, and treat at 70–75°F

(21–24°C) for germination in six to ten days for most varieties. Allow six to seven weeks from sowing for a 288-plug tray to be salable or transplantable to larger containers.

Seed can also be sown in September or October with seedlings transplanted to 1-qt. (1.1 l) or 1-gal. (4 l) containers and grown cold during the winter (once roots are established in the final container) for plants that will flower in April or May of the following year. The plants never go dormant and are kept at 35–40°F (2–4°C) night and day once roots fully develop.

Cuttings

There are a number of variegated-leaf and double-flowering selections that do not come true from seed. Unrooted cuttings brought in during the summer should be dipped into a rooting hormone, stuck into a 72 tray, and given mist for ten to twelve days. Rooting is pretty quick and a 72 tray will be ready to transplant in five to seven weeks, depending on the variety.

Cuttings are frequently promoted in December or January. These root just as easily. However, upon transplanting, the plants will not flower the upcoming spring.

Growing On

Seed

Winter or early spring sowings will produce green, salable cell packs or small pots within ten to eleven weeks when grown on at 50°F (10°C) nights. Add one to two weeks for 4.5" (11 cm) pots. These plants will not flower in spring if started in January or later.

For spring-blooming plants, sowings should be done the previous year so that the plants can be exposed to a cold treatment (vernalization) for flower bud formation. Since plants are very cold hardy, sowings can be made as late as October for late winter- or spring-flowering plants. If sown this late, however, the plants need an initial warm period to get some roots underneath them before temperatures fall below 50°F (10°C). Once roots have become established in the final pot, drop the temperature to 48°F (9°C) or lower for overwintering.

One additional point: In the cold frame the foliage frequently has a purple or bronze coloring during the cold temperatures and short days of spring. This is also evident in the garden but may not be as prevalent. Regardless, as night temperatures warm up to 50°F (10°C) and above, the plants will show less coloring.

Cuttings

Transplant rooted liners in mid- to late summer to 1-qt. (1.1 l) or 1-gal. (4 l) pots, allow to fully root, and then grow cold for the winter. These plants will flower in spring of the following year.

Related Species

A. blepharophylla 'Spring Charm' is a single-flowering, rose-colored variety that can be treated as a short-lived perennial. While plants are hardy from Zones 4 to 7, they provide two to three years of good color and performance before dying out. Sow and grow on as described for *A. caucasica.*

Argyranthemum

Argyranthemum spp. (also *Argyranthemum frutescens* or *Chrysanthemum frutescens*)
Common name: Marguerite daisy
Annual

Daisies are many gardeners' favorite. They make great spring pot plants that can be sold for the table or patio and as a component for mixed combination containers. For growers, they are easy to grow and will time right into your regular spring routine in 4" or 6" (10 or 15 cm) pots. They range in height from 12–16" (30–41 cm) tall and are available in a number of colors. However, they are not fond of hot nights and will go out of bloom when the night temperatures are 80°F (27°C) and above. The foliage will persist, and the flowers do show up again on many varieties once night temperatures cool.

Propagation

Stick cuttings in Oasis or a medium with a low starter charge and a pH of 6.0–6.5. Provide 68–75°F (20–24°C) soil temperatures and air temperatures of 70–75°F (21–24°C) days and 68–70°F (20–21°C) nights. Root cuttings under a mist set so that cuttings do not wilt. Provide at least 500–1,000 f.c. (5.4–11 klux) of light. If foliage begins to lighten, apply a foliar feed of 50–75 ppm from 15-0-15 once a week. Callus formation will take about five to seven days.

Once 50% of the cuttings have begun to form root initials, move trays to Stage 3 for full rooting. Maintain 68–75°F (20–24°C) soil temperatures and air temperatures of 70–75°F (21–24°C) days and 68–70°F (20–21°C) nights. As roots begin to penetrate the medium, begin to reduce moisture levels by reducing the frequency and duration of misting. Adjust mist frequently based on your exact conditions. Do not allow cuttings to wilt. Gradually increase light levels to 1,000–2,000 f.c. (11–22 klux). Retractable shade is ideal for this purpose. Apply a foliar feed of 100–150 ppm once a week, alternating between 20-10-20 and 15-0-15. Stage 3 will take from seven to nine days.

Harden cuttings in Stage 4 prior to sale or transplant by moving trays from the mist area. Lower temperatures to 68–75°F (20–24°C) days and 62–68°F (16–18°C) nights. Allow the medium to dry down before irrigating; however, do not allow plants to wilt. Increase light levels to 2,000–4,000 f.c. (22–43 klux). Fertilize once a week with 150–200 ppm from 20-10-20, alternating with 15-0-15. Allow cuttings to harden for seven days.

Stuck into a 105, 84, or larger liner tray, cuttings will root in three to four weeks and be ready to transplant to pots or other containers.

Growing On

Select a well-drained, disease-free, soilless medium with a high initial starter charge and a pH of 6.0–6.5. Plant 1 liner/4" or 6" (10 or 15 cm) pot or 3–4 liners/8" (20 cm). Since they are such a great addition to combinations, plan on using a part of your 4" (10 cm) production in combo patio planters for full sun. Drench with a broad-spectrum fungicide to water plants in right after planting to prevent disease problems. Do not overwater plants immediately after planting.

Marguerites will tolerate a wide range of temperatures, 28–30°F (-2–-1°C) on the low side once plants are established and as high as 85–95°F (30–35°C). However, optimum production temperatures are 65–75°F (18–24°C) days and 45–55°F (7–13°C) nights. The best-quality plants are grown cool; however, crop time will be two to four weeks longer.

Provide the highest light levels possible (i.e., over 5,000 f.c. [over 54 klux]). Give consistent moisture, but do not allow plants to be wet. Avoid severe wilting. Feed with 150–200 ppm at every irrigation, alternating between 15-0-15 and 20-10-20. If soluble salts begin to accumulate, leach with clear water at every third irrigation. Watch for interveinal chlorosis caused by magnesium deficiency if potassium rates become too high.

Once plants have rooted in (about two weeks after potting) you can pinch plants back to the fifth or sixth set of leaves if you wish to encourage branching. However, for most varieties it is not necessary. Plants in 4" (10 cm) pots may be ready to pinch in just four or five days. Florel can be substituted for a pinch; treat twice with 500 ppm.

Space plants as they grow to prevent excessive stretching. If growth regulators are needed, B-Nine (1,500–2,000 ppm) or Cycocel (1,000–1,500 ppm) can be applied from fourteen to twenty-one days after the pinch. You can reapply every two weeks as needed during production. However, growing your crop cool will be just as effective at height control as growth regulators will be.

Marguerites flower faster under long days. If you are trying to flower plants for early spring sales, use night interruption lighting.

Liners transplanted to a 4" (10 cm) pot will produce a flowering plant in the spring in six to eight weeks for most varieties.

Pests & Diseases

Aphids, fungus gnats, spider mites, thrips, whiteflies, and worms can become pest problems. Diseases that can strike include botrytis, *Phytophthora, Pythium,* rhizoctonia, and stem canker.

Troubleshooting

A medium that remains wet for extended periods of time can lead to plant collapse. High nitrogen levels can cause excessive vegetative growth, as can too much fertilizer under low light, and/or low light and overwatering. Poor branching can be due to inadequate early nutrition. Allowing plants to dry out between irrigations can result in foliage necrosis.

Varieties

There are a multitude of varieties available and most are used as combination plants in mixed containers. A few others excel as a home garden or professional landscape varieties in warm (but not hot) nights. Included in this short list is 'Butterfly', a bright yellow, flowering variety.

Armeria

Armeria maritima
Common names: Thrift; sea pinks
Perennial (Hardy to USDA Zones 4–8)

Armeria maritima is a low-growing, evergreen perennial with green to bluish-green, grassy foliage in rounded clumps. Plants grow from 6–15" (15–38 cm) tall when in bloom, although the foliage is often no more than 3–5" (8–13 cm) tall when the plants are not blooming. The plants spread from 6–10" (15–25 cm) across. The flowers are often white, rose, pink, or pastel colored. When in bloom, the flowering stalks are leafless and will wave back and forth in a light breeze. The scentless, single flowers form globes that grow up to 1" (2.5 cm) in diameter.

Propagation

Armeria is often propagated by seed or offsets. Seed can be sown anytime of the year with equal results. Clump division can be done in the spring on two- or three-year-old plants and is done on the named varieties that cannot be propagated by seed. Removal of a section of the basal rosettes (offsets) is the most common way to propagate named cultivars. This can be done spring to fall except during flowering.

Seed will germinate readily without special treatments. If using cleaned seed, use one to two seeds per 512-plug tray. If seed is raw it is better to use three seeds per 288 cell.

During Stage 1, use germination temperatures of 68–72°F (20–21°C) for best results. Cover seeds lightly or leave them exposed during germination; slightly better germination may be achieved with uncovered seeds. Seedlings emerge in four to ten days and can be transplanted to cell packs fifteen to twenty-two days after sowing.

Move trays to Stage 2 and maintain substrate temperatures at 68–72°F (20–21°C). Increase light levels to 500–1,000 f.c. (5.4–11 klux) and begin feeding weekly with 50–75 ppm 14-0-14 once cotyledons have fully expanded. Stem and cotyledon emergence will take five to eight days.

Stage 3—developing true leaves—will take from seven to ten days. Reduce the substrate temperature to 65–68°F (18–20°C) and begin to allow trays to dry thoroughly between irrigations. Increase light to 1,000–1,500 f.c. (11–16 klux). Feed with 100–150 ppm 14-0-14 weekly. Allow EC to increase to less than 1.0 mS/cm.

Harden plants in Stage 4 over seven days prior to transplant or sale by lowering the substrate temperature to 60–62°F (16–17°C) and increasing light to 1,500–2,500 f.c. (16–22 klux).

A 512-plug tray takes from three to four weeks to be large enough to transplant to larger containers. A 288 will take from five to seven weeks.

Growing On

Plants can grow slowly from seed; even two months after sowing the plants will have only a small tuft of leaves with shallow roots that will not fill out the container. To have salable, green plants in 1-qt. (1.1 l) or 1-gal. (4 l) containers for spring, sow the previous year.

Sow seed to a 512- or 288-plug tray from May to July for transplanting 50- or 72-liner cell to size up. It takes seven to ten weeks from a plug to a finished 72 tray. Transplant this to a 1-qt. (1.1 l) or 1-gal. (4 l) pot to overwinter for sales the following spring.

Another method is to sow seed to a plug tray during September, transplant to the final container and grow cool (45–50°F [7–10°C]) once the plants are fully rooted. Plants like a well-drained medium to keep them from being wet. Because of their evergreen foliage, care must be taken to avoid foliar diseases.

Sections of the basal rosette can be removed and allowed to root. Take these in spring or during active growth but not when flowering. These can be potted up once they are rooted fully. When spring propagated, these plants can be sold in late summer or overwintered for sales the following spring

If sowing seed during the fall or winter, grow the seedlings on at 50°F (10°C) and higher for the first three months after sowing. This will keep the plants actively growing so they will fill out faster in a container. Plants will tolerate temperatures down to 45°F (7°C) in this stage but are slow to develop. When grown colder, the plants are more susceptible to overwatering and often rot. Also, due to their slow nature, avoid putting small transplants in large containers; many times the plants do not live long enough to grow into the container.

Related Species

A. pseudoarmeria (A. latifolia) has broader, longer foliage giving the plant a more substantial appearance than *A. maritima.* It is hardy in USDA Zones 6 to 9.

Follow the cultural methods described for *A. maritima* for propagation, although this species is faster to grow and produce a flowering plant. Sowings can be made in November or December, grown with warm days and cool nights (60–65°F [16–18°C] days and 55–58°F [13–14°C] nights), and then kept in a cold frame above 45°F (7°C) once fully rooted in the final container.

Asclepias

Asclepias tuberosa
Common names: Butterfly weed; butterfly flower
Perennial (Hardy to USDA Zones 4–9)

Asclepias tuberosa is a premier native perennial with one of the most vivid orange flowers on the market; red and yellow colors are available as well. Butterflies, especially monarch butterflies, are attracted to this plant's rich flower nectar. The individual, crown-shaped blooms are 0.125–0.25" (3–6 mm) across and form irregular clusters measuring 1–2" (2.5–5 cm) across. In late summer as the flowers fade, an elongated, tapered, 4" (10 cm) seedpod develops from which seeds parachute to earth with silky, featherlike appendages.

Plants grow primarily upright from 1.5–3' (45–91 cm) tall and spread 12–18" (30–40 cm) across. Their stems contain a white, milky sap that exudes when the plant is wounded or cut. Plants flower in June and July and will continue through August.

Propagation & Growing On

Seed and bare-root transplants are the most common methods of propagation.

Seed

This is one of the more difficult crops to produce from seed. *Asclepias* resents transplanting, and the plant shows its distaste by dropping leaves and, in severe cases, dying if the roots become too bound in a cell or pot or if grown too cool or cold in the greenhouse or cold frame.

Seed germinates at 70–75°F (21–24°C) and can be covered or left exposed during germination. Germination takes fourteen to twenty-one days. No pretreatments are necessary. However, if germination proves difficult, chill seed in moistened peat at 36–40°F (2–4°C) for several weeks prior to germination. However, this treatment is suggested for old seed or poorly stored seed. Many companies offer enhanced seed, which allows for faster germination and greater seedling stand.

For many crops in this book, you can sow seed to a plug tray of varied sizes and transplant to the final container for plant sales in the spring or summer. *Asclepias* is a little more demanding in this early stage of production.

For plug/liner propagation sow one to three seeds per 128 or larger cell. Seed does not have to be covered with vermiculite but a light covering ensures even moisture during germination. Production time from seed to a finished liner tray is eight to ten weeks.

Another method is to sow seed two to four seeds per 50 or 72 tray in October or November and then transplant to a 1-qt. (1.1 l) pot and grow warm (58–62°F [14–17°C] nights) throughout the winter. As the plants become established and rooted in the pot, decrease the night temperatures to 55°F (13°C).

Be mindful of dropping temperatures lower than this, because the foliage will yellow and drop off. Plants will be salable green in the spring and will flower in the garden the upcoming summer. They will flower more profusely once they establish in the garden the following summer.

Bare-root transplants

Bare-root transplants are available from a number of commercial propagators. Use 1-qt. (1.1 l) or 1-gal. (4 l) containers, depending on the size of the root. Pot up in January or February with the crown 1–2" (3–5 cm) below the soil level. Grow on in a sunny location with warm temperatures (65–70°F [18–21°C] or higher during the day and no lower than 58°F [14°C] at night). Water pots thoroughly upon planting the roots, but once they establish, allow the containers to dry down a bit between watering.

Plants are salable in the spring and will flower sporadically the upcoming summer. Once established, they will flower more profusely the following year.

Asparagus

Asparagus densiflorus var. 'Sprengeri'
Common names: Asparagus fern; Sprengeri fern
Annual, foliage plant

A. densiflorus var. 'Sprengeri' is the most common asparagus fern species used in the bedding plant industry. But the plant isn't really a fern at all—it's in the lily family! Sprengeri ferns are also used as cut greens in flower arrangements. In tropical climates, they are a reliable, drought-tolerant groundcover in bright shade. Plants make terrific, low-maintenance hanging baskets and are great when used in combination pots or patio planters, providing dependable background greenery that will tolerate a wide range of conditions.

Asparagus fern is a long crop. Growing seedlings is a time-consuming process; producing a finished 4" (10 cm) pot takes about six months.

Propagation

Multisow seed in the summer or fall onto a well-drained, disease-free, germination medium with a low starter charge. Moisten the medium prior to sowing, and maintain wet conditions to germinate seed: Water should be squeezed out easily. Cover seed with vermiculite after sowing as darkness enhances germination. At 78–80°F (26–27°C), seed will germinate in four to six days.

Move trays to Stage 2 for stem and cotyledon emergence. Reduce temperatures to 72–75°F (22–24°C) and raise light levels to 500–1,000 f.c. (5.4–11 klux). Allow trays to dry somewhat between irrigations as soon as the radicles have penetrated soil. Begin a weekly feed with 50–75 ppm from 14-0-14 once cotyledons have expanded. Stage 2 will take from seven to ten days.

To develop true leaves, move trays to Stage 3. Reduce temperatures to 68–72°F (20–22°C) and increase light levels to 1,500–2,500 f.c. (16–27 klux). Feed once a week with 100–150 ppm, alternating between 20-10-20 and 14-0-14. Allow trays to dry thoroughly before irrigating. Stage 3 will take fourteen to twenty-one days for 512s (that can subsequently be transplanted into cell packs or larger plugs).

Harden plugs for seven days prior to transplanting by lowering temperatures to 65–68°F (18–20°C).

Increase light levels to 3,000–4,000 f.c. (32–43 klux) and allow trays to dry thoroughly between irrigations. Fertilize once a week with 100–150 ppm from 14-0-14.

A 288-plug tray will take from six to eight weeks from sowing to be ready to transplant to a larger container.

Growing On

Select a disease-free, well-drained, soilless medium with a pH of 5.5–6.2 and a moderate initial starter charge.

Transplant 288s into 72, 50, or similar trays and grow on for an additional eight to ten weeks. Provide moderate light levels of 2,500–4,500 f.c. (27–48 klux) and feed at every irrigation with 150–200 ppm alternating between 20-10-20 and 15-0-15. It is best to grow on at 65–70°F (18–21°C) days and 60–65°F (16–18°C) nights, although plants will tolerate temperatures that are somewhat lower and much higher. Keep temperatures above 55°F (13°C) for the fastest growth. Once the 2.25" (6 cm) pots have sized up, they can be planted into hanging baskets using 2–3 plants/basket. The crop time from a plug to a fully developed 4" (10 cm) is from twenty to twenty-six weeks.

Many southern suppliers offer asparagus fern plugs that are approximately ten to fourteen weeks old. These can be potted immediately into 4" (10 cm) pots using 1 plant/pot or 10" (25 cm) baskets using 3–4 plants/basket. Some growers prefer to establish full 4" (10 cm) pots and then put 1 pot/8" or 10" (20 or 25 cm) basket.

You can put asparagus baskets in the gutter during winter months, provided they are on drip emitters. When light intensities start to increase in the early spring, move baskets down.

Mature *A. densiflorus* 'Sprengeri' plants can flower in the summer and form red berries. They also get quite spiny as they age, so be careful if you are handling plants without gloves. The fleshy root nodules that mature plants produce help it to thrive during periods of drought—a great feature for home gardeners who sometimes forget to water their hanging baskets or patio pots!

Related Species

A. pseudoscaber and *A. setaceus* are used for cut foliage. Several other asparagus species are frequently used as foliage plants. They include *A. meyeri, A. falcatus,* and *A. pyramidalis.* While they can be germinated the same as *A. densiflorus* var. 'Sprengeri', they will take longer to germinate (in most cases) as well as take longer to finish in a 4" (10 cm) pot.

Aster

Aster novae-angliae
Common name: New England aster
Aster novi-belgii
Common name: New York aster

Europeans turned this North American native into the flowering pot plant and cut flower that it is today; although it is still widely known as a native perennial that is a stellar performer in the late summer garden.

Flower colors are available in various shades of blue, purple, rose, pink, and white. Plants range in height from dwarf-flowering selections that are only 10–15" (25–38 cm) to as tall as 3–5' (91–152 cm). Asters are a great way to incrementally add onto garden mum purchases, thus increasing total fall sales for the garden center. Garden asters will be potted about two weeks after your garden mums, but their crop culture is very similar.

Asters may be forced as a flowering pot plant in the greenhouse using photoperiod control. If this is of interest to you, pay attention to the comments on photoperiod control in the section on growing on aster as a cut flower.

Propagation

When the Danes first began working with aster as a potted plant, they treated them much like they did just about any other pot plant: They used pinched tips as the cuttings for subsequent crops. During certain times of the year, the growers discovered that the plants produced from these cuttings would flower too quickly or sporadically. Researchers later discovered that asters have a rather complicated mechanism for maintaining vegetative growth. Cutting suppliers today maintain stock plants through tissue culture rejuvenation and daylength control.

Growers can buy in rooted or unrooted cuttings. Some growers choose to direct-stick cuttings into their final growing container.

Stick cuttings into a disease-free, well-drained medium with a pH of 5.5–6.5. Remove lower leaves that would be covered with the medium. Grade cuttings by size and caliper, planting like-sized cuttings together to create a more uniform finished product.

Stick one cutting per 72-, 50-, or similar liner tray. Rooted at 70–72°F (21–24°C), liners are ready to transplant in three to five weeks.

You can also stick cuttings direct to the final container: 1–3 cuttings/4.5" (11 cm) pot and 3–4 cuttings/6" (15 cm) pot. Place cuttings 1" (2.5 cm) deep, spaced equally around the edge of pots. Make sure pots are well watered prior to sticking cuttings, and water again after sticking to settle the medium.

Place pots on propagation benches outfitted with bottom heat, mist, and movable shade screens. Maintain soil temperatures of 65–68°F (18–20°C). Time mist cycles so that plants do not wilt. Depending on nozzle size and spacing as well as the size of the propagation bench, set mist duration for ten seconds or longer. Here are some guidelines for mist frequency that you can adjust to your conditions: Days 1–3, mist every five to ten minutes; days 4–6 (callus formation), mist every twenty minutes; and days 7–10 (roots initiated), mist every thirty minutes. Night misting may be required.

During rooting, cuttings must be under long days. Provide sixteen hours of day length with night interruption lighting. Maintain ambient light levels of 3,000–3,500 f.c. (32–38 klux). Apply 200–300 ppm from 20-10-20 once roots have initiated. Rooting will take approximately ten to fourteen days.

Growing On

From a 50- or 72-liner tray, use 1 liner/4" (10 cm) pot or 3 liners/6" (15 cm) pot. Greenhouse pot plant aster growers usually use multiple cuttings per pot, while garden aster growers use less plants/pot. Select a well-drained, disease-free medium with a pH of 5.5–6.5 and a moderate initial starter charge. Make sure the medium has good moisture retention, since finished plants dry out fast.

Start plants with 300 ppm from 20-10-20, switching to 250 ppm from 15-0-15 after the last pinch. Asters are sensitive to high soluble salts, so leach with clear water periodically to avoid problems. When flowers begin to open, discontinue fertilizer altogether.

Grow pot asters on at 68–75°F (20–24°C) days and 62–65°F (17–18°C) nights. Once flowering begins, lower night temperatures to 58–60°F (14–16°C) to intensify flower color. When growing outdoors, aster flowers will tolerate a light frost.

Plants require one or two pinches to create a full pot. Pinch for the first time, leaving four to six leaves, when roots reach the bottom of the pot, about ten to fourteen days after planting. If needed, pinch a second time when plants have about 2" (5 cm) of new growth: Remove half of the new growth, leaving three to four leaves. Take care to pinch all shoots so that branches will flower evenly. Be sure to leave the necessary number of leaves after each pinch. Severely pinching or cutting back asters will encourage vegetative shoots to arise from the base of the plant. These shoots often become taller and flower much later than the upper branches, resulting in an uneven flowering pot. The last pinch date for natural-season crops should be between July 25 (northern areas) and August 10 (southern areas) to avoid any flower delay. Later pinches can be used to delay flowering if desired, but plants must be of adequate size, since little regrowth will occur after very late pinches.

Space plants to their final spacing after the second pinch. If plants are not spaced on time, lower leaves may yellow and turn brown from lack of light, or foliage diseases can occur from reduced air circulation and increased humidity from crowding. At no time during production should you allow leaves from adjacent pots to touch or overlap.

B-Nine is effective in controlling plant height. Apply at 3,750–5,000 ppm after the first pinch when new growth is 1.5" (4 cm) long. Some vigorous varieties may need a B-Nine application before the pinch. Discontinue B-Nine at visible bud to avoid clubby flower sprays. Bonzi at 5–10 ppm can also be effective: Direct the spray to stems rather than leaves.

Maintain good air circulation around the plants, as asters are highly susceptible to powdery mildew. Because they are so susceptible, avoid overhead irrigation if possible. If overhead irrigation must be used, then ensure plants are completely dry at nightfall.

If you want to grow asters as a year-round pot crop, provide long days for vegetative growth from approximately September 15 to March 15. Provide short days to trigger flowering from about March 15 to September 15. Plants need to develop vegetative growth under long days for approximately five weeks before flowers initiate. About one week after the second pinch, begin short days. Flowering should occur five to six weeks after the beginning of short days. Your supplier will be able to advise you for your specific latitude.

Aster can also be grown and sold as a perennial. These cuttings may be planted from August to October and overwintered alongside regular perennial crops in 1801s or 1-qt. (1.1 l) or 1-gal. (4 l) containers.

Cut Flowers

Using black cloth and night interruption lighting, growers can get three and a half to four and a half flushes per aster crop. Grown as an outdoor cut flower, perennial asters can be flushed once (natural season) or forced to flower twice.

Plant cut flower aster liners at a density of 1.5 plants/ft.^{2} (16/m^2) in low light; 2 plants/ft.^{2} (22/m^2) in high light. Provide one layer of support wire or netting that can be raised as the plants grow.

Vegetative growth must occur under long days. Asters require longer days than chrysanthemums to remain vegetative: Sixteen to seventeen hours are recommended. If you use night interruption lighting, make sure the total day length will be sixteen to seventeen hours. Allow plants to root in for two to three weeks and then pinch plants, leaving two to three sets of leaves on the plant. This pinch will generate two to three top shoots and will encourage additional shoots to form from the roots. Once these new shoots reach about 15" (38 cm) tall, which will take about five to eight weeks, begin short days.

Short days, provided by pulling black cloth, will trigger flowering in your aster crop. Provide plants with no more than twelve hours total light: Pull black cloth from 6:00 P.M. to 6:00 A.M., for example. You need to provide five to seven weeks of short days if night temperatures of 55–60°F (13–16°C) are maintained. Once buds are swollen and begin to show color, short days may be discontinued.

When you begin harvesting, provide long days again to ensure that the ground shoots that begin developing during harvest will maintain a vegetative growing state.

Asters tolerate most soils, but prefer medium-textured, well-drained soils that are low in salts. When asters are actively growing during long days, they are medium to heavy feeders, similar to chrysanthemums. Provide 150 ppm from 20-10-20, alternating with clear water. During bud initiation or short days, reducing nitrogen helps speed up bud set. Adding potassium nitrate is excellent during this time. Main-

Table 6. Crop Planning for Early, Shaded, Garden Aster Crops*

ACTIVITY	TIMING GUIDELINES	EXAMPLE
Plant rooted cutting	Upon receipt	May 1
First pinch	When ready; about 10–14 days after planting	May 15
Second pinch	When ready; about 10–14 days after first pinch	May 30
Short days	2 weeks after second pinch	June 14
Flower	5–6 weeks after short days	July 17–24

Note: For 6" (15 cm) or 1-gal. (4 l) pots, use 3 plants/pot. Plant May 1–June 1. Space 14" × 14" (36 × 36 cm).

* *Supplied by Yoder Bros., Barberton, Ohio. Reprinted from the 17th edition of the* Ball RedBook.

Table 7. Crop Planning for Natural-Season, Fall Garden Aster Crops*					
CROP	POT SIZE	PLANTS/ POT	PLANT DATE	NO. OF PINCHES	APPROXIMATE SPACING
Normal, pinched	8" × 5" pan, 1–1.5 gal.	1	Late May Early June	3	20" × 20"
Normal, pinched	8" × 5" pan, 1–1.5 gal.	1	Mid-June Late June	2	18" × 18"
Normal, pinched	8" × 5" pan	1	Early July	1	16" × 16"
Normal, pinched	8" × 5" pan	2	Mid-July	1	16" × 16"
6" fast crop	6–6.5"	1	Mid-July, late July	None	12" × 12"
4" fast crop	4–4.5"	1	Late July, early August	None	8" × 8"

Notes: Plant dates are based on starting with rooted cuttings. Start one week earlier if direct-sticking unrooted cuttings. Start two to three weeks earlier if rooting in 72- to 98-cell packs for transplanting to finishing container.

Plant dates are based on Midwest/eastern region growing conditions. In general, start two to three weeks earlier for the West Coast and one to two weeks later for southern production regions.

One or two plants/pot are needed as shown above when water and fertilizer are not limited, as with constant fertilization/drip-tube irrigation systems. Otherwise, one or two cuttings may be used if desired to create fuller pots.

Fast-crop plants will naturally flower about one week later than normal, pinched crops.

The above fast-crop programs could also be used with night lighting to delay flowering even more. Start lighting at planting and continue lighting until six weeks prior to desired sale date. B-Nine may be needed.

Normal, pinched 4" (10 cm) and 6" (15 cm) crops may also be produced. Plant two weeks earlier than fast-crop plants for each planned pinch. Such crops are usually bushy and compact and do not exhibit the flowering delay as in fast crops.

** Supplied by Yoder Bros., Barberton, Ohio. Reprinted from the 17th edition of the* Ball RedBook.

Table 8. Crop Planning for Greenhouse Pot Crops*			
ACTIVITY	TIMING GUIDELINES	EXAMPLE	SPACING
Plant rooted cuttings	Upon receipt	February 14	Pot-to-pot
Lights on	At planting	February 14	Pot-to-pot
First pinch	When ready, about 10–14 days after planting	February 28	Pot-to-pot
Second pinch	When ready, about 10–14 days after first pinch	March 14	Pot-to-pot
Short days	1 week after second pinch	March 21	8" × 8"
Flower	5–6 weeks after short days	April 25–May 2	8" × 8"

Notes: When direct-sticking unrooted cuttings, start one week earlier under long days.

For larger pots, use two weeks of long days after the second pinch and 4 cuttings/5" (13 cm) pot, spaced 12" × 12" (30 × 30 cm).

Desired overall height for 4.0–4.5" (10–11 cm) pot asters is 9–11" (23–28 cm). If height control is a problem, reduce the number of long days between the second pinch and the beginning of short days. For "tall" vigor varieties, start short days at the same time as the second pinch.

For Montauk daisy, use 1 plant/4" (10 cm) pot, 3 plants/5" (13 cm) pot, and 4 plants/6" (15 cm) pot. Only pinch once, start short days two weeks after pinch, and allow six and a half weeks after short days until flowering.

** Supplied by Yoder Bros., Barberton, Ohio. Reprinted from the 17th edition of the* Ball RedBook.

tain EC between 0.75 and 1.0 mS/cm. When flowering begins, discontinue all fertilizers.

Keep plants moist during the vegetative stages of growth. Once short days begin, gradually dry down beds, which will help to tone plants and set buds. Do not allow soil to become too dry, especially if salts levels are high. During finishing, allow leaves to wilt ever so slightly between irrigations.

Grow asters under full sun—7,500 f.c. (81 klux) is ideal. During long days, plants do best under cool temperatures, which will actually help them remain vegetative. Nights can be 45–55°F (7–13°C). When short days begin, raise night temperatures to 58–60°F (14–15°C) or slightly higher to speed flowering and have more complete bud set per stem. Most asters will not suffer from heat delay.

Total crop time from planting to harvest is about ten to sixteen weeks, depending on the exact growing temperatures, variety, and from which flush the plants are being forced. Generally you can flush each plant from three to five times before replanting with new liners.

After harvest, cut plants back to the ground and removing all stubble, partially cut stems, and old wood. It is important to remove old wood because it will produce prematurely budded stems of no value. Taking away the old stems will also stimulate the root system to produce underground shoots that will develop into stems and flowers for the next flush.

You can grow cut asters outdoors as well. Plants will flower naturally in September or October each year. Some growers rig lights and use black plastic on outdoor ground beds to force more than one harvest per year. After flowering, overwinter plants and then cut back all foliage and stubble in the spring to regenerate ground shoots.

Pests & Diseases

Aphids, fungus gnats, spider mites, thrips and whiteflies can attack asters. Diseases can include *Alternaria,* powdery mildew, *Pythium,* and rust. Powdery mildew is especially troublesome for greenhouse-grown crops. Provide excellent air movement, do not water late in the day, and be vigilant in heating and ventilating to avoid problems.

Cut flower growers should be careful of *Verticillium dahliae* wilt, which can build up in ground beds. Crops such as chrysanthemum, *Liatris,* and phlox can act as hosts.

Troubleshooting

Small plant size may be due to low light levels, low fertility early in the crop, improper plant spacing, and/or insufficient long days after the second pinch. Plants that look stunted may have *Pythium* root rot, low fertility, or high soluble salts/high alkalinity.

Postharvest

Cut flower asters are ready to harvest when 10–20% of the flowers are open. Large-flowered varieties are harvested when terminal flowers are open and buds show color. Place stems in water immediately. Using a bactericide in the water is mandatory. Stems also respond positively to preservatives. Store at 40°F (4°C).

Retailers should recut stems (removing 2" [5 cm] from the base), strip foliage that would be below the water line, and place stems in clean buckets with floral preservative and bactericide added. Asters should not be used in arrangements using Oasis.

Pot plant asters are ready for sale when 25–40% of the flowers are open. Asters being retailed as garden plants may be sold sooner with color picture tags—as open flowers attract bees, which can sometimes be troublesome in the retail environment. Consumers will enjoy their perennial aster for years to come in the garden.

Aster, China (see *Callistepheus*)

Astilbe

Astilbe spp.
Common name: False spirea
Perennial (Hardy to USDA Zones 3–8)

The delicate, feathery plumes of astilbe are a welcome part of early and midsummer gardens. Its foliage is equally as attractive. By selecting a range of varieties that flower early, mid-, and late season, landscapers and gardeners alike will have flowers from June through August.

Astilbe is one of the most popular perennials for the shade. Beauty combined with consistent performance will make it one of your top sellers. Also be sure to offer a range of varieties across the four main colors—pink, red, purple, and white—as well as varieties that flower at different times in the season.

Propagation & Growing On

Division (or bare-root transplants) is the most common method of propagation of astilbe. Seed (and plugs) are available as well, but they are not as popular.

Seed

Astilbe grown from seed is primarily available as mixes, but most home gardeners want separate colors. There are a few separate colors available from seed although they are of smaller stature especially useful for 4" (10 cm) or similar containers.

Sow seed in June or July to a 288-plug tray using 2–3 seeds/cell. Do not cover seed. Germinate at 72–78°F (22–25°C). Seedlings will emerge in seven to ten days, although full germination takes from fourteen to twenty-one days. Allow seven to nine weeks from sowing for a plug tray to transplant to larger containers. Transplant to 1-qt. (1.1 l) containers and grow on during midsummer for plants well rooted to overwinter cold/dormant. These can be sold in the spring as green plants that will flower in summer in the home garden. Plants will flower more profusely the following year.

Bare-root transplants

Most growers buy in bare-root transplants in the early winter and pot in January or February for green plant sales in the spring. Pot two-eye divisions into 1-gal. (4 l) pots and one- and two-eye divisions into 3-qt. (3 l) pots. Small, one-eye plants can go into 1-qt. (1.1 l) pots. Choose a well-drained, disease-free, soilless medium with a moderate starter charge and a pH of 6.0–7.0. Potted up in February, plants have well-developed roots and can be sold in mid-May as green plants. Plants will not flower profusely (if at all) when planted in the home garden in late May or June.

For fall-planted roots, establish plants in their growing containers prior to overwintering or bulk them if they are from seed-grown plugs prior to overwintering. While researchers at the University of Michigan learned that astilbe has an obligate juvenility requirement prior to receiving cold treatment, the exact size that plants need to be is unknown. To bulk plants or root them in, provide 55–65°F (13–18°C) days and 55–60°F (13–16°C) nights. Keep plants uniformly moist. Grow at 2,000–3,000 f.c. (22–32 klux) under natural days. Note that plants that have not been cooled will become dormant under day lengths of less than sixteen hours. If trying to size up plants, grow them under long days during this period. Fertilize with 100–150 ppm from 14-0-14 at every other irrigation, and keep the pH between 6.0–6.5.

Astilbe must receive a cold treatment prior to forcing into flower. Provide a minimum of twelve weeks at 41°F (5°C) in a cooler with 25–50 f.c. (269–538 lux) of light or in an unheated cold frame or greenhouse. Do not fertilize plants while they are being cooled, and do not let pots dry out.

To force plants, bring pots into the greenhouse as they are needed and gradually raise temperatures to 65–68°F (18–20°C) days and nights. Flowering can be sped up or slowed down by raising or lowering temperatures. Provide 3,000–5,000 f.c. (32–54 klux) of light, assuming growing temperatures can be maintained. As light levels increase through the year, apply shade. Fertilize at every other irrigation with 150–200 ppm from 20-10-20, alternating with 14-0-14.

Be diligent about irrigation during production. Astilbe requires consistently moist conditions. However, do not allow pots to remain wet. Use a well-drained medium high in organic matter and keep it moist.

Astilbe responds well to long days for forcing. By providing long days through four-hour night interruption lighting after you have brought plants out of dormancy, you will increase the number of flowers per plant as well as produce taller plants. Be aware, however, that late-flowering varieties may take up to four weeks longer to flower than early and mid-season varieties will.

If you would like to use growth regulators, B-Nine is effective. For producing really compact plants, two applications of 5,000 ppm can be made one week apart once flower stalks begin to show color. On some varieties, B-Nine may delay flowering by one to two weeks.

Cut Flowers

Astilbe can also be used as a cut flower. Plant in beds at a density of 1.5 plants/ft.^{2} (16/m^2) in the spring. Harvest stems when the lower halves of spikes have opened. If you plan to dry the flowers, harvest them when flowers are almost fully open. Be careful when removing flowering stalks to not damage leaves on the plants.

Varieties

Astilbe × *arendsii* varieties are the most popular astilbes. Within this group, a short list of varieties that you should consider include: 'Bressingham Beauty', 36" (91 cm) tall, early to mid-season, with pink flowers; 'Cattleya', 40" (102 cm) tall, mid-season, with airy rose-pink plumes; 'Flamingo', 20" (51 cm) tall, late season, with bright flamingo-pink plumes; 'Sister Theresa', 24" (61 cm) tall, early to mid-season, with salmon plumes; 'August Light', 28" (71 cm) tall, mid- to late season, with red plumes; 'Etna', 20" (51 cm) tall, mid-season, with dark red plumes; 'Fanal' (one of the most well-known astilbes), 20" (51 cm) tall, early season, with narrow red spikes; 'Granat', 24" (61 cm) tall, mid-season, with deep rose color that fades to rose with age; 'Bridal Veil', 28" (71 cm) tall, early season, with dusty white plumes; and 'Snowdrift', 24" (51 cm) tall, early season, with pure white plumes.

A. chinensis 'Pumila' is a dwarf, with plants reaching only 10–12" (25–30 cm) tall, making a great 4" (10 cm) crop. It thrives in well-drained yet moist, shady locations. Flowers are lavender pink, and plants are late flowering.

Other vegetatively propagated *A. chinensis* to consider include the striking 'Purple Candles', 42" (107 cm) tall, late season, with bold purple-red plumes; 'Veronica Klose', 16" (41 cm) tall, late season, with deep rose-purple flowers; and 'Visions', 12–14" (30–36 cm) tall, mid- to late season, with red-pink, sweetly scented flowers.

A. japonica varieties to consider include 'Rheinland', 24" (61 cm) tall, early season, with clear pink plumes and bold foliage; 'Red Sentinel', 24" (61 cm) tall, mid-season, with dark green leaves that set off deep red plumes; and 'Deutschland', a well-known variety, 30" (76 cm) tall, early season, with white plumes.

A. simplicifolia 'Sprite' is another well-known astilbe. Its drooping, pink plumes sit atop bronze-green foliage. Plants grow to 12" (30 cm) tall and are late-season flowering.

Postharvest

At retail, display plants under 50% shade at 65–70°F (18–21°C). Do not allow plants to dry out while they are on display, as shelf life will be adversely affected.

Aurinia

Aurinia saxatilis
Common names: Basket of gold; gold dust
Perennial (Hardy to USDA Zones 3–9)

A spring-flowering perennial with bright, golden-yellow flowers on plants from 10–12" (25–30 cm) tall and 12–14" (30–35 cm) across. Plants will often trail as their thickened, wiry stems cannot support the weight of the foliage. The single flowers are a sulfur-yellow color, 0.125" (0.3 cm) across, and bunched in terminal clusters. The gray-green foliage contrasts well with the yellow flowers. Although *Aurinia* has a short flowering season, its foliage provides continuous contrast with other garden plants throughout the summer.

Propagation & Growing On

Cuttings and seed are the two most common methods to propagate *Aurinia.*

Seed

Seed is the most common—and easiest—form of *Aurinia* propagation, but plants may vary in performance. The habit and vigor of seed-propagated plants are usually uniform, though bud set and flowering may differ.

Sow two to four seeds per cell of a 288-plug tray, leave seed exposed to light or lightly cover with vermiculite, and then germinate at 68–72°F (20–22°C). Seedlings emerge in three to eight days, and plug trays are salable in five to seven weeks after sowing. After germination, drop night temperatures to 58–60°F (14–16°C) to reduce stretching.

If sowing to an open flat, follow the same guidelines, and seedlings can be transplanted within two weeks after sowing.

For flowering plants in the spring, sow seed anytime the previous June onwards until autumn. June and July sowings can be transplanted to 1-qt. (1.1 l) or 1-gal. (4 l) pots for overwintering cold—the foliage often persists even at these cold temperatures.

Sowings made in October or November are transplanted to 1-qt. (1.1 l), 1-gal. (4 l), or 6" (15 cm) pots and grown cold (38–45°F [3–7°C]) once the plants are well rooted in their final container. These will flower the following spring.

Seed from January onwards will produce salable plants in the spring, but they will not flower until the following year.

Cuttings

Take cuttings of double-flowering or variegated-foliage varieties from early spring (before bud set) or in summer after flowering. The plants can be sheared back and fed, which will create growth of a number of shoots. Potted to 1-qt. (1.1 l) or 1-gal. (4 l) containers, the plants are overwintered cold for spring sales in flower.

Azalea (see *Rhododendron*)

B

Bacopa (see *Sutera*)

Basil (see *Ocimum*)

Begonia

Begonia × *hiemalis*
Common names: Rieger begonia; Elatior begonia; hiemalis begonia
Annual, pot plant

Hiemalis begonias have long been one of the premier flowering pot plants for gift giving. However, they also make an excellent annual for the summer shade garden. While they offer the traditional flower colors of most begonias (i.e., red, white, pink, etc.), they also offer the intermediate shades that are not commonly found in other begonias.

Flowers might appear delicate, but in reality they are stronger than you think. For growers, hiemalis begonias are 100% programmable, making them an excellent year-round producer. They are also very shippable.

Hiemalis begonias are a natural choice to include in weekly pot plant programs. Bedding plant growers will want to grow a few hundred as high-value pots to offer for Easter and Mother's Day sales. Hiemalis begonias are also a premium addition to mixed combination pots and baskets, spicing up even the most common of foliage plants with bright color and exquisite flowers.

Propagation

While seed-propagated varieties are available, the vegetatively propagated selections accentuate the market.

Cuttings

Most growers purchase liners of vegetatively propagated varieties, because hiemalis begonias are sensitive to stress, which can cause premature bud initiation and reduce growth. Additionally, sanitation during propagation cannot just be good—it must be perfect.

Some growers prefer to maintain stock plants and take cuttings. Purchase stock plants only from a reputable supplier using a clean-stock program. Maintain stock in a vegetative stage by providing a sixteen-hour day length and at least 68°F (20°C) at night. Harvest cuttings in the early morning or late afternoon. Take cuttings of uniform diameter from terminal ends, leaving one fully expanded leaf on each stem. This will require multiple passes through the plants rather than harvesting all the cuttings from each plant at once. Knives must be sterilized between each plant, as bacterial disease is a threat with this crop.

Leaf cuttings may also be used. If leaf cuttings are used, stock plants should get four weeks of short-day treatment (twelve- to thirteen-hour days). Applying short-day treatment will encourage shoot production from leaves taken off reproductive plants. Do not crush stems or leaves, as the wound can serve as a point of entry for botrytis. Cool cuttings for two hours at 48°F (9°C) to reduce their temperature. Do not store cuttings for more than twenty-four hours: If they must be stored at all, hold at 50–60°F (10–16°C).

During callus formation (first five to seven days), maintain sufficient moisture in the rooting medium (Ellepot or media); you should be able to squeeze out water. Provide root temperatures of 68–72°F (20–22°C) and air temperatures of 70–80°F (21–26°C) days and 68–70°F (20–21°C) nights. Maintain relative humidity at 75–90% at the base of the cutting. Adjust mist frequency as light and ambient air temperatures change. During callus formation, night misting may be required. However, excessive misting will delay rooting: Maintain a close watch on moisture levels and adjust according to your specific conditions. Reemay or porous plastic may help during this stage. Maintain 500–1,000 f.c. (5.4–11 klux) of light. Provide retractable shade so light levels can be increased as cuttings mature. Root cuttings under long days (sixteen hours of daylight or night interruption lighting) to maintain them in a vegetative state. (If using leaf cuttings, applying two weeks of short days can hasten shoot formation. Be careful not to use short days for too long, as flowering will be initiated.) Use 20-10-20 at 50–75 ppm applied to foliage if leaves lose color or as soon as roots start to form. Some growers direct-stick cuttings into their final growing container.

When 50% of the cuttings have initiated roots, encourage roots to develop by reducing soil moisture. This stage will take from fourteen to twenty-one days. Reduce night mist application, increase intervals between mist applications, and shorten duration. As roots form, increase light levels to 1,000–2,000 f.c. (11–22 klux) and use an alternating foliar feed of 15-0-15 and 20-10-20 at 100 ppm. Quickly increase feeding to 200 ppm as roots form.

Once fully rooted, harden cuttings for seven days prior to shipping or planting. To harden cuttings, move liners to an area with lower relative humidity, lower temperatures, and higher light (2,000–3,000 f.c. [22–32 klux]). Provide shade during the middle of the day to reduce temperature stress. Apply fertilizer twice per week, alternating 15–10–15 and 20-10-20 at 150–200 ppm.

Seed

Growers choosing to grow hiemalis begonia from seed (e.g., 'Charisma' series) can germinate seed at 75–78°F (24–26°C) in 288-plug trays. Do not cover seed. Radicles will emerge in seven to fourteen days. Follow the plug-growing guidelines for non-stop begonias in the entry for *Begonia* × *tuberhybrida* (p. 268) to grow and flower the plant.

Growing On

First, it is important to understand that hiemalis begonias initiate flowers when stressed. Stress includes but is not limited to situations such as going from long days to short days, moisture stress, chemical applications, and exposure to low temperatures. Once the plant has been stressed and blooms are initiated, more energy is put into flowering and not into growth.

Upon receiving liners, plant them immediately. Select a disease-free, sterile medium with excellent drainage, yet one that will retain water. Starting pH should be 5.5–6.0 with a balanced starter charge. Plant 1 liner/4" (10 cm) pot, 1–2 liners/1-qt. (1.1 l) or 1-gal. (4 l) pot, 2 liners/6" (15 cm) pot, and 4 liners/10" (25 cm) basket. Water in, but do not soak pots. When roots have reached the container wall, irrigate as needed, fully saturating the medium.

Begonias are one-sided: All vegetative growth and flowering will face in one direction. As you plant multiple plants in the pot, make sure to plant the flat sides of the plants facing each other in the center of the container. Avoid planting them too deeply, and be sure to plant the liner at the soil line.

After potting, maintain 68–75°F (20–24°C) days and 68–70°F (20–21°C) nights to keep plants vegetative. Closely monitor moisture levels, pH, and fertility to keep from stress.

It is vital to provide shade during summer months to prevent leaf scorch. When acclimated, begonias grown for the garden can take quite a bit of light. Slowly raise light levels to avoid damage. When able to maintain moderate growing temperatures, target 4,000 f.c. (43 klux). At higher temperatures, 70–72°F (21–22°C), reduce light levels to 3,000 f.c. (32 klux). Lower light levels to 1,500 f.c. (16 klux) or less under very warm temperatures. Too much shade will cause leggy plants and will require plant growth regulators. To maintain floral-quality begonias, moderate to lower light levels are best.

Use 150–200 ppm nitrogen (N) from 20-20-20, alternating with 15-0-15 at every other watering. As the crop is finishing during its last weeks, you may choose to reduce fertilizer levels and increase clear watering. Hiemalis begonias have roots highly sensitive to high soluble salts. Maintain EC at 1.0–1.3 mS/cm and substrate pH at 5.6–6.0.

Space pot tight for two to four weeks; do not space plants out too early. Space pots when foliage covers the container but before plants begin to undesirably stretch. The time of spacing can be an effective management tool to increase or slow the rate of growth. Allowing plants to stretch a bit allows for quick height gain. Space 4" (10 cm) pots at 7" × 7" (18 × 18 cm) and 6" (15 cm) pots at 12" × 12" (30 × 30 cm).

Always apply water from the base—either through ebb-and-flood or capillary matting or drip tubes. Do not allow foliage to become wet, as this is an invitation for disease.

Choose varieties which branch well. For larger plants or varieties with few breaks, pinch plants once above the third or fourth leaf. Crop time is increased by a pinch.

Height can be controlled with Cycocel or A-Rest. Application can begin during short-day treatment, if needed. Some growers use ultralow rates and make applications every one to two weeks until visible bud. Later applications can distort flowering. Negative DIF is also very effective for height control.

Hiemalis begonias are daylength sensitive. To remain actively growing in a vegetative state, they require fourteen-hour days. If natural day length is shorter in your area, you must light to extend the day. Many growers use night interruption lighting (minimum 12–15 f.c. [129–161 lux]) from incandescent bulbs to keep plants vegetative. Floral growers frequently use high-pressure sodium lamps and higher light levels (200–600 f.c. [2.2–6.5 klux]) to extend the day length, which also boosts plant growth.

When plants are two-thirds of the final desired height, it is time to initiate flowering. To induce flowering, hiemalis begonias must be stressed. One such stress is the short-day treatment: ten hours of daylight (fourteen hours of darkness). Pull shade from 4:00 P.M. until 8:00 A.M. the next morning. When shade/blackout is not available it is important to use more than one of the following stresses simultaneously so that you are sure to have fully initiated flowers. Some options are applying plant growth regulators, creating moisture stress, lowering fertility levels, or reducing night temperatures. However, never allow night temperatures to drop below 61°F (16°C).

For example, here is one way to schedule a typical crop. For a 4" or 4.5" (10 or 11 cm) pot, stress seven to ten days after potting. For 6" (15 cm) pots with two cuttings, stress three weeks after transplanting. As you gain experience with the crop, you can adjust the times to fit your specific circumstances. Some growers stress the plants while they are spaced pot tight. More often growers initiate when plants are two-thirds of their finished size. When using blackout, a minimum of fourteen days of short days are required during the winter and twenty-one days during the summer. Additional short days for small pots and during warmer months can help to control plant size. (See table 8 for example crop schedules.)

After short-day treatment, return plants to long days. Plants should flower within six to eight weeks after reinstituting long days.

If the plants are not stressed enough, flowering can be significantly delayed, fewer flowers may develop, and plants can grow unwieldy in height. Flowering is also not uniform; the same variety will flower over a period of four to six weeks. Growers typically have trouble keeping plants from flowering too early. Thus, the focus should be on keeping plants from pre-flowering. If plants begin to flower prematurely, simply remove the flowers. Because they are not sticky and the tissue breaks easily, they are easy to remove.

Table 8. Sample Crop Schedule for Liners from Tip Cuttings*

	4"	6"	10" BASKETS
CROP TIME	12–14 weeks	14–16 weeks	14–16 weeks
LINERS	1	2	3–4
BEGIN STRESS/LONG DAYS (TRANSPLANT)	0 weeks	2–3 weeks	4 weeks
STRESS/SHORT DAYS (FLOWER INITIATION)**	3 weeks	3 weeks	3 weeks
LONG DAYS (FORCING)	7–8 weeks	7–8 weeks	7–8 weeks

* From Ball Seed Co., West Chicago, Illinois, with additional input from Paul Hammer, editor of this section.
** A slightly longer short-day treatment (four weeks) will keep plants a little shorter and produce more flowers.

Pests & Diseases

Make sure your cuttings have been indexed and multiplied using only a clean-stock system. Hiemalis begonias can be infected by a number of bacteria, diseases, and viruses, such as *Fusarium sacchari, Xanthomonas campestris* p.v. *begoniae,* tomato spotted wilt virus, *Aphelenchoides* spp., *Steneotarsonemus pallidus,* impatiens necrotic spot tospovirus, *Pythium, Phytophthora, Erwinia,* and *Thielaviopsis.* Some of today's newer varieties are resistant to powdery mildew; however, many existing varieties are not. Under poor environmental conditions all varieties are susceptible to powdery mildew. Control *Pythium* and *Phytophthora* through irrigation control: Maintain uniform moisture, but not wet conditions. If any plants become infected with a bacterial disease, such as *Erwinia* or *Xanthomonas,* there is no cure. Remove and destroy infected plants immediately.

Insects are less of a problem, but aphids, cyclamen mites, and thrips can attack hiemalis begonias.

Troubleshooting

Reddening or darkening of leaves, cupping of leaf margins, hard growth, and stalled plants can indicate sunscald, which is caused by high light. Low fertility, high soluble salts, overwatering, poor drainage, high light, cold temperatures, insufficient root development, or preflowering may cause stunted plants. Too few short days or not enough stress, high or low soluble salts, or high or low temperatures may result in insufficient flowering. Low fertilization or *Xanthomonas* bacteria may lead to yellow spots on leaves and stem collapse. High light levels (high temperature), high soluble salts, or cold night temperatures may cause dark green, brittle leaves. *Pythium,* overwatering, cold medium, or root loss may result in gray and/or wilting foliage. Wide temperature fluctuations between day and night in combination with high substrate moisture content may lead to edema, rupturing and corking of plant cells on leaves.

Varieties

Many excellent vegetatively propagated varieties of hiemalis begonias are on the market, most originating from European breeding. Many of the traditional pot plant varieties are being successfully used in garden programs. Consult your broker sales representative for a list.

Postharvest

Pull pots for shipping when 25–75% of flowers are open. Stake 6" (15 cm) and larger pots plants at shipping (or immediately after planting) as support to prevent plants from breaking. Hiemalis begonias are highly sensitive to cold: Ship plants at 45–55°F (7–13°C) for no longer than three to five days. Flowers are ethylene sensitive; 1-MCP or silver thiosulfate can be beneficial in reducing flower and bud drop. When enjoying indoors, retailers and homeowners should display plants under at least 100 f.c. (1.1 klux) to maintain blooming. Strive for display temperatures from 65–75°F (18–24°C). Properly handled, hiemalis begonias should last two to three months.

When planting in the garden they should be put in a place where they will receive plenty of light but some shade during the hottest part of the day. Plant in good bedding soil, fertilize regularly, and keep moist. Drying out may cause buds to drop; however, plants cannot stand being overly wet. Remove dead flowers. Begonias should flower through the complete gardening season.

Begonia rex hybrids
Common names: Rex begonia; painted begonia
Foliage plant, annual

Grown for beautiful leaf markings of red, green, and silver leaves, rex begonias are enjoying a renaissance of sorts as consumers discover their versatility in partially shady garden spots, window boxes, and containers. They are mainstays for many interiorscapers and in foliage plant displays.

Because their leaves are brittle, plants are difficult to ship, making them an ideal crop addition for growers looking to provide their local market with something different.

Plants usually flower late in the year, as the days get shorter. Some of the older varieties actually go dormant, as the days grow shorter as well. However, this usually occurs with plants that are several years old and after they have flowered. Plants may also go dormant in response to short days and dry conditions. Under extremes the plants "rest" for two to three months. However, this is not common for the more current selections.

Propagation

Tissue culture and leaf cuttings are the most common forms of propagation. From cuttings, whole leaves or leaf strips made from mature leaves may be stuck in a sterile medium at a 45° angle. Maintain 75–78°F (21–24°C) substrate temperatures. Plantlets will form at the base of leaf pieces that may be separated and used as starter plants. Because this process is tedious and slow, most growers choose to buy in liners. A number of commercial propagators offer rex begonias as rooted liners.

Growing On

Pot in a well-drained, disease-free medium with a pH of 5.5–5.8 and a low initial starter charge. Use 1 liner/4" (10 cm) pot, 2 liners/6" (15 cm) pot, and 4 liners/10" (25 cm) hanging basket. Grow on at 75–80°F (24–26°C) days and 70–75°F (21–24°C) nights. Do not let day temperatures exceed 90°F (32°C) or allow night temperatures to fall below 60°F (16°C).

Light levels are critical to good leaf coloring. Grow plants between 2,200–2,500 f.c. (24–27 klux). Higher light levels reduce leaf coloring (as well as burn the leaves) and stunt growth. Too-low light levels can cause plant stretch. A-Rest at 50 ppm can provide effective height control in such instances.

Plant dormancy may occur during the short days of winter. Response is variety dependent. Reduce watering during winter months, when plants are dormant.

Maintain uniform moisture, but do not keep plants wet. Water plants with ebb-and-flood or capillary matting or drip tubes to keep foliage dry. Wet foliage encourages disease. Fertilize with 100 ppm from 20-10-20 at every other irrigation. Do not allow soluble salts levels to build up.

Maintain 50% humidity. Leaf margins will dry or brown with low humidity levels. Space plants adequately to allow for good air circulation around pots, which helps prevent disease. Pinching is not necessary.

The crop time from potting up a liner from a 72 tray is six to eight weeks for most varieties. The crop time is faster as the day length gets longer as well as warmer. There are a few varieties that are slow growing and may take a few more weeks, but commercial growers are discontinuing many of these in favor of their faster-growing counterparts.

Pests & Diseases

Watch for aphids, fungus gnats, mealybugs, thrips, and caterpillars. On occasion, whiteflies can become a problem as well. Rex begonias are susceptible to *Xanthomonas* blight. Be sure to obtain cuttings from a source that maintains a clean-stock program, as *Xanthomonas* from rex begonias can spread through worker contact to other greenhouse crops such as geranium. Other diseases that can occur include *Pythium, Fusarium,* rhizoctonia, and *Myrothecium,* which cause leaf spot and petiole lesions.

Troubleshooting

Water spots on leaves that turn light and then transparent may be due to airborne fluoride.

Postharvest

Plants are highly susceptible to cold damage, even at 35°F (2°C). Plants grown with high fertilizer rates are even more susceptible. At the retail shelf, display plants under 400 f.c. (4.3 klux) and 70–75°F (21–24°C) days and 60–65°F (16–18°C) nights. Do not overwater. Remove dead foliage and discard.

Begonia × semperflorens-cultorum
Common names: Wax leaf begonia; fibrous begonia
Annual

Bedding plant begonias are the workhorses of the bedding plant business. Year after year they perform reliably under a wide range of environmental conditions. While the popularity of other bedding plants seems to rise and fall with the trends, begonias have staying power.

For growers, begonias are a snap: Most growers buy in plugs starting with smaller plug sizes for early crops and working into larger plug sizes like 288s for later crops and summer sales. Many growers start their hanging basket begonias from 288 or 72 plugs planted in late December or January, depending on their geographic region.

Propagation

Seed is the most common method of propagation.

Begonias are an exceptionally long plug crop—eight or more weeks from sowing depending on the plug or liner size as well as the variety. Seed is small, making raw seed difficult to handle and germination is erratic and unpredictable. That is why the vast majority of growers leave plug production to specialists or buy in pelleted seed to grow their crop.

Begonias are produced as plugs grown in four stages. Stage 1, radical emergence, takes from six to seven days at a substrate temperature of 75–80°F (24–27°C). Sow pelleted seed onto a disease-free, germination medium with a pH of 5.5–5.8. Do not cover seed. If using a germination chamber, make sure to light trays. Some growers multiseed begonia plugs, especially larger sizes.

Keep the medium wet, near saturation. If it is not possible to maintain near saturation levels with mist, a very light layer of medium may help to maintain moisture levels around the seed. Reduce moisture levels once radicles have emerged. Allow the medium to dry out slightly before watering for best germination and rooting. Additional light at 100–400 f.c. (1.1–4.3 klux) may benefit germination. During germination, begonias are very sensitive to high salts, particularly ammonium. Be sure to keep ammonium levels at less than 10 ppm. Some recommend applying a fungicide drench to plug trays prior to sowing to eliminate any soilborne fungi that may be present.

Stage 2, stem and cotyledon emergence, will occur over seven to fourteen days. Maintain soil temperatures at 72–78°F (22–26°C). Begonias have a shallow root system at this stage: Keep moisture levels uniformly moist. Maintain the pH at 5.5–5.8 and EC at less than 1.0 mS/cm. Make sure ammonium levels are below 10 ppm. Fertilize with 50–75 ppm from 14-0-14 or calcium/potassium nitrate once cotyledons are fully expanded. Alternate feed with clear water.

Growth and development of true leaves will begin during Stage 3, taking twenty-one to twenty-eight days. Lower soil temperatures to 68–72°F (22–22°C). Allow medium to dry thoroughly between irrigations but avoid permanent wilting. Allowing the medium to dry down promotes root and shoot growth. Maintain a substrate pH of 5.5–5.8 and an EC of less than 1.0 mS/cm. Increase fertilizer to 100–150 ppm from 20-10-20, alternating with 14-0-14 or a calcium/potassium-nitrate fertilizer. Fertilize every second or third irrigation. If using 15-0-15, supplement with magnesium once or twice during Stage 3: Use magnesium sulfate (16 oz./100 gal. [1.2 g/l]) or magnesium nitrate. Do not mix magnesium sulfate with calcium nitrate, as precipitate will form. To control plant height during Stage 3, use DIF whenever possible, especially during the first two hours after sunrise. B-Nine and Cycocel are effective for plug height control. *Do not use Bonzi anywhere near begonias*—even the drift will adversely affect them!

During Stage 4, plugs harden and finish over seven days, becoming ready to ship or transplant. Maintain substrate temperatures of 62–68°F (16–20°C). Allow substrate to thoroughly dry between waterings. Provide light levels of 2,500 f.c. (27 klux) or higher. Continue to maintain pH at 5.5–5.8 and EC of less than 1.0 mS/cm. Fertilize as needed with 14-0-14 or calcium/potassium nitrate at 100–150 ppm nitrogen.

A 288-plug tray will take from seven to nine weeks from sowing for the majority of varieties on the market to be ready for transplanting.

Growing On

For the best-quality crops and fastest finishing times, plant plugs immediately on arrival, or as they come out of Stage 4. If you are planting multiseeded plugs, be aware that you will have more than one plant in the container or cell and depending on your specific growing conditions, the crop could become more vigorous than a single-plant plug.

Transplant begonia plugs into a well-drained, disease-free medium with a pH of 5.5–6.2 and a moderate initial nutrient charge. Because begonias comprise a significant percentage of most growers' crops, they are widely planted with automated transplanters. For maximum transplanter efficiency, make sure plants have well-developed root systems and leaves are not tangled in the plug tray.

Plant 1 plant/cell or 4" (10 cm) pot, 2 plants/6" (15 cm) pot, 3 plants/8" (15 cm) pot, and 5–7 plants/10" (25 cm) hanging basket.

Begonias are tolerant of a wide range of greenhouse conditions. Ideally, maintain night temperatures of 60–65°F (16–18°C) and day temperatures of 65–70°F (18–21°C). Provide moderate to high light levels (3,000–5,000 f.c. [32–54 klux]). Planted outdoors, plants can tolerate partial shade to full sun once established.

Fertilize at every other watering with 150–200 ppm from 15-0-15, alternating with 20-10-20. Maintain EC at 1.0 mS/cm. To harden plants, reduce nitrogen levels a couple weeks prior to sale, especially under low light conditions.

Once begonias have rooted in to container sides, you can control growing speed and plant height by withholding water and nutrients. If growth regulators are required, use Cycocel at 500 ppm for white-flowered varieties and 1,000 ppm for other colors. Begonias are not responsive to DIF. If flats become overgrown, they may be cut back and regrown. You can even bump plants into 1-gal. (4 l) pots for summer sales.

From a 288-plug tray to a flowering 32-cell pack is six to eight weeks, while a 4" (10 cm) pot takes from eight to twelve weeks depending on the variety. Due to the long crop time growers plan on sixteen to twenty weeks for total crop time for the majority of varieties from sowing to a finished 4" (10 cm) pot. Some varieties will be in flower, while others will be just starting to bloom.

Pests & Diseases

Fungus gnats and shore flies can be a problem. Be careful of botrytis, *Pythium,* and rhizoctonia. Algae growth on the medium may also be a problem, especially on flats grown on the ground in high humidity.

Varieties

Fibrous begonias are offered in green-leafed and dark-leafed varieties. The most popular color is dark-leafed red ('Vodka'), because its darker leaves and flower coloring do not fade in full-sun plantings. A wide number of excellent varieties are on the market today for pack production, as well as pot and basket production.

While older, the 'Cocktail' series is still the leading dark-leafed begonia series. No other breeder has been able to significantly improve them across the board, although newer series such as the 'Senators' are much faster to flower. 'Cocktail' includes 'Whiskey' (white), 'Vodka' (red), 'Gin' (rose), 'Brandy' (light pink), 'Tequila' (deep rose), and 'Rum' (white-edged red). 'Senators', also dark-leafed, is a great series for pack or pot production and in the garden, but it does not have the name recognition of 'Cocktail'.

For green-leafed series, if earliness is your main goal, try 'Ambassador', 'Prelude', or 'Super Olympia'. All three are available in a wide range of colors and have good outdoor performance.

In addition, breeding companies have been developing new, larger-growing series specially targeted for the landscape market. These larger varieties are best suited to basket and container production. If you are looking for great semperflorens begonias for landscape use, try the 'Party' series, the larger-growing 'Big' series, and the even-larger 'Whopper' series.

Begonia × *hybrida*
Annual

There are a number of recent introductions that are the result of various crosses between species. This list includes varieties like 'Dragon Wing', 'Big', 'Whopper', and others listed under the classification of *Begonia* × *hybrida* or as an interspecific hybrid, or that have a resemblance to wax or fibrous begonias. This group is often a larger growing, more robust selection of fibrous begonias than the traditional group.

While you can sow and grow this group as listed above for *B.* × *semperflorens* × *cultorum,* there are a few exceptions. Many of these varieties are larger-growing plants and cannot be finished for retail sales in a cell pack or liner like traditional wax begonias. Instead, an

18 tray is preferred for pack sales, if needed. However, plants are sold green and will not be in flower.

For flowering plants, use a 6" (15 cm) or 1-gal. (4 l) container and allow eight to eleven weeks from transplanting a plug or liner to a finished plant. Once plants begin to flower, they are usually well rooted to the bottom of the pot and ready to transplant to the garden.

Begonia × 'Gryphon' is classified as a *Begonia × hybrida* and can be grown like *B. semperflorens*, although it takes from two to three weeks longer to produce a plug as well as flower. In other words, a 288 plug takes from eight to nine weeks from sowing to a salable plug and five to seven weeks to finish in a 4" (10 cm) pot for spring sales. However, the plants are sold for their foliage and not their flowers. Plants will only flower under the short days of the year and not in summer. If plants are moved into the home as the nights get colder in September and placed in a room that experiences natural day length, the plants will flower during January or February.

Postharvest

Begonias are sensitive to ethylene and will drop flowers if exposed to levels over 5 ppm. If plants must be boxed for shipment, minimize transport time to less than two days. Begonias stretch in the dark, and their foliage becomes chlorotic. Display plants in partial shade (50–60%) to slow water evaporation from flats. Container displays require less shade. On most spring days, begonias will require watering twice each day. Maintain temperatures of 65–70°F (18–21°C) days and 60–65°F (16–18°C) nights. Temperatures below 50°F (10°C) for long periods can lower quality, although plants sufficiently hardened off at the nursery can tolerate temperatures in the 40s (°F [4–10°C]).

Troubleshooting

Low fertilizer rates and high light can cause bronze foliage. Leaf scorching can be a problem in regions with very high light.

Begonia × tuberhybrida
Common names: 'Non-Stop' begonia; tuberous begonia
Annual

Tuberous begonias from seed are personified by the 'Non-Stop' series, a premier selection, introduced over two decades ago, that has established itself as the brand to beat. Today, the 'Fortune' and 'Go-Go' series are gaining popularity as well. All are excellent in pots or baskets.

Tuberous begonias are available in a wide range of colors from the standard red, orange, yellow, and white to various shades of pink and salmon as well as two-tone patterns or bicolors between colors. The double-flowering blooms range in size from 3–4" (8–10 cm) across. They excel in hanging baskets or mixed combination plantings and are especially suitable for shade gardens.

Propagation

Growing tuberous begonias from seed can be difficult due to the long crop time and the exacting conditions they require. They are more challenging than fibrous begonias. Thus, many growers purchase plugs from commercial propagators.

Tuberous begonias are produced as plugs grown in four stages. Stage 1, radical emergence, takes from seven to ten days at a substrate temperature of 75–78°F (24–26°C). Do not cover seed. If using a germination chamber, make sure to light trays. If germinating on the plug bench, 100–400 f.c. (1.1–4.3 klux) of light is beneficial for germination. Most growers prefer to use pelleted seed as opposed to raw seed so seeding operators can spot-check to ensure seed placement.

Keep the medium wet, near saturation. The sowing medium should have a pH from 5.5–5.8 and soluble salts (EC) less than 0.75 mS/cm. During germination, begonias are very sensitive to high salts, particularly ammonium. Be sure to keep ammonium levels at less than 10 ppm. Some recommend applying a fungicide drench to plug trays prior to sowing to eliminate any soilborne fungus that may be present.

During Stage 2, stem and cotyledon emergence will occur over seven to fourteen days. Maintain substrate temperatures at 70–72°F (20–22°C). Keep the medium uniformly moist: Tuberous begonias will stall if seedlings dry out. The root system is very shallow at this stage. Maintain the pH at 5.5–5.8 and EC at less than 1.0 mS/cm. Ensure ammonium levels are below

10 ppm. Fertilize with 14-0-14 at 50–75 ppm or calcium/potassium-nitrate feed once cotyledons are fully expanded. Alternate feed with clear water. If fertility is too high or too low, begonias will stall out.

Growth and development of true leaves will begin during Stage 3, taking twenty-eight to thirty-five days. Lower medium temperatures to 68–72°F (19–22°C). Allow medium to dry thoroughly between irrigations, but avoid permanent wilting of the plants. Allowing the medium to dry down promotes root and shoot growth. Light is critical during Stage 3: If HID supplemental lighting is not available, light plants using night interruption lighting (50 f.c. [538 lux]) to prevent tuber formation. Provide overall light levels of 1,500–2,500 f.c. (16–27 klux) during the day; higher levels will stall growth. However, too little light will cause seedlings to be lush and stretch. Low light also encourages tuber formation. Maintain a substrate pH of 5.5–5.8 and an EC of less than 1.0 mS/cm. Increase fertilizer to 100–150 ppm from 20-10-20, alternating with 14-0-14 or a calcium/potassium-nitrate fertilizer. Fertilize every second or third irrigation. If using 15-0-15, supplement with magnesium once or twice during Stage 3, use magnesium sulfate (16 oz./100 gal. [1.2 g/l]) or magnesium nitrate. Do not mix magnesium sulfate with calcium nitrate, as precipitate will form. Be careful not to burn foliage when feeding. Rinse off foliage immediately with clear water.

To control plant height during Stage 3, use DIF whenever possible, especially during the first two hours after sunrise. B-Nine and Cycocel are effective for plug height control. *Do not use Bonzi anywhere near begonias;* even the drift will adversely affect them.

During Stage 4, plugs harden and finish over seven to fourteen days, becoming ready to ship or transplant. Maintain substrate temperatures of 60–62°F (16–17°C). Allow the medium to thoroughly dry between waterings. Provide light levels of 2,500 f.c. (27 klux) or higher. Continue to maintain a pH at 5.5–5.8 and an EC of less than 1.0 mS/cm. Fertilize as needed with 14-0-14 or calcium/potassium nitrate at 100–150 ppm nitrogen (N).

Some growers produce plugs through Stage 2 in smaller sizes and then transplant them into larger, 72-cell plugs for Stage 3 and Stage 4 production.

A 288-plug tray will take from seven to nine weeks to be ready to transplant, a 72 takes from nine to eleven weeks.

Growing On

Transplant begonia plugs into a well-drained, disease-free medium with a pH of 5.5–6.2 and a moderate initial nutrient charge. Plant 1 plant/cell or 4" (10 cm) pot, 2 plants/6" (15 cm), and 5 plants/10" (25 cm) hanging basket. Be careful not to bury the crowns.

Maintain temperatures of 68–75°F (19–24°C) days and 65–68°F (18–19°C) nights. Avoid temperatures below 50°F (10°C) as tuber formation increases at low temperatures.

Fertilize at every other watering with 150–200 ppm from 15-0-15, alternating with 20-10-20. Maintain EC at 1.5 mS/cm. Excessive ammonia will promote large leaves that show foliage necrosis. Foliage of young tuberous begonia plants can burn if watered during the brightest time of the day.

Plants are free branching and should not require pinching. However, if plants need shaping to fill out containers, they may be pinched above the fifth or sixth set of leaves once roots have reached the container edge.

Tuberous begonias need long days (days longer than twelve hours) to flower. With days less than twelve hours, tubers form at the expense of the flowers. Use HID supplemental lighting for sixteen to eighteen hours per day until April 1 to promote growth and prevent tuber formation. You may also use night interruption lighting from 10 P.M. until 2 A.M. at 50 f.c. (538 lux). However, tuberous begonias grow best under reduced light on the greenhouse bench so the foliage does not burn. As daylight lengthens in the spring, reduce light to 1,500–2,500 f.c. (16–27 klux) for variegated varieties.

Once begonias have rooted up to the container sides, you can control growing speed and plant height by withholding water and nutrients. B-Nine and Cycocel are effective growth regulators. Cycocel may be applied weekly at 150–250 ppm beginning two weeks after transplanting. Tuberous begonias are very responsive to DIF; plants are shorter under negative DIF.

Transplanting a 288 plug to a 4" (10 cm) pot will require nine to ten weeks to flower, while a basket will require eleven to fourteen weeks to be salable from a 288-plug tray.

Pests & Diseases

Mealybugs, thrips, and whiteflies may affect tuberous begonias. Diseases that can be a problem include botrytis, powdery mildew, *Pythium,* rhizoctonia, and INSV/TSWV.

Troubleshooting

Plant collapse can be due to stem canker caused by botrytis. Plants can also collapse because plants have remained wet for prolonged periods of time or because of a virus. Excessive ammonia nitrogen in the fertilizer, overfertilization under low light conditions or low light, overwatering, or wet substrate can cause too much tuber growth and a lack of flowers. INSV, iron deficiency, or low nitrogen in leaf tissue can result in interveinal chlorosis. Drying out plants between irrigations or high soluble salts in the medium can result in foliage necrosis. Poor branching or thin plants can be due to low fertilization during early stages of growth, low light conditions, or excessive light, which promote flowering and inhibit branching.

Varieties

The 'Non-Stop' series is the most popular tuberous begonia. Other varieties have been introduced that have shorter crop times, such as 'Fortune'. 'Ornament' is a dark-leafed relative of 'Non-Stop'. 'Pin-Up' is a spectacular picotee bicolor with hibiscus-sized flowers—try it in 6" (15 cm) pots. 'Illumination' is a pendulous variety that makes spectacular baskets that customers will return for year after year. It is an ultralong crop, but if you learn how to do it well, you will create a niche in the marketplace for yourself.

Other selections include the 'Go-Go' series, a wonderful 4" (10 cm) pot or basket variety. 'Fortune' can be used the same way but has a broader color range.

Postharvest

Tuberous begonias are unwieldy to ship long distances, and stems can be brittle, breaking easily with a lot of handling. Thus, this is a great crop for retail growers and smaller wholesale growers that cater to the market with value-added products. Silver thiosulfate can be beneficial in reducing flower drop. Retailers should display plants under partial shade and target 68–75°F (19–24°C) days and 65–68°F (18–19°C) nights.

Bellis

Bellis perennis
Common name: English daisy
Hardy annual

Bellis is sometimes listed as a perennial, hardy to USDA Zones 4 to 8. However, severe northern winters or the intense heat of a southern summer can quickly kill these plants.

Bellis perennis is a low-growing, rosetting plant, 6–8" (15–20 cm) tall in bloom, with an 8–10" (20–25 cm) spread. The dark green foliage is held closely to the crown. Flowers are either single or double, measuring l.5–3" (3–7 cm) across. The scentless blooms come in white, pastel pink, and carmine red with several shades in between. Without blooms, the plants are between 3–4" (7–10 cm) tall.

Propagation

Seed is the common way to propagate. Sow to a 288-plug tray, cover lightly with vermiculite, and germinate at 70–72°F (27–22°C). The root radicle emerges in a three to five days, and seedlings develop in eight to ten days. A 288 plug takes from three to five weeks from sowing to be ready to transplant to larger containers.

Growing On

Treat the bellis crop as you would pansies. In general, plants require from twelve to fourteen weeks from sow to flower, 50–100% in a 32-cell pack. Allow fourteen to fifteen weeks for a full-flowering crop in a 4" (10 cm) pot at 50–55°F (10°13C) night temperatures. This is based on late-flowering varieties. Many of the more recent varieties will flower up to seven to ten days earlier.

Varieties are cold hardy but will not tolerate severe, northern winter temperatures. For February sales, sow 'Pomponette' in October to produce flowering 4" (10 cm) pots in February and grow at 45–50°F (7–10°C) nights in northern regions. Additional lighting may not be necessary.

B. perennis could be grown in the southern states as a fall crop for winter and spring color. Due to the heat sensitivity of the crop, however, it would be best to buy in plugs available from commercial propagators as opposed to sowing seed in July for late southern summer sales. It is not recommended as a northern fall crop.

Beta

Beta vulgaris (Cicla group)
Common names: Swiss chard
Annual vegetable

Swiss chard is a leafy vegetable, and the leaf stalks vary in color from green, white, red, orange, and several in-between shades. Plants prefer full sun and grow from 18–24" (46–61 cm) depending on variety. However, if leaves are used in the juvenile stage, then the plants are notably shorter (used within twenty-five to thirty-five days after planting to the home garden). Older leaves can be brushed with a combination of garlic and olive oil and used in place of spinach once the season turns hot.

Propagation & Production

Sow seed to a 512-plug tray (one seed per cell), lightly cover with vermiculite, place under mist, and give germination temperatures of 68–72°F (20–22°C). Emergence will occur in four to eight days. Plug trays will be ready to transplant to larger containers in three and a half to five weeks. Transplanted to 32-cell packs, plants will be ready in six to eight weeks when grown at 60–65°F (16–18°C) nights once established. For 4" (10 cm) pots, allow nine to eleven weeks

Related Species

Beta vulgaris beets are a close relative of Swiss chard and can be grown in a similar fashion. However, since beets are a root vegetable and Swiss chard is a leafy one, you have to keep a few things in mind. You can use the Swiss chard crop information for beets. However, it is necessary to transplant to larger containers once the plug is ready to transplant. Do not allow the plugs or plants to become root-bound, as this will slow growth and increase other problems.

With all that said, this crop can also be sown to an open flat and have individual seedlings transplanted for a crop that is uniform in growth and yields excellent plants with minimal issues. Sow seed to an open flat in early March (following most of the other notes for Swiss Chard) and then transplant to 32 cells per flat later in the month. Plants will be ready for home gardens in mid- to late May.

Bidens

Bidens hybrids
Annual

Bidens is an up-and-coming crop. Bright yellow flowers are displayed on a plant with a delicate leaf and plant structure. The plants are fast growing with a range of habits and vigor; these include mounded compact varieties, vigorous trailing types, and upright/mounded varieties that are all readily available in the market. *Bidens* makes an excellent component in mixed containers or hanging baskets.

Propagation

Bidens is propagated from cuttings. Stick cuttings immediately upon arrival into a rooting medium. Provide soil temperatures of 68–75°F (20–24°C) and air temperatures of 70–80°F (21–28°C) days and 65–68°F (18–20°C) nights. Apply 500–1,000 f.c. (5.4–11 klux) of light. Keep medium uniformly moist so that water is easily squeezed out. Set mist frequency so that cuttings do not wilt. Apply 50–75 ppm from 15-0-15 as soon first roots emerge. Maintain a pH of 5.8–6.2. Callus will form in three to four days, with roots emerging in five to six days.

Once roots are visible, increase light levels and eliminate mist. Maintain soil and air temperatures. Begin to allow medium to dry out somewhat as roots form; however, maintain high air humidity. Gradually increase light levels to 1,000–1,500 f.c. (11–16 klux). Apply feed of 100 ppm from 15-0-15, alternating with 20-10-20. Increase the rate quickly to 200 ppm. Roots will continue to develop from root emergence until fourteen to eighteen days after sticking.

Harden plants for transplanting or shipping for seven days. Remove cuttings from the mist area. Reduce air temperatures to 68–75°F (20–24°C) days and 62–65°F (16–18°C) nights. Allow medium to dry between irrigations. Gradually increase light levels to 2,000–4,000 f.c. (22–43 klux). Fertilize with 150–200 ppm from 15-0-15.

Plan on an average of four to seven weeks for a 105-, 84-, or similar liner tray from sticking the cuttings until they are ready to transplant to a 4" (10 cm) pot.

Growing On

Plant 1 liner/ 4" (10 cm) pot, 2 liners/6" (15 cm) pot, or 3–4 liners/10" (25 cm) hanging basket. Select a disease-free, soilless medium with a high initial starter charge and a pH of 5.8–6.2. Be sure to select appropriate cultivars for the container size.

Grow on at 68–75°F (20–24°C) days and 62–65°F (16–18°C) nights under full sun (4,000–8,000 f.c. [43–86 klux]). Plant growth slows significantly when temperatures are below 50°F (10°C). *Bidens* is not day-length sensitive, although flowering increases as light levels increase. Low light levels promote stretching.

Apply 200–250 ppm at every irrigation from 20-10-20, alternating with 15-0-15. If salts begin to build up, then leach with clear water. Keep plants uniformly moist, but not wet.

Apply a pinch when plant roots have reached the container walls, in about three to four weeks. Pinch plants in 6" (15 cm) pots once and 10" (25 cm) baskets twice. For fastest crop time, do not pinch 4" (10 cm) pots. Pinch the crop back to the fifth or sixth set of leaves. Baskets can be sheared again in about two weeks. To increase branching, some growers use Florel about two weeks after the first pinch for pots and baskets.

The crop time to flowering for a 4" (10 cm) pots is from five to six weeks after transplanting one liner, while a hanging basket requires ten to twelve weeks using four to five liners.

Pests & Diseases

Insect problems that can occur include aphids, fungus gnats, thrips, and whiteflies. Disease problems can include botrytis, *Pythium,* or rhizoctonia. Avoid pesticide sprays when plants begin to flower.

Troubleshooting

Botrytis or overwatering can cause plant collapse. Excessive vegetative growth can be the result of high nitrogen concentration in the soil, overfertilization under low light, and/or wet medium in combination with low light. Low fertilizer levels can result in poor branching. Foliage necrosis can be due to plants drying out between irrigations.

Boltonia

Boltonia asteroides
Common names: Boltonia; false chamomile
Perennial (Hardy to USDA Zones 4–9)

Boltonia is a premier North American native plant for the late-summer garden with its scentless, daisy-like flowers that are either pure white or sometimes purple or lilac. Single blooms measure about 1" (3 cm) across. The plants form clumps spreading 3–4' (90–120 cm) wide and grow erect to between 5' and 6' (1.5 and 1.8 m) tall in flower. Foliage is a dull, gray-green color.

Propagation

The most common method of propagation is from cuttings, although seed can be done as well. Divisions are mostly done by the home gardener, but there are a few companies offering them.

All are equally easy to perform, depending on which method fits into your production schedule. The most common white-flowering variety, 'Snowbank', should be propagated by either cutting or division, since seed-propagated plants can be uneven or non-uniform. In addition, the seed can be difficult to find.

Cuttings

Take 1–2" (3–5 cm) cuttings any time the plants are vegetative and stick into peat moss with a 68–72°F (20–22°C) bottom heat. Rooting hormone is optional since the cuttings root pretty easily and propagation time is four to five weeks. Mist is needed for seven to eleven days, but be careful about saturation of the media—cuttings may rot. For this reason some growers use a preventative fungicide.

Potted to 4" (10 cm) or 1-qt. (1.1 l) containers, plants will not flower readily in autumn from spring- or summer-rooted cuttings.

Seed

Seed germinates in four to eight days at 70°F (21°C). Cover seed lightly with peat or vermiculite. No special treatments are necessary to obtain 70% or better germination rates on purchased seed. From sowing, allow twenty-two to thirty days before transplanting the seedlings. Plants will produce a green basal rosette with six to ten elongated leaves the first year after sowing. Green cell packs take from ten to twelve weeks from sowing at 55°F (13°C) for plants 1–1.5" (2.5–3 cm) tall. Plants will not flower the same season from seed, and the rosettes are slow growing. Instead, sow seed May or June for overwintering dormant in 1-qt. (1.1 l) or 1-gal. (4 l) pots for sale during the following spring.

Boltonia asteroides var. *latisquama*

This US native with dark-green foliage and lavender blooms is highlighted by the variety 'Jim Crockett', named in honor of James Underwood Crockett, the man who starred in the PBS broadcast of *Crockett's Victory Garden* in the 1970s.

'Jim Crockett' grows from 18–24" (46–61 cm) and flowers throughout mid- to late summer. It makes an excellent choice for the middle of a flower garden border.

Follow the boltonia cutting propagation schedule for rooting. Some additional points, cuttings root in four to five weeks in a 72 tray, when grown under short days to bulk up and given a cold treatment at 35–41°F (2–5°C) for a minimum of six weeks (but can be as long as fifteen weeks). The cold treatment helps to improve the number of blooms and increases the percentage.

Transplanted to 1-qt. (1.1 l) containers, plants will be ready for sale in five to seven weeks.

Borago

Borago officinalis
Common name: Borage
Annual herb

A mainstay of the annual herb garden, borage is noted for its edible flowers that are usually blue but are occasionally pink or white. The flowers have a mild, cucumber flavor and are used to garnish salads. Plants are large, growing from 20–30" (51–76 cm) tall.

Propagation & Growing On

Borage is propagated from seed and can be sown anytime in winter to early spring for green plant sales in the spring. Plants will flower during the summer.

Seed sown in January to early March to a 288-plug tray and lightly covered with vermiculite will germinate in five to eight days and be ready to transplant to larger containers in three to five weeks after sowing. Transplanted to 4" (10 cm) pots, the plants will be salable green in five to seven weeks for May sales. Grow on at 65–70°F (18–21°C) days and night temperatures of 55–60°F (13–16°C). Plants can be grown at warmer temperatures but usually stretch in habit, becoming floppy in appearance.

Borage is usually fast to grow and does not like to be held back at the time of transplanting. When plugs are ready, transplant to larger containers.

Bougainvillea

Bougainvillea glabra
Tropical plant; treated as an annual or tender perennial in colder areas

In the Deep South, bougainvillea is an eye-catching vine. Its colorful bracts put on an unparalleled show. It tolerates high temperatures, high light, and water stress. However, there is no reason why bougainvillea is not offered more frequently in garden centers from coast to coast. Growing the crop takes time and a little effort, but the higher prices that plants bring make the time spent worthwhile. They are great basket plants and perform fabulously in patio containers. The more root-bound plants become and the hotter it gets, the happier they are.

Propagation

Cuttings are difficult to root, which is why most growers leave the task to specialists, located mainly in Florida and California, that are able to take advantage of high heat and high light. Tip cuttings measuring 3" (8 cm) long or green stem sections with multiple leaf nodes can be rooted in Oasis or a medium under mist at 75°F (24°C).

For a 72-cell pack, using two cuttings per cell, it will take from three to five weeks to root.

Growing On

Growers have two main choices on how to grow bougainvillea: (1) buy in prefinished 4" (10 cm) pots or (2) buy in liners. Liners will yield the best-quality, largest baskets and pots, but they must be brought in around September to November. Bringing in 4" (10 cm) pots in December or early January allows you better use of your greenhouse space, but you will have less latitude in forcing the crop into flower.

Pot one 4" (10 cm) plant/6" (15 cm) pot and three 4" (10 cm) plants/12" (30 cm) basket; use four 2.5" (6 cm) liners/10" (25 cm) basket. Be extra careful when dislodging bougainvilleas. As their roots are fine, they cannot take rough handling. If using liners, it will take from eighteen to twenty-four weeks from planting to the baskets before they are ready to sell.

Use a well-drained, disease-free medium with a moderate initial starter charge and a pH of 5.5–6.2. Choose bark mixes over peatlite mixes. Plant liners and 4" (10 cm) pots at the same depth at which they are currently growing.

Immediately after planting, keep plants in a shaded area for seven to ten days to allow plants to become established if growing in a high light area. After ten days, move them to full sun. Maintain temperatures at 70–85°F (21–29°C) days and 65–70°F (18–21°C) nights. At temperatures over 90°F (32°C), flowering is retarded.

Your goal is to get the plant to root in rapidly and begin to flush with vegetation. Some believe that bougainvilleas bloom more rapidly and prolifically when they are pot-bound. Root-bound plants should flower fairly continuously. After repotting, flowering is delayed.

Use a constant liquid feed of 200 ppm from 20-10-20 and supplement with trace elements once a month. Bougainvilleas are heavy users of magnesium, iron, and manganese. If leaves yellow, occasionally add 50 ppm magnesium sulfate. However, be careful—excessive fertilizer will promote vegetative growth without flowering.

Pinch plants two to three weeks after planting and again in four weeks. If you want, you can pinch a third time for shaping. European research has shown that two applications of benzyl adenine (BA) at 50–100 ppm can help increase branching and flowering. (Use 200 ppm for more vigorous varieties.) The first application is made the day after the first pinch; apply BA a second time after the second pinch.

Drought stress and/or short days and a cool dormancy period are thought to control bougainvillea flowering; however, the picture is unclear. Plants bloom on new growth. If you brought in liners, begin dormancy in November. Pinch plants, leaving 8–10" (20–25 cm) on each break. Dormancy should last eight weeks or longer. Night temperatures can go as low as 40°F (4°C). Keep soil moist but not wet. Do not fertilize. Plants will lose part of their leaves during dormancy.

To trigger flowering, raise temperatures, move plants to full sun (you can move baskets to the greenhouse attic), and begin fertilizing. Allow four to six weeks for plants to flush out prior to sale. Once plants begin to flower, do not subject them to drought stress, as buds may drop.

Once flowering is over, you can reflower bougainvillea over and over again. As plants near the end of their bloom cycle, allow them to go into a period of moisture stress. Each day allow plants to wilt slightly and then water them sparingly. Continue this procedure for five to seven days. This will trigger the plants into reflowering.

Pests & Diseases

Aphids, fungus gnats, spider mites, and caterpillars can become problems. Botrytis and *Pythium* can infect this crop.

Varieties

'Barbara Karst' is a red and one of the best-known bougainvilleas. Good choices for basket production are 'Afterglow' (yellow-orange), 'Alabama Sunset' (orange), 'Raspberry Ice' (pink with variegated leaves), 'Purple', and 'Mary Palmer' (white).

Postharvest

Ship bougainvillea cool at 37°F (3°C) for up to six days. Bougainvillea is sensitive to ethylene, which will cause flower and leaf drop. Silver thiosulfate can be applied the week prior to shipping.

Consumers will find their bougainvilleas become vegetative during the heat of the summer, but they will reflower again in the fall. Consumers can overwinter their bougainvilleas in a garage or basement. Before temperatures drop to 40°F (4°C), they should bring the plants in and cut them back to 8–10" (20–25 cm). They should keep soil moist but not wet. When temperatures are consistently 50°F (10°C) at night, consumers should bring plants out of storage and begin watering and fertilizing to trigger growth.

Bracteantha

Bracteantha bracteata (also called *Helichrysum bracteatum*)
Common name: Strawflower
Annual

The bright, paper-thin flowers are the highlight of this crop and a must-have for any gardener who harvests from the garden to create crafts. Plants range in height from 10" to 3' (25–91 cm) and are available in a wide range of colors. Flowers are closed during cloudy or wet, rainy weather, although there are a number of recently developed varieties that flower more readily under these conditions.

Propagation

Both vegetative and seed varieties are available. Seed varieties are common on the dwarf and tall selections, while the vegetatively propagated varieties are more commonly bedding varieties that are intermediate in height.

Seed

Sow seed onto a well-drained, disease-free, germination medium with a low starter charge and a pH of 5.8–6.2. Cover lightly with coarse vermiculite to maintain moisture at the seed coat level; however, do not keep the medium saturated. Germinate at 70–72°F (21–22°C). Radicles will emerge in five to six days.

Move trays to Stage 2 for stem and cotyledon emergence, which will take from five to seven days. Provide substrate temperatures of 68–72°F (20–22°C), reducing moisture levels once radicles are emerged. Allow the medium to dry out slightly before irrigating. Fertilize with 50–75 ppm from 14-0-14 once cotyledons are expanded. Alternate feed with clear water. Always irrigate early in the day to ensure leaves are completely dry by nightfall; *Bracteantha* is susceptible to a number of foliage diseases that require freestanding water on the foliage. Increase light to 1,000–1,500 f.c. (11–16 klux).

Stage 3, growth and development of true leaves, takes from twenty-one to twenty-eight days for 288s. Lower soil temperatures to 65–68°F (18–20°C). Provide air temperatures of 70–75°F (21–24°C) days and 65–68°F (18–20°C) nights. Increase feed to 100–150 ppm, alternating between 20-10-20 and 14-0-14, fertilizing at every second or third irrigation. If needed, Bonzi can help control plant height. Increase light levels to 1,500–2,500 f.c. (16–27 klux).

Harden plugs in Stage 4 for seven days prior to transplanting by maintaining soil temperatures but lowering air temperatures to 65–68°F (18–20°C) days and 60–65°F (16–18°C) nights. Allow plugs to dry thoroughly between irrigations. Apply 100–150 ppm from 14-0-14 once a week.

If sowing to a 288 plug allow four to five weeks from sowing until plugs are ready to transplant to larger containers.

Cuttings

Stick cuttings within twelve to twenty-four hours of arrival. Cuttings can be stored overnight, if necessary, at 45–50°F (7–10°C). Maintain soil temperatures of 68–74°F (20–23°C) until roots are visible. Roots will initiate faster and more uniformly with an application of 250 ppm of indole butyric acid (IBA) applied to the base of the cutting prior to sticking. Foliar applications of IBA have also proven to be effective; however, rates are much lower for foliar applications (50–75 ppm). The most critical factor at this stage is to avoid wilting in the cuttings. Wilting of cuttings in the first five to seven days will lead to a lack of uniformity in root initiation. Frequent misting at short intervals is important in avoiding oversaturated media. Overmisting will cause leaching and delay root development. Begin fertilization with 75–100 ppm nitrogen (N) when roots become visible. Increase to 150–200 ppm N as roots develop. As the rooted cuttings develop, high light, moderate water stress, and proper air temperatures should eliminate the need for chemical plant growth regulators (PGRs). Pinching should not be necessary during propagation. Rooted cuttings should be ready for transplanting twenty-one to twenty-eight days after sticking.

Growing On

Plant one plant per 4" (10 cm) pot. Select a well-drained, disease-free medium with a high initial starter charge and a pH of 5.8–6.2. Provide full-sun conditions (5,000–9,000 f.c. [54–97 klux]). Flowering is more profuse as light levels increase. Fertilize at every other irrigation with 200–250 ppm from 20-10-20, alternating with 15-0-15. Avoid ammonium-based fertilizers to keep foliage compact. Watch for iron and boron deficiency on plants as well as iron/manganese or phosphorous toxicity. Avoid fertilizers containing phosphorous. Extended periods of excessive moisture will lead to yellowing of the foliage (i.e., poor uptake of iron). Leach pots periodically to prevent salts buildup.

Run plants as dry as possible without risking permanent wilt. Plants may collapse if the medium is wet for extended periods of time. Do not allow mature plants to dry out, or you risk plant death. Water early enough in the day so that foliage is entirely dry by nightfall to prevent foliar disease.

Grow on relatively cool at 65–72°F (18–22°C) during the day and 55–60°F (13–16°C) at night. Cool nights encourage compact growth and flowering.

Plants respond well to Bonzi if growth regulators are required. *Bracteantha* is also highly responsive to a negative DIF.

Keep plants spaced out as foliage begins to touch on the bench. Pinch vegetatively propagated varieties two weeks after planting to encourage breaks on production of larger pots of containers.

The crop time for a 4" (10 cm) pot with a 105 or similar liner to finish is six to eight weeks (budded or flowering); a 288 plug for a seed variety will be about the same. In this same crop time, dwarf-flowering seed selections like 'Chico' will be budded or in flower, while taller selections will be salable green. Taller selections will flower in the garden during late June or July.

Strawflower is also a popular field-grown cut flower. Plant plugs or direct-sow, thinning plants to 12" (30 cm) apart. Do not transplant seedlings with flowers present. Harvest flowers before they are fully open, just as centers are showing. For drying, remove all foliage, bunch and tie stems, and hang upside down in a dry, warm, and well-vented barn or warehouse. Try 'King Mix', a series with extremely large flowers.

Pests & Diseases

Aphids, caterpillars, leaf miners, thrips, and whiteflies can attack plants. Diseases that can strike include *Alternaria,* botrytis, *Pythium,* and rhizoctonia.

Troubleshooting

Botrytis stem canker or wet medium that precipitated a *Pythium* attack can cause plant collapse. Low light, short days, or warm night temperatures can result in slow or weaken flower initiation. Wet medium and/or high salts levels can lead to yellowing or dying foliage. At low light levels, plants can stretch.

Postharvest

Retailers will have the most success by displaying plants in bright, filtered light. Do not allow plants to dry out on retail shelves. Also, keep plants well spaced in displays—place pots so that leaves just barely touch. Instruct consumers to plant strawflowers in a well-drained bed or container in the full sun.

Brassica

Brassica oleracea var. *italica* (broccoli)
Brassica oleracea var. *gemmifera* (Brussels sprouts)
Brassica oleracea var. *capitata* (cabbage)
Brassica oleracea var. *botrytis* (cauliflower)
Brassica oleracea var. *acephala* (kale and collards, including flowering, or ornamental, cabbage and kale)
Brassica oleracea var. *gongylodes* (kohlrabi)
Cold-tolerant vegetable

This culture includes many of what is collectively called the "Cole Crops" and encompasses cabbage, kale, collards, Brussels sprouts, and kohlrabi. It also includes the so-called flowering (or ornamental) cabbage or kale. Flowering, or ornamental, cabbage and kale are not grown for their flowers but for their brightly colored foliage. They are often separated from the vegetable section of most catalogs and placed along with flowers, since they are commonly used as an ornamental and not eaten. However, many are used to adorn salad bars, as an edging or similar use around a plate to provide color around a salad, etc. All ornamental cabbage and kale are botanically classified under kale.

Brassica is one of the most important vegetable families in the world. Every major culture and cuisine of the world includes brassicas as dietary staples. They prefer to be grown in mild or cool weather and are grown throughout the spring and summer in the North and during fall and winter in the South.

Propagation & Growing On

Sow seed to a 512-plug tray onto a well-drained, disease-free, germination medium with a pH of 5.5–5.8 and a low starter charge. Cover seed with coarse vermiculite or medium after sowing. Keep trays uniformly moist, but not fully saturated. Seed will germinate in three to four days at 65–70°F (18–21°C).

Move trays to Stage 2 for stem and cotyledon emergence. Lower temperatures to 62–65°F (17–18°C) and increase light levels to 1,000–2,000 f.c. (11–22 klux). As radicles begin to penetrate soil, begin to allow trays to dry down somewhat between irrigations. Apply 50–75 ppm once a week from 14-0-14 as soon as cotyledons are fully expanded. Stage 2 will last about four to seven days.

Move trays to Stage 3 to develop true leaves. Maintain temperatures at 62–65°F (17–18°C) and increase light levels to 1,500–2,500 f.c. (16–27 klux). Use a weekly feed of 100–150 ppm alternating between 20-10-20 and 14-0-14. Stage 3 will take about ten to fourteen days for 512s and one to two weeks longer for 288s.

Harden plugs for seven days prior to sale or transplanting by lowering temperatures to 60–62°F (16–17°C). Allow trays to dry thoroughly between irrigations and increase light levels to 2,500–3,500 f.c. (22–32 klux). Use a weekly feed of 100–150 ppm from 14-0-14. Plugs are ready to transplant to larger containers four to five weeks after sowing.

Transplant one plug per cell or pot. Grow on at 55–60°F (13–16°C) days and 50–55°F (10–13°C) nights. Transplanted to a 32-cell flat, plants are ready in six to eight weeks; seven to nine weeks in a 4" (10 cm) pot.

Ornamental cabbage & kale

Once ornamental, or flowering, cabbage or kale is toned, it can tolerate frost during production (or when planted in the garden). Ideally, finish off your ornamental crop at low temperatures to enhance colors. Fertilize once or twice a week with 150–200 ppm from 20-10-20, alternating with 15-0-15. Do not overfertilize your ornamental crop, as leaf coloring will not be as vivid. Manage the crop dry to avoid disease problems.

If growth regulators are required for your ornamental cabbage or kale crop, Sumagic and B-Nine are effective. Sumagic provides good control at rates from 2–8 ppm (northern growers use lower rates). B-Nine at 2,500 ppm applied twice (two weeks apart) or 5,000 ppm applied once is also effective. Either can be applied beginning fourteen to twenty-one days after transplanting.

Cut foliage (cut flower) cabbage varieties

If you are a cut flower grower and are trying the new cut flower cabbages, plan on a mid-June sowing in the Midwest and Northeast. Growers elsewhere should sow about seventy-five days prior to the onset of consistently cool night temperatures (50–60°F [10–16°C]).

Pests and Diseases

Watch for aphids, cabbage loopers, flea beetles, whiteflies, and worms during production. The worst disease problems are *Pythium* and rhizoctonia.

Varieties

Vegetables

Here we offer just a short list of vegetable varieties in brassicas for you to try. Use this list combined with input from your customers to determine which varieties to offer.

Broccoli 'Premium Crop' forms 8–9" (20–23 cm) heads with small, firm buds on thick, tender stems. It is a sixty-five-day crop.

Brussels sprouts 'Jade Cross E' is an improved version of 'Jade Cross' and features densely packed sprouts that are easy to harvest. It is a ninety-day crop.

Cabbage 'Discovery' is a very sweet, standard green cabbage that makes gigantic 3–4 lb. (1.4–1.8 kg) heads! Plants are resistant to *Fusarium* yellows. It is a seventy-five-day crop. Cabbage 'Stonehead' makes nearly perfectly round 2.25 lb. (1 kg) heads that have excellent holding ability. These plants are also resistant to *Fusarium* yellows. It is a fifty-day crop.

Cauliflower 'Early Snowball A' (also called 'Super Snowball') produces smooth, pure-white, 5–6" (13–15 cm) heads that weigh 2–3 lbs. (0.9–1.4 kg). It is a sixty-day crop. Cauliflower 'Snow Crown' is a vigorous grower with pure-white heads that weigh in at about 2 lbs. (0.9 kg). It is excellent fresh or frozen and is a fifty-day crop.

Ornamental cabbage & kale

How do you tell ornamental, or flowering, cabbage and kale apart? It's easy, except both are really kale. Flowering cabbage has rounded foliage, while flowering kale has crinkly leaves and generally has two very distinct colors with a bright center that is surrounded by gray-green or purplish leaves on the outside. Some people just refer to flowering kale as wavy-leaf flowering cabbage.

Be sure to offer 'Dynasty' flowering cabbage (also called the 'Osaka' series). Available in 'Pink', 'Red', 'White', and a formula 'Mix', 'Dynasty' produces heads of tightly held, compact swirls of leaves that are semi-wavy. Plants grow 11–12" (28–30 cm) tall.

For flowering kale, try 'Emperor' (also known as the 'Nagoya' series), which is offered in 'Red', 'Rose', 'White', and a 'Mix'. Plants are extremely symmetrical and uniform, growing to 3–6" (7–15 cm) tall. For feather-leafed kale, try the 'Peacock' series in either 'Red' or 'White'. If landscapers are important to you, be sure to try the fine-leafed kale 'Flamingo Plumes', with hot pink and deep purple coloring.

If you are also into cut flowers and want to try a really cool, novel crop for fall sales, try cut flower flowering cabbage. The F1 'White Crane', with its soft cream and bridal-pink interior leaves surrounded by soft green outer leaves, looks like a thick-stemmed novel rose! It is ideal for adding interest to mixed bouquets. 'Sunset Red' and 'Sunrise White' are open-pollinated cut flower types. All grow to about 24" (61 cm) tall. Sow or plant out cut flower cabbage varieties fourteen to twenty-one days before planting ornamental cabbage bedding plants. Space plants 12" (30 cm) apart in rows. If you direct-sow, thin plants to 1 plant/12" (30 cm). Provide one to two layers of support, which is important to keep stems growing straight. You can also mound dirt around the base of stalks to keep them upright. Remove lower leaves on stems when plants reach 6" (15 cm) tall, and repeat this process several times as stems elongate. Side-dress your fertilizer early in the crop and make sure that it is depleted by the time leaves start to color up. Plants should be nearly mature by the time cool nights start. To get coloring, provide cool 50–55°F (10–13°C) nights for three to four weeks.

Bromeliaceae

The Bromeliaceae family includes the following genera: *Aechmea, Billbergia, Cryptanthus, Dyckia, Guzmania, Neoregelia, Nibularium, Tillandsia,* and *Vriesea*
Common name: Bromeliad
Pot plant

Bromeliads, members of the Bromeliaceae family, are native to tropical and subtropical areas of South, Central, and North America. Texas and Florida have many native bromeliads that are protected by conservation laws. Spanish moss and ball moss are two other bromeliads found across the southern states. The pineapple is the most familiar bromeliad. There are more than two thousand recognized bromeliad species, as well as hundreds of hybrids. Each offers something unique for the grower, the retailer, and especially the interiorscape.

Bromeliads have long been favorite houseplants in Europe, having been grown in greenhouses there for the past two hundred years. Since the early 1980s, bromeliads have become increasingly popular in North American homes and other interiors. This can be attributed to three very strong selling points. First, bromeliads are true eye-catchers, as they are available in a wide range of colors and often sport a combination of colors. Second, they require very little maintenance and thrive on neglect. Third, with reasonable care, bromeliads will bloom on the retail shelf or in the home for at least three and as long as four to six months (sometimes more).

Bromeliads range in height from several inches up to 40' (12 m). The most commonly sold varieties range in height from 10–30" (25–76 cm) and are sold in 4–8" (10–20 cm) pots. Color development is commonly rated as low, medium, and high, depending on the maturity of the flower bract.

A colorful, well-developed bromeliad retails from $4 to as much as $40, generally determined by a plant's size and desirable characteristics. A large, rare bromeliad often retails for $75 or more. Due to the wide range of colors available with bromeliads, you can tailor specific ones to the holidays: red for Christmas; orange and yellow for Thanksgiving; and pink, blue, and peach for Easter and Mother's Day.

Bromeliads are found growing naturally in diverse environments, from rain forests to cool mountains to hot, dry deserts. Most bromeliads available today are tropical, although there are a few desert types offered as well. The shape, size, and color of the flower, as well as the culture, depend on the original habitat, breeding, and species of the plant.

Most bromeliads are naturally epiphytic, clinging to tree limbs, trunks, and rocks for support. Most commercially grown bromeliads have a center rosette of leaves from which a brightly colored flower-bearing spike (bract) grows. On other varieties, a portion of the leaves near the center of the plant and surrounding the small flowers at the center change color to brilliant red, blue, or purple. These brightly colored bracts attract hummingbirds as pollinators. Many bromeliads have vase-like, watertight leaves that hold water and organic material such as leaves and bugs. As the material breaks down, the bromeliads translocate the nutrients and water through sophisticated trichomes (scales/hairs) onto the leaf surfaces in the cup.

Propagation

Only specialists propagate bromeliads. Specialist breeders, located primarily in the Netherlands and Belgium, grow plants from seed. Some varieties are also available from major commercial meristem young plant suppliers. Most growers buy in seedling or meristem liners from specialist young plant suppliers.

Plants are usually produced under contract. To receive the plants you want, you need to order at least six months in advance. Improvements in tissue culture have resulted in a great many additional varieties being introduced to the market. You can expect to see twenty to thirty new hybrid introductions every year. A few years ago, just one introduction a year was normal. This increase in breeding activity will cause bromeliads to grow in importance as a category.

Plants that cannot be grown from seed or meristem liners can often be propagated vegetatively. However, this requires considerable bench space and time. The best source of cultural information is your plant supplier.

After germinating (germination takes from two to fourteen days) the lettuce-sized seed, grasslike seedlings should be transplanted into flats after having grown to 1.5" (4 cm). From flats, plants can be moved to cell packs and eventually into individual containers as they grow larger. The more often a bromeliad is transplanted, the better it grows. For each transplant, provide more room on the bench, more light, and (to a small extent) more nutrients. Do not crowd young bromeliads on the bench, as bromeliads thrive on aeration. Bottom heating has been shown to reduce heating costs and to increase the turnover in the bench space required for commercially growing bromeliads from seed.

Growing On

While most bromeliads are truly epiphytic in nature, most adapt readily to a terrestrial or semi-terrestrial culture. The adaptable bromeliads will grow in a wide variety of media. Sufficient aeration and a slightly acidic pH are the only required conditions. Some favored mixes are: (1) 50% Canadian peat/50% coarse perlite, (2) 33% each of peat, bark, and perlite, and (3) 50% peat/50% rock wool. Coir (coconut fiber) has gained much popularity as a long-lasting bromeliad substrate. In Europe, bromeliads are grown in straight coarse peat. *Cryptanthus* and *Dyckia* are naturally terrestrial, so they will tolerate many forms of medium. Most gray-leafed *Tillandsia* species, however, will not adapt to terrestrial culture. These plants must be mounted on wood, cork, or tree fern slabs to replicate their original habitat.

Do not pot a bromeliad too deeply—just to the base of the leaves, and do not use a pot that is too large for the plant, as the danger of overwatering increases. Usually a 4–6" (10–15 cm) pot is sufficient. Use stable pots or containers, as any rocking or other motion damages the tender, developing roots. Staking may be necessary until roots are well developed.

Plants grown too close together for a long time will develop stretched leaves, and flower size and quality will be reduced. Tight spacing also reduces airflow and allows various scale insects to multiply undetected.

Exposing mature plants to ethylene gas induces flowering. You can dissolve ethylene in water and apply to the plants' water-holding cups or use the commercial product Florel. Label rates are too high, however, so you should do considerable testing to determine the correct Florel rate for your crop.

The plants flower once in their life cycle. After flowering, axial buds are stimulated to produce side shoots. These pups draw nutrients from the mother plant as the mother withers over a period of six months to a year. If the shoots are left on the mother plant, they will grow quite fast, but you will only get one to three new plants. If the pups are harvested at a size of around a quarter of the parent, it is possible to increase the yield to ten or more new plants. The pups normally bloom in one to two years.

To complete a typical bromeliad life cycle, a temperature range of 65–85°F (18–29°C) and humidity between 60–80% are usually required. Tropical bromeliads prefer high humidity. Outside of a few bromeliad types that can withstand temperatures as low as 30°F (–1°C) with protection, most bromeliads cannot withstand night temperatures below 50°F (10°C).

Bright, diffused light and genus-specific care improves growth and plant quality of bromeliads grown for commercial production. Direct sun will burn the leaves of most bromeliads. Generally, the stiffer-leaf varieties tolerate more light and may require very high light to bring out the full color of the foliage. Some soft-leaf varieties, on the other hand, will require 85% shade or more to grow properly. Once flower color is set, bromeliads will tolerate 50 f.c. (538 lux) conditions for months.

Water bromeliads like all plants: by inspection. In a greenhouse environment, the watering interval is between five and fifteen days, depending on environmental conditions affecting plant water use. Water-soluble fertilizer at 100 ppm should be applied with every other watering. The watering should be long enough for the fertilizer to reach all the roots. An additional ten minutes of clear water will dilute the fertilizer in the plant cup. Time-released fertilizer mixed in the coil or top-dressed is a good supplement to liquid feed. Bromeliads require frequent feeding of potassium. They do not store it and require it all the time.

The crop time to produce a plant is from thirty to sixty-five weeks when grown at 80°F (26°C) days and nights at 70°F (21°C).

Pests & Diseases

As epiphytes, bromeliads are very strong plants with few pest problems. Copper-based fungicides are toxic to bromeliads. Bromeliads are extremely sensitive to copper fungicides. One popular fungicide is labeled as an herbicide for Spanish moss (*Tillandsia usneoides*).

Postharvest

In the consumer environment, bromeliads require much less water. They prefer to be on the dry side inside. Instruct retailers and consumers to apply water amongst the lower leaves and soil. Allow the soil to dry between watering. Watering in the center of the cup can cause fungus or bacteria to attack the plant. Overwatering will shorten the flower life.

Bromeliads can be artificially induced to flower once the plant is mature. Florel is the most common way to treat the plants.

Browallia

Browallia speciosa 'Major'
Annual

This warm-season annual has showy, star-shaped, blue-lavender flowers. Most growers produce *Browallia* in hanging baskets, although it also makes a great 4" (10 cm) pot and does fine in larger cell packs.

Propagation

Sow seed into a well-drained, disease-free, germination medium with a pH of 5.5–5.8. Cover seed lightly with vermiculite. Maintain soil temperatures of 75–78°F (24–26°C). Keep trays uniformly moist, but not saturated. Radicles will emerge in seven to fifteen days.

After radicle emergence, move trays to Stage 2 for stem and cotyledon emergence. Lower soil temperatures to 60–65°F (15–18°C) and provide less than 2,500 f.c. (27 klux) of light. Begin fertilizing once cotyledons expand with 50–75 ppm weekly from 14-0-14. Stage 2 takes fourteen days.

Move trays to Stage 3 to develop true leaves. Lower soil temperatures to 68–72°F (20–22°C) and begin to allow trays to dry out between irrigations. Maintain light levels, but increase feed to 100–150 ppm weekly, alternating between 20-10-20 and 14-0-14. If growth regulators are needed, A-Rest, B-Nine, and Bonzi are effective. Stage 3 takes twenty-eight to thirty-five days for 384s or 288s.

Harden plugs off in Stage 4 for seven days prior to transplanting or sale by lowering soil temperatures to 62–68°F (17–20°C). Maintain light levels and water only when trays are dry, but avoid permanent wilting. Fertilize weekly with 100–150 ppm from 14-0-14.

A 288-plug tray takes from five to six weeks to be ready to transplant to larger containers.

Growing On

Transplant into a well-drained, disease-free medium with a moderate initial nutrient charge and a pH of 5.5–6.2. Use 1 plug/cell or 4" (10 cm) pot and 5–7 plugs/10" (25 cm) basket. Keep in mind that *Browallia* requires long days to flower. For early plant sales, offer the plants green with a tag.

Maintain light levels below 2,500 f.c. (27 klux). Fertilize at every other irrigation with 100–150 ppm from 20-10-20, alternating with 15-0-15. Growth regulators will likely be needed. A-Rest, B-Nine, and Bonzi are effective.

Transplant one 288 plug per 4" (10 cm) pot and allow nine to eleven weeks to be sold. Plants will be budded or in flower but will flower more profusely once planted to the garden.

Varieties

Try the 'Bells' series, available in 'Marine', 'Blue', and 'Silver' for 4" (10 cm) or hanging basket production.

C

Caladium

Caladium bicolor
Annual, tender perennial

The bright foliage of caladiums is a mainstay in the tropical southern garden and provides year-round color to malls across America in interiorscapes. Caladiums make colorful houseplants, brighten gardens, and spice up mixed containers in all climates with warm summers.

From a production standpoint, caladiums must have a lot of heat. For that reason, most growers do not begin forcing pots until later in the season to take advantage of naturally rising temperatures. When temperatures are cool, forcing can drag on for eight to twelve weeks. However, with the right temperatures, light levels, and tuber sizes, production can be an easy six weeks.

Growing On

Growers buy in tubers grown by specialist propagators. American-grown caladiums are primarily produced in Central Florida from divided dormant tubers or tissue culture plantlets. A vigorous breeding effort coordinated by the University of Florida is continuing to develop new and improved cultivars.

New crop tubers are harvested in the late fall and are available for shipping in December. Some suppliers store tubers in controlled conditions at 70–80°F (21–27°C) and 40–50% relative humidity and can ship through August. However, vigor decreases with longer storage times. Tubers are generally cured for six to eight weeks after digging and prior to shipping. Tubers that have not been stored for at least six weeks at 70°F (21°C) will sprout slowly. Because caladium tuber availability is tight in many years, it pays to order tubers as early as possible, especially to secure supply of new introductions.

Tubers are sold by size, #2 tubers are 1–1.5" (3–4 cm) in diameter; #1 tubers are 1.5–2.0" (4–5 cm) in diameter; jumbos are 2.5–3.5" (6–9 cm); and mammoths are 3.5" (9 cm) and up.

Tubers that cannot be planted immediately on arrival should be unpacked, placed on screen racks, and stored at 70°F (21°C) in a dry, well-ventilated area. Allowing exposure to temperatures below 60°F (16°C) will cause tubers to have a rubbery texture, display slow sprouting, and produce fewer leaves. Temperatures at 50°F (10°C) or lower, even briefly, can cause irreversible tuber damage.

Use a peat-based medium high in organic matter with good moisture-holding capacity, adequate aeration, and a pH of 6.0–6.5. The medium should be moist, but not saturated at planting. Recommendations for the number of tubers to use per pot are included in table 9.

Table 9. Recommended Tubers Per Pot					
BULB SIZE	POT SIZE				
	3"	4"	5"	6"	8"
#2	1	1	3	5	8
#1		1	2	3	5
Jumbo			1	1	3
Mammoth					1

New growth will come from "eyes" on the tubers. The first leaves come from the larger eyes, while smaller eyes form peripheral leaves that emerge later. Many growers remove the main bud of a #1-grade tuber or higher. Be careful not to cut or damage smaller eyes that will form peripheral leaves. By "scooping" the main bud, production time will increase by three to five days and leaves will generally be smaller. Specific recommendations on de-eyeing newer cultivars can be obtained from your supplier.

Other techniques to achieve the same result include planting the tuber upside down to allow lateral buds to develop uniformly with the center bud for more uniform finished plants. Some growers cut tubers into pieces to obtain more uniform growth; however, this practice increases the chance of disease problems.

At potting, be sure to cover tubers with 2–3" (5–8 cm) of medium because new roots develop on the upper surface of the tuber.

After potting, stack pots into pyramids and cover with plastic sheeting. Put the pots in a warm space at 80–90°F (27–30°C). If such a space is unavailable, use a bench-level heating system to provide warm soil temperatures. Maintain substrate temperatures at 75°F (24°C). Be sure to ventilate pots two to three times a week by briefly removing the plastic. In ten to twenty days, unstack pots and remove plastic sheeting permanently when the first leaf sheaths are visible.

Once leaves have emerged, grow plants warm at temperatures from 70–90°F (21–32°C). Make sure temperatures do not drop below 65°F (18°C). Provide a light intensity of 2,500–5,000 f.c. (27–54 klux) (60–80% shade). Lower light intensities cause stretched petioles, oversized leaves, and weak plants. Higher light intensities fade leaf colors and cause foliage burning. White varieties are produced in light levels at the low end of the recommended range, pink and red varieties at the upper end. Experiment to find the appropriate light level for your greenhouse and the varieties you are producing.

Be sure to keep plants well watered. Plants allowed to dry down can suffer irreversible root death and will never recover. Overwatering, however, causes tuber rot.

Top-dress 14-14-14 slow-release fertilizer at 1 tsp./6" (15 cm) pot, or with liquid fertilizer, 150 ppm nitrogen (N) and 150–200 ppm potassium oxide (K_2O) works well. Too-high nitrogen may cause excessive greening of leaves.

Caladiums are Bonzi responsive. Apply a Bonzi drench at 8 ppm (1 mg/pot for a 6" [15 cm] pot) about twenty-five days after planting (leaves are just emerged, but not unfurled) to moist medium. Pine bark in the medium will reduce Bonzi's effectiveness. B-Nine may be used as a foliar spray and is helpful in production of small containers. Applications at 2,500 ppm can be applied up to three times, spaced one week apart after plants are leafed out.

Pests & Diseases

Aphids, thrips, and, to a lesser extent, mealybugs can be problems on caladiums. As weather warms, two-spotted spider mites may also pose a threat.

Troubleshooting

Dry soil or too much light (foliage burn) usually causes the majority of caladium troubles. If temperatures are too low, tubers grow slowly and erratically. At temperatures below 50°F (10°C), tubers will rot.

Varieties

Leaf type and plant size typically classify caladiums. "Fancy-leafed" varieties have tall (18–22" [46–56 cm]), upright growth and large, fleshy, heart-shaped leaves. They account for 80–85% of the market. "Lance- or strap-leafed" varieties have heavier-textured, leathery, and narrower leaves. Plant growth of lance-leafed varieties is generally more compact with numerous leaves that do not burn or fade as easily in full sun. Lance-leafed varieties produce smaller tubers and are priced one grade above their actual size. Leaves of lance-leafed varieties may be cut and used as cut foliage. "Dwarf" varieties have heart-shaped leaves (like the fancy varieties), but plant size is overall smaller (15–17" [38–43 cm]).

While most varieties require partial shade outdoors, some varieties perform well in full sun, including 'Aaron', 'Candidum Jr.', 'Carolyn Whorton', 'Fire Chief', 'Gingerland', 'Pink Gem', 'Rosalie', and 'White Wing'. For optimum performance, be sure they are planted in a well-drained bed and kept well watered. When planted in full sun, leaf coloration will be heavily green.

White is the most popular caladium color, comprising perhaps 30% of all sales. There are several white varieties, such as 'Candidum', 'Candidum Jr.', and 'White Christmas'.

Be sure to try some of the newer varieties developed by the University of Florida. Among the varieties in commercial cultivation are 'Florida Cardinal', 'Florida Sweetheart', 'Florida Fantasy', 'Florida Elise', and 'Florida Red Ruffles'.

Postharvest

Do not expose caladiums to temperatures below 65°F (18°C) during shipping or retail display, as all plant parts are chilling sensitive. Retailers should wait to offer caladiums that will be sold as outdoor container or garden plants until night temperatures are reliably 70°F (21°C) or above.

Calceolaria

Calceolaria × *herbeohybrida*
Common name: Pocketbook plant
Spring-flowering pot plant

Bright red, orange, and yellow flowers on low-growing plants make *Calceolaria* a good choice as a cool-season flowering pot crop. In southern or West Coast markets, *Calceolaria* is used as an outdoor item in mixed combinations.

Older varieties need a combination of long days and cool nights to initiate flowering. However, some of the newer varieties bloom independent of temperature and day length. These can be grown for Valentine's Day sales or timed to flower for any holiday.

Propagation

For ease, most growers buy in plugs rather than grow their own. Buying in makes a lot of sense since most growers only produce a few hundred pots or less. Smaller growers can mix trays in a minimum order to make an assortment of cool-season crops.

Sow seed from August to early October in a disease-free, well-drained medium with pH of 5.5–5.8. Do not cover. Radicles will emerge after three to five days of 70–75°F (21–24°C) substrate temperatures.

Stage 2, stem and cotyledon emergence, takes seven days. Maintain substrate temperatures and increase light to 500–1,000 f.c. (5.4–11 klux). As cotyledons fully expand, begin fertilizing with 50–75 ppm from 14-0-14 once a week. Allow trays to dry slightly before irrigating. Move trays into Stage 3 for the growth of true leaves by lowering soil temperature to 65–70°F (18–21°C) and increasing light to 1,000–1,500 f.c. (11–16 klux). Allow trays to dry between irrigations, yet be careful to maintain adequate moisture levels, as *Calceolaria* has a very fine root system. Fertilize with 100–150 ppm weekly, alternating between 20-10-20 and 14-0-14. Apply B-Nine for height control beginning now if needed. Stage 3 takes from twenty-one to twenty-eight days. Maintain temperature, light, and fertilizer regimes. Harden plugs for sale or transplanting by lowering substrate temperatures to 60–62°F (16–17°C) and increasing light to 1,500–2,500 f.c. (16–22 klux). Fertilize weekly with 100–150 ppm 14-0-14.

A 288-plug tray sown in late August requires four to six weeks to be ready to transplant to 4" (10 cm) pots.

Growing On

Calceolaria growth can be divided into three main stages: (1) vegetative growth (weeks 1–4 after potting), (2) flower initiation (weeks 5–10), and (3) forcing (weeks 11–20). From transplant to flowering pot takes about nineteen to twenty-one weeks total for newer varieties; older varieties take twenty to twenty-seven weeks (many of these are photoperiodic, requiring long days to bloom).

Pot plugs into a well-drained, disease-free medium with a pH of 5.5–6.2 and a moderate initial nutrient charge. Pot at the same depth as in the plug tray. Do not plant too deep, as plants are subject to crown rot. Space plants as leaves begin to touch.

Maintain light levels at 2,000–3,500 f.c. (22–38 klux). Light shade may be required as light levels increase in the spring. Allow pots to dry thoroughly before irrigating, although not to the point of permanent wilting. Avoid wetting plants and flowers during irrigation, as botrytis is a constant threat. Make sure water alkalinity is below 140 ppm and EC is below 0.5 mS/cm.

Maintain temperatures as follows: vegetative growth, 65–68°F (18–20°C) days/60–62°F (16–17°C) nights; flower initiation, 48–52°F (9–11°C) days/45–50°F (7–10°C) nights; and forcing, 65–68°F (18–20°C) days/60–62°F (16–17°C) nights.

Fertilize for weeks 1–4 with 150–200 ppm 20-10-20, changing to 150 ppm 15-0-15 for weeks 5–10. Medium EC should be 1.2 mS/cm or less.

If needed, B-Nine (1,000–1,500 ppm) and Cycocel can be used for height control.

Pests & Diseases

Aphids, fungus gnat larvae, and whiteflies can attack *Calceolaria*. Botrytis, INSV/TSWV, and *Pythium* can also strike plants.

Troubleshooting

Overwatering, high pH or soluble salts, low fertility, and root rot can cause leaf chlorosis. Necrotic spots on leaves that show up after plants have been shipped may be due to INSV. To prevent INSV, control thrips, which spread the disease. Boron deficiency can result in stunted plants and strapped, hardened leaves that may also be distorted and mottled. It can occur during hot, sunny weather when growers are forced to water more frequently, thus leaching boron. Also, incorrect pH levels and high calcium can tie up boron, making it unavailable to the plant. High media pH can cause yellow upper leaves, stunted or uneven plant growth, and poor root development.

Too many weeks above 60°F (15°C) prior to bud initiation or light levels that are too high for flowering may result in large, soft, flowering plants. Not enough 60–62°F (15–17°C) nights before cold treatment or low fertility may lead to flowering plants that have not sized up. Botrytis and overhead watering can cause flowers that rot before sale. Light intensities over 5,000 f.c. (54 klux) can result in sun-scalded leaves.

Varieties

'Anytime Mixture' will flower in about eighteen to twenty weeks from sowing at 60°F (15°C), regardless of day length and without special cooling.

Postharvest

Plants are ready to sell when four to eight flowers are open. Be careful to avoid packing situations in which you will facilitate botrytis development. For example, water plants the night before packing and do not pack warm plants. Groom plants by removing dead flowers or leaves (sources of ethylene). Plants can be sleeved and boxed, provided that shipping is less than six days. Ship at temperatures from 41–60°F (5–16°C). Flowers are sensitive to ethylene; silver thiosulfate (STS) or 1-MCP may be helpful. Flower drop can occur if plants are kept in the dark for too long.

Retailers should display *Calceolaria* in 50–100 f.c. (0.5–1.1 klux) of light at 65–70°F (18–21°C). Avoid high temperatures, as plants will figuratively melt down. Do not get leaves or flowers wet when watering.

Consumers should place *Calceolaria* in a bright, cool window. Using a saucer and watering from the base will help keep plants uniformly moist, yet not wet. Plants that are allowed to dry out over and over again will die in short order. With good light and moisture, *Calceolaria* will flower for many weeks in a home environment.

Calendula

Calendula officinalis
Common name: Pot marigold
Annual

Superbright, petal-filled flowers on plants that, once established, can take a light frost make calendula a possibility for early spring sales in the Midwest and North and for fall sales in the Deep South. In southern areas with hard freezes, plants are not reliable during the coldest weeks.

Calendulas must have cool temperatures to shine; they cannot take heat and humidity. They are real stars in California, where the climate is tailored to suit them. Calendulas can reseed in the garden, so plants that died out over the hot summer in the North may spring back to life in the fall.

In the early 1900s calendula was a very important seed crop. George J. Ball, author of the first *Ball RedBook* and founder of the Ball Horticultural Company, established his seed business in the 1920s with his outstanding strains of 'Ball Orange' calendula, then an important greenhouse cut flower worldwide.

Propagation

Sow seed into a well-drained, disease-free medium with a pH of 5.5–5.8. Cover with medium or coarse vermiculite. Maintain good moisture levels around the seed and provide soil temperatures of 70°F (21°C). Radicles will emerge in three to five days. Move trays to Stage 2 for stem and cotyledon emergence by increasing light to 500–1,500 f.c. (0.5–16 klux) and beginning a weekly feed of 50–75 ppm from 14-0-14 as soon as cotyledons have expanded. Stage 2 lasts from seven to nine days. Stage 3, growth of true leaves, will take from five to seven days. Lower soil temperatures to 65–70°F (18–21°C) and increase light levels to 1,500–2,500 f.c. (16–27 klux). Feed weekly with 100–150 ppm alternating between 20-10-20 and 14-0-14. If growth regulators are needed, A-Rest, B-Nine, and Cycocel may be used beginning in Stage 3. Harden plugs for seven days prior to transplant or sale by lowering soil temperatures to 60–65°F (16–18°C) and allowing trays to dry thoroughly between irrigations. Feed weekly with 100–150 ppm from 14-0-14.

A 288-plug tray requires from four to six weeks from sowing to be ready to transplant to larger containers.

Growing On

Calendulas are large plants and will be relatively vigorous when they begin to flower. For this reason, transplant in 36-cell packs or larger or 4" (10 cm) pots. Select a disease-free, well-drained medium with a moderate initial starter charge and a pH of 6.0–6.5. Fertilize at every other irrigation with 150–200 ppm 15-0-15, alternating with 20-10-20.

Maintain high light levels (4,000–6,000 f.c. [43–65 klux]). If height control is necessary, A-Rest, B-Nine, and Cycocel are effective.

If you are growing calendulas as a cut flower crop, you will need a layer of support. The window for planting a field crop in most parts of the country will be narrow: Plants must be planted after danger of frost, yet you have to harvest before temperatures are too high. It is best to order plugs in early, grow the plants in containers in the greenhouse, and then plant into fields once the danger of frost is past. Flowers will be able to be harvested for several weeks before the heat gets them.

Plants will tolerate cold temperatures (to 45°F [7°C]), but guard against problems with disease and rots from overwatering at low temperatures. For cool-temperature production, space at a minimum of 10" × 10" (25 × 25 cm). Otherwise, space at 8" × 8" (20 × 20 cm).

If growing in 4" (10 cm) pots, plants will flower in ten to twelve weeks from a plug for dwarf varieties and twelve to fourteen weeks for taller selections. Plan on thirteen to fourteen weeks if you sow to a flat and transplant individual seedlings for 4" (10 cm) pots.

Pests & Diseases

Aphids and whiteflies can attack calendula. Plants can also succumb to root rot.

Varieties

Try the 'Bon Bon' series, available in 'Orange', 'Yellow', and a 'Mix' of the two.

Calibrachoa

Calibrachoa × *hybrida*
Annual

Calibrachoa has become one of the most significant crops in vegetative annuals. Home gardeners enjoy the vivid colors of a plant that is perfectly suited to the small spaces and container gardens that are popular today. The leading breeding companies in this crop have greatly improved its characteristics for plant habit, flower timing, and pH sensitivity, improving the appearance and production of the crop from the greenhouse to performance on the patio. The recent introduction of the double calibrachoa 'MiniFamous' series offers an exciting innovation to this growing product class.

Propagation

Choose a well-drained medium with an EC of 0.75–0.80 mS/cm and a pH of 5.4–5.8. Open boxes upon arrival. Stick cuttings within twelve to twenty-four hours of arrival. Cuttings can be stored overnight, if necessary, at 45–50°F (7–10°C). Maintain soil temperature at 68–73° F (20–23°C) until roots are visible. Avoid applying too much mist in propagation.

Begin fertilization with 75–100 ppm nitrogen (N) when roots become visible. Increase to 150–200 ppm N as roots develop. Once roots are visible, keep media moist but never saturated. This will prevent iron deficiency symptoms and chlorotic foliage, which can develop. Maintain media pH below 6.0.

As roots develop, apply appropriate water stress and moderate air temperatures to reduce the need for chemical plant growth regulators (PGRs). A spray application of B-Nine/Cycocel tank mix (with starting rates of 1,500 ppm B-Nine and 750 ppm Cycocel) applied twelve to fourteen days after sticking will reduce unwanted stretch and promote early branching. Pinch calibrachoa eighteen to twenty-four days after sticking to promote branching and improve habit. Rooted cuttings should be ready for transplanting twenty-four to twenty-eight days after sticking.

Apply a second application of B-Nine/Cycocel as laterals emerge following the pinch, usually in seven to ten days. This application can be made prior to or one week following transplant. Apply a broad-spectrum fungicide drench during the second or third week in propagation. Harden fully rooted cuttings for seven days prior to sale or transplanting by lowering air temperatures to 70–75°F (21–24°C) days and 62–68°F (16–20°C) nights. Raise light levels to 2,000–4,000 f.c. (22–43 klux).

If sticking unrooted cuttings to a 105 or 84 tray, they will be ready to transplant to 4" (10 cm) or similar containers in three to four weeks. A rooting hormone is not required.

Growing On

Use a well-drained, disease-free, soilless medium with a pH of 5.4–5.8. Routinely test the media every fourteen days to maintain this pH range throughout production. Early signs, such as yellowing foliage, can be the first indicators of the need to lower the soil pH to avoid iron deficiency.

Once roots are established, maintain nighttime temperatures of 50–58°F (10–14°C) and daytime temperatures of 71–76°F (21–24°C). Higher than recommended temperatures will cause poor branching, unwanted stem stretch, and reduced flowering. Suggested night temperatures will create maximum branching and the best possible habit. Maintain light intensities at 5,000–8,000 f.c. (54–86 klux). Low light levels cause stem stretch and poor flowering. Flowering improves as day length increases. Generally, flowering will begin in mid-spring and will be heaviest from late May to September. Crop times will increase under short day length. An autumn crop is possible if the crop is started early enough to allow for flower initiation before days shorten significantly. For fastest flowering during short day length, maintain night temperatures at 59–61°F (14–16°C) and use lighting to provide a day length greater than twelve to thirteen hours. Night-break lighting can be used.

Plants are susceptible to botrytis, so avoid high humidity and wet foliage. Calibrachoa are susceptible to root diseases if overwatered. Allow the media to dry slightly between watering, but avoid any wilt. Provide plants with adequate horizontal airflow at all times.

Calibrachoa require heavy fertilization. Use constant feed with a balanced fertilizer at 225–300 ppm N with additional iron as needed. Provide a full complement of minor elements. Use clear water with every third watering if high soluble salt problems occur.

As for crop time, one 84 liner transplanted to a 4" (10 cm) pot will be salable in six to nine weeks; although more free-flowering plants are seen eight to nine weeks after potting. For baskets, use three to five plants per 10 or 12" (25 or 30 cm) basket and plan on eight to twelve weeks for salable plants.

Media pH management

Monitor plants regularly for early, visual signs of upward pH drift (interveinal yellowing on youngest leaves). Regular soil pH tests are an excellent way to identify movements in pH before they create visual symptoms, which can be difficult to correct. Periodic application of acidic feed or drench applications of a chelated iron product will maintain appropriate pH levels and dark green foliage. An effective method of lowering pH is a soil drench of iron sulfate. Rinse foliage immediately after treatment since the iron sulfate solution can result in phytotoxicity to flowers and foliage. Early monitoring and active management of soil pH through irrigation water and fertilizer solution are a critical part of growing a successful calibrachoa crop.

Pinching

Pinch plants back seven to fourteen days after transplanting to improve basal branching. Plants can be pinched as the crop matures to improve their habit, but flowering will be delayed approximately two to three weeks. In trials, Florel has proven effective for increasing branching when applied one to three times at 250–500 ppm to stress-free, actively growing plants. Flowering will be delayed a minimum of seven to eight weeks, depending on the concentration used. Improved branching, darker green foliage, and shorter internodes will be the benefits.

Controlling growth

Use high light and cool temperatures to control growth. The 'Cabaret' series responds well to DIF in production. If necessary, growers can use one or more applications of B-Nine (1,500–3,000 ppm) starting two weeks after transplant. Calibrachoa growth can also be controlled with one to two spray applications of Topflor (5–10 ppm) or drench applications of Bonzi (3–8 ppm). Sumagic (20–30 ppm) can effectively control the growth of calibrachoa when applied once or twice as a spray. Applied when plants first reach salable size, Bonzi drenches (1–8 ppm) slow growth, maintain a tight habit, and allow normal flower development. PGRs applied late in the crop cycle can delay flowering one to two weeks. Applications should be avoided once flower buds appear. These PGR recommendations should be used only as general guidelines. Growers must trial all chemicals under their particular conditions.

Pests & Diseases

Aphids, fungus gnats, leaf miners, thrips, and whiteflies can become problems on calibrachoa crops. Disease problems can include botrytis, powdery mildew, *Pythium,* rhizoctonia, and *Thielaviopsis* as well as a number of devastating viruses. Because calibrachoa are susceptible to several viruses, it is vital to begin with cuttings supplied from clean stock.

Troubleshooting

Plant collapse can be a symptom of keeping plants wet or root rot caused by disease. High ammonium, too much fertilizer under low light conditions, and/or low light and overwatering can result in excessive vegetative growth. Poorly branched plants can be due to low fertilizer. Low light levels will cause plants to stretch. Iron deficiency/high pH levels bring on chlorosis.

Varieties

Variety selection is a key part of calibrachoa's success. Plant habits and flower timing can vary within series and from color to color. From the original 'Million Bells' to today's 'MiniFamous', a wide range of colors and habits are available. 'Callie', 'Cabaret', 'Aloha', 'Noa', and 'Superbells' are all readily available in the market.

Calla Lily (see *Zantedeschia*)

Callistephus

Callistephus chinensis
Common names: Asters; China asters
Cut flower, annual

During the late 1800s and early 1900s, asters were a staple of the American cut flower scene. George J. Ball, author of the first *Ball RedBook* and founder of the Ball Horticultural Company, first grew them in 1901 and bred single-stem asters in the 1930s and 1940s. During the 1950s, asters were grown extensively outdoors and continued to be a major cut flower. Today asters are enjoying a revival as a cut flower thanks in large part to new varieties that have come onto the market in a wide range of new colors. As a bedding plant, asters were a primary garden plant of the 1800s and early 1900s but are a diminishing crop because of their brief flowering period.

Propagation

Sow seed onto a disease-free, well-drained, germination medium with a low starter charge and a pH of 5.5–5.8. Premoisten medium prior to sowing and then manage trays dry throughout the plug stage. Do not cover. Provide 100–400 f.c. (1.1–4.3 klux) of light for more uniform germination. Seed will germinate in four to five days at 68–70°F (20–21°C).

Move trays to Stage 2 for stem and cotyledon emergence once at least half of the seed is germinated. Maintain soil temperatures at 68–70°F (20–21°C). Increase light levels to 500–1,000 f.c. (5.4–11 klux). Once cotyledons are fully expanded, begin a weekly feed with 50–75 ppm from 14-0-14. Continue to manage trays dry. Stage 2 takes about seven days.

Once you have a good stand, move trays to Stage 3 to develop true leaves. Reduce temperatures to 65–68°F (18–21°C) and increase light levels to 1,500–2,000 f.c. (16–22 klux). Fertilize once a week with 100–150 ppm, alternating between 20-10-20 and 14-0-14. Manage trays dry! Stage 3 will take about twenty-one days in 512s. *Note:* Begin long days (fourteen to sixteen hours) once four true leaves have formed.

Move trays to Stage 4 for hardening for seven days prior to sale or transplanting. Lower temperatures to 60–62°F (16–17°C). Increase light levels to 2,500–3,000 f.c. (27–32 klux). Apply a weekly feed with 100–150 ppm from 14-0-14. Manage trays dry. Continue to apply long days.

Plug production for a 200 tray will take four to five weeks.

Growing On

Transplant bedding plant crops into a well-drained, disease-free medium with a moderate starter charge and a pH of 5.5–6.2. Plant one plug per jumbo pack or 4" (10 cm) pot.

Grow the crop on at 62–65°F (17–18°C) days and 60–62°F (15–17°C) nights. Provide the highest light levels possible while maintaining these temperatures. Fertilize at every other irrigation with 150 ppm, alternating between 15-0-15 and 20-10-20. Continue to provide long days until plants are well budded (early May for spring crops in the North). A 4" (10 cm) crop will take approximately fifteen to sixteen weeks from 512s in the spring and twenty to twenty-two weeks in the fall.

Plants range in height from 10–12" (25–30 cm) for bedding selections all the way to 2–3' (61–91 cm) tall for cut flower strains.

Plants are easily susceptible to overwatering due to their fine root system. Pinching is not generally required.

Cut Flowers

Plant plugs into well-drained, pasteurized soil that has been cultivated to break up hardpans. Asters are highly susceptible to *Fusarium,* so be sure beds are disease free prior to planting. Adjust pH to 6.0–7.0. Space plugs at a density of 65 plants/yd.^{2} (78 plants/m^2). Provide one to two layers of support for field crops; greenhouse crops generally will not need it. Fertilize with 125–150 ppm from 20-10-20, alternating with 14-0-14 through a drip irrigation system. Lower rates as flower buds appear for best vase life.

Grow the crop on at 60–85°F (15–30°C) days and 50–65°F (10–18°C) nights.

Flowering is affected by the ratio of long days to short days. Increasing long days will increase stem length at the expense of a longer crop time. Provide long days (sixteen hours) until plants have received a total of ten to thirteen weeks of long days in the field or other unheated situations and six weeks in the greenhouse. After the long-day treatment, provide short days. Using this combination—long days followed by short days—will save about two weeks of forcing time compared to providing long days continually.

Some growers pinch the primary bud to develop more uniform flowering of laterals.

Growers in California and Florida can produce field-grown asters in the fall and winter. In more northern climates, plugs can be planted outdoors from April through mid-July. If planted to the field in spring, plants naturally flower in August or September.

Asters are a natural for spring sales. To quote Vic Ball from the thirteenth edition of the *Ball RedBook* (1975), "Many growers have written us to say that their spring greenhouse aster crops were the best-paying crops that they had on their place. While asters can be flowered any month of the year, the spring and early summer months seem to be the most profitable. The only important requirements are 50°F (10°C) night temperatures and mum lighting."

Pests & Diseases

Asters are susceptible to many diseases, including *Coleosporium* (rust), *Fusarium,* and *Phytophthora.* Leafhoppers can also transmit the phytoplasma *Chlorogenus,* widely known as aster yellows, one of the most difficult diseases in flower production. If aster yellows strikes, the only control is to rogue infested plants. Place them immediately in plastic bags and do not allow workers handling infected plants to touch plants not showing symptoms. Rotate beds; ideally do not plant asters back-to-back to help minimize problems. Keep leafhoppers under control to eliminate the risk of disease spread. Other insect pests include aphids, leaf miners, and spider mites.

Varieties

For bedding plant production, try 'Milady Mix'. Plants grow 10–12" (25–30 cm) tall. 'Pot 'n' Patio', great for pot production but a poor garden performer, flowers without supplemental lighting! The series is also available in separate colors. Plants grow 6" (15 cm) tall.

The leading cut flower aster series is 'Matsumoto', a medium-flowered series available in a wide color range. 'Matsumoto' is a spray form that is well suited to field or greenhouse production and has become a staple cut flower around the world. It is sold in grower bunches and mixed bouquets.

Try the 'Meteor' series for its large, 3.5–4.0" (9–10 cm) flowers and excellent plant vigor. 'Meteor' is well suited for field and greenhouse production. The series is available in separate colors as well as a mix. Production time for 'Meteor' is about two weeks longer than 'Matsumoto'.

Postharvest

Harvest cut stems when two to three flowers begin to show color. Pulse with silver thiosulfate (STS) to increase vase life. Strip the lower leaves during processing as they decay in holding solution and the resulting bacteria block stems. Store or ship at 32–34°F (0–1°C). Store stems in water treated with a bactericide and sucrose solution. Treated properly with STS and preservatives, flowers can last ten to fifteen days.

Campanula

Campanula carpatica
Common names: Carpathian harebell; Carpathian bellflower
Perennial, pot plant (Hardy to USDA Zones 3–8)

Campanula carpatica is an important summer flowering pot crop for a number of European growers. Debbie Hamrick remarked in the seventeenth edition of the *RedBook*, "We've seen acres of it in flower at Madsen's in Odense, Denmark—an unforgettable sight."

In North America the plants make an excellent perennial in the home garden. Plants are reliable in areas where the soil stays somewhat moist all summer yet is well drained. Flowering slows in the summer heat but returns in the fall when temperatures drop. Plants are low, growing to 8–12" (20–30 cm), and massed with bright blue or white, bell-shaped flowers in the early summer when days are long and temperatures mild.

Propagation

Plants can be propagated by seed or cutting.

Seed

Sow pelleted seed onto a well-drained, disease-free, germination medium with a low starter charge and a pH of 5.5–5.8. Cover seed very lightly with coarse vermiculite to maintain humidity at the seed-coat level; however, do not germinate seed in the dark. Provide 100–400 f.c. (1.1–4.3 klux) of light. Keep trays uniformly moist but not saturated. Seed will start to germinate in four to seven days at soil temperatures of 68–72°F (20–22°C). However, raw seed may not emerge all at one time and will take from ten to fourteen days after sowing to germinate more readily.

Once 50% of the seed has germinated, move trays to Stage 2 for stem and cotyledon emergence. Maintain soil temperatures at 68–72°F (20–22°C). Increase light levels to 500–1,000 f.c. (5.4–11 klux). As soon as cotyledons have expanded, begin fertilizing with 50–75 ppm from 14-0-14 once a week. Allow trays to dry down somewhat between irrigations, but be aware that campanula does not tolerate drought conditions. Stage 2 will take from seven to fourteen days.

When you have a good stand, move trays to Stage 3 to develop true leaves. Lower soil temperatures to 65–68°F (18–20°C) and increase light levels to 1,000–2,000 f.c. (11–22 klux). Fertilize once a week with 100–150 ppm from 14-0-14, alternating with 20-10-20. Again, allow trays to dry down somewhat between irrigations, but not to drought conditions. If growth regulators are needed, A-Rest, B-Nine, and Cycocel are effective. Stage 3 will take from twenty-one to twenty-eight days for 288s.

Harden plugs for seven days prior to sale or transplanting by lowering soil temperatures to 60–65°F (16–18°C). Allow trays to dry between irrigations, but do not allow permanent wilting. Increase light levels to 2,000–3,000 f.c. (22–32 klux). Fertilize weekly with 100–150 ppm from 14-0-14.

Avoid exposing plugs to long days (sixteen-plus hours of light) during the seedling stage to prevent flower initiation.

A 288-plug tray will take from eight to eleven weeks depending on variety. The 'Clips' series takes longer to fill out a plug tray, while some of the newer selections are faster to germinate and grow.

Cuttings

Plants can be vegetatively propagated from cuttings taken anytime the plants are vegetative and rooted in medium under mist. A 128- or 105-liner tray will take from four to six weeks for most varieties, especially if propagated in early summer for potting to 1-qt. or 1-gal. (1.1 or 4 l) containers. If propagated under the darker days of autumn, it might take a week or two longer.

Growing On

Select a disease-free, well-drained, soilless medium with a moderate starter charge and a pH of 5.8–6.2. Plant one plug or liner per cell, 4" (10 cm), 1-qt. (1.1 l), or 1-gal. (4 l) pot. You can use from one to three plugs per 1-gal. (4 l) or 6" (15 cm) pot, depending on the size of the plugs and the desired finished pot size.

After potting, establish plants at warmer temperatures, 70–75°F (24–29°C) days and 63–65°F (17–18°C) nights, for the first two to three weeks for liners/plugs brought in during the fall or winter. If potted up in late summer, use ambient temperatures to establish. Bulk plants that have eight to ten leaf nodes prior to forcing or overwintering. A 128 plug should have this number of leaves when fully developed and should require only two to three weeks of bulking. A 50-cell liner should have from twelve to fifteen leaves and can theoretically be put into forcing once plants are rooted in. Fertilize with 100–150 ppm from 20-10-20 at every irrigation. Discontinue fertilizer for fourteen to twenty-one days prior to sale for increased postharvest life. Grow plants under short days (less than twelve hours of daylight) to prevent plants from flowering. If you are forcing plants into flower for spring sales and still need to bulk plants on April 1, use shadecloth to provide short days. Provide 2,000–3,000+ f.c. (22–32+ klux) of light.

Once plants are bulked, lower temperatures for finishing to 60–68°F (16–21°C) days and 55–58°F (13–14°C) nights.

Campanula is a long-day plant. For winter production as a flowering pot plant, apply lighting to attain sixteen-hour days, which are required for flowering. Light for sixteen hours per day or use four-hour night interruption lighting, until buds are fully developed. Flowering is not as uniform when cyclic night lighting is used or when night lighting is less than four hours. Begin lighting ten to fourteen days after the start of forcing (plants should have a minimum of fifteen leaves); flowering should occur about thirty to forty-five days later. Do not begin applying long days until plants with eight to ten leaf nodes have been bulked up in their final containers for two to three weeks. Size your plants before initiating flowering for forced crops.

In Denmark, crops are generally potted in summer, grown outdoors, and then forced into flower for sale as a flowering pot plant beginning in the winter. These plants flower naturally in the field and are allowed to go dormant with falling autumn temperatures and shortening day lengths. If you choose to grow the crop outdoors for a portion of its production cycle, it is a good idea to drench with fungicides and/or insecticides when you bring pots indoors for forcing.

For winter-forced crops, plant quality will be higher with an additional 400–500 f.c. (4.3–5.4 klux) of supplemental light applied during the day in northern latitudes.

If you are growing *Campanula carpatica* as a perennial for spring sales, pots can be planted up from September through February. For fall planted crops, grow plants cool at 45–50°F (70–10°C) nights. Spring-planted crops that will be used in jumbo packs or 4" (10 cm) pots can be grown on at 50–55°F (10–13°C) nights. Higher production temperatures will cause plants to flower faster. Your goal is to build the plant before allowing flowering.

A cold treatment is not required for campanula flowering. However, pots can be overwintered outdoors or in unheated cold frames or greenhouses.

Do not allow the medium to dry out—campanulas do not tolerate drought. Avoid overhead irrigation if at all possible in order to prevent botrytis problems, especially once plants are in flower. If you must irrigate from overhead, do so at a time when foliage will be completely dry by nightfall.

Keep plants spaced during production, never allowing leaves to touch.

A-Rest, B-Nine, Bonzi, and Cycocel are all effective growth regulators and can be applied beginning five to six weeks after planting, usually when plants have reached one-half to three-fourths their desired finished size. Do not apply growth regulators to plants that are flowering, as it will cause spots on blooms.

If plants do not sell in the spring when they are ready, cut them back. They should reflower again in six to eight weeks.

Pests & Diseases

Aphids, fungus gnats, spider mites, and thrips can attack plants. Diseases include botrytis, leaf spot, *Pythium,* and rhizoctonia. To prevent problems with disease, be sure to use a well-drained medium and keep foliage as dry as possible. Avoid application of pesticides when plants are in flower.

Postharvest

Pots are ready to sell once they are well budded and show one open flower. Be sure pots are well watered prior to shipping to prevent flowers from inverting. For maximum postharvest life, treat with silver thiosulfate (STS) ten to twelve days prior to shipping. EthylBloc can also be used.

Ship cool; boxed campanula shipped warm will rot. Campanulas do not ship well, so they are best suited to growers catering to a local or regional market.

Campanula isophylla
Common names: Italian bellflower; Star of Bethlehem
Annual

The delightful flowers of this annual campanula make a great hanging basket or large pot. The crop time is really long: twenty-five to twenty-seven weeks from sow to sell, making it one of the longest of all seed-grown plants for spring sales. However, if you have ever seen these plants in flower in the spring, you know why growers take the time to produce them: They are spectacular! Annual campanula can also be grown in 4" (10 cm) pots, saving about three weeks of production time. However, since you will have so many bench weeks invested in it, you will want to select a container size that allows you to charge a premium for your crop.

Propagation

Sow seed from October to December for flowering pots in April and May. Seed can also be sown from February to April for crops to be sold from July to September. Sow two to three seeds per cell of a plug tray onto a well-drained, disease-free, germination medium with a low starter charge and a pH of 5.5–5.8. A 200 or other large plug is suggested over smaller sizes. Do not cover. Provide 100–400 f.c. (1.1–5.4 klux) of light. Seed will germinate in seven to ten days at 60–65°F (16–18°C). Keep trays uniformly moist but not wet.

Move trays to Stage 2 for stem and cotyledon emergence. Maintain 60–65°F (16–18°C) temperatures and increase light levels to 500–1,000 f.c. (5.4–11 klux). Fertilize once a week with 50–75 ppm from 14-0-14. Once radicals emerge, begin to allow the medium to dry down ever so slightly. However, be aware that campanula cannot tolerate drought conditions. Stage 2 will take from fourteen to twenty-one days.

Once you have a full stand, move trays to Stage 3 to develop true leaves. Lower temperatures to 60–62°F (16–17°C) and raise light levels to 1,000–2,500 f.c. (11–27 klux). Fertilize weekly with 100–150 ppm, alternating between 20-10-20 and 14-0-14. Stage 3 will take approximately thirty-five to forty-two days for 200s.

Harden plugs at 58–60°F (14–16°C) for seven days prior to sale or transplanting. Allow trays to dry down between irrigations; however, do not allow plants to reach permanent wilting. Increase light levels to 2,500–3,500 f.c. (27–38 klux). Fertilize once a week with 100–150 ppm from 14-0-14. If growth regulators are needed, B-Nine and Cycocel are effective.

A 200-plug tray will take from eleven to thirteen weeks to be ready from sowing.

Growing On

Transplant one plug per 4" (10 cm) pot or four to five plugs per 10" (25 cm) hanging basket. Select a disease-free, well-drained, soilless medium with a moderate initial starter charge and a pH of 5.5–6.2.

Grow on at 65–70°F (18–21°C) days and 58–60°F (14–16°C) nights. Because campanulas require cool production temperatures, do not hang baskets in the attic or under gutters unless they are in an unheated greenhouse that is automatically ventilated.

Allow medium to dry between irrigations; however, be aware that campanula does not tolerate drought conditions and plants that are allowed to dry down continually will yellow, brown, and eventually die. Fertilize with 150–200 ppm at every other irrigation, alternating between 20-10-20 and 14-0-14. Keep foliage dry if at all possible. If you must irrigate from overhead on occasion, do so only when it is possible to ensure leaves are fully dry prior to nightfall.

Campanula isophylla is a long-day plant, requiring days longer than twelve hours. Once fifteen true leaves appear, provide a four-hour night interruption of light for thirty evenings to initiate blooms. From bud set it will take twenty-five to thirty-five days for plants to start flowering.

If growth regulators are needed, B-Nine and Cycocel are effective. However, while plants have a "hairy" look early on, they grow out of this with increasing day lengths and cool night conditions.

From a plug to a finish 4" (10 cm) pot takes twelve to fourteen weeks. The total crop time from seed to a finished 4" (10 cm) pot is twenty-four to twenty-seven weeks.

Pests & Diseases

See the Pests & Diseases section under *Campanula carpatica* (p. 295).

Varieties

The main selection in this class has been 'Stella', but it was discontinued while this edition was being written. The culture provided here for this genus and species can be applied to other *Campanula isophylla* varieties.

Related Species

Campanula longistyla is another seed-propagated, annual, flowering campanula. 'Isabella', a blue flowering variety that grows from 6–8" (15–20 cm) tall, highlights this campanula species. Like *Campanula isophylla*, it requires long days to flower. However, the similarity in production ends there. The crop time from seed to a salable 288-plug tray is six to eight weeks, and the total crop time from seed to a finished 4" (10 cm) pot is seventeen to twenty-two weeks, including a long-day period of six to eight weeks.

Campanula medium
Common name: Canterbury bells
Annual, cut flower

Historically classified as a biennial, *Campanula medium* has undergone a number of breeding enhancements that allow it to flower as an annual or hardy annual. It tolerates light to heavy frosts but not a prolonged, cold winter. There are a few varieties that produce a crown of foliage in one year and then flower the following year after being exposed to a period of cold (or winter).

With its upright, upward-reaching stalks of bell-shaped flowers, this campanula is in a class all its own as a cut flower. The 'Champion' series has put this delightful flower on the map since it does not need cold temperatures in order to flower. The crop culture presented here pertains to the 'Champion' series.

Propagation

Sow pelleted seed onto a well-drained, disease-free, germination medium with a low starter charge and a pH of 5.8–6.2. Select larger plug trays such as 288s, 200s, etc. Cover the seed very lightly with coarse vermiculite to maintain humidity at the seed-coat level. Provide 100–400 f.c. (1.1–4.3 klux) of light during germination. Germinate at 65–68°F (18–20°C) substrate temperatures. Keep trays moist but not fully saturated. Germination will occur in seven to ten days.

Once 50% or more of the seed has germinated, move trays to Stage 2 for stem and cotyledon emergence. Increase light levels to 500–1,000 f.c. (5.4–11 klux) and maintain air temperatures of 68–72°F (20–22°C). Keep trays moist, but allow them to dry somewhat between irrigations once radicles have emerged; however, campanula does not tolerate drought conditions. As soon as cotyledons are expanded, apply 50–75 ppm from 14-0-14 once a week. To prevent foliar disease, make sure foliage stays dry. Do not allow temperatures to rise, as plants may rosette. Stage 2 will take approximately ten days.

When you have a good stand, move trays to Stage 3 to develop true leaves. Fertilize once a week with 100–150 ppm, alternating between 14-0-14 and 20-10-20. Maintain air temperatures of 68–72°F (20–22°C). Increase light levels to 1,500–2,500 f.c. (16–27 klux). Allow trays to dry somewhat between irrigations, however not to drought conditions. Stage 3 will take from ten to thirteen days in a 200 tray.

Harden plugs for seven days prior to sale or transplant by reducing temperatures to 62–65°F (17–18°C) days and increasing light levels to 3,000–4,000 f.c. (32–43 klux). Use a weekly feed of 100–150 ppm from 14-0-14.

Keep plugs under short-day conditions (eight-hour day lengths) throughout plug production to prevent premature flowering.

Plug production will take about five to six weeks. Plugs are ready to transplant when they have two to three true leaves. Plants will stall in production if they are allowed to become root-bound.

Growing On

Plant plugs for single-stem production in the greenhouse from 4–6" (10–15 cm) apart. Multistem production (i.e., pinched production) is best accomplished with outdoor plantings or in a cold frame. Plant multistem crops 10–12" (25–30 cm) apart. Yield will be about eight to ten stems per plant. Provide one to two layers of support.

Once plants have sized up (eight to ten leaves, four to five weeks after transplanting), begin long days. (*Note:* Prior to long days, be sure plants are growing under short days of eight hours so they are sized up prior to flowering. Flowering response will be more uniform.) Apply sixteen-hour days by using four hours of night interruption lighting for forty to forty-five days. If you are flowering your crop in the late spring through the early summer, lights are not required.

Keep 'Champion' well irrigated: Campanula does not tolerate drought conditions. However, make sure that the medium is very well drained, as campanulas do not like wet conditions.

The ideal production temperatures are constant 55–60°F (13–16°C) days and nights. Avoid night temperatures lower than 50°F (10°C). Growers in mild climates who are able to maintain these production temperatures can flower 'Champion' year-round using black cloth.

If you are in a temperate to mild winter area, for a niche product in the market, try planting plugs outdoors in the fall and overwintering them. Sakata's Russ Gillum reports that some growers have produced 40" (1 m) spikes that are spectacular. However, trial this procedure first before using it on a large crop.

Pot Plants

Plant plugs from a 200 tray into 6" (15 cm) pots. Select a disease-free, well-drained, soilless medium with a moderate starter charge and a pH of 6.0–7.0. Establish the plants at 68°F (20°C) days and nights. Fertilize with 100–150 ppm from 20-10-20 or another balanced fertilizer.

Begin forcing plants by dropping temperatures to 50–55°F (10–12°C) and giving long-day treatment (sixteen hours) with night interruption lighting. Apply long days for two weeks to initiate budding.

About seven weeks after potting, apply a 2 ppm drench with A-Rest and raise average temperatures to 59°F (15°C). Pots should begin to flower about five weeks later.

Pests & Diseases

See the Pests & Diseases section under *Campanula carpatica* (p. 295).

Troubleshooting

Tip distortion and abortion may be caused by boron deficiency. Iron chlorosis will show as leaf tip burn. Tip burn can also show up if plants are allowed to dry out. Stems that suddenly become crooked during production may be showing a response to drying out

Varieties

The series that has created this class as a cut flower is 'Champion'. Both 'Champion Blue' and 'Champion Pink' are Fleuroselect Gold Medal award winners. The series is high yielding and quick to finish.

Postharvest

As a cut flower, 'Champion' is ready to harvest when two to three lower buds are open. Place stems in tepid water immediately after harvest. Pulse for twenty-four hours with a sucrose postharvest solution prior to sale.

While stems have the best quality and vase life when they are moved immediately into the market, stems can be stored upright in water at 36°F (2°C) for three weeks, or stored dry at 36°F (2°C) for up to two weeks. Stored flowers will last from six to nine days. Flowers not treated with postharvest solution have a fourteen-day vase life if they are not stored.

'Champion' buds continue to open after harvest, even in the cooler! However, flowers will shrivel up if they are exposed to ethylene gas. Treating with silver thiosulfate (STS) or EthylBloc is highly beneficial. A treatment should render stems protected from ethylene for up to seven days.

Retailers should recut stems and treat with a sucrose postharvest solution. Ideally, do not use 'Champion' in floral foam, as stems take up large amounts of water.

Canna

Canna × *generalis*
Common names: Canna; canna lily; Indian shot
Tender perennial (Hardy from USDA Zones 7–11)

Long admired as one of the tallest plants in the annual garden in the northern United States and as a perennial in more southern locations, cannas are one of the few upright plants with various foliage colors (including green, red, green/white striped, and green/yellow striped) and with a number of flower colors (including red, yellow, orange, pink/rose, and white). While the green-leafed varieties come in all of these flower colors, the variegated- or bronze-foliage types are more limited. Plants can range in height from 2.5–6' (76–183 cm) tall by mid- to late summer.

Propagation & Production

Seed, tubers, and tissue culture are the preferred ways to propagate cannas. However, with the recent virus issues plaguing tuber varieties, an increasing number of commercial propagators are moving toward either seed or virus-indexed plants through tissue culture.

Seed

The popular 'Tropical' series is the only seed-propagated canna. Seed can be sown in early winter for flowering plants in late spring and summer.

Soak seed for ten to fourteen hours prior to sowing. Pour warm water (as warm as your own skin can handle) over the seed late in the day and let stand overnight. Sow the following morning. Soaking helps to soften the seed coat and improve overall germination.

Sow seed to a large plug tray (128 to 72 cells per flat) using one seed per cell with a soil temperature of 70–75°F (21–24°C). Cover seed with peat moss. Seed germinates in eight to twelve days. A 128-plug tray is ready to transplant to larger containers in four to six weeks.

After being transplanted to 6" (15 cm) or 1-gal. (4 l) containers, plants are salable in mid-May to early June when grown at 65–75°F (18–24°C) daytime and 60–65°F (16–18°C) nighttime temperatures. In addition the plants will have developed foliage but no flowers. 'Tropical' needs from ninety to one hundred thirty days from sowing to flowering, but this depends on warm conditions (70–75°F [21–24°C] days).

Tissue culture/liners

A number of commercial propagators offer rooted liners of plants produced from tissue culture. These are initially disease free and sold as rooted liners in 32, 52, and other liner sizes.

Potted up to 1-gal. (4 l) during winter or early spring, these plants will be rooted and salable in seven to ten weeks when grown at 65–75°F (18–24°C) days and 60–64°F (16–18°C) nights. Plants will not be in flower but, after planting to the garden, will be in bloom in three to four weeks.

Tubers

Using two- or three-eyed tubers, pot up to 6" (15 cm) or 1-gal. (4 l) pots. Plant the tubers 2–3" below the soil line and grow at 68–75°F (20–24°C) days and 65–68°F (18–20°C) nights until shoots begin to emerge. Once the plants begin to develop, grow at 65–75°F (18–24°C) days (for rapid growth) and 60–65°F (16–18°C) nights.

Plants will be ready for sale in seven to nine weeks and will be salable green with no foliage color. Tuber-grown plants that have been grown warm will flower in eleven to thirteen weeks. However, they are upright in stature and will fall over, so provide support to keep the plants vertical. Green plants are salable seven to nine weeks.

Troubleshooting

Warm temperatures are key for cannas. Equally, long days help to enhance growth as well as keep internodes short. Warm temperatures with cloudy conditions cause stretching. Growing on with cooler night temperatures helps to keep plants in check.

Capsicum

Capsicum annum
Common names: Pepper; ornamental pepper
Annual, fruiting pot plant, vegetable

This culture covers ornamental peppers that are used as pot, garden bedding, or landscape plants as well as the traditional vegetable pepper. Both are from the same genus and species. In all fairness, ornamental peppers can be used as edible peppers, if preferred. However, while they are as hot in taste to a vegetable pepper, they are more desired as an ornamental in the United States and Canada because of their multicolored fruit and smaller size.

Peppers are increasing in importance as a bedding plant with the new breeding of ornamental varieties, which thrive in the summer and make a colorful splash of fruit color in the garden; a few also have colorful foliage. Ornamental varieties are also great as a novel pot plant, especially in the fall.

As a bedding plant vegetable, they are already an important crop and becoming more so as America's love affair with hot peppers heats up. Be sure to offer a range of edible peppers to appeal to a discerning American appetite.

Propagation

Germinate seed on a disease-free, well-drained medium with a pH of 5.5–5.8. Single-sow to a 288-plug tray and cover seed. The root radicle will emerge in five to seven days at a substrate temperature of 70–75°F (21–24°C). Many pepper varieties are treated—some with fungicide and others with a hot water dip—to help prevent fungus attacks in the seedling stage.

Move trays to Stage 2, which will last seven to ten days, by increasing light levels to 1,000–2,500 f.c. (11–27 klux) and allowing trays to dry between irrigations. Use a weekly application of 50–75 ppm 14-0-14.

Stage 3, growth of true leaves, will last from ten to fourteen days. Lower soil temperature and increase feed to 100–150 ppm once a week from 14-0-14. Maintain light levels and continue to allow trays to dry between irrigations. Harden trays prior to sale or transplant for seven days by lowering substrate temperatures to 60–62°F (16–17°C). Maintain light and moisture levels and continue weekly feeds.

A 512-plug tray takes from four to five weeks to be ready to transplant to larger cells, while a 288 takes five to six weeks to be ready. One additional point: Unlike tomatoes and other fast-germinating crops, some pepper varieties germinate quickly, while others take another week or two to be ready.

Transplant into a disease-free, well-drained medium with a pH of 5.5–6.3 and a moderate initial starter charge. Grow on at 60–70°F (16–21°C) days and 60–65°F (16–18°C) nights. If transplanting a 288 plug to a 32-cell pack, plants will be salable green in eight to ten weeks. For fruiting plants allow fourteen to eighteen weeks from transplanting. Fertilize at 150–200 ppm with 15-0-15 at every third irrigation. Manage plant height by withholding fertilizer and water and running a negative DIF.

The crop time for a 288-plug tray varies. Most will be ready in four to six weeks from sowing in winter for spring sales. However, dwarf or slow-growing selections will take a little longer.

Pot Plants

Transplant plugs into 4 or 6" (10 or 15 cm) pots filled with a disease-free, well-drained medium that has a pH of 5.8–6.2 and a moderate initial starter charge. Use 3 plants/6" (15 cm) pot. Keep medium uniformly moist: Avoid plant wilting, especially during flowering and fruit set, when plants are extremely sensitive. Water stress at the wrong time can cause fruit to abort. Because plants require so much water, be very mindful of disease development.

The total crop time for a fruiting 4" (10 cm) pot takes from eighteen to twenty-four weeks when sown in spring for late summer/fall sales. Growth can be separated into two distinct parts: (1) transplant to flowering (eight to twelve weeks) and (2) flowering to fruit (five to six weeks). Grow on at 70–75°F (21–24°C) days and 62–65°F (17–18°C) nights. Maintain light levels at 4,000–6,000 f.c. (43–65 klux).

Fertilize three to four times a week with 150 ppm from 15-16-17 and maintain EC below 0.5 mS/cm. High fertilizer levels can inhibit flowering and fruit set, so reduce fertilization several weeks prior to sale.

Ornamental peppers can be top or crown pinched when there are four to five true leaves, but this is more common for older varieties. The newer varieties do not require pinching, although as a pot plant some growers feel that a top pinch helps to make a more mounded plant.

Pests & Diseases

Phytophthora blight is a serious problem for seedlings, while botrytis can attack plants in production. Pest problems can include aphids, mealybugs, spider mites, thrips, and whiteflies. Worms and grasshoppers can infest plants grown outdoors.

Table 10. How Hot Is Hot?*

Below is a ranking by hotness of hot peppers sold by Ball Seed. In descending order, from very hot (a rating of 4) to mild (a rating of 1). Peppers grown under stress and/or high heat and humidity will be hotter than peppers grown under moderate climatic conditions.

VARIETY	HOTNESS
'Caribbean Red Hot'	4
Habanero	4
'Hottie'	4
'Ancho San Martin'	3
'Balada (Kung Pao)'	3
'Serrano'	3
'Super Chili'	3
'Tabasco'	3
'Apache'	2
'Big Bomb'	2
'Burning Bush'	2
'Cayenne Long Thin'	2
'Cheyenne'	2
'Chichimecca'	2
'Early Jalapeño'	2
'Holy Molé'	2
'Hot Portugal'	2
'Hungarian Yellow Wax'	2
'Ixtapa'	2
'Mucho Nacho'	2
'Nainari'	2
'Red Cherry Hot–Large'	2
'Tajin'	2
'Tiburon'	2
'Alma Paprika'	1
'Biggie Chili'	1
'Garden Salsa'	1
'Mariachi'	1

** Information supplied by Ball Horticultural Co. from the Ball Seed Company catalog, 2010–2011.*

Troubleshooting

For vegetable bedding plants, low moisture or insufficient calcium levels can cause blossom end rot. Too much moisture may lead to fruit abortion. Peppers have shallow roots and are very sensitive to moisture extremes after fruit has set. Fruit can become sun scalded—light-colored and wrinkled—after too much intense sunlight.

For ornamental peppers, fruit that does not develop adequately for the consumer can be due to plants being sold too early or low light levels in the interior environment.

Varieties

Vegetable peppers are classified into several groups: bell peppers, which include green, yellow, red, and pimiento peppers; Anaheim chili peppers, which include red chili, paprika, and cayenne peppers; jalapeño peppers; cherry peppers; wax peppers; and tabasco peppers (*C. frutescens*). If you are a larger bedding plant grower, ask your plug supplier to custom sow special pepper varieties for you: Your customers will love you for it! In addition to growing a couple of bell and sweet pepper varieties, literally spice up your bedding plant assortment by selecting some hot peppers from table 10.

For standard bell peppers, 'California Wonder Select' is the old standby. For variety, try 'Better Belle', 'Lady Bell', and 'Red Beauty'. 'Fat 'n' Sassy' is one of the largest bell peppers ever!

For ornamental peppers, try 'Black Pearl' with black foliage on plants up to 18" (46 cm) tall. 'Purple Flash' has plants growing up to 15" (38 cm) tall and features flashes of bright purple, ivory, and black on its foliage. 'Explosive Ember' grows up to 12" (30 cm) tall and is a striking plant with purple leaves and purple fruit.

Prefer a chill to these hotter selections? 'Medusa' or the All-America Selections Award winner 'Chilly Chili' are just the thing. While many ornamental peppers have fruit that is too hot to handle, these varieties have fruit that is mild, making them child-safe and suitable for outdoor plantings in public spaces. Fruit starts out ivory and then turns yellow, then orange, and finally red. One plant can display the entire range of colors at once!

Related Species

Solanum pseudocapsicum (commonly called Jerusalem cherry or Christmas cherry) used to be a common sight at Christmas. However, the popularity of poinsettias has helped to crowd them off retail shelves.

Sow seed from January to early March for transplanting to 6" (15 cm) pots. Grow in cold frames or a similar environment to allow for crosspollination to get fruit set. By mid- to late summer plants will be dotted with green, 0.5" (1 cm) berries. Bring back into a warm greenhouse or cold frame as outside temperatures cool and grow on at 58–62°F (14–17°C) nights. Fruit will start to turn reddish orange, and by late November plants are salable.

These make excellent houseplants and can be grown year-round in the home and placed outside in the summer. (This helps the flowers to pollinate and produce fruit.) As plants age, they become woody at the base. The only negative is that fruits are not edible, which has contributed to this plant's reduced favor as a pot plant.

Postharvest

Once fruit has set on ornamental peppers, lower temperatures to 65–75°F (18–24°C) days and 60–65°F (15–18°C) nights. Take care during packing and shipping not to break fruit from stems. Partial shade in the retail display may help slow down plant drying and preserve quality. Do not ship boxed ornamental peppers for longer than three days.

Carex

Carex comans
Common name: Leatherleaf sedge
Grass, tender perennial (Hardy to USDA Zones 7–9; treated as an annual in cold-winter climates)

Ranging in height from 10–16" (25–40 cm) tall, *Carex comans* has fine, elongated leaves that grow out of clump into an arching habit. Plants prefer full-sun locations. Foliage colors are available in bronze tones (brown to some) as well as green or green/chartreuse.

Propagation & Growing On

Seed

Seed sown to a 72-liner tray using several seeds per cell (or a multiseeded pellet), lightly covered with vermiculite in January or early February, and germinatedat 70–72°F (21–22°C) will emerge in five to twelve days. The liner trays will be ready to transplant to larger containers in seven to ten weeks.

Transplanted to 18-deep cell trays or 4" (10 cm) pots when ready, plants will be salable in May to early June.

If using a 288-plug tray, use a multiseeded pellet with one pellet per cell and cover lightly with coarse vermiculite. Germinated at 70–72°F (21–22°C) seedlings will emerge in six to ten days. Trays will be ready to transplant to larger containers in five to seven weeks.

Transplanted to 18-deep cell trays, plants will be salable in five to eight weeks when grown at 55–58°F (13–14°C) nights and 65–72°F (18–22°C) days.

Varieties

This species is highlighted by varieties like 'Bronco', with bronze-brown foliage and an arching habit on plants that grow from 10–14" (25–36 cm) tall and spread from 13–16" (33–41 cm) across. Also, 'Amazon Mist', a variegated, light and medium green variety with a twisting habit to its leaves. Plants grow from 10–14" (25–36 cm) tall.

Related Species

Carex morrowii is a variegated variety growing from 12–16" (30–41 cm) tall in flower and blooms in spring. It is hardy to USDA Zones 5–9 but needs to be protected in Midwestern gardens to survive. One of the more popular varieties in this class is 'Ice Dance'. It is commonly available from a number of professional propagators from autumn to spring in liner trays of thirty-two to seventy-two plants per tray.

Propagated by division of the crown, plants are separated during late summer and early fall using a clump of shoots with a root to the tray sizes noted above. Plants can be propagated in the spring as well.

If purchased from a professional propagator in winter, plants are potted to 4" (10 cm) pots for spring sales.

Carex flagellifera
Common name: Carex
Grass, tender perennial (Hardy to USDA Zones 7–9; treated as an annual in cold-winter climates)

Ranging in height from 14–16" (35–40 cm) tall, *Carex flagellifera* has fine, elongated leaves that grow out of clump into an arching habit. Plants prefer full-sun locations.

Propagation & Growing On

Seed sown to a 72-liner tray using several seeds per cell. Lightly cover with vermiculite in January or early February and germinate at 70–72°F (21–22°C). Seedlings will emerge in five to twelve days, and the liner trays will be ready to transplant to larger containers in seven to ten weeks.

Transplanted to 18-deep cell trays or 4" (10 cm) pots when ready, plants will be salable in May. For a 288-plug tray, follow the culture under *Carex comans* for similar results.

Varieties

This species is highlighted by the popular variety 'Bronzita' with bronze-brown foliage and a semi-arching habit. Plants are slightly taller and have a broader leaf than 'Bronco.'

Carnation (see *Dianthus*)

Catharanthus

Catharanthus roseus
Common names: Vinca; Madagascar periwinkle
Annual

Few bedding plants have experienced as many advancements as *Catharanthus*, commonly called vinca (but not to be confused with vinca vine), in the past ten years. While notable improvements were detailed in the seventeenth edition of the *Ball RedBook*, the amount of achievements has continued to upgrade the class.

While growers in the South had always offered vinca as a standard part of their bedding plant assortment, disease had continued to increase and limit their garden performance, until the introduction of 'Cora'. F_1 hybrids, which have added to the vigor of garden performance, are now available in a number of series including 'Cora', 'Titan', 'Boa', and others. In flower colors, 'Titan Dark Red' now personifies red; other series also define the colors they represent. As for a trailing habit, seed varieties have seen improvements in existing varieties such as 'Mediterranean' as well as in the introduction of 'Cora Cascade'. All these improvements mark a milestone that few other classes have seen in such a short period of time.

When vinca is grown under the right conditions, you cannot find a better crop. And that is the catch: You have to provide the right conditions. Plants prefer a full-sun, warm location in the garden and produce single blooms on plants from 8–12" (20–30 cm) tall. While most varieties are upright in performance, there are a number of trailing selections as well.

Propagation

Sow seed onto a disease-free, germination medium with a pH of 5.5–5.8 and an EC of less than 0.75 mS/cm. Use only new trays for vinca, as using old trays can spread *Thielaviopsis.* Cover seed with coarse vermiculite or medium to maintain 100% humidity around the seed coat. The medium should be nearly saturated. Maintain soil temperatures of 75–80°F (24–26°C). Keep ammonium levels at less than 10 ppm. Radicles will emerge in four to six days.

Once root radicles have emerged, move trays to Stage 2 for stem and cotyledon emergence. Lower soil temperatures to 72–75°F (22–24°C) and begin to allow trays to dry down before watering. Keep the media pH at 5.5–5.8 and the EC below 0.75 mS/cm. Provide ambient light levels of 450–700 f.c. (4.8–7.5 klux). As soon as cotyledons are expanded, fertilize with 50–75 ppm from 14-0-14 once a week. Stage 2 will take from seven to ten days. Many plug growers apply a preventative fungicide drench during this time to guard against disease.

Move trays to Stage 3 to develop true leaves, which will take from fourteen to twenty-one days for 512 trays. Increase light levels to 1,000–2,500 f.c. (11–27 klux) and lower soil temperatures to 68–72°F (20–22°C). Continue to allow trays to dry down between irrigations: Vinca prefers warm, dry conditions. Fertilize once a week with 100–150 ppm from 20-10-20, alternating with 100–150 ppm from 14-0-14. Supplement with magnesium sulfate at 16 oz./100 gal. (1.2 g/l) once or twice during Stage 3. Maintain a soil EC of less than 1.0 mS/cm. Maintain a pH of 5.8–6.2 during this stage. If growth regulators are needed, A-Rest, B-Nine, and Bonzi can be used at this time.

Harden plugs during Stage 4 for seven days prior to sale or transplanting. Allow soil temperatures to drop to 65–68°F (18–20°C) and increase light to 2,000–3,500 f.c. (22–38 klux). Allow trays to dry thoroughly between irrigations. Fertilize with 100–150 ppm from 14-0-14 once a week. Apply a fungicide drench again as a preventative. Avoid ammonium and phosphorus in Stage 4.

Most growers buy in plugs for their vinca crops because it is more convenient. Smaller plug sizes such as 512s or 384s are suitable for pack production, and larger sizes such as 288s are best for pot production. For hanging baskets, 72s are better, as it is easier to manage water relations with larger plugs.

A 288-plug tray will take from five to seven weeks to be ready from sowing to transplant to larger containers, while a 200-plug tray will take from six to eight weeks.

Growing On

Select a very well-drained, disease-free medium with a pH of 5.5–6.3 and a moderate initial starter charge. Plant 1 plug/36- or 18-cell; 1–2 plugs/4" (10 cm) pot; 3–5 plugs/6" (15 cm) or 1-gal. (4 l) pot; and 5–7 plugs/10" (25 cm) basket. Only use new pots and trays for vinca to prevent spreading *Thielaviopsis.*

Do not plant vinca too deep: Plants should be placed at the same depth as they were growing in the plug tray. Water plugs in well after planting.

Fertilize at every other irrigation with 150–200 ppm from 15-0-15, alternating with 20-10-20. Maintain an EC of 1.0 mS/cm. Maintain a pH of 5.5–6.3 to avoid micronutrient problems. Grow vinca warm—target temperatures are 70–75°F (21–24°C) days and 65–68°F (18–20°C) nights. Hanging baskets can be grown on drip tubes in the attic. Before attempting to grow plants at lower temperatures, make sure they are well established first. At temperatures of less than 65°F (18°C), yellow foliage can develop and growth is stunted.

If you need to control height with growth regulators, A-Rest (16–33 ppm), B-Nine (2,500–5,000 ppm), and Bonzi (10–15 ppm) can be used. (*Note:* Bonzi may cause black spotting on leaves.) Do not treat with growth regulators until plugs have rooted in fully.

Transplanting a 288-plug tray to a jumbo pack will take from five to eight weeks for spring sales, while a 4" (10 cm) pot will require from six to nine weeks. Plants will usually have one to two blooms in more northern locations and be freer flowering in the South or in warmer greenhouses. Plant timing is dependent on warm temperatures and high light levels.

Pests & Diseases

Diseases are the most serious production threat, especially *Thielaviopsis,* better known as black root rot. Symptoms include yellow lower leaves, loss of root hairs, brown discolored roots, slow plant growth, or even plant death. *Thielaviopsis* is most likely to occur when plants are under stress: high pH, high soluble salts, and low growing temperatures. Remove infected plants from the greenhouse immediately. Other diseases that can strike vinca include *Pythium* and rhizoctonia. Insects such as aphids and thrips can also attack vinca.

Troubleshooting

High media pH can lead to yellow upper leaves, poor root development, and/or stunted or uneven growth. When media pH rises above 6.5, iron can become deficient. For a quick temporary treatment to use while you lower your media pH, apply iron sulfate at 33 oz./100 gal. [2.7 g/l]). Be sure to rinse plants off immediately afterwards to prevent leaf burn. To lower pH 0.5, use Cleary's liquid sulfur ($FeSO_4$) at 16 oz./100 gal. (1.2 g/l)

Low production temperatures can cause leaf rolling. Before attempting to grow vinca cool, allow plugs to fully root into their growing containers.

Postharvest

Retailers should display vinca in full sun and warm air temperatures. Protect plants from cool temperatures, less than 65°F (18°C), and under no circumstances should vinca be exposed to freezing temperatures. As much as possible, retailers should try to water plants from their base to keep leaves dry. This can be done using a Dramm water breaker and directing the stream along the sides of trays and pots rather than allowing the water to fall through the foliage from above. If this is not possible, then make sure all irrigation is completed in enough time so that foliage is completely dry well before nightfall. Do not overwater vinca; however, trays that are allowed to dry down frequently to the point of plant wilt will quickly become shopworn.

Once established in the landscape, vinca is drought tolerant. Landscapers should avoid overhead irrigation to prevent aerial *Phytophthora*. Ideally, incorporate micronutrients into beds at planting and adjust the pH to 5.5–6.0 prior to planting to avoid micronutrient problems.

Celosia

Celosia argentea

Note: Depending on your botanical reference, *Celosia* species can be classified a number of ways. Here both plume/feather and crested types are included under *Celosia argentea.*

Common names: Cockscomb; plumed cockscomb; feathery amaranth

Annual, cut flower

For landscape impact and bright colors, it is tough to beat *Celosia.* For growers, they are relatively easy, provided you keep plants moving through production: *Celosia* does not perform well when growth is checked through cool/cold weather, it becomes root-bound, or it is allowed to flower in the plug. Under extremes, plants will never perform well in the garden.

Celosia is one of those bedding plant crops that just does not shine as a flowering bedding plant in packs. As a matter of fact, it is best to sell *Celosia* in packs green. That said, well grown in 4 or 6" (10 or 14 cm) pots, *Celosia* looks fabulous and will walk out the door. If you are a retail grower, providing *Celosia* at the right stage to your customers is just one more way you can set yourself apart from competition in the market.

Propagation

Sow seed onto a well-drained, disease-free, germination medium with a low starter charge and a pH of 5.5–5.8. Cover seed with coarse vermiculite. Maintain moist germination conditions, but do not allow trays to become fully saturated. At 75–78°F (24–26°C), seed should germinate in four to five days.

Move trays to Stage 2 for stem and cotyledon development. Lower temperatures to 72–75°F (22–24°C) and increase light levels to 500–1,000 f.c. (5.4–11 klux). Once radicles have penetrated soil, begin to allow trays to dry down somewhat between irrigations. Apply 50–75 ppm from 14-0-14 once cotyledons have fully expanded. Stage 2 will take about seven days.

Move trays to Stage 3 to develop true leaves. Lower temperatures to 65–68°F (18–20°C) and increase light levels to 1,500–2,000 f.c. (16–22 klux). Allow trays to dry thoroughly between irrigations. Apply a weekly feed of 100–150 ppm, alternating between 20-10-20 and 14-0-14. If growth regulators are required, A-Rest, B-Nine, Bonzi, and Cycocel are effective. Stage 3 will take from fourteen to twenty-one days for 512s.

Harden plugs for seven days prior to transplanting or sale by lowering temperatures to 60–62°F (16–18°C) and increasing light levels to 2,500–3,000 f.c. (27–32 klux). Allow trays to dry thoroughly between irrigations. Use a weekly feed of 100–150 ppm from 14-0-14.

Most growers buy in *Celosia* plugs. Once plugs are ready to transplant, do not hold them! Transplant them on time, as plants that have their growth checked will not fully recover. Never allow *Celosia* to flower in the plug tray.

A 288-plug tray will take from four to five weeks from sowing to be ready to transplant to larger containers.

Growing On

Plant one plug per cell pack or 4" (10 cm) pot, and use up to three plugs per 6" (15 cm) or 1-gal. (4 l) pot. Select a disease-free, well-drained, soilless medium with a moderate initial starter charge and a pH of 5.5–6.2.

Grow on at 65–70°F (18–21°C) days and 65–68°F (18–20°C) nights. *Celosia* needs heat and does not like wide temperature fluctuations during production. Plants grown too cool will flower prematurely and will never attain full size when planted out. If you must hold plants once they are ready for transplanting, hold them at 50–55°F (10–13°C).

Provide the highest light levels possible while maintaining production temperatures.

Fertilize at every other irrigation with 150–200 ppm, alternating between 20-10-20 and 15-0-15.

Growing 'Bombay' *Celosia* Cut Flowers

With a vase life of close to twenty-one days, a planting density of 6–7 plants/ft.^{2} (64–80 plants/m^2) and only eleven to thirteen weeks' crop time, 'Bombay' *Celosia* has become a wildly popular cut flower among greenhouse cut flower growers. They are excellent sold as grower bunches or in mixed bouquets.

Here, 'Bombay' breeder, Kieft Seeds, provides cultural tips. 'Bombay' may be direct-seeded (using pelleted seed) into the greenhouse or grown from plugs. Provide one to two layers of netting for support. 'Bombay' was developed for greenhouse production: Outdoor cultivation is generally not advised.

Celosia 'Bombay'

Do not allow 'Bombay' to become water stressed during culture, as plants will flower prematurely. Overhead irrigation is fine until flowers appear. However, make sure foliage is dry by nightfall to prevent disease. Just before flowering, discontinue overhead irrigation and allow soil to dry out, irrigating only when foliage begins to wilt. Applying too much water at this stage of production can cause flowers to become top heavy and fall over.

Low light intensity under short days and cool growing temperatures may cause stems to be flat and flower combs to shatter. Do not pinch plants during culture.

Allow flowers to ripen: Stems need three to four more weeks from the moment they "look" to be mature. What is ripe? According to Kieft the lower third of the "neck" should be past flowering, the middle third should be flowering, and the upper third should not be in flower yet. After harvest, treat water with Florissant 500 and keep water fresh to avoid bacteria buildup.

If growth regulators are required, A-Rest, B-Nine, Bonzi, and Cycocel are effective. B-Nine may be applied two weeks after transplanting and reapplied every seven to ten days as needed. Cycocel may cause leaf damage, so do not apply it late in the production cycle.

After transplanting a 288 plug to a cell pack, it will be salable in five to six weeks for spring sales, and plants will be green or just starting to flower. 4" (10 cm) pots require six to eight weeks for spring sales with first bloom.

Related Species

Celosia spicata (commonly called wheat celosia) is highlighted by the 'Flamingo' series, a tall-growing selection used as a landscape or field-grown cut flower. Grow as noted above as a pack or 4" (10 cm) plant. However, the 'Flamingo' series, regardless of selling in a pack or pot, is sold green and not in flower. Heights range from 18–26" (45–64 cm) on 'Flamingo Feather' (pink flowering) to as tall as 48" (120 cm) on 'Flamingo Purple'. Plants will flower in the garden once established, although 'Flamingo Purple' will take from three to four weeks longer to flower than 'Flamingo Feather'.

Pests & Diseases

Aphids, spider mites, and whiteflies may become pest problems. Diseases that can strike include botrytis, *Pythium,* and rhizoctonia.

Postharvest

Retailers should advise consumers not to plant *Celosia* until after the danger of frost has passed. When established early in the season, plants flower throughout the summer heat!

Centaurea

Centaurea cyanus
Common names: Bachelor's button; cornflower
Annual, cut flower

Centaurea's blue flowers have been an American garden staple since Colonial times. Its common name, bachelor's button, originates from a bygone tradition: In times past, bachelors would place centaurea in their buttonhole when they went courting.

Today, centaurea has expanded beyond its humble beginnings with annual varieties in red, pink, and white. Its fringed flowers are a welcome, casual addition to any modern garden or cut flower bouquet.

Most growers offer annual centaurea in 4" (10 cm) pots or 36-packs, sold green in the spring. Cut flower growers can offer centaurea year-round with photoperiod control. Natural-season flowering peaks in the spring for California and Florida crops and in the summer months for local production in northern regions.

Propagation

Centaurea is generally produced as seed-sown plugs grown in four stages. Stage 1, root radical emergence, will take from five to seven days at 65–70°F (18–21°C). Keep medium uniformly moist. Light is not required for germination until radicle emergence. Medium should have a pH of 5.5–5.8 and soluble salts (EC) should be less than 0.75 mS/cm.

Stage 2, stem and cotyledon emergence, will occur over the next four to seven days. Maintain soil temperatures at 62–65°F (17–18°C). Keep media pH at 5.5–5.8 and EC at less than 0.75 mS/cm. Make sure ammonium levels are below 10 ppm. Begin fertilizing with 50–75 ppm from 14-0-14 or a calcium/potassium-nitrate feed once cotyledons are fully expanded. Alternate feed with clear water. To prevent disease, be sure to irrigate early in the day so that foliage is dry by nightfall.

Growth and development of true leaves will begin during Stage 3, taking fourteen to twenty-one days. Maintain soil temperatures at 62–65°F (17–18°C). Allow medium to dry thoroughly between irrigations, but avoid permanent wilting; this promotes root and shoot growth. Maintain a substrate pH of 5.5–5.8 and an EC of less than 1.0 mS/cm. Increase fertilizer to 100–150 ppm from 20-10-20, alternating with 14-0-14 or a calcium/potassium-nitrate fertilizer. Fertilize every second or third irrigation.

During Stage 4, plugs harden and finish over seven days, becoming ready to ship or transplant. Lower medium temperatures to 60–62°F (16–17°C). Allow medium to thoroughly dry between watering. Maintain pH at 5.5–5.8 and drop EC to less than 0.75 mS/cm. Fertilize as needed with 14-0-14 or calcium/potassium-nitrate at 100–150 ppm nitrogen (N).

Some cut flower growers sow centaurea seed directly in the field during the spring once soil temperatures are warm enough (April to mid-May). Seed can be direct-sown (0.25–0.50 oz. [8–16 g] makes about a thousand plants). Crops will flower from June to September. Florida and California growers may also sow seed in September for flowering from February to June. (If using plugs for fall-started field crops, expose plants to four to six weeks of short days prior to planting.) Greenhouse-grown crops planted in the fall will need long days provided by night interruption lighting (10 P.M.–2 A.M.) from mid-January until mid-March. Centaurea flowered in the greenhouse in May can be sown from December until mid-January without lighting.

A 288-plug tray requires from four to five weeks to be ready to transplant to larger containers. Transplant to a jumbo or similar pack or to 4" (10 cm) pots. Grow under short days (twelve hour day lengths or less) to bulk up and then under long days to encourage blooming. Seed sown in late February or March produces salable green plants for May. Planted to the garden these flower, depending on variety, from mid- to late June. Plants will not flower throughout the summer, succumbing to high heat and humidity when both are in abundance. Plants prefer cool temperatures to perform their best.

Growing On

Transplant bedding plant crops into 4" (10 cm) pots or 36-packs using one plant per container. Use a well-drained, disease-free medium with a moderate initial nutrient charge and a pH of 5.5–6.2. Fertilize at every other irrigation with 15-0-15, alternating with 20-10-20 at 150–200 ppm. Maintain EC at 1.0 mS/cm. Fertilize cut flower crops with a rate of 100–150 ppm to prevent weak plants. Avoid fertilizers high in ammonium.

Grow plants under high light levels while maintaining moderate temperatures; ideal light levels are from 3,000–5,000 f.c. (32–54 klux). Bedding plant crops may be grown at 55–60°F (13–16°C) days and 50–55°F (10–13°C) nights. Cut flower yield is highest at 55–60°F (13–16°C) under long days.

Plant cut flower crops in raised beds on 12" × 12" (30 × 30 cm) spacing or in the field on 6" × 9" (15 × 23 cm) centers. Plants will require one or two levels of support as they grow in the greenhouse. Support for field crops is dependent on wind conditions.

Withholding water once plants have rooted to the sides of the container effectively controls centaurea bedding plant height. Alternatively, A-Rest or B-Nine can be used to control the height of bedding plant centaurea.

Centaurea is a long-day plant: Plants must be exposed to at least three weeks of long days for flowering to initiate. Growers can use night interruption lighting for four hours each night or cyclic on-off lighting—two minutes on, eight minutes off. Once flowering has initiated, it will continue through short days.

Cut flower growers planting in the summer must provide short days for four to six weeks prior to planting to bulk up plants prior to planting. Otherwise plants will be short, flower quickly, and have low yield.

Transplanting a 288 plug to 4" (10 cm) requires seven to eight weeks to be ready to sell for planting to the gardening. Few varieties (if any) will be in flower on the retail shelf but will flower in two to four weeks once transplanted to the garden. Plants grow and develop best under cool night conditions. Hot and humid temperatures aid in plant decline.

Cut Flowers

One source reports that some Israeli growers have experienced good results forcing November and December flower crops by sowing seed beginning week 33 (mid-August). Shoots can be elongated using gibberellic acid sprays in week 39 (late September) and harvested for the Thanksgiving and Christmas holidays.

University of Kentucky researchers have tested a containerized cut flower production regime as follows.

Day-old seedlings germinated in 288-plug trays are moved into the greenhouse where they receive twenty-four-hour supplemental light for ten to eighteen days. Plugs are transplanted to 3 or 4" (8 or 10 cm) pots (two plants per pot) and spaced pot tight for a plant density of 15–20 stems/ft.^{2} (161–215/m^2). Pots are put in a greenhouse with long-day lighting and grown at 55°F (13°C) and fertilized with a constant liquid feed at 100–150 ppm. One layer of mesh support is used. The crop requires eight to nine weeks from sowing to harvest. Individual stems have four flowers and are uniformly 20–26" (51–66 cm) tall.

Pests & Diseases

Aphids, leafhoppers, and spider mites can be problems. Aster yellows can be spread by leafhoppers, making insect control a priority. Just about every disease attacks centaurea. Make sure to provide excellent air circulation around plants and avoid wet foliage when temperatures are low or nightfall is approaching. Among the diseases to watch for, there are botrytis, downy mildew, *Fusarium,* powdery mildew, rust, *Sclerotinia, Verticillium,* and root rots caused by *Pythium* and rhizoctonia.

Postharvest

Harvest cut flowers when they are 25–50% open. Place in water with a floral preservative solution and cool to 35–41°F (2–5°C). Long-term storage is not recommended, making centaurea an excellent locally grown cut flower. Properly handled, flowers should have a vase life of ten days.

Retailers displaying bedding plant centaurea should provide temperatures of 55–60°F (13–16°C) in either full sun or partial shade.

Chives (see *Allium*)

Chrysanthemum

Chrysanthemum × *morifolium*
(also *Dendranthema* × *grandiflora*)
Common names: Chrysanthemum; mum; dendranthema; daisy
Flowering potted plant, fall bedding plant, cut flower

Chrysanthemums may very well be the world's most-grown flower. There are three primary types of chrysanthemums: (1) pot mums, (2) cut mums (spray mums and standard mums, known as football mums to some), and (3) garden mums. In Asian cultures, chrysanthemums have special cultural and spiritual meanings. On All-Saints Day (November 1) in Europe, chrysanthemums are an honored tradition. In the United States and Canada, chrysanthemums not only help form the foundation of the day-in and day-out weekly floriculture business but are also a popular holiday plant for American and Canadian Thanksgiving Day sales. They are a staple pot plant for supermarket and florist sales. Spray chrysanthemums imported from the Medellín area of Colombia are utilized in just about every mixed bouquet sold in the United States. Florists rely on cut chrysanthemums for the structure of many floral arrangements. Garden mums are a major crop for North American bedding plant and pot plant growers to produce for summer and fall sales.

Chrysanthemums play a key role in floriculture for several important reasons. First and foremost, they are highly programmable. Flowering is precisely timed by manipulating day length. Long day lengths promote vegetative growth, while short days (twelve or more hours of complete darkness) stimulate the initiation and development of flowers. Chrysanthemums can be produced year-round in virtually any quantity needed. Breeders have developed a wide array of varieties well suited for modern greenhouses and production practices. Second, chrysanthemums are available in an excellent array of colors, including white, yellow, pink, purple, bronze, orange, coral, salmon, red, bicolors, and two-tone color combinations. There are also many flower forms, including anemones, daisies, cushions (decoratives), spoons, Fuji (spider) types, pom-poms, and buttons. Third, chrysanthemums are adaptable to a wide range of climatic conditions, making them suitable to grow just about anywhere. And fourth, they offer a tremendous value to the consumer. Through superior genetics, the postharvest life of most varieties is counted in weeks, not in days. For the average person buying a chrysanthemum bouquet, garden mum, or pot mum, it is easy to be successful. You will not find all of these attributes rolled up into any other single flower species.

Because scheduling is the most important aspect of growing a chrysanthemum crop—and it is similar for all types of chrysanthemum production—we are going to break from the *RedBook* format to discuss the how's and why's of mum scheduling before diving into crop specifics.

Scheduling Mums to Flower

In 1920 when USDA researchers first discovered the secret to why mums flower, they began a revolution: programmed crop production. The discovery that mums flower based on exposure to short days allowed growers to begin producing chrysanthemum crops year-round. It also paved the way toward understanding why scores of different crops seemed to flower naturally at various times of the year—they were acting in response to either the lengthening or the shortening of day lengths.

Chrysanthemums, left to flower under natural conditions, will flower in the fall when day length shortens. Armed with this information, growers began "forcing" chrysanthemums into flower by pulling black cloth over benches to block out light for at least twelve hours between 5 P.M. and 8 A.M. After successfully flowering a "forced" crop and earning the higher prices that marketing early flowers bring, it was only a matter of time before growers began to flower mums year-round, turning them into the staple they are today.

As an aid to scheduling, chrysanthemums are classified based on their response group (i.e., how many weeks from the beginning of short-day treatment until the crop flowers). Varieties can take as few as six weeks for some garden mums or as long as fourteen to fifteen weeks for some cut mums.

Most chrysanthemum growers utilize "mum lighting" both as a scheduling and a cultural tool throughout the year.

Before beginning short-day treatment, growers allow plants to grow and develop for up to several weeks of "long days" to prepare for floral initiation. For example, if you were to grow an eight-week- response pot chrysanthemum from a cutting, you would grow the crop under two to three weeks of long days and then begin eight weeks of short days for a total crop time of ten to eleven weeks from planting to flowering.

The length of time required for long days to keep plants vegetative will vary based on the variety being grown, the time of year, and the location. For instance, northern winter cut spray chrysanthemum crops might need four weeks of long days to develop enough stem length and substance for the low-light, slow-growing winter crop. A summer or fall flowering crop in the same northern greenhouse would only require one or two long-day weeks, because the crop grows so much faster in higher light intensities and the longer days of summer and fall.

Growers in the Deep South experience, in effect, high light and relatively longer days year-round. Their winter schedules tend to follow the schedules and variety selections used by northern growers in the summer and early fall.

In general, garden mums are early-response plants (six-, seven-, and eight-weeks) and normally flower outdoors from early September to mid-October. In addition, the majority of commercial pot and cut mum varieties are becoming faster thanks to intensive breeding. Most cut and pot varieties have a seven- to eight-week response time. Few, if any, commercial varieties used today have the ten-week or longer response times that were common in the past.

Propagators generally provide complete schedules prescribing a precise number of weeks of long days and short days for crops to flower each week of the year. Schedules are also available for varying degrees of latitude from north to south. Typically, these schedules also recommend specific crop varieties for each latitude. Your supplier should be able to provide this information when you are placing your cuttings order.

Mums initiate buds and develop their flowers when day lengths are twelve hours or less. A few varieties will flower with day lengths of twelve and a half hours. In fact, research has pointed out that it is the length of the night, not the day that matters, but the results are the same. During short days, it is important to ensure plants are in total darkness during the night. Avoid stray light from outside sources, or the exposed plants may be delayed or display poor uniformity in flowering.

To encourage vegetative growth and prevent mums from initiating flower buds, growers need to provide long-day treatments. This occurs naturally in most locations in North America between March 15 and September 15. Buds will not form as long as the uninterrupted darkness periods are less than seven hours. To maintain vegetative growth during the short days of fall and winter, growers must extend the day length with either supplemental lighting or night interruption lighting. For example, assume midwinter natural light is from 8 A.M.–4 P.M. Lights provided from 10 P.M.–2 A.M. will produce two six-hour periods of darkness, 4 P.M.–10 P.M., and 2 A.M.–8 A.M. No buds will form. (*Note:* Most varieties will rosette, making short, spreading growth with no stems or flowers, if the temperature is 50°F [10°C], even if exposed to long days).

Night interruption lighting can be used intermittently to save electrical costs. Use the same hours per night as with continuous light, but during the hours of light, turn the lights on for six minutes out of each thirty-minute period. In other words, six minutes on, twenty-four minutes off. Some growers use twelve minutes on, eighteen minutes off.

If you are flowering successive crops in the same greenhouse, be sure that light from long-day areas does not leak onto crops under short days. Most growers use black fabric or plastic (which is carefully checked for holes and patched, if needed).

At latitudes near the equator where natural day length is close to twelve hours year-round, winter days are longer and fewer hours of light are needed to keep plants vegetative. In fact, in Colombia, where most of the cut mums are produced for the North American market, growers are very close to the equator. They find that no buds occur on some varieties even with no lights at all in the midwinter. The same circumstance happens with certain varieties in Hawaii.

Pot Mums

If you are a new grower considering pot mums as a crop, do your homework before planting the first cutting. In North America, pot mum production is flat and highly consolidated in the hands of pot plant specialists who cater to supermarkets. The consolidation has occurred because it is easier to gain production efficiencies on a larger scale, and pot mums are easily shipped long distances. This does not mean there are not good market opportunities, but it does mean that you need to know how and where you will market your crop before planting.

What size pots will you produce? Standard pot sizes in the market are 4, 4.5, 5, 5.5, and 6" (10, 11, 13, 14, and 15 cm), but 4 and 6" (10 and 15 cm) are especially popular. Many growers that do not cater to the supermarket trade choose to grow in larger containers or focus on upscale forms (disbuds). How many cuttings per pot will you use? What spacing will you use? Upscale 7" (18 cm) pots may get six or seven cuttings and be grown at a 16" × 16" (41 × 41 cm) spacing, but a supermarket pot may measure 5.5 or 6" (14 or 15 cm) and get three to four cuttings and be spaced much closer on the bench.

Approach customers and potential customers and ask them what they are looking for and what kind of price they are willing to pay. Then grow the crop on paper. Your objective is to determine if the crop you are planning at the spacing and price you are planning will make a profit. Include packing costs, selling costs, overhead expenses, and shrink. Play with different spacing and prices, and compute a net profit on each variable before you grow the crop. Plan to make money.

Pot mums can be grown year-round. Weekly production is often augmented with considerable peaks for major holidays. A typical year-round grower will bump his average weekly production by three to five times for Easter, Mother's Day, Thanksgiving, and other holidays. In some cases, pot mums are only produced seasonally, such as for Thanksgiving, or at times when the market makes them profitable.

Scheduling

Pot mum varieties are classified as short, medium, or tall in terms of vigor. In addition to a cultivar's response time, whether or not that variety is short, medium, or tall will also affect scheduling. A sample commercial pot mum growing schedule is shown in table 11. Note that all varieties of the "tall treatment" group are potted on January 28 and handled the same way. The seven-week variety flowers on March 25, eight-week variety flowers on April 1; and the nine-week variety flowers on April 8. These flowering dates are from a starting short-day date of February 3. Shading in all cases does not start until March 15 because days are naturally short enough until then.

Short-treatment varieties need several weeks more time to develop strength and substance, compared to a naturally tall variety. Therefore, from the same potting date (January 28), the short treatment varieties have a later short-day date (February 17 versus February 3). The extra two weeks from plant until short days allows the plant to develop more substance. Therefore, the short-treatment, nine-week variety will flower two weeks later than the nine-week tall treatment (April 8 versus April 22). When you order cuttings, the catalog will categorize varieties as tall, medium, or short and also provide the response time (in weeks).

Propagation

Growers can order either unrooted cuttings or rooted liners. Purchase cuttings only from a reputable specialized propagation company that is maintaining a rigid clean-stock program. Chrysanthemums can harbor a number of latent diseases.

Most large-scale pot mum growers bring in unrooted cuttings and plant them into their final growing container. Stick cuttings immediately. Stick cuttings into a disease-free, well-drained medium with a pH of 5.8–6.2. Removal of lower leaves that will be covered with medium is optional. Grade cuttings by length and caliper and plant like-sized cuttings together to create a more uniform finished product. Treat the base 0.5" (1.3 cm) of cuttings with 1,500 ppm IBA rooting hormone and stick 1–3 cuttings/4.5" (11 cm) pot and 3–5 cuttings/6" (15 cm) pot. Stick cuttings 1–2" (3–5 cm) deep—deep sticking is preferable to shallow sticking. Space the cuttings equally around the edge of the pot. Stick cuttings into a moist, well-drained medium and water them in after sticking. Many good pot mum growers will also fertilize the cuttings while rooting with 300 ppm of 20-10-20 fertilizer one, five, and ten days after sticking.

Place pots on propagation benches outfitted with bottom heat, mist, and movable shade screens. Maintain substrate temperatures of 70–74°F (21–23°C) and air temperatures of 70–85°F (21–29°C). Time mist cycles so that plants do not wilt. Depending on nozzle size and spacing, set the mist duration for ten seconds or longer. Here are some guidelines for frequency that you can adjust to your conditions: Days 1–3, mist every five to ten minutes; days 4–7 (callus formation), mist every twenty minutes; and days 8–15 (roots initiated), mist every thirty minutes. Night misting is usually not required except in exceptionally hot regions. Too much misting will turn the plants yellow and should be avoided. Applying 100 ppm of fertilizer in the mist is beneficial. Consult your media scientist for a recommendation that will take into account the quality of your water.

During rooting, cuttings must be under long days year-round. Maintain ambient light levels of 3,000–3,500 f.c. (32–38 klux). Begin a constant liquid fertility program with 250–300 ppm from 20-10-20 once roots have initiated.

Most varieties will be rooted and ready to pinch in twelve to eighteen days.

Growing On

If you are buying in liners, pot them in a well-drained, disease-free medium with a pH of 5.8–6.2 and a moderate initial nutrient charge. Mums are heavy feeders; however, as the crop matures and finishes, plants require less fertilizer. For weeks 1–5, start with 250–300 ppm from 20-10-20 at every irrigation. From weeks 6–9, lower rates to 150–200 ppm and switch to 15-0-15 at every irrigation. From week 10 until sale, discontinue feed and just use clear water. You will produce high-quality plants that will have outstanding postharvest life.

Drip tubes irrigate most pot mums in the United States. Mum production also works very well with ebb-and-flood bench and floor systems. Maintain uniform moisture levels and do not allow wilting. Plants that are water-stressed during flower initiation may have reduced flower size and delayed flowering.

After potting, you will need to provide long days to maintain plants in a vegetative state. Provide at least two to three weeks of long days with night interruption lighting from 10 P.M.–2 A.M. Maintain ambient light levels at 4,500–7,500 f.c. (48–81 klux). Many northern growers use supplemental lighting (500 f.c. [5.4 klux]) to help boost winter mum quality. Also during winter months, spacing plants farther apart will help more light to reach each leaf. Allow an extra seven to ten long days in the schedule (compared to summer) to allow plants to develop more substance during the winter months.

Provide day temperatures from 72–80°F (22–26°C) days and 65–68°F (18–20°C) nights for the first five weeks. From weeks 6–8, lower night temperatures to 62–65°F (16–18°C). For the last week or two of a crop, consider lowering the night temperature to 58–60°F (14–15°C). Cooler temperatures at finishing can intensify flower color. Note that production temperatures of less than 60°F (15°C) can delay flowering or cause uneven flowering of many varieties. High temperatures (90°F [32°C]) for one to three weeks may cause heat delay (late flowering).

Pinch the cuttings once roots have reached the bottom of the pot and six to eight leaves have formed, which is about ten to fourteen days after potting liners and sixteen to eighteen days after sticking unrooted cuttings. Remove the complete tip of the plant, and be sure not to pinch too softly. As a guide, try to allow five to seven leaves remaining on the cutting after the pinching.

What is a "delayed pinch"? A delayed pinch refers to pinching the plant in short days. The opposite would be to pinch the plant in long days. Delayed pinching causes a number of very beneficial changes to occur in the growing pot mums. If a well-growing plant is placed into short days with the growing tip still on the plant, the induction of the reproductive stage is started. The plant is stimulated by short days to change from a vegetative plant to a flowering plant. This short-day initiation also causes branching to occur in the plant. Because branching is induced by the short-day treatment, it occurs very evenly and most of the potential lateral branches develop. Pot mums produced using a delayed pinch will be fuller because they have more branches and, therefore, more flowers. Plants typically will also be shorter and have fewer leaves. If the objective is to grow a short- to medium-sized plant, this technique is excellent. When growing zero bud removal (ZBR) plants, this also helps to produce excellent quality.

Height can be controlled with B-Nine or Alar and scheduling. Rates vary by variety and time of year, but are generally from 2,500–5,000 ppm. Begin applications after pinching when breaks are 1.5–2.5" (4–6 cm) long (about fourteen to sixteen days after pinching) and repeat every ten to fourteen days if needed. Short-treatment varieties may need no or only one B-Nine application, medium-treatment varieties from one to two applications, and tall-treatment varieties two or three applications. B-Nine use also depends on the time of year. In the winter, some growers delay application until shoots are longer (3" [8 cm]) or do not apply at all. Note that some growers use a B-Nine dip (1,000 ppm) for unrooted cuttings prior to sticking to minimize stretching during propagation. Dip the unrooted cuttings when received, shake off the excess solution, place the cuttings in a plastic bag with a label, put the cuttings in a cooler at 35–40°F (2–4°C) overnight, and stick the cuttings the next morning. As a bonus, the cuttings will be very turgid and very easy to stick.

Crowd plants carefully—for profit. The more space you give a plant, the better the end quality, but the lower the profit tends to be. The traditional 14" × 14" (36 × 36 cm) spacing for 6" (15 cm) pots with four or five cuttings has given way to 12" × 13" (30 × 33 cm) or even 12" × 12" (30 × 30 cm). The effect on profit by moving from 14" × 14" (36 × 36 cm) to 12" × 12" (30 × 30 cm) is monstrous: You get 27% more plants on the bench. It takes skill to crowd plants and maintain quality—you will not be growing a deluxe plant, but a slightly more restricted plant with fewer flowers.

Disbudding pot mums has been the traditional way to produce the crop for many decades. In this process, each one of the eight to twenty-five stems is disbudded—all lateral buds are removed, leaving only the top bud on each stem to flower. The result is one large terminal flower per stem. This is a very labor-intensive and, therefore, costly procedure. Growers frequently pay laborers on a piecework rate to do this, but that requires careful follow-up inspection. Disbudded crops feature extra-large flowers and have become the

Table 11. Sample Commercial Pot Mum Schedule

TREATMENT GROUP	POT	LONG DAYS	PINCH	SHORT DAYS	VARIETY FLOWERING DATE		
					7-WEEK	8-WEEK	9-WEEK
Tall	Jan. 28	Jan. 28–Feb. 4	Feb. 11	Mar. 15	Mar. 25	Apr. 1	Apr. 8
Medium	Jan. 28	Jan. 28–Feb. 11	Feb. 11	Mar. 15	Apr. 1	Apr. 8	Apr.15
Short	Jan. 28	Jan. 28–Feb. 18	Feb. 11	Mar. 15	Apr. 8	Apr. 15	Apr. 22

"florist" mum since they rarely appear on supermarket shelves. Because of the expense, disbudding has been declining each year. A supermarket buyer may ask a grower to send a case of disbudded pot mums, and if the grower decides to grow this type of mum, he should be paid extra for this kind of quality

Center bud removal (CBR) of mums was developed many years ago to save labor compared to the disbudding process. This technique involves removing only the terminal bud from the top of each stem. This allows a spray of flowers to develop. This is much quicker and results in more flowers per stem in bloom, although they are smaller than disbudded flowers. Growers have decided to grow this type of mum to save money, and breeders select varieties for this type of culture.

Multiple bud removal (MBR) involves a second pinch of the side shoots of the plants approximately fourteen days after the start of short days. This somewhat involved technique is very variety and climate specific. When done correctly, it will produce spectacular results. It is costly and takes time and focus to master.

More recently, breeders are releasing zero bud removal (ZBR) varieties. ZBR is a new term that describes an exciting breakthrough. As growers have worked to find ways to save money, many have resorted to no disbudding with varying results. Royal Van Zanten has developed new varieties that are ideally suited for this type of culture and that produce plants of extremely high quality. The actual look of the finished product when done correctly can resemble that of a disbudded or a CBR pot mum.

Pests & Diseases

A number of pests routinely attack chrysanthemums. The most notorious of which include leaf miners and thrips. Aphids, caterpillars, fungus gnats, spider mites, and whiteflies can also become problems.

Disease problems include bacterial leaf spot, crown gall, TSWV/INSV, *Fusarium, Pythium,* rhizoctonia, and leaf and flower blights. White rust, a devastating disease to chrysanthemums, is not indigenous to North America; however, it has been discovered on pot mums in California in years past. The affected growers were quarantined with economically devastating results. Since then no reports of white rust on pot mums have been publicized.

Troubleshooting

Plants that do not size up or break well could be experiencing poor root development, low fertility, low light levels, or low production temperatures. Root loss can be due to high soluble salts, poor drainage, or disease. Poor root development may be due to lack of sufficient aeration in the medium, low calcium and phosphorus, or insufficient water. Flower buds that do not develop on schedule may be due to temperature extremes (below 60°F [15°C] or above 90°F [33°C]), poor fertilization, or insufficient long or short days. High humidity, poor air movement, or lack of preventive controls can cause Flower blight or foliar disease.

Varieties

Two main pot mum breeders supply the North American market: Royal Van Zanten and Syngenta Yoder. The variety picture changes rapidly as new and improved varieties are introduced each year. Your supplier can help you develop a list of varieties that are suitable for production in your climate and the holidays or months of the year that you wish to produce plants.

Postharvest

Plants are ready for sale when one-third to three-fourths of the flowers are open. If plants are sold prematurely, the flowers will not open to their full potential. Store and ship pot mums at 35–40°F (2–4°C) for no more than seven days. Ethylene does not adversely affect chrysanthemums.

Retailers should display flowers under a minimum of 50 f.c. (538 lux) of light to maintain flower color. At retail, maintain plants at temperatures from 65–75°F (18–24°C) if possible. Cool-white fluorescent lighting can distort the color of chrysanthemums, especially reds, purples, and bronzes. Try using incandescent spotlights to enhance flower colors on retail displays.

Consumers can plant pot mums purchased in the spring and summer into the garden. If cut back, plants will bloom again in the fall (usually in late October to early November), provided temperatures do not get too cold. Pot mum varieties are marginally hardy and may or may not survive the winter. Pot mums are very vigorous when compared to garden mums and do not have good growth habits for the garden. Garden mums are much better suited for growing in the garden.

Cut Mums

Spray chrysanthemums are the workhorses of the cut flower marketplace. Most spray chrysanthemums sold in the United States are imported, with a majority coming from the Rio Negro (Medellín) region of Colombia. There are a limited number of North American growers producing spray chrysanthemums today.

In the Netherlands, spray chrysanthemums are the most-grown commercial floriculture crop and the second most important crop in value. However, spray chrysanthemum growers in the Netherlands are under pressure: Prices have been stable or declining, while costs have been rising. Fewer spray chrysanthemum growers remain in business each year. In the Westland area, where most spray chrysanthemum production is located, older growers often sell out and retire without an heir. Often houses are built where there used to be greenhouses. Younger growers entering the business for the first time often choose higher profit crops.

If you are thinking about adding spray chrysanthemums or other cut mums to your product assortment, make a plan for the crop that includes a market and profitability study before you plant the first cutting. Spray chrysanthemums are plentiful in the market and their long postharvest life means they can be shipped over long distances. With that said, spray chrysanthemums are a lightweight flower sold at low prices in many markets worldwide, which makes them a less-than-ideal candidate for airfreight since the airlines prefer heavier cargoes. You may be in a position where a limited amount of spray chrysanthemum production could be a profitable venture. Grow your crop on paper first to find out.

Chrysanthemums can be scheduled to flower every week of the year if you choose. If you opt for year-round production, you will be able to turn the same space at least three times—maybe even four times—during the year. Your supplier will have year-round schedules developed specifically for your climate.

Most commercial spray chrysanthemums are eight-week varieties and trending to seven-week responses. Summer and fall crops in the North will typically be seven- and eight-week varieties with one to two long-day weeks. The winter crop will be eight-week varieties, often with three to five weeks of long days, which increases the crop time. In Southern California, seven-, eight-, and nine-week varieties are used year-round. If possible, simplify your operation and plan to grow similar response group varieties together.

Propagation

Growers may order either unrooted cuttings or liners to start the crop. Purchase cuttings only from a reputable, specialized propagation company that is maintaining a rigid clean-stock program. Chrysanthemums can harbor a number of latent diseases. Ideally, handle unrooted cuttings immediately; however, if you must store cuttings prior to sticking, they can be stored at 35°F (2°C) for up to one week.

Unrooted cuttings should be stuck into a disease-free, well-drained, peat/perlite or similar medium with a pH of 5.8–6.5. Removing lower leaves that would be covered with media prior to sticking is optional. Grade cuttings by length and caliper and plant like-sized cuttings together. This helps to create a uniform finished product. Treat the base 0.25" (6 mm) of cuttings with 1,500 ppm IBA rooting hormone. Stick cuttings 1.5" (4 cm) apart and water in well. Many growers water cuttings in with a fungicide drench as a preventative treatment against disease. Maintain substrate temperatures of 70–74°F (21–23°C) and air temperatures of 70–85°F (21–29°C). Do not allow temperatures less than 60°F (15°C) or higher than 95°F (35°C). Time mist cycles so that plants do not wilt. Depending on nozzle size and spacing, set mist duration for ten seconds or longer. Here are some guidelines for mist frequency that you can adjust to your conditions: Days 1–3, mist every five to ten minutes; days 4–7 (callus formation), mist every twenty minutes; and days 8–14 (roots initiated) mist every thirty minutes. Night misting is usually not required except in hot, low-humidity environments. During rooting, cuttings must be provided with mum lighting to ensure long-day photoperiods. Maintain ambient light levels of 3,000–3,500 f.c. (32–38 klux); however, during the first week, use shade over cuttings so they do not burn.

Overmisting will turn plants yellow and, thus, should be avoided. Applying 100 ppm of fertilizer in the mist is beneficial. Consult your soil scientist for a recommendation that will take into account the quality of your water.

Cuttings will be ready for planting in fourteen to sixteen days.

Growing On

Plant cuttings in well-worked, steam-sterilized ground beds. Work soil to a depth of 16" (41 cm) making sure there is no hard pan. Ideally, the pH is between 6.0 and 7.0 and the starting EC is 1.0 mS/cm. Incorporate magnesium, if there are deficiencies, in the form of magnesium sulfate at a rate of 8.8–17.6 lbs./120 yd.^{2} (4–8 kg/100 m^2) and phosphorus in the form of triple superphosphate at a rate of 8.8–17.6 lbs./120 yd.^{2} (4–8 kg/100 m^2) prior to planting. If no other fertilizer is in the soil, also add 17.6 lbs. (8 kg) of a balanced nitrogen-phosphorus-potassium (NPK) fertilizer. It is best to take soil samples and provide starting nutrient charges to beds based on the recommendations of your soil laboratory. Plant cuttings at a density of about 6/ft.^{2} (64/m^2) or approximately 4" × 6" (10 × 15 cm). For winter production or under low light, be more generous in spacing and reduce density to 4–5 plants/ft.^{2} (48–56 plants/m^2), approximately 5" × 6" (10 × 15 cm).

Irrigate plants overhead for the first three to four weeks. Once leaves begin to touch, switch to drip irrigation. Fertilize at each irrigation with 200–250 ppm from 20-10-20 or a similar fertilizer, adjusting nitrogen sources to your specific conditions and the time of year. Finish the crop for the last two to three weeks with clear water only, which will increase postharvest life and reduce salt levels in the planting beds for the next crop.

Chrysanthemums may be produced under a wide range of temperatures, the ideal being 65–77°F (18–25°C) days and 62–65°F (17–18°C) nights. The most important thing is to maintain consistent temperatures. Do not expose plants to prolonged temperature extremes (below 60°F [15°C] or above 100°F [38°C]). Lower temperatures, if possible, to 62–64°F (17–18°C) days and 58–60°F [14–16°C] nights during finishing for more intense flower color.

Grow plants under long days to achieve stem length: Long-day treatment takes about one to two weeks in the summer, two to three weeks in the spring and fall, and four to five weeks in the winter. If you plan to pinch the crop, add two additional weeks. Use night interruption lighting, applying at least 10 f.c. (108 lux) at crop level. Light for two hours per night in June and July; three hours in August, September, April, and May; and four hours nightly from October to March.

When plants are 10–12" (25–30 cm) tall, it is time to begin short-day treatments. Do not apply short days any sooner or stem length may be reduced and the spray formation may be of lower quality. Use blackout shading for at least twelve hours per night from March 15 to September 15 from approximately 7:00 P.M. until 7:30 or 8:00 A.M. From the start of short days, an eight-week variety should be ready to harvest in approximately fifty-two to fifty-six days.

B-Nine at 2,500 ppm can be used to get a more desirable spray formation on spindly growing varieties. Gibberellic acid (GA) can be used to stretch peduncle length, which certain markets demand. Apply 10 ppm at disbudding.

Normally cut mum crops are not pinched—most growers produce one stem from one cutting. This allows for the most efficient crop times. However, if maximizing crop turns is not a priority, some growers pinch the cuttings. If you are going to pinch your crop, plant cuttings at a density of 3 cuttings/ft.^{2} (32/m^2), decreasing density to 2 cuttings/ft.^{2} (24/m^2) in the winter. To get more blooms, you can pinch two to three weeks after planting. However, pinching the cutting will extend crop time and reduce final stem quality. Pinch off the top 1.0–1.5" (3–4 cm), leaving the two strongest breaks (three on outside rows). If more side shoots are present, remove them.

Practically no one disbuds cut mums anymore, as the practice is too labor intensive and prices do not justify the expense. For spray mums, however, the center bud is typically removed, leaving five to nine flowers on the spray, depending on the variety.

Pests & Diseases

See Pests & Diseases in the Pot Mums section (p. 314).

Varieties

See Varieties in the Pot Mums section (p. 314).

Postharvest

Harvest spray chrysanthemums over the course of a week, taking those stems with flowers that are half to three-quarters open. There should be five to eight open flowers per spray. It is better to harvest flowers as they are ready and store them in the cooler than to let them remain on the plant. Hydrate stems for four hours prior to packing and shipping. If you must store stems, place them in buckets of water and hold them in coolers with temperatures at 35–38°F (2–3°C).

Harvested at the right stage and handled properly, stems should have more than a two-week postharvest life. Generally, leaves will deteriorate before flowers do.

Retailers should recut stems, strip foliage that would be below the water level, and hydrate for two hours. Chrysanthemums respond well to commercial postharvest preservatives.

Garden Mums

Garden mums used to be the chrysanthemum stepchild—an afterthought planted only because there were no other alternatives. That all changed in the 1990s. New varieties helped to fuel a revolution in culture and production. Faster crop schedules, more flower forms, and a broader color assortment all combined to stimulate the market. Simultaneously, more and more growers started to produce pansies in the East and South. Growers and retailers everywhere started to realize the potential that is today's fall marketplace. Garden mums play a large role in that market.

Creativity in containers has helped too: While 6" (15 cm) pots are definitely the most-grown size, you will also find garden mums in all sorts of containers, from 4" (10 cm) up to ½-bushel (18 l) baskets. Retail growers organize special fall harvest events complete with rides, food vendors, and pumpkin and gourd sales around their garden mum crop. Halloween is America's second largest holiday! By thinking creatively, there is every reason to believe that you can tap into consumers' predisposition to spend money at that time of year.

Garden mums are categorized in two ways. For natural-season crops there are very early, early, midseason, late-season, and very late or "season extender" varieties. In catalogs, you will see flowering dates for varieties, such as September 15 or October 5. Early-season mums initiate and develop flowers under the long days of summer. Temperatures can significantly influence them. Cool summer temperatures promote early flowering, while excessively hot summers can create heat delay. Midseason mums can flower under long days at cool temperatures, but they will not flower under long days at high temperatures. Late-season mums need short days to flower. Most natural-season growers select an assortment of mums to grow so they can offer fresh pots each week.

The second category is black cloth, or shaded, garden mums. These are forced into flower for sales in July and August by creating artificial short days using black cloth or black plastic. Here you will see varieties categorized by their black cloth response times (i.e., 6-, 7-, or 8-week varieties).

Propagation

Both unrooted and rooted cuttings are available. While most growers bring in rooted cuttings, many growers purchase unrooted cuttings. Unrooted cuttings can be direct-stuck, if space is available, into 4–6" (10–15 cm) pots. Cuttings that will go into 8" (20 cm) and larger pots are usually rooted in propagation trays. (See information on rooting pot mum cuttings beginning on p. 312.)

Growing On

Most growers receive rooted cuttings from late May until early July, earlier for August and early September crops of early-season varieties, later for late bloomers such as season extenders. Although some growers prefer to plant all varieties (early, midseason, and late) on the same date, other growers believe that it's easier to manage plant growth when plantings are somewhat staggered.

Plant liners or stick unrooted cuttings into moist soil when they arrive. If cuttings must be held, put them in a cooler at 35–40°F (2–4°C) for up to a few days. Garden mums are very reproductive. It is highly likely that you will receive some cuttings with terminal buds. If that is the case, plant up as normal, keep the plants warm and moist, and fertilize with 300 ppm of 20-10-20 immediately after potting and at each irrigation. You may give the cuttings an optional pinch four to six days after planting. (These buds may also be left on, which will cause the plant to pinch itself.)

If direct-sticking into a pot, select a well-drained medium with a pH of 5.8–6.2 and water pots well prior to sticking cuttings. Water again immediately after sticking cuttings to form a good connection between the medium and the cutting base. Mist pots frequently during the first few days to keep cuttings turgid.

Ideally, garden mums are grown outdoors (temperatures inside structures in the summer get too hot for the plants) with pots set out in fields equipped with automatic irrigation (preferably drip tubes and/or drip tape) and covered with permeable black ground cloth for weed control. Drip irrigation is preferred so that foliage stays dry, which helps prevent disease. Many growers start pots off in the greenhouse, where it is easier to maintain environmental conditions conducive to rooting or starting the cuttings. Northern growers may need to begin the crop indoors to avoid the chilly days and nights that can encourage premature crown buds in early to mid-June.

Garden mums, like all chrysanthemums, are considered to be heavy feeders. Maintain a constant liquid feed between 250–300 ppm. Proper nutrition, adequate spacing, and genetically superior varieties will provide exceptionally well-shaped plants. Begin fertilizing as soon as cuttings have roots and then fertilize at every irrigation with 250–300 ppm 20-10-20. If you are using a bark mix, increase the rate to 300–350 ppm. To help control plant growth and final height, switch fertilizers through the crop. Once plants are actively growing, rotate in 15-0-15 at every third irrigation. Once flower buds appear, switch entirely to 15-0-15, which will harden and tone plants. Supplement with magnesium to develop dark foliage and prevent any deficiency. Maintain a media pH from 5.5–6.2.

Some growers like to use slow-release fertilizers; however, because plants will be outdoors and exposed to the elements, release can be faster or slower than required. More than one application of a slow-release fertilizer may be necessary to ensure adequate crop nutrition. Once a slow-release fertilizer is added, it cannot be removed. Be careful not to apply too much slow-release fertilizer.

If you experience heavy rains for extended periods, be aware that nutrients may leach from pots. If this occurs, simply liquid feed plants as soon as there is a break in the rain. Discontinue feeding when buds are pea-sized and larger, switching to just clear water. By doing so, you will increase the postharvest life of your crop.

Do not let garden mums experience water stress, especially early in the crop during the first few weeks. Allow 10% to drain through at every irrigation. As the crop matures, plants can be allowed to wilt very slightly. Be sure your field is level. Low spots, where plants may stand in water after rains, are invitations to disease problems, especially *Pythium.*

Give plants plenty of space: The economics of serving your market will determine how much. For example, if you are shipping to the mass market, it is likely that you will not want oversized plants. However, if you are producing for garden centers, your own retail sales, or other higher-priced markets, your plants will be mounded and as much as 24–30" (61–76 cm) across. Six-inch (15 cm) pots can be spaced 12" × 12" (30 × 30 cm) to 15" × 15" (38 × 38 cm). Larger pots are generally spaced from 18" × 18" (46 × 46 cm) to 24" × 24" (61 × 61 cm). Bushel baskets may go all the way to 36" × 36" (91 × 91 cm) on center.

Temperature affects garden mums in the same way it affects other mum culture: High temperatures delay flowering. Flowering is fastest at 65°F (18°C). The ideal temperatures for growing garden mums are 65–75°F (18–24°C) days and 65°F (18°C) nights. If plants receive several nights of cool temperatures in June, the crop may bud prematurely. However, when it gets warm, the plant will generally grow around these buds. Cool temperatures at the end of the production cycle will cause pinking in some white varieties.

Provide high light, 4,500–7,500 f.c. (48–81 klux). If you will be flowering garden mums under natural day length for fall sales, there is no need to provide long days (unless rooting unrooted cuttings) or short days. Garden mums are not entirely photo-responsive: Many growers have discovered this when their crop has budded up, no matter how many hours they were given light.

Most natural-season growers do not pinch their plants any more, no matter what the size of the container. If pinching, some growers pinch at planting; many pinch ten to fourteen days after cuttings are planted, when new growth is 1.0–1.5" (3–4 cm) long and roots are at the bottom of the pot. Remove the top 0.5" (13 mm) of new growth. Virtually no growers pinch their garden mums a second time. In areas with cool nights, such as the Pacific Northwest and coastal California, pinch no later than July 4 for mid-August flowering. Southern growers should pinch no later than August 1. Other growers should target July 15 as the last pinch date.

Another natural-season production practice is to use two cuttings per 8" (15 cm) pot, plant from approximately July 15–20, and not pinch. The result is a beautiful 8" (15 cm) pot that looks like one planted up back in mid-June or so.

Height control is a topic of conversation for every garden mum grower. When buds are pea-sized, plants should be 75–80% of their desired finished height. B-Nine at rates from 2,500–5,000 ppm can be used early in the growing season, approximately early to mid-July, and repeated if needed two weeks later. Do not spray B-Nine too late, as flowering could be delayed and flowers could become discolored.

Once a mum has reached its final height, a 1–2 ppm Bonzi drench can control height in the final weeks of production. Be sure to irrigate the day before applying Bonzi so that pots are uniformly moist.

Some growers use Florel (300–500 ppm) to help minimize early-forming crown buds and perhaps increase branching. Be aware that response is variety dependent. Florel can also delay flowering if used after mid-July.

Early shaded program

Markets often want early-flowering garden mums in July and August. Plant cuttings for these early programs from April to early June. To schedule crops for specific dates, you will have to provide short days (black cloth) after you have finished the vegetative production phase. Plants can be scheduled as you would a pot mum.

Spring garden mums

Some growers produce garden mums in 4" (10 cm) pots and/or 18- to 36-cell packs. They make a nice spot of bright color, especially in combination pots. The best part is that consumers can cut them back and put them in the garden, and they will flower again in the fall.

Spring garden mums can be grown in the greenhouse without lighting or black cloth. For crops flowering from early March to April, plant rooted cuttings from January to early March (seven weeks prior to sale date). Pinch ten to fourteen days after planting a liner and follow the temperature and fertilizer recommendations above. B-Nine at 2,500 ppm can be used three weeks after pinching to create a stockier plant. No light/no shade programs are great for 18- and 36-pack programs and are generally grown from unrooted cuttings produced as a direct-stick crop—total crop time can be as fast as seven to eight weeks.

If you prefer to build up more vegetative growth prior to flowering or are growing 4" (10 cm) pots, provide crops with two weeks of long days after planting. For crops scheduled to flower after May 1, provide short days (twelve hours of complete darkness each night after March 15) after the completion of long days. Shade can be discontinued after buds begin to show color. Plants will be ready to sell about seven weeks after the start of short days.

Some varieties perform better in the spring than others; be sure to ask your supplier for a list.

Pests & Diseases

See Pests & Diseases in the Pot Mums section (p. 314). Because garden mums are an outdoor crop, most growers treat with preventive fungicide drenches and avoid overhead irrigation.

Varieties

See Varieties in the Pot Mums section (p. 314).

Postharvest

Plants are ready for sale as soon as flowering begins. Garden mums can even be sold budded, with color picture tags showing the color. Generally speaking, garden centers and retailers appreciate a few open flowers per plant so consumers can see what the flowers look like. Landscapers often prefer just a hint of color in the plants. Suppliers often can help you pick the best varieties for landscape jobs with superior color retention, rain tolerance, etc.

Due to their bulk, size, and wholesale price, most garden mums are sold very near to where they are grown. Most growers ship plants on rolling carts with liftgate trucks or in vans or short trucks fitted with shelves. Some growers sleeve plants to protect them, especially large premium plants

Retailers should be sure plants in flower do not dry out. They should display plants in cool temperatures (60–65°F [15–18°C]) and protect them from frost. Let garden mums be the foundation of your bustling fall season!

Cilantro (see *Coriandrum*)

Cineraria (see *Pericallis*)

Citrullus

Citrullus lanatus
Common name: Watermelon
Vegetable, annual

Watermelons are a vining plant and one of the highlights of the mid- to late summer ripening vegetables preferring warmth and full sun. Watermelons have enlarged fruits with mostly juicy, red innards; although yellow also exists. Fruits weigh up to several pounds.

Sow seed to the final container, such as a peat pot, rice hull pots, or Ellepots using three to four seeds per container and push down 0.25" (0.64 cm) below the surface of the media. Germinate at 75–80°F (24–27°C), and seedlings will emerge in eight to twelve days. Of all the vining crops (i.e., squash, pumpkin, cucumber, etc.), watermelon is the slowest to germinate and grow so many sow it earlier than the other crops mentioned.

Grow on at 65–70°F (18–21°C) nights and days at 70–75°F (21–24°C) and fertilize from 100–150 ppm nitrogen (N) from 20-10-20 alternating with 14-0-14, 15-5-15, or other calcium–potassium nitrate fertilizer.

From sowing to salable pot depends on the grower as well as the retailer's expectations. It will take from five to six weeks from sowing for pots to leaf out to the first or second true leaf stage (cotyledons develop first, true leaves next). However, the plants can be sold as early as two weeks after sowing, although signage should be included with instructions for home gardeners on how to plant. When sold at this young stage, home gardeners tend to bury pots with the cotyledons planted too deep, causing damage and eventual rotting. Others tend to leave the pot too high in the garden and, when not watered until the roots are established, the pots dry out and the plants die. Upon planting, the consumer should leave the top of the pot flush with the soil surface and cover the surface of the container with soil so all but the very base of the seedlings are under garden soil. If watered in, the plants will perform well.

One final point: Do not hang on to pots in the greenhouse if the plants start to stretch and vine. As the cotyledons yellow, the plants should be planted immediately to the garden or discarded. Sowings done every two weeks help to keep the crop looking fresh.

Clarkia (see *Godetia*)

Clematis

Clematis spp.
Common names: Clematis; virgin's bower
Perennial vine (Hardy to USDA Zones 3–8)

Clematis is the leading perennial vine, having no rival for flower color, performance, and number of varieties available on the market. In the United States, clematis is characterized by upright vining varieties like 'Jackmanii', which can stand alone in a yard or be mixed in with other plants. These large-flowered varieties are often grown up a trellis, arbor, lattice, or fence; although there are some varieties that will look their best when allowed to sprawl across the ground or over rocks or terraces. English gardeners have trained the plants onto the outer limbs of trees and over shrubs, a practice not common in the United States.

Plants climb by twisting their petioles (small stems between leaves and stems) around any stationary object onto which they can hold. Unlike ivies, whose modified roots along the plant stems can affix to the sides of walls, clematis needs poles, wire, or other structures for petioles to wrap around. Clematis is deciduous and, in Midwestern winters, die back to the ground. The stems are primarily, but not always, woody and possess large, single flowers from 4–6" (10–15 cm) across; although when grown under ideal conditions, there are some selections with flowers up to 10" (25 cm). The flowers are not a combination of petals and calyx, as is the case of many other flowering plants, but instead are composed of sepals, since there is only one set of floral appendages instead of two. Flower colors include red, white, yellow, purple, blue, pink, and many bicolors. Plants can vine from 6–25' (1.8–7.5 m) long, depending on the variety.

There are a number of other hybrids and species that do not fit this description. Most commercially available or wild species will have a flower size of 3" (7 cm) or less and often sweetly scented blossoms. Some hybrids have a slight scent, but most lack it altogether. Double-flowering varieties are also available but are not as common as their single-flowering counterparts.

Clematis is divided into three groups based on the time of year they flower and pruning requirements: (1) early-flowering selections that are not pruned until after they have flowered in spring, (2) early to midseason cultivars with vines that are pruned in late winter or early spring, and (3) late-flowering cultivars that bloom on new growth beginning in summer with vines can be cut down to the crown of the plant. Plant catalogues will indicate what time of year their various varieties will flower.

Propagation

Cuttings and bare roots are the most common methods of propagation.

Cuttings

Most growers buy in liners or small pots as opposed to starting from their own cuttings; cuttings must be made at the height of the spring season and clematis is difficult to root, with high losses on the propagation bench. Softwood cuttings must be used. Harvest internodal cuttings with a leaf pair, dormant buds at their base, and a stem below the leaves. Stick cuttings into a well-drained, sterile, rooting medium and maintain 75–78°F (24–26°C) soil temperatures until cuttings root four to six weeks later. Maintain a media pH of 7.0–7.5. Do not keep cuttings too wet; spritzing during the day is preferred to frequent misting, depending on ambient air temperatures. Under hot conditions, misting may be necessary. Harden cuttings off at 60–65°F (16–18°C) and hold at these temperatures until potting. Cuttings can be potted as soon as four weeks after sticking, although some selections take two to four weeks longer. Cuttings rooted in the spring form the basis for crops sold the following year.

Bare roots

A number of firms offer bare roots that can be potted up in winter into a 1-gal. (4 l) pot and grown for spring sales.

Growing On

For the methods below, one trellis per pot is recommended to keep the plants upright. Once actively growing, you will need to do redirect the vines from other pots and wrap around the trellis.

Cuttings

Upon receiving liners or small pots, unpack them immediately and water. Pot one plant per 4.5" (11 cm), 1-gal. (4 l), or 1.5-gal. (6 l) pot. For larger pot sizes, plant in smaller containers and step plants up as they obtain size.

More and more growers are purchasing clematis liners in the summer, growing them on through the fall and overwintering them to sell the following year. The advantages to such a crop are that plants can be frequently pruned to encourage branching. The resulting final product is a specimen plant that your customers will not find anywhere else.

Irrigation is critical to success with clematis. Be sure to water early in the day so that foliage is completely dry by nightfall. Make sure the medium is completely dry (but not to the point of wilting) before watering. Be sure to water pots thoroughly so that 15% of the water applied drains as leachate.

Fertilize with 20-20-20 or 20-10-20 at 150 ppm. If plants are growing quickly, increase the rate to 200 ppm. Leach with clear water every third or fourth irrigation. Avoid using slow-release fertilizers, as clematis goes through periods when it does not feed and you can quickly develop soluble salts problems with a continually releasing fertilizer.

Maintain the pH at 5.8–6.5. When the pH rises above 6.5, uptake of minor nutrients is inhibited, which can result in leaf yellowing. Some varieties are more sensitive and show symptoms of high pH earlier, such as 'Nelly Moser', 'Arctic Queen', 'Pink Champagne', 'Sugar Candy', and 'Henryii'.

Grow clematis in full sun during the spring; however, 30–50% shade is needed for production during summer months to moderate light levels and help maintain lower temperatures.

Clematis must pass through a dormant period before it will flower. A combination of short days and cool temperatures bring on dormancy. During production, begin lowering temperatures in November gradually over a period of seven to twenty days. Maintain 45–52°F (7–11°C) days and 36–42°F (2–5°C) nights. Cut plants back to 8–12" (20–30 cm). For good sanitation, remove leaves from the cold frame once plants have defoliated. Maintain dormancy for a minimum of six weeks. Plants can be held dormant until February or March. Then raise temperatures to 70°F (21°C) days and 60°F (15°C) nights. Do not pinch. Let them grow out, and they will flower in six to nine weeks.

When you receive clematis liners after December, they should have already passed through their required dormancy period. Containers started in early January from liners can be pinched about January 20, leaving two sets of nodes. Pinch again about February 20, leaving one or two sets of nodes. Do not pinch plants after this time if you want plants to bloom in early May. Keep an eye on plants after pinching to make sure new shoots emerge. If they do not emerge within seven days, leach with clear water to lower soluble salts levels. Shoots should then emerge within a few days.

Plants started after February 15 should not be pinched if you want them to bloom in early May.

In general, for a winter-potted liner, it will take from fourteen to sixteen weeks for flowering early or midseason varieties.

Bare root

Roots potted up in January or February and started out at 50–55°F (10–13°C) will be salable in May. Plants will not be in flower when sold and will bloom sporadically during their first summer with a greater amount of flower color the following year.

Pests & Diseases

Aphids, whiteflies, spider mites, and thrips can be problems. Botrytis, *Fusarium,* and clematis wilt (*Ascochyta*) are the most common diseases. To prevent disease problems, manage humidity levels, make sure plants are dry before they are watered, and maintain vigorous growth.

Troubleshooting

If you use a slow-release fertilizer, be sure to watch plants closely after pruning. Plants that do not sprout within seven to ten days may have too high soluble salts levels. Leaf yellowing can be caused by a pH that is too high. If you see leaf yellowing, test media for pH and soluble salts levels.

Postharvest

Clematis plants in bud and flower will hold up fairly well in shipping and can take boxing for brief periods of time. Flowers are sensitive to ethylene, so make sure to groom pots of all dead and dying foliage and old blooms prior to boxing, and do not store plants near sources of ethylene like ripening fruits and vegetables.

Retailers should display plants in a bright, cool, frost-free area. A display under lath or light shade will help keep temperatures moderate and reduce irrigation frequency. Treated well, clematis flowers will last several weeks.

Cleome

Cleome hassleriana
Common name: Spider flower
Annual

Cleome is one of the few tall, flowering annuals. Growing from 3–5' (91–152 cm), they are available in flower colors of rose pink, cherry, violet purple, and white. However, they have been difficult to germinate. Today, the 'Sparkler' series is available in primed seed, making germination headaches an issue of the past. If you have never tried *Cleome,* now is the time: Consumers love their exotic-looking flowers and bold plants. Landscapers love them, too, because they deliver.

Propagation

Sow seed into large plug trays (288s or larger). Select a sterile, well-drained, coarse, germination medium with a pH of 5.5–6.0. Cover with coarse vermiculite and germinate on the bench. (*Note:* Do not presoak plug trays with potassium nitrate prior to sowing, as this will inhibit germination of primed seed.)

Once cotyledons have expanded, begin fertilizing with 50–75 ppm from 14-0-14. Gradually increase light levels and fertilizer rates as seedlings grow. When at least one set of true leaves has formed (from fourteen to sixteen days after sowing), treat plugs with Bonzi at 2–4 ppm or B-Nine at 5,000 ppm. Treat plants before they are 1" (3 cm) tall.

Plugs will be ready for sale or transplant in three to six days after treating with growth regulators.

A 288-plug tray will be ready in three to four weeks.

Growing On

Keep *Cleome* seedlings actively growing by transplanting them before they are root-bound. Select a disease-free, well-drained medium with a moderate to high initial starter charge and a pH of 5.6–6.0.

Grow on at 75–80°F (24–27°C) days and 60–65°F (15–18°C) nights. High temperatures combined with low light cause plants to stretch.

Keep *Cleome* well irrigated: Do not let plants wilt, but do not keep media saturated either. Fertilize with 300 ppm from 20-10-20 at every irrigation. Leach with clear water every fourth irrigation. Media EC should be 1.8–2.0 mS/cm. High fertility levels promote branching while preventing leaf curl and yellow lower foliage.

Apply Bonzi at 20–30 ppm seven to fourteen days after transplanting. Make additional applications as needed based on your conditions. B-Nine is also effective at 4,000–5,000 ppm; however, it may delay flowering by four to seven days.

As plants grow, space them so that their branches will fully develop.

Upon transplanting 288 plugs, plants require ten to twelve weeks for 6" (15 cm) or 1-gal. (4 l) pots to be sold. Plants are sold green

Varieties

The 'Sparkler' series is great for 4" (10 cm), 6" (15 cm), or 1-gal. (4 l) pot production. Separate colors available include 'Lavender', 'Rose', and 'White'. 'Sparkler Blush' is an All-America Selections Award winner. 'Sparkler' is dwarf, growing to about 24" (61 cm) tall and just as wide. Primed seed is available, which means that 'Sparkler' is a snap to grow.

'Linde Armstrong' is a new vegetatively propagated, dwarf, pink selection that is thornless.

If you want to offer some of the old-fashioned, ultra-tall 3–5' (0.9–1.5 m) *Cleome,* try the 'Queen' series in 'Cherry', 'Rose', 'Violet', 'White', or 'Mix'. 'Queen' makes a bold landscape presence, capable of providing color in the most massive of settings and all summer long in most climates. It is a great plant for highway plantings. Sell 'Queen' green in April or May for planting out once the danger of frost has passed.

Postharvest

Plants will flower profusely through the summer, thriving in the heat. Stems will become somewhat prickly with spurs forming along them. Do not plant *Cleome* in areas where small children might handle them. *Cleome* will reseed itself in most locations.

Coleus (see *Solenostemon*)

Consolida

Consolida ambigua
Common name: Larkspur
Annual, cut flower

Larkspur is widely available as a cut flower in grower bunches or mixed bouquets throughout the spring. It is an ideal field cut crop that can be planted en masse. Plants thrive in cool temperatures and will die out in the heat and humidity of summer.

Plants grow from 2–3' (61–91 cm) tall in the field, but as a cut flower in a cold frame, they can reach much taller heights. Bedding plant growers can grow larkspur too. Offer 4" (10 cm) pots of direct-sown larkspur to customers for planting in the fall if you live in California or south of Washington, D.C., or St. Louis, or for planting in the early spring if you live in the Midwest or Northeast. It is also a great flower for gardeners to sow themselves!

Propagation

Sow seed onto a well-drained, disease-free, germination medium with a low starter charge. Choose larger plug trays and cover heavily with coarse vermiculite. Seed will germinate over ten to twenty days at 60–65°F (16–18°C). Keep germination temperatures below 70°F (21°C). Acclimatize plugs at 40–50°F (4–10°C) prior to planting outdoors in the spring. Transplant plugs on time—do not allow plants to become root-bound. Allow approximately twenty-eight to thirty-five days for plug production.

Growing On

In mild-winter areas (i.e., south of 38° latitude) and California, growers can sow seed directly to the field in September and October for harvest in March. In the Midwest and North, transplant plugs to the field in April and May, or sow seed in April and May for harvest ten to thirteen weeks later. For greenhouse crops, seed sown in November or December will flower in April in the Midwest. California growers can crop larkspur year-round, except during the heat of the summer.

If you are direct-sowing, calculate about 25 oz. of seed/acre (175 g/ha) for double 8" (20 cm) rows spaced 3' (91 cm) apart. Sow six to seven weeks prior to the ground freezing to give plants enough time to become established prior to the onset of cold temperatures. You can also wait to sow seed late enough to avoid germination before the ground freezes. This seed will germinate quickly in the spring and flower two to three weeks earlier than if you waited to sow until the ground dries in the spring. Cover seed after sowing. Seed will germinate over fourteen to twenty-five days at a rate of 60–80%.

Select only well-drained fields for larkspur production. Prior to planting, incorporate a preplant fertilizer. Apply a sidedress of fertilizer once or twice during the growing season.

If you are planting out plugs, space plants at a density of about 2 plants/ft.^{2} (20/m^2) and provide two to three layers of support.

Pests & Diseases

Fusarium and powdery mildew can become problems.

Varieties

Try the 'Giant Imperial' series, which grows from 36–48" (91–120 cm) tall. The series comes in eight separate colors from deep blue to rose, lavender, pink, and white as well as a mix. The 'QIS' series is excellent for fresh and dried cut flower production. The series is exceptionally uniform in flower color and stem quality and is available in seven separate colors.

Postharvest

Stems are ready for harvest when two to five of the lower flowers are open, or as much as a third of the stem is open. If you will be drying larkspur, harvest stems in near-full flower, when just a couple of buds have not opened. Some growers sell the central leader as a fresh cut flower and harvest laterals for the dried market.

Place fresh stems in water treated with floral preservative as soon as possible. Larkspur is sensitive to the effects of ethylene and responds well to STS or EthylBloc. Store in a cooler at 36–41°F (2–5°C) for only one or two days, not longer. Ideally, ship larkspur upright in water.

Retailers should recut stems and keep flowers away from sources of ethylene and heat.

To dry, hang stems upside down in a well-ventilated, dark, dry place at 70–80°F (21–27°C). Cure for two to four weeks.

Cordyline

Cordyline indivisa
Common names: Dracaena spike; spikes
Annual, foliage plant

Bedding plant growers should be sure to have dracaena spikes on hand in the spring: They are the vertical focal point in thousands and thousands of mixed combination planters sold coast to coast featuring a red zonal geranium and variegated vinca vine. Without this slice of true Americana sitting on the front stoop of home after home, it just wouldn't be summertime!

While ornamental grasses are stealing some of the limelight for use in combination pots, dracaena is still very popular.

Propagation

Most growers buy in plugs or liners because dracaena is a long crop from seed, taking twenty to twenty-two weeks for a 128 plug. Sow seed in summer into a well-drained, germination medium with a pH of 5.5–5.8. Most growers produce dracaena in larger plug trays such as 128s or smaller trays, and then bump plants up as they grow. Germinate at a soil temperature of 75–78°F (24–26°C). Do not cover seed. Make sure to supply 150–400 f.c. (1.6–4.3 klux) of light. Keep medium uniformly moist but not saturated. Germination is slow, so be patient. It takes from thirty to forty days at 75–78°F (24–26°C).

Gradually increase light levels as plugs grow and develop. Once cotyledons are expanded, apply 50–100 ppm from 14-0-14 at every other irrigation, increasing to 100–150 ppm as plants develop roots.

Once seedlings are established, they can be grown in full sun and warm temperatures through the summer and fall.

Growing On

Plant a 128 liner in a 4" (10 cm) pot or 18-cell packs. From a well-grown liner, you will be able to sell pots as soon as they are rooted in; eight to ten weeks from liners for 4" (10 cm) pots. Grow in high light and ambient temperatures: They tolerate a wide range of conditions. If growth regulators are required, A-Rest may be used.

Troubleshooting

Cordyline is an easy crop to grow; however, sometimes when media pH is too high (above 6.5), upper leaves may become yellow from iron deficiency. Plants may also be stunted or grow unevenly. High pH can also cause poor root development. Under cool, wet growing conditions, iron uptake is inhibited. If the problem is due to true iron deficiency, use iron sulfate as a drench. Rinse plants immediately to prevent burning. If wet conditions are the problem (and iron levels are fine), increase soil temperatures to dry out plants.

Coreopsis

Coreopsis grandiflora
Common name: Tickseed
Perennial (Hardy to USDA Zones 4–9)

This reliable performer is native in much of the United States. Double and single, daisy-like flowers are bright golden-yellow and appear throughout the summer. For growers, coreopsis is an easy crop: It can be forced into flower for spring sales and will subsequently deliver summer color for consumers who are dedicated to deadheading.

Propagation

Seed

Sow seed onto a disease-free, germination media with a pH of 5.5–5.8 in a 288-plug tray. Cover lightly with vermiculite. Keep media evenly moist but not saturated. Germinate seed at 70–75°F (21–24°C) soil temperatures. Radicles will emerge in four to seven days.

Move trays to Stage 2 for stem and cotyledon emergence, which will occur over five to nine days. Maintain substrate temperatures at 65–75°F (18–21°C). Begin to dry trays down once radicles have emerged, allowing media to dry slightly before irrigating. Provide 500–1,000 f.c. (5.4–11 klux) of light. As soon as cotyledons have expanded, apply 50–75 ppm from 14-0-14 at every other irrigation.

Once the stand is up, trays are ready for Stage 3, growth and development of true leaves, which takes eighteen to twenty-four days for 288s. Maintain soil temperatures. Increase feed to 100–150 ppm from 14-0-14, fertilizing at every second or third irrigation. Increase light levels to 1,000–2,000 f.c. (11–22 klux). Grow plugs under natural day lengths (less than twelve hours). If height control is needed, B-Nine and Sumagic are effective.

Harden plugs prior to transplanting or sale for seven days. Lower soil temperatures to 60–62°F (16–17°C) and allow the medium to dry thoroughly between irrigations. Increase light levels to 2,000–3,000 f.c. (22–32 klux). Grow plugs under natural day lengths. Fertilize with 100–150 ppm from 14-0-14 as needed.

A 288-plug tray with one seed per cell takes five to six weeks to be ready to transplant to larger containers.

Cuttings

There are a number of varieties that are propagated vegetatively by cutting. Patented varieties are protected, and a license is required.

Direct-stick one cutting per cell of a 105-liner tray (or similar tray) with or without a rooting hormone. Rooting hormones are better during the colder months of the year. Place the tray under mist for four to six days at 70–72°F (21–22°C) soil temperature to allow roots to begin to initiate. Do not pinch. Allow four to five weeks to root fully. One key aspect during propagation is to avoid saturated media during rooting.

Growing On

Seed

Plant one 288 plug per 4" (10 cm) pot and one to three plugs per 6" (15 cm) or 1-gal. (4 l) pot, depending on the season and plug size. Select a disease-free medium with a pH of 5.5–6.2 and a moderate initial starter charge.

The key to seed-propagated varieties is to understand which varieties will flower in May or June from a winter sowing and which ones only flower with a vernalization (cold treatment). For instance, varieties like 'Early Sunrise', 'Rising Sun', and others will flower in spring or summer when sown in January or February. There are other varieties, however, that will not flower unless they are sown from June to September for potting up August to November and are overwintered in a cold environment.

In order to flower, plants must be bulked up to a minimum size and then receive a cold treatment. Bulk plants to eight nodes (sixteen leaves) in pots prior to providing cold treatment. To bulk, grow on at 60–65°F (16–18°C) days and 55–60°F (13–15°C) nights. Allow pots to dry thoroughly between irrigations. Provide 2,000–3,000 f.c. (22–32 klux) of light and natural day lengths (less than twelve hours). Fertilize at every other irrigation with 100–150 ppm from 14-0-14, alternating with 15-0-15. You can also transplant a 288 plug to a liner tray (32, 50, or 72 cells per flat) to bulk up as well prior to the cold treatment.

Once plants have sized up (sixteen leaves), provide a cold treatment in a lit cooler (25–50 f.c. [269–538 lux]) or cold frame. Plants need eight to twelve weeks of cold treatment at 35–41°F (2–5°C).

Bring pots or trays out of cold treatment. (Pot plugs if necessary). For 6" (15 cm) or 1-gal. (4 l) pots, use multiple cooled plugs per pot for fastest crop times. Fertilize cooled plants with 150 ppm from 20-10-20 at every other irrigation. Grow on at 65–68°F (18–20°C) constant day/night temperatures. Allow plants to thoroughly dry between irrigations. Coreopsis is a rapid grower, as plants grow they fill in pots quickly. Be careful to keep small pots irrigated as needed as plants wilt quickly once they have fully rooted in. Provide 3,000–5,000 f.c. (32–54 klux) of light and long days (more than fourteen hours). To time pots for flowering, provide three weeks of long days by extending the day length or using night interruption lighting. If growth regulators are needed, B-Nine and Sumagic are effective. Flowering will be somewhat delayed with growth regulator use.

Some growers buy in cooled plugs in liner trays from specialist perennial suppliers.

Note: You can bring plugs of 'Early Sunrise' into the greenhouse in March with your normal bedding plant shipments for flowering 4 or 6" (10 or 15 cm) pots by mid-April to May. Based on Michigan State University recommendations, 'Early Sunrise' is best timed if you are forcing pots by providing plants with six weeks of short days (less than twelve hours) followed by long days until sale.

Cuttings

Transplant one liner (105 or larger sizes) per 1-qt. (1.1 l) or 1-gal. (4 l) container and grow on at daily temperatures of 62–67°F (16–19°C) to root in the final container. Once rooted, drop temperatures to 35–41°F (2–5°C) to provide a cold treatment (vernalization). For those varieties requiring this treatment for flowering, the exposure to cold requires eight to twelve weeks. There are a number of varieties that benefit from a cold treatment but do not require it to flower. Most companies note this in their culture information.

After the cold treatment, plants will flower naturally as days lengthen and temperatures rise in late spring. Plants in 1-qt. (1.1 l) containers take seven to nine weeks to flower, while 1-gal. (4 l) containers take eight to ten weeks. If plants are sold green in May, the plants will flower in either June or July.

Pests & Diseases

Whitefly can show up in the greenhouse or closed cold frame. Aphids can attack flowering plants. Watch for powdery mildew and INSV/TSWV.

Varieties

The flowers of 'Baby Sun' are golden yellow with red flecks at the base of each petal. Plants grow 12–16" (30–41 cm) tall. 'Early Sunrise' is a popular seed-grown variety. 'Early Sunrise' will flower the first year it is sown from seed, so plants do not require a cold treatment or overwintering for flowering. 'Early Sunrise' is so fast to flower that crops can be sold in just one hundred days from sowing in the spring.

Try the tall-growing 'Sunburst'. With a height of 3' (91 cm), it is perfect for gardeners looking for cut flowers. Its habit is open and floppy. 'Sunray', one of the most popular coreopsis varieties, grows to 24" (61 cm) tall and has a tidy, compact plant habit with golden-yellow flowers.

'Tequila Sunrise', a vegetatively propagated variety is patented, making unlicensed propagation illegal. Leaves are variegated with cream and yellow. New foliage in the spring is reddish pink, turning to mahogany in the fall. Flowers are yellow-orange sporadically through the summer.

'Sunny Day', another protected, vegetatively propagated variety, is a pure yellow, flowering variety with green foliage. It is an excellent garden performer.

Postharvest

Plants are ready to sell green with color tags once pots have filled in or in flower once flowering has started. Note that coreopsis will not flower all at once, making bench runs of flowering pots difficult in the spring. Retailers should display pots in partial shade to reduce watering requirements, as flowering plants allowed to wilt frequently will quickly become shopworn.

Instruct consumers to keep plants deadheaded for continual flowering from late May to late fall. Plants may last as long as four years in the garden and oftentimes will reseed themselves.

Coreopsis verticillata
Common name: Threadleaf coreopsis
Perennial (Hardy to USDA Zones 4–9)

The single, yellow, daisy flowers of this drought-tolerant perennial sit atop soft, needlelike leaves. It is a mainstay for summer flowering.

Propagation & Growing On

C. verticillata is more commonly propagated by cuttings than by bare-root transplants. Plants are propagated in the spring or fall.

Cuttings

Direct-stick two cuttings per cell of a 105-liner (or larger) tray with or without a rooting hormone. Rooting hormones are better during the colder months of the year. Place the tray under mist for six to nine days at 70–72°F (21–22°C) soil temperature to allow roots to begin to initiate. Pinch the plants eighteen to twenty-four days after sticking the cuttings. Allow five to six weeks to root fully.

Bare-root transplants

C. verticillata is more commonly propagated by cuttings than by bare-root transplants. Liners or bare-root plants purchased in the fall will make the largest, highest-quality pots. Plant No. 1 transplants into 6" (15 cm) or 1-gal. (4 l) pots and liners into 4 or 5" (10 or 13 cm) pots. The larger the plants you start with, the larger the finished product. Be sure to only use large plants in 1-gal. (4 l) pots in order to fill the pots.

Select a well-drained, disease-free medium with a moderate initial starter charge and a pH of 5.8–6.2.

While plants do not require a cold treatment in order to flower, overwintering in a cold frame for ten to fifteen weeks at 41°F (5°C) will increase the flower number as well as speed flowering by one or two weeks. But since a cold treatment is not necessary, you can also wait to bring in liners or bare-root transplants; be aware, however, that crops grown from spring plantings will not be as full.

If you choose to overwinter plants, allow them to fully root into the pot and become established prior to exposing them to cold treatments. Bulk plants under natural short days. Allow pots to dry somewhat between irrigations. Fertilize at every other irrigation with 100 ppm of 14-0-14. Grow at 55–60°F (15–16°C) days and 50–55°F (10–15°C) nights. Root in plants under short days (less than twelve hours).

Provide the cold treatment in a cooler (under 50 f.c. [538 lux] of light) or cold frame at 41°F (5°C) for ten to fifteen weeks. During this time, do not fertilize plants.

Bring pots out of the cold treatment and force them in a cool greenhouse at 65–68°F (20°C) days and nights. Provide 3,000–5,000 f.c. (32–54 klux) of light. Keep plants dry to somewhat moist, being careful to allow them to dry out between irrigations but not to reach permanent wilting. Consistent moisture is ideal. Feed with 100 ppm from 14-0-14 or 20-10-20 at every other irrigation. Maintain pH at 5.8–6.2 and EC at 1.0 mS/cm. Provide long days (more than fourteen hours) by extending the day length or using night interruption lighting until flowering. Flowering will take about seven weeks.

If you are bringing plants in during the winter for potting and sale in the spring and summer, size plants up under short days prior to allowing plants to flower. Bare-root transplants will not require sizing up prior to flowering, but make sure that pots are well rooted in prior to long days. Conversely, you can provide a pinch to liners at planting to force plants to develop bulk.

B-Nine and Sumagic can be used to control height two to three weeks after the end of cold treatment, but flowering may be somewhat delayed.

Pests & Diseases

Be careful of powdery mildew and root rots on threadleaf coreopsis.

Varieties

'Moonbeam', with its familiar soft-yellow flowers, is the most popular threadleaf coreopsis variety. Plants grow to 18" (46 cm) in the garden and flower throughout the summer. It is a great perennial in just about every corner of the country. 'Zagreb' is also very popular and has golden-yellow flowers on more dwarf-growing plants (12" [30 cm]). It is especially good under difficult conditions in the garden.

Related Species

Coreopsis rosea, commonly called pink coreopsis or pink threadleaf coreopsis, is usually propagated by cuttings. The above information under *Coreopsis verticillata* can be followed for its propagation and production. One additional point: *Coreopsis rosea* can be an aggressive variety and can fill in an area in a few short years.

Postharvest

Ship plants as soon as they begin to flower. Plants can get floppy in the pot if held after flowering starts. Cut plants back once flowering is finished to encourage rebloom.

Coriandrum

Coriandrum sativum
Common names: Cilantro; coriander
Annual herb

Mexican cuisine has become haute couture in the United States in the past two decades, and coriander is one of the herbs that complement many of these dishes. The foliage is called cilantro, and the leaves are mandatory for the best-tasting salsas, guacamoles, tacos, and more. The seed is called coriander, and it is an equally important ingredient in many Asian and Middle Eastern dishes as well. It is somewhat ironic that the iconic plant is not a Mexican native but instead was brought to current-day Mexico by Spaniards.

Propagation & Growing On

Cilantro/coriander is easily propagated by seed. However, it does not appreciate transplanting since it forms an enlarged root that can become root-bound in a pot, which spoils later performance. For this reason it is best to sow it directly into the home herb garden.

That being said, you can sow seed directly into the final 4" (10 cm) or similar-sized container using three to six seeds per pot and allowing ten to twelve weeks from sowing for a salable pot for retail sales. Grow on at 55–65°F (13–18°C) days and slightly cooler nights. Since this is a cool-season crop, it is understandable why it bolts in warm to hot summers; it bolts just as quickly when it is transplanted to the garden as opposed to sown directly.

If needed, multiseed two to three seeds into 288-plug trays and cover lightly with coarse vermiculite. Germinate at 65–70°F (18–21°C). Seed will germinate in about seven to ten days. Once cotyledons have expanded, begin fertilizing once a week with 50–75 ppm from 14-0-14, increasing the rate to 100–150 ppm as seedlings mature. Provide 500–1,500 f.c. (5.4–16 klux) of light, gradually increasing light levels to 2,000–2,500 f.c. (22–27 klux) as seedlings develop. Maintain cool growing conditions. Seedlings that dry out too much bolt prematurely. Allow three to four weeks for plug crops.

Transplant one plug per cell or 4" (10 cm) pot. Grow plants on at 62–68°F (17–20°C) days and 55–58°F (13–14°C) nights. Fertilize with 150–200 ppm from 20-10-20 at every other irrigation and grow under full sun, while maintaining cool temperatures.

Plants will bolt, flower, and then die out in warm summers. Once plants have flowered, gardeners can pull them out and plant another crop in the fall, enjoying fresh cilantro leaves until hard frost.

Cortaderia

Cortaderia selloana
Common name: Pampas grass
Perennial or tender perennial grass (Hardy from USDA Zones 7–10)

This vigorous and beautiful grass grows from 7–12' (2–4 m) tall in warm-winter locations with large, feather duster–like plumes in white or light pink. The plumes can be as large as 15–24" (38–61 cm) long, depending on the growing season, and are excellent in dried flower bouquets.

Propagation & Growing On

Cortaderia is a warm-season grass and performs best during warmer parts of the year. For this reason seed is sown in spring or summer for plants to overwinter for sales the following year. Temperatures above 60°F (16°C) help to stimulate growth.

Seed sown in December or early January and covered lightly with vermiculite or peat moss will germinate in about a week or less (at 72°F [22°C]) and can be transplanted to larger containers. You can sow several seeds per cell of a 48-cell pack, allow eight to ten weeks, and then transplant to 1-qt. (1.1 l) or 1-gal. (4 l) containers. Quart containers will be ready in ten to fourteen weeks after transplanting. Keep in mind that this reflects growing under late winter and spring conditions, the cooler parts of the year. For the best performance, sow seed during summer and transplant to larger containers (1-gal. [4 l] pots), overwinter, and then sell the plants the following year.

Seed sown in December or January produces well-rooted 1-qt. (1.1 l) pots in the spring. However, these plants grow to only 3–4' (91–122 cm) by late summer with limited or no blooms by autumn. For flowering plants, sow the year before and overwinter for spring sales the following year.

Corynephorus

Corynephorus canescens
Common names: Clubawn grass; grey hair grass
Grass, tender perennial (Hardy to USDA Zones 7–9)

This low-growing, blue-green grass produces spiky tufts, grows from 8–20" (20–25 cm) tall, and prefers full-sun to part-shade locations in the garden. Some references suggest that this plant is hardy to USDA Zone 4. However, in cold-winter areas the plants have not been reliably hardy. Plants perform best in a well-drained location.

Propagation & Growing On

Sow several seeds (or use one multiseeded pellet, if available) to a 128- or 288-plug tray in January. These will be ready to transplant in six to eight weeks to 18 trays per flat or 4" (10 cm) pot for salable plants in May. The total crop time from seeding to salable plants is sixteen to eighteen weeks.

Varieties

'Spiky Blue' is the most common variety available in this species. It makes a wonderful border or edging grass but only spreads 8" (20 cm) across so space plants accordingly.

Cosmos

Cosmos bipinnatus
Common name: Cosmos
Annual

This summer annual is easy to grow and ranges in height from 24–36" (61–91 cm) tall. Flowers in colors of pink, white, carmine, and rose are mostly single, although semi-double and tubular flower forms are also available.

Propagation

Seed is the most common method of propagation. Seed sown to a 288 plug, lightly covered with vermiculite, and germinated at a temperature of 68–72°F (20–22°C) will produce seedlings in three to seven days. A 288-plug tray will be ready to transplant to larger containers in four to five weeks.

If temperatures are warm, seedlings can grow rather quickly. A 15 ppm Bonzi spray can be used to control growth in Stage 1 (seedling emergence).

Growing On

Transplant one plug per 18-cell pack or 4" (10 cm) pot, and plants will grow quickly. They grow mostly upright, many times without branching. After transplanting into 4" (10 cm) pots, allow six to seven weeks for budded to flowering plants on more dwarf selections. For taller varieties, plants are sold green in large packs (18 cells per flat) or 4" (10 cm) pots and will flower later in the summer.

Varieties

The 'Sonata' series is a more dwarf variety and can be sold in color, unlike the majority of cosmos selections, which will not flower until later in the summer and are sold green.

Crossandra

Crossandra infundibuliformis
Flowering potted plant, annual

Crossandra is an unusual, attractive pot plant that originally came from India and is commonly used in the warm/hot areas of North America. Its bright salmon-orange flowers appear atop gardenia-like glossy foliage, making a bold, tropical statement.

It is a long crop and takes a lot of heat, which limits its use to more southern locales. In northern climates, *Crossandra* can take as long as seven to nine months to finish from sowing. Crop time is about three months if using cuttings.

Propagation

Sow seed onto a well-drained, disease-free medium with a pH of 5.5–5.8. Cover seed with coarse vermiculite. Maintain total saturation of the germination medium: You should be able to easily squeeze water from the medium. Alternate day and night temperatures to improve germination: Grow at 80°F (27°C) days and 70°F (21°C) nights. Radicles will begin to emerge in ten to fourteen days. Germination is slow and sporadic.

Move plants to Stage 2 for stem and cotyledon emergence when you have about 50% or more germination. Maintain the same alternating day and night temperatures as above. Begin to allow the medium to dry somewhat between irrigations once radicles have emerged. As soon as cotyledons appear, fertilize with 50–75 ppm weekly from 14-0-14. Provide 1,000–1,500 f.c. (11–16 klux) of light. Stage 2 will take about ten to fourteen days.

Move trays to Stage 3 to develop true leaves. Continue to alternate day and night temperatures as outlined above. Allow the medium to dry somewhat between irrigations and maintain 1,000–1,500 f.c. (11–16 klux). Increase feed to 100–150 ppm weekly, alternating between 20-10-20 and 14-0-14. If growth regulators are needed, B-Nine may be used. Stage 3 takes from seven to fourteen days for 512s.

Harden plugs prior to transplanting them into 2.5" (6 cm) liners for seven days. Continue the same alternating day and night temperatures. Fertilize once with 100–150 ppm from 14-0-14.

Transplant 512 plugs into 2.5" (6 cm) liners. Select a well-drained, disease-free medium with a pH of 5.8–6.5. Plugs will be about seven weeks old at this stage. Provide day temperatures of 75–80°F (24–27°C) and night temperatures of 70°F (21°C). Do not allow temperatures to drop below 45°F (7°C). If temperatures drop too low, leaves will turn black and fall off. Allow the medium to dry between irrigations. Increase light levels to 1,500–1,800 f.c. (16–19 klux); light levels must remain below 1,800 f.c. (19 klux) to prevent flowering. Fertilize weekly with 150–200 ppm from 20-10-20, alternating with 14-0-14. Time to grow liners out will be about ten to fourteen weeks.

Crossandra can also be vegetatively propagated from tip cuttings. Rooting takes about twenty-one to twenty-eight days at a 75°F (24°C) substrate temperature and using the alternating day/night air temperatures outlined above.

Growing On

Plant liners into 4" (10 cm) pots using one plant per pot. Use a disease-free medium with a pH of 5.8–6.5.

Plants will spend their first six weeks in the vegetative phase of production. During this time, grow at 75–80°F (24–27°C) days and 70°F (21°C) nights. Maintain light levels at 1,500–1,800 f.c. (16–19 klux). Keep plants uniformly moist but not wet. Fertilize at every other irrigation with 150–200 ppm from 20-10-20, alternating with 14-0-14. Space plants as leaves begin to touch.

Four to six weeks after transplanting, pinch all shoot tips that emerge to encourage branching. Pinching will delay flowering by up to fourteen days. B-Nine at 2,500–5,000 ppm can be used after the pinch when new growth is 1.5" (4 cm) long. Bonzi at 50 ppm is also effective two weeks after the pinch.

For production weeks 7–12, your goal is to flower plants. Do so by raising day temperatures to 78–82°F (26–28°C) days and lowering night temperatures to 65°F (21°C). Increase light levels to 2,500–3,500 f.c. (27–38 klux) to encourage flowering. Fertilize at every other irrigation with 150–200 ppm from 20-10-20, alternating with 14-0-14. Space plants as leaves begin to touch.

Crossandra likes it hot: Do not grow at temperatures less than 65°F (18°C). It takes twelve to fourteen for flowering pots from liners.

Pests & Diseases

Aphids, mites, and whiteflies can become problems on *Crossandra.* Diseases that can strike include bacterial leaf spot, *Pythium,* and rhizoctonia.

Troubleshooting

Plants that dry out or wilt will show leaf edge burn. High soluble salts levels can cause hard plant growth and/or stunted plants. Slow-growing, slow-flowering plants may be due to low production temperatures, low light levels, or low fertility. Light levels that are too high early in the crop, failure to pinch off early flowers, and/or low fertility can lead to flowering on small plants. Lack of fertilizer, plants not being pinched, or low potassium levels can result in small, weak-stemmed plants. Leaves turning black and dropping off means that plants have been exposed to temperatures that are less than 50°F (10°C).

Varieties

'Tropic Flame' is fast growing and heat tolerant, making it a regular in the southern landscape. Plants grow to about 10" (25 cm) tall and produce salmon-orange flowers.

The 'Florida' series is vegetatively propagated and more cold tolerant than 'Tropic Flame'. Available in 'Florida Flame', 'Florida Passion', 'Florida Summer', and 'Florida Sunset', flowers are shades of red, orange, and yellow. To turn foliage glossy, treat the series with B-Nine at 5,000 ppm when new breaks are 1.5" (4 cm) after the pinch. A-Rest and Bonzi will also cause leaves to gloss up.

Postharvest

Plants are ready to sell once they are 50% in flower. Store and ship *Crossandra* warm (50–55°F [10–13°C]). Never ship plants below 45°F (7°C). *Crossandra* is sensitive to ethylene, so avoid high temperatures, do not box plants for long periods of time, and keep plants away from ripening fruits and vegetables. Retailers should display plants at 65–70°F (18–21°C) under 300–1,000 f.c. (3.2–11 klux) of light.

Cucumis

Cucumis melo
Common names: Melon; cantaloupe; rockmelon
Annual vegetable

Actually from the tropical regions of India and Africa, we may not think of cantaloupe as a tropical plant but its heat-loving requirements suggest otherwise. Plantsare trailing if left on the ground but will vine up a fence as well. Cantaloupe prefers full sun.

Propagation & Growing On

Sow seed to the final container, such as a peat pot, rice hull pot, or Ellepot, using three seeds per container. Push seeds down 0.25" (0.6 cm) below the surface of the media. Germinate at 70–75°F (21–24°C), and seedlings will emerge in seven to ten days.

Grow on at 65–70°F (18–21°C) nights and 70–75°F (21–24°C) days and fertilize from 100–150 ppm nitrogen (N) from 20-10-20 alternating with 14-0-14, 15-5-15, or other calcium/potassium-nitrate fertilizer.

From sowing to salable pot depends on the grower as well as the retailer's expectations. It will take from four to five weeks from sowing for pots to leaf out to the first or second true leaf stage (one set of cotyledons develop first, true leaves next). However, the plants can be sold as early as two to three weeks after sowing with signage for home gardeners on how to plant.

When sold at this young stage, some home gardeners tend to bury pots with the cotyledons planted too deep, causing damage and eventual rotting. Others tend to leave the pot too high in the garden and, when not watered until the roots are established, the pots dry out and the plants die. Upon planting, consumers should leave the top of the pot flush with the soil surface, cover the surface of the container with soil so all but the very base of the seedlings are under garden soil, and water in. If home gardeners follow these instructions, plants will perform well.

One final point: Do not hang on to pots in the greenhouse if plants start to stretch and vine. As the cotyledons yellow, the plants should be planted immediately to the garden or discarded. Sowings done every two weeks help to keep the crop looking fresh.

Cucumis sativus
Common name: Cucumber
Annual vegetable

This vining plant is often left as a trailing plant across the garden, although it can be planted to vine up a fence or similar structure. Fruits are classified in two ways, slicers or pickles. Cucumbers prefer full sun.

Propagation & Growing On

Follow the same culture information for cucumbers as for *Cucumis melo.*

However, there is one major difference: Cucumbers are earlier to germinate and grow. You can easily take one week off of the production noted for melons. As for germination, seedlings emerge in four to seven days, several days ahead of melons.

Cuphea

Cuphea hyssopifola
Common name: Mexican heather
Annual

This tough-as-nails annual becomes semi-woody in some regions, performing under even the hottest and most humid of conditions and flowering all summer long. In mild coastal climates and many parts of the Deep South, it can overwinter, serving as a low-growing backdrop all year long.

Cuphea is a great addition to your spring pot or hanging basket program, so be sure to have some on hand when the season hits. If the landscape trade is important to you, you will also want to plan on including *Cuphea* in your product mix.

Propagation

Plants may be propagated by seed or cuttings. Cover seed lightly after sowing with coarse vermiculite and germinate at 70–75°F (21–24°C). Seed will germinate over twelve to fifteen days. A 288-plug tray will take from six to seven weeks to be ready to transplant to larger containers.

Growing On

Select a well-drained, disease-free, soilless medium with a moderate starter charge and a pH of 5.6–5.8. Bark-based mixes are excellent. Use 1 liner/4 or 5" (10 or 13 cm) pot and 1–2 liners/6" (15 cm) pot or 10" (25 cm) basket.

Do not allow plants to become water stressed, as this promotes leaf drop. Maintain uniform moisture levels. Fertilize with 100–150 ppm from 20-10-20, alternating with 15-0-15 at every other irrigation. *Cuphea* is highly sensitive to soluble salts damage. If salts become a problem, leach with clear water. Avoid applying excessive ammonia, as this will cause overgrown, large leaves that show marginal necrosis.

About ten to fourteen days after potting, pinch plants to promote branching. Pinch back to the fifth or sixth pair of leaves, leaving 1.0–1.5" (3–4 cm) above the substrate line. Pinch 4 or 5" (10 or 13 cm) pots once; 6" (15 cm) pots and 10" (25 cm) baskets twice. If you need growth regulators, Cycocel is effective. Reduce growth regulator rates once light intensity increases.

While *Cuphea* is day neutral, flowering is affected by total light accumulation. Plants grown during the winter will take much longer to finish than plants grown in the spring or early summer will.

Pests & Diseases

Aphids, fungus gnats, spider mites, and whiteflies can become pest problems. Diseases that can strike include botrytis, *Cylindrocladium, Phytophthora, Pythium,* and rhizoctonia.

Troubleshooting

Stem canker, overwatering, drought stress, or high soluble salts can cause plant collapse. Lack of flowering on vegetative plants can be due to excessive ammonia fertilizer, too much fertilizer under low light conditions, and/or low light and overwatering combined. Plants drying out between irrigations or high soluble salts can lead to foliage necrosis. Low fertilization during early stages of growth can result in poorly branching plants that are spindly.

Varieties

'Allyson', the industry standard, is a fabulous plant. Its pink-lavender flowers appear on vase-shaped, 12" (30 cm) plants. 'Alba' is a white selection.

Related Species

Cuphea ignea, commonly known as cigar plant, features tubular, red-orange flowers that are highlighted with black. Plants are more upright, reaching 12–24" (30–61 cm) tall, and become quite woody at the base. 'Dynamite' is a dwarf selection, reaching only 8–10" (20–25 cm) tall. Foliage is deep, dark green, and plants become covered with small, tubular, bright-scarlet flowers.

Sow seed to a 288-plug tray and cover lightly with vermiculite. Germinate at 70–75°F (21–24°C). The root radicle emerges in four to six days, and plug trays are salable after five to six weeks.

Transplanted to a 4" (10 cm) pot, allow six to eight weeks for a salable plant. Plants do not need to be pinched and will be sold green or at first flower, as opposed to fully flowering.

Cuphea ignea

Cyclamen

Cyclamen persicum
Flowering pot plant

Cyclamen have gotten a bad rap over the past few years. Their long crop time (particularly when grown from seed) and rather exacting growing needs can be frustrating and turn growers toward easier, less-demanding crops. But cyclamen commands more money, both at wholesale and retail, than many easier crops. They can also be sold from November to May (including December 26 and after, when those last few poinsettias may not be selling so well).

Cyclamen are also becoming popular as landscape plants in mild-winter areas. They may be more expensive than other bedding plants but usually last twice as long in the landscape, so they can actually cost less in the long run.

Cyclamen do not like stress, and they do not like abrupt changes to their environment. Cleanliness and consistency are the keys to growing this crop. Rapid fluctuations in temperature, fertilization, and watering regimes; constantly moving plants from one place to another; and unsanitary growing conditions make this crop a lot harder to grow than it needs to be. Keep careful records when growing so you can repeat what went right. Remember that cyclamen grows slowly and any changes you make in production today may not show up for two or three months.

Propagation

Most growers buy in plugs (large cells such as 50s, 72s, or 128s) to increase crop turns, as growing cyclamen from seed is a very long-term proposition.

Many growers will start their cyclamen crop in 288s and transplant later to a larger cell, such as a 50 or 72. This method conserves germination chamber and bench space.

Sow seed in a disease-free, peat/perlite, germination medium with excellent porosity, a pH of 5.5–5.8, and an EC of 0.5 mS/cm. Take care not to compact the media when filling the trays. Cover seed lightly with coarse vermiculite after sowing to maintain humidity around the seed. Water trays prior to sowing and again after sowing and covering with vermiculite to help obtain 100% humidity at the seed level. The seed coat of cyclamen is very hard; it takes about three weeks for the seed coat to soften and for the radicle and cotyledon to emerge. Cyclamen require total darkness for germination; this is critical. The easiest way to do this is in a germination chamber with the lights off. Some growers place germinating flats on a cart and wrap the whole cart in black plastic.

It is best to germinate cyclamen in a temperature- and humidity-controlled germination chamber, providing a constant temperature of 64–68°F (18–20°C) and 100% humidity. Varying temperatures or humidity below 100% will reduce germination significantly. By maintaining 100% humidity you will not have to water the plug trays while in the chamber and you will avoid "sticks"; a stick is what happens when the cotyledon is completely enclosed inside the seed coat. If this is happening, it is a definite sign that humidity is too low. If the seed coat does not come off the cotyledon, the plant will be set back in its growth considerably or may not even grow at all. If after removal from the germination chamber you have a lot of sticks, it is best to lightly mist the seedlings frequently (not so much mist that you are keeping the soil too wet). This will soften the seed coat and promote it to fall off the cotyledon. Stage 1, radicle emergence, takes from nineteen to twenty-four days.

While the cyclamen seeds are germinating, prepare the greenhouse by sanitizing all hard surfaces with a disinfectant such as quaternary ammonium. Starting clean will help you maintain a healthy crop and will reduce the need for expensive fungicide applications.

Once radicles have emerged, move seedlings to the greenhouse for Stage 2 for stem and cotyledon emergence. Provide 500–1,000 f.c. (5.4–11 klux) of light and begin to allow the medium to dry down slightly between irrigations. This is very important to prevent algae growth on the soil surface, which will reduce oxygen entry into the media and will result in poor root growth. Maintain temperatures of 65–68°F (18–20°C) days and 62–65°F (16–18°C) nights and apply 50–100 ppm from 15-5-15 once a week after cotyledons are fully expanded. If needed, apply a fungicide drench as plants start to get their first true leaf—about two to three weeks after leaving the germination chamber. This should not be necessary if proper sanitation measures have been taken. Stage 2 takes from seven to fourteen days.

Move plants to Stage 3 and maintain soil temperatures of 65–68°F (18–20°C) days and 62–65°F (16–18°C) nights. If these temperatures can be maintained, increase light levels to 2,500–3,000 f.c. However, reduce light levels if temperatures are too high. Feed at 100–150 ppm from 15-5-15 every two to three irrigations. Let the plugs dry down moderately between irrigations to promote healthy root growth. If needed, a broad-spectrum fungicide drench can be applied when two true leaves have developed. If you used 288-plug trays, about eight weeks after sowing (as leaves begin to touch) and plan to transplant to a 6" (15 cm) pot or larger, transplant seedlings to 72- or 50-cell plugs. However, 288s are fine to transplant directly to a 4" (10 cm) (but not larger) pot. If you started seedlings in larger containers, there is no need to transplant. Grade plants as you transplant to make watering easier. Plants will remain in these containers for another four to six weeks.

Transplant to the final growing container when seedlings have six to eight true leaves.

Growing On

As with growing the plugs, when finishing cyclamen it is critical to practice strict sanitary measures. Prior to transplanting be sure to disinfect all hard greenhouse surfaces. Always use new pots and clean media.

By the time plants are ready for transplanting they will have started to form a corm—a round bulbous structure from which stems, roots, and leaves arise. Plant so that the bottom of the corm is flush with the top of the medium; it will settle into the medium when you irrigate. Use a porous, well-draining media. This is critical for cyclamen because it allows enough oxygen to reach the roots and promotes rapid drying between irrigations.

Place plants pot tight on the bench immediately after planting and then space plants as leaves begin to touch (four to eight weeks later). Space 4" (10 cm) pots on 8" (20 cm) centers and 6" (15 cm) pots on 12" (30 cm) centers.

Cyclamen crops for Thanksgiving or Christmas sales are normally sown in April to June, meaning that the production cycle in the final containers begins during the hottest time of the year. Both purchased and homegrown plugs will have been happily settling into their routine, and then they become stressed by transplanting and high temperatures—this is the most critical time of production.

For the first seven weeks after potting, maintain 65–68°F (18–20°C) nights to promote root development. For the next seven weeks, maintain 62–65°F (17–18°C) nights. Day temperatures can be 5–10°F (3–6°C) higher. How can you maintain cool temperatures? Pull out all the stops—pad and fan cooling, ridge vents, sectioning off greenhouse space, shading, and, above all, air movement at the plant level. Simply keeping air flowing through the dense canopy can be a big help.

During irrigation, keep foliage dry and avoid watering late in the day to help reduce disease problems. The plant canopy is so dense that drip tube or ebb-and-flood irrigation is easiest. Do not overwater cyclamen; allow the plants to dry to almost wilting between irrigations.

Fertilize at 100–150 ppm from a calcium-based feed such as 15-5-15 every second or third watering. Leach with clear water at every third or fourth irrigation to prevent salts buildup. An ammonium-based feed such as 21-5-20 at 150–200 ppm can be used to promote leaf expansion if plants are too hard. About four weeks prior to shipping, feed with potassium nitrate alone at an EC of 2.0 mS/cm every other irrigation to promote flowering and reduce flower stem stretch.

Cyclamen prefers bright light, but requires shade during the hottest times of the year. Maintain 2,500–4,000 f.c. (27–43 klux). Cyclamen requires sixteen to twenty weeks from transplanting until a 4" (10 cm) pot begins to flower.

Pests & Diseases

Diseases such as *Fusarium, Erwinia,* rhizoctonia, and TSWV/INSV can strike cyclamen. TSWV/INSV can be especially devastating; monitor thrips levels and maintain a strict control program. Cyclamen or broad mites can also infest plants around the corm. Aphids, fungus gnats, mealybugs, spider mites, and thrips can also be problems.

Sanitation is the best disease prevention. Keep walkways and benches clear of debris. Do not leave the water breaker on the ground—hang it up. Use plastic liners in trashcans and use a new liner each time. Remove dead or diseased plants from the greenhouse. Instruct employees to wash their hands before handling the crops. When you are removing the old blooms and leaves, take away the entire petiole or peduncle. Grasp the bottom end of the peduncle or petiole at the point where it attaches to the corm, give it one or two turns clockwise, and pull it off. Any part that is left behind serves as an entry point for disease.

Troubleshooting

High temperatures can cause bud drop and small flowers. Low light, excessive drying, and high salts can also cause bud drop or yellow leaves. Low feed toward the end of the crop will promote earlier flowering but small flowers on stretched stems.

Weak or stretched growth can be due to high temperatures, low light, low or high fertility, fertilizer containing too much ammonium, crowding, or excessive soil moisture.

High EC/high soluble salts, poor root development, excessive drying out, a lack of aeration, transplanting root-bound plants, or excess light levels can result in stunted plants. Plant wilting or yellowing that is not due to moisture stress can be a symptom of either a disease or fungus gnat larvae. Note that older leaves will turn yellow on plants that are allowed to dry out severely.

High EC or soluble salts, dry media, or sudden environmental changes can cause blasting of flower buds or small flowers. Delayed flowering can be due to variety selection, high or low temperature extremes, faulty nutrition, or poor root development.

Varieties

All cyclamens grown today are F_1 varieties. Series for 2, 4, and 6" (5, 10, and 15 cm) pot production are available in a wide range of colors. The 'Sierra Synchro' (6" [15 cm] production) and 'Laser Synchro' (4" [10 cm] production) series from Goldsmith are very popular as pot plants and for the landscape. Both come in a wide color range with excellent habits and good bloom uniformity across the colors.

The 'Halios' (6" [15 cm] production) and 'Latinia' (4" [10 cm] production) series from Morel are also popular, offering wide color choices and uniform habits. Mini-cyclamens very often have wonderfully scented flowers and can be grown in pots ranging from 2–4" (5–10 cm) in size. Popular mini series include 'Metis' or 'Miracle'.

Many unique cyclamen varieties have come to market in the past decade, including silver-leafed types such as 'Silverado' (mini) and 'Winter Ice' (standard). The 'Halios Select' series includes such novelties as 'Fantasia' with a bicolor flower and 'Victoria 50', which sports beautiful, fringed flowers.

Postharvest

Plants are ready for sale when at least three to five flowers are open. With today's modern hybrids you can bench run most cyclamen, unlike in the past with the old, open-pollinated varieties, which could experience up to a month of flower timing difference between plants.

Ship plants cool at 35–40°F (2–4°C) and keep plants boxed and sleeved for no longer than three to five days. Select paper sleeves over plastic, as botrytis can be a problem in shipped plants. To help reduce the chance of botrytis infection, water the day before shipping so foliage and flowers are completely dry. Retailers should unpack and unsleeve plants immediately and place in a light, airy display environment. Cool retail display temperatures (65°F [18°C]) will also prolong shelf life. Retailers should provide 300–500 f.c. (3.2–5.4 klux) of light to maximize postharvest life. Do not allow plants to wilt; flower longevity decreases and leaves will yellow. Flowers are ethylene sensitive. Applications of 1-MCP prior to or during shipping may aid in preventing flower drop.

Consumers should place plants near windows and away from heat vents. Make sure each plant is sold with a saucer so that plants can remain moist (not wet) at all times. Once plants are finished flowering, they make excellent foliage plants for months, especially the silver-leafed varieties. Many consumers in Texas and California are discovering how versatile cyclamens are as outdoor plants on the patio or in the garden. They plant cyclamen in beds in November, where plants last until May or whenever temperatures become too warm.

Cyperus

Cyperus alternifolia
Common name: Umbrella plant
Sedge, annual (also grown as a houseplant)
(Hardy to USDA Zones 10–11)

Growing from 18–24" (46–61 cm) tall the first year from seed, the plant can grow as tall as 3' (91 cm) during the second season, if overwintered warm. Narrow fronds or leaflets sit atop singular stems that grow upright and can arch over. Plants prefer full sun to part shade and moist conditions. Cyperus is an aquatic in most locations.

Propagation & Growing On

Sow three to six seeds per 50- or 72-liner tray and lightly cover with vermiculite. Seeds do not need to be covered for germination, but coarse vermiculite helps to increase the moisture around the seed. Germinate at 70–72°F (21–22°C), and seedlings should start to emerge in seven to twelve days. The best performance is from a germination chamber or similar managed environment. If germinating on the greenhouse bench, some growers have reported a higher success by wrapping trays with plastic to increase the humidity and heat around the seed. This is especially true when germinated during times when outside temperatures are cool to cold. Under these conditions, germination has occurred ten to fourteen days after sowing.

Liner trays will be ready six to ten weeks after sowing, especially if sown under late summer or early autumn conditions. The best production has been from seed sown from September to early November. Sowings done in late January or later tend to be ready by July or later.

Transplant one 72 liner to a 4" (10 cm) pot and allow another ten to fourteen weeks when grown at 70–75°F (21–24°C) days with nights only 5–8°F (3–4°C) cooler. Total crop time from seed to a finished green plant is eighteen to twenty-four weeks in a 4" (10 cm) pot. Higher light and warmer temperatures hasten production, while the opposite is true under cooler, lower light conditions.

One additional point: If sowing a minimal number of seeds (two to four) per cell, then the resulting plants have broader leaves and are taller. If sowing eight to ten seeds per cell, then the leaves are thinner, the plants are shorter, but the overall effect is still attractive.

Varieties

'Wild Spike' is the only variety currently available from seed.

Related Species

Cyperus papyrus is the famed sedge along the banks of the Nile River in Egypt. The word *paper* is derived from *papyrus*. Upright plants grown from 2–7' (0.6–2 m) (depending on variety) in one season. Plants are propagated vegetatively by dividing the crown. Plants are also readily available in liner trays of varying size.

Rooted liners (usually in a 32-, 50-, or 72-cell tray/flat) can be brought in from commercial or professional propagators in January or early February. Stick one liner per 1-gal. (4 l) pot. These are grown warm (70°F [21°C] days and 60–62°F [16–17°C] nights), and plants grow moderately during these low-light winter days. As day length increases and warmer temperatures prevail, plants start to grow readily.

Some growers in the northern United States and Canada have reported that if the weather is cloudy for a prolonged period during the early months of the year, the plants can slow to the point of rotting. If these conditions express themselves, pot the liner to a 4" (10 cm) pot (instead of a 1-gal. [4 l] pot) and allow it to root before transplanting to a 1-gal. (4 l) pot. You can also bring the liners in later and transplant to smaller pots (4 or 6" [10 or 15 cm]), but the plants take longer to reach their final height.

D

Dahlia

Dahlia × *hybrida*
Pot plant, annual

Dahlia's many supersized flowers make a bold statement wherever they're found, from decks and patios in containers to the dining room table as flowering potted plants.

They are a Mother's Day tradition for many, which is good because plants must have cooler temperatures to thrive. Once the heat of summer hits, flowering is sporadic at best until temperatures fall off again in the fall. For growers, early crops require daylength control: Long days (night interruption lighting) to prevent plants from forming tubers and "forces" them to "focus on flowering." Plants are heavy feeders and a bit finicky with their irrigation requirements. Nevertheless, they are a great impulse-sale item for the spring. When you have done a great job growing full-sized plants topped with large, boldly colored flowers, you won't be able to keep them in stock.

Dahlias may be grown from bought in plugs started from seed, unrooted cuttings or liners, or purchased tubers (actually they are tuberous roots, an important distinction when it comes to gardening and propagation). All three methods have pros and cons. Seed-grown dahlias tend to be grown in jumbo 36s, 1801s, and 4" (10 cm) pots—they are mass-market items. For most growers, dahlias are a small part of their total bedding plant crop.

Buying in tuberous roots is the most traditional way to grow dahlias for 6" (15 cm) indoor pots, as a nursery item for planting out, or for larger patio containers. The crop has a certain retro, old-fashioned appeal and a lack of uniformity that some find endearing. Flower form is the reason why you should take a look at growing tuberous dahlias. While seed- and cutting-propagated dahlias are generally single or double and usually dwarf to compact in size, tuberous dahlias have a wide range of flower types: singles, decoratives, cactus-flowered, and even formal, totally symmetrical, pompon types. Flower size varies from 1" (3 cm) to a whopping 10" (25 cm) across, with plant height equally as diverse, making growth regulator applications critical (see the tuberous section).

By far the most modern way to grow dahlias is to buy in unrooted cuttings or liners of the new dwarf plants, such as the 'Dahlietta' or 'Dahlinova' lines. Crop response is uniform, and flowers are more fully double across all plants. Plus, the best part is that crops are really fast: You can turn 4" (10 cm) pots in just six to eight weeks. Today, many greenhouse growers make a dahlia crop a regular part of their spring program.

Propagation

Seed, unrooted cuttings or liners, and tubers are the common methods of propagation. Seed and cuttings are discussed in the Propagation section, while tubers are presented in Growing On.

Seed

Sow seed onto a well-drained, disease-free, germination medium with a pH of 5.5–5.8 and cover with coarse vermiculite. Germinate at 68–70°F (20–21°C) substrate temperatures. Keep medium uniformly moist, but not saturated. Keep ammonium levels at less than 10 ppm. Radicles will emerge in three to four days.

Move trays to Stage 2 for stem and cotyledon emergence. Maintain substrate temperatures at 68–70°F (20–21°C) and begin to allow trays to dry down slightly between irrigations. Provide 1,500–3,000 f.c. (16–32 klux) of light. Begin fertilizing with 50–75 ppm from 14-0-14 as soon as cotyledons have fully expanded. Stage 2 will last about five to seven days.

Move trays to Stage 3 to develop true leaves. Lower substrate temperatures to 65–68°F (18–20°C) and allow trays to dry down between irrigations. Maintain light levels. Increase fertilizer to 100–150 ppm weekly from 20-10-20, alternating with 14-0-14. If growth regulators are needed, A-Rest, B-Nine, Bonzi, or Cycocel are effective at this time. Stage 3 will last about fourteen to twenty-one days for 384s and twenty-one to twenty-eight days for 288s.

Harden plugs in Stage 4 for seven days prior to sale or transplanting. Lower substrate temperatures to 60–62°F (16–17°C) and increase light levels to 2,500–4,000 f.c. (27–43 klux). Feed weekly with 100–150 ppm from 14-0-14.

Note that providing long days during Stages 2, 3, and 4 with night interruption lighting will prevent tuber formation, important for growers using transplanters to plant dahlia plugs. Plugs grown under long days also tend to flower faster based on Cornell University trials with the seed variety 'Sunny Yellow'. However, plugs grown under long days will be taller than plugs grown under short days. Cornell researchers had the best results growing plugs under long days, switching to short days (ten hours) for two weeks immediately after transplanting, and then providing long days (more than fourteen hours) to finish the crop.

A 288-plug tray will take from three to five weeks from sowing to be ready to transplant to larger containers.

Unrooted cuttings

Root cuttings of 'Royal Dahlietta' in a well-drained, disease-free medium with a pH of 5.8–6.2. Oasis wedges may also be used. Stick cuttings 0.5" (1.3 cm) below the substrate line so that the leaf canopy touches the mix. Use IBA rooting hormone at 2,500 ppm to speed rooting and increase rooting uniformity. Root cuttings under long days to prevent tuber formation by giving night interruption lighting from 10 P.M.–2 A.M. Provide substrate temperatures of 65°F (18°C) and air temperatures of 65–70°F (18–21°C). Mist only as much as absolutely required to keep foliage turgid and reduce mist as soon as possible to reduce disease risk (usually seven to ten days). Rooting will take about three weeks for a 105-, 84-, or similar-sized liner tray. Cuttings are not pinched.

Growing On

Bedding plant crops

Plant one plug per cell or 4" (10 cm) pot. Select a well-drained, disease-free medium with a moderate initial starter charge and pH of 6.0–6.5. Fertilize at every other irrigation with 250–300 ppm nitrogen (N). Grow under as high light as possible while maintaining moderate production temperatures [see the Vegetative (liner) crops section]. Immediately after transplanting, grow dahlias under two weeks of short days, and then begin to provide long days (more than fourteen hours) with night interruption lighting to achieve the fastest flowering.

A soft pinch about two weeks after transplanting is beneficial to bulk up plant size and "build the chassis," especially for 4" (10 cm) pots. B-Nine can be used for height control, beginning in the plug tray at the first true leaf stage and throughout production as needed. A-Rest, Bonzi, and Cycocel may also be effective.

The crop time from transplanting a 288 plug to a 4" (10 cm) pot is six to eight weeks for the majority of the seed-propagated varieties on the market. Plants tend to only have one flower per plant at this crop time.

Vegetative (liner) crops

Plant rooted cuttings into a well-drained, disease-free medium with a low initial starter charge and a pH of 5.8–6.2. Use one liner per 4" (10 cm), one or two per 6" (15 cm), and three per 8" (20 cm) pot. Plant cuttings deep to encourage stem roots. Deep planting, at least up to the first leaf, is recommended; burying one to three leaves with medium will not hurt plants at all. Because dahlias become so top heavy, encouraging stem root development will help to stabilize the pot down the road.

If you bought in rooted cuttings in Oasis wedges, be sure to cover the entire wedge with medium and be careful with watering in the first weeks. Your goal is to encourage roots to grow out of the Oasis and into the pot—to do that, the pot and Oasis wedge must remain uniformly moist.

Grow 'Royal Dahlietta' and similar varieties on at 65–70°F (18–21°C) days and 60–65°F (16–18°C) nights. Cool temperatures (less than 65°F [18°C]) when combined with short days (less than fourteen hours) promote tuber formation: This is not your goal! You want plants to flower; therefore, be diligent in maintaining temperatures and grow plants under long days by using night interruption lighting. Make sure to provide 10 f.c. (108 lux) at crop height for four hours each night to prevent tuber formation, encourage lateral shoots (and thus more flowers), and shorten crop time. Do not grow plants too warm, as plant habit will open up and spread.

Dahlias prefer to grow under high light (4,500–6,000 f.c. [48–65 klux]), provided moderate production temperatures are maintained. If temperatures become too high, provide shade to cool plants down after May 1. Always provide 2,000 f.c. (22 klux) or more of light for best growth: Northern growers will find benefits from 400 f.c. (4.3 klux) of assimilation lighting during the dark winter months (light for fourteen hours total). When grown under high light, 'Royal Dahlietta' does not need growth regulators. If growth regulators are needed, B-Nine at 1,500 ppm can be applied two to three weeks after potting. Bonzi and Sumagic are also effective.

Some growers prefer to pinch plants to increase plant size and flower number. Pinching two weeks after potting (once roots have reached the container sides) will delay flowering by a week. Remove the topmost leaf pair. Some growers disbud dahlias by removing the center bud to encourage laterals to develop and become larger.

Fertilize at every other irrigation with 250–350 ppm from 20-10-20, alternating with 14-0-14 or 15-0-15. Dahlia is very sensitive to high soluble salts under short photoperiods. Be careful at all times to avoid excessive ammonia fertilizer, as this encourages soft growth.

Be sure to space plants adequately to encourage good airflow around pots. Space 4" (10 cm) pots at 3–4/ft.^{2} (33–44/m^2), 6" (15 cm) pots at 2/ft.^{2} (22/m^2), and 8" (20 cm) pots at 1/ft.^{2} (11/m^2).

Spring crop times are six to ten weeks for 4" (10 cm) pots (longer crop time if you pinch) and eight to ten weeks for 6 and 8" (15 and 20 cm) pots.

Tubers

Tubers will arrive beginning in late January, after the supplier has provided the requisite six- to eight-week resting period. Consult closely with your supplier about specific planting dates and schedules. Tubers planted in late January to early February will be flowering from mid- to late April. Tubers planted from early to mid-February will be flowering for early May sales.

Plant dahlias brought in as No. 2 tubers into 8" (20 cm) or 1-gal. (4 l) pots and No. 1 tubers into 2-gal. (8 l) pots, using a well-drained, disease-free medium that has significant substance and weight. Composted bark mixes work very well. Smaller pots can be used; however, you will have to be aggressive in height control. Make sure crowns are 1" (3 cm) below the medium surface. Some growers water pots in with a preventative fungicide drench.

Many growers prefer to add a slow-release fertilizer to the medium, while others prefer to use a constant liquid feed of 250–300 ppm nitrogen (N) from fertilizer sources determined by your specific water quality. Begin your liquid fertilizer program after tubers have sprouted. Leach pots weekly to ensure soluble salts do not build up. As buds begin to develop, dahlias have a significantly higher irrigation requirement.

Grow on at the temperatures and light levels outlined above. Do not allow day temperatures in excess of 80°F (27°C), as flowering can be delayed.

Provide adequate space to avoid stretched plants. Immediately after potting, you can space plants pot tight. As they begin to grow, space 1-gal. (4 l) pots on 14" (36 cm) centers and 2-gal. (8 l) pots on 16" (41 cm) centers.

Pinch when three to four leaf pairs have formed, and throughout production remove "wild" growth and side buds to keep plants shaped. Some general pinching rules for tuberous dahlias are: Pinch all plants that develop only a single shoot, and do not pinch plants that produce two shoots, unless one is strong and the other is weak, in which case pinch the strong shoot.

If plants make three shoots, do not pinch. Pinching will delay flowering by five to ten days and will produce taller plants than non-pinched plants. Pinch only those plants that need to be pinched, not the entire crop. Some growers also remove the center bud, which allows laterals to more fully develop. This will add an additional five to ten days to the total crop time.

Grow plants under long days (more than fourteen hours) using night interruption throughout the entire crop cycle. Growth control is essential for many cultivars. Topflor (fluprimidol) drenches at approximately 0.5–1.5 mg/pot (4–2 ppm, if using 4 oz. per pot) are very effective, but rate depends on cultivar and initial local trials are needed to develop the correct dose. You will be able to sell the crop about ten weeks after planting.

Tuberous dahlias also make a great local cut flower crop. Harvest stems when flowers are three-fourths to fully opened and place immediately in clean buckets with a sugar-based cut flower food. Sear the bottom of stems in 160°F (71°C) water prior to shipping. Retailers should maintain cut flower food treatment and refrigerate at 35–40°F (2–5°C). Recent research from North Carolina State University has shown that dahlia cut flowers do not benefit from protection against ethylene. Vase life is about seven to ten days, provided preservative solutions have been used.

Pests & Diseases

Aphids, fungus gnats, thrips, and whiteflies are all insects that can attack dahlias. In retail and home conditions, spider mites can become a problem. Botrytis, *Pythium,* and rhizoctonia can also attack plants.

Troubleshooting

If plants collapse or fall over, the problem may be botrytis, physical damage to the liner during transit, or liners not planted deep enough to encourage stem root development. When dahlias produce too much vegetation and no flowers, the problem can be excessive ammonia in the fertilizer, overfertilization under low light conditions, short days/cool temperatures, and/or low light combined with overwatering or wet medium. Premature flowering on small plants can be due to high light and high temperatures, inadequate fertilizer, or growth that has been checked because plants were allowed to dry too much between irrigations. Too much light or excessive B-Nine use can cause excessive leaf curl and crinkled leaves.

Varieties

In seed-grown bedding plant dahlias, 'Figaro' with flowers is the mainstay. The series comes in separate colors and a mix. Even though 'Figaro' has been around for a while, it is still the most uniform series available, although plants are famous for throwing a significant percentage of single flowers. Other seed varieties include 'Fresco', a mix of double-flowering plants from 10–12" (25–30 cm) tall, and 'Harlequin', a mix of uniquely colored, bicolor blooms growing from 12–14" (30–36 cm) tall.

For vegetatively propagated dahlias in 4" (10 cm) or 6" (15 cm) pots, try the 'Dahlietta' or 'Dahlinova' series. Both are available in a full range of colors and suitability for different size pots. The 'Gallery' series of vegetatively propagated dahlias also make fine large pots.

Additionally, a wide range of tuberous dahlias is available. Consult your sales representative for a listing of currently offered varieties.

Postharvest

Dahlias are ready to sell when the flower buds begin to show color. Ship at 45–48°F (7–9°C) and keep boxed for no longer than two to three days, as flowers are sensitive to ethylene. Open flowers last about nine days.

Retailers should display dahlias under cool (50–70°F [10–21°C]), bright conditions (500 f.c. [5.4 klux] minimum). Do not allow plants in flower to dry out. Keep dead flowers groomed from plants to encourage subsequent flowering.

Daylily (see *Hemerocallis*)

Delphinium

Delphinium spp.
Perennial, cut flower (Hardy to USDA Zones 4–7)

Delphinium blue is a color—actually it's a series of colors, from the surrealistic blue of the waters off the coast of Hawaii to the bright periwinkle blue of Provence, France. Just about every shade in-between is featured, as are lavenders, white, and even pink. Flowers are at once stately and casual, evoking the grandeur of a Victorian garden and the call to nature of the mountains at the same time. This popular perennial is a joy in any garden, albeit a short-lived one. Delphiniums are often treated as annuals in many parts of the country such as the South, with dedicated gardeners replanting them each year.

For growers, producing and selling delphinium as a green plant in 4 or 6" (10 or 15 cm) pots is easy. Growing larger sizes and flowering those plants is more of a challenge, although one for which the market will reward you handsomely! Cut flower growers routinely offer field-grown delphinium in season, with California growers cropping plants for successive flushes throughout the year. Some southern field cut flower growers plant delphiniums in the fall for harvest the following spring.

For those who grow in containers, the F_1 series from various companies are the most economical. Because they are F_1 hybrids, they are more programmable and growers can bench-run their production.

Propagation

Sow only fresh delphinium seed. The best germination occurs with seed that is only four to six months old. If you must store your seed for summer sowings, keep it refrigerated in its sealed packet. Sow onto a well-drained, disease-free germination medium with a pH of 5.5–5.8. Cover lightly with coarse vermiculite and germinate at 70–75°F (21–24°C). If you are using old seed or are experiencing poor germination rates, try alternating day/night temperatures of 80°F (27°C) days and 70°F (21°C) nights to boost germination rates. This is usually more for 'Pacific Giant' since it generally shows poorer germination than belladonna types. D. × belladonna can be germinated at 65–70°F (18–21°C). The F_1 hybrids, however, germinate readily at 68–70°F (20–21°C).

Move trays to Stage 2 for stem and cotyledon emergence once 50% of seed has germinated. Provide soil temperatures of 75–80°F (24–27°C) and increase light levels to 500–1,000 f.c. (5.4–11 klux). Fertilize with 50–75 ppm from 14-0-14 weekly once cotyledons have fully expanded. Delphinium seed will germinate over twelve to eighteen days—Stages 1 and 2 may very well run together.

Once you have a full stand, move trays to Stage 3 to develop true leaves. Lower substrate temperatures to 65–70°F (18–21°C) and increase light levels to 1,000–2,000 f.c. (11–22 klux). Fertilize once a week with 100–150 ppm from 14-0-14, alternating with 20-10-20. Stage 3 will last fourteen to twenty-one days for 288s; allow longer for larger plug sizes.

Prior to sale or transplant, harden plugs for seven days at 60–62°F (16–17°C) substrate temperatures and increase light to 1,500–2,000 f.c. (16–22 klux). Fertilize once a week with 100–150 ppm from 14-0-14.

For a 200-plug tray, the crop time from sowing to transplanting to a larger container is six to seven weeks. However, this applies mainly to newer selections on the market; some older strains require another two to three weeks.

Growing On

Most growers buy in delphinium plugs or plants for their perennial crops in the fall and overwinter the crop. Cut flower growers also bring in plants early enough so they will receive cold treatment. Some perennial suppliers offer delphinium as a cooled 50 plug that is ready to pop into 1-qt. (1.1 l) or 1-gal. (4 l) containers for rooting in and forcing for spring sales either green or in flower. While delphinium will flower its first year from seed without cold, flowering is more prolific and uniform after plants receiving a five- to six-week cold treatment at 35–40°F (2–4°C). Delphinium naturally flowers in the early summer, but you can force cooled plugs into flower for sale in the spring. However, you do not have to overwinter plants or purchase cooled plugs in order to produce a flowering crop.

Plant one plug per 4" (10 cm), 6" (15 cm), or 1-qt. (1.1 l) pot. Some growers transplant plugs to a liner (84, 72, or 50 tray) to bulk up plants to transplant to 1-gal. (4 l) or use three plugs per pot. Do not grow delphinium in cell packs for retail sales, as plants are highly susceptible to overwatering, which is far too easy to do in a shallow bedding plant tray. Select a long-lasting, disease-free medium with a pH of 5.5–6.2 and a moderate initial starter charge. Bark-based mixes are ideal if plants will be overwintered. Prior to overwintering or cooling, bulk plants to at least five nodes by growing them at 60–65°F (16–18°C) days and nights. Fertilize lightly with 75–100 ppm at every other irrigation, alternating between 15-0-15 and 14-0-14. Provide 2,000–3,000 f.c. (22–32 klux) of light. If growth regulators are required, A-Rest, Bonzi, and Sumagic are effective. However, allowing plants to thoroughly dry between irrigations is the best way to control plant height.

Cool plants for five to six weeks at 35–40°F (2–4°C) in an unheated cold frame or greenhouse or in a cooler with 25–50 f.c. (269–538 lux) of light. Do not water or fertilize plants during the cold treatment.

Cooled plants can be forced into flower in eight to ten weeks. Provide 65–70°F (18–21°C) days and 60–65°F (16–18°C) nights. If you are just trying to bring the plants out of dormancy for green plant sales, use lower temperatures, which will also encourage more stocky growth. Grow under the highest light levels possible (3,000–5,000 f.c. [32–54 klux]) while maintaining production temperatures. Flowering is not daylength sensitive, but long days increase stem length and quality for cut flower production. With that said, Michigan State University research has shown that flower number is higher under short days for *D. grandiflorum*. Fertilize with 150–200 ppm from 15-0-15 at every other irrigation. Keep pots uniformly moist and do not allow wilting, especially once plants have budded. Plants are naturally basally branching and do not require pinching.

Pots can be sold green, in spike, or in flower, depending on your market. Plants in flower will benefit from EthylBloc, which will lessen petal shattering.

Sowings done from late summer to November produce the best overall plants. Sowings can be done later, although plants often grow erect, without branching, and are not as fully rooted as their counterparts that have been sown earlier and grown cool during winter in their final container for spring. This is primarily true for the taller types. The shorter selections (*D. grandiflorum*) can be sown in December or January for spring flowering.

Cut Flowers

In coastal California, plugs are generally transplanted into the field from August to October and February to early May. Fall transplants will flower the following spring (beginning in February). Spring transplants will flower in late spring and fall. Do not allow plugs to become root-bound, or growth will be checked. Space plants in well-drained, raised ground beds outdoors at 12" × 12" (30 × 30 cm) if you will be replanting annually, or 12" × 18" (30 × 46 cm) if you will be harvesting plants as perennials. Do not irrigate plants from overhead, as plants are highly susceptible to aerial disease. Keep plants moist, but not wet. Provide one level of support for D. × belladonna types and two levels of support for taller-growing D. elatum hybrids.

Use night interruption lighting on plantings made in the late summer or fall to speed flowering. Use low to moderate fertilizer rates (150–200 ppm from a complete fertilizer) and keep temperatures at 63–65°F (17–18°C) days and 45–55°F (7–13°C) nights. Avoid temperatures in excess of 75°F (24°C). Once delphinium plants begin to flower, they will tolerate higher production temperatures; however, the best-quality stems are grown at cooler temperatures. Discontinue nitrogen fertilizers once buds have initiated, and increase applications of fertilizers containing potassium. To keep disease problems to a minimum, reduce fertilizer and water applications during flowering. Delphinium dislikes both humidity and high temperatures, which is why southern growers treat it as an annual. It takes generally thirteen weeks to go from planting plugs to harvest.

Pests & Diseases

Aphids, cut worms, and leaf miners can attack plants. Crops can also become infested with a number of diseases due to poor air circulation or wet conditions. These include pseudomonas black spot, crown rot due to *Sclerotium, Pythium,* and powdery mildew, among others. Delphinium can also become infected with INSV/TSWV.

Varieties

'Belladonna' grows to 30–36" (76–91 cm) and produces light blue flowers. 'Bellamosum' also grows to 30–36" (76–91 cm), but produces deep blue flowers. Both 'Belladonna' and 'Bellamosum' may be direct-sown to the field for cut flower production. 'Casablanca' grows to 42" (1.1 m) and produces white flowers (a small percentage is pale blue).

D. grandiflorum produces loose spikes of flowers in lavender, blue, or white, depending on the variety. Plants are very short lived in the garden. 'Blue Mirror' has deep, electric-blue flowers that are spurless and have a more rounded shape. Because plants are so short lived, most in the trade sell it as an annual. Plants grow to 16" (41 cm).

D. elatum produces tall, elegant spikes. While *D. elatum* as a species is rarely produced, the hybrids in the market with *D. elatum* blood are a mainstay. The most famous *D. elatum* are the 'Pacific Giant' hybrids, with tall plants that stretch to 5' (1.5 m). This series from seed has plants with a range of colors, each flower with a white bee, except 'Blue Jay' and 'Black Knight', which have a black bee. 'Pacific Giant' may be sown directly in the field for cut flower production: Add eight to ten weeks to the production schedule if you direct-sow.

With the advent of F_1 hybrids, germination is easier and production more uniform to finish. Series like 'Aurora', 'Guardian', and others have reinvented the *D. elatum* group. Flower colors are the traditional blue and lavender with shades in between.

Postharvest

Flowers are ready to harvest when 25–30% of flowers have opened. Cut stems at least 12–15" (30–38 cm) below the flowers and be sure to leave at least 3–4" (8–10 cm) of stem on the plant to prevent disease from entering the crown. Plants yield stems (flush) over several weeks if temperatures are ideal. Some varieties flush three or four times during the year.

Remove leaves on portions of the stem that will be in water, and pulse stems with STS solution or treat with EthylBloc to prevent petal shattering caused by ethylene. Stems that are not treated will shatter immediately! Store stems in water with floral preservative and bactericide at 34–38°F (1–3°C).

Dendrobium

Dendrobium hybrids

Flowering pot plant, cut flower

Dendrobium orchids have become an important pot plant crop recently. They are often considered the "starter orchid" for the consumer because of their relatively low retail price, from $10 to $20, and their hardiness once grown to flowering size. Flowers can last six to ten weeks, making them a real value for consumers.

Dendrobium is the largest genus within the orchid family, Orchidaceae, consisting of thousands of extremely variable plant forms. The information provided here considers only the very small group of commercial plants in the general market today. The dominant plants within *Dendrobium* are the phalaenopsis types (*D. bigibbum* ssp. *phalaenopsis*). These plants were so named due to their flower shape, which strongly resembles the flowers of orchids in the genus *Phalaenopsis*.

Young plants are obtained primarily from specialist breeders and propagators in Southeast Asia. Since orchids are protected under Convention for International Trade of Endangered Species (CITES) laws, they require special permits to import. Most commercial *Dendrobium* are propagated from tissue culture because seedlings are extremely variable and not suitable for commercial production. Today's commercial pot plant *Dendrobium* originates from the large-scale cut flower production of this plant. Tens of millions of *Dendrobium* sprays are shipped worldwide each year. The production center for *Dendrobium* as a cut flower in the United States is Hawaii. There is also some young plant production available from Hawaii.

Because *Dendrobium* is such a long crop and has high heat and light requirements, large-scale production is restricted to southern and tropical areas, where plants can be produced with low input costs. Most growers in northern climates purchase prefinished plants in spike for finishing or finished pots for resale.

Propagation

Seed and tissue culture are the two ways this crop is propagated, although tissue culture is more common. Growers purchase prefinished plants from commercial propagators or as flasked plantlets. Flasks are the result of tissue-cultured plants that are propagated in Agar and shipped to the grower.

Growing On

One of the most debated topics among commercial orchid growers is in what medium to plant. Virtually every grower uses something different. However, the rules are simple enough. Commercial *Dendrobiums* are epiphytic plants—they evolved growing on tree branches. The most popular media simulate this natural occurrence. Media such as large chunks of hardwood bark, mixes of foam rubber and cork, and solid slabs of coconut have all made successful substrates. The rule is, in actively growing seasons (warm to very warm), the plants want to get wet and then dry. It is a common practice in Asia to water the plants in the morning and late afternoon, allowing them to dry in between. We cannot do this in greenhouse production, so we compromise by trying to achieve the same goal of plants getting water but being allowed to dry between watering. It is very difficult to underwater *Dendrobium,* yet it is very easy to overwater.

Dendrobiums grown in straight bark mixes require higher levels of nitrogen in a 3-1-1 ratio to make up for the nitrogen-binding properties of bark. If you are using other media, fertilizer in a 1-1-1 ratio is fine. *Dendrobiums* like to be fed while growing and prefer less or no fertilizer during their dormant periods. When night temperatures drop below 55°F (13°C), leaf drop may occur.

Dendrobiums have very high light requirements, preferring 5,000 f.c. (54 klux) and long days for vegetative growth. Plants store starch and water in their canes (psuedobulbs) to live through periods of drought. This is what makes the plants so hardy for consumers.

Shorter day lengths induce flowering. However, modern *Dendrobium* clones have been selected to freely flower year-round. Even so, there is a surge of flowering during the naturally occurring short days from October to January.

Dendrobiums do not like temperatures below 60°F (16°C). Low temperatures can cause bud blast as well as leaves to turn yellow and drop off.

It will take from seventy to one hundred twenty weeks to finish if using flasked plants.

Pests & Diseases

Keeping plants too wet can cause algae that will allow fungus gnats and shore flies to breed. Fungus gnat and shore fly larvae can damage roots and introduce soilborne disease. Aphids and thrips are attracted to flowers. Spider mites can attack foliage and flowers. Botrytis can occur on flowers in high humidity and in an environment with stagnant air conditions.

Troubleshooting

There are three common problems a grower may encounter when receiving flowering or prefinished plants:

1. **Ethylene damage.** Orchids are extremely sensitive to ethylene gas in very small amounts. Keep them away from all sources of ethylene.
2. **Botrytis spots on the flowers.** High humidity and poor air circulation cause this.
3. **Overwatering.** *Dendrobium* are easy to overwater, and subsequently received plants may have rotted roots.

Postharvest

Since *Dendrobium,* like all orchids, is highly sensitive to ethylene damage, keep plants away from ripening fruits and vegetables. Also, maintain plants under sanitary greenhouse and shipping conditions: Rotting plant debris and old flowers can also serve as sources of ethylene. Under reasonable conditions, *Dendrobium* is a strong plant with long flower life on the shelf. A common mistake for retailers is to put these beautiful flowering plants near the door, where they are subjected to blasts of cold outside air.

After the consumer has taken plants home and once the flowers are finished, plants should be placed in a south or east window with direct sunlight. A new cane will grow from the base and flower the following year.

Dianthus

Dianthus barbatus
Common name: Pinks; sweet william
Perennial, hardy annual (Perennial varieties are hardy to USDA Zones 5–8.)

Many consumers recognize the flowers of sweet william because they commonly appear in late spring in many regions; although once naturalized throughout some areas in North America, there are many patches that have died out. It is a tender perennial in most parts of the country, reseeding itself freely in northern gardens. In the South, it is usually a short-lived perennial. Some varieties are better categorized as hardy annuals and, while many return the following year, they do not tolerate as much cold as the true perennial varieties do.

There are a number of varieties that are true *D. barbatus,* however there are a larger number of varieties on the market that are crossed with other species, resulting in *Dianthus barbatus interspecific*. While this is not a recognized botanical name, it is frequently used throughout the wholesale industry.

Propagation

Dianthus barbatus can be propagated by seed or cuttings.

Seed

For the perennial varieties, seed and seedling stands differ in quality between the varieties. For the traditional or older selections, like 'Midget Mix' or 'Indian Carpet Mix' use two to four seeds per 288-plug tray and cover lightly with vermiculite and germinate at 70–72°F (21–22°C).

Dianthus barbatus is one crop that can be germinated at cooler temperatures to get a more uniform stand that is less prone to stretched seedlings. Germination temperatures of 64–68°F (18–20°C) are suggested but are difficult to achieve without a germination chamber when sowing during warm summer months.

Seedlings will emerge in seven to ten days. Trays of 288 plugs will be ready to transplant to larger containers in five to six weeks. This is recommended if using raw seed. If seed is enhanced, use only one to two seeds per cell. Seed should be sown in late spring or early summer to overwinter in 1-qt. (1.1 l) pots for sale the following spring.

The newer selections, mostly offered under *Dianthus barbatus interspecific,* include such varieties such as 'Bouquet', 'Amazon', and 'Diabunda'. These selections have better overall germination and performance. They can also be treated as hardy annuals, since they flower in late spring or summer from an early to midwinter sowing. These plants often overwinter up to Zone 4. Sow one seed per 288-plug tray, cover lightly with vermiculite, and germinate using the guidelines listed above. Seed germinates in three to seven days, and plugs are transplantable in five to six weeks.

Cuttings

Cuttings are taken from stock plants in late spring to late summer and stuck one cutting per liner tray. If cuttings are supple and taken in midsummer, they do not need rooting hormone. Cuttings taken later, especially as plants become older and the stems become more hardened, a rooting hormone is suggested. Also, plants grown under a portion of shade produce soft tissue that is easily propagated. However, avoid elongated stems taken as cuttings as they tend to be weak in habit and produce an inferior liner tray.

From a cutting to a rooted 105-liner tray takes from four to six weeks, depending on variety.

Growing On

Provide a disease-free medium with a neutral to slightly basic pH.

Seed

The traditional varieties must be vernalized with a cold period (40–45°F [4–7°C]) before they will flower. When they are at least twelve to sixteen weeks old, pot up plants to larger containers to be sold in flower the following spring. When plants are too young, they are not receptive to cold temperatures. Once potted, plants may be put in an unheated cold frame or cooler for the twelve-week vernalization period.

For newer selections, seed is sown to plug trays from January to early March to transplant to cell packs or 4" (10 cm) pots. Plants are salable green in April or May and will flower (in pot or garden) in late June onwards.

Use 1 plant/4" (10 cm) pot and 2–3 plants/6" (15 cm) pot. Allow twelve to fourteen weeks from sowing for sale of green 4" (10 cm) pots in the fall or spring.

Cuttings

While a 105 tray was noted above under Propagation, commercial propagators offer dianthus in a number of liner sizes (from 32 to 50 and so on up to a 105). These are available in winter to early spring to be potted to larger containers. For the larger liners transplanted during the colder parts of the year, pot to 1-qt. (1.1 l) or 1-gal. (4 l) containers. For smaller liners, especially if planted as the days get longer and the temperatures warm, transplant to 1-qt. (1.1 l) or 6" (15 cm) pots.

Under cooler conditions plants develop more evenly and are salable in April or May from an early to midwinter potting. Plants may or may not be in flower at the time they are sold.

Seed & Cuttings

Flowering naturally occurs in May and June. Plants can be forced into flower for early sales at 70–74°F (21–23°C) days and 65–70°F (18–21°C) nights and supplemental lights to increase winter light levels.

Do not overwater *D. barbatus*. Allow pots to dry between irrigations. Use 100–150 ppm from 20-10-20. Growth regulators are not required.

D. barbatus may also be produced as a cut flower crop. Harvest flowers when at least 10% of flowers in the cyme are open.

Pests & Diseases

Spider mites can be a problem during production. *Fusarium* and rhizoctonia may also be problems.

Varieties

'Double Mix' and 'Indian Carpet Mix' are the true perennial varieties that require a cold treatment to flower. Crop time is about twenty-six weeks from sowing. The 'Sweet' series is more annual in its performance and sowings made from January to early March will flower the same summer. Growing cold is beneficial and keeps a uniform habit. It also helps to increase flowering. However, they last usually for two to four years at best in the garden before dying out.

'Amazon', Bouquet', 'Diabunda', 'Dulce', 'Supra', and 'Dynasty' are just a few of the excellent garden performers that are the result of various crossbreeding. For this reason sowings can be made in winter or spring for flowering plants in summer. All of these varieties will flower in summer from a winter to early spring sowing. Equally, these varieties, and all the newer selections, differ in their hardiness tolerance. See the seed catalogs for details.

Postharvest

Display flowering pots of *D. barbatus* in cool temperatures and high light. Since flowers are relatively short lived and plants will not reflower throughout the summer, many growers prefer to sell plants green with a color picture tag.

Cut flowers are highly sensitive to ethylene and benefit from silver thiosulfate (STS) treatment.

Dianthus chinensis (includes *D. chinensis* × *D. barbatus* and *D. hybrida*)
Common names: Garden dianthus; dianthus
Annual, hardy annual, or tender perennial

No other bedding plant class features the bold, bright, electric hot-pink, red, and white flowers of dianthus. These cool-season performers are a must for gardeners when temperatures are cool. In the past, most bedding plant dianthus was *D. chinensis*, an extra-large-flowered species. However, in the heat and humidity, plants went out of flower and melted. New breeding has expanded the range with varieties that are much more heat tolerant and flower all summer long. Many are *D. chinensis* × *D. barbatus* crosses; others are hybrids selected for their heat tolerance.

Northern growers offer dianthus during the spring, while southern and many western growers can offer them in both the spring and fall: Planted in the fall, dianthus stops flowering in the coldest times of late December and January and returns to flower as light levels and temperatures rise. Plants may overwinter in well-protected locations but deep, continuous freezing around the roots kills plants. In mild climates, bedding plant dianthus can perennialize.

Propagation

Sow one pelleted or one to two standard seeds on a germination medium with a pH of 5.5–5.8 and an EC of less than 0.75 mS/cm in a 288-plug tray. Maintain soil temperatures of 70–75°F (21–24°C) and keep medium evenly moist, but not saturated. Light is not necessary for germination, but seed can be covered with vermiculite. Radicles should emerge in three to five days.

During Stage 2, stem and cotyledon emergence, maintain 70–75°F (21–24°C) substrate temperatures. Reduce substrate moisture levels once the radicle emerges, allowing the soil to dry out slightly before watering for the best germination and rooting. Continue to keep ammonium levels below 10 ppm. Once cotyledons have fully expanded, begin fertilizing with 14-0-14 or a calcium/potassium nitrate at 50–75 ppm. Alternate between fertilizer and clear water. Apply 50 f.c. (538 lux) of night interruption lighting from 10 P.M.–2 A.M. to promote flowering if HID supplemental lights are not available. Irrigate early in the day to ensure that foliage is dry by nightfall to prevent disease.

During Stage 3, growth and development of true leaves, reduce soil temperatures to 65–70°F (18–21°C). Allow medium to thoroughly dry between irrigations, but avoid permanent wilting. This promotes root growth and controls shoot growth. Keep the medium pH at 5.5–5.8 and the EC at less than 1.0 mS/cm. Increase fertilizer to 100–150 ppm from 20-10-20, alternating with 14-0-14 or another calcium/potassium-nitrate source. Fertilize every two to three irrigations. If using 15-0-15, supplement with magnesium once or twice during Stage 3 using magnesium sulfate (16 oz./100 gal. [1.2 g/l]) or magnesium nitrate. Choose one or the other: Do not mix these two magnesium sources, as precipitate will form. Continue to provide night interruption lighting if supplemental lights are not available. Use DIF whenever possible, especially during the first two hours after sunrise to control plant height. A-Rest, Bonzi, and Cycocel can also be used to control plant height. Plants will remain in Stage 3 for about twenty-one to twenty-eight days.

Stage 4, preparing plants for sale or transplanting, takes seven days. Reduce soil temperatures to 60–62°F (16–17°C) and continue to allow soil to dry thoroughly between watering. Fertilize with 14-0-14 or a calcium/potassium-nitrate feed at 100–150 ppm, as needed.

A 288-plug trays takes from five to seven weeks from sowing to transplant to larger containers.

Growing On

Select a well-drained, disease-free medium with a moderate initial nutrient charge and a pH of 5.5–6.2. Use 1 plant/cell for bedding plant packs, 1 plant/4" (10 cm) pot, and 2–3 plants/6" (15 cm) pot.

Dianthus performs best under cool temperatures and long days (day lengths of fourteen hours and longer). Many growers produce them in the late winter or early spring in unheated greenhouses or cold frames. For fall sales, keep plants as cool as possible during production, providing light shade and moderate temperatures. Ideal growing temperatures are 55–65°F (13–18°C) days and 50–55°F (10–13°C) nights. Provide as much light as possible, while moderating temperatures. Plants grown under low light become grassy, spindly, and weak. If you are trying to produce flowering plants during the winter months, provide supplemental lighting or night interruption lighting (50 f.c. [538 lux] from 10 P.M.–2 A.M.) to stimulate flowering and improve crop uniformity. However, lighting may cause some stretch that will require additional growth regulators.

Fertilize at every other irrigation with 15-0-15, alternating with 20-10-20 at 150–200 ppm.

Height can be controlled once plants have rooted to the sides of containers by allowing plants to wilt before irrigation. Withholding phosphorus and ammonium-nitrogen fertilizers will also help in height control. DIF (reverse day/night temperatures) will help product shorter plants. A-Rest, Bonzi, and Cycocel can also be used to control height.

Transplanting a 288 plug to a 32-plug tray takes from four to six weeks to finish for spring production. Plants will not be in flower. Upon transplanting to the garden plants will flower in three to four weeks.

Pests & Diseases

Thrips and aphids may be problems during production. Leaf spot caused by rust or *Alternaria* may also be a problem.

Varieties

D. chinensis 'Super Parfait' and 'Valentine' make fabulous 4" (10 cm) pots for Mother's Day sales. Some growers in milder climates with high light even produce them for Valentine's Day sales. Be warned that these varieties have no tolerance to heat and humidity. Under extremes the plants will die quickly when planted in the garden. With that being said, they are excellent impulse-sale pot plants

Interspecific crosses such as 'Floral Lace', 'Ideals', and 'Telstar' are the choice for growers looking to provide reliable performance to retailers and gardeners across a wide range of climatic conditions. All come in a wide range of colors, including eye-catching bicolors, and are great in the garden.

Postharvest

Provide cool temperatures (50–65°F [10–18°C]) during shipping and retail display. Provided temperatures are cool, display dianthus outdoors in full light. Potted dianthus will hold better at retail than pack dianthus, as plants become robust and put on growth spurts once they begin to flower.

Dianthus caryophyllus
Common name: Carnation
Cut flower, flowering pot plant

Cut Flower Crops

Carnations are a staple in the world's cut flower supply. Their reliability when properly handled after harvest makes them a must in most cut flower arrangements and mixed bouquets. Some colors are also pleasantly clove scented.

The production of carnations has moved from traditional consumer markets such as the United States, Germany, Japan, and the Netherlands to flower-exporting nations such as Colombia, Israel, Kenya, Spain, and China. Lower labor costs, higher light levels, and, in the case of Colombia, perfect growing conditions allow growers there to produce stems at lower costs. When combined with the magic of silver thiosulfate, which can increase vase life to thirty days, the carnation has become the poster child for the ideal cut flower to grow offshore.

While locally grown carnations are considered to be of superior quality worldwide, imported stems dominate. If you are a grower in a developed nation looking to grow and sell carnations, do market research first to determine what angle you can pursue to carve out a market niche for your crop.

Propagation

Carnations are susceptible to a wide range of bacteria and viruses, not to mention diseases. Because these pathogens are devastating to crops in production, it is best not to propagate carnations; instead purchase liners taken from virus-indexed mother stock. Make sure your supplier has the means to develop and implement a clean-stock program because you do not want to even think about buying cuttings that have not been maintained in a rigorous clean-stock system.

Growing On

Prepare soil in ground beds by tilling it before planting so it is loose and able to retain sufficient air and water. Such conditions stimulate rooting. Soil must be well drained, so break up impermeable layers.

As a basic fertilizer, a mix of 50% composted manure and 50% peat should be mixed into the soil before disinfection. Add 3–4 yds.^{3}/100 yds.^{3} (2.3–3.1 m^3/76 m^3) of greenhouse area. Before planting, and on a regular basis during growing, analyze the soil to measure its nutrient content. Table 12 lists target values based on a 1:2 volume extract.

Table 12. Target Values for Nutrients for Carnations

	TARGET	RANGE
EC	1.0 mS/cm	0.8–1.6
pH	6.0	5.5–6.5
NH_4	<0.2 mS/cm	0.1–0.5
K	2.1	2.0–4.0
Na	1.8	1.0–3.5
Ca	2.0	1.7–2.8
NO_3	4.3	3.3–6.0
Cl	1.0	0.5–3.5
SO_4	1.8	1.0–3.5

Disinfect the soil before planting. You will obtain the best results by steaming. Make sure the soil is watered well after the disinfection (also when steaming) in order to rinse out remaining chemicals. Have the soil analyzed. If the EC, especially the sodium and chloride, content is over 3.5 mS/cm, leach with water again. Always disinfect the soil before planting. Virgin soil may be clean, but when reused, soil may contain many pathogens. *Fusarium,* in particular, can build up in soil over time, making some ground beds unsuitable for carnation cultivation, even after disinfecting.

Some carnation growers have switched to hydroponic systems using coir, rice hulls, or other soilless medium. These growers remove 6–8" (15–20 cm) of soil from the ground beds, line the excavated area with plastic pond liner, and fill in with the artificial medium. While more expensive than growing in the ground, for growers who have been repeatedly cropping carnations on the same land for years, it is the only alternative to *Fusarium* buildup.

Plant only clean, healthy plant material from a reputable supplier. Keep the base of the cutting above the soil line. Never plant well-rooted cuttings deep: Deeply planted cuttings are at risk for root rot.

Plant at a density of 3 plants/ft.^{2} (31/m^2). Carnations eventually build up a huge plant mass, so a strong support system is required. The distance between poles should be no more than 10' (3 m). Use four to five layers of support nets, preferably made of wire. Insufficient support will result in flowers with bent stems. Openings in the support netting should be 6" × 8" (15 × 20 cm); space layers 8" (20 cm) apart.

Carnations perform best in relatively cool climates. When beginning a new crop, the soil temperature should be about 60°F (16°C), and the greenhouse

Hydroponic carnation production in Colombia

temperature 61–68°F (16–20°C). High temperatures combined with low light intensity will result in flowers with inferior quality. Do not allow the greenhouse temperature to exceed 77°F (25°C). Ventilation, shade cloth, whitewash, or a cooling system such as fog, roof sprinklers, or evaporative pad-and-fan cooling will be required in most regions.

The optimum greenhouse humidity for producing carnations is 65–90%. When humidity is higher, plant activity drops and plants become more susceptible to disease. When condensation occurs, there is a high risk of diseases such as *Alternaria* or botrytis. High light levels and too low humidity are harmful to a young crop and may cause leaf burning, weak foliage with closed stomata, and drought-related infections from red spider mites. However, while maintaining humidity levels is vital to prevent foliar disease problems, do not allow free water to stand on plants.

Pinch plants two to four weeks after planting, leaving four to six leaf pairs. Basically, each leaf pair will give one shoot. If too many leaf pairs remain, stem quality and growing speed may be dramatically reduced. If you leave too few leaf pairs, production is insufficient.

To spread out production, a 1.5 pinch is possible. About four weeks after the first pinch, 50% of the shoots should be pinched a second time. To push production for a particular sales window, a double pinch might be considered: Pinch all shoots a second time.

Grow plants at 55–60°F (13–16°C) days and 50°F (10°C) nights. Watch for boron deficiency during production.

Carnation flowers are induced when the plant has developed seven full-grown leaf pairs and when day length is thirteen to fourteen hours of sufficient intensity. Night interruption lighting from 10 P.M.–2 A.M. is commonly used to force flower induction during periods of low light intensity and short days.

Under low light levels, carnations produce more leaves and flower initiation is slow. Plants flowered under low light show low-quality, weak stems. During low light periods, avoid ammonium forms of nitrogen to prevent soft winter growth. Be careful, as too much nitrate nitrogen can also cause weak flower stems.

Once buds have initiated, temperature influences flowering. Warmer temperatures (above 60°F [16°C]) reduce flower size. Maintain cool temperatures (55–60°F [13–16°C]) and flower plants slowly to produce the highest-quality, large flowers.

The side buds of standard carnations must be removed (disbudded). This can be done in three phases. Buds should be taken away before they get too big, as they will negatively affect flower size and will take energy away from the plant. All buds and shoots within 12–16" (30–41 cm) below the terminal bud should be removed. Be careful when disbudding close to the main bud. This should be done just before it flowers, or the flower head may become deformed. With miniature carnations, the center bud should be removed as soon as the bud shows color.

Be careful of calyx splitting, which is caused when the calyx (flower) develops too quickly. Provide uniform night temperatures and a difference between day and night temperatures to prevent the problem. Using nitrate nitrogen and making sure plants have enough boron also helps. Some growers prevent the problem by placing rubber bands around buds when workers are disbudding.

There are several cropping options: 1-year, 1.5-year, 2-year, and crown-flower culture. The choice of culture depends on the climate (temperature), the quality perception of the flowers in the market, crop planning, and personal preference. Two-year culture is most common.

Pests & Diseases

Aphids, red spider mites, and thrips are the most frequently encountered insect pests. *Alternaria,* botrytis, *Fusarium,* rhizoctonia, and rust can all cause problems. Scout for problems frequently and control diseases immediately and aggressively to prevent major outbreaks.

Postharvest

Harvest flowers at the correct stage. For standard carnations, that is when they are 50% open and the first petal lies horizontally. For miniature carnations, harvest when two or three flowers are open. Do not cut flowers too deeply, as this may negatively affect production during the next flush. When harvesting standard carnations, leave two to three nodes on the shoot. On miniature carnations, leave four to eight nodes on the shoot. Place flowers in fresh, clean water as soon as possible and handle them with care.

Carnations are very sensitive to ethylene. Silver thiosulfate (STS) preservative solutions such as Chrysal AVB or Florissant 100 can be used. Preservatives work best at temperatures from 59–68°F (15–20°C). After STS treatment, stems may last up to thirty days.

After pretreatment, store flowers in a cooler at 41°F (5°C). To avoid condensation in the flowers and stem breakage, do not put the flowers in a cooler immediately after arrival from a "hot" greenhouse. Allow them to cool down slowly.

Maintain cool temperatures during transport. When shipping flowers in boxes, make sure the insides of the boxes have been cooled thoroughly, either by precooling or by keeping the open boxes in a cooler for at least eight hours. Transport should be as quick as possible and under cool conditions.

If stems have not been treated with STS, they will be much more susceptible to the ill effects of ethylene, which worsen at higher temperatures. For such stems, maintain storage temperatures of 33–35°F (1–2°C) in a cooler equipped with an ethylene scrubber.

Retailers should unpack boxes as soon as possible and hydrate flowers that have been dry-packed for two to twelve hours in deionized warm water with a pH of 3.5. Once hydrated, move stems to a sucrose floral preservative solution. Bacteria can also block water vessels and reduce longevity, so keep solution fresh or choose a preservative with a germicide.

Varieties

Standard varieties include 'America', 'Glacier', 'Tasman', 'Dover', 'Pleasure', and 'Shannon', to name just a few. Mini varieties include 'Elsy', 'Goldstrike', 'West Diamond', 'Malea', 'Scarlette', and 'Lior'.

Pot Plant Crops

Pot carnations are great for Valentine's Day, Easter, and Mother's Day sales. Grown well, plants will provide excellent value throughout the chain: from grower to retailer to consumer.

Propagation

Pot plant culture is similar to culture for rooting cut flower crops. Root cuttings at 64–68°F (18–20°C). Most growers buy in liners for two reasons. First, the absolute best varieties to grow are protected by plant patents and breeders' rights laws that prohibit vegetative propagation. Second, due to the susceptibility of diseases such as pseudomonas, corynebacterium, and viruses specific to carnations that can infect stock plants and not show up as a problem until you are half-finished with the crop or hit cuttings in the rooting bed, many growers leave the job of maintaining a clean-stock system and the rigorous sanitation required for propagation to specialists. It is best to buy clean, unrooted cuttings or liners from a reputable supplier.

Rooted cuttings will flower in 4" (10 cm) pots in about twelve to sixteen weeks when planted from week 50 (end of December) to week 18 (early May). Using pinched liners will shorten crop time about five weeks.

Growing On

Select a disease-free medium with a pH of 5.5–6.0 and a moderate nutrient charge. Plant 1 liner/4" (10 cm), 2–3 liners/6" (15 cm), and 3 liners/8" (20 cm) pot. Be sure to keep the crown above the soil line at planting. Due to their upright growth pattern, plants can be grown almost pot tight.

Grow plants on at mild temperatures, 70–72°F (21–22°C) days and 56–65°F (13–18°C) nights. Temperatures below 56°F (13°C) can delay flowering; every 2°F (1°C) drop adds one week of production time. Provided cool temperatures can be maintained, grow plants under as high light levels as possible: 4,000–7,500 f.c. (43–81 klux).

Early high fertilization is important for a shorter crop time. For the first five weeks after potting, fertilize with 250 ppm nitrogen (N) 20-10-20 plus 100 ppm ammonium nitrate at every irrigation. From weeks 6 to 13, drop fertilizer back to 15-10-15 at 200 ppm at every irrigation. During weeks 14 through 16, reduce fertilizer levels even further before flowering. Add boron at a rate of 0.25 oz./100 gal. (0.02 g/l) after weeks 14 through 16. The EC of fertilizer runoff should be 1.5–1.8 mS/cm until after bud development. Once buds are developed, lower it to 1.2–1.5 mS/cm. Maintain the pH at 6.2–6.5.

Do not overirrigate! Carnations have a higher tolerance for drier conditions than most other crops do, so allow pots to dry thoroughly between watering. Ideally, do not use overhead irrigation, as it can spread disease. Keep soluble salts levels in the mix low: Maintain an EC of 0.2–0.5 mS/cm. Irrigate with clear water weekly or as needed.

Pinch plants six to eight weeks after planting. The lead stem buds should be well developed. Pinch down to the first branch shoot on the stem below the bud. Grooming may be necessary to remove premature lateral buds.

Pot carnations do not need growth regulators.

Pests & Diseases

Pot carnations can get fungus gnats, aphids, thrips, spider mites, and caterpillars. They are among the most susceptible crops to a wide range of diseases including *Alternaria,* botrytis, *Fusarium,* powdery mildew, *Pythium,* and rust. Because carnations are susceptible to a range of diseases and viruses, it is vitally important to purchase liners grown from indexed stock that has been maintained in a clean-stock program.

Troubleshooting

Low night temperatures can cause slow plant development and uneven flowering. Plants can also grow slowly with uneven flowering due to low fertility under low light, poor root development, or *Pythium* caused by poor potting mix aeration. While plants are not daylength sensitive, flowering is delayed when plants receive a combination of low temperatures and short days. Providing night interruption lighting (50 f.c. [538 lux] from 10 P.M.–2 A.M.) can help to alleviate the problem but will cause internode stretch. For crops grown during long days, it may be beneficial to provide short days for the first five weeks of crop development to encourage plant growth and development.

Low initial fertility, premature pinching, and/or low light or low temperatures may cause poor branching. Boron or calcium deficiency, low light, constant day/night temperatures, and low day temperatures can lead to low flower petal count. Leaf tip burn may be a phytotoxic response to insecticide sprays, high soluble salts, excessive drying of medium, or high light/high temperatures combined.

Postharvest

Ship plants cool at 36–41°F (2–5°C). Retailers must keep pot carnations away from sources of ethylene, as plants are highly susceptible. Silver thiosulfate (STS) or EthylBloc may be applied to pot carnations prior to shipping to inhibit the flowers' response to ethylene and prolong flower life. Display plants under high light conditions with cool temperatures. Consumers who place their blooming pots in sunny windows should enjoy flowers for weeks to come since plants have not been disbudded and buds will flower.

Dianthus deltoides
Common names: Pinks; garden pinks; maiden pinks
Perennial (Hardy to USDA Zones 6–8)

The dainty, delightful flowers of garden pinks are a joy in the spring. Plants bloom for two to three weeks and scatter bloom for a week or two more. During summer the foliage persists and is best used as a border or edging plant.

Propagation

Sow two to four seeds onto a well-drained, disease-free medium with a low starter charge and a pH of 5.5–5.8 in a 288-plug tray in summer. You can lightly cover with vermiculite to help moderate humidity levels at the seed coat, or you can leave it exposed. Keep trays uniformly moist, but not fully saturated. At 60–70°F (16–21°C), seed will germinate in five to seven days.

Move trays to Stage 2 for stem and cotyledon development. Maintain temperatures at 60–70°F (16–21°C) and raise light levels to 500–1,000 f.c. (5.4–11 klux). Keep substrate moist, but once radicles begin to penetrate soil, allow trays to dry down somewhat but not letting seedlings wilt. Provide 50–75 ppm once a week from 14-0-14, once cotyledons have fully expanded. Stage 2 will take from four to seven days.

Move trays to Stage 3 to develop true leaves. Lower temperatures to 60–62°F (16–18°C) and increase light levels to 1,000–2,000 f.c. (11–22 klux). Increase the weekly feed to 100–150 ppm, alternating between 20-10-20 and 14-0-14. Allow trays to dry down somewhat between irrigations, but do not allow seedlings to wilt. Stage 3 will take from twenty-one to twenty-eight days for 288s.

Harden plugs for seven days prior to sale or transplanting by lowering temperatures to 55–60°F (13–16°C) and increasing light levels to 2,000–3,000 f.c. (22–32 klux). Provide short days (less than thirteen hours). Feed once a week with 100–140 ppm from 14-0-14. If growth regulators are needed, Bonzi and Sumagic are effective.

Using two to four seeds per cell, a 288-plug takes from four to six weeks, while a 128-plug tray takes from seven to nine weeks to be ready to transplant to larger containers from a June or July sowing. A late- spring or summer sowing that has been overwintered in its final container will produce flowering plants the following year. Most sowings made in January or later will produce either mediocre flowering plants or no flowers at all.

Growing On

Select a medium that is well drained and disease free with a moderate initial starter charge and a pH of 5.8–6.5. If you will be overwintering plants, select a long-lasting medium, such as a bark medium, and adjust pH levels.

You can use your own or purchase plugs or liners for planting in the late summer or fall and overwinter. Potted up to 1-qt. (1.1 l) or 1-gal. (4 l) pots, allow pots to fully root and then overwinter cold for flowering sales the following spring.

Commercial propagators offer cooled 50-cells that arrive in January or February for potting up and fast cropping.

If you bring in liners during the fall, bulk up plants prior to overwintering. Ideally, use larger plants, 128 plugs or 72-cell liners. Make sure plants are bulked up to at least eleven or twelve leaf nodes and are completely rooted into pots prior to the onset of cooling. Grow at 55–65°F (13–16°C) days and 50–55°F (10–13°C) nights. Provide short days (less than thirteen hours) to keep plants vegetative if you are establishing plants prior to mid-September. Fertilize at every other irrigation with 100–150 ppm from 14-0-14.

When you are ready for cooling, dianthus can be cooled in unheated cold frames or greenhouses or in a cooler with 25–50 f.c. (269–538 lux) of light. Cool for five to ten weeks at 41°F (5°C). *Note:* Cool 'Brilliant' for ten to fifteen weeks.

Bring plants out of dormancy for forcing, or allow them to flower naturally. If you are buying in cooled 50-cell liners, pot them up in January or February and grow on as outlined above. Keep night temperatures below 65°F (18°C). You can flower plants under natural day lengths. If growth regulators are needed, Bonzi and Sumagic are effective.

Plants prefer more alkaline conditions, so keep an eye on pH levels.

Pests & Diseases

Aphids and thrips can become problems during production, as can *Alternaria* and rust.

Varieties

'Brilliant' is rose red, while 'Flashing Light' is deep, bright red with densely packed bronzy foliage.

'Zing Rose' is probably the best known and best performing of the garden pinks. Its large (1" [3 cm]), deep rose-red flowers appear in the spring and then sporadically all summer long. It is not as well adapted to areas with hot humid summers as other varieties are, however, and may be short lived in those regions. 'Zing Rose' may be sown in the winter for green sales in packs in the early spring. Plants will flower in May. This variety also makes an excellent companion plant for sale along with pansies in the fall or early spring.

Diascia

Diascia barberae
Annual

Diascia is a cool-season crop with pink, coral-orange, scarlet, and white flowers. It makes an ideal cottage-garden plant, almost custom-made for a border along a white picket fence. Northern growers will want to be sure to offer *Diascia* as part of their spring plant mix. Southern and West Coast growers can offer plants in the fall with pansies for an unrivaled spring flower show. Plants grow from 4–6" (10–15 cm) tall and prefer morning sun and afternoon shade (especially in warm-summer climates). Full sun is fine for cooler areas.

Propagation

Diascia is commonly cutting propagated, although seed is available as well.

Seed

For a 288-plug tray, use one seed per cell and cover lightly with coarse vermiculite. Germinate at 65–70°F (18–21°C) and seedlings will emerge in four to six days. The crop time for a plug tray from sowing to transplanting is four to five weeks.

Cuttings

Stick two stem cuttings immediately upon arrival into Oasis foam or rooting medium. Provide soil temperatures of 68–75°F (20–24°C) and air temperatures of 70–80°F (21–28°C) days and 65–68°F (18–20°C) nights. Apply 500–1,000 f.c. (5.4–11 klux) of light and mist. Keep medium uniformly moist so that water is easily squeezed out. Set mist frequency so that cuttings do not wilt (six to ten days of mist on average). Apply 50–75 ppm from 15-0-15 as soon as foliage begins to lose any color. Use tempered water in mist lines as cold water can lower soil temperatures. Maintain a pH of 5.5–5.8. Callus will form in five to seven days.

Once 50% of the cuttings have begun to form root initials, move cuttings to the next stage. Maintain soil and air temperatures. Begin to allow medium to dry out somewhat as roots form; however, maintain high air humidity. This can be accomplished by adjusting the mist duration and frequency. Gradually increase light levels to 1,000–1,500 f.c. (11–16 klux). Apply a foliar feed of 100 ppm from 15-0-15, alternating with 20-10-20. Increase the rate quickly to 200 ppm. Roots will develop over nine to fourteen days.

Harden plants for transplanting or shipping for seven days. Remove cuttings from the mist area. Reduce air temperatures to 68–75°F (20–24°C) days and 62–65°F (16–18°C) nights. Allow medium to dry between irrigations. Gradually increase light levels to 2,000–4,000 f.c. (22–43 klux). Fertilize once with 150–200 ppm from 15-0-15.

A 105 tray requires three to four weeks to root.

Growing On

Plant 1 liner or plug/4" (10 cm) pot, 2/6" (15 cm) pot, or 3–4/10" (25 cm) hanging basket. Select a disease-free, soilless medium with a high initial starter charge and a pH of 5.5–5.8. Be careful not to bury the crown at planting, as this can cause crown rot.

Grow on at 62–75°F (16–24°C) days and 50–55°F (10–13°C) nights under full sun (4,000–8,000 f.c. [43–86 klux]). Plant growth slows significantly when temperatures are below 50°F (10°C), but plants will tolerate temperatures as low as 35°F (1°C) without damage. *Diascia* is not daylength sensitive, although flowering increases as light levels increase. Low light levels promote stretching.

Apply 200–250 ppm at every irrigation from 20-10-20, alternating with 15-0-15. If salts begin to build up, leach with clear water.

Apply a pinch when plant roots have reached the container walls, in about three to four weeks. Pinch plants in 6" (15 cm) pots once and 10" (25 cm) baskets twice. For the fastest crop time, do not pinch 4" (10 cm) pots. Pinch the crop back to the fifth or sixth set of leaves. Baskets can be sheared again in about two weeks. To increase branching, some growers use Florel (as a spray at 200–300 ppm) about two weeks after the first pinch for pots and baskets. If a plant growth regulator is needed, use B-Nine as a spray at 3,000–5,000 ppm.

Keep plants uniformly moist, but not wet. Overwatering can result in chlorosis of the growing point. If this happens, apply iron chelate. Water stress can also cause leaf-edge damage. Ideally, keep plant foliage dry during irrigation to reduce chances for foliar disease.

The crop time to a finished 4" (10 cm) pot is five to seven weeks from a 105 liner and seven to nine weeks from a 288-plug tray. For larger containers (6 and 10" [15 and 25 cm]) add another week or two with two plants per pot.

Pests & Diseases

Insect problems that can occur include aphids, fungus gnats, thrips, and whiteflies. Disease problems can include botrytis, *Pythium,* or rhizoctonia. Avoid pesticide sprays when plants begin to flower.

Troubleshooting

Botrytis or overwatering can result in plant collapse. Excessive vegetative growth can be due to high nitrogen concentration in the soil, overfertilization under low light, and/or wet medium in combination with low light. Low fertilizer levels can cause poor branching. Plants drying out between irrigations can lead to foliage necrosis.

Dicentra

Dicentra spectabilis
Common name: Bleeding heart
Perennial (Hardy to USDA Zones 3–9)

With arching stems decorated with heart-shaped flowers, *Dicentra* is an old-time favorite, one that consumers adore. Plants naturally bloom in May and June. Bleeding heart is a great plant to force into flower for Valentine's Day sales if you are a retail grower with good winter traffic or a pot plant grower selling locally.

Propagation

Dicentra is propagated by seed, stem cuttings, tissue culture, and root divisions. Named cultivars have to be propagated vegetatively to maintain the integrity of the variety, while seed can reproduce the species. *D. spectabilis* can also be propagated by root cuttings taken during the winter.

Tissue culture, root divisions, and cuttings are the most common methods of propagation, and most commercial propagators offer either liners or root divisions (bare root), which are both easier and quicker for the grower to produce a crop for spring sales. Most growers buy in precooled bare-root plants that were produced by bare-root perennial specialists in the United States or the Netherlands. Roots are sold based on how many growing points, or eyes (crown buds), they have: two to three, three to five, etc.

Growing On

If you are not able to pot up plants immediately on arrival, store roots at 35°F (2°C) until planting. Pot bare-root transplants in 1-gal. (4 l) containers. Use a well-drained medium with a moderate initial starter charge and a pH of 5.5–6.2. Trim roots, if necessary, at potting. Plant crowns of bare-root plants a bit high so that when you are watering pots in after planting, the medium will pack in around roots but crowns will not be buried. One plant is required per pot.

Commercial propagators have bare roots available during late fall and winter. Many times these are already precooled so the roots can be planted to produce flowering pot plants for the winter and early spring holidays or potted up during mid- to late winter for spring perennial plant sales.

Since *Dicentra* requires a vernalization before plants will flower, you will need to provide eight to ten weeks at a temperature of 40°F (4°C) or less if you do not have precooled roots.

The roots are potted up into 3- or 4-qt. (3 or 4 l) containers and grown on at 50–55°F (10–13°C) nights. *Dicentra spectabilis* will develop quickly, and roots potted in February or March will yield flowering plants in five to seven weeks. If grown under cooler night temperatures or under prolonged, cloudy springs, the plants will require up to one to two weeks longer to flower.

If you are flowering the crop for winter or spring sales as a pot plant, apply B-Nine at 2,500–3,000 ppm just as sprouts begin to grow (six to eight days after planting). Spray again five days later since emergence is not uniform. Bonzi and Sumagic are also effective. The crop time for a flowering pot plant for an early Easter or for Valentine's Day is from six to seven weeks at 50–55°F (10–13°C) nights.

Plants that are being forced into flower can be grown at 70–75°F (21–24°C) days and 55–65°F (13–18°C) nights. However, cooler temperatures (50–55°F [10–13°C] days and nights) produce a higher-quality plant, although these temperatures will increase production time by two to three weeks. Once plants have sprouted, apply a 200 ppm application of 20-10-20. Plants do not need a lot of fertilizer; they require only one or two applications during production. Plants will flower about twenty-one to twenty-eight days after potting.

If forcing plants for Mother's Day, maintain short days throughout forcing (less than fourteen hours) to keep plants compact. Plants forced for sale earlier may be grown under natural day lengths.

Varieties

D. spectabilis var. *alba* is a white-flowering form. Plants are not as vigorous

Related Species

D. eximia is a related species native to North America. Leaves are fringed, and flowers are mostly pink to lavender. If forcing *D. eximia* as a winter pot plant, you will likely not need to treat with growth regulators, since plants are shorter growing. *D. eximia* may take one or two weeks longer to flower than *D. spectabilis.*

Postharvest

Plants that have been forced into flower are ready to sell when they are 10–14" (25–36 cm) tall and have half of their flowers open. *Dicentra* does not ship well, as flowers are fragile. If you must sleeve plants, do so for the least amount of time possible. Once flowering is finished, consumers can plant *Dicentra* in a well-drained, yet moist shady spot in the garden, where it will perennialize. *D. spectabilis* will go dormant in the heat of the summer, disappearing until the following year. In contrast, *D. eximia* will not go dormant in the summer. Make sure *Dicentra* stays watered in the landscape, as drought may case permanent dieback.

Dichondra

Dichondra argentea
Annual

Dichondra is a low-growing plant that is perfect as a groundcover or as a trailing plant in mixed combination containers or hanging baskets. Plants can grow 6' (1.8 m) across (or down) in one season in some regions and no more than 2–3" (5–7 cm) tall. Dichondra's foliage makes the perfect backdrop for flowering plants, and its draping habit adds a sense of elegance and style wherever it is planted. For growers, dichondra is very easy. You can simply add a few plug trays onto your existing bedding plant order.

Propagation

Sow seed onto a disease-free, germination medium with a pH of 5.5–6.3 and a moderate starter charge. Cover lightly with coarse vermiculite to maintain humidity. Germinate at 72–76°F (22–24°C). Dichondra can be germinated in a chamber. Once 50% of the seed has cracked (about four to five days), move trays to Stage 2 for stem and cotyledon emergence. Reduce moisture levels once radicles have penetrated the medium: Keep seedlings moist, but not wet. Lower temperatures to 65–72°F (18–22°C) and gradually increase light levels to 1,000–2,500 f.c. (11–27 klux). Use 50–75 ppm from 15-0-15 as soon as cotyledons have expanded.

Once you have a full stand (after about ten to fourteen days), move trays to Stage 3 to form true leaves. Provide 65–70°F (18–21°C) temperatures and increase fertilizer to 100–150 ppm. Increase light levels to 3,000–4,000 f.c. (32–43 klux). To promote branching, spray B-Nine at 2,500 ppm about one week before transplanting. Prior to sale or transplanting, harden plugs off for a week by lowering temperatures to 62–65°F (16–18°C) and increasing light levels to 5,000 f.c. (54 klux). A 406 tray takes about five weeks. Dichondra is very sensitive to salts in the medium higher than EC 1.2 mS/cm. Clear-water irrigations between feeds will help minimize salt buildup in the medium.

A 288-plug tray will take from five to seven weeks from sowing to be ready to transplant to larger containers.

One application of B-Nine 5,000 ppm spray two weeks after transplant increases branching, controls stem length, and prevents plants from becoming tangled (left pot). This application also makes the foliage more silver at retail.
Photo courtesy of PanAmerican Seed

Growing On

Transplant plugs into a well-drained, disease-free medium with a moderate initial starter charge and a pH of 5.5–6.5. Use 1 plant/4" (10 cm) pot, 3 plants/6" (15 cm) pot, or 3–5 plants/12" (30 cm) hanging basket.

Grow on at 65–75°F (18–24°C) days and 62–65°F (16–18°C) nights. Provide full-sun conditions: High light levels promote shorter internodes and more silvery foliage. Keep plants on the dry side; plants can be allowed to wilt slightly. Dichondra grows most quickly under dry conditions. Feed once a week with 200 ppm from 20-10-20 or another complete fertilizer. Dichondra is very sensitive to salts in the medium higher than EC 1.2. Clear-water irrigations between feeds will help minimize salt buildup in the medium.

Increase branching and help to control plant growth by using B-Nine (5,000 ppm) as a foliar spray two weeks after transplanting. The treatment also causes foliage to become more silvery. Using growth regulators will also help keep plants from becoming entangled on the growing bench.

The crop time for a 4" (10 cm) pot is seven to nine weeks, while a basket is nine to ten weeks.

Varieties

Try 'Silver Falls', which changed the industry when it was introduced. Its leaves and stems are silver in color. 'Emerald Falls' is a solid green-leaf form. 'Emerald Falls' has a tighter habit than 'Silver Falls', but it is not as popular as its silver-leafed counterpart.

Dieffenbachia

Dieffenbachia spp.
Common name: Dumb cane
Foliage plant

Dieffenbachia is a star in the interiorscapes of offices, malls, and other public spaces. Its large, boldly patterned leaves brighten up and add interest to even the dullest interiors. Plants are offered in both small pots, such as 4 and 6" (10 and 15 cm), and as large-stemmed, freestanding, floor specimens in 10" (25 cm) or larger pots.

Most growers in the North buy in finished plants as part of their regular Florida foliage shipments, since dieffenbachia is a long-term crop requiring several months for large pots and must have heat in order to thrive.

Propagation

Tissue culture is becoming the standard way to propagate dieffenbachia. By purchasing tissue-cultured liners, growers are assured they are bringing in clean stock that is free of devastating diseases such as *Erwinia* or dasheen mosaic virus. Crops finished from tissue-cultured liners are also more uniform. In addition, dieffenbachia varieties are being protected more and more by plant patents and breeders' rights, thus unlicensed propagation is illegal. To access these superior new varieties, growers must buy in liners from the plant developer.

Dieffenbachia may also be grown from tip cuttings, but the risk of spreading disease is high. Some growers direct-stick tip cuttings into the final growing container for larger dieffenbachias such as 'Tropic Snow'.

Growing On

Use 1 liner/6" (15 cm) pot and 3 liners/8 or 10" (20 or 25 cm) pot. Select a disease-free, sterilized, peatlite medium with a moderate initial nutrient charge and a pH of 5.8–6.2. Make sure your medium choice does not hold too much moisture. Plant liners of suckering varieties a bit deep to promote suckering. Also, provide these varieties with more space, so that light reaches all leaves of the plant.

Provide a constant liquid feed of 200 ppm from 20-10-20. Leach with clear water regularly to prevent soluble salts problems. Rinse foliage after applying fertilizer. Watering is critical to your success: Dieffenbachia needs uniform moisture but does not like to be continually wet, so allow pots to dry between irrigations. Be careful not to overwater plants, as this can contribute to disease development. But if plants experience moisture stress, they will lose their lower leaves. Make sure foliage is dry by nightfall to prevent disease.

Provide moderate light levels of 2,000–3,000 f.c. (22–32 klux). For the best leaf coloration and suckering, grow plants at the high end of this range.

Grow dieffenbachia warm; 85–90°F (29–32°C) days are ideal. Maintain night temperatures above 65°F (18°C). Between fourteen and seventeen weeks are required for finished 6" (15 cm) pots.

Pests & Diseases

Disease can be a serious problem, especially *Erwinia* bacterial blight and *Xanthomonas campestris* pv. *dieffenbachiae.* Dasheen mosaic virus shows up as streaks in leaves, which may also be distorted and have ring spots. Plants infected with the virus may also lack vigor. The virus is not uncommon in plants propagated by cuttings. *Colletotrichum, Fusarium, Myrothecium,* and *Phytophthora* can also be problems. Be aware that Subdue can cause toxicity problems. Insect problems can include aphids, mealy bugs, scale, spider mites, and thrips.

Troubleshooting

Stunted growth and poor root development could be due to low production temperatures. Severe water stress can cause leaves to yellow and then drop. Nitrogen deficiency can lead to small, poorly colored leaves on plants that are growing too slowly.

Postharvest

Hold plants under lighted conditions, 100 f.c. (1.1 klux) minimum. Plants may be sleeved, boxed, and shipped at 60–68°F (16–20°C). Ship plants for no longer than seven days. Ethylene affects dieffenbachia: You will see signs to exposure when leaves become chlorotic. Holding plants in dark conditions for too long will cause lower foliage to yellow.

Retailers should unpack and unsleeve plants as soon as possible. Move plants to 100 f.c. (1.1 klux) of light and display them in temperatures from 70–80°F (21–26°C).

Consumers will find their plants perform best in 150–250 f.c. (1.6–2.7 klux) of light.

Varieties

Most production is in two species: *Dieffenbachia amoena* and *D. maculata.* Additionally, a number of hybrids have been developed. The most popular variety of the tall-growing *D. amoena* is 'Tropic Snow'. Well-known *D. maculata* varieties include 'Camille', 'Exotica Compacta', and 'Perfection Compacta'. Other well-known hybrid dieffenbachias are 'Paradise', 'Star Bright', and 'Tropic Marianne'.

Digitalis

Digitalis purpurea

Common name: Foxglove

Annual or biennial (Plants tolerate cool to cold winters from Zone 4 south.)

With 3–5' (91 cm–1.5 m) spires of gently colored spikes of flowers, foxglove is a favorite of spring cottage gardens. *Digitalis purpurea* is the most common of the foxgloves. Often classified as a perennial, digitalis usually only lives for one to two years in cold winter/wet climates. The older varieties are traditionally biennial in performance (plants rosette the first year from seed and then flower the following year after vernalization). They are considered perennial because many of them will reseed and provide color for several years.

However, the introduction of 'Foxy' changed the species' biennial status. 'Foxy' won an All-America Selections Award in 1967 as the first digitalis variety that would flower the same year as sown from seed. In the mid-2000s 'Camelot', 'Dalmatian', and others upped the ante with the first F_1 hybrids available in separate colors. Like 'Foxy', they all flower in summer from early to midwinter sowing. The hybrids will also reseed but, due to their genetic makeup, will not exhibit the same color, habit, and/or performance of the plant from which the seed dropped.

Propagation

Sow one seed per 288 cell onto a well-drained, disease-free medium with a low initial starter charge and a pH of 5.5–5.8. Seed does not need to be covered for germination. Keep trays uniformly moist, but not fully saturated. Provide 100–400 f.c. (1.1–4.3 klux) of light during germination. At 60–65°F (16–18°C), seed will germinate in two to three days on the more recent introductions.

Move trays to Stage 2 once about half have germinated. Maintain 60–65°F (16–18°C) temperatures. Increase light to 500–1,000 f.c. (5.4–11 klux) and apply 50–75 ppm from 14-0-14 fertilizer once a week as soon as cotyledons have fully expanded. As soon as radicles have begun to penetrate soil, allow trays to begin drying down somewhat, but do not allow plants to wilt. Stage 2 will take from four to seven days.

Stage 3, development of true leaves, will take from twenty-one to twenty-eight days for 288s. Reduce temperatures to 58–62°F (14–17°C) and begin to allow trays to dry between irrigations. Apply a weekly feed of 100–150 ppm from 20-10-20 fertilizer, alternating with 14-0-14. Increase light levels to 1,000–1,500 f.c. (11–16 klux).

Harden plugs for seven days prior to sale or transplanting at 55–60°F (13–16°C). Increase light levels to 1,500–2,500 f.c. (16–27 klux) and feed once a week with 100–150 ppm from 14-0-14 fertilizer.

On the more recent F_1 hybrid introductions, especially those from pelleted or enhanced seed, use one seed per cell, and a 288-plug tray will take from four to six weeks from sowing. Older varieties, especially sown from raw seed, may be less uniform in growth and stand and replugging may be needed.

Seed can also be sown to a 288 plug as noted above and then transplanted to a 72, 50, or 32 liner to bulk up for larger inputs. These liners can be transplanted to 1-gal. (4 l) or 2-gal. (8 l) containers depending on the size you want to finish in spring. The crop time to a finished liner is three to five weeks from transplanting to the liner from a plug, depending on the size of the final liner or tray size.

Transplanting a 288 plug to a final container (6" [15 cm] or 1 gal. [4 l]) is more commonly done in early to midwinter, while a liner is transplanted from a plug sown during summer to late autumn to pots (1 gal. [4 l] or larger) for overwintering in a cold house during winter for sales in the spring.

Growing On

Select a disease-free, well-drained, soilless medium with a moderate initial starter charge. Do not allow plugs to become root-bound. While digitalis can be transplanted to cell packs, this is only for sizing up as a liner to move into larger containers. Avoid allowing small cell packs to become root-bound. This will spoil later performance.

If using a 288-plug tray, transplant one plug to a 1-gal. [4 l] container in late summer in order overwinter the pots in a cold environment. Plugs transplanted to a 72, 50, or 32 liner can be transplanted to a 1-gal. [4 l] or larger container from late fall or early winter for plants sold green in the spring but will flower in late spring or summer.

Establish plants in their growing containers at 60–65°F (16–18°C) days and 50–55°F (10–13°C) nights. Once plants are fully rooted in their final containers, grow on cold at 45–55°F (7–13°C) days and 40–45°F (4–7°C) nights. Most of the newer varieties do not require a chilling period to set flower, but cooler temperatures produce a more compact plant and increase the number of flowering stalks.

Plants will flower in late May or June in their growing containers or in the garden, if they are sold green as long as they get high light. Long days (more than twelve hours long) enhance flowering.

Fertilize with 150–200 ppm at every other irrigation with 15-0-15. Grow plants under the highest light levels possible while maintaining cool production temperatures. Manage plants dry to avoid disease problems.

Varieties

The varieties introduced during the past ten years have unquestioningly changed the course of digitalis since the last major improvement (i.e., the All-America Selections Award-winning 'Foxy' in 1967). Many of these are F_1 hybrids and are easier to produce—many flowering in summer from an early to midwinter sowing.

Varieties like 'Camelot', 'Dalmatian', and 'Virtuoso' (all F_1 hybrids) are recent introductions that are offered as an enhanced-seed product that growers can purchase pelleted or in other forms. All can be sown in summer for overwintering in 1-gal. (4 l) or similar containers or sown in December or early January for flowering plants the upcoming summer. They vary in heights, from 28–48" tall (70 cm–1.2 m), but all enjoy morning/afternoon sun or all day sunny locations as well as cool nights and warm days with little humidity. Treated as annuals in the Midwest, they flower in summer from a winter or autumn sowing but seldom overwinter for more than one year.

'Foxy' is a mix of pastel colors in rose, lavender, pink, yellow, and white. Plants should be regarded as annuals, as they will generally not survive the winter. The groundbreaking variety is becoming less and less available each season.

'Giant Excelsior' is a well-known, traditional foxglove. Plants are huge, growing to 6' (1.8 m) tall! However, as a true biennial, 'Giant Excelsior' will not flower the first year from seed. Plants can be sold green to experienced gardeners in cell packs or 4" (10 cm) pots for planting out in the spring and flowering the following spring/summer. You can also grow it as you would a perennial that requires cold treatment (provide fifteen weeks). These plants will flower the following spring/summer.

Doronicum

Doronicum orientale
Common name: Leopard's bane
Perennial (Hardy to USDA Zones 4–7)

One of the earliest perennials to flower, *Doronicum orientale* is a hardy, but often short-lived, plant, lasting only two to four years in Midwestern gardens. The scentless flowers are single to semi-double, golden-yellow with a daisy-like form. They measure up to 2.5" (6 cm) across. Flowers are held terminally on plants to 2' (60 cm) tall. Foliage is kidney shaped and deep green in color and spreads to 15" (38 cm) across.

Propagation

Seed is the most common method of propagation. The best time to sow *Doronicum* is June to August for overwintering plants in their final containers for spring sales since vernalization (cold treatment) is required to develop flowers.

Sow two to three seeds per cell of a 288-plug tray using a well-drained, disease-free, germination medium with a low starter charge and a pH of 5.5–5.8. Cover seed very lightly with coarse vermiculite to maintain humidity at the seed-coat level; however, do not germinate seed in the dark. Provide 100–400 f.c. (1.1–4.3 klux) of light. Keep trays uniformly moist, but not saturated. The root radicle will start to emerge in two to four days at 68–72°F (20–22°C) soil temperatures. However, they do not emerge all at one time and will take from ten to fourteen days after sowing to germinate more readily.

Move trays to Stage 2 for stem and cotyledon emergence. Reduce soil temperatures to 64–68°F (18–20°C). Increase light levels to 500–1,000 f.c. (5.4–11 klux). As soon as cotyledons have expanded, begin fertilizing with 50–75 ppm from 14-0-14 once a week. Allow trays to dry down somewhat between irrigations. Stage 2 will take from four to seven days.

Move trays to Stage 3 to develop true leaves. Lower soil temperatures to 60–65°F (16–18°C) and increase light levels to 1,000–1,500 f.c. Fertilize once a week with 100–150 ppm from 14-0-14, alternating with 20-10-20. Stage 3 will take from twenty-one to twenty-eight days for 288s.

Harden plugs for seven to fourteen days prior to sale or transplanting by lowering soil temperatures to 55–60°F (13–16°C). Allow trays to dry between irrigations. Increase light levels to 1,500–2,500 f.c. and fertilize weekly with 100–150 ppm from 14-0-14.

In general, a 288-plug tray takes from seven to eight weeks to be ready to transplant to larger containers.

Growing On

Transplant a 288 plug to a 1-qt. (1.1 l) container in midsummer and allow to bulk in the final container until the start of cold weather. Overwinter dormant in a cold frame or similar environment at 35°F (2°C). Plants can be sold green in spring and will flower in the pot or garden.

Plugs from seed sown in September or October can be transplanted to 1-qt. (1.1 l) pots and overwintered cold in a pansy or similar greenhouse or cold frame at 40–45°F (4–7°C) nights once fully rooted in the final container. These will also flower the following spring. However, they will not be as free flowering as those sown earlier.

Sowings made in January and later seldom produce flowering plants until the following year.

Pests & Diseases

Aphids, powdery mildew, botrytis and *Pythium* are common issues for *Doronicum orientale.*

Dracaena

Dracaena spp.
Foliage plant

If you're not a grower in a tropical climate, don't even think about producing dracaena as a foliage plant. The exception to that would be highly mechanized 4" (10 cm) production, and even then the economics of crop production in the North would be questionable. Most growers outside of tropical areas who sell dracaena buy in finished plants: The quality is higher, and they are inexpensive, even after paying for shipping.

Propagation

Stock plants in Central America are the source for most dracaena cuttings coming into the United States. *Dracaena deremensis* and *D. marginata* are rooted from tip cuttings, a process that lasts about three weeks. *D. fragrans* is grown from tip cuttings or as a single- or multiple-sprouted cane. Again, cane is brought in from Central America. Make sure to buy only fresh, mature cane. Tip cuttings and cane are generally rooted in the final growing container. Rooting takes three to six weeks.

Growing On

D. deremensis and *D. fragrans* are typically grown with three tip cuttings per 10" (25 cm) pot. One tip can be used for 4" (10 cm) pots. *D. fragrans* is also grown with multiple canes per pot, such as two or three staggered canes in a 10" (25 cm) pot.

Grow *D. deremensis* at moderate temperatures: 75–85°F (24–29°C) days and 65–70°F (18–21°C) nights. Grow plants under light levels up to 4,500 f.c. (48 klux), provided temperatures can be maintained. Apply shade in areas with high light to reduce temperatures. Plants are very susceptible to boron and fluoride: Use only low-fluoride or fluoride-free water for irrigation.

D. fragrans grows best at 3,000–3,500 f.c. (32–38 klux): Be sure to space so that all canes receive good light. Grow on at 75–95°F (24–35°C) days and 65–70°F (18–21°C) nights. Do not allow temperatures below 55°F (13°C).

Grow *D. marginata* on at 3,000–6,000 f.c. (32–65 klux). At lower light levels, plants may drop foliage. Plants are moderately heavy feeders, so use 200 ppm at every irrigation, leaching with clear water occasionally.

The crop time for a 6" (15 cm) pot is from ten to fourteen weeks.

Pests & Diseases

Root rots caused by *Fusarium* or *Pythium* can become problems. *Erwinia* and rhizoctonia can also attack. Insect problems include mealybugs, scale, and thrips.

Growing out mass cane in southern Florida

Troubleshooting

Boron or fluoride toxicity can result in burned leaf tips. Poorly developed root systems can lead to strappy leaves that are narrow. In *D. fragrans,* high soluble salts and/or moisture stress can cause tip burn.

Varieties

Dracaena deremensis 'Warneckii', with its gray-, green-, and white-striped leaves, and *D. deremensis* 'Janet Craig', with its dark green leaves, are very well known. Both tolerate low light in interiors and make great floor plants.

D. fragrans 'Massangeana' is also known as the corn plant and/or mass cane: Leaves are striped with yellow and green. 'Massangeana' also tolerates lower light levels in interiors and makes a great floor plant. *D. marginata* has red- and green-striped leaves.

Postharvest

Display plants at retail under a minimum of 150 f.c. (1.6 klux) of light. *D. marginata* needs a minimum of 200 f.c. (2.2 klux). Plants grown under very low light for long periods of time will produce narrow new leaves. Ship at 60–65°F (16–18°C). Do not expose plants to temperatures below 50°F (10°C). Chilling injury may show up two weeks following exposure when new leaves emerge and are chlorotic or leaves become water-soaked. When shipping 'Massangeana' canes, pack trucks with pot spacers so that pots are held tight to prevent canes from leaning or becoming dislodged. Allow plants to dry between irrigations. Once *D. fragrans* has flowered, plants will not grow any taller.

Draceana Spike (see *Cordyline*)

Dusty Miller (see *Senecio*)

E

Echinacea

Echinacea purpurea
Common names: Coneflower; purple coneflower
Perennial (Hardy to USDA Zones 3–9)

This iconic North American native flower is beloved for its reliable performance as a perennial under a wide variety of circumstances. Beginning in midsummer, *Echinacea* begins its flower show. Common to this species are oversized bright, golden-orange discs placed atop a ring of downward-curving purple rays; although through the breeders' art, there are a myriad of other colors and forms available. Herbalists and natural medicine practitioners use *Echinacea* roots/rhizomes to treat all manner of infections. It also attracts a wide variety of bees and butterflies. With plants ranging in height from 24–40" (61–102 cm) with mostly single blooms (double-flowering varieties are available as well), *Echinacea* is a flower of national treasure.

Propagation

Most *Echinacea* is propagated from seed or tissue culture. Tissue-cultured plants are propagated by specialists and sold by commercial propagators in various tray or liner sizes, such as 50 or 72 plants per liner. While propagating by seed is noted below, tissue culture is a more exacting process best left to the commercial propagators. Information is provided from a rooted liner to a finished plant under Growing On.

Seed

Sow seed onto a disease-free, well-drained, germination medium with a low starter charge and a pH of 5.5–5.8. Do not cover. Germinate at 70–75°F (21–24°C) and provide 100–400 f.c. (1.1–4.3 klux) of light during germination. Keep medium moist, but not fully saturated. Stage 1 takes two to three days.

Move to Stage 2 for stem and cotyledon emergence once 50% of the seed has germinated. Maintain 65–70°F (21°C) soil temperatures and increase light levels to 500–1,000 f.c. (5.4–11 klux). Once cotyledons have fully expanded, apply 50–75 ppm from 14-0-14. Stage 2 takes four to seven days.

Move trays to Stage 3 to develop true leaves by increasing light levels to 1,000–2,000 f.c. (11–22 klux). Maintain 65–70°F (18–21 C) soil temperatures. Fertilize once a week with 100–150 ppm from 14-0-14, alternating with 20-10-20. Stage 3 will take about fourteen days.

Harden plugs for seven days prior to transplant or sale by lowering soil temperatures to 60–65°F (16–18°C). Increase light levels to 2,000–3,000 f.c. (22–32 klux) and short days (ten to twelve hours).

Fertilize once a week with 100–150 ppm from 14-0-14. Stage 4 will take seven days.

In total, a 512-plug tray takes from four to five weeks, while a 288-plug tray takes six to seven weeks from sowing to a salable or transplantable tray.

Growing On

While plants will flower without it, providing a cold treatment will hasten flowering and increase uniformity. However, flowering is primarily a long-day response. If you plan to force the crop for flowering in the spring, you need lights. If long days are not provided, any sowings done in late summer or fall and transplanted to their final container to keep cool during winter will flower naturally in July and August in the Midwest.

Seed

Select a long-lasting, well-drained, disease-free, soilless medium with a pH of 6.0–6.8. Use 1–2 plugs/4" (10 cm) or 1-qt. (1.1 l) pot and 3–4 plugs/6" (15 cm) or 1 gal. (4 l). When using seed-propagated plugs, you can also sow two to four seeds per plug instead of one or more multiple plugs per pot. Because each plant only produces one flower in its first year, using multiple plants for pots that will be sold in color creates the best presentation.

If you will be overwintering your *Echinacea* crop from a late-summer or fall planting, bulk plants to a minimum of four to five true leaves prior to providing the cold treatment. Grow on at 60–65°F (16–18°C) day and night temperatures. Make sure plants receive short days during the bulking process to keep them vegetative. If you are potting plugs from mid-September to mid-winter, this should not be a problem. Keep plants uniformly moist, but not wet. Fertilize at every other irrigation with 100–150 ppm from 14-0-14, alternating with 15-0-15. If growth regulators are needed, A-Rest, B-Nine, and Cycocel are effective.

Provide a cold treatment that lasts from ten to fifteen weeks at 41°F (5°C). Overwintering can be done in a cold frame or unheated greenhouse or in a cooler under 25–50 f.c. (269–538 lux) of light.

To bring plants out of dormancy, raise temperatures to 65–68°F (16–18°C) days and nights and increase light levels to 3,000–5,000 f.c. (32–54 klux). Run the crop somewhat dry during forcing to avoid disease problems. Fertilize with 150–200 ppm from 15-0-15 at every other irrigation. Maintain the pH at 5.8–6.2. If growth regulators are needed, A-Rest, B-Nine, and Cycocel are effective. For fastest flowering, grow plants under long days (more than fourteen hours) using night interruption lighting. Establish plants in the pot prior to beginning lighting. Because night interruption lighting will cause plants to stretch, make your first growth regulator application before lighting begins and apply again two weeks after lighting begins. Discontinue lighting once plants bolt (roughly seven to eight weeks later). Plants that do not receive night interruption lighting in the spring will flower naturally in July.

Tissue culture

Tissue-cultured plants are available in a number of liner sizes (128 or 105 plants per tray and larger), and these are commonly available for double-flowering selections, different flower colors, and unique flower forms.

Use one cell from a 128 liner (received from commercial propagators) per 1-gal. (4 l) container in the fall or winter. Potted up from December to early February, plants are grown under warmer temperatures to allow for root development in the final container (65–68°F [18–20°C] days and nights) and then moved to the cold frame in March for green plants in May. For flowering plants, follow the temperatures and other recommendations noted for seed-propagated plants.

Echinacea can be grown as a fresh or dried cut flower. Space out in fields at a minimum of 15" (38 cm) centers. If flowers will be sold dried as disks only, allow flowers to mature on the plant before harvesting: Petals will be easier to remove. Dry by hanging stems upside down in a well-ventilated, dry area. Fresh *Echinacea* has a short, seven-day vase life.

Troubleshooting

Avoid stressing plants during flower initiation, as flowers can become distorted.

Varieties

'Magnus' (from seed), a basally branching variety, grows to 26–36" (66–91 cm) tall and produces large 4.5" (11 cm) blooms. Petals are held flatter than the species or other varieties. It was chosen as the Perennial Plant Association's perennial variety of the year in 1988.

While there are a number of other varieties from the 1990s, it is the wealth of the introductions from the late 1990s and beyond that have changed the face of the varieties available in this class. The list of these varieties is too long to go through each selection. However, the vegetatively propagated 'Big Sky' series and the seed-propagated 'Primadonna', 'PowWow', and 'Prairie Splendor' have all helped, along with other seed and tissue-cultured varieties, to change the course of improvements in this class.

Eragrostis

Eragrostis elliotii
Common name: Love grass
Annual grass

Love grass grows upright at first and then, as the summer progresses, arches with narrow, blue-green leaves that change to beige in the autumn as temperatures cool. Plants prefer full sun and grow from 3–4' (0.9–1.2 m) tall once in bloom.

Propagation & Growing On

This grass is easily sown as seed to plugs or liner trays. Seed sown to a 406-plug tray, left uncovered, will germinate in two to three days at 72–74°F (22–23°C), and plug trays will be ready to transplant to larger containers in three to four weeks. 'Wind Dancer' is available as multiseeded pellets, so you need only one pellet per cell, or you can use three to four seeds per cell. Grow on at 68–74°F (20–23°C) days and nights at 64–66°F (18–19°C). Transplant to 18-cell flats, and plants will be ready in six to eight weeks.

Seed can also be sown directly to a 105- or larger liner tray. Seed sown to 96 trays in February, using six to eight seeds per cell and left exposed to light will germinate and grow as above. Trays can be transplanted to 4 or 6" (10 or 15 cm) pots in early April, and plants will be ready in mid-May.

Warmer growing temperatures help to develop this grass quickly so move flats or pots to cold frames where nights can be maintained at 55°F (13°C) or higher to hold.

Varieties

'Wind Dancer' is the only variety in this species currently available on the market. Plants grow from 3–4' (0.9–1.2 m) tall and have an arching effect in the garden.

Erysimum

Erysimum linifolium
Common names: Wallflower; Alpine wallflower
Perennial (Hardy to USDA Zones 6–8)

Wallflowers get their names as they grew out of rock or stone walls, preferring rocky soils that were well drained in locations with warm days and cool nights. Plants grow from 12–18" (30–45 cm) in the northern part of North America, but taller in areas where they are truly perennial. Plants persist in cool-winter areas but will not survive the winter in areas where the ground freezes. Flowers are single, measure from

0.75–1.5" across (2–4 cm) and come in bright lavender, mauve, yellow, and white colors.

Propagation

The most common method of propagation is through stem cuttings, although seed propagation, while not common in North America, can also be done.

Seed

Sow to a 288 plug using three to four seeds per cell and cover lightly. Final germination rates can often be lower than other crops, and the additional seeds help to excel plug performance. Germinate at 68–70°F (20–21°C). Stage 1 is two to three days for root radicle emergence, and Stage 2 is four to eight days for cotyledon expansion. From sown seed to a finished plug in a 288-plug tray takes five to seven weeks from an autumn to early-winter sowing.

For best performance in the garden, sow seed from November to early January for transplanting to larger containers for spring color sales. Sowings made in February or later will not be in flower when sold but will flower if planted green to the garden. Plants usually flower in two to three weeks here in the Midwest. However, since these are cool-season plants, flowering is best in spring or late summer.

Cuttings

Unlike plants started from seed, there is a myriad of varieties available from stem cuttings. Stick cuttings from autumn to early winter for spring sales. Cuttings can be taken later, but plants prefer cooler temperatures to flower profusely.

Stick one cutting per 50-, 72-, or 105-liner tray and treat with rooting hormone to encourage faster rooting. Provide seven to nine days of mist to root at 72–74°F (22–23°C). Liner trays are ready to transplant in four to seven weeks, depending on liner size.

Growing On

Seed

Transplant one plug from a 288-plug tray to a 32-cell pack or 4" (10 cm) pot. Grow on at 65–68°F (18–20°C) days and 55–58°F (13–14°C) nights. Plants will tolerate colder night temperatures once they are fully rooted but will "hold." That is, they will not grow but stay about the same height and size until temperatures warm up.

A 32-cell pack will be ready in six to eight weeks but should be sold green at retail. Once planted to the garden the plants will flower in another two to four weeks. Plants that flower in the cell pack are upright in habit and take a while to fill in after planting to the garden.

A 4" (10 cm) pot will be in flower in seven to nine weeks.

Cuttings

Transplant to 4 or 6" (10 or 15 cm) pots or in warm-winter areas (e.g., the Pacific Coast and in the Deep South), 1-qt. or 1-gal. (1.1 or 4 l) containers.

Plants prefer a pH of 5.8–6.2 and light levels of 5,000–9,000 f.c. (54–97 klux). Grow on at 62–67°F (16–19°C) days and 58–61°F (13–19°C) nights and feed at 175–225 ppm nitrogen (N).

Use one liner per 4" (10 cm) pot and allow seven to ten weeks to flower. For 6" (15 cm) pots, use one to two liners per pot and allow eight to twelve weeks to flower. For larger pots, especially gallons, use three liners per pot and allow fourteen to seventeen weeks to flower.

Pests & Diseases

Aphids and whiteflies are the common insect issues, while root and stem rots are the prevalent diseases.

Euphorbia

Euphorbia hypericifolia
Annual

'Diamond Frost' hit the market with great fanfare for gardeners and floral growers alike. Its delicate-looking white flowers are tough and add a unique, fancy look to mixed containers. Shortly after its introduction, it was marketed as an accent plant for poinsettias and other indoor floral plants. Since then many breeders have come to market with *Euphorbia hypericifolia* varieties that offer a variety of different performances both vigorous and more compact. There are even varieties with bicolor leaves and flowers that give a darker foliage look.

Sales have recently slowed and stabilized, securing a place for *Euphorbia hypericifolia* in all grower programs. Its wide tolerance to varying temperature, sun/shade, and moisture levels once it is established into a planting make it a season-long performer for the end consumer.

Propagation

Euphorbia hypericifolia is propagated by cuttings. Cuttings are relatively easy to root. If cuttings are wilted, place in a clean cooler at 50–55°F (10–13°C). Provide a relative humidity of near 100% by wetting the floor and misting open bags with clean water. Stick cuttings in Ellepots or medium. The use of a hormone such as Hormodin #1 will help with uniformity. Root cuttings under mist at a substrate temperature of 70–72°F (21–22°C) and air temperatures of 70–72°F (21–22°C) days and 68–70°F (20–22°C) nights. Provide only enough mist to keep cuttings turgid. Apply CapSil at 4 oz./100 gal. (0.3 ml/l) on the first day after sticking. This should help reduce the frequency of misting. Provide 500–1,000 f.c. (5.4–11 klux) of light.

If cuttings develop yellow leaves after sticking, it is important not to panic. Cuttings typically will continue to root even if all of the leaves drop. The dropped leaves also do not usually cause any disease issues. The most important thing to do is significantly reduce the mist, as the plants need very little mist when no leaves are present to transpire. The loss of leaves will extend the rooting time to five to six weeks.

When the cuttings are callused and begin to develop roots as early as five days after sticking, move to Stage 2. Begin to apply an overhead feed of 50–75 ppm from 20-10-20.

In Stage 3 maintain soil and air temperatures. As root initials form, begin to dry out soil between irrigations. Increase light levels to 1,000–2,000 f.c. (11–22 klux) as cuttings root. Apply a foliar feed of 100 ppm from 15-0-15, alternating with 20-10-20. Increase the rate to 200 ppm as roots form. Flowering in propagation can be expected and will not inhibit rooting. Do not apply Florel, as it will cause defoliation.

Stage 4 consists of hardening the cuttings for seven days prior to sale or transplant. At the beginning of Stage 4 you can pinch or trim the liners to help create lower breaks. Do not apply Florel. Reduce air temperatures at night to 65–68°F (18–20°C). Move liners from the mist area and increase light levels to 2,000–4,000 f.c. (22–43 klux). Shade plants during midday to reduce temperature stress. Fertilize with 200 ppm from 15-0-15, alternating with 20-10-20.

A 105 tray will take from three to four weeks at 68–72°F (20–22°C) to be ready for transplanting to larger containers.

Growing On

Transplant 1 plug or liner/4" (10 cm) pot; use 1–2 plugs or liners/6" (15 cm) pot; use 4–6 liners in 10" (25 cm) or larger. Select a disease-free, well-drained medium with a pH of 5.8–6.3. Grow plants on at 70–75°F (21–24°C) days and 65–68°F (18–20°C) nights. Temperatures below 60°F (16°C) will significantly slow or even stop growth. Fertilize with 200–250 ppm constant feed with a complete fertilizer. Ammonium-based fertilizers will cause softer growth as opposed to nitrate-based fertilizers, which will provide harder growth.

Early in production plants are sensitive to moisture extremes. As plants develop and mature, they are much more forgiving. Grow under high light levels (4,000–7,000 f.c. [43–75 klux]). Plants are day neutral and can be expected to flower throughout the growth process. This will not affect growth.

Pinching can be used to control height and shape the plant. Pinch as early as the plants have roots to the side of the finish container. You can pinch multiple times to create the desired shape. Do not pinch within three to six weeks of the sale date (less time later in the spring season when the weather is good).

Many varieties will not need any chemical plant growth regulators. However, when needed B-Nine at 1,500–2,500 ppm or a Bonzi drench at 0.5–1 ppm is effective for height control. Be careful not to overapply as it can severely stunt new growth.

Some growers may choose to grow *Euphorbia hypericifolia* as an accent plant with the traditional poinsettia or other indoor crops. This is done by planting *Euphorbia hypericifolia* in a larger container, while leaving an empty pot in the center. Later in the production the empty pot is removed and the poinsettia or other plant is placed into the center. Growing is the same as above. *Note: Euphorbia hypericifolia* tends to shed its flowers in low light conditions such as a consumer home.

The crop time with one plug per 4" (10 cm) pot is five to seven weeks, while twelve to fourteen weeks are required for a 10" (25 cm) basket.

Pests & Diseases

Aphids, spider mites, thrips, and whiteflies can attack plants. Disease problems that can strike include botrytis and *Pythium*. Like poinsettias, *Euphorbia hypericifolia* can be prone to root issues.

Troubleshooting

Plants may collapse due to wet medium or botrytis. High ammonium levels in the medium, overfertilization under low light conditions, and/or low light and overwatering can cause excessive vegetative growth. Plants drying out between irrigations, high soluble salts, or low substrate pH can lead to foliage necrosis or leaf drop. Production temperatures that are too low or an overdose of growth regulators can result in a lack of growth.

Varieties

'Diamond Frost', 'Silver Fog', 'Breathless', and 'Hip Hop' all have moderate vigor. 'Star Dust Sparkle' is more compact and displays profuse flowering. 'Breathless Blush' and 'Stardust Pink Glitter' have bicolor foliage and white and pink flowers.

Postharvest

Display plants in high light with moderate temperatures. Do not allow plants to wilt, yet avoid keeping pots wet. Plants are retail tough but should be protected from large swings in temperatures and moisture, which could cause the flowers to drop. Plants sell well as a component in a mix.

Consumers should plant in well-drained soil in full sun to part shade. Once established, plants will flourish in moist soil and moderate temperatures. Plants will withstand lower temperatures later in the fall and continue to flower. When displaying inside, consumers should provide as much natural sunlight as possible.

Euphorbia pulcherrima
Common name: Poinsettia
Flowering pot plant

> *Editor's note:* Jack Williams of the Paul Ecke Ranch in Encinitas, California, edited the poinsettia culture for the eighteenth edition of the *Ball RedBook* and passed away before the book went to press. In one of his emails, after completing the culture, he wrote: "Thanks for the opportunity to continue to represent poinsettias in the *Ball RedBook.* Your publication and Ecke both have such a great history of being a source of information that it is awesome to be part of continuing that legacy! Jack"
>
> That was Jack. Our industry will miss this universal friend.

Poinsettias have been the top-selling flowering potted plant in America based on United States Department of Agriculture data for many years; although in the most recent data they have dropped to number two behind orchids, based only on "value" and not volume. US growers produce more than 65 million pots, valued at over $240 million each year. They are one of the most important crops for several reasons. First, for bedding plant growers, they are countercyclical to bedding plants in the greenhouse. About the time bedding plant growers have moved out their summer annual crops and are more than midway through hardy mums, poinsettias begin. They take up greenhouse space in the summer and fall, when bedding plant demand in many areas is low. As the poinsettia crop finishes in December, it is time to receive early-season bedding plants and hanging baskets and time to clean up in preparation for the major liner and plug shipments arriving from January to March.

Poinsettias are also an important crop for the retail and consumer market based on their visual attributes, which is why you will find growers who specialize in pot plants growing them as well. Poinsettias are available in a range of shades and tones of red as well as different designer colors such as pink and white as well as complex bicolors that include marble, "glitter," 'Jingle Bells', orange, and "peppermint." Red has been, and still is, the most important color based on demand during the Christmas holidays, and poinsettias deliver the traditional holiday look better than any other crop available. In addition to novelty colors, there are many varieties that are anything but traditional in form and shape; 'Avantegarde', 'Carousel', 'Jester', 'Valentine', and 'Winter Rose' are some of the most significant examples. Breeding work in the past twenty years has developed plants with significantly stronger production characteristics to reduce shrink and enhance postharvest life. Walk into the average American home in mid-January and the poinsettia they purchased or were given for Christmas will still be there looking good, still in color, and with most of their leaves still attached. They are a great value for the consumer.

For most growers producing poinsettias, it is the most significant monoculture crop they will ever produce. You will get to know the crop well while you spend several months with poinsettias filling just about every square foot of your greenhouse. For a new grower that is a scary proposition: Do it right, and you will get it right on a big scale; do something wrong and you have the opportunity for a major blunder. In the career of some growers it is possible to produce anywhere from twenty-five to thirty crops, and these growers can tell you there are no normal years for poinsettia production. One year it's hot, the next it's cloudy, and the next something new and unexpected happens and impacts the crop. Whiteflies are still the primary insect pest to control worldwide, and botrytis is the most common disease problem to prevent. Height control is a perennial challenge. Be sure to keep good notes on the crop and plenty of digital pictures as reference points for crop development on a week-by-week basis. Jot down the date, your activities for the week, problems encountered, and solutions tried with successes and failures, and be sure to note the weather—sunlight levels and temperatures. Take notes variety by variety. Each variety performs differently under different circumstances and in different regions. These notes will serve as an invaluable record for you through the years as you make decisions about different treatments needed.

The cultural information provided here is but a scratch on the surface of the body of knowledge that exists on how to grow poinsettias. There are books (e.g., *The Ecke Poinsettia Manual)*, pamphlets, and flyers available as well as a plethora of information available on the web written by everyone from breeders to university researchers. With the vast amount of knowledge that is out there to pull from and the technical support provided by the major breeders, there is no reason that growers cannot produce poinsettias with minimal problems. Please look at what you find here as a beginning point to help you get started. By all means, read everything available and attend seminars on poinsettia production. Ask your supplier for information. If you are a new poinsettia grower, take time during the crop to visit one or two other poinsettia growers. It is always great to compare notes and to see how someone else's crop compares in development to your own. Most growers are happy to share information and cultural tips, especially if they do not compete directly with you.

Marketing

In North America there are still no organized marketing programs outside of a few variety-specific promotions organized through the varieties' breeders; the 'Polar Bear' poinsettia is one such example. "Stars for Europe" is a marketing and public relations program used in Europe where additional money collected from each cutting sold is pooled to support advertisement and promotion efforts throughout the European Union. The focus of this program has been to cultivate younger consumers of poinsettias and to help increase the total use and value of poinsettias. It is hard to say if this program has been effective or not, but certainly every effort made to promote the use of this holiday favorite is of value to our industry. Maybe one day a similar program will be instituted here in North America.

Plants start showing up on retail shelves earlier and earlier each year, as early as October in some mass markets. The bulk of plants are sold from Thanksgiving until about mid-December. Most years there are no shortages of plants, although in the past few seasons there has certainly been fewer in the market as greenhouses have discontinued growing poinsettias due to pressure from competition and rising energy costs. However, a well-managed crop is still possible and profitable! Before planting the first cutting it is important to understand the expectations of your customers so that finished plants can be designed to meet their specifications. Even within a specific retailer there may be multiple product specifications required, making it even more difficult for the producers of these plants. Without a clear knowledge of what the end product needs to be, growers are at risk of having their shipments rejected

Know your market and produce for that market. If you choose to grow 6–6.5" (15–17 cm) pinched plants, you will be competing with the 1,000,000 $ft.^2$ (93,000 m^2) grower two counties away. On the other hand, if you choose to produce 8" (20 cm) pots with three pinched plants or five to six straight ups in each pot, you will be competing with mid-sized growers in your area, some of which may be selling direct to the public through their own retail stores. Certainly the greater the volume of plants produced by competitors the less flexibility you will have in getting your desired price for this product. Specialty forms and varieties can help differentiate product and price even in large retail outlets.

Fundraisers are good outlets to consider as more and more growers are working with groups to sell their plants. Schools and churches are always looking for ways to raise money, and poinsettias are an excellent opportunity. The grower sells to the group at wholesale prices, and the group sells at retail prices, keeping the markup.

Just as you should visit other greenhouses during the production cycle, get out and see six or seven retail outlets from Thanksgiving until December 5 to see what is being offered at retail. How many cuttings does each size pot have? How developed are the bracts? What about plant height—too tall, too short? What kinds of add-ons are being sold with the plants? Are consumers buying the plants or just picking them up and putting them down? Stand and watch consumers at the poinsettia displays. Ask one or two of them why they selected the plants they did. Look to see how plants are being displayed and cared for and what's selling. With the emphasis placed on managing "shrink" in most retail outlets, it is important to do everything possible to produce and deliver plants that can hold up in these conditions and make it to the consumer's shopping cart! In the long run, being closely attached to the product and the buyers/customers will separate growers who make money from their poinsettia crop from the growers who break even or lose money.

Propagation

Today, most growers buy in unrooted poinsettia cuttings from a few large producers; mostly from offshore farms. Few growers still grow poinsettia stock plants from which they take their own cuttings. When considering the economics of each option, purchasing unrooted cuttings (URCs) usually pencils out much better than producing your own cuttings. Running the numbers, if a grower does not get at least twenty cuttings from each mother plant, the costs of this production are not being met. With good prices and effective transport options available, more and more growers have switched to purchasing in rather than growing their own mother plants. This also allows growers to use the space otherwise required by mother plants to produce and sell other crops from this same space, making money rather than costing them resources and missed opportunities.

For growers who do still produce their own poinsettia stock, these plants are generally shipped in March, April, or May. For stock plants pinched three times, plant in March; two pinches, plant in April; and one pinch, plant in May. The more pinches, the more cuttings possible per plant. However, since most growers producing poinsettias are bedding plant growers, they tend to wait to receive stock plants until bench space is opened up. (If you receive stock plants in March or April, be sure to provide long days with night interruption lighting to keep them in a vegetative state. See Growing On.)

The majority of cuttings are harvested or delivered from July through August each season, although we have seen a shift toward earlier crop start dates by producers using cold-growing programs to compensate for the slower rate of development due to lower average greenhouse temperatures (more on cold-grow programs below). If you are a new poinsettia grower, do not attempt to grow stock plants or root your own cuttings the first year in production. It is safer to buy in liners for your first crop; if you find that easy, buy in unrooted or callused cuttings. Evaluate the cost of growing your own stock plants before committing to this option.

Producing poinsettia liners begins with the stock plant. The quality of your finished crop will in part be determined by the quality of the cuttings harvested by you or your supplier. The objective is to develop the stock plant so it yields high-quality cuttings that will perform in production.

Regardless of the number of pinches given to the stock plant (one, two, or three), the stock-plant program consists of three phases: (1) developing the chassis, (2) producing cuttings, and (3) propagation.

Developing the chassis

During this phase, plants are pinched based on a set schedule and their rate of development to ensure the sufficient development of nodes for future cutting harvest. Pinch plants back to four to six nodes about three weeks after potting. Pinch again every four to six weeks later. Stock plants are grown on through the summer. Using plant growth regulators (Cycocel, Florel, or other PGRs) will reduce stock plant height and maximize stem strength to prevent stem breakage and improve plant tone. Be cautious to not overapply these chemicals as the result on rooting and uniformity of the cuttings can impact the crop.

Employ integrated pest management (IPM) to monitor and control insect and mite pests to prevent outbreaks on stock plants, because once you start taking cuttings, control is difficult. Monitor and control diseases to reduce the incidence of botrytis, *Pythium*, powdery mildew, and *Thielaviopsis*.

Fertilize to optimize lateral branching, leaf expansion, and shoot development. During this phase, maintain a higher level (35–50%) of ammoniacal/urea nitrogen to improve lateral branching, leaf expansion, and leaf color. Excessive levels of ammonia can cause oversized leaves, stretched growth, greater susceptibility to disease, and delayed rooting. It is critical to maintain calcium levels in the medium to maximize calcium uptake. (See Fertilizer.)

Producing cuttings

After the final pinch, fertilize stock plants to improve tone, color, and rooting. Optimize internode elongation of the developing cuttings by modifying the growth regulator rate or application frequency. Cuttings will be ready to harvest about five weeks after the pinch, depending on the variety, to result in quality cuttings that callus and root uniformly. This maturity factor is very important to the quality of cuttings produced so be sure you or your supplier provides proper attention to this quality-control factor. One-pinch stock plants will yield about eight to nine cuttings over a three-week harvest. Two-pinch stock plants will yield fifteen to twenty cuttings over a four- to five-week harvest. Cuttings should have an internode that is about 1" (25 mm) at the base of the cutting, with each subsequent internode a little bit shorter.

After the last pinch, reduce the ratio of ammoniacal to nitrate nitrogen to less than 15% to improve the tone, reduce the leaf size, and improve rooting of the developing lateral shoots (future cuttings). Do not eliminate ammonia/urea if you plan to harvest additional cuttings. A small amount of ammonia is recommended to improve the lateral branching on the return flush of cuttings. The ratio of ammoniacal to nitrate nitrogen can be changed by modifying the fertilizer used or increasing the number of nitrate feeds when alternating between ammonia and nitrate feeds (i.e., three applications of 14-0-14 or similar calcium-magnesium (Cal-Mg) feed, then one application of 20-10-20). (See Fertilizer.)

Harvest cuttings by snapping them or preferably by cutting them with a knife. Use only sharpened knives to ensure clean cuts without damaging plants or cuttings. Take only cuttings that are uniform in diameter and length (2.5–3" [6–8 cm]) to ensure uniform rooting and subsequent uniformity in the finished crop. Disinfect the knife periodically. Harvest in the early morning or very late in the afternoon to avoid heat stress. Do not crush stems, as the damage can serve as a point of entry for botrytis.

Store cuttings in a cooler with a small amount of moisture provided for a minimum of two hours at

48°F (9°C) to reduce the temperature and improve turgidity. Maintain 75–90% humidity in the cooler to prevent cuttings from drying out.

Propagation

For growers purchasing unrooted cuttings it is important to unbox and put cuttings into propagation as quickly as possible. These cuttings have been in transit for anywhere between twenty-four and thirty-six hours, so delays in sticking can result in significant losses. For cuttings harvested off stock plants and cooled as suggested above, remove from the cooler and stick within twenty-four hours. Cuttings can be successfully rooted in many different mediums: Oasis wedges, Preforma plugs, Jiffy-Pots, Ellepots, etc. Some growers use a rooting hormone (IBA or mixes of IBA and NAA) to improve rooting uniformity. Make sure the medium is uniformly moist so that water is easily squeezed out. Orient cuttings as you stick them so that leaves from neighboring cuttings do not cover the terminals and do not overcrowd the cuttings. Typically, densities of about twelve cuttings per square foot (120/m^2) work well. Maintain soil temperatures at 76–78°F (24–26°C) and air temperatures at 70–75°F (21–24°C). Use tempered water (70°F [21°C]) in the mist lines so soil temperatures are more easily maintained. During callus formation (taking five to seven days) it is acceptable to have free water in the Oasis wedges, but other mediums like Jiffy-Pots should not be allowed to stay this wet. Mist enough to maintain turgidity without being excessive: Too much mist will leach nutrients from the leaf. You might start at ten seconds of mist every four to six minutes and adjust to your conditions and as cuttings root. If leaching is a problem, apply a foliar feed of 50–75 ppm from 14-0-14 or 15-0-15. Adjust mist frequency based on your ambient light and temperature conditions. Maintain light at 800–1,000 f.c. (8–11 klux). Higher light levels will stress cuttings due to warming. Retractable shade is ideal since you will be able to increase light levels gradually as cuttings mature. Once 50% of the cuttings begin differentiating root initials, move them to the next phase.

By the second phase, development of roots (which takes seven to fourteen days), moisture levels should be lower. As roots develop, begin drying the medium down and gradually increasing light levels to 1,000–1,200 f.c. (11–13 klux). Foliar feed with 100 ppm 14-0-14, alternating with 20-10-20, increasing rapidly to 200 ppm. Rinse cuttings lightly after any fertilizer treatments that contain phosphorous (P) to prevent damage to the growing tip and young leaves (more on this under Leaf Distortion). Monitor the pH and EC of leachate daily. Maintain a substrate pH of 5.8–6.2 and an EC between 1.0–2.0 mS/cm. If needed, growth regulators (Cycocel or combinations of Cycocel and B-Nine) may be used to prevent unwanted cutting stretch. (*Warning:* Do not apply growth regulators to low-vigor varieties such as 'Winter Rose' in the cutting stage.)

During the final rooting phase (seven days to harden before sale or transplanting), watering the medium is more important than misting the foliage. Allow medium to dry thoroughly between irrigations. Increase air temperatures to 72–75°F (22–24°C). Increase light intensity to 1,500–1,800 f.c. (16–19 klux) or higher, depending on temperatures and conditions. Using shade during the heat of the day will reduce stress on the crop. Fertilize weekly with 150–200 ppm of 14-0-14. Apply a foliar fungicide twenty-four hours prior to shipping to minimize botrytis during shipping. Before shipping, precool cuttings to 48°F (9°C) to remove heat from the boxes and, if possible, put ice packs in the boxes.

Note that it is possible to direct-stick cuttings to their final growing containers, saving valuable labor. However, growing a direct-stick crop requires more greenhouse space and mist lines. The best success for direct stick is done with smaller pots such as 4" (10 cm) up to 6.5" (17 cm).

In general allow, twenty-one to twenty-eight days for rooting in medium or Oasis foam.

Growing On

Select a disease-free, well-drained medium with a pH of 5.8–6.2 and a moderate initial nutrient charge. If starting with rooted cuttings, make sure the young plant medium is moist before transplanting. At potting, be sure to plant the cuttings level with or slightly above the potting mix. Water immediately after potting to settle the potting mix firmly around the young cutting and its roots. Make sure the Oasis foam stays moist until roots are firmly established in the potting medium by irrigating lightly each day for the first week to prevent drying out of the foam, other mediums may not require this additional step at this stage. To ensure good root development, water pots thoroughly and then allow the growing medium to dry down before another full watering. Excess moisture at this stage contributes to problems with fungus gnats, which can cause significant damage to the crop. During this stage, plants are susceptible to heat delay from high soil temperatures (greater than 85°F [30°C]). Moderate soil temperatures by irrigating at the hottest time of the day, which will cool temperatures.

During the first phase of production (from planting until October 1), maintain temperatures at 75–85°F (24–27°C) days and 66–68°F (19–20°C) nights. Grow at 4,000–6,000 f.c. (43–65 klux) of light. Fertilize at every irrigation at 250 ppm from 20-10-20 or a complete, balanced fertilizer that is compatible with your water quality. Fertilizing early in the crop is critical for strong lateral development. Maintain the pH at 5.8–6.2 and soil EC at 1.0–1.5 mS/cm.

During the second phase of production (from October 1 to bract color) when producing at traditional temperatures, maintain temperatures at 70–75°F (21–24°C) days and 64–66°F (18–19°C) nights. Cold-grow temperatures are described in more detail for those who choose to use this method of production in order to save energy and costs (see the Cold-Growing Poinsettias sidebar). Grow at 4,000–6,000 f.c. (43–65 klux) of light. Continue fertilizing at 250 ppm, alternating at every other irrigation between a general-purpose fertilizer and a calcium-based fertilizer. Maintain the pH at 5.8–6.2 and EC at 1.0–1.5 mS/cm.

For the third phase of production (from bract color to sale), maintain temperatures at 70–75°F (21–24°C) days and 64–66°F (18–19°C) nights. Reduce light levels to 1,000–3,500 f.c. (11–38 klux) to protect colored bracts. Lower fertilizer to 150–200 ppm from 14-0-14 once a week. Failure to reduce the feed at the end of the crop will increase bract-edge burn. Maintain the mix pH at 5.8–6.2 and EC at 1.0–1.5 mS/cm.

Poinsettia growth and development is a function of average daily temperatures. For an explanation, see Height control (p. 383).

Poinsettias are short-day plants; they require a minimum of twelve hours of uninterrupted darkness to stimulate flowering. Plants will initiate flowers naturally when the temperature is less than 72°F (22°C) and day length is less than twelve hours (after September 21, when days and nights are twelve hours each). So from about October 1 until early March, natural daylight will stimulate flower initiation. Some varieties may initiate flowers earlier (around September 15), resulting in early maturity and sale dates.

In the South, flower initiation can be delayed when the night temperature remains above 72°F (22°C) for several nights, resulting in "heat delay."

To grow high-quality plants, you must produce the vegetative portion of the plant under long days prior to giving short days. If the days are short, supplemental lighting or night interruption lighting (10 f.c. [108 lux] at plant level from 10 P.M.–2 A.M.) can be used to keep plants vegetative so they grow to the proper size before initiating bract coloration. The general rule of thumb is to begin lighting on September 5 to prevent floral initiation and keep plants vegetative.

Conversely, for plants that need to be colored early in the season, shadecloth can be used to provide short days and promote floral initiation for any of the varieties that do not set bud early (fortunately there are now many good varieties available that can be in flower by early November without any manipulation). When using shade- or black cloth, the general rule is to begin short days before September 15. To grow "natural season" poinsettias, you do not light or black cloth plants. Such plants (with the exception of varieties with early initiation characteristics) will remain vegetative until September 21 and then begin to initiate flowers. If you have questions about the initiation characteristics of the varieties you produce, your breeders should be able to provide you with details that allow you to properly schedule your crop.

Whether or not you are pulling black cloth or using naturally short days, once short-day treatment begins, ensure there is no stray light hitting plants. Even just 1–2 f.c. (11–22 lux) of light will cause problems. For example, streetlights or lights from adjacent greenhouses or buildings can cast enough light to cause flowering problems. The rule of thumb many growers have used successfully is if you can read a newspaper in the greenhouse at night, then too much light is present and can result in delay.

It is important to have the plant built before initiating flowering. When plants initiate at the time of the pinch, they will only have five to seven leaves on the stem, and the finished crop will be very short. The number of days of vegetative growth provided between the pinch and the start of short days affects the number of leaves and the finished height potential of the crop. Schedule properly to get the best results and best quality.

Cold-Growing Poinsettias

Another option for production is to use an approach referred to as cold growing. Under this regime, plants should be started and pinched about ten to fourteen days earlier than they would be under a traditional growing program in order to compensate for slower growth due to lower average temperatures. To compare this program to the traditional production method, during the first phase of growth (planting to October 1) temperatures are managed cooler than described, with a target of 65–75°F (18–24°C) days and 60–64°F (15–18°C) nights. During the second phase of growth (October 1 to November 1) targeted temperatures are dropped to 60–65°F (15–18 C) days and 58–62 F (14–17 C) night. In the final phase of production (November 1 to sale) average daily temperatures of 60 F (15 C) are used.

Each of the major poinsettia breeders offer genetics that are well adapted to cold production, along with support information on the ideal temperatures and conditions needed to be successful with these varieties. Also, a visual tool called the Ecke Poinsettia Bract Meter is available from the Ecke Ranch. Growers can use this tool to visually track development of the poinsettias as they develop color on the bracts, along with detailed temperatures to manage and adjust timing based on the percentage of color development. Using a cold-grow program like this one, growers can save up to 21% of the energy required to produce this crop without compromising quality!

Response groups

Poinsettias are classified by the number of weeks between flower initiation (short-day treatment) and pollen shed. (Poinsettias are sold when the first flowers, or cyathia, begin to show pollen. This is also called anthesis. Pollen shed occurs about two weeks later, depending on environmental conditions.) Each variety is different. You can select varieties from different response groups to time your crop to flower successively with early-, mid-, and late-season varieties. Or, you can use black cloth and lighting to time an individual variety to be in flower early, midseason, or late.

To time different response groups, the vegetative period from planting to the start of short days should be shifted earlier by the appropriate number of days for each of the flowering periods. For example, a nine-week response group variety will start short days one week earlier than an eight-week response group variety in order for them to both be in flower and ready to market on the same date.

Schedule your crop by counting backwards from your desired sale date. Plants are ready to sell at visible pollen (anthesis) on the first cyathia. The bracts are actually not flowers; they are specialized leaves. The flowers (cyathia) are in the center of the bracts. If plants are shipped before pollen shows on the first cyathia, they will never develop correctly in postharvest. If pollen has begun to shed, the plants are too old to sell. You have roughly a fourteen-day window to move plants once they start flowering.

With a flowering date in mind, determine when short days will begin. If you are growing an eight-week variety and want them in flower the week before Thanksgiving (week 47), you would begin short days eight weeks earlier (week 39) in late September. You will need to schedule the pinch date next, which varies by variety. In general, allow two to three weeks between potting and pinch.

Scheduling the time between pinching and the start of short days varies by pot size and the region of the country from two to five weeks. See the Schedule Samples sidebar for examples of pinched and straight-up crops from nine-week varieties by pot size.

If you would like to offer single-stem plants (unpinched), more commonly known as straight-ups, cropping will be slightly different. Scheduling single-stem poinsettias is similar, but the time between short days and potting is shorter. For a three-cutting crop in 6.5" (17 cm) pots, pot liners about two to three weeks later than a pinched crop. Growers that are harvesting their own cuttings from stock plants may find it easy to use the last "pinch" for straight-up production. Use three to four cuttings in a 6.5" (17 cm) pot, seven in an 8" (20 cm) pot, and ten in a 10" (25 cm) pot.

Your cutting supplier and your poinsettia varieties' breeders will also be able to supply you with scheduling information.

Fertilizer

Managing the fertility of poinsettias is critical to the success of the crop. With so many of the varieties sold today having dark-green leaves, fertility management is different from the programs of fifteen or twenty years ago. As a rule, the dark-leaf varieties use about 20% less feed overall than light-leaf varieties. Since most growers have both in the greenhouse production, developing an effective program that works for all can be a challenge, especially if everything is supplied from a common injector system. Many have found it helpful to feed the entire crop at a level appropriate for the dark-leaf types (200 ppm constant liquid feed) and then supplement their dark- and light-leaf varieties

Schedule Samples

These schedules are for a nine-week response plant in various pot sizes. Adjust the schedule according to the response group, growth habit, and region of the country in which you are located. For an eight-week response group, start the process one week later. For a ten-week response group, start the process one week earlier.

4–4.5" (10–11 cm) pot
Pinched
No. plants/pot: 1
No. leaves on liner at pinch: 4–5
No. nodes on the lateral at flowering: 5

FINISH WEEK	PLANTING WEEK	PINCH WEEK	START OF SHORT DAYS*
46	34	36	37
47	35	37	38
48	36	38	39**
49	37	39	40
50	38	40	41
51	39	41	42

4" (10 cm) pots
Straight-up
No. plants/pot: 1
No. nodes on the lateral at flowering: 7–9

FINISH WEEK	PLANTING WEEK	PINCH WEEK	START OF SHORT DAYS*
46	35		37
47	36		38
48	37		39 **
49	38		40
50	39		41
51	40		42

* Use four hours of night interruption lighting prior to short days to promote vegetative growth. Use a shadecloth at the start of short days to promote flowering.

** Natural season, no shadecloth or night interruption lighting needed.

(Continued)

with fertilizers that provide the additional nutrition each variety requires. You can influence the growth of your poinsettia crop based on the type of fertilizer you use by varying the amount of ammonia versus nitrate-nitrogen in your fertilizer (table 13).

Immediately after planting, ammonia nitrogen (recommended at 25% of the total fertilizer applied) helps to improve branching, expand leaves, and color bracts. Once plants reach the transition bract stage (the first bract color), reduce ammonia nitrogen to less than 15% and increase nitrate fertilizer to improve tone, decrease upper leaf size, increase root development, and improve postharvest life.

Poinsettias use fertilizers differently as they grow and develop, which is why you need to check soluble salts regularly. A soluble salts reading is a measurement of the amount of fertilizer left in the medium after the plant has taken up what it needs. If you apply a fertilizer with a reading of 1.5 mS/cm and check the soluble salts before the next irrigation, the reading

Schedule Samples *(Continued)*

6" (15 cm) pots
Pinched plants
No. plants/pot: 1
No. leaves on liner at pinch: 5–7
No. nodes on the lateral at flowering: 7–9

FINISH WEEK	PLANTING WEEK	PINCH WEEK	START OF SHORT DAYS*
46	31	33	37
47	32	34	38
48	33	35	39**
49	34	36	40
50	35	37	41
51	36	38	42

6.5" (17 cm) pot
Pinched plant
No. plants/pot: 1–2
No. leaves on liner at pinch: 5–7
No. nodes on the lateral at flowering: 7–9

FINISH WEEK	PLANTING WEEK	PINCH WEEK	START OF SHORT DAYS*
46	31	33	37
47	32	34	38
48	33	35	39**
49	34	36	40
50	35	37	41
51	36	38	42

* Use four hours of night interruption lighting prior to short days to promote vegetative growth. Use a shadecloth at the start of short days to promote flowering.

** Natural season, no shadecloth or night interruption lighting needed.

(Continued)

Table 13. How Does Fertilizer Choice Affect Plant Growth?		
	AMMONIA FERTILIZERS	NITRATE FERTILIZERS
Leaf size/tone	Large, soft	Small, hard
Leaf color	Dark green	Light green
Stems	Stretched	Compact
Overall plant tone	Soft, floppy	Hard
Root growth	Average	Strong, branched
Postharvest life	Poor	Excellent

Schedule Samples *(Continued)*

6.5–7.5" (17–19 cm) pot
Straight-up
No. plants/pot: 3–5
No. nodes on the lateral at flowering: 9–11

FINISH WEEK	PLANTING WEEK	PINCH WEEK	START OF SHORT DAYS*
46	33		37
47	34		38
48	35		39**
49	36		40
50	37		41
51	38		42

7–7.5" (18–19 cm) pots
Pinched plants
No. plants/pot: 3
No. leaves on liner at pinch: 5–7
No. nodes on the lateral at flowering: 7–10

FINISH WEEK	PLANTING WEEK	PINCH WEEK	START OF SHORT DAYS*
46	30	32	37
47	31	33	38
48	32	34	39**
49	33	35	40
50	34	36	41
51	35	37	42

* Use four hours of night interruption lighting prior to short days to promote vegetative growth. Use a shadecloth at the start of short days to promote flowering.

** Natural season, no shadecloth or night interruption lighting needed.

(Continued)

will tell you how much fertilizer was removed from the medium. For example, going from 1.5–0.5 mS/cm tells you that the plants removed most of the salts.

Poinsettias take up nutrients at different rates during the production cycle. Just after potting and before roots reach the edge of the pot, plants do not need much fertilizer because there is limited plant growth and the growing medium is relatively large in comparison to the extent of roots, allowing for ample nutrition in the soil. When plants begin to grow side shoots after pinching, the need for fertilizer increases to support the active and rapid growth stage of the plants. Fertilizers through this stage of development should be balanced and contain the full range of elements required by the plants. It is important to have your water quality tested and to understand the mineral content and chemical properties of your water in order to choose the proper fertilizers for your crop.

Monitor soluble salts weekly during production, paying particular attention to key developmental phases such as pinching, flower initiation, and transition bract development. Through regular monitoring, you can change fertilizer concentration (ppm) to prevent salts damage.

Schedule Samples *(Continued)*

8–8.5" (20–22 cm) pots
Pinched plants
No. plants/pot: 4
No. leaves on liner at pinch: 5–7
No. nodes on the lateral at flowering: 9–11

FINISH WEEK	PLANTING WEEK	PINCH WEEK	START OF SHORT DAYS*
46	28	31	37
47	29	32	38
48	30	33	39**
49	31	34	40
50	32	35	41
51	33	36	42

8–8.5" (20–22 cm) pots
Straight-up
No. plants/pot: 5–7
No. nodes on the lateral at flowering: 13–15

FINISH WEEK	PLANTING WEEK	PINCH WEEK	START OF SHORT DAYS*
46	31		37
47	32		38
48	33		39**
49	34		40
50	35		41
51	36		42

* Use four hours of night interruption lighting prior to short days to promote vegetative growth. Use a shadecloth at the start of short days to promote flowering.
** Natural season, no shadecloth or night interruption lighting needed.

If micronutrient problems are going to occur, they will likely show up by October. Be ready to adjust the pH or apply a soluble trace element mix (S.T.E.M.) to correct any deficiencies. By the way, poinsettias have a high need for molybdenum, which can be added at every irrigation at a rate of 0.1 ppm.

Pinching

By far, the majority of poinsettias sold are pinched plants. Varieties vary in their branching characteristics, or the number of "breaks" they produce. If your goal is to produce a multibract plant, pinching at least once is required. Before pinching, be sure roots have developed all the way to the pot edge, otherwise there will not be enough root mass to support shoot development and you will slow the crop down.

How to pinch: Remove the leaf that is beginning to unfold and all the smaller leaves around it. Remove young, rapidly expanding leaves below the pinch to enhance lateral shoots and increase the light getting to the interior of the plant. You should get as many breaks as you have leaves left on the plant after pinching.

Many growers use Florel to enhance branching and provide a mild level of growth control. The rule of thumb for using this plant hormone/growth regulator is to make an application of between 350–500 ppm one week before and again one week after pinching. Using this chemical should maximize branching and mildly control the shoot elongation that results. If using Florel, monitor the development of the crop height prior to applying any additional PGRs. Do not use Florel past flower initiation and the start of short days.

Height control

Use graphical tracking to monitor growth so you can correct problems quickly. (For more information, see the Graphical Tracking sidebar.)

The best height control is achieved when cuttings are treated before pinching and the plants are treated regularly through production. A-Rest, B-Nine, Bonzi (paclybuturol), Cycocel, and Sumagic are all effective growth regulators for poinsettias. Topflor is a relatively new PGR introduced to the market, and growers are still learning how to use this chemical effectively. Rates for all these PGRs vary by variety, stage of development, and climatic conditions. General rates are: 0.25–0.5 mg A-Rest a.i. as a drench; a 2,500 ppm B-Nine/1,500 ppm Cycocel tank mix; 10–30 ppm Bonzi as a spray; 0.1–1 ppm Bonzi as a drench; 1,000–3,000 ppm Cycocel; and 2–10 ppm Sumagic. Many seasoned poinsettia growers have their own PGR recipes based on years of experience, notes, and developing a feel for their exact climatic conditions.

You can use growth regulators to effectively manage plant height and crop quality. PGR applications should be completed by mid-October to minimize any risk of reducing bract size and delaying flower maturity, especially in northern regions. However, it is well documented now that late applications at very low concentrations with Bonzi or similar materials can be used during the late stages of development with excellent results. When considering these applications, information on the process, rates, and important details can be found online or in research reports available from the University of Florida, Gainesville (contact Dr. Jim Barrett). Without proper guidance and appropriate rates to follow, these programs should not be attempted.

Poinsettia height can also be controlled using DIF. Today, most poinsettia growers use a combination of growth regulators and DIF. The best ones are choosing what to do based on how each crop's height is tracking graphically.

Positive DIF (warm days, cool nights) increases stem elongation, while negative DIF (cool days, warm nights) decreases stem elongation. When using DIF, it is important not to change average daily temperatures, as that will affect timing and bract size. In some parts of the country, it is impossible to run a negative DIF due to high daytime temperatures. In such cases, focusing on lowering temperatures for one to two hours prior to sunrise will work. North Carolina State University researchers recommend running a minimum night temperature of 64°F (18°C) and then dropping the temperatures for one to two hours just prior to sunrise to as close to 58–60°F (14–16°C) as possible. Simultaneously, control daytime temperatures and reduce the positive DIF as much as possible.

If you are using DIF to help manage plant height, it is vital to be aware that poinsettia growth and development is entirely dependent on average daily temperatures. If you use day and night temperature combinations that reduce the average daily temperature, you will slow your crop's development. The use of negative DIF is not suggested if growing poinsettias with the "cold-growing" program (see the sidebar).

How do you determine the average daily temperature? As an example, if plants are grown at 75°F

(24°C) for ten hours and 65°F (18°C) for fourteen hours, the average growing temperature would be 69.2°F (20.7°C). Average daily temperature is calculated with the following formula (you can substitute Celsius for Fahrenheit):

$$\text{Average daily temperature} = \frac{(\text{day °F} \times \text{hrs.}) + (\text{night °F} \times \text{hrs.})}{\text{24 hours}}$$

$$\text{Example average} = \frac{(75°F \times 10\text{ hrs.}) + (65°F \times 14\text{ hrs.})}{\text{24 hours}} = \frac{750 + 910}{24} = \frac{1{,}660}{24} = 69.2°F$$

DIF is most effective once shoots have developed after pinching, but discontinue negative DIF at least three weeks before pollen is showing (your scheduled flowering date).

Spacing

Your plant quality will be directly related to the space given to grow plants. Most growers determine the financial results they expect from the crop and space accordingly. If you are selling into the mass market, you will be at 10" × 10" (25 × 25 cm) centers, although some mass-market growers are pushing toward 9" × 9" (23 × 23 cm), but at such a close spacing they are getting less than five bracts per plant. Plants produced so close together not only have poor bract count, but they generally get tall and straggly, which is in many cases exactly what lands on too many mass-market shelves. There are some varieties available today that are more efficient in the use of space due to a strongly upright branch position. These poinsettias can be used with high-density production to yield better quality than varieties with a broader branch angle. Breeders can provide the list of varieties especially well suited for this style of production.

If you are selling to your own retail customer base, upscale outlets, or even fundraisers, give 6.5" (17 cm) plants 15" × 15" (38 × 38 cm). A general guideline for pot spacing would be: 4" (10 cm) pots 9" × 9" (23 × 23 cm); 5" (13 cm) pots 12" × 12" (30 × 30 cm); 6" (15 cm) pots 13" × 14" (33 × 36 cm); and 7" (18 cm) pots 17" × 17" (43 × 43 cm).

Graphical Tracking

Use graphical tracking to make informed decisions throughout the crop. Graphical tracking is based on the fact that plants follow a specific growth pattern regardless of variety. The grower monitors the crop over time by graphing the actual height compared to the desired height. The rate of development changes depending on the average daily temperature, while stem elongation is dependent on the DIF (the difference between day and night temperatures). A free, web-based program for height tracking named OnTarget is available at the Ecke Ranch website (www.ecke.com).

How to collect data

Collect specific temperature and crop development information for each variety by container size and flowering date. Most growers monitor three to five major varieties and graphically track these crops. Measure plant height at least weekly; however, collecting data more frequently will improve accuracy.

Interpreting data

The following table is useful in calculating a desired graphical track for a poinsettia crop. Determine the response group for the variety you are growing. Subtract the height of the pot and the cutting height at pinch from the desired final height. Multiply the relative height by the final desired height. If the height of the plant is lower than the calculated height, it will be necessary to push the crop. If the actual height is higher than the calculated height, it will be necessary to hold back the crop. In the chart, 'Freedom', an eight-week response variety, is used. The desired final plant height is 24" (61 cm). So, fourteen days after the pinch, plants should be 2.8" long: 24" final plant height minus 6.5" pot minus 4" cutting times 0.2122, equals 2.8". Using this chart, you can develop a graphical track for any variety you are growing.

(Continued)

Pests & Diseases

Whiteflies are the most serious insect pest. Prevent outbreaks by maintaining a rigorous scouting program. Fungus gnats, mealybugs, mites, and thrips can also be problems.

Botrytis and/or powdery mildew can be important disease problems when bracts are colored up, as spores land on the bracts, grow, and produce brown necrotic spots, which are unsightly and harm sales. *Erwinia,* powdery mildew, *Pythium,* and rhizoctonia can also be problems. Preventing and controlling all of these pests require diligent integrated pest management practices throughout production.

Troubleshooting

Leaf distortion

Leaf deformity can occur at many different stages of a poinsettia crop in reaction to chemical applications, fertilizers, and environmental stresses. There is no one reason for these distortions, so it is important to investigate what factors might cause the distortion upon observing the symptoms in the crop. The symptoms are extremely variable based on the actual cause.

On Christmas-season plants, leaf distortion frequently occurs in late September and early October

Graphical Tracking *(Continued)*

	RESPONSE GROUP			EXAMPLE
	8 Week	9 Week	10 Week	6.5" single-stem, pinched Final height 24" tall (6.5" pot, 4" liner) 'Freedom' (eight-week response)
Relative Height	**Days from pinch to flower**			
	8 weeks	9 weeks	10 weeks	
	Days after the pinch			
0.00%	0.0	0.0	0.0	
2.32%	2.8	3.2	3.5	0.3" = (24 – 6.5 – 4) x 0.0232
4.88%	5.6	6.3	7.0	0.65" = (24 – 6.5 – 4) x 0.0488
8.87%	8.4	9.5	10.5	1.2" = (24 – 6.5 – 4) x 0.0887
14.38%	11.2	12.6	14.0	1.9" = (24 – 6.5 – 4) x 0.1438
21.22%	14.0	15.8	17.5	2.8" = (24 – 6.5 – 4) x 0.2122
29.05%	16.8	18.9	21.0	3.9" = (24 – 6.5 – 4) x 0.2905
37.41%	19.6	22.1	24.5	5.1" = (24 – 6.5 – 4) x 0.3741
45.88%	22.4	25.2	28.0	6.2" = (24 – 6.5 – 4) x 0.4588
54.08%	25.2	28.4	31.5	7.3" = (24 – 6.5 – 4) x 0.5408
61.75%	28.0	31.5	35.0	8.3" = (24 – 6.5 – 4) x 0.6175
68.71%	30.8	34.7	38.5	9.3" = (24 – 6.5 – 4) x 0.6871
74.9%	33.6	37.8	42.0	10.1" = (24 – 6.5 – 4) x 0.749
80.29%	36.4	41.0	45.5	10.8" = (24 – 6.5 – 4) x 0.8029
84.91%	39.2	44.1	49.0	11.5" = (24 – 6.5 – 4) x 0.8491
88.84%	42.0	47.3	52.5	11.9" = (24 – 6.5 – 4) x 0.8884
92.13%	44.8	50.4	56.0	12.4" = (24 – 6.5 – 4) x 0.9213
94.88%	47.6	53.6	59.5	12.8" = (24 – 6.5 – 4) x 0.9488
97.15%	50.4	56.7	63.0	13.1" = (24 – 6.5 – 4) x 0.9715
99.02%	53.2	59.9	66.5	13.4" = (24 – 6.5 – 4) x 0.9902
100.00%	56.0	63.0	70.0	13.5" = (24 – 6.5 – 4) x 1.0000

after the plants have been moved from propagation to the finishing area. Branches that develop after pinching may have two to three misshapen and distorted leaves as well. In most instances these leaves remain distorted but green throughout the forcing period. Leaves that expand later are usually normal and hide the damaged leaves by market time. Potential causes of these early-season distortions can include environmental stresses, but many times are also linked to phosphorous damage from overhead fertilization with materials that contain phosphorous. Once the crop is hooked up to individual irrigation tubes or placed on ebb-and-flood benches or flood floors, where overhead feeding is no longer required, new growth generally develops free of this disorder.

The causes of leaf distortion are not well understood. It seems that when cells in very young leaf tissues are ruptured or killed, the leaf becomes misshapen as it expands. Drying of tissue, burn from fertilizer, nutrient deficiency, and chemical burn have all been suspected of damaging young leaf cells. Plants under stress from bright light, extremely warm temperatures, or moving air often have more leaf distortion. It is helpful to provide shade and to syringe (mist) the foliage until roots are well established and the side branches begin to develop. Leaf distortion may also result from insects, such as thrips, feeding on the young leaf tissues. As leaves mature, damage from the pest becomes apparent and prevents normal expansion.

Many plants, including poinsettias, have leaf structures that include hydathodes, or vein endings opening along the edges, tips, and sometimes leaf surfaces. Under cool, humid conditions, with ample growing mix moisture supply and elevated growing mix temperature, high fluid pressure in the conducting system may occur. If a rapid rise in temperature and drop in humidity occur simultaneously, as frequently happens during mornings of bright days, dissolved contents will become more concentrated. Sudden use of air-conditioning fans or natural air movement from wind can cause the same effect. This concentrated solution may be strong enough to cause cell damage, and when sudden stress on the plant occurs simultaneously, the concentrated fluid may be drawn back into the vein endings and cause damage to cells in and around the area. Since the phenomenon occurs only on immature leaves still undergoing expansion, subsequent growth in areas of cell injury will be inhibited and developing leaves will be distorted.

Control of such leaf-edge damage can best be achieved by maintaining low humidity at night and avoiding conditions of rapid drying in the morning. Syringing of foliage in the early morning may also help by slowing transpiration. A complicating factor is frequently that of infection of injured tissue by botrytis.

Growers have also experienced leaf dehydration of dark-leaf cultivars at various times in production. This damage is usually evident on expanded, mature foliage. Symptoms begin with darkening of the leaf tissue, followed by one edge of a leaf rolling up. The affected tissue eventually dries with no further progression of symptoms. Damage is usually isolated to a few leaves per plant and does not continue beyond this. This damage is most likely a result of sudden changes in the environment. Conditions typically noted prior to leaf dehydration include extended cloudy weather followed by bright, warm days. Leaf dehydration is most evident on plants that are well fertilized or where the growing medium is allowed to dry. The best prevention for this disorder is to maintain uniform substrate moisture with low soluble salts. Whenever rapid changes in growing conditions are noted, mitigate the conditions by syringing plant foliage.

Bract burn

A condition or disorder that affects blooming poinsettias, bract-edge burn (BEB) first appears as small, brown, necrotic spots at the tips or along the edges of mature bracts. As the condition progresses, entire bract margins may die and turn brown, giving a burned appearance. This injured tissue is also ideal for botrytis to establish on and cause further damage to the plant.

Bract-edge burn has been virtually eliminated in the past decade as most varieties grown and sold are dark-leaf types that are less susceptible to the disorder.

Severe bract burn has been encountered where extreme rates of fertilizer have been used. Under these conditions the leaves may show no damage. One theory is that during growth there is a diluting effect of plant-absorbed fertilizer, but at flowering, new tissue development has virtually ceased and the fertilizer salts accumulate in the youngest mature and most sensitive tissue, the bract. This accumulation causes cell damage, usually starting on the bract edges. With slow-release fertilizers, the usage rate should be modest and application should be early enough to ensure almost complete depletion at time of flowering.

Calcium seems to be the most important nutrient associated with BEB. Plants use calcium to build strong cell walls. Without strong cell walls, plant tissues are vulnerable to damage from soluble salts, drying, sunburn, or invasion by diseases or insects. The young, developing bract margin cells are the kind of tissue most likely to be damaged under any of these stress conditions. Providing sufficient calcium levels is critical to protecting these young cells from damage. Calcium is moved from the root solution through the plant in the water stream. Any condition that restricts water movement in the plant will also limit the uptake of calcium required for strong bract cell walls. Factors that affect the uptake of water and calcium include loss of roots to disease or burn; high relative humidity and poor air circulation, which limit transpiration; and low soil temperatures. Among substrate chemistry factors are high soluble salts in the growing medium; nutrient imbalances that favor the uptake of elements other than calcium; low pH in the potting mix, affecting the availability of calcium for uptake; and low calcium availability in the mix as a result of fertilization programs.

Monitor calcium levels in the substrate and tissue throughout the crop. If deficiencies are noted prior to or during bract formation, foliar applications of calcium can be beneficial in preventing BEB. Use of laboratory- or technical-grade calcium chloride at 300 ppm may provide the calcium required to overcome this problem.

Another factor that influences the occurrence of BEB is the age or maturity of the bracts. The disorder is more likely to happen on plants that are being held for sale in the greenhouse. The conditions used to hold a crop are the same conditions that limit water movement in the plant. As the plants sit waiting to be sold, bract-edge burn, botrytis, root rot, or cyathia abscission is likely to occur. To help avoid these conditions, schedule crops for precise timing and improved shipping quality.

Reduce bract necrosis by:

- Selecting varieties that are not susceptible to this disorder.
- Avoiding high fertility rates and heavy watering practices during the final four weeks of the production period. It may be advisable to discontinue fertilization and use water only, beginning two weeks before the flowering plants are to be sold.
- Avoiding use of fertilizers that contain 50% or more of their nitrogen in the ammoniacal form. Fertilizers that contain mostly nitrate nitrogen are readily available or easily formulated.
- Modifying the greenhouse environment to reduce humidity and increase air circulation throughout production. Chemical disease-control methods may be used to minimize the spread of botrytis on the bracts.
- Using calcium chloride as a foliar spray on developing poinsettia bracts whenever tissue and soil analysis indicate calcium deficiency in the crop.

Premature cyathia drop

During some Christmas flowering seasons, the true flowers, or cyathia, may drop from the center of the bract presentation before the flowers reach maturity. This may occur before the plants are ready for market, particularly in northern climates with low light conditions. Because this detracts from the appearance of the poinsettias and makes them appear to be overly mature, it also reduces their value.

Michigan State University researchers determined that low light levels or high forcing temperatures cause premature cyathia drop. Water stress exacerbates the problem. These conditions allow the food reserves of the plant to become depleted. As the food reserves become low, the plant reacts by dropping the cyathia.

Low light levels may result from dark or cloudy weather conditions or simply from spacing the plants too tightly on the bench. It is not uncommon for skies to become overcast during late October and November in northern states. Lower light levels and cooler temperatures accompany this cloudy weather. These are not ideal conditions for high-quality poinsettias. With lower light levels, plants do not make as much food. Also, as poinsettia bracts develop, they shade the green leaves below and further restrict the amount of light available for photosynthesis.

If poinsettias are grown at lower-than-optimum temperatures during the early part of the production period, flower development may be delayed when cloudy weather begins. To speed up the rate of flower development, it then becomes necessary to raise greenhouse temperatures. As temperatures increase, the plants' food reserves are used at a faster rate.

To produce high-quality poinsettias and reduce the possibility of premature cyathia drop, it is important to follow the old adage: "Make hay while the sun shines." Especially in northern areas, it is essential to maintain optimum greenhouse temperatures for rapid poinsettia development during the early part of the production period. When cloudy weather begins, it may then be possible to start lowering greenhouse temperatures and slowing the depletion of food reserves.

The Michigan State University research also demonstrated that if the growing medium dries to the point that the poinsettias begin to wilt after the time the flower buds become visible, the chances of premature cyathia drop become greater.

We have also noticed that cold-grow poinsettia programs have resulted in less cyathia drop, perhaps due to both the longer production time, which results in more carbohydrates being accumulated, and the reduced depletion of these food reserves due to cooler greenhouse temperatures.

Lessen the possibility of premature cyathia drop with the following cultural guidelines:

- Schedule your poinsettia program early enough so that the plants can do most of their "growing" early in the fall, while good light intensities are available.
- Do not attempt to grow a poinsettia crop at lower-than-optimum temperatures during the early part of the production period in an attempt to save energy and reduce fuel costs.
- Do not allow the growing medium to become excessively dry. This only hastens the start of premature cyathia drop.
- Grow your crop under a clear greenhouse cover to admit as much light as possible during October and November.

If the crop develops properly, it should be possible to reduce temperatures late in the production period, when light levels are low, thus preserving part of the plants' food supply.

Latex eruption

Plants belonging to the *Euphorbia* genus contain latex, which is exuded upon cell injury. This became a problem in poinsettia production when the variety 'Paul Mikkelsen' and its sports first became popular. The malady is sometimes termed crud. The mechanism is one of bursting cells resulting from high turgor pressure, with latex spilling over the tissue and, upon drying, creating a growth-restricting layer. When this occurs at developing stem tips, growth is distorted or stunted. The exuding of latex has also been observed on fully expanded leaves, sometimes giving the appearance of mealybug infestation due to the white splotches scattered over the leaf surfaces.

All contributing factors have not been clearly defined, but several obvious ones include high moisture availability and high humidity, both of which result in high fluid pressure within the cells. Low temperature is an important contributing factor. Suddenly lowering temperatures can trigger the reaction, but, fortunately, most varieties are not highly sensitive to the problem. Mechanical injury from rough handling or from excessively vigorous moving air may also increase cell injury. High rates of photosynthesis may contribute by building up a high osmotic pressure in cells from carbohydrate accumulation.

Using a growing medium that dries out in a reasonable length of time is the best way to attain control. Also, avoid humidity extremes, particularly at night. Moderate shading in extremely bright weather might also be helpful.

Stem splitting

Under certain conditions poinsettias will suddenly produce stem branches at the growing tip. Careful examination will reveal that the true stem tip has stopped growth or aborted. This phenomenon is known as splitting.

Splitting is actually the first step in flower initiation. The stimulus to flower increases with the age of stems, exposure to cold temperatures, and lengthening of nights. Even with short nights and normal growing temperatures in the 60–70°F (16–21°C) range, splitting can be expected if the stem is permitted to grow until twenty or more leaves are present.

Keeping stock plants pinched back on a regular basis will help prevent splitting of harvested cuttings. Stem tips that are continuously propagated carry an increasing tendency to flower. To ensure against this, light stock plants until May 15. Plants propagated prior to July 15 should be grown as multiflowered or branched plants *only* with tips discarded.

Another cause of splitting can be cold temperatures. Lowering night temperatures to around 60°F (16°C) was an old trick to help initiate flowering at the end of September in the 1950s and 1960s. Growers have experienced early flower initiation and splitting with today's cultivars, however, when growing temperatures become cool and heaters are not yet operating. Avoid greenhouse temperatures that drop below 60°F (16°C) whenever plants should be maintained in a vegetative state.

Stems heavily shaded by a canopy of higher foliage may experience such a reduction in light as to split even in periods when day length would be considered adequately long to keep apexes vegetative. If daytime conditions are unusually cloudy, plants may not receive enough total light to maintain vegetative growth. During weather patterns like this or under fluctuating day lengths, it is prudent to use night lighting to prevent premature bud set.

There is a clear connection to splitting and specific varieties in the market. While most are not highly susceptible to this disorder, be sure to check with your supplier or breeders to be sure you understand which poinsettias are more subject to this problem, especially when considering growing larger product forms, like trees, that need to grow to significant heights without risk of splitting.

Leaf drop

Older varieties were much more prone to a sudden loss of leaves than are modern varieties. There are several indirect causes of leaf drop. Under conditions of moderate to severe stress, it is not uncommon for older leaves to form an abscission layer at the juncture of the petiole and the supporting stem. It is believed this is due to loss of auxin from the leaf blade under stress conditions. Once started, the reaction is irreversible, and the leaf petiole is virtually severed from the stem. Also, when plants are kept under very low light intensity for a period of several days, lower leaves will turn yellow and drop.

Before better sanitation procedures reduced or eliminated disease problems, leaves of the older varieties would frequently drop in the greenhouse as root disease reduced the plants' ability to supply water to the top. A parallel contributing factor was the deliberate attempt by growers to keep the growing medium dry in order to restrict disease organism activity. Even with healthy roots, many of the cultivars would drop leaves within a day or two after being moved from the humid glasshouse to a warm, dry home or office. The change in environment caused more water stress than the leaves could tolerate. The moisture loss exceeded the ability of the roots to supply water.

The Paul Ecke Ranch website (www.ecke.com) has an excellent Diagnostic Center where you can access a plethora of data and photographs to help you troubleshoot problems you may be experiencing with your crop.

Varieties

The most popular poinsettia variety in the North American market today is 'Prestige Red', although a number of other varieties have significant market share. 'Prestige' is easy for growers to produce and is highly resistant to stem breakage; it is the first, and still the only, family of poinsettias that is genetically resistant to stem breakage! For growers producing other varieties it has become common to use pot rings that force branches to grow upright and help prevent breakage during transit and retail.

Red varieties have always been the majority of production and are likely to continue as the most popular color because poinsettias are sold during the December holiday season. That said, the percentage of nonred varieties sold changes each year based on the latest trend in decorating and fashionable colors influencing the market. By far, the biggest news in poinsettias in recent years has been the explosion of novelties—in plant form and in colors (table 14).

Table 14. 2010 US Poinsettia Color Trends

Color	Share
Red	80%
White	6%
Pink	5%
Marble	3%
'Jingle Bells'/Glitter	2%
Other Novelty*	4%

* Includes 'Monet Twilight', 'Miro', 'Picasso', 'Ice Crystal', 'Ice Punch', 'Orange Spice', and 'Burgundy'.

Source: The Paul Ecke Ranch, Encinitas, California

Novel reds such as 'Carousel', with its fun, crinkly bracts; 'Jester', with its upright, spiky leaves and bracts, or 'Winter Rose', with its tight, stately bracts, have provided the market with niche products that offer consumers choices.

More recently new breeding is giving the industry non-traditional colors such as burgundy, orange, and yellow. In addition, there are lots of complex color combinations such as marble, with pink centers surrounded by white on the bracts, and "peppermint," as well as varieties that have a solid-bract background color, usually pink, and small, red flecks of color that are evenly spread across the bracts or concentrated on the edges, such as 'Monet', 'Miro', and 'Picasso'. Also the various "glitter" varieties, which are red with white spots and complement the already popular red-and-pink-spot types, known as 'Jingle Bells'. By far some of the most popular and interesting new novelty colors include 'Ice Punch' and 'Ice Crystals'; both display a strong white center surrounded by red ('Ice Punch') or a dark peppermint ('Ice Crystals'). And of course, there is always a range of poinsettias with variegated foliage. Truly there is a poinsettia for every taste!

Postharvest

Plants are ready for sale when pollen is showing on the first cyathia (flower). By this time, bracts should be

Poinsettia Varieties by Breeder

DUMMEN*	SYNGENTA	SELECTA	PAUL ECKE RANCH	
'Arctic White'	'Carousel Dark Red'	'Christmas Beauty'	'Advent Red'	'Marblestar'
'Avantgarde Marble'	'Carousel Pink'	'Christmas Carol'	'Autumn Red'	'Maren'
'EarlyGlory Red'	'Cinnamon Star'	'Christmas Carol Pink'	'Chianti'	'Max Red'
'EuroGlory Red'	'Cortez Burgundy'	'Christmas Carol White'	'Classic Marble'	'Monet Twilight'
'Flame'	'Cortez Early Red'	'Christmas Day'	'Classic Pink'	'Orange Spice'
'Infinity Fire'	'Cortez Electric Fire'	'Christmas Eve'	'Classic Red'	'Peppermint Twist'
'Infinity Marble'	'Cortez Pink'	'Christmas Feelings'	'Classic White'	'Peterstar Marble'
'Infinity Pink'	'Cortez White'	'Christmas Feelings Cinnamon'	'Early Joy Pink'	'Peterstar Pink'
'Infinity Polar'	'Da Vinci'	'Christmas Feelings Dark Salmon'	'Early Joy Red'	'Peterstar Red'
'Infinity Red'	'Marblestar'	'Christmas Feelings Marble'	'Eggnog'	'Peterstar White'
'Marblestar'	'Maren'	'Christmas Feelings Merlot'	'Enduring Marble'	'Polar Bear'
'Maren'	'Mars Marble'	'Christmas Feelings Pink'	'Enduring Pink'	'Polly's Pink'
'Marco Polo'	'Mars Pink'	'Christmas Feelings Red Cinnamon'	'Enduring Red'	'Prestige Early Red'
'Merlot'	'Mars Red 09'	'Christmas Feelings White'	'Enduring White'	'Prestige Maroon'
'Pink Cadillac'	'Mars White'	'Christmas Feelings Select'	'Freedom Early Marble'	'Prestige Red'
'Premium Apricot'	'Mira Red'	'Christmas Season'	'Freedom Early Pink'	'Red Angel'
'Premium Early Red'	'Mira White'	'Christmas Season Fire'	'Freedom Early Red'	'Red Glitter'
'Premium Ice Crystal'	'Novia'	'Christmas Season Marble'	'Freedom Early White'	'Red Velvet'
'Premium Lipstick Pink'	'Olympus'	'Christmas Season Pink'	'Freedom Bright Red'	'Red Velveteen'
'Premium Marble'	'Orion Early Red'	'Christmas Season White'	'Freedom Fireworks'	'Salmonstar'
'Premium Miro'	'Orion Red'	'Crazy Christmas'	'Freedom Jingle Bells'	'Shimmer Pink'
'Premium Picasso'	'Puebla'	'Happy Christmas'	'Freedom Marble'	'Shimmer Surprise'
'Premium Polar'	'Pink Elf'	'Marbella'	'Freedom Peppermint'	'Solstice Red'
'Premium Red'	'Red Elf'	'Merry Christmas'	'Freedom Pink'	'Snowcap'
'Premium White'	'Ruby Frost'	'Merry White'	'Freedom Red'	'Strawberries N' Cream'
'Redlight Bright Red'	'Silverstar Marble'	'Noel'	'Freedom White'	'Tapestry'
'Redlight White'	'Silverstar Red'	'Pink Candy'	'Gala Red'	'Visions of Grandeur'
'Scandic Early'	'Sonora Jingle'	'Valentine'	'Gala White'	'Winter Blush'
'Viking Red'	'Sonora Marble'	'Vintage Red' (new 2011)	'Ice Punch'	'Winter Rose Early Marble'
	'Sonora Pink'	'Wintersun'	'Independence Red'	'Winter Rose Early Pink'
	'Sonora Red'		'Jester Red'	'Winter Rose Early Red'
	'Sonora White'		'Jester White'	'Winter Rose Dark Red'
	'Sonora White Glitter'		'Jubilee Red'	'Winter Rose White'
	'Whitestar'			

* Varieties based on the company's 2010 catalog.

fully colored. When plants are sold prematurely, they generally will not have good postharvest life. Remove dead and damaged plant parts before sleeving. Always use a permeable sleeve to prevent mechanical and cold damage. Poinsettias can be shipped at 55–60°F (13–16°C). Do not expose plants to temperatures below 50°F (10°C) or ship above 70°F (21°C). Keep poinsettias boxed for the least amount of time possible—definitely not more than three days. Plants held in the dark will drop leaves.

Unbox plants and remove sleeves immediately. Poinsettias are sensitive to ethylene: It causes leaves and bracts to droop. The act of sleeving plants or otherwise damaging stems and leaves causes the plants themselves to generate ethylene. While plants will "bounce" back after they are unsleeved and placed in an inviting retail environment or home interior, the longer they are sleeved (thus exposed to ethylene), the longer it will take for them to recover.

At retail, provide 100–500 f.c. (1.1–5.4 klux) of light. Poinsettias are best displayed under incandescent light or under lamps emitting the full light spectrum. Red plants especially will look dull and washed out when displayed under fluorescent light. Do not allow plants to dry out at retail. Plants should not be "standing" in water either—be especially mindful of this if plants are in foil or pot covers without drainage. Do not wet foliage or bracts when watering.

Display plants in areas away from hot or cold drafts and keep them out of aisles and doors so that people do not bump up against them, thereby damaging bracts and breaking stems.

Consumers should place their poinsettia in a bright spot in the home interior away from warm or cold drafts. Light levels should be a minimum of 100 f.c. (1.1 klux). Provided plants do not dry out too often and are receiving a reasonable amount of light, they will last well into January or February. Poinsettias can even be enjoyed as a foliage plant on the deck or patio over the summer. Once the danger of frost is past, they can trim stems back to remove spent bracts, repot into fresh medium, and keep them watered and fertilized. If the consumer chooses to try to reflower the plant the following year, they will have to provide them the same short days as outlined previously.

Eustoma

Eustoma grandiflorum

Common names: Lisianthus; Texas bluebell; prairie gentian

Annual, cut flower, flowering pot plant

Retailers and consumers love lisianthus. This Texas native is as versatile as its flowers are lovely. Lisianthus makes an excellent heat-loving annual, an unusual flowering pot plant, and a durable addition to any cut flower arrangement or bouquet. Its casual prairie flowers have just the right cottage-garden appearance wherever they are found, lending an air of casual class.

So, what's the catch? Production. Lisianthus is a challenging crop. First, lisianthus plugs take a minimum of ten weeks to grow, and seedlings are finicky. They are not available on the spec market, so you have to order plants well in advance of when you will need them: A minimum three months (twelve weeks) lead time is normal. Second, during production, temperatures have to be right or plants will rosette rather than flower. Third, some cut flower varieties will respond to high heat and light levels by flowering prematurely. Fourth, growers must be vigilant in controlling disease and insects. And, finally, crop time from plug to harvest for 4" (10 cm) pots is fourteen weeks and for cuts it is thirteen to eighteen weeks. To justify tying up so much valuable greenhouse space and consequent overhead costs, market prices must be high. You will not find lisianthus widely available at retail for these reasons. So, if you like a technical challenge and you are confident that you can get the prices you need to make lisianthus profitable in your market, go for it!

Propagation

Lisianthus is propagated from seed. Because plugs take so long to produce and can easily succumb to disease pressures if environmental conditions are not right, most growers buy in plugs. Start with larger plugs, such as 288s as the minimum size.

Single-sow pelleted seed into a disease-free medium. Do not cover, as seed needs light to germinate.

Germinate at a soil temperature of 72–77°F (22–25°C) and maintain air temperatures of 70–75°F (21–24°C) days and 60–65°F (16–18°C) nights. Seedlings emerge in ten to twelve days. Provide 100–400 f.c. (1.1–4.3 klux) of light. Maintain the substrate pH at 6.2–6.4 to ensure calcium availability. Sakata Seed recommends placing seed trays on capillary mats or plastic to keep uniform substrate moisture and to ensure uniform emergence. Most commercial plug growers cover trays with Reemay or plastic to moderate moisture levels from above. Just as soon as seed has germinated, remove covering or take trays off of plastic/capillary matting. Stage 1 takes from ten to twelve days.

Move trays to Stage 2 for stem and cotyledon emergence. Maintain daytime air temperatures, but lower soil temperatures to 68–72°F (21–24°C). Do not allow day temperatures to rise above 75°F (25°C) or night temperatures to go below 59°F (15°C) to avoid rosette problems. Sakata recommends using cool-night temperatures to avoid rosettes: After germination, run 63–65°F (17–18°C) nights and days at 75–80°F (25–27°C). Maintain cool-night temperatures for twelve hours. Using cool-night temperatures should help avoid rosettes even in hot growing conditions. Reduce moisture levels as soon as the radicle has emerged: Allow medium to dry slightly between irrigations and ensure excellent air circulation around trays. Increase light to 1,000–1,500 f.c. (11–16 klux) and apply 50–75 ppm of 14-0-14 foliar feed after cotyledons are fully expanded. Stage 2 can take fourteen to twenty-one days: You may feed two times. If so, be sure to maintain ammonium levels below 10 ppm; lisianthus is very sensitive to ammonium toxicity. Supplemental light for nine to twelve hours during the fall and winter can benefit seedling growth.

Stage 3, growth and development of true leaves, takes twenty-eight to thirty-five days. Maintain temperature regimes, being careful to avoid high and low temperatures to prevent rosetting. Increase light levels to 1,200–1,500 f.c. (13–16 klux) and allow trays to dry down between irrigations. Avoid excessive humidity. Use calcium-based fertilizers. Fertilize with 100–150 ppm from 15-15-15 or 11-5-19, alternating with 14-0-14 or another calcium/potassium fertilizer. Fertilize every second or third irrigation, as needed.

By Stage 4, hardening off, seedlings should have four true leaves. Harden plugs for seven days prior to planting or selling. Continue to maintain air temperature regimes; reduce medium temperatures to 62–65°F (17–18°C) and increase light to 2,500–4,000 f.c. (27–43 klux). Continue fertilization as needed.

Lisianthus plugs must be transplanted as they are ready, otherwise growth is checked and plants will not branch as readily. Transplanting plugs on time ensures roots are always actively growing. Overgrown lisianthus plugs can be identified by encircling masses of root growth within the plug-tray cell and elongated internodes on the developing shoot. Checked plugs are likely to flower prematurely.

Note: Some researchers recommend refrigerating winter-blooming cut flower lisianthus varieties at 33–34°F (1–2°C) for thirty days after sowing. After refrigerating, move to a greenhouse at 70–75°F (21–24°C) days and 58–60°F (14–15°C) nights and grow normally. Growers in cooler plug production locations do not need to do this, and more vigorous varieties also do not require this treatment.

A 288-plug tray will take from nine to eleven weeks to be ready to transplant to larger containers.

Pot Plant & Bedding Plant Crops

Growing On

Pot liners or plugs into a disease-free medium with a moderate initial nutrient charge and a pH of 6.5–7.0. Irrigate lisianthus when medium dries slightly. Immediately after transplanting, plants will be slow growing, so be careful not to overwater. As stems begin to elongate, do not allow plants to become moisture stressed. The goal is to maintain moist but not wet conditions. When plants are in flower, do not allow the medium to dry out!

Maintain high light levels at 4,000–5,000 f.c. (43–54 klux). When plants begin to flower, provide light shade. Fertilize with 100–200 ppm from 14-0-14, alternating with 15-15-15 or 17-5-17 at every other irrigation. Maintain the substrate EC at 1.5 mS/cm.

Height control can be accomplished using DIF: Lisianthus will be shorter with a negative DIF. A-Rest and B-Nine are also effective: Apply two weeks after pinching when breaks are 1–2" (3–5 cm) long, repeating the application two to three weeks later. As days get longer, lisianthus requires less growth retardant.

Lisianthus is not long-day obligate; however, long days will shorten the time to flower. Long days also help to prevent rosetting. Begin lighting after transplanting. If you do not have supplemental lighting to extend day length to more than thirteen hours, use four hours of night interruption lighting. To see significant results in shortening the crop time, plants must receive ninety days of long days. Light from transplant (three sets of true leaves) until terminal bud flowering begins. You can stretch stems by extending the day length: Apply lighting from 6 P.M. to midnight, or from midnight to 6 A.M. Night interruption lighting will not stretch stems. Some growers do not apply light, but instead plant when the crop will experience increasing days. For example, John Kister at Sunlet Nursery, Fallbrook, California, plants lisianthus plugs from January 15 to June 1.

It takes from twelve to fourteen weeks from plugs to flowering 4" (10 cm) pots

Pests & Diseases

Aphids, leaf miners, thrips, and whiteflies can be problems with lisianthus.

Botrytis, *Fusarium, Pythium,* and rhizoctonia can be severe disease problems. Control diseases by maintaining proper growing temperatures and irrigation levels and by watching humidity levels.

Troubleshooting

Rosetted growth is a serious problem resulting from high temperatures. When seedlings show rosetted growth, it is very difficult to overcome. You will recognize rosetted growth by clustered leaves with very short internodes. The best cure is to avoid the problem by maintaining the proper temperature regimes throughout the seedling stages and growing on phases. Lisianthus does not like high temperatures (greater than 80°F [27°C]), especially during the seedling stage. Gibberellic acid applications have been known to encourage plants to grow out of rosettes. If rosettes appear three to six weeks after planting, begin gibberellic acid treatments. Although rosetted plants will eventually begin to grow, stem quality will be reduced and the time to flower will be unacceptably long. Another source shows that providing four to five weeks of 50°F (10°C) will reverse rosetting, but at the expense of lost time in production cycles. If you are buying in plugs, purchase from a plug producer with a good reputation for lisianthus plugs because when temperatures are correct in plug production, rosetting is little or no problem.

Low substrate pH can result in chlorotic lower leaves and checked growth. Unfortunately, once symptoms of pH difficulties begin to show in lisianthus, growth has already been checked. Be sure pH is correct *before* planting. To avoid problems, conduct regular medium pH tests.

If, during visible bud, young leaves fail to expand and/or terminal buds become trapped inside the leaves, calcium may be deficient, especially under high light conditions. Calcium can become deficient even if adequate calcium levels are available but the air is so humid that translocation of calcium cannot occur. A foliar application of calcium may be needed.

Poor aeration and high fertility can cause algae on the potting mix.

When plugs are transplanted too late, flowering can be delayed or premature.

Varieties

Try the 'Florida' series for 6" (15 cm) pots. Plants grow to about 10" (25 cm) and are extremely well branched for pot plant production or bedding plants. The series comes in 'Blue', 'Sky Blue', and 'Pink'. 'Lisa' is great for 4" (10 cm) pots. Plants are somewhat resistant to rosetting. The series comes in four separate colors: 'Blue', 'Pink', 'Lavender', and 'White'. 'Forever Blue', an All-America Selections Award winner, is great for northern growers, because it performs well under lower light conditions. Plants can be grown in 4 or 6" (10 or 15 cm) pots. The 'Mermaid' series, in 'Blue', 'Lilac Rose', 'Pink', and 'White', is extremely early to flower and is great in 4" (10 cm) pots. 'Sapphire' (4" [10 cm] pots) and 'Forever' (6" [15 cm] pots) were both bred for basal branching.

Postharvest

Retailers displaying lisianthus outdoors should put plants under light shade to help modify temperatures and slow drying out. Do not allow plants in flower to dry out! Ideally maintain 68–70°F (20–21°C) day and 60–65°F (16–18°C) night temperatures.

Cut Flower Crops

Growing On

Transplant plugs immediately into new or steam-sterilized beds high in organic matter that have been cultivated to a depth of 18" (46 cm). A pH of 6.5–6.8 is recommended. Plant "high" to help avoid stem rots. Space plants at 4" × 6" (10 × 15 cm) for single flowering during fall and winter; 5" × 8" (13 × 20 cm) if the same plants are to be reflowered. Space plants at 3.5" × 4.5" (9 × 11 cm) for single flowering during the spring and summer. Do not allow plant growth to be checked by failing to transplant on time: Keep roots active. When five pairs of true leaves appear, the plant and root system will begin to develop rapidly.

Black plastic mulch placed over the beds in the winter can help to raise winter mix temperatures, while white plastic in the summertime can help to lower mix temperatures. For optimum results, maintain substrate temperatures of 55°F (13°C) absolute minimum and 73°F (23°C) maximum.

One to two levels of support netting, typically with squares of 6" × 6" (15 × 15 cm) or 6" × 8" (15 × 20 cm), will be needed. You can pinch the growing stem once to improve branching and yield; however, crop time will be delayed three to four weeks. Most commercial cut flower growers have determined that the extra cost in labor, additional growing time, and thinner stems from growing pinched plants does not justify the cost of pinching lisianthus.

Water beds thoroughly and uniformly after planting. However, do not maintain wet conditions, as that will encourage high humidity and could spur diseases. Some growers like to bury irrigation lines at 2–3" (5–8 cm) deep. Supplying water below the surface helps to mimic the natural habitat of lisianthus (desert areas along riverbeds and in low areas). As the rains become less frequent, the plants push their roots down deep in the soil to obtain water.

Grow on at 68–78°F days (20–25°C) and 60–65°F (15–18°C) nights. Increase ambient light levels to 4,000–5,000 f.c. (43–54 klux) and begin providing long-day treatment after transplanting (see comments in the Pot Plant & Bedding Plant Crops section). As plants begin to flower, provide light shade to maintain flower color.

Fertilize with 150–200 ppm 20-10-20, alternating with 15-0-15 at every other irrigation. Since plants can become susceptible to calcium deficiency at visible bud stage, apply calcium nitrate regularly. Maintain the soil pH at 6.5 and EC at less than 1.5.

Pests & Diseases

See Pests & Diseases under the Pot Plant & Bedding Plant Crops section.

Troubleshooting

See Troubleshooting under the Pot Plant & Bedding Plant Crops section.

Varieties

Basically, cut flower lisianthus may be divided into single- and double-flowering forms. Single-flower forms have four or five petals, while double-flower forms can range from six or seven to as many as twenty-five. Double-flower forms predominate in popularity in the Americas, and their popularity in Asia and Europe continues to rise each year. A further division in lisianthus is flowering response time. Optimal performance comes from selecting the right variety for the time of year the plants will be flowering. They are divided into three categories: (1) winter flowering, (2) spring and fall flowering, and (3) summer flowering.

Be aware that seed-breeding companies are actively engaged in developing and trialing new cut lisianthus varieties on a year-round basis. It is important for growers to trial both experimental and new varieties under their own growing conditions to determine not only which varieties grow well but also have the greatest market potential.

Postharvest

Flowers are ready for harvest after the terminal bud is open and at least one major lateral is flowering. If stems are harvested too early, color development in dark colors is poor and flowers will not reach their potential final size. Indoors, buds from stems harvested too early open to a faded whitish color. Harvesting one week after the first flower has opened is beneficial for stems that will be transported long distances. Harvest can be delayed even longer for local markets: Wait until five or six flowers are open for a more impressive display of color. The delay is possible because each flower may last up to fourteen days. The extra boost in flower color and size will make locally grown stems stand out.

Harvest will occur over two weeks. If you plan to recut the same plants, harvest the first cuts above the third or fourth leaf internode. The second flush will come six to eight weeks later.

Immediately after harvesting, place stems in water with 10% sucrose floral preservative. When stems are stored dry after harvest, leaves dry out quickly but the flowers remain turgid for two to three days. Young buds halt development. Provide continuous light (fluorescent lamps are ideal) for twenty-four hours to extend vase life. Refrigerate at 35–40°F (2–5°C).

Evolvulus

Evolvulus nuttallianus
Annual

The blue flowers of *Evolvulus* sit atop silvery, gray-green foliage. It is a striking combination that consumers find especially appealing in hanging baskets. It is also one of the few plants that is salt tolerant for coastal plantings. As a bonus, *Evolvulus* is perennial in many parts of the Deep South.

Propagation

Ideally, stick cuttings immediately upon arrival. If that is not possible, store them at 50–60°F (10–15°C) for no longer than twenty-four hours. Stick cuttings in Oasis foam or a soilless medium with a low starter charge and a pH of 5.5–6.5. Provide soil temperatures of 68–75°F (20–24°C) and air temperatures of 70–80°F (21–28°C) days and 65–68°F (18–20°C) nights. Provide light levels of 500–1,000 f.c. (5.4–11 klux). Keep medium fully saturated, applying mist so that cuttings remain fully turgid. If you notice any loss in leaf color, apply 50–75 ppm from 15-0-15 once a week. Cuttings should form calluses within five to seven days.

Once 50% of cuttings have begun to form root initials, move trays to Stage 3 for full rooting. Provide substrate temperatures of 68–75°F (20–24°C) and air temperatures of 70–80°F (21–28°C) days and 65–68°F (18–20°C) nights. Increase light levels to 1,000–2,000 f.c. (11–22 klux). Apply 100–150 ppm of fertilizer weekly, alternating between 20-10-20 and 15-0-15. As soon as roots begin to penetrate the medium, reduce mist frequency and duration, allowing the medium to dry down somewhat. Cuttings should be fully rooted in within seven to nine days.

Harden cuttings prior to sale or transplanting by lowering substrate temperatures to 65–70°F (18–21°C) and air temperatures to 68–75°F (20–24°C) days and 62–65°F (16–18°C) nights. Move trays from the mist area and increase light levels to 2,000–4,000 f.c. (22–43 klux). Allow trays to dry down somewhat before irrigating; however, do not allow cuttings to wilt. Apply a twice-weekly feed of 20-10-20, alternating with 15-0-15 at 150–200 ppm.

Evolvulus is easily brought in as a liner from a number of excellent suppliers. A 105 tray takes from three to four weeks to be ready to transplant.

Growing On

Select a well-drained, disease-free medium with a high initial starter charge and a pH of 6.0–6.5. Plant 1 liner/4" (10 cm) pot, 3 liners/6" (15 cm), and 3–4 liners/10" (25 cm) basket. Allow eight to nine weeks for 4" (10 cm) pots in the early spring.

Grow on at 65–75°F (18–24°C) days and 55–60°F (13–16°C) nights. Grow the crop at the highest light levels possible, provided you can maintain production temperatures. Keep plants moist, but not wet. Once plants are established, they can tolerate drying out. Ideally, do not allow foliage to get wet. Fertilize with 150–200 ppm at every irrigation, alternating between 20-10-20 and 15-0-15. As plants mature, you can increase the rate to 200–250 ppm. If salts build up, leach at every third irrigation with clear water. Excessive ammonia can promote leaf stretch and soft growth.

Pinch 4" (10 cm) pots once, 6" (15 cm) pots once or twice, and 10" (25 cm) baskets two to three times. Pinch for the first time fourteen days after potting and again two weeks later. Pinch hanging baskets again when stems reach the edge of the basket. If growth regulators are needed, Bonzi and Cycocel are effective.

Plants flower faster as day length increases and will seem to take forever to flower during winter or early spring.

When transplanting one liner per 4" (10 cm) pot, the crop time to finish is six to eight weeks. A 10" (25 cm) basket takes ten to twelve weeks.

Pests & Diseases

Fungus gnats, leaf miners, thrips, and whiteflies can all become insect problems. Diseases that can strike include botrytis and *Pythium*.

Troubleshooting

Watch for yellowing of foliage under cold or wet conditions. Raise temperatures and/or dry out pots to remedy this situation.

Varieties

'Blue Daze' is the most common variety available.

Exacum

Exacum affine
Flowering pot plant

Blue, a rare color in flowering plants, is one of the main reasons why *Exacum* is a popular pot plant. Plants with white or pink flowers and even large-flowered types with flowers measuring 0.5" (1 cm) across are also available. Double-flowered types are newer to the market.

Exacum's delicate scent makes it even more desirable. However, it is not the easiest flowering pot plant crop growers can choose to grow. For success, they have to pay attention to details and be meticulous in preventing disease.

Propagation

Most growers purchase liners or plugs rather than propagating themselves. Seed is slow to germinate, taking fourteen to twenty-one days. After sowing onto a lightweight, germination medium, cover seed lightly and mist; do not water. A 128-plug tray will take from five to seven weeks.

Double-flowering varieties are propagated by cuttings, because plants are sterile. Cuttings root in three to four weeks at 72–75°F (22–24°C) in a 105- or 92-liner tray.

Growing On

Pot liners or plugs into a very light, loose, well-drained medium. Good root aeration is important. Use one plant per pot. Initial watering during the first ten days should be very light to encourage root action. Alternatively, some growers plant high to avoid stem disease and then water normally and stake plants.

Immediately after potting, treat with Rovral at 1 lb./100 gal. (1g/l). Spray to wet the foliage of the plant using 6–8 oz. (170–227 g)/4 or 6" (10 or 15 cm) pot. Plants not treated for two to three weeks after potting may be 50% infected with stem canker botrytis, and treatment with Rovral fungicide will not prevent dieback.

Space pot tight for four weeks, then increase pots incrementally to final spacings of 11" × 11" (28 × 28 cm) for 4" (10 cm) pots and 14" × 14" (36 × 36 cm) for 6" (15 cm) pots. Overcrowding plants on the bench will result in decreased flowering and higher disease incidence.

Grow on at temperatures of 75–80°F (24–27°C) days and 60–65°F (16–18°C) nights. Grow under full sun in winter months and at 4,500–6,000 f.c. (48–65 klux) (50–65% shade) in the summer. In spring and early fall, grow plants in full sun and apply light shade when plants begin to flower. Shading at this time will produce darker-colored blooms. Excessive light and heat will fade flowers; however, adequate light is required for premium plants and rapid flowering.

Fertilize by alternating 16–16–17 Peatlite Special and calcium nitrate at the rate of 2 lb./100 gal. (2 g/l) every third watering. During the summer, supplement liquid feed with a slow-release fertilizer top-dressed at 0.25–0.5 teaspoon (1–2 ml) per 6" (15 cm) pot. High ammonium-nitrogen levels delay flowering. Be careful not to overfertilize as plants are very sensitive to high soluble salts levels. High fertilizer rates can also delay flowering; result in soft, vegetative plants; and decrease postproduction longevity.

Exacums do not require pinching, as they are self-branching plants. In the case of premature budding on small plants, remove the earliest flowers to grow larger plants.

Height can be controlled with B-Nine at 2,500 ppm applied one week after potting. If needed, a second application can be given two to three weeks later for plants that are being grown under low light. *Exacum* is also responsive to A-Rest or a Bonzi spray or drench. Regulating the amount of water plants receive can also control height. If small plants are desired, allow plants to dry out more between watering.

Winter production tends to have more growing problems than summer production does. Lower light levels and shorter days make softer, vegetative plants that are more easily injured and attacked by disease. This must be compensated by lower fertilizer levels and reduced watering to make a harder plant. Water early in the morning so that foliage is dry by late afternoon. Provide good air circulation around plants and reduce fertilizer levels by half. Overwatering and high nutrient levels, besides promoting disease, cause delayed flowering. Attempt to grow the plant "hard."

Exacum requires much less fertilizer and soil moisture than chrysanthemums, lilies, or poinsettias. They respond poorly if fed with high levels of constant liquid feed fertilizer. To accelerate winter flowering, lower fertilizer levels and make sure plants dry out between watering.

Exacum growth is based on total light energy. Flower bud initiation and development are not affected by day length, but plant growth is increased by longer days. Therefore, supplemental lighting in winter is very beneficial. High intensity discharge (HID) lights or even providing long days through night interruption lighting (10–20 f.c. [108–215 lux] of incandescent lights four to six hours per night) can speed up production time in the winter by two or more weeks.

Timing *Exacum* is variable by season. In the summer, marketable 6" (15 cm) pots can be grown in seven to eight weeks from a liner; in the winter production may require up to twelve to fourteen weeks. Smaller plants in 4.5 or 5" (11 or 13 cm) pots for mass-market sales can be produced in six to seven weeks from a liner in the summer and nine to ten weeks in the winter. Production during November and December is not recommended for northern growers.

Pests & Diseases

Broad mites may be found on the upper parts of the plant and can cause the leaves and growing tips to become yellow and distorted and the buds to fail to open. Worms or thrips may also be problems.

Stem canker caused by botrytis is a serious problem for *Exacum*. Most of the infection occurs within one to two weeks after potting. Prevent problems by drenching at planting, as outlined above. *Pythium* and *Phytophthora* can also become disease problems. By taking the following important steps, disease problems on *Exacum* can be virtually eliminated.

- Disinfect benches, walls, and floors before putting plants in the greenhouse. This is especially important if *Exacum* will be following a poinsettia crop.
- Provide good horizontal air circulation.
- Use a well-aerated medium.
- Lower nutrient levels during production from January through April.
- Make sure fungus gnat populations are low.
- Avoid overhead irrigation and water early in the day.
- Avoid high light, excessive watering, and high nutrition.

Troubleshooting

Excessive leaf curl or crinkle may be related to excessive light or B-Nine and possibly a low copper level in the leaf. A foliar spray using Tri-Basic Copper at 1 lb./100 gal. (1 g/l) applied two weeks after potting has been very successful in reducing crinkle. Soil applications of copper on *Exacum* have not been useful.

Too much vegetative growth and a lack of flowers may have several causes, including excessive nitrogen balance in the fertilizer, overfertilization and low light conditions, plants being produced on benches next to chrysanthemums under short days, and low light and overwatering.

Premature flowering on small plants may be caused by checked growth due to low fertilizer, excessive drying between watering, and/or high light and high temperatures.

Postharvest

Plants are salable when 10–20% of the flowers are open. Store and ship plants at 55–60°F (13–16°C). Low temperatures (40°F [4°C]) cause black spots on foliage and can kill plants.

In the retail display and interior home settings, plants should receive bright light (50–100 f.c. [0.5–1.1 klux]) to ensure continued flowering and high-quality foliage. Do not overwater. Allow pots to dry between irrigations; however, if plants wilt, flowers will die. Treated well, *Exacum* will continue to flower indoors for three to four weeks. If placed outdoors, provide semi-shade. Plants are hardy to freezing.

F

Fern, Boston (see *Nephrolepis*)

Festuca

Festuca glauca (*Festuca ovina* var. *glauca*; *Festuca cinerea*)
Common names: Blue fescue; glaucus fescue
Perennial grass (Hardy to USDA Zones 4–8)

Blue fescue is a blue-green, narrow-leafed grass with dense, compact tufts growing from 10–14" (25–36 cm) when in flower. Flowering stalks turn brown to beige as they age.

Propagation

Seed and division are the common methods of propagation.

Division

Named cultivars are often propagated by division of the crown. The plantlets are stuck to 32, 50, or other liner sizes and more commonly sold by commercial propagators. A 72 liner takes from seven to ten weeks from sticking, while a 32 can take twelve to sixteen weeks, depending on the variety.

Seed

Sow three to four seeds per cell of a 288-plug tray, lightly cover with vermiculite, and germinate at 68–72°F (20–22°C) soil temperature. Seeds germinate in four to six days.

Plugs can be transplanted to larger containers in six to seven weeks.

One final note: The more recent introductions have given excellent germination. However, if collecting your own seed or if germination is slow or erratic, there is another method that works well. Upon sowing to a plug tray, place in a cool environment for one to two weeks at 45°F (7°C) and then 70–75°F (21–24°C) until germination begins (usually in two to four weeks). However, seedlings usually germinate over a period of time and not all at once, so several transplantings to larger containers may be necessary.

Growing On

Division

A 32 liner transplanted to a 1-gal. (4 l) pot requires seven to nine weeks when grown at 55–65°F (13–18°C).

Seed

When transplanting a 288 plug to a 4" (10 cm) pot, plants will be ready to sell or transplant in seven to ten weeks for a spring crop. Seed sown after the first of the year will not produce flowering plants the upcoming summer.

Seed can be sown summer to mid-autumn in the northern United States or Canada to liner trays (50–105 cells per flat) ready to be transplanted to 1-qt. or 1-gal. (1.1 or 4 l) containers in fall. These produce larger plants overall, and there is a greater degree of flowering than those started later.

Ficus

Ficus benjamina
Common name: Weeping fig
Foliage plant

As one of the world's leading foliage plants, *Ficus benjamina* is well known by consumers and growers. Plants grace malls, office buildings, homes, and other interiorscapes nationwide. Most gardening consumers have purchased a ficus at some point. Because ficuses are relatively inexpensive to buy and readily available, growers in northern locations generally buy in plants for resale. In tropical climates, ficus is a popular tree, shrub, and hedge plant.

Propagation

Common ficus roots easily from cuttings under mist or fog at a bottom heat of 80–82°F (27–28°C). Rooting hormones enhance rooting. Cuttings root in three to five weeks.

Some growers propagate via air layering to grow standard trees and braided forms. Some varieties are propagated via tissue culture as well.

Growing On

Use a well-aerated medium that can retain uniform moisture. Peat and bark combinations work very well. Maintain good light levels, 50–70% shade in tropical regions (3,500–5,000 f.c. [32–54 klux]). Keep soil temperatures below 95°F (35°C).

Keep plants well watered: Water stress causes leaf drop and reduced growth. Use a constant liquid feed of 20-10-20 at 200 ppm nitrogen. Growers producing larger containers may prefer granular fertilizers. Maintain a set nutritional program to protect plants against leaf-spot disease such as *Xanthomonas* and pseudomonas. Maintain a substrate pH of 6.0–6.5.

The crop time for a 6" (15 cm) pot to finish is from thirteen to sixteen weeks grown warm.

Pests & Diseases

Mealybugs, scale, thrips, and spider mites can attack ficus plants during production. Make sure plants are free of spider mites before shipping, as often a mite infestation will not be apparent in the nursery but runs rampant once the plants are in the interior. A number of diseases can affect plants. *Phomopsis* causes twig dieback and can be especially problematic in interiorscapes, where plants do not receive adequate water. *Pythium, Fusarium,* and rhizoctonia can be problems if plants are kept too wet. Leaf spot caused by *Xanthomonas campestris* pv. *fici* can cause problems. Maintain good nutrition and keep foliage dry to aid in control.

Troubleshooting

High light intensity can cause pale leaves with margins that curl upward. Moisture stress can lead to reddish spots on the undersides of leaves. High soluble salts can result in stunted plants that have excessive leaf drop and wilting. Excessive leaf drop can also be due to plants being improperly stored or exposed to mercury-containing compounds, such as some paints. Nutritional deficiency can cause pale older leaves. Manganese deficiency can result in terminal leaves with interveinal chlorosis, while magnesium deficiency can cause lower leaves with chlorotic margins. Boron toxicity, showing up first as leaf chlorosis and later as leaf death, can be also a problem.

Postharvest

Ficus plants that have been improperly acclimated will drop an excessive amount of leaves, especially if plants were grown in full sun in Florida and then sold to an unsuspecting consumer or retailer in the North. Most growers produce plants under full-sun conditions for a portion of the crop to create well-shaped, full plants. However, depending on the size of the plants, they require as long as nine months of lower light levels (4,000–6,000 f.c. [43–65 klux]—50–70% shade) before sale. Shade-grown plants have a looser growth habit and hold leaves horizontally. Make sure to provide an absolute minimum of 150–250 f.c. (1.6–2.7 klux) of light for retail display and interior settings—higher levels are preferred.

Moisture stress can also cause leaf loss. Lynn Griffith Jr. of A&L Southern Agricultural Laboratories says that what actually happens is that the plants themselves begin to generate ethylene when they go into moisture stress. Because ficus originated in a climate where cycles of moisture and drought are normal, lack of water triggers the plant to begin working on its own survival by dropping leaves: It does so by generating ethylene that causes leaves to drop, thus enabling the plant to survive a drought. During postharvest, allow plants to dry slightly before rewatering. Avoid very dry or very wet conditions. Since plants are ethylene sensitive, avoid shipping them with fruit or produce.

Maintain shipping temperatures of 55–60°F (13–16°C). Established plants in landscape situations can tolerate light frost; however, during shipping, exposure to fluctuating temperatures lowers plant quality with chilling injury or desiccation.

Other ficus species that are commercially important include *F. elastica, F. lyrata, F. maclellandii, F. pumila,* and *F. retusa nitida.*

Foeniculum

Foeniculum vulgare
Common names: Fennel; leaf fennel
Annual herb (Hardy in mild-winter areas)

Fennel is known in two ways. One is leaf fennel, which is covered here. The other is bulb fennel, which is highlighted under Related Species. Both are closely related but vary in certain aspects.

Like dill (a relative), fennel has edible seeds and foliage and is used where the flavor of anise (licorice) is preferred. Leaf fennel has feathery leaves on plants that grow from 3–4' (0.9–1.2 m) tall.

Propagation & Growing On

Fennel is propagated from seed and is an easy crop to grow. Seed can be sown to a plug tray or open flat.

Sow seed to a 288-plug tray and cover lightly with vermiculite. Germination begins in five to eight days in a germination chamber or seven to ten days on a greenhouse bench at 70–72°F (21–22°C). Plugs are ready to transplant to cell packs or pots in four to six weeks.

However, once ready to transplant, don't delay. Many annual plants will tolerate being held in a plug tray for a period of a week or two or three, etc. depending on genus. On fennel, like dill and others in this family, they often bolt (stretch) in habit once they are transplanted to the final container if they are held too long in a plug tray. This is especially true when the crop is sown later in the spring for sales in May or June and day temperatures soar to the 80s (°F or 27–32°C) for a period of time.

While open-flat sowing is an outdated method, it still serves a purpose. A sowing in February or March can be transplanted as individual seedlings to cell packs and planted out of doors in late April or May—whenever the last frost-free date has arrived.

Upon transplanting to the final container, regardless of the method, grow on at 55–60°F (13–16°C) nights with days 8–10°F (4–5°C) higher. Plants prefer a cooler temperature during their production season but tolerate warm to hot growing temperatures in the garden during the summer.

Varieties

Fennel is one of those herbs of which seed companies carry only a few varieties because there is no need for a wide range. With fennel, you only need one or two varieties to select from to serve all of your needs.

However, there is one difference: Fennel has one or two bronze-leafed varieties available that, while used as an herb, can also be used as an ornamental.

Foeniculum vulgare dulce
Common names: Sweet fennel; bulb fennel; finocchio

This species differs from common fennel in producing a bulb. In all other regards, it is very similar: It tastes the same, grows to similar heights, etc. However, it is not a crop that has been perfected by greenhouse growers for sale in retail stores in the spring.

Sweet fennel plants do not tolerate being rootbound when grown in a plug tray, cell pack, or pot. For this reason this crop is sown directly into the garden for its best performance.

Fragaria

Fragaria × *hybrida*
Common name: Strawberry
Perennial, fruit

If you have only thought of strawberries as that delightful fruit to top your cereal, allow yourself to expand your thinking. Strawberries make fabulous hanging baskets. Their low-growing, bushy plant habit is also attractive in combination pots. The fact that your customers will be able to literally enjoy the fruits of their labor is just a bonus.

Propagation

Seed and bare-root transplants are the most common methods of propagation. Bare roots are available from a number of commercial propagators.

Seed

Sow seed onto a well-drained, disease-free, soilless medium with a pH of 5.5–5.8. Select larger trays, such as 288s, using one to two seeds per cell. Wet trays and maintain them at full saturation during germination. Cover lightly with coarse vermiculite to maintain humidity at the seed-coat level but do not germinate in the dark. Provide 100–500 f.c. (1.1–5.4 klux) of light. At 70°F (21°C), seed will germinate in five to seven days.

Move trays to Stage 2 for stem and cotyledon development. Maintain temperatures at 70°F (21°C) and increase light levels to 500–1,500 f.c. (5.4–16 klux). Allow trays to dry somewhat between irrigations as soon as radicles begin to penetrate soil. Once cotyledons are fully expanded, fertilize with 50–75 ppm from 14-0-14 once a week. Stage 2 takes from sixteen to twenty-one days.

Move trays to Stage 3 to develop true leaves. Reduce temperatures to 65–68°F (18–20°C) and increase light levels to 1,500–2,500 f.c. (16–27 klux). Fertilize once a week with 100–150 ppm, alternating between 20-10-20 and 14-0-14. Allow trays to dry down somewhat between irrigations, but do not allow seedlings to wilt. Stage 3 takes from seven to fourteen days for 288s.

Harden plugs for seven days prior to sale or transplanting. Lower temperatures to 60–62°F (16–17°C) and increase light levels to 2,500–3,500 f.c. (27–38 klux). Fertilize once a week with 100–150 ppm from 14-0-14. Allow trays to dry down somewhat between irrigations, but do not allow seedlings to wilt. Total crop time from sowing to a transplantable 288-plug tray is five to seven weeks.

Growing On

Seed

Transplant plugs to cell packs, 4" (10 cm), baskets, or other containers using 1 plug/cell or 4" (10 cm) pot and 3 plugs/10" (25 cm) hanging basket. Grow on at 60–65°F (16–18°C) days and 55–60°F (13–16°C) nights. Provide the highest light levels possible while maintaining production temperatures.

Keep plants uniformly moist, not allowing the medium to ever fully dry out. Feed at every irrigation with 150–200 ppm, alternating between 20-10-20 and 15-0-15. Monitor pH levels closely and make sure the range stays between 6.5–7.5.

4" (10 cm) pots take seven to nine weeks to finish for spring sales, while fruiting baskets require eleven to thirteen weeks. Provide fourteen-hour days through night interruption lighting or by extending the day length for early flowering.

Bare roots

A number of commercial propagators have bare-root plants available during the winter. These are potted up to 6" (15 cm) or 1-gal. (4 l) pots in January or February, grown on at 60–65°F (16–18°C) nights until rooted and top growth begins to develop, and then grown on at 55–58°F (13–14°C) nights with days 5°F (3°C) warmer. Plants will be well rooted by May. Plants may not fruit well the first year but will do so the following year.

Pests & Diseases

Watch for spider mites. Disease problems can include leaf spot and powdery mildew.

Varieties

From seed 'Berri Basket' is an F_1 hybrid ever-bearing variety that is fabulous in 8", 10", or 12" (20, 25, or 30 cm) baskets. Plants create a full, lush look quickly. Plants will make berries until frost. 'Berries Galore', another F_1 ever-bearing variety, puts out so many runners that even the most timid of gardeners will feel like a success. Again, plants will bear fruit until frost.

Freesia

Freesia hybrids
Cut flower, flowering pot plant

Freesias originate in the Cape of South Africa. Because of their native habitat, they require dry storage at very warm temperatures, followed by cool, moist precooling and/or growing conditions in the greenhouse. Today's commercial varieties are the result of extensive breeding efforts. These improvements have helped to increase demand for freesias. Modern varieties come in a wide range of colors, have single or double flowers, are oftentimes fragrant, are highly suitable for low-temperature forcing, and have good keeping quality. They can be forced as either fresh cut flowers or flowering potted plants.

For growers, the keys to success with freesia are: (1) closely coordinating the market and time of arrival of the corms with the supplier; (2) using a well-drained, fluoride-free, pathogen-free planting medium with a pH of 6.5–7.2; (3) using low forcing temperatures and high light intensities; and (4) marketing each product at the proper stage of floral development. To force freesia year-round, you will need to use a soil-cooling system or be located in a favorable freesia climate.

Propagation

Growers buy in corms that normally measure 2.0–2.75" (5–7 cm) in circumference. Specialists in the Netherlands produce most of the corms. Order the corms at least four months prior to the date you need them. Flower development is regulated primarily by temperature and light intensity. Because of the need to precisely control the growth and development of the corms and plants, the forcer must closely coordinate the precooling and planting schedule with the supplier. Corms must be stored at 86°F (30°C) for at least three months before being shipped to the forcer. Because the quantities are generally small and the transport period must be short (less than seven days), corms are always shipped by air.

After they have flowered, the corms can be harvested and stored for forcing the next season. However, to do this requires proper storage facilities that meet specific temperature, relative humidity, and ventilation requirements and is not recommended in most cases.

The production information that follows covers greenhouse production from corms. It is also possible to produce flowers from seed.

Growing On

Cut flower crops

In most areas of North America, planting can begin in September in northern areas and be as late as December in the South. It can, however, be year-round, especially when a soil-cooling system is used. Depending on the varieties and greenhouse temperatures, flowering starts 110–120 days (fourteen to seventeen weeks) after planting and lasts about four weeks. The forcer who wants flowers for several months must stagger the plantings.

On arrival, inspect the corms to be certain they are free from serious diseases or physical damage. Be prepared to plant the corms on arrival. If they must be stored, place them in open trays at 55°F (13°C) under nonventilated conditions, but only up to three weeks. *Caution:* Do not return the corms to 86°F (30°C).

The planting medium must be well drained and free of pathogens and fluoride-containing additives and have a pH of 6.5–7.2. Corms can be planted in either ground beds or raised benches that are at least 10" (25 cm) deep. They can also be started in special propagating trays and subsequently transplanted. The ground bed or bench must have a mesh support system for the growing plants. Average plant heights are 20–30" (51–76 cm), but the actual fresh cut flowers are usually 14–16" (35–40 cm) long.

Plant corms 2" (5 cm) deep and use about 80–100 corms/yd.^{2} (97–120/m^2). The exact planting date depends on prevailing soil temperatures, which must be in the range of 55–60°F (13–16°C). After planting, keep the planting medium moist but not wet.

Freesias require medium to high light intensities (2,500–5,000 f.c. [27–54 klux]). Use 50–55°F (10–13°C) night temperatures and avoid day temperatures over 63°F (17°C), especially during the short days of winter. During warm-temperature months, soil-cooling systems can be used to maintain a planting medium temperature lower than 63°F (17°C).

In the greenhouse, freesias can be forced with 1,000 ppm carbon dioxide (CO_2) during daylight hours. After plants have roots and begin to grow, fertilize them with 200 ppm of 20-20-20 every other week.

Because freesias are sensitive to fluoride toxicity, do not use superphosphate or other fluoride-containing amendments to the medium or use water that contains fluoride either during forcing or after cutting the flowers.

Pot plant crops

Forcing freesias as flowering potted plants requires some experience. Before beginning, determine how the plants will force under your conditions. The objective is to produce a marketable plant in sixty to eighty days (nine to eleven weeks) from planting, with an average total plant height of 10–16" (25–41 cm). At present, these goals are not always achieved with the cultivars, average daily temperatures, and plant growth regulator treatments available. Royal Van Zanten, Risenhout, the Netherlands, has released five dwarf cultivars called the name 'Easy Pot'. Under most forcing conditions, they do not require plant growth regulators to reduce the height of the marketable plants.

Always inspect the corms for your potted freesia crop on arrival to be certain they are free from serious diseases or physical damage. Since they do not normally require a precooling treatment or plant growth regulator, corms of 'Easy Pot' varieties should be planted immediately upon arrival. Other varieties should be stored in open trays at 55°F (13°C) for forty-five to forty-nine days with a high relative humidity and good air circulation, but no ventilation. *Caution:* Do not return the corms to 86°F (30°C) and do not store the corms more than forty-nine days. This can cause the corms to pupate, that is, form a new corm instead of a shoot. Sometimes, corm suppliers will provide the 55°F (13°C) precooling treatment. If this is done, the transport period must be very short.

A preplant soak of Bonzi will reduce the height of flowering potted freesias. The concentration required varies with each variety. Always follow the basic guidelines and test new varieties for the concentration needed. Corms should be dipped for one hour immediately after the 55°F (13°C) dry-storage treatment. Also, it is important that the corms be planted immediately after dipping. Do not allow them to dry out!

Plant corms 1" (3 cm) deep. Use 4–6 corms/4" (10 cm) standard-depth pot, 6–10 corms/6" (15 cm) three-quarter-depth pot, or 10–15 corms/8" (20 cm) bulb pan. Use a well-drained, fluoride-free, sterilized medium with a pH of 6.5–7.2. After planting, keep the medium moist but not wet. For plants that require staking, special rings are available from suppliers.

Potted freesias require a greenhouse with medium to high light intensities (2,500–5,000 f.c. [27–54 klux]). Use 55–60°F (13–16°C) night temperatures and avoid temperatures above 63°F (17°C), especially during the short days of winter. Forcing times range from fifty-five to ninety days, depending on the variety. After plants begin to grow, use either 200 ppm of 20-20-20 every other week or top-dress with 14-14-14 Osmocote.

Pests & Diseases

Other than viruses, which originate with the corms, *Fusarium* is the most common disease. Varieties demonstrate variable susceptibility to it. The most common insect problem is aphids.

Troubleshooting

Freesias can exhibit flower abortion caused by low light intensities and/or high temperatures during the period of rapid flower development in the greenhouse.

Freesias are sensitive to fluoride toxicity and care must be taken to avoid fluoride (F) in the production of the cut or pot crop. Plants are susceptible to very low fluoride levels. Symptoms of fluoride toxicity are "leaf scorch," dead/brown leaf tips or marginal areas, usually with a halo of yellow between the dead area and the remaining healthy green leaf. Common fluoride sources are superphosphate fertilizers, certain other fertilizer sources including some resin-coated fertilizers, and a number of commonly used potting mix components. Perlite, an oft-cited fluoride source, is probably not a primary fluoride source based on research performed by Paul Nelson at North Carolina State University.

Varieties

There are many freesia varieties and, in general, they change every seven to ten years. Freesia varieties are available as double and single flowering, some with fragrance and some without. Some varieties are used both as fresh cut flowers and flowering potted plants, while others have only one basic use. To reiterate, close communication with a reputable supplier is crucial to success for both pots and cuts, and forcing programs and varieties must be chosen with the supplier.

Postharvest

Harvest cut freesias when the first (lowermost) floret begins to open. Potted freesias are ready to sell when the first floret begins to color. Although the open flowers are not highly sensitive to ethylene, the floral buds are sensitive and can abort. Thus, the flowers respond to a pretreatment of 1-MCP (EthylBloc). Little or no storage is advised for either cuts or pots. However, for short-term storage, hold cut flowers dry at 32–35°F (0–2°C) and 95% relative humidity; for long-term storage, place flowers in water at 33–35°F (0–2°C). Some flower preservatives can aid in the bud opening of cut freesias. Do not use water that contains fluoride!

Retailers should display potted freesias under cool, bright conditions and away from sources of ethylene, such as ripening fruit and vegetables. Consumers should be advised to place pot plants in the coolest, best-lit area of the home in order to obtain maximum flower life.

Fuchsia

Fuchsia × *hybrida*
Annual

Fuchsias are a traditional hanging basket crop. Perhaps few other spring plants have the sheer grace and elegance of a well-grown fuchsia hanging basket. Many times, flowers are striking bicolors: red with blue; purple and pink; or pink and white. For some, fuchsias are a standard Mother's Day gift purchased year after year.

If you have grown fuchsias before, it may be time to expand your horizons. While trailing fuchsias are the leading sellers, extensive breeding has developed a number of new fuchsia varieties that are easy to grow and perform well. Some newer varieties have improved heat tolerance, others have upright habits great for pots, and some are even hardy outdoors to

Zone 7. Take a new look at fuchsias and commit to spicing up your assortment: Your customers and sales manager will love you for it.

Propagation

Tip cuttings and, to a minor degree, seed are used to propagate fuchsia.

Cuttings

Most fuchsias are vegetatively propagated from tip cuttings. Harvest uniformly sized cuttings measuring about 3" (8 cm) long from actively growing plants. Root cuttings in Ellepots or a medium with a pH of 5.0–5.5. Provide substrate temperatures of 68–72°F (20–22°C) and ambient air temperatures of 75–80°F (24–26°C) days and 68–70°F (20–21°C) nights. Provide mist using tempered water in the mist lines. Adjust the frequency based on your specific light, relative humidity, and temperature conditions. Depending on nozzle size, spacing, and the size of the propagation bench, set mist duration for ten seconds or longer. Here are some guidelines for frequency that you can adjust to your conditions: Days 1–3, mist every five to ten minutes; days 3–7 (callus formation), mist every twenty minutes; and days 7–14 (roots initiated), mist every thirty minutes. Fertilize with 50–75 ppm from 20-10-20 if you notice loss of leaf coloration or as callus forms. Provide 500–1,000 f.c. (5.4–11 klux) of light. Night misting is usually required. Cuttings will callus in five to seven days.

When 50% of cuttings have differentiated root initials, move trays to the next stage for root development. Maintain soil and air temperatures and light levels. Increase fertilizer to 100–200 ppm once a week from 20-10-20, alternating with 15-0-15. Roots should develop over seven to fourteen days. Harden cuttings off before sale or transplanting by lowering air temperatures to 70–75°F (21–24°C) days and 62–68°F (16–20°C) nights. Fertilize twice a week with 150–200 ppm, alternating between 20-10-20 and 15-0-15.

When growing stock plants keep them actively growing in a vegetative state. Day-sensitive types should be kept under short days. Both day-sensitive and -neutral types should be grown soft with regular cutting maintenance. Most growers buy in rooted cuttings in January and begin baskets upon arrival, because it is more convenient.

Seed

Fuchsia can also be grown from seed. Plug crops will take from eight to ten weeks for 288s and 128s. Use a medium with a pH of 6.0–6.5. Germinate seed at 70–75°F (21–24°C) soil temperatures. Cover seed after sowing. As cotyledons emerge, provide 1,000 f.c. (11 klux) of light and begin fertilizing with 50–75 ppm from 14-0-14. Increase fertilizer rates to 100–150 ppm and begin alternating in 20-10-20 as plugs develop true leaves. Once true leaves begin to develop, start to allow trays to dry down between irrigations. If growth regulators are needed, A-Rest, Bonzi, and Florel can be used in the plug stage.

Growing On

Cuttings

Plant 5 cuttings/10" (25 cm) hanging basket or 7 cuttings/14" (36 cm) basket. Use 1 cutting/4" (10 cm) pot or 2 cuttings/6" (15 cm) pot. Select a well-drained, disease-free medium with a high initial starter charge and a pH of 5.0–5.5. Pinch plants at planting (or wait and pinch one week later, pinching back to four or five sets of leaves). To help with uniformity, some growers prefer to establish plants in 4" (10 cm) pots and bump them into hanging baskets.

If you plant liners directly to baskets, be extra careful in watering the first weeks after planting, as it will be easy to overwater plants in these conditions. The medium must be moist enough so that cuttings root in, but not too moist as to inhibit rooting in.

Fuchsias like fertilizer: Once plants are fully rooted in, use a constant liquid feed of 200–300 ppm from 15-0-15, alternating with 20-10-20. Prior to that, use 150–200 ppm at every other irrigation. Avoid fertilizers high in phosphorus.

Do not grow fuchsia hanging baskets in the attic! Begin plants at 65°F (18°C) days and nights to promote rapid vegetative growth. Later, grow on at 65–75°F (18–24°C) days and 60–65°F (15–18°C) nights. Do not expose plants to temperatures below 59°F (15°C) or above 86°F (30°C).

Grow plants under lower light intensities (1,000–2,000 f.c. [11–22 klux]) in the beginning weeks to promote vegetative growth, and then increase light intensity to 4,000–7,000 f.c. (43–75 klux) once plants reach their desired size in order to promote flower initiation.

When growing day-sensitive fuchsia, provide long days (more than twelve hours) to initiate flowering:

Plants require twenty-five long days to ensure flowering. Once flowering is induced, plants will continue to flower regardless of day length. For hanging basket crops grown from cuttings received in January and sold for Mother's Day in May, manipulating day length to speed flowering should not be an issue for most growers. However, if you want to bring cuttings in earlier for April basket sales, extending the day length of the crop will allow you to produce flowering baskets faster. Extend the day length with incandescent lights (chrysanthemum lighting) for four hours at the end of the day rather than using night interruption lighting. Before long days, make sure plants are sized up. Fuchsia will flower 6.5–7.5 weeks after the start of long days. *Note:* Most fuchsias flower from auxiliary buds. Terminal buds are vegetative under all photoperiods. However, some varieties bloom on both terminal and axillary buds, regardless of photoperiod.

New breeding has developed day-neutral, well-branched varieties. There are numerous varieties on the market so be sure to check with your broker or breeder rep to confirm that the plants you are growing are day neutral. Day-neutral varieties will respond to the same growing regimen as day-sensitive varieties. Blooming plants as early as propagation is possible and does not typically inhibit growth. Using plant growth regulators such as Florel will most likely only slow growth. Daylength extension is not required to flower, but it can quicken the growth and shorten the weeks to flower.

For less-branching cultivars, pinch fuchsia again about seven to eight weeks into the production cycle to adjust plant shape. Some growers also prefer to pinch a third time. Allow eight to nine weeks from the final pinch to sale. For Mother's Day baskets, you will be giving the last pinch about mid-March. Florel at 500 ppm increases branching and is especially helpful to control growth under short days and/or low light conditions. To produce fuchsia topiaries, some growers use gibberellic acid (GA) on upright-growing varieties at 200–400 ppm to stretch stems.

To control plant height, fuchsia responds very well to negative DIF (nights warmer than days). B-Nine at 3,000 ppm can also be used to control plant height. Cycocel sprays are also effective. Drenches of Bonzi at 5–10 ppm or Sumagic at 2–5 ppm can also be applied. *Note:* A-Rest at 25–75 ppm has been shown to increase flowering. Spray five to six days after pinching and reapply every four weeks, as required.

Seed

Transplant one to two plugs per 4" (10 cm) pot. Unlike their vegetative counterparts, seed-propagated selections are best in pots and not hanging baskets since they do not trail. Grow on at 65–75°F (18–24°C) days and 60–65°F (15–18°C) nights. The crop time to a finished plant is seven to nine weeks.

Pests & Diseases

Aphids, fungus gnats, spider mites, thrips, and whiteflies can become problems. Be on alert for whiteflies especially.

Diseases that can attack include botrytis, *Fusarium, Pythium, Phytophthora,* rhizoctonia, and leaf rust. Fuchsia rust is highly infectious. If plants are infected, you will see orange-brown pustules on the undersides of leaves.

Troubleshooting

Excessively high light and low relative humidity can cause leaf scorch. Excessive vegetative growth may be due to high nitrogen concentration in the planting mix, overfertilization under low light, and/or low light and wet planting mix. Low fertilization or lack of nitrogen can result in poor branching. Low light conditions can lead to stretched plants. High light intensity, long days, and/or high production temperatures can cause excessive flowering. Root rot, high salts, or cold production temperatures can stunt plant growth. Reddish spots can be due to a phosphorous deficiency, especially noticeable at temperatures below 55°F (13°C).

Varieties

Most fuchsias are vegetatively propagated; however, one variety, 'Swing', is available from seed. Its large, white flowers have red sepals and are 75% semi-double and 20% fully double. 'Swing' can be grown without pinching.

In vegetative varieties, 'Swingtime' (pink/white), 'Dark Eyes' (purple/pink), and 'Southgate' (pink) are among the most well-known varieties. Expand your assortment by trying 'Applause' (light pink/dark pink), 'Flying Cloud' (bridal pink), 'Lisa' (purple/pink double), 'Pink Marshmallow' (light pink), 'Sophisticated Lady' (double, white edged in pink), and 'White Eyes' (red/white).

In upright types, try 'Dollar Princess' (pink/purple), 'June Bride' (red), 'Lord Byron' (red/purple),

'Tom Thumb' (mini single, pink/purple), and 'Winston Churchill' (double pink/lavender). 'Aretes' is a day-neutral, self-branching series.

Postharvest

Plants are ready to sell when they begin flowering. Most growers like to sleeve upright-growing varieties to protect blooms in transit. Trailing varieties are harder to ship and tend to be grown and sold in local markets. Fuchsia is sensitive to ethylene and responds to silver thiosulfate (STS) treatment. Ship plants cool at 38°F (3°C).

Retailers can display flowering fuchsias in full sun, provided temperatures are moderate (70–75°F [21–24°C]). Keep pots and baskets moist to prevent wilting and prevent soluble salts problems. Fuchsias that are allowed to continually wilt will drop their flowers and become shopworn rapidly.

Consumers can display plants in full sun, provided temperatures are moderate and baskets are watered religiously. Under most conditions, they will have greater success putting the basket in spots that are shaded from the heat of the day.

G

Gaillardia

Gaillardia aristata
Gaillardia × *grandiflora*
Note: Depending on the botanical references you use, these two gaillardia species are often used interchangeably. However, *Gaillardia aristata* is actually one of the parents of *G.* × *grandiflora*. Most of the varieties on the market today are the result of crosses between species. For this reason most recognize *G.* × *grandiflora* as the botanical name.
Common name: Blanket flower
Perennial (Hardy to USDA Zones 3–8)

Gaillardia is a real garden performer: It is one of the few perennials that flowers all summer long from June through frost. However, its first bloom in June is fuller than the color seen in July or August. Gaillardia thrives in high light and well-drained soils. The daisy-like, single flowers come in apricot, burgundy, and crimson as well as shades of yellow atop gray-green foliage. Mounding plants grow from 12–30" (30–76 cm) high. Gaillardia has naturalized in some parts of the western United States. Although gaillardia is typically hardy from Zones 3–8, some of the newer cultivars are only hardy to Zone 6, so review your catalogs for the hardiness of individual varieties.

Propagation

Cuttings

Take tip cuttings from new shoots when plants are actively growing but without flowers. A rooting hormone is not necessary on supple growth. However, cuttings dipped into a solution of indolebutyric acid (IBA) at 1,000 ppm have been shown to be more uniform to root.

Stick cuttings into a 105, 96, or larger liner size using one plant per cell. Give mist for five to eight days but be careful about overwatering, which will lead to rotting. If possible, high humidity with limited misting is more desirable. Cuttings will root from three weeks up to five or six weeks. During cooler temperature and lower light times of the year, it may take a week or two longer. Pinch off any flower buds that start to develop.

Seed

Sow seed onto a well-drained, disease-free, germination medium with a pH of 5.5–5.8. Seed can be left exposed during germination, especially if germinated in a plug or germination chamber. Otherwise, cover lightly with coarse vermiculite. Germinate at 70–75°F (21–24°C) medium temperatures. Maintain uniformly moist, but not saturated, medium. Provide 100–400 f.c. (1.1–4.3 klux) of light in the germination area. Keep ammonium levels below 10 ppm and sodium below 40 ppm. Radicles should emerge in four to five days.

Once radicles have emerged, move trays to Stage 2 for stem and cotyledon emergence. Maintain substrate temperatures and increase light to 500–1,000 f.c. (5.4–11 klux). As soon as cotyledons are fully expanded, begin to fertilize once a week with 50–75 ppm from 14-0-14. Alternate feed with clear water. Stage 2 will take three to ten days.

Move trays to Stage 3 to develop true leaves. Reduce soil temperatures to 65–70°F (18–21°C) and raise light levels to 1,000–2,000 f.c. (11–22 klux). Fertilize once a week with 100–150 from 14-0-14, alternating with 20-10-20. Keep the EC below 1.0 mS/cm. Stage 3 takes seven to fourteen days.

Harden plants for seven days prior to sale or transplanting by reducing soil temperatures to 55–60°F (13–16°C). Allow trays to dry thoroughly between irrigations. Increase light levels to 2,000–3,000 f.c. (22–32 klux) and begin to grow plants under short days to encourage vegetative growth. Fertilize once with 100–150 ppm from 14-0-14.

Plant growth regulators are generally not needed. However, B-Nine/Alar (daminozide) at 2,500 ppm can be applied at Stage 3, if necessary.

A 288-plug tray takes from five to six weeks to be ready to transplant to larger containers.

Growing On

Pot one plant per 4" (10 cm), 6" (15 cm), 1-qt. (1 l), or 1-gal. (4 l) pot. Use a well-drained, disease-free medium with a moderate initial starter charge and a pH of 5.5–6.2. Fertilize with 100–200 ppm from 14-0-14 at every other irrigation. Alternate feed with 20-10-20. Provide 2,000–3,000 f.c. (22–32 klux) of light.

There are several ways you can choose to produce gaillardia: (1) Sow your own seed, or receive plugs, liners, or No. 1 transplants in the fall, then allow them to root, and move them to a cold frame to sell in the spring; (2) receive plants in January, pot them up, root them up, and sell them green in the spring; or (3) buy in vernalized 50-cells in January or February that can be flowered for sale in the spring.

From your own sowings, it is best to sow seed in early to midsummer for overwintering plants in their final container. Sowings done in November or after will produce a flowering plant in late spring. Allow eighteen to twenty weeks with high light (4,000 f.c. [43 klux] of light at 60–70°F [15–21°C] days and 50–60°F [10–15°C] nights).

If you purchased rooted liners during the winter months, pot up to 1-qt. or 1-gal. (1.1 or 4 l) pots for green sales in the spring. Planted to the garden, they will be in flower in three to four weeks.

The following cultural program was developed at Michigan State University based on tests conducted with gaillardia 'Goblin'. For plants received in the fall, bulk plugs or liners up under short days for six to seven weeks. Grow plants at 72–76°F (22–24°C) days and 60–65°F (16–18°C) nights. Allow pots to dry thoroughly between irrigations. To develop vegetative growth, provide short days (less than twelve hours) by pulling black cloth. Prior to going into cool treatment, plants should have sixteen leaves. Large plugs, such as 50-cells or 2.25" (6 cm) cell packs, can be put into cool treatment in the plug cell. Plants started from No. 1 transplants can move into cool treatment once they have rooted into the pot, generally after two to four weeks.

Provide ten to fifteen weeks of cool treatment at 41°F (5°C) in either a cooler (25–50 f.c. [269–538 lux] of light) or cold frame.

Bring pots out of cool treatment and force into flower under long days (more than sixteen hours) with night interruption lighting and constant 65–68°F (18–20°C) day and night temperatures. Fertilize with 150–200 ppm from 15-0-15 at every other irrigation. Growing plants cool at less than 65°F (18°C) will increase crop time, but it will also keep plants compact.

Pests & Diseases

Watch for aphids. Disease problems can include aster yellows and powdery mildew.

Varieties

'Baby Cole' is vegetatively propagated, grows 6–8" (15–20 cm) tall, and has red flowers with yellow margins. 'Fanfare', the most popular gaillardia, has fluted red flowers with yellow margins. Plants grow taller, up to 18–24" (24–61 cm). 'Goblin', also popular, has red flowers with yellow margins as well and grows up to 12" (30 cm) tall. While 'Goblin' is also available from seed and cuttings, plants are more uniform when they are vegetatively propagated. If you are selling large lots to the same customer for planting landscape beds, be sure to grow vegetatively propagated gaillardia, as plants from seed are not uniform.

However, a recent introduction, the All-America Selections Award-winning 'Mesa Yellow' is both uniform and perennial to USDA Hardiness Zone 6, although it has been reported as far north as Chicago (Zone 5).

Other seed-propagated gaillardias include 'Dazzler', with red petals tipped in yellow; 'Golden Goblin', a pure yellow; and 'Monarch Strain', an upright group of red and yellow combinations on plants growing to 30" (76 cm) tall.

G. aristata 'Burgundy' is a seed-raised gaillardia with large, 3" (8 cm) diameter deep wine-red flowers on 24–30" (61–76 cm) plants. From seed, flowers can be dull colored and frequently get lost in foliage.

Postharvest

Plants are ready for sale as soon as they begin flowering. The crop will come into flower over a period of weeks. Growers, retailers, and homeowners should keep gaillardia deadheaded to encourage continuous flowering. In many climates, gaillardia is a short-lived perennial. Be sure to advise homeowners to select only well-drained locations for planting.

Gaillardia pulchella
Common name: Blanket flower; annual blanket flower
Annual

As you would expect, annual gaillardias flower all summer long and tolerate a range of temperatures and locations. They are not as common as their perennial-growing counterparts nor are there as many varieties available. Plants grow from 10–12" (25–30 cm) tall and have globed flower heads that are 2" (5 cm) wide and available in various colors. Plants are heat and drought tolerant.

Propagation

Plants can be propagated by either cuttings or seed.

Cuttings

Take tip cuttings from new shoots when plants are actively growing but without flowers. Follow the culture under perennial gaillardia noted above. Exceptions include that the cuttings root in three to four weeks in a 72 tray.

Seed

Follow the culture under perennial gaillardia for information on pH, EC, etc. Seed is sown from January to March to 288-plug trays, lightly covered with vermiculite, and germinated at 70–75°F (21–24°C) medium temperatures. Maintain uniformly moist, but not saturated, medium. Provide 100–400 f.c. (1.1–4.3 klux) of light in the germination area. Radicles should emerge in three to four days.

A 288-plug tray will take from four to five weeks to be ready to transplant to larger containers.

Growing On

Pot one plant per 4" (10 cm), 6" (15 cm), 1-qt. (1.1 l), or 1-gal. (4 l) pot. Use a well-drained, disease-free medium with a moderate initial starter charge and a pH of 5.5–6.2. Fertilize with 100–200 ppm from 14-0-14 at every other irrigation. Alternate feed with 20-10-20. Provide 2,000–3,000 f.c. (22–32 klux) of light.

From rooted cuttings in a 72 cell, plants will be salable in six to seven weeks with one plant per 4" (10 cm) pot. Plants are sold green or with one flower in the greenhouse but will flower more profusely once transplanted to the garden. B-Nine is effective at 1,500–2,000 ppm, and Cycocel is effective as a 750–1,000 ppm tank mix and used as a spray. Bonzi is also effective at 20–30 ppm spray.

From a seed-sown, 288-plug tray, plants are salable in eight to ten weeks in a 4" (10 cm) container. Plants are sold green or with early color. Like their vegetative counterparts, they flower much better once planted to the garden. For more dwarf plants, follow the plant growth regulator comments in the previous paragraph.

Gardenia

Gardenia jasminoides
Common name: Cape jasmine; gardenia
Flowering pot plant, cut flower

Few other flowers have the sweetly scented flowers of gardenia. Just one flower can add fragrance to an entire room. In the Deep South, gardenias are treasured outdoor shrubs that flower naturally in the spring and grow up to 5–6' (1.5–1.8 m) tall.

Most growers should not consider growing gardenias from liners. Instead, bring in a few cases of prefinished plants in the late winter, force them to flush into flower, and sell them to your local market. Plants in bud and flower do not ship well at all, making gardenias excellent niche pot plants.

Propagation

Gardenias are propagated by cuttings. Terminal cuttings measuring 4" (10 cm) long can be harvested from actively growing plants at any time of the year. Root cuttings in propagation medium at a temperature of 70–75°F (21–24°C) with bottom heat. Provide both mist and shade. Rooting will take from six to eight weeks. From a cutting, plants take eighteen to twenty-four months to reach flowering size. This long crop time is why most growers buy in a few cases of prefinished or finished pots grown in Florida.

Growing On

If you have purchased prefinished plants, such as 4" (10 cm) pots, you can bump them into a larger pot size. Some growers buy in liners and then use 3 liners/6" (15 cm) pot. Select a well-drained, disease-free medium high in organic matter with a pH of 5.0–5.5.

Fertilize with 150 ppm from 20-20-20 at every irrigation. Keep nutrient levels lower (an EC less than 1.0 mS/cm) to increase flower and plant quality. Keep the potting mix pH below 6.0 to avoid iron chlorosis. Maintain high relative humidity and keep pots moist but not wet. Grow gardenias under full light; in the summer in southern regions, shade is required.

Short days enhance flower induction, and night temperatures of 60–62°F (16–17°C) are required for flower bud formation and development. Be sure to provide day temperatures higher than night temperatures to ensure good bud development. Flowering is continuous at night temperatures below 65°F (18°C). If well-budded plants are kept at warmer temperatures, abscission occurs.

To speed flowering and cause a flush of flowers prior to sale in the spring, you can receive prefinished 4 or 6" (10 or 15 cm) plants in December or January, bump them into 6" (15 cm) pots, and grow them under naturally short days and as high light as possible for four to eight weeks at 60–62°F (16–17°C). Then provide long days with night interruption lighting from 10 P.M.–2 A.M. to speed flowering. Be sure to provide as much light as possible during the winter months. All the while, maintain 60–62°F (16–17°C) nights. In general, they require eight to ten weeks to finish.

Maintain adequate spacing between pots so that leaves and stems do not overlap. Proper spacing is vital to ensure good air circulation, which will reduce disease problems.

If you are growing your crop from liners, pinch two to three times during production to encourage branching. Pinch the first time when roots have reached the edge of the container, about two to four weeks after potting; thereafter, pinch as required to shape plants. Do not pinch plants after October 1 if you will be flowering them the following spring.

Bonzi and Cycocel are effective growth regulators and also increase flower quantity.

Pests & Diseases

Gardenias are highly susceptible to insects and disease. Bacterial leaf spot and root knot nematodes can attack plants, as can a number of root rots. Mealybugs, scale, spider mites, and whiteflies can also attack. Plants can become black with sooty mold when insect populations are high.

Troubleshooting

Iron chlorosis can show as yellowing on leaves when the pH drops below 6.0. If chlorosis becomes a problem, you can drench with iron sulfate at 1 lb./100 gal. (1.2 g/l). A number of issues can cause flower drop, including overly moist conditions, high fertilization, high temperatures, rapid temperature drops, and/or insects or root damage.

Varieties

'August Beauty' and 'Mystery' are two tall, large-flowered varieties. 'Veitchii' produces an abundance of smaller flowers.

Postharvest

Flowers are ready to harvest as cut flowers when they are almost fully open. Cut stems so that blooms can be handled by the stem. Most growers place stems into perforated cardboard, mist the flowers with water, and then place the tray in plastic bags or wax boxes to maintain humidity. Stems do not need to be put in water tubes. Plants should be stored in a cooler at 32–36°F (0–2°C). Gardenia petals should not be touched—doing so will brown them.

Gardenia pot plants can be sold as soon as buds are swollen and puffy. Ensuring that plants have flowers at point of sale when they are shipped long distances has not proven practical, making gardenias an excellent product for local growers to supply. Retailers must display plants cool (65°F [18°C]) and under high light to ensure continuous flowering. Retailers should not allow gardenias in flower to dry out and should avoid touching open flowers, as they brown each time they are touched.

Instruct consumers purchasing gardenias as potted plants to place them in the sunniest window in the home and to keep them away from heat vents for continued flowering. Keep ambient air temperatures cool (65°F [18°C]). Once the danger of frost has passed, consumers can move plants outside to the patio or deck for flowering in the summer.

Gaura

Gaura lindheimeri
Common name: Whirling butterflies; wandflower
Perennial (Hardy to USDA Zones 5–9)

In the past, *Gaura lindheimeri* was a plant with white flowers. It was not commonly grown until Siskiyou Nursery in Oregon developed 'Siskiyou Pink', which introduced pink flowers to the species as well as an extended flowering season. 'Siskiyou Pink' has changed the face of *Gaura* in North America. Now, there are a multitude of varieties.

Gaura, in general, grows from 2–4' (0.6–1.2 m) when in full bloom, depending on variety. Plants like full sun and well-drained locations in the garden. Some of the newer varieties are only hardy as far north as Zone 6 or 7, so be sure to refer to each company's catalog for a clearer understanding of a variety's hardiness.

Propagation

Gaura can be propagated by both seed and cuttings. However, cuttings are the most common propagation method.

Seed

Seed germinates in five to eleven days at 70–72°F (21–22°C). Do not cover or cover lightly with coarse vermiculite to maintain moisture around seeds during germination. If sown to a 288-plug tray allow five to six weeks to transplant to larger containers.

Cuttings

Vegetative tip cuttings can be stuck to a 105, 72, or similar liner tray. Cuttings do not require a rooting hormone to root, but plants often appear more uniform when treated with it. Provide mist for six to ten days and pinch the crop fourteen to twenty-one days after sticking. A 72 tray requires five to six weeks from sticking to be ready to transplant to larger containers.

Growing On

Seed

Gaura sown from seed will flower the same summer as when sown in early February or earlier. After transplanting a 288 plug to a 32-cell flat, plants are ready to transplant to the garden or larger containers in five to seven weeks. A 6" (15 cm) pot is salable in flower in ten to thirteen weeks, as the days get longer. For larger containers, such as a 1-gal. (4 l) container, sow in November for transplanting to the final container in December or early January and grow on at 65–70°F (18–21°C) days and 55–60°F (13–16°C) nights, once roots are established in their containers.

Cuttings

Growers can produce *Gaura* in two main ways: (1) Bring in liners or small pots in the fall, overwinter plants, and force them into flower early or sell them green; or (2) buy in liners or small pots in the winter, pot them, and sell them green in the spring and/or flowering beginning in the summer. *Gaura* is a really easy and fast plant to force into flower early on, and, unlike a lot of perennials forced into flower, it will continue flowering through the summer once the consumer plants it outdoors. To force flowering for late March, April, and early May sales, plants require a cold treatment.

Pot liners or plugs into a well-drained, disease-free medium with a moderate initial starter charge and a pH of 5.8–6.2. If you choose to overwinter plants, bulk plants up to at least seven to ten leaves prior to beginning the cool treatment. Bulk plants at 55–60°F (13–15°C) days and 50–55°F (10–13°C) nights. Keep plants watered well, but allow pots to dry between irrigations. Provide moderate light levels of 3,000 f.c. (32 klux) and natural day lengths. Fertilize with 100–150 ppm from 14-0-14 at every other irrigation.

Once plants are bulked up, begin cool treatment. Overwinter pots in a cooler (25–50 f.c. [269–538 lux] of light) or cold frame at 41°F (5°C) for fifteen weeks or more.

When plants come out of cold treatment, they will force quickly under long days. Light with incandescent lighting from 10 P.M.–2 A.M. *Gaura* is a light to moderate feeder; use 100–150 ppm from 15-0-15 at every other irrigation. Do not overwater, and allow pots to thoroughly dry between irrigations. Grow plants under as high light as possible, 3,000–5,500 f.c. (32–59 klux), while maintaining cool production temperatures of 65–68°F (18–20°C) days and nights. You can force flowering quickly—in just six weeks—or you can choose to bulk plants up under short days for a few weeks first to fill pots out, and then begin long-day treatment for forcing. If you simply bring pots out of overwintering and allow them to flower under natural days, they will begin to flower as days naturally get longer in late May and June.

Postharvest

Plants with color picture tags are ready to sell as soon as they are rooted in well. Retailers should be sure to instruct consumers to plant *Gaura* in a superbly drained location with well-amended soil. In its northernmost reaches (Zone 5), *Gaura* is marginally hardy and does better in a protected spot. In the South, *Gaura* goes great in the hottest and most humid summers. Keep plants deadheaded to encourage continuous flowering.

Gazania

Gazania igens
Hardy annual, tender perennial

The bright, colorful flowers of this drought-tolerant bedding plant have been getting a lot more play these past few years in areas where rainfall has fallen short of normal. This tough plant can handle full-sun conditions and a wide range of soils, including soils with salt incursion. However, garden performance suffers in areas with high humidity and/or excessive rain. Gazania likes it hot, but not hot and humid. In fact, gazania's biggest downfall as a bedding plant is that flowers close up in the rain or under dark, cloudy weather.

Even with its downfalls, gazania makes a great addition to pack and pot programs and is an underused flower capable of being a focal point in combination pots. Newer breeding is focusing as much on gazania's silvery foliage as on its bright flowers, making an attractive plant even out of flower. In frost-free areas, gazania makes a great perennial.

Propagation

Using a 288-plug tray, sow seed onto a well-drained, disease-free, germination medium with a low starter charge and a pH of 5.8–6.2. Cover lightly with vermiculite. Light is not needed until radicles emerge. The root radicle will emerge in one to two days at 68–70°F (20–21°C) in the plug chamber.

Move trays to Stage 2 for stem and cotyledon emergence. Maintain substrate temperatures at 68–70°F (20–21°C). Increase light levels to 1,000–2,500 f.c. (11–27 klux). As soon as cotyledons expand, provide 50–75 ppm from 14-0-14. Keep trays on the dry side to improve germination and plant growth. Stage 2 lasts for about five to seven days.

Move trays to Stage 3 to develop true leaves. Reduce substrate temperatures to 65–70°F (18–21°C) and run air temperatures of 70–75°F (21–24°C) days and 60–65°F (16–18°C) nights. Maintain light levels at 1,000–2,500 f.c. (11–27 klux). Increase fertilizer to 100–150 ppm once a week, alternating between 14-0-14 and 20-10-20. Keep trays on the dry side, allowing them to dry thoroughly between irrigations, but avoid permanent wilting.

Harden plugs for seven days prior to sale or transplanting by reducing soil temperatures to 60–65°F (16–18°C) and air temperatures to 70–75°F (21–24°C)

days and 60–65°F (16–18°C) nights. Fertilize once a week with 100–150 ppm from 14-0-14.

A 288 plug takes from four to six weeks to be ready to transplant to larger containers.

Growing On

Select a well-drained, disease-free medium with a moderate initial starter charge and pH of 5.8–6.3. Transplant 1 plant/cell or 4" (10 cm) pot. Use 3–4 plants/6" (15 cm) pot. Grow on at 65–75°F (18–23°C) days and 60–62°F (16–17°C) nights. Manage plants dry, making sure pots are thoroughly dry before watering them. Provide high levels, 4,000–5,000 f.c. (43–54 klux). Fertilize with 150–200 ppm from 15-0-15 at every other irrigation.

Transplant one 288 plug per 4" (10 cm) pot. Plants will be in flower for spring in nine to ten weeks for the majority of varieties. If cell packs are desired, plants will be in flower in a 32 tray in eight to nine weeks

Pests & Diseases

Aphids, thrips, and whiteflies can become insect problems. Diseases that can strike include botrytis, *Pythium,* and rhizoctonia.

Varieties

The 'Daybreak' series is an F_1 hybrid that has won numerous Fleuroselect awards for its garden performance. The series comes in eight separate colors, the most striking of which is 'Red Stripe', a bright yellow with medium red stripes. The series is fast to flower.

For something unique, try the F_1 'Tiger Mix', a blend of unique colors and flower patterns. Colors range from creamy white with rose stripes to bright yellow with red stripes.

The 'Talent' series features silvery-white foliage that is lovely with or without flowers! The series is available as a mix or in four separate colors.

The large-flowered, F_1 'Kiss' series is available in separate colors and boasts overlapping petals. The series was selected for uniformity of blooming among colors. There is also a 'Big Kiss' series with larger flowers than standard 'Kiss'. It is also an F_1 hybrid.

A new series called 'New Day' will be coming in 2012. At the time of this writing the variety is just being introduced. This new series will offer compact plants with short flower peduncles and large, overlapping pedals in the flowers.

Geranium

Geranium sanguineum
Common name: Hardy geranium; cranesbill
Perennial (Hardy to USDA Zones 4–8)

The bright magenta flowers in the spring and early summer that are followed by bright red foliage in the fall have made this hardy geranium a popular perennial. For growers, starter plants are readily available in the fall or spring, so it is an easy addition to your normal perennial assortment.

Propagation

Plants may be propagated through division, cuttings, and tissue culture. Commercial propagators preform the tissue culture, and plants are sold to growers in small pots or cell packs (32s, etc.). Cuttings are sold the same way.

No. 1 transplants are available from a number of sources. Potted in winter, these will be fully rooted in the container in nine to eleven weeks. Plants will not be in flower when sold if grown at 65–68°F (18–20°C) days and 65–68°F (18–20°C) nights until rooted and then grown in a cold frame or similar environment at 58–60°F (14–16°C) until sold.

You can receive plants in the fall or spring: According to Michigan State University tests, cold is beneficial, but not necessary, for flowering. After fifteen weeks of cold treatment, plants flower in eight weeks under long days (when provided with night interruption lighting) or eleven weeks under short days. Grow plants under as high light as possible (3,000–5,500+ f.c. [32–54+ klux]) while maintaining moderate production temperatures of 65–68°F (18–20°C) days and nights.

Pests & Diseases

Please note that hardy geraniums can act as a Typhoid Mary in your greenhouse by harboring the dangerous bacterium *Xanthomonas campestris* pv. *pelargonii.* You or your workers could spread *Xanthomonas* from your hardy geraniums to your zonal geranium crop without being aware of it. If you choose to grow both crops at the same production facility, keep plants isolated from each other and do not allow employees who handle your perennial geraniums to also work in your zonal crop.

Varieties

G. sanguineum produces bright magenta-pink flowers in season and, as a bonus, bright red foliage in the fall! 'Alpenglow' has rose-red flowers on 8" (20 cm) tall plants. 'John Elsley' has clear, carmine flowers on 12–18" (30–46 cm) spreading, drought-tolerant plants with fine, lacy foliage. Flowering lasts from early to midsummer. *G. sanguineum striatum* 'Lancastriense' grows to 8" (20 cm) tall and produces pink flowers with darker veins.

Related Species

Related species include the following: *G.* × *cantabrigiense* 'Karmina' has bright raspberry-pink flowers on 6–8" (15–20 cm) tall plants. *G.* × *oxonianum* 'Bressingham's Delight' has light pink flowers with dark pink veins atop deep green foliage. 'Claridge Druce' has bright magenta-pink flowers above gray-green foliage on 18" (46 cm) plants. However, after flowering, the foliage is often detracting because of foliar disease, such as rust. *G. psilostemon* 'Patricia' is a long-blooming selection growing to 4–5' (1.2–1.5 m) in England. Flowers are magenta. *G.* × 'Johnson's Blue' is an interspecific cross with bright blue, cup-shaped flowers. The flower season is long; however, its performance in the garden in most parts of the country is marginal.

Postharvest

Keep plants deadheaded and sheared to encourage rebloom.

Geranium, annual (see *Pelargonium*)

(see *Pelargonium*)

Gerbera

Gerbera jamesonii
Annual, flowering pot plant, cut flower

The electric colors of gerbera and their rayed, daisy flower form have catapulted them into being one of the world's top-ten flowers. Bedding plant growers produce gerbera in 4" and 6" (10 and 15 cm) pots for sale in the spring. In the South, some gardeners make a ritual out of planting "gerber daisies" each year.

Gerberas also make showy flowering pot plants and tend to pop up in supermarket floral department displays throughout the spring and summer and into the fall. Cut flower gerberas are a mainstay: Their extra-large blossoms grace mixed bouquets and brighten up florists' arrangements worldwide.

Most pot plant and bedding plant gerberas are started from seed-raised plugs, while the cut flower varieties are often propagated by tissue culture and sold as liners.

Growing gerberas can be tricky. They are famous for "blinds," or plants that just will not flower. This problem is usually associated with planting too deeply and/or covering the crown.

Bedding and Pot Plant Crops

Propagation

Seed

Gerbera bedding and pot plants are produced as plugs grown in four stages. Stage 1, radical emergence, takes five days at a substrate temperature of 70–75°F (21–24°C). Keep the medium uniformly moist, but not saturated. Sow standard or coated seed and cover with coarse vermiculite to maintain humidity levels around the seed coat. Light at 500 f.c. (5.4 klux) benefits germination, increases seedling bulk, and speeds time to visible bud. Sowing medium should have a pH of 5.5–5.8, and soluble salts (EC) should be less than 0.75 mS/cm. Reduce moisture levels once the radicle emerges. During germination, gerbera are very sensitive to high salts, particularly ammonium. Be sure to keep ammonium levels at less than 10 ppm.

During Stage 2, stem and cotyledon emergence will occur over seven to ten days. Maintain soil temperature at 70–75°F (21–24°C). Keep the medium pH at 5.5–5.8 and the EC at less than 1.0 mS/cm. At pH levels above 6.0, leaves may become chlorotic from iron or manganese deficiency. Make sure ammonium levels are below 10 ppm. At every other watering, begin a liquid fertilizer of 14-0-14 at 50–75 ppm or calcium/potassium-nitrate feed once cotyledons are fully expanded. Alternate feed with clear water. Be sure to irrigate early in the day so that foliage is dry by nightfall. Continue to provide supplemental light.

Growth and development of true leaves will begin during Stage 3, taking fourteen to twenty-one days. Lower soil temperatures to 65–70°F (18–21°C). Allow medium to dry thoroughly between irrigations, but avoid permanent wilting. Be very careful not to overwater at this stage. Overwatered seedlings develop into hard, distorted young plants. Allowing medium to dry down promotes root and shoot growth. Maintain a medium pH of 5.5–5.8 and an EC of less than 1.5 mS/cm. Increase fertilizer to 100–150 ppm from 20-10-20, alternating with 14-0-14 or a calcium/potassium-nitrate fertilizer. Fertilize every second or third irrigation. If using 15-0-15, supplement with magnesium once or twice during Stage 3, using magnesium sulfate (16 oz./100 gal. [1.2 g/l]) or magnesium nitrate. Continue to provide supplemental light. To control plant height during Stage 3, use DIF whenever possible, especially during the first two hours after sunrise. A-Rest, B-Nine, and Bonzi may also be used to control plant height.

During Stage 4, plugs harden and finish over seven days, becoming ready to ship or transplant. Maintain substrate temperatures of 62–65°F (17–18°C). Allow the medium to thoroughly dry between watering. Continue to maintain the pH at 5.5–5.8 and drop the EC to less than 1.0 mS/cm. Fertilize as needed with 15–16–17 at 100–150 ppm.

A 128-cell plug takes from six to eight weeks from sowing until it is ready to be transplanted to larger containers.

Growing On

Transplant plugs immediately into pots from 4" (10 cm) up to 6" (15 cm). If you are not able to transplant plugs immediately, then remove every other plug and place them into another plug tray: This will ensure that plants receive high enough light so that flowering will not be delayed. Do not hold plugs for more five days. Ideally, sort plugs by size at planting to help even out the crop times of the finished crop. Select a disease-free medium with a moderate starter charge and a pH of 5.8–6.0. Make sure the medium has a low or zero bark content. When planting, be sure that the crown is placed slightly above the soil line: Buried crowns are an invitation for disease and/or blind plants.

Pots can be spaced pot tight in the beginning; be sure to space plants before the leaves cover the crowns of other plants. Gerbera plants must have adequate sunlight at the crown in order to initiate flowers. When plants are grown in crowded conditions, flowering is delayed and inhibited and leaves are longer than normal. Space 4" (10 cm) pots to 6" × 6" (15 × 15 cm) after four weeks and 6" (15 cm) pots to 10" × 10" (25 × 25 cm) after six weeks.

Grow plants at 72–75°F (22–24°C) days and 65–70°F (18–21°C) nights. Temperatures below 60°F (15°C) or above 90°F (32°C) will delay flowering. During the summer months, producing plants under 25–30% shade may be beneficial in lowering temperatures.

The exact flowering mechanism for potted gerbera is not pinpointed. Research has shown that potted gerberas form buds after they have developed from ten to fourteen leaves in the primary leaf whorl and after two to six leaves have developed in the secondary whorl. Once this happens, the next (secondary) flower will become visible in eleven days under short days or eighteen days under long days. However, other research has shown a correlation between light levels and temperature, with high light intensity during seedling growth being most important to speeding flowering.

In general, grow gerberas under moderate to high ambient light levels (3,000–4,000 f.c. [32–43 klux]). During the summer months, shade may be beneficial. Some European growers that produce plants during the winter months apply supplemental light at 40 watts/m^2 (543 f.c. [5.8 klux]). If lighting during the winter, do not use lights for more than fourteen hours each day, or stretching may occur. Gerberas are slightly photoperiodic. Some Europeans believe that for the first three to four weeks after potting, gerberas should receive long days (until plants are spaced). Thereafter, provide short days (ten to twelve hours). They believe this results in the shortest growing time and the most uniformity. Longer days increase crop time and may promote blind buds.

Before watering, make sure plants are thoroughly dry. Overwatering causes plant stretch and makes plants susceptible to disease. However, do not allow plants to wilt down just after they are potted. Ideally apply a calcium-based fertilizer for fall and winter production and an ammonium fertilizer during the spring and summer. Use 150–200 ppm. Maintain an EC of about 1.0 mS/cm. Maintain the pH at 5.5–6.0. Do not allow the pH to rise above 6.5. If iron or magnesium deficiency occurs, apply magnesium sulfate at 16 oz./100 gal. (1.2 g/liter) and chelated iron at 0.5 oz./100 gal. (37 mg/l).

Once plants begin to flower, make sure relative humidity levels are below 80% to prevent botrytis.

Also, ensure plants have good air circulation and ventilation at night to prevent botrytis.

Gerberas grown dry normally do not require growth regulators. When plants are rooted to the container sides, they can be allowed to wilt before irrigation. B-Nine is effective at height control if needed. After plants have been planted for twelve to fourteen days, B-Nine may be used at 1,000–1,500 ppm. If required, apply again ten to fourteen days later. A-Rest (200 ppm) and Bonzi are also effective. Discontinue all growth regulator treatments when buds have visible stems.

The total crop time from sowing to a finished flowering 4" (10 cm) pot is sixteen to twenty weeks, depending on variety. When transplanting a 128 plug to a 4" (10 cm) pot, plan on eight to ten weeks until flowering when grown at temperatures of 66–68°F (19–20°C) day and 62–66°F (17–19°C) nights.

Pests & Diseases

Leaf miners, thrips, whiteflies, cyclamen mites, and aphids can attack gerberas. Diseases to watch for include *Alternaria,* botrytis, *Phytophthora,* and powdery mildew. Maintaining good sanitation, providing excellent air circulation, and allowing plants to dry between irrigations are vital for gerbera production.

Troubleshooting

Low light levels from plants spaced too closely can cause delayed flowering. Low light, high ammonium levels, and B-Nine rates that are too low can also cause long leaves and stretched flower stems.

Short flower stems can be caused by excessive B-Nine, running plants too dry, high soluble salts, or temperature extremes. Insect damage from cyclamen mites or thrips, excess salts, or temperature extremes may result in distorted flowers. Fasciated flowers—or two flowers that have fused—is inherent in gerbera, and there is little you can do to control it. Lack of uniform flowering may be the result of a multitude of factors, including variety selection, low light levels, and long days.

Varieties

'Festival' has been the long-time favorite. The series is available in a wide color range in mini as well as standard forms with single, semi-double, and spider flower types. Plants grow from 8" (20 cm) for mini types to 12" (30 cm) for standard types.

Two series that have continued to increase in popularity are 'Jaguar' and 'Revolution'. Both are available in a wide range of colors.

Postharvest

For maximum postharvest life, decrease fertilizer to 100–150 ppm during the last three weeks of production and market plants when they are 50% open. Flowers are considered "open" when the two outer rows of disc florets are open and pollen can be seen. Plants may be shipped at 35°F (2°C) for up to three days. Advise retailers and consumers to display plants under at least 100 f.c. (1.1 klux) and temperatures of 70°F (21°C). At higher temperatures, shelf life is reduced. Gerberas are sensitive to ethylene, so do not store plants with or near fruits or vegetables, and be sure to keep plants groomed, as dead leaves are a source of ethylene production.

Cut Flower Crops

Propagation

Specialist breeder/propagators produce gerberas for cut flower production from tissue culture. Most of the time, growers receive liners produced from tissue-cultured plantlets. However, some growers may receive Stage 3 plants in jars or bags. Handle plants immediately. Wash away agar and plant plantlets into a well-drained medium in 2.25" (6 cm) cell packs or Jiffy strips. Once planted, put plants under 50% shade and intermittent mist. Set mist cycles dependent on light levels and temperature. Drench with a broad-spectrum fungicide such as Subdue. Maintain 77°F (25°C) days and a minimum of 60°F (16°C) nights. Gradually reduce mist prior to planting. Establishing plants will take about six weeks.

Begin feeding 100 ppm of nitrogen (N) and potassium (K) two to three weeks after plants are established. Do not allow fertilizer to stand on new growth, as it can burn foliage.

Establishing plants from Stage 3 plantlets can be tricky with many losses. For that reason, most growers prefer to buy in established liners.

Growing On

The largest Dutch gerbera growers and some American growers produce plants on rock wool using hydroponic systems. However, in many other locations, gerberas are primarily produced in ground beds, which will be discussed here.

Make sure ground beds are well worked and have an excellent air-water balance. Soil must be disinfected prior to planting. Gerberas are very susceptible to disease: Crops begun in improperly disinfected soil will be a problem from the start. After steaming, be sure to leach soil to remove any excess dissolved manganese.

Make sure soil beds have been watered to field capacity prior to planting. If you incorporate organic medium into the soil to improve aeration, avoid using pine bark. Also avoid using fresh manure or compost, unless it has been sterilized. Fully composted manure or compost does not have to be disinfected. When working in organic medium, be careful not to compact the soil. If a hardpan is present, soil must be harrowed, busting it up before planting. The hardpan will inhibit root growth and prevent water from draining quickly. However, be careful not to compact soil when working with heavy equipment over soil beds: Work to maintain the porosity of the soil.

If your water table is high, install drains at an 8" (20 cm) depth (10.5–27.5" [27–70 cm] depth for sandy soils) to ensure plants will not have wet feet.

Soil heating benefits gerbera growth by speeding plant establishment, increasing winter production, speeding spring production, and lowering infection by disease. Flower length and diameter are also greater on plants grown with soil heating. Lay pipe 8.0–19.5" (20–50 cm) deep on 27.5–31.5" (70–80 cm) spacing. To prevent root damage, water circulating through pipes should be no hotter than 104°F (40°C) to maintain a soil temperature of 64–68°F (18–20°C). Higher temperatures are detrimental. Turn off soil heating in the late spring and summer.

In principle, gerberas may be planted year-round. However, in practice, there are two primary planting seasons: spring (January, February, and March) and summer (June, July, and August). Due to high heating costs and lower light levels, planting in the fall and winter is not as profitable. Spring-planted crops are best started from plants established in 4" (10 cm) pots. Spring planted crops are generally harvested for one and a half years. Summer-planted crops going into heavier soils are ideally planted in May or June. On lighter soils, delay planting until week 30 (third week of July) as a rooted cutting; week 33 (mid-August) for 3" (8 cm) pots. Summer-planted crops will have a reasonable fall production. These plants may be harvested for one, one and a half, or two years.

In general, gerbera is not cropped for two years, because during the second year winter production levels and quality fall dramatically. Crops grown for one and a half to two years also require peak demand for labor as the crop cycles. This is not the case for one-year crops, which tend to be steadier.

For plants cultivated longer than one year, leaves must be pruned in order to maintain ventilation and light levels around the plant. Remove old leaves judiciously throughout the production of a two-year crop, only occasionally for a one-and-a-half-year crop, and never for a one-year crop. When culling leaves, do not pick too many at once, as the wounds can be a point of entry for disease. Discard old leaves and leaf debris in the production area, as they serve as hosts for botrytis.

Transplant plants into a two-row system measuring 40" (1 m) across, including the aisle. Space rows 14–18" (36–46 cm) apart and plants within the row 10" (25 cm) apart. In areas with high light and low humidity, plants can be spaced closer together. Most growers mound beds to a height of 10–14" (25–36 cm), depending on the soil type, to allow leaves to droop and improving air circulation and facilitating harvest. If day temperatures are 86°F (30°C) or higher, plant in the morning or evening. Do not plant too deeply. The crown should be level with the soil line.

Immediately after planting, run temperatures at 72–77°F (22–25°C) days and 68–72°F (22–25°C) nights for three to four weeks. Afterward, for spring plantings run 57–60°F (14–16°C) nights and for summer plantings 54–57°F (12–14°C) nights. Shade is necessary during summer months to moderate light levels, lower air temperatures, and prevent plants from drying out.

During winter months, maintain minimum temperatures of 57–62°F (14–16°C) days and 54°F (12°C) nights. Too-high temperatures during the winter will result in high heating bills, weak plants, and low-quality flowers.

Maintain humidity levels at 80–85%. Humidity levels higher than 90% will cause deformed flowers. Carbon dioxide at 700 ppm is beneficial. (*Note:* A very few varieties find carbon dioxide levels above 300 ppm toxic).

Lansbergen "Moves" Gerbera

The future of cut flower automation can be found in Pijnacker, the Netherlands, near the Westland. This is home to Lansbergen Gerberas' Moving Flowers system, which has been attracting the attention of cut flower growers around the world. Lansbergen's 9,000 tables of cut gerbera plants move slowly up and down the 1,378'–long (420 m–long) sections of the ultra-modern greenhouse while workers stand in one place to harvest the flowers as they roll past. It is probably the most automated cut flower range in the world today, and future plans are to automate the harvest operation, cutting Lansbergen's labor by 80%.

This empty table lets you see the trough design and gravity-feed drip irrigation system. Orange funnels on each table collect water from outlets (the tables stop periodically for filling—they are not filled during movement). Note the length of each bay—1,365' (420 m)—and the traveling grow lights overhead.

Lansbergen is a world leader in cut gerbera production. They handle every aspect of the crop, from breeding and tissue culture production of young plants, to producing millions of cut flowers per year, all grown hydroponically on rock wool. But while Lansbergen was expanding rapidly, labor expenses were soaring and labor availability was shrinking.

"In the future, we see big problems in Holland finding workers," Hein Lansbergen says. "My brothers said to me five years ago that we want the company built to be automatic in ten years' time." Half way into that time schedule, Lansbergen has cut labor by 40–50%, he says.

Harvesting is the most labor-intensive part of gerbera production. Hein felt that the best way to cut labor was to bring the plants to the workers, rather than the workers walking up and down the aisles to harvest. "Like a car factory," he explains. "People stand still and the cars come by." Except at Lansbergen, the gerberas come by.

Hein saw potential in the moving tables used by pot plant growers. With the help of an internal transport specialist, he designed aluminum tables equipped with two gutters to hold the plants and a self-contained, gravity-flow irrigation system that feeds the plants through drip tubes.

The Venlo-style greenhouse for the new system covers 721,000 ft.^{2} (67,000 m^2) and is divided into ten two-bay sections. Each bay is 26' (8 m) wide and an incredible 1,365' (420 m) long! Gutters are 16' (5 m) high, and the glass is 3.75' (1.25 m) wide, making for a very open, bright greenhouse. It is equipped with a "traveling" light system where HID assimilation lights move over the crop. This lets Lansbergen provide supplemental lighting to the entire greenhouse with fewer fixtures.

Each of the ten two-bay sections has two rows of roller track, with nine hundred tables per section. The tables roll toward the back of the greenhouse in one bay, cross over to the next bay, and then roll back toward the front of the greenhouse, all automatically. The 2,600' (800 m) round trip takes two days.

The irrigation reservoirs are filled with 1.3 gal. (5 l) of water and fertilizer as needed, by outlets along each bay. Nine hundred tables can be filled in two minutes. Hein says designing this was the most difficult part of the project, as pot plant systems flood the entire bench.

To harvest flowers in the old stationary system, workers walk up and down between the plants to take flowers. With the new system, the plants come to the workers, who stand on plywood platforms to easily reach the flowers. Each table's two-day trip through the greenhouse is long enough for new flowers to develop. Hein has been able to cut his harvesting force from forty workers to twenty (two per section) because they are more productive, each harvesting 800 stems/hour compared with 600/hour the old way. The workers say they appreciate the clean, comfortable work environment.

Such a system is expensive—about \$4.3 million, Hein says. It costs him about \$3.00/ft.^{2}/year (\$31.20/m^2/year) to operate the system (not counting plants). But he adds that he is saving \$0.63/ft.^{2}/year (\$6.70/m^2/year) on labor costs, so in actuality he is making an annual profit from the system of \$0.12/ft.^{2} (\$1.25/m^2). An added cost benefit is realized in space savings: Hein says he gets 8% more space in the greenhouse because workers do not need so much aisle space to harvest flowers. That equates to about 2.5 million more stems per year.

Reprinted from the seventeenth edition of the Ball RedBook.

Overhead irrigation is possible in the first weeks after planting until leaves begin to touch; afterwards use drip lines to keep water off plants and flowers. Water thoroughly rather than frequently while establishing plants.

Base fertilizer applications on monthly soil samples. Maintain a soil pH of 6.0–7.0. Watch for iron and manganese deficiency during growth. In general gerbera are light feeders. Maintain an EC of 1.2–1.6 mS/cm.

Pests & Diseases

See comments under the Bedding and Pot Plant Crops section. Gerbera are one of the few cut flowers harvested without leaves. For this reason, a large number of growers are successful using biological controls. High disease susceptibility is one reason many cut flower gerbera growers have switched from soil culture to hydroponic rock wool culture in troughs. Plants get better ventilation since troughs are raised above the ground. Rock wool is also sterile. Yields are also about 20% higher.

Troubleshooting

Yellow leaves may be an indication of iron deficiency, especially if yellowing begins on younger leaves and veins remain green. Thick, crispy leaves with yellowish leaf margins may be an indication of magnesium deficiency. Older yellow leaves may signal manganese deficiency.

Postharvest

Flowering will begin seven to twelve weeks after planting. Begin harvesting when two to three whorls of stamens have entirely developed. Double types are picked at a more developed stage; flatness and how open the flowers are determine harvest. If picked too soon, the postharvest life is shortened.

Gerberas are harvested by pulling on stems: Be sure to harvest the first flowers carefully in order not to dislodge the plants. Harvest stems by pulling them from the plant using an up-and-down twisting motion to locate the natural breaking point. Depending on the variety, harvest between two and four times per week.

Place stems in water immediately upon harvest. To stimulate water uptake, cut the stem end (heel) off. By removing the woody lower 1–2" (3–5 cm), the stem can take up water. Pulse with sucrose and silver nitrate or 1% bleach. Tween 20 (surfactant) after harvest helps prevent bacterial blockage.

Gerberas are sensitive to ethylene; however, rates detrimental to many flowers cause little damage to gerbera.

Cool stems after harvest to 35°F (2°C). Stems may be stored for up to five days in a preservative solution in a cooler. If flowers must be stored dry, be sure to cut off the heel of stems first so stems will later take up water. Avoid using fluoridated water—fluoride will cause brown spots.

To prevent stem and flower damage and provide maximum presentation at the wholesale level, growers use special gerbera boxes or Procona buckets that allow flowers to be transported without damage. Be sure to use a bactericide in the water. Make sure buckets have been sterilized between each use.

Gerbera can be held in a refrigerator and shipped at 35°F (2°C).

At retail, use only clean water and sterilized buckets. Leave flowers in the perforated cardboard trays while they rehydrate to prevent permanent bending of stems that may be limp. Once flowers are turgid, they can be removed from trays. Hydrate for one to two hours with warm water (pH of 3.5 with citric acid). After hydrating, move stems to a silver-nitrate postharvest solution at 25 ppm.

Gerbera stems can elongate if they are stored in a sucrose solution due to sugar uptake. If this becomes a problem, change the solution to either a bleach or silver-nitrate solution. Store at 35–40°F (2–4°C).

Geum

Geum spp.
Common name: Avens
Perennial (Hardy to USDA Zones 4–8)

Geums appear in the garden in a blaze of red, yellow, and shades in-between for a couple of weeks in May or early June. In flower, they are luxuriously spectacular. Plants grow to a height of 14–30" (36–76 cm) depending on variety.

Propagation

Plants are grown from seed, bare-root transplants (divisions), tissue culture, or cuttings. Many items that are propagated by tissue culture and cuttings (softwood and root) require a license to propagate. Professional propagation companies perform tissue culture, and the resulting plants are often available as a rooted liner tray. (For more information on tissue cultures, see Cuttings.)

Seed

Seed should be sown from June to July for transplanting to final containers and overwinter dormant. Seed can also be sown later (August to October), although these plants are grown cold (40–45°F [4–7°C]) once well rooted in the final container. All of these plants will flower in spring. If seed is sown in January or later, most varieties will either not flower at all or flower sporadically with limited color. Seed varieties often require vernalization (cold treatment) to produce flower buds, although some of the newer selections (especially those propagated from cuttings) do not require a vernalization but benefit from cool temperatures to improve overall flowering.

Sow fresh seed onto a disease-free, germination medium. Cover lightly with coarse vermiculite to maintain 100% humidity at the seed coat. Fresh seed will yield from 65–80% germination; however, if you have saved seed from the prior year, germination rates will be low and germination will be difficult. You can increase the rate from old seed by alternating day and night temperatures by 8–10°F (4–5°C variation in day/night temperatures).

Fertilize with 50–75 ppm from 14-0-14 once cotyledons are fully expanded, and provide 500–1,000 f.c. (5.4–11 klux) of light. Begin to allow trays to dry down somewhat between irrigations once cotyledons have expanded. Increase fertilizer to 100–150 ppm once a week from 14-0-14, alternating with 20-10-20 as plants grow. Plugs (288s or 128s) will be ready to transplant into pots about eight to ten weeks after sowing.

Bare-root transplants

These are commonly available in the late fall and winter for planting one root to a 1-gal. (4 l) pot. Allow to root at 65–70°F (18–21°C) days and nights 5–7°F (3–4°C) cooler.

Cuttings

Softwood and root cuttings can be done on the cultivars. However, most of these are protected varieties and asexual propagation is illegal without a license. Liner trays are available in the fall and winter as 20, 30, 50, etc. plants per 11" × 22" (28 × 56 cm) flat. This is the same with tissue-cultured plants. They are available in similar liner sizes and can be treated in the same manner.

Growing On

Seed

Because most geums will not flower the same year sown, most growers buy in plugs or liners in autumn so plants can be overwintered and sold the following spring either green in April or flowering in May and June. While there are a few newer selections that are touted as first-year flowering, they are still best sown the previous year up to November to grow cool or cold over the winter. Some suppliers also offer cooled plants that can be received in January or February, potted up, and sold either green in April or in flower in May and June.

Bare-root transplants

The best performance is from roots that are potted up in fall or early winter. Grow on in the greenhouse at 65–70°F (18–21°C) until plants fill out the crown of the pots. Within five to seven weeks the roots will be filling out the container and the plants can be moved to a cold frame and grown on at 58–60°F (14–16°C) days and 50–55°F (10–13°C) nights.

Roots can be potted up as late as March for plants that are still salable in late April or May. However, the roots will not be all the way to the bottom of the container and the media will fall away from the root-ball when planted to the garden.

Cuttings

Liner trays should be received in the late summer or fall to pot to 1-qt. or 1-gal. (1.1 or 4 l) containers and then overwintered cold or dormant once the plants are fully rooted in the containers. These plants are sold in the spring either green or in bloom.

Postharvest

The best way to sell geums is green with a large picture tag, that way the consumer will be able to best enjoy all two weeks of the plants' glory. Plants that have flowered in the spring will not reflower again until the following year.

Gladiolus

Gladiolus spp.
Common names: Gladiolus; glads
Cut flower, garden annual

The production of gladioli as a cut flower requires a high degree of specialization, not only in knowledge but also in equipment and facilities. Thus, the information presented is intended for those outdoor growers who want to supplement their production programs on a limited scale.

Gladioli species and cultivars produce a multi-flowered inflorescence that can contain ten to twenty-five florets. Cultivars are available in most colors and can be up to 6' (1.8 m) tall. In coastal areas of California, they can be produced year-round, while in Florida flowers can be produced from November to June. In other areas of the United States, flowers can be produced from early summer to fall.

Growing On

Purchase high-quality corms from a gladiolus specialist. Commercial corm sizes are: large (Jumbo and No. 1) and medium (No. 2 and No. 3). Smaller sizes are usually used as planting stock. Staggered planting is the basis of continual harvesting throughout the season.

On arrival and before planting, corms should be stored at 35–41°F (2–5°C) under highly ventilated conditions. Growers should select open fields with well-drained soils with a pH 5.8–6.5. They should not use amendments that contain fluoride; using fresh soils or utilizing a minimum five-year field rotation with fumigation is advised.

To fertilize for flower production, use three applications of 2.2 lbs. (1 kg) of 5-10-10 per 100' (30 m) linear: (1) prior to planting, (2) as a side dressing one month after planting, and (3) when the floral spike becomes visible in the foliage. If the soil is very sandy or the season is very rainy, an additional application may be needed between applications two and three. For gladioli, watering must be carefully controlled. Gladioli should never be allowed to dry out, especially after the floral spike begins to grow out of the leaves. Thus, drip or overhead irrigation should be available.

Two planting systems are available. The first is planting the corms about 5–9" (13–23 cm) deep at a density of about 4–6 corms/1' (30 cm) linear. The second is planting the corms 2–3" (5–8 cm) deep, and, when they have produced six to seven leaves, hill the rows to provide additional support. The rows should be about 3' (91 cm) apart.

Gladioli are quantitative long-day plants; therefore, flowering is more prolific and faster as the days lengthen.

Plants flower in ten to fifteen weeks in ground beds in outdoor fields.

Pests & Diseases

Gladioli are susceptible to many soilborne diseases, e.g., *Fusarium, Stromatinia,* and nematodes. Therefore, the planting stock must be properly treated (a hot-water treatment in combination with fungicides) and either fresh soil or minimum five-year rotations of fumigated soils must be used. There are many other fungal, bacterial, or viral diseases that can affect corms, leaves, and/or flowers. In the mid-2000s, gladiolus rust was discovered in commercial production areas in Florida and subsequently in Southern California. Gladiolus rust (*Uromyces transversalis*) is a disease of international quarantine significance, and all gladiolus growers must be watchful for it and control weeds in fields. In addition, there are many insects, for example, armyworms, aphids, cutworms, loopers, and thrips, which can affect gladioli.

Varieties

There are many gladiolus cultivars available. The top twenty varieties in (corm) production in Holland are: 'White Prosperity', 'Priscilla', 'Traderhorn', 'Plum Tart', 'Peter Pears', 'Nova Lux', 'Hunting Song', 'Green Star', 'Jessica', 'Fidelio', 'Wig's Sensation', 'Advance', 'Prinses Margaret Rose', 'Jester', 'Spic And Span', 'Rose Supreme', 'Joyeuse Entrée', 'Bangladesh', 'White Friendship', and 'Oscar'.

Postharvest

Harvest stems with at least two to three leaves and with the florets in the tight bud stage of development (i.e., with one to five florets emerging and showing color). Gladiolus flowers are not extremely sensitive to ethylene, but exposure can cause the developing floral buds to abort. They can be treated with 1-MCP (EthylBloc). Gladioli are sensitive to fluoride, so only high-quality water should be used. In the field, the cut flowers should be immediately placed upright in either fresh water or a solution prepared with floral preservative containing 20% sugar (sucrose or glucose). As quickly as possible, they should be placed at 33–35°F (0.5–2°C) until graded and packed. Subsequently, they should be stored upright in fresh water to prevent stem bending and whenever possible transported at 33–35°F (0.5–2°C).

Gloxinia (see *Sinningia*)

Godetia

Godetia whitney (also called *Clarkia amoena*)
Common name: Satin flower
Annual, flowering pot plant, cut flower

Godetia, or satin flower, has been transformed from a unique and uncommon garden plant into a dependable cut flower for greenhouse and field production and for pot plant production. *Godetia* is a cool-season crop that is native to the North American West Coast from California to British Columbia, growing naturally within a few miles of the Pacific Ocean.

Propagation

Seed is the common method of propagation, but most growers buy in *Godetia* plugs in the late fall for pot plants and in the early spring for cut flowers. If you are planning to have plugs arrive for planting cut crops at Thanksgiving or Christmas, make sure they are coming from a cool climate, such as California, because seedlings do not tolerate warm temperatures in the plug stage.

Plug production is four weeks for cuts and six weeks for pots. Germination is usually uniform and yields high percentages. Select 288 trays or larger cells (128s or 72s) for pot production. A late December or early January sow date should result in plants ready to sell for Mother's Day in most of the Northeast and Midwest.

Single-sow seed into germination medium with little or no starter charge. Lightly cover seed with medium or vermiculite. Maintain uniform substrate moisture with intermittent mist, and maintain substrate temperatures of 70°F (21°C) for seven to ten days, until radicles have emerged.

For Stage 2, cotyledon and stem emergence, move trays to a cool, bright (500–1,000 f.c. [5.4–11 klux]), well-ventilated greenhouse; optimum temperatures are 55–60°F (13–16°C). At this stage, fertilizer applications directly affect the number of lateral branches. Use almost no fertilizer (beyond a small starter charge in the germination medium) if the plants will be grown as single-stem cut flowers. Fertilize pot crops and multiple-stem cut flower crops one or two times during plug production to get four to eight lateral branches per plant. Apply 50–100 ppm fertilizer from 14-0-14 one time. Do not use constant liquid fertilizer applications or the crop will be too soft. Do not use ammonium- or urea-based fertilizers. Supplemental lighting for six to twenty-four hours each day will greatly enhance plug growth in low-light areas. Stage 2 takes about three to four days.

For Stage 3, development of true leaves, continue to maintain cool temperatures and low fertilizer applications. Increase light to 1,000–2,000 f.c. (11–22 klux). The use of a negative DIF (55°F [13°C] days and 65°F [18°C] nights)—or a two-hour temperature drop of 5–10°F (3–5°C) at daybreak, followed by moderate day temperatures of 60–65°F (16–18°C)—is ideal for *Godetia* production. Weekly applications of B-Nine (2,500–5,000 ppm) can also be used to help control plant height for pot production. Cool temperatures and the wise use of water and fertilizer will make the best plugs. Stage 3 takes about six to seven days.

Harden plugs for one week prior to sale or transplanting by continuing to maintain temperature, fertilizer, and irrigation regimes. Increase light to 2,000–3,000 f.c. (22–32 klux).

Transplant plugs as soon as possible when they are ready.

Growing On

Cut flowers

Most field-grown *Godetias* are produced on the West Coast because uniform cool temperatures are common during production. Field production is possible in the upper Midwest and New England and at higher elevations with some experience. California growers harvest sequential crops from May into July from fall, winter, and spring sowings. Crop time and stem length decreases as average temperature, light intensity, and day length increases from spring into summer.

Outdoors, plant plugs in rows 5' (1.5 m) apart with in-row spacing of 2.5–3' (76–91 cm). This will yield large, individual plants with twenty to fifty branches and cut stem lengths of 14–20" (36–51 cm). For longer stems (20–36" [51–91 cm]), plant in double rows with a spacing of 9–12" (23–30 cm) between plants. Transplants are pinched at transplanting to remove the primary stem, leaving four to eight lateral branches. These rows must be supported in the field.

Outdoor production requires more fertilizer than greenhouse production, but still much less than most crops. Use low-fertilized soils with good drainage. Keep the soluble salts below 0.6 mS/cm. Organic fertilizers have been successful in field production in California. Be careful not to damage plugs at transplanting and do not plant too deeply.

In the greenhouse, *Godetia* is a spectacular cut flower crop with long stems and large flowers. *Godetia* must be grown with almost no fertilizer for the twelve to sixteen weeks the crop is in the greenhouse. Before planting, make sure ground beds are nearly fertilizer free. *Godetia* planted in beds following other crops or where bedding plants or pot crops were grown will probably fail due to the relatively high fertilizer residue in the soil. Overfertilized plants are soft, their stems are crooked, many lateral branches are along the stem, and plants are easily knocked down with watering and are difficult to support.

Space crops that will be pinched in ground beds at 1–2 plants/ft.^{2} (11–22/m^2). Pinched plants require two to three weeks more production time than single-stem plants. The crop will require one to two layers of support.

Take care when watering newly planted crops: Younger plants, eight weeks old to when buds are 1" (3 cm) long, are soft and easy to knock down with careless overhead watering. Always water the soil rather than the foliage. It is best to keep the plants on the dry side; however, wilted plants will have permanently crooked stems when they recover.

Godetia is a great candidate for cut flower production in pots. You can grow stems from 24–40" (61–102 cm) long from single-stem plants grown as 1 plant/4" (10 cm) pot or 2–4 plants/6" (15 cm) pot, with a plant density of 8–10 plants/ft.^{2} (86–108/m^2). Try not to set these pots on ground beds, where they will root into the soil below and deteriorate because of the fertilizer residue in the soil.

Supplemental lighting will speed flowering in the variety 'Grace' by four to ten weeks, depending on the color and time of year. Best winter greenhouse production will come from plugs grown under four to six weeks of HID supplemental lighting for twenty-four hours per day. After transplanting, provide long days by extending the day length (4 P.M. to midnight) with incandescent light.

Crop scheduling depends on temperature, supplemental lighting treatments, time of year (total light accumulated by the crop), and variety. In Kentucky, four-week-old plugs transplanted in late September flowered in mid-December (eleven to twelve weeks, for Christmas); in mid-October, they flowered in early February (fifteen to sixteen weeks, for Valentine's Day); and in early November, they flowered in mid-March (seventeen to eighteen weeks). These results all came when supplemental incandescent lighting was used for the whole crop in a greenhouse with 50°F (10°C) night and 60°F (16°C) day temperatures.

Pot plants

Ideally plant plugs when they arrive, however if that is not possible, hold plugs in a well-lit greenhouse at 40°F (4°C) until space is available for planting. Plant 1 plug/4" (10 cm) pot and 2–3 plugs/6" (15 cm) pot. Use a well-drained, sterile medium with a low starter charge. *Godetia* plants are sensitive; dislodge them from the plug tray by pushing up from the bottom. Avoid pulling the plants out of the tray by hand, which may damage the stem. Avoid planting the plug below the substrate line, to guard against stem rot and ensure a healthy transition.

Grow on in a cool greenhouse at 60–65°F (16–18°C) days and 50–55°F (10–13°C) nights, preferably with negative DIF, after transplanting. *Godetia* is as tolerant of cold temperatures as pansies, snapdragons, and dianthus are, making it an excellent candidate for outdoor pot production on rolling tables for some climates.

Although the growth retardants B-Nine, Cycocel, Bonzi, and Sumagic are effective, they are not necessary when *Godetia* is grown cool, dry, and with very low nutrition. Growing with a negative DIF (see comments under Propagation) is ideal.

Irrigation and nutrition practices are critical to pot *Godetia* production. Careless overhead watering will weaken the plants and open the plant canopy: Use subirrigation or drip irrigation, especially when plants are in flower. Overhead watering with strong water pressure will weaken the plants and open up the plant canopy.

Use very little fertilizer on the plants after transplanting. A single application of a quarter teaspoon (1.2 ml) of Osmocote 14-14-14 at transplanting is sufficient for 4" (10 cm) pots. If you prefer liquid feeding, use 50–100 ppm from 15-0-15 or 20-10-20 every fifteen days. (*Note:* Plants that appear undernourished during the middle of the production cycle will have stronger stems. Plants can be greened up in the last two weeks as the flowers begin to open.) Maintain a pH of 5.5–6.5.

Be sure individual plants always have sufficient space as they develop, to prevent stretching.

Godetia 'Satin' is responsive to day length, and using night interruption lighting from 10 P.M.–2 A.M. or extending the day length to midnight (six hours), using ordinary mum lighting (10 f.c. [108 lux] from incandescent bulbs placed on 6' [1.8 m] centers), will hasten flower development. Plants will flower in thirteen to fifteen weeks from sowing or nine to eleven weeks from transplanting a plug for spring production.

Pests & Diseases

Aphids, thrips, whiteflies, and spider mites can infest *Godetia* when these insects are problems on other plants in greenhouse or field production. Root- and stem-rotting diseases, such as *Fusarium* and *Pythium,* can be a problem when these plants are grown in poorly drained soils or when they are overwatered. Botrytis can cause leaf and stem damage on tightly packed, overfertilized plants in the greenhouse.

Varieties

The 'Grace' series of hybrid cut flower *Godetia* has 'Salmon', 'Red', 'Pink', 'Rose/Pink', 'Shell Pink', 'Lavender', 'Lavender Eye', 'White', and 'Mixed'. The 'Flamingo' series of hybrid cut flower *Godetia* has 'Lavender', 'Lavender Pink', 'Pink', 'Red', 'Salmon', and 'White with Rose Eye'. A number of open-pollinated varieties may be found from some suppliers, as well.

The 'Satin' series is a dwarf hybrid suitable for 4" (10 cm) and 6" (15 cm) pot production with 'Deep Rose', 'Lavender', 'Lilac Rose', 'Red', 'Red with White Edge', 'Salmon', 'Pink', 'White', and 'Mixed'.

Postharvest

Harvest when the first flowers on each stem open. Flowers do not open when stems are harvested prematurely in the tight bud stage. Because they ship poorly, *Godetia* is an excellent locally grown cut flower.

Cut stems have a vase life of fourteen to eighteen days, and all flower buds open to normal size and color when the flowers receive proper care. Research has shown that floral preservatives with sucrose will damage the leaves and reduce vase life. Cut stems in tap water performed equally or better than in preservatives in vase life trials.

Acclimated plants of *Godetia* can tolerate light frosts, permitting spacing outdoors in late April or early May, when greenhouse bench space is at a premium. For garden performance, *Godetia* does best under mild weather conditions. In areas where summer temperatures regularly exceed 80°F (27°C), plants will benefit and perform better if given shade during the hot afternoon.

Gomphrena

Gomphrena globosa
Common name: Globe amaranth
Annual, cut flower

Gomphrena is a great plant for those locations that have heat and humidity and is also drought tolerant. Globe-shaped flowers on medium-sized plants add interest planted en masse or individually. For growers, it is an excellent plant for 4" (10 cm) pots in the spring and 4 or 6" (10 or 15 cm) pots in the summer. *Gomphrena* is an easy-to-grow field cut flower that is very popular dried.

Propagation

Sow seed onto a well-drained, disease-free medium with a pH of 5.5–5.8. Cover lightly with coarse vermiculite after sowing to maintain 100% humidity at the seed coat. Germinate at 75–78°F (24–26°C) soil temperatures. Medium should be fully saturated. Provide ambient light levels of 100–400 f.c. (1.1–4.3 klux). Keep ammonium levels below 10 ppm. Radicles should emerge in two to five days. (*Note:* Some suppliers may offer scarified seed that will germinate more readily.)

Move trays to Stage 2 for stem and cotyledon emergence by lowering soil temperatures to 68–72°F (20–22°C). Increase light levels to 1,000–2,500 f.c. (11–27 klux) and continue to keep the medium fully saturated. Once all seedlings are up, begin to allow trays to dry slightly between irrigations. Once cotyledons are fully expanded, apply 50–75 ppm from 14-0-14 once a week. Stage 2 will take from seven to ten days.

Move trays to Stage 3 to develop true leaves by lowering soil temperatures to 60–65°F (16–18°C).

Provide air temperatures of 70–75°F (21–24°C) days and 60–65°F (16–18°C) nights. Allow trays to dry slightly between irrigations. Fertilize weekly with 100–150 ppm from 14-0-14, alternating with 20-10-20. Maintain EC below 1.0 mS/cm. If growth regulators are needed, B-Nine and Cycocel are effective. Stage 3 takes from twenty-one to twenty-eight days for 384s and twenty-eight to thirty-five days for 288s.

Harden plugs for seven days prior to sale or transplanting by lowering soil temperatures to 60–62°F (16–17°C). Allow trays to dry thoroughly between irrigations. Continue to maintain air temperatures. Fertilize once or twice with 50–75 ppm from 14-0-14.

Most growers purchase plugs, as *Gomphrena* is a minor crop and a few trays are all that most growers need.

The crop time from sowing to a finished 288-plug tray is five to seven weeks.

Growing On

Use 1 plug/1801 cell or 4" (10 cm) pot and 2–3 plugs/6" (15 cm) or 1-gal (4 l) pot. Plant into a well-drained, disease-free medium with a moderate initial starter charge and a pH of 5.5–6.2. Grow plants under full sun at 70–75°F (21–24°C) days and 62–65°F (17–18°C) nights.

Fertilize at every irrigation with 100–150 ppm from 15-0-15, alternating with 20-10-20. Overfertilized *Gomphrena* will be oversized and soft. If growth regulators are needed, B-Nine, Bonzi, and Cycocel are effective. The crop time from transplanting a 288 plug to a 4" (10 cm) is eight to ten weeks. Dwarf varieties will start to flower, but taller selections are sold green and will flower in the garden once temperatures warm.

Cut flower growers should establish plugs into 3" or 4" (8 or 10 cm) pots for six to eight weeks prior to planting in the field. Space 6–9" (15–23 cm) apart in rows 16" (41 cm) apart. Support is not required.

Pests & Diseases

Aphids, leaf miners, spider mites, thrips, and whiteflies can attack plants. Disease problems can include *Alternaria, Cercospora,* and rhizoctonia.

Varieties

For bedding plant and pot plant sales, try 'Gnome', a dwarf series available in 'Pink', 'Purple', 'White', and a 'Mix'. 'Buddy', available in 'Purple' and 'Rose', is also an excellent choice; however, 'Gnome' is much more uniform in production.

Try the 'Woodcreek' series available in 'Lavender' and 'Red' for cut flowers. 'Woodcreek Red' is very similar to the variety 'Strawberry Fields'. 'Woodcreek' grows to 24" (61 cm) tall. Also for cuts, try *G. haageana* 'QIS Carmine' with dark leaves, 10–18" (25–46 cm) stems, and very large flower heads.

Other taller-growing selections include the 'Audray' and 'Las Vegas' series; both growing to 20" (50 cm) tall. They are available in a number of colors. However, the tallest variety 'Fireworks' can be as tall as 4' (1.2 m). It has bright pink-purple flowers that are tipped with yellow.

Postharvest

Harvest cut flower stems that will be sold fresh when flowers are colored, but before they are fully open. Flowers that will be dried should be harvested when they are open. To dry, remove leaves, bunch and tie stems, and hang them upside down.

Gypsophila

Gypsophila paniculata
Common name: Baby's breath; gyp
Cut flower

Gypsophila, commonly called gyp throughout the trade, is the traditional filler flower for mixed bouquets and arrangements. It is ubiquitous: You will find stems of gyp everywhere flowers are sold worldwide.

For growers, gypsophila is a relatively easy crop, provided you get your start with clean plants. Pot plant and

Outdoor gypsophila production in Ecuador

bedding plant growers are increasingly taking advantage of gypsophila's popularity by offering it as a flowering pot plant and as a mixed combination plant in some areas.

Propagation

Gypsophila may be produced from cuttings or seed. Most gypsophila is started vegetatively, with the best gypsophila liners being produced from tissue culture, because crown gall (*Agrobacterium tumefaciens*) can be a serious problem of gypsophila. To produce clean plants, gypsophila suppliers must rigorously index their stock, multiply their plants through tissue culture, and maintain their lines in a clean-stock program. Plants grown from clean stock are healthy and vigorous. Some suppliers offer liners and plants grown from Stage 2 tissue culture.

Growers buy in liners or bare-root transplants from a reputable supplier. Cuttings can be rooted as you would carnations. Cuttings respond to an IBA dip prior to sticking. Root under mist at 70–72°F (21–22°C) substrate temperatures for ten to fourteen days.

Some gypsophila is propagated by seed. Sow seed into a well-drained, disease-free, germination medium with a pH of 6.5. Cover lightly with coarse vermiculite and germinate at 70–72°F (21–22°C). Germination will occur over five to ten days.

Growing On

Cut flowers

Most gypsophila grown as a cut flower is produced outdoors in fields. When gyp is grown in the greenhouse, flowering can be scheduled using lighting. In theory, gypsophila can be planted year-round; however, for field production summer and fall are the two major planting periods. Outdoor crops will flower most profusely with flushes in the summer and fall, with a smaller flush in the spring. Using rolling hoop houses to help warm temperatures and incandescent mum lights to extend the day length from the end of February until flowering, outdoor plants can be manipulated into flowering earlier. Indoor growers can also force a winter flush through manipulating light and temperature. For crops planted in the summer for fall flowering, provide a hard pinch as soon as possible after plants are rooted in.

Select only well-drained sites. Soils with a coarse, gravely texture are ideal. Adjust the pH to 6.5–7.5.

Choose fresh fields that have not been planted with gypsophila in the past, if possible, as the Danziger "Dan" Flower Farm in Israel warns that gypsophila plants themselves secrete toxic material into soil that can build up over time and harm crops. If you plan on harvesting for multiple years (gyp can be harvested for two to three years), space plants at 4' (1.2 m) apart. For plantings that will be renewed yearly, plant plants at 12–24" (30–61 cm) apart. Support is not required for greenhouse crops, although some outdoor growers in areas with frequent storms may use one layer.

Establish plants for seven to ten days in growing beds, by carefully monitoring irrigation and light levels to protect the new crop. Once established, gypsophila passes through various stages of growth: Vegetative (twenty to thirty days), flower induction and stem elongation/flower initiation (twenty to fifty days), and flowering (twenty to fifty days). In general, plants flower eleven to eighteen weeks after planting.

Day length, temperature, and plant age all interact to affect gyp flowering. Gypsophila is an obligatory long-day plant if plants were not previously cold treated. At higher temperatures under the long days and warm temperatures of summer, plants can flower quickly, but the resulting stems will be of low quality. The best quality is produced when plants are grown more slowly at cooler temperatures. Cold-treated cuttings or plants at 41°F (5°C) will flower regardless of photoperiod during forcing.

Vegetative growth, to bulk up plants to ten to twelve nodes for flowering, requires short days, high light, and cool temperatures (50–55°F [10–13°C]). This takes about five weeks. Plants that are not sized up prior to exposure to long days will flower erratically and stem quality may be poor. Plants induce flowering based on exposure to long days (thirteen to eighteen hours), each variety responds to differing day lengths. However, plants flower with the highest quality and most quickly at longer photoperiods (more than sixteen hours).

To force a spring-flowering crop, provide long days (more than sixteen hours) using night interruption lighting or continuous lighting from the end of February until flowering begins. Start using lights four to six days after pruning, or when the first new growth appears. To force crops planted or pruned in October, begin lighting seven to ten days after shoots start to grow. To force fall crops planted in September, provide long days (more than sixteen hours) using night interruption lighting or continuous lighting two to four weeks after planting. To force plantings made in July and August, begin lighting three to five weeks later. Shoots should be at least 12–16" (30–41 cm) long before discontinuing long days. However, ideally, continue to light until flowers have fully matured. Grow at night temperatures of 61–68°F (16–20°C) and in full-sun conditions. At lower light levels, reduce night temperatures.

Some growers prefer to use precooled plants to start their cut flower crops, planting from January to March. (Plants may also be dug from fields and stored in a cooler through the summer and replanted in the fall for winter-flowering crops in southern climates, such as Florida.) Plants that have been cooled for at least seven weeks at 32–34°F (0–1°C) flower independent of day length, although flowering is heavier and on stockier stems under long days. Using cooled plants, growers can produce crops indoors or outdoors year-round, regardless of day length. In warm climates, growers replace these plants each year.

During production, do not overwater gyp. However, some varieties, such as 'Million Stars', have more shallow root systems that require more frequent irrigation. Maintain the pH at 6.5–7.0 and provide 150 ppm from a balanced fertilizer at every irrigation. Reduce fertilizer to 100 ppm when flowering begins.

Some growers apply gibberellic acid from one to three times at 250–300 ppm to stimulate stem elongation. Plants are treated at one-week intervals in the field once the rosette is developed.

After harvest, dry plants down and prune back to encourage reflowering and provide long days as outlined above. Prune shoots back to 1–2" (3–5 cm), but remove thin stems down to the ground. Some growers leave plant debris in fields for several days to help shade newly pruned plants during hot weather, while others provide temporary shade netting.

Pot plants/perennials

Gypsophila also makes a great perennial in most parts of the United States, provided plants are put in a well-drained location and beds have been adequately limed.

Buy liners in the fall or winter and pot them into 1-qt. (1.1 l) or 1-gal. (4 l) containers. Select a well-drained, disease-free medium with a moderate initial starter charge and a pH of 6.5–7.0. Use larger pots as opposed to smaller ones so that the finished product will be proportional.

Plants flower faster after a cold treatment. Before cold treatment, bulk plants to eight to ten nodes under 65–70°F (18–21°C) days and nights. Grow under moderate to high light, providing 3,000 f.c. (32 klux) and short days (ten to twelve hours). Fertilize lightly with 75–100 ppm from 15-0-15. Allow pots to dry thoroughly between irrigations.

Once plants are bulked up to ten nodes, provide twelve to fifteen weeks of cold treatment at 41°F (5°C) in a cooler (25–50 f.c. [269–538 lux]) or cold frame. After cooling, plants will force in about twelve weeks at 65–68°F (18–20°C) days and nights. Grow under full sun (4,000–6,000 f.c. [43–65 klux]) and light continuously using chrysanthemum lighting. Feed with 150–200 ppm from 15-0-15 at every other irrigation. If height control is needed, A-Rest, Bonzi, and Sumagic are all effective.

Gypsophila can also be grown without forcing and flowered under natural days in early summer. In such a case, you can receive liners in the winter and sell pots green or flower in the spring, depending on the variety.

Pests & Diseases

A number of pests attack gypsophila. Crown gall (*Agrobacterium tumefaciens*) can become a serious problem. Plants are also susceptible to *Pythium* and rhizoctonia. Aphids, armyworms, leaf miners, spider mites, and thrips may attack plants.

Troubleshooting

Flowers may become brown when they are too mature under the long days and high temperatures of summer. Plants may fail to flower if they were not cold treated or received too few long days. Growing gypsophila under low light can also cause lack of flowering as well as weak, spindly stems. Flowers and branches may be distorted if they were produced under high temperatures and high humidity.

Varieties

'Million Stars', an extremely prolific small-flowering selection from Danziger in Israel, has swept the gypsophila community off its feet. As a cut flower filler, the variety is so much improved over previous selections that it is often sold by name in the trade.

'Perfecta' has large, 0.25" (6 mm), full, white flowers with somewhat lower production than the old mainstay, 'Bristol Fairy'. Flowering is also a little later and takes place during a shorter time period. However, its large flower size generally brings higher prices. 'Bristol Fairy' has small, full, white flowers. In decades past, it was *the* gypsophila until the larger-flowered 'Perfecta' ousted it.

'Flamingo' is a double, pink variety, while 'Pink Fairy' is a light pink double. 'Snowflake' is a seed-grown gypsophila. However, plants are not as uniform or robust as vegetatively propagated varieties, so it is rarely grown.

G. repens 'Rosea' grows to 5" (13 cm) tall, making it a wonderful groundcover perennial. The annual gyp, *G. muralis,* is available as 'Garden Bride' with pink, single flowers and 'Gypsy' with double, pink flowers.

Postharvest

Harvest stems when 25–50% of the flowers are open, but not too mature. Some growers shipping long distances harvest when only 5% of flowers are open—retailers then open these flowers at 70°F (21°C) under a minimum of 100 f.c. (269 lux) using special gypsophila-opening postharvest solutions.

Immediately after cutting, place harvested stems in water. Gypsophila is very sensitive to ethylene. Grade, bunch, and place stems in buckets with a postharvest solution containing silver thiosulfate (STS), sucrose, and a bactericide. Refrigerate open flowers at 35°F (2°C). Gypsophila may be dried (harvest flowers in a more open state for drying). Stems also readily take up dyes.

Retailers should unpack gypsophila as soon as possible, recut stems, and place them in clean buckets filled with preservative solution. They should refrigerate cut flowers at 35–40°F (2–5°C).

Pot plant gypsophila is ready to sell once flowering has begun. Plants are sensitive to ethylene, so boxing and long-distance transportation may not be practical, making gypsophila a great niche product for local growers. Plants are responsive to 1-MCP, which will reduce the harmful effects of ethylene.

Retailers should keep plants in flower well watered, although not wet, and display them in cool to moderate temperatures (65–70°F [18–24°C]) under high light. Instruct consumers to plant gypsophila in limed soils that are slightly alkaline and have excellent drainage. Consumers should cut plants back after flowering to encourage reblooming.

H

Hedera

Hedera helix
Common name: English ivy
Perennial (Hardy to USDA Zones 5 or 6–9, depending on variety)

As the quintessential groundcover and wall covering, hedera covers everything and grows from 3–5" (8–13 cm) tall but trails with vigor. Its flowers are insignificant, while its foliage personifies the species. Hedera enjoys shade or morning sun to afternoon shade in well-drained locations.

Propagation

Tip or stem cuttings are the most common method of propagation. Stick one to two cuttings per 50-liner tray or one cutting per smaller liner-cell size. Dip cuttings into rooting hormone and provide mist for seven to ten days. However, reduce the amount of mist as you get closer to day 7. Upon rooting, pinch off tip cuttings and allow six to eight weeks to sale or transplant to larger containers.

Growing On

Rooted liners can be transplanted to 1-qt. or 1-gal. (1.1 or 4 l) containers and will be salable in eight to ten weeks for 1-qt. (1.1 l) containers and ten to twelve weeks for 1-gal. (4 l) containers. Grow on at 65–70°F (18–21°C) days and 56–60°F (13–16°C) nights. Maintain a pH of 5.8–6.5 and light levels of 2,000–8,000 f.c. (22–86 klux).

One to two additional pinches will help to increase the number of branches. However, if individual stem length is more important, then do not pinch at all. When growing in pots (instead of hanging baskets), growing branches start to weave through serrated greenhouse benches and have to be pulled or ripped out if the pots have not been regularly picked up and repositioned.

Pests & Diseases

Mites are a common issue, while leaf spot is the most common disease that affects hedera.

Helenium

Helenium amarum
Common name: Sneezeweed; tick weed
Annual

Highlighted by the variety 'Dakota Gold', this North American native grows 12–14" (30–35 cm) tall and 15–16" (38–40 cm) wide with bright, golden, daisy-like flowers and fine-textured, green foliage. With a short crop time, particularly under high temperature and long day conditions, each plant has a compact, mounded habit, making it great for mass plantings or as a border plant. Tough and vigorous, this annual does well in high heat, with either high or low humidity, but also tolerates cool conditions. It's a perfect choice for landscape plantings!

Propagation

Sow seed into a well-drained, disease-free, soilless medium with a pH of 5.5–6.3 and a medium initial nutrient charge (EC 0.75 mS/cm with a 1:2 extraction). Recommended tray size is 288–128 cells (with a multiseeded pellet). Cover the seed with vermiculite for added humidity around the seed for germination. Seed germinates in three to five days at 65–75°F (18–22°C). Light is not required for germination. Maintain 95% relative humidity until cotyledons emerge. Keep soil moisture high until radicle emergence, and then reduce moisture levels after the radicle penetrates the medium. Do not allow the seedlings to wilt.

Drop the temperature in Stages 2 and 3 to 68–72°F (20–22°C). Feed with 15-0-15 at 50 ppm in Stage 2. Increase feed to 100–150 ppm during Stages 3 and 4. Alternate with 20-10-20 and clear water to deliver a balance feed program in the plug development. No growth regulators are required.

A 288-plug tray is ready to transplant to larger containers three to five weeks after sowing.

Growing On

Transplant into a well-drained, disease-free, soilless medium with a pH of 5.5–6.5 and a medium initial nutrient charge. Grow on the warm side until up to size. Then cool down and finish. Provide 65–70°F (18–21°C) days and 64–66°F (18–19°C) nights.

Plants grow better in heat. Higher temperatures result in a faster crop time. They also like high light. Grow under as high light as possible. Plants will flower regardless of day length, but the growing habit is quite related to day length. Plants grow slowly when grown under day lengths shorter than twelve hours and become very flat or even rosette when grown under day lengths shorter than ten hours. Growing plants under long days (twelve hours or more) is recommended.

Avoid both excess watering and drought. Feed plants weekly with 200 ppm nitrogen (N) in a

Table 15. Weeks from Transplant for Different Container Sizes*

CONTAINER SIZE	PLANTS PER POT (multiseeded plug)	WEEKS FROM TRANSPLANT
Premium 306 pack	1	5–7
4.5" (11 cm) pot	1	6–8
6" (15 cm) pot or 1-gal. (4 l) container	3	6–8

* This crop time is based on a 68°F (20°C) daily average temperature. When plants are grown in warm tem peratures, crop time can be two or more weeks shorter.

complete fertilizer. Plant growth regulators are generally not required. If necessary, a B-Nine spray at 5,000 ppm can be used about two weeks after transplant for height control. This treatment can also prolong shelf life. Pinching is not required.

Pests & Diseases

Aphids and powdery mildew can be problems for this crop.

Troubleshooting

Grow crop on the warm and dry side.

Helenium autumnale
Common name: Sneezeweed
Perennial (Hardy to USDA Zones 3–8)

Vigorous and erect, *Helenium autumnale* grows semi-columnar, reaching 4–5' (1.2–1.5 m) in height and spreading no more than 2' (60 cm) wide. The daisy-like 2" (5 cm) flowers are single, scentless, and yellow, apricot, reddish-brown, orange, or combinations of these colors. Plants flower in August or September.

Propagation

Seed, tissue culture, cuttings, and bare-root transplants are commonly available.

Seed

Seed can be sown in late spring or summer and transplanted to larger containers for overwintering dormant or sown in late autumn to early winter for growing on cold for spring sales. Seed-sown plants, regardless of whether sown the previous year or in January, will produce flowering plants in the garden late in the upcoming summer.

For a 288-plug tray, sow one to two seeds per cell and leave seed exposed or lightly covered with vermiculite. Germinate at 70–75°F (21–24°C). Seedlings emerge in eight to twelve days.

From sowing to a salable or transplantable plug takes four to six weeks.

Cuttings

Stick one cutting per 50, 72, or similar tray and root at 70–72°F (21–22°C). Cuttings are taken anytime the mother stock is vegetative and not in flower. 72-liner trays are available six to eight weeks after sticking (depending on variety).

Bare-root transplants

Bare roots potted up to 1-gal. (4 l) containers in February or earlier and grown at daytime temperatures of 65–68°F (18–20°C), with 5°F (3°C) cooler nights, will root and begin to grow on the greenhouse bench during late February and March. By early to mid-April they can be moved to a cold frame and kept at 50°F (10°C) nights. Plants are sold green in the spring.

Growing On

In all cases, seedling-sown plugs, vegetative liners, or bare roots received in January or February, potted up to 1-qt. or 1-gal. (1.1 or 4 l) containers can be sold as green plants in the spring. When planted to the garden in May or June, these plants will flower naturally in late summer.

Helianthus

Helianthus annuus
Common name: Sunflower
Annual, cut flower, pot plant

There was not one person in the floriculture industry that believed that the sunflower craze, which began in the mid-1990s, was more than a flash in the pan. But sure enough, it has lasted. Today sunflowers are permanent fixtures in garden centers beginning in May and wherever cut flowers or pot plants are sold almost year-round.

To give credit where credit is due, new breeding that introduced fun bicolors, pot plant varieties, and multiple-flowering stems have helped to keep interest heightened.

If you are a retail grower, plan on having a few fresh pots of sunflowers every week from spring until midsummer. They are easy to grow, and your customers will love them.

Pot Plant Crops

Propagation

Sunflowers have a deep taproot. You will have the most success by direct-sowing seed into the final growing container. If you choose to grow plugs, be sure to use large, deep cells, such as 200s, but be sure to transplant to larger containers when ready. As the seedlings develop, the lower foliage yellows once the root has established. Transplant the plug as soon as possible, or the final plant will be shorter in overall height and plant performance in pots, gardens, or flowerbeds will be diminished.

Sow one to five seeds per 6" (15 cm) pot. Select a well-drained, disease-free medium with a moderate initial starter charge and a pH of 6.5–7.0. Cover seed lightly after sowing. Germinate at 70–75°F (21–24°C) substrate temperatures. As seedlings germinate, gradually increase light levels from 400–2,500 f.c. (4.3–27 klux). Begin fertilizing weekly with 50–75 ppm from 14-0-14 once cotyledons are fully expanded. As true leaves emerge, increase the rate to 100–150 ppm weekly and begin alternating with 20-10-20. Thin seedlings to three plants per pot. Some growers prefer to sow seed into cell packs and then transplant into pots, but your fastest crop will come from direct-sowing. Grow seedlings at 70–75°F (21–24°C) days and 65–68°F (18–20°C) nights.

Growing On

Maintain growing temperatures as previously outlined; warmer temperatures can cause stems to stretch. Increase light levels to 3,000–5,000 f.c. (32–54 klux), as seedlings become five weeks old.

Keep plants well irrigated to promote active growth. Fertilize with 200–250 ppm from 14-0-14, alternating with 20-10-20 once seedlings are fully established. Discontinue fertilizer seven days prior to flowering to extend postharvest life.

Growth regulators can be helpful in controlling height. B-Nine at 4,000–8,000 ppm and Sumagic at 16–32 ppm are effective as sprays. A Bonzi drench at 2–4 mg of active ingredient (a.i) is also effective.

The crop time for flowering is ten to fourteen weeks, depending on variety. Single-flower selections flower the earliest, while dwarf selections flower first and taller ones later. The double-flowering selections take the longest to flower, requiring from twelve to fourteen weeks depending on height.

Pests & Diseases

Diseases such as *Alternaria,* botrytis, *Pythium, Phytophthora, Septoria,* and *Verticillium* can be problematic for sunflowers.

Insects such as caterpillars, thrips, and whiteflies can attack plants.

Varieties

For pot plants, try 'Miss Sunshine', 'Ballad', 'Big Smile', 'Pacino', 'Sundance Kid', 'Sunspot', and 'Teddy Bear' (double). 'Pacino' makes an especially great pot, because it produces four to six secondary flowers in addition to the main flower.

Postharvest

Sell plants once roots have reached the bottom of the pot and plants are budded. Once buds appear, plants will be in full flower in five to ten days. Sunflowers will bloom for fourteen to twenty-one days. One test has shown that plants can be held when flowers begin to show color at 42°F (5°C) for a week. However, before attempting on a wide scale, conduct your own test.

After flowering, sunflower seeds ripen and flowers dry up and fall away. Many birds, especially goldfinches, love sunflower seeds. Because sunflowers do not like to be transplanted, consumers should plan on enjoying the flowering pots as they are, placing them in a spot where birds can enjoy them too once the flowers are spent.

Cut Flower Crops

Propagation

Sow seed directly into growing beds, as plants do not like to be transplanted and those transplanted from plugs will not reach a tall height. Place seed 9–10" (23–25 cm) apart in rows spaced 18" (46 cm) apart. If you choose to transplant container-grown plants, use deep plug trays and sow no earlier than three to four weeks prior to planting out.

Soil beds should have a neutral pH from 6.5–7.5. Based on soil tests, incorporate a starter fertilizer and nitrogen, phosphorous, and potassium, as required. Place fertilizer 5" (13 cm) to the side of seed and 2–3" (5–8 cm) below seed so that seedlings are not burned at germination. Seed can also be sown outdoors to fields once soil temperatures are 46–50°F (8–10°C). Be sure to buy enough seed for subsequent weekly sowings to ensure fresh stems continuously each week.

Growing On

Provide one or two levels of support that can be raised as plants grow. Use drip irrigation so that foliage and flowers will remain dry. Do not cause water stress in plants, as this will slow flowering.

Sunflowers require good fertilization. In addition to the starter nutrients, every four to six weeks apply most of the nitrogen fertilizer (10-10-10) as a side dressing while plants are actively growing, beginning when they are 1' (30 cm) tall. Greenhouse-grown crops can be fertilized with constant liquid feed at 200–250 ppm from 14-0-14, alternating with 20-10-20.

Sunflowers grow and flower fastest at warm temperatures: 70–75°F (21–24°C) days and 65–68°F (18–20°C) nights. Avoid temperatures below 50°F (10°C).

Varieties

For cut flwer varieties, try 'Full Sun', a golden-yellow F_1 with 6–8" (15–20 cm) wide flowers! 'Ring of Fire' is an All-America Selections Award winner with bicolored yellow and mahogany blooms. Its stems are heavily branched and flowers measure 5–6" (13–15 cm) across! 'Sunbright Supreme' is a bright yellow, single-stemmed variety. The 'Sunrich' series, available in a number of separate colors, is single-stemmed and grows 5–6' (1.5–1.8 m) tall.

Postharvest

Stems are ready to harvest when petals are just beginning to show color. Harvest stems in the early morning or evening, when temperatures are cool. Cut stems as long as possible. Strip excess foliage in the field and place stems in clean buckets filled with water and a commercial preservative solution. Refrigerate at 36–41°F (2–4°C). Select pollen-less varieties for the longest vase life.

Helichrysum

(see also *Bracteantha, Plectostachys*)

Helichrysum petiolare
Annual

What did the world do before the bedding plant industry discovered the silvery foliage of *Helichrysum*? This versatile plant provides the perfect backdrop for just about any colorful flower in hanging baskets and combination pots and baskets. *Helichrysum* is an easy greenhouse crop.

Propagation

Root cuttings in Ellepots or rooting medium under mist at 68–72°F (20–22°C) soil temperatures and air temperatures of 75–80°F (24–26°C) days and 68–70°F (20–21°C) nights. Provide 500–1,000 f.c. (5.4–11 klux) of light and apply a foliar feed of 50–75 ppm from 20-10-20 as soon as there is loss of foliage color. Keep rooting medium uniformly moist. Callus will form in five to seven days. Root cuttings under mist.

Once half of the cuttings have begun to develop root initials, move cuttings to the next stage for root development, which takes from seven to fourteen days. Maintain soil and air temperatures. Increase light levels gradually to 1,000–2,000 f.c. (11–22 klux) as cuttings foot out, and begin drying out medium by reducing mist frequency. Apply a foliar feed of 100 ppm from 15-0-15, alternating with 20-10-20, and increasing the rate to 200 ppm as roots develop. Harden cuttings for seven days prior to transplanting by increasing light levels to 2,000–4,000 f.c. (22–43 klux) and removing cuttings from mist. Shade plants during the heat of the day to lower temperature-related plant stress. Fertilize with 150–200 ppm from 15-0-15, alternating with 20-10-20 once a week. Some growers treat with Florel as cuttings harden off to encourage subsequent branching.

Sticking one cutting per 72 tray takes from three to four weeks to be ready to transplant to larger containers.

Growing On

Use 1 liner/4" (10 cm) pot, 1–2/6" (15 cm) pot, and 3–4/10" (25 cm) hanging basket, although plants are most often used in mixed combination containers and not singly in a basket. Pinch 4" (10 cm) pots once when roots have reached the container sides. *Helichrysum* is a heavy feeder. As plants mature, increase fertilizer rate to 200–300 ppm from 15-0-15, alternating with 20-10-20. If iron chlorosis becomes a problem, apply iron chelate. Florel is effective in causing branching, but the last application should occur eight weeks prior to sale. Bonzi applied as a drench is also effective.

A 72-liner tray transplanted to 4" (10 cm) pots takes from seven to eight weeks to be ready for sale.

Pests & Diseases

Aphids, caterpillars, leaf miners, thrips, and whiteflies can attack plants. Diseases that can strike include *Alternaria,* botrytis, *Pythium,* and rhizoctonia.

Troubleshooting

Plants may collapse if the medium is kept wet for an extended period of time. High ammonia concentration, low light and/or overwatering, and excessive or late applications of Florel can cause excessive vegetative growth. Poor plant branching may be due to low fertilizer. Phosphorous toxicity may result in chlorotic yellow or red older foliage or bleached-out young foliage. Low light can cause plants to stretch.

Varieties

'Licorice' comes in 'Splash' with variegated leaves, 'Petite' with smaller leaves, and 'White' with more silvery leaves, as well as plain 'Licorice'. 'Minus', like 'Petite', is a small-leafed version. 'Limelight' is bright green.

Related Species

Helichrysum thianschanicum is better known in the trade as 'Icicles', 'Spike', and 'Silver Spike'. Plants feature soft, needlelike foliage on mounding plants that grow to 12" (30 cm) in the garden.

Postharvest

Retailers must keep *Helichrysum* properly watered: Avoid wilting, but do not keep plants wet! Keep foliage dry in the retail setting to avoid foliar disease problems.

Heliopsis

Heliopsis helianthoides
Heliopsis helianthoides var. *scabra*
Note: Most of the cultivars sold today are listed under *H. helianthoides.* However, the seed variety sold as 'Summer Sun' is *H. helianthoides* var. *scabra.*
Common names: False sunflower; hardy sunflower; sunflower heliopsis
Perennial (Hardy to USDA Zones 3–9)

This excellent, native perennial has a long flowering period, blooming from June to September here in the Midwest. It grows between 3 and 5' (90 and 152 cm) tall and has a 15–24" (38–61 cm) spread. Stems grow erect, without branching, from a central crown. As many as ten stems develop, and each one produces a flower canopy at its terminal point. The dark-green foliage measures up to 4" (10 cm) long. The scentless flowers are yellow, single to semi-double, and 3–4" (7–10 cm) wide.

Propagation

Seed

Heliopsis will flower during summer from sowings made anytime during the winter (up until late March), although earlier sowings produce larger plants and more profuse flowering. Sowings after April 1 will flower during the late summer, however flowering is so late that the color cannot be enjoyed for long before the start of cool weather.

If using 288-plug trays, use one to three seeds per cell. Either do not cover or lightly cover with vermiculite to maintain moisture. Plugs will be ready to transplant or sell four to six weeks after sowing.

Bare-root transplants

There are a number of cultivars of *Heliopsis* that are available. Pot up to 1-gal. (4 l) pots no later than early February for sales in May.

Growing On

Seed

Heliopsis grows upright without side or basal branching. If the plants start growing too fast and become too tall, pinch them back to the second node. Usually two side shoots develop, which helps to make a larger crown and a more uniform plant as well as one that performs better in the garden. The habit and flower power are much improved the second year after planting in the garden. Do not grow the plants below 58°F (14°C) as this will cause yellow foliage, which will fall off as cool conditions continue. Keep night temperatures above 60°F (16°C) for best performance.

Transplant 288-plug trays to 1-gal. (4 l) pots in January or February for well-rooted plants by April or May. These will flower upon planting to the garden in summer.

Bare-root transplants

Grow on warm as noted for seed-propagated crops. Pinching and plant growth regulators are not necessary. Pots are salable in eight to ten weeks with well-rooted, 1-gal. (4 l) containers.

Troubleshooting

Warm temperatures are key for *Heliopsis*. Plants do not appreciate cool temperatures and go dormant when grown below 50°F (10°C)

Heliotropium

Heliotropium arborescens
Common name: Heliotrope
Annual

The sweetly scented flowers of heliotrope are well known among long-time gardeners but are not seen nearly enough at retail or in greenhouses. This easy-to-grow crop is perfect for retail and wholesale growers catering to the landscape trade or upscale garden centers.

Propagation

Heliotrope may be propagated by seed or cuttings.

Seed

Sow seed onto a well-drained, disease-free, germination medium with a pH of 5.5–5.8 and an EC of less than 0.75 mS/cm. Cover lightly with coarse vermiculite. Media temperatures should be 68–72°F (20–22°C). Light is not necessary for germination but is essential once radicles emerge. Keep the medium moist, but not saturated. Radicles will emerge in three to seven days.

Once 50% of seed has germinated, move trays to Stage 2 for stem and cotyledon emergence. Reduce substrate temperatures to 65–68°F (18–20°C). Begin fertilizing with 50–75 ppm from 14-0-14 as soon as cotyledons have expanded.

Move trays to Stage 3 to develop true leaves by maintaining soil temperatures and increasing fertilizer to 100–150 ppm once per week from 20-10-20, alternating with 14-0-14. Allow trays to dry slightly between irrigations. Stage 3 will last from twenty-eight to thirty-five days for 288s. If growth regulators are required, Cycocel is effective.

Harden plugs off for seven days prior to sale or transplant. Lower soil temperatures to 60–65°F (15–18°C) and increase light levels to 2,000–3,000 f.c. (22–32 klux). Fertilize once per week with 100–150 ppm from 14-0-14 as needed.

For 288-plug trays allow six to eight weeks from sowing for plug trays to be ready to transplant to larger containers.

Cuttings

Heliotrope cuttings are relatively easy to root. Stick cuttings in medium or Oasis wedges. Root cuttings under mist at substrate temperatures of 70–72°F (21–22°C) and air temperatures of 70–72°F (21–22°C) during the day and 68–70°F (20–22°C) at night. Provide 500–1,000 f.c. (5.4–11 klux) of light. Apply a weekly foliar feed of 50–75 ppm from 20-10-20 once leaves begin to lose color. If growth regulators were not used during stock plant production, apply Cycocel as soon as cuttings are turgid. Once half of the cuttings have differentiated root initials (about one week), move them to Stage 3 for root development.

Maintain the same soil and air temperatures for root development. As root initials form, begin to dry out soil, but do not allow air humidity levels to drop. Increase light levels to 1,000–2,000 f.c. (11–22 klux) as cuttings root. Apply a foliar feed of 100 ppm from 15-0-15, alternating with 20-10-20. Increase the rate to 200 ppm as roots form.

Harden cuttings for seven days prior to sale or transplant by reducing air temperatures to 70–80°F (21–26°C) days and 65–68°F (18–20°C) nights. Move liners from the mist area and increase light levels to 2,000–4,000 f.c. (22–43 klux). Shade plants during midday to reduce temperature stress. Fertilize with 150–200 ppm from 15-0-15, alternating with 20-10-20 once a week.

The crop time for a cutting from sticking to a 72 tray to a rooted liner is seven to eight weeks.

Growing On

Transplant 1 plug or liner/4" (10 cm) pot and 1–2 plugs or liners/6" (15 cm) pot. Select a disease-free, well-drained medium with a moderate initial starter charge and a pH of 5.5–6.2. If you are growing vegetatively propagated heliotrope, use a medium with a high initial starter charge. Monitor pH levels, as heliotrope can decrease the media's pH. If leaves begin to roll, increase pH levels.

Grow plants on at 70–75°F (21–24°C) days and 65–68°F (18–20°C) nights. Fertilize with 150–200 ppm at every other irrigation with 15-0-15, alternating with 20-10-20. As plants mature, rates can be increased to 200–300 ppm. Do not add iron to the fertilizer solution. Excessive ammonium fertilizers can lower pH and promote iron uptake. Do not allow the medium to remain wet, as plants may show marginal leaf necrosis.

Grow under high light levels (4,000–7,000 f.c. [43–75 klux]). Heliotrope flowers faster as days lengthen naturally in the spring and as light intensity increases. At photoperiods of less than ten hours, flowering is inhibited. Low light intensities increase crop time and cause stems to stretch.

Pinch pots once plants reach the six-leaf stage to promote more uniformity. Leave 1–1.5" (3–4 cm) of growth above the substrate line. Many of the newer varieties do not require pinching, although a pinch will produce a more uniform, rounded plant. Cycocel at 500–1,000 ppm is effective for height control and may be used two or three times during production, depending on light intensity and plant growth.

The crop time for a 288-plug tray to finish in a 4" (10 cm) pot is eight to nine weeks. The crop time for a 72-liner tray to finish in a 4" (10 cm) pot is six to eight weeks. In both cases plants will be budded and a few will be in flower, depending on variety. A few more weeks are required for all plants to be in flower, if that is desired.

Pests & Diseases

Aphids, spider mites, thrips, and whiteflies can attack plants. Disease problems that can strike include botrytis, powdery mildew, *Pythium,* rhizoctonia, and leaf spot.

Troubleshooting

Plants may collapse due to wet medium, botrytis, or iron toxicity from low pH. High ammonium levels in the medium, overfertilization under low light conditions, and/or low light and overwatering can lead to excessive vegetative growth. Low fertilization during early growth stages can result in poor branching. Plants drying out between irrigations, high soluble salts, iron toxicity, or low substrate pH can cause foliage necrosis. Leaf curl may be due to low substrate pH or iron toxicity.

Varieties

For seed varieties, 'Marine' (also available as rooted cuttings) is the standard selection, with fragrant, deep blue flowers on plants that can grow from 15–20" (38–51 cm) tall, although some plants can grow as tall as 24" (61 cm).

There is a broader range of vegetative varieties. Be sure to review plant company catalogs for further details.

Postharvest

Display plants under light shade and cool temperatures. Do not allow plants to wilt, yet avoid keeping pots wet. Advise consumers to plant heliotrope in well-drained beds that are in partial shade to full sun.

Hemerocallis

Hemerocallis hybrids
Common name: Daylily
Perennial (Hardy to USDA Zones 3–9)

Daylilies are the second best-selling perennial, right after hostas. Their foolproof performance for gardeners and landscapers make them a base perennial in many landscapes. For growers, daylilies are easy.

Although typified with yellow flowers, there are a multitude of other available colors. These include russets, reds, roses, pinks, apricots, oranges, whites, purples, and various bicolored combinations of all these colors.

Propagation

Most growers buy in bare-root divisions, sometimes called fans, which are dug in late summer or early fall by specialist propagators. Daylilies are propagated primarily by dividing fans in early spring or fall.

Your bare-root supplier may have started its crop with tissue-cultured plants instead of dividing plants. Ask. Reliable daylily plant suppliers will grow out tissue-cultured plants in the field for a couple of seasons, rouging off-types and ensuring that only plants true to name are supplied.

The fans most often sold are one or two years old. One-year old plants will not flower readily their first year. (Daylilies coming from Europe will be smaller, one-fan divisions because they must be washed to have all dirt removed due to phytosanitary regulations.) If you are looking to sell plants in flower, buy two-year old plants (one to two fans), which will fill out a 1-gal. (4 l) pot nicely. Because of the time involved in propagation, the expense of maintaining clean stock, and the field space required to grow out plants (not to mention the labor and equipment involved to dig, sort, and divide plants), most growers leave starter plant production to specialists.

With all this said, there are a handful of perennial producers that have plenty of land and are growing their own divisions. But for most growers it is too expensive and time consuming. If you are not trying to offer the latest, greatest varieties, fans can be very inexpensive to buy in from a number of reputable suppliers.

Growing On

Ideally, plant fans immediately upon arrival. If that is not possible, open boxes and store them dry and cool at 35°F (2°C) for up to seven to ten days. If you buy in daylily liners, do not hold plants so long that they become root-bound. Most daylilies will require a 1-gal. (4 l) pot, due to their size. Some dwarf varieties may be able to go into 1-qt. (1.1 l) or 5" (13 cm) pots if you prefer.

Use a well-drained, disease-free medium with good moisture-holding capacity and long-lasting components. Media pH should be 5.5–6.5 with a moderate to high initial starter charge. Spread roots out in the pot and plant fans deep enough so that 0.5–1.0" (1–3 cm) of the medium will cover the crown. Water plants in immediately after planting so that the medium settles in around the roots.

For the first four to six weeks, do not fertilize: Wait for plants to sprout, then use 100–150 ppm 15-0-15 once or twice as roots start to wrap around the inside of the container. Until plants are well rooted in, maintain uniform moisture in the pots. Once plants are established, they can wilt slightly between irrigations. When plants begin to flower, maintain consistent moisture to increase flower size and substance.

If you are buying in bare roots and planting pots to overwinter in an unheated cold frame, be sure plants are well rooted in for six to eight weeks before freezing temperatures begin. Fall-planted pots will be much fuller and flower much better than winter-planted pots will.

However, plants brought in and potted up in the winter can be grown warm until established (60–65°F [16–18°C] nights) in the pot and then moved to a cold frame once days get longer with cool 50–55°F (10–13°C) nights. Roots potted up in February will produce salable green plants in late May. However, the pots will not be fully rooted. Daylilies hold well in containers. If not sold in May, the plants can be left in the pot, and during the warmer days of June and July, the plants will grow more quickly and begin to flower. The crop time from potting roots to salable plants is ten to twelve weeks.

Note that evergreen daylilies do not overwinter well in containers, and losses can be significant. Even then, plants that make it through the winter look unattractive in the early spring, as leaves look beat up and yellow, giving the impression that the plants are unhealthy to retailers and consumers alike. Conversely, dormant varieties produce fresh, new foliage each spring, giving a nicer presentation for retail sales.

A cold treatment may be necessary for the newer, ever-blooming varieties, which appear to form their flowers the year before and require cold temperatures followed by high light and long days to flower. Provide a minimum of fifteen weeks at 41°F (5°C) with at least 25–50 f.c. (269–538 lux) of light. Plants may be cooled in a cooler with lights or an unheated cold frame.

Pests & Diseases

Aphids, mites, and thrips can become pest problems. Diseases rarely attack daylilies.

Varieties

Daylily varieties are classified as evergreen, semi-evergreen, and dormant, as well as by dwarf, low, medium, and tall sizes. Bloom time is another classification: early, midseason, or late.

Evergreen varieties have foliage that persists through the winter months in southern locations. The foliage on dormant and semi-evergreen varieties turns brown as winter approaches and needs to be hand pulled or trimmed. Foliage type is a rough indicator of daylily hardiness, with evergreen types performing better in southern climates and dormant types performing better in northern climates, where they receive enough cold during the winter months. Do not attempt to grow dormant varieties in the Deep South or evergreen varieties in far northern regions. Exceptions to this general rule are 'Chorus Line', 'Lullaby Baby', and 'Pandora's Box'; all three perform well in the North or South.

There are thousands of daylily varieties: Daylily enthusiasts and devotees follow their introduction with cult-like devotion. The American Hemerocallis Society lists more than 46,000 varieties. Performance of each daylily variety differs greatly by region of the country.

Probably the most famous daylily variety for greenhouse growers is 'Stella d'Oro', an ever-blooming, low-growing variety that flowers from June through frost. Commonly known as Stella, this reliable performer is literally everywhere, especially in commercial landscape plantings. First introduced in 1975, Stella pioneered continuous flowering for daylilies: Prior to its introduction, daylilies flowered for their requisite three weeks during their bloom window. Mary Walters of Great Garden Plants Inc. offers the following suggestion in answer to the one of the most common questions she receives: "Why doesn't 'Stella d'Oro' flower all summer?" She responds, "In the landscape, it's important to divide plants about every three years. As clumps get bigger, their performance may dwindle." Moral of the story: Even the best perennial varieties need care in the garden.

New breeding has matched Stella's outdoor reblooming performance. Many of the best reblooming varieties tend to be yellow and gold. One variety that is a reliable rebloomer is 'Brocaded Gown'. Among colored daylilies, some varieties will have at least one rebloom. Some varieties to look for are: 'Baby Bear', 'Bric-a-Brac', 'Bridgeton Brass', 'Pardon Me', 'Pastel Classic', 'Siloam John Yonski', 'Siloam Plum Tree', 'Strawberry Candy', and 'Sue Rothbauer', a deep rose, large-flowered dormant variety.

Most daylily varieties will flower for the three weeks around their bloom window. If the varieties you are offering are not ever blooming, be sure to have an assortment so that you will be able to offer plants in flower all season long.

Note that new daylily varieties tend to be very expensive when they first come onto the market. Ramping up production of plants is time consuming and takes years. As a variety is introduced, its price is generally a reflection of how widespread the availability is: The hottest new color is likely to be two, three. or even four times more expensive than an old standby. If your market will absorb the extra expense, work in some of the newer breeding, especially of ever-blooming types, which have a longer bloom window (six to eight weeks) and offer characteristics such as better growth habit, stronger flower scapes, and better branching. However, be warned that some in the trade rename old varieties and reintroduce them as "new." Also, know your supplier: Will enough stock of the super-new variety shown in the catalog photograph actually be available to meet demand, or will the supplier be subbing?

Postharvest

Daylily flowers only last for a day; however, plants produce so many of them that the flower show is ongoing. Retailers should keep old blooms picked off to prevent plants from making seedpods. New reblooming varieties require full sun all day long, combined with good soils that hold moisture.

Heuchera

Heuchera hybrids
Note: Heuchera has a number of true species that are commonly propagated including *H. villosa* and *H. sanguinea.* However, most of the varieties available today are hybrids between various species. The culture below can be used on all true species and hybrids with similar results.
Common names: Coral bells; alumroot
Perennial (Hardy to USDA Zones 3–9)

Heuchera is one of the most versatile perennials, and gardeners love them for both their wonderful foliage and their delicate flowers. Recent introductions have added new life to the genus, with varieties that have even more interesting foliage and are more prolific bloomers. Anyone with a perennial program has already made *Heuchera* a mainstay.

Its foliage grows from 10–18" (25–45 cm) tall with flowers in upright to arching spikes. Flowers are scentless, light in color, and usually not as popular as the foliage—but that is not always the case. *Heuchera* prefers morning sun and afternoon shade, but many of the newer varieties will tolerate more sun than those offered twenty years ago.

Propagation

The best *Heucheras* are vegetatively propagated, but they can also be seed propagated on a few selections.

Seed

Sow seed in May, June, or July onto a well-drained, disease-free, germination medium with a pH of 5.5–5.8. Do not cover. Seed will germinate over fourteen to twenty-one days at 65–70°F (18–21°C). Keep trays moist, but not wet. Once 50% of the seed has germinated, move trays to Stage 2 for stem and cotyledon emergence.

Stage 2 will take from seven to fourteen days. Increase light levels to 500–1,000 f.c. (5.4–11 klux). Maintain soil temperatures at 65–70°F (18–21°C). Once radicles have penetrated the medium, begin to allow trays to dry somewhat between irrigations. As soon as cotyledons are expanded, apply 50–75 ppm from 14-0-14 fertilizer once a week.

When you have a good stand, move trays to Stage 3 to develop true leaves. Reduce soil temperatures to 62–65°F (17–18°C). Increase light levels to 1,000–2,000 f.c. (11–22 klux) gradually and allow trays to begin to dry completely before irrigating. Use 100–150 ppm from 14-0-14 once a week. For 128s, Stage 3 will take about twenty-eight to thirty-five days or longer. If growth regulators are needed, B-Nine and Sumagic are effective.

Harden plugs off for seven days prior to sale or transplanting. Reduce substrate temperatures to 60–62°F (16–17°C) and increase light levels to 2,000–3,000 f.c. (22–32 klux). Fertilize weekly with 100–150 ppm from 14-0-14.

Growing *Heuchera* plugs from seed is time consuming—seedlings are slow growing. For 1-qt. (1.1 l) plants that will be large enough to sell in the spring, seedlings should be started the year before. For that reason, most growers buy in plugs. Also, seed-raised *Heuchera* has drawbacks. The biggest downfall of seed-raised *Heuchera* is its variability. From seed, *Heuchera* is not stable, although the newer varieties are helping to change that.

From sowing to a finished or salable 288 takes from six to eight weeks for most sowings. Using two to three seeds per cell helps to diminish the number of empty cells within a plug tray.

Bare roots

Many varieties are propagated by divisions taken in the spring or fall. Crown divisions can be made in the late summer or early fall, potted into 1-qt. or 1-gal. (1.1 or 4 l) pots, and overwintered for sale the following year. Allow six to eight weeks from potting before the onset of cold weather for plants potted up in late summer.

Bare roots can be purchased and potted in February or early March to 1-gal. (4 l) containers. Grow on at 65–70°F (18–21°C) days with nights no lower than 58°F (14°C) for active growth. Plants can be moved to a cold frame and grown on with nights no lower than 50°F (10°C) until sold—ten to twelve weeks after potting.

Cuttings

Treat tip cuttings with a rooting hormone, stick to a 105, 72, 50, or similar tray, and provide soil temperatures of 70–72°F (21–22°C). Give mist for six to ten days and do not pinch. Liner trays will be ready to transplant to larger containers in four to six weeks, depending on cell size.

Tissue-cultured plants are available on a number of varieties and are sold by commercial propagators as rooted liners in various sizes.

Growing On

Heuchera needs a cold treatment to flower. Plants require a minimum of eight to twelve weeks at 35–40°F (2–4°C). In general, flower number will be greater with more cold weeks.

Seed

Sowings made in late spring or early summer are transplanted to 1-qt. or 1-gal. (1.1 or 4 l) containers, grown through summer, and overwintered dormant or cold. Plants are sold green in the garden center the following spring and will flower in the garden in a few weeks.

Many suppliers offer vernalized plugs in a 50-liner that can be potted into a 1-qt. or 1-gal. (1.1 or 4 l) pot in the winter and grown at 55°F (13°C) nights. Plants will be rooted in about eight to ten weeks in 1-qt. (1.1 l) pots and eleven to fourteen weeks for 1-gal. (4 l) pots.

Bare roots

Some growers buy in bare-root plants and schedule them to arrive in the fall or winter. No. 1 bare-root transplants are preferred for 1-gal. (4 l) production and are potted in the fall to early February.

If you are potting your *Heuchera* crop in the late summer or fall in the North and overwintering it, select a disease-free, long-lasting, soilless medium with a pH of 5.8–6.2. Bulk plants under natural day lengths at 55–60°F (13–16°C) days and 50–55°F (10–13°C) nights for eight to ten weeks. Provide 2,000–3,000 f.c. (22–32 klux) of light and keep pots irrigated while plants are actively growing. However, allow plants to be moist but drain thoroughly between irrigations. If growth regulators are needed, use B-Nine and Sumagic. Before overwintering, bulk plants to a minimum of sixteen leaf nodes.

Overwinter *Heucheras* in a cooler, unheated greenhouse, or cold frame for eight to twelve weeks at 35–40°F (2–4°C) or throughout the winter in the northern part of North America. Do not fertilize during this time. Irrigate very carefully, if at all, since *Heucheras* keep a lot of their leaves when dormant, botrytis can become a problem. A fungicide drench as a preventive prior to overwintering is recommended and is often applied in the northern United States during the third week of November. Make sure plants receive minimal light, 100–200 f.c. (1.1–2.2 klux), if you are overwintering in a cooler.

When brought out of the cold, force plants for six to eight weeks at 65–68°F (18–20°C) days and with night temperatures 5°F (3°C) cooler. Irrigate when pots are thoroughly dry and provide 3,000–5,000 f.c. (3.2–5.4 klux) of light. At higher light levels, plants will require additional moisture. *Heucheras* are not daylength sensitive, so you can force plants under natural day lengths. Fertilize at every other irrigation with 150–200 ppm from 15-0-15. Growth regulators such as B-Nine and Sumagic can be used for height control once flowers begin to appear in foliage.

If you are forcing plants into flower for early spring sales in the North or other low-light areas, consider applying supplemental lights (400 f.c. [4.3 klux]) to increase foliage color and flowering quality and color.

If potting bare roots in early February, grow at 65–70°F (18–21°C) days and 58–60°F (14–16°C) nights. Once rooted and days become longer, move to a cold frame and grow on at no less than 50°F (10°C) nights. Days should be 10°F (5°C) warmer. Plants are salable green ten to twelve weeks after potting the roots. They will not be in flower but should be sold green.

Cuttings

Cuttings and tissue-cultured plants are commonly available as liner trays (82, 50, or 32) from commercial propagators. For those received in February, plants require eight to ten weeks in 1-qt. (1.1 l) containers to be salable, while 1-gal. (4 l) containers require eleven to fourteen weeks. If the liners arrive vernalized (already subjected to a cold treatment), the plants will be ready to sell in 1-qt. (1.1 l) containers in five to seven weeks and will flower four to six weeks after planting to the garden.

Pests & Diseases

Insect pests are rare in production, but plants can succumb to root rot and/or become infested with botrytis, especially during overwintering.

Varieties

For seed-raised *Heuchera,* 'Bressingham Hybrids' is one of the most common *H. sanguinea* varieties sown. Plants produce flowers in a mixture of scarlet and light crimson. Plants grow to 18–22" (46–56 cm) tall with an upright habit. Also from seed, 'Splendens' (also called 'Spitfire' or 'Firefly') has deep red flowers on uniform plants.

H. micrantha has deep purple-bronze leaves with yellow flowers. Plants flower in June and July and then sporadically afterwards until frost. 'Palace Purple', the best-known *Heuchera* variety of all, has dark red foliage that gets bronzy and darker at higher light levels. Flowers are creamy yellow on plants that reach 18–22" (46–56 cm) tall. 'Palace Purple' can be propagated by seed or division; however, about 3–5% seedlings will be off-type (green foliage), while vegetative propagation yields plants that are true to type.

But the jewels of this genus are the varied cultivars that have resulted from various crosses and are most commonly known as hybrids. There has been and continues to be a never-ending introduction of these selections to the market. The hybrids began with Dan Heims and Terra Nova's innovated introductions but continue today with a number of companies offering up new selections.

For something truly unusual, try some of the "heucherellas" on the market from Terra Nova. These plants are the result of crosses between *Heuchera* and *Tiarella.* The result is a sterile plant that just flowers and flowers and flowers. × *Heucherella* 'Silver Streak' has palm-shaped leaves overlaid with silver and purple.

Postharvest

Heucheras prefer moist, semi-shady conditions in the garden. Instruct consumers to place plants in a well-drained site that receives afternoon shade. Cut plants back after their first flowering for a second flush of flowers.

Hibiscus

Hibiscus rosa-sinensis
Common name: Tropical hibiscus
Flowering pot plant, foliage plant

Hibiscus is the plant most of us think about whenever we picture a tropical paradise. Their large, bright flowers on deep green, lush foliage is the ideal backdrop to daydream about swaying in a hammock by warm, turquoise waters. They are exotic and make great container plants on patios and decks across North America. In frost-free southern and tropical areas, hibiscus is a shrub.

Growers can produce hibiscus in two ways: from rooted cuttings or prefinished 4–4.5" (10–11 cm) pots. Pencil out which works best for you by taking a look at your available space and the price you can earn from the finished pots. If you are a retail grower or a small- to medium-sized wholesale grower focusing on bedding plants, you will probably find that prefinished pots are your best bet: They finish quickly, and since they are produced in sunny Florida, they are full and bushy and will make great flowering pot plants that you can turn quickly at a profit alongside your regular bedding crops. Some northern growers do a combination of prefinished pots 8" (20 cm) and larger pots they buy in from Florida producers.

If you are looking for a crop for late-season sales, hibiscus offers you even further opportunities as a summer plant: Hibiscus loves the heat and humidity.

Propagation

Hibiscus is vegetatively propagated from stem cuttings. However, the most exciting new colors and varieties are patented and available only from licensees or liner suppliers. Most growers buy in liners rather than root their own cuttings.

Harvest tip cuttings with two or three mature leaves from stock plants every two to three weeks. Rooting hormone (IBA) is beneficial. (In Europe, hibiscus specialists take cuttings from plants in production using the portion that is pinched off.) Stick cuttings in a loose, peat/perlite blend or Oasis wedge and mist for five seconds every ten minutes. Cuttings require shade in high light areas and during the summer elsewhere. Maintain substrate temperatures of 79–86°F (26–30°C). Keep dead leaves and debris removed from the propagation area to reduce disease. Cuttings will be rooted in thirty-five to forty days and ready for transplant. Cuttings may also be direct-stuck into the final growing container.

Growing On

Transplant liners into a well-drained, disease-free medium with a moderate initial nutrient charge and a pH of 6.0–6.5. Plant 1 liner/4" (10 cm) pot and 2–3/6" (15 cm) pot. Plant Oasis cuttings deep enough so that the top of the wedge is covered with medium. Plant prefinished pots at the same depth as they are currently growing.

Hibiscus love the heat, producing their greatest growth and flowering in the summer. Maintain temperatures at 75–85°F (24–29°C) days and 65–70°F (18–21°C) nights. Plants will become damaged at temperatures below 50°F (10°C). Grow hibiscus in full sun: 1,000 f.c. (11 klux) is the absolute minimum. Under low light, crop times will be extended. Some varieties tolerate lower light levels better than others. If this is a concern for you or you are developing a year-round pot program, ask your supplier to develop a list of varieties that are suited to low-light production.

Do not allow plants to dry out. Provide constant liquid feed at 200 ppm from 20-10-20. If needed, increase the fertilizer rate during periods of rapid growth.

Pinch plants when new shoots are 1.25–2" (3–5 cm) long, approximately one to two weeks after planting. Pinch as often as required to shape plants, but do not pinch back into the hardwood portion of the stem. Remove any premature buds that may form in laterals for the first month after potting. Pinch for the last time nine to twelve weeks prior to sale in flower. Generally, 4" (10 cm) pots will require one pinch combined with Cycocel sprays.

Cycocel is very beneficial for hibiscus. Plants become shorter, leaves darken up, and flowers become more plentiful. Begin application two weeks after the pinch, once shoots have begun to grow again. Rate recommendations vary from 200–1,000 ppm. For example, the Aris Horticulture Company recommends two to three total Cycocel applications at 460 ppm as needed after each pinch for its variety 'TradeWinds'. Discontinue Cycocel once buds are pea sized. A-Rest and Bonzi are also effective in controlling hibiscus height. Some of the newer hibiscus varieties do not require growth regulators.

Plants may be spaced while they are in production as leaves begin to touch.

Pests & Diseases

Aphids and spider mites can affect hibiscus, and outdoor hibiscus plants sometimes act as magnets for whiteflies. *Xanthomonas* angular leaf spot can affect hibiscus: Infested plants should be removed from the premises and destroyed.

Troubleshooting

Loss of older foliage can be due to plants spaced too closely or uneven watering. Plants drying out, insects, or low light can cause bud drop. Interveinal chlorosis can be due to iron deficiency caused by too-wet conditions, disease, or high pH. Flowers only last one day, so do not be alarmed when they die by nightfall.

Postharvest

Plants are ready to sell when two or more flowers have opened. Hibiscus is sensitive to ethylene: Silver thiosulfate (STS) sprays ten days before sale may help to reduce bud and leaf drop. Plants can be sleeved and boxed, provided transit is four days or less. Ship plants at 50–60°F (10–16°C).

Retailers should unpack and unsleeve plants immediately upon arrival. If needed, water plants immediately. Do not allow plants to dry out at retail, as this causes bud drop. Put pots in high light so that subsequent buds will continue to flower. Do not display at temperatures below 50°F (10°C) or above 80°F (27°C).

Hibiscus moscheutos
Hibiscus hybrids
Note: Both of these botanical names are used throughout the industry. There are a number of varieties that are known botanically as *Hibiscus moscheutos,* especially those that are seed propagated. *Hibiscus* hybrids are cultivated varieties that are the result of hybrid crossing between hibiscus species.
Common names: Swamp rose mallow; rose mallow; hardy hibiscus
Perennial (Hardy to USDA Zones 4–9)

Fabulous and underutilized are the two adjectives that come to mind for hardy hibiscus. These North American–wetland natives are showstoppers in the landscape. Their performance is unbeatable. Bring on hot nights and humid days: They love 'em.

Plants grow from 2–6' (61–183 cm), depending on variety, and are highlighted with single, large, almost plate-like blooms that grow up to 12" (30 cm) across. The wide range of flower colors includes white, white with eye, pink, rose, red, and shades of plum.

Because this plant is from America's wetlands, it prefers moist, but not wet, conditions. It can be grown in standing water (make sure the water is warm) or as a

marginal (along the edge of a stream or pool of water). However, it is versatile and can be grown in a mixed flowerbed with well-drained soils.

Propagation

Hibiscus moscheutos is propagated by seed, cutting, tissue culture, and bare-root transplants.

Seed

Highlighted by varieties like the 'Southern Belle' mix and the 'Disco Belle' and 'Luna' series, hardy hibiscus seed can be sown anytime of the year depending on the size of the plant you want to sell and when as well as whether you want to overwinter plants or start from inputs in winter for spring sales. In all cases, the plants will flower in June or July their first year.

Pour warm (not hot) water over the seed and let stand from a few hours to no more than overnight in a warm location in the greenhouse or office but out of direct light. This process is not necessary for the 'Luna' series; however, it is beneficial for the remaining seed-propagated selections.

Pat the seed dry in a paper towel, sow one seed per 200, 162, 144, or larger plug tray, and cover with plug media. Seeds germinate over a wide range of temperatures, from 68–80°F (20–27°C); although on the greenhouse bench the seedlings should be germinated at 75°F (24°C) or less. When using a germination chamber, warmer temperatures can be used. However, remove the plug tray once the root emerges (usually in two to four days), or the seedlings will stretch. Regardless, lower germination temperatures take longer but produce a more uniform plug tray.

If the plugs are starting to stretch, apply a Cycocel spray at 300 ppm ten days after sowing. The foliage will become darker green, and seedling stretch will be stalled. Do not pinch if plants are stretching. It adds more time to finish to a flowering crop.

On the 'Luna' series, a 200-plug tray takes from three to five weeks from sowing to be salable or transplantable to larger containers. If sowing to a 72 tray, from sowing to transplanting, is five to six weeks. On all other seed varieties, add another one to two weeks.

Cuttings

Commercial propagation companies offer a number of varieties in 72, 50, or similar tray sizes in the winter and spring. These can be from stem cuttings or the result of tissue-cultured plants.

Bare roots

Bare-root transplants are available from commercial propagators and are potted up in winter for green plant sales in the spring. Bring in the roots in early February and pot up to 1-qt. or 1-gal. (1.1 or 4 l) pots.

Growing On

Seed

Seed sown in January will produce flowering plants in April or May, but require long days to flower. Use three plants per 1-gal. (4 l) pot, since hibiscus flowers with only one bloom per plant at a time.

Growers can purchase seed-grown plugs or 2.5" (6 cm) liners. Pot liners or plugs into a well-drained, disease-free medium with a pH of 5.5–6.2 and a moderate initial nutrient charge. Provide constant liquid feed of 150 ppm from 20-10-20 and maintain substrate EC at 1.0 mS/cm. Plants flower from fifteen to seventeen weeks after potting, depending on variety.

Hardy hibiscus does not require vernalization for flowering, so seed sown in January will flower the following summer. If you want to force plants into flower for sale in April or May, you will have to start them early and provide long days and warm temperatures (above 75°F [24°C]) to force flowering. Provide extended days by lighting from 4–10 P.M. with incandescent lights or supplemental lights. If daylength extension lighting is not practical, night interruption lighting also works.

Potting plugs or liners from mid-December to

mid-January and extending the day length should yield flowering plants twelve weeks later, provided you grow them warm. Consumers can buy hibiscus forced to flower early and enjoy subsequent flowers on those same plants through frost.

Hibiscuses are vigorous growers. Liners may be pinched at planting. Pinch again when new growth is 2" (5 cm) long, and apply Cycocel about two weeks after the pinch to control plant height. Spray at 460 ppm every fourteen days for a total of two or three applications. Bonzi is also effective. A drench at 5 ppm can be made either three weeks after planting or once new growth is 0.5" (13 mm) long after pinching. A second application can be made at bud set.

Grow hibiscus under high light: 3,000 f.c. (32 klux) or above is ideal. During production, do not allow plants to dry out; although once they are established in the landscape, they can handle drought situations.

Cuttings

For pots sold green in May or flowering in June or July, bring in vegetatively propagated 2.25" (6 cm) liners in February or March and pot into 1- or 2-gal. (4 or 8 l) pots. Pinch plants once or twice and/or use Cycocel or Bonzi as outlined above to control height.

Bare roots

Potted up in February, grow warm at 60°F (16°C) nights and days 5–10°F (3–5°C) warmer. Unlike many perennials that are cold hardy, hibiscus prefers to be finished warm (nights no lower than 55°F [13°C]). Otherwise the foliage can turn yellow, drop off, or both.

Pests & Disease

In the greenhouse, hibiscus is a whitefly haven. Outdoors, Japanese beetles will be pests if they are a problem in your region.

Varieties

From seed, 'Disco Belle' is a well-known variety. Another favorite, 'Southern Belle' is a taller selection and one of the author's personal favorites. In field trials, it is always guaranteed to turn heads. However, seed is perennially in short supply for both of these selections, so put your orders in early. Plants grow from 24–30" (60–75 cm) tall on 'Disco Belle' and up to 5' (1.2 m) on 'Southern Belle.'

The 'Luna' series grows from 24–30" (60–75 cm) and is another favorite. It is easy to produce and makes a wonderful statement in the garden.

Hippeastrum

Hippeastrum hybrids
Common name: Amaryllis
Flowering pot plant, cut flower

Found everywhere during the Thanksgiving and Christmas holidays with its bold, large, and bright lily-like flowers, amaryllis is a stately, elegant flowering pot plant that is as easy for growers to produce as it is for consumers to enjoy. Native to Central and South America, amaryllises thrive in warm-cool growing temperatures (warm days, cool nights) and high light. Bulbs can be flowered as late as June, and they make excellent patio plants or cut flowers.

Propagation

Amaryllises are vegetatively propagated by either mother bulb offsets or twin scaling. Growers purchase bulbs from specialists in the Netherlands, Israel, Brazil, and South Africa. South African crops are generally forced for very early pot plant sales (in October or November); Brazilian, Israeli, and Dutch bulbs are grown for later crops.

After harvest, roots and leaves are cut off and the bulbs are quickly dried. During this and subsequent processes, it is critical that the old root system remains viable. Normally bulbs are cured for two weeks at 73–77°F (23–25°C) with high ventilation. They are then stored at 48–55°F (9–13°C) at 80% relative humidity for eight to ten weeks. Bulbs stored for longer are held at 35–48°F (2–9°C). They should be held at very low temperatures if the leaves or flower stalks emerge from the bulbs.

Bulb size and cultivar influence the number of floral stalks produced. Commercial bulbs are sold by circumference in cm: 20/22, 24/26, 28/30, and 32/up. The number of flower stalks is a cultivar response and ranges from two to six; most produce four flowers per stalk. Larger bulbs tend to produce two or more floral stalks.

Bulbs should be transported at 48°F (9°C). They must be protected from freezing and drying out.

Growing On

Inspect bulbs on arrival for signs of disease or mechanical damage. Do not plant seriously affected bulbs. Bulbs should have the roots intact. Bulbs with all roots removed to the root base are inferior: They may bloom, but the quality is impaired because water and nutrients cannot be taken up readily and stored food reserves will be depleted. Plant bulbs immediately upon arrival. If they cannot be planted immediately, hold at 35–48°F (2–9°C): The exact temperature will depend on the sprouting condition of the bulb. If bulbs have started to sprout, store at 35°F (2°C); if not, store at 41°F (5°C).

Normally, one bulb is used per 6" (15 cm) standard pot. From three to nine bulbs can be used for larger containers and patio planter production. Bulbs may also be planted in ground beds or cut flower production. Select a disease-free, well-drained, potting medium with a pH of 6.0–6.5. Leave about a quarter of the bulb above the soil line. Never use fresh manure or bark as part of the medium. Some growers add sand to the mix, which will add stability to the pot. Place about 1" (3 cm) of medium in the bottom of the pot and plant the bulb, leaving a third of the bulb exposed. Force bulbs pot to pot on the bench.

Bulb size, temperature, and moisture influence flowering. Water in pots thoroughly after planting. Irrigate pots when they dry out, keeping the medium only slightly moist. In order to stimulate regrowth of the basal root system, it is important not to overwater the plant. Normally watering once a week is satisfactory. Use tepid water, and do not water over the bulb noses. Bulbs do not need fertilizer during forcing.

Grow the crop on at 70–86°F (21–30°C) days and 65–75°F (18–24°C) nights. Temperatures less than 60°F (16°C) are detrimental. Grow in full-sun conditions. It is possible to start bulbs under dark, temperature-controlled conditions before the bulbs are placed in the greenhouse. Force plants in a well-ventilated greenhouse, and do not allow humidity to build up.

The crop time from a programmed bulb to a salable plant ranges from three to seven weeks.

Pests & Diseases

The primary disease that affects amaryllis is *Stagonospora,* also called fire or red spot, which can attack leaves, stems, and flowers and is occasionally caused by overwatering. There is unfortunately no real treatment for this disease once forcing has started.

Also watch for mealybugs, bulb mites, and thrips. While insects are not common, maintain a scouting program anyway.

Postharvest

Pots are ready to sell when floral stalks are 12" (30 cm) long. Leaves should be 6–12" (15–30 cm), and the second stalk should be starting to grow. If plants must be held, hold them at 48°F (9°C). Do not hold plants in the dark for more than a few days, as stems will continue to elongate.

Harvest cut flower amaryllis when the floral buds are fully colored but not open. To prevent the splitting and outrolling of cut stems, hold flowers in a sucrose preservative solution (6 oz./gal. [45 g/l]) for twenty-four hours at 72°F (22°C) before shipping. This treatment will lengthen flower life and help all buds to open on the stem.

Retailers should unpack potted amaryllis immediately, place them in a bright, cool display at (65–70°F [18–21°C]), and irrigate with tepid water.

Advise consumers to place pots in a bright, cool spot in the interior. Flowers should last from fourteen to twenty-one days. Consumers should fertilize plants once or twice a month when plants are actively growing.

When the danger of frost is past, plants can be placed outside. Plants can be reforced, if they maintain active growth during the summer. Full sun to light afternoon shade is best in addition to plenty of water and fertilizer. The objective is to grow as many healthy leaves as possible. Before frost, consumers should take plants back inside in the fall, allow them to dry thoroughly, and store them for at least eight weeks at 50–60°F (10–16°C). Then, dried leaves can be cut off and pots can be watered and put in a warm area to begin the forcing process. To avoid the storage process, grow plants in the light at 50–60°F (10–16°C) for eight to ten weeks and then place in a warm area to reflower.

In many climates (Zones 7–11), amaryllises are perennial, returning year after year and blooming in the late spring.

Hosta

Hosta spp.
Common name: Hosta; plantain lily; funkia
Perennial (Hardy in USDA Zones 3–9)

Hostas are the top-selling perennial. If you already grow or are planning to add perennials to your product assortment, your customers will expect you to offer hostas. In most parts of the country, hostas are great performers: They tolerate a wide range of soils and moisture conditions, and, unlike just about every other plant, they thrive in the shade. The world loves hostas for their foliage: ranging from large and bold to small and delicate to brightly variegated. Plants can be a petite 6" (15 cm) or a stately 3' (91 cm) tall.

To the uninitiated, hostas are a confusing group of plants: Which species do you offer? What new varieties are hot right now? Why don't the shoots emerge and grow uniformly so I can just sell them all right away? Hostas almost always look better in the garden than they do in the pot because they grow slowly, taking several years to attain full size.

From a marketing perspective, there are two roads you could take: If you are a wholesale grower with an excellent landscaper trade or selling to the mass market, you will want to offer plenty of standard hostas, focusing on the basics and being competitive with price. Buying in plants or liners in the winter, rooting them in well, and moving them out will be your goal. If you are a wholesale grower who sells to garden centers or you are selling direct to the public, be more selective in choosing your hosta offerings. You can buy your plants in the fall, overwinter them, and produce large pots that command premium prices. The mass merchandisers tend to offer thousands of pots of standard hostas at rock-bottom prices throughout the summer. You will never be able to compete with that. But, if you have rarer foliage combinations on supersized plants, your customers will be able to justify the price difference.

Propagation

Hostas are propagated primarily through division by specialist propagators. Most bare-root hostas sold in the United States come from suppliers in the States or the Netherlands. Plants may be divided in spring or fall. One-year old transplants with either one or two growing points are replanted for production of two-year-old plants or are sold.

Tissue culture is another popular propagation method. Stage 2 tissue-cultured plants are brought in by propagators and grown into 2" (5 cm) liners over the course of nine to twelve months. However, a 2" (5 cm) liner may require up to two additional years to finish in a 1-gal. (4 l) pot. Larger liners can be purchased that will finish in a 1-gal. (4 l) pot in the same season. If you are buying in tissue-cultured liners such as these for spring, make sure to buy plants that have received a cold treatment.

Like propagation for most perennials, hosta propagation is a long-term process. The bare-root plants that most growers buy in for potting into 1- or 2-gal. (4 or 8 l) pots are two years old. Because of this, most growers leave propagation to the specialists.

H. ventricosa is the only hosta that comes true from seed.

Growing On

Hostas can be potted beginning in August and September from freshly harvested bare-root plants. These plantings are overwintered in a cooler or unheated cold frame for sale the following spring. Some growers may choose to pot plants in the winter, buying in bare-root plants that have been held in cold storage since harvest. Note that plants potted in the fall and overwintered will be significantly fuller and more robust and have more well-developed roots compared to plants potted in the winter and sold the same spring. Know your market to decide which route is best for you.

Hostas may also be planted year-round from actively growing liners (many of these are the result of tissue culture). Liners planted in the winter and spring can be sold in smaller pots (4" [10 cm] or 1 qt. [1.1 l]) eight to ten weeks after potting (provide long days and temperatures above 65°F [18°C]). Liners can be planted later in the spring and summer and sold the same year once plants have rooted in. Finishing times become significantly shorter as temperatures warm up.

Select a long-lasting, well-drained medium high in organic matter with a pH of 5.8–6.2. Bark mixes are great. If you are using a bark-based medium, consider incorporating a slow-release fertilizer. Hostas require uniform moisture, so don't skimp on watering. This is especially true as plants begin to emerge from winter-planted bare-root divisions: When roots have not developed adequately yet leaf production has begun, plants are very susceptible to drought stress. Any stress at this time may cause marginal leaf burning, which will delay sale until new, replacement leaves can be produced.

Six weeks of cold temperatures at 32–41°F (0–5°C) increases emergence uniformity. If you are bringing plants out early to force, provide long days (more than fourteen hours). Hostas require long days for continual leaf production. Flowering is also triggered by long days. Many growers overwinter their hosta crop in an unheated cold frame covered with white plastic and use microfoam to cover pots. Others prefer a minimally heated greenhouse (35°F [2°C]).

Finishing is about nine to eleven weeks in the spring. However, you will be able to finish pots much faster as temperatures warm up. It is possible to schedule receiving hostas year-round, but plants need six to eight weeks of cold treatment to force well.

A spray application of 3,000–4,000 ppm benzyl adenine (BA) to established, single-eye divisions will drastically increase the number of shoots the plants produce—you will see the results within one month. The BA stimulates lateral and latent buds to grow. Using BA, researchers have taken a one-year-old bare-root plant and grown a 1-gal. (4 l) that is as large as a comparable pot planted from a two-year-old plant. Plants can also be dipped in 1,000 ppm BA prior to planting. BA can be applied to increase lateral formation anytime plants are actively growing.

Space pots of larger-growing varieties as leaves begin to touch to avoid plant crowding. For forcing plants to leaf out for spring sales, provide growing temperatures of 70–75°F (21–24°C) days and 68–70°F (20–21°C) nights. Avoid temperatures above 80°F (27°C), as plants can lose their distinctive varietal look: Leaves can become smaller and strappier, plants may rosette, and colors may be faded.

Provide constant liquid feed from 15-0-15 at 150–200 ppm. Be sure to leach 10% of the applied volume so that soluble salts don't build up.

Grow green and variegated varieties under 50% shade and blue-leafed varieties under 70% shade, which will help plants retain their blue foliage longer.

Pests & Diseases

Hostas are magnets for slugs and a delicacy to deer, while voles tunnel right through hosta roots. Be prepared to fend off these problems before your crop begins to leaf out. Be sure to use rodent bait around pots that are overwintered.

Aphids, mites, and thrips may also become pest problems. Rhizoctonia may attack roots of newly planted pots. Bacterial crown rot can attack plants that are injured by surface salts on their leaves, so be sure to rinse off fertilizer.

Troubleshooting

Burned leaf tips can result from water-stressed plants, a situation made worse if soluble salts are also high. High levels of salts in the upper portion of pots can cause plants to be unstable in the pot because they have not adequately rooted in. Leaching, adding fresh medium to pots, or repotting plants can solve this. Severely water-stressed plants may go dormant. Leaves may be showing signs of sunburn if the topsides of upper leaves have brown patches.

Blue-leafed varieties may lose their blue cast, becoming greener as the summer progresses. According to the American Hosta Society, this is because they have lost a layer of epidermal wax. High summer temperatures cause this condition, which is exacerbated by overhead watering or rains during the heat of the day.

Varieties

Growers in hosta havens know that having this year's hot new variety can spell temporary windfall, even if the young plants are exorbitantly expensive. Hosta collectors and devotees will drive miles to acquire the ideal new plant for their collection. The problem is guessing what that variety will be and then timing the market with the right quantities in a ready-to-sell size.

For most of us mere mortals and everyday gardeners, sticking to offering a basic product assortment that is beyond the basic green *H. fortunei* and *H. undulata* 'Albo-marginata', which consumers see by the millions at the mass market, is the road to take.

If you are going to offer such an assortment, consider this list as a starting point from which to build. Consult your supplier and regular customers and take a market survey of what competitors offer to develop a hosta assortment tailored to your customer base.

The twenty most popular varieties, according to the *AHS Hosta Popularity Poll 2009* conducted by members of the American Hosta Society, are: 'Sagae', 'June', 'Liberty', 'Sum and Substance', 'Blue Angel', 'Paradigm', 'Regal Splendor', 'Krossa Regal', 'Paul's Glory', 'Guacamole', *H. montana* 'Aureomarginata', 'First Frost', 'Striptease', 'Guardian Angel', 'Whirlwind', 'Halcyon', 'Great Expectations', 'Stained Glass', 'Niagara Falls', and 'Praying Hands'.

If you are interested in diving into hostas, there are entire books devoted to hostas and their nomenclature. Contact the American Hosta Society at http://www.americanhostasociety.org for details.

Postharvest

Retailers should display hostas in shade. If customers wish to divide their plants, instruct them to do so in the spring, just as leaves are sprouting and when they are 1" (3 cm) tall. Hosta flowers attract hummingbirds and butterflies.

Hyacinth

Hyacinthus orientalis
Common name: Hyacinth
Pot plant, cut flower, garden perennial

Hyacinths are prized for their sweet fragrance and quintessential, spring-like qualities. They are most commonly forced in 4 or 6" (10 or 15 cm) pots but can also be grown in larger containers in monochromatic colors or mixes or as sprouted growing plants in various bedding plant containers. Essentially 100% of the world's hyacinths come from the Netherlands. Cooling facilities (temperature-controlled rooting rooms) are essential for proper cooling of the bulbs after planting. For early forcing (flowering before early January), *prepared* bulbs are necessary. Preparation is a sequence of specific temperature treatments performed in Holland to help ensure early forcing potential. For approximately mid-January and later forcing, *regular bulbs* are used. These bulbs have no special temperature treatments given before export.

The keys to success are: (1) proper cultivar selection; (2) obtaining bulbs from reliable suppliers who use controlled-temperature shipping containers for transport; (3) inspecting bulbs upon arrival for mechanical damage or the presence of molds or bacteria; (4) using a clean, well-drained, planting mix with a pH of approximately 6.3 and low soluble salts; (5) after planting, providing the proper number of cold weeks per cultivar and forcing period; (6) using the correct greenhouse forcing temperature; and, (7) when needed, giving the correct PGR treatment.

"Lifting" of hyacinths occurs as a result of too-shallow planting into lightweight potting mixes. This problem is solved by planting deeper or using a mix with 5–15% sand.

These hyacinth bulbs have been grown in 2.5" (6 cm) pots in the Netherlands. After using a temporary foam-rubber pad above the bulbs during the first two to four weeks of rooting, the bulbs have been placed on top of the planting mix, yielding a product in which the bulb is an integral part of the "experience."

Growing On

Bulbs should be planted with at least 1" (3 cm) of mix on top of the bulb to prevent bulbs from lifting out of the pot during rooting. After planting, pots are placed into coolers initially at 48°F (9°C). Decrease temperatures as rooting and shoot growth proceeds (similar to tulip and daffodil production). The total cold duration is about ten weeks for prepared bulbs and thirteen weeks or more for regular bulbs. It is important to promptly reduce temperatures to restrict shoot growth so that plants do not grow up into the above rack or crate. For cut flower production, it is best to leave temperatures at 48°F (9°C), since this temperature allows the slow stem growth needed for cut flowers.

Hyacinths are very vigorous rooters, and as such "lifting" out of the pot is a common problem (see photo). In the Netherlands, growers have developed systems whereby pieces of foam rubber are placed on top of the freshly planted bulbs with another crate on top of that in an alternating manner. After a few weeks of rooting, the pads are removed. Since the bulbs are already fully rooted, they do not lift out (i.e., the pads act as a counterforce to allow rooting without bulb lifting). This system actually results in a totally different bulb product than what we have in North America, as the bulb is clearly visible and becomes part of the entire product (see photo).

After cooling, pots are moved into the greenhouse at temperatures that vary depending on the season. For earliest (mid-December) flowering, prepared bulbs are needed and forcing temperatures should be set at 73–77°F (23–25°C). Later in the season, temperatures of 63°F (17°C) are good. Forcing times vary by season from two to three weeks for very early forcing to just a few days for very late forcing.

Growth Regulation

Traditionally, hyacinths are sprayed one or two times with Florel (an ethylene-releasing chemical) at 500–2,000 ppm, depending on cultivar and time of season. Hyacinth growth regulation does not benefit the grower as much as it benefits the consumer, as stems are heavy and can topple. While Florel can work very well, sometimes inconsistent results can occur. Over the past several years, extensive trials with Topflor drenches and preplant dips have been conducted. Excellent postharvest growth control can be obtained with 20–40 ppm preplant bulb dips for five to thirty minutes. Topflor drenches are also effective at 1–2 mg/pot. More detailed information by cultivar can be found online at the Flower Bulb Research Program website at www.flowerbulbs.cornell.edu.

Postharvest

Pot hyacinths must be marketed promptly as floral development is rapid at or above room temperature. Pots should be marketed in the green bud stage. A common problem at retail is marketing plants too late, which does not give enough postharvest life to the consumer. Cut hyacinths are harvested when one or two florets are ready to open. For cuts, it is critical to harvest in such a way that a portion of the base plate remains attached to the stem, as this allows continued water uptake by the stem. Hyacinths are best kept at 33–38°F (0–3°C) as much as possible after harvest. Ethylene exposure to open flowers at room temperature can cause premature floret wilting and discoloration.

Pests & Diseases

Hyacinths suffer from relatively few major problems. Since they are in the greenhouse such a short time, insects are usually not a major issue. The main diseases include botrytis, *Fusarium*, and *Pythium* as well as the bacteria known in the industry as "snot" (*Xanthomonas hyacinthi*).

Troubleshooting

Inadequate cooling can result in "green tips" with florets that are not fully colored. "Spitting" is a random problem in which the shoots are literally severed from the base plate inside the bulbs. This problem is commonly ascribed to radical temperature changes coming out of the cooler (that is, going from cold to warm temperatures too quickly), and seems to be cultivar dependent. However, after several years of experiments, Cornell researchers were unable to cause spitting, even when immediately moving plants from the cooler to 90°F (32°C) greenhouses; so the true cause of this condition remains unknown.

Hydrangea

Hydrangea macrophylla
Flowering pot plant

Who could have guessed that the wave of popularity hydrangeas enjoyed throughout the 1990s would show no signs of slowing down in the new millennium? They are showing up everywhere: in floral prints, on magazine covers, and in fresh, dried, and silk flower arrangements. They are being grown on patios and decks in containers and are showing up in all sorts of retail outlets as flowering pot plants.

Hydrangeas grown outdoors make vegetative growth in July and August, with initiation of terminal flowers in September and October, after which the flower buds are in a resting state and resume growth after normal winter chilling and leaf shedding. Overwintered flower buds are usually in flower by late June. The usual method of greenhouse forcing simply mimics the natural sequence, with propagation in May or June and the substitution of a controlled cold period of six to eight weeks for winter chilling, followed by twelve to fourteen weeks of forcing in the greenhouse. The period of availability for "summer production" of hydrangeas as blooming pot plants extends from January to early June. With freezing techniques or environmental manipulation, this season can be extended.

Propagation

Commercial greenhouse growers buy in dormant plants for forcing because a successful plant for forcing depends on the quality of growth made the previous summer. Producing plants under ideal circumstances yields a better end result. Plants grown and shipped dormant are generally produced in 4 or 6" (10 or 15 cm) pots. Plants from 4" (10 cm) pots can become very root-bound, making them tricky to bump into larger pot sizes. Inexperienced growers or northern growers often use 6" (15 cm) plants and produce a better crop as a result.

Stem cuttings can be taken from April to June from the tips of vegetative shoots and the lower nodes of long shoots used for two-eye (butterfly) cuttings or single-eye (leaf-bud) cuttings. Cuttings may also be made of mid-stem pairs of leaves. Treat the base of cuttings with rooting hormone (IBA) prior to sticking. Cuttings root in three to five weeks.

Cuttings are generally stuck directly into pots filled with a well-drained, peatlite medium. For blue-flowering varieties, pH should be 4.5–5.5; for pink-flowering varieties, pH should be 5.5–6.5. During rooting, provide light at 2,500–3,000 f.c. (27–32 klux). Shade, up to 50%, may be needed, depending on local conditions. Use mist to maintain cutting turgidity. Maintain substrate temperatures of 72°F (22°C).

Make sure that leaves of adjacent cuttings do not excessively shade each other. Cuttings should be rooted in three to four weeks. Begin applying 100–200 ppm from 20-10-20 as soon as cuttings make roots. If you are rooting your own cuttings, you will have to develop the plant over the summer. Begin programming plants to be pink or blue immediately. (See the fertilizer discussion under Growing On). Increase rates quickly up to 200 ppm as cuttings develop, peaking in July and August; discontinue fertilizing in September.

Pinch cuttings after three nodes have formed, leaving two nodes. Make sure all pinches are completed by early to mid-July. Plants usually require three mature sets of leaves on each shoot to produce a flower bud.

Growing On

Hydrangea flowers are initiated in the late summer to early fall. They develop in cool temperatures, either outdoors or in refrigeration, during the required 1,000–1,200 hours of cooling at 40–45°F (4–7°C). Before cooling, plants are defoliated chemically or by hand. It is important not to move plant debris into the cooler, as it will serve as a launching pad for disease. During cooling, it is vital that plants do not dry out. After cooling, plants can then be grown at warmer temperatures to allow the stem with the flower buds to elongate and the buds to further develop.

The resting hydrangea bud contains five to eight sets of leaves in addition to the initiated and partially developed flower bud. The flowering shoot must unfold these leaves and the flower by flowering time; therefore, a relatively long forcing period is required—ten to twelve weeks from dormant plants.

Most growers buy in precooled dormant plants. Thoroughly soak dormant plants coming out of the cooler or that have arrived just prior to potting: Adequately irrigating plants once they are transplanted is difficult. It is best to start with watered plants and then water in again after potting. Pot hydrangeas in a medium with a pH of 5.5–6.5, except when plants are to be "blued." A source of slowly available minor elements is also necessary; a medium with compost usually needs no further minor elements added. When potting up dormant plants (shifting plants to a larger pot for forcing), a soilless mix containing peat moss or peat moss alone can be packed around the original mutilated soil ball. Add gypsum to soils for both pink and blue plants in order to supply calcium without affecting pH. Use a medium with a low initial nutrient charge. Some suppliers recommend that you use a medium with at least 25% topsoil. Be sure to obtain high-quality loam that is herbicide free.

Hydrangeas can be flowered year-round. Many growers routinely force hydrangeas for late January and Valentine's Day sales. Forcing hydrangeas for sales later than Mother's Day has always been possible by holding plants in refrigerated cold storage. Much of the heating costs of winter forcing are avoided, but market potential must be developed for any out-of-season production. Freezing dormant plants, after their cold-storage requirement has been satisfied, may offer a means of having plants at other periods of the year, although many varieties are not suitable.

Dormant plants are placed in the greenhouse at forcing temperatures of 60–65°F (16–18°C) immediately after removal from cold storage. Plants can be placed in the greenhouse prior to potting and then potted as time permits, provided potting is done within a reasonable time.

Table 16. Hydrangea Blooms and Pot Sizes		
BLOOMS	INCHES	CENTIMETERS
1	4.5–5.5	11–14
2	6	15
3, 4	6.5	16.5
5	8	20

Should plants show signs of insufficient cold storage—as evidenced by slow development, short internodes, small leaves, or a general rosetted appearance—consider an application of gibberellic acid (GA). To overcome cold-storage deficiency, GA at 2–6 ppm is used in the forcing period. A single foliar application may be adequate, but weekly applications may be made if plants do not respond. Careful observation is the only means to determine the number of applications necessary to restore growth. *Note:* Do not apply GA if flower buds are larger than 0.375" (0.95 mm) in diameter.

One serious problem can develop when the plants are repotted in fresh medium into larger pots than are used for summer growth (e.g., repotting a 4" [10 cm] plant into 6.5" [17 cm] pot): failure to root into the new soil, with a subsequent lack of stem elongation and flower expansion. This is especially true when plants are forced, particularly in the North. Plants should root in within three to four weeks. If they do not, there are several ways to overcome this problem: (1) Use 6" (15 cm) dormant plants instead; (2) scarify the root-ball rather severely to damage and expose root tips, thus encouraging them to grow into the fresh medium; (3) use a medium low in fertilizers; (4) place the lead-weight watering tube directly over the original root-ball, ensuring that the original root-ball does not shrink away from the new media; and, (5) as suggested by some experts, start forcing in trays out of the pot until root growth begins. In any event, understanding the nature of the problem is necessary in solving it and avoiding further problems.

Throughout production hydrangeas need to be supplied with adequate moisture: Avoid wilting plants. However, do not allow the medium to remain saturated.

Space pots closely for the first two to four weeks to save on heat and space. Thereafter, provide adequate space to plants: Plants with three flowers need nearly 1.25 ft.^{2}/plant (8 plants/m^2); plants with four to five flowers need 1.5 ft.^{2}/plant (7 plants/m^2). Calculate about 0.3–0.5 ft.^{2}/flower as a space requirement (21–32 flowers/m^2), depending on the variety and your target market.

During forcing, make every effort to prevent plant growth from becoming soft and subject to excessive water loss or desiccation injury when removed from the greenhouse. Maximum sunlight, adequate space, and low humidity are important. Do not wet the leaves. Tube watering is practical during the forcing period, but mat or capillary watering is acceptable. Growth will be more vigorous with a constant supply of moisture to the roots.

Growth regulators are frequently used to prevent excessive height and to reduce space requirements. Plants forced for Mother's Day, particularly, may need a growth regulator. Apply a spray of A-Rest at 25–50 ppm or B-Nine at 1,250–2,500 ppm during the third week of forcing, when leaves are 1.5–2" (4–5 cm) long. Some varieties do not need growth regulators. 'Sister Therese' should have a "delayed" application for toning up plants. 'Rose Supreme' and 'Blaumeise' require higher and/or repeated applications.

Photoperiod may affect the rate of development and type of growth during forcing. Plants placed in cold storage early and forced under the long nights of November, December, and January and those that have not had an adequate storage period will benefit from night interruption lighting from 10 P.M.–2 A.M. with 10 f.c. (108 lux) of incandescent light. Additional light will have little effect on plants forced late in the season, as these plants have had longer bud development and rest periods.

Forcing temperatures regulate not only the rate of development but also the ultimate height, size of cymes, intensity of sepal color, and quality of the finished plant. Basically, hydrangeas are cool-temperature plants, making their best growth at night at temperatures below 60°F (16°C), although the rate of development will be faster at a higher temperature. Night temperatures at 52–57°F (11–14°C) will produce taller stems, larger leaves, and larger flower heads with more intense color than growing at 62–65°F (17–18°C). Representative forcing periods at different night temperatures are sixteen weeks at 54°F (12°C), twelve weeks at 60°F (16°C), and ten weeks at 65°F (18°C). At a temperature of 60–62°F (16–17°C), buds are visible eight weeks before bloom; they measure 0.75" (2 cm) at six weeks and 1.5" (4 cm) in diameter at four weeks before flowering. The old rule of buds being pea-sized eight weeks before sale, nickel-sized six weeks before sale, and half-dollar-sized four weeks before sale is still valid. Plants showing sepals are approximately eighteen days away from sale.

Mature hydrangea plants can be held in refrigeration at 35–40°F (2–4°C) for several weeks, if necessary. Watch for botrytis if you store plants.

Grow on under 5,000–7,500 f.c. (54–81 klux). As the sepals enlarge and become pigmented, it may be necessary to reduce light intensity to 2,500–3,000 f.c. (27–32 klux) to prevent fading and injury to sepals from excessive transpiration. Harden the plants as the sepals approach maturity by giving cooler night temperatures and ample ventilation. If plant growth has been restricted through environmental manipulation and growth regulators, the staking and tying of flower heads should not be required. Multiflowering plants usually need no support.

Gibberellic acid (GA, Pro-Gibb), sprayed at 2–5 ppm on young growth out of storage, is a powerful growth stimulant. Early forced plants often benefit from GA applications, but applications must be made before the flower bud reaches the size of a dime. Later applications can cause distorted, open flower heads.

Understanding the effects of fertilization on color change in pink-blue hydrangeas is essential. The sepals of hydrangeas contain a red anthocyanin pigment that becomes blue upon reacting with certain metals, including aluminum. The relative availability of aluminum, which is abundant in most field soils, is thus the principal factor in determining the color of the forced hydrangea. Unless steps are taken to prevent aluminum uptake, the pink sepals gradually become blue. Just as important, unless enough aluminum is present to react with all the anthocyanin completely, an intermediate color will be produced instead of the desired clear blue color. Intermediate colors are not attractive in most varieties.

Aluminum becomes more available to plant roots as soil acidity increases (pH values become lower). Growers usually lime to a pH of 6.5 to produce pink hydrangeas and acidify to a pH of 5.5 or less for the production of blue flowers. (Do not use phosphoric acid to reduce the pH of irrigation water for production of blue plants.) In soilless media, these pH values are often a half to one unit lower.

Phosphorus will also render soil aluminum unavailable—high phosphorus and high nitrogen

Avoid "Blurple" Hydrangeas

The key to ensure clear pink or blue colors is first to order plant material that is programmed to develop the desired color. After that, you need to continue to control the availability of aluminum sulfate, which produces the blue color. Keep aluminum away, and you will get pink flowers. White varieties will be white regardless of aluminum availability.

Putting the Pink in Hydrangeas

Keep pink hydrangeas pink by not using mineral soils in your medium. Do not use fertilizers that contain aluminum. Make sure your fertilizer has a relatively high level of phosphorus, which helps prevent aluminum uptake. You can incorporate triple superphosphate (0-45-0) at 3–4.5 lbs./yd.^{3} (2–3 kg/m^3) of potting mix or rotate monoammonium phosphate (11-53-00) into your feed program. A sample fertilizer program for pink hydrangeas is continuously feed with 150 ppm from 20-10-20, rotated with 100 ppm from 11-53-00 every third feeding.

It is also important to keep the medium pH at least 6.0–6.2, because aluminum becomes more available at lower pH levels. Do not let the pH rise above 6.4, or you will get iron-deficiency-induced chlorosis. If your irrigation water pH is above 6.5, consider adding acid to reduce it to 6.3. Use phosphoric acid to increase the phosphorous level in the medium. Supply only low to moderate levels of potassium as high levels will increase the bluing of hydrangeas.

Creating Hydrangea Blue

Dormant blue hydrangeas will have received aluminum-sulfate treatment from the plant producer prior to shipment. You need to keep supplying aluminum while you are forcing them into bloom.

Transplant blue hydrangeas into a phosphorus-free medium and use a phosphorus-free fertilizer. When feeding, apply high levels of potassium. A sample fertilizer program would be nitrogen at 150 ppm and potassium at 300 ppm for each irrigation, supplied with ammonium nitrate plus potassium nitrate.

Start drenching with aluminum immediately after transplanting. Drench with an aluminum-sulfate solution. Make sure the soil is moist before you drench to avoid damaging the roots. Apply the drench every ten to fourteen days.

Check your medium pH ten days after application. If the pH is greater than 5.6, drench again. The aluminum sulfate not only provides aluminum, it also keeps the pH low (5.2–5.5), which is desirable for blue hydrangeas. If you need to use acid to lower the irrigation water pH, use sulfuric acid because it will not add unwanted phosphorus and is not as caustic as more concentrated sulfuric or nitric acid.

Of course, leaving a few blue hydrangeas untreated is not always a bad thing: Sometimes those weird "blurple" plants are the first to sell!

Information adapted from the North Carolina State University Cooperative Extension website.

during flower development promotes clear pink sepals. Low phosphorus and nitrogen, but an abundant supply of potassium, promotes clear blue sepals when the medium contains plenty of aluminum. (Do not add or use superphosphate when producing blue hydrangeas.)

Hydrangea growth during forcing requires a relatively high nitrogen ratio: A 2-1-1 or 3-1-1 ratio is adequate. Plants are not fertilized in cold storage, nor are they heavily fertilized during the late phases of summer growth. For this reason, give particular attention to the effects of fertilization in both the potting medium and the liquid feed program in the greenhouse. Raise fertilization with liquid fertilizers or top-dressed dry fertilizers after root growth is initiated in the medium. If plants have been cold-stored in the pots in which they are to be forced begin fertilization immediately. In a nutshell: The rule is for pink hydrangeas, grow at a pH of 6.0–6.5 with high phosphorus, high iron, low aluminum, and low potassium. For blue hydrangeas, grow at a pH of 5.0–5.5 with low phosphorus and high aluminum.

For production of pink sepals, alkaline-residue fertilizers, such as calcium nitrate, are used. When pink color is difficult to attain, ammonium phosphate (either mono- or di-ammonium) at 700 ppm nitrogen can be used on alternate weeks after flower buds are visible. The phosphate fertilizer ties up aluminum in the soil, preventing the blue color from developing. Replace fertilization when sepals are in full color and plants are hardening.

In the production of blue-flowering plants, the dormant (summer) plant grower needs to have made summer applications of aluminum sulfate. Summer applications of aluminum sulfate alone, however, are not sufficient to produce a reliable blue color. Four or more applications are necessary in the greenhouse, in addition to planting in a medium low in lime. Aluminum sulfate at the rate of 10 lbs./100 gal. of water (4.8 kg/400 l) is satisfactory. This material should be applied only to a moist medium.

Fertilization for plants being forced for blue sepals should be lighter than for pink sepals. Use low phosphorus and high potassium levels for the clearest blue color. Additional applications of aluminum sulfate, made on several alternate weeks after flower buds are visible, should ensure complete bluing of the sepals. Additional applications may be required if the soil or water is alkaline.

Low nitrogen rates are usually used on plants being forced as blue to produce clear blue colors. For blue, use 100–200 ppm nitrogen, and use 200–400 ppm nitrogen for pink varieties. White varieties are best fertilized on the pink sepal program for best plant appearance. Reduce fertilizer rates by half as soon as plants begin to show color to improve postharvest life.

Pests & Diseases

The most common problems are aphids during the forcing period and two-spotted spider mites during summer growth. Plants should undergo continuous inspection. Slugs and snails may be present on plants as they are brought from storage. They can also be particularly troublesome where plants are forced on soil or on solid-bottom beds.

Botrytis can occur in situations with high humidity and high moisture. Powdery mildew is the most prevalent disease on outdoor plants in the fall and is also a problem in the greenhouse under conditions of high humidity and crowding. Hydrangea ring-spot virus has been found in most present-day commercial varieties. Typical symptoms show only during winter growth, and the effect on susceptible varieties is generally weakened or smaller growth. Roguing is difficult, and virus-free plants of commercial varieties are not currently available. Green sepal mycoplasma complex has been responsible for a series of problems. With a severe infection, extreme stunting, small leaves with vein yellowing, and dwarf, green cymes are followed by death of the plant. An intermediate case presents a reduction in vegetative growth, but with normal leaf expansion and continued vein yellowing. The cymes will contain both green or bronzed sepals and normal-colored sepals.

Troubleshooting

Iron deficiency can show up in pink varieties being forced at a pH of 6.0–6.5 because iron is less available at higher pH levels. If you are growing in soilless medium, this is especially true. Applying chelated iron every three to four weeks through the production cycle can alleviate the problem. Chlorosis can also be due to alkaline soil, overfertilization, or overwatering during summer growth. Plants grown for pink are more likely to show iron chlorosis than those being blued because of the greater availability of iron at lower pH. An application of 0.5 lb. (227 g) of iron sulfate and 0.5 lb. (227 g) of ammonium nitrate is also very helpful.

Failure to initiate flower buds or evidence of crippled buds during forcing may be due to poor culture during summer growth, frost injury during storage, or bud rot (gray mold) in storage or shipping. Initiation of flowers early in the summer may result in fewer leaves than normal at forcing, causing poor flower development because of the lack of leaf area. Cymes containing leaves are also associated with early initiation. Removing leaves from the cymes early in forcing usually permits the cyme to develop normally.

Plants can show signs of chemical burn. The young growth of hydrangea leaves and flowers at forcing is very susceptible to injury from insecticides, fungicides, and growth regulators. Use caution with any chemical spray and confine dosages to the lower recommended rate. Hydrangeas do not tolerate herbicides of any kind. The use of bleach in close proximity to hydrangeas can cause severe damage.

Varieties

Most hydrangea varieties originated in Germany, France, Belgium, and Switzerland. Some have retained their original names, while others have been renamed by the introducers and are known by different names in the United States. The outdoor types with flat cymes and staminate flowers only at the outer edge are known as lace caps; they have sparked new interest in hydrangea as a flowering pot plant. Only one currently popular variety, 'Rose Supreme', is of US origin.

In the North, early varieties are expected to force in twelve weeks, midseason varieties in thirteen weeks, and late varieties in fourteen weeks under usual forcing conditions. Sepal colors are always more intense when forced at cooler temperatures.

At one time, 'Rose Supreme' was the most popular hydrangea variety (table 17). In the United States, 'Merritt Supreme', a very different plant, has taken its place. 'Rose Supreme' is very vigorous and produces extra-large, light pink or light blue heads. Intermediate colors are also good. It is tall, requiring B-Nine applications during summer growth and usually two to three B-Nine applications during forcing to reduce height. 'Rose Supreme' is a late variety, requiring about one week longer to force than most, and two weeks longer than 'Merritt Supreme'. It stands up well in the heat and is used extensively in Texas, Florida, and other southern areas. 'Rose Supreme' may be difficult to bring to flower for the earliest Easters. It is always suitable for Mother's Day, though.

Also difficult to grow in the summer, 'Kuhnert' is an excellent forcer if grown cool. It blues easily, giving clear, light blues. 'Kuhnert' is not particularly adapted for southern forcing, although some use it for its color. At Dahlstrom and Watt, all 'Kuhnert' plants are blued in the summer. It branches well and has smaller flowers. 'Kuhnert' typifies the often-discussed "new European" varieties, but it has been grown in the United States for years.

Most whites are not of high quality in regard to hardiness of flower and compactness of growth. However, 'Sister Therese' is as good as any and better than some. It forces early and readily with large heads. Provide extra protection against sunburn during forcing. 'Sister Therese' will not usually need B-Nine early, but it will benefit from a "toning" application later, after its height is nearly established. 'Sister Therese' flowers ten days ahead of 'Merritt Supreme'. When it is received in shipment at the same time as

Table 17. Hydrangea Characteristics

VARIETY	COLOR	FLOWER-HEAD SIZE	DAYS TO FLOWER AT 60°F (16°C)	PLANT HEIGHT	HEAT TOLERANT
'Firelight'	Red	Large	85	Medium	Yes
'Jennifer'	Red	Small	85	Short	Yes
'Kuhnert'	Light blue	Medium	90	Medium	Somewhat
'Mathilda Gutches'	Blue	Medium	90	Medium tall	Yes
'Merritt Supreme'	Medium pink	Large	85	Medium	Yes
'Rose Supreme'	Light pink	Very large	95	Very tall	Yes, very
'Sister Therese'	White	Large	78	Short	Not very

'Merritt Supreme', it should be delayed by cold storage or in a cold greenhouse for ten days before forcing.

Another white, 'Regula', has a more symmetrical habit than 'Sister Therese'. It also flowers later, which as a result may offer advantages. 'Regula' is not often offered, though, because of its rank growth and extremely soft nature.

Among the reddest hydrangeas available, 'Jennifer' has light green foliage, and its sepals are greenish in the center, especially in the early stages. It makes an outstanding addition to the color selection. Growth is not especially vigorous with respect to stem diameter, and flowers are generally smaller than on 'Merritt Supreme'. Flower numbers, however, make up for flower size. Some growers use up to 10% of finished plant production in red.

'Mathilda Gutches' (also called 'Mathilda Gutges'), a well-formed European variety with excellent steel-blue flowers, branches profusely. Its flowers are smaller, but the show is comparable because of flower numbers. It makes an excellent blue as well as a medium pink. 'Mathilda Gutches' is also one of the most popular varieties for blue. It can be tall, requiring B-Nine because of its smaller-diameter stems.

'Firelight' (also called 'Leuchtfeuer') is dark pink and more rose-colored than 'Merritt Supreme'. Under very cool temperatures (50–55°F [10–13°C]) at finishing, it can be nearly red. 'Firelight' breaks freely, has sturdy stems, and features pretty, large flowers. Some say it is susceptible to mildew, but that has not been everyone's experience. We believe it forces slightly faster than 'Merritt Supreme'. It apparently performs well in the South. 'Firelight' is a vigorous variety and will require B-Nine.

'Masja' is a pleasing, red-pink color and forces along with 'Merritt Supreme'. It has good growth habit and light-green, attractive foliage and grows well. Its vigorous growth will likely require B-Nine.

The large, diverse group of plants known as lace caps has one row of sterile flowers surrounding a cluster of fertile flowers. The fertile flowers are usually an attractive blue. Lace caps are novelties. Forcing time is about two weeks slower than 'Merritt Supreme'. The color of the flowers improves with age with the sepals being very light when they first expand. White lace caps are of great interest. However, powdery mildew can be a problem with all lace caps.

'Blaumeise' is a versatile variety that can be blued or grown pink and is the most prominent variety for these colors. It is a vigorous grower.

'Sara', a light pink of very good substance, has a forcing time about one week slower than 'Merritt Supreme'. It is worthy of trial.

Postharvest

Plants are ready for shipping when they are fully colored. Do not keep plants in the dark for more than two or three days, as leaves can drop. Ideally store and ship plants cool (35°F [2°C]) for the least amount of time possible. Prior to boxing, cool plants down in the cooler to avoid condensation on sepals, which can cause botrytis.

Retailers should unpack and unsleeve plants immediately on arrival. Provide at least 250 f.c. (2.7 klux) of light and cool temperatures 60–70°F (15–21°C) in the retail display. Do not allow plants to dry out, as it will cause leaf and flower drop.

The hydrangea, like the poinsettia, has the capability of being a long-lasting flowering plant because the showy parts are not petals that rapidly fade and fall, but instead are sepals. Treated well, plants will flower for three to four weeks. Once they are finished flowering, consumers in Zones 6–9 can enjoy them for years in the garden as flowering shrubs.

Hydrangea flowers may be dried using glycerin. Harvest stems for drying when they are fully colored. After treating, generally for three to four days, rinse stems with clear water, cut off the portion of the stem that was in the glycerin, and hang them to dry for one to two weeks.

Hypericum

Hypericum spp.
Cut flower

In the fall, consumers cannot get enough of the bright red, orange, and purple berries of *Hypericum.* At other times of the year, this relatively new cut flower shows up in mixed bouquets. Worldwide, *Hypericum* has caught on as a great cut, with Ecuadorian, Israeli, European, and African growers producing stems by the thousands. *Hypericum* is novel without being obtrusive and is versatile in arrangements. Growers can harvest bushes for years and years, provided rust does not become a problem.

Since the best *Hypericum* cut flowers are patented and propagation is rigidly controlled, we will skip the propagation section for this crop.

Growing On

Choose only well-drained soils for *Hypericum* culture. Perform a soil test prior to planting and adjust nutrients according to the results. Adjust the pH to 5.5–6.0.

Keep the integrity of the soil ball intact when you plant and do not plant too deeply. Begin new plantings at a higher density and remove plants after they have been cropped for a couple of years. Start off new plantings at 13 plants/yd.^{2} (16 plants/m^2), later removing plants as they mature to a density of 3 plants/yd.^{2} (4/m^2) for fully mature plants.

If you are growing your crop outdoors, provide 30% shade. Pinch cuttings when they begin to stretch to encourage fuller plants, leaving three leaf pairs.

Grow the crop at 60–65°F (16–18°C) days and 55–60°F (13–16°C) nights, although plants will tolerate a wide temperature range (from 24–90°F [–4–32°C]). *Hypericum* is hardy to about Zone 5; however, it does not like extreme heat combined with high humidity.

Adjust your fertilizer composition as plants move through their development. When plants are vegetative, maintain nitrogen at a higher ratio than potassium. When stems reach 10" (25 cm) long (beginning of long days), increase the ratio of potassium to nitrogen. Constant liquid feeding through drip irrigation is also suitable. To prevent disease, avoid overhead irrigation.

Hypericum is a long-day plant. Provide fourteen-hour days until stems are 10–12" (25–30 cm) long, increasing to eighteen hours until the end of flowering and then returning to fourteen hours until harvest. *Hypericum* responds better to extending the day length than to night interruption lighting. If you do not light plants, expect one flush per year when growing under natural day lengths.

Back off on irrigation when berries are ripening to prevent them from bursting.

From a new planting, expect two stems per plant for the first flush, six stems on the second flush, and fifteen stems per plant on subsequent flushes. The first flush will occur about sixteen to eighteen weeks following planting, with subsequent flushes every sixteen weeks thereafter. You can time harvests in order to target holidays or special programs by pinching plants.

Pests & Diseases

The most serious pest problem is rust, which causes yellow spots on the top of leaves and orange spots on the undersides of leaves. The disease is easily spread and very difficult to control. Keep beds clean and treat any infestation immediately. Breeders are working on rust-resistant varieties. Nematodes can also be a problem, as can aphids and thrips.

Troubleshooting

If flower set is poor, try a twenty-hour day length. You can also stress the plant by withholding water to force it into reproductive growth.

Postharvest

Stems are ready for harvest when all the berries have matured. Cut stems back hard, leaving only 0.5" (1 cm) above ground. You will be harvesting from ready plants over the course of one to two weeks. Treat stems with a floral preservative solution that contains a bactericide immediately after harvest to improve water uptake.

Hypoestes

Hypoestes phyllostachya
Common name: Polka dot plant
Annual, foliage plant

Hypoestes is one of those old-fashioned plants that is just right for today's garden. Grown for its foliage rather than its flowers, leaves generally have green mottling that is pierced with pink, red, or white coloring. It is a great 4" (10 cm) foliage plant and be sold either as a stand-alone pot plant or used in dish gardens. *Hypoestes* also makes an ideal backdrop in color bowls and combination pots. Some landscapers love it as a solid mass of color planted in shady beds or used as an accent with solid-colored flowers, such as impatiens. For growers, *Hypoestes* is an easy crop. Retail bedding plant growers can add a few trays to their production to brighten up their product offerings, ordering it along with regular bedding plant plugs.

Propagation

Sow seed onto well-drained, disease-free, germination medium with a pH of 5.5–6.5 and a low starter charge. If desired, cover seed lightly with coarse vermiculite. Multisow plugs for the shortest finishing times. Some growers put as many as eight or ten seeds per cell to produce 4" (10 cm) pots. Germinate at 68–75°F (20–24°C). Seedlings will emerge in four to seven days. 288s require four to five weeks from sowing to transplanting.

Some growers sow seed directly to the final growing container, using as many as seven to fifteen seeds per pot. Direct-sowing will result in a slightly faster crop time, with pots finishing in seven to nine weeks in late spring or summer.

Growing On

Plant 1 single or multiplant plug/cell pack, 3 plugs/4" pot, or 5 plugs/6" pot. Select a disease-free, well-drained medium with a moderate starter charge and a pH of 5.5–6.0.

Provide 50–60% shade under high light. When *Hypoestes* is produced under light levels that are too high (more than 2,500 f.c. [27 klux]) and high temperatures (more than 85°F [29°C]), leaves curl.

Fertilize at every other irrigation with 150–200 ppm from 20-10-20 or 15–5–15. Maintain uniform moisture levels.

Using Cycocel for height control will intensify leaf coloring and keep the central stem in check, which creates a uniform, mounded plant shape. Apply at a rate of 700–1,000 ppm once the second set of leaves develops. Use again in two weeks. B-Nine at 1,000 ppm is also effective for height control, as is a B-Nine/Cycocel tank mix. Apply growth regulators to leaf coverage, not runoff.

Eight to ten weeks is required for 4" (10 cm) pots from plugs.

One additional note: *Hypoestes* flowers under short days, and many consider the flowers to be unattractive detractors from the foliage. Sowings made in winter or spring will not flower until late summer. However, if you have to keep plants from flowering, provide long days (sixteen hours or greater), especially for winter crops when light levels are low. Mum lighting from 10 P.M.–2 A.M. works well. Also, be sure to maintain good light levels at 1,500–2,000 f.c. (16–22 klux), as low light levels cause plants to produce flowers rather than leaves.

Varieties

Try 'Confetti', which is available in 'Burgundy', 'Red', 'Rose', or 'White', or 'Splash Select', which is available in 'Pink', 'Red', 'Rose', or 'White'. 'Confetti' grows to 18–24" (46–61 cm) in the garden, while 'Splash Select' grows 8–10" (20–25 cm).

Postharvest

Instruct landscapers and consumers to keep plants trimmed to their desired size in the garden. Established plants can tolerate higher light levels outdoors. Keep flowers pruned for best appearance.

I

Iberis

Iberis sempervirens
Common names: Candytuft; hardy candytuft; evergreen candytuft
Perennial (Hardy to USDA Zones 3–9)

A perennial favorite for the home garden, *I. sempervirens* is a low-growing, woody-stemmed, evergreen perennial. It grows about 9–12" (23–30 cm) tall and spreads 10–15" (25–38 cm) across. The flowers are single, pure white, and scentless and develop in clusters measuring 2–3" (5–7 cm) across in early spring. Plants are dependably hardy and are one of the first returning perennials to show color in the spring.

Propagation

Iberis can be propagated by seed, cutting, or bare-root transplant.

Seed

The first thing to know about seed-propagated plants is that they are not as uniform in habit as their vegetatively propagated counterparts. Plants can be up and down in performance.

Seed is sown from May to July for transplanting to 1-qt. or 1-gal. (1.1 to 4 l) containers to overwinter for flowering the following spring. Any sowings done in January or beyond may not flower or will not flower profusely. Plants need to be kept cold during the winter at 32–34°F (0–1°C) and amply protected since their foliage is evergreen and rots under wet, cold conditions in the cold frame.

Sow two to three seeds per 288 or larger cell, seed does not need to be covered (although some growers state they get higher germination if they cover the developing seedlings lightly with vermiculite as the root radicle develops—four to six days in the plug tray after sowing in summer). Allow six to seven weeks from sowing to be ready to transplant to larger containers.

Cuttings

Tip cuttings are taken anytime the plants are green and not flowering (summer to winter) or as long as the plants are kept warm in a vegetative state.

Tip cuttings stuck to a 105, 84, or larger liner tray, dipped in rooting hormone, and given mist for nine to fourteen days will salable in five to seven weeks. Pinching can be done at this time although it is better in the growing on stage to make plants more uniform in appearance.

Bare-root transplants

Potted up in winter to 1-qt. or 1-gal. (1.1 or 4 l) containers, plants are kept on the greenhouse bench until shoots emerge. As roots develop and reach the side of the container, plants can be moved to a cold frame at 58–65°F (14–18°C) or higher days and 50–55°F (10–13°C) nights until sold green.

Growing On

Candytuft must have excellent drainage and high light. Use a well-drained, disease-free medium with a moderate initial starter charge and a pH of 5.5–5.8. If you overwinter, be sure the medium will not break down quickly: Bark mixes are great.

Seed or cuttings

Transplant plugs or liners to 4", 6", or 1-qt. (10 cm, 15 cm, or 1.1 l) containers and grow on at 65–68°F (18–20°C) days and 58–60°F (14–16°C) nights to root plants once received in late summer or early autumn. Provide 2,500–3,000 f.c. (27–32 klux) of light and fertilize with 75–100 ppm from 14-0-14 at every other irrigation. Stop fertilization by early September.

To induce flowering, expose plants to ten weeks at 35–41°F (2–7°C) in an unheated cold frame or lighted cooler (25–50 f.c. [269–538 klux] of light). Cooled mature plants will flower fast—in just three to five weeks after bringing them out of the cold and growing at a minimum of 55°F (13°C) nights and days at 6–10°F (3–5°C) warmer. Make sure plants are well developed and bulked up before vernalization. If you are growing plug seedlings, plants need to be at least a 50-cell size or have fifty leaves to overcome juvenility and flower consistently. No. 1 bare-root liners are generally a sufficient size already. Encourage branching by trimming plants back after they come out of vernalization. Maintain high light levels (4,000–6,000 f.c. [43–65 klux]). Keep plants on the dry side and fertilize at every other irrigation with 150 ppm of 20-10-20.

Bare-root transplants

Potted up in late January or February to 1-qt. or 1-gal. (1.1 or 4 l) containers will be in flower, in most cases, by mid-April to early May. Grow on at the temperatures noted above under Growing On for seed or cuttings.

Pests & Diseases

Insects rarely bother candytuft. *Pythium* can be a problem, however, so watch watering and allow plants to dry thoroughly between irrigations.

Troubleshooting

If you see some pinking in flowers, that's normal: Cool day and night temperatures bring out pink in the flowers.

Postharvest

Keep plants cool (65–68°F [18–20°C] days and nights 10–15°F [5–8°C] cooler) and out of the elements to prolong flower quality. If needed, you can trim back mature plants to induce a second flowering, although it is not as profuse as the first flowering. Indeed, in most locales it does not rebloom well.

Impatiens

Impatiens hawkeri
Common name: New Guinea impatiens
Annual

New Guinea impatiens have become one of the most important bedding plants in the market, rising meteorically in their forty-year history. The crop was introduced to the commercial trade through the United States Department of Agriculture (USDA). Their positive impact in the industry is a success story demonstrating that government and industry can indeed work together. When New Guineas were first released to the trade in the early 1970s, commercial plant breeders went to work to develop today's modern introductions. The plants that you are able to grow today are the result of decades-long breeding programs instituted by some of the leading plant supply companies in the world. New Guinea impatiens are one of the most important spring plants, comprising of high percentages of pot plant and hanging basket sales at wholesale.

Modern New Guinea impatiens are a joy for homeowners: Their oversized, bright flowers and fun foliage are a treat in the garden. Provided they receive good light and adequate moisture, they are reliable garden performers. Breeding has also lead to more sun tolerance, which has opened new possibilities for their use. For growers, New Guineas practically grow themselves as long as growers remember that too much fertilizer will inhibit flowering, are careful to irrigate as the plants need it, and have a zero-tolerance policy for thrips.

You will want to offer a wide range of New Guinea colors and leaf patterns in 4 and 6" (10 and 15 cm) pots and hanging baskets. They also make great additions to combination container designs.

Propagation

New Guinea impatiens are primarily propagated by cuttings, although seed is also available.

Cuttings

New Guinea impatiens can harbor a number of diseases, the most notorious of which is tomato spotted wilt virus/impatiens necrotic spotted wilt virus (TSWV/INSV). It is imperative that you begin with clean cuttings taken from mother plants maintained in a rigorous clean-stock system. Know your supplier!

Some growers prefer to buy in rooted cuttings (liners); others prefer to buy in unrooted cuttings. Most growers do not attempt to maintain stock plants of New Guineas: The threat of disease is just too great. Besides, it is easy to get cuttings in, and if you are careful with whom you work, you can be certain they are clean.

It is always best to stick cuttings right away, but if you must store cuttings, store them for no longer than twenty-four hours at 50–60°F (10–16°C). If your cuttings are wilted, place open bags on a clean surface in a cooler with a relative humidity near 100% for at least four hours. Mist them to increase turgidity. Stick cuttings into Ellepots or a disease-free rooting medium with a pH of 5.5–6.0. If you choose to use a medium, limit the peat moss component, as peat moss will hold too much water. Some growers prefer to use rooting hormones to unify rooting across all varieties. Cuttings (tips) should measure 0.75–1.5" (2– 4 cm) long with two expanded leaves. Place cuttings so that 0.5" (13 mm) of the stem is below the soil line: The leaf canopy should touch the soil.

Root cuttings at a substrate temperature of 70–75°F (21–24°C) and air temperature of 70–75°F (21–24°C) days and 65–70°F (18–21°C) nights. Make sure that the medium is moist: You should be able to squeeze water out easily. Provide low ambient light levels of 500–1,000 f.c. (5.4–11 klux); retractable shade is ideal, as you will increase light levels as cuttings mature.

The only reason to mist is to maintain cutting turgidity. Only apply as much as is needed. While you should adjust mist frequency and duration to your exact greenhouse and climate conditions, some starting guidelines for misting are: week 1, five seconds every fifteen minutes in winter (five seconds every ten minutes in spring); week 2, five seconds every thirty minutes in winter (five seconds every twenty minutes in spring); and week 3, mist only as needed to keep humidity at 80–90%. Excessive misting will cause disease. Be diligent in evaluating your timing and mist duration to ensure cuttings do not wilt, yet stay dry enough so that disease cannot become established.

As soon as you notice any loss in foliage color, apply a foliar feed of 50–75 ppm from 20-10-20. Once 50% of the cuttings have begun to show root initials (about five to seven days after sticking), move cuttings to the next stage of development.

The next stage, root development, will last from nine to eleven days. Reduce substrate temperatures to 68–72°F (20–22°C) and maintain air temperatures of 70–75°F (21–24°C) days and 65–70°F (18–21°C) nights. Cold temperatures during propagation will cause leaves to reflex downward. As soon as cuttings begin to develop roots, allow the rooting medium to begin drying out somewhat (see the mist guidelines above). Remember that roots will grow when they have to search out water. However, avoid drying out the air, as this will increase evapotranspiration, which will reduce the root zone temperatures. Increase light levels to 1,000–1,500 f.c. (11–16 klux) as the cuttings begin to root out. Apply a foliar feed of 100 ppm from 15-0-15, alternating with 20-10-20; increase rate rapidly to 200 ppm as cuttings root in. The majority of fertilizer should be nitrate fertilizer 15-0-15. Maintain the substrate pH at 5.5–6.0.

As cuttings finish rooting, harden them off prior to planting or sale for seven days. Lower ambient air temperatures to 68–75°F (20–24°C) days and 65–68°F (18–20°C) nights. Cuttings should no longer be under mist. Increase light levels to 1,500–2,500 f.c. (16–27 klux). Provide shade during the heat of the day to reduce crop stress. Fertilize once per week with 150–200 ppm from 20-10-20, alternating with 15-0-15.

Cuttings rooted in Ellepots or medium will root in five to seven days at 70–75°F (21–24°C) and require three to four weeks to be ready to transplant to larger containers from a 105-liner tray.

Seed

Sow the seed into a 288- to 128-size plug using a well-drained, disease-free, seedling medium with a pH of 5.5–6.0 and an EC about 0.75 mS/cm (1:2 extraction). Water adequately after sowing. Cover seed lightly with coarse vermiculite to maintain high moisture around the seed coat. Germination takes six to eight days.

Seed may be germinated either on the bench or in a germination chamber. Germinate at 72–78°F (22–26°C). After germination, keep the air temperature at 70–75°F (21–24°C) and soil temperature at 70°F (21°C) until transplant. Light (up to 2,500 f.c. [27 klux]) appears to improve germination. In Stages 3 and 4 increase light levels up to 5,000 f.c. (54 klux), if temperature can be controlled. Supplemental lighting is not required but will decrease total crop time.

Maintain 100% relative humidity (RH) until radicles emerge. RH can be reduced gradually to approximately 50% as plugs mature.

Keep plug trays in high moisture until late Stage 2, and then start reducing moisture. Avoid wilting: New Guinea impatiens cannot tolerate wilt.

At radicle emergence, apply 50 ppm nitrogen (0.4 mS/cm EC) from low phosphorous-nitrate form fertilizers, such as 13-2-13. As cotyledons expand, increase to 100–150 ppm nitrogen (0.9–1.3 mS/cm EC). If growth is slow, apply 20-10-20 with every other fertilization.

Plant growth regulators are not needed in the plug stage for the 'Divine' series.

Plugs are ready for transplanting when "pullable" from the plug tray. Do not allow plugs to get root-bound. A 288-plug tray takes from five to six weeks from sowing to be ready to transplant to larger containers.

Growing On

Cuttings

Plant 1 cutting/4" (10 cm) pot; 1–2 cuttings/6" (15 cm) pot; 3 cuttings/8" (20 cm) hanging basket; and 4–5/10" (25 cm) hanging basket. Pot cuttings into a well-drained, disease-free medium with a moderate initial starter charge and a pH of 6.0–6.5. Be sure to maintain higher pH levels, as low pH can cause minor nutrient toxicity.

Paying attention to irrigation immediately after potting is critical, especially in hanging baskets, when it will be easy to apply too much water. Excessive water can inhibit rooting because the roots do not need to search for water. Overwatering can also cause disease problems. Some growers prefer to establish liners in 3 or 4" (8 or 10 cm) pots first, and then plant their hanging baskets. Later, as plants root into their containers and become established, it is important not to water-stress New Guineas: Keep the medium uniformly moist to keep plants continually growing.

Grow on at 68–75°F (20–24°C) days and 65–68°F (18–20°C) nights. As plants come into flower, reduce night temperatures to 62–65°F (17–18°C) to intensify flowering. Southern growers target 70–80°F (21–26°C) days and 68–72°F (20–22°C) nights. New Guinea impatiens are very sensitive to DIF: Avoid high day temperatures and low night temperatures (positive DIF), as plants will stretch rapidly.

Grow plants on at moderate light intensities of 2,500–4,000 f.c. (27–43 klux). Maintain the higher range of light levels for best leaf coloration, provided growing temperatures can be moderated. Never hang New Guinea impatiens in baskets in the greenhouse "attic": Both light levels and temperatures are too high. Most growers produce their New Guinea impatiens hanging basket crops on the floor or bench. Do not hang other crops that can drip onto your New Guinea baskets.

New Guinea impatiens are daylength neutral; however, as light levels naturally increase through the year, flowering also increases. During winter months, New Guineas benefit from supplemental lighting. In the South, shade may be necessary by the end of the spring and into the summer to moderate temperatures.

New Guineas are highly sensitive to soluble salts and have a low to moderate fertilizer requirement. Immediately after transplanting, do not fertilize for a couple of weeks. As plants begin to root in, use 50–100 ppm at every other irrigation, alternating between 15-0-15 and 20-10-20. As plants mature, root in, and actively grow, increase rates to 200–250 ppm. Leach with clear water every third or fourth irrigation. *Note:* Excessive ammoniacal fertilizers will cause large leaves and poor flowering. Maintain soil EC below 2.0 mS/cm.

Space pots once leaves begin to touch. For 4" (10 cm) pots, space at 6" × 8" (15 × 20 cm); 6" pots, 10" × 10" (25 × 25 cm); for 8" (20 cm) hanging baskets, 14" × 14" (36 × 36 cm); and for 10" (25 cm) hanging baskets, space at 18" × 18" (46 × 46 cm). Crop times vary, but, in general, in the spring you can count on eight to ten weeks for 4" (10 cm) pots in the North and six to eight weeks for 4" (10 cm) pots in the South. Ten-inch (25 cm) hanging baskets will require twelve to fourteen weeks in the North and ten to twelve weeks in the South.

If growth regulators are required, Bonzi (25–30 ppm) or Sumagic (5–10 ppm) can be applied as a spray after plants are actively growing and before stretch. Reapply as needed. Reduce rates under higher light conditions.

Using one plant per pot, the crop time to a flowering 4" (10 cm) pot is seven to nine weeks.

Seed

Seed-propagated New Guinea impatiens are best suited to 306 premium packs, 1801 flats, 4" (10 cm) pots, and hanging baskets.

Maintain air temperature at 65–80°F (18–26°C) from transplant to sale. The warmer the temperature, the faster the plant will flower. Lowering the temperature to 61–65°F (16–18°C) in the weeks during flower development will encourage larger flowers.

Maintain light levels as high as possible, while maintaining the appropriate temperatures. Flowering is also related to light accumulation—low light will delay flowering.

No major disease problems will arise if using good cultural and integrated pest management practices. Thrips are the most common insect pests.

Total crop time from transplanting a liner to a flowering 4" (10 cm) pot is seven to eight weeks.

Pests & Diseases

Aphids, fungus gnats, spider mites, and thrips can attack New Guinea impatiens. Disease problems can include botrytis, *Pythium*, rhizoctonia, stem canker and TSWV/INSV.

As thrips are the vector that spreads TSWV/INSV, it is vital to maintain a zero-tolerance policy for thrips. Monitor insect populations rigorously with blue or yellow sticky cards. If TSWV/INSV symptoms appear, collect samples to be sent in for positive disease identification. In the meantime, treat that block of plants as if it is infected: Restrict access to only the most trusted employees. Irrigate as needed, being careful not to touch plants. If test results are positive, collect infected plants, place them in plastic trash bags, and remove them from the premises. There is no cure or treatment for TSWV/INSV, and your goal is to prevent it from spreading to the rest of your New Guineas, other impatiens, or other greenhouse crops. Most growers remove all plants showing symptoms; in most instances that would be all of the plants within a variety color for the same ship date. Some growers like to remove the immediately adjacent plants as well, just to be safe.

The best way to control TSWV/INSV is to prevent it in the first place. You can do that by adhering to a strict policy not to carry over any impatiens plants from one year to the next. Some floricultural crops can harbor TSWV/INSV without showing symptoms, serving as a Typhoid Mary in your greenhouse. Ideally you should bring in new plant material each year for all the crops you grow; material should be purchased from a propagator and/or breeding company dedicated to maintaining a clean-stock program. Continue to monitor thrips populations and create a fully integrated pest management approach for long-term control.

Troubleshooting

Plant collapse could be caused by a medium that has been wet for extended periods of time and/or botrytis (stem canker). Excessive nitrogen in the fertilizer; overfertilization and/or low light; overwatering and/or low light; and excessive or late Florel applications can result in too much vegetative growth and a lack of flowering. Drying out plants between irrigations or excess minor nutrient levels in the soil may lead to foliage necrosis. Low fertilization during the early stages of growth can lead to poor branching and thin plants. Leaves that appear dark, shiny and wavy, or rippled and/or cupped may be showing signs of overfertilization.

Varieties

Try the 'Celebrette' and 'Petticoat' series for 4 and 5" (10 and 13 cm) pots and the 'Celebration', 'Riviera', and 'Sonic' series for 6" (15 cm) pots and hanging baskets. All these series come in a full color range and feature a wide selection of leaf colorings, from light green to dark green to variegated types.

Other vegetatively propagated varieties include 'Fanfare', 'Paradise', and 'Harmony'. 'Baby Bonita' and 'Sweeties' are small-flowered series. The 'Super Sonic' and 'Magnum' series are a result of breeding for very large flowers.

'SunPatiens' are a newer series that has been bred for full sun. The original breeding came to market with very vigorous plants. The most recent releases have corrected this with more compact plants.

The 'Java' and 'Divine' series are a seed-propagated New Guinea impatiens. Plants make great 36s, 1801s, or 4" (10 cm) pots for mass-market sales. All colors in these series flower within one week.

Postharvest

New Guinea impatiens are very sensitive to ethylene gas. Before shipping, remove leaf and flower debris by grooming plants, which will remove a source of ethylene gas production. Also, do not ship or store New Guineas near ripening fruits or vegetables. Some growers have found 1-MCP helpful as a treatment applied at shipping to prevent flower and bud drop during transport. Ship plants for the shortest time possible: No longer than three days and ideally on racks, not boxed. Maintain shipping temperatures at 55–60°F (13–16°C).

Retailers should display New Guineas under partial shade, about 1,000 f.c. (11 klux) of light. They should keep plants moist at all times: Allowing regular wilting can cause flower and bud drop. Plants need to be groomed by removing dead leaves and flowers to limit ethylene gas exposure.

Consumers should plant New Guineas in a well-drained, partially sunny location and keep them well watered. New Guinea impatiens can tolerate full-sun conditions in the landscape, provided temperatures are moderate and plants have adequate moisture.

Impatiens walleriana
Common names: Bedding plant impatiens; impatiens
Annual

Impatiens are one of the main reasons for the explosion of bedding plant sales through the 1980s and 1990s. They are reliable, both for the grower and the consumer. Intensive breeding efforts by flower seed companies have created a rainbow of colors, flower sizes, and plant types. And best of all, they are easy to grow. Seed treatments and improvements have lead to nearly 100% germination and germination speeds and uniformity unimaginable just a few years ago. Such advancements have made impatiens the model bedding plant crop.

Impatiens are so reliable that in some parts of the country they are overproduced, especially in cell packs or flats and 4" (10 cm) pots. From plugs, flats of finished impatiens can be turned in three weeks during the late spring.

Propagation

Impatiens are produced as seed-grown plugs in four stages. Stage 1, radical emergence, takes from three to five days at a substrate temperature of 75–78°F (24–26°C). Keep the medium moist, near saturation. If it is not possible to maintain near saturation levels with mist, a very light layer of medium may help to maintain moisture levels around the seed.

Reduce moisture levels once radicles emerge. For best germination and rooting, allow medium to dry out slightly before watering. Additional light at 100–400 f.c. (1.1–4.3 klux) may benefit germination. The sowing medium should have a pH of 6.0–6.2 and an EC less than 0.75 mS/cm. During germination, impatiens are very sensitive to high salts, particularly ammonium. Be sure to keep ammonium levels at less than 10 ppm.

During Stage 2, stem and cotyledon emergence will occur over ten days. Maintain soil temperatures at 72–75°F (22–24°C). Provide ambient light levels of 500–1,000 f.c. (5.4–11 klux). Supplemental light at 450–700 f.c. (4.8–7.5 klux) for twelve to eighteen hours per day can be provided for two weeks once cotyledons expand. Keep the substrate pH at 6.0–6.2 and the EC at less than 1.0 mS/cm. At every other watering, begin a liquid fertilizer of 14-0-14 at 50–75 ppm or calcium/potassium-nitrate feed once cotyledons fully expand. Alternate feed with clear water.

Growth and development of true leaves will begin during Stage 3, taking fourteen to twenty-one days. Lower soil temperatures to 68–72°F (22–22°C). Allow the medium to dry thoroughly between irrigations, but avoid permanent wilting. Provide ambient light levels of 1,000–1,500 f.c. (11–16 klux). Continue supplemental lighting for twelve to eighteen hours per day at 450–700 f.c. (4.8–7.5 klux) for two to three weeks after the cotyledons have expanded. Maintain substrate pH of 6.0–6.2 and an EC of less than 1.0 mS/cm. Increase fertilizer to 100–150 ppm from 20-10-20, alternating with 14-0-14 or a calcium/potassium-nitrate fertilizer. Fertilize every second or third irrigation. If using 15-0-15, supplement with magnesium once or twice during Stage 3, use magnesium sulfate (16 oz./100 gal. [1.2 g/l]) or magnesium nitrate. Do not mix magnesium sulfate with calcium nitrate, as precipitate will form. To control plant height during Stage 3, use DIF whenever possible, especially during the first two hours after sunrise. A-Rest, Bonzi, and Sumagic may also be used to control plant height.

During Stage 4, plugs harden and finish over seven days, becoming ready to ship or transplant. Maintain substrate temperatures of 62–65°F (17–18°C). Allow the medium to thoroughly dry between watering. Provide light levels of 1,500–2,500 f.c. (16–27 klux). Continue to maintain the pH at 6.0–6.2 and drop the EC to less than 0.75 mS/cm. Fertilize as needed with 14-0-14 or calcium/potassium nitrate at 100–150 ppm nitrogen. Avoid fertilizers with ammonium or phosphorus during Stage 4.

Most growers buy in impatiens plugs rather than grow them—it's just easier. Because finishing an impatiens crop is so easy, many retail growers also buy in plugs rather than buying them in as finished flats. Common impatiens plug sizes run the gamut, including 512, 384, 288, and 180. Orders for spring bedding plant crops are generally placed in the fall. Growers plant smaller plug sizes into cell packs for flat sales and larger sizes into pots and hanging baskets.

A 288-plug tray takes four to five weeks from sowing to be ready to transplant to larger containers.

Growing On

For the best crops and fastest finishing times, plant plugs immediately upon arrival or as they come out of Stage 4. If plugs must be held prior to transplanting, store at 45°F (7°C).

Transplant impatiens plugs into a well-drained, disease-free medium with a pH of 6.2–6.8 and a moderate initial nutrient charge. Because impatiens comprise a significant percentage of most growers' crops, they are widely planted with automated transplanters. For maximum transplanter efficiency, make sure plants have well-developed root systems.

Plant 1 plant/cell or 4" (10 cm) pot; 2 plants/6" (15 cm); 3 plants/8" (20 cm), and 5–7 plants/10" (25 cm) hanging basket.

Impatiens are tolerant of a wide range of greenhouse conditions. Ideally, maintain night temperatures of 62–65°F (17–18°C) and day temperatures of 65–75°F (18–24°C). Temperatures below 60°F (15°C) will delay flowering and slow growth. Higher temperatures, above 85°F (29°C), can stretch plants and lessen postharvest life. At visible bud, reduce night temperatures to 60°F (16°C). Do not grow impatiens at less than 55°F (13°C), as plants can become damaged due to chilling injury.

Maintain strong light levels of 2,500–3,500 f.c. (27–38 klux). Plants can tolerate light levels as low as 2,000 f.c. (22 klux) or, if moisture and temperatures are favorable, light levels as high as 8,000 f.c. (86 klux). However, high light combined with high temperatures can cause leaf scorch. Crops produced during the summer months require shade to reduce light levels and temperatures. Impatiens produced under high light and high temperatures show poor postharvest life.

Fertilize plants as needed with 15-0-15, alternating with 20-10-20 at 150 ppm. Maintain an EC of 1.0 mS/cm. Excessive fertilization will result in lush growth with few flowers or cause flowers to develop underneath the foliage canopy. During flat production, you may fertilize only two or three times, depending on your exact situation.

Height control is vital to producing a quality impatiens crop. Once plants are rooted to the sides of containers, they can be allowed to wilt prior to irrigation to provide some height control. However, prolonged wilting harms plant quality: Maintaining the line between allowing wilting to control plant height and growth, while not causing permanent damage, requires vigilant grower attention. Withholding fertilizer, especially phosphorous, can also control height and ammonium-based nitrogen. DIF temperature regimes are very effective in impatiens height control as are A-Rest, Bonzi, and Sumagic growth regulators.

A 4" (10 cm) pot takes from five to six weeks to start flowering when potting up a 288 plug.

Pests & Diseases

Rhizoctonia stem rot can cause dark-brown stem lesions near the substrate line, causing plants to collapse. On leaves and stems, rhizoctonia causes foliage to melt and collapse, often spreading from one location in a plug tray or flat. Usually rhizoctonia is only a problem with temperatures above 75°F (24°C) and high relative humidity. Botrytis can cause tan or brown leaf or flower spots.

Pseudomonas bacterial leaf spot is easy to confuse with botrytis, impatiens necrotic spot virus (INSV), or tomato spotted wilt virus (TSWV). Leaves develop black, brown, tan, or purple spots, sometimes becoming water-soaked and infecting the petiole. Splashing water or handling the plants can spread the disease. Sanitation is the key to control. Disinfect anything that contacts the plants. Be especially careful when patching plug trays, as this is an excellent way to spread the disease. Workers should wash their hands frequently. Keep foliage dry as much as possible, especially at night. Drought stress in the seedling stage may also increase pseudomonas susceptibility. Pseudomonas develops very well under wet conditions at moderate temperatures. The disease does not progress well when temperatures increase; therefore, the problem may be worse early in the season. Some varieties are more susceptible, especially stars and bicolors and those with apricot, coral, salmon, or lavender flowers. There are no chemical controls; however, some growers have reduced spread using Kocide 101 or 606 or Phyton 27.

INSV/TSWV symptoms include black, brown, tan, or purple leaf spots, often in concentric rings. Portions of the stem may be affected and turn black. Plants may be infected with the virus and not show symptoms until the plant is under stress, at which time the plant will collapse rapidly. The disease is spread by thrips feeding on infected plants and then transmitting the disease to other plants as they feed throughout the crop. There is no control; affected plants must be discarded.

Insects such as aphids and thrips may be problems during production and cause distorted leaves. Thrips, as mentioned above, vector INSV/TSWV.

Troubleshooting

Excessive fertilizer, especially ammonium-form nitrogen fertilizers, will result in excessive foliage and few flowers. Impatiens that are moderately drought-stressed will flower more profusely than those with uniform moisture levels will. Plants grown in low light intensities (less than 2,000 f.c. [22 klux]) will not flower profusely. Flower buds can also abort in response to ethylene in the air or the use of Florel. Low temperatures stunt plants and cause leaf yellowing. Boron deficiency or high substrate pH can also cause problems.

Varieties

Excellent impatiens varieties are abundant in the marketplace. Several series to consider include 'Impreza', 'Accent', 'Dazzler', 'Xtreme', and 'Super Elfin' for pack, 4" (10 cm), and 6" (15 cm) production, and 'Show Off' for 4" (10 cm), 6" (15 cm), and hanging basket production.

Postharvest

Flats are ready to sell when they are 20% in color. Impatiens can be shipped in the dark for up to thirty-six hours. During storage and shipping, keep temperatures above 55°F (13°C). Ethylene gas, even at very low concentrations, can cause leaf curl and bud drop.

At retail, make sure plants are displayed with at least 700 f.c. (7.4 klux) of light to maintain flowering. Do not display plants in full sun or on asphalt, as they will dry out too fast and need to be watered several times a day. Shade is mandatory for an impatiens display to maintain plant quality. Make sure temperatures stay above 55°F (13°C) to avoid stressing plants. Because impatiens are very ethylene sensitive and because dead and dying plant parts, as well as cars, generate ethylene gas, keep displays groomed, free of plant debris, and away from parking lots.

Gardeners should be sure to plant impatiens in part shade for best performance. While plants may perform for a time in full sun, they will be in a nearly constant state of wilting for much of the summer.

Impatiens walleriana
Common name: Double-flowering impatiens
Annual

Double-flowering impatiens have been around for years and years. In the past decade, the new breeding that has been released has sparked a craze for this fabulous plant. Even though they too are *Impatiens walleriana,* just as bedding plant flat impatiens are, they are important enough to need their own cultural guidelines. There are several important differences between the two. First, the double-flowering impatiens that growers and consumers have come to know and love are vegetatively propagated. For growers, these plants are exceptionally easy to grow, as long as they start with clean stock and focus on proper nutrition and irrigation. For consumers, they are practically guaranteed to deliver a summer of show-stopping waves and flushes of flowers. Be sure to offer a big assortment for Mother's Day sales.

Propagation

Cuttings

Most growers start with rooted cuttings. Some, however, prefer to buy in unrooted cuttings. As with all impatiens, TSWV/INSV is a devastating disease. Be sure to start off your crop with only cuttings that come from a reputable supplier that adheres to a strict disease-indexing program and maintains its stock in a clean-stock system. The best double-flowering impatiens are protected with plant patents and breeder's rights, so unlicensed propagation is illegal.

Plant cuttings immediately on arrival. If that is not possible, store them at 50–60°F (10–16°C) for no longer than twenty-four hours. Stick cuttings into Ellepots, Oasis wedges, or a disease-free rooting medium with a pH of 5.5–6.0. Place cuttings so that 0.5" (1.3 cm) of the stem is below the substrate line.

Root cuttings at a substrate temperature of 70–75°F (21–24°C) and air temperatures of 70–75°F (21–24°C) days and 65–70°F (18–21°C) nights. Make sure the medium is moist: You should be able to squeeze water out easily. Provide low ambient light levels of 500–1,000 f.c. (5.4–11 klux). Retractable shade is ideal, as you will increase light levels as cuttings mature.

Maintain cutting turgidity with mist. However, excess moisture will delay rooting: Only mist three to four times per day. If you are not using mist, then cover rooting benches with white plastic over hoops and securely attach the sides to increase humidity around the cuttings. Once roots are visible, remove the plastic at night. Adjust mist frequency and duration to your exact greenhouse and climate conditions.

As soon as you notice any root initials, apply a foliar feed of 50–75 ppm from a fertilizer low in phosphorus and ammoniacal nitrogen. This will help reduce unwanted stretch. Once 50% of the cuttings have begun to differentiate root initials (about five to seven days after sticking), move cuttings to the next stage of development.

The next stage, root development, will last from nine to eleven days. Reduce soil temperatures to 68–72°F (20–22°C) and maintain air temperatures of 70–75°F (21–24°C) days and 65–70°F (18–21°C) nights. As soon as cuttings begin to develop roots, allow the rooting medium to begin drying out somewhat. However, avoid drying out the air, as this will increase evapotranspiration, which will reduce the root zone temperatures. Increase light levels to 1,000–1,500 f.c. (11–16 klux) as the cuttings begin to root out. Apply a foliar feed of 100 ppm from 15-0-15, alternating with 20-10-20; increase rate rapidly to 200 ppm as cuttings root in. The majority of fertilizer should be nitrate fertilizer (15-0-15). Maintain the substrate pH at 5.0–6.0.

As cuttings finish rooting, harden them off prior to planting or sale for seven days. Lower ambient air temperatures to 68–75°F (20–24°C) days and 65–68°F (18–20°C) nights. Cuttings should no longer be under mist. Increase light levels to 1,500–2,500 f.c. (16–27 klux). Provide shade during the heat of the day to reduce crop stress. Fertilize once per week with 150–200 ppm from 20-10-20, alternating with 15-0-15.

A 105-liner tray is ready in three to four weeks after sticking the cuttings.

Growing On

Plant 1 cutting/4" (10 cm) pot, 2 cuttings/6" (15 cm) pot, and 4–5/10" (25 cm) hanging basket. Select a well-drained, disease-free medium with a moderate initial nutrient charge and a pH of 5.8–6.2.

Double-flowering impatiens, like all impatiens, have a low fertilizer requirement. Fertilize only at every other irrigation with 15-0-15, alternating with 20-10-20. Do not fertilize plants immediately after transplanting; instead wait a couple of weeks and then begin with 100–150 ppm. As plants root into their growing containers and begin actively growing, increase the rate to 200–250 ppm. Impatiens are very sensitive to high soluble salts, so leach with clear water at every third or fourth irrigation. *Note:* Excessive ammoniacal fertilizer promotes leaf growth and inhibits flowering.

Irrigation is critical: Wet medium will delay flowering and cause plants to stretch. Excessive humidity will also lower flowering and encourage soft growth. Immediately after planting, be very careful not to overwater plants until they become rooted in. Once plants reach their final size, water stress will promote flowering. However, do not allow plants to wilt severely, as this will cause yellow leaves and flower drop. To prevent salts buildup, be sure to water thoroughly when you irrigate.

Grow plants under moderate light intensities of 2,500–4,000 f.c. (27–43 klux); provided you can moderate production temperatures, grow plants at light intensities on the higher end of the range. Under high production temperatures, reduce light intensity to prevent flower and leaf burning. At light intensities below 2,000 f.c. (22 klux), plants will stretch.

Double-flowering impatiens can be shaped with pinching. Pinch plants one to two weeks after potting, once roots have reached the edge of their growing container. Pinch plants back to 2–3" (5–8 cm) above the substrate line, above the fifth or sixth set of leaves. Pinch plants again in two weeks.

Do not crowd plants on benches. As leaves begin to touch, space pots.

Both Bonzi and Sumagic can effectively control height. Apply five to six days after the pinch and again as needed. As light intensity increases, reduce rates. University of Florida trials have found that rates from 1–3 ppm of Bonzi applied as a drench when plants are at their desired sale height slows growth without inhibiting flowering. Midwestern growers should halve rates for their own trials.

Note that if plants become overgrown due to a slow-moving market, they may be cut back and reflowered in four to six weeks.

Using one liner per pot, a 4" (10 cm) pot takes from five to seven weeks to flower, while a 10–12" (25–30 cm) basket takes from nine to twelve weeks to flower.

Pests & Diseases

See Pests & Diseases for *Impatiens walleriana* (bedding plant impatiens) and *Impatiens hawkeri.*

Troubleshooting

See Troubleshooting for *Impatiens walleriana* (bedding plant impatiens) and *Impatiens hawkeri.*

Varieties

The 'Fiesta' series has turned the American bedding plant scene on its head. These reliable performers make fabulous pot plants and hanging baskets. 'Fiesta Olé' is more compact. 'Fiesta' is available in 'Apple Blossom', 'Blush', 'Burgundy Rose', 'Coral Bells', 'Deep Orange', 'Lavender Orchid', 'Olé Cherry', 'Olé Frost', 'Olé Salmon', 'Olé Stardust Pink', 'Orange Spice', 'Pink Ruffles', 'Purple Piñata', 'Salmon Chiffon', 'Salmon Sunrise', 'Salsa Red', 'Sparkler Red', 'Sparkler Rose', 'Sparkler Salmon', 'Stardust Lavender', and 'White'.

Other double-flowering impatiens series include 'Rosebud' and 'Tioga'.

Impatiens hybrida
Common name: Exotic impatiens

The term "exotic" is applied to the 'Fusion' series of impatiens. This series is available in brighter colors than standard impatiens, including shades of yellow, peach, and others. Propagate and grow as you would a 'Fiesta' impatiens.

Postharvest

See Postharvest for *Impatiens walleriana* (bedding plant impatiens) and *Impatiens hawkeri.*

Ipomoea

Ipomoea batatas
Common name: Sweet potato vine
Annual

Ipomoea batatas has become a highly popular foliage accent for mixed containers and annual landscape beds. It thrives in warm temperatures and is very tolerant to heat and humidity. Almost all varieties are very vigorous and are best combined with vigorous flowering plants or used sparingly, otherwise they tend to outgrow other plants. Lower temperatures slow growth.

Propagation

Ipomoea can easily be propagated from cuttings. Propagation may occur year-round, but maintain warm temperatures during winter months for year-round production. Most standard tray sizes for propagation are between 58 and 105 cells per tray. Stick one cutting per cell. Use a standard propagation medium with a pH of 5.5–6.2. Temperatures of at least 75–80°F (24–26°C) for the rooting process are best, ideally with bottom heat. Keep the medium moist at all times, but not saturated. Until the plants have developed roots of their own, mist several times a day to avoid wilting. The amount and frequency of misting depend on the temperature and air circulation in the greenhouse. After cuttings have rooted, move plants to a hardening area with slightly increased light levels and slightly lower temperatures.

More and more growers choose to buy in liners instead of self-propagating due to the often higher quality available from specialized young-plant producers and the fact that the best new varieties are patented, making propagation illegal without a license.

Cuttings take from four to five weeks to root and be ready to transplant to larger containers.

Growing On

Transplant liners into a well-drained, soilless medium, such as a peat/perlite mix, with a pH of 5.5–6.2 and a moderate initial starter charge. Use 1 liner/4 or 6" (10 or 15 cm) pot and 3–4/10" (25 cm) basket. *Ipomoea* is excellent as a component in mixed container gardens: You can plant the liner right in the pot if temperatures are warm enough.

Ipomoea grows well in partial to full sun. A light range of 4,000–7,000 f.c. (43–75 klux) is appropriate. At light intensities below 1,500 f.c. (16 klux), leaf color is less intense. It grows best in warm temperatures. Growth is halted at temperatures around 40°F (4°C) and should not drop below 35°F (1°C).

Keep the medium moist, but not saturated. Apply a constant feed of 150–200 ppm 20-10-20, alternating with 15-0-15 and an average amount of micronutrients with slightly higher levels of iron.

Pinching is optional, but one soft pinch within one to two weeks of liner planting is recommended. *Ipomoea* responds well to cutting back if it gets too big. *Ipomoea* also responds well to B-Nine sprays at 2,500 ppm.

If transplanting a liner to a 4" (10 cm) pots, it takes from four to six weeks to be ready to sell.

Pests & Diseases

Scout for whiteflies and aphids. Sweet potato weevil is a problem in the southeastern United States, where there are restrictions on imports of sweet potato vines from some states. If you are located in the Southeast, consult your broker sales representative for a list of suppliers that can ship plants from unrestricted, weevil-free areas.

Drench with a broad-spectrum fungicide at liner planting to minimize the danger of soilborne diseases.

Varieties

There are several standard *Ipomoea* selections. 'Blackie' has dark, purplish-black foliage that is deeply lobed. 'Black Heart' is also known as 'Ace of Spades' and features the same color and vigor as 'Blackie' but has heart-shaped leaves. 'Marguarita' (also known as 'Marguerite' or 'Limelight') has chartreuse foliage and a very vigorous habit.

More recent introductions include the 'Sweet Caroline', 'Illusion', and 'Sidekick' series. These are as vigorous as 'Blackie', 'Black Heart', and 'Marguarita' but offer a few more unique foliage colors and leaf shapes.

Postharvest

Sell plants when they are ready; pots of *Ipomoea* will quickly become a tangled mess on the bench if you let the crop get away from you. However, don't start selling too early, especially in northern climates, because plant performance is best once outdoor temperatures have warmed. If plants become overgrown on the bench, you can pinch them back as you would coleus; they will regrow quickly.

Ipomoea tricolor
Common name: Morning glory
Annual

Also in the genus *Ipomoea* is the morning glory. Morning glories have a vining growth habit and showy, trumpet- or bell-shaped flowers that are often blue in color, although varieties with white, pink, purple, or bicolored blooms are available. Plants are fabulous on trellises and fences or bounding over the edges of containers, providing privacy when grown upright and elegance when trailing. Morning glory flowers open every morning and stay open until the heat of the day saps them. Like their relative, the sweet potato vine, morning glories need warm temperatures for optimum growth.

Propagation

Morning glories are started from seed. It is easiest to sow seed direct to the final growing container, such as 4" (10 cm) pots or hanging baskets. If you choose to grow morning glories in plugs, use large trays, such

as 98- or 128-cell trays. Sow scarified seed for the highest germination rates. If you do not have access to scarified seed, soak seed overnight in water prior to sowing. Sow onto a disease-free, germination medium or growing medium with a pH of 5.5–6.2. Cover seed with growing medium. Germinate at 70–75°F (21–24°C). Keep the medium moist but not wet. Seed will germinate in four to six days.

Growing On

Grow seedlings on at 75–80°F (24–27°C) days and 55–65°F (13–18°C) nights to maintain steady growth. Feed lightly with 50 ppm from 15-0-15, alternating with 20-10-20 to maintain green leaf color; however, fertilizer is not necessary. Too much fertilizer will delay flowering and produce extra-large foliage. Grow plants under full-sun conditions.

Be sure to sell plants as soon they are ready or stick to hanging basket production: Vines will become a tangled mess on the bench!

From sowing, 10" (25 cm) baskets require seven to nine weeks to be salable. Plants will not be in flower, but they will have lush foliage.

Varieties

The 'Good Morning' series has a compact, mounding habit without the vigorous vining habit of most morning glories. Its foliage is marbled green and white, and the series comes in 'Red', 'Violet', and 'Pink'; all of these are edged in white, which adds additional color and interest.

'Heavenly Blue' is a very popular morning glory variety, with the plant's famous sky-blue flowers that brighten up summer mornings. 'Happy Hour Rose' is great in baskets or sold in pots as a groundcover. Plants have rose flowers with a white throat and can spread up to 3' (0.9 m). Foliage is marbled green and white.

Iresine

Iresine herbstii
Common name: Iresine
Annual

This annual with red-purple—or yellow on some varieties—foliage is fabulous as a mid-ground plant, as a centerpiece in mixed plantings, or alone en masse in a bed. For growers, its culture is really easy and crop times are relatively fast: just nine to eleven weeks for 4" (10 cm) pots and thirteen to fourteen weeks for 10" (25 cm) baskets from seed.

Propagation

Plants can be propagated by seed or cuttings.

Seed

Sow seed onto well-drained, disease-free, germination medium with a moderate starter charge and a pH of 5.5–6.3. Cover seed lightly with coarse vermiculite. Provide 100–400 f.c. (1.1–4.3 klux) during germination. At 72–76°F (22–24°C), seed will germinate in three to four days.

Move trays to Stage 2 for stem and cotyledon emergence. Lower substrate temperatures to 65–72°F (18–22°C) and gradually raise light levels to 1,000–2,000 f.c. (11–22 klux). As soon as cotyledons expand, fertilize with 50–75 ppm from 15-0-15. Once radicles penetrate the medium, begin to allow trays to dry down somewhat. However, do not allow seedlings to wilt at any point during plug production.

Once you have a good stand, move trays to Stage 3 to develop true leaves. Lower temperatures to 65–70°F (18–21°C) and slowly increase light levels to 2,500–3,500 f.c. (27–38 klux). Increase feed to 100–150 ppm applied once a week, alternating between 20-10-20 and 15-0-15. Allow trays to dry between irrigations, but do not allow seedlings to wilt.

Harden plugs at 62–65°F (16–18°C) for seven days prior to sale or transplanting. Increase light levels to 4,000–5,000 f.c. (43–54 klux), provided temperatures can be maintained. Apply 100–150 ppm once a week from 15-0-15.

Allow five to six weeks for a 288-plug tray.

Cuttings

Stick cuttings within twelve to twenty-four hours of arrival in a well-drained medium with an EC of 0.75–0.80 mS/cm and a pH of 5.6–5.9. Cuttings can be stored overnight, if necessary, at 45–50°F (7–10°C). The soil temperature should be maintained at 68–74°F (20–23°C) until roots are visible. Begin fertilization with 75–100 ppm nitrogen when roots become visible.

A media drench with Bonzi (0.5–1.5 ppm) prior to sticking the cuttings has proven to reduce stem elongation in propagation. Apply to already-moist media and adjust the volume and rate of application for local conditions. A Bonzi spray (1–3 ppm) applied eight to ten days after sticking is also effective in reducing stretch in propagation.

The timing of the first pinch is critical to successful production. Apply a soft pinch fourteen to eighteen days after sticking. Pinching too low will result in very prostrate growth in the plant. A soft pinch will encourage lateral breaks that will grow upright, contributing to the mounded habit that is best suited for retail and landscape application.

Iresine rooted cuttings should be ready for transplanting twenty-one to twenty-four days after sticking.

Growing On

Seed

Select a well-drained, disease-free medium with a pH of 5.5–6.5 and a moderate initial starter charge. Use 1 plant/4" (10 cm) pot, 1–2/6" (15 cm) pot, and 3–4/10" (25 cm) basket.

Grow on at 65–75°F (18–24°C) days and 62–65°F (16–18°C) nights. Allow plants to dry slightly between irrigations. Feed weekly with 150–200 ppm from 20-10-20 or another complete fertilizer. Higher fertilizer rates will result in faster-growing plants. Growth regulators are not required. If you need to slow plants down, withhold feed and reduce water.

A 288-plug tray transplanted to 4" (10 cm) requires from nine to eleven weeks to finish.

Cuttings

Select a well-drained, disease-free medium with a pH of 5.5–6.5 and a moderate initial starter charge. Use one plant for a 4" (10 cm) pot to as large as a 1-gal. (4 l) pot. Maintain temperatures in the range of 60–70°F (16–21°C) nights and 72–85°F (22–29°C) days. Cool night temperatures will extend crop time.

Keep light intensities at 4,000–10,000 f.c. (40 to 100 klux). Extremely low light levels result in poor branching and stem stretch. Extremely high light levels will have a bleaching effect on the leaves.

The media should dry slightly between watering. Avoid severe wilting. Use constant feed with a balanced fertilizer at 150–200 ppm. Leach regularly to avoid soluble salts buildup. Pinch plants seven to fourteen days after transplanting, as needed, to improve basal branching.

Use high light and the recommended temperatures to control growth and produce the best possible habit. A Cycocel (1,000–1,500 ppm) and B-Nine (2,500–4,000 ppm) tank mix applied one to three times or a Bonzi (5–10 ppm) spray are both effective. These recommendations should be used only as general guidelines. Growers must trial all chemicals under their particular conditions.

A 4" (10 cm) pot takes from seven to nine weeks from potting to be salable, while a 1-gal. (4 l) pot takes from nine to eleven weeks.

Troubleshooting

If foliage becomes reddish, increase fertilizer. Leaves can become crinkled under a combination of high light and high fertilizer rates.

Varieties

The low-growing 'Purple Lady' can spread as wide as 3–4' (0.9–1.2 m), growing to about 6–8" (15–20 cm) tall. 'Purple Lady' prefers partial shade but is able to handle higher light levels as humidity levels increase.

Vegetatively, both green/yellow- and red-foliage varieties are available. The 'Blazin" series offers a selection of colors and is excellent in morning sun and afternoon shade locations. 'Blazin' Rose' grows from 18–30" (45–90 cm), while 'Blazin' Lime' grows from 12–16" (30–40 cm).

Iris

Iris × hollandica
Common name: Dutch iris
Cut flower

You find them everywhere cut flowers are sold in the winter and early spring: Dutch irises. Their elegant blue, purple, yellow, white, and bicolored flowers are integral in arrangements as well as mixed and single-species bouquets. Growers located in areas where temperatures are 70–75°F (21–24°C) year-round offer Dutch irises throughout the year, making them a regular for consumers.

For growers, Dutch iris is a moderately easy crop if you follow the basics and are diligent about preventing and controlling diseases and keeping plants free of light and temperature stress. High density planting and the ability to be cut tight and stored cold prior to sale make this one of the most programmable cut flowers, even though its vase life at room temperatures is just five days.

While commonly grown as a cut flower and hardy to USDA Zones 8–10, Dutch iris has been reportedly hardy in the home garden as far north as USDA Zone 5. However, it is not as common a garden plant as it is a cut flower.

Propagation

Propagation of Dutch iris as a cut flower is left to specialists located primarily in the Netherlands or Washington state. Bulbs require several weeks of heat treatment and/or ethylene treatment immediately after they have been harvested. Dutch-grown bulbs are usually dug in late July and immediately given a ten-day curing period at 90°F (32°C) or exposed to ethylene at 500 ppm for twenty-four hours, which accelerates flower bud formation. All subsequent storage conditions are dependent on when the bulbs will be programmed to flower, early to year-round. Dutch irises that will be shipped to customers for planting after mid-November are put into a retarding room at 86°F (30°C) with 50–60% humidity. They are stored there until they are precooled at 41–48°F (5–9°C), which occurs about six to eight weeks before planting, depending on the bulb size, variety, and time of year.

Precooled bulbs force more quickly and uniformly than those that are not precooled. They must be cooled at 41–48°F (5–9°C) at the proper time for six to eight weeks and then removed just prior to planting. Precooled bulbs must be planted immediately. Some growers have the facilities to do their own precooling and can use heat-treated (retarded) bulbs. Generally these are shipped six weeks prior to planting. Trials have shown that a two-week temperature treatment of 65°F (18°C) either before or after the precooling treatment reduces the amount of foliage and improves flower size. This treatment also reduces the probability of flower abortion. Unless you have the facilities to properly handle bulbs, buy in bulbs that have been programmed with heat treatment and cool treatments.

Purchase large-sized bulbs such as 9/10 or 10+ cm and up for early crops; you can use smaller bulbs for later crops.

Growing On

Plant programmed bulbs immediately upon arrival. If you must store them prior to planting, then store at 35°F (2°C). Do not store bulbs for more than two weeks. Holding bulbs at higher temperatures extends the bulb treatment and can reduce flowering. Consult your bulb supplier for advice.

Prior to planting, make sure ground beds or planting medium is free of disease and fluoride. Always allow a two-week period following steam pasteurization or chemical treatment prior to planting. Adjust pH to 7.0. Do not use superphosphate, as irises are very sensitive to fluoride. The soil EC should be 1.0–1.5 mS/cm. If it is any higher, you will need to leach salts out prior to planting. Do not add preplant fertilizer, as rooting will be delayed.

Provide a thirty-minute fungicide dip prior to planting to control diseases. Plant bulbs 1" (3 cm) deep at a density of about 90–100 bulbs/$yd.^2$ (108–120/m^2) for bulbs 10 cm and up; 100–120 (120–144/m^2) bulbs/$yd.^2$ for 9/10s; and 110–130/$yd.^2$ (132–155/m^2) for 8/9s. Plant outdoor crops 4"

(10 cm) deep. Dutch irises may be grown in either 8" (20 cm) deep, raised or ground beds. If you are using flats, make sure they are at least 4" (10 cm) deep. Some growers use a layer of support to keep foliage upright, although it is not necessary.

Immediately after planting, keep the medium moist but not wet. Once the crop is rooted and begins to actively grow, keep plants well irrigated. In flats, iris forms a thin, matted layer of roots on the bottom: Thorough watering is a must. Do not place flats too near heating pipes or plants will dry out too quickly and flower buds may abort as a result.

Force at 60–63°F (15–17°C) days and 55°F (13°C) nights. Maintain temperatures above 47°F (8°C), but no higher than 68–73°F (20–23°C). Do not grow at temperatures above 63°F (17°C) during the short days of winter or during low-light periods. As a rule, lower forcing temperatures when light intensity is low. Dutch research has shown this is especially critical two weeks prior to flowering (i.e., when the flower stalk is rapidly growing out of the leaves). Produce irises under moderate light levels, 2,500–4,000 f.c. (27–43 klux). The total crop time from a programmed bulb to a flowering plant is six to eight weeks.

Once bulbs are rooted and leaves have emerged, fertilize weekly with calcium nitrate at 250 ppm.

Pests & Diseases

Disease problems can be significant with a Dutch iris crop. *Pythium* and rhizoctonia root rots can attack plants. *Penicillium* (blue mold) and *Fusarium* (bulb rot) can cause serious losses. Both are present on the bulbs as spores. The spores can infect the bulbs through wounds. These diseases often appear when bulbs are exposed to high humidity and/or mechanical damage. Aphids and thrips can become problems.

Troubleshooting

Using small bulbs or improper heat temperature treatments can cause blind plants, ones that produce three leaves but no flowers. Flower abortion may occur at any stage of flower development and may be due to low light, high temperatures, incorrect nutrition, and/or moisture stress. The critical period for flower abortion is about fourteen days before flowering, when the floral stalk is rapidly elongating. Always avoid stresses, especially during this period of development.

Calcium deficiency can cause stem topple. Leaf tips may become burned (scorched) from fluoride in water used for irrigation.

Varieties

'Blue Diamond' and 'Professor Blaauw' (also called 'Blue Ribbon') are dark blue. 'Telstar' is blue-violet. 'Ideal' is light blue. 'Blue Magic' is purple. 'Apollo' is yellow/white. 'Casablanca', 'Madonna', and 'White Wedgwood' are white. 'Golden Beauty', 'Golden Harvest', and 'Yellow Queen' are yellow.

Postharvest

The ability to store cut stems in a cooler is imperative for growers, wholesalers, and retailers. The International Flower Bulb Centre in the Netherlands recommends 10 ft.^{3} of cooler space for every 1,000 ft.^{2} (0.3 m^3/93 m^2) of bulb production. It recommends harvesting stems with bulbs attached and placing stems in the cooler at 35°F (2°C). Precool stems prior to grading, bunching, and removing bulbs.

Normally, flowers are ready to harvest when the flower has emerged from the sheath but before falls begin to open. The exception being 'Professor Blaauw', which must be harvested as the falls begin to open. STS is beneficial in helping this variety to open.

Dutch irises are generally between 20–25" (51–64 cm) in length when they are cut. To store, place stems in distilled water (water must be fluoride free) with floral preservative and hold at 32–35°F (0–2°C).

Subsequent steps in the distribution chain such as wholesalers and retailers should recut stems by 0.25" (6 mm), place them in distilled water (*caution:* tap water may contain fluoride) with a floral preservative, and refrigerate at 32–35°F (0–2°C). Once removed from the cooler, flowers last from two to five days at room temperatures.

Iris hybrids
Common names: German iris; bearded iris; fleur-de-lis
Perennial (Hardy to USDA Zones 4–8)

Fleetingly beautiful, bearded iris provides a spectacular show with which millions of gardeners have fallen in love.

For growers, your perennial iris offerings will be limited only by your available space and imagination. There are dozens of irises for all growing conditions: from tall, upright iris specimens in the border to Japanese irises for water gardens and dwarf, bulbous species that peek out of the ground just after daffodils. By far the most popular iris is the bearded iris, loved for its oversized flowers.

Iris aficionados plan their gardens to have irises in flower continually from spring to midsummer. If you are a retail grower with a customer base looking for irises, offering an excellent selection is one way to set yourself apart from the competition in the marketplace. Irises are sold green most of the time because their flowering window is so brief. Thus, most mass marketers avoid stocking them in pots. The culture that is presented here relates primarily to bearded iris, but it can be adapted to other types as well.

Propagation

Bearded irises are propagated by dividing rhizomes in mid- to late summer. Cut rhizomes up into smaller sections that have roots attached, trim back leaves to 5–6" (13–15 cm) long, and plant them into pots for fall sales or overwintering.

While propagation is extremely easy, most growers buy in rhizomes sold as one- or two-fan divisions from commercial suppliers in either the fall or winter so they can get an assortment of the best varieties.

Growing On

Select a disease-free, well-drained medium with a pH of 6.5–7.0 and a low starter charge. Walters Gardens recommends soaking irises in a 10% bleach solution for fifteen to twenty minutes when they arrive. Allow rhizomes to dry prior to planting. Place rhizomes so the roots will be spread out in the pot and the rhizome will lie horizontally on the media surface. Gently press the rhizome into the medium, leaving the tops of rhizomes uncovered. Place one rhizome per 1-qt. (1.1 l) pot. However, bearded iris should generally be potted in 1-gal. (4 l) containers, so that its rhizome size is proportional to the pot.

For the best flowering the following spring, plant up rhizomes from late summer to fall and overwinter pots in a cold frame to keep pots dry. Plants that do not receive cold treatment will flower sporadically their first year. Before overwintering, make sure rhizomes have rooted all the way to the bottom of the container. Water irises only after pots are thoroughly dry, as too much moisture can cause rot. Cut foliage back to 4–5" (10–13 cm) and spray plants with a 20% bleach solution. Do not cover pots with an overwintering material that will come in direct contact with the plants.

Provide potted irises with at least ten weeks of cold treatment at 41°F (5°C). After cold treatment is over, grow plants on at 55°F (13°C) nights. If plants are developing too quickly, lower temperatures to 48°F (9°C). Plants will be salable in seven to twelve weeks and will flower as they would normally in the garden, depending on the variety.

For those planted in January or February for spring sale in 1-qt. or 1-gal. (1.1 or 4 l) containers you might have to use two roots per 1 gal. (4 l), if the roots are small. Grow on the greenhouse bench at 68–70°F (20–21°C) days with nights 5–10°F (3–5°C) cooler. Plants will be ready to sell green in mid- to late May, although the roots may not be fully massed to the bottom of the 1-gal. (4 l) pots. Plants will flower sporadically the first year but more profusely the following year in the home garden.

Varieties

There are scores of bearded iris from which to choose. Following are the top twenty varieties for 2010, as voted on by members of the American Iris Society (www.irises.org).

1. 'Dusky Challenger' (dark purple)
2. 'Queen's Circle' (white with blue border on falls)
3. 'Conjuration' (white and violet with white horns)
4. 'Jesse's Song' (white and violet plicata)
5. 'Paul Black' (dark purple, orange bd.)
6. 'Gypsy Lord' (white standards, violet falls with white center, red bd.)
7. 'Decadence' (apricot standard, plum falls with apricot edge)
8. 'Sea Power' (ruffled blue)

9 (tie). 'Thornbird' (ecru, tan, violet horns)

9 (tie). 'Silverado' (light silver blue)

11. 'Golden Panther' (gold/bronze)
12. 'Drama Queen' (cyclamen/capucine plicata)
13. 'Stairway to Heaven' (off-white standards with medium blue falls)
14. 'Florentine Silk' (peach standards, violet falls with pink edge)

15 (tie). 'Mesmerizer' (white with white flounces)

15 (tie). 'Happenstance' (pink)
17. 'Beverly Sills' (pink)
18 (tie). 'Stepping Out' (white and violet plicata)
18 (tie). 'Before the Storm' (near black)
20. 'Daughter of Stars' (purple with white edge)

Related Species

In addition to bearded (German) iris, also be sure to offer related irises.

I. ensata, the Japanese iris, has rather flat flowers from 4–6" (10–15 cm) across on plants growing to 20–40" (51–102 cm) tall. Plants flower in June and July and come in blue, lavender, red, violet, and yellow. Some flowers are marbled with gray or white. They bloom about one month later than bearded irises. Grow Japanese irises a bit wetter and in a medium with more acid than bearded irises. In the landscape, they like being planted near water features such as lakes, streams, or water gardens, tolerating wet areas and acid soils very well.

I. sibirica, the Siberian iris, likes moist conditions. Its foliage grows 24–36" (61–91 cm) tall. Plants are freer flowering than bearded iris, with blooms appearing in June. Flowers are primarily blue, violet, and white.

I. pallida 'Variegata' has fabulous variegated foliage and lavender-blue flowers in the summer. Foliage is attractive enough to stand on its own in the garden.

Louisiana irises are native to the Gulf Coast region. They prefer acidic soils and higher moisture conditions than bearded iris. Showy flowers are in red, lavender, white, and purple.

Postharvest

Since iris flowers last for such a brief period of time, it is important to market plants with color tags. Develop a system to ensure that you always know which plants are which by permanently affixing tags to the pot.

Instruct consumers to divide plants every two or three years. Do not mulch irises in the garden or landscape.

Isolepsis

Isolepsis cernua
Common name: Fiber-optic grass
Annual grass

This dwarf-growing grass grows from 6–8" (15–20 cm) with each radiating leaf blade tipped with a tiny, "fiber-optic flower spike—hence the common name.

Propagation

Seed is the most common method of propagation. Use a well-drained, disease-free, soilless medium with a pH of 5.5–6.1 and a medium initial nutrient charge (i.e., an EC less than 0.75 mS/cm with a 1:2 extraction). Each multiseed pellet will generally yield three to four plants. Plug tray sizes from 406 to 288 cells are recommended. If desired, 128 and 72 cells are also suitable.

Seed can be sown without cover. Light is required for germination and early seedling growth. For best germination have a soil temperature of 64–68°F (18–20°C). Seed germinates most uniformly around 65°F (18°C). Germination takes six days. Keep soil wet during Stage 1. Maintain 100% relative humidity (RH) until radicles emerge.

After germination, maintain a soil temperature of 64–68°F (18–20°C) and reduce soil moisture slightly to allow roots to penetrate into the media. Apply fertilizer at less than 100 ppm nitrogen and an EC less than 0.7 mS/cm from nitrate-form fertilizers with low phosphorous. Keep light levels up to 2,500 f.c. (27 klux)

Once the plants are actively growing, maintain the same temperatures and increase light levels up to 5,000 f.c. (54 klux), if temperature can be controlled. Increase the fertilizer rate to 100–175 ppm nitrogen and EC of 0.7–1.2 mS/cm. Allow media to dry until the surface becomes light brown before watering. However, do not allow the seedlings to wilt.

The total plug time in a 288 is approximately five weeks. Add one more week for 128- and 72-cell plug trays, but reduce post-transplant crop time by one week. Plugs can be held for about two weeks without affecting the subsequent growth after transplanting, provided they are adequately watered.

Growing On

Use a well-drained, disease-free, soilless medium with a pH of 5.5–6.2 and a medium initial nutrient charge. Grow on at 64–66°F (18–19°C) nights with 66–74°F (19–23°C) days. Plants can be grown under temperatures as low as 50°F (10°C), but the crop time will increase significantly. Light levels should be 3,000–5,000 f.c. (32–54 klux). Isolepsis can tolerant light shade and likes moist to wet medium. Plants can even grow in standing water. Do not allow plants to dry out, as this will cause the foliage to become yellow.

Applying fertilizer with 175–225 ppm nitrogen (EC 1.2–1.5 mS/cm) once per week is adequate, but plants can tolerate higher rates or more frequent fertilization as well. The heavier the fertilization is, the faster the plants will grow. Because isolepsis is naturally compact, growth regulators and pinching are not necessary.

Varieties

'Live Wire' is a long-day plant and will flower when day length is longer than twelve hours. If you get your plugs in early (before mid-March), chances are it will not have the fiber-optic look. Once days are over twelve hours, the plants will start to look as expected.

Table 18. *Isolepsis* Schedules from Transplant to Saleable Size

CONTAINER SIZE	PLANTS PER POT/BASKET	WEEKS FROM TRANSPLANT	TOTAL WEEKS
306 premium pack	1	6–7	11–12
2.5" (6 cm) pot	1	4–5	9–10
4–4.5" (10–11 cm) pot	1	6–7	11–12
6–6.5" (15–16 cm) pot	3	6–7	11–12
1-gal. (4 l)	3	6–7	11–12

J

Juncus

Juncus inflexus
Common names: Juncus; blue arrows
Perennial grass (Hardy to USDA Zones 5–9)

This blue-green, upright grass grows from 24–30" (61–76 cm) tall, even taller in warmer areas. These versatile, long-lived plants are excellent in wet to moist locations. In drier locations, they are shorter and not as vigorous in habit, but are still long lived.

Propagation & Growing On

Sow seeds to a 288-plug tray with either three to six seeds per cell or a multiseeded pellet and leave uncovered. At temperatures of 70–74°F (21–23°C), seed germinates in seven to eight days and plug trays are salable in six to eight weeks after sowing. When transplanted to 18-cell flats and grown on at 68–74°F (20–23°C) days with nights 10°F (5°C) cooler, plants are salable in another six to eight weeks, while 4" (10 cm) pots for spring production are about the same.

Varieties

'Blue Arrows' is the standard variety on the market and available from a number of suppliers. It is offered as raw seed or multiseeded pellets with four to six seeds per pellet, depending on the producer.

Juncus pallidus
Common name: Juncus
Annual grass (Hardy to USDA Zones 7–10)

This green, upright grass grows up to 4' (1.2 m) tall and 18–22" (46–56 cm) across. As versatile as *Juncus inflexus*, plants are best suited to moist/wet locations but also excel in well-drained locations.

Propagation & Growing On

The scheduling is very similar to the crop culture noted under *Juncus inflexus*. However, since it is an annual, *Juncus pallidus* is usually one to two weeks earlier in plug production as well as total crop time. Review the information under *Juncus inflexus* and adjust accordingly. Sow four to six raw seeds or one multiseeded pellet per cell. Plants of *Juncus pallidus* will also be taller in habit when sown along with 'Blue Arrows.'

Varieties

'Javelin' is the most common variety available at the time of this writing and is offered as single seeds or multiseeded pellets.

Juncus effuses spiralis
Common names: Twisted juncus; spiraled juncus
Perennial grass (Hardy to USDA Zones 5 or 6–9)

A tender crop in prolonged, cold winters, this juncus has a unique feature: curling leaves that shoot from the crown upright as well as off the sides. These curling leaves make the plant a rather interesting addition to mixed or combination containers. Plants grow from 12–14" (30–36 cm) tall with a similar spread. Plants do not flower in the same season they are seeded.

Propagation & Growing On

Twisted juncus is slower growing than *Juncus inflexus* and *Juncus pallidus*. Using the culture under *Juncus inflexus*, add another one to two weeks to the plug culture and another two to three weeks to finish.

Varieties

There are a number of varieties available under this species. 'Twister' is available as a multiseeded pellet, which is the easiest seed form to grow and achieve the best results.

K

Kalanchoe

Kalanchoe blossfeldiana
Flowering pot plant

Kalanchoe is a succulent herb and shrub that is native to temperate and tropical regions. It is a short-day plant, flowering in January or February in temperate regions under protection.

Kalanchoe can be flowered year-round on a scheduled program very similar to the pot mum program: long-day requirement for vegetative growth, followed by short-day response for flowering, using a long-day period and growth regulators to control plant habit and flowering.

There are many new commercial varieties with bright and pastel colors, pleasing foliage, and long shelf life—all of which increase customer satisfaction. Sales have been increasing for 4" (10 cm) pots flowered for mass-market outlets. Plants are easy to care for in mass-market displays and have good postharvest life in consumer homes.

Propagation

Kalanchoes are vegetatively propagated from cuttings. While kalanchoes have been propagated from seed in the past, it is not practical for commercial growers because of long crop times. Most growers purchase either liners or, more commonly, unrooted cuttings from specialist propagators. It is vital that suppliers index their stock to ensure it is disease-free and then maintain their stock in a clean-stock program. Kalanchoes can be infected with a number of diseases, such as the bacteria *Erwinia,* and the several viruses, including yellow spotted virus, kalanchoe latent virus, potato-Y (Poty) virus, kalanchoe mosaic virus (KMV), tobacco mosaic virus (TMV), carnation mottle carmovirus (CMC), impatiens necrotic spot virus (INSV), and tomato bushy stunt virus (TBSTV). The only way to control these is to not get them, and the best way to ensure that is to work only with clean cuttings from indexed stock and to immediately discard any plants that show signs of bacteria or viral disease.

Large growers producing a weekly crop can receive unrooted cuttings or liners to ensure a steady supply of flowering plants. The best unrooted cutting is either a two- or a three-node cutting that measures 1.5–2.5" (4–6 cm) long, from a clean-stock plant growing under long days with a minimum of thirteen hours of light. These cuttings will root within fourteen to twenty-one days with soil temperatures of 70–74°F (21–23°C) provided by bottom heating. You can direct-stick cuttings into their final growing containers. Make sure the medium has a high starter charge and a pH of 5.5–6.0. Use 1 cutting/4" (10 cm) pot, 1 cutting/5" (13 cm) pot, and 3 cuttings/6" (15 cm) pot.

Root cuttings under about 2,000–4,000 f.c. (22–43 klux) of light and provide 40% shade during the summer. You do not have to use mist. If you do, use very little mist, or preferably use overhead syringing a couple times per day for the first ten days or so. Some growers prefer to cover pots with plastic for the first ten to fourteen days rather than misting or syringing to maintain turgidity. (Use milky-white plastic in summer and clear plastic in winter.) Other growers like to cover pots with cheesecloth and syringe a couple times per day. No rooting hormone is required, although it will speed rooting of difficult-to-root varieties by a couple days. Some growers like to water-in pots with a preventive fungicide drench immediately after sticking. During the winter months, it is very beneficial to root cuttings under supplemental lighting and continue to use HID lighting during long days (vegetative period).

About fourteen days after sticking, or just as cuttings are beginning to form roots, begin fertilizing with 20-10-20 at 50–75 ppm. Kalanchoes have a high fertilizer requirement during their vegetative growth phase, so these early feeds are important.

Cuttings will root in about two to three weeks during the summer and three to five weeks in the winter. Root cuttings under long days, using night interruption lighting. (See more comments on providing long days under Growing On.)

Growing On

Kalanchoes are daylength sensitive, which means they can be scheduled for year-round flowering by providing long days to keep plants vegetative followed by short days to initiate flowering. The minimum number of short-day weeks for flowering is six (forty-two days). Why the long-day and short-day periods? Kalanchoe flowers when exposed to short days (normally under ten-hour day lengths). If short days are applied on cuttings that are just planted, the finished plant will be too short. Prior to flowering, build up (bulk) the plant under long days before flowering them under short days. Day length is discussed in depth below.

Kalanchoes have a very fibrous root system. The medium should be well aerated, but also able to retain moisture. The pH should be 5.8–6.5. A peatlite mix with a pH close to 5.8 is ideal. In any medium, add dolomite to increase calcium levels. Also add superphosphate and micronutrients, if a complete water-soluble fertilizer is not used. If you are planting liners, be sure to plant them deep enough so that the lower leaves touch the soil. This will result in a sturdier, more compact plant.

After cuttings are fully rooted in, kalanchoe should not be watered from overhead on a regular basis. Smaller pots work well on a mat or an ebb-and-flood system. Kalanchoe does not have a high transpiration rate; plants are drought resistant. However, their root systems are sensitive to rapid changes in moisture and soluble salts. Young plants in the early stages of growth through bud initiation should not be stressed by high temperature or lack of water. It is very important to maximize the growth of the young plants through flower initiation in order to size them up in relation to their pot size. After flower initiation, overwatering can soften and stretch flower stems. Thus, it is a good practice to tone the plants at the finishing stages.

Kalanchoes require less water than chrysanthemums. Since you are watering less frequently, you need to increase the fertilizer concentration. Apply 300–400 ppm from 20-10-20 through visible bud. After that stage, alternate liquid feed with clear water, or reduce fertilizer rates to 150–200 ppm. Maintain the substrate pH at 5.8–6.3. Discontinue fertilizer when flowers begin to show color for better postharvest performance.

Temperature control is critical to your success with kalanchoes. Low night temperatures slow down growth and disrupt schedules. The minimum night temperature until visible bud should be 65°F (18°C), but 65–68°F (18–20°C) is optimal. Bud initiation and development is slowed at night temperatures higher than 75°F (24°C), and heat delay can occur. Should the crop be developing too quickly, the night temperature can be reduced to less than 60°F (16°C) after the buds are initiated. Prior to sale, drop temperatures to 62°F (16°C) to increase postharvest life.

Kalanchoes do not usually do as well under high temperatures and high light. Leaf temperature is everything in growing a good kalanchoe. Under high temperatures and high light, the leaves will bleach out and harden, even developing a red color. When temperatures exceed 75°F (24°C), reduce the sunlight to about 2,500–4,500 f.c. (27–48 klux). Reduce the light to lower the leaf temperature. Under low light levels, kalanchoes will stretch and not flower as heavily and flowers will be thin and weak. In the winter or other low light periods, use supplemental (HID) lighting to start plants.

Knud Jepsen, Denmark's largest kalanchoe grower and a leading kalanchoe breeder, recommends a morning temperature drop to 60–65°F (15–18°C), which will help to make plants more compact. If these temperatures are not possible during the summer, then lower morning temperatures as much as possible to create a greater DIF, he says.

B-Nine and Cycocel are effective for height control. Begin controlling vigorous varieties early on, during long days or at the start of short days. Some vigorous varieties may require their first application during propagation. In warm temperatures and high light, when plants are growing quickly, you will be making more applications than in cooler times of the year. B-Nine can be scheduled as often as every three weeks during the plant development period to produce a more compact plant. Usually no more than two applications of 5,000 ppm are necessary; 2,500 ppm is also effective for some growers, depending on how vigorously plants are growing. Using the lower rate, B-Nine can be applied weekly in the summer, every two weeks in the fall, every two to three weeks in the winter, and every two weeks in the spring. Some markets prefer very short kalanchoes with flower heads right on top of the foliage. B-Nine can shorten peduncle stretch on tall-growing varieties. Use 2,500 ppm when the buds are clearly visible. Not all varieties need B-Nine to shorten peduncle stretch, and some growers have also used Bonzi as a growth regulator.

Kalanchoes can be grown either pinched or unpinched, depending on the size of the pot and the number of cuttings used. As a rule, with the smaller single plants (2.5–4.5" [6–11 cm]), no pinch is necessary if the minimum number of long days and growth regulators are used to develop a compact plant habit. If growing a single plant in a larger pot, apply more long days and a pinch to increase plant size. If growing three plants in a larger pot, crop time can be reduced by four weeks by not pinching.

There are two ways to pinch plants. When pinching kalanchoes, it is better to take the tip with one set of leaves to get a plant with more basal breaks. Leave two to three leaf pairs in the summer and three to four leaf pairs in the winter. The first way to pinch is to make an early pinch at the start of short days when laterals are 0.25" (6 mm) long.

The second type of pinch is a late pinch that can be used to reduce growth regulator use. This pinch is given one week after the start of short days. This type of pinch increases flower number and shortens the response time by five to six days. Applying a late pinch from November–January is not advised.

Because of their compact growth habit, kalanchoes can be grown closer to each other than most pot plants. They can be grown pot to pot until the foliage touches, usually at the beginning of short days. Place plants at final spacing when short days begin (table 19).

Table 19. Kalanchoe Final Spacing

POT SIZE	SPACING
4" (10 cm)	4–6 plants/ft.^{2} (43–45/m^2)
5" (13 cm)	7" × 7" (18 x 18 cm)
6" (15 cm)	10" × 10" (25 x 25 cm)
6.5" (17 cm)*	12" × 12" (30 x 30 cm)

* Three plants per pot; spacing is between pot centers. All other sizes have only one plant.

Long-day treatment

The number of long-day weeks is critical to kalanchoe vegetative development, determining plant size and plant habit. Insufficient long-day weeks reduces the size and height of the plants. Too many long-day weeks stretches the plants. The long-day weeks on pinched plants should be balanced. For example, if three long-day weeks come before a pinch, three long-day weeks should follow the pinch.

If too many long-day weeks are given, the result is a tall plant. Given good culture programs, the schedules in table 20 should produce balanced plants for any size. There are varietal differences, and experience will fine-tune the results. Remember that one week of long days in the winter does not equal one week of long days in the summer. To consistently grow high-quality kalanchoes throughout the year, adjust long days or change your varieties as the seasons change.

Long-day treatment consists of 10 f.c. of light (108 lux) at plant level. Use two hours per night from March 1 through October 31 and four hours per night from November 1 through February 28. Using this schedule takes some of the details out of changing schedules every month. Mum lighting is compatible with the light levels and timing of kalanchoe lighting.

Here are some guidelines for the number of weeks of long days to provide prior to providing short days to flower plants when starting from unrooted cuttings: two weeks for minis (2" [5 cm]); four weeks for 3.5" (9 cm) pots; four to five weeks for 4–5" (10–13 cm) pots; and five to eight weeks for 6" (15 cm) pots.

One of the most common mistakes made with kalanchoes is premature budding during long days. This happens when light bulbs burn out or timers malfunction. To avoid premature budding, check your lights and timers daily to ensure they are working properly.

Short-day (black cloth) treatment

Kalanchoes are short-day plants. They require a longer dark period than chrysanthemums do. Once plants are sized up, begin the short-day treatment to force flowering. Pull black cloth until buds are visible; generally short days last six to seven weeks. Some growers pull black cloth through flowering, as they may often have plants at different growth stages in the same greenhouse; however, during the summer months it is best to restrict short days to six weeks. The total time from the start of short days until sale is nine to twelve weeks, depending on the variety and time of year.

When night temperatures are high, pull black cloth to provide fourteen to fifteen hours of total darkness from 7:30 P.M.–9:30 A.M. instead of 5:00 P.M.–7:00 A.M. During the summer months, black cloth can be opened from 10:00 P.M.–2:00 A.M. to let heat escape.

The consistency of applying black cloth is more critical to kalanchoes than to chrysanthemums. Since most of the larger pots require long crop time, do not miss any nights of black cloth application or the crop will be delayed. Holes in the black cloth and stray light will cause delays in flowering. The short-day treatment should start on March 1 and end on October 1. Be mindful to also prevent any lighting spillover from adjacent areas from October 1 to March 1. Be aware that heat delay can occur in summer production if there is high temperature buildup (84°F [29°C]) under the black cloth.

Pests & Diseases

The most common insect problem is aphids at the later stages of maturity, when buds and flowers develop. Some other insect problems include worms, mealybugs, thrips, and whiteflies. During propagation, fungus gnats can be a serious problem.

Kalanchoes are very sensitive to many spray materials. Emulsifiable oils often burn the leaves. This burning happens because many of the kalanchoe leaves cup up and hold the spray material. The best materials for kalanchoes are wettable powders or water-soluble pesticides. Oil-based sprays should always be tested before application. If an oil-based spray is used, wash it off after a short period; do not let spray material remain on the leaves. Also, be cautioned against fog materials that have an oil base and can settle on the leaves.

Table 20. Kalanchoe Crop Schedule[a]

	UNROOTED CUTTINGS					LINERS				
Pot size	2–3"	4–4.5"	5–5.5"	6–6.5"		2–3"	4–4.5"	5–5.5"	6–6.5"	
Plants per pot	1	1	1	1	3	1	1	1	1	3
Long-day weeks	0–2	3	5	8	4	0	2	3	6	3
B-Nine spray[b]	Week 3	Week 4	Week 4	Week 5	Week 5	Week 2	Week 4	Week 4	Week 6	Week 4
Approx. weeks plant to pinch	No	No	Week 5	Week 6	Optional	No	Optional	Week 2	Week 3	Optional
Short-day weeks	6	6	6	6	6	6	6	6	6	6
B-Nine spray[b]	Week 6	Week 7	Week 8	Week 9	Week 9	Week 5	Week 7	Week 7	Week 9	Week 7
B-Nine spray[c]	Yes	Yes	Varietal	Varietal	Yes	Yes	Yes	Varietal	Varietal	Yes
Weeks to salable plants[d]	9–15	12–16	14–18	16–20	13–17	9–13	11–15	12–16	15–19	12–16

Note: Short-day treatment is not needed in northern areas from October 1 through March 1. Days are naturally short enough to induce flowering.

[a] Use a minimum temperature of 65°F (18°C) at night. This schedule can be used from March through September in the North and year-round in the South. Longer crop times may be required in the winter and in lower light areas.

[b] B-Nine is applied for height and foliage size control at 5,000 ppm, for example, during the third week after plants are potted.

[c] B-Nine is applied for peduncle length control at 2,500 ppm when the bud is visible, only if needed. Some varieties will not need this third application.

[d] Two long-day weeks plus six short-day weeks totals eight weeks, not eleven to fifteen weeks. Reason: You do not need to shade kalanchoe all the way to maturity. Also, a range of weeks is given because varieties differ in their earliness.

The most difficult disease problem of kalanchoes is bacterial soft stem rot, which can spread at any stage of development. The crop should be started from disease-free cuttings, and sanitation is important in culture. The benches should be cleaned and treated for disease with bleach. If growing on the ground, the soil should be sterilized between crops. Other disease problems include *Myrothecium* and powdery mildew.

Troubleshooting

Plants that are growing too slowly may have salts buildup, may have not receiveed enough long days prior to beginning short days, may have received too much growth regulator, may be reacting to low production temperatures (below 60°F [16°C]), or may have been treated with phytotoxic chemicals. Fasciations, where two or more stems or flowers fuse together, may be caused by high production temperatures or may be genetic. Blind shoots or delayed flowering can be caused by temperatures below 61°F (16°C). At high production temperatures—greater than 84°F (28°C)—kalanchoes will experience heat delay. At temperatures over 93°F (35°C), plants will not flower. Red leaves may be due to high light intensities, cool temperatures (below 63°F [17°C]), moisture stress, or the genetics of the variety. Runaway side shoots, when the center buds are surpassed by side shoots, are the result of the plant changing from a flowering stage and reverting to vegetative growth. This condition can be due to low temperatures during short days, high temperatures (heat delay), too few short days, light leakage in short days, or a seasonal influence such as too many dark days in the winter. Cells that rupture due to high moisture, cool temperatures, and high relative humidity can cause corky lesions and scabs on stems and leaves.

Varieties

Varieties differ in their response times by season, sometimes by as much as two to three weeks. A variety that may flower in the spring after eight weeks of short days may take eleven weeks in the winter. If you are growing a year-round program, choose varieties with shorter response times or add extra production time to your schedule during the winter months. In the North, the overall crop time for kalanchoes can vary from eleven weeks in the summer to eighteen weeks in the winter.

For an interesting new product, try five cuttings of 'Santorini', 'Light Jacqueline', 'Timor', or 'Kerinci' in a 10" (25 cm) hanging basket.

Kalanchoes also make a great garden plant. Once established, they are relatively drought resistant, and they will flower multiple times outdoors with flowers getting larger and larger at each successive flowering.

Postharvest

Kalanchoes are ready for sale when about 30–50% of the flowers are open. Shipping the plants before terminal flowers open will set the plants back and the flowers will never develop fully, so be sure the plants have mature flowers before shipping. Plants shipped promptly at this stage will have six to eight weeks of shelf life. Plants may be sleeved, boxed, and shipped at 40°F (4°C). Plants can be stored in flower for as long as three weeks, but postharvest life is significantly reduced. Kalanchoes are sensitive to ethylene, so be sure to groom them prior to packing to remove any dead leaves or flowers, which can generate ethylene. At higher temperatures, plants are more susceptible. Do not ship kalanchoe with ripening fruits or vegetables.

Retailers should unsleeve and display pots in a bright (100 f.c. [1.1 klux]) area at 60–75°F (15–24°C). Plants exposed to ethylene and showing damage when unpacked may never recover. Retailers should water when plants are dry.

Homeowners can place kalanchoes in a bright window. Kalanchoes can tolerate a wide range of temperatures with ease.

Kniphofia

Kniphofia uvaria
Common names: Red hot poker; torchlily
Perennial (Hardy to USDA Zones 5–9)

Kniphofia uvaria is an impressive plant with long, thin, coarse leaves radiating out of a central crown that grows from 15–20" (38–51 cm) tall when not flowering. The foliage will develop out of the crown and arch back over to the ground, but not necessarily gracefully. A large group planting may appear unkempt, since the long, linear leaves sometimes bend to the ground.

When in flower, the plants will top out at 3–4' (90–120 cm) tall. The flowers are small and tubular and droop or hang from the crown of an erect 3' (90 cm) stalk. Collectively, a mass of blooms occupies the top 4–6" (10–15 cm) of the spike. The blooms are often two-tone or bicolor in yellow and red or orange. If transplanted to the garden the previous year, one to three spikes of color can be expected with more in proceeding years as the plants become established.

Propagation

Seed and bare-root transplants are the two primary propagation methods. Seed can be sown year-round, but read the germination notes in the Seed section to achieve the best performance. Divisions are best taken during the spring and early summer when the anchor roots are forming and will establish readily after transplanting.

Seed

Sow seed to a 288-plug tray and leave seed exposed during germination or lightly cover with coarse vermiculite. Germinate at 65–75°F (19–24°C) soil temperatures under mist.

Seed will develop a root radicle but will then form a small bulbil, or round growth, at the root crown. The growing shoot will actually develop from this. Seed starts to germinate in six to eight days after sowing, but a high percentage of seedlings will not be seen until fourteen to twenty days after sowing when exposed to 65–75°F (19–24°C) temperatures. Seed will germinate whether it is covered or not and has the same germination results in either case.

As seedlings emerge, they will be erect, 3" (7 cm) long, grass-like spikes approximately twenty days after sowing. A 288-plug tray will be salable in seven to eight weeks from sowing.

If sowing to an open flat, seedlings can be transplanted thirty to forty-five days after sowing. When transplanting, be sure that seedlings are strong enough to be handled, but do not allow them to establish a strong root in the sowing flat or tray prior to transplanting. Although they have a taproot, they are not as prone to transplant shock as other tap-rooted perennials. However, there is also no need to slow down the plants' growth and development by letting roots become restricted within the germination tray or plug cell.

Bare roots

Bare roots are available from commercial propagators. Potted up in winter to 1-gal. (4 l) containers and placed on the greenhouse bench, plants will start to root and develop green shoots as days are kept at 65–68°F (18–20°C) or slightly warmer and nights at 55–58°F (13–14°C).

Growing On

Seed

Seed sown to a 288-plug tray in June or July and transplanted to the final 1-qt. (1.1 l) or 1-gal. (4 l) containers can be overwintered cold or dormant for sales the following spring. Seed can also be sown to plug trays in late summer or autumn and overwintered in a cold greenhouse or cold frame in the final containers once the roots are fully developed in the final containers. However, the plants do not develop so quickly as to allow them to go dormant.

In the spring, as the days grow longer and the temperatures warm, these plants develop and can be sold green in May. In the home garden they will be in flower in July.

For seed sown in the winter (January), plants develop slowly; cloudy weather or cool temperatures will accentuate this condition. Sowings made after the New Year will not be salable the same year from seed unless you are planning late summer sales in the North. However, there are exceptions, including the All-America Selections Award-winning 'Flamenco'. Sowings made in January to early or mid-February will produce 4 or 6" (10 or 15 cm) pots by May or June, and plants will flower in the garden in late July or August. Granted, they will not be as profuse in flowering as those sown the previous year, but 'Flamenco' offers you the opportunity to sow late and still have a flowering crop.

Bare roots

From an early February potting to a 1-gal. (4 l) pot, plants are rooted well enough by the first of April here in the Midwest to be moved to a cold frame or similar environment and grown on at no less than 48°F (9°C) nights and days at 10–15°F (5–8°C) warmer. Plants are salable in late May. However, while the top growth will be developing well, the roots will not be fully massed throughout the container. However, if the plants are held on in the nursery or garden center until summer and sold then, the roots will be fully massed. Roots potted up the previous year, overwintered dormant, and sold in spring are more fully rooted than those potted up in winter for spring sales. Regardless, potting roots in the winter is commonly done and is not a problem.

Koeleria

Koeleria glauca
Common name: Blue hair grass
Perennial grass (Hardy to USDA Zones 4–9)

Tufted crowns of this blue-green grass grow from 6–8" (15–20 cm) tall and, when in bloom, from 12–16" (30–41 cm) tall. Plants prefer a full-sun location in well-drained soil and from May through June.

Propagation & Growing On

Sow several seeds (from two to six) or one multiseeded pellet, if available, per cell to a 288-plug tray. Seeds do not need to be covered. Germinate at 70–72°F (21–22°C), and seedlings emerge in four to six days. Plug trays are ready to transplant to larger containers in five to seven weeks. Grow on at 65–72°F (18–22°C) days and 60–64°F (16–18°C) nights, once established. A 4" (10 cm) pot will take another six to eight weeks to be ready to sell.

L

Lamium

Lamium maculatum
Common names: Dead nettle; spotted dead nettle
Perennial (Hardy to USDA Zones 3–8)

Lamium maculatum is a premier plant for the shade that, under ideal conditions, can spread rapidly. The foliage has silver highlights mottled with green next to the midrib and across the leaf surface. The plants grow upright, are less than 12" (30 cm) tall in bloom, and fill in well when given a moist but drained location.

Lamium's lipped flowers are 0.75–1" (2–2.5 cm) long, pink to rose-red or purple in color, and single in form. The plant's flower quality is secondary to its speckled foliage and ability to fill in readily.

Propagation

Lamium is easily propagated from cuttings. Propagation can occur year-round, as long as active growth is maintained. Maintain high light levels and warm temperatures during winter months for year-round production. Most standard tray sizes for propagation are between 32 and 128 cells per tray. One cutting per cell is sufficient, as *Lamium* cuttings root easily. Rooting hormone is not necessary, but can be used if desired.

Use a standard propagation medium with a pH of 6.0–6.5. Temperatures of at least 70–75°F (21–24°C) are best for the rooting process, ideally with bottom heat. Keep the medium moist at all times, but not saturated. Until the plants have developed roots of their own, mist several times a day to avoid wilting. The amount and frequency of misting depends on the temperature and air circulation in the greenhouse, although six to nine days of mist are usually sufficient for rooting. After cuttings have rooted, move plants to a hardening area with slightly increased light levels and lower temperatures. From sticking to rooting takes three to four weeks for a 105- or 84-cell pack/liner tray.

More and more growers are choosing to buy in liners instead of self-propagating due to the often higher-quality available from specialized, young plant producers.

Growing On

Transplant liners into a well-drained, soilless medium, such as peat/perlite mix, with a pH of 6.0–6.5 and a moderate initial nutrient charge. For winter to early spring potted plants, use one liner per 4 or 6" (10 or 15 cm) pot. *Lamium* can also be used as a component in mixed container gardens: Many growers choose to establish plants in smaller pots before moving them into combination containers.

Lamium grows best under moderate light levels, although some varieties are quite sun tolerant. A range of 3,500–7,000 f.c. (32–75 klux) is appropriate. *Lamium* is an excellent plant for shady areas.

Keep medium moist, but not saturated. Apply a constant feed of 150–200 ppm from 20-10-20, alternating with 15-0-15 and average amounts of micronutrients with slightly higher levels of iron.

To encourage branching, give one soft pinch within one to two weeks of planting. Some varieties can get quite leggy over the course of the season and may be cut back as desired. Other varieties are naturally more compact.

The crop time for a 4" (10 cm) is five to seven weeks when potted up in late winter or early spring for May sales. Plants will be salable green and not in flower.

Pests & Diseases

Watch for whiteflies. Drench with a broad-spectrum fungicide at liner planting to minimize the danger of soilborne diseases.

Varieties

'Aureum' has soft, golden-yellow foliage and pastel, lavender-pink flowers. 'Beacon Silver' is a very popular variety, with silver leaves, green margins, and pinkish-purple flowers. 'Golden Anniversary' is a more recent, patented introduction with golden-yellow, green, and white variegation and purple flowers. It has a compact, self-branching habit. 'Orchid Frost' is another recent, patented introduction with silver leaves, green margins, and orchid-pink flowers. 'Pink Pewter' is a low-growing variety with soft pink blooms and creamy silver and green variegation. 'White Nancy' is similar to 'Beacon Silver', but has pure white flowers.

Related Species

Lamium galeobdolon is more commonly referred to as *Lamiastrum galeobdolon* within the industry, even though the correct botanical name is actually *Lamium galeobdolon.* (At least this was true at press time—the name has changed quite frequently.)

L. galeobdolon 'Hermann's Pride' has striking silver leaves with dark green veins. *L. galeobdolon* 'Variegatum' has silver-variegated leaves with green veins and margins and a quite vigorous habit. Propagate and grow *L. galeobdolon* on as noted for *Lamium maculatum.*

Postharvest

Lamium is an excellent plant for often-difficult shady locations. The cooler the outdoor temperatures, the more sun *Lamium* tolerates.

Lantana

Lantana camara
Annual

Whether they are in hanging baskets or 4" (10 cm) pots, lantana is one of those die-hard performers that consumers turn to year after year. Once established, plants can take heat, full sun, and moderate drought. Depending on the variety, they will grow to the size of a small shrub over the summer, continuously producing flush after flush of flowers, provided temperatures are warm enough. To be honest, at the point of sale, plants rarely have the instant color appeal of petunias, geraniums, or New Guineas, but once a consumer has experienced lantana, they are repeat buyers. Fortunately, breeders have been paying attention to lantana, because new varieties that make a more colorful presence on the retail shelf have been showing up in the market.

For growers, lantana is easy: Plant up baskets, throw them in the second tier of the greenhouse attic on drip tubes, and practically forget about them until sale.

Propagation

Lantana is primarily propagated from softwood tip cuttings. Harvest uniformly sized cuttings measuring about 2–3" (5–8 cm) long from actively growing plants. Root cuttings in Ellepots or medium with a pH of 5.5–6.2. A rooting hormone is beneficial, but is not required. Provide substrate temperatures of 68–72°F (20–22°C) and ambient air temperatures of 75–80°F (24–26°C) during the day and 68–70°F (20–21°C) at night. Provide mist using tempered water in the mist lines. Adjust the frequency based on your specific light, relative humidity, and temperature conditions. Depending on nozzle size and spacing as well as the size of the propagation bench, set the mist duration for ten seconds or longer. Maintain medium at or very near full saturation so that cuttings do not lose turgidity. Fertilize with 50–75 ppm from 20-10-20 if you notice loss of leaf coloration. Provide 500–1,000 f.c. (5.4–11 klux) of light. Night misting is usually required. Cuttings will callus in five to seven days.

When 50% of cuttings have differentiated root initials, move trays to the next stage for root development. Maintain substrate and air temperatures. Increase light levels to 1,000–2,000 f.c. (11–22 klux) and increase fertilizer to 100–200 ppm once a week from 20-10-20, alternating with 15-0-15. Roots should develop over seven to nine days.

Harden cuttings off for seven days before sale or transplanting by maintaining air temperatures at 75–80°F (24–26°C) days and 68–70°F (20–21°C) nights. Increase light levels to 2,000–4,000 f.c. (22–43 klux). Fertilize twice per week with 150–200 ppm, alternating between 20-10-20 and 15-0-15.

Some growers buy in unrooted cuttings of unpatented varieties and root their own liners. Others harvest a tip cutting of nonpatented varieties when they pinch their basket crops and root liners for 4" (10 cm) production. Increasingly, more and more growers are buying in liners as newer, superior varieties are protected with plant patents and/or breeders' rights, and unlicensed propagation is illegal.

An 84 tray will root in four to five weeks for most varieties.

Growing On

Plant 3–4 cuttings/10" (25 cm) or 7 cuttings/14" (36 cm) basket. Use 1 cutting/4" (10 cm) or 1–2 cuttings/6" (15 cm) pot. Plan to offer 1-gal. (4 l) pots in the summer for landscape sales. Select a well-drained, disease-free medium with a high initial starter charge and a pH of 5.5–6.2.

If you have planted liners directly to baskets, be extra careful when watering during the first two weeks after planting, as it will be easy to overwater plants in these conditions. If you have rooted cuttings in Oasis, it is even more critical that you maintain near-perfect moisture conditions in the basket, as the medium must be moist enough so that cuttings root in.

Grow plants warm: 75–80°F (24–26°C) days and 62–65°F (16–18°C) nights. Do not allow temperatures to peak over 95°F (35°C), as flowers will abort.

Grow under high light (4,000–7,000 f.c. [43–75 klux]), while maintaining production temperatures. You can place baskets in the greenhouse attic on drip tubes in late winter and early spring, but move them down to the second or third tier as the season progresses and outdoor temperatures rise. Keep plants thoroughly watered; however, allow pots to dry down between irrigations. Do not allow permanent wilting because leaves will burn. Botrytis and powdery mildew can become problems under humid conditions where foliage is allowed to remain wet.

Lantana flowers more profusely as light levels improve through the season. Under the low light of winter (below 1,000 f.c. [11 klux]), crops take longer to finish and stems can stretch. In the summer, provide some shade in the greenhouse.

Fertilize twice a week with 200–300 ppm from 20-10-20, alternating with 15-0-15. Pinch plants back once liners are established (about two weeks after potting). Pinch plants back to the fifth or sixth leaves above the substrate line. Pinch 4" (10 cm) pots once, 6" (15 cm) pots two to three times, and 10" (25 cm) baskets two to three times. Cycocel (3,000 ppm) or Florel can be applied five to six days after pinching and every two weeks after that, as needed. Florel causes bud abortion, so do not use it after buds set. As light levels increase, reduce Cycocel rates to 1,500 ppm.

With one or two liners per pot, a 4–5" (10–13 cm) pot takes six to seven weeks to root and flower. A 10–12" (25–30 cm) basket with three to six liners will take from ten to twelve weeks to flower. *Note:* These crop times are based on the newer selections (see the Varieties section)

Pests & Diseases

Aphids, fungus gnats, spider mites, thrips, and whiteflies can attack plants. Diseases include *Alternaria,* botrytis, *Fusarium, Pythium,* and rhizoctonia.

Troubleshooting

Excessive vegetative growth can be due to high ammonia levels, too much fertilizer under low-light conditions, low light and overwatering, and too much or a late application of Florel. Low fertilizer (lack of ammonia), low light, and excessive flowering can cause poor branching. Plants drying out too much between watering and/or high soluble salts in the medium can lead to foliage necrosis.

Varieties

Lantana breeders have been improving this class with zeal for the past ten years. While 'New Gold', 'Dallas Red', and others have dominated the market, there are a multitude of additional varieties to consider. Some of these are patented, so propagation is illegal unless you are licensed. Be sure to visit plant company websites for details.

'Athens Rose' grows 3–4' (0.9–1.2 m) in height in one season and is hardy in most parts of the South. 'Miss Huff' grows to 5–6' (1.5–1.8 m) in a season and continually produces orange and yellow flowers. Plants are hardy in most parts of the South. 'Samantha' has variegated foliage and bright yellow flowers. For basket production, it is faster to stay with the unpatented varieties like these and pinch.

The 'Patriot' series is compact and is excellent for 4 or 6" (10 or 15 cm) pots. 'Patriot Rainbow' is a more compact version of 'Confetti'. 'Patriot Bouquet' and 'Patriot Desert Sunset' have orange-gold, pink, and coral flowers. 'Patriot Firewagon' has coloring similar to 'Radiation'. Of the series, be sure to have pots of 'Patriot Rainbow' and 'Patriot Honeylove' (pale pink, pastel yellow, and cream white).

The 'Lucky' series is available in a multitude of flower colors on plants that make excellent compact plants in 4" (10 cm) pots and grow to 12–16" (20–40 cm) tall. Be sure to try 'Sunrise Rose' with its unique, deep purple-rose color and yellow center.

The 'Landmark' series is a larger-growing variety specially selected for the landscape market. Plants grow from 15–20" (38–50 cm) tall, and the series has been listed highly in regional trials across North America.

The 'Bandana' series is another excellent choice for landscapes and large containers, growing from 16–20" (41–50 cm) tall. 'Lemon Zest' is a particularly showy variety that makes a great component in mixed container designs.

The 'Bandito' series is the more compact line from the 'Bandana' series but is similar in height (12–14" [20–30 cm]) to the 'Lucky' series.

Related Species

L. montevidensis, a trailing lantana with lavender flowers, is available from some suppliers. These plants make excellent baskets.

Postharvest

Instruct retailers to make sure pots dry thoroughly between irrigations. They should display plants under full sun.

Larkspur (see *Consolida*)

Lathyrus

Lathyrus odoratus
Common name: Sweet pea
Annual, cut flower

At one time sweet pea was one of the most popular greenhouse-grown cut flowers. Equally, it was one of the most common items for home gardeners to sow to the soil for late spring flowers. It is well known that the soft, sweet scent and delicate flowers of sweet peas were once a favorite of June brides for wedding bouquets. Today, consumers still love them, but because they are such a difficult crop to grow from a temperature standpoint and very labor intensive, most growers have turned away from them. In the Netherlands alone, the supply of sweet peas to the Dutch auctions dropped 40% from 1995 to 2000 (to 9.7 million stems per year). Like in the United States, most Dutch production hits the market in the spring (i.e., April, May, and June).

Bedding plant growers can offer sweet peas in color in pots in the spring as an impulse item; however, these plants will not perform well when planted out in the garden.

Propagation

Sweet peas do not like to be transplanted. Sow seed for cut flowers directly to the location where vines will grow. Likewise for pot plants, sow the seed to the final container.

Pot plant growers can sow seed directly into the final container in December or January. Cover seed and germinate at 65–70°F (18–21°C). Thin seedlings to three or four per 4" (10 cm) pot or five or six per 12" (30 cm) pot. Grow seedlings on at 58–60°F (14–16°C) days and 50–55°F (10–13°C) nights. Fertilize with 150–200 ppm from 15-0-15 at every other irrigation. Maintain good moisture levels to keep plants actively growing, but do not keep plants wet.

Growing On

There are two classes of sweet peas: summer flowering and winter flowering. Summer-flowering sweet peas branch close to the ground and produce a number of active shoots. Winter-flowering varieties produce one vegetative shoot, which first flowers and eventually branches along the stem. Of the two, summer-flowering varieties are long-day plants and will not flower under the short days of winter. Since most cut flower material is sown in the summer for winter flowering, use only winter-flowering varieties in northern greenhouses for the best performance. Crops sown in June through September will flower in October through March or April.

As a cut flower, sow seed onto well-drained soils with a slightly alkaline pH. Ground beds are best since sweet peas are deep rooting and require 6–8' (1.8–2.4 m) of headroom. In the past, many cut flower growers would sow a few rows of winter-flowering sweet peas along end or sidewalls. Sow seed directly onto steam-sterilized beds in trenches that are 1.0–1.5" (3–4 cm) deep, space seed 1" (3 cm) apart, and later thin to one seedling every 2" (5 cm). You can sow into double rows that are 6" (15 cm) apart with 3' (91 cm) walkways between each set of rows. Cover with moist soil and then a removable mulch (newspaper works well) or perforated plastic that will prevent drying out until seedlings have emerged.

When seedlings are 12" (30 cm) high, they must be supported by string and wire. Wires are run at ground level with a parallel row at 7–8' (2.1–2.4 m) high for the length of the bed. Lace strings vertically between the two every 2–3" (5–8 cm) apart to provide a support for the growing tendrils.

Maintain good moisture levels, but be especially careful not to overwater plants in the winter months when the threat of disease is highest. Fertilize with 150–200 ppm from 15-0-15 at every other irrigation.

Cut flower varieties require from fourteen to sixteen weeks to flower.

For pot plants, growers select the dwarf-growing (bedding plant) varieties and sowings made in December directly into the final container will be in flower in twelve to fourteen weeks.

Pests & Diseases

Aphids, cabbage loopers, and leaf miners can attack sweet peas. Diseases such as botrytis, bacterial wilt, fungal leaf spots, and powdery mildew can afflict plants.

Varieties

'Knee High Mix' is a bush-type sweet pea that grows 24–30" (61–76 cm) tall and is well suited for pot production. 'Winter Elegance Mix' is a multiflora blend for flowering under short days from winter to early summer. 'Mammoth Mix' is a cut flower sweet pea with extra-large flowers and long stems.

The 'Winter-Flowering' series does not revert to summer flowering, as other series do. Plants produce four to five fragrant flowers per stem on trellises. Sowings from August to October begin flowering in December and continue to produce until the onset of warm weather. The series comes in 'Crimson', 'Deep Blue', 'Lavender', 'Pink', 'Rose Crimson', 'Rose Pink', 'Salmon Pink', 'Scarlet', and 'White'.

Postharvest

Sweet peas are easily damaged during postharvest. Harvest flowers when the first flower is showing good color and other flowers are in the bud stage. Flowers are highly sensitive to ethylene gas. Do not store them near ripening fruits and vegetables, and keep coolers and work areas clear of dead flowers and other plant debris. Place in floral preservative and silver thiosulfate (STS) solution to extend its vase life to seven days. Store and ship stems at 50°F (10°C). Retailers should recut stems immediately and place them in a floral preservative with STS.

Lavandula

Lavandula spp.
Common name: Lavender
Herb, perennial (Hardy to USDA Zones 5–8)

Lavender is the grande dame of herbs. Cherished for the soothing scent of the oil in its leaves, lavender is an ingredient in many home-scent products, which range from candles to air fresheners, and in personal care products, such as lotions and hand soaps. On the retail shelf, lavender plants are one of the first herbs to sell out, as it seems lifestyle magazine editors and television gardening personalities cannot seem to feature it enough.

Lavender needs mild Mediterranean or Pacific temperatures to thrive: cool to warm days and cool evenings with low humidity in sunny locations make the plant excel. Plants are perennial, but there are only a few that are reliably hardy outside of the Pacific Coast and similar climates. They are not long lived in warm to hot, humid summers and cold-winter environments. However, a few varieties have demonstrated good performance abilities in protected locations across the northern part of North America and live from year to year. These are highlighted in the Varieties section.

Propagation & Growing On

Seed

There are a wide number of varieties of lavender available from seed. Many of these are *Lavandula angustifolia* (or similar parentage) and are excellent garden, bedding, or border plants, and a few are even hardy, but they are not commonly used for culinary purposes. They possess only a limited amount of the essential oils preferred in the kitchen.

For spring sales, sow in December or early January for green plants (not flowering) in 4" (10 cm) pots or 18 trays (18 cells per standard 11" × 22" flat). For a 512-plug tray, sow seed and leave exposed to light and germinate at 68–72°F (20–22°C). Seedlings emerge in six to ten days on some selections (including 'Lavender Lady'), while others will require as much as fourteen to eighteen days from sowing to emerge. While the seed catalogs do not necessarily state it, there are a few lavenders that undergo seed treatments to allow them to germinate more readily with a greater germination percent. If these treatments were not done, seed will require more time to emerge as well as have a lower germination percentage.

Commercial propagating firms will sow lavender seed in the summer for plants sold in 50-liner trays or similar for sales later that year. These are often pinched once to allow for more uniform growth and can be potted up to 1-qt. or 1-gal. (1.1 or 4 l) pots for overwintering or the liner tray can be exposed to 35–44°F (2–7°C) for ten to fifteen weeks to vernalize the plants. This is especially helpful on varieties like 'Hidcote' and 'Munstead' that will keep them more uniform and help them to flower.

Cuttings

The preferred method of propagation for culinary varieties of lavender is from cuttings.

Cuttings can be taken from stock plants or received from commercial companies, dipped in rooting hormone, and stuck in summer to early autumn in a 105- or similar liner or plug tray. Propagated under high humidity (preferably through overhead mist), keep under low mist or place cuttings under a tent and mist once per hour for seven to eleven days. It takes about three to five weeks, depending on variety, to root with soil temperatures of 68–74°F (20–23°C).

Transplant the 105 liners to 32 or 50 cells per flat and allow the plants to bulk up (develop roots and top growth) for six to ten weeks, depending on variety. Generally, plants do not need to be pinched. However, if they are growing erect with no basal branching, a soft pinch will help to correct this.

These can then be either transplanted to 1-qt. (1.1 l), 1-gal. (4 l), or similar containers or vernalized at 35–44°F (2–7°C) for ten to fifteen weeks for plants that will bloom after they have been sold. (Flowering is not a crucial matter, but this culture provides what is needed to ensure a blooming plant the same year as the plant is sold to the home gardener.) Plants that have been transplanted to 1-qt. or 1-gal. (1.1 or 4 l) containers are overwintered in a cold greenhouse or cold frame with special consideration for their evergreen foliage. Take care not to allow these plants to rot.

The vernalized plants in the 32 or 50 cells are often sold during the winter to early spring from commercial propagators to greenhouse or nursery companies to pot up to 1-qt. (1.1 l) containers for spring sales. Potted to 1-qt. or 1-gal. (1.1 or 4 l) containers, these plants are ready for spring sales in eight to twelve weeks.

In general, it takes eight to twelve weeks from transplanting a plug to a salable 4.5" (11 cm) pot. Grow on a 60–70°F (16–21°C) days and nights 10°F (5°C) cooler.

Forcing

Lavender in flower is an especially attractive impulse item. The following forcing information has been compiled in order to educate growers on how to grow fabulous flowering lavender. This forcing information is based on research conducted at Michigan State University with *Lavandula angustifolia.*

Lavender must receive cool treatment (vernalization) for flower induction. Bulk up plants at temperatures of 70–75°F (21–24°C) days and 65–70°F (18–21°C) nights. Allow plants to dry out before irrigating. Maintain light levels at 2,000–3,000 f.c. (22–32 klux) and natural daylight conditions. Lavender is a light feeder; fertilize every other irrigation with 20-10-20 at 150 ppm. Maintain a substrate pH at 6.5–7.0 and an EC of 1.0 mS/cm. Once plants have developed forty to fifty leaves, they are ready for vernalization.

Prior to vernalizing, pinch or trim plants to encourage branching and a denser plant at finishing. Vernalize plants for fifteen weeks at 41°F (5°C) in a cooler or unheated cold frame. Maintain 25–50 f.c. (269–538 lux) of light. Plants in 50-cell liners with forty to fifty leaves are large enough to be vernalized prior to potting. They can then be potted in the winter and sold that spring. Plants may also be vernalized in the final growing container.

Trim plants again when they emerge from vernalization. After vernalization, plants must receive more than sixteen hours of daylight to flower; begin using supplemental lighting or night interruption lighting to provide light until visible bud. Grow plants at constant day/night temperatures of 65–68°F (18–21°C). Fertilize with 100–150 ppm from 20-10-20, alternating with 15-0-15 at every other irrigation. Do not overwater. A-Rest and Sumagic can be used on lavender for height control. However, if the plants may be consumed, avoid plant growth regulators.

Pests & Diseases

Rhizoctonia can be a problem. Foliage is sensitive to rotting in high humidity. *Phytophthora* can be a problem during summer months. Insect problems are very rare.

Varieties

The most common lavender is *Lavandula angustifolia* (commonly called English lavender). *L. angustifolia* is very fragrant and showy in flower. 'Munstead' and 'Hidcote' are named varieties. 'Hidcote' grows approximately 12" (30 cm) tall, while 'Munstead' ranges from 12–24" (30–61 cm) tall. 'Lavender Lady' is a 'Hidcote'-type seed variety that produces uniform, true-to-type plants growing to the same height, while the 'Potpourri' series, also similar in height to 'Hidcote', is available in a number of flower colors. As for hardiness, 'Hidcote' has shown greater winter hardiness in the Midwest than 'Munstead', while 'Potpourri' has also proven good in the Midwest, lasting from two to three years.

Lavandula dentata is French lavender. It has a different leaf shape than English lavender. Its foliage is green or gray with a deep serration on the leaf margin. Plants are more sensitive to cold temperatures and thus less winter hardy. Plants can reach up to 3' (91 cm) tall and are suitable for use as topiaries. French lavender will not tolerate a cold, wet winter and is not considered winter hardy outside of the Pacific Coast.

Lavandula stoechas has the showiest flowers of all the lavenders and is a vigorous grower. Selections of *L. stoechas* swept through Europe in the late 1990s as the "in" plant.

Lavandula pinnata is an excellent annual that can be used in mixed combination containers and is a very tender species that can be used in the landscape only in very mild southern climates. As for winter hardiness, they will not survive a midwestern or northern winter but excel along the Pacific Coast.

Intermediate hybrids such as *L. intermedia* add confusion to lavenders. These are crosses between hybrids of *L. angustifolia* and *L. latifolia.* 'Provence' is similar to *L. angustifolia* in fragrance but is larger and taller with a lighter-colored flower. It is used in the garden as a hedge. 'Goodwin Creek Gray' is a cross of *L. dentata* and woolly lavender. It looks like *L. dentata* with gray-green foliage and deep blue flowers. In warm climates, it will bloom year-round and thus is a good southern variety. *L. latifolia* is prized for its pungent oil in household products. While not a garden species, hybrids between *L. angustifolia* and *L. latifolia* are used in the perfume industry. Like *L. dentata* and *L. stoechas,* these will not survive a cold, wet winter but excel along the Pacific Coast and similar climates.

Postharvest

Plants can be sold green or when one to three flower spikes are in color. Retailers should display lavender in high light areas and maintain cool temperatures. Warn retail customers not to overwater lavender.

Winter-hardy lavenders do not like to be mulched. This would keep moisture levels too high and the plants would rot. In northern areas, overwintering success is determined by the size of the root system before the plants become dormant.

Leucanthemum

Leucanthemum × superbum
(also *Chrysanthemum × superbum*)
Common name: Shasta daisy
Perennial (Hardy to USDA Zones 5–8)

Shasta daisy is one of the most popular perennials, a harbinger that summer has truly arrived. With one of the purest white flowers (dotted with a yellow center), Shasta daisies are a result of the work of renowned American breeder, Luther Burbank.

Plants flower naturally in the garden from June to August. By providing cold treatment and long days, you can force them into flower for sales of flowering pots in April and May. Many popular Shasta daisies will then continue to flower through the summer.

Propagation

There are a number of varieties that are propagated by seed, although both cutting and tissue culture is common as well. Most commercial propagators offer tissue-cultured plants as rooted liners in various cell or liner sizes. Some have been vernalized (given a cold treatment to enhance flowering) as well.

Seed

Sow seed into a well-drained, sterile medium with a pH of 5.5–6.2 and cover lightly with vermiculite to maintain humidity levels at the seed coat. Germinate at substrate temperatures of 65–75°F (18–24°C). Keep trays uniformly moist. Once radicles have emerged, in approximately three to five days, move trays into Stage 2 for stem and cotyledon emergence. Drop substrate temperatures to 62–70°F (17–21°C) and increase light levels to 500–1,000 f.c. (5.4–11 klux). Once cotyledons have expanded, apply 50–75 ppm of 14-0-14 weekly. Stage 2 will take from four to nine days.

Move plugs into Stage 3 for true leaf development by lowering substrate temperatures to 60–65°F (16–18°C) and raising light levels to 1,000–2,000 f.c. (11–22 klux). Begin to allow trays to dry down between irrigations. Increase fertilizer rates to provide 100–150 ppm weekly from 14-0-14, alternating with 20-10-20. Stage 3 will last from thirty-five to forty-two days for 128s. Harden plugs in Stage 4 for seven days prior to sale or transplant by lowering soil temperatures to 55–60°F (13–16°C) and increasing light levels to 1,500–2,500 f.c. (16–27 klux). Continue to allow trays to dry down between irrigations. Fertilize weekly with 14-0-14 at 100–150 ppm.

From sowing to a finished 128-plug tray will take from eight to ten weeks using one seed per cell.

Cuttings

Stick tip cuttings into a 105 tray and then provide seven to ten days of mist and 70–74°F (21–23°C) bottom heat. Cuttings do not need to be treated with a rooting hormone, although they often root more uniformly when dipped into one prior to sticking into the media. Plants will root shortly, and the tray will be ready to transplant in four to six weeks.

Smaller liner trays (such as a 105) are fine for winter rooting and potting in January to their final container for spring sales. For late summer liner trays, consider a 72, 50, or similar size. These will take from six to eight weeks to root during this time of year.

Growing On

Seed

Plugs can be transplanted in early or late fall as well. These plants are overwintered cold or with minimal heat in cold frames. Some suppliers offer vernalized plugs that can be brought in from late January through March and sold in flower from late March through May.

Seed sown in January or early February will produce flowering plants of a few varieties the upcoming summer, but the bloom is not as profuse as plants sown the previous year.

Shasta daisies are very sensitive and do not like to be too wet, so select a well-drained medium with a moderate initial starter charge and pH of 5.8–6.2. Do not bury the crown at transplanting.

Shasta daisies flower based on a combination of cold temperatures (although beneficial, a cold treatment is not required for flowering but it does enhance it) and photoperiod. Plants should be bulked (grown on to fully develop roots and crown) under short days for four to eight weeks.

Seedlings must have at least twelve to sixteen leaves before vernalization begins. (If you choose to vernalize in a propagation tray, ideally use 50-cell liners or 2.5" [6 cm] pots.) Bulk plants up by maintaining day and night temperatures of 60–65°F (16–18°C). Allow pots to dry thoroughly between waterings. Fertilize at every other irrigation with 100–150 ppm from 14-0-14. If height control is an issue, Bonzi and Sumagic have proven to be effective. Bare-root transplants may be potted, rooted in for a couple of weeks, and moved into vernalization.

Shasta daisies can then be forced into flower by providing ten weeks of cold temperatures at 41°F (5°C) under 25–50 f.c. (269–538 lux) of light and then bringing plants out into warmer temperatures and long days. Plants do not require vernalization to flower; however cold is beneficial as it helps to increase the percentage of flowering and earlier flowering.

Bring plants out of cooling and into a greenhouse with 65–70°F (18–21°C) day and night temperatures. Provide long days by either extending the natural day length (lighting from 4 P.M.–10 P.M.) or night interruption lighting (10 P.M.–2 A.M.). Shasta daisies need more than sixteen hours of daylight to induce flowering. Maintain ambient light levels of 3,000–5,000 f.c. (32–54 klux). Fertilize with 100–150 ppm from 15-0-15 at every other irrigation. Allow pots to thoroughly dry down between irrigations. Plants should be ready to sell in seven weeks.

Cuttings

Rooted liners are available in late August or September. They can be potted and grown with cool nights

(50°F [10°C]) until rooted and then overwintered in a cold greenhouse or cold frame. Plants can be allowed to go dormant if well rooted in the final container.

A 50-liner tray, potted in August to a 1-gal. (4 l) pot, will be rooted by mid-November for overwintering. Plants can be grown dormant or cold during the winter and sold green in the spring. Plants will flower in summer.

Fifty-liner trays are also available from commercial propagators and are sold in winter, many as vernalized liner trays. These can be potted up to 1-gal. (4 l) containers, allowed to root in a cool greenhouse (55°F [13°C] nights and 60°F days [16°C]), and then moved to a cold frame to finish for spring sales. These plants will also be sold green and will flower in summer.

As for a 105-liner tray, use 1–3 liners per 1-gal. (4 l) container, when transplanted in summer for overwintering. One liner can be used for a 6" (15 cm) pot, if transplanted in winter for spring sales. Summer- or fall-planted containers will always be fuller, with more flowers, than their winter-potted counterparts for spring sales.

Pests & Diseases

Crown rot can be a problem if plants stay too wet.

Varieties

'Alaska' is the one variety that helped to define Shasta daisies. Developed by Luther Burbank, 'Alaska' grows to 36" (90 cm) tall with single, white flowers. It is seed propagated.

The vegetatively propagated 'Becky' is another famed variety. It has received a Perennial Plant Association (PPA) Perennial Plant of the Year award and has helped to renew the interest in Shasta daisies. Plants grow to about 24" (61 cm). 'Becky' is an especially good variety for the South.

The list of Shasta daisy varieties is broad for both seed- as well as vegetatively propagated varieties. There are many notable varieties from which to choose. Be sure to review the list from each perennial plant company to get the most up-to-date selections available.

Postharvest

Plants are ready to sell once buds are ready to open. To preserve flower quality, ship and display plants cool. Display temperatures should be no higher than 70°F (21°C). Remove old flowers to encourage plants to rebloom. Shasta daisy is a short-lived perennial, generally dying out after two to three years, especially when planted in an area that gets a lot of winter rains.

Liatris

Liatris spicata
Common names: Gay feather; blazing star
Perennial, cut flower (Hardy to USDA Zones 3–9)

Today, *Liatris* is used in floral arrangements year-round. However, this hasn't always been the case. This native perennial was "discovered" by the Dutch, taken back to Holland, turned into a cut flower, and then exported back to the United States. Before then, Americans enjoyed *Liatris* only as a tall, stately perennial that flowers in July and August. Today, it is appreciated as both a perennial and a cut flower.

Propagation

Growers generally begin their cut flower crops with corms that are produced by specialists. Plants may also be propagated from seed or bare-root transplants.

Corms

After dividing corms, allow them to cure in a warm, well-ventilated area for three to seven days.

Seed

Sow seed to a 128-liner tray using two to three seeds per cell in summer for plants that are overwintered dormant for spring sales the following year. Seeds do not need to be covered, but a light covering of vermiculite helps to maintain moisture around the seed as the root emerges. Germinate at 68–72°F (20–22°C). Seedlings emerge in twenty-one to twenty-eight days. They appear similar to grass seedlings.

Growing On

Perennial crops can be started from corms, plugs, or bare-root transplants. Most growers start with 2.25" (6 cm) liners or bare-root plants. Either can be potted in early winter for green sales in May. If you would like to force plants for flowering sales, you will need to provide ten to fifteen weeks of cold temperatures (33–40°F [1–4°C]) and long days (see the cut flower production information for guidelines). Before cold storage, make sure plants are well rooted into containers.

Select a well-drained medium with a pH of 6.5–7.0. Place one 2.25" (6 cm) liner or bare-root transplant per 1-qt. or 1-gal. (1.1 or 4 l) container; corms can be spaced 2" (5 cm) apart, using three per 6" (15 cm) pot.

Liatris is very sensitive to overwatering, so the medium must be well drained and must dry between irrigations. Provide 150–200 ppm from 15-0-15 at every other irrigation.

Grow under high light levels (4,000–6,000 f.c. [43–65 klux]), while maintaining cool temperatures (55–60°F [13–16°C] days with nights 10°F [6°C] cooler).

Pot plants forced into flower early for spring sales may be planted in the garden; however, they will not reflower for consumers again until the following summer.

Cut Flowers

Purchase 6/8 or 8/10 grade corms. Freshly harvested corms require ten to fifteen weeks of cold storage at 32–35°F (0–2°C). In the long-term, corms can be held frozen at 28–30°F (–2– –1°C) in moist peat and brought out for year-round forcing. Do not refreeze corms. After freezing, corms may be held for up to two weeks at 45°F (7°C) prior to planting. Gibberellic acid (GA) may substitute for cold treatment. Soaking corms in GA at 500 ppm for one hour after removing them from five weeks of cold storage ensures 100% flowering. *Liatris* may be planted year-round; however, corms planted in February and March have the highest yields, based on University of Georgia trials.

Plant only in well-pasteurized beds; *Liatris* is highly susceptible to *Verticillium*. Substrate pH should be from 6.0–7.0. Plant corms 2–3" (5–8 cm) deep. Depending on corm size, space at 2" (5 cm) apart in rows, using 6–8 corms/ft.^{2} (63–84/m^2). Fertilize with 100–150 ppm from 20-20-20 once plants have begun to emerge.

Both cold storage and long days accelerate *Liatris* flowering. Be careful, however: If corms receive short cooling treatment, then short days will enhance flowering. Temperature also affects flowering. At cool temperatures (55°F [13°C]), long days enhance flowering; while at warmer temperatures, day length has little effect. Research has shown that the greatest effect on flower acceleration comes from using long days in the first five weeks after foliage emergence. So provide night interruption lighting from 10 P.M.–2 A.M. for five weeks after foliage emergence. (Prior to using long days, research has shown that three weeks under short days will boost total flower production. Use short days for two to three weeks after emergence, then provide five weeks of long days.) Flowers will be ready to harvest sixty to seventy days after emergence.

Maintain substrate temperatures below 68°F (20°C) for the first four weeks after planting. Afterward, force at 65–68°F (18–20°C).

Plants may be cropped for three years, becoming more productive in the second and third years. After the first year, use two layers of wire supports for stems.

Pests & Diseases

Plants are highly susceptible to *Verticillium*. Botrytis and *Sclerotinia* can also be problems. Aphids or thrips may attack, but it is rare.

Troubleshooting

Leaves can become sunburned, especially at the early stages of flower development. Sometimes sunburn causes flower abortion.

Varieties

The 'Floristan' series, which is available in 'Violet' and 'White', is the most-grown *Liatris* and makes an excellent cut flower. 'Kobold' is more compact, grows to 24" (61 cm), and is the first *Liatris* to flower in the summer.

Note that seed-grown varieties will flower sporadically when sown the same season they are sold.

Postharvest

Harvest lasts two weeks. Stems are ready to cut when three to four florets have opened. Flowering occurs from the top down. Remove the lower stem foliage before placing stems in water. Pulse for twenty-four to seventy-two hours with 5% sucrose for tight flowers and 2.5–5.0% solution for normal stems. Store in water at 32–35°F (0–2°C). Using postharvest preservatives drastically improves vase life. If stems are harvested tight and shipped dry immediately, recut stems, remove lower foliage, and pulse as above with 5% sucrose before sale.

Liatris sold as a perennial can be sold when flowering begins. Be sure to ship and display plants in cool temperatures (60–70°F [16–21°C]) to preserve flower quality.

Ligularia

Ligularia dentata
Common names: Bigleaf goldenray; bigleaf ligularia
Perennial (Hardy to USDA Zones 5–8)

L. dentata is a clump-forming perennial with elongated stalks that each terminate in a rounded, kidney-shaped leaf. The flowers are daisy- or rudbeckia-like in their form, golden or yellow-orange in color, single or semidouble in shape, and 2.5–4" (6–10 cm) across.

Ligularias are shade-loving plants that often grow to 2–3' (60–90 cm) tall in full bloom and spread 2–3' (60–90 cm) across when established in the garden. In gardens farther south than our Chicago-area trialing station, however, the plants have been up to a foot (30 cm) taller or broader, depending on location. Ligularias appear robust but are not invasive.

Propagation

Bare-root divisions and tissue culture are the most common form of propagating ligularias, since clones are preferred over seed-grown selections. Both of these forms of material are available from commercial propagating companies. *Note:* Many of the tissue-cultured plants are protected, and asexual propagation is prohibited.

Growing On

Divisions

Bare-root divisions can be potted up to 1-gal. (4 l) or larger containers in the summer or winter. Plants potted up in summer are grown outside under ambient day and night temperatures and shaded to avoid leaf scorch. Plants are overwintered dormant in a cold frame or cold greenhouse. Plants are sold leafed out in spring and will flower in June.

Bare-root divisions potted up to 1-gal. (4 l) containers in January or February are started on the greenhouse bench. Pot so that the root crown is flush with the soil, water in thoroughly, and grow on with 60–62°F (16–17°C) soil temperatures during the day and nights 5–10°F (3–5°C) cooler. After plants start to establish and roots reach the side of the container (in approximately four to five weeks), they can be grown cooler (no lower than 45°F [7°C]). Shoots will emerge within ten days, and pots will be well rooted in twelve to fifteen weeks. However, from February-potted roots, the container is often not fully rooted for April sales. The plants are still salable but need up to twenty weeks to be fully rooted when potted during winter.

Tissue culture

Professional labs perform tissue culture, and liners are commonly available from commercial propagation firms. These are commonly available in cell packs or liner trays in 32, 50, or 72 plants per tray.

Potted up to 1-gal. (4 l) containers in January, the plants are ready for green plant sales in May. Read the comments regarding temperatures and culture for growing on bare-root divisions. Many of these comments relate to liners produced from tissue-culture plants as well.

Varieties

Unless otherwise specified, all of the following varieties feature rich, vivid purple stems, petioles, and foliage when young. This coloring will persist as the plants emerge in the spring, though as summer approaches, the surface of the leaf will be green in color with a dark purple highlight. The undersides and leaf edges will remain purple, however.

'Britt-Marie Crawford' has orange-yellow, daisy-like flowers with chocolate-purple foliage that has a dark purple below. Plants grow from 3–4' (36–48 cm) and are excellent performers.

'Desdemona' is the most vigorous in spread and the tallest in height of the varieties listed here. The plants grow from 3–4' (90–120 cm) tall but are closer to 4' (120 cm) in full bloom. The flowers tend to have more yellow in them than the other cultivars listed here, but they are still golden yellow, rather than pure yellow. Flowers measure up to 3.5" (8 cm) across, and plants spread to 3.5' (1 m).

'Osiris Café Noir' emerges purple black and then changes to olive green as the summer progresses. Flowers are yellow, and plants grow from 20–24" (51–61 cm) tall.

'Othello' is related to 'Desdemona' but often grows into a mirror image of 'Dark Beauty' in appearance. This is a good cultivar with a 3–4' (90–120 cm) height and a 3' (90 cm) spread. Its golden-orange flowers grow 2.5–3" (6–7 cm) across. 'Othello' has a dark leaf coloring similar to 'Desdemona' and 'Osiris Café Noir'.

Related Species

The species noted below develop spike-type flowers as opposed to the daisy-like blooms on the *L. dentata.* Also, though they possess a dark stem coloring as seen on some *L. dentata* varieties, the coloring is not as rich in color or as persistent.

Ligularia przewalskii has yellow flowers on plants that grow to about 3' (90 cm) tall in bloom and even taller in more fertile and moist locations. Its flowers are held in spikes from 10–15" (25–38 cm) long and will bloom before *L. dentata.* Plants will spread from 2–2.5' (60–75 cm) across.

L. stenocephala is an interesting plant, but it is a specific clone, 'The Rocket', that has gained the market's attention. This variety is named for its elongated flower spikes, broader at the base and tapered at the tip. 'The Rocket' is often confused with *L. przewalskii* and is sometimes listed as its clone. There is one point of difference, however: 'The Rocket' often has more flowering stems than does *L. przewalskii* in trials—as much as three times more. Alan Bloom of England developed this cultivar. It grows to a 3–4' (90–120 cm) height with a 3' (90 cm) spread. Its yellow flowers are held in upright spikes from 10–15" (25–38 cm) long and, like *L. przewalskii,* will flower earlier than *L. dentata* varieties.

For both of the above species, hardiness, propagation methods, and cropping are the same as for *L. dentata.*

Lilium

Lilium × *hybrida*
Common names: Asiatic lilies; Oriental lilies; LA (or *L. longiflorum*–Asiatic), LO (or *L. longiflorum*–Oriental), OT (Oriental-trumpet) hybrids
Flowering pot plant, cut flowers

Lily, especially as a cut flower, has been one of the fastest-growing crops worldwide. Its clean, straight stems appear perfectly swirled in clear vases. Its perfect flowers sit high on stems with fully saturated color and adorn hotel lobbies, restaurants, and homes across Europe. The scent of fresh lilies is practically everywhere.

As pot plants, Asiatic and Oriental lilies are frequently offered in the spring for Easter and Mother's Day sales, and in more recent years lilies have "crossed over" to become a "nursery perennial," sold as growing plants for patio use or for planting directly into the garden. In many regions (Zones 5–7), the bulbs will perennialize when planted in the garden, flowering naturally in midsummer year after year.

Asiatic and Oriental lilies are a diverse group of plants that can be grown year-round. They are relatively easy to grow, have a high crop value per square foot of greenhouse space used, and are an excellent value for the consumer. Ever-increasing variety selections, better height control through improved genetics and growth regulators, the potential to be grown in a perennial context, and new types of lilies that can be forced at low greenhouse temperatures are all pluses.

Propagation

Most Asiatic and Oriental lilies are produced in Europe (the Netherlands and France), South America (Brazil and Chile), and the United States (the West Coast, from California to Washington). Plants can be propagated by scales or tissue culture. Seed can also be used. Growers purchase bulbs to force from commercial bulb suppliers, who generally contract their production. Most bulbs are harvested in the fall and shipped to growers after vernalization.

Most Asiatic hybrids are vernalized for at least six weeks at 34–36°F (1–2°C). Orientals are cooled eight to ten weeks at 34–35°F (1–2°C). LA hybrids (hybrids of Easter lily and one or more Asiatic hybrids) need at least six weeks at 35°F (2°C). Additional time at these temperatures can be used for short-term holding. For later plantings (early January and after), bulbs are frozen in. Asiatics are frozen at 28°F (–2°C) and Orientals and LAs at 30–31°F (–1°C). Bulbs must be properly cooled before they can be frozen in for storage. Freezing allows year-round planting, while lessening the danger of sprouting. Invariably, the bulb supplier, not the forcer, performs the freezing process. It is very much a specialist's job. In fact, some lilies are also held in ultra-low oxygen atmospheres while frozen (so-called ULO storage) to further reduce growth (that very slowly occurs even at freezing temperatures used) and to allow for even longer storage. To minimize water loss, bulbs are packed in moist peat moss and wrapped in polyethylene before freezing. If bulbs arrive frozen, thaw them slowly over one to three days at temperatures below 55°F (13°C), then plant as soon as possible. If you must store bulbs before planting, do so for no longer than three weeks at 34°F (2°C). Bulbs that have been frozen in must never be refrozen.

Growing On

Select a well-drained, pasteurized, fluoride-free (no superphosphate) medium with a pH of 6.2–6.5. Keep in mind that mixes containing pine bark tie up growth regulators, thus reducing their efficacy. Many growers dip bulbs in an approved fungicide before planting, but many do not as well. Most hybrid lilies are heavily stem-rooted, which means that deep planting in the pot is essential. Deep (standard) pots are also important for proper rooting depth.

Typically, plant three 12/14 cm bulbs per 6" (15 cm) pot and five per 8" (20 cm) pot. Water in well after planting, but be careful not to overwater pots in the days just after planting. Run plants on the dry side for the first two to three weeks or until stems are 1–3" (3–8 cm) tall to improve stem root growth.

Do not fertilize until stem roots are growing to prevent soluble salts buildup in the upper level of the pots. A program of 200 ppm from calcium-potassium-nitrate alternated with 20-10-20 is sufficient. Some growers also add a slow-release fertilizer. 'Star Gazer' requires supplemental iron, usually as a foliar spray.

Asiatics need to be grown cooler than Orientals. Grow Asiatics at days 70°F (21°C) or lower days and 55–62°F (13–17°C) nights. Do not let temperatures rise above 85°F (29°C). Depending on the schedule, Asiatics can be grown much colder than this. Orientals respond better to 75°F (24°C) days and 65–67°F (18–19°C) nights. As with Easter lilies, timing is mainly controlled by the twenty-four-hour average temperature, and some level of height control can be achieved by using DIF (warmer nights than days).

The average temperature is simply not the average of the day and night temperatures. Calculate it as follows:

$$\frac{\text{Day temperature} \times \text{hours} + \text{Night temperature} \times \text{hours}}{24} = \text{Average daily temperature}$$

You can moderate temperatures with shade from overhead screens during warm periods.

As a general rule, the number of days from planting to emergence is based on the length of cold storage or freezing in. Days to emergence vary from two to three weeks for early crops to three to four days for crops planted in May and beyond. After that, timing depends on variety and season. For example, timing to visible bud may vary from two to four weeks after emergence. The number of days from visible bud to flowering is also highly variety specific but does not vary with the length of cooling. In general, the number of days from visible bud in the foliage to sales is about thirty days for Asiatics, thirty-five days for LA hybrids, and forty-five to fifty days for Orientals.

As with Easter lilies, bud sticks can be used to time the crop from visible bud on to flowering. Unfortunately, bud development rates vary widely by variety, and bud sticks must be developed for individual varieties. The chart below provides bud sticks for the Orientals 'Star Gazer', 'Dimples', 'Melody d'Amor', 'Mona Lisa', and 'Ready', based on greenhouse temperatures of 75°F (24°C) days and 65°F (18°C) nights. Warmer temperatures will speed bud development and cooler temperatures will slow development.

Lighting—providing long days—can speed flowering of many Oriental varieties. Crop time of 'Sans Souci', an old variety, was reduced by three to four

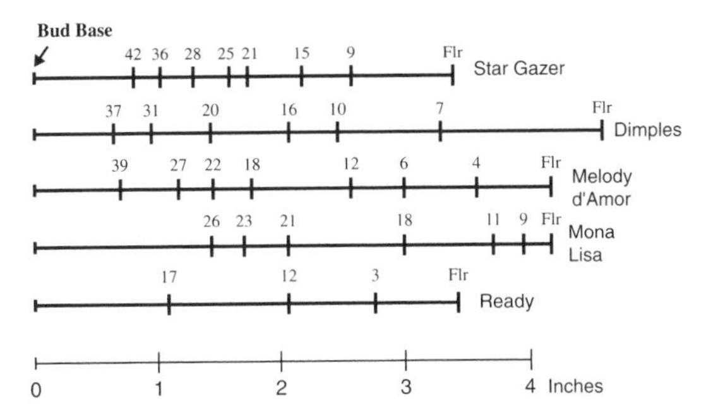

Use bud sticks to time flowering during later production stages. Hold the end of the stick up to the bud and read the number of days for the bud to open at a temperature of 65°F (18°C) nights and 75°F (24°C) days. Warm temperatures speed bud development, while cool temperatures slow development.

weeks by providing at least 10 f.c. (108 lux) (60-watt bulbs placed 4' [1.2 m] above benches and spaced 4' [1.2 m] apart) at plant height from 10 P.M.–2 A.M. 'Star Gazer' responds in a similar manner. In general, lighting Orientals for three to four weeks beginning at emergence is useful, although not required.

To control lily height for pot plant production, most growers start with a one-minute dip into one of three plant growth regulators (PGRs): Sumagic (2.5–10 ppm), Topflor (less than 10–25 ppm), or Bonzi (50–200 ppm). Asiatics require the lowest rates, LA hybrids and intermediate-growing hybrids need higher rates, and Orientals require the highest rates. Specific response to PGRs is highly cultivar dependent. Specific information on the PGR requirement for more than sixty hybrid lily cultivars can be found on the Flower Bulb Research Program's website at www.flowerbulbs.cornell.edu. Later sprays (at visible bud) with 2.5–5.0 ppm Sumagic can help to control late-season stretch.

Cut Flowers

In addition to the general guidelines above, following are more specific guidelines for forcing Asiatic and Oriental lilies as a cut flower crop. Lilies are extremely susceptible to root diseases: You must begin the crop in disease-free medium. Create raised beds, working the soil to a depth of 12" (30 cm), if possible. Do not use perlite or superphosphate as amendments in beds to avoid leaf scorch problems. The medium must be very well drained. Some growers prefer to grow their cut lily crops in pots or plastic bulb crates, which are easily moved around. For a higher-quality crop with longer stems, plant the bulbs first in bulb crates and then allow a two- to three-week rooting period at 40–45°F (4–7°C) before placing them in the greenhouse.

Plant Asiatics at a density of 5–6/ft.^{2} (50–60/m^2) and Orientals at a density of 4–5 bulbs/ft.^{2} (40–50/m^2) depending on the bulb size, type, variety, and ambient light levels. Your supplier can advise you of the optimum spacing, since the ideal density varies by cultivar and your specific growing climate. If you are planting bulbs that have been frozen in, check a few of the bulbs for damage prior to planting the crop: Slice through the bulb and look at the growing meristem and basal plate to ensure healthy tissue.

If possible, plant bulbs 6–8" (15–20 cm) deep to allow good stem root development. Provide at least one layer of support netting that can be raised as the crop grows. If you are forcing the crop during warm months, mulching beds will help to moderate soil temperatures.

In areas with low winter light levels, cut lily crops respond well to assimilation lighting.

Pests & Diseases

Many growers apply preventive fungicide drenches every four weeks to avoid problems with root rot such as *Pythium*, rhizoctonia, and *Fusarium*. Apply the first drench within one to three days of planting. For more pest and disease information, refer to the Pests & Diseases section for *Lilium longiflorum* (p. 521).

Troubleshooting

See the Troubleshooting section for *Lilium longiflorum* (p. 522) for general lily troubleshooting guidelines.

A problem specific to Oriental hybrid lilies deserves mention here, and that is upper leaf necrosis (ULN). With ULN, the edges of the top several upper leaves turn brown and the leaves curl and become disfigured. This is a calcium deficiency problem that, in a practical sense, is not solvable by foliar calcium sprays. The best solution is to have significant airflow on the plants, such that the young leaves actually move from the airflow. This must be done in the first forty days of the crop. (The injury occurs very early, but is not seen until one to three weeks later). For pot crops of 'Star Gazer', it is possible to treat the plants with two sprays of 1,000 ppm Florel at twenty and thirty days after planting. This will cause the very young leaves to bend away form the spindle and increase calcium movement into the leaves because of higher transpiration. It also gives significant height control, such that preplant bulb dips are not needed, but there is the danger of flower abortion from the ethylene! While used successfully in many operations, this is a highly advanced technique and should be approached with due caution.

Classes & Cultivars

Since about 2000, the LA hybrid class has overtaken Asiatics in the white/orange/red/yellow/pink segment of the industry. LA hybrids (for *L. longiflorum*–Asiatic crosses) have several advantages over the Asiatic hybrids they are slowly replacing. Compared with Asiatics, LAs typically are more resistant to botrytis, *Fusarium,* and *Pythium,* and buds open rapidly. Newer LAs also have much brighter colors and generally have nicer foliage and beefier buds than Asiatics. However, based on acreage of bulbs produced, the queen of the crop is, by far, the Oriental hybrid. Another class of hybrid lilies is coming onto the market: OTs, crosses between Oriental hybrids and trumpet lilies. The flower size and forcing schedule are similar to Orientals, but at slightly cooler temperatures. Leaves are thick and leathery, lasting a long time after harvest. Breeders are also working on eliminating pollen from the flowers (which stains permanently) and on scent reduction.

A discussion of cultivars is very difficult as lily breeding progresses on a massive scale, and dozens or more cultivars are introduced each year. Cultivar selection should proceed on an individual grower level and will change throughout the forcing season. It is important to realize that most lily breeding has been done for the cut flower industry and adaptation of varieties to pot culture has been mainly through the wise use of PGRs. More effort has been put into pot lily breeding and marketing, however, with one end result being the 'Looks' series of pot lilies introduced in the United States recently. The 'Sunny' series of pot Orientals also has potential. Overall, the need in pot lilies is for uniform, fast-growing varieties that are tolerant of low temperatures.

Postharvest

Potted hybrid lilies for indoor use are ready to sell when the first bud is swollen and colored, but before it opens. Harvested before this stage and placed indoors, younger buds often fail to open properly. When harvested too late, the plant and flowers are correspondingly older and open flowers are damaged. As with Easter lilies, there is the temptation to cold-store budded plants prior to sale. In general, Oriental lilies store poorly. When they are removed from the cooler, plants often drop a large portion of their leaves, and quality is drastically reduced. However, research at Cornell and other universities in the late 1990s and early 2000s has indicated that gibberellin (GA_{4+7}) is extremely effective as a postharvest aid on all types of lilies, with the ultimate result being the registration of Fascination as a pot lily postharvest aid. Fascination may be sprayed onto the lower leaves of LAs and Asiatics at the rate of 25 ppm (each, of BA and GA). For Orientals, you can treat the entire plant. In both cases, spraying should occur within two weeks of harvest or cold storage. LAs and Asiatics can be stored at 35–38°F (2–3°C); Orientals should be stored at 40–43°F (4–6°C). Do not store any lily for more than two weeks. More information can be found in *Research Newsletter 16* at www.flowerbulbs.cornell.edu.

Most growers sleeve and box pot lilies. Small lily buds are sensitive to ethylene, which can cause bud blasting. The condition is exacerbated when plants are shipped at high temperatures.

Harvest lily stems by cutting them rather than pulling them up. Harvest in the morning to prevent drying out, and make sure lilies are dry when they are harvested to prevent botrytis. Cut stems at the proper stage: For five to ten buds, harvest when there are at least two colored buds; for stems with less than five buds, harvest when one bud is showing color. When flowers are harvested too mature (open flowers), petals can be discolored by pollen. Petals can also become bruised, and aging will accelerate for the remaining flowers and buds due to the production of ethylene by open flowers.

Cool stems after harvest to 34–36°F (1–2°C). Once cooled, stems can be bunched. During bunching, remove the leaves from the lower 4" (10 cm) of the stem. To prevent stems from drying, limit the sorting and bunching process to one hour or less.

Store bunched stems upright in clean, fluoride-free water in the cooler to hydrate them. Once lilies are hydrated, they can be stored dry in cold storage. If stems must be stored for any length of time, provide temperatures of 34–36°F (1–2°C). In cold storage, some Oriental hybrid cultivars are susceptible to a problem termed "bud necrosis," where brown spots or blotches appear on the petal edges of unopened buds. This problem is variety specific and is caused mainly by too-low cold storage, especially with hot and high light production environments. To avoid "bud necrosis," initially place stems in a somewhat warmer cooler. If possible, initially store at 46°F (7°C) for two days before reducing temperature to 40°F (4°C) for longer storage. Shading plants at least one week before harvest under high light and high temperature seasons can also solve this problem. More information on "bud necrosis" can be found in *Research Newsletter 14* at www.flowerbulbs.cornell.edu.

Pack lilies in boxes perforated with air holes to prevent ethylene buildup. Ship at 34–36°F (1–2°C). For long-term shipping, precool boxes prior to loading the truck.

Overall, leaf yellowing is a major postharvest issue with hybrid lilies and will continue to be an issue as we push the postharvest chain. Leaf yellowing is exacerbated by cold storage after harvest. Research at Cornell University is showing that gibberellin (specifically, GA_{4+7}) increases the vase life of Oriental lilies by dramatically improving leaf quality and lengthening flower life. A rather wide range of concentrations, pulse durations, and treatment temperatures may be used. No products are yet labeled for this use, but researchers have found that 25 ppm of GA_{4+7} pulsed for three to four hours works well. Lower concentrations may also give good results.

Florists and consumers should recut stems, remove any leaves that will end up below the water line, and place in fluoride-free water with floral preservatives. Florists should hold stems in a cooler at 35–41°F (2–5°C).

Lilium longiflorum
Common name: Easter lily
Flowering pot plant

The Easter lily, with its white trumpet, is the most popular lily for greenhouse pot lily production. It is a major holiday pot plant crop that has maintained its popularity over the years. It is a traditional plant, with a core of demand centered on its religious and traditional significance. Although the lily of religious paintings and writings is the Madonna lily (*L. candidum*), *L. longiflorum* types have replaced it for practical cultural purposes. Tradition has also played a role quite apart from religious considerations. Probably the most important factor contributing to the continued popularity of the Easter lily has been the profitable nature of the crop, for both the producer and the seller, and the perceived value by the customer. The Easter lily was one of the first plants to be sold through mass-market outlets.

Easter lilies are the most profitable major-holiday pot plant crop produced. This is true even though the initial cost of a lily bulb is high. When combined with the cost of the pot, medium, and labor of potting, growers often have 30–35% of the selling price invested initially. The space occupied by the crop compensates for this, so the return per square foot is high. Consider that a chrysanthemum, poinsettia, or hydrangea may occupy 0.75–1.5 ft.^{2} (0.07–0.14 m^2) of bench space at finished spacing. Assuming an average for a poinsettia of 1.25 ft.^{2} (0.12 m^2) and a price of \$4 per plant, then the return per square foot of finish space is \$3.20 (\$34.43/m^2). Lilies are grown at a density of 2–3.5 pots/ft.^{2} (22–38/m^2). If 2.4 lilies are grown per square foot (26/m^2) and sold at \$3.50 each, then \$8.40 per square foot (\$90.38/m^2) is realized. This is more than two times the per-square-foot income to be realized from other holiday pot plant crops. Depending on marketing area, lilies may be even more dramatically profitable.

However, they are a challenge for the grower. Easter dates move from year-to-year, meaning that in some years when Easter is late, the lily crop competes for greenhouse space alongside regular spring crops.

In years when Easter is early, weather can affect the crop. Fortunately for growers, lilies are one of the most programmable crops, and there is a lot of basic knowledge about how and why they grow and the precise environmental conditions that are needed. Armed with the right knowledge, you can track your crop's progress and manipulate its development to come in right on time. If you are a new grower, do not attempt Easter lilies for a couple of years. Buy them in from someone else to begin developing your market before you attempt to produce your own crop.

Propagation

Most Easter lilies are produced on the West Coast, around the Oregon-California border area (many fields overlook the Pacific Ocean) between Harbor, Oregon, and Smith River, California. Lily bulb production requires two to four years' growth in the field, depending on size and whether scale production is used or plants are started from bulblets (small bulbs formed around the belowground stem above the bulb). From scales (modified leaves broken from "mother bulbs"), bulbs called scalets can be produced in one year. Scaling, combined with tissue culture, offers a means for a more rapid buildup of desirable stock. Other factors being equal, scalets produce more uniform crops than bulblets.

Bulblets or scalets are graded and planted to produce 4–8" (10–20 cm) circumference bulbs, called yearlings, in the first year. These produce "commercials" in the second year. Because of demand for larger-sized bulbs, some smaller commercials are replanted for still another year.

Bulbs are harvested in late September and October, usually being completely packed by October 20. Rain is a determining factor in some years. The rainy season can begin on the West Coast around September 15, thus delaying the harvest. Forcers should be aware of this and prepare to adjust their procedures should harvest be delayed. Lily case packing standards are presented in Table 20.

Table 20. Easter Lily Case-Packing Standards

BULB CIRCUMFERENCE	NUMBER PER CASE
6.5–7.0" (16–18 cm)	200
7–8" (18–20 cm)	150
8–9" (20–23 cm)	125
9–10" (23–25 cm)	100
10–11" (25–28 cm)	75

Excerpted from *Lilies*, edited by D.C. Kiplinger and Robert W. Langhans, Ithaca, NY: Cornell University, February 1967.

Bulbs are packed in peat moss with a standardized moisture content. The ratio of bulbs, peat moss, and moisture is of critical importance: Bulbs must not dry out during the vernalization period or afterwards.

Vernalization (precooling)

Easter lily bulbs of today's varieties have similar, though not exactly the same, vernalization requirements. Vernalization is the cold treatment that lasts several weeks and precedes initiation of flower buds. While vernalization is the proper term, the process is often referred to as precooling, cooling, chilling, cold treatment, and others. It is important to remember that not only is the cold treatment critical, but it must be given under moist conditions. Cold received in late October, November, and December is "remembered" by the stem to cause flowers to be initiated in January. If plants are not exposed to cold (or long days), the stem will eventually grow—perhaps indefinitely—and not initiate flowers. Stems with over three hundred leaves have been recorded. Thus, cold causes the bulbs to cease making leaves and to begin forming flowers.

Pot cooling

Pot cooling is a broad term meaning that plants are given their cold treatment after potting. Pot cooling has many variations, such as cold framing, which refers to early potting and placing pots in a cold frame or another location that will prevent freezing but otherwise is not temperature controlled. Outdoor cooling refers to potting and placing pots outside, with perhaps a straw cover to prevent drying, protect from heating during the day, and ward off frost at night. Controlled temperature forcing (CTF) is a popular, preferred system of pot cooling. The CTF system, or a variation, allows more definite control of vernalization. When the investment in bulbs, soil, pots, and labor is considered, it is prudent to perform all pot cooling under controlled conditions.

A variation in pot cooling is to pot bulbs in 4.4" (11 cm) pots for vernalization. This saves space in expensive refrigeration facilities. After vernalization (and before emergence), the knocked-out bulb and root-ball are carefully placed in the bottom of an inverted pot (to pot deeply). The pot, turned right side up, is then filled with soil. This entails two pottings but saves space in coolers and allows potting in a fresh, non-compacted medium. Root growth is explosive after final potting.

Vernalization is a process that takes place under cool, moist conditions. Drying during vernalization, either in the case or in the pot, can prevent the bulb from receiving cold temperatures and result in unevenly or only partially vernalized bulbs. If vernalization is not complete, exposure to temperatures near 70°F (21°C) or higher can erase the cold treatment and can cause growth anomalies. If vernalization has been completed (that is, by having six weeks at temperatures near 40°F [4°C] under moist conditions), temperatures at or slightly above 70°F (21°C) will not cause devernalization.

Case cooling

Case-cooled bulbs are precooled in the packing case, wherever it takes place, either with the bulb grower or with the finished producer. Bulbs are shipped from the production area to either commercial cold storage facilities or to greenhouse growers, who place them into refrigerators in the cases in which they have been shipped. It is important that temperature and time are carefully controlled. Table 21, though reflecting some variation in the data, illustrates the effect of too little or too much vernalization or precooling.

A vernalization time of six weeks (1,000 hours) is almost always recommended. With six weeks expect forcing times of 100–110 days. Longer vernalization results in faster forcing but lower bud count. This is a trade-off. Times longer than six weeks are not suggested. Bulbs probably vary from year to year either in the amount of cold they have accumulated in the field or in the time requirement for vernalization, and possibly in the optimum temperatures as a result of seasonal changes. It has not been necessary to vernalize longer than six weeks or less than four weeks. For practical purposes, never vernalize less than five weeks, and do that only in the years when bulbs have received some cold in the field.

Table 21. Case Cooling and Flowering

WEEKS OF STORAGE	DAYS TO FLOWER	NUMBER OF BUDS
0	196	10.0
1	176	9.7
2	160	9.1
3	135	7.1
4	123	6.4
5	114	6.5
6	109	5.6
7	112	5.6
8	110	5.2
9	103	5.0
10	100	4.9
11	98	4.4
14	103	4.5

CTF cooling

Bulbs are potted immediately after being received in October. After potting, many schedules call for three weeks of 63°F (17°C) for root growth and six weeks at 40–45°F (4–7°C) for vernalization. This is a total of sixty-three days. When bulbs are received early and Easter is relatively late, this is no problem. When Easter is early or bulbs are received late, there is not enough time to accommodate the entire CTF nine-week schedule. The six weeks of vernalization are the most important.

Table 22 shows (after six weeks' vernalization) forcing times and days to flower and demonstrates that at least 110 days should be allowed for bringing the pot into the heated greenhouse until shipping. The 120-day figure is a good one because of slower forcing. Usually a third of the crop is shipped up to two weeks before Easter.

Consider the forcing times shown in table 22 for a crop that is to be shipped seven days before Easter. For the two earliest Easter dates, plants must be brought into the greenhouse by December 1–8 or earlier to allow time for forcing. On the latest Easter dates, enough time is available from a December 15–21 date.

Table 22. Forcing Times to Easter

	FORCING DAYS FROM		
Easter	Dec. 1	Dec. 8	Dec. 14
March 26	109	101	95
April 7	121	113	107
April 14	128	120	114
April 21	135	127	121

Most practically, whatever the date of Easter, we strongly suggest that plants be in the greenhouse no later than December 15. On a very early date, December 1 is much preferred. This allows time for the plant to develop. If December 15 is used as a date to begin forcing (no matter what the Easter date), then the schedule before forcing, the forcing time, or both must be adjusted.

Note that for the earliest Easter, a vernalization date of October 26 is suggested (table 23). Earlier is better. If bulbs are shipped from the West Coast on October 10, then plan five days for transit and three days to get them potted—to approximately October 18. At most, there are only seven to eight days for 63°F (17°C) rooting treatment. The movement to the greenhouse at the proper time and a full vernalization treatment are more important than three weeks of rooting at 63°F (17°C); therefore, the rooting period should be cut short. Note also that in many cases, weather and other factors prevent early bulb shipments. Also, as more forcers elect pot cooling, more early shipments are requested, but all shipments cannot be made at once. Further, bulbs are not always out of the ground to honor all early shipment requests.

In summarizing pot-cooling techniques, remember that Easter lilies need 100–110 days from the start of forcing to flowering. Many troubles in forcing result from bringing pots into the greenhouses too late and "starting from behind." Pot cooling is recommended for those who can use and understand it. The 63°F (17°C) rooting period should be adjusted (eliminated, if necessary) to allow for a full six weeks of vernalization and getting pots into the forcing greenhouse in time. In many cases, a slower start in the greenhouse at relatively low temperatures helps early root growth. It is also critical that forcers understand that they must keep pots moist to ensure that bulbs can perceive the proper cold temperatures. Many problems are blamed on the bulbs when, in fact, the problem has been caused by drying during pot cooling.

With today's earlier poinsettia-shipping schedules, lilies can be brought into the greenhouse at the proper time. While this may require special management, the extra trouble will be more than repaid by the crop quality that results. Raising the temperature of the storage after precooling and prior to moving the pots to the greenhouse is a method to start forcing at the proper time if some problem prevents moving pots to the greenhouse on schedule and if sprouting has not started. Growth in storage is probably slower than in the greenhouse because there is no solar radiation. There is also danger of sprouting in dark storage, so use care.

Modifications of pot and case cooling

In the past some growers have requested that bulbs be shipped to them early, prior to the finish of case cooling, in order to be potted in late November. These growers then run cool temperatures (50–55°F [10–13°C]) until late December, when they raise temperatures to 60–65°F (16–18°C) to start forcing. This is a workable system. The cool temperatures during December allow for some rooting and some vernalization. The four weeks at 50°F (10°C) equates to nearly two weeks at 40°F (4°C). This is a system that was widespread before long-lasting poinsettia varieties became common.

In some warm areas, such as the South and the inland valleys of California, a modified pot-cooling program works, although controlled conditions are preferred. Some forcers allow bulbs to have two to four weeks of vernalization in the case. Then they pot them up and allow the balance of the vernalization to proceed under natural conditions. This procedure, too, allows some rooting to occur.

Providing long days can substitute for vernalization on a day-for-day basis. Long days also have the same effect on reducing bud count, as does increased vernalization. From a practical view, lights can be very beneficial if combined with sorting. Provide 10 f.c.

Table 23. Controlled Pot-Cooling Schedule

	VERNALIZATION		FORCING	
Easter	**Days**	**Start**	**Start**	**Days to Easter**
March 26	42	October 26	December 7	102
April 7	42	November 3	December 15	114
April 14	42	November 3	December 15	121
April 21	42	November 3	December 15	128

(108 lux) at the crop level from 10 P.M.–2 A.M. nightly. Begin lighting immediately on emergence (have lights on one to two days prior to emergence) and continue to light for the number of days desired.

There is one potential dilemma: Early-emerging plants could receive more long days than they need, and late-emerging plants not enough. Observations indicate, however, that slow-emerging plants often have fewer leaves than early-emerging ones, thus they flower in nearly the same time with slightly less lighting, eliminating the worry. Remember that lighting for too long reduces bud count and stretches plants.

Bud count and bulb size

Bud count is controlled by bulb size, vernalization, and growing factors. Table 24 is idealized, but it gives a rough idea of the number of flowers to be expected from a given size of bulb treated properly. Note that pot cooling can increase bud count. Bud count is at least partially controlled by the meristem area at the start of the flower initiation period (from January 7 to February 7). Thus, larger bulbs, which have larger meristems, have more flowers. Similarly, anything that promotes vigorous growth of the new stem can increase flower count. The most recognized of these growth factors is temperature. Other factors, such as sunlight, good fertilizer, high carbon dioxide (CO_2) levels, good roots, and proper watering, all have definite effects on bud count.

By reducing the night temperature during the flower initiation period, growth slows and the meristem apparently expands, allowing more flowers to form. Reducing temperatures to 55–58°F (13–14°C) for seven to fourteen days can increase bud count appreciably. More importantly, it must be emphasized that raising the temperature during this period can cause a severe loss of flowers. In no instance is it suggested that the grower use a temperature dip unless the leaf count method of timing is being used to monitor crop development.

The negative effects of oververnalization have already been covered. The most beneficial effect of pot cooling is that roots are established when flower initiation occurs. This allows lots of water and nutrients to be absorbed to promote vigorous growth. Similar to how a pot mum cutting fattens after planting, a lily stem also expands. The better the growth, the higher the bud count. A further effect of rooting prior to stem emergence is the apparent control of leaf elongation by the root system. Plants that emerge prior to rooting will have shorter lower leaves than those well rooted prior to emergence.

Growing On

Pot bulbs either upon receipt or completion of cooling (see the Vernalization section). Delaying potting is a serious potential problem. Plant bulbs deep in standard pots (6" × 6" [15 × 15 cm], for example) to protect against early emergence in controlled storage and to allow adequate room for the development of stem roots above the bulb. One inch (3 cm) of medium in the bottom of the pot is adequate, but 2" (5 cm) of medium over the bulb is preferred.

A few bulbs may sprout in the case. Sprouted bulbs are not hurt. The critical factor is to bury the entire etiolated (white) stem below the medium. Sometimes planting the bulb on its side can accomplish this. If the entire stem is covered, growth will be normal on emergence. If a portion of etiolated stem remains aboveground, leaves will not elongate and small stem bulblets will form. If this is 1" (2 cm) or less, it will not be noticed at flowering.

Table 24. Easter Lily Bulb Size and Flower Number

BULB CIRCUMFERENCE	CASE VERNALIZED		POT VERNALIZED	
	'ACE'	'NELLIE WHITE'	'ACE'	'NELLIE WHITE'
6.5–7.0" (16–19 cm)	3–4	2–3	4–5	3–4
7–8" (18–20 cm)	4–5	3–4	5–6	4–5
8–9" (20–23 cm)	5–6	4–5	6–7	5–6
9–10" (23–25 cm)	6–7	5–6	7–8	6–7
10–11" (25–28 cm)	7–8	6–7	8–9	7–8

Note: Figures are the average number of flowers per plant, not the maximum or minimum. Pot cooling, in many of its variations, will usually produce one or more buds more than indicated in each category here.

Some forcers still prefer gravel in the bottom of pots, mostly for weight to prevent tipping. Gravel serves no drainage purpose; in fact, because it reduces the water column and contributes to a wetter root area, it is a practice that should be discarded.

The potting medium should have high water-holding capacity, be well drained, and have good fertilizer-holding capacity. It is unlikely that the keeping quality of lilies grown in lightweight peat mixes is as good as that of lilies grown in heavier soil or compost-based mixes. Further, lily media should have enough weight to prevent tipping should lilies get taller than desired. Depending on physical characteristics, 15–50% of the mix can be soil. Vermiculite is excellent to increase nutrient exchange capacity. Peat moss (at 25% or more) and bark (with added nitrogen to correct for bark decomposition) are suitable. Bark can be a problem if A-Rest drenching is used because of absorption. Remember that on a volume basis, peat moss has very little fertilizer-holding capacity. A useful mix is 33% soil, 33% peat moss, 17% vermiculite, and 17% perlite. You could replace half the peat with bark, giving 17% of each in the mix.

Adjust the pH to 6.2–6.5 with calcium carbonate, using 2 lbs./yd.^{3} (1.2 kg/m^3) as a basis for soil-based mixes, but nearly a 0.5 pH unit lower for artificial mixes. Certain limestone deposits have high fluoride content, and known concentrated sources of this material should be avoided. For this reason, perlite should not be used in lily mixes. Calcine clay is an excellent aggregate where available.

Fertilize with a complete fertilizer for example, 12–12–12 added at the rate of 1 lb./yd.^{3} (0.6 kg/m^3). Do not use superphosphate (read about leaf scorch in the Troubleshooting section). If bark is added to the mix, add an extra 0.5 lb. of fertilizer per cubic yard (0.3 kg/m^3) to compensate for nitrogen tie-up in bark decomposition. If the medium is to be stored, organic nitrogen should not be used. Trace elements are probably best added in liquid form.

Easter lilies need fertilizing early in their development to produce vigorous stem expansion and a large leaf canopy. For this reason, the medium should contain adequate fertilization. Good nutrition begins with a growing mix of high initial fertilization but not exceptionally high in total soluble salts. High soluble salts could cause erratic sprouting and in extreme cases could prevent sprouting completely. Soluble salts should be kept below 2.0 mS/cm using a 1:2 soil water dilution.

Provide adequate nitrogen to prevent lower leaf yellowing and subsequent leaf loss. Lower-leaf loss caused by nitrogen deficiency most likely starts at bud initiation. Lilies have a high nitrogen requirement at this time, and organic matter decay can lower nitrogen levels in the medium. At this point, it is often difficult to make an adequate number of liquid fertilizer applications because of constantly wet soils due to dark weather and other liquid applications. Frequently, the yellowing of a few lower leaves is attributed to root loss or drying, and nothing is done. Nitrogen deficiency then becomes progressively more pronounced until it is too late to correct it. The best way to prevent nitrogen deficiency and ensure a dark green, shiny, healthy leaf surface is to topdress with or incorporate at preplant a dry, slow-release nitrogen source.

For best keeping quality, reduce fertilizer in the greenhouse during the last week. Clear water applications in the last two or three irrigations before shipping will leach out any excess salts. Lilies, like foliage plants, use much less fertilizer and are more susceptible to fertilizer injury when they are moved from the growing environment to a non-growing environment. This is especially true when lilies are boxed and cold-stored prior to shipping.

Irrigation

In most cases, lilies should not be automatically watered by mat or drip tube systems because plants will grow taller—although current PGR and temperature management strategies can accommodate this. Overhead sprinkling is satisfactory. As with all crops, water management is tied to the medium. Heavier, soil-based media are harder to manage with respect to water relations than lighter media, but the extra customer satisfaction resulting from soil-based mixes makes the extra effort worthwhile. Early in the life of the crop, in the dark days of December and January, fungicide drenches, liquid fertilizer applications, and liquid growth regulator applications are difficult to time. Our theory has been "wet is wet," so needed fertilizer or growth regulator applications are made mostly on schedule.

Timing

A great amount of research has been devoted to making lilies more predictable. In spite of the progress made in recent years, lilies are a difficult crop to grow to perfection. Until A. N. Roberts of Oregon State

University devised the leaf-counting method of timing lilies, most growers relied on height to time the crop until flower buds were visible. From visible flower buds, the use of "bud sticks" aided timing. The pattern can be traced onto a pot label to make a bud stick. The pointed end is sharpened and then aligned with the base of the small developing bud (where the peduncle ends and the petals and sepals begin). The tip of the flower is then aligned with three numbers that show the number of days at three temperatures required to bring that bud into flower. Note that a bud nearly 0.5" (1 cm) long—about the visible stage—can be brought to flower in as little as twenty days by "hard forcing" at near 70°F (21°C) night temperatures (with day temperatures 10–20°F [6–11°C] higher) or as many as thirty-six days at 54°F (12°C).

Table 25. Inflexible Old Easter Lily Development Schedule

WEEKS TO EASTER	DAYS	SUGGESTED STAGE OF DEVELOPMENT
17	December 3	Potted, 60°F (16°C)
16	December 10	Making roots
15	December 17	Making roots
14	December 24	Making roots
13	December 31	Growth coming through
12	January 7	Growth 2–3" (5–8 cm)
11	January 1	Growth 4–6" (10–15 cm)
10	January 21	Growth 6–8" (15–20 cm)
9	January 28	Growth 10" (25 cm)
8	February 4	Growth 15" (38 cm)
7	February 11	You can feel the buds
6	February 18	You can see the buds
5	February 25	Buds 0.5–1.0" (1–3 cm) long
4	March 4	Buds 2–3" (5–8 cm) long; few bending down
3	March 11	Buds 3–5" (8–13 cm) long
2	March 18	Buds fully developed
1	March 25	Buds whitish; cooled
1/2	March 28	Some opening; cooled
0	April 2	Easter

Time of emergence and height are still useful guides for early development. In table 25, these criteria are presented in an old schedule based on an April 2 Easter.

Note that the essential problem of such a height schedule is that it completely fails to take growth factors that influence height into account. It is rigidly fixed in allowing six weeks for development from visible bud and is also relatively rigid upon emergence. The only flexibility is in the middle of the schedule. Since it does not allow any method to determine how fast development should be in the middle of the schedule, it really is quite useless.

If buds are not seen by a desired time, temperatures can be raised. Obviously, it would be advantageous if an earlier measure were available. Such a system was devised by Oregon State University researchers and was widely publicized and refined by Harold Wilkins, who is now retired from the University of Minnesota.

Using the leaf-counting procedure, a grower can determine by mid- to late January exactly how many leaves the crop has (remember, the number of leaves varies with the amount of cooling). Since the rate of leaf unfolding is determined by temperature, a grower can figure the best forcing temperature as early as the leaf count is made. By monitoring the rate of leaf unfolding, development can be continuously monitored. The number of leaves the crop has must be determined after flower initiation has occurred. To count leaves, use the following procedure:

1. From January 15–20, select three to five representative plants in each major lot to be monitored.
2. With a felt tip pen or by notching a leaf, select the uppermost "unfolded" leaf.
3. Start at the bottom and count all the leaves that have unfolded, up to the notched or marked leaf (in step 2). Write this number down.
4. Start with the notched or marked leaf and remove and count leaves toward the growing point. This is easy until the leaves get to be 0.25–0.5" (0.6–1 cm) long. At this point, a mounted hand lens and needle will be necessary. Count the leaves right into the growing point; buds should be visible. Write down the number of leaves. As an example, the numbers written would appear as follows:

Leaves unfolded	50
Leaves not unfolded	45
Total leaves	95

5. Average the number of leaves for three to five plants. If, for example, the count is made on January 15 and forty-five leaves are left to unfold, then compute the number of leaves per day that must unfold to make your schedule.
6. Counting back from Easter, compute the number of days before Easter that buds should be visible. Usually six weeks is used, but less can suffice. So, if Easter falls on April 19, then six weeks earlier would be March 8.

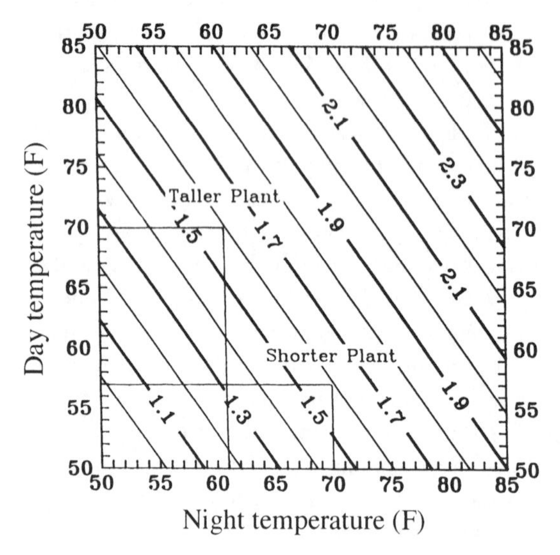

Lily leaf unfolding rate at various day and night temperatures

7. There are fifty-two days from January 15 to March 8. Divide forty-five leaves not unfolded by fifty-two days. The answer is that 0.87 leaves must unfold per day in order to see buds on March 8.
8. Now the question is: How many leaves per day can be unfolded? After the leaves per day are determined, the data in the figure below can be used as a starting point to pick a forcing temperature.
9. It is nice to read in a reference book that for a "large bulb," 60°F (16°C) will cause one leaf per day to unfold. However, it is even better to actually measure the average number of leaves unfolding per day after the indicated temperature change has been made. To accomplish this, follow this procedure.

1. Select another three to five plants from each major lot. Put a label in each pot and use a tall, flagged stake to mark it so that it will be easy to find.
2. Write the date on the label and notch or mark a leaf in the same relative position as indicated in step 2 of the previous procedure.
3. Wait four to five days and again mark a leaf in the same relative position as in step 2.
4. Then count the number of leaves that have actually unfolded between the most recently marked leaf and the earlier marked leaf. Write this on the label.
5. Compute the leaves per day as follows: days ÷ leaves unfolded = number of leaves per day. For example, five unfolded leaves in four days means 1.25 leaves per day.
6. The computation in step 5 of the previous procedure indicated that 0.87 leaves must unfold per day. This current computation shows that 1.25 leaves per day are actually unfolding. Therefore, temperatures need to be slowly lowered to decrease the rate of unfolding to 0.9–1.0 leaf per day. The figure gives typical leaf unfolding rates for different temperatures.

By counting leaves every four to five days, crop progress can be followed. This is a workable system. All lily growers should use it. Remember that leaf counting allows timing to start on January 15 rather than when buds are visible. Growers have four to six weeks longer to manipulate the temperature using this system.

Leaf counting also allows more reasoned decisions regarding temperatures, which can result in fuel savings. When coupled with graphical tracking, keeping consistent records of the rate of leaf unfolding along with major cultural events—periods of high or low light intensity, irrigation and fertilizations, CO_2, temperature changes, root loss, and fungicide applications—makes it possible to determine the effects of these changes on the growth rate. It builds a body of knowledge that can be used to judge the effects of future cultural changes.

Height control

Temperature management should be looked upon as the first source of height control. A zero DIF not only makes determination of the average daily temperature easy, it will reduce stem elongation relative to a more traditional warm day/cool night scenario. Strong negative DIFs should be avoided due to exacerbated leaf yellowing and potentially catastrophic leaf yellowing problems in postharvest.

Both A-Rest and Sumagic can be used. If roots are present, drenches may be made early, before the leaf surface is expanded to absorb a spray. The treatment may have a more lasting effect than a foliar spray, but only if a root system is present. Shredded pine bark absorbs A-Rest (and probably other plant growth regulators), thus reducing its effectiveness; fir bark is less absorptive than pine. More growth regulator must be used to counter either loss. The greater material expense incurred may offset the advantage of soil application. Rates of A-Rest near 0.5 mg/pot as an application to the medium and 33 ppm as a spray are effective. Sumagic is commonly used as a foliar spray at 3–7 ppm. One notable problem is that Sumagic has been associated with "rubber stems" (stems that lose strength and literally topple over). Very little is known about this problem. Growth regulators are less effective during periods of low light. In higher light areas or if bark is not used (these rates presume use of bark media), the lower rates probably should be used. Spraying is gaining favor over media drenches because of flexibility, especially in view of negative DIF.

Growth regulators should be applied in relation to leaf number in order to have a more predictable response from year to year. First spray applications can be made when twenty-five to thirty leaves have unfolded. Sprays any earlier have less dwarfing effect, apparently because of lack of leaf surface. Usually two or more applications are necessary, the second application should occur seven to fifteen days after the first. There may be some advantage to a first spray application, which gives a rapid effect, followed by a drench, which seems slower to take effect but lasts longer.

When night temperatures are higher than day temperatures, it is referred to as a negative DIF situation. Conversely, with higher day temperatures than night temperatures, a positive DIF results. The greater the DIF, positive or negative, the greater the height difference will be, taller or shorter. Unless the negative DIF has been too excessive, a few days of positive DIF will correct the problem. As a practical matter, many growers use a zero DIF (equal day and night temperatures).

The tools to control height with graphical tracking are widely available. The challenge is to apply them.

The graph shown below illustrates tracking a crop for an April 7 Easter, with planned flowering one week earlier. The graphical tracking method is based on the assumption that the height of a lily (from the top of the pot to the top of the plant) will double from buds visible to flowering.

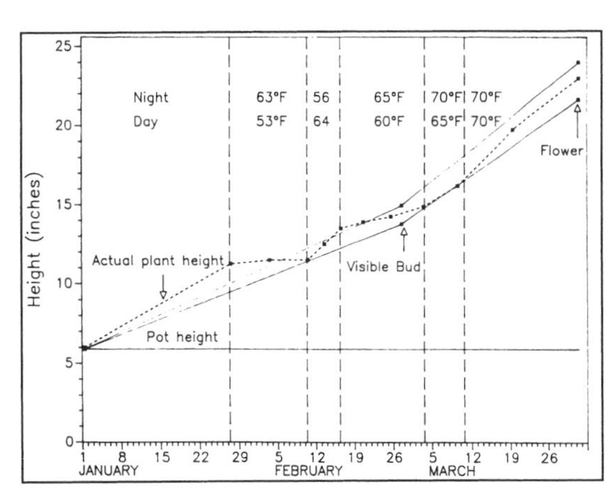

Crop tracking for an April 7 Easter

To follow the graphical tracking procedure, record the date of emergence on a graph. This is the starting point. The next step is to determine the desired final height—for example, 20–22" (51–56 cm). Third, specify the date buds should be visible. In the case illustrated, this is February 28, forty-two days before Easter. Assuming a height of one-half the final height at buds visible, a buds-visible range of height should be 10–11" (25–27 cm). Draw lines connecting the emergence date through the upper and lower height allowed at the visible bud date and then to the upper and lower final heights. This defines the envelope of goal height over time.

Height can be monitored closely by measuring test plants at least twice per week. The graphs should be filled in while in the greenhouse so that any modifying actions, such as temperature changes or A-Rest applications, can be entered at the same time, thus creating a record for future crops. A potential problem of graphical tracking is false information that could result from handling the plants to get measurements. More than one uniform plant should be selected at the beginning of the tracking. This allows switching plants, should stunting from handling occur.

Pests & Diseases

Root rots are probably the most serious disease problem of Easter lilies. *Pythium,* rhizoctonia, and, to a lesser extent, *Fusarium* appear to cause the most problems. A well-drained medium and attention to good irrigation practices are important.

Two major viruses affect lilies. When cucumber mosaic virus and lily symptomless virus are at high levels, fleck symptoms appear. These viruses can be crippling. Besides fleck spots on the leaves, plants can sometimes become distorted. All lilies in commercial production have lily symptomless virus. Clones made free of this virus grow much taller than infected clones, but flower count and other characteristics are apparently not much affected. Poor growing conditions can magnify virus symptoms. Low temperature at starting, low humidity, and other conditions can cause more symptom expression. Due to rouging and aphid control in the field, viruses are usually not a problem, however.

Under low-light, high-humidity conditions, botrytis can be a problem on unopened buds and flowers. The best control is to add heat while venting to reduce relative humidity. Under severe humidity and free-water situations, botrytis can also occur on leaves.

Bulb mites consume the sloughed-off outer scales of the bulbs as they grow from the inside out. Mites will tunnel into stems, apparently entering between the bulb and the surface of the medium. Most severe bulb mite damage has been seen in pot-cooled lilies. Plants are often bent because the lesions on one side of the stem stop growth while the opposite side elongates. Dwarfed, stunted growth with thickened leaves and crippled growing points has been attributed to mites.

Aphids should be rigorously controlled in greenhouses. A few aphids early in the crop can cause considerable damage due to honeydew deposition, sooty mold growth, and bud disfiguration. Fungus gnats can build up on lilies that are kept especially moist or when much algae is present.

Troubleshooting

Leaf scorch is a serious disorder, and many younger growers have not seen the extreme loss that can result. Leaf scorch symptoms are very characteristic: half-moon-shaped areas, often with concentric rings of varying colors of brown. These scorched areas are almost never located at the tip of the leaf but are usually 0.25" (6 mm) or more from the tip. The cause of this scorch is fluoride toxicity. Develop your fertilization program to eliminate fluoride sources. High calcium levels and low phosphorus can control leaf scorch. This is the reason why lime should be used for pH control, and why no superphosphate should be used in the initial mix. After potting, calcium nitrate at 200–300 ppm nitrogen, coupled with soluble trace elements and potassium at 200 ppm, is satisfactory, depending on media mixes and irrigation schedules. Calcium nitrate can be applied at 200–400 ppm regularly, and at up to 750 ppm to boost fertilization. Phosphorus at 20 ppm in irrigation water does not cause leaf scorch and certainly provides adequate phosphorus.

No-shows, or lilies that never emerge, can be a problem. Lily bulbs are a field crop, and by the very nature of production, all bulbs are not perfect and each year a percentage of the crop may not emerge. There are several causes of this phenomenon, although some non-emergence cannot be explained. Among the most common causes are broken sprouts. If lily bulbs summer-sprout in the field—if the stem that normally emerges in the greenhouse emerges in the field—the sprout may be broken off during the harvest operation. Bulbs are inspected prior to packing, but the broken sprout may not be detected. A bulb with a broken sprout will eventually make good roots and sprout, but the plant will be off cycle and will not bloom for Easter.

No-shows can also be caused by stem dieback or sometimes by *Phytophthora* and other soilborne diseases. This may occur on a prematurely sprouted bulb in the field or in a sprouting bulb after potting. Such bulbs are very difficult to detect at packing. At other times, the dieback occurs after packing or potting. Such bulbs eventually make good roots and sprout, but they, too, will be off cycle.

During the harvesting process some bulbs are broken. This is evidenced by bulbs that do not emerge and, when dug up and scaled down, show excessive amounts of bulblet formation.

In addition to these determinable causes, there are still unexplained no-shows. Some bulbs make roots but do not sprout. This is seldom a problem, but it does occur. Other bulbs do not sprout or make roots. Non-emerging bulbs of this kind can occur in the field, during vernalization, or in the greenhouse. Severe drying, anaerobic conditions from a medium high in readily decomposable organic matter or water, and probably other factors can cause rest in the bulb. That is, even though the current environment is favorable, internal factors prevent sprouting and growth.

Bud blasting can be seen when buds stop growing, then shriveling begins at the base of the bud, and further browning of the bud follows. Most often the cause is high-temperature forcing, especially if low humidity has been allowed. Lack of water is also critical. This situation may be made worse by root rot at the time.

Sample Pot Cooling Schedule Controlled temperature forcing schedule for an April 15 Easter date		
WEEKS BEFORE EASTER	**DATE**	**ACTION**
25	Bulb arrival	Pot bulbs. Water in and drench with Terraclor and Subdue. Put in 63°F (17°C) cooler to initiate roots.
24	October 30	Continue warm rooting.
23	November 6	Start vernalization. Lower temperatures to 40°F (4°C). Make sure pots are moist.
22	November 13	Week 2 of cooling. Water if needed.
21	November 20	Week 3 of cooling. Watch for sprouting. If present, lower temperatures a few degrees; do not go below 35°F (2°C).
17	December 18	Move plants into the greenhouse at 68°F (20°C) days and 65°F (18°C) nights. If you have had problems with inadequate cooling, provide long days with night interruption lighting from 10 P.M.–2 A.M. for two weeks after emergence.
16	December 25	Maintain temperatures until emergence. Begin 200–300 ppm fertilizer from 15-0-15.
15	January 1	Continue temperatures. Check root development. Plants should begin to emerge.
14	January 8	Most plants should be emerged. When 50% or more of crop is up, lower temperatures slightly.
13	January 15	Crop is fully emerged. Move no-shows to warmer areas. Begin graphical tracking and leaf counting. Shoots should be 1" (3 cm) tall.
12	January 22	Buds are probably initiated. Complete leaf counting to determine leaves to unfold before visible bud.
11	January 29	Consider height control options depending on graphical tracking information. Shoots should be 3" (8 cm) tall.
10	February 3	Space plants. Be careful of high day temperatures, which will cause stretching. Consider application of Fascination (GA_{4+7}) at 10–25 ppm. Spray only lower leaves, and do not spray excessively to avoid dripping into the medium. Shoots should be 4" (10 cm) tall.
9	February 12	Adjust temperatures in relation to leaf-counting and leaf-unfolding rates. Watch for aphids. Plants are 11" (28 cm) tall.
8	February 19	Since half of the final height is determined by visible bud, track and adjust height with graphical tracking. Keeping plants short now helps later. Check leaf-unfolding rate. Plants should be 12" (30 cm) tall.
7	February 26	Height can get away from you now. If plants are too short, run positive DIF. Plants are 13" (33 cm) tall.
6	March 5	First buds are visible. Plants are 14" (36 cm) tall. Switch to 200–250 ppm from calcium nitrate.
5	March 12	All plants should be at visible bud. Maintain graphical tracking and implement height control measures as needed. Plants are 15.5" (39 cm) tall.
4	March 19	Buds are 1" (3 cm) long. Use a bud stick to measure them. Plants are 16.5" (42 cm) tall.
3	March 26	Buds are 2" (5 cm) long. Plants are 18.5" (47 cm) tall.
2	April 2	Buds are 4" (10 cm) long. Watch for botrytis. The first plants are shipped or go to the cooler at 35–40°F (2–4°C). If plants go into the cooler, make sure the medium is moist, foliage is dry, plants had a root rot drench within the last fourteen days, and you have used gibberellic acid on lower foliage at 25 ppm (where legal). Plants are 20.5" (52 cm) tall.
1	April 9	Shipping is in full swing. Buds are more than 5" (13 cm) long, and plants are 22" (56 cm) tall.
0.5	April 10/11	Last ship dates. Plants are 22" (56 cm). Buds are cracking open.
0	April 15	Easter Sunday

Buds can also be lost by abortion, which occurs just after bud initiation. A grower will see signs that buds were present but then were lost. Again, high temperatures just after the initial bud set period are most often responsible. Drying or any other growth factor can affect this. Small scars, small "pimples," and bract-like leaves (which are always present below a bud) are all signs that buds were there, or were potentially there, but were lost.

The third main bud problem is the loss of potential. In severe situations, buds may be lost but leave no telltale signs. This phenomenon is best observed by comparing bulbs from the same case that were grown under different conditions, such as pot and case cooling. Again, high temperatures during the bud-set period is most frequently responsible, but poor growing conditions from potting though the bud initiation period can also be the cause.

In most cases, greenhouse forcing problems can lead to excessively low bud counts. So many factors affect bud count that simple answers are often not possible. Most problems, however, relate to temperatures during the bud initiation period.

The cause of greenhouse "twist" on lilies is debated. Some believe an organism—perhaps a bacterium—is the cause. Whatever its cause, twist can be serious in limited situations. Symptoms are circle-shaped leaves appearing at the growing point, often with brown, necrotic areas bordering the inside of the circle. The plants may produce only one or two such leaves or several and then may outgrow the problem and be salable.

Leaf yellowing

Fast, catastrophic leaf yellowing from the bottom up is usually associated with a carbohydrate shortage in combination with, or as a result of, late growth regulator applications, negative DIF, high-temperature forcing, or long-term storage. All of these can cause nearly complete yellowing of the plant. It has recently been established that ethylene exposure to lilies *after two weeks in cold storage* can dramatically increase leaf yellowing and may well be the ultimate "cause" of catastrophic postharvest yellowing. The second type of yellowing, gradual yellowing at the bottom of the plant, usually starts at bud initiation and often progresses to an unsightly appearance at sale time. Avoiding late growth regulator applications, late high-temperature forcing, long box storage, ethylene, and long delivery times are essential in controlling catastrophic yellowing.

Gradual bottom leaf yellowing is a complex problem. Control it by preventing nitrogen deficiency in the lower leaves. This usually requires extra applications of dry fertilizer or high rates of liquid fertilizer. Stress of any kind is also likely to be important. Research has shown that high-density planting (more than 2.5 plants/ft.^{2} [27/m^2]) contributes to the problem. In addition, excessive drying, root rot, and high-temperature forcing are all critical. Dry fertilizer applications to supplement liquid feeding are very important, especially when it is difficult to sequence liquid feeding with fungicide or growth regulator drenching.

Both kinds of leaf yellowing are controllable by applying GA_{4+7}, now available to greenhouse growers as Fascination. Details of its use are on the product label. Due to its ability to cause cell elongation, Fascination can also stretch lilies. To minimize stretch, it is crucial to avoid spraying upper, actively growing regions of the plant.

Varieties

'Ace' first gained widespread popularity around 1953, and 'Nellie White' was slowly accepted from about 1964, replacing 'Ace' in popularity around 1979. At present, 'Nellie White' is the only important cultivar, and by present floricultural standards is considered to be an old variety. Although persistent efforts to breed superior varieties continue, 'Nellie White' has many good characteristics and remains preeminent. Great hopes were held for several newer selections, but they have not met expectations. A fundamental issue is the changing Easter date where it is impractical to ramp up or down bulb numbers of two or more varieties to meet the holiday.

Postharvest

Pots are ready for sale when buds reach the "white puffy" stage. Most retail florists receive plants unboxed, while most mass-market lilies are delivered in boxes. In many instances, plants are packed when they reach the "white puffy" stage and are placed in a cooler. However, excessively early packing is being abused. Packing too early can reduce plant quality for stores and ultimately for consumers by decreasing flowering time.

If plants must be stored, they can be stored under lights (50 f.c. [538 lux]) at 35°F (2°C) for up to three days. Sleeve, box, and ship plants at 38–42°F (3–5°C). A lack of keeping quality in stores and homes is potentially the most severe problem in the industry. Long box time, high-temperature forcing from late potting, media with no soil base, high growth regulator rates, and other factors may also be responsible for this problem. Flowers are sensitive to ethylene: Do not store lilies near ripening fruits or vegetables, and keep plants groomed by removing dead or dying plant debris to remove sources of ethylene production.

Retailers should unsleeve plants upon arrival. Display pots in a cool (65–70°F [18–24°C]), bright area under 100 f.c. (1.1 klux) of light. Some retailers remove pollen; others do not. If plants with open flowers will be handled, it is best to remove pollen and thus keep flowers clean. (See the Postharvest section in *Lilium hybrida*.) Keep plants moist: Drying out can cause buds and leaves to drop.

Consumers can enjoy lilies in a cool, bright window. Be sure to instruct them to keep dead flowers groomed away. In USDA Hardiness Zones 4–7, consumers can plant their bulb outdoors once the danger of frost is past and enjoy flowers in the summer for years to come.

Limonium

Limonium sinuatum
Common names: Statice; common statice; garden statice
Annual, cut flower

This popular cut flower filler is a component of many mixed bouquets sold in the United States. Its bright purple, pink, yellow, or white flowers complement just about everything. Statice is so ubiquitous, why would anyone even think of using it in a new way? Perhaps it is because statice is under our very noses that few have taken advantage of it as a bedding plant. Annual statice makes a fine garden flower and offers consumers the opportunity to harvest their own stems right out of a cutting garden. If you are a retail grower or a grower catering to the landscape trade, give it a try.

For growers, annual statice is easy to produce as either a cut flower or in pots to be sold as a bedding plant. Plugs are readily available year-round for greenhouse or field cut crops or for production as a flowering pot plant or bedding plant. Production is not at all difficult.

Propagation

Sow seed onto a disease-free, germination medium with pH of 5.5–5.8. Cover seed lightly with coarse vermiculite to maintain humidity at the seed coat and encourage more uniform germination. However, do not germinate in the dark; provide 100–400 f.c. (1.1–4.3 klux) of light during germination. Keep trays moist, but not fully saturated. Seed will germinate in three to five days at 70–75°F (21–24°C).

Once 50% of seed has germinated, move trays to Stage 2 for cotyledon emergence, which will occur over four to seven days. Maintain soil temperatures at 70–75°F (21–24°C) and increase light levels to 1,000–1,500 f.c. (11–16 klux). As soon as cotyledons have expanded, provide 50–75 ppm from 14-0-14 once a week. Begin to dry trays down somewhat between irrigations when radicles penetrate the medium.

When you have a stand, move trays to Stage 3 to develop true leaves. Reduce substrate temperatures to 65–70°F (18–21°C). Increase fertilizer rates to 100–150 ppm and apply weekly, alternating between 20-10-20 and 14-0-14. Stage 3 will last about twenty-one days for 288s.

Harden plugs for seven days in Stage 4 prior to sale or transport. Lower temperatures to 60–65°F (16–18°C) and increase light levels to 2,500–3,500 f.c. (27–38 klux). Allow trays to dry thoroughly between irrigations. Use a weekly feed of 100–150 ppm from 14-0-14.

A 288-plug tray takes from six to seven weeks from sowing to transplant to larger containers for spring sales.

For cut flowers, plugs sown in January can be harvested as cut flowers in the greenhouse or cold frame in the North starting in May and later.

Growing On

For bedding plant or pot plant production, transplant plugs into a well-drained, disease-free, soilless medium with a pH of 5.5–6.2 and a moderate initial starter charge. Grow on at 55–60°F (13–16°C) days and 50–55°F (10–13°C) nights. Ideally, do not grow plants at higher temperatures. Allow plants to dry thoroughly between irrigating. Fertilize with 150–200 ppm at every other irrigation, alternating between 20-10-20 and 15-0-15. Growth regulators are not required. Four-inch (10 cm) pots should be ready for sale in about eight weeks in the spring.

To grow annual statice as a cut flower, space plugs at 10" × 10" (25 × 25 cm) in the greenhouse or 12" × 12" (30 × 30 cm) in the field. Do not plant outdoors until the danger of frost is past, but plant out before nights rise above 55–60°F (13–16°C). Established plants will flower until they are killed by a hard frost. Ensure ground beds are well drained, as statice is highly susceptible to a number of soilborne diseases. Adjust pH to 6.5. Plants do not require staking. While statice is not daylength sensitive, plants flower more freely under long days. Using night interruption lighting on actively growing plants will help to speed up flowering.

Grow *Limonium* dry. While it is important to irrigate freely while plants are establishing themselves, sharply reduce water once stems form. Mature *Limonium* requires very little water. Overwatering causes elongated, weak, thin stems. Do not overfertilize *Limonium.* Excess fertilization results in tall plants, weak stems, and flower abortion.

For cut flower growers in mild climates, annual statice can be cropped as a cut flower from December to June from successive plantings.

As a bedding plant, plugs that are transplanted to an 18-cell tray will be salable green in spring in seven to nine weeks when grown at 55°F (13°C) nights with days 5–10°F (3–5°C) warmer. Grow plants cool in the greenhouse or cold frame. While they grow under warmer conditions, they produce a stronger flowering plant when given an extended period of cool temperatures. Plants are sold green in the spring but will flower in the garden in June or July.

Pests & Diseases

Aphids can attack plants. Diseases are more serious however and include botrytis, *Fusarium,* powdery mildew, and rust.

Varieties

Growers producing cut flowers or bedding plants can use the same varieties, all of which are readily available as seed or from a number of plug suppliers specializing in cut flowers. Cut flower growers will want to experiment with some of the newer vegetatively propagated varieties on the market, which offer greater uniformity, yield, and color range than seed types do.

'Fortress' is probably the best-known annual statice and comes in a mix. The 'Qis' series, however, is available in separate colors and flowers more freely in both heat and cold. It also performs as a greenhouse or field cut.

Related Species

Also an annual statice, *Psylliostachys suworowii* (also called *L. suworowii*), is commonly known as Russian statice or rat-tail statice. Russian statice was once a favorite greenhouse-grown cut flower in the North. Russian statice prefers cool production temperatures and can be grown outdoors in the Deep South and in California during the winter. You germinate Russian statice as you would *L. sinuatum*; however, germination will take about twice as long. Russian statice produces lavender flowers that are excellent for drying. Plants that become root-bound flower more readily.

L. peregrinum is a popular cut flower. Try 'Ballerina Rose', a vegetatively propagated variety with large, striking flowers that are bold enough to be used as a center flower in arrangements or bouquets. Plants are tender perennials.

L. perezii, an annual in most parts of the country but also sold as a tender perennial, is commonly known as seafoam statice. This tender perennial is a real standout in coastal California, where it thrives as a wildflower. In other parts of the country, seafoam statice is a short-lived perennial that is best treated as an annual. Flowers are deep blue on 36" (91 cm) stems. Grown from seed, *L. perezii* flowers in its first year and is a fast grower. 'Violet', great for cut flower production, was selected for its deep color, earliness, and high production. Plant only in disinfected soils, as *L. perezii* is susceptible to *Sclerotinia.* Manage plants dry and use raised beds for cut flower production to encourage good drainage. Keep fertilizer levels low because excess nitrogen softens plants and makes them more susceptible to disease. The vegetatively propagated 'Cosita' grows to only 9–10" (23–25 cm) tall and makes an attractive 6" (15 cm) flowering pot plant, flowering for up to eight weeks. Each plant produces eight to ten flowering spikes. Total crop time from tissue-cultured liners is twelve to fourteen weeks in the spring and summer and fourteen to eighteen weeks in the fall and winter.

Postharvest

Harvest stems when both stems and flowers are dry, as disease can become a problem during storage and transport. Stems are ready for harvest when most of the flowers are showing color. Store in clear water or water treated with postharvest solution. Fresh statice lasts seven to fourteen days. Instruct retailers to recut stems and remove any damaged plant parts and as much lower foliage as possible. Keep refrigerated in a preservative solution at 40°F (4°C).

Yellow leaves can be due to stems being held out of water. Do not allow statice to dry out if it is being used fresh.

Statice is easily dried. Simply hang stems/bunches upside down in a well-ventilated area for two weeks. Dry stems can last twelve months.

Limonium latifolium
Common names: Sea lavender; wideleaf sea lavender
Perennial (Hardy to USDA Zones 4–9)

In the most northern climates of its hardiness zones, *L. latifolium* is a short-lived perennial. However, in areas where it thrives, plants produce massive clumps of spreading basal leaves, and, once established after several years, the summer flower show is spectacular. Like its annual relatives, plants are excellent cut flowers used either fresh or dried.

While our propagation information centers on growing *Limonium* from seed, be sure to take a look at the expanding new varieties from Europe and Japan—many of these are from tissue culture. The cut flower industry has taken note of these new varieties, revitalizing the *Limonium* market.

Propagation

L. latifolium is commonly propagated by seed for home garden use and by tissue culture for cut flower growers and the commercial market.

Seed

Sow seed in June or July onto a well-drained, germination medium with a pH of 5.5–5.8 in a 288-plug tray. Seed does not need to be covered to germinate, but a light covering of coarse vermiculite is helpful. Provide 100–400 f.c. (1.1–4.3 klux) of light during germination. Keep the medium moist, but not fully saturated. Seed will start to germinate in five to seven days at 65–75°F (18–24°C) but can be uneven to emerge and it may take another four or five days for a more uniform appearance of seedlings.

Once 50% of seed has germinated, move trays to Stage 2 for stem and cotyledon emergence. Increase light levels to 500–1,000 f.c. (5.4–11 klux). Allow trays to dry down somewhat between irrigations once radicles have penetrated soil. Apply 50–75 ppm from 14-0-14 once a week when cotyledons are fully expanded. Maintain temperatures.

Move plugs to Stage 3 to develop true leaves. Reduce temperatures to 60–65°F (16–18°C) and increase light levels to 1,000–1,500 f.c. (11–16 klux). Fertilize weekly at 100–150 ppm, alternating between 20-10-20 and 14-0-14. Allow trays to dry thoroughly between irrigations as seedlings mature. Stage 3 will take from twenty-one to thirty-five days for a 288. Harden plugs for seven days prior to sale or transplant at 55–60°F (13–16°C). Increase light levels to 1,500–2,000 f.c. (16–22 klux) and allow trays to dry thoroughly between irrigations. Fertilize weekly with 100–150 ppm from 14-0-14.

A 288-plug tray will take from eight to nine weeks to be ready from sowing to transplant.

L. latifolium will not bloom when sown in January or later for same-year flowering. Sowings need to be done the previous year.

Tissue culture

Tissue-cultured plants are produced by a number of firms in Europe and Japan. These are transplanted to liner trays in North America by commercial propagators and sold to greenhouse and cut flower growers to pot up to larger containers or ground beds for sales primarily as cut flowers, although there are a few sold as bedding plants. However, considering the price of the inputs and the crop time, cut flowers are the most profitable way to go.

Growing On

Seed

Perennial *Limonium* plants must be well established and receive a cold treatment before they will flower. If plants are not established, they will not flower, even under low temperatures. If plants are mature enough but do not receive cold temperatures, they will continue to grow vegetatively.

Select a well-drained, disease-free, long-lasting, soilless, growing medium with a pH of 5.5–6.2 and a moderate initial starter charge.

Transplant plugs from September to February for flowering plants in the summer. Plugs planted in the winter (January and February) can be grown cold at 30–35°F (–1–2°C). If you transplant plugs in the fall, be sure to fully establish a root system prior to overwintering; this will ensure a higher survival rate.

Place pots in an unheated or minimally heated cold frame or greenhouse and overwinter at 30–35°F (–1–2°C). Do not fertilize when plants are being held dormant. Some growers root plants into cell packs for overwintering and then plant up pots in the winter for subsequent sale in the spring.

Plugs transplanted in the spring from a winter sowing may be sold green in packs or pots for flowering the following year. Grow these plants at 60°F (16°C) days and 55°F (13°C) nights.

When you are establishing plants prior to overwintering or are bringing plants out of dormancy in the spring, fertilize with 150–200 ppm from 15-0-15 at every other irrigation. Too much fertilization produces tall, lanky plants.

Force plants at 60–65°F (16–18°C) days and 55–60°F (13–16°C) nights. Provide full sun during the spring months (3,000–5,000 f.c. [32–54 klux]).

Perennial statice may also be started from bareroot plants brought in for fall potting.

Tissue culture

Commercial propagators have rooted liners from tissue-cultured plants available for sale. Most of these are marketed to cut flower growers and will require three to four months to begin flowering after transplanting to a ground bed.

Pests & Diseases

See the Pest & Diseases section for *Limonium sinuatum*.

Varieties

'Stardust', from seed, produces small, white and yellow flowers on multistemmed, 24" (61 cm) plants.

Related Species

L. altaica is propagated either by seed, division, or cuttings. Flowers are lavender to light blue in color and are great as fresh or dried cuts. The vegetatively propagated cut flower varieties 'Emille', 'Pink Emille', and 'Lavender Emille' actually prefer higher fertilizer rates than most other statice. In warm climates, plants produce a flush in early summer and again in the fall, taking a break from flowering in the summer heat. In regions with mild summers, plants flower continuously.

L. gmelinii, commonly known as Siberian statice, produces lilac plumes. Plants are vigorous and produce 24–36" (61–91 cm) long stems. Siberian statice is also suitable as a cut flower. The interspecific cross between *L. latifolium* and *L. gmelinii,* 'Charm Blue', has a deep purple calyx. Stems are 30" (76 cm) in their first year and can be 40" (1 m) in their second year of production! Unlike other statice, keep plants well irrigated for best results, even when flower stalks appear and begin to elongate.

L. tataricum (also called *Goniolimon tataricum*) is referred to as German statice and makes a great dried cut flower. Flowers are small and pale blue to white with red inner petals. German statice is also suitable as a cut flower, but the flowering window is brief. Plants are hardy to USDA Hardiness Zone 5. Sow in summer for overwintering plants in their final container. Plants will flower the following spring or summer.

Linaria

Linaria hybrida
Common names: Linaria; baby snapdragon; spurred snapdragon

The sweet-smelling flowers of *Linaria* perform well in cool seasons and offers small, snapdragon-like flowers that continue to bloom in cool conditions, making it a great winter bedding plant in freeze-free climates. Upright and well branched, *Linaria* grows to heights of 12–16" (30–40 cm) tall. *Linaria* is best suited to spring and mid- to late-autumn premium-pack and 4" (10-cm) pot programs for bedding and mixed containers.

Propagation

Seed is the most common method of propagation for *Linaria.*

Sow multipelleted *Linaria* to a 288-plug tray with a well-drained, disease-free, soilless medium with a pH of 5.5–6.2 and a medium initial nutrient charge (EC ≤0.75 mS/cm with a 1:2 extraction). Leave the seed uncovered or cover the seed lightly with vermiculite (to maintain moisture around the seed). Germinate at 65–68°F (18–20°C) and remove from the germination chamber immediately at radicle emergence to avoid excessive seedling stretch. Light is not required for germination but significantly improves seedling quality, since hypocotyls can elongate rapidly under dark conditions. Light levels should be 1,000–2,500 f.c. (11–27 klux). Keep soil moisture high until radicle emergence, then reduce moisture levels after the radicle penetrates the medium.

The root radicle emerges in two to three days and germination is complete within a week to twelve days. Do not allow the seedlings to wilt. At radicle emergence, 50–75 ppm nitrogen from 14-0-14 or 13-2-13, alternating with a 20-10-20–type fertilizer. As cotyledons expand, increase to 100–150 ppm nitrogen. Plant growth regulator application is key to producing high-quality plugs. Spray Bonzi at 2 ppm right after radicle emergence to control rapid hypocotyl elongation. A second Bonzi spray at 5 ppm should be applied about seven to ten days later.

Finish the plugs at 65–70°F (18–21°C) with light levels up to 5,000 f.c. (54 klux), if temperature can be controlled. Plugs can be held at cooler temperatures for a week or two (62–65°F [16–18°C]). Temperatures cooler than this will result in slow down and longer crop time.

Plug crop time in a 288 is approximately four to five weeks. Add a week or two for larger plug sizes.

Growing On

Use a well-drained, disease-free, soilless medium with a pH of 5.5–6.2 and a medium initial nutrient charge. Growing on temperatures should be 50–60°F (10–16°C) nights with days at 55–65°F (13–18°C). Cool temperatures give the best-quality plants and more intense flower color and fragrance. Light levels should be kept as high as possible, while maintaining the recommended temperatures. Avoid both excessive watering and drought. Do not allow plants to wilt. Feed with 150–200 ppm nitogren once a week, alternating between 15-5-15– and 20-10-20–type fertilizers, while keeping the EC at less than 1.75 mS/cm with a 1:2 extraction.

Trials at a research facility in Northern Illinois found that using a 10 ppm Bonzi spray one week after transplanting to control main shoot elongation was effective while allowing secondary shoots to develop. Once secondary shoots reach about 1.5" (4 cm), spray with Bonzi at 20–30 ppm. Repeat the Bonzi spray if necessary. The Bonzi spray will strengthen the stems and make the flower color more intense. Researchers also found that a spray tank mix of B-Nine at 2,500 ppm and CCC at 300–500 ppm can significantly reduce plant height. However, the plant stems are not as strong as those with Bonzi treatment.

Pinching is not necessary to promote branching. However, since *Linaria* grows rapidly, pinching can be done if plant growth regulators (PGRs) are not applied on time, especially under warmer conditions. If the plants get too big prior to PGR application, pinching or shearing the main stems back will yield fuller plants without significant flower delay. Growing cold is still the best growth control.

For fall sowing in warm climates, sow in mid- to late autumn, when temperatures begin to moderate. Sowing in late summer or early fall, when temperatures are very high, will make plants grow too quickly and the height will be difficult to control.

Pests & Diseases

Common pests include aphids. *Linaria* does not have any common disease issues.

Varieties

This culture is based on the variety 'Enchantment', which is the first *Linaria* F_1 hybrid available on the market. It has intense, magenta flowers highlighted with a gold "bee," or "nose." An interspecific hybrid, 'Enchantment' has better heat tolerance and gives longer season performance than other cool-season crops.

Related Species

Linaria maroccanna is the more common species of this genus. One of the best varieties of this class is the 'Fantasy' series. This series is available in a number of separate colors and grows from 12–14" (30–35 cm) tall. Its crop time is similar to the culture noted for 'Enchantment'.

Table 26. *Linaria* Transplant to Salable Pot

CONTAINER SIZE	PLANTS PER POT	WEEKS FROM TRANSPLANT	TOTAL WEEKS
306 pack	1	5–6	9–10
4–4.5" (10–11 cm) pot	1	6–7	10–11
6–6.5" (15–16 cm) pot	3	6–7	10–11
1-gal. (4 l) container	3	6–7	10–11

Lobelia

Lobelia erinus
Common name: Lobelia
Annual
Anyone who has ever been to California knows how spectacular lobelia looks when it is grown under perfect conditions—bright, diffused light and cool to warm (not hot) days and cool nights. Plants create blue, pink, lilac, or white walls of color. Up and down the West Coast, across Canada, and in other parts of the United States with cool temperatures, lobelia is a real star. Most growers, even those living outside lobelia country, offer plants in early spring. They are easy to grow, and blue lobelias are real eyecatchers.

Propagation

Seed has dominated this class for decades, but during the past ten years there has been an infusion of vegetatively propagated selections available as well. These tend to be more robust and have a higher heat tolerance.

Seed

Multisow six to ten seeds per plug or buy multiseeded pellets. Select a disease-free, well-drained medium with a pH of 5.5–5.8, and do not cover. Maintain soil temperatures of 75–80°F (24–27°C) and keep trays moist, but not saturated. Lobelia can be germinated in a chamber. Radicles will emerge in four to six days.

Move trays to Stage 2 under 500–1,500 f.c. (5.4–16 klux) of light. Lower soil temperatures to 68–72°F (20–22°C) and provide 50–75 ppm from 14-0-14 once cotyledons have fully emerged. Stem and cotyledon emergence will take seven days. Move trays to Stage 3 by lowering soil temperatures to 65–68°F (18–20°C) and increasing light to 1,000–1,500 f.c. (11–16 klux). Fertilize once per week at 100–150 ppm from 20-10-20, alternating with 14-0-14. If growth regulators are needed, A-Rest, B-Nine, or Bonzi are effective. Allow trays to dry thoroughly between irrigations. Stage 3 will take from fourteen to twenty-one days for 288s.

Harden plugs for shipping or transplant for seven days at soil temperatures of 60–62°F (16–17°C). Increase light levels to 1,000–2,000 f.c. (11–22 klux) and fertilize weekly with 100–150 ppm from 14-0-14.

A 288 plug will take from four to six weeks from sowing to a finished plug tray. This culture is based on multiseeding the plug tray or using a multiseeded pellet and not using a single seed per plug tray.

Cuttings

Stick cuttings of uniform length and diameter into rooting hormone (hastens rooting) and then into Ellepots or medium in a liner tray (105, 84, 72, etc.). Maintain substrate temperatures of 68–72°F (20–22°C) and air temperatures of 75–80°F (24–26°C) days and 68–70°F (20–21°C) nights. Provide mist (for an average of six to nine days) to keep cuttings turgid, adjusting frequency based on light levels and temperatures. Maintain 500–1,000 f.c. (5.4–11 klux) of light. Fertilize with 50–75 ppm from 20-10-20 once a week. Cuttings should form callus within five to seven days.

Move cuttings to the next stage for root development, which will take from seven to fourteen days. Maintain temperatures and increase light levels to 1,000–2,000 f.c. (11–22 klux). Increase fertilizer to 100–150 ppm once a week from 20-10-20, alternating with 15-0-15. Begin to cut back on mist applications. Harden cuttings for seven days prior to sale or transplant by lowering air temperatures to 70–75°F (21–24°C) days and 62–68°F (16–20°C) nights. Take cuttings out of the mist area and increase light levels to 2,000–4,000 f.c. (22–43 klux). Increase fertilizer to 150–200 ppm once a week from 20-10-20, alternating with 15-0-15. Pinch cuttings eight to twenty-one days after sticking. This will help to increase branching.

The average propagation time from sticking the cutting to a salable liner tray is four to five weeks.

Growing On

Seed

Most growers buy in plugs. Transplant plugs into a well-drained, disease-free medium with a moderate initial starter charge and a pH of 5.5–6.0. Use 1 plug/cell; 1–2 plugs/4" (10 cm) pot; and 7–10/10" (25 cm) basket, depending on the size of the plug.

Lobelia loves cool temperatures: 55–60°F (13–16°C) days and 50–55°F (10–13°C) nights. Once the summer heats up in the East, Midwest, and South, lobelia dies out.

Fertilize at every irrigation with 150–200 ppm from 20-10-20, alternating with 15-0-15. If height control is an issue, A-Rest, B-Nine, and Bonzi are effective.

Crop time from transplanting a plug to a finished container is five to six weeks for a cell pack and ten to twelve weeks for a 10" (25 cm) hanging basket.

Cuttings

Transplant 1 liner/4" (10 cm) pot; 2 liners/6" (15 cm); and 3–4/10" (25 cm) basket. Use a well-drained medium with a pH of 5.6–5.8 and a moderate initial nutrient charge. Provide a constant liquid feed of 200–300 ppm from 20-10-20, alternating with 15-0-15. Irrigate with clear water every third watering to prevent salts buildup. Water stress causes foliar necrosis, so keep the medium moist.

Once plants have rooted in, pinch plants above the fifth or sixth leaves, leaving 1.0–1.5" (3–4 cm) above the potting mix line. One to two pinches help to produce a uniformly shaped plant.

After plants are established, reduce temperatures to 40–45°F (4–7°C) for six weeks to promote flowering. Finish at 68–75°F (20–24°C) days and 55–65°F (13–18°C) nights.

Provide light levels of 4,000–9,000 f.c. (43–97 klux) (2,000–4,000 [22–43 klux] for *L. trigonacaulis*). Lobelia is day neutral but will flower profusely as light conditions improve after the cooling treatment.

The total crop time from transplanting to a flowering 4" (10 cm) pot is eight to ten weeks using one liner per pot; nine to eleven weeks for a 6" (15 cm) pot using two liners per pot; and up to thirteen weeks for a hanging basket with four to five liners per basket. This crop time includes one pinch.

Pests & Diseases

Spider mites, thrips, and whiteflies can become problems. Diseases that affect lobelia can include botrytis, *Pythium,* and rhizoctonia.

Varieties

For seed varieties, try the 'Moon', 'Palace', or 'Riviera' series for early upright types and 'Fountain' and 'Regatta' for trailing types in baskets. 'Crystal Palace' is one of the oldest variety names in floriculture, being named for the world-famous Crystal Palace greenhouse in England in the late 1800s. Flowers are deep blue on bronze foliage. ('Palace Royal' has similar flower and foliage color on earlier plants). Note that white varieties will have 5–10% off-types that are blue.

For cutting varieties, try the 'Waterfall', 'Techno', 'Hot', and 'Laguna' series for excellent 4" (10 cm) pots, baskets, and combination containers. Flower color is similar to bedding plant lobelias, but many appear to have better heat tolerance than the seed-propagated selections. In addition, cutting-propagated varieties have a more robust and better trailing plant habit, filling in baskets better and faster than their seed-grown counterparts.

Related Species

Highlighted by the variety 'Periwinkle Blue', *Lobelia hybrida* is the result of crosses between species. For this annual, vegetatively propagated lobelia, follow the same cultural guidelines for propagating and growing *L. erinus* cuttings to flowering 4" (10 cm) plants.

Insects & Diseases

Aphids, fungus gnats, spider mites, thrips, and whiteflies can attack plants. Disease problems can include botrytis, *Pythium,* and rhizoctonia.

Troubleshooting

Plant collapse may be due to extended periods of wet medium or botrytis infection. High nitrogen concentration in the medium, overfertilization under low light conditions, and/or low light and overwatering may cause excessive vegetative growth. Low fertilization during early crop stages may result in poor branching.

Lobelia cardinalis
Common names: Cardinal flower; red lobelia; scarlet lobelia; Indian pink
Perennial (Hardy to USDA Zones 3–9)

Lobelia cardinalis is a native perennial predominant in moist woodland settings and stream sides from eastern Canada through the United States to the Gulf of Mexico. Plants are erect, taller than they are wide, from 2–4' (60–120 cm), with a 2' (50 cm) spread when fully mature. The flowers are a bright cardinal-red on plants with medium to dark green foliage. Blooms are from 1–1.5" (2.5–3 cm) long and more open in appearance than *L. siphilitica.* Flowers are held out and slightly up on racemes to 14" (35 cm) long, sometimes longer if the plants are self-sown or if they sprout within a naturally existing clump. Blooms appear from July to September. Although obviously hardy, the plants are often short lived, lasting from two to four years depending on soil and exposure.

Propagation

Seed is the common propagation method, although plant clumps can be divided in spring or fall as well. Many lobelias produce a cluster of new basal growth in late summer after the central stem dies. These can be removed from the plant—along with their roots—and potted up or moved to the propagation bench to fully root for sales after the pots become fully established. Plants can be overwintered as well for sales the following year if they are not subjected to below freezing temperatures that could cause die out.

Growing On

Refer to the Growing On section for *L. siphilitica*; similar results will be obtained for *L. cardinalis.*

Varieties

L. cardinalis can be readily found at native wildflower companies and seed exchanges as well as commercial seed houses in the United States. Commercial propagators often offer this as a 32, 50, or other liner-tray size in the winter or spring.

Related Species

L. splendens (or *L. fulgens*), commonly called Mexican lobelia, is similar to *L. cardinalis.* The major difference is the bronze leaves on *L. splendens* varieties. The hybrids of the group, rather than the species themselves, have gained the most amount of attention. They are propagated both vegetatively and from seed.

These hybrids are the result of *L. splendens, L. cardinalis,* and *L. siphilitica* crosses. *(L. amoena* is considered to be a parent as well by some authorities.) The hybrids are known by the botanical name of *L.* × *hybrida.* The plants are not hardy in cold-winter locations and readily winter kill. So they are often grown as annuals or "protected perennials."

Seed is available and can be sown in winter for summer flowering. November sowings potted to 1-gal. (4 l) pots will be salable as vigorous growing plants in May. These, once planted to the garden before Memorial Day, will often be in flower in June or early July.

The 'Fan' series is also young by comparison—it was introduced to the trade in the late 1980s. Plants are often 2' (60 cm) tall, and separate colors of orchid, rose-scarlet, or deep red are available. The leaves are deep green, except on the scarlet-flowered plants, which have bronze leaves.

For summer-flowing plants, sow from November to early January. Plants can be sold green in either 6" (15 cm) or 1-gal. (4 l) pots in the spring.

'Queen Victoria' is often listed as a *L. cardinalis* variety and may have been listed as *L. cardinalis atrosanguinea* in older texts, referring to the bronze foliage. The plants possess brilliant, red flowers similar to *L. cardinalis,* although the foliage is a consistently vivid bronze-red. Plants can grow as tall as 4.5' (1.3 m) but seldom reach over 3' (90 cm) in the first year after sowing.

One final note regarding these lobelias: Don't pinch the plants back as they are growing. Lobelias have a central stem with a number of basal side shoots that develop. By pinching back this primary stem, the secondary ones develop, increasing your crop time to first flower by as much as four weeks.

Lobelia siphilitica
Common names: Blue cardinal flower; great lobelia; great blue lobelia
Perennial (Hardy to USDA Zones 4–8)

Lobelia siphilitica is a native perennial found in moist woodland settings in a sweeping range from Maine westward and south from the Dakotas to Louisiana. It is an erect and stately plant from 2–3' (60–90 cm) tall with a spread from 15–20" (38–51 cm) across. Plants will appear leafy in form and habit, either when growing in a pot or in the garden. The flowers are single, with a purple-colored upper lip and a lower set of blue petals highlighted in white. Its flowers are tightly held, appear tubelike, and grow to about 1" (2.5 cm) long. They develop in upright racemes that measure from 6–9" (15–23 cm) long. Flowers appear from July or August until September.

Propagation

L. siphilitica is primarily propagated by seed. It germinates in a range of 65–85%, and some companies offer enhanced seed for both higher and easier germination as well as plug performance.

If you are not using an enhanced seed product, sow several seeds per 288-plug tray in June or July. The seed does not need to be covered for germination, but a light covering of coarse vermiculite helps maintain moisture and promote germination uniformity. The root radicle emerges in five to eight days, but emergence will happen over a period of time. Germinate with a substrate temperature 70–72°F (21–22°C).

If the seed does not germinate readily, take the plug tray and expose it to cold conditions between 28–35°F (–2–2°C) for three to four weeks. Then put back in a cool greenhouse at 58–65°F (14–16°C) until germination is complete.

Seedlings will develop in Stages 3 and 4, over a period of several weeks. Expect a 288 plug to be ready seven to nine weeks after sowing, while a 128-liner tray will be ready in ten to eleven weeks after sowing.

Another method is to sow up to five seeds per 50-, 72-, or similar-sized cell pack in summer. These are ready to transplant to larger containers in eleven to twelve weeks after sowing in June or July.

Note: L. siphiliticas will flower the first summer after sowing in winter or spring but are one-third the size of the full-grown specimens in the garden. They look, for lack of a better term, puny. They will also vary in color from light to medium blue with a few white-flowering plants as well.

Growing On

For flowering plants in summer the same year as the plant is sold, seed should be sown to plug trays from June to July of the previous year and can be transplanted to the final container, such as a 1-qt. or 1-gal. (1.1 or 4 l) container.

However, if these plugs are transplanted up to a 32, 50, or larger liner or sown directly to this cell size (as noted under Propagation), these cell packs can be kept in a cold greenhouse or cold frame during fall and early winter to vernalize and then transplanted in early January or February to 1-qt. or 1-gal. (1.1 or 4 l) containers for spring sales.

Conversely, a 32 or 50 cell can be transplanted in late August or September to a 1-gal. (4 l) container, and the container can be overwintered cold or dormant for sales the following spring once the plants are well rooted.

Lobularia

Lobularia maritima
Common names: Alyssum; sweet alyssum
Annual

The sweet scent of alyssum permeates the air at any bedding plant grower's operation during the height of the spring season. While these low-growing annuals are a minor crop, everyone offers a few trays during the bedding plant season.

Propagation

Multiseed plugs with standard or multipelleted seed; do not cover. Germinate at 78–80°F (25–27°C) substrate temperatures and under 100–400 f.c. (1.1–4.3 klux) of light. Maintain a substrate pH of 5.5–5.8 and keep the substrate uniformly moist. Radicles will emerge in two to three days.

Move trays to Stage 2 for stem and cotyledon emergence, which will take from five to seven days. Lower the substrate temperatures to 72–75°F (22–23°C) and allow trays to dry down some between waterings. Increase light levels to 1,000–2,000 f.c. (11–22 klux) and fertilize weekly with 50–75 ppm from 14-0-14 once cotyledons have expanded.

Move trays to Stage 3 to develop true leaves by lowering substrate temperatures to 65–68°F (18–20°C) and increasing light levels to 2,000–2,500 f.c. (22–27 klux). Allow trays to dry out between irrigations and fertilize with 100–150 ppm weekly from 20-10-20, alternating with 14-0-14. Harden plugs in Stage 4 for seven days before sale or transplant. Reduce substrate temperatures to 60–62°F (16–17°C) and continue to maintain light levels. Fertilize weekly with 100–150 ppm from 14-0-14. If growth regulators are needed, B-Nine and Bonzi are effective.

In general, a 288-plug tray requires from four to five weeks from sowing until ready to transplant to larger containers. However, some growers seed alyssum directly into the final growing container.

Growing On

Transplant one plug per cell or 4" (10 cm) pot. Select a well-drained, disease-free medium with a moderate initial starter charge and a pH of 5.5–6.2. Fertilize once every other week with 100–150 ppm from 15-0-15, alternating with 20-10-20.

Alyssum prefers cooler temperatures. Grow on at 55–60°F (13–16°C) days and 50–55°F (10–13°C) nights. Deep South growers can offer alyssum in the fall for continuous color outdoors until the following summer. Plants will take temperatures to 28°F (–2°C).

A 32-cell pack requires from five to seven weeks from transplanting a plug until ready to sell in flower. A 4" (10 cm) pot requires from seven to eight weeks from transplanting until ready to sell.

Pests & Diseases

Whiteflies can attack alyssum. Disease problems can include downy or powdery mildew.

Varieties

The 'Clear Crystal', 'Wonderland', and 'Easter Bonnet' series are the most well-known varieties today. 'Clear Crystal' is the most recent of the group and is durable and vigorous in habit and performance.

Postharvest

Alyssum will perform best in outdoor plantings in the cool temperatures of spring, early summer, or late summer and fall. Plants stall out in the heat of summer; oftentimes white varieties will bounce back once night temperatures cool off again. Alyssum will reseed itself in some parts of the country. Consumers can keep plants looking fresh and reflower them by shearing plants back after they've been in flower for one month. Bees like alyssum.

Lupinus

Lupinus polyphyllus
Common name: Lupine
Perennial (Hardy to USDA Zones 4–6)

Breathtakingly spectacular is the only way to describe what it is like to see a garden of lupines in flower. Finding them in the trade happens, but not with great frequency, especially in flower. As a matter of fact, it is likely that no one is going to beat your doors down looking for lupines: Their fabulous flowers are one of the best-kept secrets in ornamental horticulture. However, if your customers have ever seen lupines in their native habitat in the Pacific Northwest, California, or western prairie, they will request them from you. They will also request them if they are the type to peruse gardening catalogs: Photographs of the flowers are as spectacular as real life. If you are a retail grower and you are planning on offering perennials for sale in the fall, add a few trays of lupine plugs onto your orders. Your experienced gardening customers will appreciate it, and it is one of those touches that can set your business apart from your competition.

Propagation

Sow seed onto a well-drained, disease-free, germination medium with a low starter charge and a pH of 5.5–5.8. Cover seed, but germinate on the bench or in a chamber under 100–400 f.c. (1.1–4.3 klux) of light. In most cases, seed will germinate readily.

However, if seed is of an unknown age, some recommendations state that in order to increase germination rates, scarify seed and/or soak it overnight prior to sowing. Other recommendations include alternating day and night temperatures of 80°F and 70°F (27°C and 21°C) to increase germination. Conversely, you can also choose to germinate at a constant 65–75°F (18–24°C). Keep trays moist, but not totally saturated. Seed should germinate in two to four days.

Move trays to Stage 2 for stem and cotyledon emergence. Maintain soil temperatures of 65–75°F (18–24°C) and increase light levels to 500–1,000 f.c. (5.4–11 klux). Apply 50–75 ppm from 14-0-14 once cotyledons are fully expanded. As radicles emerge, begin to allow trays to dry down somewhat between irrigations. Stage 2 will take from four to eight days.

Once you have a good stand, move trays to Stage 3 to develop true leaves. Lower temperatures to 60–65°F (16–18°C) and increase light levels to 1,000–1,500 f.c. (11–16 klux). Increase feed to 100–150 ppm applied once per week, alternating between 20-10-20 and 14-0-14. Allow trays to thoroughly dry between irrigations. Stage 3 will take from fourteen to twenty-eight days for 288s.

Harden plugs for one week prior to sale or transplant at 55–60°F (13–16°C). Increase light levels to 1,500–2,500 f.c. (16–27 klux). Apply a weekly feed of 100–150 ppm from 14-0-14. Allow trays to thoroughly dry down between irrigations.

Plugs seeded in January can be potted in March for green plant sales in May but for flowering the following year. These can be grown in 1801s or 4" (10 cm) pots and sold green in the spring. However, it is highly likely that these plants will not survive a hot, humid summer in many regions.

Plugs seeded in June through August can be potted in September, sold in October and November, and overwintered in an unheated cold frame or greenhouse. These plants will flower the following year.

Plugs seeded in September or October can be transplanted in January or February and grown on cold in an unheated greenhouse or cold frame at 32–40°F (2–4°C). If you choose this route, grow plugs in large trays such as 50s or 72s or plan to step plugs into cell packs before planting into their final growing containers (1-gal. [4 l] pots or containers). These plants will flower the year they are sold.

Growing On

Select a well-drained, disease-free medium with a moderate starter charge and a pH of 5.5–6.2. If you will be overwintering plants, a bark-based soilless mix is excellent. Ideally, grow lupines in 1-gal. (4 l) containers only, as plants become quite large. Throughout production, be careful of overwatering: Lupines do not tolerate a wet medium and are highly susceptible to root rots. Manage plants dry. Be especially careful about overwatering when transplanting smaller plugs to large pots.

For plugs planted in the fall, establish and bulk plants at 55–60°F (13–16°C) days and 50–55°F (10–13°C) nights. Fertilize lightly with 50–75 ppm from 15-0-15 at every other irrigation. Prior to sale or overwintering, make sure that plants have fully rooted into their growing container. For overwintering, place plants pot tight. Ideally, overwinter in a covered area—an unheated greenhouse or cold frame—as lupines are highly susceptible to root rot when plants become wet for long periods of time. Mulch plants in to help moderate conditions in the pot. Do not fertilize during dormancy and provide water only if needed.

Bring plants out of dormancy in the spring naturally with 55–60°F (13–16°C) days and 50–55°F (10–13°C) nights. Grow plants under the highest light levels possible (e.g., 3,000–5,000 f.c. [32–54 klux]) while maintaining cool temperatures. Fertilize at every other irrigation with 150–200 ppm from 15-0-15. For crops planted in February from an October sowing, grow on at 45–48°F (7–9°C) nights initially for their first six to eight weeks, then allow ambient day/night temperatures. Plants should flower in May or June. Colorado State University researchers have shown that lupines that have not had a cool treatment can be flowered under long days (plants flower seven to ten weeks later). However, long days cause significant stretching. Cooled plants can also be forced in eight to twelve weeks in the greenhouse.

Lupines that become root-bound in the plug or pot are more susceptible to root rots.

Pests & Diseases

Aphids can become a problem, as can botrytis, powdery mildew, and root rots.

Varieties

'Russell Hybrids', well known for years among perennial growers, comes as a mix of eight colors or as separate colors. Plants grow 3–5' (0.9–1.5 m) tall and produce 12–15" (30–38 cm) long spikes!

'Gallery' is a dwarf version with plants growing to 20" (51 cm) tall and producing 6–8" (15–20 cm) tall spikes. 'Gallery' is available in five separate colors and a mix. However, seed is in notoriously short supply.

Postharvest

Gardening customers will have the best success with lupine when it is planted out in the fall in most regions. Plants prefer very well-drained, acidic soils rich in organic matter. Mulch plants in to protect them from fluctuating temperatures through the winter. Plants will flower the following spring and early summer. In most parts of the country, lupines are short-lived perennials and are annuals in the South. Sometimes, when conditions are right, they reseed themselves. Lupines will not survive hot, humid summers.

Lycopersicon

Lycopersicon esculentum (also known as *Solanum lycopersicum*)
Common name: Tomato
Annual vegetable

If you are a bedding plant grower, tomatoes will be your most important vegetable crop and one of your biggest bedding plant sellers. Tomatoes are one of the world's most important vegetables—serving as the base for several different culinary styles. The taste of "real" tomatoes harvested from the garden on top of a grilled hamburger is an unparalleled slice of Americana. For lovers of Italian cuisine, a Caprese salad with tomatoes, mozzarella cheese, and basil is the epitome of a midsummer picnic delight. Baby boomers have discovered fresh tomatoes as a great source of lycopene, vitamin A, and other antioxidants.

Fresh, vine-ripened greenhouse tomatoes have become a standard in the past decade in just about every American supermarket. Newer breeding has developed medium-sized fruits that can be sold "stem attached" in clusters. Fruit can be harvested ripe, so that the fruit is flavorful. Specialized US, Canadian, and Dutch growers produce these crops year-round to sate a market where "fresh" is the byword. *Note:* Producing greenhouse tomatoes is specialized and is not the focus of this section.

Propagation

Sow seed into 288-plug trays. Select a disease-free, germination medium with a pH of 5.5–6.0. Cover seed with vermiculite for germination. Maintain substrate temperatures of 70–72°F (21–22°C) and ensure trays do not dry out. The substrate pH should be 5.5–5.8. Radicles will emerge in two to three days.

Move plug trays into Stage 2 and reduce irrigation once the radicle has emerged, allowing the medium to dry slightly before rewatering. Drop substrate temperatures to 68–72°F (20–22°C). Increase light levels to 1,000–1,500 f.c. (11–16 klux) and begin fertilizing with 50–75 ppm from 14-0-14 once cotyledons have expanded. Stage 2 will take about seven days.

Stage 3, growth and development of true leaves, will also take seven days. Drop substrate temperatures to 60–65°F (16–18°C) and allow the substrate to dry thoroughly between irrigations. Increase light to 1,000–2,500 f.c. (11–27 klux).

Prepare plugs for sale or transplanting in Stage 4 over seven days by lowering soil temperatures again to 60–62°F (16–17°C) and increasing feed to 100–150 ppm from 14-0-14 once a week.

In general, seedlings emerge in four to nine days. Allow five to six weeks from sowing to a transplantable 288 plug. There are a few dwarf and pot plant varieties that may take one more week to size up.

Growing On

Transplant into a well-drained, disease-free medium with a moderate initial starter charge and pH of 5.5–6.2. Grow on at 55–65°F (13–18°C) days and 50–60°F (10–16°C) nights. Transplanted to a 32-cell pack, plants are ready in four to six weeks for spring sales; 4" (10 cm) pots are ready in six to seven weeks.

Tomatoes respond to increased fertilization with increased growth. The earliest yield will come from plants that are not stressed from insufficient nitrogen. Fertilize every third irrigation with 15-0-15 at 100–150 ppm. Maintain the EC at 1.0 mS/cm.

If you are not able to sell plants as they are ready, do not try to hold plants back with moisture stress, as you may adversely affect results for an unsuspecting customer later on. Tomatoes are very sensitive to moisture stress and may never recover if they are stressed early in their development.

If you need to slow plants down, use DIF by dropping temperatures for a couple of hours just before sunrise.

Pests & Diseases

Aphids, thrips, and whiteflies can become pests on tomatoes in production. Diseases such as botrytis, tomato spotted wilt virus (TSWV), *Pythium,* and rhizoctonia can also be problems. While a number of other diseases can potentially be problems, many varieties have been bred for disease resistance.

Varieties

There are two types of tomatoes: indeterminate and determinate. Indeterminate plants (often called stake tomatoes) grow tall, flower, and make tomatoes through the growing season.

Indeterminate tomatoes make fruit on many stems and are ideally pruned and staked to keep fruit off soil. To prune indeterminate plants, pinch out suckers that develop between the main stem and a branch. Provide stakes and/or strings because plants can grow to 6–7'(1.8–2.1 m).

Determinate plants are more compact and produce bushy plants to 3–4' (0.9–1.2 m). Plants grow to a set height, flower, set fruit, and ripen in a short time, so harvest is concentrated. Cages are ideal to keep determinate plants upright.

Shape is also used to categorize tomatoes. There are five basic shapes of tomatoes: plum, pear, cherry, standard, and beefsteak (largest). Further, tomatoes are characterized by their maturity: the number of days from planting outdoors to harvesting fruit. For example, early tomatoes can be harvested in fifty-five to sixty-five days; midseason tomatoes at sixty-six to eighty days; and late types at more than eighty days.

'Better Boy' and 'Early Girl' are probably the two most well-known tomato varieties. Plants are indeterminate and fruit midseason. Fruit weights 16 oz. (454 g). 'Burpee Big Girl' is an indeterminate midseason with 8-oz. (227 g) fruit. 'Jet Star' is an indeterminate midseason with 6–8-oz. (170–227 g) fruit.

'Bush Early Girl' is a determinate, early season producer with fruit weighing 6.5–7.5 oz. (184–312 g).

For cherry tomatoes, try 'Sugar Snack', an indeterminate early-season with 1-oz. (28 g) fruit as well as America's favorite 'Sweet 100', an indeterminate early-season tomato with 1" (3 cm) clusters of tomatoes on long branches.

Postharvest

Displaying plants in partial shade at retail may help ease watering requirements.

Instruct consumers to plant tomatoes only after the danger of frost is past. If the gardener chooses to produce tomatoes in containers, recommend larger containers: A minimum diameter of 12" (30 cm) will help to maintain moisture around roots. Blossom end rot, caused by a lack of calcium from moisture stress and/or plants that grew too rapidly, can be controlled by adding a calcium soil amendment at planting and mulching to help maintain uniform moisture. Tomatoes are especially sensitive to moisture stress: Do not allow them to dry out.

Lycopene is responsible for giving tomatoes their red coloration. Ideal temperatures for reddening tomatoes are 70–75°F (21–24°C)—lycopene production declines above 80°F (26°C) and is almost nothing above 90°F (32°C). Orange pigmentation, carotene, is easily developed at these high temperatures, which is why fruit ripened in the height of the summer has an orange color. Southern gardeners may choose to plant early-maturing varieties again in August for a fresh fall crop. Tomatoes may be harvested until hard frost.

Lysimachia

Lysimachia congestiflora (also called *L. procumbens*)
Annual

Lysimachia congestiflora is a popular combination pot plant. Bright, golden-yellow flowers appear clustered profusely above round, 2" (5 cm) leaves. For growers, the crop is easy, and when plants start to flower, they will walk off the bench.

Propagation

Stick cuttings immediately upon arrival. If they must be stored, place them at 40–50°F (4–10°C) for no more than twenty-four hours.

Stick cuttings in Oasis wedges or a peat-based medium with a pH of 6.5–7.0. Use a rooting hormone during winter months. Maintain soil temperatures of 68–72°F (20–22°C) and air temperatures of 75–80°F (24–26°C) days and 68–70°F (20–21°C) nights. Temperatures below 68°F (20°C) inhibit rooting. The medium should be moist, but not saturated. Apply mist for six to nine days, adjusting the frequency to ensure cuttings remain fully turgid. However, *Lysimachia* should be rooted on the dry side, so reduce mist as soon as possible to reduce stretch. Provide from 500–1,000 f.c. (5.4–11 klux) of light. Apply a foliar feed of 50–75 ppm from 20-10-20 as soon as foliage shows any loss in color. Cuttings should callus in five to seven days.

Once half of the cuttings have formed root initials, move cuttings to the next stage to develop roots, which will take from nine to eleven days. Maintain soil and air temperatures. Once cuttings begin to root in, begin drying down the medium by reducing the frequency and duration of mist. Increase light intensity gradually from 1,000–2,000 f.c. (11–22 klux). Apply a foliar feed of 100 ppm from 15-0-15, alternating with 20-10-20. Increase the rate gradually to 200 ppm. Maintain pH levels of 6.5–7.0.

Harden cuttings for transplant or shipping for seven days by lowering air temperatures to 72–75°F (22–24°C) days and 62–68°F (16–20°C) nights. Move the cuttings from the mist area into a hardening greenhouse and increase light levels to 2,000–4,000 f.c. (22–43 klux). Shade during the height of the day to help moderate temperature stress on plants for cuttings stuck in late spring to early autumn. Fertilize once a week with 100–150 ppm from 15-0-15, alternating with 20-10-20.

Cuttings stuck to a 105 tray will take from three to four weeks to be ready to transplant to larger containers.

Growing On

Pot rooted liners into a well-drained, disease-free, soilless medium with a high initial starter charge and a pH of 6.5–7.0. Make sure the medium is low in phosphorus to prevent leaf chlorosis. Use 1 plant/4" (10 cm) pot; 2–3/6" (15 cm); and 4–5/10" (25 cm) or 12" (30 cm) hanging basket. Ideally establish plants in 4" (10 cm) pots before planting up combination planters, although, by carefully monitoring irrigation frequency and amounts, you can plant liners directly into large combination pots. Just be careful not to overwater. *Lysimachia* is quite drought tolerant and will survive regular wilting episodes. Conversely, if plants are kept too wet, root rot can quickly become a problem.

Lysimachia has a light fertilizer requirement. Fertilize once or twice a week with 100–150 ppm from

15-0-15, alternating with 20-10-20. Use the lower rate until plants become established. Avoid excessive ammonium-based fertilizers as pH can lower, which will promote uptake of boron and iron. Be sure to maintain the pH range through production to reduce minor nutrient uptake. If you notice marginal leaf necrosis and/or chlorosis, salts could be building up: Leach with clear water to reduce salts.

Grow the crop warm at 75°F (24°C) days and 65–68°F (18–20°C) nights. Temperatures above 95°F (35°C) are not recommended. Plants are also sensitive to temperatures below 45°F (7°C).

Pinch plants at least once to improve basal branching, leaving about 1.0–1.5" (3–4 cm) above the soil. Since plants flower after the seventh leaf pair, if you delay the pinch, your plants will not be well branched. Plants in 4" (10 cm) pots do not need to be pinched, and growth regulators are not needed during production.

Lysimachia flowers better under long days: In winter, crops take longer to produce. At photoperiods of less than nine hours, plants remain vegetative. While plants will tolerate lower light levels, low light promotes stretch. The best growth is obtained at more than 5,000 f.c. (54 klux).

The crop times for 4" (10 cm) pots with one liner per pot is four to five weeks to finish; a 6" (15 cm) pot with two to three liners takes from five to eight weeks to finish; and a 10–12" (25–30 cm) basket takes from eight to ten weeks to finish when using with thrre to four 105 liners.

Pests & Diseases

Aphids, fungus gnats, thrips, spider mites, and whiteflies can attack plants. Disease problems can include botrytis, powdery mildew, *Pythium,* and rhizoctonia.

Troubleshooting

Exposure to cold temperatures, medium that remains wet for too long, and/or botrytis infection can lead to plant collapse. High ammonia levels in the medium, overfertilization under low light conditions, low light combined with a wet medium, or short day lengths can result in excessive vegetative growth. Low light causes poor branching. Foliage necrosis can be due to plants drying out between irrigations, high soluble salts, or excess phosphorus, boron, or iron in the medium.

Varieties

Try 'Sunset' for its deep-green and lime-green variegated leaves—a great combination with its yellow flowers.

Related Species

Lysimachia nummularia, commonly called creeping Jenny, has dark green, rounded leaves on plants with a trailing habit. Plants grow to no more than 8" (20 cm) tall in bloom with a 2' (60 cm) spread. Flowers are bright yellow, 1" (2.5 cm) across, and fragrant.

Lysimachia nummularia

L. nummularia 'Aurea' is the lime-green or light-yellow foliage variety for shadier locations. 'Goldilocks' is a commonly available variety. It is prized for its foliage since its yellow blossoms get lost in the foliage. *L. nummularia* 'Aurea' is hardy from USDA Zones 3–8 and can be propagated by divisions or stem cuttings. Propagate as noted under *L. congestiflora.*

L. punctata, commonly called yellow loosestrife, is an upright plant with 0.75–1" (2–2.5 cm) wide, bright yellow flowers. The flowers are in whorls around the leaf axis. Blooms are close to the stem but are large and colorful enough to make yellow loosestrife a useful landscape plant or cut flower. The plants are hardy from USDA Zones 4–8 and, like the other species described here, will be vigorous in fertile and moist locations. *L. punctata* also prefers the same sunny location as *L. clethroides.* Propagate as noted for *L. congestiflora.*

M

Marguerite Daisy (see *Argyranthemum*)

Marigold (see *Tagetes*)

Matricaria (see *Tanacetum*)

Matthiola

Matthiola incana
Common name: Stock
Annual, cut flower

This old-fashioned cut flower and garden favorite adds a touch of class to any bouquet or arrangement. The scent of cloves permeates the air whenever stock is around, adding to its consumer appeal. Stock shows up in today's market as a seasonally offered cut flower, a flowering potted plant during the cooler months, or a bedding plant in jumbo packs or 4" (10 cm) pots in some markets.

At one time, stock was a major greenhouse cut flower crop in the Midwest, South, and eastern United States. George J. Ball, the original author of the *Ball RedBook,* offered stock to the United States cut flower market from the early 1910s to the 1930s. Today, the domestic cut flower crop is largely produced outdoors in Arizona and California.

For growers, stock is a relatively long crop and requires cool temperatures in order to flower. If you are planning to add it as a pot plant or cut flower, work with enough plants so that you can justify providing the specialized environment that plants require in order to perform. If you are a bedding plant grower, it is easy enough to add a few plug trays onto an existing order to try it out in jumbo packs, 1801s, or even 4" (10 cm) pots. Stock requires cool temperatures in production and in the garden; plants will not perform outdoors in the summer heat. Temperatures above 65°F (18°C) inhibit flowering. Cool nights at 50°F (10°C) are ideal. While plants will tolerate freezing temperatures and light frosts, stock is killed at temperatures too much below freezing.

If you offer stock as a bedding plant, be sure to offer plants early enough so that gardeners can enjoy them. Growers in mild climates, where lows rarely go below 32°F (0°C) in the winter, can offer stock in the fall. Outdoor plantings in mild climates will flower continuously from November to April.

Cut flower stock can be grown in the greenhouse or outdoors. In general, stock is a greenhouse cut flower harvested from January to June in areas north of 35° latitude (i.e., north of Bakersfield, California; Oklahoma City, Oklahoma; and Charlotte, North Carolina). Stock can also be grown in the greenhouse or at other times of the year in the South at high altitudes or in other cool climates. Field stock is harvested in the late winter and spring in California. In northern Pacific areas and in isolated cool pockets in the rest of the country, it is grown as a summer cut flower.

Propagation

Sow seed onto a well-drained, disease-free, germination medium with a pH of 5.5–5.8. Cover seed very lightly with coarse vermiculite to help maintain humidity levels at the seed coat. Do not place trays in the dark. Provide 100–400 f.c. (1.1–4.3 klux) of light for germination. Keep trays moist enough so that water can easily be squeezed from the medium. Seed will germinate in three to five days with substrate temperatures of 65–75°F (18–24°C).

Once 50% of the seed has cracked, move trays to Stage 2 for stem and cotyledon emergence. Lower the substrate temperatures to 62–75°F (17–24°C) and increase the light levels to 1,000–1,500 f.c. (11–16 klux). Provide 50–75 ppm from 14-0-14 once cotyledons are fully expanded. As radicles penetrate the medium, begin to allow the medium to dry down somewhat between irrigations. Stage 2 takes from five to seven days.

When you have a stand, move trays to Stage 3 to develop true leaves, which will take about fourteen to twenty-one days for a 288. Reduce substrate temperatures to 60–72°F (16–22°C) and increase light levels to 1,500–2,500 f.c. (16–27 klux). Feed weekly with 100–150 ppm, alternating between 14-0-14 and 20-10-20. You will not need to apply growth regulators on stock.

Harden plugs for seven days prior to sale or transplant by lowering temperatures to 55–65°F (13–18°C) and increasing light levels to 2,500–3,500 f.c. (27–38 klux). Feed once with 100–150 ppm from 14-0-14.

Sow bedding plant plugs from August to February for flowering from October to March in mild climates. See the Cut Flower section below for times to sow cut flower crops.

Growing On

Transplant bedding plant plugs in to a well-drained, disease-free medium with a moderate initial starter charge and a pH of 5.5–6.5. In order to avoid root rots, be careful not to bury the crown of the plant. Use one plant per jumbo pack or 4" (10 cm) pot and three plants per 6" (15 cm) or 1-gal. (4 l) pot. From a 288, 4" (10 cm) pots can be ready to sell in just five to seven weeks in the spring. Provide the highest light levels possible, while still maintaining growing temperatures.

Warm temperatures inhibit flowering. Flowering is optimal at 50°F (10°C), although plants can survive at freezing temperatures. Grow plants on at 55–60°F (13–16°C) days and 50–55°F (10–13°C) nights. Once plants have sized up and initiated flowers, lower night temperatures to 45–50°F (7–10°C) to tone and/or hold plants.

Fertilize with 150–200 ppm, alternating between 15-0-15 and 20-10-20.

Cut Flowers

Grow stock in well-drained soils with a pH of 5.5–6.5. For outdoor plantings, incorporate 10-10-10 prior to planting, depending on your soil test results. Add additional potassium, if needed.

Stock plugs are ready to transplant when the second true leaf has unfolded. If you are buying in plugs, allow seedlings to acclimate to greenhouse conditions for twenty-four hours prior to planting, but do not hold plug trays any longer, as flowering may be delayed. If you must hold plugs, hold them in a cooler (36–40°F [2–4°C]) under fourteen hours of 250 f.c. (2.7 klux) lighting for as brief a time as possible. Treat with fungicide prior to storage to prevent botrytis.

Plant plugs at a density of 54–59 plants/yd.^{2} (64–70/m^2). Nonbranching plants sown in the field can be sown in rows 18" (46 cm) apart. Thin plants, leaving about 3" (8 cm) between them once you have a full stand and seedlings are established. If you have sown a selectable stock, you can thin out plants to increase the percentage of doubles. Otherwise, remove the smallest plants. For greenhouse plantings, space plants at 3" × 6" (8 × 15 cm). Support should not be needed, provided you give plants the correct temperatures.

Plants can be irrigated from overhead in the early stage, switching to drip irrigation once plants have filled in. Use clear water to water in the new planting. Afterwards, provide a constant liquid feed of 150–200 ppm from 14-0-14, reducing feed as flowers mature. Light, sandy soils require 200–300 ppm. Stock is a moderate feeder and will produce inferior cut flowers if it is given excessive nitrogen. Use fertilizers low in ammonium.

Note that temperature control is critical to success. When average temperatures are higher than 65°F (18°C) for more than six hours a day, flower initiation is inhibited. Plants will grow vegetatively until temperatures drop below 65°F (18°C). Normally flower buds are initiated three to four weeks after planting.

After planting, grow on at 60°F (15°C), but do not allow average temperatures to go above 65°F (18°C). Once plants have reached the proper size to form buds (ten to fifteen leaves), they need a minimum of twenty-one days at 50–60°F (10–16°C) to initiate buds on all plants under natural winter conditions. However, low temperatures before plants have sized up properly will prevent buds from forming on columnar stock.

Once buds have formed with the cool treatment, provide 60°F (15°C) for flowering.

It is interesting to note that stock leaf margins are determined by temperature as well. Once plants have reached the critical ten-to-fifteen-leaf stage, temperatures below 60°F (16°C) cause smooth margins. Plants grown at higher temperatures will develop lobed leaves but will once again produce smooth margins when they are exposed to lower production temperatures.

For winter crops, once you have initiated buds with cool temperatures, provide four hours of night interruption lighting to speed flowering by two weeks.

Scheduling your stock crop is also vital to success. Sowings for greenhouse stock made between July 15 and February 15 will flower from January to early June. However, if you sow seed outside of this time frame, plants may be blind because of high temperatures.

Plantings made between November 15 and May 15 may be grown under natural day lengths. January and February plantings can be harvested in about fourteen weeks; March, April, and May plantings in eight to ten weeks. To get good results from plantings made from May 15 to August 15, keep young plants actively growing by keeping the greenhouse as cool as possible.

Stock is sensitive to high temperatures, and flower initiation is inhibited at temperatures above 65°F (18°C). If you can provide the right temperatures, summer plantings can be harvested in ten to fourteen weeks. However, as previously stated, producing a salable crop is very difficult in high temperatures. For plantings made from August 15 to November 15 in northern latitudes, use supplemental lighting for fifteen to sixteen hours per day, beginning three to five weeks after planting. Without the supplemental lights, spikes can become stretched and "gappy." Stock planted from October to December will be ready for harvest in about seventeen weeks.

Pests & Diseases

Aphids can spread viruses to stock, which will appear as streaked or blotched flowers. Leaf rollers and thrips can also attack plants. Diseases can include *Phytophthora* and rhizoctonia. Bacterial blight can appear as sudden wilting and collapse of cotyledons during the seedling stage. On older plants, it appears as water-soaked areas around leaf scars, while the main stem is soft and water-soaked in appearance.

Troubleshooting

Potassium deficiency appears as a dying of lower leaves from the tips and margins inward, progressing up to the plant nearly to the top. Leaf loss is most serious at flowering.

When cut flower stock plants are too short, temperatures may have been too low when the seedlings were developing. Sowings made after December usually produce short plants. A cooling period that is too short causes blind plants. Be sure to provide the full twenty-one days of temperatures below 60°F (16°C). Avoid growing stock during the wrong time of year for your climatic zone.

If stems fail to take up water, they may be low in carbohydrates due to low light intensity or high temperatures. Plants with hard stems because of old age or having been grown in dry soil also often fail to take up water.

Varieties

The 'Midget' series is early and dwarf (8–10" [15–25 cm]). You can "select" seedlings to keep only the double-flowering plants by keeping only the plants with serrated leaves; plain-leafed types will be single. 'Harmony' is also dwarf and grows from 9–12" (23–30 cm) tall. It is not selectable as a seedling but produces the same amount of doubles as a selectable stock. 'Vintage' is like 'Harmony' and 'Midget' in earliness but is much more strongly branched. 'Vintage' is 60% double.

'Trysomic 7-Week Mix' is a dwarf, double, seven-week stock that is used as a bedding plant in some regions, especially California. Plants grow to a height of 12" (30 cm). 'Trysomic' is the only bedding plant strain that will flower under high temperatures. The dwarf, double plants usually put out a central spike in May and then produce numerous side shoots that will flower later. Sow in February and move plugs to a cold growing house as soon as seedlings are established.

For cut flowers, most growers choose to grow columnar, or nonbranching, stock. The 'Cheerful' series is 90% double. Plants set buds without a cold period, so an August sowing can be harvested in November, even in warmer regions. 'Cheerful' is recommended for greenhouse production only. The series is available only in white and yellow colors. For field production, the 'Katz' series is one of the latest selections in breeding and is available in ten separate colors. Crops can be easily finished in plastic tunnels in the field. The 'Column' series is the most common variety for field production. Plants grow 2.5–3.0' (76–91 cm) tall, with 50–60% of the plants bearing double flowers.

Postharvest

Cut flower stock is ready for harvest when seven to ten flowers are open. You can cut stems or pull out entire plants and remove the roots later. Premature harvest leads to poor color development and reduced flower size as flowers continue to open. Strip leaves from the lower portion of stems and place them immediately in water. After grading and bunching, place stems in preservative solution. Stock responds well to a sucrose-based preservative solution with bactericide to facilitate water uptake and inhibit stem plugging. Hold stems at 45–50°F (7–10°C) for at least six hours or overnight prior to shipping.

Stock can be stored dry in a 40°F (4°C) cooler for no longer than two days. If stored dry, rehydrate and condition flowers in a preservative solution. For longer-term storage, wrap the highest-quality spikes in plastic to prevent them from drying out and hold them in a preservative solution at 32–40°F (0–4°C). However, be aware that storing stems too long can negatively affect flower fragrance. Since stock has a relatively short vase life after harvest, storage is not advised.

Instruct retailers to recut stems and place them in a clean bucket with a floral preservative/biocide. Stock should be stored in a lit cooler at 40°F (4°C).

Bedding plant stock should be displayed cool at 55–60°F (13–16°C) days and 50–55°F (10–13°C) nights. Light shade may be beneficial in moderating temperatures.

Mentha

Mentha spp.
Common name: Mint
Herb, perennial (Hardy to USDA Zones 4–9)

Mint is a very useful herb. It can be used fresh or dried to flavor teas, jellies, candies, meats, and vegetables. It is the perfect garnish for fresh fruit as well as many meat dishes. Mint is easy to grow for both gardeners and growers. The trick for growers is to offer true-to-type varieties. However, mint can be invasive. Many varieties are aggressive and take over a portion of the garden in which they have been planted. For that reason, give it plenty of space, as it will take it over anyway.

Propagation

Seed

Mint seed is readily available, but it is not a preferred method of propagation. Even though spearmint and peppermint are commonly sold from seed, they are not true to type and will not produce the best plants for culinary use. Plants propagated from seed vary in their essential oils, and most are better used as ornamentals (for potpourri and sachets) than for cooking.

Sow seed to a 512-plug tray and lightly cover with coarse vermiculite. Germinate at 70–75°F (21–24°C).

The root radicle emerges in two to three days, and plug trays are ready to transplant three to four weeks after sowing. Transplanted to 4" (10 cm) pots, the plants are ready in nine to twelve weeks for spring sales. Grow on at 64–75°F (18–24°C) days and 55–60°F (13–16°C) nights.

Cuttings

Harvest stem or root cuttings from actively growing plants any time of the year. Cuttings can range from 2–3" (5–8 cm) long. Stick cuttings into a soilless rooting medium with a low starter charge and a pH of 5.5–5.8. Rooting hormone is not usually necessary. Place cutting trays under mist and shade. (*Note:* Using mist, you can also direct-stick two unrooted cuttings per 4" [10 cm] pot and save two weeks of production time.) Root cuttings at substrate temperatures of 72–77°F (22–25°C). Gradually reduce the mist duration and frequency as root initials form. Cuttings should be rooted within seven days.

Once cuttings have formed root initials, begin fertilizing with 50–75 ppm from 14-0-14 at every other irrigation. Gradually reduce the mist duration and frequency as the cuttings mature. Once trays are fully rooted, remove them from the mist area and, over the course of a week, gradually increase light levels and lower substrate temperatures to 65–72°F (18–22°C). Harden cuttings off for seven days prior to sale or transplant by lowering temperatures to 62–65°F (17–18°C) and increasing the fertilizer to 100–150 ppm from 14-0-14. Liner production takes from twenty-one to twenty-eight days.

If purchasing plugs or liners, transplant a 105- to 72-cell tray to a 4" (10 cm) pot. When grown on at 64–75°F (18–24°C) days and 58–64°F (14–18°C) nights, the plants will be ready for spring sales in five to seven weeks (depending on plug size). Provide 4,000–5,000 f.c. (43–54 klux) of light. Fertilize at every other irrigation with 100–150 ppm from 20-10-20, alternating with 14-0-14. Do not allow plants to dry out, as foliage will yellow. Some growers prefer to pinch out the terminal leader to encourage branching. This is especially effective on plants potted up during the darkest days of the winter. For plants potted up in March or later, pinching plants is more to shape them than to get them to branch. Plants will not be in flower but will start to spread or runner at the time they are sold. If they are getting away from you, shear them back slightly and they will fill out again within two weeks, if done in late March or later.

Pests & Diseases

Aphids, spider mites, thrips, and whiteflies can attack plants. The best control is to avoid pest problems. It is illegal to use conventional pesticides on an herb crop. When insect problems are caught early, products such as Safer's Insecticidal Soap should offer good control.

Diseases that strike include root rots and foliar rust. If rust strikes plants, discard them immediately, as there is no control.

Species & Varieties

The most famous mint is spearmint, *M. spicata.* However many consumers will demand a broader selection, and there are many to consider. Start with curly spearmint, *M. spicata* 'Crispa', which offers spearmint flavoring on highly ornamental leaves. Orange mint, *Mentha aquatica* 'Citrata', has a fruity flavor. *M.* × *piperita* 'Citrata' is commonly known as lemon mint and features bronze-purple leaves. *M.* × *gracilis* 'Variegata' is known as ginger mint, which is grown for its attractive, yellow and green foliage. Pineapple mint (*M. suaveolens* 'Variegata') is grown for its attractive, creamy white and green leaves. *M. requienii,* or Corsican mint, is a creeping groundcover with highly pungent leaves and grows only 1–4" (3–10 cm) tall. And do not forget to offer peppermint, *M.* × *piperita.* It is important to note that all of these mints are vegetatively propagated.

Postharvest

Mints can become invasive in the garden, so they are best planted in an area where they can be contained. Leaves may be used fresh or dried. Some mints have very attractive flowers.

Mimulus

Mimulus × *hybridus*
Annual

Mimulus is the ideal plant for damp, shady areas and for early spring sales. Tolerant of frost, this short, spreading plant is quickly covered in masses of brightly colored blooms that tolerate rain and cool sun very well. A long-day plant, *Mimulus* responds well to daylength manipulation and growth regulators. It is the ideal cool crop.

Propagation

Single-sow pelleted seed onto a disease-free, germination medium with a pH of 5.5–5.8. Cover lightly. Germinate at substrate temperatures of 60–65°F (15–18°C). Radicles will emerge in two to three days. Maintain substrate temperatures throughout Stage 2, which will last three to five days. A light spray with Cycocel as soon as the first true leaves have expanded will help to prevent stretching in the plug. Maintain light levels at 450–700 f.c. (4.8–7.5 klux).

Stage 3, developing true leaves, will last from seven to ten days. Lower soil temperatures to 50–55°F (10–13°C) and ensure that the trays never fully dry out. Fertilize weekly with 100–150 ppm 7-0-14. Low doses of Cycocel or B-Nine can be used as required. Increase light levels to 1,000–2,500 f.c. (11–27 klux).

Harden plugs in Stage 4 for seven days prior to transplant or sale. Maintain soil temperatures of 50–55°F (10–13°C) and keep the trays moist at all times. Fertilize weekly with 7-0-14 at 100–150 ppm and raise light levels to 2,500–4,000 f.c. (27–43 klux).

A 288-plug tray takes from four to six weeks to be ready to transplant to larger containers.

Growing On

Transplant plugs into a well-drained, disease-free medium with a pH of 5.5–6.2 and a low initial starter charge. Postproduction life of *Mimulus* is best with cell packs, but it also makes an excellent pot bedding plant. Grow on as cool as possible (frost protection is all that is needed), but 55–60°F (13–16°C) days and 40–45°F (4–7°C) nights will maintain quality while promoting strong growth. Maintain light levels as high as possible (3,000–4,000 f.c. [32–43 klux]).

Fertilizer will not normally be needed, but if the plants do show signs of deficiencies, then a well-balanced fertilizer can be applied with the irrigation (20-10-20) at 150 ppm; maintain EC at 1.0 mS/cm. If height control is necessary, Cycocel and B-Nine are effective.

The crop time from transplanting a 288 plug to a salable 4" (10 cm) pot is six to seven weeks for most varieties.

Pests & Diseases

The short cultivation period of *Mimulus* means that few pests or diseases have the chance to take hold, so sprays of insecticides and fungicides are rarely necessary. However, keep an eye out for aphids and whiteflies. If the plants have been allowed to grow too quickly, then botrytis can infect the plants.

Varieties

'Magic' is available in a range of colors and grows to about 10" (25 cm). The majority of its flowers are solid colored, but there are several bicolors, and some are being developed with deeply contrasting markings.

'Mystic' grows from 10–12" (25–30 cm) tall and, like the 'Magic' series, is a clear mix of colors—its flowers lack the speckling that is common in *Mimulus*.

Postharvest

Do not ship plants at temperatures above 60°F (16°C), or they will stretch in transit. Ship when the first color is showing, as the flowers are sensitive to ethylene. Avoid boxing plants if at all possible and do not store near ripening fruits or vegetables.

Mimulus is a plant of shady water margins (marshes), and so it is important not to expose the plants to full sun. At retail, 50% shading is beneficial to prevent plants from drying out and to prolong the life of the flowers. Display at 50°F (10°C) day and night. Temperatures can drop to below 40°F (4°C) without harming the plants. To avoid disease problems, do not water plants late in the day. Groom plants to remove dead flowers and leaves so that sources of ethylene are not present.

Plantings in color bowls are particularly effective, and the color display rivals impatiens. *Mimulus* will flower continuously from late spring through midsummer, if they are placed in a cool, shady spot and are watered daily. Deadheading plants in the garden will encourage flowering for a longer season.

Morning Glory (see *Ipomoea*)

Muehlenbeckia

Muehlenbeckia axillaris
Common name: Creeping wirevine
Annual

Muehlenbeckia, commonly known as creeping wirevine, is a strong plant to add to your assortment and mixed containers. Its aggressive trailing habit works great as an accent plant in mixed containers and tiered gardens. It can take drought conditions and will hold up to some of the hottest summer temperatures. Its unique, wire-sized stem and small, round leaves add fantastic texture to any combination planting.

Propagation

Muehlenbeckia is propagated by cuttings. If cuttings are wilted, place in a clean cooler at 45–50°F (7–10°C). Provide a relative humidity of near 100% by wetting the floor and misting open bags with clean water. Stick cuttings in Ellepots or medium that drains well and has a pH of 5.8–6.2. Most suppliers ship two-for-one cuttings because cuttings are so small. If this is the case, then stick two in one rooting cell. Be sure to provide good soil-to-stem contact by lightly squeezing the soil or running it through a high-volume water tunnel. Do not to dibble the hole too deep. Root cuttings under mist at a substrate temperature of 70–72°F (21–22°C) and air temperatures of 70–72°F (21–22°C) days and 68–70°F (20–22°C) nights. Because the cuttings are small, it is extremely important to only provide enough mist to keep cuttings turgid. Too much mist will inhibit rooting and increase disease. Provide 500–1,000 f.c. (5.4–11 klux) of light. It should take around twelve days to reach the callus stage.

Once roots begin to form root initials move to Stage 2. Reduce and quickly cease the use of misting and begin to apply an overhead feed of 50–75 ppm from 20-10-20.

In Stage 3 maintain soil and air temperatures. As roots form, begin to let the soil dry down between irrigations. Increase the light levels to 1,000–2,000 f.c. (11–22 klux). Apply a foliar feed of nitrogen at 100 ppm from a balanced fertilizer. Increase the rate to 150 ppm as roots form. Stage 4 consists of hardening the cuttings for approximately seven days prior to sale or transplant. At the beginning of Stage 4, you can pinch or trim the liners to help create lower breaks. Reduce air temperatures at night to 65–68°F (18–20°C). Liners can be held, if needed, at 54–58°F (12–14°C). Move liners from the mist area and increase light levels to 2,000–4,000 f.c. (22–43 klux). Shade plants during midday to reduce temperature stress. Fertilize with nitrogen at 200 ppm from a balanced fertilizer.

The crop time for an 84-liner tray is five to six weeks using two cuttings per cell.

Growing On

Transplant 1 plug or liner/4" (10 cm) pot; use 1–2 plugs or liners/6" (15 cm) pot; and use 4–6 liners/10" (25 cm) or larger container. Select a disease-free, very well-drained medium with a pH of 5.8–6.2. Grow plants on at 65–75°F (18–24°C) days and 58–62°F (14–17°C) nights. Fertilize with a constant feed of 150–250 ppm nitrogen with a complete fertilizer. Monitor and maintain a pH of 5.8–6.2.

Because plants have such little leaf surface, they should be started on the dry side. Too much water will reduce root growth leading to a greater production time and potential loss. As plants mature and build a root system, they will require moderate irrigation. Grow under high light levels (4,000–7,000 f.c. [43–75 klux]). Pinching or trimming plants can help shape and create fuller plants. Growth regulators should not be needed.

The crop time from transplanting to a salable plant is six to eight weeks for a 4" (10 cm) pot.

Pests & Diseases

Regular greenhouse insects attack this plant. Botrytis, rhizoctonia, and *Pythium* can be problems when soil is overly saturated.

Troubleshooting

Plants may collapse due to moderate to severely wet medium or excessive botrytis. High ammonium levels in the medium, overfertilization under low light conditions, or high production temperature can cause excessive vegetative growth.

Varieties

Muehlenbeckia is available from many cutting suppliers. Also try a tricolor variety for a dose of red, light green, and light gray.

Postharvest

Display plants in high light with cool temperatures. Do not allow plants to wilt, but keep them on the dry side.

Consumers should plant in well-drained soil in full sun. They should fertilize regularly. Plants are great when used for mixed containers, retaining walls, and tiered gardens.

Myosotis

Myosotis sylvatica
Common name: Forget-me-not
Pot plant, biennial

Forget-me-nots. There is probably no one other plant that has such a distinctive common name—and none whose name has quite the same emotional impact: never forget me. *Myosotis sylvatica* is a low-growing plant that reaches about 10–12" (25–30 cm) tall and is dotted with 0.125–0.25" (3–6 mm) flowers in light to mid-blue. Additional cultivars are available in rose or white. It is more common, however, to find a blue-flowering selection than any other color at a seed house or garden center. The flower is five-lobed with a yellow (or sometimes white), star-shaped eye in the bloom's center. The foliage is usually coarse and somewhat hairy. Plants are often short lived in the garden.

Myosotis sylvatica has traditionally been considered a biennial plant. That is, seed sown in January and after will produce green foliage plants throughout the summer and autumn and then flower the following year after a vernalization period (cold treatment). However, this is truer in Europe than in North America. Typically, *Myosotis* is treated as an annual in North America, where it excels under cool climates but will die in locations with hot summers and cold or freezing winters. It can live for several seasons in areas along the Pacific West Coast, but it suffers inland and frequently won't live beyond one season.

Plants prefer shaded, moist locations: This is probably the primary factor in their short life in the home garden. Native to Europe, *Myosotis* has naturalized itself in the eastern North American woodlands, where the flowers grow larger (up to a 0.5" [l cm]) and plants can be as tall as 16" (41 cm).

Propagation

Sow one to three seeds to a 288-plug tray and leave exposed or lightly covered with coarse vermiculite. Germinate at a substrate temperature of 68–74°F (20–23°C) and maintain a 95–97% relative humidity until cotyledons emerge. The root radicle emerges in three to six days.

In Stage 2, drop the temperatures to 65–75°F (18–24°C) days and 60–65°F (15–18°C) nights. Light levels should be 2,500 f.c. (27 klux) or less, and keep the media moisture level at medium or medium wet during both Stages 2 and 3. Fertilize at 50–75 ppm nitrogen with a nitrate-form fertilizer and low phosphorous. Stage 2 takes ten to fourteen days

During Stage 3, maintain the same temperatures as noted in Stage 2. Increase fertilizer to 100–175 ppm nitrogen. Stage 3 takes two to three weeks.

In Stage 4, provide 60–70°F (15–21°C) days and 55–60°F (13–15°C) nights and light levels at 5,000 f.c. (55 klux) or less, as long as moisture can be maintained.

A 288-plug tray takes from five to six weeks to be ready to transplant from sowing.

Growing On

Transplant one 288 plug per 4 or 5" (10 or 15 cm) pot. Maintain a soil pH of 5.6–5.8, and grow on at 60–70°F (15–21°C) days and 50–55°F (10–13°C) nights. Once established in the final container, plants can be grown at 40–45°F (4–7°C) nights with days 5°F (3°C) warmer.

Sowings should be made from late summer to autumn for overwintering plants cold to vernalize for flowering. Plants sown as late as October and grown cold in their final container can be sold in flower or as green plants in the spring. They will flower in several weeks.

Starting a week after transplant, apply fertilizer at a rate of 175–225 ppm nitrogen (with an EC of 1.2–1.5 mS/cm), using a nitrate-form fertilizer with low phosphorus. If needed, alternate with an ammonium- and nitrate-based fertilizer to encourage growth as well as a balanced media pH. Maintain the media EC at 1.5–2.0 mS/cm and the pH at 5.6–5.8. Avoid high media pH, as this will cause interveinal chlorosis of the developing foliage caused by iron deficiency.

For the varieties that require vernalization in order to flower, sowings made in October or November will flower in April or May. Most varieties do require this vernalization period; however, there are exceptions. One of these is 'Mon Ami' (or 'My Friend'). This annual flowering selection does not require a prolonged cold period to flower. Sowings made in January or February will flower in a 4" (10 cm) pot thirteen to fifteen weeks after sowing

Pests & Diseases

Aphids are one of the biggest issues with this crop.

N

Narcissus

Narcissus hybrids
Common name: Daffodil
Cut flower, flowering pot plant

Temperature-controlled rooting rooms are essential to properly program daffodils forced as fresh cut flowers and for growing and flowering potted plants. Generally, the flowering season is from mid-December to April. Narcissus bulbs are now sold by circumference, with 15/17 being a common size for cultivars such as 'Ice Follies' and 'Carlton' and 12/14 being the common size for 'Tete-a-Tete'.

The keys to success are: (1) select proper cultivars for marketable products; (2) use temperature-controlled reefers and trucks to transport the bulbs; (3) upon arrival, always inspect bulbs for the presence of *Fusarium*; (4) use a clean, well-drained, planting medium with a pH of 6.0–7.0; (5) use proper temperature sequences and number of cold weeks in the rooting room; and (6) use optimal greenhouse temperatures for specific market products.

Daffodils are propagated by offsets or twin scaling, which is done by professional propagators or specialists. Bulbs are commonly received from August to November.

Growing On

Always inspect the bulbs upon arrival and immediately before planting and discard diseased bulbs. *Fusarium* is the major disease of daffodils and appears as a basal soft rot. If present, it normally comes with the bulbs, which are infected before they are harvested. Otherwise, daffodils have few other disease or insect problems that affect forcing.

Daffodils begin forming the flowers in the field and the immature flower bud is usually fully developed when the bulbs are harvested. Thus, bulbs can be immediately precooled at 48°F (9°C) or stored at 63°F (17°C) for later planting.

Always use a well-drained, disease-free, planting medium with a pH of 6.0–7.0. Daffodils should be planted bulb-to-bulb in any container. In the rooting room, they must be well rooted at 48°F (9°C) before the temperature is lowered to 41°F (5°C) and then lowered further to 33–35°F (0.5–2°C). The lower temperatures are used to keep the shoots from becoming too long—they should not exceed 4" (10 cm) in the rooting room. This prevents shoots and leaves from growing up into crates stacked above, and this is also important, especially for potted plants, because plants of a marketable height and quality are produced in the greenhouse. Keep the planting medium moist in the rooting room.

Normally, cultivars for flowering potted plants require fifteen continuous cold weeks and those used for fresh cut flowers require seventeen to eighteen continuous cold weeks.

Pot and cut daffodils should be forced at 60–63°F (16–17°C) day and 50–55°F (10–13°C) night temperatures. By growing at these temperatures, the bulb will produce either short potted plants or long cut flowers. Force cut daffodils at low light intensities and pot daffodils at high light intensities. Daffodils do not need to be fertilized in the greenhouse, but the planting medium must be kept moist.

Height control in daffodils has long been achieved through Florel sprays. Depending on the cultivar and time of season, one or two sprays at 500–2,000 ppm are recommended. At the time of spraying, the plants should be dry, the shoots should be about 4" (10 cm) tall, and the foliage should not be wetted for a day after being treated. Unfortunately, Florel sprays are not always effective on daffodils and the reasons for this are not well understood. In addition, the main need for daffodil height control is actually *postharvest* . . . during and after retail. Research from Cornell University is leading to new concepts in daffodil height control, especially with preplant Topflor bulb soaks. While cultivars vary, excellent height control can be obtained from Topflor soaks at 25 ppm for five to ten minutes. Some tested cultivars, such as 'Ice Follies' and 'Pink Charm', benefit from longer soaks (e.g., thirty minutes). A key benefit from these dips is superior control of postharvest leaf and stem stretching. Additional information can be found on the Flower Bulb Research Program at Cornell University's website at www.flowerbulbs.cornell.edu.

Postharvest

Fresh cut daffodils should be cut and marketed in the "gooseneck" stage of flower development. To store the flowers, place them dry and upright at 32–35°F (0–2°C). Flowering pot plants should be marketed in the "pencil" stage. For maximum post-greenhouse life, they must be stored at 32–35°F (0–2°C) until purchased by the consumer. Advise consumers to place plants and flowers in the coolest part of the home in order to maximize their post-harvest life. After they flower, forced potted daffodils can be placed in home gardens in USDA Zones 3–8 in the spring.

The proper stage at which to cut narcissus

Narcissus tazetta

Common name: Paperwhites

Cut flower, flowering pot plant

Most paperwhite narcissus bulbs used as fresh cut flowers and flowering pot plants in the United States and Canada are produced in Israel. Normally, the marketing season extends from mid-November until April. Bulb sizes available are (in cm circumference) 13/14, 14/15, 15/16, and 16/17 and 17/+, with the most consistent results being obtained with 15/16 and 17/+ bulbs. All cultivars have very fragrant flowers.

Paperwhites are very easy to force. Generally, they have no serious disease or insect problems and only require temperature-controlled storage rooms.

When paperwhite narcissus bulbs are harvested, the apical meristem is vegetative. The flowers begin forming under high temperatures. Thus, paperwhites should be transported from Israel at 77–86°F (25–30°C) under highly ventilated conditions. Upon arrival, they should be stored under well-ventilated conditions at 77–86°F (25–30°C) until either shoots or roots begin to grow out of the bulb. Then they should be stored at 35°F (2°C) to retard development.

In general, it is advisable not to place paperwhites at 35°F (2°C) before November 1. Prior to shipping to forcers or planting, bulbs must be placed at 63°F (17°C) for two to three weeks. This is known as an "end treatment," which stimulates uniform root and shoot growth. This end treatment is required regardless of whether the bulbs were previously stored at 77–86°F (25–30°C) or 35°F (2°C).

Growing On

Plant bulbs immediately after the end treatment. Prior to planting, check bulbs carefully. Discard any damaged or diseased bulbs; occasionally a bulb will have *Fusarium*. Use a well-drained, planting medium with a pH of 6.0–7.0. If bulbs must be stored after the end treatment, place them at 48°F (9°C) under well-ventilated conditions, but only for a few days. Take care not to damage the shoot if it has emerged from the nose of the bulb.

Cut Flower Crops

Growing On

'Inbal', 'Sheleg', and 'Ziva' are the major cut flower cultivars. Plant thirty-five to fifty bulbs in a minimum of 4" (10 cm) deep flats or trays. The flats or trays do not need to be spaced out on the forcing bench.

Use a low light intensity (2,500 f.c. [27 klux]) in a well-ventilated greenhouse. No fertilization is required in the greenhouse. Keep the growing medium moist. Maintain 60–63°F (16–17°C) night temperatures. At these temperatures, forcing normally takes two to five weeks. Lower temperatures can be used, but the plants will take longer to reach the market stage.

Pests & Diseases

Generally, diseases and insects are not a problem during forcing. However, always scout for botrytis and aphids.

Postharvest

Cut the flowers when the first floret is fully colored. If they must be stored, place them upright and store dry at 32–35°F (0–2°C). Advise wholesalers and retailers to market paperwhites when the first flower is fully colored. Stems readily take up floral dyes, and excellent colors can be obtained from the normally white flowers. Consumers should be advised to place cut flowers in the coolest part of the home in order to maximize their vase life.

Pot Plant Crops

Growing On

'Galilee', 'Ariel', and 'Ziva' are the major pot plant cultivars. 'Ariel' and 'Ziva' are both early cultivars. Use a well-drained, planting medium with a pH of 6.0–7.0. Plant 3 bulbs/5" (13 cm) standard-depth pot; 4–5 bulbs/6" (15 cm) standard-depth pot; or 7–9 bulbs/8" (20 cm) bulb pan. Grow plants pot-to-pot on the greenhouse bench. Keep the growing medium moist.

Use a low light intensity (2,500 f.c. [27 klux]) in a well-ventilated greenhouse. No fertilization is required in the greenhouse. Maintain 60–63°F (16–17°C) night temperatures. Lower temperatures can be used, but the plants will take longer to reach the market stage. Forcing takes two to five weeks.

The proper stage to market pot paperwhites

To reduce excessive elongation of the flower stalk and leaves of 'Ziva', plants can be sprayed to runoff with ethephon at 2,000 ppm when the shoots are 4" (10 cm) long. It is important that the foliage be dry when the plants are sprayed. Also, do not wet the foliage for twelve hours after treatment. Late afternoon is the best time to spray. This treatment can delay flowering by two to four days.

Consumers purchasing dry bulbs can be advised to try the so-called "Cornell Gin Method" for paperwhite height control! Basically this entails irrigating paperwhites growing in pebbles with a sevenfold dilution of gin (or vodka, whiskey, or tequila) to water. The dilute alcohol can significantly reduce stem growth, resulting in plants that do not flop over. More information is available at www.flowerbulbs.cornell.edu.

Pests & Diseases

Generally, diseases and insects are not a problem during forcing. However, always scout for botrytis and aphids.

Postharvest

Potted paperwhites should be marketed when shoots are 8–10" (20–25 cm) tall and flowers are visible. Do not wait until they begin to show color. If plants must be stored, place them at 41°F (5°C) with moist planting medium. Advise wholesalers and retailers to market paperwhites when they are in the bud stage of development. Consumers should keep the plants in the coolest area of the home in order to obtain maximum satisfaction. In USDA Zones 8–11, consumers can plant paperwhites in the garden after they have finished flowering.

Nemesia

Editor's note: Nemesia was a crop that was featured in the *Ball RedBook* in earlier editions but was removed in recent editions due to a lack of interest. However, with the rise of the vegetatively propagated selections, it has returned to this edition. In those earlier editions it was strictly seed propagated; interest in seed propagation has also returned, probably due to the increase of the vegetative selections.

Nemesia foetans
Common name: Nemesia
Annual

While we may only think of nemesia as a vegetative crop, there are seed-propagated selections available as well. The following culture is based on the seed-propagated variety 'Poetry', although any nemesia from seed will finish in a similar way. Because it is from seed, it offers efficiencies in production when used with today's automation.

Propagation

'Poetry' is available in pelleted form; it is recommended to multisow using four pellets per plug cell for better-quality liner production. 'Poetry' can be produced in a 288-cell or similar-sized plug trays.

Use a well-drained, disease-free, soilless, plug media with a pH of 5.5–6.2 and an EC of 0.75 mS/cm (1:2 extraction). Germination takes approximately three to five days at a germination temperature of 68–70°F (20–21°C). Light is not required for germination. Keep the media moisture at medium wet (level 3 to 4) during Stage 1. Maintain 95–97% relative humidity until cotyledons emerge.

After germination, maintain soil temperatures 68–72°F (20–22°C) during the day and 62–65°F (17–18°C) at night. Light levels should be on the low side (up to 2,500 f.c. [27 klux]) until the plant is more mature.

Keep the media moisture at medium (level 3) to medium wet (level 4). As the plug matures, lower the moisture levels to medium dry (level 2) to medium (level 3) during Stages 3 and 4. This will help develop a stronger root system.

In Stage 2, apply fertilizer at rate 1 (less than 100 ppm nitrogen and less than 0.7 mS/cm EC) with a nitrate-form fertilizer with low phosphorus. During Stages 3 and 4, increase the fertilizer to rate 2 (nitrogen at 100–175 ppm and an EC of 0.7–1.2 mS/cm). Maintain a media pH of 5.8–6.2 and an EC at 0.7–1.0 mS/cm (1:2 extraction).

Generally plant growth regulators (PGRs) are not required for plug production. However, if necessary, you can apply a foliar spray of B-Nine/Alar (daminozide) at 2,500 ppm (3.0 g/l of 85% formulation or 4.0 g/l of 64% formulation) once at about three weeks after sowing to tone the plugs. Do not use PGRs before radicle emergence as this can delay or stop germination.

In finishing the plug, it is best to lower temperatures to 65–68°F (18–20°C) days and 60°F (16°C) nights. Light levels can be up to 5,000 f.c. (54 klux) if temperatures can be maintained. Maintain the same soil moisture and fertility levels.

It takes approximately four weeks to finish a 288-cell plug.

Growing On

Most nemesia are grown for accent plant programs and uses. Plugs are commonly transplanted into 306-packs as well as 4–4.5" (10–11 cm), 6" (15 cm), and 1-gal. (4 l) pots.

Transplant into a well-drained, disease-free, soilless media with a pH of 5.8–6.2 and a medium initial nutrient charge. Plants like it on the cool side with day temperatures of 62–70°F (17–21°C) and cooler nights at 55–60°F (13–16°C). Keep light lev-

Table 27. Nemesia Crop Scheduling from Transplant to Finish			
Container size	Plugs per pack or pot	Weeks from transplant	Total weeks from sow
306-pack	1	5–7	9–11
4–4.5" (10–11 cm) pot	1	6–8	10–12
6" (15 cm) pot	3	6–8	10–12
1-gal. (4 l) pot	3	6–8	10–12

els as high as possible, while maintaining the optimal temperatures.

Starting a week after transplant, apply fertilizer at rate 3 (nitrogen at 175–225 ppm and an EC of 1.2–1.5 mS/cm) using predominately nitrate-form fertilizer with low phosphorus. Maintain the media EC at 1.5–2.0 mS/cm and the pH at 5.8–6.2. Nemesia does not like extreme fluctuations in the soil moisture. Maintain optimal media moisture (not too wet nor too dry).

PGRs are generally not required, especially when grown under cool temperatures, as temperature can be the best natural growth-controlling factor. But when producing the crop under warmer temperatures and, if needed, you can apply a foliar spray of B-Nine/Alar (daminozide) at 5,000 ppm (5.9 g/l of 85% formulation or 7.8 g/l of 64% formulation) after transplant. The first application can be done at seven to ten days after transplant with a second application a week later.

Pinching is not required, as the multisown plugs will finish naturally well.

Pests & Diseases

Check and monitor for thrips on nemesia as they can spread impatiens necrotic spot virus (INSV).

Nemesia fruticans
Common name: Nemesia
Annual

Commonly propagated by cuttings, *Nemesia fruticans* is a great plant to include in your cool-season assortment. Great colors, form, and pleasant fragrance make this a unique plant to add to your offerings, and low-cost production makes this a profitable crop. Use it as a stand-alone crop or mix it in planters. Be sure to devote effort into planning space for it, because it needs cool, dry conditions to flourish.

Propagation

If cuttings are wilted, place in a clean cooler at 45–50°F (7–10°C). Provide a relative humidity of nearly 100% by wetting the floor and misting open bags with clean water. Stick cuttings in Ellepots or a well-drained medium with a pH of 5.8–6.2. Some suppliers ship two-for-one cuttings. If this is the case, then stick two cuttings in one rooting cell. Be sure to provide good soil-to-stem contact by lightly pinching the soil while sticking or running through a high-volume water tunnel. A rooting hormone might help increase rooting uniformity. Root cuttings under mist at a substrate temperature of 70–72°F (21–22°C) and air temperatures of 70–72°F (21–22°C) days and 68–70°F (20–22°C) nights. Because cuttings are small, it is extremely important to only provide enough mist to keep cuttings turgid. Too much mist will inhibit rooting and increase disease. Provide 500–1,000 f.c. (5.4–11 klux) of light.

After about seven days, the cuttings should be callused and begin to develop roots. Quickly move to Stage 2. Significantly reduce or cease the use of misting and begin to apply an overhead feed of 50–75 ppm nitrogen from a balanced fertilizer.

In Stage 3, maintain soil and air temperatures. As roots form, begin to let the soil dry down between irrigations. Increase light levels to 1,000–2,000 f.c. (11–22 klux). Apply a foliar feed of nitrogen at 100 ppm from a balanced fertilizer. Increase the rate to 150–200 ppm as roots form. Stage 4 consists of hardening the cuttings for approximately seven days prior to sale or transplant. At the beginning of Stage 4, you can pinch or trim the liners to help create lower breaks. Reduce air temperatures at night to 65–68°F (18–20°C). Liners can be held, if needed, at 54–58°F (12–14°C). Move liners from the mist area and increase light levels to 2,000–4,000 f.c. (22–43 klux). Shade plants during midday to reduce temperature stress. Fertilize with nitrogen at 200 ppm from a balanced fertilizer.

Cuttings stuck to a 105 tray will be ready to transplant to larger containers in four to five weeks.

Growing On

Transplant 1 liner/4" (10 cm) pot; use 1–2 liners/6" (15 cm) pot; use 4–6 liners in 10" (20 cm) or larger containers. Select a disease-free, very well-drained medium with a pH of 5.6–6.0. Grow plants on at 65–75°F (18–24°C) days and 55–60°F (13–15°C) nights until roots have developed. You can then choose to lower temperatures further, which will help plant form and growth control. The highest-quality plants will be grown as low as 50°F (10°C), but this will add to the production time. Fertilize with a 200 ppm nitrogen constant feed with a complete fertilizer. Too much ammonium-containing fertilizer can cause unwanted stretch and limited branching. Supplemental iron may be provided with an application of chelated iron. Monitor and maintain a pH between 5.6 and 6.0.

Nemesia should be grown on the dry side, particularly very early in production. Plan the space and companion plants beforehand and make sure it happens. It is critical that plants are not overwatered as this can cause problems that would lead to significant loss. *Remember:* With lower temperatures the plants will use less water. Grow under high light levels, 4,000–7,000 f.c. (43–75 klux). Pinching or trimming once plants have reached the edge of the container can be done to shape larger containers. Depending on the time of season, do not pinch within three to six weeks of the sale date.

Cooler production temperatures, moderate regular water stress, and high light will probably keep nemesia from needing any growth regulators. However, when any of these are not controlled as well as for vigorous varieties, spray applications of B-Nine (2,500 ppm), Sumagic (3–5 ppm), or Florel (350 ppm) can provide effective control. Florel should not be applied four to six weeks before sale.

Liners transplanted to a 4" (10 cm) pot will be salable in five to six weeks.

Pests & Diseases

Aphids, thrips, whiteflies, leafminers, and fungus gnats can all attack nemesia. Scout regularly. Botrytis, rhizoctonia, and *Pythium* can all become problems when the soil is overly saturated and plants are wet overnight.

Troubleshooting

Plants may collapse due to moderately to severely wet medium or excessive botrytis. High ammonium levels in the medium, overfertilization under low light conditions, or high production temperatures can cause excessive vegetative growth. Low fertilization and low light conditions can lead to poor branching. Late transplanting, late spacing, or excessive phosphorous can result in stretched plants.

Varieties

'Aromatica', 'Confection', and 'Innocence' are a few series on the market. Check with your broker or breeder representative for the best varieties for your production.

Postharvest

Display plants in high light with cool temperatures. Do not allow plants to wilt, but keep them on the dry side. Do not let plants remain wet at night.

Consumers should plant in well-drained soil in full sun and fertilize regularly. Nemesia is great when used for mixed containers and boarder plantings.

Nepeta

Nepeta cataria
Common name: Catnip
Perennial herb (Hardy to USDA Zones 3–9)

The quintessential herb for cats—although humans like it, too! These upright plants range from 2–3' (61–91 cm) with a gray-green appearance to foliage. Flowers are less than 0.5" (1.2 cm) and usually pink or white in color. Plants flower from late June or July to August.

Propagation & Production

Seed

Seeds germinate quickly when sown directly to the final cell pack or an open flat. Lightly cover seed with vermiculite and at a media temperature of 70–75°F (21–24°C). Seedlings emerge in five to seven days. There are many times when catnip simply does not germinate well. In these cases, use of "primed," or treated, seed can be advantageous. Please note that primed seed can *never* be used for organic production.

If sown to an open flat, seedlings can be transplanted in fourteen to twenty days to an 804-cell pack (32-cell flat), and total crop time for green plant sales is ten to twelve weeks after sowing.

If sowing to a 512-plug tray, follow the same guidelines as those for an open flat. Allow three to five weeks for the plug tray to be ready to transplant to larger containers. Transplant to cell packs or 4" (10 cm) pots. Plants will be salable in six to eight weeks for cell packs and in eight to ten weeks for 4" (10 cm) pots grown for spring sales.

Sowing multiple seeds into a 105 tray allows a shorter time from transplant to sales. Allow four to five weeks from sow to transplant. Pinching as soon as root initials are evident will greatly improve the transplant cell. Allow four weeks from transplant to be ready for retail. Grow in high light on the dry side, pinch once, and grow cool for best quality.

Warm temperatures encourage rapid development and stretching. Use night temperatures of 55–60°F (13–16°C) and days from 60–62°F (16–17°C). If the plants start to yellow due to warmer temperatures and dry out due to being root-bound, then pinch the plants back, pot up to larger containers, and sell later in the season.

Use a 15-0-15 or 15-5-15 fertilizer, alternating with 20-10-20 at 150–200 ppm nitrogen.

Cuttings

Cuttings can help eliminate some of the strong variation from seed propagation. Stock plants are better if they are grown under high light, pinched or cut often, and grown cooler. Be aware of insect and disease issues at all times. Allow four weeks of rooting in 105s with a media temperature of 72°F (22°C). Lowering the temperature and cutting mist as soon as callus and initials are seen will reduce the tendency to stretch. Pinching as early as possible will greatly improve the crop. Allow four weeks after transplant for a finished 4" (10 cm) pot.

Nephrolepis

Nephrolepis exaltata
Common names: Boston fern; sword fern
Foliage plant

This old-fashioned plant was once a fixture in parlors. Today you are more likely to find Boston ferns in 4 and 6" (10 and 15 cm) pots used as a houseplant or in hanging baskets gracing decks, porches, and backyards of homes across America.

Growing Boston ferns is time consuming but relatively easy. While some growers prefer to bring in finished baskets from Florida or other local growers, many more find that Boston ferns are a great crop to grow. They are one of the most popular hanging baskets in the spring months—buying Boston ferns each year is a ritual for many homeowners. Since you can sell pots and baskets into the summer, having pots going into June is a relatively low risk. If you choose to grow your own fern baskets, you will have to weigh the

Table 28. Finishing Times for Fern Liners*

STARTING SIZE	FINISH SIZE			
	4" (10 cm)	6" (15 cm)	8" (20 cm)	10" (25 cm)
	Weeks to finish (# of plants/pot)			
Boston fern/'Porter's Roosevelt'				
72-tray	12–14 (1 liner)	14–16 (1 liner)	14–16 (3 liners)	16–20 (3 liners)
4" (10 cm) prefinished	—	4–6 (1 pot)	8–10 (1 pot)	10–12 (1 pot)
'Boston Compacta'				
72-tray	12–16 (1 liner)	16–18 (1 liner)	16–18 (3 liners)	18–20 (3 liners)
'Dallas'				
72-tray	8–12 (1 liner)	12–16 (1 liner)	16–20 (3 liners)	18–22 (3 liners)
3" (8 cm) prefinished	4 (1 pot)	10 (1 pot)	14–18 (1 pot)	16–20 (1 pot)
4" (10 cm) prefinished	2–3 (1 pot)	4–6 (1 pot)	8–12 (1 pot)	10–14 (1 pot)

* From Ball Seed Co., West Chicago, Illinois

pros and cons of running a warm greenhouse to keep plants actively growing in the winter months. Most growers find that the pricing works out in their favor, and Boston ferns are one of the few foliage plants that growers nationwide routinely produce.

Plants are propagated by tissue culture, which is done by specialists, and various liner sizes are sold; 72s are commonly available.

Growing On

Most growers receive fern liners propagated using tissue culture. Typically plants are sold in 72 trays. In the North, growers receive plants in late spring for fall and winter production in 4 and 6" (10 and 15 cm) pots. Use 1 liner/4" (10 cm) pot and 2 liners/6" (15 cm) pot. Growers wishing to grow plants for spring sales in 10" (25 cm) baskets bring in the liners the summer or fall before, pot them into 4" (10 cm) pots, and repot plants into hanging baskets ten to twelve weeks later. Liners can be planted directly into baskets, in which case irrigation for the first few weeks becomes critical (see below). Some growers choose to buy in 4" (10 cm) prefinished pots for 10" (25 cm) basket production. These can be received from January to March, potted one to a basket, and sold in April and May.

Pot ferns into a well-drained, disease-free, peatlite medium with a pH of 5.0–5.5. Do not use a soil-based medium. Avoid fertilizer for the first three weeks after potting to allow plants to root in. Just after potting, be careful with watering: Since plants have not yet rooted into their new larger containers, it is easy to overwater them. A lot of growers like to remove the basket saucer when they bump pots into baskets, reattaching it at sale. Allow Boston ferns to dry slightly between irrigations.

Once plants are rooted in, fertilize with 100–200 ppm from 20-10-20 at every other irrigation. Reduce fertilizer frequency during the winter months when plants are not growing as actively. Water with clear water about once a week to avoid soluble salts problems. Maintain a lower substrate pH, because high pH will adversely affect growth.

Ferns love humidity; some types require as much as 90% humidity. Ferns also need moisture, but they do not like to stand in water. Runner growth is a good indicator of how well you are doing: Few runners mean too much water.

Ferns can tolerate a wide temperature range, but the fastest growth and best quality will be produced at a constant day/night temperature of 72°F (22°C). For highs and lows, keep day temperatures above 60°F (16°C) and below 80°F (27°C). Grow Boston ferns at light levels of 1,000–3,500 f.c. (11–38 klux). Use higher light in the spring and fall, when it is easier to maintain moderate production temperatures, and lower light levels in the summer. Providing 50% shade in the summer months will help moderate production temperatures. To encourage fronds to lengthen during the short days of winter, provide night interruption lighting from 10 P.M.–2 A.M.

For 1-gal. (4 l) pot sales in April, bring in 72s from June to October. Use a well-drained medium, yet one that also retains consistent moisture. Many growers apply a broad-spectrum fungicide at planting. Do not use slow-release fertilizers; instead use 100–150 ppm from 20-10-20 and discontinue feeding in September. Irrigate with clear water only from September to March. Begin fertilizing again when plants start to grow the following spring. Overwinter pots above 25°F (–3°C). Bring plants out of dormancy the following spring as new growth emerges by maintaining temperatures above 35°F (2°C).

You can also bring in 72s in October for a 4" (10 cm) crop. Root plugs into pots for four weeks and then overwinter these at 45°F (7°C). You can force them to finish in six to ten weeks by raising temperatures to 65–70°F (18–21°C).

Pests & Diseases

Aphids, mealybugs, scale, and whiteflies can attack ferns. Since ferns are sensitive to many insecticides, make test applications first before treating the entire crop. Foliar nematodes can also be a problem on some species. *Pythium* and rhizoctonia may attack plants that remain too wet during production.

Troubleshooting

Brown fronds indicate overwatering or an environment that is too dry. Water stress can lead to graying foliage. Too much light can result in lightly colored fronds.

Varieties

Folklore has it that the Boston fern is a cultivar of the sword fern, *Nephrolepis exaltata.* The story goes that in 1894 a Philadelphia grower shipped 50,000 sword ferns to a Boston distributor. The plants were different from traditional sword ferns and thus became known as *Nephrolepis exaltata* 'Bostoniensis'. Today, there are a wide range of Boston ferns grown commercially.

In addition to being able to purchase regular Boston ferns, a number of named selections are available. 'Dallas' is a great, well-known variety and with its low-growing habit, is excellent for 4 and 6" (10 and 15 cm) pot production. 'Boston Compacta' makes a tidy, symmetrical 10" (25 cm) basket. It takes about two weeks longer to grow than a regular Boston, but its uniformity is worth the wait. 'Porter's Roosevelt' is an extra-frilly Boston fern that is great in baskets. Grow it with slightly less fertilizer than typical and at a lower pH. 'Kimberly Queen' makes a fabulous 1- or 2-gal. (4 or 8 l) pot. Its habit is upright, bold, and stately. Add a couple of weeks to your production schedule for 'Kimberly Queen'.

Related Species

Nephrolepis biserrata, the fishtail fern, is a striking 10" (25 cm) basket. Plants tolerate higher light levels (above 3,000 f.c. [32 klux]). Grow as you would a typical Boston fern, but at slightly lower fertilizer rates.

Asplenium nidus, Japanese bird's nest fern, thrives in 4, 6, or 8" (10, 15, or 20 cm) pots. Allow plants to dry thoroughly between irrigations. Their root systems are fairly weak, so use shallow pots. Grow under 1,200–1,700 f.c. (13–18 klux) of light.

Davallia trichomanoides, rabbit's foot fern, is a true novelty. Its hairy rhizomes give this fern its name. Be sure to allow pots to dry thoroughly between irrigations. Grow them in 4, 6, or 8" (10, 15, or 20 cm) baskets. Grow under 1,100–2,500 f.c. (12–27 klux). Rabbit's foot fern prefers warmer temperatures.

Pellaea rotundifolia, the button fern, is great in 4 or 6" (10 or 15 cm) pots or 8" (20 cm) baskets. Do not let plants get totally dry between irrigations. Plants that are significantly water stressed will not recover. Fertilize at 100 ppm from 24-8-16 and keep the pH at 6.0.

Platycerium bifurcatum, staghorn fern, makes a great specimen fern and can be mounted on all sorts of decorative wood, wire, and poles or even grown and sold potted. Consumers love its oversized fronds for the summer patio or in a well-lit interior location. Do not overwater staghorn ferns. Feed at 100–200 ppm from 24-8-16 and grow under 2,000–3,000 f.c. (22–32 klux) of light. Staghorns will tolerate temperature dips to 34°F (1°C).

Among the most popular hardy ferns that you can offer are: *Athyrium filix-femina* (Miss Vernon's crested lady fern) 'Vernoniae Cristatum'; *Athyrium nipponicum* (Japanese painted fern) 'Pictum'; *Cyrtomium falcatum* (Japanese holly fern) 'Rochfordianum'; *Dryopteris erythrosora* (autumn fern); *Dryopteris* × *complexa* (robust male fern) 'Robust'; *Osmunda cinnamonea* (cinnamon fern); *Polystichum polyblepharum* (tassel fern); and *Thelypteris kunthii* (southern wood fern).

Postharvest

If your customers will be using your crop as houseplants, acclimatize pots for one week under 750 f.c. (8.1 klux) of light prior to sale. Water plants prior to packing. Ferns may be sleeved, boxed, and shipped at 55–60°F (13–15°C). Do not transport for more than two weeks; when shipping for such a long time, be sure to maintain humidity levels. Boston ferns are somewhat sensitive to ethylene, so avoid exposure by keeping plants away from ripening fruits and vegetables.

Retailers should display plants under 200 f.c. (2.2 klux) of light to maintain the best appearance. Garden centers displaying plants outdoors should place baskets in shade houses and keep temperatures at 60–75°F (16–24°C) during the day and above 50°F (10°C) at night. Be sure to maintain uniformly moist conditions at retail.

Nicotiana

Nicotiana alata
Common name: Flowering tobacco
Annual

Underused is the best adjective to describe nicotiana. Year after year they look so good in the California Pack Trials at the breeder and in field trials from coast-to-coast, yet the amount of nicotiana actually produced by growers is miniscule. If the landscape trade is important to you, it's likely you have already discovered nicotiana. If you are a retail grower and you are looking for a practically bulletproof bedding plant for your customers, nicotiana just may be your ticket.

Nicotiana culture is easy, and plants tolerate a wide range of conditions. The biggest trick in production is to keep plants actively growing. For that reason, nicotiana is best grown in pots or sold green in packs.

Propagation

Sow pelleted seed onto a disease-free, germination medium with a pH of 5.5–5.8. Do not cover seed. Provide 100–400 f.c. (1.1–4.3 klux) of light and maintain moderate moisture conditions: The medium should be uniformly wet, but not saturated. Seed will germinate at 75–78°F (24–26°C) in three to five days. Ideally use larger plug trays so that roots do not become restricted.

Once 50% of the seed has cracked, move plugs to Stage 2 for stem and cotyledon emergence. Reduce soil temperatures to 68–75°F (20–24°C) and increase light levels to 500–1,200 f.c. (5.4–13 klux). Once cotyledons have expanded, fertilize weekly with 50–75 ppm from 14-0-14. Once radicles have penetrated the soil, begin to allow the medium to dry out somewhat between irrigations. Stage 2 lasts for seven to ten days.

After you have a full stand, move trays to Stage 3 to develop true leaves. Reduce temperatures to 65–70°F (18–21°C) and increase light levels to 1,200–1,500 f.c. (13–16 klux). Allow trays to dry thoroughly between irrigations. Apply a weekly feed at 100–150 ppm, alternating between 20-10-20 and 14-0-14. If growth regulators are needed, B-Nine is effective. Stage 3 takes from twenty-one to thirty-five days for 288s.

Harden plugs for seven days prior to sale or transplant by reducing temperatures to 60–62°F (16–17°C) and increasing light to 1,500–3,500 f.c. (16–38 klux). Feed once a week with 100–150 ppm from 14-0-14.

Some companies suggest that providing fourteen hours of supplemental light (50 watts/m^2 or 679 f.c. [7.2 klux]) is beneficial for plug production.

A 288-plug tray takes from five to six weeks to be ready for transplant to larger containers.

Growing On

Transplant plugs on time, as nicotiana does not like to have its roots restricted. Root-bound plants exhibit stunted growth. Use 1 plug/cell or 4" (10 cm) pot and up to 3 plugs/6" (15 cm) or 1-gal. (4 l) pot. Select a well-drained, disease-free, soilless medium with a moderate nutrient charge and a pH of 6.0–6.5.

Grow on at 65–70°F (18–21°C) days and 60–65°F (16–18°C) nights. At cool temperatures, white spots may appear on young foliage. High humidity and fluctuating temperatures can cause leaf mottling. Provide full-sun conditions, if moderate production temperatures can be maintained. Feed at every other irrigation with 150–200 ppm, alternating between 15-0-15 and 20-10-20. If growth regulators are needed, B-Nine is effective. Do not overwater nicotiana. For best height control, grow plants on the dry side.

Flowering is more uniform under long-day conditions (more than ten hours). If you are growing crops that will be sold from late September to February, use night interruption lighting throughout production to stimulate flowering. However, some of the newer varieties do not require lighting.

Transplanting a 288 plug to a 4" (10 cm) pot requires five to seven weeks to finish for dwarf bedding varieties. Taller selections require from three to five weeks longer.

Varieties

The 'Saratoga' series comes in a number of separate colors and a mix. Plants grow 10–12" (25–30 cm) in the garden, making 'Saratoga' one of the most compact series on the market. It is suitable as a pack or pot item. For 4" (10 cm) pots, try the 'Whisper' series in three colors. Plants grow from 32–40" (81–102 cm) tall. For something truly unusual, try the scented 'Perfume' series. Available in separate colors, plants grow 18–20" (46–51 cm) in the garden.

Postharvest

Sell nicotiana plants on time. Do not allow plants to become root-bound in their containers, as their growth may become so stunted that they will not perform for the end user.

O

Ocimum

Ocimum basilicum
Common names: Basil; sweet basil
Annual herb

For any cook, summer is not summer without the sweetly aromatic leaves of basil permeating the air. Far too often, growers and retailers offer basil plants at the same rate as other herbs, selling out of basil pots well before even a fraction of other herbs have been sold. Basil seed is cheap, germination is a snap, growing on is easy with enough heat, and it is popular. There is really no reason why you should not offer a fresh crop of direct-sown pots from biweekly sowings during the spring season and into midsummer.

Several growers around the country earn a good living growing and supplying fresh basil stems to local restaurants and wholesale produce distributors year-round. Weekly sales of basil form the basis for these businesses that can then also offer a full range of other fresh culinary herbs freshly cut or sown weekly throughout the year.

Propagation & Production

Seed is the most common method of propagation, although cuttings can be done as well—especially on variegated-leaf varieties (such as 'Pesto Perpetuo').

Seed

For a 288-plug tray, use one seed per cell. The seed can be left exposed or lightly covered with coarse vermiculite for germination. Seed germinates in four to seven days at a media temperature of 70–75°F (21–23°C). Plug trays are ready to transplant to 4" (10 cm) pots or cell packs four to five weeks after sowing. During northern United States winters, it can take five to seven weeks.

Transplant one plug per 4" (10 cm) pot and allow five to seven weeks to finish, depending on variety. The critical factor is temperature. Warmer temperatures accelerate growth, while cooler ones slow it down. Grow on at 70–75°F (21–24°C) days in the greenhouse and keep night temperatures above 60°F (15°C). Plants will tolerate warmer conditions but medium- to tall-growing varieties will start to stretch. Shorter varieties, especially those that are more round in shape, will become more open in habit.

You can also multiseed basil into a 4" (10 cm) pot using eight to ten seeds per container. Cover lightly with vermiculite or peat. Germination is 80–85%. Begin fertilizing direct-sown pots once seedlings have reached the true leaf stage, about ten to fourteen days after sowing. Use 100–150 ppm 20-10-20, alternating with 14-0-14 or another calcium/potassium-nitrate fertilizer. Fertilize at every second or third irrigation.

Alternatively, you may grow fast-finish liners in a 105 plug, sowing five to six seeds in each. Allow four weeks for plug growing and anywhere from two to four weeks for retail-ready finish, depending on the season and temperatures. For a mid- to late May 4"

(10 cm) pot, a mature, well-grown 105 plug can finish in less than two weeks.

Cuttings

There are currently only a small number of basils that are propagated by cuttings. Unrooted cuttings received during the winter months can be stuck into a 105 or similar-sized tray, using one cutting per cell. Allow four to five weeks to transplant to larger containers. When propagating under mist on a greenhouse bench, maintain early-winter production temperatures of 65–70°F (18–21°C). If kept too cool, with too much mist, the plants may easily rot. There are more losses under the cold, cloudy days of early to midwinter in the North than at any other time of the year. Pinching is critical to make good, strong, attractive pots.

Pests & Diseases

Fusarium causes yellow foliage and discolored stems. Downy mildew has become a very serious problem for commercial basil production and must be scouted and controlled ruthlessly. Bacterial and fungal leaf blights are also common. Avoid overwatering plants to control disease. Aphids, slugs, spider mites, and whiteflies may also attack plants.

Troubleshooting

Basil is a warm-season crop. Under cool or cold temperatures plants stall in growth and the leaf edges turn black. If the plants are also exposed to excess watering at this time, they will rot. This usually occurs when the plants are finished in cold frames and on the retail bench in the garden center in early spring, when night temperatures can still be cool. Temperatures under 58°F (14°C) are detrimental at all times.

Branches that begin to flower can stop making leaves: Keep plants pinched. Leaves harvested before plants flower have the highest concentration of essential oils. Harvest whole stems (as opposed to separate leaves) by cutting stems above a leaf pair. The plant will branch at that point. If plants are allowed to become leggy in the pot, they will stall. Cut overgrown plants back and bump them into larger pots for sale two weeks later. Be aware that basils that have been cut back are far more susceptible to foliar diseases until the new growth flushes, and it is important to make cuts just above nodes—the "sticks" are a major source of disease growth.

Varieties

There is a wealth of basil varieties available today, and many of the seed catalogs carry a broad selection of heights, scents, and flavors. Most customers will be happiest with varieties of *O. basilicum,* sweet basil. In warm climates, plants can grow in excess of 40" (1 m), developing woody stems by the end of the summer. In cooler climates, plants grow to 2' (61 cm) or less.

O. basilicum 'Minimum' is known as bush basil. Plants remain compact and grow to no more than 10–12" (25–30 cm) tall. 'Spicy Globe' and 'Green Bouquet' are dwarf types.

O. basilicum purpurescens are purple-leafed varieties. 'Amethyst', 'Purple Ruffles', and 'Red Rubin' are all varieties of dark-leafed basil. However, leaves frequently revert to green in 5–10% of the plants, although 'Amethyst' is a superior selection. Pinch purple-leafed varieties to encourage full plants.

Scented-leaf basils include the lemon-scented *O. basilicum citriodorum* 'Sweet Dani' and 'Mrs. Burns Lemon'. Pot two or three seedlings per pot and shear once to encourage full plants. 'Sweet Dani' is the first basil to flower, so be sure to keep plants pinched to prevent flowering.

'Sweet Genovese' and 'Nufar' are good choices for regular basils, while 'Magic Michael' and 'Siam Queen' are Thai basils. 'Boxwood' and 'Aristotle' are the best miniatures by far. If you would like to try a novelty basil, 'Cinnamon', 'Lime', 'Holy Red', and 'Green' can provide something different to your selection.

'Pesto Perpetuo' is one of few basil varieties that is propagated vegetatively. It has a light green leaf with a cream-white edge on upright plants that grow to 18–24" (46–61 cm) in the garden (even taller in warmer locations). It is best used as fresh garnish and not for culinary applications.

There are other great vegetative varieties that you can add to your basil offerings. 'African Blue' is a superb garden plant with excellent flower power. This tough-as-nails plant is wonderful for pollinators. 'Columnar' is the origin of 'Pesto'. This superb garden plant has no flowers and is ideal for culinary use. Finally, 'Dwarf Purple' is a gorgeous basil that looks great in beds, combination plantings, and more.

Postharvest

Basil plants will be among the first to die out in the garden with frost. Cut basil should be kept in a glass of water *outside* the refrigerator and is always best when used shortly after cutting.

Oenothera

Oenothera macrocarpa (formerly *Oenothera missouriensis*)
Common names: Ozark sundrop; Missouri evening primrose
Perennial (Hardy to USDA Zones 4–8)

A perennial native to the central United States, *Oenothera macrocarpa* is well known for grand displays of funnel-shaped, bright canary-yellow blossoms. The blooms range from 3–4" (7–10 cm) across and can have a light fragrance. As the plants grow and develop, growing tips, sepals, and even the aging blossoms can be highlighted or brushed with red. Plants grow from 6–12" (15–30 cm) tall—although in the Midwest they may only reach 8" (20 cm) in flower. However, in drier climates the plants can reach heights of 15–20" (38–51 cm) tall.

Plants are often short lived but can reseed—but not to a nuisance. Plants can produce a clump from 1–2' (30–60 cm) across. However, in the Midwest plants usually have a closely held crown and a spread of only 10–12" (25–30 cm) across. These clumps are considerably larger on plants in the wild. Plants flower in June and July.

Propagation

Sow two to three seeds per 288-plug tray and lightly cover with a coarse vermiculite or not at all. Vermiculite helps to maintain moisture around the seed as it germinates. However, some growers do not cover the seed at all at first and then upon the emergence of the root radicle, cover the seed tray with a light layer of vermiculite.

Give the seed tray a substrate heat of 70–72°F (21–22°C) or slightly lower, but no lower than 65°F (18°C). The root radicle will emerge in four to seven days, especially when bench-sown in a greenhouse. When seed is germinated in a germination chamber, where the temperature is uniform and the humidity higher, some growers have achieved a more uniform stand of seedlings when germinated at lower temperatures of 62–65°F (17–18°C).

From sowing to a finished 288-plug tray will take six to seven weeks.

Growing On

Seed

Transplant one 288 plug to a 1-qt. (1.1 l) pot, or transplant to a 32-, 50-, or similar cell pack and grow on for six to eight weeks and then transplant to 1-gal. (4 l) containers. This culture reflects a sowing made in June or July and overwintered dormant in the final container. However, sowings can be done from August to October, and the resulting plants can be kept in a cold greenhouse or cold frame during winter. Both sowings will produce flowering plants.

A 32-cell pack is ready to transplant to larger containers ten to twelve weeks after sowing. Seed can be sown to plug trays as noted above and then transplanted to the cell pack; or two to four seeds can be sown directly to the final cell pack.

Sowings made in January or February will produce plants large enough for 4" (10 cm) containers for late spring sales, but the plants will only flower sporadically that same summer when planted to the garden in June.

Overwatering is a possibility because plants are not equal in their ability to dry out between waterings. If plant crowns are kept constantly moist, water rots tend to become established and can kill a number of plants.

Cuttings

For cuttings, avoid the woody rootstock at the plant's center and concentrate on the trailing roots just below the soil surface. These can be cut, removed, and potted up. Cuttings can also be taken from shoots as they emerge from the crown in spring. These are often called basal cuttings. Remove and root as you would a terminal tip cutting. Some growers wait until after flowering to take cuttings, but some have achieved better performance from basal cuttings.

Origanum

Origanum majorana
Origanum vulgare
Common names: Marjoram; sweet marjoram *(O. majorana)*; Oregano *(O. vulgare)*
Herb (Oregano is a perennial herb from USDA Zones 5–8, while culinary sweet marjoram is best treated as an annual in cold-winter areas.)

Marjoram and oregano are closely related but differ in their aroma. Sweet marjoram is sweeter (hence the common name) and is used in medicinal preparations, aromatherapy, and cooking as a milder form of oregano.

Oregano, on the other hand, has a stronger aroma and is used in Italian cooking in pizza, sauces, and the like. It has other uses as well, but its culinary use takes center stage. While available from seed, most culinary types are propagated vegetatively.

The entire plant of both these *Origanum* species may be used in cooking or medicinal preparations.

Propagation & Growing On

Seed

Oregano from seed is almost *never* true to type and is not considered a culinary cultivar.

Origanum germinates in four to eight days at 70°F (21°C) media temperatures. Cover lightly with vermiculite. Seed can be sown directly to final growing containers or into plug trays. Germination is about 75–80%, and seedlings emerge in four to eight days.

Seed sown to a 512-plug tray and lightly covered with vermiculite is transplanted in three to five weeks into a 4" (10 cm) pot. Plants will finish in seven to nine weeks at temperatures of 65–70°F (18–21°C) days and 55–62°F (13–16°C) nights.

Cuttings

Cutting propagation pertains specifically to *Origanum vulgare.* There are a number of varieties of oregano that are propagated vegetatively. Most have a stronger flavor than any of the seed-sown strains and maintain true-to-type performance.

Cuttings received in the late fall and winter should be dipped in a rooting hormone and stuck in a 128-, 105- or similar-sized plug/liner tray. Place under mist and allow four to five weeks to root at a 72–74°F (22–23°C) media temperature. However, due to the fact that leaves are pubescent (hairy), be careful of heavy mist during dark periods when propagating on the greenhouse bench. The plants may rot.

Plants benefit from a hard pinch in plugs or at transplant, and pinching as needed in pots. Grow cool and in high light. Run dry, just before wilt, and feed lightly with 14-4-14. Allow five weeks to finish a 4" (10 cm) pot.

Varieties

Greek oregano is considered the true culinary type. Italian oregano is less hardy and more subtle in flavor. There is a nice variegated Greek oregano, and there is also a beautiful golden form from Israel. All of these types are worth growing. Grow the Italian form because consumers equate oregano with Italy and will ask for and purchase it. But grow and *recommend* Greek oregano for its great culinary flavor.

Marjoram seed is acceptable, however there is a very nice, vegetative cultivar named 'Marjoram Compacta'. It has good flavor and is much hardier and far more attractive than seed varieties. 'Gold-Tip Marjoram' is a fabulous variety with brilliant chartreuse leaves and pendulous, crinkled texture. It is winter hardy and makes a stunning combination plant.

The Encyclopedia of Herbs and Their Uses (DK, 2001) points out that commercially dried oregano is made from a range of different plants, including *Lippia graveolens* (Mexican oregano), *L. palmeri, O. vulgare hirtum* (Greek oregano), and *O. syriacum.* If your customers are looking for the kind of oregano they are accustomed to seeing on the supermarket shelf, steer them toward Mexican or Greek oregano.

Osteospermum

Osteospermum ecklonis
Common name: African daisy
Annual

Osteospermum hit the US market with a big splash in the 1990s. Its bright, daisylike flowers generated interest in the hungry grower and retailer communities from coast-to-coast. Many Americans who frequently traveled to Europe knew that Danish growers had been growing *Osteospermum* as a flowering pot plant with great success for years.

After a few years in the United States, we discovered that, while *Osteospermum* is a great plant for spring sales, its flower power would disappear in the summer heat. Lack of through-the-summer performance left many growers, retailers, and consumers with a bad taste.

Breeders have worked hard over the last decade to develop series with greater heat tolerance. *Osteospermum* can have a place in your product mix no matter your location when one keeps the following factors in mind. The most common time to sell it is early spring when night temperatures are cool. In mild climates in the South and West, osteospermum can be planted in the landscape in the fall for color over the winter months, because cool nights promote flowering and plants tolerate a light frost. As a grower, one needs to understand that night temperatures affect flower performance. Typically, consistent night temperatures above 60–65°F (16–18°C) will reduce the amount of flowers. The mountainous regions along the Pacific Coast have the most promising climates for this product. Consumers love the plant's innocent flowers, now available in a wide range of colors. Spend some time trialing plants in your local area so you can promote them properly. Many new varieties are showing excellent promise for summer performance.

Propagation

While commonly propagated by cuttings, *Osteospermum* can also be propagated by seed.

Cuttings

Dip cuttings into a rooting hormone and stick into Ellepots or medium at 75°F (24°C) substrate temperatures and 68–74°F (20–23°C) air temperatures.

Apply mist or fog only as needed to keep cuttings fully turgid. Once roots are visible, quickly reduce mist frequency and duration. Cease misting except when needed, and begin overhead fertilizing with 75 ppm nitrogen (N) using a well-balanced fertilizer. Quickly increase to 150–200 ppm N. Gradually increase light levels to 4,000 f.c. (43 klux), when roots are visible to help control growth. Only apply plant growth regulators if needed. The goal should be to keep cuttings as soft as possible to later aid in branching. A fungicide drench at half strength after sticking will help to prevent disease during propagation.

A 105-liner tray is ready to transplant to larger containers in three to five weeks, depending on the variety.

Seed

Osteospermum seeds are relatively large and thus are fairly easy to use. Sow seed into a well-drained, disease-free, germination medium with a pH of 5.8–6.2. Cover seed after sowing with coarse vermiculite, but do not germinate in the dark. Water the trays in well.

Osteospermum does not need light to germinate, but its hypocotyls (stems) stretch very quickly in a dark germination chamber. Use at least 50–100 f.c. (0.5–1.1 klux) of light in the chamber if you think it is likely plants will remain in there for more than four days. Maintain 100% humidity for optimum germination. At 70–75°F (21–24°C) substrate temperatures, seed should be almost fully germinated in four to five days but can be removed from the chamber in three to four days to prevent stretch. Germination will continue after being removed from the chamber.

Move seedlings to Stage 2 for stem and cotyledon emergence. Lower soil temperatures to 68–72°F (20–22°C) and increase light levels to 500–1,000 f.c. (5.4–11 klux). Begin to allow the medium to dry down somewhat once radicles have penetrated the soil. Apply 50–75 ppm from 14-4-14 weekly once cotyledons have fully expanded. Stage 2 will last about seven days.

Move trays to Stage 3 to develop true leaves by increasing light to 1,000–1,500 f.c. (11–16 klux) as well as ramping up fertilizer to 75–100 ppm and alternating with clear water. Continue to maintain the Stage 2 temperatures and allow trays to dry thoroughly between irrigations. Stage 3 takes about seven days for 288s. Once the first true leaves are almost fully expanded, a spray application of B-Nine at 2,500–3,500 ppm will help prevent stretch. It is imperative that the hypocotyl (stem) does not stretch, so that finished plants are not floppy at the base.

Harden plugs for seven to ten days prior to sale or transplanting by lowering temperatures to 60–65°F (15–18°C) days and 55°F (13°C) nights. Allow trays to dry thoroughly between irrigations. Increase the light levels to 1,500–2,500 f.c. (16–27 klux). Use 100 ppm from 14-4-14 every other irrigation.

A 288 plug requires from four to five weeks from sowing to transplant to larger containers, while a 128 requires from five to seven weeks.

Growing On

Cuttings

Select a disease-free, well-drained, soilless, potting medium with a pH of 5.5–6.3 and a high starter charge. Plant 1 liner/4" (10 cm) pot; 2–3 liners/6" (15 cm) pot; and 3–4 liners/10" or 12" (25 or 30 cm) basket. Be careful to avoid overwatering the crop immediately after planting.

Start the crop at 60–65°F (16–18°C) to establish plants for two weeks, then grow on at 50–70°F (10–21°C) days and 50–60°F (10–15°C) nights. While *Osteospermum* generally flowers at "normal" greenhouse temperatures, low night temperatures (45–55°F [7–13°C]) promote profuse, uniform flowering. Growers can provide this environment in an unheated or minimally heated greenhouse through the winter months. It is important to keep the crop from freezing. Growing the crop under low night temperatures for the entire crop cycle will increase the time to flower but will result in more heavily budded plants. If you choose, after flower buds are initiated, you can raise night temperatures to 50–55°F (10–13°C) to speed up the finish of the crop.

If cool growing temperatures can be maintained, grow plants under full light (5,000–6,000 f.c. [54–65 klux]). High light levels are best for branching and height control.

Though branching has also improved with breeding, there are still plenty of poorly branched plants making it to retail. Pinching is very important to overcome this. The sooner the plants can be pinched the better. Some growers are pinching sooner than the traditional two weeks after potting. It is possible to pinch the cuttings in the liner tray when they are soft. This can also greatly improve uniformity, but they should then be planted immediately. Locations further south, which experience very high light, will find branching occurs easily. Leave six to seven nodes above the substrate line for 6" (15 cm) pots and four to five for 4" (10 cm) pots.

The best way to control height is to provide as much light as possible and to space on time. When this is not enough, a Sumagic or Topflor drench is the next best control. Be sure to anticipate growth and apply before the plants stretch. Depending on your soil and local conditions, 0.5–1.0 ppm at labeled volumes will provide good control. It is very important to apply the correct volume (twice the recommended volume is double the rate). A second application can be made at or after bud set. When temperatures warm up expect an additional one-third growth in height and be sure to apply plant growth regulators if your crop is already tall.

B-Nine at 2,500 ppm can be applied after the pinch but is less effective for height control. It can be applied weekly, as needed, for three to four weeks after the pinch. Cycocel, at 3,000 ppm as a drench, can also be used. However, Cycocel sprays can cause yellowing that will not disappear over time, so never apply it as a spray.

Always test growth regulators before making broad applications, as different varieties have varying responses. Do not spray growth regulators once plants are budded. To help prevent plant stretch, avoid warm day temperatures as much as possible. Keeping plants spaced during production so that the foliage does not touch also helps to keep height in check. For high light southern locations, Florel can be used at 500 ppm about two weeks after potting, instead of a pinch.

While *Osteospermum* is a moderate to high feeder, withholding feed and stressing the plants during the flower initiation stage may result in more profuse flowering. Quickly return to a constant liquid feed of 250 ppm from 20-10-20 or 15–16–17. *Osteospermum* is sensitive to high soluble salts, so be sure to monitor salts levels and leach if salts become a problem. Maintain an EC below 1.0 mS/cm.

Osteospermum is not daylength sensitive, however for earlier spring flowering, use night interruption lighting starting only in the flower initiation stage. Be sure not to use this too early, as it will inhibit growth if flowers are set too early.

The crop time from planting a 105 liner to a 4" (10 cm) pot is eight to thirteen weeks, depending on the variety. Check the varieties from each of the companies you are considering. Early-flowering selections bloom in eight to eleven weeks, while mid- and late-flowering selections require ten to thirteen weeks.

Seed

Modern *Osteospermum* from seed does not require pinching. If grown cool, it will also not require any growth regulator treatments. If grown later in the spring in warmer climates, B-Nine at 2,500–5,000 ppm or Cycocel at 750–1,000 ppm may be used to control stretch. Always test growth regulators before making broad applications, as different varieties have varying responses. Do not apply any growth regulators once plants are budded. To help prevent plant stretch, avoid warm day temperatures as much as possible. Keeping plants spaced during production so that the foliage does not touch also helps to keep height in check.

Follow the cultural notes for cuttings in terms of temperatures, fertilizer requirements, and timing.

Transplanting a 288 plug to a 4" (10 cm) pot requires from ten to fourteen weeks to flower.

Pests & Diseases

Aphids, caterpillars, fungus gnats, mites, and thrips can attack plants. Thrips are the most common pests feeding on the pollen and disfiguring the flowers.

The most serious diseases that can strike are botrytis, *Pythium, Phytophthora,* and rhizoctonia. Keep air circulating around plants and do not overwater. Viruses can also infect *Osteospermum,* so be careful to keep plants clean and control vectors such as aphids and thrips.

Troubleshooting

Remaining wet for extended periods of time can cause plant collapse. Excessive nitrogen in the fertilizer, overfertilization under low light, and/or low light and overwatering can lead to a lack of flowering and too much vegetative growth. Yellow leaf margins may be due to excess sodium in the water supply or fertilizer. Pinching liners that are too hard and/or are under too low of light levels can result in poor branching.

Varieties

Contact your broker or breeder representative for the most current cultivars propagated from cuttings that show the best summer performance. Try the 'Serenity', 'Summertime', 'Astra', 'Margarita', or 'Zion' series for general bedding crops. 'Symphony', 'Swing', and 'Cascadia' are all great, spreading, landscape or basket types. 'Voltage' in yellow is an interspecific cross that performs through the summer in many locations.

Three seed *Osteospermum* series are commercially available. 'Asti' is available in 'White' (an All-America Selections winner), 'Lavender Shades', 'Purple', and a mix. Since the white is larger than the purple, it is best to grow 'Asti' only as separate colors in order to simplify the uniform application of PGRs. Plants grow from 17–20" (43–50 cm). The newest variety on the market is 'Akila'. It is available in three separate colors plus a mix and grows from 16–20" (40–50 cm) tall. The other seed variety is 'Passion Mix', an All-America Selections winner that comes in a mix of pink, rose, purple, and white and grows to 12–18" (30–45 cm) tall.

Postharvest

Plants are ready to sell when the first two to three flowers open and buds are visible for subsequent blooms.

Display plants for sale under cool conditions (60–65°F [15–18°C]) and bright light. Applying light shade may be beneficial to moderate temperatures.

Retailers should instruct consumers to cut plants back lightly if flowering stops during the heat of the summer. Consumers should fertilizer regularly. The new growth that results will produce another flower display once temperatures cool down for fall.

Oxalis

Oxalis sp.
Common names: Oxalis; shamrock; wood sorrel
Pot plant, annual, tender perennial (Hardy to USDA Zones 8 and higher)

Oxalis sp. are members of the Oxalidaceae. The name *oxalis* is derived from the Greek meaning "sour" or "acid," due to the characteristic presence of oxalic acid in the plant. There are reportedly between 600 and 900 species of oxalis with only a few species currently cultivated for ornamental purposes. *Oxalis regnellii,* also known as the "shamrock plant," is grown and marketed in the spring, primarily for the St. Patrick's Day holiday. *Oxalis triangularis* (purple shamrock) and other new, red-leaf oxalis cultivars are primarily produced for use in the landscape and mixed containers. *Oxalis tetraphylla* (lucky shamrock) and *O. adenophylla* (silver shamrock) are typically grown for landscape settings.

Aside from its reputation as a weedy species, oxalis has great potential as an ornamental pot plant for use in individual pots, mixed containers, or the landscape. Abundant diversity in leaf shape and flower color exists in oxalis. Future breeding could yield interesting and novel oxalis as well, providing ornamental greenhouse producers with even more options.

Propagation

Some oxalis varieties readily propagate from seed, lending themselves to an invasive, weedy nature; however most commercially produced oxalis species are asexually propagated from rhizomes, bulbs, and tubers. *Oxalis regnellii* and *O. triangularis* develop underground scaly rhizomes (1–2" [2–5 cm]), while *O. adenophylla* and *O. tetraphylla* develop small, tunic-covered bulbs. *O. triangularis* has also been propagated through tissue culture using different explants.

Growing On

Greenhouse forcing parameters are species dependent and are summarized in table 29. Typically, one to three rhizomes are planted 0.4" (1 cm) deep, per 4" (10 cm) pot and should be planted upon arrival, unless otherwise indicated by your supplier. The media should be well drained, and the pH should be maintained between 5.8 and 6.2 for most species. Forcing temperatures for *O. regnellii* and *O. triangularis* are significantly warmer (70–75°F [21–24°C]) than for other oxalis species. A warm start is essential for *O. regnellii* and *O. triangularis.* Light intensities are species specific, but in general when forcing in early to midwinter, medium to high light intensities are beneficial (1,000–2,500 f.c. [11–27 klux]). Many oxalis species are not heavy feeders and will produce quality, marketable plants with 200–250 ppm nitrogen. Fertilizer formulations with added calcium, magnesium, and iron enhance growth for *O. regnellii.* Plants are typically marketable after six weeks, when one to two flower stalks are beginning to show color.

Pests & Diseases

Oxalis are relatively free of pests and diseases. Thrips (*Thrips tabaci*) can be very problematic, causing foliar distortion and blemishes. Fungus gnats (*Bradysiai* spp.) have also been observed during production. Spider mites (*Tetranychus urticae*) can also be problematic with oxalis. Mold (*Penicillium*) can be a major problem with oxalis during the storage and shipment of rhizomes and can be reduced with a fungicide dip.

Table 29. Storage Organ, Greenhouse Forcing Temperatures, Weeks to Force, and Flower Color for Common Oxalis Species					
SPECIES	**COMMON NAME**	**STORAGE ORGAN**	**FORCING TEMPERATURE (°F)**	**WEEKS TO FORCE (10 CM POT)**	**FLOWER COLOR**
O. adenophylla	Silver shamrock	Bulb	54–59° (early), then 64°	4–5	Pink
O. regnellii	Shamrock plant	Scaly rhizome	70–75°	6–8	White or pink
O. tetraphylla (syn. *O. deppei*)	Lucky clover, Iron cross shamrock	Scaly bulb	55–61°		Rose or rose-pink
O. triangularis	Purple shamrock	Scaly rhizome	70–75°	6–10	Pink
O. versicolor	Candycane sorrel	Bulb	50–61°	7–9	White with rose/pink edge
O. vulcanicola	Volcanic sorrel	—	70–75°	6–8	Yellow and orange

* Adapted and modified from John M. Dole and Harold F. Wilkins. *Floriculture: Principles and Species.* Upper Saddle River, NJ: Prentice Hall. 2005.

Powdery mildew (*Microsphaera russellii*) and rust (*Puccinia* sp.) have also been reported to affect oxalis.

Shamrock chlorotic ringspot virus has been observed and reported in *O. regnellii*. Virus symptoms may look very similar to interveinal chlorosis (resulting from a nutritional deficiency) with the most identifiable virus symptom being a characteristic chlorotic ring spot surrounding an island of green tissue (see photo). As the viral infection progresses, the chlorotic ring spots fade into chlorotic blotches and streaks, while infected rhizome scales become dark brown or black. The virus is thought to be transmitted via aphid feeding; through mechanical contact between diseased roots and healthy plant roots; and from propagating infected plants. Careful and immediate rouging of symptomatic plants is the most effective control of the virus.

Troubleshooting

Oxalis typically does not require growth regulator applications when grown under ideal conditions. However, low light situations often occur during typical *O. regnellii* forcing periods (December to March), especially in northern latitudes, and can lead to increased petiole lengths, or "legginess." Bonzi sprays

Symptoms of virus on oxalis, likely shamrock chlorotic ringspot virus.

of 1–4 ppm or drench applications 0.05–0.1 mg of active ingredient (a.i.) along with A-Rest spray applications of 33 ppm have been effective in controlling plant height in *O. regnellii* and *O. triangularis.*

Careful attention must be paid to irrigation practices when producing oxalis, so as not to overwater. Higher light intensities can reduce petiole elongation; however leaves will typically fold down under higher intensities (and during darkness). Thus, caution should be exercised in order to prevent overwatering as oxalis roots are fine and very sensitive to excess water in the media and will easily decay. This can ultimately lead to nutrient deficiencies and chlorosis.

Oxalis behaves very similar to other greenhouse ornamental crops, in that, when the pH rises above the generally accepted upper threshold of 6.2, iron (Fe) can become limited, resulting in interveinal chlorosis. If iron deficiencies occur, iron chelate (Sprint-138 and Fe-EDDHA) media drenches are successful at 5 oz./100 gal. (374 mg/l). Foliar iron chelate applications have been reported to be successful, but limited information exists.

Varieties

The number of oxalis cultivars continues to increase. Some of the cultivars on the market include: *O. regnellii* 'Fanny', 'Birgit', 'Irish Mist', and 'Charmed Jade'; *O. tetraphylla* 'Iron Cross'; *O. triangularis* 'Charmed Wine', 'Charmed Velvet', and 'Myke'; *O. versicolor* 'Grand Duchess'; and *O. vulcanicola* 'Molten Lava' and 'Zinfandel'.

Postharvest

Little information exists about the postharvest care of many oxalis species. *O. regnellii* and *O. triangularis* have been grown as long-lived indoor plants. Plants often become etiolated and benefit from cutting back foliage and/or repotting. Placing plants in bright, indirect light conditions will keep plants more compact.

P

Paeonia

Paeonia spp.
Common name: Peony
Perennial, cut flower (Hardy to USDA Zones 2–7)

Chances are, if you live in an older home, you have inherited peonies in your garden. These long-lived perennials survive from generation to generation, thrilling gardeners each May with oversized blooms, many of which are lightly scented.

For retail growers, peonies are a no-brainer. Your customers will demand them once they see them flowering all over town. Be sure to have flowering pots on hand when the rush hits.

Propagation

Plants are propagated by root division in late summer or early fall by specialists. It is a time-consuming task that requires large amounts of available field space. Most growers produce their finished crops from purchased roots. Growers buy in plants based on the number of "eyes," or growing points, they have. Roots are commonly offered with two to three eyes or three to five eyes.

Growing On

While peony roots can be potted in the autumn or winter, most growers buy in peonies in autumn, pot them up to 1-gal. (4 l) or larger containers (depending on the size and shape of the root), and overwinter them in cold frames or a cold greenhouse for sale the following spring. Most of the varieties offered will flower readily. For roots potted up in February, plants will be salable in May, but few will flower profusely.

Select a well-drained medium with a pH of 6.5–7.0. Plant roots with eyes up and deep enough so that the eyes are covered with no more than 2" (5 cm) of medium. Allow the plants to root for four to six weeks before dropping the temperatures to freezing (32°F [0°C]).

Peonies require cold treatment for flowering. The exact requirements are not clear, but plants will break dormancy after six weeks at 34°F (1°C). Once plants have passed through cold treatment, you can force them into flower if you choose or allow them to flower naturally in May. Force them cool by maintaining production temperatures at 64–68°F (18–20°C) days and 64°F (18°C) nights. At high temperatures, flowering slows and flowers age faster.

Do not overfertilize because flowering will be inhibited. Fertilizing with 100–150 ppm of 20-10-20 two to three times once plants are vegetative will be sufficient. Make sure to keep nitrogen levels low. Overfertilized plants may become burned. Peonies need good drainage to control disease, but be careful not to water-stress plants when they are actively growing or flowering.

At one time, peonies were a very popular cut flower. Some growers today are once again producing it as a field-grown cut flower. In cold climates, plants are extremely long lived and can be cropped for seven years. Yields peak after three years. Peonies prefer heavier, clay soils with good drainage and good moisture-holding capacity. Plant roots with three to five eyes directly into fields in the fall and space them at 1 plant/4 ft.^{2} (2.5 plants/m^2). Flowers can be disbudded several weeks prior to flowering, when buds are pea-sized. By removing laterals, one extra-large flower will be produced; by removing the center bud, a spray of flowers will be produced.

Some Dutch growers force peonies into flower several weeks earlier using rolling polyethylene greenhouses. They roll the house over portions of the field beginning in March to raise air temperatures. If forcing this way, be sure to maintain excellent ventilation for disease control and to maintain cool air temperatures below 70°F (21°C).

The best time to fertilize cut flower peonies is after they have flowered, when they are in the process of forming flowers for the following year. A light sidedress of a complete dry or slow-release fertilizer is fine.

After plants die back in the fall, remove all debris, as it serves as a source of disease the following year.

Pests & Diseases

Nematodes can attack plants in the ground. Ants are frequently seen on flower buds, but they are not a problem. Disease problems can include botrytis, *Cladosporium, Sclerotinia,* or root rots.

Troubleshooting

Peonies that fail to flower may have received too much fertilizer or have been planted too deep.

Varieties & Species

Most garden peonies are crosses between *Paeonia lactiflora* and *P. officinalis.* Dozens of peony varieties are available, and the plants are so popular that there is a society of peony enthusiasts and breeders. Most peonies are categorized by their flower type: single, Japanese, anemone, semi-double, and double. Additionally, varieties are categorized by their bloom season: early, mid, or late. Since flowering lasts only two weeks, offer a range of varieties to extend the sales season. One of the most famous peony varieties is 'Sarah Bernhardt', a medium pink.

P. tenuifolia, an early-blooming species, and *P. officinalis,* a midseason flowering peony, are two species that are worthwhile to offer if you find plants available. *P. tenuifolia* (commonly called fern leaf peony) has lacy, cut foliage and comes in either single or double flowers. *P. officinalis* 'Mollis' is an early, single pink.

P. suffruticosa is a woody species known as the tree peony. Stems remain over the winter and reflush with leaves and flowers the following spring. Tree peonies are produced from grafted plants. According to Song Sparrow Perennial Farm, tree peonies may not flower until the second or third year after grafting. If you offer tree peonies, be sure to use color picture tags so consumers know what they are buying.

Postharvest

Cut stems when tight buds are showing true color; by this time the calyx will be loose. Allow more time for doubles and red varieties. If stems are not going to be stored or shipped, harvest when blooms are partially open. When harvesting cut flowers, allow as many leaves as possible to remain on the plant; the more leaves left after harvest, the better plants will grow. Be sure to leave at least three leaves after harvest. If flowers are harvested too open, vase life is five days or less. Store and ship stems dry at 34°F (1°C). Retailers should hydrate stems in a citric acid solution at 70°F (21°C) and then treat with preservative solution. Hold open flowers at 35–40°F (2–5°C). Average vase life is seven days.

Pansy (see *Viola*)

Papaver

Papaver nudicaule
Common name: Iceland poppy
Annual, tender perennial (Not hardy in cold, severe winter locations)

Poppies have showy, paper-thin, brightly colored, single flowers that are eye-catchers for consumers on the retail shelf or in the garden. Flowers range in size from 3–4" (8–10 cm) across. Plants prefer cool conditions to excel and are one of the brightest of spring offerings for the garden with colors of yellow, orange, light scarlet, and pink as well as shades in between. Plants grow to 15" (38 cm) tall but have a small habit, measuring only 6" (15 cm) across.

Propagation

Sow one seed per 288-plug tray and lightly cover with vermiculite or peat moss. Germinate seed at 64–68°F (18–20°C) media temperature. The root radicle emerges in three to six days.

Seedlings develop rapidly and can become stretched if grown under too warm an environment. Upon emergence, lower substrate temperatures to 50–55°F (10–13°C).

Maintain a soil pH from 5.5–5.8 throughout propagation, allowing pH to rise as high as 6.2 during forcing and finishing. Maintain soil EC levels of less than 0.75 mS/cm during germination, raising it to 0.75–1.0 mS/cm as seedlings mature and into finished production.

During the final stages of propagation and throughout forcing and finishing, maintain relatively dry conditions, allowing the medium to fully dry between irrigations. Always allow foliage to dry by nightfall to avoid disease.

Plug trays will take from four to six weeks to be ready to transplant to larger containers.

Growing On

Seed can be sown anytime in winter up to February for flowering from sixteen to twenty weeks after sowing for most varieties. While seed can be sown later, the plants are not tolerant of heat or high humidity and are short lived in either extreme.

If using a 288 plug, transplant one to two plugs per 4" (10 cm) or three per 6" (15 cm) pot and grow on at 50–55°F (10–13°C) days with nights a little cooler until plants are established in the final container. Then drop the media temperature down to 40–45°F (4–7°C) to keep plants from stretching.

When transplanted from a plug in late February or March, plants will start to bud up in April and flower soon after. As natural day length increases and night temperatures stay above freezing, plants respond by growing faster and flowering.

Varieties

Papaver nudicaule, often sold as a perennial, is an annual or very tender perennial. Plants are not tolerant of hot summers in the central or southern United States and perform best sold in pots for planting in the garden or to be enjoyed indoors. In southern regions, plants may be purchased and planted in the fall for a profusion of flowers the following spring. 'Champagne Bubbles', an F_1 mixture of scarlet, bronze, orange, pink, and cream blooms, is the most popular Iceland poppy variety. Plants will flower in about seventeen weeks from seed.

Papaver orientale
Common name: Oriental poppy
Perennial (Hardy in USDA Zones 3–7)

Similar in description to Iceland poppies, Oriental poppies often have equally large flowers (3–4" [8–10 cm]) across but usually in darker colors of red, scarlet, orange, burgundy, and purple in addition to lighter shades including white, salmon, or pink. The main color absent in Oriental poppies is yellow. Oriental poppies have paper-thin, almost crepe-paper-like blooms. Plants grow to 15" (38 cm).

Propagation

Seed and divisions (bare-root transplants) are the common methods of propagation among growers and commercial propagators.

Seed

Seed is sown from May to July for overwintering dormant in 1-qt. or 1-gal. (1.1 or 4 l) pots. Sow two to four seeds per 288-plug tray and leave the seed exposed to light or lightly cover with vermiculite. In addition, some growers leave the seed exposed to light upon sowing but then cover it lightly with vermiculite as the root radicle emerges and just as the cotyledons first develop. Germinate at 70–75°F (21–24°C), and seedlings emerge in seven to ten days.

A 288-plug tray takes from seven to nine weeks from sowing to be ready to transplant to larger containers, while a 128 liner/plug takes from nine to twelve weeks from sowing to transplant.

Seed can also be sown from October to early November for overwintering plants cold but not dormant. However, once established in the final container the plants are grown on cold to allow for vernalization so that plants will still flower in the spring.

Bare-root transplants

Available from commercial propagators, bare roots can be planted in October or in spring. Autumn-potted roots tend to produce larger overall plants than those planted from roots in the winter.

Growing On

Transplant one 128 liner per 1-gal. (4 l) pot in summer to early autumn. As for bare roots, use one root per 3-qt. or 1-gal. (3.3 or 4 l) pot. Allow the liners or bare roots to root and then overwinter the plants dormant in the cold frame or cold greenhouse. When sown from seed, *Papaver orientale* will not flower unless it receives cold-temperature vernalization. Seed sown in January will not flower its first year.

Plants potted in September should be overwintered in an unheated greenhouse or cold frame. Plants potted in November can be held at 40–45°F (4–7°C) for twelve to fourteen weeks.

In the spring, finish plants at 50–55°F (10–13°C) nights and 55–60°F (13–16°C) days. Provide moderate light intensities of 1,500–3,000 f.c. (16–32 klux). Plants can be allowed to wilt between irrigations for height control. Other height control measures include withholding fertilizer and using a negative DIF.

Flowering is relatively short; plants will go dormant during the summer and many times will not survive southern summers. Stems can be harvested for cut flowers: After cutting, sear stem ends to prevent the milky stem latex from "bleeding" before placing stems into water. Postharvest life of poppies is just a few days.

Varieties

'Allegro' is an 18" (46 cm) tall, scarlet-red variety. Plants require two years from seed to flower but can be sold green in packs ten to twelve weeks after sowing. 'Beauty of Livermore' is deep red; 'Brilliant' is bright scarlet; 'Oriental Mix' has red, scarlet, pink, and salmon; and 'Princess Victoria Louise' is salmon.

Parsley (see *Petroselinum*)

Pelargonium

Pelargonium × *domesticum*
Common names: Martha Washington geranium; regal geranium
Flowering pot plant

Regals are one of the most spectacular flowering plants growers can produce. That they are a long crop and require cool temperatures and daylength treatments have given them the image of being too hard for most growers to tackle. However, both traditional types and new hybridized lines are relatively easy to produce if you follow the protocols set forth in this chapter. Creating a new market, new hybridized types have shown great summer performance even in some of the hottest trial sites in the South.

It is important to note whether you are growing a traditional regal or a new hybrid and then follow the appropriate protocol. With traditional regals, most growers start with precooled plants; otherwise, pots need four weeks of constant cool temperatures to initiate flowers. Cooled plants are available year-round, making regals a great holiday pot plant opportunity if you can produce them at night temperatures of 58°F (14°C) or less. Newer hybridized series do not require the traditional cooling period. However, they do need specific conditions (listed below) to flower to their full potential.

Propagation

Root cuttings in Ellepots or a media substrate. Maintain 72–77°F (22–25°C) substrate temperatures and air temperatures of 68–74°F (20–23°C) days and 65–67°F (18–19°C) nights while cuttings form callus. Use light levels of 500–1,000 f.c. (5.4–11 klux) and apply a 50–75 ppm foliar feed of 20-10-20 if there is any loss in foliage color. Maintain the substrate pH at 5.5–5.8. Use tempered water in mist lines and mist only as much as necessary. Excessive mist can cause disease. Once 50% of the cuttings have formed root initials, move them to the next stage (root development) by reducing humidity and substrate moisture, increasing light to 1,500–2,000 f.c. (16–22 klux), and increasing fertilizer to 100–200 ppm, alternating between 20-10-20 and 14-0-14. Once cuttings are fully rooted, begin hardening them off for planting by maintaining air temperatures of 58–62°F (14–17°C), increasing light to 3,500–4,000 f.c. (38–43 klux), and allowing the medium to dry down between irrigations.

For traditional regals, propagators provide precooled cuttings, making the job of finishing regals much easier. Newer hybridized series typically can be planted directly after rooting.

In general, regal geraniums root in three to five weeks in a 105 tray.

Growing On

Pot liners into a well-drained, disease-free medium with a pH of 5.8–6.0 and a moderate initial nutrient charge. For traditional regals, as soon as cuttings are established in the pot, begin floral initiation cool temperature treatment. Provide four full weeks at 45–48°F (7–9°C) days and nights. Lighted coolers can be used (2,000 f.c. [22 klux]). Low light levels may increase the cooling period. If you are going to pinch plants, do so prior to cooling (see below). Most growers purchase precooled liners, which have been cooled for six weeks, from their supplier, so they can skip this step.

For newer hybridized series, once roots have grown to the edge of the pot, you can lower the night temperatures to 54–58°F (12–15°C) for four to six weeks. Although this isn't required, it can increase uniformity and the bud count. Day temperatures should be kept below 75°F (24°C). *Note:* Excessively cool night temperatures can add weeks to the production time. After the cool-night period, you can raise night temperatures to 65–67°F (18–19°C) to speed flowering.

Regals must receive the proper light levels, day length, temperatures, and humidity to flower to their full potential. Between mid-September and April, plants must receive sixteen to eighteen hours of light per day from supplemental lamps (400 f.c. [4.3 klux]) or night interruption lighting from 10 P.M.–2 A.M. However, there is a strong correlation between the total amount of accumulated light and flower quality. For this reason, when possible, use HID lights for daylength control. Grow plants under as high light levels as possible while maintaining cool production temperatures. Maintain a minimum of 2,000 f.c. (22 klux) of light at all times. A regal grown under high humidity and no moisture stress will remain vegetative, even under proper temperatures and light. Maintain humidity below 70% for the best quality. Even a bud-initiated liner will revert to a vegetative state when humidity levels are too high.

Regal geraniums are sensitive to drying out: Maintain a uniformly moist, but not wet, medium. Fertilize with 150–300 ppm from 20-10-20, alternating with 14-0-14 at every other irrigation. For newer hybridized types, flowering seems to be highly affected by the fertilizer levels. Higher levels have been shown to keep plants vegetative, while lower levels promote flowering. In general, feed 150 ppm. To help initiate flowers, you can irrigate with clear water twice six to eight weeks from finish.

Maintain cool production temperatures of 60–65°F (16–18°C) days and 55–58°F (13–14°C) nights.

To encourage full plants, make a hard pinch two to three weeks after transplant, leaving five leaf nodes above the substrate line. If you pinch plants, wait to begin cooling until the smallest shoots are 0.75–1.0" (2–3 cm) long. However, know your crop—many newer cultivars branch well without a pinch. Pinching can add weeks to your production schedule.

If needed, Cycocel can be used to control height. Apply three applications (1,000 ppm) beginning when roots reach the edge of the pot (about the time cool treatment begins). Apply the second application after cooling is completed, and the third application one week later. If you are not cooling plants, you may begin the first application as roots reach the edge of the pot, and then treat two more times, spaced one week apart, one week after the first application. Bonzi, applied weekly at relatively low rates until plants show color, is also effective. Use Bonzi cautiously and trial it before treating your entire crop.

Give regals enough space on the bench to develop full plants: Optimally space plants one 6" (10 cm) pot/ft.^{2} (10–11/m^2).

For traditional regals that require a vernalization period, the total crop time from sticking a cutting to a flowering plant is twenty to twenty-three weeks or ten to twelve weeks for 6" (15 cm) pots from a precooled liner. For the newer hybrids, the total crop time from unrooted cutting to flowering plant is from sixteen to nineteen weeks.

Pests & Diseases

Spider mites and thrips can become problems, as can botrytis and *Pythium*. *Xanthomonas* can be present in regal geraniums and not display symptoms. It is recommended to grow this crop and your regular geranium crops separately.

Postharvest

Regal geraniums are extremely susceptible to ethylene. To help prevent petal drop, use silver thiosulfate (STS) at the first sign of color or 1-MCP (1-Methylcyclopropene) (see label for guidelines). Groom plants by removing dead flowers and leaves prior to shipping. This will also remove sources of ethylene production. If plants must be boxed, do so for only as long as necessary. Ship plants at 42°F (5°C).

If plants have been shipped in boxes, retailers should open boxes immediately and unpack plants. Display regal geraniums in cool temperatures (60–70°F [16–20°C]), under high-light conditions (minimum of 100 f.c. [1.1 klux]). Do not allow plants to dry out, but avoid wet conditions. Do not expose regal geraniums to any sources of ethylene, such as ripening fruits and vegetables or areas with decaying leaves and flowers.

Consumers should place plants in a minimum of 50 f.c. (538 lux) of light to ensure continuous flowering.

Pelargonium × *hortorum*
Common names: Geranium; zonal geranium
Annual

Geraniums are an American tradition. Front stoops, patios, decks, mailboxes, and window planters provide the stage, and their bright, bold flowers provide the enjoyment. The traditional combination planter with dracaena spike surrounded by red geraniums and vinca vine is as American as apple pie and baseball. Often called "your grandmother's plant," the geranium is easy to grow for the consumer and performs in virtually all climates. Consumer success is why it continues to be a leading plant at retailers.

The American consumers' love affair with geraniums is big business too: US growers produce enough geraniums for each household to have nearly two plants each. *Pelargonium* × *hortorum* is by far the most popular geranium, while the *P. peltatum* (ivy geranium) sales continue to be strong. New to the market are cultivars that have been crossed between zonal and ivy geraniums. This combines the dark-red flowers and dark, attractive foliage of ivies to the garden performance of zonals. Vegetatively propagated geraniums, commonly known as zonal geraniums or zonals, are grown twice as frequently as seed geraniums are. Ivy geranium production is covered under *Pelargonium peltatum,* starting on page 597.

For growers, geraniums are a deceptive crop: They are easy to grow; however, geraniums are subject to a host of serious disease problems, the most notorious of which is *Xanthomonas. Xanthomonas* is a serious disease. The only way to make sure crops do not get *Xanthomonas* is to start with clean, disease-indexed stock that has been produced using a defined, clean-stock program. Far too often, problems arise when growers get a few good years of crops under their belts and relax their internal procedures. (See Propagation and Pests & Diseases for more about *Xanthomonas.*)

If you are a bedding plant grower, geraniums will be an important part of your crop mix. There is plenty of room in most markets for a wide variety of approaches: From oversized, spectacular patio containers to 4" (10 cm) mass-market pots of seed geraniums or fast-cropped zonals. The trick is to know your market and to know your greenhouse so you can maximize your profit potential.

Seed-propagated geraniums are mainstay bedding plants in North America, although they are not as popular as those that are vegetatively propagated. They are typically produced in 4" (10 cm) pots for mass-market sales and planted in ground beds, as opposed to containers, which are the mainstay for their vegetative cousins—zonal geraniums. Seed geraniums differ from zonals in that they have single flowers that are "self-cleaning," that is they lose their petals (shatter) as they age so they don't require deadheading like zonal geraniums. Thus, their popularity as bedding plants.

Entire books have been written about growing geraniums. The following is an overview. With every crop you grow, you will learn more and delve deeper. Be sure to keep notes on each crop so you can learn from year to year.

Propagation

Pelargoniums are commonly propagated by vegetative cuttings, although seed can be used as well.

Cuttings

The world's top cutting producers have invested enormous resources in offshore cutting facilities. Because of the risk of *Xanthomonas*, only the very best producers are certified by the USDA and able to ship to North America. Though a few growers still choose to do their own stock, most now rely on cutting producers. With the recent addition of real-time inventory and the oversupply of cuttings, growers can feel confident that they will receive what they ordered. The risk of a serious disease should be enough to discourage most growers from attempting to maintain their own stock. In general, there are much easier and less risky plants with which to self propagate.

Growers who choose to take their own cuttings purchase geranium stock plants and take cuttings beginning in December for hanging baskets through March for fast-cropped 4" (10 cm) pots. Generally, stock plants should arrive in the summer. If you choose to maintain your own stock, please pay particular attention to the sections on *Xanthomonas* and on clean-stock programs in this book. Finally, be sure to get a license from the breeder, as unlicensed propagation is illegal.

Order stock plants from a reliable supplier adhering to rigid protocols regarding their clean-stock program. Ask questions. You will be bringing in several hundred or even thousands of plants from this supplier and keeping them for eight to nine months, so you have a right to know the supplier's procedures. The plants must be planted in an isolated, clean greenhouse.

Most growers choose not to take their own cuttings at all and purchase unrooted cuttings. Unpack unrooted cuttings immediately upon arrival. Open bags and allow air to circulate to disperse ethylene. For best results, stick cuttings immediately. If they cannot be planted immediately, store them in opened boxes or in the cutting bags on shelves in a clean cooler with nearly 100% humidity and 40–45°F (4–7°C) for up to forty-eight hours. If cuttings are wilted, mist with clean water and cool for at least four hours. Some foliage yellowing may result if cuttings are stored.

Stick cuttings in Ellepots or a medium. Some growers dust cuttings with a powdered rooting hormone prior to sticking. (Never dip geranium cuttings in a liquid rooting hormone, as you will spread any diseases that may be present in the plant sap.) Provide enough spacing so as not to allow leaves to overlap. Provide bottom heat, maintaining a root temperature of 68–72°F (20–22°C). Mist only enough to maintain turgidity. As soon as possible, cease the use of mist at night. Too much mist will cause disease and nutrition problems. First adjust moisture and then use fungicides only when needed. Use tempered water so that the mist does not cool down the rooting medium. Maintain air temperatures of 68–72°F (20–22°C) days and 65–70°F (18–21°C) nights. Provide 1,000–2,000 f.c. (11–22 klux) of light and apply a light foliar feed of 20-10-20 at 50–75 ppm if needed. Maintain the substrate pH at 5.5–6.0. When 50% of the cuttings have begun to differentiate root initials, they may be transferred to the next rooting stage.

Once cuttings have callused and root initials appear (after seven to ten days), increase light levels to 2,000–2,500 f.c. (22–27 klux) and increase fertilizer to 100–200 ppm, alternating between 20-10-20 and 15-0-15. Raise the substrate pH to 5.8–6.0 and maintain the EC at less than 1.0 mS/cm. If height control is a concern, A-Rest, Bonzi, and Cycocel can be used. When possible the best height control is higher light, lower temperatures, and space.

Harden cuttings for shipping or planting for seven days by allowing the medium to moderately dry down between irrigations, increasing light levels up to 4,000 f.c. (43 klux) and increasing fertilizer to 150–200 ppm from 20-10-20, alternating with 15-0-15. At the beginning of the hardening stage, you can apply 350 ppm Florel to abort flower buds, promote branching, and control growth. Follow label instructions as factors such as the time of day, amount of spray volume, and pH of the spray water affect its potency. Flower buds should quickly turn yellow when used correctly.

Apply Florel as soon as cuttings have roots and before buds are able to mature.

For the skilled grower, direct-sticking unrooted or callused cuttings is possible when he is able to provide conditions close to those previously mentioned in a larger production area. This can save labor and decrease production time by one to three weeks. Generally, callused cuttings will root five to eight days faster than unrooted cuttings and provide security in losses when conditions are less than ideal. As you become more skilled at the process, direct-sticking unrooted cuttings will provide the greatest return.

The crop time from sticking a cutting until ready to transplant in a 105 tray is three to four weeks.

Seed

Single-sow seed into a germination medium with a pH of 6.4–6.5 and cover lightly with coarse vermiculite. The media EC should be 0.75–1.0 mS/cm; high EC levels will discourage roots from penetrating into the media. Maintain soil temperatures of 73°F (23°C). Keep the medium uniformly moist. Light is not necessary for germination, but if you think it is likely the plug trays will remain in the chamber for more than four days, then providing 50–100 f.c. (0.5–1.1 klux) of light will help prevent stretch. Keep ammonium levels at less than 10 ppm. Radicle emergence will occur in one to three days.

Stage 2, stem and cotyledon emergence, takes five to ten days. Reduce temperatures to 70–75°F (21–24°C) days and 65–70°F (18–21°C) nights and provide 1,000–2,500 f.c. (11–27 klux). When cotyledons are fully expanded, begin fertilizing with a calcium-based feed such as 14-4-14 or 13-2-13 at 50–75 ppm of nitrogen once per week. Maintain an EC of 1.0–1.2 mS/cm. If possible add supplemental light during Stages 2–4, using HID lighting with a minimum of 400 f.c. (4.3 klux). Geraniums are light accumulators, so adding supplemental light during the plug stage will significantly speed up flowering later in the crop at a reasonable cost.

During Stage 3, growth and development of true leaves, maintain light levels and temperatures and begin allowing the substrate to dry between irrigations. Increase feed to 100–150 ppm twice a week from 14-4-14 or 13-2-13. If growth regulators are required during this stage, apply Cycocel at 750 ppm when three to five true leaves are present. Stage 3 takes about ten days.

Harden plants for seven days during Stage 4 before shipping or planting. Reduce temperatures to 70–75°F (21–24°C) days and 60–65° (15–18°C) nights and allow soil to dry thoroughly between irrigations. Monitor the pH to ensure it stays above 6.4. Maintain the feed program. Be sure not to apply ammonium-nitrate fertilizers since they will drop the pH below the desired range.

A 288-plug tray requires from four to five weeks from sowing to be ready to transplant to larger containers.

Growing On

Cuttings/liners

Select a disease-free medium with a moderate initial starter charge and a pH of 6.2–6.5. The initial pH of the media is crucial to prevent problems with iron or manganese toxicity during the growing of the crop. Most geranium growers use a special media for this crop that contains extra lime.

Pot liners even with the surface of the potting medium, being careful not to compact the soil when planting. Remove any flower buds that might be present on cuttings at potting to obtain the best vegetative growth. Take care to regularly disinfect all that comes in contact with the geraniums. For maximum growth, grow zonals on at 65–70°F (18–21°C) days and 62–65°F (17–18°C) nights. Slightly cooler temperatures with high light will add some production time but will produce the best-quality plants.

Geraniums love fertilizer. Feed 15-5-15 constant liquid feed and alternate with 20-10-20. Use as much calcium nitrate as possible while maintaining a pH of 5.6–6.0. Alternate with ammonium only to control pH or push size. Calcium nitrate when supplied at rates of 200–300 ppm will keep plants tight and healthy. Use 100–150 ppm during low-light periods, increasing to 250–300 ppm in high-light periods. If soluble salts build up, water with clear water every third watering. Maintain the EC at 1.0–1.5 mS/cm. When possible, grow zonals with a drip irrigation system to limit the potential spread of diseases.

Though not common with today's genetics, you may choose to pinch vegetative geraniums that are to become specimen plants or are in large pots to encourage better branching and higher flower number. Each pinch will delay finishing by two weeks. Do not pinch fast-cropped 4" (10 cm) pots.

Once liners have rooted into their pots, allow plants to dry between irrigations just to the point before wilt, which will help control plant height and maintain healthy roots. Cycocel and Cycocel/B-Nine tank mixes can also be used to control plant height. For less vigorous cultivars, 750–1000 ppm Cycocel can be enough to control growth. Multiple applications are better than one high rate. For more vigorous cultivars higher rates are necessary (1,000–1,500 ppm). A tank mix with B-Nine can reduce the risk of yellowing and increase the results. It is important to apply growth regulators before leaves expand and shoots stretch. Bonzi is effective but can cause undesirable changes to the plant structure and flower peduncles. Florel is quite effective in producing high-quality geraniums, especially in conjunction with well-fertilized plants. Florel may be applied as a 250–500 ppm spray after plants have become established and are not stressed by high light or temperatures. A second application can be applied fourteen days later. Discontinue using Florel at least seven weeks prior to sale.

Geraniums also respond to a negative DIF. Several hours before sunrise, temperatures may be dropped 5–10°F (3–5°C).

One 105 liner potted up to a 4" (10 cm) pot will finish in eight to eleven weeks, depending on variety.

Seed/plugs

Select a disease-free medium with a moderate initial starter charge and a pH of 6.2–6.5. The initial pH of the media is crucial to prevent problems with iron or manganese toxicity during the growing of the crop. Most geranium growers use a special media for this crop that contains extra lime.

Grow seed geraniums slightly warmer than vegetative varieties at 70–75°F (21–24°C) days and 65–70°F (18–21°C) nights. Maintain high light at 4,000–7,000 f.c. (43–54 klux). Shade may be needed during summer months to control heat.

Seed geraniums don't require as much fertilizer as zonals do. Use a constant liquid feed at 200 ppm nitrogen (N) with a calcium-based feed such as

Florel: A Different Perspective

Bob Frye

Florel is a chemical that, by itself, can do more to increase geranium quality than any other single factor I know of. Generally, growers should use lower rates of about 300–400 ppm and make multiple applications about ten to fourteen days apart to maximize this chemical's safety and effectiveness. Some really vigorous cultivars can handle one-time applications of 500 ppm.

On geraniums, growers will notice some unique things happening as the plant reacts to Florel, compared with other plant growth regulators. The angle of the leaf pedicel will rapidly incline vertically about 15–20° compared with an untreated check. There are a couple of schools of thought to explain this, but this generally happens because internode distances decrease and the pedicel has no choice but to incline toward vertical. This is normal with Florel. In my scenario, where I maximize the chemical and keep internodes *really* compressed, I'll even get what I call "leaf abscission," where the nodes are so compressed that the lifeline of the leaf pedicel is cut off completely. This is also normal under this specific situation.

In addition, the plant will exhibit a more-yellow-than-normal color. This is merely cosmetic and has nothing to do with nutrition, so don't go chasing nutrient deficiencies. Leaf size will be smaller, which is also good. The dark-leafed cultivars aren't affected as much as the lighter-leafed cultivars are.

One of the tricks I use to maintain what I consider a more attractive foliage color: I'll back off my predetermined Florel rate by about 50–70 ppm and add 1,000–1,250 ppm of B-Nine, which almost always offsets the yellowing caused by the Florel. I feel this also helps the plants absorb solar energy, since dark colors absorb more than light colors do.

The first applications are the most important because they establish a base line of quality early. I don't feel most growers acknowledge this importance. My *very most important* application is the first one, before the cuttings are even planted: 200 ppm while the cuttings are still in the plastic Oasis trays, right after unpacking but prior to planting. The caution here is that they must be stress free, turgid, and kept from the beating sun until the Florel is dry.

13-2-13 or 14-4-14. Maintain the EC at 1.2–1.5 mS/cm. Leach with clear water if EC rises above 1.6, since high salts can lead to leaf-edge burn or spotted foliage.

Once plugs have rooted in to their pots, allow plants to dry down between irrigations, which will help control plant height and promote healthy roots. Cycocel is the preferred growth regulator for seed geraniums. Several applications at 750 ppm two weeks apart will produce compact, well-toned plants that flower earlier. Do not apply after flower buds are pea sized or larger since you will reduce the size of the flowers.

Geraniums also respond to a negative DIF in lieu of growth regulator applications. Several hours before sunrise, temperatures may be dropped 5–10°F (3–6°C).

How Bob Frye Grows Premium 7" Geraniums

*Bob Frye**

The production and marketing of premium geranium plants is all about five factors: (1) selecting proper cultivars, (2) maintaining plant health, (3) maximizing branching, (4) controlling plant height and form, and (5) properly marketing the product. Don't forget these five items, especially maximizing branching—it's often the most critical, most neglected, and least understood of all the production objectives.

First, determine your destination. The following is a short list of our 7" (18 cm) fancy zonal production specs. If you have specifications, you can score your progress as you grow the crop. The ideal specs can generally be described as:

- Height: 14–15" (36–38 cm), including 7" (18 cm) azalea pot
- Width: 13–15" (33–38 cm)
- Branching: Fifteen to thirty-five breaks (cultivar dependent)
- Maximum internode length: 0.75" (2 cm)
- Flowering: Five to seven fully open; six to eight unopened or breaking buds
- Foliage: Dark green, no yellow leaves, no chemical residues
- Canopy: Extremely dense
- Form: Symmetrical
- Health: No deficiencies, no disease
- Roots: White, clean, and slightly root bound
- Growing Time: 110 days

The Rooted Cutting

Production at The Plantation starts with a clean, healthy rooted cutting that should meet the following specifications: disease free; 1.5–2.5" (4–6 cm) tall from the top of the Oasis wedge to the terminal end; basal stem caliper of 0.25–0.30" (6.4–7.6 mm) or larger; minimum of three to four clean, fully functional, dark green yet small leaves; short petioles and four to five stacked nodes; clean, white roots; and a turgid and soft appearance with no shipping or frost damage.

We have pots spaced, saturated with fertilizer solution, and prepared for planting prior to the arrival of the cuttings. The greenhouse environment has been warmed to 72°F (22°C) for forty-eight to seventy-two hours.

When cuttings arrive, they are immediately unpacked, with each Oasis strip being separated to provide plenty of air circulation. We consider these first steps and procedures critically important to maximize branching potential later. Once the cuttings are laid out, the Oasis wedges are saturated, and the plants are fully turgid, we make the first application of 100–225 ppm of Florel to the unplanted cutting. We make the application in the evening to prevent burning tender foliage. Properly done, this application will yield significant branching. Apply it improperly or at the wrong rates, and you'll regret making this early application. Make trial applications before trying this unconventional Florel application on your whole crop.

Pot cuttings properly and in a timely fashion. Keep plants stress free so that branching is maximized.

Continued

How Bob Frye Grows Premium 7" Geraniums *(Continued)*

Days 1–10

Once planted, we immediately drip irrigate with Subdue Maxx. We maintain 72°F (22°C) days and nights for the first ten days. We feel it takes four full days for the roots to become functional and another six before new roots begin to emerge from the Oasis.

During this time we expose plants to high levels of carbon dioxide (CO_2), as we feel this aids rapid rooting. We rarely use CO_2 for more than two weeks. After the first six to seven days, we remove any less-than-desirable leaves, which are not contributing to photosynthesis. At the first sure signs of new growth, we shift to a –6°F (–3°C) DIF regime (72°F nights and 66°F days [22°C nights and 29°C days]). On days of bright, intense light, we increase the negative DIF to 10°F (6°C) or more. Precise temperature control is a constant priority throughout the entire growing season. Deviation of 2°F (1°C) or more is unacceptable. The only exceptions are the warm, sunny days of spring, when full greenhouse cooling proves impossible.

Days 11–60

This is what we consider the most intense phase of production. At about fourteen days after sticking rooted cuttings, we begin the exacting process of regulating with chemicals. During this stage we use Florel. If you've never used it, realize that it is a wonder chemical when used properly. Our rates can range from 225–800 ppm, with frequencies ranging from two to eight times and the intervals from once a week to once every three weeks. For all applications, mix Florel with distilled water. Coverage can range from a light mist to wetting to drip off. Some would say we live on the edge with our unconventional use of this chemical. We feel we survive these calculated risks by meeting the following: superior cultivar selection and knowledge of the traits of that cultivar; near luxury feeding; high night temperatures (72°F [22°C]); preapplication irrigation (twelve to twenty-four hours in advance of spraying); and evening and nighttime applications.

"Proceed with extreme caution" might be appropriate words of advice before using Florel in our unconventional manner. The three variables of growth regulator application—rate, frequency, and interval—are based on cultivar knowledge and what we see happening physiologically prior to application. We stop using Florel seven to eight weeks prior to scheduled bloom dates.

Days 61–110

When the calendar forces us to stop applying Florel, we continue the benefits of chemical growth regulation with a mix of Cycocel and B-Nine. We have not reached a point in the plant's development that I call "the point of critical mass." This is a point where there seems to be much less risk in overregulating. At this time and under our conditions, the plant is so dynamic and becoming so massive in its steep development curve that it seems almost immune to growth regulators at normal rates.

Our rates range from 750 ppm of Cycocel and 1,000 ppm B-Nine to 1,000 ppm Cycocel and 3,000 ppm B-Nine. Frequencies range from two to seven times and intervals from five to fourteen days. We make our applications with a carrier of Capsil-30 using only distilled water. Coverage ranges from a light mist to heavy runoff. Late applications are accomplished by holding the spray nozzle below any breaking color and reducing the pressure, because splatter of the mix would scar bloom petals.

Disbudding

When we're in the Cycocel/B-Nine mode, the bud-aborting effects of Florel wear off and buds begin to flush. Twice a week we remove these buds manually and simultaneously clean any leaves that need to be removed. Disbudding enhances our quality and gives us an opportunity to inspect every plant's internode length and pinpoint exacting schedules for future Cycocel/B-Nine applications. Generally, disbudding continues until four weeks prior to first scheduled sales dates.

Continued

Potting up a 288 plug to a 4" (10 cm) pot takes from six to eight weeks to bloom on early-flowering selections, while the majority of selections take from eight to ten weeks to flower.

Pests & Diseases

Cuttings

No other crop that you will grow in your greenhouse can become devastated as fast as vegetatively propagated geraniums can in warm weather by bacterial diseases such as *Xanthomonas campestris* pv. *pelargonii* and *Ralstonia solanacearum.* You cannot treat plants for these pathogens; infected plants must be safely removed from your property and destroyed. An infestation is so serious that prevention is the only control—see the Sanitation section for more information. Other diseases that can affect geraniums include *Alternaria,* botrytis, *Pythium,* rhizoctonia, and rust.

Seed

Fortunately, *Xanthomonas* and *Ralstonia* are not seedborne so infections will not arise if all you grow are seed geraniums. If you also grow any geraniums from cuttings, you should separate these by a considerable distance from your seed geranium crops, and sanitation measures should be put in place to reduce any chance of spreading disease from the cutting crops to the seed crops. A preventative fungicide drench program is common with seed geraniums to prevent *Pythium* infections. Insects that can attack seed geraniums include aphids, caterpillars, slugs, and thrips, but all of these are uncommon problems.

Sanitation

The best way to grow clean geranium plants from cuttings is never to have a problem. Plants cannot be treated for the two most serious diseases, *Xanthomonas campestris* pv. *pelargonii* and *Ralstonia solanacearum.* Prevention is the best defense.

Always order cuttings from a reputable supplier who is working with indexed mother stock and is rigidly adhering to a clean-stock program. Indexing stock is an expensive proposition. All reputable geranium suppliers have had problems in the past, and the

How Bob Frye Grows Premium 7" Geraniums *(Continued)*

DIF and DIP

During production there's only about a two-week period (during cutting establishment) when we're not operating some form of negative DIF. In addition, about March 10 we begin to use a substantial DIP. Our DIP practices consist of an early-morning ventilation of cool air (48°F [9°C]) starting fifteen to twenty minutes before first light and lasting for a total duration of two hours. This is not a slow, gradual ventilation—it's a blast of cold air. We maintain the cold DIP temperature for no longer than two hours because we don't want to appreciably lower root temperatures. At The Plantation, this practice is continued until the crop is gone. In addition to promoting compact growth and branching, it also helps to acclimate plants to realistic cold-morning conditions when they leave the greenhouse. After the two-hour DIP, normal DIF temperatures are resumed, and then right back to 72°F (22°C) for the night warm up. The exact DIP and DIF regimes are adjusted daily, being fine-tuned based on how the crop is developing.

Irrigation

Geraniums should not be grown on the dry side for any length of time if you want to produce lush, well-branched, premium plants. Moderate wet/dry cycles should be established to encourage clean, healthy, expanding root systems, but not to the point that it starts to harden soft, green tissues and retard or eliminate branching potential. We have the ability to automatically or semiautomatically water, fertilize, or apply fungicide to our entire production of pots and baskets in three hours or less. We segregate cultivars into their own irrigation zones so we can manage them correctly and independently.

Our process has been designed to maximize our quality. The process directs our endeavors and production schedules. Product revenue is elastically tied to the quality of geraniums as perceived by our retail public. The most secure way we can protect ourselves in "our" market is to produce the highest-quality plants possible.

* *Excerpted from* GrowerTalks, *September 1999, pp. 51, 56.*

best suppliers have extensive protocols for clean plant production. While no supplier is immune to the occasional problem, the best ones can isolate the source stock plants from which a cutting in an individual shipment originated. In this way, even if a problem slips through the cracks, it can be quickly tracked and the plants discarded before a problem arises.

The best zonal geranium growers think in terms of managing the risk posed by bacterial diseases on geraniums and develop strategies to limit that risk. You can begin preventing geranium diseases by establishing a plan for how you receive your cuttings. When you unpack boxes, make sure to keep cuttings of different varieties from different sources and different rooting stations in separate greenhouses or Quonsets. It is a good idea to keep the supplier tag with the plants and indicate origin on a pot stake for the lot. Then, once the cuttings are potted, keep plants from different sources separated (or quarantined) until you are certain they are clean. Always keep the supplier tag with the lot of plants. That way, if you have a problem later, you will be able to isolate it.

Routinely scout your geranium crop to catch suspicious plants quickly. Scout first thing in the morning before doing anything else. Warm weather combined with overhead irrigation can quickly spread an infection. Be diligent in scouting: Plants caught very early can be discarded before they have a chance to affect other geranium varieties.

Restrict traffic in your greenhouse to minimize problems. When workers move from one greenhouse to the next, they can spread bacterial disease on their hands and clothing or through insects that may be hitchhiking. Assign certain employees to the geranium crop and ask them to take note as to where they have been in the greenhouse and where they are going. When they go from one geranium greenhouse to another, they must wash their hands. Do not allow any employee that has visited another production facility—or even a retailer that markets geraniums—to reenter your geranium production area afterward.

Avoid hanging any baskets over your geranium crop. Do not hang ivy geraniums overhead: Splashing water spreads disease from one plant to another. Ideally, irrigate geraniums with drip tubes.

Do not grow any perennial *Geranium* species if you are also growing zonals. Be very careful with specialty geraniums such as scented or fancy-leafed varieties. Make sure your source is working with indexed stock. Do not grow seedling geraniums near zonal

Southern Bacterial Wilt

P. Allen Hammer and Karen Rane

Ralstonia solanacearum, formerly *Pseudomonas solanacearum,* is a serious disease of geraniums that results in plant death. *Ralstonia* invades plants primarily through the roots, and it can be introduced to a greenhouse crop through the same methods as *Xanthomonas campestris* pv. *pelargonii.* Both pathogens invade the vascular system of the plant, block water transport, and result in wilt and disease.

Why is it important to determine which bacterial wilt disease is responsible for wilting geraniums when both diseases are fatal and the pathogens spread in a similar manner? There are two major differences in these pathogens, and they have a potentially significant impact on disease management.

The first difference is host range. *Xanthomonas* affects only *Pelargonium* and *Geranium* species. *Ralstonia* is reported to have a wide host range that includes numerous vegetable and herbaceous ornamental crops. It is possible for *Ralstonia* to spread to other crops in the greenhouse, especially if irrigation water from infected geraniums runs down the bench and comes in contact with roots of other plants.

The second difference is pathogen survival. *Xanthomonas* cannot survive for long periods of time outside of host tissue. When *Xanthomonas*-infected plant debris decays, the pathogen population declines dramatically. This is why cleaning a greenhouse of all geranium plants and plant debris effectively reduces the chances of a recurrence of the disease to near zero. In contrast, *Ralstonia* is known to survive for several years in the soil of warm climates without the presence of susceptible host tissue. Severe disease outbreaks have occurred in susceptible field crops after long rotations with non-host crops. This means that the potential exists for *Ralstonia* to survive in greenhouses on solid floors or benches with contaminated soil, sand, or other substrates.

Growers should send any wilted geranium plant to a diagnostic laboratory for confirmation of bacterial disease.

geraniums. Seed geraniums are highly susceptible to *Xanthomonas* and die quickly.

Do not keep any geraniums in the garden on or near your premises. Ask employees to put in fresh plants each year, discarding plants at the end of the summer—even if geraniums will overwinter in your area. Under no condition should you ever hold a geranium over from one year to the next.

Some growers conduct random testing for *Xanthomonas* through the production cycle. Portable test kits are available from Agdia Inc. Additionally, your supplier or floriculture extension agent can provide you with a list of testing laboratories that will accept plant samples.

If you have a bacterial disease outbreak, remove all plant debris from the infected areas. Treat benches and all greenhouse surfaces with a quaternary ammonium compound before introducing the next crop. *Xanthomonas* only survives in plant tissue, dead or alive. It cannot live on concrete or metal. Quaternary sprays, however, will not kill the bacterium in the plant tissue. Be sure to remove all debris from the growing area, propagation area, material handling equipment (carts), and production lines. One dried up leaf can be all it takes to start a new outbreak. Don't dump your diseased plants on your property, bag them and send them to the landfill. Use only new pots and new, sterile soil.

Troubleshooting

Chlorotic, reddish lower leaves combined with a flat, gray-green foliage color can indicate iron and/or manganese toxicity. Symptoms first appear in March or early April and continue through the season. The problem is due to manganese/iron interaction. Keep the iron-to-manganese ratio between 1:1 and 2:1 and maintain a pH of 5.8. Use fertilizers with low manganese levels and keep the water pH at 6.5 or higher. Ammoniacal nitrogen should never exceed 25–30% of the total nitrogen. Supply most nitrogen from nitrate fertilizers.

Low temperatures (below 45°F [7°C]) or severe phosphorus deficiency can cause reddish foliage, while high temperatures (above 85°F [20°C]) can lead to chlorotic leaves.

A pH drop can result in sudden lower-leaf yellowing and death. Monitor pH levels throughout the crop and adjust as needed by switching nitrogen sources or adding lime. Excessive stretch and leaf expansion can be due to too much ammoniacal nitrogen.

Varieties

Four main companies in the US supply zonal geraniums: Ball FloraPlant, Dummen, Oglevee, and Syngenta. These are the leading suppliers, and each has excellent genetics. Growers have preferences, as some varieties perform better in certain environmental

What If You Suspect a Bacterial Disease?

1. Send samples of affected plants to a reputable testing lab. Your supplier or floriculture extension agent can provide you with a list. Select a wide range of plants showing symptoms—some showing serious effects and others just beginning to show symptoms. Wrap the soil ball in a plastic bag to prevent soil from mixing with leaves during transit. Do not wrap the foliage in plastic—use newspaper.
2. Keep any suspect plants away from the rest of your geranium crop. Place the affected crop on quarantine and make the area off-limits. Allow access only by the most trusted employee to water and fertilize as needed. Irrigate only as needed and do so without splashing.
3. If tests confirm a bacterial infection, contact your supplier immediately. Note the date and name of the person with whom you spoke.
4. Unless otherwise instructed, immediately rogue the affected plants and those within 3' (0.9 m) of them. When discarding affected plants, place them in plastic trash bags and remove them from the greenhouse. Burn or bury plants: Never compost plants suspected of being infected with bacteria. Sometimes it's prudent to discard all of the plants of the variety, especially if symptoms seem to be limited to one variety. Keep a record of how many plants and of which varieties you dispose.
5. Remove all leaves and other debris and disinfect the bench with GreenShield, Physan 20, or ZeroTol. Ideally, do not put another geranium crop in this space for the season.
6. Treat any remaining geraniums with a copper-based pesticide to help reduce the spread of bacteria.
7. Throw out any geraniums remaining at the end of the production season. Do not compost these plants and use the compost for later geranium production.

conditions than others, but most of the time growers will order varieties from several sources to get the best selection for their specific market.

Geraniums in the 'Allure', 'Designer', 'Rocky Mountain', and 'Survivor' series are the most versatile plants on the market. Used primarily for 6" (15 cm) pots and larger, these versatile plants can fill many sizes. Series such as 'Fantasia' and 'Savannah' have dark foliage and are more reactive to growth regulators, lending them to also work well in smaller pots. Interspecific geraniums (i.e., crosses between ivy and zonal geraniums) have shown great summer performance. Varieties such as 'Calliope' combine the dark foliage and dark-red flower colors of ivies and the summer performance of zonals into one great plant.

Seed geraniums have become a somewhat mass-market item in the past decade but are making a comeback at retail nurseries as an affordable way to have geraniums in the garden. Breeders are now offering more unique colors to stimulate interest. Seed geraniums are also generally more heat tolerant than zonals, so garden performance is better in hot climates. 'Horizon', 'Maverick', and 'Pinto' for larger pot sizes, such as 4.5" or 6" (12 or 15 cm); 'Orbit Synchro' and 'Ringo 2000' for medium pot sizes, such as 4" (10 cm); and 'Elite' and 'Multibloom' for packs are all seed geranium series to consider.

Postharvest

Geraniums are sensitive to ethylene. Seed and single ivy geraniums are subject to petal drop (shattering). Plants under stress produce more ethylene, which in turn creates more petal drop. Lower temperatures inhibit ethylene production and help to reduce petal fall. If plants must be held once they are ready for sale, hold them at 50–55°F (10–13°C). In the past, some growers have treated plants with silver thiosulfate (STS). The newer product, 1-MCP, is also an ethylene inhibitor and may be helpful in reducing petal shatter (see the 1-MCP label for usage guidelines). Some growers ship plants before flowers open to avoid ethylene problems.

Zonal geraniums do not have problems with petal drop; however, all shipped geraniums can suffer from yellow leaves. Do not ship plants for more than forty-eight hours and ship at lower temperatures, 40°F (4°C).

Retailers should provide a high-light environment and maintain uniform moisture. Since geranium leaves are thick, plants can be in significant water stress when they are wilted. Be sure to remove dead flowers and foliage to remove sources of ethylene production.

Consumers should deadhead geraniums through the summer to ensure continuous flowering. While in most areas of the country geraniums should be planted in full sun, in hot, humid southern regions, shade during the hottest four hours of the day is ideal. Fertilize regularly.

Pelargonium peltatum
Common names: Ivy geranium; balcony geranium
Annual

Anyone who has ever visited northern Europe in the summer has seen window boxes, patio planters, and balconies everywhere ablaze with color from ivy geraniums. The wave of new balcony plants like bacopa, *Bidens, Scaevola,* and verbena are now planted in combination with ivy geraniums in masterpiece combinations befitting the finest European painters.

In the United States, ivy geraniums have not caught on to the same degree. In the garden, they perform best when night temperatures are cool—the 50s and 60s (°F [10–20°C]) are ideal, bringing out their best color. In the United States, that restricts them to mountainous areas, New England, and the

Pacific Coast. Elsewhere plants will struggle through the summer, returning to their glory with a final blaze of color in the fall before frost.

For growers, ivy geraniums are tricky. They are subject to a condition known as edema (also spelled oedema in some texts), which causes cells in the leaf to burst and cork over, making unsightly spots. However, with proper water management this problem can usually be avoided. The good news here is that there are many new zonal and ivy interspecific geraniums now on the market that resemble ivies, especially with their flower form, but resist edema and are tolerant of higher temperatures. You may want to steer your customers into using these newer plants since consumers will experience better performance from these.

Propagation

Ivy geraniums are commonly propagated by cuttings, but seed varieties are also available.

Cuttings

While ivy geraniums are not susceptible to *Xanthomonas*, they can harbor the pathogen without symptoms. For that reason, it is vital that you begin your vegetative crop with only indexed stock that has come from a reputable supplier with a rigorous clean-stock program. Most growers buy in unrooted cuttings or liners to ensure they are getting off to the best start.

Root ivy geraniums as you would zonals with these exceptions: Run EC levels at 0.5 mS/cm or less until plants are ready to be hardened for sale or planting. Ivy geraniums do not tolerate light levels as high as zonals. In hardening, raise light levels to 2,000–3,000 f.c. (22–32 klux). For really full plants, make a first application of Florel at 150–200 ppm one week prior to transplant.

The crop time from sticking unrooted cuttings until a finished 105-liner tray is four to five weeks.

Seed

In the past twenty years or so, seed companies have made great strides with ivy geraniums from seed. Seed is relatively expensive, so again, rather than taking risks losing seedlings, many growers buy in plugs—125s or larger.

Sow seed into a disease-free, germination medium with a pH of 5.8–6.2 and a low starter charge. Cover with vermiculite to maintain moisture around the seed. Expect radicle emergence in about four days at 73°F (23°C). Germination temperature is a critical factor for germinating ivy geraniums since temperatures about 76°F (24°C) can cause thermodormancy, which will reduce germination percentages dramatically. Maintain a high substrate moisture level at all times to ensure fast emergence. Germination chambers and full light provide ideal conditions, but the trays must be moved to the greenhouse bench as soon as full emergence is reached to prevent the hypocotyls from stretching. Ivy geraniums from seed are bred to branch freely and maintain a compact habit but under low-light conditions Cycocel at 300 ppm can be applied when the first three to four leaves have expanded.

A 128 plug/liner takes from seven to eight weeks from sowing until it is ready to transplant to larger containers.

Growing On

Most growers produce ivy geraniums in hanging baskets. Use baskets that are 10" (25 cm) or larger. The plugs can be transplanted directly into a cell pack or 4" (10 cm) pot and sold in 4" (10 cm) pots or can later be repotted to hanging baskets. Plant 1 liner or plug/4" (10 cm) pot; 2 liners/6" (15 cm) pot; and 3–5 liners/10" (25 cm) hanging basket. If flower buds are present, remove them before planting. Establish plugs in pots first before transplanting them to baskets. Grow ivy geraniums planted into hanging baskets on the bench first until plants are firmly established before hanging them. Since ivy geraniums can carry *Xanthomonas* without showing the disease, never hang them over a zonal geranium crop.

Grow on at a night temperature of 68°F (20°C) until the plants are established, after which the thermostat can be set at about 63–65°F (17–18°C) for the remaining crop time. Sufficient soil temperatures are critical during establishment. Most ivy crops are lost at the beginning by growers starting with substrate temperatures that are too low and soil that is too wet, leading to problems with *Pythium*. In bright, sunny weather, allowing the daytime temperatures to rise to 70–75°F (21–25°C) before venting will promote faster growth and earlier flowering.

Some ivy geranium varieties branch better than others. For varieties that do not branch freely, a Florel application two weeks after planting at 350 ppm is recommended. Make sure plants are well irrigated prior to the Florel application and plan the application for late in the day so the spray does not dry on the foliage too fast. To increase branching even more, a soft pinch can be made one week after applying the Florel.

Ivy geraniums require only moderate light levels, 2,500–3,500 f.c. (27–38 klux), but will benefit from as much light as possible as long as greenhouse temperatures are controlled. Do not grow ivy geranium hanging baskets in the attic or on a gutter hanging next to glass or plastic, as edema will become a problem (see Pests & Diseases) due to rapid heating and drying out of the medium. A high light level also hardens foliage, setting the stage for edema.

Keep plants well irrigated to promote active growth. Fertilize with 150–220 ppm and alternate with a calcium-based feed such as 14-4-14 and an ammonium-based feed such as 21-5-20. Maintain the pH at 5.5–5.8 to avoid iron problems. Under higher temperatures or if the pH is too high, you may see iron chlorosis, which usually starts to show as a yellowing of the newest growth. Use iron chelate at 3–4 oz./100 gal. (22–29 g/100 l) to correct this problem.

Growth regulators can be helpful in controlling height, and Cycocel is the most effective. Not only will it control stem elongation, but it will also promote earlier flowering. Apply Cycocel when new growth is 1.0–1.5" (3–4 cm) long at 750–1,000 ppm. Several applications at lower rates will produce a more even crop with less risk of phytotoxicity from the Cycocel. Many growers pinch and then treat with Cycocel immediately afterward.

Ivy geraniums flower according to how much light they have accumulated, so as days naturally lengthen in the spring and light levels improve, crop times will shorten.

For vegetative varieties, with one 105 liner per 4" (10 cm) pot, the crop time is eight to ten weeks, while a 10–12" (25–30 cm) basket with four or five liners per pot will take from eleven to thirteen weeks.

For seed varieties, a 4" (10 cm) pot with one 128 plug/liner will flower eight to nine weeks after transplanting, while a 12" (30 cm) basket with five liners will flower in ten to eleven weeks.

Pests & Diseases

Pythium is the major disease of seed-raised geraniums, and a preventative program of sprays and/or drenches should be followed. However, rhizoctonia can also be troublesome, especially in high-density crops. More recently *Thielaviopsis* has also become a problem on ivy geraniums that are grown stressed, especially by high salts.

Edema looks like a disease but is actually a physiological problem caused when leaf cells burst and subsequently cork over. It occurs when leaves are not able to transpire rapidly enough, namely under high humidity conditions. To reduce the incidence of edema, choose varieties that are more resistant, such as the new interspecific varieties. Also make sure that ivy geraniums are growing in a well-ventilated portion of the greenhouse. During early crop stages, be careful to apply only as much water as plants need and, as they age, do not allow the medium to completely dry out before rewetting. Provide light shade under high-light conditions. Your goal is to maintain uniform moisture levels in the medium, which will help to avoid wide fluctuations in humidity at the plant level. Growing well-branched plants that form a dense canopy over the tops of baskets and pots will also help to control edema.

Whiteflies and aphids are the main pests; but if the plants are being grown in a particularly dry environment, then spider mites could be a problem. Monitoring the crop and applying suitable insecticides in a timely manner will help to prevent problems from arising.

Troubleshooting

Too much ammoniacal nitrogen in the fertilizer and overfertilization under low-light conditions can cause excess vegetative growth and lack of flowers. Plants drying out between irrigations, low pH, high soluble salts, or *Pythium* can cause foliage necrosis. Poorly branched, thin plants can be the result of low fertilizer in early growth stages, low light, or poor variety choice.

Varieties

Only a few ivy geraniums from seed are currently on the market. 'Summer Showers' comes in three separate colors and a mix. 'Tornado' is a well-branched, compact series that does not require a pinch nor growth regulators. It has eight separate colors and a mix.

A large number of vegetatively propagated ivy geranium series are available. All of them include a wide range of colors. Among the series on the market today are 'Precision', 'Contessa', 'Freestyle', 'Temprano', 'Global', 'Pacific', and 'Atlantic'. New interspecific geraniums that resist edema and are more heat tolerant but still retain a somewhat ivy look include 'Caliente' with eight colors; 'Galleria' with seven colors; and 'Sonata' with seven colors.

Postharvest

Sell ivy geraniums when flower heads are well above the foliage and just showing color. Ivy geraniums are a great addition to mixed hanging baskets and color bowls. If they are being used in this way, they should be planted up about four weeks before the intended time of sale.

If night temperatures are high, instruct consumers to place ivy geraniums in partially shaded spots to help moderate daytime temperatures and light levels.

All geraniums are highly sensitive to ethylene petal drop, so if shipping long distances treat plants with EthylBloc (MCPP) either prior to or during shipping.

Pennisetum

Pennisetum glaucum
Common name: Ornamental millet
Annual, ornamental grass

Ornamental millet is a tall, grass-like plant that adds dramatic height to gardens and large containers. A relative newcomer to home gardens, millet is upright and stately. While the variety 'Purple Majesty' was the first garden selection in this species as well as a winner of the 2003 All-America Selections Gold Medal, there are a number of other varieties available also.

Propagation

Because millet is a larger seedling and plant, most growers sow into a larger plug cell sizes like 200s, 128s, and 72 liners. Larger cells result in shorter overall crop times. Multisowing two to three seeds per plug results in fuller, more attractive plants at retail. Sow millet into a well-drained, disease-free, soilless medium with a pH of 5.5–6.3 and a medium initial nutrient charge (EC 0.75 mS/cm with a 1:2 extraction).

Millet germinates best when the seed is covered or sown into an indentation, or dibble hole, and then covered with sowing media. Light is not required for germination. Germinate at 72–78°F (22–25°C). Keep moisture levels constant at medium (level 3) to medium wet (level 4). Seed germinates in two to three days at the recommended temperatures. Temperatures below 68°F (20°C) will significantly delay germination.

As long as the soil is kept evenly moist, high air humidity is not required for germination. Therefore, seed can be germinated directly on the bench. Keep soil moisture high until radicle emergence, and then reduce moisture levels after the radicle penetrates the medium. The moisture should be medium (level 3) to stimulate the root system. Do not allow the seedlings to wilt. After germination the temperatures can be lowered to 68–72°F (20–22°C).

Millet is a relatively heavy feeder. Apply 50–75 ppm nitrogen (N) from 15-0-15 at radicle emergence. Increase the feed levels to 100–150 ppm N as leaves develop. As the plug matures, increase the fertilizer to rate 2 (100–175 ppm N and an EC of 0.7–1.2 mS/cm). Maintain a media pH of 5.8–6.2 and EC at 0.7–1.0 mS/cm (1:2 extraction).

Plugs that become root-bound or are stressed by drought or nutrient deficiency will not perform well after transplanting.

When growing millet plugs, no plant growth regulators are required. If needed for toning in the plug, a Bonzi spray at 2 ppm during Stages 3–4 will work.

The crop time from sown seed to a transplantable 128 cell is two to three weeks. Seed can also be sown directly to the final container (see Growing On).

Growing On

Millet is normally grown in or transplanted into 6" (15 cm) to 1-gal. (4 l) containers. Some growers will direct-sow into a 4" (10 cm) pot, which reduces crop time by two weeks. Use a well-drained, disease-free, soilless medium with a pH of 5.5–6.5 and a medium initial nutrient charge. Growing on at temperatures of 68–85°F (20–30°C) days and 64–66°F (18–19°C) nights is optimal. Millet is a warm-season crop. Higher temperatures result in faster growth and taller plants. An average temperature below 64°F (18°C) will significantly delay crop time; below 60°F (16°C) will stop plant growth.

Keep light levels as high as possible. Higher light results in stronger, thicker stems and better basal branching. Young plants are green. The stem and midrib of the foliage first turn purple after about eight leaves have developed. Foliage coloration occurs when the plants are moved from the greenhouse outside to full sun. Because the plants will be mostly green when sold, a color picture label is recommended to help consumers understand what the plant looks like after it is planted in the garden.

Millet likes a uniform, constant moisture level (level 3). Maintain even moisture. Do not allow plants to wilt. Feed plants weekly with 150–200 ppm N in a complete fertilizer.

Applying Bonzi at an early stage results in bushier plants with more side shoots and does significantly affect the final plant height. If seeds are sown directly into final containers, apply a 6–8 ppm Bonzi drench four weeks after sowing. If using plugs, a 3–5 ppm Bonzi drench can be applied one week after transplanting.

You can also apply Florel as an optional growth regulator treatment. Apply two applications of Florel spray at 500 ppm. The first application can be done one week after transplant or four weeks after sowing. The second application can be done ten to fourteen days later. These treatments can result in bushier plants with more side shoots. However, Florel is not as strong as Bonzi in height control.

If seeds are sown directly into final containers, two applications of a 6–9 ppm Bonzi drench can be used to control plant height. The first application can be done four weeks after sowing. Repeat ten days later.

If seeds are sown into plug trays, apply a 6–8 ppm Bonzi drench one week after transplanting into final container. Only one application is needed.

These treatments result in plants with the first flower spike approximately 2–2.5' (60–75 cm) above the top of the container for 'Purple Majesty' and 1.8–2' (55–65 cm) for 'Purple Baron' and 'Jester'.

Millet's response to PGRs varies with container size and different environmental conditions. Be sure to run an in-house trial to determine the best rate or method for your conditions.

Based on the PanAmerican Seed research trial at Elburn, Illinois, transplanted plugs require fewer

Table 30. Millet Crop Scheduling[a]

CONTAINER SIZE	PLANTS PER POT	WEEKS FROM TRANSPLANT
1801s, 4–4.5" (10–11 cm) pot	1–2[b]	4–5 (green)
1-gal. (4 l) container or 8" (20 cm) standard pot	3[b]	5–6 (green)
1-gal. (4 l) container or 8" (20 cm) standard pot	3[b]	11–13[c] (flowering)

[a] Assumes a sow-to-transplant time of two to three weeks for a 288-cell plug tray. Crop time is based on a 68°F (20°C) average daily temperature. When plants are grown in warm temperatures, the crop time can be two or more weeks shorter.
When selling plants "green," the crop time is for plants with roots established enough to hold the soil ball together and with a height of 12–16" (30–40 cm). Allowing plants to become excessively root-bound or to flower prior to planting in the landscape will result in shorter plants.

[b] For multisown plugs, only one plug is needed per pot. For single-sown plugs, plant the plugs close together in the center of the pot.

[c] The crop time for a 1-gal. (4 l) container is for plants with flower spikes emerging. See the growth regulator recommendations for producing shorter plants with flower spikes.

PGRs and make bushier plants after PGR applications, but crop timing is one to two weeks longer than direct-sown plants. Do not pinch.

Also, do not allow the plants to be stunted from water stress or inadequate fertilizer, and do not allow the plants to become root-bound. Plants that are stunted in a young stage may produce only single, short stems and not reach their full potential.

The crop time can be reduced by two weeks if seed is direct-sown into the final container. If directly sown, the seed can be easily germinated in the finished area. *Note:* 'Jester' does not perform as well in cooler temperatures.

Varieties

'Purple Majesty' reaches 4–5' (1.2–1.5 m) tall in the garden with 12–14" (30–35 cm) flower plumes. At the time of flowering, 'Jester' is 3–4' (90–120 cm) tall, while 'Purple Baron' is 2.5–3.5' (75–110 cm) tall. 'Jade Princess' offers chartreuse foliage in contrast with a burgundy-bronze flower plume and grows 24–30" (60–75 cm) tall.

When grown under favorable conditions each plant produces one to three main stems, with some secondary branches from the base. 'Jester' has the most branching, followed by 'Purple Baron'. 'Purple Majesty' produces a taller, statelier plant. It is recommended that these varieties be sown with multiple seeds per plug cell because of this natural habit. Multiseed pellet is also available.

'Jester' flowers seven to ten days later than 'Purple Baron' and 'Purple Majesty', and therefore stays compact longer into the season than the other two. These F_1 hybrids are suitable for growing in 4" (10 cm) and larger containers. 'Purple Majesty' is a 2003 All-America Selections Gold Medal winner.

Pennisetum setaceum

Common names: Fountain grass; pink fountain grass

Grass, annual in cold-winter locations, perennial (Hardy to USDA Zones 8–10)

This green-leafed, upright, arching grass lives up to its common name of fountain grass as it foliage arches all the way back down to the soil. Growing from 24–36" (61–91 cm) tall (in flower), plants have light pink flowers and prefer to be planted in full-sun locations.

Propagation & Growing On

This is a very easy grass to germinate and grow. Sowings made to a 288-plug tray, lightly covered with vermiculite or left exposed to light, will germinate in three to six days when germinated at 70–72°F (21–22°C) soil temperature. These can be transplanted to 4" (10 cm) pots or an 18 tray (18 cells per standard flat/tray) in five to seven weeks. Allow another six to seven weeks to finish for spring sales at 70°F (21°C) days and nights at 60–62°F (16–17°C).

You could also sow five to seven seeds directly into a 32-cell tray in early to mid-February following the germination guidelines above (do not cover the seed). Transplant plugs to 6" (15 cm) or 1-gal. (4 l) pots in six to eight weeks. Plants in 6" (15 cm) pots should be ready in another six to eight weeks for spring sales. At six to eight weeks, plants in 1-gal. (4 l) pots may be filled out on top, but the pots may not have sent roots to the bottom of the container and the soil ball may fall apart. It is better to allow 1-gal. (4 l) pots eight to ten weeks from transplanting for better rooting. Grow on at 55–60°F (13–16°C) nights, if the plants are starting to grow too fast.

Pests & Diseases

Few problems exist with this crop.

Pennisetum setaceum 'Rubrum'

Common name: Purple fountain grass

Grass, tender perennial or annual in cold-winter locations

Purple fountain grass seems to show up everywhere these days, being tucked into corners, crevices, and planted en masse by creative landscapers or home gardeners. Purple fountain grass is treasured for its dark burgundy foliage, arching flower heads, and pink-red flowers.

For growers, the hardest part about purple fountain grass is obtaining the young plants. Once you have your plants, they are slow to grow during the winter

months. But you will not be able to keep the resulting filled-out 1-gal. (4 l) pots in stock, so do whatever you have to do to make sure you get your plants.

Propagation

Sometimes not readily available, *P. setaceum* 'Rubrum' is sterile: Plants do not produce seed. If they did, plugs would be widely available. Stock plants have to be held over in a frost-free greenhouse, which requires a lot of space.

Cuttings

Purple fountain grass can be propagated by cuttings. Harvest 2.5–3.5" (6–9 cm), first-node (closest to the crown) cuttings from flowering stalks (culms). The node should be about 0.25" (6 mm) from the end of the cutting. Stick cuttings in the medium and place under mist. Cuttings will root in about twenty-one days. Provide short days during propagation.

Division

Plants can also be divided in December. Cut plants back to about 6" (15 cm) tall. Trim roots back to leave about 3" (8 cm) of actively growing roots. Divide plants into two- to three-shoot plants and place in 36- or 50-cell trays. Root out under short days at 73°F (23°C).

Growing On

Transplant to a disease-free, well-drained, soilless medium with a pH of 5.8–6.2 and a moderate initial starter charge. Plants will finish in eleven weeks from a 36- to 50-cell liner. The larger the starter plant, the faster plants finish.

Grow on at an average temperature of 68°F (20°C). Do not let temperatures drop below 50°F (10°C). Plants are easily killed by frost. Fertilize with 150–200 ppm with 15-0-15 at every other irrigation. Grow plants under full sun: 4,000–6,000 f.c. (43–65 klux).

Provide long days (more than thirteen hours) for ten weeks after planting through night interruption lighting or by extending the day length to force plants into plume.

If growth regulators are needed, 5 ppm Sumagic can be used twenty-one days after planting. Reapply, if needed, two weeks later.

Plants transplanted in 1-gal. (4 l) containers require from nine to eleven weeks from transplanting to sales.

Pests & Diseases

Aphids, leaf miners, thrips, and whiteflies can be problems.

Pentas

Pentas lanceolata
Common name: Egyptian starflower
Annual

Pentas has been a long-time favorite for southern gardens, but its popularity with growers has really grown since the advent of dwarf varieties such as 'Graffiti'. *Pentas* is an especially strong landscape performer in hot, humid climates but is becoming a summer favorite for more northern areas with the introductions of strong garden performers such as 'Butterfly', 'Northern Lights', and 'Starla'. Typically grown for spring sales in 4" (10 cm) or 1-qt. (1.1 l) pots, *Pentas* is also a great season-extending crop when grown in 1-gal. (4 l) containers.

In the garden, *Pentas* provides flowers all summer long in a rainbow of colors—from red to pink to white to purple. Its flowers attract hummingbirds and butterflies. You can find tall and dwarf seed-propagated *Pentas* series as well as vegetatively propagated varieties. Be sure to grow *Pentas* warm and under high-light conditions for the best quality and fastest crop time. For northern growers, sales should be scheduled to start around Mother's Day; for markets in the Deep South, the sales season can be almost year-round.

Propagation

Use only pelleted seed, since *Pentas* seed is quite small. Sow seed onto a premoistened, well-drained, disease-free, germination medium with a pH of 6.5–6.8. *Pentas* is very sensitive to low pH levels, and germination, will be slow and erratic at pH levels below 6.4. Do not cover. Light is not necessary for germination, but providing 10–100 f.c. (108–1,076 lux) of light will improve germination uniformity and reduce stretch potential. It is best to germinate in a chamber that provides 100% humidity. Germinate at 70–72°F (21–22°C) substrate temperatures. Expect radical emergence in seven to ten days.

Move trays to Stage 2 for cotyledon expansion. Drop the substrate temperatures to 68–72°F (20–22°C) day and night. At this point reduce humidity to about 50%; this will improve germination and encourage root penetration into the media. As soon as cotyledons expand, begin fertilizing with 50–75 ppm from 14-0-14 or 14-2-14. Alternate feed with clear water. Be sure foliage is completely dry by nightfall to prevent disease problems. Stage 2 lasts approximately seven to ten days.

Stage 3, development of true leaves, takes thirty-five to forty-two days for 288s. Maintain soil temperatures at 68–72°F (20–22°C). Allow trays to dry down between irrigations, but not to the point of wilting. Under high light conditions increase feed to 150 ppm nitrogen (N) using 17-5-17; under lower light feed at 100 ppm N with a calcium-based feed such as 14-2-14. Fertilize every second or third irrigation to maintain an EC of 1.0 mS/cm.

If growth regulators are required to prevent stretch in the plug tray, apply B-Nine at 2,500 ppm or Cycocel at 500 ppm.

Harden trays for seven days prior to transplant by continuing to maintain substrate temperatures and allowing trays to dry thoroughly between irrigations. If growth is not as expected, the pH may be at too low a level. Check the pH to be sure it is above 6.4. If it is below, apply liquid lime at a low rate since plugs do not have a high buffering capacity.

Most growers buy in *Pentas* plugs, as they are not only a long crop in propagation, but they require relatively high production temperatures and high light levels. If you are in the North, your quickest finishing times will come from plugs grown in high-light southern locations.

A 288 plug takes from seven to nine weeks from sowing to be ready to transplant to larger containers.

Growing On

Use one plug per 4 or 6" (10 or 15 cm) pot. Select a well-drained, disease-free medium with a moderate initial nutrient charge and a high lime charge to adjust the media pH to 6.5–6.8. High lime soil used for geraniums works well for *Pentas.*

Provide as much light as possible during finishing. Plants flower much quicker with higher light levels. *Pentas* are facultative long-day plants; that is, long days are not required for flowering, but the plants will flower earlier under long days. During winter months when light levels are low and the photoperiod is less than ten hours, crops take much longer to finish.

Maintain minimum temperatures for growing on of 72–75°F (22–24°C) days and 62–65°F (13–16°C) nights, or an average daily temperature of 68°F (29°C). *Pentas* can tolerate much higher temperatures once plants are established; however, day temperatures above 100°F (38°C) are not recommended. Under very high day temperatures provide shade to reduce stress to the crop.

Fertilize at every other irrigation with 150–200 ppm using a calcium-based fertilizer such as 15-5-15. *Pentas* requires relatively high levels of calcium. If calcium is deficient, the plants will exhibit some leaf distortion. *Pentas* is not a heavy feeder. Leach at every fifth irrigation to prevent soluble salts buildup. Avoid excessive ammonia fertilizers (20-20-20), as they may drop the pH and promote iron uptake, leading to iron toxicity and slow growth. An occasional fertilization with 20-10-20 at 200 ppm N can be used to promote leaf expansion if necessary. Do not add iron to the fertilizer solution. *Pentas* will decrease the pH of the medium over time—monitor pH closely. (See the Troubleshooting section.)

Keep *Pentas* on the dry side; do not overwater. Allow plants to dry thoroughly between irrigations; however, plants that are severely water stressed will show damage to leaf margins and/or yellow leaves. To prevent foliar disease, keep foliage as dry as possible, especially going into the night hours.

If growth regulators are required, Bonzi, Cycocel, and B-Nine are effective. Apply growth regulators prior to flower bud formation to avoid reduction in bloom size. Some growers like to pinch their 6" (15 cm) *Pentas* crop to make a fuller crop, but with modern genetics this is usually not necessary. You can pinch plants once they have rooted into their final growing container. Pinch to the third or fourth leaf node. Pinched crops require two to three more weeks to finish.

The total crop time from sowing to a flowering plant is sixteen to eighteen weeks. If transplanting a 288 plug, the crop will take eight to nine weeks until plants flower for early to mid-spring production. For late spring or early summer crops, plants flower six to seven weeks after transplanting a 288 plug.

Pests & Diseases

Aphids, fungus gnats, spider mites, thrips, and whiteflies can attack plants. Botrytis or powdery mildew can also affect flowers in high humidity and wet conditions, as can a variety of root rots such as *Pythium* and rhizoctonia if plants are grown too wet.

Troubleshooting

Iron toxicity and magnesium deficiency can occur at a pH below 6.0. Plants in flower may show signs of low pH, as *Pentas* can release hydrogen into the soil much like a geranium. Leaf rolling is one symptom of low pH and is usually accompanied by low calcium levels. Be sure to monitor pH closely throughout the crop and maintain pH levels at or close to 6.5. To increase pH, apply 12 oz. of hydrated lime/100 gal. (90 g/100 l) of water as a soil drench. Follow up with 1 tbsp. (1.4 cl) of limestone/pot. Do not apply hydrated lime ifthe medium's ammonium levels are above 10 ppm. Be sure to rinse foliage after applying liquid lime.

Poorly branched plants may be due to poor fertilization during early growth stages. Excessive drying of media between irrigations, high soluble salts, iron toxicity, and/or low soil pH can lead to foliage necrosis. High ammonia fertilizers such as 20-20-20, overfertilization under low light, low-light conditions, and/or overwatering can cause excessive vegetative growth.

Varieties

Leading *Pentas* seed varieties include the 'Butterfly', 'Graffiti', 'Kaleidoscope', and 'Starla' series. All are available in separate colors.

'Butterfly' is especially well suited to northern production. Plants produce larger flowers and umbels, are able to withstand stressful garden conditions, and grow larger and fuller, filling out pots faster. 'Butterfly' is suitable for 4" (10 cm), 6" (15 cm), or 1-gal. (4 l) pots. Plants grow to 12–14" (30–36 cm) tall in northern climates and 18–22" (46–56 cm) in southern climates. The series is available in 'Blush', 'Red', 'Deep Pink', 'Deep Rose', 'Lavender Shades', 'Light Lavender', and 'White'.

'Graffiti' is dwarf, growing to 12–14" (30–36 cm) tall in the garden. Plants do well in 4" (10 cm) production and require little to no PGRs. Colors include 'Pink', 'Bright Red', 'Red Lace', 'Rose', 'Violet', 'White', 'Lavender', and 'Lipstick'.

The 'Kaleidoscope' series is a more vigorous, upright series and is great for 6" (15 cm) or 1-gal. (4 l) production. Plants grow 18–24" (45–60 cm) tall in the garden, making them an excellent background plant or landscape item. The series is available in 'Lilac', 'Carmine', 'Appleblossom', 'Deep Red', 'Deep Rose', and 'Pink'.

'Starla' is a moderately to vigorously growing series, reaching a garden height of 14–18" (35–45 cm). Uniform plants are good for 4 and 6" (10 and 15 cm) production. Colors include 'Appleblossom', 'Deep Rose', 'Pink', 'Red', 'Lavender Shades', and 'White'.

'Northern Lights Lavender' is probably the most cold tolerant of the *Pentas* on the market, making it a favorite for northern climates. Garden height for this plant is about 20" (50 cm).

Pepper (see *Capsicum*)

Pericallis

Pericallis cruenta (also known as *Senecio cruentus*)
Common name: Cineraria
Flowering potted plant

Cineraria—the common name sounds like a dreadful disease and belies this colorful plant's bright, daisy flowers. Cineraria is a great cool-season pot crop and brightens up retail displays of flowering pot plants from January through April. In milder climates, it also makes a striking window box or front stoop attraction in planters.

Propagation

Sow seed onto a disease-free, germination medium with a pH of 5.5–5.8. Do not cover. Maintain uniform moisture in germination trays, but do not keep trays wet. Germinate at substrate temperatures of 70–75°F (21–24°C). Provide 100–400 f.c. (1.1–4.3 klux) of light. Radicles will emerge in five to seven days.

Move trays to Stage 2 for stem and cotyledon emergence, which occurs over seven to ten days. Maintain substrate temperatures. Increase light to 500–1,000 f.c. (5.4–11 klux) and begin fertilizing with 50–75 ppm from 14-0-14 once a week when cotyledons are fully expanded. Begin to allow the medium to dry slightly between irrigations once cotyledons have expanded. Make sure foliage is dry by nightfall.

Stage 3 takes about twenty-one to twenty-eight days for 288s. Lower substrate temperatures to 65–70°F (18–21°C) and allow trays to dry thoroughly between irrigations. Drop ambient air temperatures to 65°F (18°C). Increase light to 1,000–1,500 f.c. (11–16 klux). Fertilize at every second or third irrigation with 100–150 ppm from 20-10-20, alternating with 14-0-14. Keep pH at 5.5–5.8; high pH will cause leaf chlorosis. If growth regulators are required at this stage, B-Nine is effective and can be used once true leaves have expanded.

Harden plugs in Stage 4 for seven days prior to sale or transplant. Lower substrate temperatures to 60–62°F (16–17°C). However, do not drop air temperatures below 50°F (10°C); flower initiation will be delayed if the temperature is held too low for too long. Fertilize as needed with 100–150 ppm from 14-0-14. Do not use ammonium fertilizers at these low growing temperatures.

Some growers sow seed into open seed flats and transplant seedlings to 50-cells. Most growers buy in plugs of cineraria because it is easier and more convenient.

Be sure to transplant plugs on time: Do not allow plants to stretch or become overgrown.

A 128-plug tray takes from five to seven weeks to be ready to transplant to larger containers.

Growing On

Use one plant per 4" (10 cm) pot. Plant plugs a little deeper in their final growing container than they were growing at in the plug tray; this will add stability to the plant.

Select a disease-free medium with a moderate initial starter charge and a pH of 5.5–6.2. Do not use coarse mixes, which may dry out too quickly.

Irrigation is critical to cineraria. Plants require uniform moisture; however, their fine root systems cannot tolerate being wet. Be sure to place pots on benches that allow good drainage—expanded metal or wire mesh is great. Make sure that water alkalinity is above 140 ppm and EC is greater than 0.8 mS/cm. When under stress, plants benefit from a light syringing. Cineraria has a high transpiration rate and dries down quickly on warm, sunny days.

Cineraria is a cool-season plant. For the first two weeks (weeks 1–2) after transplant, grow plants on at 65–68°F (18–20°C) days and 60–62°F (16–17°C) nights to root plants into their growing containers. For the next four weeks (weeks 3–7), lower temperatures to 50–58°F (10–14°C) days and 45–55°F (7–13°C) nights to initiate flowers. Force plants for the next eight to ten weeks (weeks 8–18) at 62–65°F (17–18°C) days and 60–62°F (16–17°F) nights. *Note:* Cineraria can tolerate temperatures from 40–68°F (4–20°C). At lower production temperatures, growth is slow and stunted. However, running lower daily temperatures during finishing will create more compact plants, lessening the need for growth regulators.

Grow plants under 3,000–4,000 f.c. (32–43 klux) of light. At higher light levels, plants may wilt. As plants begin to flower, be sure to provide good air movement to prevent disease on flowers.

Start the crop off with 150–200 ppm from 20-10-20. When plants are at the bud stage (initiation), switch to 150–200 ppm from 15-0-15 and alternate with clear water. During forcing, increase fertilizer in relation to the amount of water used. Cinerarias are fast growers—if you overfeed, growth will become excessive, plants will take up too much bench space, and profitability will decline. Many growers leach prior to sale to improve postharvest performance but maintain adequate fertility to avoid yellow foliage.

Do not pinch cinerarias, as plants are naturally branching. However, be sure to space pots before leaves from adjacent plants begin to touch and the stems stretch. During the cool treatment (flower initiation), plants can remain pot tight. When you raise temperatures for forcing, plants will need spacing. For height control, plants respond to negative DIF or B-Nine. B-Nine (2,500 ppm for most varieties) can be applied as soon as plants become established in their growing containers. Apply subsequent sprays as needed.

The crop time from transplanting a 128 plug to a 6" (15 cm) pot until flowering is twelve to fourteen weeks.

Pests & Diseases

Aphids, spider mites, thrips, and whiteflies may attack plants. Disease problems can include botrytis, powdery mildew, *Pythium,* and *Verticillium* wilt. Control powdery mildew by maintaining good air movement around plants and spacing plants on time. Plants can also become infected with INSV/TSWV. Be sure to isolate your bedding plant crops, especially impatiens, from your cineraria crop to avoid spreading problems. INSV may show itself as black necrotic spots on leaves after the plants are stressed during the shipping process.

Troubleshooting

Plants can become deficient in boron at either high or low pH levels, which causes stunting and strapped and hardened leaves that are distorted and mottled. The problem may show up during hot, sunny weather when you have to water heavily, thus leaching boron from the pot. Maintain the substrate pH to prevent problems. Supplement with S.T.E.M. (soluble trace element mix) at one-half the recommended rate twice during the cropping cycle to prevent problems. Solubor or Borax can also be used. Leafy plants that grow large may have received too much fertilizer or may be growing under low light/high temperature conditions. High light levels can cause leaf burning and sun scald, which are exacerbated by high temperatures at the same time. Plants that fail to flower may not have received adequate cool treatment for flower initiation. Wilting plants combined with loss of roots can be due to a poorly drained medium, overwatering, high soluble salts, or disease.

Varieties

There are a number of varieties available, most as mixtures. However, separate flower colors are available in a few selections as well. Catalogs often describe varieties as 4 or 6" (10 or 15 cm) pot plants. Large-flowered selections are best in larger containers, while smaller-flower types are well suited for smaller containers.

Postharvest

The crop is ready to sell when five to six flowers are open and subsequent buds are present. Plants can be boxed and shipped at 40–45°F (4–7°C). Pack plants that have been precooled so that water does not form on leaves, which could cause disease. Cineraria is sensitive to ethylene: Groom plants to remove dead foliage and flowers as sources of ethylene. Do not place cineraria near ripening fruits or vegetables.

Retailers should display plants in a cool (55–60°F [13–16°C]), bright (500 f.c. [5.4 klux]) area. Do not allow plants to dry out at retail. Consumers will achieve the best results by placing cineraria in a bright window with indirect light and making sure the plant does not fully dry out.

Perilla

Perilla frutescens
Common name: Perilla

Perilla is a striking, multicolored, accent plant that is tolerant of sun and shade. Perilla fills in mixed containers fast and delivers great heat tolerance and summer performance in the landscape. Plants grow from 24–36" (60–90 cm) tall.

Propagation

Choose a well-drained medium with an EC of 0.75–0.80 mS/cm and a pH of 5.6–5.9. Stick cuttings within twelve to twenty-four hours of arrival. Cuttings can be stored overnight, if necessary, at 45–50°F (7–10°C). Soil temperature should be maintained at 68–74°F (20–23°C) until roots are visible. Once roots are visible, the media should be moderately moist but never saturated. Begin fertilization with 75–100 ppm nitrogen (N) when roots become visible. Increase to 150–200 ppm N as roots develop. As the rooted cuttings develop, high light and moderate air temperatures reduce the need for chemical plant growth regulators (PGRs). Several PGRs are effective in reducing elongation on propagation, including Topflor, Bonzi, and B-Nine/Cycocel tank mix. For finishing in small containers, 'Magilla Perilla' does not require pinching during propagation. To improve branching and habit and for larger containers (6" [15 cm] or larger), plants can be pinched seven to ten days before transplanting. Perilla rooted cuttings should be ready for transplanting twenty-one to twenty-four days after sticking into a 105 tray.

Growing On

Use a well-drained, disease-free, soilless medium with a pH of 5.8–6.0. For active growth, maintain night temperatures of 60–70°F (16–21°C) and day temperatures of 75–85°F (24–29°C). Cool temperatures will slow crop times dramatically. Maintain light intensities of 4,000–10,000 f.c. (43–100 klux). Low light levels promote poor branching, stem stretch, and poor foliage color. Allow the media to dry slightly between watering, but avoid severe wilting. Use constant feed with a balanced fertilizer at 150–200 ppm. Leach regularly to avoid the buildup of soluble salts.

Managing the pinching and growth regulation on perilla is dependent on the type of finished plant needed. Pinch plants seven to fourteen days after transplanting to improve branching. A 4" (10 cm) crop can be produced with no pinch if necessary. Use high light and the recommended temperatures to control growth and produce the best possible habit. A Cycocel (1,000–1,500 ppm) and B-Nine (2,500–4,000 ppm) tank mix, applied one to three times, is effective in controlling growth. Low rate and high volume drenches on Bonzi and Topflor have demonstrated effective control. Recommendations for PGRs should be used only as general guidelines. Growers must trial all chemicals under their particular conditions.

The crop time from sticking a 105 liner to a finished 4" (10 cm) pot is six to seven weeks. The plants will crop faster as days become warmer and brighter in the spring.

Pests & Diseases

The most common insect pressures on perilla come from aphids, whiteflies, and mites. Maintaining a healthy root system and avoiding condensation on the foliage will alleviate many foliar diseases. Rhizoctonia and *Pythium* will attack weakened or stressed plants.

Troubleshooting

Undersized plants may be a result of growing under excessively cool conditions, poor system (*Pythium*), or excess moisture. Warm temperatures, low light, and high levels of ammonia in feed can lead to excessive growth. Perilla is a tough plant and rarely shows signs of environmental stress.

Table 31. Perilla Crop Schedule & Uses*

	4" (10 CM) POT 1 ppp	6" (15 CM) POT 1–2 ppp	10" (25 CM) POT 3–4 ppp
Unrooted cuttings	8–10	9–11	11–14
Rooted cuttings	5–7	6–8	8–11

ppp = Plants per pot
* Crop schedule listed in weeks.

Perovskia

Perovskia atriplicifolia
Common name: Russian sage
Perennial (Hardy from USDA Zones 5–9)

A striking plant noted for its color and form as well as its fragrance, *Perovskia atriplicifolia* is a clump-forming perennial from 4–5' (120–150 cm) tall, though its seldom seen over 3.5' (105 cm) in Chicago-area gardens. The stems are covered with a glaucous (grayish-white) waxy layer, and the foliage is light green or gray green on the top and grayish white on the underside. The flowers are light blue, approximately 0.25" (6 mm) in size and arranged in panicles, or branching flower clusters. *Perovskia* grows from 3–4' (90–120 cm) across once established. Plants flower from July to September.

Propagation

While commonly propagated by cuttings, seed can also be sown. However, use only enhanced seed products that will give you increased and more uniform germination and seedlings.

Seed

The following is based on the variety 'Taiga', which is offered as an enhanced seed. Sow one to two seeds per 288-plug tray in summer and lightly cover with coarse vermiculite. Germinate at 70–72°F (21–22°C), and seeds will germinate in six to ten days. A 288-plug tray will be ready to transplant to larger containers in six to eight weeks.

Note: Until the introduction of 'Taiga', germination of *Perovskia* was irregular and seedling stands were uneven. Seed was not a preferred method of propagation. In addition, if seed was slow to germinate, protocols included sowing seed to a plug or liner tray, exposing the tray to 35°F (2°C) for up to three or four weeks, and then bringing the trays back into the greenhouse to a bench or germination chamber at 65°F (18°C) to allow for germination. The media was kept moist but not wet.

Cuttings

Stick one cutting per 105-, 84-, or larger liner tray with a peat-based media. Cuttings taken in summer tend to root readily, but a rooting hormone is helpful during other times of the year. Give mist to encourage rooting for six to ten days and reduce or eliminate mist as soon as possible. A preventative fungicide may be necessary. A 105-liner tray will take from four to six weeks. As plants root, pinch out the terminal tip of the shoot.

Growing On

Seed

For sowings made in spring or early summer, transplant a 288 plug to a 32-cell pack and grow on for six to nine weeks. Then transplant to a 1-gal. (4 l) pot to allow to root. Overwinter the plants dormant as long as the roots have filled the final container. Plants are sold the following spring and will flower in summer.

Seed sown in January or February can be transplanted to 4 or 6" (10 or 15 cm) containers for spring sales. However, while plants will flower sporadically, they are not as free flowering as plants sown from seed the previous year and are thought to be inferior. In addition, plants will not fully root out a 1-gal. (4 l) pot for May sales.

Cuttings

Cuttings rooted one year are often bulked up and then vernalized in a 84-, 72-, or 50-liner tray and made available from commercial propagators in the winter for potting up to 1–3-qt. (2–3 l) containers. These containers are sold in the spring for planting to the garden in May and flowering in summer.

Petroselinum

Petroselinum crispum
Common name: Parsley
Herb, annual, biennial

This versatile herb is more than just a garnish for restaurants. For some ethnic cuisines, parsley is the main ingredient in salads. Parsley is an excellent source of vitamins A and C and is high in calcium. Its chlorophyll is a great breath freshener.

In mild-winter climates, this biennial's leaves look lovely through its first winter, providing a splash of green in a texture not often found in herbaceous plants and making an excellent backdrop for pansies and violas in fall combination pots. For butterfly lovers, parsley is a garden staple: Butterflies such as the monarch and black swallowtail find parsley to be an especially fine feast. For that reason, do not kill caterpillars or worms eating parsley in the summer months. As a matter of fact, some retail growers are using this as a marketing tool and selling parsley as butterfly food.

Propagation

Parsley may be sown directly to its final growing container or an open flat or started as a plug. If sown to a 512-plug tray, the seed lightly covered with vermiculite and given 70–72°F (21–22°C) media temperatures will germinate in twelve to sixteen days. Plug trays will be ready to transplant to larger containers four to five weeks after sowing.

When transplanted to 4" (10 cm) pots, plants will be ready for spring sales in six to eight weeks after being grown at 65–70°F (18–21°C) days and 60–65°F (16–18°C) nights. Once the plants are established in their final containers, the night temperatures can be dropped another 5–10°F (3–6°C) to hold or finish the plants. Parsley is tolerant of cool conditions with little problems.

If you are sowing directly to the final growing container on the greenhouse bench, sow three to four seeds per 4" (10 cm) pot and cover lightly with medium or coarse vermiculite. Maintain uniform moisture around the seed and substrate temperatures of 70°F (21°C). Seed will germinate in fourteen to twenty-one days. Allow twelve to fourteen weeks from sowing to a salable plant for spring sales.

Plants make deep taproots, so if you are going to offer larger sizes, make sure pots are deep. Use a well-drained, peat-based medium with a pH of 6.0–6.5. Parsley performs best in cooler weather, making an excellent garden accent in spring and fall. It is a biennial: Once plants have flowered their second year, the foliage is bitter and no longer suitable for culinary uses.

Varieties

Curly-leafed parsley, or common parsley, is deep green and grows about 12" (30 cm) tall; it is best used as a garnish. Italian parsley is lighter green with flat leaves, making a somewhat coarser texture and can grow to 15–24" (38–61 cm).

Petunia

Petunia × *hybrida*
Common name: Petunia
Annual

Virtually every American gardener has petunias in her garden. They are a mainstay for bedding plant growers from coast to coast, thriving in all kinds of conditions. Superior breeding and continual renewal of varieties has kept the petunia flame fanned since the 1940s and 1950s. Just when you think petunias might be getting a bit stale, breeders introduce new forms, new habits, and new colors.

In many ways the evolution of petunias has paralleled the development of the industry. They easily made the leap from being grown in fields and dug bare root to being grown from seedlings in cell packs. They leaped from open seed trays to plug production—the transition was bumpy at first, but now plug production techniques are well refined. As the industry has turned more and more toward segmenting the market and niching products into segments, petunias are once again making the leap from seed-grown varieties for packs to vegetatively propagated landscape and garden varieties bred for 4 and 6" (10 and 15 cm) production, hanging baskets, window boxes, and even specifically for combinations. Today, growers routinely produce both seed-grown bedding plant petunias as well as vegetative petunias. Each type has different cultural requirements, but all are wildly popular with consumers.

For growers, petunias are a major part of their bedding plant assortment. With such a pallet of growing types, colors, and requirements you must invest the time to understand the different series and their best uses. Your broker and breeder rep can help direct your purchases to help you build a successful program.

Here is an overview of the offerings on the market and how they can be best utilized.

- Stores are filled with packs from late spring to early summer. Most of these offerings are seed-grown varieties, which make it easier to provide this product for value. Gardeners, rather than decorators, typically use this product, because these plants must be planted and cared for to get the best performance.
- Four-inch (10 cm) and 1-qt. pots have become an important segment in petunias. More vegetative varieties are used in this arena than in packs. These are grown full and in color. They are typically intended for container gardening, although unique varieties can be incorporated into flower plantings.
- Utilizing the latest breeding efforts, growers put special colors and performers into 6" (15 cm) and 1-gal. (4 l) pots. These are definitely value-added products and become an impulse purchase. Many times growers will choose to also use a pot which is printed with special POS (point of sale) marketing information for the specific color or series grown in the pot.
- Growing in a large, decorative container or basket with petunias can be done easily when you comply with the following production protocols. New breeding has developed plants that require a lot less maintenance. This segment is marketed to both gardeners and decorators. Gardeners buy these products to put the finishing touches into their landscapes, but more importantly this becomes an

impulse buy for decorators. Someone who sees the beautiful color when going into the supermarket or the local garden center for their list will be sure to buy a couple.
- Mixed containers are again marketed for the impulse buy. Many petunias bred for mounding habits work well with many other plants. They are used as the filler plant that provides color all summer long, while other plants are used for accents, the centerpiece, or structure. Many breeders have recipes to help get you started. Pay particular attention to the combinations you put together because vigorous plants can cover the other plants in the planter and ruin the mix.

Consumers buy petunias year after year because they are reliable. With such a predisposed buying attitude, you can let your imagination run wild to create new products that entice them to buy even more than they had planned.

The Crop from Seed

Propagation

Petunias are grown from seed for bedding plant pack and pot production of grandiflora, multiflora, and spreading petunias (highlighted by the 'Wave' series). Most growers buy in their petunia plugs when they receive other bedding plant plugs, because it is just easier to do so.

Sow pelleted petunia seed onto a well-drained, disease-free medium with a pH of 5.5–5.8. Do not cover. Provide 100–400 f.c. (1.1–4.3 klux) of light to prevent watery-looking seedlings and improve germination. Provide substrate temperatures of 75–78°F (24–26°C). Maintain moist substrate conditions, but do not saturate the medium. ('Wave' petunias require above-average amounts of substrate moisture and light during Stage 1.) Radicles will emerge in three to five days.

Once radicles emerge, move trays to Stage 2 for stem and cotyledon emergence. Reduce substrate temperatures to 68–75°F (20–22°C). Allow soil to dry out slightly between irrigations in Stage 2. Provide fourteen to eighteen hours of supplemental light (450–700 f.c. [4.8–7.5 klux]) for five weeks in the plug stage. Ambient light should be 1,000–2,500 f.c. (11–27 klux) to promote early and uniform flowering. Keep ammonium levels below 10 ppm. As soon as cotyledons have expanded, apply 50–75 ppm from 14-0-14, alternating with clear water. Once the stand is full, apply a preventative fungicide drench to control *Pythium* and *Thielaviopsis.* Stage 2 will take from seven to ten days.

Move trays to Stage 3 to develop true leaves by reducing substrate temperatures to 65–70°F (18–21°C). Allow soil to dry thoroughly between irrigations, but avoid permanent wilting. Continue to provide 450–700 f.c. (4.8–7.5 klux) of supplemental lighting for fourteen to eighteen hours a day (you need to light for five weeks). Ambient light should be 1,000–2,500 f.c. (11–27 klux). Increase fertilizer to 100–150 ppm, alternating between 20-10-20 and 14-0-14. Alternate feed with clear water. If you need growth regulators at this time, A-Rest, B-Nine, and Bonzi are all registered. 'Purple Wave Classic' will require more growth regulator than other 'Wave' colors. If you are using Bonzi on 'Wave', conduct trials to determine the best rates and timing for your specific conditions. A rate of 6 ppm is a good starting point. Stage 3 will take from twenty-one days to twenty-eight days.

Harden plugs for sale or transplanting by lowering substrate temperatures to 60–62°F (16–17°C) for seven days. Flowering may be delayed if the temperature is held too low for too long and the plants have not initiated flowers (thus the necessity of long days in Stages 2 and 3). Allow trays to fully dry out between irrigations. Increase ambient light to 2,500–3,000 f.c. (27–32 klux). Feed once a week with 100–150 ppm from 14-0-14. Potassium nitrate has been shown to promote flowering in petunias. Do not use ammonium nitrate at these temperatures. Consider an additional preventative fungicide drench in Stage 4.

A 288-plug tray will take from four to five weeks to be ready to transplant to larger containers.

Growing On

Be sure to schedule multiple crops so you will always have fresh petunias available. Transplant petunia plugs promptly. Hanging onto plugs only makes them more difficult to dislodge from the tray and will slow rooting. Transplant 1 plug/cell per 4" (10 cm) pot, 2–3 plugs/6" (15 cm) or 1-gal. (4 l), and 5–7 plugs/10" (25 cm) basket. The growth of 'Wave' petunias is so vigorous that it is best not to grow them in packs at all; 4" (10 cm) pots is a minimum size. One 'Wave' plug will fill a 4 or 6" (10 or 15 cm) pot. For baskets, use three 'Purple Wave' plugs and four for other colors in the series. 'Easy Wave' petunias are well suited for 1801s or jumbo 306s.

Select a well-drained, disease-free, peatlite medium with a high starter charge and a pH of 5.5–6.3. Peatlite and peat-bark mixes are excellent for petunias. Water in immediately after planting, but be aware that petunias are very sensitive to overwatering. Allow the medium to dry slightly between irrigations. 'Wave' petunias grow quickly and may require more water than your regular bedding plant crop.

Basic production temperatures are 65–75°F (18–24°C) days and 62–68°F (17–20°C) nights. Petunias can tolerate temperatures a lot higher and lower than that. However, you will want to stay on the lower end of the ranges to produce the best-quality crop and to make the crop more manageable. Once plants are budded, you can reduce nights to 50°F (10°C) to hold them. Lower temperatures on flowering flats also increases the postharvest life of finished trays.

Feed at every other irrigation with 150–200 ppm, alternating between 20-10-20 and 15-0-15. 'Wave' petunias require more feed than other petunias: Use 200–250 ppm. When 'Wave' petunias are fertilized with less than 200 ppm, nutrient deficiencies are likely to show up. 'Wave' petunias can handle rates as high as 400 ppm, but plant growth will be exceptionally lush. To boost performance at retail and for the consumer, apply a topdress of Osmocote or another slow-release fertilizer to 'Wave' baskets ten days before shipping. For your regular bedding plant petunia crop, cut fertilizer rates in half at visible bud to increase postharvest life.

Grow plants under the highest light levels possible (i.e., 4,000–7,000 f.c. [43–75 klux] or more). Provided your petunia bedding plant plugs were lighted in the plug stage, they should have initiated flowers. Note that 'Wave' petunias flower fastest at day lengths longer than thirteen hours. 'Purple Wave Classic' shows a stronger response. For early crops (sold in March and April), you can significantly reduce crop time and increase the number of flowers per plant by providing long days. After planting, supply 100–200 f.c. (1.1–2.2 klux) of light daily from 10 P.M.–2 A.M.

Height control is an important issue with petunias. Seasoned growers know the fine line between keeping petunias actively growing with water and fertilizer and simultaneously holding back the reins with judicious irrigation, well-timed growth regulators, and low production temperatures. As soon as plants are rooted into containers, you can allow petunias to wilt between irrigations. You can also hold them back by withholding feed. Combine the two and you have the "run 'em dry and hungry" technique that most growers use to produce their petunia crops. Add cool or even cold production temperatures and you will be "growin' 'em hard." However, it takes time to develop a feel for withholding water and fertilizer and not destroying the crop. Once your crop is rooted in, do not irrigate them unless they are a bit wilted. Petunias can actually handle fairly significant wilting before flowering begins. Cool production temperatures will be one of your greatest allies in controlling height and growth: A lot of growers produce fine petunia crops with as little heat as possible in Quonsets or cold frames. Many more growers use growth regulators to help with the process, especially if the petunias are grown in the same structure with a number of other bedding plant crops and temperatures have to be compromised for the good of the whole. Temperatures above 75°F (24°C) will significantly decrease branching and cause plants to stretch.

Florel does an excellent job on petunias: An application once plants are established will cause plants to develop many laterals. Some growers of 'Purple Wave Classic' treat plugs with Florel at transplanting and then again when plugs have rooted in. The resulting plants are densely branched and can measure 8" (20 cm) or more across before runners are produced. Florel applications will delay flowering by as much as four weeks after each application.

As a general rule, apply growth regulators such as B-Nine (2,500–5,000 ppm), Bonzi (15–50 ppm), and Sumagic (10–30 ppm) for the first time when plants are established. B-Nine can be sprayed, while Bonzi may be sprayed or drenched (only once). Petunia varieties respond differently to growth regulators, so you will be experimenting. Test with your first application at lower rates and adjust as needed. Do not apply any growth regulators after visible bud. 'Wave' petunia size will be reduced slightly with growth regulator application, but plants grow out of treatment quickly once they are transplanted to the field. Note that 'Purple Wave Classic' is more vigorous and will require more growth regulators than other petunias. B-Nine at rates as high as 7,000–10,000 ppm once a week are not unheard of. Bonzi as a 5–7 ppm drench applied once is also effective on 'Purple Wave Classic'. For other 'Wave' colors, start out with B-Nine at 3,000–5,000 ppm; a Bonzi drench can be applied once at 2–4 ppm. To produce excellent 'Easy Wave' petunias in packs or pots, spray weekly with B-Nine at 5,000 ppm beginning seven days after transplant.

For production in small containers, weekly sprays of B-Nine at 5,000 ppm control plant size and promote branches. These plants have been sprayed four times.

Keeping daily notes that you can look back on from year to year will help you perfect your petunia height control techniques. It is an important crop, and learning how to keep them actively growing in a highly controlled way is critical to your success.

The crop time from transplanting a 288 plug to a flowering cell pack is four to six weeks; a 4" (10 cm) pot is five to seven weeks; while a 10" (25 cm) basket requires seven to eight weeks.

Pests & Diseases

Aphids, thrips, and whiteflies can attack plants. Botrytis can attack open flowers. Be diligent in managing water once your petunias start to flower by making sure to water early enough in the day so that foliage and petals will be dry by nightfall. Keep air circulating through greenhouses. If you have poor airflow in one corner of the greenhouse, that's the spot where you will first find botrytis spots on petunia flowers.

Troubleshooting

Tan or brown spots on flowers can be due to botrytis. Be sure to maintain excellent air circulation and reduced humidity in the greenhouse to control it. Keep dead or diseased plants rouged to remove sources of spores. A pH levels above 6.8 can cause interveinal chlorosis from iron deficiency. Some varieties are more sensitive than others are. If the pH is above 6.5, lower it with iron sulfate. Iron or manganese toxicity at extremely low pH levels can result in brown or tan lesions on the foliage. Switch to a base-forming fertilizer such as 15-0-15. If symptoms don't improve or if the pH is below 5.5, irrigate with hydrated lime. Boron deficiency can lead to distorted foliage, tip dieback, and proliferation of side shoots below the meristem. Maintain pH between 5.5–6.3 and use a boron supplement (borax dissolved in hot water at 0.5 oz./100 gal. [4 g/100 l]), if necessary. Solubor works well too.

Varieties

Single grandifloras are one of the most popular petunias on the market. Characterized by large flowers that grow to 3.5–5" (9–13 cm) across, the single grandifloras have dominated the market since the early 1950s, when 'White Cascade' was introduced. Grandifloras are suitable for pack, pot, and basket production. Continually renewed through improvements by breeders, they are real stars for growers in production and for consumers in the garden. Flowers are offered in a range of colors, including stars and picotees, and with frilled or rounded petal edges. In the South, plants are hardy in areas with mild or no winters, doing well in full sun. However, a hard frost will kill them.

There are a number of excellent grandiflora varieties on the market today. You will want to make 'Storms' or 'Dreams', with their improved weather tolerance, your mainstay series. Both series come in various standard colors plus mixes. 'Supercascade' and 'Ultra' make excellent hanging baskets. Be sure to spice up your assortment with a few colors of the veined 'Daddy' series, the picotee-flowered 'Frosts' or 'Hulahoops', the star patterns of the 'Ultra' series, and the two 'Falcon Morn' colors, 'Red' and 'Pink'. Try a few doubles too, such as 'Purple Pirouette' and 'Cascade', with greatly improved germination.

While grandifloras have dominated for more than fifty years, the multifloras came on strong in the 1980s, unseating a significant market share from the grandifloras for good reason: weather performance. Multiflora petunias do not become battered by the rain and are more disease tolerant. Flower size is comparable to a medium-sized grandiflora for 'Madness', smaller for other series. For superior greenhouse and garden performance and wide color range, you will want to make 'Madness' your mainstay multiflora series for packs, pots, and baskets. The series comes in twenty-one colors encompassing solids from blue to red to white as well as stars, veins, and designer mixes. 'Madness' is also available in six double colors. For smaller-flower multiflora series, try 'Carpet', 'Celebrity', and 'Merlin'; all are excellent in packs and 4" (10 cm) pots, especially for landscape sales.

'Wave' petunias were a breath of fresh air into bedding plant petunias. Anyone can be successful with them, as they thrive in all kinds of soils and under a wide range of climatic conditions. 'Purple Wave Classic' and 'Pink Wave' can spread 3–4' (0.9–1.2 m) in the garden, hugging the ground with masses of color. There is also 'Purple Wave Improved', which is up to thirteen days earlier to flower than the original 'Purple Wave Classic'—the All-America Selections winner. The 'Wave' series has been expanded to also include 'Blue', 'Lavender', 'Misty Lilac', and 'Rose'. Additionally, a number of other series are available to compete with 'Wave'. The most notable of these is 'Avalanche', which has 'White' and 'Red'. 'Avalanche' is less daylength sensitive than 'Purple Wave' is. 'Plush', 'Explorer', and 'Ramblin'' are other entrants into the spreading-petunia-from-seed class.

'Easy Wave' petunias are great for pack or pot production. Plants are less daylength sensitive and are great for early-season sales. 'Easy Wave' colors include 'Cherry', 'Pink', 'Shell Pink', and 'White'. Another series for pack or pot is 'Shock Wave', available in a number of colors plus several mixes. It is the earliest to flower spreading series and the only small-flower spreading series from seed.

If your customer base includes quite a few landscapers, try the 'Tidal Wave' series. Plants literally make small hedges of color in the garden that are 16–22" (41–56 cm) tall. The closer the spacing outdoors, the taller they will grow. Flowers are resistant to botrytis, and plants bounce back quickly following rain. The series comes in 'Cherry', 'Hot Pink', 'Purple', and 'Silver'. Try them in 6" (15 cm) or 1-gal. (4 l) pots.

Postharvest

Petunias are ready to ship when they reach the size you want and just as soon as flowering begins. If you hold your seed plants cold and hungry to time the market, give them a shot of light feed when you water trays before shipping. Do not ship plants in the dark for long periods, as leaves will yellow. You can ship trays cool (at or above 50°F [10°C]).

Retailers should display petunias under filtered sun conditions and in cool temperatures—50°F (10°C) is ideal. Do not allow plants to freeze, and do not allow baskets or flats to dry out completely. Ideally, feed petunias weekly in the retail setting to maintain the highest plant quality.

Consumers will find that petunias are tolerant of a wide range of conditions as long as they get enough light, water, and fertilizer. Be sure to provide excellent drainage, especially for vegetative petunias, which will become chlorotic in waterlogged soils. Consumers should regularly pinch back one-third of the plant to encourage continued blooming and control size and shape. Most established petunias are able to handle a light frost when they are hardened. Many petunias may prove to be winter-hardy in the South, taking occasional nights with temperatures to 25–28°F (–4–-2°C). They can also tolerate temperatures well into the 90s (°F or 32–37°C).

The Crop from Cuttings

Propagation

Vegetatively propagated petunias, such as 'Suncatcher', 'Surfinia', or 'Potunias', are propagated by cuttings. Plant patents and breeder's rights protect the best of these varieties, so unlicensed propagation is illegal. Not only are they protected, which makes taking cuttings illegal, but petunias can also harbor a number of latent diseases. Among the viruses that can infect plants are potato Y virus, tobacco mosaic virus (TMV), cucumber mosaic virus (CMV), and TSWV/INSV. If you bring in cuttings infected with these diseases, you may infect other crops in your greenhouse. The best breeders and suppliers routinely index and clean their stock, continually refreshing mother blocks with clean material. They then supply their own stock production and subsequent rooting stations with clean plant material that has been maintained in a rigorous clean-stock program. Your best route is to work with a reputable supplier that is maintaining a clean-stock program so you are assured of receiving only clean stock.

Under no circumstances should you ever maintain vegetative petunias from one year to the next as stock plants. Growers of trailing petunias in the United Kingdom learned an expensive lesson on the value of clean stock, when in 1994 about 5 million trailing petunias were dumped due to the widespread infection of TMV and potato Y virus. Due to the exploding demand, suppliers elected to maintain their mother blocks a second season to increase supply faster. The growers did not know that the heldover stock was infected. The viruses soon spread to the new elite stock. The incident received broad publicity—even to the consumer press, with headlines blaring the message of diseased plants for sale.

Suppliers ship both unrooted and rooted cuttings. Many smaller and medium-sized growers buy in liners for convenience. Workers who smoke cigarettes should not handle or work in propagation, particularly with petunias because of the ease of TMV transmission. TMV can be transferred even from clothing when the employee is using gloves.

Unpack boxes and inspect cuttings upon arrival. Stick cuttings immediately. If that isn't possible, they may be stored in a clean cooler. Unpack from the box or at the very least open the box. The cooler temperature should be 40–45°F (5–8°C) with close to 100% humidity, and cuttings should be in there no longer than twenty-four hours.

Stick cuttings in Ellepots or a rooting medium with a pH of 5.5–6.2. Consider a half-strength drench of Daconil after sticking, if you consistently have problems with botrytis. Provide substrate temperatures of 68–74°F (20–24°C) and air temperatures of 70–75°F (21–22°C) days and 68–74°F (20–24°C) nights. Ensure the medium is uniformly moist so that water can easily be squeezed out. Maintain ambient light levels at 500–1,000 f.c. (5.4–11 klux). Retractable shade is ideal, as you will be able to gradually increase ambient light levels as cuttings root in and mature. Provide mist only to maintain cutting turgor. During the first two to three days, frequent night misting may be required. Adjust mist timing and frequency to your specific conditions. The use of Capsil at 4 oz./100 gal. (30 g/100 l) of water can help reduce the amount of mist needed. Always remember that the more mist applied, the greater chance of disease. Mist also leeches critical nutrients from the plants. Begin a foliar feed of 50–75 ppm of 20-10-20 as soon as the plants callus. This will help build nutrients into the soil so the plants can immediately take up nutrients once roots grow. Early yellowing, particularly at the growing tip, can give the finish grower trouble through the growing cycle. This can easily be avoided with early fertilization. For some older varieties, which stretch wildly in propagation, light growth regulator applications can be started once cuttings are turgid. Once 50% of the cuttings have differentiated root initials (after five to seven days), move them to the next stage.

Develop roots on cuttings for nine to fourteen days. Maintain substrate and air temperature regimes. As cuttings begin to develop roots, allow the soil to dry down, but avoid drying out. Begin drying liners down by reducing and then as quickly as possible eliminating mist applications at night and reducing the duration and frequency of mist during the day. Increase ambient light to 1,000–2,000 f.c. (11–22 klux). Foliar feed with 100 ppm from 15-0-15, alternating with 20-10-20 and increasing the rate rapidly to 200 ppm as cuttings root in. Monitor the substrate pH and EC daily. Keep the pH at 6.0 and the EC at 0.5–1.0 mS/cm.

Harden cuttings for seven days prior to sale or planting by lowering air temperatures to 70–75°F (21–24°C) days and 62–68°F (16–20°C) nights. Move cuttings from the mist area to an area with lower relative humidity and higher light (2,000–4,000 f.c. [22–43 klux]). Maintain the pH at 5.5–6.2 and the EC below 0.5 mS/cm. Fertilize with 150–200 ppm from 15-0-15, alternating with 20-10-20. For "harder" plants, you can use more 15-0-15 but be sure to closely monitor your pH. The use of a trimming machine can be very useful in creating uniform, branched liners that are ready to flourish for the finish grower. Trimming or pinching is the preferred method of controlling height and form but can be combined with growth regulators on vigorous, stretching cultivars. Be sure to maintain your cutting devices and regularly disinfect the cutting implements before, during, and after operation.

A 105-liner tray takes from three to four weeks from sticking the cutting until ready to transplant to larger containers.

Growing On

Growing petunias doesn't have to be hard. Now that you have selected the best and newest breeding, you can utilize the following techniques to finish high-quality plants. Be sure to understand the plant or series you are growing and what the desired look of the finished crop is.

Plant 1 cutting/4 or 6" (10 or 15 cm) pot and 3–5 cuttings/10" (25 cm) hanging basket. Overplanting can cause bald centers, since plants will stretch to compete with each other. Select a disease-free, well-drained medium with a high starter charge and a pH of 5.6–6.2. A medium with high pH will cause chlorosis due to iron deficiency. Plant the liners at the same level as they are currently growing. Deep planting can invite disease troubles.

Watering is a big challenge with vegetative petunias: Plants are highly susceptible to overwatering and can get root rots as a direct result. Avoid keeping plants too wet, especially after planting. Plants are also highly susceptible to botrytis and powdery mildew when humidity is too high, so be extra careful to avoid wet foliage. The best way to grow them is in hanging baskets on drip tubes. Run them dry as opposed to wet. If your water is naturally high in salts, leach regularly to avoid salts problems. Do not be afraid to hang them up in an attic with roof vents once plants are rooted in; they will like the extra light and you will be able to open vents to provide cool temperatures—it is an unbeatable combination for producing high quality.

Provide the highest light levels you can while maintaining production temperatures; 4,000–7,000 f.c. (43–75 klux) or more is ideal. Low light levels promote stem stretching. Grow on at slightly cooler temperatures than your bedding plant crop: 65–72°F (18–22°C) days and 62–65°F (17–18°C) nights. Temperatures of 65°F (18°C) or higher promote the most rapid growth, while the best-quality plants are grown at temperatures less than 60°F (16°C). In low light levels and/or high production temperatures, growth regulators will be required.

Though many vegetatively propagated petunias must have long days to flower, in recent years breeding has developed fully day-neutral petunias. Be sure to research the plants you choose to grow and note the differences between day neutral and early varieties. In the winter, under short days, most crops will take longer to finish. To speed flowering for early spring crops, provide long days by lighting crops from 10 P.M.–2 A.M. with incandescent lights.

Vegetative petunias are very high feeders. Fertilize with 200–300 ppm constant liquid feed, alternating between 20-10-20 and 15-0-15. As described in liner production, you can use a nitrate-based feed to grow "harder" plants. High levels of ammonium and overall low salt levels (EC) in the soil can cause undesirable soft growth. For vegetative petunias it is not recommended to grow lean. This creates lower salt levels in the soil, which allow the plants to absorb more water and creates soft, "watery" growth.

If iron chlorosis becomes a problem, an iron-sulfate drench or iron-chelate spray or drench can be used. Some growers apply iron a couple of times during the production cycle just to make sure they don't encounter problems. Many growers supplement their liquid feed program with a slow-release fertilizer such as Osmocote. A topdress of slow-release fertilizer applied in the weeks before shipping may improve performance for the retailer and consumer. Fertilizer companies have released commercial feeds specific to chronic iron-hungry crops such as petunias and calibrachoa. These feeds are designed to maintain a stable pH while providing three forms of iron, which are available at different pH values, thus widening your pH window. These should not be used on crops such as geraniums.

For plants that tend to stretch, pinch once plants are established in their final growing containers. Subsequent pinches can be made to shape plants, but flowering will be delayed. An application of Florel, applied once plants are established, can help reduce stretch and promote lateral branching. Apply Florel under long-day conditions instead of a pinch. Do not apply Florel too late, as flowering can be delayed. B-Nine (1,500–5,000 ppm) or Bonzi (30–45 ppm) sprays can be applied to control plant height. A one-time Bonzi drench (up to 4 ppm) about one week after the pinch can hold aggressive cultivars, but be careful later in the crop because of its potential to significantly delay flowering. Multiple low-rate Bonzi drenches (0.5–1.5 ppm) made two to four weeks apart can be a strategy to shape and control growth on mounding types. When applied just before shipping, it can slow vegetative growth enough and will allow the flowers to be better displayed, giving the perception of increased flowering. Response to growth regulators is variety specific—they may delay flowering significantly. Vegetative petunias are very responsive to DIF and morning DIPs; plants will be more compact with these strategies. When using a morning DIP, it is important that the temperature is dropped two hours before sunrise and held for two hours after sunrise.

Warning: If you are growing vegetative petunias in 4 or 6" (10 or 15 cm) pots, be prepared to space and/or sell plants as they are ready. A 4" (10 cm) crop can be ready in as little as four to five weeks, depending on your exact growing conditions. If you let them get away from you on the bench, you will have a tangled mess that will keep your employees busy for hours trying to extract something salable. One plant can have runners that eventually grow 4' (1.2 m) in the garden. If you find yourself in such a predicament, you can cut pots back, reflower them, and/or put them into combination pots. Hang baskets and use drip emitters as soon as they are rooted in for this same reason. Hanging baskets can take as few as eight weeks and as many as sixteen weeks, depending on how you choose to grow them. Be sure to take into account things such as how many times you pinch, if and when you apply Florel, the growing temperatures, and whether or not you light.

As for crop times, a 4" (10 cm) pot with one liner will flower in five to seven weeks; a 6" (15 cm) with two to three liners will take from six to eight weeks; while a 10–12" (25–30 cm) basket takes from eight to eleven weeks with three to five liners.

Pests & Diseases

Aphids, caterpillars, fungus gnats, leaf miners, thrips, and whiteflies can attack plants. Botrytis, powdery mildew, *Pythium,* rhizoctonia, and several viruses can attack vegetatively propagated petunias. Be diligent in managing water to control disease and buy only cuttings that have been produced in a clean-stock program to ensure that you do not introduce viruses into your greenhouse. Limit any contact between employees who smoke and your petunia crops.

Troubleshooting

Plant collapse can be due to wet potting mix for an extended period or an attack by *Pythium* or rhizoctonia on plants that were planted too deep. High ammonium in the medium or low light combined with overwatering can cause excessive vegetative growth. Low fertilizer or a lack of nitrogen can lead to poor plant branching. Low light levels can cause stretched plants. A high pH that induces iron deficiency can result in leaf chlorosis. Short days or late applications of growth regulators can delay flowering. A weak lower stem at the base of the plant, which makes the plant floppy, is the result of the liner being grown too soft. See also Troubleshooting in The Crop from Seed section for nutrient toxicity and deficiency symptoms.

Varieties

'Surfinia' is the petunia that created the vegetative petunia craze back in the early 1990s. In the United States, the Proven Winners group introduced it with flair. Overtime, the series has taken off, and when combined with a range of really cool, new plants like *Bacopa, Calibrachoa,* and *Scaevola,* it has helped to establish the line as a value-added container and hanging basket crop.

While 'Surfinia' continues to perform, other breeders have come into the ring with petunia series of all types, sizes, and uses. The 'Potunia' series broke new ground when it hit the market in 2006. Fully day neutral, it is easy to grow with its rounded, tight growing habit. It performs well in mixed containers, staying controlled and not overtaking combinations.

Doubles and mini-flowered series add to mix-designing pallets. Bicolor, veined, and extreme colors like 'Black Velvet' keep the petunia world changing each year. While there are many tried and true varieties, it is worth the investment each year to visit one or two independent trial sites located throughout the country. Investigate the latest varieties and give them a try in your own garden. Be sure to utilize your broker's and breeder reps' advice, and because there are so many habits and types, carefully follow specific breeder production recommendations for that variety or series.

Postharvest

See Postharvest in The Crop from Seed section.

Phalaenopsis

Phalaenopsis hybrids
Common name: Moth orchid
Flowering pot plant, cut flower

Phalaenopsis orchids have become important American pot plants in the last fifteen years because of their relatively low price at retail, $15–30, and their adaptability once they have grown to flowering size. Flowers can last ten to fourteen weeks, making them a real value for consumers.

Phalaenopsis is the most widely propagated and most commercially important genus within the Orchidaceae family.

Propagation

Orchids are typically propagated through tissue culture at specialized laboratories or from seed. In the early 1990s there were very few sources for *Phalaenopsis* young plants. Today, there are many providers. The most energetic breeding activity is in Taiwan. Taiwan is also the top supplier of seedling plants. Seedlings that are stable and line bred are available at reasonable prices. New hybrids are in demand, but they are risky to grow. The Netherlands is the leading supply source for tissue-cultured plants. *Phalaenopsis* are difficult, slow, and expensive to tissue culture. Consequently, young tissue-cultured plants are four to ten times the cost of seedlings. The advantage of starting off with tissue culture is the reliability of the resulting crop. Buying in flasked plants from orchid laboratories or prefinished plugs will save six months of growing time.

Orchids are protected under CITES (Convention for International Trade of Endangered Species) laws and require special permits to import.

Growing On

The very long crop time and high heat and light requirements necessary during production render large-scale production to areas with a low cost for these inputs. Plants grow much faster in warmer areas. *Phalaenopsis* plants do prefer a 15–20°F (8–11°C) difference between day and night temperatures. They require this temperature differential in order to flower. Because of that, most growers in northern climates purchase prefinished plants in spike for finishing or finished pots for resale. It is still possible to grow the crop profitably in the North, but as production increases, the prices are going down for finished plants, making buying them in a much more viable option.

If you choose to give production a go or wish to repot plants for reflowering, the following cultural tips will help you.

One of the most debated topics among commercial orchid growers is what medium to plant in. Virtually every grower uses something different. The rules are simple enough. Commercial *Phalaenopsis* plants are derived from epiphytic plants: They evolved growing on tree branches. The most popular media mimic this and stimulate their natural tendencies. Successful substrates that have been used include large chunks of hard wood bark or even mixes of foam rubber and cork. The rule is this: In active-growing seasons (warm to very warm), the plants want to get wet and then to get not quite dry. When *Phalaenopsis* plants are cultured in pots, the roots become very soft, fleshy, and delicate. The strength of the plants lies in the starches stored in their roots. Overwatering easily damages these roots.

Phalaenopsis plants grown in straight bark mixes require higher levels of nitrogen in a 3:1:1 ratio to make up for the nitrogen-binding properties of the bark. If you are using other media, using fertilizer in a 1:1:1 ratio is fine. *Phalaenopsis* likes to be fed while growing and prefers less or no fertilizer during dormant periods.

Phalaenopsis is a medium-light plant, preferring 1,500–3,000 f.c. (16–32 klux). Flowering is induced by a 15–20°F (8–11°C) difference between day and night temperatures. Ideally, the daytime temperature will not exceed 80°F (27°C) during spike and bud formation. The natural flowering season is late winter until early spring, with the peak about March 15. Growers, using artificial cooling to induce flowering, produce many off-season *Phalaenopsis*. This requires at least three weeks of cold treatment.

Plants require seventy to one hundred twenty weeks from flask to a finished plant. For this reason, most growers order in prefinished plants to sell.

Pests & Diseases

Erwinia is the most significant problem for this crop. During summer months, inspect the crop daily, removing and destroying any infected plants, as there is no control. Botrytis can occur on flowers in high humidity and with low air circulation. Overly wet conditions can cause algae that will serve as breeding grounds for fungus gnat and shore fly larvae, which can damage roots and introduce soilborne diseases. Aphids and thrips are attracted to flowers. Spider mites can attack foliage and flowers.

Postharvest

Orchids are extremely sensitive to ethylene gas, even in very small amounts. Do not place orchids near ripening fruits and vegetables or decaying plant debris, all of which are sources of ethylene gas. EthylBloc may be beneficial in prolonging shelf life.

Store and ship plants at 45–50°F (7–10°C).

Many *Phalaenopsis* plants are sold with flowers staked for support during handling. This helps to prevent damage to the flower stalk in shipping and at retail.

Orchids are easy to overwater. Under reasonable conditions, *Phalaenopsis* is a strong plant with a long shelf life. Instruct retailers to keep these beautiful, flowering plants away from the door, where blasts of cold outside air are likely to kill them.

After the consumer has taken the plants home and the flowers are finished, the plants should be placed in a southern or eastern window with direct sunlight. Once the last flower is spent, it is possible to cut the flower spike above the second node and a new secondary spike will form with fewer flowers than the original spike. It is also possible for the home hobbyist to reflower *Phalaenopsis* by placing it in a shaded southern window or eastern or western window.

Philodendron

Philodendron scandens oxycardium
Philodendron selloum
Foliage plant

The philodendron is a mainstay foliage plant. Most growers in the North buy in finished pots or baskets because it is more cost effective to produce them in Florida, Texas, and California, where temperatures are warm. The most popular, *P. scandens oxycardium,* with its vine-like stems and heart-shaped leaves, makes an excellent addition to combination pots and baskets when mixed with annuals for spring sales in the North.

Propagation

Vining species such as *P. scandens oxycardium* (heart-leafed philodendron) are propagated from stem or leaf and eye cuttings, which root easily at 80°F (27°C) under 2,000–3,000 f.c. (22–32 klux) of light. *P. selloum* is commonly grown from seed. Sow six or more seeds per 72 cell and cover very lightly. Seed needs 300–600 f.c. (3.2–6.5 klux) to germinate. Germinate warm at 75–80°F (24–27°C). Many self-heading and hybrid philodendrons are propagated through tissue culture and sold as liners.

Allow three to six weeks from sticking cuttings until liners are rooted.

Philodendron selloum

Growing On

Pot cuttings, seedlings, or liners in a well-drained medium with a pH of 6.0. Maintain minimum growing temperatures of 65°F (18°C), with ideal growth occurring at 70°F (21°C) or above. Grow *P. scandens oxycardium* at 1,500–3,000 f.c. (16–32 klux) and maintain a temperature of 75°F (24°C) or above. For best plants, be sure to keep the substrate temperature at 65°F (18°C) or above. When growing *P. scandens oxycardium* on totems or in hanging baskets, be sure to space plants as soon as vines touch, otherwise the production bench will become an entangled jungle.

P. selloum can be grown at 3,000–6,000 f.c. (32–65 klux) of light. Some Florida growers produce these plants under full-sun field conditions, bringing them indoors for acclimatization prior to sale as an indoor foliage plant. *P. selloum* is more tolerant of lower growing temperatures than other species are, even tolerating brief periods below freezing as a mature landscape specimen in southern regions.

Use a 3:2:1 NPK fertilizer and provide a constant liquid feed at 200 ppm. At feed levels that are too low, philodendrons produce smaller leaves and shorter internodes. Plants may show marginal chlorosis on older foliage when magnesium is deficient. Calcium deficiency can cause root tips to die. Some experts say a magnesium/calcium foliar spray is essential to produce the best quality foliage.

Allow ten to twelve weeks from transplanting a liner for 10" (25 cm) baskets to be salable.

Pests & Diseases

Erwinia, a bacterial pathogen that shows up as water-soaked lesions, can be a serious problem on philodendrons. Also serious is *Xanthomonas campestris* pv. *dieffenbachiae,* which causes older leaves to show a characteristic red edge. Aphids, mealybugs, and scale can also be problems.

Varieties & Related Species

There are a number of *P. selloum* selections available. Pothos (*Epipremnum aureum*), a popular variegated vine sold in hanging baskets, is often mistaken to be a philodendron.

Postproduction

Provide at least 150 f.c. (1.6 klux) of light to hold plants prior to sale and at retail. While *P. scandens oxycardium* will tolerate very low light levels in the home, down to 50 f.c. (538 lux), higher light levels will encourage better plant growth. *P. selloum* requires minimum light levels of 250 f.c. (2.7 klux). Maintain temperatures between 65–85°F (18–29°C) on the display table and in the interior. Philodendron is ethylene sensitive and will show chlorosis and leaf drop when exposed to ethylene gas for several days. To help prevent damage, hold or ship plants cool at 55–60°F (13–16°C).

Phlox

Phlox drummondii
Common name: Annual phlox
Annual

The breeding efforts going into annual phlox over the past decade are paying off in spades—the new strains are real showstoppers. Ideal in 4" (10 cm) pots for spring sales, annual phlox is a great impulse sale item and features the same bold colors of petunias on smaller flowers and low-growing plants. While phlox is native to Texas, outdoors its performance is less than stellar once temperatures heat up in southern regions of the country, where it is enjoyed as an early spring or fall-planted flower. In milder and northern climates, plants flower through the summer from spring plantings. In parts of California, phlox is a great fall/winter plant.

Propagation

Single-sow seed onto a well-drained, disease-free, germination medium with a pH of 5.5–5.8. Cover the seed with vermiculite or germination medium: Phlox germinates best in the dark. Keep trays on the dry side. Germinate at 60–65°F (16–18°C). Keep ammonium levels below 10 ppm. Radicles will emerge in five to seven days.

Move trays to Stage 2 for stem and cotyledon emergence, which will take about seven days. Maintain 60–65°F (16–18°C) soil temperatures. Allow trays to dry between irrigations. As cotyledons expand, begin fertilizing with 50–75 ppm from 14-0-14 once a week. Provide 500–1,000 f.c. (5.4–11 klux) of light.

Stage 3, development of true leaves, will take about twenty-one to twenty-eight days for 384s. Continue to maintain substrate temperatures at 60–65°F

(16–18°C). Increase feed to 100–150 ppm once a week from 20-10-20, alternating with 14-0-14. Allow trays to dry thoroughly between irrigations. Increase light to 1,000–1,500 f.c. (11–16 klux).

Harden plugs for seven days prior to sale or transplant. Maintain soil temperatures of 60–65°F (16–18°C) and increase light to 1,500–3,000 f.c. (16–32 klux). Feed once a week with 14-0-14 at 100–150 ppm.

If growth regulators are required in the plug stage, B-Nine is effective.

Most growers buy in plugs rather than worry about maintaining cool substrate temperatures during germination. Transplant plugs on time: Do not stress annual phlox, as plants may stall.

A 288-plug tray takes from five to six weeks from sowing to be ready to transplant to larger containers.

Growing On

Use 1 plug/4" (10 cm) pot and 3 plugs/1-gal. (4 l) pot. Select a well-drained, disease-free medium with a moderate initial starter charge and a pH of 5.5–6.2. Some growers like to apply a fungicide drench after planting as a preventative precaution.

Allow plants to dry out between irrigations, but do not allow severe wilting. As plants begin to flower, do not allow them to wilt. Fertilize with 150–200 ppm from 15-0-15, alternating with 20-10-20 at every other irrigation.

Grow plants cool under full-light conditions at 55–60°F (13–16°C) days and 50–55°F (10–13°C) nights. Annual phlox responds very well to negative DIF for height control. B-Nine is also effective.

A 4" (10 cm) pot requires seven to nine weeks to finish, while a 1-gal. (4 l) requires ten to twelve weeks to fully root to the bottom of the container when using one 288 plug per 4" (10 cm) pot and three per 1-gal. (4 l) container.

Pests & Diseases

Aphids and thrips can attack plants. Diseases can include *Pythium* and rhizoctonia.

Troubleshooting

High media pH can cause leaf chlorosis.

Varieties

'21st Century' is an F_1 that comes in a range of separate colors plus a formula mix. Plants are early and well branched. Consumers love their flower power.

Postharvest

Plants may be sold in flower or green with color picture tags. Display plants under partial shade to moderate temperatures, which will reduce watering and lower temperatures. Display plants in cool temperatures: 60°F (16°C) days and 50°F (10°C) nights. Keep flowering plants well watered.

Consumers should select a well-drained area in full sun and water in plants well. Mulch will help to moderate soil temperatures and reduce watering needs later on.

Phlox maculata
Common name: Annual phlox
Annual

This class of phlox is highlighted by the varieties 'Intensia' and 'Phloxy Lady'. *Phlox maculata* is a great plant, which flourishes with sun and drier conditions, making it a great bedding plant. Newer breeding has developed self-branching, mounded habits that are easy to grow and control. However, early production issues with too much moisture will still cause large losses with this crop. With some effort and commitment to be sure that the crop gets a cool, dry start, you can have a fantastic plant for a wide range of customers.

Propagation

Phlox maculata is propagated by cuttings. If cuttings are wilted, place in a clean cooler at 50–55°F (10–13°C). Provide a relative humidity of close to 100% by wetting the floor and misting open bags with clean water. Stick cuttings in well-drained Ellepots or medium at pH 5.6–5.8. A rooting hormone might help increase the uniformity of rooting. Root cuttings under mist at a substrate temperature of 70–72°F (21–22°C) and air temperatures of 70–72°F (21–22°C) days and 68–70°F (20–22°C) nights. It is extremely important to only provide enough mist to keep cuttings turgid. Provide 500–1,000 f.c. (5.4–11 klux) of light. Too much mist will inhibit rooting and increase disease.

When the cuttings are callused and begin to develop roots, quickly move them to Stage 2. Significantly reduce or cease the use of misting and begin to apply an overhead feed of 50–75 ppm from 20-10-20. Monitor and treat for fungus gnat larvae.

In Stage 3, maintain soil and air temperatures. As root initials form, begin to dry out soil. Increase light levels to 1,000–2,000 f.c. (11–22 klux) as cuttings root. Apply a foliar feed of 100 ppm from 15-0-15, alternating with 20-10-20. Increase the rate to 200 ppm as roots form. Stage 4 consists of hardening the cuttings for approximately seven days prior to sale or transplant. At the beginning of Stage 4, you can pinch or trim the liners to help create lower breaks. Reduce air temperatures at night to 65–68°F (18–20°C). Liners can be held, if needed, at 54–58°F (12–14°C). Move liners from the mist area and increase light levels to 2,000–4,000 f.c. (22–43 klux). Shade plants during midday to reduce temperature stress. Fertilize with 200 ppm from 15-0-15, alternating with 20-10-20.

A 105 tray will be ready to transplant to larger containers in four to five weeks.

Growing On

Transplant 1 plug or liner/4" (10 cm) pot; 1–2 plugs or liners/6" (15 cm) pot; and 4–6 liners/10" (25 cm) or larger. Select a disease-free, very well-drained medium with a pH of 5.6–5.8. Grow plants on at 65–68°F (18–20°C) days and 60–65°F (15–18°C) nights until roots have developed. Then lower temperatures to 54–58°F (12–14°C), which will help plant form and control growth. The highest quality plants will be grown as low as 50°F (10°C) but this will add weeks to the production time. Fertilize with 200 ppm nitrogen (N) constant feed with a complete fertilizer. Supplemental iron may be provided with an application of chelated iron. Monitor and maintain a pH between 5.8 and 6.2

Phlox must be grown on the dry side, particularly very early in production. Plan the space and companion plants beforehand and follow this plan. It is critical that plants are not overwatered as this can cause problems, which will lead to significant loss. Consider applying a preventative fungicide treatment to protect the sensitive root system. Remember that with lower temperatures, the plants will use less water. Grow under high light levels (4,000–7,000 f.c. [43–75 klux]). Pinching can be used to shape the plant. Pinch as early as the plants have roots to the side of the finish container. You can pinch multiple times to create the desired shape. Depending on the time of season, do not pinch within three to six weeks of the sale date.

Newer breeding and cooler production temperatures will probably keep phlox from needing any growth regulators. However, for vigorous varieties spray applications of Topflor at 10–15 ppm, or B-Nine or Dazide (Daminozide) at 2,500–5,000 ppm, are effective for control. Continue to monitor for fungus gnat larvae and treat as needed.

Transplanting one liner per 4" (10 cm) pot will be in flower in five to seven weeks.

Pests & Diseases

Major greenhouse pests can attack phlox. Fungus gnat larvae can be devastating pests. Powdery mildew and botrytis can attack this plant when conditions permit. Root rots and fungi can be a problem as well when plants are grown too wet.

Troubleshooting

Plants may collapse due to moderately to severely wet medium or excessive botrytis. high Ammonium levels in the medium, overfertilization under low-light conditions, or high production temperature can cause excessive vegetative growth.

Varieties

'Intensia' and 'Phloxy Lady' series are both mounding-type phlox. Check with your broker or breeder rep for the best varieties for your production.

Postharvest

Display plants in high light with cool temperatures. Do not allow plants to wilt, but keep on the dry side. Keep plants from being wet at night.

Consumers should plant in well-drained soil in full sun. Once established, plants will flourish in moist to dry soil and high temperatures. Fertilize regularly throughout the summer for continued performance. Plants are great when used for mixed containers and border plantings.

Phlox paniculata
Common names: Perennial phlox; garden phlox
Perennial, cut flower (Hardy to USDA Zones 4–8)
Phlox paniculata is a widely grown perennial commonly available in colors of white, pink, lavender, and salmon; red, purple, and light blue shades are also available. Many have dark-colored centers or bands around the flower center. The flower clusters grow from 8–10" (20–25 cm) across, with individual flowers measuring 1–1.5" (2.5–3 cm). Some cultivars have scented flowers. Plants grow 2–3' (60–90 cm) tall and from 15–24" (38–61 cm) across and flower from July to September.

Bright colors and an old-fashioned country garden appeal make phlox one of the most popular perennials on the market. New varieties that are resistant to powdery mildew are sure to boost its popularity even more.

Propagation

Bare roots

Perennial phlox is primarily vegetatively propagated from divisions or bare roots potted up in late summer or during winter. Leave three to five crowns to make good-sized plants to fill in 1 gal. (4 l) pots.

Stem cuttings

Stem cuttings at 3–6" (8–15 cm) long can be harvested anytime plants are actively growing but not in flower. Terminal cuttings root fastest. If cuttings are taken from older growth, use a rooting hormone. Stick cuttings into a peatlite media within a 105, 84, 72, or larger liner or cell and provide 70–72°F (21–22°C) bottom heat. In addition, provide mist for about a week or so, and then remove from mist and place on the greenhouse bench. Plants will fully root in four to five weeks but need to bulk up for another two or three weeks for a total of six to eight weeks.

Growing On

Select a well-drained, bark-based or peat medium with a pH of 5.8–6.2 and a moderate initial nutrient charge.

While a cool treatment is not mandatory for flowering, plants that have been cooled have flowered more profusely, more uniformly, and with longer stems in Michigan State University tests.

Phlox plants must attain a minimum of eight to twelve leaf nodes before given a cold treatment. "Bulk" plants up to meet this maturity requirement before the cold treatment begins. Grow on at 65–70°F (18–21°C) days and 60–65°F (16–18°C) nights. Keep plants moist but not saturated. Fertilize at every other irrigation with 14-0-14 at 100–150 ppm. Provide short days (less than twelve hours) if day length is not already short enough when you start the crop. Maintain a soil EC of 1.0 mS/cm.

Either bare roots or rooted liner trays can be potted up to 1-gal. (4 l) containers during late summer to early fall, and the plants can be bulked up prior to cold treatment. Plants can be sold green the following spring and will flower during the summer once planted to the garden.

Growers potting plants in the fall can provide ten to fifteen weeks at 41°F (5°C) in a cold frame or cooler (25 f.c. [269 lux] of light) to meet cooling requirements. Use only clear water to irrigate plants, as needed, during the cool treatment.

For forcing plants into flower for spring (April or May) sales, the plants must receive long-day treatment (more than sixteen hours) with supplemental light (400–500 f.c. [4.3–5.4 klux]) or night interruption lighting from 10 P.M.–2 A.M. To force plants for spring flowering, move pots into a cool greenhouse from the cooler or cold frame during winter and begin long-day treatments immediately. Pinch plants as you begin forcing to encourage branching. Provide constant day/night temperatures at 65–68°F (18–20°C) and provide high light at 3,000–5,000 f.c. (32–54 klux). Low light levels cause weak plants. During the winter months, supplemental lighting at 400–500 f.c. (4.3–5.4 klux) is beneficial to plant quality. *Note:* While plants may be forced into flower early for spring sales in color, the postharvest life of flowering plants is very short. For maximum consumer value, it is best to grow full, bushy plants and sell them green with color tags.

Fertilize with 15-0-15 at 150–200 ppm at every other irrigation and maintain a pH of 5.8–6.2 and an EC of 1.0 mS/cm. Bonzi, Cycocel, and Sumagic can be used for height control.

For bare roots or liners transplanted to 1-gal. (4 l) pots in winter, grow plants on the bench until roots are established at 55–60°F (13–16°C) nights with days 5–8°F (3–5°C) warmer. Once roots reach the side of the container, four to five weeks later, the containers can be moved to a cold frame with 45–50°F (7–10°C) nights and days 5–10°F (3–6°C) warmer. These plants can be sold green in the spring, although smaller than their fall-potted counterparts. Plants will flower sporadically in summer and will bloom more freely in following summers.

Cut Flowers

Plant at a density of 2 plants/ft.^{2} (22/m^2) for indoor or field production. Make sure ground beds have been properly steamed and are free of disease. Phlox prefers well-drained loam but can tolerate heavier soil. Use one level of support. Adjust the pH to 6.0–7.0. Incorporate a preplant fertilizer of 20-20-20 or 12-10-18 and add calcium nitrate as the growing season progresses. Make sure to apply micronutrients once. Ensure that plants get adequate nutrition while they are actively growing and up until bud set. After budding, switch to potassium nitrate.

Irrigation is critical: Keep soil moist, but not wet at all times. Dry soil can reduce stem length and overall quality.

Grow under 4,000–7,000 f.c. (43–75 klux) of light. Temperatures can fluctuate between 60–80°F (16–27°C) during the day and 40–60°F (4–16°C) nights. If good nutrition and irrigation practices are used, plants will tolerate higher light levels and temperatures.

Field plantings in regions where plants are hardy can be harvested for two to three growing seasons. Southern growers can plant field crops from September until November; California growers can plant in January and February; and growers in the Midwest and North can plant in May or after the danger of heavy frost has passed. Depending on your exact conditions, you can get two to three and a half flushes per year. After harvest, cut back outgrowth to the ground to begin initiating the next flush.

Greenhouse plantings made in high-light areas in January can be flowered two to three times: early spring, early summer, and early fall—early winter flowerings are even possible. In lower-light areas, spring, summer, and fall flushes are normal. Production in low-light conditions will not develop properly. In the North, growers can drop greenhouse temperatures to 20°F (–7°C) so plants go dormant for the winter. In late January to February, cut plants back to the ground and begin fertilizing and raising temperatures to normal production temperatures.

Pests & Diseases

P. paniculata is notorious for its susceptibility to powdery mildew. The best control is to grow only mildew-resistant varieties. Viruses can also infect stock plants: Know your source and buy only plants propagated from clean stock. *Verticillium* and root rots can also affect plants. Aphids, thrips, and whiteflies can also be problems on plants.

Varieties

For the best outdoor performance for consumers, select only mildew-tolerant phlox. The Chicago Botanic Garden evaluated more than twenty phlox varieties for mildew resistance. Only one *P. paniculata* received a four-star rating: 'Katharine', a July-to-September bloomer with white-eyed, lavender flowers. Plants grow 36–43" (0.9–1.1 m) tall and nearly as wide.

Other mildew-resistant *P. paniculata* include 'Laura', a lavender-purple with a white star center and a dark eye; 'Bright Eyes' with clear, pink flowers and a red eye; and 'Robert Poore', a medium pink with a light fragrance. 'David', a pure white and a Perennial Plant Association Plant of the Year, is very mildew resistant and delightfully fragrant. 'David' grows to 36–40" (0.9–1.0 m) tall, shorter under dry soil conditions. Somewhat mildew resistant are 'Franz Schubert' (pale lavender with dark lavender and a white eye) and 'Nicky' (magenta purple), which is very fragrant.

Postharvest

Harvest cut flowers when two to three flowers are open. Use an STS (silver thiosulfate) preservative with bactericide after harvest. Handled properly, stems can last ten to fourteen days.

In their gardens, consumers can thin established plants, removing some of the stems in a clump to increase bloom size and count on the remaining stems. Divide plants every three to five years.

Phlox subulata
Common names: Moss pink; creeping phlox; ground phlox
Perennial (Hardy to USDA Zones 2–9)

Phlox subulata is an excellent evergreen groundcover with needlelike, dark green foliage. Plants reach 4–6" (10–15 cm) tall and spread 15–24" (38–61 cm) across. The single, open flowers measure 0.5–0.75" (1–2 cm) wide in colors of pink, blue, and white and flower in April and May.

Propagation

Plants are easily propagated by tip cuttings. Cuttings may be taken from actively growing plants that have been sheared after flowering. Harvest semi-mature cuttings, not woody stems, in late spring and mid- to late summer and stick into a peat moss–based media within a 105-, 84-, or larger liner tray. For larger liners, some growers stick two cuttings to get faster rooting.

If cuttings are supple, then a rooting hormone is not necessary, but it is useful if you allow the stems to harden. The longer you wait after the plants flower or if you have not kept the stems actively growing, the more difficult it is to get a more uniform stand of rooted liners.

Place the rooting trays under mist, allowing six to ten days and cuttings will root at 72–75°F (22–24°C) in two to three weeks. However, allow another three to five weeks to fully fill out the liner tray.

Growing On

Liners may be potted up in the summer or fall and overwintered in cold frames. You can also receive plants in the winter for potting up as well. Creeping phlox may be sold in flower or green with color tags.

Use 1 plant/1-qt. (1.1 l) pot or 2 plants/1-gal. (4 l) pot. Select a long-lasting, bark/peat medium with a pH of 6.0–6.5 and a moderate initial starter charge. Some growers like to drench with a fungicide immediately after planting to prevent disease.

The best plants are grown from liners received in summer or fall, bulked up, and given a cool treatment. Grow 1-gal. (4 l) pots from a 32, 50, or similarly large liner for twelve weeks to bulk plants prior to providing cool temperatures. Quarts (1.1 l) can be bulked in about eight weeks. Michigan State University researchers recommend providing a minimum of eight weeks at 41°F (5°C) in a cooler (50 f.c. [538 lux] of light for nine hours daily) or cold frame. Plants may be held in the cooler until you are ready to force them for sale. Creeping phlox has a very short shelf life once in flower; be careful in scheduling so you will have a steady flow of fresh plants through the season.

After cooling, temperature is the next determining factor for when plants will flower: The warmer the temperature, the faster flowering. Once you are ready to force plants into flower, grow them on at constant 65–68°F (18–20°C) day/night temperatures in the greenhouse or uncover cold frames and grow them with natural outdoor temperatures (50 [10°C] days and 35–40°F [2–4°C] nights). Fertilize with 100–150 ppm from 20-10-20 once a week. Do not keep plants wet. Allow them to dry slightly between irrigations, but do not allow them to wilt. Irrigate plants early in the day so that foliage is totally dry by nightfall to help prevent disease. Grow under full light in the spring.

Some liner suppliers offer cooled 50-cells that you can receive in January and grow on in a cool greenhouse at 50–55°F (10–13°C) nights. No. 1 transplants can also be received in March or April, potted, and forced for almost immediate sale. Walters Gardens recommends growing a crop in this way outdoors under white, permeable row cover. Irrigate as needed, removing the row cover once plants are budded. Protect plants from hard freezes.

Pests & Diseases

Creeping phlox is susceptible to a number of diseases, including *Alternaria,* botrytis, *Colletotrichum, Fusarium,* rhizoctonia, and *Thielaviopsis.* For the cleanest crops, be sure to start with liners produced from stock maintained in a clean-stock system to ensure that you are selling only virus-free plants. Spider mites and thrips may also attack creeping phlox.

Varieties

The most popular creeping phlox varieties include 'Candy Stripe' (pink and white striped), 'Crimson Beauty' (rose), 'Emerald Blue' (light blue), 'Emerald Pink' (pink), 'Millstream Daphne' (dark pink/dark eye), 'Red Wings' (crimson), and 'White Delight' (solid white).

Postharvest

Plants are ready to sell just before they begin to flower. Ship budded plants cool (40°F [4°C]) on racks. Retailers should display plants under bright, cool conditions. Once flowering begins, keep plants as cool as possible and well watered (but not wet) to prolong shelf life.

Consumers should shear plants back in midsummer to encourage branching. Do not shear after mid-August, since flower buds may have already formed for the following season.

Pimpinella

Pimpinella anisum
Common name: Anise
Annual herb

Valued for its seeds and leaves, anise looks similar to either *Ammi majus* or *Daucus carota* with plate-like umbels that have small, white flowers. Plants grow from 2–3' (61–91 cm) tall.

Propagation & Growing On

Like a number of plants in the Umbelliferae family, anise has a long taproot that makes it difficult to transplant. It is not recommended for open-flat sowing and singular seedling transplant or plug/liner production. Instead, it can be either sown to the final pot (3 or 4" [8 or 10 cm]) or direct to the home garden (this latter method is the most common).

For pots, sow seed directly to the final container, such as a Jiffy pot, Ellepot, or other pot can be planted directly into the garden. Use three to six seeds per pot and cover lightly. Germinate at 68–70°F (20–21°C), and seedlings will emerge in six to ten days. Thin as needed and grow on at 58–60°F (14–16°C) nights. Pots will be salable in five to eight weeks from sowing. Growing slightly cooler helps to improve overall appearance but adds a week or so to the crop time.

Platycodon

Platycodon grandiflorus
Common name: Balloon flower
Perennial (Hardy to USDA Zones 3–9)

Platycodon grandiflorus is a clump-forming perennial noted for its inflated, balloon-like flower buds in white, pink, and, especially, blue. These buds will eventually split open, expanding to 3" (7 cm) when fully developed. The scentless flowers are single or sometimes double and emerge terminally at the stem ends. Stems are erect and basal branching; they develop out of the center of the plant.

The stems, when broken, will exude a white, milky latex; sear them with a flame before placing in a vase as a cut flower. The roots are fleshy, so dividing the plants requires care. Dwarf varieties grow 10–12" (25–30 cm), while the cut flower strains reach 2–3' (60–90 cm) in height. Plants spread from 10–15" (25–38 cm) across and flower from late June to August.

Propagation

Sow one seed per 288-plug tray in a well-drained, disease-free, germination medium with a pH of 5.5–5.8. Do not cover or cover lightly with coarse vermiculite (covering lightly with vermiculite helps to maintain moisture around the seed during germination). The root radicle emerges in two to five days at 68–72°F (20–24°C). Provide 100–400 f.c. (1.1–4.3 klux) of light during germination. Keep the medium uniformly moist but not saturated. Once the root radicle emerges, move trays to Stage 2 for stem and cotyledon emergence, which will take five to ten days. Provide substrate temperatures of 60–65°F (16–18°C) at night with days 4–5°F (2–3°C) warmer. Begin to dry trays down a bit once radicles are emerged. Gradually raise light levels to 500–1,500 f.c. (5.4–16 klux). Once cotyledons expand, begin fertilizing with 50–75 ppm from 14-0-14, alternating with clear water.

Once you have a stand, move trays to Stage 3 for growth and development of true leaves. Stage 3 will take twenty-one to thirty-five days for 288s. Reduce substrate temperatures to 58–62°F (14–17°C) and allow the medium to dry thoroughly between irrigations. Increase light levels to 1,500–2,500 f.c. (16–27 klux). Increase fertilizer to 100–150 ppm from 20-10-20, alternating with 14-0-14. Feed at every second or third irrigation.

Harden plugs in Stage 4 for seven days prior to sale or transplanting by reducing soil temperatures to 55–60°F (13–16°C) and allowing soil to thoroughly dry out between irrigations. Increase light levels to 2,500–3,500 f.c. (27–38 klux), and feed once with 100–150 ppm from 14-0-14.

A 288-plug tray requires five to six weeks from sowing to be ready to transplant to larger containers. A 128 tray requires from eight to nine weeks from sowing.

Growing On

While most of the current dwarf selections will flower in July from a January sowing, the best performance in the garden is from sowings done in the summer, potted to 1-qt. (1.1 l) or larger containers, and kept cold during the winter for green plant sales the following spring. Plants can be stored during the winter in an unheated greenhouse at 30–35°F (–1–2°C). These plants will flower more readily than those sown in January or later. However, today's *Platycodon* varieties do not require cold temperatures in order to flower.

Select a disease-free, commercial, potting medium with a moderate starter charge and a pH of 5.5–6.2. Plant 1 plug/4" (10 cm) pot and 2–3 plugs/6" (15 cm) pot. *Platycodon* has a taproot, so choose deep pots. Provide light shade immediately after potting to lessen transplant shock when transplanted in summer. Once plants have rooted in, grow in full sun (4,000–6,000 f.c. [43–65 klux]).

Provide day temperatures of 60–65°F (16–18°C) and nights of 55–60°F (13–16°C) for the fastest crop time. Fertilize with 150–200 ppm from 15-0-15 at every other irrigation. Run plants dry; however, plants that dry out too much will have leaf yellowing.

If you are forcing the crop from a winter or spring planting, you can speed flowering by providing night interruption lighting (four hours). Additionally, growers producing flowering pot plants will benefit from using twelve to fourteen hours of supplemental lighting (400–500 f.c. [4.3–5.4 klux]), which will increase both the shoot and flower number. If you choose to light, watch plant height, as plants may stretch.

Pinching plants once they are rooted in and actively growing increases both the branching and flower number. Plants that have been overwintered will generally not require pinching. Once flowering begins, pinch old flowers off unsold pots, because plants stop growing when flowers are left on the plant.

B-Nine can be used for height control. When you treat with a growth regulator, be sure to water plants first so leaves do not yellow.

Prior to overwintering pots planted in the summer or early fall, make sure roots have fully filled the container/medium. The plant's fleshy taproot can be highly susceptible to root rots, so overwinter pots under cover rather than outdoors. *Platycodon* is extremely slow to emerge after dormancy, taking until June to emerge in the garden in some parts of the country. Do not plan to sell overwintered pots early in the season.

Platycodon can also be grown as a cut flower. Plant outdoors in rows in the fall for harvest the following summer. Space plants at least 12" (30 cm) apart and provide one to two layers of support. Flowering is not prolific the first year but will be profuse in subsequent years.

Transplanting a 288 cell to a 4" (10 cm) pot requires nine to eleven weeks to flower.

Pests & Diseases

Aphids and whiteflies may attack plants. *Pythium* and rhizoctonia can become disease problems.

Varieties

The variety that got the bedding plant world's attention is 'Sentimental Blue', a dwarf variety (6–8" [15–20 cm]). 'Sentimental Blue' tends to produce one flower at a time as a pot plant. The 'Astra' series was introduced next and, more recently, the 'Balou' series. Developed for the 4.5" (11 cm) pot market and flowering profusely, 'Astra' is available in white, pink, traditional blue, and semi-double flower forms in blue and lavender. 'Balou' is currently available in both blue and white selections.

Plectranthus

Plectranthus spp.
Plectranthus coleoides
Plectranthus amboinicus
Common name: Swedish ivy
Annual, tender perennial

Plectranthus is valued as an ornamental for its leaves, which may be green, variegated, or colored. However, flowering varieties like 'Mona Lavender' are equally prized. It used to be considered primarily a foliage plant and its use was limited to hanging baskets. But now you find *Plectranthus* turning up in a variety of combination baskets, planters, and window boxes. Its foliage is a great backdrop for bright-colored flowers. Growers find it very easy to produce, and consumers love the fact that it is virtually maintenance free.

P. coleoides is widely used as an ornamental, especially in hanging baskets and mixed planters. Known commonly as Swedish ivy, leaves are variegated, and plants are an outstanding addition to combination plantings for the sun or shade. The popularity of *P. amboinicus* as an ornamental is more recent, but it has been used as an herb throughout history. You'll recognize this plant as soon as you break off a leaf: The pungent odor is reminiscent of very strong oregano.

Plectranthus coleoides

Propagation

Plectranthus can easily be propagated from cuttings. Propagation may occur year-round as long as active growth is maintained. Maintain high light levels and warm temperatures during winter months for year-round production. Most standard tray sizes for propagation are between 84 and 128 cells per tray. One cutting per cell is sufficient. Use a standard propagation medium with a pH of 5.5–6.5. Temperatures of at least 70–75°F (21–24°C) for the rooting process are best, ideally with bottom heat. Keep the medium moist at all times, but not saturated. Until the plants have developed roots of their own, mist several times a day to avoid wilting. The amount and frequency of misting depend on the temperature and air circulation in the greenhouse. After cuttings have rooted, move plants to a hardening area with increased light levels and lower temperatures.

Due to the often higher quality available from specialized young-plant producers, more growers are choosing to buy in liners instead of self-propagating.

A 105 tray will be ready to transplant to larger containers three to four weeks after sticking cuttings.

Growing On

Transplant liners into a well-drained, soilless medium such as a peat/perlite mix with a pH of 5.5–6.5 and a moderate initial starter charge. Use 1 liner/4" (10 cm), 2 liners/6" (15 cm) pot, and 3–5 liners/10" (25 cm) basket. *Plectranthus* is excellent as a component in mixed container gardens and can be planted directly into combination pots as a liner.

Plectranthus grows best under moderate to high light levels. A range of 3,500–8,000 f.c. (38–86 klux) is appropriate. Some varieties require partial shade, while others can tolerate full sun.

Keep the medium moist, but not saturated. Apply a constant feed of 150–200 ppm from 20-10-20, alternating with 15-0-15, and average amounts of micronutrients with slightly higher levels of iron. *P. amboinicus* requires slightly less fertilizer: Use a rate of 100–150 ppm. Watch for salts buildup—especially if you note salts buildup on the top of the pot—and leach with clear water if necessary.

To encourage branching, give one soft pinch within one to two weeks of planting. Larger pots and baskets may require additional pinches for shaping throughout production.

Transplanting a 105 liner to a 4" (10 cm) pot requires from six to eight weeks for the plant to be ready to sell.

Pests & Diseases

Watch for whiteflies. Drench with broad-spectrum fungicide at liner planting to minimize the danger of soilborne diseases.

Varieties

P. coleoides 'Variegatus' or 'Variegata' is the most common type. Also commonly called variegated Swedish ivy, it features creamy, yellow variegation on trailing, fountain-like foliage. Plants can take full sun and are a low-cost, bright addition to combination pots.

P. amboinicus varieties include 'Variegatus' and 'Athens Gem', both brought to the market by Allan Armitage from the University of Georgia. Both are very tolerant of heat and humidity and have fragrant foliage similar in scent to oregano. 'Variegatus' has green leaves with white margins and an upright growth habit of up to 18–24" (46–61 cm). 'Athens Gem' has green and yellow variegation and a more spreading/trailing habit.

Plectranthus coleoides 'Variegata'

Postharvest

Avoid bruising leaves during shipping and handling. Consumers can move plants indoors to a bright window and enjoy them as houseplants through the fall and winter months.

Plumbago

Plumbago auriculata
Common names: Plumbago; leadwort

Propagation

Sow seeds to 200, 128, and larger cells onto a well-drained, disease-free, soilless medium with a pH of 5.8–6.2 and a medium initial nutrient charge (EC less than 0.75 mS/cm with 1:2 extraction). Seeds are very large and must be covered lightly with a coarse-grade vermiculite at sowing to retain moisture during germination. Germinate at 72–75°F (22–24°C) with optional light. Keep soil wet (level 4) during Stage 1. Maintain 95% or more relative humidity (RH) until radicle emergence. Germination takes approximately seven to ten days.

After germination, maintain a soil temperature of 70–72°F (21–22°C) with light levels up to 2,500 f.c. (27 klux). Reduce soil moisture slightly (level 3) to allow the roots to penetrate into the media. Apply fertilizer at rate 1 (less than 100 ppm nitrogen and less than 0.7 mS/cm EC) from nitrate-form fertilizers with low phosphorous. Plugs will finish in four to five weeks for 200-cell plugs.

Growing On

Most varieties are finished in 4" (10 cm), 6" (15 cm), or 1-gal. (4 l) containers. They are vigorous growers that can be finished in 4.5" (11 cm) pots up to 2-gal. (20 cm) containers. Use 1 plant per 4.5" (11 cm) pot and 1–2 plants in 1- or 2-gal. (18 or 20 cm) containers.

Transplant into a well-drained, disease-free, soilless medium with a pH of 5.8–6.2 and a medium initial nutrient charge. Maintain air temperatures at 62–80°F (17–26°C) from transplant to sale. High light levels and warm temperatures promote the best branching. Avoid both excessive watering and drought. Do not allow the plants to wilt. Starting one week after transplant, apply fertilizer at rate 4 (225–300 ppm nitrogen [N] and 1.5–2.0 mS/cm) once a week, predominately using a nitrate-form fertilizer with low phosphorus and high potassium. Maintain the media EC at 1.5–2.0 mS/cm and the pH at 5.8–6.2. For a constant fertilizer program, you can apply fertilizer at rate 3 (175–225 ppm N and 1.2–1.5 mS/cm) while maintaining the same recommended EC and pH ranges.

For fall flowering, the plants may need long-day treatment after September 1.

Plumbago 'Escapade' produces bushier plants when pinched once or twice. The first pinch should be done when the plants have developed fifteen leaves. A hard pinch removing five leaves is recommended. If pinched at the tip level, or a soft pinch, the plant will redevelop a main leader and not branch. Pinching will delay flowering by about two weeks.

Not many plant growth regulators work on *Plumbago*. Applying a Florel spray at 1,000 ppm one week before pinching will produce a bushier, more compact plant.

The total crop time is location dependent. In the South, crop time is twelve to sixteen weeks, while in the North it is sixteen to twenty weeks. When sown in January, 'Escapade' will flower in May under natural days.

Pests

Insect problems on *Plumbago* include thrips and aphids.

Varieties

'Escapade' is the most commonly propagated variety from seed in North America. Available in separate colors of blue and white, plants grow from 12–18" (30–45 cm) upwards to 6' (1.8 m) in southern locations.

Postharvest

Home gardeners will see the best results when they place 'Escapade' plants in full-sun locations. These moderately drought-tolerant plants grow vigorously during hot weather. A perfect choice for the warm South, the versatile plants can grow up to 6' (1.8 m) tall to create a bushy, informal hedge and also work well in mixed containers. In the North, *Plumbagos* 'Escapade' are best suited for patio planters, growing to about 12–18" (30–45 cm) tall, with a slightly greater spread.

Poinsettia (see *Euphorbia*)

Porphyrocoma

Porphyrocoma pohliana
Common name: Brazilian fireworks

This relative newcomer to the bedding plant industry is highlighted by the variety 'Maracas.' Plants grow from 6–8" (15–20 cm) and perform best as shaded ground or container plants.

Propagation

Sow one seed per cell in a 400- to 288-cell plug tray. Cover the seed with coarse vermiculite. Use a well-drained, disease-free, soilless medium with a pH of 5.5–6.1 and a medium initial nutrient charge (EC less than 0.75 mS/cm with a 1:2 extraction).

In Stage 1 (germination), provide soil temperature of 65–75°F (18–24°C). Seed germinates slightly better under dark conditions. Keep soil wet (level 4) during Stage 1 and maintain 100% relative humidity (RH) until radicles emerge. Stage 1 takes approximately four to five days.

In Stage 2, keep temperatures at 68–70°F (20–21°C) and provide light levels up to 2,500 f.c. (27 klux). Reduce soil moisture slightly (level 3) to allow roots to penetrate into the media. Apply fertilizer at rate 1 (less than 100 ppm nitrogen [N] and an EC less than 0.7 mS/cm) from nitrate-form fertilizers with low phosphorous.

In Stage 3, provide soil temperatures of 66–68°F (19–20°C) and light levels up to 2,500 f.c. (27 klux). Allow media to further dry so the surface becomes light brown (level 2) before watering. Do not allow the seedlings to wilt. Increase the fertilizer to rate 2 (100–175 ppm N and an EC of 0.7–1.2 mS/cm). Plant growth regulators are not needed.

In Stage 4, keep soil temperatures at 68–70°F (20–21°C) and light up to 5,000 f.c. (54 klux) if temperature can be controlled. Maintain the same moisture and fertilizer requirements as in Stage 3.

A 288-plug tray takes from five to six weeks from sowing to be ready to transplant to larger containers.

Growing On

Use a well-drained, disease-free, soilless medium with a pH of 5.5–6.2 and a medium initial nutrient charge with temperatures of 72–80°F (22–27°C) days and 66–68°F (19–20°C) nights. Brazilian fireworks is a heat-loving crop. The warmer the temperature is, the shorter the crop time. The crop time will be significantly longer when grown in temperatures below 66°F (19°C). In addition, temperatures below 66°F (19°C) can cause the leaves to turn yellow and pucker. Provide shade when the light level is above 5,000 f.c. (54 klux). Brazilian fireworks is a shade crop that can tolerate light levels as low as 1,500 f.c. (16 klux). High light can cause leaf puckering.

Avoid both excessive watering and drought. Do not allow plants to wilt. Apply fertilizer at rate 3 (175–225 ppm N and an EC of 1.2–1.5 mS/cm) once a week from a nitrate-form fertilizer with low phosphorus, alternated with a balanced ammonium-and-nitrate-form fertilizer as needed.

Growth regulators are not needed for height control, but Bonzi (paclobutrazol) spray at 3–5 ppm (0.75–1.25 ml/l, 0.4% formulation) will result in more silver-colored foliage and flowering about two weeks earlier than without treatment. Using a higher rate than 5 ppm (1.25 ml/l, 0.4% formulation) can induce puckering of the leaves. To determine the best rate for your conditions, conduct an in-house trial.

Crop time for transplant to flower is ten to fourteen weeks, depending on temperature. The warmer it is, the shorter the crop time. The total crop time is similar to *Crossandra.*

Table 32. Crop Time for Brazilian Fireworks

The number of weeks from transplant to salable foliage plant without flower

CONTAINER SIZE	PLANTS PER POT OR BASKET	WEEKS FROM TRANSPLANT	TOTAL WEEKS
306 premium pack	1	7–8	12–14
4 or 4.5" (10 or 11 cm) pot	1	7–8	12–14
6 or 6.5" (15 or16 cm) pot	3	7–9	12–15
1-gal. (4 l) basket	3	7–9	12–15

Portulaca

Portulaca grandiflora
Common name: Moss rose
Annual

Most landscapers wouldn't be caught without portulaca for those difficult beds with both poor soil and full sun. Brightly colored, double flowers on low-growing, almost succulent plants characterize this summertime bedding plant star.

Propagation

Most growers buy in multiseeded plugs. For bedding plant packs, 512s are fine; but for later-season pot production and hanging baskets, buy in 288s or 72s.

Growing plugs takes a minimum of four weeks. Stage 1, radicle emergence, occurs in two to three days at substrate temperatures of 78–80°F (26–27°C). Maintain uniform moisture levels. Multiseeded plugs fill out faster, so most plug growers sow multiseeded pellets, especially on larger plug sizes.

Stage 2, stem and cotyledon emergence, takes seven days at soil temperatures of 72–75°F (22–24°C). During Stage 2, begin to allow soil to dry slightly between irrigations. Maintain the soil pH at 5.5–5.8 and the EC at less than 0.75 mS/cm. Start fertilizing with 50–75 ppm from 14-0-14 once cotyledons are fully expanded. Alternate feeding with clear water.

Stage 3 takes from fourteen to twenty-one days. Maintain substrate temperatures at 68–72°F (20–22°C) and allow the medium to dry between irrigations. Portulacas prefer warm, dry conditions. Fertilize with 100–150 ppm 20-10-20, alternating with 14-0-14 every second or third irrigation. Supplement with magnesium once or twice at this stage. If needed, B-Nine can be used to control plant height.

Stage 4, hardening plants for transport or transplant, takes seven days. Cool substrate temperatures to 65–68°F (18–20°C) and allow the medium to dry thoroughly between irrigations. Fertilize with 100–150 ppm from 14-0-14 as needed.

Ideally, provide more than thirteen hours of day length throughout the plug stage for fastest flowering.

A 288-plug tray takes from four to five weeks from sowing to be ready to transplant to larger containers.

Growing On

Transplant plugs into cells or pots filled with a well-drained, disease-free medium with a moderate initial nutrient charge and a pH of 5.5–6.2.

Grow on under full-sun conditions while maintaining moderate air temperatures. Fertilize at every other irrigation with 150–200 ppm 15-0-15, alternating with 20-10-20. Maintain an EC at 1.0 mS/cm.

PanAmerican Seed advises that portulaca is sensitive to short days. Under day lengths less than eleven hours, the variety 'Margarita' can rosette, and flowering will be delayed. Extend the day length using mum lighting to hasten flowering in early crops.

One 288 plug transplanted in a 4" (10 cm) pot requires six to seven weeks to flower.

Pests & Diseases

Spider mites and thrips can become pest problems. Diseases that can attack portulaca include *Pythium* and rhizoctonia.

Varieties

Portulaca requires high light levels for flowers to remain open. The Holy Grail of every seed breeder working in the crop is to develop strains that flower rain or shine. For the best flowering, choose F1 varieties. Two F1 series are 'Sundial', a spreading type, and 'Tequila', a mounding type. Both come in a wide color range. If you are in a hot climate or if landscapers are an important client base for you, be sure to offer separate colors of either series in addition to trays of 'Mix'.

Portulaca oleracea
Common name: Purslane
Annual

New breeding revived this traditional hanging basket crop. When 'Yubi' was introduced in the late 1990s, growers began to once again pay attention to this die-hard performer. Finally large flowers in bright, electric colors were available on plants that love the heat. The new vegetative portulacas make great hanging baskets and are finding their way onto retail shelves from coast to coast. Not to be outdone, seed-propagated selections have been improved as well.

Propagation

Seed and cuttings are the two common methods of propagation for these portulacas.

Cuttings

Stick callus cuttings in Ellepots or a medium for five to seven days at substrate temperatures of 68–72°F (20–22°C) and air temperatures of 75–80°F (24–26°C) days and 68–72°F (20–22°C) nights. Maintain uniform moisture with mist and provide light at 500–1,000 f.c. (5.4–11 klux). Apply a foliar feed of 50–75 ppm 20-10-20 once a week. Maintain the substrate pH at 6.5–7.0 and soluble salts (EC) below 0.5 mS/cm.

When 50% of the cuttings have begun to differentiate root initials, move them to the next stage. Maintain air and substrate temperatures as cuttings develop roots over the next nine to eleven days. Increase light levels to 1,000–2,000 f.c. (11–22 klux) as cuttings begin to root out. Apply foliar feed of 100 ppm 15-0-15, alternating with 20-10-20. Increase the rate rapidly to 200 ppm as roots form and cuttings mature. Maintain the same soil pH and EC levels.

To harden cuttings for sale or planting, reduce night temperatures to 65–68°F (16–20°F) and increase light levels to 2,000–4,000 f.c. (22–43 klux). Increase the target EC level to below 1.0 mS/cm.

A 105 liner will root in three to four weeks.

Seed

Sow seed to 288-cell trays using one multiseeded pellet or three to four individual seeds per cell for the most uniform results. Do not cover seed. Use a well-drained, disease-free, soilless media with a pH of 5.8–6.2 and a medium initial nutrient charge (an EC of 0.75 mS/cm).

Stage 1 (germination) takes from three to four days at 68–74°F (20–23°C). Light is not required for germination. Keep media wet (level 4) during Stage 1. Maintain 95% or more relative humidity (RH) until radicles emerge.

In Stage 2, increase to temperatures to 72–75°F (22–24°C) and light levels up to 2,500 f.c. (27 klux). Reduce soil moisture slightly (levels 3–4) to allow the roots to penetrate into the media. Apply fertilizer at rate 1 (less than 100 ppm nitrogen [N] and an EC of less than 0.7 mS/cm) from nitrate-form fertilizers with low phosphorous.

For Stage 3, decrease temperatures to 64–68°F (18–20°C) and maintain light levels up to 2,500 f.c. (27 klux). Allow media to dry further until the surface becomes light brown (level 2) before watering. Keep the moisture to wet-dry cycle (moisture level 4–2). Increase the fertilizer to rate 2 (100–175 ppm N and an EC of 0.7–1.2 mS/cm). If growth is slow, apply a balanced ammonium-and-nitrate-form fertilizer with every other fertilization. Maintain a medium pH of 5.8–6.2 and an EC between 1.0 and 1.5 mS/cm (1:2 extraction). Plant growth regulators are not needed.

In Stage 4 maintain soil temperatures at 65–68°F (18–20°C) and light levels up to 5,000 f.c. (54 klux) if the temperature can be maintained. Keep the same moisture and fertility levels as in Stage 3.

A 288 plug takes from four to five weeks to be ready to transplant to a larger container.

Growing On

Cuttings

Plant liners into a well-drained, disease-free medium with a pH of 6.5–7.0 and a high initial nutrient charge. The medium must dry thoroughly between irrigations to prevent root rot. Use 1 liner/4" (10 cm) pot; 2–3 liners/6" (15 cm) pots; and 5–7 liners/10" (25 cm) hanging basket.

Plants are day neutral, but flowering is affected by total light accumulation. As light intensity increases, the number of leaves formed prior to initiation of the terminal flower decreases. Under winter light conditions, plants take longer to finish. Low light also promotes stem stretch and thin, weak stems. Grow in light levels as high as possible, targeting 4,500–6,000 f.c. (48–65 klux). You can hang portulaca baskets in the greenhouse attic.

Fertilize with 200 ppm of 15-0-15, alternating with 20-10-20. Increase to 300 ppm as plants mature. Depending on light intensity, purslane has moderate to high fertilizer requirements.

Grow at 75–80°F (24–26°C) days and 65–68°F (18–20°C) nights. Do not overwater.

Pinch plants in pot sizes larger than 4" (10 cm) at least two times during production to improve branching and basket shape. Once liners are established, pinch plants to about 2.0–2.5" (5–6 cm) above the substrate line. Unpinched plants growing in pots will become tangled; make sure to maintain adequate space around plants and/or to pinch.

Cycocel can improve branching and flowering. Apply 5,000 ppm five to six days after pinching.

Potting up one 288-cell plug to a 4" (10 cm) pot requires from five to seven weeks to flower and be ready for retail sale. Hanging baskets require eight to nine weeks.

Seed

Transplant one 288 plug per 4" (10 cm) pot using a well-drained, disease-free, soilless media with a pH of 5.5–6.2 and a medium initial nutrient charge (an EC of 0.75 mS/cm). Grow on at temperatures of 68–75°F (20–24°C) days and nights at 65–68°F (18–20°C). Maintain light levels as high as possible, if temperature can be controlled. 'Toucan' can flower under any day length but will flower slightly faster under shorter days.

Grow on the dry side. Apply fertilizer at rate 2 (100–175 ppm nitrogen [N] and an EC of 0.7–1.2 mS/cm) using predominantly nitrate-form fertilizer with low phosphorus and high potassium. Maintain the media EC at 1.5–2.0 mS/cm and the pH at 6.0–6.5.

Grown from multiseed pellets, plants generally do not need a PGR treatment if produced under low feed, dry watering, and high-light conditions. However, if necessary, a Topflor (flurprimidol) 30 ppm (7.9 ml/l, 0.38% formulation) spray can be applied one week after transplant. Repeat the spray two weeks later. Alternatively, a Bonzi (paclobutrazol) 5 ppm (1.3 ml/l, 0.4% formulation) drench can be used one week after transplant. Pinching is not needed.

Potting up one 288-cell plug to a 4" (10 cm) pot requires from five to seven weeks to flower and be ready for retail sale. Hanging baskets require eight to nine weeks.

Table 33. Crop Times for Portulaca

CONTAINER SIZE	NUMBER OF PLANTS	WEEKS FROM TRANSPLANT	TOTAL WEEKS
1801 flats or 306 packs	1 plug per cell	6–7	11–12
4" (10 cm) pot	2–3 plugs per pot	6–7	11–12
6" (15 cm) pot	2–3 plugs per pot	6–7	11 –12
10" (25 cm) basket	3–4 plugs per basket	8–9	13 –14

Pests & Diseases

Aphids, thrips, oligochaetes, spider mites, fungus gnats, and mealybugs can be pest problems. Diseases that can attack purslane include botrytis, rhizoctonia, *Pythium,* INSV, and potato virus Y.

Troubleshooting

Plant collapse may be the result of the medium being kept wet for extended periods or the presence of botrytis, fungus gnats, or oligochaetes. High ammonia concentration in the soil, overfertilization under low light, or low light and overwatering can cause excessive vegetative growth. Low fertilizer during early growth stages can result in poor branching. Drying the plant out between irrigations, spray damage, or viruses can all cause foliage necrosis.

Varieties

Vegetatively, the 'Yubi' series has revolutionized hanging basket purslane. They look nothing like the old varieties everyone grew for years, and they have become the building block on which many of the new varieties have built themselves. Check with your plant distributor for the latest in selections.

From seed, the 'Toucan' series is available in a number of separate colors plus a mix.

Primula

Primula acaulis
Common name: Primrose
Flowering pot plant, annual

Spring just wouldn't be as bright without the bright, fragrant flowers of primrose. For growers, *Primula* is a moderately easy-to-grow crop, requiring low production inputs. You can grow plants in a cool to cold greenhouse, and once they are put out, they are practically labor free until sale. While wholesale prices are modest, production costs are too. You see their bright flowers showing up on retail shelves across the country from supermarkets and mass markets to the most upscale garden centers and gardening stores beginning in February after Valentine's Day and continuing into the spring. In the Pacific Northwest and along the California coast, *Primulas* are a season unto themselves.

Propagation

Primula is seed propagated. Its crop time is relatively long and requires cool temperatures during the summer months, impossible for most growers to provide. For these reasons and convenience, most growers buy in plugs that originate from cool-summer areas. If you are germinating your own seed under suboptimal conditions, consider using primed seed to increase germination and uniformity. To extend your sales period with fresh plants, be sure to offer early-, mid- and late-season types, all of which can be planted at the same time.

Since *Primula* seed is short-lived, store the seed prior to sowing in a refrigerator at 45°F (7°C). Sow seed onto a well-drained, disease-free, germination medium with a pH of 5.5–5.8 and a low starter charge. Cover with a thin layer of coarse vermiculite to maintain 100% humidity at the seed coat. Provide substrate temperatures of 64–68°F (18–20°C). Do not allow substrate temperatures to rise above 68°F (20°C), as total germination will be reduced and seedlings will not be uniform. Keep trays evenly moist but not saturated. Light during germination is not necessary, but providing 100–400 f.c. (1.1–4.3 klux) of light will help prevent seedling stretch. Keep ammonium levels below 10 ppm. Radicles will emerge in seven to ten days.

Move trays to Stage 2 for stem and cotyledon emergence. Reduce soil moisture and humidity levels once the cotyledons emerge, allowing the medium to dry slightly before watering. Drop substrate temperatures to 62–65°F (17–18°C) and increase light to 1,000–1,500 f.c. (11–16 klux). As soon as cotyledons expand, begin fertilizing weekly with 50–75 ppm from 14-0-14. Stage 2 takes about fourteen days.

Move trays to Stage 3 to develop true leaves. Increase light levels to 2,500–3,000 f.c. (27–32 klux). Reduce substrate temperatures to 60–62°F (15–17°C) and begin allowing the soil to dry thoroughly between irrigations, but avoid permanent wilting. Increase fertilizer to 100–150 ppm from 14-0-14 and fertilize at every second or third irrigation. Do not use ammonium-nitrate fertilizers. Stage 3 will take about thirty-five days for 288s, longer for larger plugs.

Prior to sale or transplanting, harden plugs off for one week at soil temperatures of 60–62°F (16–17°C). Allow the soil to dry thoroughly between irrigations. Fertilize with 100–150 ppm from 14-0-14 as needed.

A 288-plug tray takes from six to seven weeks from sowing to be ready to transplant to larger containers.

Growing On

Pot 1 plug/4 or 4.5" (10 or 12 cm) pot and 2–3 plugs/6" (15 cm) pot. Be very careful not to bury the crown of the plant; set plugs out slightly higher than the substrate line of the plug. Select a disease-free, well-drained medium with a moderate initial starter charge and a pH of 5.5–6.0. Select a peat-based mix with peat comprising 60–80% of the mix.

Drainage is vital to success with *Primula*; set pots on surfaces that will not allow water to pool underneath them. Allow pots to dry thoroughly between irrigations. However, make sure that pots are uniformly moist after watering, since *Primula* can develop dry, brown leaf edges when irrigation is inconsistent. Be especially careful not to allow plugs to dry out immediately after transplanting. Always be sure that foliage is dry well before nightfall to prevent disease problems.

Grow *Primula* under light levels of 3,000–4,000 f.c. (32–43 klux). At lower temperatures, plants can take higher light levels (full sun), but most growers in the South will need to shade plants down to as low as 2,000 f.c. (22 klux). Fertilize with 150–200 ppm from 21-5-20 at every other irrigation for the first four weeks, switching to 15-0-15 after the cold treatment. Keeping adequate nutrition to the crop in early stages of growth is critical to later production success. Maintain the soil pH below 6.0 to prevent nutrient deficiencies, especially iron deficiency.

Primula culture can be divided into three stages: (1) establishing plants, (2) initiating flowers, and (3) forcing plants. Each stage has its own temperature regime. Establish plants at 60–65°F (16–18°C) days and 55–60°F (13–16°C) nights for the first four weeks (production weeks 1–4). Do not allow temperatures during this phase of growth to drop below 47°F (8°C) if possible. Before beginning cool treatment to initiate flowering, plants should be fully rooted in and have eight to ten leaves. To initiate flowers, drop temperatures to 45–48°F (7–9°C) days and 35–40°F (2–4°C) nights for the next six weeks (production weeks 5–11). Once you see flower buds, raise temperatures to force flowering. Force plants into flower at 54–58°F (13–14°C) days and nights. Temperatures above 58°F (14°C) result in lower-quality flowers. You can also hold plants at 40–45°F (4–7°C) for later forcing and in so doing schedule pots for successive flowerings over a period of weeks or months.

Some of the newer series do not require cold treatment in order to flower. For these series, providing long days (sixteen hours) with 10 f.c. (108 lux) of night interruption lighting is beneficial for earlier flowering.

Under short-day conditions, running a negative DIF (warmer nights than days) has been shown to speed flowering. Negative DIF is also beneficial in helping to keep plants compact and reduce stem length.

Be sure to maintain adequate air circulation around plants and to keep foliage dry during flowering to prevent botrytis. Space pots as leaves begin to touch to a density of about 4/ft.^{2} (43/m^2).

PGRs are rarely used on *Primula*. If necessary, B-Nine at 2,500 ppm may be used.

The crop time for 4" (10 cm) pots to flower is from thirteen to sixteen weeks after transplanting a plug.

Pests & Diseases

Aphids, cutworms, fungus gnat larvae, leaf miners, thrips, and whiteflies can become pests. Fungus gnat larvae can feed on the crown of the plant, allowing crown rot diseases to enter.

Disease problems can include botrytis, *Pythium*, and rhizoctonia. TSWV/INSV can attack *Primula*, showing up as browning along leaf veins and yellow mottling of foliage. Since these tospoviruses are spread only by thrips, controlling these pests will minimize these diseases. Also, there is no cure for these viruses so any infected plants must be collected and removed from the premises to prevent spread to other crops. *Primula acaulis* are very susceptible to their flowers being infected with botrytis, so it is crucial to keep the flowers as dry as possible, especially at night.

Troubleshooting

Plants may become chlorotic due to medium that is too wet or poorly drained; an iron, magnesium, and/or nitrogen deficiency caused by a high pH; or ammonium toxicity. High ammonia fertilizers can cause large, leafy plants, especially during periods of cloudy weather. Flower stems that are long and weak may be due to light levels that are too high or high day/night temperatures (above 70°F [21°C]). Cold growing temperatures or low fertility can cause premature bud set on small plants. Be sure to bulk plants up prior to providing the cool treatment. Plants drying out and/or inconsistent watering or soluble salts damage can result in necrotic leaf margins. Blind plants can be due to low production temperatures during the bulking up phase of growth. Flower stems that are too short on *P. acaulis* may be the result of temperatures remaining low (less than 40°F [4°C]) for too long or nighttime forcing temperatures being above 65°F (18°C).

Varieties

If you are not able to provide cool temperatures, go with early-season varieties that force easily, such as 'Danova' or 'Primera'. Midseason series will force well under cool, frost-free conditions provided buds are set prior to warm temperatures, while late-season varieties do best with cold temperatures. Most *Primula* series will come into flower over a two-to-four-week flowering window.

The 'Danova' series is probably the most widely grown *Primula* series. It's available in a broad range of separate colors from yellows to white, pinks, reds, oranges, and blues. Most growers buy in plug trays of 'Mix' so they can offer an assortment. 'Danova' is early, uniform, and programmable.

'Dania' is a midseason series with plants that are slightly larger than 'Danova', while 'Daniella' is a mid- to late-season series.

'Primera' is an early-season *Primula* that doesn't require cool temperatures to initiate flowers. 'Orion' is a midseason series that flowers very uniformly. Both series come in a broad assortment of colors.

'Primlet' can bring that special touch to your *Primula* assortment with its unique double flowers that look like small bouquets of roses. Like all *Primula acaulis*, 'Primlet' has scented blooms that make them great gift plants.

Related Species

Primula × *polyantha* (also called *P. veris*) is similar to *P. acaulis*, but plants produce flower stalks that are 8–12" (20–30 cm) tall with clusters of flowers. They are great outdoors in beds when planted in regions that experience only light frosts. Because they hold their flowers high above the foliage, they are much less susceptible to botrytis in the landscape than other *Primula* species.

'Pacific Giants' is the best-known polyanthus series and is very popular in California. This series is a vigorous grower and thus has great landscape performance. 'SuperNova' also has great landscape performance and can be grown to flower in the fall. Try a fast-crop of this polyanthus *Primula* with these production tips. 'SuperNova' can be sown in November in larger plug sizes such as 128s or 72s, which helps to shorten the crop cycle. Germinate at 68°F (20°C) under long days (fourteen hours). Provide supplemental lights and long days, transplanting plugs at ten weeks. Transplant and establish plants at 60°F (15°C) for rooting in. Grow plants at 50–54°F (10–12°C) under natural day lengths. Plants sown in mid-November should be ready to sell in mid-April.

Try some hardy *Primulas* such as *P. vialii, P. denticulata, P. japonica*, or *P. auricula* for a botanical-print look that your customers will love.

Postharvest

Plants are ready to ship as soon as flowering begins. Plants can be boxed and shipped cool at 40°F (4°C). Foliage and flowers must be dry prior to shipping to prevent severe botrytis infections. Flowers are also sensitive to ethylene; if shipping for a long period of time, use EthylBloc (MCPP) to greatly increase shelf life.

If the crop flowers early, you can pinch plants back and they will rebud in about six weeks as long as the weather remains cool.

Retailers should unpack plants and place pots in a bright, cool location, below 70°F (21°C). If *Primulas* are to be displayed outdoors, protect them from rain and frost. Consumers can enjoy their *Primulas* indoors in a cool, bright window, or they may enjoy them planted in the garden or patio containers in frost-free areas of the country.

Primula obconica
Flowering pot plant

This flowering pot plant *Primula* makes a wonderful gift or spring container plant. In the past, *P. obconica* was infamous for the rash the primine in its leaves produced on just about anyone who handled the plants. Working with the crop meant wearing gloves and thick, long sleeves. Today, new primine-free strains of this brightly colored pot plant are flowing into the market, making *P. obconica* a great pot for winter and spring sales. Unlike for its relatives, *P. vulgaris* and *P. malacoides*, you won't have to provide a cold treatment for induction of flowering of *P. obconica*. Bright pink, purple, blue, red, orange, bicolored, and white flowers are held upright in trusses. A well-grown *P. obconica* is a mound of color atop a swirl of flat, lime-green leaves. You can even plant them outdoors in frost-free areas in the spring, especially in areas protected from rainfall.

Propagation

Primula obconica is not as temperature sensitive as *P. acaulis,* so it is easier to start from seed. Sow seed onto a well-drained, disease-free, germination medium with a pH of 5.5–5.8 and an EC of 0.50–0.75 mS/cm. Cover with a thin layer of coarse vermiculite to maintain 100% humidity at the seed coat. Provide substrate temperatures of 72–75°F (22–24°C) until radical emergence, then drop to 65–68°F (18–20°C). Keep trays uniformly moist, but not saturated. Light during germination is not necessary, but providing 100–400 f.c. (1.1–4.3 klux) of light will help prevent seedling stretch. Radicles will emerge in seven to ten days.

Move trays to Stage 2 for stem and cotyledon emergence. Reduce the soil moisture and drop humidity levels to 40–70% once the cotyledons emerge, allowing the medium to dry slightly before watering. Maintain substrate temperatures at 65–68°F (18–20°C) and increase light to 1,000–1,500 f.c. (11–16 klux). As soon as cotyledons expand, begin fertilizing weekly with 50–75 ppm from 14-0-14. Stage 2 takes about fourteen days.

Move trays to Stage 3 to develop true leaves. Increase light levels to 2,500–3,000 f.c. (27–32 klux) Reduce substrate temperatures to 62–65°F (17–18°C) and begin allowing the soil to dry thoroughly between irrigations, but avoid permanent wilting. Increase fertilizer to 50–100 ppm from 14-4-14 and fertilize at every second or third irrigation. Do not use ammonium-nitrate fertilizers. Stage 3 will take about thirty-five days for 288s and longer for larger plugs.

Prior to sale or transplanting, harden plugs off for one week at soil temperatures of 60–62°F (16–17°C). Allow soil to dry thoroughly between irrigations. Fertilize with 100–150 ppm from 14-0-14 as needed.

A 288-plug tray takes from six to seven weeks from sowing to be ready to transplant to larger containers.

Growing On

Transplant 1 plug/4" (10 cm) or 6" (15 cm) pot and 3 plugs/8" (20 cm) pot. Cool temperatures are not necessary to induce flowering. Establish plants at 68–70°F (20–21°C) days and 60–65°F (16–18°C) nights until the roots are well established. Initiate flowers and finish plants under cool temperatures (60–70°F [16–21°C] days and 50–55°F [10–13°C]) nights. Since *P. obconica* does not require cool production temperatures in order to flower, you will be able to offer it for sale in a wider market window. Shading during high-light times of the year will be necessary to moderate production temperatures.

Another major difference in *P. obconica* is irrigation: Plants have a higher moisture requirement. Do not allow plants to dry out. *P. obconica* is also a light feeder; use 100–150 ppm nitrogen (N) only as needed. Use a calcium-based fertilizer such as 14-4-14 or 15-5-15, since ammonium-based fertilizers will promote leafy plants. Plants are highly sensitive to high salts.

If growth regulators are required, B-Nine at 5,000 ppm is effective.

Space 4" (10 cm) pots at a final density of 2–3 pots/ft.^{2} (22–32/m^2) once leaves begin to touch.

From transplanting a 288 plug to having a finished 6" (15 cm) pot is ten to twelve weeks.

Pests & Diseases

Thrips are the primary cause of problems in *P. obconica.* The primine that was bred out of this plant is a natural insect repellent, so the newer primine-free varieties are very susceptible to thrips' feeding damage on flowers and foliage and are also very susceptible to INSV/TSWV. Whiteflies can also be an occasional problem. Botrytis flower blight can be problematic, especially under cool, wet conditions.

Troubleshooting

Primula obconica does not suffer from too many problems. High pH can lead to iron chlorosis. Low light and warm temperatures, especially coupled with high ammonium fertilizers, can lead to very leafy plants with small flowers.

Varieties

Don't even bother growing a series with primine anymore. There are so many good primine-free series on the market today that you don't have to grow the older series.

Try the 'Libre' series, available in separate colors and a mix. 'Libre' can be germinated at higher temperatures (72°F [22°C]), along with other bedding plant crops. It makes a great early spring bedding plant in frost-free areas. Planted outdoors, 'Libre' can be used as a bedding plant in zones where there are fewer than thirty days above 86°F (30°C).

From Schoneveld Twello in the Netherlands is the 'Twilly Touch-Me' series of primine-free *P. obconica.* It comes in a broad range of separate colors of blue, pink, red, white, and bicolors. This series is all the rage among *P. obconica* growers in Europe.

Postharvest

Plants are ready to ship as soon as flowering begins. Plants can be boxed and shipped cool at 40°F (4°C). Foliage and flowers must be dry prior to shipping to prevent severe botrytis infections. Flowers are also sensitive to ethylene; if shipping for a long period of time, use EthylBloc (MCPP) to greatly increase shelf life.

Retailers should unpack plants and place pots in a bright, cool location, below 70°F (21°C). If *Primulas* are to be displayed outdoors, protect them from rain and frost. Consumers can enjoy their *Primulas* indoors in a cool, bright window, or they may enjoy them planted in the garden or patio containers in frost-free areas of the country.

R

Ranunculus

Ranunculus asiaticus
Common name: Buttercup
Flowering pot plant, cut flower

Ranunculus is one of those crops that screams spring. Flowers look surreal, as bold as an Andy Warhol work of art and just as brightly colored. Ranunculus's flower petals are substantial and tremendously abundant, since flowers are semi- or fully double. Their beauty and fullness make you want to take the day off and just enjoy life.

For growers, ranunculus fits a couple of production spots. The most convenient niche is for late winter/early spring sales. That's when you have some days with warmer temperatures, and few other crops are available to sate consumers in their quest to experience spring. Ranunculus can also be grown and sold as a fall crop for outdoor planting in areas with a Mediterranean climate. When plants are overwintered in cool, frost-free conditions, ranunculus will put on a flower show the following early spring that is unparalleled. As a matter of fact, The Paul Ecke Ranch in California named its spring plant program The Flower Fields after the famous ranunculus flower fields in Encinitas, a Southern California tourist hot spot each spring.

For bedding plant growers, the hardest part about ranunculus is remembering to order your plugs when you are still in the height of the bedding plant sales season. Make a date to do so on your calendar—your customers will appreciate it.

Propagation

Ranunculus can be propagated from seed or tubers. Most recently, some tuber suppliers have begun working with tissue culture. These tissue-cultured liners are being used mainly for cut flower production. Two notable features are that the resulting plants finish faster and are more uniform in flowering response and color. For simplicity, seed propagation is covered in this section.

Sow seed into a disease-free, well-drained, germination medium. Select larger plug sizes: 288s would be the minimum size, but 128s are preferred. Cover lightly with coarse vermiculite. Maintain substrate temperatures of 50–55°F (10–13°C) and a pH at 5.5–5.8 with an EC reading below 0.75 mS/cm. For success in germinating ranunculus a cooler is necessary. Light is not necessary for germination.

Ranunculus is a slow-growing seedling; maintain moisture at the seed coat by providing 100% humidity until germination is finished. Radicle emergence will take about seven to fourteen days.

Move trays to the greenhouse for Stage 2, stem and cotyledon emergence, which will take from seven to fourteen days. Continue to maintain soil temperatures, but reduce irrigation to promote root penetration into the media. Increase light to 500–1,500 f.c. (5.4–17 klux) and begin a weekly 50–75 ppm feed with 14-4-14. Stage 3 will take from twenty-eight to thirty-five days for the development of true leaves. Raise substrate temperatures to 55–60°F (13–15°C) and increase light levels to 1,500–2,500 f.c. (17–27 klux). Begin to allow trays to dry somewhat between irrigations. Increase weekly feed to 100–150 ppm from 14-0-14.

Harden plugs in Stage 4 for seven days prior to sale or transplant. Maintain soil temperatures, but allow trays to thoroughly dry out before irrigating. Increase light to 2,000–3,000 f.c. (22–32 klux) and continue the weekly feeding schedule. If plant growth regulators are needed, B-Nine (1,500–2,500 ppm at the second true leaf) is effective. Continue feeding with 14-4-14; do not use ammonium-nitrate fertilizers.

A 288-plug tray takes from eight to ten weeks from sowing to be ready to transplant to larger containers.

Growing On

Transplant well-watered plugs into a well-drained, disease-free medium with a moderate initial nutrient charge and a pH of 5.5–6.0. Ranunculus grows slowly for the first weeks: It is critical that the medium is disease-free and very well drained to control disease.

Plant 1 plug/4" (10 cm) pot and 1–3 plugs/6" (15 cm) pot. Plant plugs at the same level as the plug's substrate line. Provide light levels up to 5,000 f.c. (54 klux). When flowering begins, reduce light to 2,000–2,500 f.c. (22–27 klux). Fertilize at every other watering with 150–200 ppm from 15-5-15. Do not use ammonium-nitrate fertilizers when the average temperature is less than 65°F (18°C).

To help height control, you can allow plants to wilt just slightly before watering once plants are rooted to the sides of containers. Ranunculus is DIF responsive: Maintain a 60°F (15°C) average daily temperature and work with a negative 5–10°F (3–6°C) DIF. B-Nine (2,500 ppm) can be used when buds appear; however, overuse will dwarf flower stalks. When ranunculus is grown under the proper temperatures, growth regulators are not needed. Space plants throughout production as leaves begin to touch to prevent stretching.

Grow at 55–60°F (13–15°C) days and 40–50°F (4–10°C) nights. For fastest flowering, maintain temperatures at the upper end of the range. Temperatures below 50°F (10°C) encourage plants to branch, while temperatures above 65°F (18°C) can cause stretching, leggy growth and plant dormancy or premature flowering. Many growers in some parts of the country produce ranunculus in unheated greenhouses.

Ranunculus flowers naturally under winter's short days (eight to ten hours of daylight). Under long days (more than ten hours), plants are slow to develop and flowering is delayed. Note that under prolonged long days, plants will form tubers and go dormant, especially if long days are combined with high temperatures.

Plants form a rosette and initiate flower buds in the center. Stems elongate from the rosette. Before they flower, plants will produce from nine to sixteen leaves.

Some growers grow their crops from tubers. Take precooled tubers measuring from 1.5–2.25" (4–6 cm) in diameter and plant them, with prongs facing down (eyes up), 2" (5 cm) deep. Use 1 tuber/4" (10 cm) and 2–3 tubers/6" (15 cm) pot. Maintain soil temperatures at 65°F (18°C) night temperatures. Roots will begin to sprout in two to three weeks.

Cut flower growers can plant tubers from September to January, provided cool temperatures are possible. Flowers can be harvested sixty to ninety days after planting.

Transplanting a 288 plug to a 4" (10 cm) pot requires from fourteen to sixteen weeks to flower depending on variety.

Pests & Diseases

Thrips, aphids, leaf miners, and spider mites can become pest problems. Diseases such as botrytis, *Fusarium*, powdery mildew, and *Pythium* can also attack plants. Ranunculus is highly sensitive to INSV/TSWV, so thrips prevention is imperative to keep these diseases out of your crop.

Varieties

Both the 'Bloomingdale' and 'Mache' series are popular, seed-propagated, potted plant varieties. Both come in a wide array of separate colors. 'Magic' is the only genetically dwarf series on the market and is excellent for high-density production. 'Magic' should only be grown in 4" (10 cm) pots and requires about two more weeks to flower.

Few cut flower ranunculus varieties are grown from seed; most are grown from tubers. Popular cut flower varieties include 'Tecolote', 'La Belle', 'Gigi', and 'Super Greens'.

Postharvest

Harvest cut flowers in the morning while they are still closed (flowers close at night). Immediately after harvest, store stems in water at 36°F (2°C). Transport should be short, no longer than two to three days to the final market. Make every effort not to store ranunculus, but if longer-term storage (ten days distribution time) is needed, pulse stems in water for five to six hours in a cooler at 36°F (2°C) then store dry. Prior to sale at retail, stems should rehydrate for seven to eight hours.

Pot plants can be harvested when one or two flowers per pot are ready. Ship cool. Retailers should display plants in a cool, well-lit location. Maintain temperatures no higher than 70°F (21°C) and at least 100 f.c. (1.1 klux) of light. When consumers place ranunculus in a cool, bright window, they will enjoy them for one to two weeks.

Rhododendron

Rhododendron × *obtusum*
Rhododendron simsii
Common name: Azalea
Flowering pot plant

Azalea is one of the premier flowering pot plants. This flowering shrub draws attention wherever you see it. It is the prize of the South and West, planted by the hundreds of thousands in gardens around homes, offices, schools, and parks. As a flowering pot plant, it is an important player in spring plant sales, available in an array of flower forms, sizes, and colors. For growers, potted azalea is a high-value crop that commands a premium price. Therein lies its real beauty: It is a bit too expensive for the average, run-of-the-mill, mass-market shelf, especially in larger sizes and specialty forms. That makes it great for both retailers and wholesale growers who are looking to target a marketplace niche. If you really want to make a statement, try some azalea trees or pyramids—just be sure to put your order in a couple of years early so you can get in line to receive plants.

Propagation

Azaleas are vegetatively propagated from stem or tip cuttings harvested from actively growing plants. While rooting is relatively easy, growing a "florist" azalea takes as long as two to three years. Most growers purchase dormant plants sized up for finishing or liners from specialist growers in the Pacific Northwest, Alabama, Florida, and New York.

Terminal cuttings measuring 3–4" (8–10 cm) can be taken any time of the year from healthy stock plants that are maintained in a protected environment. Shoots for cuttings should "break with a snap." Limber, succulent cuttings do not root as readily. Dust the basal end of the cutting with rooting hormone (IBA) and direct-stick cuttings into peat blocks (Jiffy) or a peat/perlite/sand mixture. Root cuttings with bottom heat, maintaining 70°F (21°C) rooting mix temperatures and mist for twenty-four hours a day during the first three to four days to keep cuttings turgid. Thereafter, mist only during the day. You can make the first pinch (very soft) seven weeks after sticking. Cuttings should root in about eight to twelve weeks.

Growing On

Dormant or precooled plants

A few growers buy liners and grow their crops from rooted cuttings that have been sheared (see the Liners section). However, most growers bring in dormant or precooled azaleas. Dormant plants require cooling in order to break dormancy, while precooled plants are ready to force. Note that most azaleas are cooled outdoors by the supplier. Cooled plants arriving from the West Coast in October for December sales may require additional time in your cooler to ensure that they have received enough cooling and are ready to force.

Sizes that are available include 2" (5 cm), 4" (10 cm), 5" (13 cm), 6" × 6" (15 × 15 cm), 6" × 8" (15 × 20 cm), 6" (15 cm), trees, and pyramids. Unpack plants immediately upon arrival. If they look frozen, put them in a cooler at 32–35°F (0–2°C) until the root-ball becomes soft. Then pot them into 4, 6, or 8" (10, 15, or 20 cm) pots, etc., depending on the plant height and bloom count. Many growers use clay pots for a more upscale look. Select a peat-based medium; 50% peat and 50% bark is ideal. Azaleas like acidic conditions, so make sure the medium has a pH of 4.5–5.5. Water three times after potting, allowing each previous watering to fully percolate, to ensure the root-ball is fully saturated.

If you choose to force plants in their original growing containers and unpack them straight into the greenhouse, take care to keep the crop shaded in the beginning to acclimate them to their new environment. Cold pots coming off a refrigerated truck and going straight into a hot greenhouse will transpire faster than the plants can take up water. The result can be burned and dropped leaves.

If you're not ready to force plants right away, you can hold them in the cooler at 35–40°F (2–4°C) for several weeks. Note that plants held at colder temperatures (35–37°F [2–3°C]) will become dormant and flowering will stop, so hold them at the higher end of the range if you're going to be forcing them quickly once they are out of the cooler.

You will need to provide cool temperatures to dormant plants. Budded plants require four to six weeks of cool temperatures to break dormancy. Provide temperatures of 38–40°F (3–4°C) in a cooler, cold frame, or greenhouse. Kurume varieties like the Dogwoods require four weeks of cooling; other types require six weeks. The Vogels will flower without cooling, but flowering is more easily controlled with cooling. If cooling temperatures are higher than 40°F (4°C), provide 120 f.c. (1.3 klux) of light for twelve hours a day to prevent leaf drop. At 35°F (2°C), lights are not required. Provide 80–90% relative humidity in the cooler. Some growers cool plants naturally outdoors; however natural cooling is not as precise or as predictable as cooling plants in a refrigerator. If you choose to use an unheated greenhouse to cool plants, as many growers in milder parts of the country do, cover it with shade to prevent high leaf temperatures and moderate water loss from the medium. Water plants in cool storage regularly to keep them from drying out, but do not fertilize. Plants cooled at 40–50°F (4–10°C) with less than 120 f.c. (1.3 klux) of light may require watering twice a week.

Gibberellic acid (GA) can be used to break azalea bud dormancy in substitution for cooling. Using four to six sprays at 1,000 ppm at weekly intervals can substitute for cold. Buds must be fully developed prior to application. Alternately, three sprays at 250–500 ppm can be used following three to four weeks of chilling. Response to GA is variety specific and varies depending on growing conditions. Test plants first to see how GA works in your exact situation. When treating with GA, spray plants to runoff; do not apply GA after buds are showing color. GA has been shown to reduce postharvest life in one research study, so be aware of that possibility.

Force plants slowly using cooler temperatures to increase postharvest life. Flowering speed after the cooling treatment is temperature dependent. For example, to flower a crop at Easter, the crop time at 50°F (10°C) is seven weeks; at 60°F (15°C), six weeks; and at 70°F (21°C), five weeks. To flower a crop for Mother's Day that was received as cooled plants in March, keep the plants as cool as possible (35–40°F [2–4°C]), as forcing will only take two to three weeks at 60°F (10°C) at that time of year.

There is no need to fertilize plants during forcing. Plants forced without fertilizer have more intense flower color and last longer in postharvest without it. Plus, fertilizing during forcing may cause plants to generate vegetative growth. Some growers do, however, provide a light feed to keep the foliage deep green.

Irrigation is critical: Azaleas budded and ready to force cannot dry out. Inconsistent irrigation practices at this time will cause flower buds to drop and flowering to be irregular. Since the best medium to use for azaleas is peat, keeping pots moist but not wet at all times is very important, as peat that dries out is very difficult to rewet. Many growers choose to drip irrigate their plants, which keeps foliage and flowers dry, thus reducing foliar disease problems.

Maintain high light levels to ensure high flower number and good flower color on reds, pinks, and bicolors. During the spring, provide full-sun conditions.

Liners

Growers can also finish 6" (15 cm) pots from liners. In general, the propagator will have pinched liners as many as three times. The finishing grower can then pot these up during the summer, pinch them once, provide a cool treatment, and force them into flower for winter and spring sales.

Place liners into pots at the same depth the cuttings are currently growing. Select a well-drained, disease-free, peat-based medium with a low starter charge and a pH of 4.5–5.5.

Grow plants at 77–86°F (25–30°C) days and 68°F (20°C) nights for vegetative growth. Grow plants under 2,000–4,000 f.c. (22–43 klux) of light; shade will be required during the summer months in most parts of the country. Since vegetative growth is stimulated under long-day conditions for many varieties, providing 10–20 f.c. (108–215 lux) of light with night interruption lighting from 10 P.M.–2 A.M. should be used in a year-round growing program. If you are bringing liners in over the summer, you won't have to worry about a long-day treatment. To shorten crop time and keep plants vegetative, provide long days (sixteen hours) continuously for twenty to twenty-four weeks. For growers with year-round forcing programs, that means you will be lighting from around September 1 until March 31.

Azaleas have a fine root system that is susceptible to underwatering, overwatering, and soluble salts buildup. Water stress retards vegetative growth and promotes premature flower bud initiation. Use a well-drained medium to ensure against overwatering, and then do not allow plants to dry out.

Feed once a week with 21–7–7 at 300 ppm. Supplement with trace elements at every fourth irrigation. If you encounter problems from iron deficiency because pH is too high, apply iron chelate. Discontinue fertilizer prior to providing cool treatment, which will reduce the number of brown leaves during and after cooling.

Pinching plants during the vegetative production phase promotes lateral branching and increases the number of flowers per plant. In a small-pot program (e.g., 4" [10 cm]), plants should be pinched three times. Pinch for the first time at potting and two more times at six-week intervals. (During dark winter months, increase the interval between pinches to eight weeks). Provide a soft pinch, removing about 0.5" (13 mm) of growth of all shoot tips.

Azaleas may be pinched chemically with Atrimmec, which will save a lot of time and labor. However, follow label directions and be sure it is applied correctly.

Azaleas flower on old wood. Allow plants to grow for six to eight more weeks after the final pinch prior to initiating flower buds. Allow about twenty-six weeks between the last pinch and flowering date under controlled conditions. (If you plan to cool the plants under natural conditions, this span will be longer.)

Maintain adequate space between plants in production. Space plants as the leaves begin to touch. When leaves overlap each other, they act as filters. Shoots on plants exposed to the far-red end of the spectrum will be more elongated than those exposed to the red band as the leaves absorb more far-red light. Lateral branching will also be reduced when far-red light is absorbed, so plant shape and size are adversely affected by crowding.

Azaleas initiate flower buds when they are given short days, warm temperatures, and/or a B-Nine treatment. Under natural conditions, azaleas develop flower buds during the days of late summer, when days are shortening and nights are still warm. To promote bud development, give plants eight weeks of warm temperatures with a minimum night temperature of 65°F (18°C). Combine this with eight weeks of short days (ten hours). If you are forcing pots for year-round programs, you will need to provide short days from around April 1 until August 31. Flowering should occur about eight to twelve weeks after the end of short days.

B-Nine (2,500–3,500 ppm) also promotes azalea buds. Apply five weeks after the final pinch. Azaleas treated with B-Nine will be smaller than nontreated plants and flowering may be slightly delayed, but B-Nine will increase the number and uniformity of buds. Providing both short days and B-Nine treatment results in higher uniformity of flowering following cooling. Similar responses can be expected from Bonzi (25 ppm) and Cycocel (2,500 ppm).

Budded plants require four to six weeks of cool temperatures to break dormancy. Following cool temperatures, plants may be forced into flower (see the section on dormant or precooled plants for more information).

Pests & Diseases

Azaleas are susceptible to a wide range of insect, mite, and disease problems. Botrytis, leaf gall, powdery mildew, and *Septoria* can all afflict leaves. *Cylindrocladium* and *Phytophthora* can be problems in roots.

Aphids, lace bugs, leaf miners, nematodes, red mites, spider mites, and whiteflies can attack plants. Most growers maintain an active preventative program to deter problems before pests can become established.

Troubleshooting

Azaleas that lose their leaves in the cooler could be reacting to the presence of ethylene, a lack of humidity (below 70%), or a lack of light at cooling temperatures above 40°F (4°C). Water stress at any time during production, cooling, or forcing can also cause leaf drop.

Uneven flowering may be due to late pinching, forcing too early after cooling by not providing enough cooling, and low light and/or irrigation stress during forcing. Be sure to allow at least fourteen to sixteen weeks after the final pinch until beginning cooling. A high pH can cause foliage to become chlorotic.

Azaleas can develop bypass shoots that surround the flower bud during forcing. They must be removed, as bypass shoots may cause bud blasting. Remove them using a sideways, twisting motion, when they are 0.5" (13 mm) long. You will encounter trouble with bypass shoots when you wait too long between the final pinch and the start of cooling. University of Florida researchers have shown that a Bonzi application applied seven weeks before plants went into the cooler caused the bypass shoots to be so short that they were not noticed.

Varieties

Azaleas are one of the most hybridized flowering plants in ornamental horticulture. In the South, entire cities dedicate festivals to them when they are flowering. In these areas, most consumers can spout off the names of at least a half dozen varieties. Just about every azalea can be forced into flower in the greenhouse. Some are easier than others, especially for timing early, midseason, and late forcing. Suppliers who specialize in florist azaleas are working with hybrids that force especially well. Follow their recommendations for your best success. You will want to choose early varieties for Christmas sales, midseason varieties for Valentine's Day sales, and late-season varieties for Easter and Mother's Day sales.

Postharvest

Plants are ready to ship when 25% of the flowers are open and/or in the "candle" stage. Some growers ship plants earlier, but the color on flowers opened indoors is pale and washed out. Plants that are still in the tight bud stage but are showing some color when planted in a low-light environment may never open as they should. Water plants well prior to packing so that the medium is moist, but do not pack plants with wet leaves or flowers, as this invites botrytis to attack during shipping.

Plants may be sleeved, boxed, and shipped dark for no longer than one week. Ship plants cool at 35–40°F (2–4°C). Azaleas are sensitive to ethylene and can drop leaves when they are exposed. Be sure to groom plants, removing any dead and dying plant debris prior to sale, as debris serves as a source of ethylene gas. Also, do not ship or store azaleas near ripening fruits or vegetables.

Retailers should unpack plants immediately upon arrival and place them in a bright, cool location (100 f.c. [1.1 klux] of light). Temperatures of 68–72°F (20–22°C) are ideal. If plants are dry, water them prior to putting them in displays. Keep dead flowers and foliage groomed from plants in order to remove these as a source of ethylene production.

Instruct consumers that it is critical the pot does not dry out, as the medium will be very difficult to rewet. In many areas of the country, consumers will be able to plant their azaleas outdoors after the danger of frost has passed and enjoy their plants for years to come as a spring flowering shrub.

Rosa

Rosa hybrids
Common name: Rose
Flowering pot plant, cut flower, garden flower

Roses are the most important floriculture crop in the world. While color trends affect demand of individual varieties, roses are always at the top of production charts in any country where flowers are grown. They are surrounded by a regal aura and never seem to go out of style.

Consumers love their velvety petals and rich flowers. For most, roses are deeply symbolic since they have evolved with mankind through the ages. What other flower expresses the deepest emotions as well as a rose? It is the worldwide symbol of love and the most thought-of flower.

Roses are a crop that many producers choose to specialize in. They are demanding and rarely perform up to par when given the backseat to other crops or are an afterthought when a product mix is expanded. Often, cut rose growers specialize in roses. While domestic cut rose growers frequently also produce other crops, their rose greenhouses are considered the most elite of all their production. Pot plant growers frequently add miniature roses into their product assortment, but because pot roses require a specific environment, pot rose plants are often grown in an isolated growing zone where the required light and temperature regimes can be delivered. Many retail growers and bedding plant growers offer garden roses in the spring that have been forced in containers. Plants are either grown from rooted liners, prefinished plants, or dormant bare-root plants that are containerized. Subsequently, the plants are grown into leaf and flower to sale during peak season.

The cultural information presented here is but a fraction of the information available for growing roses. If you are planning to grow roses, plan to seek out additional information and visit other rose producers.

We will divide this section into four distinct parts: (1) mini roses grown as flowering pot plants, (2) cut flower roses, (3) roses grown outdoors for cut flowers, and (4) garden roses forced to sale in containers during the spring.

Pot Roses

In many countries, specialized growers produce millions of direct-stuck pots a year in weekly programs that are sold in supermarkets, mass markets, retail florists, and independent garden centers.

In North America, pot roses are popular and are still regarded as premium pot plants, especially when the grower or retailer takes the time to dress the pot up. Market demand in the United States and Canada is principally in 4 and 6" (10 and 15 cm) pots, which parallels other greenhouse potted flowering crops. Currently, the principal channel of distribution is through supermarkets and chain stores, but with more 6" (15 cm) products on the market, the number of plants in retail florists is expanding. Overall demand for colors seems to be about 25–30% red and 70–75% other colors. Demand for red is higher during some holidays, such as Christmas and Valentine's Day. Overall, the trend in varieties is toward larger flowers.

Many bedding plant growers and retail growers buy in liners or prefinished pots to force into flower for holiday sales.

Propagation

If a grower has the specialized equipment, available environment, and proper licensing, rose cuttings are direct-stuck using cuttings taken from the pinch from a production crop. Only disease-free cuttings taken from HID-lit stock will produce uniform flowering plants. Select a medium with a pH of 5.5–6.2 and a moderate initial starter charge of fertilizer. Cool cuttings in a plastic bag overnight at 34°F (1°C) prior to preparing them for sticking. Trim cuttings that will be stuck to 1" (3 cm) long, leaving one five-leaflet node. Discard any cuttings that do not have one five-leaflet node or have begun producing axillary shoots. Some growers dip cuttings in IBA rooting hormone prior to sticking, but plants will root without it. Stick four cuttings for a 4" (10 cm) pot or five or six cuttings for a 5 or 6" (13 or 15 cm) pot into a moist medium, placing them around the edge of the pot. Drench with an all-purpose fungicide to guard against root rots during propagation. Root cuttings under mist or fog at a substrate temperature of 74°F (23°C). Maintain 800–1,000 f.c. (8.6–11 klux) of light. Apply a foliar feed of 150 ppm of 20-10-20 twice during the rooting process, leaching with clear water between feedings. Specialized pot rose growers often place pots on rolling ebb-and-flood tables that are moved through a specialized propagation house for rooting. Once plants are rooted, the tables are moved out of the propagation area and into production greenhouses. The entire production should be under HID lighting.

Cuttings will initiate roots in seven to ten days and will be sufficiently rooted in fourteen days. At this point, you can begin to decrease humidity and increase light.

Growing On

Select a well-drained medium with a pH of 5.5–6.2 and a moderate initial starter fertilizer charge. Liners generally have multiple cuttings per cell. Plant liners as soon as they arrive for best results. If you must delay planting, unpack boxes and place liners in a cool, bright area of the greenhouse. Many times the supplier has also provided one pinch to plants; be sure to ask about how the plants have been handled prior to shipping so you can adjust your schedule accordingly.

Plant rooted liners or prefinished pots at the same level as plants are currently growing. Drench with a general fungicide immediately after planting.

Grow plants on at 70–75°F (21–24°C) days and 62–65°F (16–18°C) nights. Pot roses will grow well at 80°F (27°C). Growing plants at the lower range of nighttime temperatures will slow development. Start the crop at higher temperatures, gradually lowering temperatures approximately two weeks after final spacing. This will increase crop time but improve postharvest life. Plants can tolerate higher production temperatures at lower light levels. Avoid wide fluctuations in temperature throughout the entire crop.

This pot rose production greenhouse in Denmark is growing grass for biological control of aphids.

Provide high-light, full-sun conditions, as long as you can maintain moderate production temperatures. During the winter, most pot rose growers provide supplemental lighting (400 f.c. [4.3 klux] for thirteen hours per day).

Fertilize with 200 ppm from 20-10-20 at every irrigation. Slow-release fertilizers can be used at half rate. Avoid fertilizers high in ammonium or urea. Nitrate fertilizers provide the best results. Keep plants uniformly moist to avoid wilting. Roses drop leaves when they dry out. Mini roses are a great crop for ebb-and-flood irrigation systems, which are best since they keep foliage dry throughout production. If you will be using ebb-and-flood irrigation, you may need to reduce fertilizer rates. Apply fertilizer at half rate or leach regularly during the final three weeks of production to increase shelf life. Apply supplemental iron at every irrigation at 5–10 ppm or drench with 30–50 ppm every week.

Pinch plants two to three times by shearing plants with hand shears or hedge trimmers. Apply the first pinch from two and a half to three weeks after rooting; pinching plants back to three-to-five leaf nodes. (*Note:* Most liners have already been pinched once.) A second pinch can be made about three weeks later. (For pinching bought in liners, pinch plants once they are fully rooted into pots). Additional pinches can be made in three weeks. Pinch plants back to about 0.5" (1.25 cm) above the previous pinch.

During a cutback, a significant portion of the leaf canopy is removed. Irrigation should be adjusted accordingly. The irrigation immediately prior to the cutback should be done one day before and without fertilizer in order to ensure that the salinity is not increasing. Subsequent to the cutback, irrigation should be given sparingly to avoid overwatering and root lost.

Provide high-light, full-sun conditions if you can maintain moderate production temperatures. Roses require high, uniform light: Keep plants spaced and never let leaves touch. Until pinching, pots are generally pot tight and then spaced. A rule of thumb to follow is to space plants one pot-width apart. Plants spaced too closely together after the pinch will grow upright, stretch, and lose lower leaves. Use shade in the summer to produce the best foliage and flower quality. Shade young plants when light exceeds 3,000 f.c. (32 klux). Shade from week 4 or 5 (after planting liners) until finish, if light levels exceed 4,100 f.c. (44 klux). During the winter, many pot plant growers provide supplemental lighting (400 f.c. [4.3 klux] for thirteen hours per day).

Bonzi is effective and can be applied from two to three times during production. Apply for the first time about two weeks after the first pinch. Rates vary by location and growing conditions, so experiment under your specific conditions. A starting point is 20–25 ppm, with applications spaced one week apart. Make applications before buds are visible for maximum effectiveness. Be sure to apply Bonzi sprays and any other chemical sprays in the morning to allow foliage to be totally dry by nightfall.

Roses are also responsive to DIF to control height. Grower literature usually recommends not using greater than 5°F (3°C) negative DIF to avoid weak stems and flowers.

Pests & Diseases

Insects that can attack roses include aphids, thrips, spider mites, and fungus gnats. The best insect control is to grow clean. Use screens on vents and doors, and be rigorous about scouting to detect problems early. Disease problems include black spot, botrytis, downy mildew, powdery mildew, and *Pythium*. Be sure to keep plants well ventilated and rogue suspect plants.

Troubleshooting

Slow growth and slow flowering can be the result of low temperatures, low light levels, or poor air circulation. High soluble salts can cause root necrosis and leaf drop. Moisture stress, overwatering, or low fertility can lead to limited plant growth and chlorotic foliage. Check roots regularly on direct-stuck crops to make sure roots are white, well branched, and uniformly spaced and that the roots have active root hairs. Discolored roots can be a sign of disease or soluble salts problems. Interveinal chlorosis on upper leaves can be a symptom of iron deficiency. Some growers apply chelated iron at 5–10 ppm at every irrigation to avoid problems. Ethylene, water stress, root loss, or low light can result in bud drop. Low light, uneven pinching, or poor nutrition can cause uneven flowering. Low light levels and/or temperatures can lead to blind shoots.

Varieties

There are a limited number of rose hybridizers that specialize in commercial pot rose varieties worldwide. The principal hybridizing companies are Kordes Roses ('Kordana'), Poulsen ('Parade'), and Roses Forever.

Postharvest

Plants are ready to sell when at least two flowers are open and an additional two to three buds show color. Pot roses may be sleeved and boxed, but make sure foliage is completely dry before packing. Store and ship plants at 35–40°F (2–4°C). Do not keep plants in boxes or storage/transit for longer than six days. Plants are sensitive to the harmful effects of ethylene, so never ship plants with fruit or vegetables.

Retailers should unpack boxes immediately and place them in a cool (68°F [20°C]), well-ventilated area with 300 f.c. (3.2 klux) of light. Do not display roses near ripening fruits or vegetables. Water

Table 34. Grower's Guide to Finishing Pot Roses

ITEM	TRANSPLANT POT SIZE[a]		SHEAR[c] WEEK[d]	TARGET SEASON[e]	FINISH SPACING			GROWTH TIME	
	IN.	CM			WEEK[d]	IN.	CM	AFTER FINISH SPACING	TOTAL WEEKS[d]
50-cell liner	4	10	2	Winter Spring	5 3	6.5	17	4–6 4–5	9–11 7–8
32-cell liner	5–6	13–15	3	Winter Spring	6 4	10–12	25–30	6 6	12[f] 10[f]
4" (10 cm) prefinished pot	N/A[b]		Do not shear.	Winter Spring	2 1	6.5	17	4 4	6 5

[a] In all cases, one transplant per pot. Also, the initial spacing is equal to the pot size.
[b] Not applicable
[c] To promote branching, cut back plant to 2–3" (5–8 cm) above the pot top.
[d] After potting
[e] Winter finish time refers to crop finishing up to and including Easter. Spring finish time applies to any crop from Mother's Day through November 1. Use spring finish time year-round if plants are grown under HID lights with at least 450 f.c. (4.8 klux).
[f] Grow out times for 32-cell roses are estimates only.

plants carefully to avoid wetting foliage. Keep plants groomed, removing dead flowers and leaves to reduce sources of disease problems.

Consumers can plant pot roses in patio containers as well as the garden for subsequent enjoyment.

Cut Greenhouse Roses

If you decide to take a look at cut roses, really think your decision through before acting. Know exactly where and to whom you will be selling your crop before talking to breeders. Once you have your niche hammered out, then talk to breeders. You may have to wait a long time to receive plants, especially if you are looking to grow newer varieties. The lag time to purchase grafted bushes can be nine months or more. Look at the delays in starting as the time to gather market intelligence, line up potential customers, and learn your competition. Buy plants from the best breeder and propagator that you can afford. As you do your market research, you will quickly discover that roses are bought and sold by variety name more than any other flower. Having the right variety with yields and crop performance suited to your environment at the right time and place can be very profitable. However, growing a variety ill-suited to your climate or market is equally as unprofitable.

Propagation

Growers buy in rose plants from commercial propagators, who, in turn, work with various breeders. Do your homework. For growers located in temperate or tropical climates where weather extremes are common, it's important to start your rose plantation off with plants that are certified clean of viruses and *Agrobacterium.* Ask questions of the companies with which you plan to do business. The best ones are happy to talk to you about NAK Tuinbouw certification programs and

Here roses are being inspected at the Aalsmeer auction.

how they maintain clean stock. If a supplier is offering you inexpensive rose plants, there's a reason! Buy such plants at your own risk.

Most growers buy in grafted rose plants because grafting the desired variety onto an understock suited to your production situation generally provides higher yields, disease resistance, and better flower quality. The understock you choose depends on your growing conditions and market. Bench grafting takes place in the winter and early spring. Generally cuttings of the desired variety are grafted onto understock grown outdoors the year before. Plants are joined under propagation tents at 82–86°F (28–30°C) and ready for planting out in fourteen to twenty-one days. Among the most common understocks used are *Rosa indica* 'Manetti', *R.* 'Natal Briar', *R. indica* 'Major', *R. canina* 'Inermis', and interspecific crosses between them.

R. canina 'Inermis' is the most common understock in northern Europe. Flower color and bud shape are

Table 35. Suggested Pot Rose Color Mixes

SALES SEASON	% RED	% WHITE	% YELLOW	% PINK	% CORAL, SALMON
General	30–35	5–10	20	35	10–15
Christmas	75–85	15–25			
Valentine's Day	50	5–10		20–30	10–15
Easter, Secretary's Day	25–30	5–10	10–15	15–25	10–15
Mother's Day	30	10–15	20-25	25-30	10–15

Source: Newflora

the best of any understock. Roots are deep and heavy, and it is the best understock for clay soils. 'Inermis' is only multiplied from seed, meaning the resulting plants can be highly variable. Plants are slower starting than with other understocks. 'Inermis' is winter hardy and less sensitive to powdery mildew than other roses are. However, it is declining in popularity as newer propagation methods, such as stentling production, take over for crops grown on artificial substrate systems.

R. 'Natal Briar', native to South Africa, is a heavy water user. Since 'Natal Briar' is vegetatively propagated, make sure your supplier has disease-free mother stock. Production is strong on 'Natal Briar'. The long-stemmed plants are fast to root and recover. Plants are sensitive to high boron levels. 'Natal Briar' provides good top growth but can yield lighter flower color.

R. indica 'Major' (Chinensis type) is the traditional understock used in the Mediterranean region. Plants are vegetatively propagated, so make sure your supplier has disease-free mother stock. *R. indica* 'Major' is the most sensitive understock to *Agrobacterium.* On *R. indica* 'Major', the resulting flower colors are palest compared to other understocks. Plants are not winter hardy, but production declines as temperatures decline.

R. indica 'Manetti' (Chinensis type) is the traditional understock for South American production. Plants are vegetatively propagated, so make sure your supplier has disease-free mother stock. Plants are sensitive to *Agrobacterium.* 'Manetti' prefers warm, semi-humid climates, but can tolerate low night temperatures. Plants are productive on 'Manetti'. Its fine root system is best in light, well-aerated soils.

Which understock should you use? Consult with the breeder, your rose propagator, and other growers in your area before making a decision. Understock can play a significant role in yields, especially for ground culture; however, the interaction between understock and environmental factors, understock clones, light, and vigor of plants is complicated and little hard data exists in this area.

While most growers use grafted plants, in the Netherlands, "stentlings" have taken over for most rose production. Stentlings are also grafted, yet take much less time to produce than traditional grafted plants since they are produced by placing a cutting from the desired variety on an unrooted cutting of the desired understock. The "graft" is held together with a clothespin. Plants are then rooted and established in high-tech greenhouses under precise environmental conditions. Generally, stentlings can be produced in seven to nine weeks. Bad cultural practices in stentling production, such as when cuttings of varying thickness are used, will cause knots to form at the graft.

Another recent trend in the Netherlands is for growers to use roses on their own roots. Rose crops produced from cuttings are increasing very quickly among Dutch growers. The turnaround from ordering until plant delivery is just six to eight weeks, and plants begin producing flowers quickly.

Growers producing their crop in artificial substrate systems such as rock wool or coco fiber generally choose to buy stentlings or roses growing on their own roots. Cuttings for these young plants consist of stem sections 2" (5 cm) or more below buds (bud is removed) with one five-leaflet leaf present. IBA rooting hormone and/or a fungicide dip aid rooting and prevent disease. Cuttings are rooted under tents or fog at 72–82°F (22–24°C).

Root-grafted plants and half-year rose bushes are other type of plants you can purchase.

Please note that many rose plant suppliers cite production numbers for varieties based on Dutch conditions under supplemental lights and in hydroponic growing systems. Not every rose that performs well in an artificial substrate system is suited to ground culture.

Growing On

When planting a new rose plantation, remove all broken stems, shoots, or roots before planting. For grafted plants and bushes, some growers also trim canes to leave three to four buds.

If you are planning soil cultivation, take soil samples and have them analyzed by a lab that specializes in horticultural production. Tell the lab running the test that you will be planting a commercial cut rose crop and they will provide you with guidelines on how to amend the soil. Prior to planting, make sure soil is free of weeds, insects, disease, and nematodes. Adjust the pH of soil to 5.5–6.0. Soil should be worked to a depth of 18–24" (46–61 cm). Make sure soil is well drained, adding drain tiles if needed. Ground bed growers typically provide three or more layers of support.

Newly planted rose plants require constant care until they are well established. Slowly increase growing temperatures and be careful not to overwater plants, as roots will not be actively growing immediately. Grafted plants may be delivered in a dormant state coming out of a cooler, so treat them gently.

If you purchased plants rather than bare-root grafts, plant only actively growing plants with white roots showing.

Determine your planting density based on your market. Plants spaced closer together will yield a higher number of lower-quality stems. Defer to your research in determining the best plant density to use for your market.

Start your crop off at 68–72°F (20–22°C) nights and higher humidity levels for the first couple of weeks. Syringe plants with a hose or mist attachment frequently during the day if light levels are high. Once plants put out new growth, gradually lower humidity and temperatures to normal: 65–75°F (18–24°C) days and 62°F (16°C) nights. Overhead shade curtains are ideal, since they will allow you to moderate temperatures. (For more information, read the Commercial Production of Outdoor Cut Roses section.)

Focus on developing your plants in the beginning of a new crop at the expense of harvesting stems, which will cause higher yields in later years. The overall amount of bottom breaks your plants produce will be directly related to subsequent yield. That's why you remove the first flowers.

Many rose growers in South America and, to some extent in Africa, grow their roses the traditional way, by establishing the plant's structure, or "chassis." To establish the plant chassis that will provide your cut flower crops, provide two soft pinches to your new plantation. Pinch shoots off new plantings once the flower bud is almost the size of a garden pea (about four weeks after planting). Pinch shoots back to the second five-leaflet leaf on the main stem. Pinch again when flower buds are almost the size of a garden pea, about five to six weeks later, depending on your exact growing circumstances and the variety you are growing. You will then be able to make your first harvest forty-two to fifty-six days later.

In general, the total time to cutting the first flowers from a planting can vary from 70 to 90 days (ten to thirteen weeks) in a one-pinch program and from 108 to 127 days (fifteen to eighteen weeks) in a two-pinch program. While growers in some regions use this traditional program to establish the chassis, others prefer to use more modern production techniques, such as the bent-stem technique. (See the section on the bent-stem technique for more information)

Ground bed growers will irrigate with perimeter irrigation tubes. Growers in hydroponic systems such as rock wool or coir will provide a constant liquid feed of 150–200 ppm. Many growers in developed nations such as the Netherlands prefer hydroponics, as yields are much higher than with ground bed production. Ammonium fertilizers help to stimulate stem elongation.

Many rose species naturally go dormant in the winter. Maintaining warm production temperatures and providing as much light as possible can prevent dormancy. In the past, some soil growers chose to prune plants back and allow them to go dormant for a period of time, "resting" at cool temperatures just above freezing. Growers would bring plants out of dormancy gradually by raising temperatures and lowering humidity. However, growers today rarely allow their plants to go out of production.

Some growers, however, do choose to let plants go out of flower in mid- to late summer, when prices are lower and temperatures are high. To do so, cut plants back to an active bud, leaving enough leaf canopy and hardwood behind to generate a harvest in about four weeks.

Light levels are critical to good rose production, as they directly determine shoot (bottom breaks) and leaf production. In areas with high light and high temperatures, many growers produce their crops under shade varying from 15–50%. Whitewash, white screens, or black Saran plastic can be used. Removable or automatic screening systems are ideal, as you will be able to adjust the shade based on climatic conditions.

Modern rose production at Zuurbier, Heerhugowaard, the Netherlands

Building the plants

One way to build up newly planted roses, or to rebuild reserves in established plants, is a method known as deheading and deshooting. The technique is labor intensive, but it does develop basal shoots, which aid in the future production of cut flowers.

When growers dehead, they generally follow one of two options: (1) They remove all developed flower heads that show color by snapping them off immediately below the flower bud (i.e., at the peduncle). Growers should dehead all stems that reach this phase during the program. (2) They allow all flowers in a bed to completely flower out and drop their petals. At this time the flower head is usually removed in the same way, snapping flowers off at the base (at the peduncle).

Deshooting follows deheading. Rose plants will respond to deheading by initiating numerous lateral shoots. These new lateral shoots need to be removed as they develop. This usually involves two to three passes per week through the growing bed. The program will not work unless this portion is performed continuously. Normally, the only shoots removed are those that initiate on what would have been cut flower stems. Do not remove lower or ground shoots during this operation, since the purpose of the program, after all, is to encourage basal development. Dehead and deshoot for six to eight weeks to renew your plantation.

Harvesting

Harvesting your rose crop is the most critical area of production. Not only will most of your labor be spent in harvesting (50% or more), but also your future success is determined by how well workers cut stems. Stem length and flower quality determine price—both are determined to a great degree by the quality of the plant you've built, which is directly affected by how workers harvest.

Subsequent crops will be produced from growth flushes after harvest. Leaf nodes on the upper portion of plants produce flowers faster and with fewer leaves than lower nodes. Therefore, for the fastest flowering cycle, choose to harvest stems high. Flowering shoots will regenerate quickly, although resulting stem lengths will be shorter. Harvesting at the third node or lower will slow down shoot regeneration but yield stems with longer length. Harvesting below the juncture of the shoot and stem (the knuckle) will slow down regeneration even more but yield very long stems. However, if you need to reduce plant height, consistently cut below the knuckle for several harvests. Lower cuts rejuvenate the plant. If you choose to harvest at lower cutting points to pursue the quality market, be sure plants have enough leaf mass to regenerate shoots. Bending a few stems (see the section on the bent-stem technique) can help to build leaf mass.

In practice, growers use a combination of upper and lower cuts to manage plant growth, the quality of stems, and plant yields, depending on their precise situation.

With experience in your conditions and good notes on yields by greenhouse and by variety, you will be able to predict the number of stems you can harvest and at what stem length. Keep records. Your exact notes based on your specific growing conditions will be your most valuable crop-planning assets.

Removing flowering shoots from plants five to eight weeks before harvest for a holiday will cause a flush of production. The time for plants to flower depends upon light levels and duration as well as temperature. During times of the year when rose demand is low, you can cut older plants back hard (to 24" [61 cm]).

Dutch researchers have shown that the more young shoots present at harvest, the longer the vase life of the roses harvested. At the end of a flush, plants tend to have more shoots than at the beginning of a flush, which implies that stems harvested at the end of the flush should last a bit longer in postharvest.

As with harvesting, grading and packing also consume massive labor requirements. The benchmark in the Netherlands for whether or not a grading machine pays is an annual volume of 7.5 million stems.

The bent-stem technique

Japanese growers first began growing their rose plants using bent stems in the 1980s. Rose plants used for the bent-stem technique are started as single node cuttings, which are rooted and grown frequently in rock wool cubes. Rooting takes from twenty to forty days depending on the time of year. Rooted cuttings are then pinched and put on rock wool slabs in raised growing gutters placed in one- or two-row gutter growing systems. Many growers are also using stentlings for bent-stem culture.

No matter how the plants get started, growers utilizing the bent-stem technique use growing benches to facilitate delivering water and nutrients. Benches are about knee high and also make plants easier to access for workers. Planting density is from 0.5–1.0 plants/ft.^{2} (7–10 plants/m^2).

After planting, remove buds from the shoots that have formed. These stems will be bent gently downward in an open arch that sweeps over the side of the bench. Do not force bending. The bend is above the second leaf node. Wait to bend until the plant has around sixty or seventy leaves (counting three- and five-leaflets). These stems will provide a photosynthetic factory that supplies the rest of the plant, enabling higher production of shoots that will arise from axillary basal buds. These buds develop into shoots that are later harvested as cut flowers. Blinds that arise from axillary basal buds are bent down during harvest to provide shoots for the next flush of flowers, and so on through the year. Some growers pinch flower buds and then wait a few days to bend stems, because this lessens the total shock to the plant, as photosynthesis drops by 10–20% in the leaves of a bent shoot, taking three weeks to bounce back. The goal is to have a balance of leaf area in bent stems and developing shoots for flower production. Once you have developed a good plant chassis and canopy, bend only one out of every two or three shoots.

The bent-stem technique has a number of advantages, especially since stems are longer using the technique and many markets pay based on stem length. Workers have better access to plants since the plants are generally raised up, making maintenance and harvesting much easier. Flowers are harvested at the stem base without a leaf node, thus simplifying harvest, formerly a critical process. Raised gutters also greatly increase air circulation around the plant, which lowers disease. While total production can be 20% less because flowers are cut before plants are established, plants yield a higher number of No. 1–quality stems, thus offsetting lower production numbers. Stems are of more uniform length as they are harvested, which also decreases labor needed for grading.

In the Netherlands, research has shown that the highest yields on 'Frisco' and 'First Red' were obtained (although stem weight was lowest) when the primary shoot was bent at the beginning of cultivation. Subsequently, only stems that grew from this shoot were bent. The other bending techniques compared to this showed less yield yet higher stem weights in this order: (1) bending primary shoot and the first flush of bottom breaks plus lateral shoots; (2) bending primary shoot and bottom breaks of inferior quality and other stems of inferior quality; and (3) primary shoot, bottom breaks of inferior quality bent at a rate of one shoot bent per three to four branches of 'First Red' or one shoot bent per five to six branches of 'Frisco'.

Most growers who have used the technique for several seasons do not bend stems in the winter months (November through January), as light levels are too low.

The bent-stem technique, also known as arching, is especially well suited to spray and mini-spray roses.

Pests & Diseases

See Pests & Diseases in the Pot Roses section. If you choose to grow cut roses, botrytis and/or powdery mildew will strike the minute you let up on environmental control. Keep air circulating and avoid humidity buildup. Keep freestanding water off plants and flowers.

Troubleshooting

There are numerous reasons why rose plants may suddenly drop old leaves. Most of these are related to air and the medium. Conditions that can induce leaf drop include excessive humidity; changes in temperature, especially low night temperatures; and changes in light levels, especially overcast or other low-light conditions. A rose plant may have leaf drop from ethylene exposure or may suffer phytotoxicity from a pesticide or from sulfur burning. Undesirable medium qualities, such as excess salinity or a low nitrogen level, can also be responsible for leaf drop. Low soil pH can cause manganese or aluminum toxicity or magnesium or calcium deficiency, but high pH can also be very unfavorable. If the problem seems to be high bicarbonates in the soil or irrigation water, use water with less than 122 ppm (2 meq/l) of bicarbonates.

A rose plant could lose roots for various reasons, which might lead to leaf drop. Nematodes may be causing trouble. Downy mildew is a more noticeable problem. Finally, rose leaves do not live forever, so there could be leaf senescence from simple old age.

Varieties

Cut roses are divided into three categories: hybrid teas (flower diameter greater than 3.5" [9 cm]), spray roses (three or more flowers/buds per stem), and sweethearts (flower diameter below 3.5" [9 cm]). Within the hybrid tea category, there are large-, medium-, and small-flowering varieties. Some small-flowering hybrid tea roses may be smaller than the 3.5" (9 cm), but yield long stems, more than 36" (91 cm), which allows them to be classified as hybrid teas.

Red is thought to be the most important flower color since it is preferred for Valentine's Day and Christmas holiday sales; however, yellow is coming on strong right behind it. Dutch auction statistics show that the market in the Netherlands is about 25% red, 25% yellow, 10% orange, 10% pink, 10% white, 5% salmon, and 15% all others. In the early 1990s, red used to comprise as much as one-third of the supply. Superior breeding, combined with competition from new production centers such as Ecuador, have woken growers up around the world to the possibilities and market demand for colored roses.

When growing garden roses as outdoor cut flowers, correct variety selection is a key factor in the success of an operation. Not just any garden rose variety will work, since the variety must be able to perform as a cut flower after harvest. This means the flowers must have a satisfactory vase life and ability to be shipped, attractive foliage, and acceptable color under commercial postharvest conditions.

Since most greenhouse cut roses have little or no fragrance, growers should make fragrance an important criterion in variety selection. However, in general, extremely fragrant roses often have very soft petals, which can cause a less-than-ideal vase life.

How do you select a rose variety? Nancy Laws, industry rose expert and consultant, contributes the rest of the rose variety discussion, outlining important considerations for cut rose growers in today's marketplace.

What is your market? There are two distinct and mutually exclusive markets. The first is large, metropolitan markets that have florists who demand big roses, cut open, ready to include in flower arrangements. These roses are usually delivered directly to florists in buckets on the same day of harvest. Florists expect them to last a week. Many of these roses are scented. Some must be cut open in order to develop fully. Since they are cut open, they do not fit densely in a shipping box. The second are the major export markets—either shipped long distance or airfreighted to another country. These roses are sold in the bud, not open, and must look good as a bud and have a tight cut point so that as many roses as possible can be packed in a box. Flowers should not bruise in transport. They must have a long vase life.

There are also two major types of buyers. The first type of buyer is largely interested in size, beauty, new colors, elegant shape, flowers that "spiral" open, color and luster of the foliage, and so forth. Generally these buyers are large metropolitan wholesalers. The second type of buyer is interested in having a standard array of colors at a cheap price, consistent quality, and a rose that lasts twelve days (six days to get through the distribution chain, and six days for the end consumer).

Purchasers of long-stemmed roses tend to be in the first group. Purchasers of medium roses tend to be in the second group. The supermarket buyers who dominate purchasing of 20" (51 cm) roses have been very effective in negotiating lower rose prices for imported roses. The average price of medium roses has dropped 12–30% from 1998–2002 in most major import markets. If you choose to compete in this market, know your costs.

You must pick your market. If you choose to ship your stems, all of the roses grown on the farm should be similar in length. A grower can specialize in 20" (51 cm) roses for supermarkets or 28" (71 cm) for retail florists, but not both. If your stems are to be sold locally, they can be big roses with eight to a hundred petals harvested at an advanced stage of opening. Ideally, do not choose rose varieties that require different cut points.

Choose your breeder. Once you have identified your main market and customer base, research the breeders. There are less than two-dozen rose breeders specializing in commercial cut flower varieties. Seven of them are in the Netherlands (Olij, Interplant, Terra Nigra, De Ruiter, Schreurs, Preesman, and Lex +). Germany has two large breeders (Rosen Tantau and Kordes Roses); France has four (Meilland, NIRP, Delbard, and Fazarri); the United States has two (Jackson & Perkins and Hills); New Zealand has one (Franko Roses), and smaller breeders are located in Japan and Israel.

Choose your variety. Characteristics to look for in a variety:

- **Size of the bloom.** Recent rose breeding has aimed to increase the size of large hybrid teas up to 4.5" (11 cm). Few supermarkets will accept a hybrid tea bloom size under 3.5" (9 cm).
- **Vase life.** The vase life of some cut roses has been extended up to twenty-one days. A minimum of twelve days is essential. Spray roses are expected to have a minimum twelve-day vase life.
- **Productivity.** Productivity should approach 240 blooms/yd.^{2} (200/m^2) per year for large roses and 360 blooms/yd.^{2} (300/m^2) per year for medium roses.
- **Thorns.** Stems should have few thorns. Thorns rip and tear the leaves during shipping.
- **Stem length.** Few supermarkets will accept roses under 20.5" (52 cm) in length. This is very important in climates with hot summers, because the rose length and head size cannot be under the minimum required by the client, even in the hot summer

months. When in doubt, pick a rose that is longer than you believe your clients will need.

- **Bud shape.** Tall, cylindrical hybrid tea roses that spiral open slowly over several days are preferred. On the other hand, buyers demand that roses open fully. They do not like roses that stay permanently cup-shaped.

Currently there are more than one thousand roses traded around the world. Assume that 80% of them are not for you. Don't even be tempted to buy them, except in small quantities, for experiments.

Also, do not limit yourself to just new varieties. Just because a variety is old does not mean it isn't good. Ecuadorian rose growers are very successful selling roses in both the United States and Europe, and most of their export varieties are not "new." 'Aalsmeer Gold' has been on the market so long that its royalty period has expired (though use of the trademark requires compensation). In fact, many of the roses grown in Ecuador, Colombia, Kenya, and Zimbabwe are "old" varieties. In some cases, they are excellent and there is no better replacement. In other cases, the new replacements show significant improvements. There are now reds that feature outside petals that do not turn black, which has been a drawback of 'First Red'. There are yellows that do not fade and turn pale as they open. There are roses that produce 240 stems/yd.^{2} (200/m^2) instead of 144 (120). There are roses that last fourteen to sixteen days instead of only eight to ten. There are excellent new bicolors. A new rose introduction may command a short-lived price premium, but certainly do not count on this being a long-term effect.

When you are discussing varieties with your salespeople, be sure to ask key questions such as: How well does this rose travel? What is its vase life? Does it blacken on the outside of the petals? Does the head get smaller in the heat? What about vase life, flower color, and stem length in high-temperature production? Is it better than any other rose hybrid on the market in its category? If your salesperson evades these questions, don't buy. Wait a year or look for a substitute from another breeder.

Postharvest

The two factors most important to long postharvest life of roses are variety and temperature. Long-lasting cut flower varieties such as 'Prophyta', 'Frisco', and 'Vivaldi' can last twenty days when handled correctly. Ask prospective breeders about vase life and ask them to tell you how they determine the vase life of their varieties: You may discover that it will be impossible to compare vase life numbers from one breeder to the next because they are measured in different ways.

During production, strive to produce greenhouse roses at lower humidity levels. Dutch research has shown that the higher the relative humidity during production, the shorter the vase life. For the variety 'First Red', produced in the winter, each 1% rise in humidity decreased vase life by 0.1–0.25 days.

Field-grown cut roses are generally cut in the morning and the evening. This practice may not be necessary in field production during cooler times of the production cycle, but will be during the warmest times of the year if the best-quality blooms are to be harvested.

As soon as possible place cut stems in a hydration solution such as water acidified (with citric acid) to a pH of 3.5. Use a bactericide in all postharvest solutions to prevent bacterial growth, which will clog water vessels. Use cold water (32°F [0°C]) for fastest water uptake into stems. Store roses in a postharvest solution in a cooler with 80% or less humidity at 35–39°F (2–4°C). High humidity in the cooler may cause problems with botrytis. One researcher has shown that providing 100–300 f.c. (1.1–3.2 klux) of light in the cooler increases solution uptake. The rule of thumb in the Netherlands for how much cooler space you will need at the greenhouse is about 5 yd.^{2}/1,200 yd.^{2} (4 m^2/1,000 m^2) of greenhouse. After a few hours of cooling, when roses are full of water, they can be graded, bunched, and returned to the cooler.

After grading, water used in buckets for storing roses should be treated with chlorine to keep bacteria from developing. Using chlorine at a rate of 100 ppm will eliminate bacteria from rose buckets. Sanitation is important for good flower quality.

Roses are generally graded according to stem length. Most growers grade roses in 4" (10 cm) increments. Standard grades are:

- Short: 10–14" (25–36 cm),
- Medium: 14–18" (36–46 cm),
- Long: 18–22" (46–56 cm),
- Extra long: 22–26" (56–66 cm),
- Fancy: 26–30" (66–76 cm), and
- Extra fancy: 30" (76 cm) or more.

All broken, crooked, or bent stems are culled, as are those roses with poor head shape, bent necks, or disease.

After grading, most cut roses are bunched in bundles of twelve to fifteen stems. Roses are often bunched in a round pack and then wrapped with commercial plastic or cellophane. For large-headed varieties, many growers use a spiral pack, while others nest roses so that twelve flowers are arranged below the other thirteen. The wrap must never be tight, as roses grow during storage and shipping. If packed too tightly, the growing heads will actually bruise each other or even snap off. Also, extend the wrap 2" (5 cm) above the flower heads in the bunch in order to help avoid bruising the tops of the flowers. Label all bunches by grade and variety. After wrapping, stems are usually tied together with rubber bands, string, or twist ties.

If you keep stems dry, store at 33–35°F (1–2°C). Recut stems under clean water each time bunches are out of water to remove air embolisms. Rehydrate each time stems are cut (i.e., after harvest, arrival at wholesaler, arrival at retailer, etc.) using a commercial postharvest formulation for roses.

Roses are sensitive to ethylene and may benefit from a 1-MCP or silver thiosulfate (STS) treatment. Do not store near ripening fruits or vegetables.

Growers who produce garden roses outdoors and are close to the final market often deliver stems in water and refrigerated trucks. If you are selling at roadside markets or through local florists, you should stress that the flowers are locally produced, garden roses.

Retailers should unpack boxes, cut stems under water using only clean water, and place them in disinfected buckets with a commercial postharvest solution designed for roses. Do not use a solution with more than 2% sugar. Remove leaves and thorns below the water line.

Commercial Production of Outdoor Cut Roses

If you are looking for a great seasonal cut flower crop with good profitability, consider producing cut roses outdoors or under high tunnels. Florists and consumers love the casual, old-fashioned appearance, romantic colors and blends, improved vase life, and fragrance of many of today's rose varieties that have been hybridized with an eye for use in outdoor commercial production. The production of roses as an outdoor commercial crop has evolved significantly in the last decade. Countries such as Germany have as many as 600 acres (250 ha) in outdoor and tunneled production. Many new varieties offer a range of characteristics such as flowers with quartered and rosette blooms, large sprays of buds and blooms, longer stems, fragrance, productivity, improved water uptake, and longer vase life.

Climate, location, and physical access and proximity to available labor and markets are all important considerations in determining the potential of a commercial cut rose operation. Water availability, water quality, soil, and drainage are also essential considerations when determining if an area is suitable for a commercial plantation.

Although ideal conditions for growing garden roses as cut flowers are cool nights coupled with warm, sunny days, excellent quality cut roses can be grown under many environments. The best spring, summer, and fall temperatures for outdoor rose production are minimum night temperatures in the range of 45–65°F (7–18°C) and maximum daytime temperatures in the range of 75–90°F (24–32°C). However, there are considerable differences in how varieties respond to temperature. Some varieties will still grow well with nights as cold as 40°F (4°C), while others need 50°F (10°C) or above to form well-shaped flower heads. In addition to improving quality under rainy and wet conditions, tunnels can aid production under some of these lower temperatures. Production of commercial outdoor roses is generally best in USDA Hardiness Zones 5–9. In regions with USDA Hardiness Zones of 5 and less, some type of protection of the dormant plants in winter is a best.

Climates best suited for outdoor cut flower production are those with moderate to low humidity and with somewhat limited rainfall (less than 30–45" [76–114 cm] annually) during the growing season. The timing of rainfall is important. Winter rains while plants are dormant are not as problematic as summer rains while plants are in bloom are. In areas with higher rainfall, roses will also grow well; but as with any floral crop, you should pay special attention to the potential for problems associated with water standing in opening flowers and on foliage. In rainier locations, some type of polyethylene cover over the crop ensures more consistent quality, earlier flowering in the spring, as well as continued bloom into the fall. Windbreaks should be considered in areas with strong winds during the growing season. Many simple systems are effective. Trees, bamboo lattice, snow fencing, and commercial plastic fabrics are commonly used.

In certain high light and high temperature areas, producing under shadecloth enhances flower and foliage quality. Generally, stem length is improved, flower bud size increases, flower colors fade less, foliage expansion (leaf size) is enhanced, and foliage gloss under high tunnels or shade. Conduct trials before committing to one shade density for the entire plantation. Additionally, you should install shadecloth in a manner that allows it to be bunched and fastened out of the way during lower light periods or in the event of high winds and excessive snow or hail. Densities in use range from 15–50% shade. White and reflective-type shadecloths have been used successfully in some warmer areas. Shade can also aid in screening out some insects pests.

In selecting a site for roses, make air circulation a prerequisite. Freely circulating air will promote the rapid drying of foliage and help reduce the potential of fungal diseases. Perhaps the most significant change in production methods over the last decade has been the move to lower density plantings to aid in air circulation. Previously, beds were planted with multiples of plants placed across the beds. Currently, outdoor plantings are usually planted as single rows. If planting on a slope, locate roses toward the top of the rise. This will encourage free air movement. Low-lying areas can be more susceptible to diseases and frosts.

Other considerations for a production site should take into respect sources of dust, deer, rabbits, and other animals. If deer or rabbits are present, usually the only effective way to avoid plant damage is by fencing.

In areas that require irrigation, the quality and quantity of available irrigation water is critical to success.

Planting Stock

Most growers plant dormant one- to two-year budded plants from a supplier that has commercial varieties suitable for commercial production and marketing. Budded plants help to ensure the maximum stem length and productivity from commercial varieties. In addition to dormant plants, bench-grafted "stentlings" are available from some specialist propagators and are well suited for outdoor production. One benefit to using stentlings is their shorter lead time.

Companies that specialize in rose plants for commercial outdoor cut rose production offer growers a number of options for plants. The most commonly available options are one- to two-year rose bushes, started-eye rose plants (one year), June-budded rose plants, and/or bench-grafted plants (stentlings). Varieties should be budded or grafted on a suitable understock for the climate and soil of the region of production. Many excellent understocks are used around the world, and the benefits of their use should be discussed with your plant supplier. A few popular understocks include *Rosa* 'Dr. Huey', *R.* 'Manetti', *R. multiflora, R. canina* 'Natal Briar', and *R. laxa.*

Establishing a Plantation

Fertility is a key factor to a successful plantation. Always use a soil analysis and recommendations from a laboratory experienced in floriculture crops to guide fertilization rates and practices. Amendments to pH and basic elements are always best done ahead of time. Also, know your water quality and adjust your fertilization and irrigation accordingly. Water sources high in boron (above 0.5 ppm) and bicarbonate (above 2 meq/l [approximately 130 ppm]) should be avoided as they can lead to long-term production problems.

Roses will thrive in many types of soil, from pure sand to heavy clays, if soil preparation and irrigation are managed to match the rose plants' needs. Prior to adding soil amendments, any supplemental drainage requirements, such as underground tiles or plastic drains, should be addressed. Ideally, the area selected for production should be ripped with a tractor and ripper attachment. The best soil for roses is deep, well-drained, and fertile.

If soil is sandy or gravelly or has low water-holding or cation-exchange capacity, amendments should be chosen that improve water and nutrient retention. Peat moss, coco peat, composted pine bark, composted sawdust, many types of compost, and vermiculite are beneficial in improving rose growth in these types of soils. If soil is high in clay or silt, consider soil amendments that will improve soil aeration and drainage. Medium-ground composted pine bark, composted rice hulls, coarsely ground corncobs, composted peanut hulls, coarse washed sand, perlite, and volcanic scoria are just a few of the materials used worldwide.

If possible, the soil should be worked to a depth of 12–18" (30–45 cm). In addition to any of the above soil amendments, well-composted manure can be considered as a soil amendment and fertilizer. It should be added in judicious amounts, since excessive amounts can and will cause problems, such as excessive soil salinity. Let your soil lab be your guide. It should also be noted that in most regions, there are local rosarians in a local rose society who have ample experience in growing roses under local conditions.

Another approach is to not grow in the soil at all, but to grow the plants in containers. In Italy and other parts of the world, producers grow some of the best outdoor roses by transplanting two plants in a 5–10 gal. (20–40 l) or larger container. The media within these containers can include a variety of components, such as volcanic pumice, perlite, coco coir, vermiculite, or other amendments commonly used in hydroponics. Drip tubes provide brief, frequent irrigations throughout the day. Containers are set out in rows on nursery ground cloth for weed control. Growers in extremely cold-winter areas should consider additional winter protection for their production system.

Always base your fertilizer program on a soil analysis done by a competent soil laboratory with experience in floriculture soil fertility management. This will reveal potential problems that can be alleviated before planting. For example, roses grow best in soils that are slightly acidic; the ideal soil pH is 6.2–6.5. The lab can also recommend preplant soil amendments and any fertilizers that should be incorporated prior to planting.

Once the new rose plants arrive, open the boxes to inspect the contents and then store in a cool, shaded place or in refrigeration. The plants should be sprayed with water, and the boxes and liners reclosed in order to avoid dehydration. It is generally best that the rose plants are transplanted immediately. Dormant rose plants can be stored for a limited period of time prior to planting. The deal storage temperatures is 34–36°F (1–2°C), and plants should be left sealed in their original shipping cartons. At these temperatures, dormant rose plants may be stored for two or three weeks. Do not store them with fruits, vegetables, or any other sources of ethylene. If the roses cannot be planted in the field for some time, it is best to containerize until you are ready to plant them.

Row spacing and planting densities often depend on the size of the grower's equipment for cultural activities and harvest. Sufficient space should be allowed at the end of the rows for turning equipment. Commonly used distances between rows are 48–100" (120–260 cm), depending on the width of equipment. For example, a grower with equipment that has a width of 43" (110 cm), the distance between rows should be approximately 71" (180 cm). This distance is advisable in new projects. Decreasing the planting density will improve the habit and production of each plant; however, the total production of stems per acre or hectare may decrease. Single-file rows are now the norm in cultivation. The spacing between plants in the row should range from 10–12" (25–30 cm). Planting in beds or double-row planting systems are no longer advisable because of reduced air circulation around plants and increased disease risk. Overall, the rows generally have the look of a hedge as one looks down a planted row.

If the area to be planted is dry, it should be irrigated approximately one week prior to planting in order to have the soil in a workable condition but with good soil moisture.

High-quality cut roses are produced in many planting scenarios—raised rows, raised beds, flat field plantings, and even in the bottom of the furrow. Experience, soil type, drainage, and irrigation method should all be considered when choosing the placement of the plants. One successful way of preparing the soil for planting is to simply trench or plow out a row of desired length in a field with prepared soil using a small tractor. Plants are then placed in the furrow and covered with soil to give the proper planting depth.

Dehydrated plants can greatly impact initial performance. Dormant plants should be hydrated by immersing them in water from two to twenty-four hours prior to planting. Plant roses one at a time, ensuring that the roots are spread out uniformly and fully covered with soil. *At all costs, avoid allowing the plants to dehydrate.* The planting depth depends upon the understock that the roses are budded to and the severity of winter in the region in which the roses are planted.

- With 'Dr. Huey' understock, plant so the bud union is at or just above the soil line.
- With 'Dr. Huey' understock in areas with more severe winters, plant so the bud union is well below the soil line.
- With *R. multiflora* and *R. laxa*, plant so the bud is 1–2" (2–5 cm) below the soil line.

Make a basin around the roses to ensure that irrigation water penetrates the area around the roots.

If canes are longer than 6" (15 cm), cut them back to an outside eye, leaving not more than 6" (15 cm) of cane—4" (10 cm) is even better.

After planting, it is advisable (especially under dry and warm conditions) to mound some type of organic matter, such as pine bark, rice hulls, mulch, chopped straw, or aged sawdust around each plant to protect it from dehydration, direct sun, drying winds, and/or frost. After two or three weeks, this organic matter can be spread out around the base of the plant to serve as mulch.

To encourage maximum bud emergence, maintain high relative humidity around the newly planted roses. If the weather is rainy and cool, this will occur naturally. However, if drier conditions exist, then syringe, sprinkle, or mist the plants briefly during the sunnier periods of the day to reduce the chance of dehydration. However, care should be taken to not wet (waterlog) the soil excessively, which can result in poor soil aeration and subsequent root initiation and growth. Syringing should be terminated early enough so as to ensure that foliage is completely dry by late afternoon; otherwise the risk of disease increases. Once new, white roots have fully emerged to support the plant, discontinue syringing in order to encourage root development.

As the eyes on the original canes force and begin to produce new canes, several options are available to build a chassis on the plants for future production. The traditional method is to pinch the new growth on stems when the flower buds have developed a diameter of 0.25" (4–5 mm or roughly the size of a pea). The pinch is accomplished by cutting back the stem slightly above the second five-leaflet leaf (roughly translating to removing the upper half of the stem). This pinch will force one or two eyes below the pinch to break and force new flowering shoots. Half of the resulting shoots from the first pinch should be pinched in the same fashion a second time. Half of the resulting shoots from the first pinch can be allowed to flower for harvest. The stems that are harvested should also be cut above the second five-leaflet leaf.

In many regions, roses often require some type of irrigation to ensure quality flower stems throughout the season. Drip irrigation, with the potential for fertilizer injection, is a popular choice. Avoid overhead irrigation in order to reduce the incidence of disease. If you must irrigate from overhead, the water quality must be excellent to avoid foliage residue and to generally enhance foliage quality. As a rule of thumb, approximately 1–2" (25–50 mm) of water should be applied per week.

Use a soil probe to check the penetration of the irrigation water. After irrigating, the soil should be wet to a depth of 12–18" (30–46 cm). Winter soil moisture is also important. Plants should not be allowed to dehydrate in the winter in drier areas.

Roses require moderate to high fertilization levels. The best pH range for roses is 6.2–6.5. Soil salts should be kept in the moderate to low range for best yield. Established roses require dry granular fertilization in the spring, in late June, and in mid-August. As always, the amount and type should be based upon a soil analysis. However, as a general rule, each fertilizer application using a dry granular fertilizer should amount to about 4 oz. (115 g) per plant per feeding of a 10–10–10 or 10–15–10 fertilizer. Put fertilizer into the soil about 6–8" (15–20 cm) around the base of each plant. Simply bury it below the soil surface and irrigate. Minor elements can also be applied, as indicated in the soil analysis, by using dry fertilizers.

Constant liquid feed is a very efficient method of fertilization. Drip systems, in addition to reducing water usage, have the advantage of allowing fertilizer to be injected into the irrigation lines with the objective of giving a small amount of fertilizer at every irrigation. Use the results of your soil analysis as a guide. Most growers will feed a balanced nitrogen-phosphorus-potassium (N-P-K) fertilizer at 100–150 ppm for two irrigations, then irrigate with clear water, and then repeat.

Outdoor roses used for cut flower production should be harvested when the calyx loosens and reflexes and the outer petals have begun to unfold. Generally, the stems are cut above the second five-leaflet leaf above the previous cut, slightly above the node (point of leaf attachment). If extra length is desired, stems may be undercut (below the last cut) or cut to the first five-leaflet leaf. Undercutting must be done infrequently or not enough foliage will be left on the plant to ensure rapid and productive growth of the next crop.

Most commercial operations use "cut and hold" shears, which allow the individual harvester to carry the harvested stems in one arm while cutting with the other.

Pruning is generally done in late winter. A guideline often used for determining the best date to prune in any region is to prune when the forsythia is in bloom. For most varieties, prune to strong canes about 18" (46 cm) above the bud union.

Pests & Diseases

Disease problems can be seasonal, with black spot, powdery mildew, and botrytis being the three most common problems. Air circulation is an important consideration in laying out a plantation. Maintain clean growing conditions and keep debris removed to reduce carryover disease. Generally, insect problems include western flower thrips, aphids, and spider mites.

All outdoor rose plantations should be sprayed at approximately four-week intervals during late fall, winter, and early spring with a dormant oil and copper spray. Dormant oil sprays should occur on clear days when the temperatures are expected to remain above 50°F (10°C) for at least twenty-four hours. For the best results, spray these oils before buds begin to swell in the spring.

Troubleshooting

There are numerous reasons why rose plants may suddenly drop old leaves. Most of these are related to the environment and the medium. Conditions that can induce leaf drop include excessive humidity; changes in temperature, especially low night temperatures; and changes in light levels, especially overcast or other low light conditions. A rose plant may experience leaf drop from ethylene exposure or from phytotoxicity due to a spray. Excess soil salinity or excessively high or low nitrogen levels can also be responsible for leaf drop. Low soil pH can cause manganese or aluminum toxicity or magnesium or calcium deficiency, but high pH can also be very unfavorable. If the problem seems to be high bicarbonates in the soil or irrigation water, use irrigation water with less than 122 ppm (2 meq/l) of bicarbonates.

A rose plant could lose roots for various reasons, which might lead to leaf drop. Nematodes may be causing trouble. Sudden leaf drop could be from downy mildew, and corrective action should be taken immediately. Rose leaves do not live forever, so there could be leaf senescence from simple old age.

Varieties

There are a few rose breeders that are focusing on creating roses that are suitable for commercial cut flower production. The leading breeders in this new variety development are European rose breeders Kordes Roses, Meilland Roses, and Rosen Tantau. Some criteria for the new selections are nostalgic flower shapes (often quartered, cupped, or rosetted flowers with an "English garden" look), large spray production, high production, long upright stems, fragrance, ease of production, and prolonged vase life. The breakdown of colors for a commercial plantation has shifted to a range of pink, white, yellow, and novelties. Red is far less important in outdoor cut rose production as compared to greenhouse production. Discuss your variety selections with your customers prior to ordering and planting. Budded or grafted plants are best for production, because the understock is often more adaptable to the variable soil conditions encountered around the world and helps ensure vigorous, productive plants.

Postharvest

A number of factors determine the potential postharvest life of any cut flower. These factors include species, flower variety, production method and conditions, time of year, flower cut point, postharvest hydration treatment, presence of ethylene, ambient light, relative humidity percentage, and temperature. Plant suppliers should have vase life trial results in their literature. Although only general guides, this data should be used as a guide in selecting varieties for production, taking into consideration that the data is often not comparable between breeders.

Field-grown cut roses are generally cut during the coolest part of the day, such as morning and evening. This practice may not be necessary in field production during cooler times of the production cycle, but is essential during the warmest times of the year if best-quality blooms are to be harvested. If the variety is one that produces sprays (multiple inferences per stem), most of the uppermost floral lateral flower buds and blooms should be left on the cut stems and not removed when grading and bunching. This will yield a more "natural, garden-like" flower as well as yielding a fuller bunch of blooms. For the floral buyer, it is vital that stems look as if they came from the garden and look natural.

As soon as possible place cut stems in a hydration solution such as water acidified (with citric acid) to a pH of 3.5. Use a bactericide such as chlorine (i.e., 50 ppm chlorine bleach, or NaOCl) in all postharvest solutions to prevent bacterial growth, which can clog water vessels in the cut stems. Use cold water (32°F [0°C]) for the fastest water uptake into stems. Store roses in a postharvest solution inside a cooler with 80% or less humidity at 35–39°F (2–4°C). High humidity in the cooler may cause problems with botrytis. After a few hours of cooling, roses should be full of water, they can then be graded, bunched, and returned to the cooler.

After grading, water used in buckets for storing roses should be treated with chlorine to keep bacteria from developing. Using chlorine at a rate of 100 ppm will eliminate bacteria from the buckets. Sanitation is important for good flower quality.

Roses generally are graded according to stem length. Most growers grade in 4" (10 cm) increments.

Standard grades are:

- Short: 10–14" (25–36 cm),
- Medium: 14–18" (36–46 cm),
- Long: 18–22" (46–56 cm),
- Extra long: 22–26" (56–66 cm),
- Fancy: 26–30" (66–76 cm), and
- Extra fancy: 30" (76 cm) or more.

All broken, crooked, or bent stems are culled, as are those roses with poor head shape, bent necks, or disease.

After grading, most cut roses are bunched in bundles of twelve to fifteen stems and any thorns are removed from the lower portions of the stem by a thorn stripper. Roses are often bunched in a round pack and then wrapped with commercial plastic or cellophane. For large-headed varieties, many growers use a spiral pack, while others nest them so that twelve flowers are arranged below the other thirteen. The wrap must never be tight, as roses grow during storage and shipping. If packed too tightly, the growing heads will actually bruise each other and even snap off. Also, extend the wrap 2" (5 cm) above the flower heads in the bunch in order to help avoid bruising the tops of the flowers. Label all bunches by grade and variety. After wrapping, stems are usually tied together with rubber bands, string, or twist ties.

If you dry-store stems, store them at 33–35°F (1–2°C). Then, recut stems under clean water each time their bunches are out of water in order to eliminate air embolisms. Rehydrate each time stems are cut (i.e., after harvest, arrival at wholesaler, arrival at retailer, etc.) using a commercial postharvest formulation for roses.

The postharvest behavior of roses is distinguished from that of other flowers mainly by their low sensitivity to ethylene and the absence of flower bud and leaf abscission. However, both flower bud and leaf abscission may occur on exposure of flowers to elevated levels of ethylene. Roses should not be stored near ripening fruits or vegetables. To ensure ethylene does not build up in coolers, relatively inexpensive activated charcoal air filters are available for flower coolers.

Growers who are producing cut flower roses outdoors and are located close to the final market often deliver stems in water and refrigerated trucks. If selling at roadside markets or through local florists, stress that the flowers are locally produced.

If retailers receive dry-shipped roses, they should unpack boxes, ideally recutting the stems under water using only clean water, and place them in disinfected buckets with a commercial postharvest solution designed for roses. Do not use a solution with more than 2% sugar. Remove leaves and thorns below the water line.

Consumer and florist usage of today's outdoor 'garden-like' commercial cut roses with their casual, old-fashioned appearance, romantic colors and blends, improved vase life, and fragrance continues to expand. The production of roses as an outdoor commercial crop has evolved significantly in the last decade, and their production can yield a good commercial return to the grower.

Forced Garden Roses

The forcing of roses in containers is experiencing a renaissance of sorts. More growers are producing plants that are "patio ready" for consumers who wish to dress up their outdoor living space for the entire growing season with such long-blooming types as floribunda, shrub, and landscape roses. Often, growers are producing these plants in more decorative containers that can be used directly on the patio or terrace. To ensure good sell-through, growers should focus on producing plants with eye-catching sales appeal.

There are two basic approaches to producing container roses. The more traditional approach involves "potting up" or "containerizing" bare-root, field-grown rose stock into 2–3 gal. (8–12 l) nursery containers. Alternatively, more and more roses are grown from start to finish as own-root plants in containers. This method expedites the production process, but is only possible when suitable varieties are grown under the proper conditions.

Over 35 million bare-root roses are shipped annually from California and Arizona. The majority of these are used for potting up the traditional "bud and bloom," retail-ready pot rose plant. Many retailers buy in "bud and bloom" versus doing it themselves at the garden center, because the garden center is better able to use its resources on other endeavors. However, growing roses from dormant, bare-root plants is still very popular in some areas of the country among retailers who use their growing area as a sales tool. Dormant bare roots also allow growers a very quick turnover.

There are several basic classifications of plants that are commonly used for finishing garden roses in containers.

1. Two-year-old, budded rose plants. These are the most commonly and widely used plants. 'Dr. Huey' is the most popular understock in the United States. Worldwide there are many popular understocks, each with its own vigor, winter hardiness, ease of production and storage, ability to ship, and disease resistance to be considered.

2. One-year, own-root, field-grown plants. This method of production is increasing in popularity at a brisk pace. Only varieties that have been shown to grow well as own-root plants should be considered, as only a small percentage of all rose varieties perform well in this manner. Initially, the list consisted principally of landscape and miniature types. The category has broadened as certain rose breeders have focused on developing varieties suitable for growing as own-root plants.

3. "Standard" tree or patio tree roses. These vary in age from one to three years old and can be from 18–60" (0.5–1.5 m) in length.

4. Rooted rose liners (dormant or actively growing). Although they are also gaining in popularity, this type of plant is used mostly for smaller container finishing and does best in more temperate climates.

The large number of rose types can be confusing to a grower just starting a container rose program. As with any new product, perform market research with customers and suppliers to determine which type is best suited for your market. Most growers produce a mix of hybrid tea, grandiflora, floribunda, miniature, and landscape roses (hedge types or groundcovers), with some patio and landscape trees. Your geographical location also plays a big role, as northern growers will emphasize winter-hardy, own-root cultivars, while growers in moderate climates will have more choices.

The quality, size, and price of bare-root material vary quite widely. Be sure to buy your plants from a reputable supplier of virus-indexed material from certified sources such as the University of California-Davis. This is becoming more and more critical as more state agricultural departments are inspecting roses for the presence of viruses. Virus-free roses also grow more vigorously and give a better overall performance for you and your customers. Most growers use USDA No. 1–grade plants. Grade No. 1.5 can offer a more affordable option, although plants produce a smaller finished product.

Growing container roses from liners is very similar to growing pot roses from liners, so it will not be discussed further here, except to say that it will take longer to reach salable size from a rooted liner when compared to a bare-root plant.

Handling Bare-Root Roses

Most plants grown in North America are produced in California and Arizona, which together account for about 85% of the total production of bare-root roses. Other small production areas exist on the West Coast and in Texas and Canada.

Harvest begins as early as October and is usually finished by mid-February, depending on the weather and the location. Roses are shipped from mid-December to the mildest areas (Southern California, Texas, Arizona, and Florida) until late March or even early April to the upper Midwest.

Because the time between the first harvest and last shipment is so long, storage of dormant plants is critical. Unless you have the proper refrigeration facilities, do not time your roses to arrive any more than a week to ten days prior to planting. Dormant roses are very sensitive to ethylene, which may cause delayed start, malformation of the breaks, or even plant death.

If you are storing bare-root roses, always use a refrigeration unit with good ventilation and never store fruits or vegetables with roses. Sealed boxes are no protection against ethylene, since the gas can easily penetrate boxes and plastic. The optimum storage temperature is 34°F (1°C).

As soon as you receive your bare-root shipment, open the boxes and examine plants for possible damage in transit. Bare-root roses will tolerate freezing to some extent without detrimental effects if they are thawed out slowly. If plants arrive frozen, close the boxes and place them in a cool, dark location or in a cooler at 35–40°F (2–4°C) to thaw out slowly. Do not water them at this time nor try to thaw them quickly in a warm room. This could result in severe damage and possibly the loss of the plants.

If plants are to be grown in a warm environment (65°F [18°C] or higher), a common method called "sweating" can be used to ensure a better start. To sweat your plants, open the box and the plastic bag containing the roses. Spray plants liberally with plain water and then close the plastic liner as tightly as possible before closing the box again. Keep boxes at ambient temperatures of 60–70°F (15–21°C). The combination of high humidity and warm temperatures will help the plants begin to break. Check the boxes daily until the eyes have forced shoots that are 0.125–0.25" (3–6 mm) long and some white roots begin to show. This process can take as little as three days and usually no longer than one week.

When the plants are to be grown in an unheated growing environment (40°F [5°C]) in cold frames or outside, it is best not to use the sweating procedure. For these plantings, remove plants from the box and soak the roots overnight to rehydrate them prior to planting. Once potted, place containers directly in the growing environment. This enables the plants to initiate roots prior to the bud breaking and ensures a more balanced growth afterwards.

For container roses that will be grown outside or in a cold frame, you may need to protect plants from frost. Sprinkling with overhead water in mild climates or covering pots with a plastic blanket in colder climates will control any damage, as long as the temperature does not drop below the mid-20s (–5– –2°C). If temperatures drop below 25°F (–4°C) for any length of time, you may need to apply a minimum amount of heat.

Growing On

You can choose from a wide range of container sizes for your crop. The most popular are 2- and 3-gal. (8 and 12 l) containers. Miniature roses, own-root liners, and one-year, own-root, field-grown roses can be grown in 1-gal. (4 l) containers. However, growing in small pots may require excessive root trimming and very frequent watering as the plants grow. Most tree roses are grown in 5-gal. (20 l) or larger pots.

Roses may be produced in a wide range of media. The mix should be well drained and high in organic matter (optimally around 80%). Provide a starter charge with calcium, phosphorus, and micronutrients mixed into the medium. The optimum pH for growing roses is 6.0–6.5.

Prior to planting, prune the canes to 6" (15 cm) above the bud union for two-year roses or 4.5–5" (11–13 cm) for one-year plants. Such pruning encourages uniform breaks and a better finished plant. Pruning also reduces the chance of desiccation. Remove any weak or damaged canes at the same time as well. Always use sharp, clean pruning shears to avoid cane dieback. If you use the sweating procedure, the canes should be cut after the plants are in the container. Root pruning is not recommended, unless the roots are too big to fit in the container.

It is extremely important to keep the plants from drying out during the entire planting operation. Do not leave roses for very long without water between planting and spacing.

When planting roses, the bud union (graft area) should be at or slightly above the soil line. Firming the soil around the roots of the plant during potting is important. If the soil is not firmly tamped around the plant roots, air pockets may dry out the roots. Fill the pot to within 1" (3 cm) of the top. It is a good idea to stake tree roses to keep them straight in the containers.

The plants are now ready to place in the growing area. The growing area will vary depending on your geographical location, the size of your nursery or greenhouse, and whether the roses are grown for direct retail sales or for shipping by a wholesale grower. However, since most container roses are grown in nurseries using cold frames (hoop houses), most of the information given here will relate to that particular situation.

In most nurseries, roses are spaced directly at the final spacing. Starting plants off with pot-to-pot spacing can save space initially and is especially useful if roses are started in a heated/warm area with space limitations, but it is usually not very cost effective. Furthermore, spacing too late may result in significant breakage. This is why pot to pot is mostly used by retail growers or by growers who do a lot of shipping when the plants are still in the early stages of development.

The best spacing will allow for the plants to have 150% of the space occupied by the container. For example a 2-gal. (8 l) container with a 12" (30 cm) diameter will need 18" (46 cm) spacing.

Once spaced, water pots thoroughly, until the soil mix is totally saturated. A drench with an all-purpose fungicide will prevent soil diseases and dieback on the fresh cut canes.

In this interesting new system being used in Europe, pots sit atop a mat consisting of three layers, and small sprinklers are used for frost control. The system grows very nice plants, and all the water and fertilizers are recycled.

It is absolutely critical to maintain very high humidity during the first two to three weeks of growing. Failure to do so will result in poor bud break, drying out of the canes, and possibly plant death.

There are different ways to keep the humidity high. In warm climates where roses are grown outdoors, this can be achieved by using overhead sprinklers. In hoop houses, growers usually cover the plants with clear polyethylene film. The film should be pulled off when the uppermost shoots come in contact with it, when they are about 1" (3 cm) long. The poly blanket also helps protect from deep frost. When grown outdoors, newly potted roses should also be protected from wind and sun until plants are rooted in. Most growers in this situation—mostly in California and other areas of the Sunbelt—use wet burlap to cover the canes instead of plastic to avoid sunburn.

Begin fertilization when shoots are 1.0–1.5" (3–4 cm) long, usually two to three weeks after potting. The first leaves will be fully expanded.

In a greenhouse, with sweated plants, the plants should be started at 50°F (10°C) nights. Fertilizer applications can start one week after potting, and temperatures can be raised to 60–65°F (16–18°C) nights and 75°F (24°C) days for finishing. A slow-release fertilizer such as 14-14-14 or 18-9-9 can be used, although most large growers are using liquid feed at 150 ppm from 20-10-20. If you choose constant liquid feed, use clear water only at every third irrigation to avoid salts buildup.

Drench with iron chelate to avoid iron deficiency about five to six weeks into the growing period. Make sure to do it on a cloudy day to avoid leaf and tip burn. Add macronutrients to the fertilizer program during the third or fourth week of production.

Throughout production, be sure to keep plants moist at all times. Drying out at any time will result in a poor finished product and severe leaf drop.

Roses grown in greenhouses as opposed to cold frames are softer and more succulent and, therefore, will need to be gradually acclimated to outdoor conditions at finishing.

Crop scheduling and the timing of finished plants vary drastically with geographic location and environmental conditions. Most container rose sales occur from early to late spring, with a large percentage being finished for Mother's Day. The following guidelines are given as examples only. These are for No. 1–grade plants without pinching:

- Outdoor production in Southern California for early spring sales: Production time is nine weeks. Pot in mid-December and ship late February.
- Outdoor production in Northern California for Mother's Day sales: Production time is twelve weeks. Pot in late January.
- Cold frame production in the Northeast for Mother's Day sales: Production time is ten to eleven weeks. Pot in early February and ship from late April to early May.
- Outdoor production in Texas for Mother's Day sales: Production time is eight to ten weeks.
- Greenhouse production in the Midwest for Easter sales: Production time is ten weeks.
- Greenhouse production in the Northwest for Mother's Day: Production time is seven to eight weeks.

In general, smaller plants such as grade No. 0.5 plants will usually require a pinch, adding three to four weeks to the total crop time, unless they are grown in very mild climates. Also, miniatures will usually finish faster, followed by floribunda and hedge roses, and then hybrid teas and grandifloras. Climbers and groundcover/landscape roses are typically the last to flower, but there are differences between cultivars.

Pests & Diseases

Downy mildew is probably the most common disease encountered when growing in cold frames, where controlling moisture on the leaves at night is not always easy. Use preventative sprays to control it before you develop a problem. Botrytis—because of poor air circulation, cool temperatures, and high levels of moisture on the plants—is the other main disease encountered in cold-frame production. Begin preventative sprays when plants leaf out, after three to four weeks in the cold frame or two to three weeks in the greenhouse. Black spot and rust, the other major diseases encountered with roses, are usually not a problem in production but can become an issue if the roses stay too long on the retail lot.

Aphids and spider mites are the most common insect and mite pests, although whiteflies and thrips can be problems as well, especially late in the season when temperatures are warmer. Borers, midges, caterpillars, and beetles can also be encountered, especially at the retail level.

Troubleshooting

Failure to break and leaf out is the most common problem with container roses. The problem is usually due to desiccation or drying out of the canes and/or the roots during the planting process or during the initial weeks after planting. Be sure to maintain a high enough humidity environment before the canes leaf out.

Postharvest

Plants may be sold as soon as they are fully leafed out when being shipped or in leaf or flowering for direct retail sales.

Fully leafed out, "bud and bloom" container roses can be prone to breakage during shipping. This is why growers use large paper sleeves and ship on decked trucks or on rolls, not stacked. Temperatures in the truck should never be allowed to go over 85°F (29°C) because long exposure to high temperatures and darkness will result in severe leaf drop due to ethylene buildup. It is better to use refrigerated trucks for late-season shipments.

Because most consumers like to know what they are buying, it is always helpful to have a few plants showing color as well as waterproof, 5" × 7" (13 × 18 cm), color picture cards at the point of sale to show what the green plants will look like in flower. Instruct consumers that they must keep the root-ball intact when transplanting.

Rosmarinus

Rosmarinus officinalis
Common name: Rosemary
Herb, tender perennial (Hardy to USDA Zones 6–9)

No herb garden is complete without at least one rosemary plant. In many parts of the country, rosemary is perennial (hardy to USDA Zone 7). Dedicated herb gardeners adore the soothing scent of rosemary. Rosemary is a symbol of friendship and loyalty, and its fresh branches have been used through the ages in ceremonies such as weddings and funerals. Today, rosemary—both fresh and dried—finds its way into all sorts of culinary dishes as well as potpourri and aromatherapy treatments. Some growers have made quite a name for themselves growing rosemary topiaries and standards offered as attractive table decorations for the fall and Christmas seasons.

Propagation

While rosemary may be germinated and grown from seed, the plants vary in their amount of essential oils. Vegetatively propagated plants are higher in these oils, which are valued for their culinary uses among other things. In addition, seed germination is notoriously low. Thus, most rosemary is vegetatively propagated by cuttings, and most growers buy in liners or take their own cuttings.

Seed

If you choose to propagate by seed, sow seed in a well-drained, sterile, germination medium with a low starter charge and a pH of 6.0–6.5. Do not cover. Germinate seed at 70°F (21°C). Seedlings will emerge in two to three weeks. Some experts recommend freezing seed and/or alternating day and night temperatures at 70° and 55°F (21° and 12°C) to increase germination rates. In addition to low germination rates, plants grown from seed are slow and erratic, taking a year and a half to two years until plants are large enough from which to harvest cuttings.

Cuttings

Propagate rosemary by harvesting 2–4" (5–10 cm) cuttings from new growth. Strip leaves away from the base of the cuttings and dip in a rooting hormone prior to sticking. Root under light mist for six to ten days under shade at 70–75°F (21–24°C). It takes from three to six weeks to root and another two to three weeks to bulk up the liners. Rosemary cuttings can be harvested year-round; however, rooting is poor through the hottest summer months, especially under high humidity. Because rosemary is highly susceptible to root rot in warm, humid conditions, transplant cuttings as soon as they are ready.

Plants may be layered by placing lower limbs on the ground and covering them with sand or medium. Roots will form at points where stems touch the ground, at this point you can cut the new plant away.

Growing On

Pot cuttings into a well-drained medium with a low starter charge and a pH of 5.8–6.2. Using one to two plants per 4–5" (10–13 cm) pot, it will take from six to eight weeks to be ready for spring sales. A 1-qt. (1.1 l) pot with two to three plants will take from seven to nine weeks. Grow at 67–75°F (19–24°C) days and 62–68°F (17–20°C) nights.

Irrigation is critical to crop success. Do not allow plants to become waterlogged at any time during production. Strive to find the balance between keeping plants consistently moist, yet not overwatered. Fertilize with 150–200 ppm from a complete feed at every other irrigation.

Rosemary responds well to shearing and pinching, which increase lateral breaks but also increase crop time. In general, pinch only once or twice, which will be enough for container sales.

Varieties

'Barbeque', an evergreen upright, grows to 4' (1.2 m) tall. Edible flowers appear from April to September. 'Tuscan Blue', perhaps the most well known of all rosemary, has deep blue-green leaves and dark blue flowers and grows up to 36" (91 cm). 'Albus' has white flowers. 'Arp', with a more open plant habit, has the reputation of being the most cold-hardy variety. 'Prostratus' is low growing, making an excellent plant in hanging baskets or combination pots.

Postharvest

In areas where rosemary is hardy, plant it in a well-drained, sunny location. Other gardeners will find rosemary a great container plant that can be brought indoors for overwintering and enjoyed every year on the deck or patio. Instruct consumers to prune plants after they have finished flowering to encourage branching.

Rudbeckia

Rudbeckia fulgida
Common names: Black-eyed Susan; rudbeckia
Perennial (Hardy to USDA Zones 3–9)

Rudbeckia fulgida is a premier, mid- to late-summer-flowering perennial. It signals summer's official arrival since it is a long-day plant, requiring more than fourteen hours of day length in order to flower. Single, 3" (7 cm), scentless, golden-yellow flowers highlighted with a dark, contrasting center dot its crown. Plants grow upright with each stem terminating in a flower. The dark green foliage is uniform and reaches 2.5–3.5' (75–105 cm) tall. *R. fulgida* spreads 15–18" (38–46 cm) across. Plants flower from July to September.

Propagation

Seed is the common method of propagation but also can serve as the primary source of problems for the propagator. Raw seed germinated at 70–72°F (21–

22°C) will seldom achieve high germination rates. In addition, seedling emergence will occur over a period of weeks instead of at one time.

A key to faster germination and more uniform seedling stands is to increase the media temperatures upon sowing. Sow two to three seeds per 288-plug tray, leave the seed exposed to light or lightly cover with coarse vermiculite, and then place in a plug chamber at 82–85°F (28–29°C) days and 72–75°F (22–24°C) nights. Some growers have reported excellent results at germination temperatures of 86–88°F (30–31°C) with germination results in five to seven days but with higher overall emergence than at cooler media temperatures.

If germination still appears to be slow, chill seeds at 40°F (5°C) for two to four weeks. In addition, enhanced seed is available from a number of seed companies. These offer higher germination rates and are especially recommended for commercial plug propagators.

Stage 1 of plug production takes from five to nine days. Maintain uniformly medium substrate moisture levels and a pH of 5.5–5.8 with an EC below 1.0 mS/cm. Rudbeckia is very sensitive to high salts, especially ammonium and sodium, so keep ammonium below 10 ppm and sodium below 40 ppm.

Stage 2, stem and cotyledon emergence, takes from ten to fourteen days at 70–72°F (21–22°C) substrate temperatures. Provide light levels of 500–1,000 f.c. (5.4–11 klux) and begin feeding at 50–75 ppm 14-0-14 once a week.

Stage 3, development of true leaves, will last from fourteen to twenty-eight days for 288 and 128 plugs. Reduce substrate temperatures to 65–70°F (18–21°C) and increase light levels to 1,000–2,000 f.c. (11–22 klux). Increase fertilizer to 100–150 ppm from 14-0-14 once a week. Growth regulator treatments for height control may begin during this time. A-Rest, B-Nine, and Bonzi are all effective.

Stage 4, hardening, takes seven days. Lower temperatures to 60–65°F (16–18°C) and reduce moisture levels, allowing soil to dry thoroughly between irrigations.

A 288-plug tray will take from seven to eight weeks from sowing to transplant to larger containers, while it will take a 128-liner tray takes from nine to ten weeks from sowing to transplant.

Growing On

Sowings made in June and July can be transplanted to larger containers later on in the summer. Transplant one 128 liner per 1-qt. or 1-gal. (1.1 or 4 l) container or one 288 per 32-cell pack.

A 128 liner transplanted to a 1-qt. or 1-gal. (1.1 or 4 l) pot in August or early September is overwintered in a cold frame or greenhouse for green plant sales the following spring. Plants will flower in July or August.

A 288 plug transplanted in August to a 32-cell pack can be transplanted to a 1-qt. or 1-gal. (1.1 or 4 l) pot in September or October (allow five to seven weeks) are grown cool until the roots are established in the final container. Plants can then be grown cold for the remainder of the winter. Plants are sold green in the spring and will flower in the garden in July or August.

For sowings made in December or early January, grow plants warm to establish and then cool to finish off for green plant sales in the spring in 1-qt. (1.1 l) containers. Plants will flower in summer but not profusely. The best flowering is made from sowings the previous summer.

Rudbeckia plants must attain a minimum of ten leaf nodes before flower induction begins. Bulk plants up to meet this maturity requirement before cool treatment and daylength treatments begin. Grow on at 65–70°F (18–21°C) days and 60–65°F (16–18°C) nights and maintain relatively dry growing conditions. Fertilize at every other irrigation with 14-0-14 at 100–150 ppm. Maintain a soil EC of 1.0 mS/cm. During this stage, plants should form a densely packed rosette of foliage. In general, growers starting with field-grown plants will be able to begin cool treatment and then be followed by photoperiod treatments once plants have rooted in.

While cold treatment is not mandatory for flowering, plants that have received cool treatment flower three weeks sooner than nonvernalized plants. Growers potting plants in the fall can provide ten to twelve weeks at 41°F (5°C) in a cold frame or cooler to meet cooling requirements.

If flowering plants are required for spring, plants must receive long-day treatments with supplemental light (400–500 f.c. [4.3–5.4 klux]) for fourteen hours and/or four hours of night interruption lighting from 10 P.M.–2 A.M. so that plants perceive fourteen hours or more of day length.

To force plants for spring flowering, move pots into the greenhouse from the cooler or cold frame and begin long-day treatments immediately. Provide constant day/night temperatures at 68–70°F (20–21°C) and provide high light (3,000–5,000 f.c. [32–54 klux]). During the winter months, supplemental lighting at 400–500 f.c. (4.3–5.4 klux) is beneficial to improving plant quality. Fertilize with 15-0-15 at 150–200 ppm at every other irrigation and maintain a pH of 5.5–6.2 and an EC of 1.0 mS/cm. A-Rest, B-Nine, Bonzi, and Cycocel have a slight effect on height, based on Michigan State University research.

Pests & Diseases

Aphids and powdery mildew may be problems during production or in the garden.

Varieties

'Goldsturm' is the most widely grown perennial rudbeckia. It should be. It was the winner of the Perennial Plant Association's Perennial Plant of the Year in 1999 and is a stellar performer. It can be propagated vegetatively or from seed. 'Goldsturm' flowers from July to September. 'Goldstrum' is a commonly misspelled form of 'Goldsturm'.

Rudbeckia hirta
Common names: Gloriosa daisy; black-eyed Susan
Annual

This species is hardy to USDA Zone 4, but the seed strains, especially any of the dwarf varieties, are best treated as short-lived perennials or annuals. They may live from one year to the next, but they have to be in a protected location and under cover. The crop can reseed, however.

Rudbeckia hirta is a native plant, growing up to 2.5–3' (75–90 cm) tall in flower. While the wild forms are widely seen growing across the Great Plains eastward, the seed strains are commonly grown in gardens today. These range in height from 10–40" (25–102 cm) tall and spread 10–36" (25–91 cm) across. The single and semi-double flowers are shades of yellow or orange and measure 3–5" (7–13 cm) across, depending on the variety. Notorious for their sensitivity to powdery mildew, gloriosa daisies often succumb to the disease, and many die by the end of summer. Plants flower from June until frost.

Propagation

Sow one seed to a 288-plug tray, cover lightly with vermiculite, and germinate at 70–72°F (21–22°C). The root radicle emerges in three to five days. Move plants to Stage 2 and maintain substrate temperatures at 70–72°F (21–22°C). Stage 2 takes from five to seven days.

Move to the greenhouse bench for Stage 3 and grow on at 65–70°F (18–21°C) days with nights 5°F (3°C) cooler. Maintain relatively dry growing conditions. Fertilize at every other irrigation with 14-0-14 at 100–150 ppm. Maintain a soil EC of 1.0 mS/cm. During this stage, plants should form a densely packed rosette of foliage. Stage 3 takes from seven to fourteen days. Stage 4 takes another seven days and is best at 60–65°F (16–18°C) media temperatures.

Depending on variety, a 288-plug tray takes from five to seven weeks from sowing to transplant to larger containers.

Growing On

Plants flower beginning in June through the summer. Like its relative, *R. fulgida, Rudbeckia hirta* requires long days to initiate flowering. January sowings flower in June, making annual rudbeckia a natural addition to just about every bedding plant program. Plugs are widely available, which is how most growers start.

Transplant plugs into a well-drained medium with a pH of 5.5–6.2 and a moderate initial nutrient charge. Plants are salable green in an 18-cell pack in six to seven weeks and in a 4" (10 cm) pot in eight to nine weeks. Although plants are sold green, they will flower in the garden several weeks later.

Pests & Diseases

Powdery mildew, rust, and aphids can affect annual rudbeckia. To control disease, do not overwater and maintain moderate humidity levels.

Varieties

'Becky' and 'Toto' are the two shorter-growing selections. Both are available in separate colors as well as a mix, and both are very dwarf, growing only to 8–10" (20–25 cm) tall. They make excellent impulse items in 4" (10 cm) pots.

Later-flowering, tall-growing varieties include the spectacular 'Indian Summer', 'Irish Eyes', the All-America Selections Award-winning 'Prairie Sun', and 'Rustic Mixture'. 'Tiger Eye', the first F_1 hybrid in the group, is an excellent performer. These bold plants can be sold green in packs in the spring or flowering in 1-gal. (4 l) pots in the summer.

S

Saintpaulia

Saintpaulia ionantha
Common name: African violet
Flowering pot plant

In the days when windowsills were large and every kitchen sink had a window, African violets were a favorite of housewives and budding horticulturists. Treated properly in the home environment, they are a showy, long-lasting pot plant. For growers, they are a bit trickier, being subject to a host of diseases and finicky on irrigation requirements.

Most growers who produce African violets do so on a large scale, using capillary mat or ebb-and-flood bench systems. However, they are a great item for any grower with a solid, market niche for pot plants. Prefinished pots that finish in just a few weeks are readily available. For growers wanting to get into bigger-scale production, liners are plentiful. African violets are easily programmed for year-round production and are a mainstay in dish gardens.

Propagation

Stick leaf petiole cuttings at 0.5–2.0" (1–5 cm) long into a medium that is 80% peat and 20% perlite. Be sure leaves do not touch. Root cuttings at a substrate temperature of 70–72°F (21–22°C) and air temperatures of 75–80°F (24–26°C) days and 70–72°F (21–22°C) nights. Provide low light levels of 100–300 f.c. (1.1–3.2 klux) during rooting. Root only mature yet juvenile leaves that are solid green and firm from non-flowering plants. Maintain moderate moisture levels in the medium. Use heated 70°F (21°C) water in mist lines, since cold water will lower substrate temperatures and discolor leaves. Cover rooting trays with a plastic tent to maintain humidity levels. Plantlets will form in twelve to sixteen weeks. Plantlets can be harvested up to three times, taking the largest plantlets each time. Uniformity of plantlets is important to the quality of the finished crop.

Transplant plantlets to 1" (3 cm) liner cells and reduce the medium's temperature to 68–72°F (20–22°C). Do not plant plantlets too deeply, as this will result in multiple crowned plants with a lower market value. As root initials begin to form, start drying down the medium. Avoid lowering the humidity, as this will increase evapotranspiration and thus reduce the root zone temperature. Increase light to 500–900 f.c. (5.4–9.7 klux) as cuttings root out. Apply a foliar feed at 75 ppm from 14-0-14. Keep the substrate pH at 5.8–6.0 and the EC below 0.5 mS/cm. Grow plants into 1" (3 cm) liners for seven weeks.

Harden plants for shipping or planting for one week. Maintain growing temperatures and increase light to 900–1,200 f.c. (9.7–13 klux). Fertilize with 100 ppm 14-0-14 once a week.

Plants root in twelve to sixteen weeks.

Growing On

Irrigate liners well before potting if they are dry. Pot liners in a well-drained, sterile, peat-based mix at the same depth they were growing in propagation. Again, be careful not to plant liners too deeply. Press the liner firmly into the pot to encourage a tighter plant structure. The medium should have a low or no starter charge and a pH of 5.8–6.0. Space pots in a pot-tight, staggered format after potting. Cover pots with Reemay or a light-gauge fabric for the first week to help liners root in faster.

Irrigation is critical. Do not allow plants to wilt. Maintain moist medium at all times; however, the medium must not be wet. Plants that dry out lose roots. While plants are spaced pot tight, overhead irrigation may be used. If using overhead irrigation, use tempered water that it is the same temperature as the leaves, otherwise leaves will become spotted. Just a 5°F (3°C) difference in leaf and water temperature is all that's required to cause damage to the leaves. Some growers like to water violets from overhead as long as possible to avoid the soluble salts buildup that can occur with subirrigation. As plants mature and flower once they are spaced, do not irrigate from overhead. Because irrigation is so important, most growers produce violets on capillary mats or ebb-and-flood benches. These make it easy to ensure that the medium has the right amount of moisture and that the foliage stays dry later in the production cycle, thus avoiding spotted leaves.

Provide a consistent 900–1,200 f.c. (9.7–13 klux) of light. As plants begin to flower and mature, they can handle higher light levels, up to 1,300 f.c. (14 klux). In the low light of winter, violets will take longer to finish. Four to six hours of supplemental light (300–500 f.c. [3.2–5.4 klux]) is beneficial during winter months. During high-light periods, shade is necessary. A two-layer shading system is best, because it allows continual light adjustment as the crop matures. Holtkamp, a leading African violet breeder, says that the light level is perfect if during full sun at noon you can barely see the shade of your extended hand as you hold it out over the plants.

Grow on at temperatures of 75–80°F (24–26°C) days and 70–72°F (21–22°C) nights. Night temperatures below 70°F (21°C) will delay flowering; production temperatures above 85°F (29°C) are detrimental to flowering and foliage color.

African violets are very sensitive to high soluble salts levels. A weekly feed with 100 ppm of 15–10–15 is all that is needed. Do not apply fertilizer from the bottom, as nutrients will not be equally distributed throughout the pot. If soluble salts do build up, apply a top irrigation every third watering to leach salts. Maintain an EC of 0.5 mS/cm. Do not use a slow-release fertilizer with African violets.

Space pots as leaves begin to touch, about five to six weeks after potting. Space to a density of about 6 pots/ft.^{2} (65 pots/m^2). Do not provide too much space at this stage, as it will encourage leaves to become hard and grow downward. Space pots again two weeks later (week 8 after potting) to about 4 plants/ft.^{2} (43 plants/m^2). Flowering 4" (10 cm) pots can be finished in eight to ten weeks from a liner and in one to three weeks from a prefinished liner. Flowering 6" (15 cm) pots (2–3 plants/pot) take from ten to twelve weeks. Six-inch (15 cm) pots may also be produced by repotting 4" (10 cm) pots eight weeks after potting. At repotting, remove any flower buds and space one 6" (15 cm) pot per ft.^{2} (11/m^2).

Pests & Diseases

Cyclamen mites can become a problem on foliage, causing gnarled, twisted leaves. Botrytis, crown root rot, petiole rot, and powdery mildew can also become problems.

Troubleshooting

Chlorotic or bleached leaves may be due to high light intensities, high leaf temperatures, or a lack of fertilizer. Overexposure to fluorescent light can cause tight centers with small, bunched leaves, so increase the distance between the light and the plants. Chilling can also cause this condition; maintain 70°F (21°C) minimum night temperatures. Overfertilization or standing water on crowns can lead to brown centers or excessively hairy leaves. Cold water on foliage from overhead irrigation can result in brown spots or rings on leaves. Lack of magnesium in the medium, excess sulfur, or excess fertilizer can cause brittle leaves. Plants that bud but do not bloom may be due to a pest infestation, low humidity, or swift changes in temperature. Low light, low temperatures, high moisture, or high nutrients can result in solid colors on bicolor varieties.

Postharvest

Plants are ready for sale when five or more flowers are open. Most growers use special paper sleeves that are wrapped around pots to protect flowers during shipping. *Note:* Leaves break easily when sleeving or unsleeving traditional sleeves.

African violets are sensitive to low temperatures during shipping: Do not expose plants to temperatures below 50°F (10°C). Unpack plants immediately upon arrival, if they are boxed. African violets are very sensitive to ethylene, so do not store plants with ripening fruits or vegetables.

Violets respond very well to fluorescent lights in retail displays, which will encourage continuation of flowering. Retailers should provide constant moisture, watering from the base. Avoid overhead irrigation, as using water that has a temperature colder than the leaf temperature permanently spots leaves. To keep plants fresh, pinch old flowers or flower stalks.

Salvia

Salvia farinacea
Common name: Mealycup sage
Annual

Salvia farinacea is one of those unassuming flowers that provides an unrivaled performance, upon which landscapers especially have come to rely. With lavender-blue or white flowers atop silvery, gray-green foliage, plants make a classy, understated statement in beds and simply refuse to go out of flower all summer long, right up to the hardest frost. It has an old-fashioned appeal that is perfect for today's home garden look. Plants do great in the heat and in dry locations. In the South, it can be a tender perennial.

Propagation

Sow seed onto a well-drained, disease-free medium with a pH of 5.5–5.8. Cover lightly with coarse vermiculite to maintain 100% humidity at the seed coat. Maintain wet germination conditions: The medium should be fully saturated. Germinate at soil temperatures of 75–78°F (24–26°C). Keep ammonium levels below 10 ppm. Radicles should emerge in five to seven days.

Move trays to Stage 2 for stem and cotyledon emergence, which will take about seven days. Reduce soil temperatures to 72–75°F (22–24°C) and begin to allow trays to dry down somewhat between irrigations. As soon as cotyledons expand, begin fertilizing with 50–75 ppm from 14-0-14. Maintain light levels at 1,000–1,500 f.c. (11–16 klux).

Move trays to Stage 3 to develop true leaves. Drop soil temperatures to 68–70°F (20–21°C), and begin to allow soil to dry thoroughly between irrigations. Fertilize weekly with 100–150 ppm from 20-10-20, alternating with 14-0-14. Increase light levels to 1,500–2,500 f.c. (16–27 klux). If growth regulators are required, A-Rest, B-Nine, and Bonzi can be used at this time.

Harden plugs for seven days prior to sale or transplanting by increasing light levels to 2,500–3,000 f.c. (27–32 klux) and reducing soil temperatures to 65–68°F (18–20°C). Feed weekly with 100–150 ppm from 14-0-14.

Most growers buy in a few plug trays of *S. farinacea* with their normal bedding plant shipments rather than growing the plugs themselves. Transplant salvia plugs as soon as they arrive. If plugs are held past the transplantable stage, plant growth will be severely checked and root growth will quickly decline.

A 288 plug is ready to transplant to a larger container five to seven weeks after sowing, depending on variety.

Growing On

S. farinacea plants grow pretty tall, so most growers pot them into 4" (10 cm) or 1-gal. (4 l) pots rather than growing them in packs (although jumbo packs can be used). If you prefer to grow flats, use 1801s and plan to sell plants green. Plant one plug per cell or pot.

Select a disease-free, well-drained medium with a moderate initial starter charge and a pH of 5.5–6.2. Grow on at 60–65°F (16–18°C) days and 55–60°F (13–16°C) nights. Fertilize at every other irrigation with 150–200 ppm from 20-10-20. Be especially careful of soluble salts buildup, especially early in the production cycle. Discontinue feed one to two weeks prior to sale to increase shelf life.

If growth regulators are needed, A-Rest, B-Nine, Bonzi, and Cycocel are effective.

After transplanting one 288 plug per 4" (10 cm) pot, plants are salable in seven to nine weeks. Plants may be budded but are mostly green with limited or no flower color. Plants will flower in the garden in about three to four weeks.

Pests & Diseases

Aphids, leafminers, spider mites, thrips, and whiteflies can attack plants. Disease problems can include botrytis and leaf spot.

Varieties

For *S. farinacea,* try the 'Victoria' series in 'Blue' or 'White' for plants growing to 18–20" (46–51 cm) in the garden. For more dwarf plants that grow 12–14" (30–36 cm), try 'Rhea' with mid-blue spikes or the All-America Selections Award winner 'Strata' (blue and white bicolor spikes). 'Evolution Deep Blue' is another All-America Selections Award winner. It has deep blue flowers and grows to 18" (45 cm).

Postharvest

Salvia is highly sensitive to ethylene gas. Do not expose plants to ripening fruits and vegetables or even automobile and truck exhaust. Keep plants groomed by removing old flowers and dead leaves to remove this source of ethylene.

Retailers should display salvia under partial shade and cool temperatures (70–75°F [21–24°C]). Consumers should plant salvia in partial shade in the South and keep plants deadheaded in the garden to encourage continuous flowering. Salvias will often put on their best flower show in the fall, so instruct your landscape customers not to pull plants out early.

Salvia nemorosa (formerly known as *S. superba*)
Common names: Perennial salvia; perennial sage
Perennial (Hardy to USDA Zones 4–8)

Bushy growth, excellent basal branching, and a superb garden performance highlight this tried-and-true perennial. Plants grow from 15" to over 2' (38–60 cm) tall and spread from 15–18" (38–46 cm) across, depending on the cultivar. The flowers have violet-blue to white to sometimes rose spikes from 6–10" (15–25 cm) long. Tubular in shape, they are small, single, and 0.125–0.25" (3–6 mm) across. As a class, salvias are easy to propagate and grow. They provide reliable performance year after year. Plants flower in May or June and, if cut back, will reflower once more during the summer.

Propagation

Plants are commonly propagated by seed, stem cuttings, or bare-root transplants.

Seed

While perennial salvia can be grown from seed, the crop is more variable in habit and flowering than cultivars propagated from cuttings or bare roots.

Sow one or two seeds per 288-plug tray onto a well-drained, disease-free medium with a pH of 5.5–5.8. Maintain uniform moisture: The medium should not be saturated. Germinate at soil temperatures of 68–72°F (20–22°C). Keep ammonium levels below 10 ppm. Light levels should be at 500–1,000 f.c. (5.4–11 klux). Radicles should emerge in two to three days.

Move trays to Stage 2 for stem and cotyledon emergence, which will take from two to five days. Maintain substrate temperatures at 68–72°F (20–22°C) and keep trays uniformly moist. As soon as cotyledons have expanded, begin fertilizing with 50–75 ppm from 14-0-14. Maintain light levels at 500–1,000 f.c. (5.4–11 klux).

Move trays to Stage 3 to develop true leaves. Drop substrate temperatures to 65–68°F (18–20°C), and begin to allow medium to dry slightly between irrigations. Fertilize weekly with 100–150 ppm from 14-0-14. Increase light levels to 1,000–2,000 f.c. (11–22 klux). Stage 3 will take from twenty-one to twenty-eight days for 288s and from twenty-eight to thirty-five days for 128s.

Harden plugs for seven days prior to sale or transplanting by increasing light levels to 1,500–2,000 f.c. (16–22 klux) and reducing substrate temperatures to 60–65°F (13–16°C). Feed weekly with 100–150 ppm from 14-0-14.

A 288-plug tray takes from five to seven weeks to be ready to transplant to larger containers. While many varieties are listed as first-year flowering from seed, the best-performing containers and plants come from sowings done in summer for plugs transplanted to larger containers for overwintering dormant. Sowings made as late as November, potted to 1-qt. or 1-gal. (1.1 or 4 l) pots and grown cool once rooted in the final pot were salable in flower in May. However, the 1-gal. (4 l) containers were not as fully rooted when done from a November sowing as those overwintered dormant from the summer sowing.

Stem cuttings

Most of the well-known salvia cultivars are vegetatively propagated. Tip cuttings measuring 2–3" (5–8 cm) long can be taken any time the plants are actively growing and stuck into a 32, 50, 72, or 105 tray. A rooting hormone and a preventative fungicide are both recommended. Provide mist for seven to eleven days, although it's imperative that mist be reduced as soon as possible to prevent disease. A pinch is not necessary to promote branching.

An 85 tray takes from three to six weeks to be ready to transplant to larger containers.

Bare roots

Bare-root divisions are available from a number of companies and can be potted up in summer and fall for overwintering dormant or potted up in winter to 1-gal. (4 l) pots to force for the spring. Using one bare root per 1-gal. (4 l) container, pot plants up in February, water in, and then grow on initially at 55–65°F (13–18°C) days until rooting begins.

Growing On

Seed

Bulk plugs up under long days for two to four weeks prior to transplanting. Plant one plug per container. Use 1-gal. (4 l) pots when potting up in summer to fall for containers to overwinter dormant for spring sales the following year.

When transplanting 288 or smaller plugs in mid- to late winter, use one plug per 6" (15 cm) pot for May sales. If larger pots are used, February transplanted plugs will not fill out the container by April or May for planting to the garden. They will be more completely rooted by mid-June or July.

Grow plants at 60–65°F (16–18°C) days and 55°F (13°C) nights. Allow pots to dry thoroughly between irrigations. Provide long days (sixteen hours) with night interruption lighting for early flowering. If growth regulators are needed, B-Nine is moderately effective. Fertilize with 100–150 from 20-10-20 at every other irrigation.

While many varieties are listed as first-year flowering from seed, the best performance is usually from sowings done in summer for plugs transplanted to larger containers for overwintering dormant. Sowings made as late as November, potted to 1-qt. or 1-gal. (1.1 or 4 l) containers, and grown cool once rooted in the final pot are also salable in flower in May. For those sown in January or later, the plants flower but are neither as uniform nor as large as those sown the previous year.

Cuttings

Liner trays are available in various sizes by professional propagators. A 70 cell that has been overwintered cold from cuttings stuck the previous year can be potted up to 1-qt. or 1-gal. (1.1 or 4 l) containers in January or February. Grow on in the greenhouse at 55–65°F 13–18°C) days until the roots begin to establish and then move to a cold frame once well rooted (about three to four weeks). These will be salable green or in flower in the container in spring.

If using smaller liner sizes, potting to a 1-gal. (4 l) pot in February will not fully root to the bottom of the container by spring if desired for May sales. For this reason, 6" (15 cm) or 1-qt. (1.1 l) containers are preferred.

Bare roots

If roots are potted up in winter to 1-gal. (4 l) containers, start out with 55–65°F (13–18°C) days and 55°F (13°C) nights. Once roots have established in the upper half of the container and green shoots are massed above the crown (about three to four weeks later), move the plants to a cold frame or similar environment and grow on at 50°F (10°C) nights with days 5–8°F (3–4°C) warmer.

If the transplants you are using have not been vernalized (kept cold to encourage more uniform flower bud set), you can move containers into a cool treatment once they have rooted into the pot, generally after three to four weeks. Provide ten to fifteen weeks of cool treatment at 41°F (5°C) in either a cooler with 25–50 f.c. (267–538 lux) of light or a cold frame.

Bring pots out of cool treatment and force into flower under long days (more than sixteen hours) with night interruption lighting and constant 65–68°F (18–20°C) day/night temperatures. Fertilize with 150–200 ppm from 15-0-15 at every other irrigation. B-Nine (1,500–2,500 ppm) can be used to achieve some height control. Growing plants cool (below 50°F [10°C]) will increase crop time but will also keep plants compact.

Lights are not necessary unless you want early flowering. Plants will naturally flower in May or June using the cold treatment but not with the added lights. Plants grown under long-day treatment are used for flowering outside of their normal period in mid-spring.

For 'May Night', follow the same recommendations as for *Salvia nemorosa*, but provide fifteen weeks of cool treatment. 'May Night' can be bulked under natural day lengths. When providing long days to force, provide more than fourteen hours.

Pests & Diseases

Botrytis and root and stem rots are the most common among the diseases. Be watchful of prolonged exposure to moisture, which aggravates these conditions.

As for pests, aphids and whiteflies are the most common.

Varieties

The best-known perennial salvia is 'Mainacht' (or 'May Night') with its midnight blue flowers. Plants are vegetatively propagated from tip cuttings or division. This reliable performer is a real winner across multiple climatic zones and a garden staple. 'Oostfriesland' ('East Friesland') is another popular vegetatively propagated variety. Plants grow to 24" (61 cm) tall, and flowering is extended from June to August.

Seed-grown perennial salvia includes 'Blue Queen' (also called 'Stratford Blue'), which may also be vegetatively propagated. 'Blue Queen' is bushy and grows to 24" (61 cm) tall. The 'Select' series is new and comes in 'Blue' and 'Rose'. This variety has been selected for commercial production and to have a long flowering period. 'Select' grows from 12–14" (30–36 cm) tall.

Postharvest

Consumers can cut back plants that have flowered, as some may flower again in the late summer.

Salvia officinalis
Common names: Sage; garden sage; common sage
Perennial, herb (Hardy to USDA Zones 4–8)

Salvia officinalis, or common sage, is one of the most popular culinary herbs, but it has also gained appreciation for its ornamental value in mixed landscape plantings and containers during recent years. Its use as an herb has been recorded in history, and it is traditional in Mediterranean cooking. Its ornamental value stems mostly from its foliage, with varieties ranging in color from gold to purple. Its blue flowers are attractive but are often secondary next to its colorful foliage.

While there are many species of salvia, ranging from annuals to hardy perennials, common sage is hardy from USDA Zones 4–8. Understanding the Mediterranean climate, where summers are hot and dry, helps clarify the ideal growing conditions for common sage.

Growers in milder climates will be able to offer sage twice during the year: once in the spring alongside regular herbs and again in the fall in 'Fall Magic' combination pots and 4" (10 cm) pots as a complement to pansies.

Propagation & Growing On

Seed

Common sage can easily be propagated by seed, although choice selections are more readily propagated by cuttings. For a 288-plug tray, sow one or two seeds per cell and lightly cover with vermiculite. Germinate at 70–72°F (21–22°C), and seedlings will emerge in five to seven days. Allow five to seven weeks for the plug trays to be ready. Transplant one plug per 4" (10 cm) pot and allow another six to eight weeks for plants to be salable when growing at 55–60°F (13–16°C) nights with days 10–15°F (5–8°C) warmer.

Cuttings

For those selections that need to be propagated by cuttings, you can either maintain your own stock plants or purchase in unrooted cuttings. Most standard tray sizes for propagation are between 84 and 128 cells per tray. One cutting per cell is sufficient. Use a standard propagation medium with a pH of 5.5–6.5. Maintain substrate temperatures of at least 70–75°F (21–24°C) for the rooting process, ideally with bottom heat. Keep the medium moist at all times, but not saturated. Be careful not to have excess moisture on the leaves. Due to their hairy foliage, too much mist during rooting will rot the leaves.

Until the plants have developed roots of their own, mist several times a day to avoid wilting. The amount and frequency of misting depend on the temperature and air circulation in the greenhouse. After cuttings have rooted, move plants to a hardening area with increased light levels and lower temperatures. Depending on the variety, salvias will root and will be salable in six to eight weeks in a 128-liner tray.

Transplant liners into a well-drained, soilless medium with a pH of 5.5–6.5. Use one liner per 4 or 6" (10 or 15 cm) pot and grow on at 60–65°F (16–18°C) nights until rooted and then grow on at cooler night temperatures of 55–58°F (13–14°C), which helps to tone the plants. *Salvia officinalis* is tolerant of cool growing conditions. Vigorous varieties in 4" (10 cm) pots are ready to sell in five to seven weeks, while slower ones take six to nine weeks.

Provide high light levels. Sage can be grown outdoors once the danger of frost has passed. Direct sunlight enhances foliage color and fragrance, especially under cooler conditions. Keep the medium moist, but not saturated. Take care not to overwater. Apply a constant feed of 100–125 ppm 20-10-20 and average amounts of micronutrients.

To encourage branching, give one soft pinch within one to two weeks of planting the liners. Trim as desired for shaping, but trimming is not required for plant performance.

Varieties

'Albiflora' is a selection with attractive white flowers. 'Berggarten' has broad, silver-colored leaves that are soft to the touch as well as a compact growing habit. 'Icterina' has gold and green, variegated foliage and is also referred to as golden sage. 'Purpurascens' has purplish foliage with a silver sheen to the leaves. Under high temperatures and low-light conditions, the foliage is more green than purple. 'Tricolor' shows a variety of colors, ranging from white, pink, purple, and silver-green. Colors are especially prominent on young growth. Cool temperatures and high light levels intensify the colors.

Postharvest

Give consumers ideas about how to use sage for its culinary value by providing tags with usage ideas or recipes. Instruct retailers to display plants in full sun or partial shade. Watering is critical: Allow pots to dry between irrigations, but do not allow plants to permanently wilt, as foliage will brown and plant growth will become stunted.

In the garden, like most Mediterranean natives, common sage prefers a sunny location. Good drainage is critical; avoid heavy soils. Plant performance is often diminished in areas with very hot, humid summers or a high amount of rainfall during the summer months.

Salvia splendens
Common name: Scarlet sage
Annual

Salvia splendens is one of the most requested salvias for use in the home garden. While available in various colors, it is the scarlet hues that garner the most attention. It is frequently used in landscaping to highlight gas stations and fast food restaurants that use red in their logos and signs. Plants grow from 10–24" (25–60 cm) tall when flowering.

Propagation

Sow seed onto a well-drained, disease-free medium with a pH of 5.5–5.8. Cover lightly with coarse vermiculite to maintain 100% humidity at the seed coat. Maintain wet germination conditions: The medium should be fully saturated. Germinate at soil temperatures of 75–78°F (22–23°C). Keep ammonium levels below 10 ppm. Radicles should emerge in four to five days.

Move trays to Stage 2 for stem and cotyledon emergence. Reduce soil temperatures to 72–75°F (22–24°C) and begin to allow trays to dry down somewhat between irrigations. As soon as cotyledons are expanded, begin fertilizing with 50–75 ppm from 14-0-14. Maintain light levels at 1,000–1,500 f.c. (11–16 klux).

Move trays to Stage 3 to develop true leaves. Drop soil temperatures to 68–70°F (20–21°C) and begin to allow soil to dry thoroughly between irrigations. Fertilize weekly with 100–150 ppm from 20-10-20, alternating with 14-0-14. Increase light levels to 1,500–2,500 f.c. (16–27 klux). If growth regulators are required, A-Rest, B-Nine, and Bonzi can be used at this time. B-Nine can be applied at 2,500–5000 ppm to tone plugs.

Harden plugs for seven days prior to sale or transplanting by increasing light levels to 2,500–3,000 f.c. (27–32 klux) and reducing soil temperatures to 65–68°F (18–20°C). Feed weekly with 100–150 ppm from 14-0-14.

A 288-plug tray takes from five to six weeks from sowing until it is ready to transplant to larger containers.

Growing On

S. splendens may be produced in jumbo packs, 1801s, or 4" (10 cm) pots. Use one plug per cell or pot. Taller-growing varieties such as 'Bonfire' are great in 1-gal. (4 l) pots for the landscape trade; use two or three plugs per pot. For all salvia, if you must use packs, use deep packs to enhance shelf life, as plants will dry out less frequently.

Select a disease-free, well-drained medium with a moderate initial starter charge and a pH of 5.5–6.2. Grow on at 60–65°F (16–18°C) days and 55–60°F (13–16°C) nights. Fertilize at every other irrigation with 150–200 ppm from 20-10-20. Be especially careful of soluble salts buildup, especially early in the production cycle. Discontinue feed one to two weeks prior to sale to increase shelf life.

If growth regulators are needed, A-Rest, B-Nine, Bonzi, and a B-Nine/Cycocel mix are all effective.

The crop time from transplanting a 288 plug to cell packs is from five to six weeks, and a 4" (10 cm) pot requires from six to seven weeks, based on if plants are early- or midseason flowering. Late-flowering varieties are sold green with no flower color and will bloom later in the garden.

Pests & Diseases

Aphids, leafminers, spider mites, thrips, and whiteflies can attack plants. Disease problems can include botrytis and leaf spot.

Varieties

S. splendens varieties are offered in three primary types: pack, garden, and large, landscape varieties. For a mainstay pack variety, try the 'Vista' or 'Sizzler' series for a fast turn. Both series come in a wide range of colors, including bicolors. In 1801s or 4" (10 cm) pots, 'Vista' makes a great garden salvia. Its color range is not as large as 'Sizzler', but its plants are superior outdoors. 'Red Hot Sally' and 'Fuego' are top-selling reds. These older varieties perform reliably in packs and pots for growers and in the garden for consumers. 'Flare', growing to 18" (46 cm) outdoors, is also an excellent garden performer. For landscape salvia, you can't beat 'Bonfire', which grows to 26" (66 cm) outdoors and puts on a bright red flower show that will dazzle passersby.

Postharvest

Salvia is highly sensitive to ethylene gas. Do not expose plants to ripening fruits and vegetables or even automobile or truck exhaust. Keep plants groomed by removing old flowers and dead leaves to remove this source of ethylene.

Retailers should display salvia under partial shade and cool temperatures of 70–75°F (21–24°C). Consumers should plant salvia in partial shade in the South. Keep plants deadheaded in the garden to encourage continuous flowering. Salvias will often put on their best flower show in the fall, so instruct your landscape customers not to pull plants out early.

Santolina

Santolina chamaecyparissus
Common name: Lavender cotton
Perennial (Hardy to USDA Zones 6–8)

Prized for its evergreen, gray-green, finely divided foliage, *Santolina chamaecyparissus* is considered a subshrub. It develops a woody base the first year regardless of how it is propagated. Plants are aromatic but not overpowering, with the scent more prevalent after a rain shower or when touched. Santolina grows 12–24" (30–61 cm) tall and spreads about the same. Its golden-yellow flowers are rather unattractive but are appreciated by some for their brightening effect in herb gardens. Blooms are single, 0.5–0.75" (1–2 cm) across, and globed shaped. However, its foliage is preferred.

Propagation

Santolina can easily be propagated from cuttings. Stock plants should be kept vegetative and in active growth so that cuttings are not too hard. Avoid using woody cuttings because they take longer to root and result in higher losses.

The best time to propagate is in spring and summer, when the plants are actively growing. However, propagation can occur year-round as long as active growth is maintained. Most standard tray sizes for propagation are between 84 and 128 cells per tray. Multiple cuttings per cell (usually two) are recommended to achieve a fuller liner. A rooting hormone is not necessary if the cuttings are supple, although an application doesn't hurt the cuttings either.

Use a standard propagation medium with a pH of 5.5–6.5. Temperatures of at least 70–75°F (21–24°C) for the rooting process are best, ideally with bottom heat. Keep the medium moist at all times, but not saturated. Until the plants have developed roots of their own, mist several times a day to avoid wilting. However, excessive mist will cause rotting, and botrytis can become an issue as well. The amount and frequency of misting depend on the temperature and air circulation in the greenhouse. After cuttings have rooted, move plants to a hardening area with increased light levels and lower temperatures.

Liners are available from commercial propagators, and greenhouse growers more commonly buy them in rather than root their own. The convenience of receiving plants when desired and not having to maintain stock plants are two key reasons for buying in liners.

An 84-liner tray, with two cuttings per cell, will take from five to seven weeks to fully root.

Growing On

Transplant liners using a well-drained, soilless medium such as a peat/perlite mix with a pH of 5.5–6.2. Use 1 liner/4" (10 cm) pot or 1–3 liners/1-gal. (4 l) pot. Santolina can also be used in hanging baskets or mixed container gardens.

Provide high light levels. Santolina can be grown outdoors once the danger of frost has passed. Direct sunlight enhances foliage color and aroma. Keep the medium on the dry side, as santolina is sensitive to overwatering. Apply a constant feed of 100–125 ppm from 20-10-20.

To encourage branching, give one soft pinch within one to two weeks of planting. Santolina responds well to cutting back: Trim plants as desired for shaping, however trimming is not required for plant performance.

From potting, 4" (10 cm) plants can be sold in five to seven weeks and ten to twelve weeks in a 1-qt. (1.1 l) container.

Pests & Diseases

Fungal diseases can become a problem under very hot, humid conditions and will result in rotting plants. As a precautionary measure in all climates, drench with a broad-spectrum fungicide at liner planting. Watch for aphids during summer months.

Varieties

Although there are a few different varieties available, the *Santolina chamaecyparissus* species is probably the most commonly grown. 'Nana' is a more compact selection. *S. rosmarinifolia,* also known as *S. virens,* is similar to *S. chamaecyparissus,* but with narrower, green leaves.

Postharvest

Although santolina is hardy to USDA Zone 6, it might die in very hot, humid summer climates.

Sanvitalia

Sanvitalia speciosa
Common name: Sanvitalia
Annual

Sanvitalia is widely popular in Europe because of its great habit in mixed plantings, utility in border plantings, and hard-to-find, yellow-colored flowers. Newer breeding is improving its plant form and significantly increasing the size of the flowers. Great summer performance in tough, high-light summer conditions should make this a candidate for increased production in the United States.

Propagation

Sanvitalia is propagated from cuttings. If cuttings are wilted, place in a clean cooler at 50–55°F (10–13°C). Provide a relative humidity of nearly 100% by wetting the floor and misting the open bags with clean water. Stick cuttings in Ellepots or medium.

Root cuttings under mist at a substrate temperatures of 70–72°F (21–22°C) and air temperatures of 70–72°F (21–22°C) days and 68–70°F (20–22°C) nights. Provide only enough mist to keep cuttings turgid. Provide 500–1,000 f.c. (5.4–11 klux) of light.

When the cuttings are callused and begin to develop roots, move to Stage 2. Significantly reduce or cease the use of misting and begin to apply an overhead feed of 50–75 ppm from 20-10-20.

In Stage 3 maintain soil and air temperatures. As root initials form, begin to dry out soil in between irrigations. Increase light levels to 1,000–2,000 f.c. (11–22 klux) as cuttings root. Apply a foliar feed of 100 ppm from 15-0-15, alternating with 20-10-20. Increase the rate to 150 ppm as roots form.

Stage 4 consists of hardening the cuttings for approximately seven days prior to sale or transplanting. At the beginning of Stage 4, you can pinch or trim the liners to help create lower breaks. If growth control is needed B-Nine at 2,500 ppm should provide control. Reduce air temperatures at night to 65–68°F (18–20°C). Plants can be held at 50°F (10°C) assuming dry, high-light conditions. Move liners from the mist area and increase light levels to 2,000–4,000 f.c. (22–43 klux). Shade plants during midday to reduce temperature stress. Fertilize with 150 ppm from 15-0-15, alternating with 20-10-20.

A 105 tray will be ready to transplant to larger containers in four to five weeks.

Growing On

Transplant 1 plug or liner/4" (10 cm) pot; use 1–2 plugs or liners/6" (15 cm) pot; use 4–6 liners in 10" (25 cm) or larger. Select a disease-free, well-drained medium with a pH of 5.6–6.2. Grow plants on at 65–68°F (18–20°C) days and 60–65°F (15–18°C) nights. Plants can be held at 50°F (10°C). Fertilize with a constant feed of a complete fertilizer at 150 ppm. Sanvitalia is not a heavy feeder. If higher rates are used, alternate with fresh water.

Sanvitalia should be grown on the dry side. Grow under high light levels (4,000–7,000 f.c. [43–75 klux]). Pinching can be used to shape the plant. Pinch as early as the plants have roots to the side of the finish container. You can pinch multiple times to create the desired shape. Do not pinch within three to six weeks of the sale date (less time later in the spring season when the weather is good).

Withholding water is the best growth control. However, when needed spray applications of B-Nine or Dazide at 2,500 ppm is effective control. Do not apply Truban, Banrot, or azole-containing fungicides as they can stunt plant growth.

The crop time to a flowering 4" (10 cm) is six to nine weeks after transplanting a 105 liner.

Pests

Aphids, spider mites, thrips, and others can attack this plant. Include this crop in general scouting of bedding plant crops.

Troubleshooting

Plants may collapse due to wet medium or botrytis. High ammonium levels in the medium, overfertilization under low-light conditions, and/or low light and overwatering can cause excessive vegetative growth. Production temperatures that are too low or an overdose of growth regulators can result in a lack of growth.

Varieties

'Sanvi Super Gold' and 'Sunbini' are a couple of proven performers. Varieties are changing quickly because breeders are working hard on this crop. Contact your broker or breeder rep for the most up-to-date varieties.

Postharvest

Display plants in high light with moderate temperatures. Do not allow plants to wilt, but keep on the dry side.

Consumers should plant in well-drained soil in full sun. Once established, plants will flourish in moist todry soil and high temperatures. Fertilize throughout the summer for continued performance.

Scabiosa

Scabiosa caucasica
Common name: Pincushion flower
Perennial (Hardy to USDA Zones 4–7)

While *Scabiosa* is usually winter hardy, stress from summer heat and humidity usually results in short-lived plants in Chicago. They last two to four seasons at best.

Scabiosa caucasica is an erect-growing perennial with long, linear, basal leaves that arch and curve toward the ground, creating a wispy appearance. Leaves that develop along the stems are more lobed than the basal leaves but look similar. Plants don't have a full habit in Chicago's climate, although in less hot and humid areas of the country, they perform better. *Scabiosa* flowers are blue or white, single, and usually scentless and open on long stems above the plant. The striking flower heads are 2–3" (5–7 cm) across and are composed of large, slightly cupped or flat petals at the outer margin with a tufted center cluster. The prominent, pin-like stamens give this perennial the name of "pincushion flower." In bloom, they grow from 18–24" (46–61 cm) tall and 10–14" (25–35 cm) across. Plants flower in May or June.

Propagation

Seed is the most common method form of propagation. Sow one or two seeds per 288-plug tray, using a peat-based medium with a pH of 5.5–5.8. Cover lightly with vermiculite and germinate at a soil temperature of 68–72°F (18–22°C). The root radicles emerge in four to six days.

Upon moving the plug trays to the greenhouse bench, maintain soil temperatures at 65–70°F (18–21°C) and light levels from 500–1,500 f.c. (5.4–16 klux). Start fertilizing at 50–75 ppm 14-0-14 once a week. Stage 2 takes from eleven to fourteen days.

In general, a 288-plug tray takes from five to six weeks from sowing to transplanting, depending on variety.

Growing On

While sowings in January or February will flower sporadically the upcoming summer, sowings done the previous year up until mid-autumn (November) and transplanted to 1-qt. (1.1 l), 4" (10 cm), or 6" (15 cm) pots for May sales have a higher bud count and more free-flowering plants than those sown in February. While plants do not need a cold treatment (vernalization) to flower, growing them at 50–55°F (10–13°C) nights (45–50°F [7–10°C] once fully established in the final container) improves the bloom count and uniformity of flowering. Containers are sold green in May, and flowering will occur from June to early August, depending on the sowing.

Sowings are also done in summer (June or July) for transplanting to 1-qt. or 1-gal. (1.1 to 4 l) containers. Plants are grown outside during the summer and then moved to a cold frame for overwintering dormant.

Commercial propagators offer *Scabiosa* in small pots or as plugs/liners. Primarily available during the autumn and winter, either of these can be potted up into 1-qt. and 1-gal. (1.1 or 4 l) containers for late spring and summer sales.

Varieties

'House Hybrids' ('House Mix') is a common variety found across North America through a number of seed companies. It is known outside the United States as 'Isaac House Hybrids' in honor of its selection at the Issac House in England. The hybrids are a mixture of blue- and white-flowering varieties that grow 2–3' (60–90 cm) tall. 'Fama' has lavender-blue flowers on plants that are 2–3' (60–90 cm) tall. Its habit is more uniform than 'House Hybrids', but it is available in only one color. 'Compliment' is a dark-lavender variety that grows to 2' (60 cm) tall. It's darker in color than 'Fama'. The 'Perfecta' series is another notable *Scabiosa*. 'Perfecta' is a lilac-flowering variety to that reaches 2' (60 cm) tall, while 'Perfecta White' is its white-flowering counterpart. All of these varieties are seed propagated.

Scabiosa columbaria
Common name: Pincushion flower
Perennial (Hardy to USDA Zones 5–7)

Like *Scabios caucasica, Scabiosa columbaria* is usually winter hardy, but stress from summer heat and humidity usually results in short-lived plants in Chicago. They last one to two seasons at best. While there is a range of varieties available, most of the current, popular selections are dwarf growing from 8–15" (20–38 cm) tall and are highlighted with mostly blue or purple flowers and an occasional white. Their foliage is uniform in appearance and has a compact habit. Plants spread from 10–15" (25–38 cm) across.

Propagation

S. columbaria can be propagated by seed, tissue culture, or cuttings.

Seed

Follow the propagation notes for *Scabiosa caucasica* for similar results.

Cuttings

Cuttings are available from a number of commercial companies. Stick one cutting per 32, 50, 72, 84, or 105 tray, dipping each into a rooting hormone for uniform rooting. Give mist for nine to twelve days, but avoid saturating the media. Do not pinch. Cuttings root in four to five weeks.

Growing On

Seed

Follow the growing on notes for *Scabiosa caucasica* for similar results.

Cuttings

Select a disease-free, well-drained medium with a moderate initial starter charge and a pH of 5.8–6.2. Pot up one liner per container.

While a cold treatment is beneficial, it is not required for flowering. Start by bulking up the plants under short days for four to eight weeks. Then provide a six- to fifteen-week cool treatment in a cooler (under 25–50 f.c. [269–538 lux] of light) or cold frame at 41°F (5°C). This helps to increase overall flowering.

Fertilize with 100–150 ppm from 14-0-14 at every irrigation. Provide high light levels, from 4,000–6,000 f.c. (43–65 klux). Grow on at 65–70°F (18–21°C) days and 60–65°F (16–18°C) nights.

In general, a 1-qt. (1.1 l) container takes from five to six weeks to be ready for sale from a rooted liner, while a 1-gal. (4 l) container can take from seven to twelve weeks, depending on variety. If liners are transplanted in late winter, the pots may not be well rooted for May sales. While the plants will be salable from a visual point of view, liners should be potted up in December or January for 1-gal. (4 l) pots for mid-spring sales.

If you don't want to worry about cooling plants, just bring in large liners in January or February and pot them into 4 or 5" (10 or 13 cm) pots. Plants will be a lot smaller at flowering, but 'Butterfly Blue' doesn't have to be bulked up or cooled in order to flower.

As for height control, *Scabiosa columbaria* is responsive to B-Nine sprays at 2,500–4,000 ppm alone or as a tank-mix spray with Cycocel 1,000–1,500 ppm. Sumagic sprays at 10–20 ppm are also effective.

Varieties

Varieties from cuttings include key selections like 'Butterfly Blue' and 'Pink Mist'. They are the two cornerstones to this species, although others like 'Harlequin Blue' are also popular. All are good garden plants but also make wonderful pot plants as well.

From seed, the variety 'Blue Note' makes an excellent 4" (10 cm) pot item, growing from 8–10" (20–25 cm) tall.

Postharvest

Plants are ready to sell when flowering begins. If you miss your target sales window, trim off the flowers, and they will rebloom in four to five weeks.

Retailers should display plants in full sun, provided temperatures are cool (60–70°F [15–21°C]) and plants do not become water stressed.

Scaevola

Scaevola aemula
Common name: Fan flower
Annual

Commonly known as fan flower, *Scaevola* is one of the wave of plants that hit the United States from Australia in the 1990s. Plants thrive in hot conditions, providing a continuous color show throughout even the worst summers.

Growers generally offer *Scaevola* in 4" (10 cm) pots, hanging baskets, and combinations. While it is a more expensive item than most seed-propagated bedding plants, some landscapers have discovered how great *Scaevola* is planted in landscape beds as well. Outdoors *Scaevola* creates mounds of green punctuated with blue flowers that seem to glow in the sun. *Scaevola* plants have even brightened the mall in Washington, DC, with large plantings outside the Smithsonian.

Propagation

While most growers purchase rooted liners, *Scaevola* is also available as an unrooted cutting from some suppliers.

Unpack and plant cuttings immediately upon arrival. If planting must be delayed, store cuttings for no more than twenty-four hours at 48–50°F (9–10°C). Stick cuttings into Oasis wedges or a rooting medium. Irrigate the medium so that it is uniformly moist. Callus formation takes place in five to seven days at a substrate temperature of 68–72°F (20–22°C). Maintain relative humidity at 75–90% at the base of the cuttings. Use mist to decrease wilting; increase and decrease frequency as light and air temperatures change during the day. Provide 500–1,000 f.c. (5.4–11 klux) of light during this stage. Higher light intensities will stress plants due to warming. Retractable shade is ideal to allow light intensity to increase as cuttings mature. Use 50–75 ppm 20-10-20 as a foliar feed as soon as there is loss of foliage color. Maintain a substrate pH of 5.0–5.5 with an EC less than 0.5 mS/cm.

When 50% of the cuttings begin differentiating root initials, move them to the next stage for root development, which will take place over seven to fourteen days. Maintain substrate temperatures of 68–72°F (20–22°C) and air temperatures of 75–80°F (24–26°C) days and 68–70°F (20–21°C) nights. As soon as cuttings start to form root initials, begin to dry down the medium by reducing the duration and frequency of mist and moving mist start and end times. Avoid drying out the air, since this will increase evapotranspiration, which reduces root zone temperatures. Begin increasing light intensity to 1,000–2,000 f.c. (11–22 klux) as cuttings root. Apply growth regulators as needed. Begin using a foliar feed at 100 ppm 15-0-15, alternating with 20-10-20, and then increasing rapidly to 200 ppm. Increase the frequency and rate at each application to prevent salt problems. Keep the soil EC below 0.5 mS/cm.

In the final stage of rooting to get plants ready to ship or transplant, maintain 70–75°F (21–24°C) days and 62–68°F (16–20°C) nights. This final stage will last about seven days.

Move cuttings from the mist area to a bench with lower humidity, lower temperatures, and higher light levels (2,000–4,000 fc. [22–43 klux]). Provide shade during midday to reduce temperature stress. Maintain the substrate pH at 5.0–5.5 and the EC below 1.0 mS/cm. Fertilize twice per week using 150–200 ppm 15-0-15, alternating with 20-10-20.

A 105 tray is ready to transplant to larger containers four to five weeks after sticking cuttings.

Growing On

Pot liners into a well-drained, disease-free medium with a pH of 5.0–5.5. Make sure the medium has a high initial nutrient charge. Pot 1 plant/4" (10 cm); 1–2 plants/6" (15 cm); and 3–5 plants/10" (25 cm) hanging basket. If you are planning to use *Scaevola* in combination pots, root plants into 3 or 4" (8 or 10 cm) pots first, and then transplant to baskets. *Note: Scaevola* is sensitive to being overwatered, which is easy to do with liners in a large pot. Pinch plants as they are potted, repeating three weeks later. If needed, plants may require a third pinch to shape them.

Florel at 500 ppm helps produce good branching. Apply about two to three weeks after the first pinch. Subsequent pinches may not be needed after a Florel application.

Cool night temperatures (60–65°F [15–18°C]) are ideal for *Scaevola,* with days at 65–75°F (18–24°C). Warm nights cause plant stretching. For late season/ early summer crops, *Scaevola* is an excellent candidate for full-sun outdoor production in containers.

Scaevola is a heavy feeder. Provide constant liquid feed using 15-0-15, alternating with 20-10-20 at 200–300 ppm once plants are established. Avoid fertilizers high in phosphorus. A yellow-red chlorosis and bleached-yellow chlorosis of young leaves is an indication of phosphorus toxicity. Slow-release fertilizers may also be used. An iron supplement at 5 ppm is recommended periodically. Maintain an EC of about 1.0 mS/cm.

Keep the medium moist to avoid soluble salts problems. Use clear water at every third irrigation if salts problems arise. Do not allow *Scaevola* to wilt once flowering begins, as flowering will be detrimentally affected.

Provide high light intensities at 4,000–7,000 f.c. (43–75 klux), while maintaining moderate temperatures. *Scaevola* is day neutral, but flowering is stimulated as light conditions improve. Note that some varieties are less sensitive and flower faster in the spring than others. By summer, there is no difference. During the winter and dark periods in the spring, crops take longer to finish. Low light promotes stem stretch.

Cycocel and Bonzi sprays have been shown to be effective at controlling height. A one-time Bonzi drench is also effective.

A 105 liner that is transplanted to a 4" (10 cm) pot requires from six to eight weeks to flower.

Pests & Diseases

Botrytis and powdery mildew can become a problem if humidity is too high during production. Avoid wet foliage to prevent problems. *Pythium* root rot can also occur. Fungus gnats, leaf miners, thrips, and whiteflies can all attack *Scaevola.* When selecting pesticides for insect control, avoid insecticidal soaps or chemicals dissolved in petroleum-based solvents when spraying crops in flower, as flowers will be damaged.

Troubleshooting

Do not keep the medium wet for extended periods, as plants may simply collapse. Too-high nitrogen levels, overfertilization and low light, and/or low light and overwatering can all result in excessive vegetative growth. Low light can cause plants to stretch. Check for phosphorus toxicity if you find yellow-red older foliage or bleached younger foliage.

Varieties

'New Blue Wonder', one of the Proven Winners, is the variety that established *Scaevola* as a spring plant. There are now a number of selections on the market including ones that feature additional colors of white, pink, and even yellow. Contact your broker for more details.

Postharvest

Scaevola performs best in full sun. However, it is especially important at retail to provide moderate temperatures (65–72°F [18–22°C] days) while providing high light. To ensure that soluble salts are not a problem, keep plants moist.

Planted in the garden, *Scaevola* can tolerate full sun and, once established, high temperatures (up to 95–100°F [35–38°C]). Plants can also tolerate temperatures slightly below freezing. Established plants are relatively drought tolerant.

Schizanthus

Schizanthus × *wisetonensis*
Common names: Poor man's orchid; butterfly flower
Flowering pot plant, annual

Schizanthus loves cool temperatures, making it an excellent candidate for early spring pot plant sales or as a spring bedding plant in areas with prolonged cool weather. If you have never tried *Schizanthus* and are looking for an item to excite customers before the onset of the main season, give it a try. *Schizanthus* can be grown with primula and other cool crops. Newer breeding that has come out in the past several years has created plants with a more compact habit and shatter-resistant flowers. If you are a retail grower, *Schizanthus* will offer you a product that your customers will not find at your competitors, because it doesn't ship well. Flowers truly do resemble orchids, as its common name, poor man's orchid, implies.

Propagation

Sow seed into 288- or 125-cell trays and cover with vermiculite. Germinate seed in the dark at 64–72°F (18–22°C) substrate temperatures. Radicles emerge over two to five days. Keep plug trays in the dark for Stage 2, stem and cotyledon emergence, which takes from five to nine days. Maintain substrate temperatures and begin fertilizing weekly with 50–75 ppm from 14-0-14. Maintain moist, but not wet, plug trays during Stages 1 and 2.

Move trays to 500–1,000 f.c. (5.4–11 klux) of light for Stage 3, growth and development of true leaves. Drop substrate temperatures to 62–65°F (17–18°C) and begin allowing trays to dry out between irrigations. Apply fertilizer once a week using 100–150 ppm, alternating between 20-10-20 and 14-0-14. If growth regulators are required, B-Nine can be applied beginning in Stage 3. Plugs will stay in Stage 3 for seven to fourteen days.

Move plugs to Stage 4 for hardening for seven days prior to sale or planting. Increase light levels to 1,000–3,000 f.c. (11–32 klux) and fertilize weekly with 100–150 ppm from 14-0-14.

Schizanthus may be a custom-sown item from most plug producers, which means that you will need to be sure to get orders in early and be willing to order the minimum quantity for a custom sowing, which is usually around twelve trays.

A 288-plug tray takes from four to six weeks to be ready to transplant to larger containers.

Growing On

Do not allow plugs to become root-bound before transplanting. Pot 1 plug/4" (10 cm) or 3 plugs/6" (15 cm) pot into a well-drained medium with a moderate initial starter charge and a pH of 5.5–6.3.

Temperature affects growth, with the fastest flowering at higher temperatures but the most compact and uniform plant habit at lower temperatures. Grown at 47°F (8°C), the plants require 165 days to flower, while at 65°F (18°C) plants flower in ninety days. Plants grown at lower temperatures are shorter with a better growth habit. Since you will be able to grow an acceptable plant at either temperature, weigh the pros and cons of a longer crop time against fuel cost. An average temperature of 60°F (16°C) will increase days to flower to one hundred, yet will produce an acceptable plant habit, a good compromise.

Light intensity also affects days to flower. At 53°F (12°C), *Schizanthus* needs 143 days to flower at 800 f.c. (8.6 klux), yet only 125 days at 1,100 f.c. (12 klux). Maintain light levels as high as possible (3,000–6,000 f.c. [32–65 klux]), while maintaining cool temperatures.

While newer varieties do not require pinching, plants will develop a more uniform habit from a pinch. Days to flower will increase slightly.

Height can be controlled with A-Rest, B-Nine, Bonzi, or Sumagic. Use A-Rest or Bonzi as a drench and B-Nine or Sumagic as a spray. Apply growth regulators only after plants have rooted out to the edge of the pot. At higher temperatures or under low-light conditions, more growth regulators may be needed.

Space plants through production so that leaves do not overlap.

Irrigation is critical. Allow the medium to dry out between watering; however, drought stress will cause lower foliage to yellow and not recover. To prevent leaf spots, do not water plants from overhead.

Do not withhold fertilizer from *Schizanthus* for any reason. Fertilize with 150–200 ppm from 15-0-15 at every other irrigation. Switch to a fertilizer with more potassium than nitrogen (13–0–44) when buds start to show color. Do not stop fertilizer prior to sale. Withholding fertilizer may cause lower leaves to turn yellow. Plant foliage color is naturally light green, although plants treated with B-Nine may have darker green foliage than will untreated plants.

The crop time to a flowering 4" (10 cm) pot plant is from seven to nine weeks.

Pests & Diseases

Aphids, spider mites, thrips, and whiteflies may become pest problems. Botrytis or leaf spots can become disease problems.

Troubleshooting

Lower leaf yellowing may be due to a number of causes: plants that are too closely spaced, keeping light from reaching lower leaves; low nutrient levels; low magnesium levels; drought stress; or leaf-spot diseases.

Varieties

'Royal Pierrot' and 'Atlantis' are both available as mixes and are programmable, so you will be able to more precisely schedule production.

Schlumbergera

Schlumbergera × *buckleyi*
Schlumbergera truncata
Common names: Christmas cactus; holiday cactus; Thanksgiving cactus
Flowering potted plant

Just about anyone who's remotely good with plants has a Christmas cactus hanging around their house. Home gardeners delight in the fact that they can grow plants outdoors on the deck in the summertime, bring them inside when weather cools, and enjoy a spectacular holiday flower show year after year.

While during the holidays Christmas cactus shows up in all the familiar retail locations, it is not as widely grown among growers. A few growers specialize in Christmas cactus; some pot plant growers put in significant programs, but for the majority of typical bedding plant growers, it's an afterthought. And that's too bad because *Schlumbergera* is one of the most highly sensitive plants to ethylene. It is virtually impossible for a grower shipping plants any distance to put plants in the market with significant color. Maybe there are a few buds and a flower here or there that opens on the retail shelf, but nothing compared to the beauty that a grower selling direct to the public can produce. The majority of pot plants today are sold in supermarkets, which means Christmas cacti are shipped in. There is a whole generation of Americans who do not know what a "real" Christmas cactus is supposed to look like. While plants can be treated to prevent bud drop, they look nothing like the gorgeous plants that can be grown and sold locally.

Schlumbergera is a great niche opportunity for local growers who market pot plants to garden centers or florists or sell direct to the public. Several suppliers offer prefinished programs so that you can even buy in a few cases to spice up your holiday sales. If you are one of those retail growers today who is holding poinsettia production volumes steady, expanding your offerings at the holidays by doing a fabulous job with spectacular plants that your customers just don't see in the supermarket or mass market is one way you can boost sales numbers. Think about *Schlumbergera* in that regard.

Propagation

Plants are easy to root by cuttings; however, since the best varieties are patented, you need to be licensed. Most growers buy in liners or even prefinished 4" (10 cm) plants.

Single-segment leaf cuttings are twisted off the upper portion of plants and stuck in a medium (72-cell or 1.5–2.25" [38–57 mm] size rooting trays). Select only mature leaf segments, and only use segments from disease-free stock. Stick two to four segments per cell. Use a well-drained, sterile, propagation medium. Maintain substrate temperatures of 70–75°F (21–24°C). Do not use mist. Cuttings can be propagated until March to produce a flowering crop the following November and December. To keep cuttings vegetative, place cuttings taken from early September to late April under long days (daylength extension or night interruption lighting). Use 50–75 ppm from a well-balanced fertilizer as soon as roots have formed. As cuttings mature, increase the rate to 100 ppm and apply weekly. Pinch off the new shoots that cuttings generate after about six to eight weeks to encourage the development of multiple shoots. Pinching takes time, but the full plants you will grow are worth it. Benzyl adenine (BA) can be applied at this time to further encourage branching. However, BA application can be touchy and is very variety dependent.

Cuttings propagated from November to March will be ready to transplant in April, May, or June.

Growing On

Use a well-drained, disease-free medium with a pH of 5.8–6.2. Pot 1 cell (with 3–5 plants/cell)/4" (10 cm) pot; 2–3 cells/6" (15 cm); and 8 or more cells/8" (20 cm) basket. Plant 6 and 8" (15 and 20 cm) pots earlier to enhance their size.

Grow on at 70–76°F (21–24°C) days and 62–65°F (17–18°C) nights. Maximum day temperatures should be 80–85°F (27–29°C). Under ideal conditions, plants will produce one tier (one leaf segment) every six to eight weeks. Maintain light intensity at 1,500–3,000 f.c. (16–32 klux). When light levels and/or temperatures are too high, some varieties become chlorotic. Also, at high light levels you may find yellow foliage, hard growth, or small, brown spots on leaves.

Provide long days until plants are ready for flower initiation. Normally, if you are growing a crop for holiday production, you will not have to provide night interruption lighting to do so.

Irrigate plants lightly just after potting. As they become established, irrigate more frequently, keeping the medium moist but not saturated. *Schlumbergera* are succulents—do not overwater them. If necessary, run them dry rather than too wet.

By September, 4" (10 cm) pots should be between four and five segments long. Plants that are too tall can be shortened, or "twisted" as they say in Europe, by removing the upper leaf segment(s). In the United States, this pinch is called leveling. Level plants off approximately seven to ten days after the start of short days (see below). Correctly done, this will yield heavily budded plants that are uniform in height. Leveling also does two other things to improve the finished plant quality. First, it encourages more upright growth, which means plants will sleeve and ship better. Second, it causes flower buds to initiate on more mature leaf segments, which means the part of the plant that's supporting them has more structure to bear the weight. After leveling, treat with BA to encourage heavier bud set.

Flowering is controlled by photoperiod and temperature. When temperatures are below about 59°F (15°C), flower initiation will occur under any photoperiod, including continuous light. When temperatures are between 60 and 75°F (16 and 24°C), plants will initiate flowers under short days but will remain vegetative under long days. Thus, holiday cactus is a short-day plant when grown from 60–75°F (16–24°C). The critical day length is between twelve and a half and fourteen hours for plants grown at 64–65°F (18°C) nights and 70–72°F (21–22°C) days. Growers can use either natural flowering or controlled flowering to grow the crop. However, temperatures below 50°F (10°C) will prevent flowering.

In the North, a naturally flowered cactus will flower mainly in mid-November at 62–65°F (17–18°C) nights. In even cooler areas, where temperatures could dip into the 50s (°F or 10–16°C) during August and September, plants will flower earlier. Plants in the South will flower mainly in early to mid-December. However, flowering plants under natural conditions is limited: Each variety has a limited flowering period, and crop scheduling is difficult because the exact flowering date will vary from year to year by variety.

It's better to schedule flowering. To schedule plants, grow them at 68–72°F (20–22°C) days and 62–65°F (17–18°C) nights. During natural long days (late April to early September), induce flowering by reducing the day length to less than twelve hours. It's important to maintain short-day conditions on a daily basis for at least three weeks. Poor or uneven bud set can occur if temperatures under the black cloth exceed 75°F (24°C). Finish plants under short days at 75°F (24°C) days and 62°F (16°C) nights.

Flowering can also be delayed by daylength manipulation. Beginning the first week of September, light plants with night interruption lighting to delay flowering.

Schlumbergera has a low water requirement; therefore, it is important to feed early in the production program. For the first eighteen weeks after potting, provide a constant liquid feed of 150 ppm from 20-10-20 and alternate with 15-0-15 at every third or fourth irrigation. From weeks 18–21, the floral initiation phase (short days), use clear water to avoid damaging buds. Do not stress plants from a lack of water during floral initiation. Then for weeks 21–30 (finishing), use 150 ppm from 15-0-15 at every other irrigation. Do not allow the pH to drop below 5.5, as micronutrient toxicity can occur.

BA can increase branching in vegetative plants and increase the number of flower buds on reproductive plants. To increase flower buds, apply 50–100 ppm as a spray when pinpoint buds are visible (natural flowering) or about ten to twelve days after starting short days (controlled flowering). Using BA for a natural-season crop is more difficult because each variety will bud differently.

The crop time from a transplanting a 2.5" (6 cm) liner to a flowering 4" (10 cm) pot is from twenty-one to twenty-four weeks.

Pests & Diseases

Caterpillars, fungus gnat larvae, mealybugs, spider mites, and thrips can be pest problems. Diseases can be a serious problem because *Schlumbergera* is such a long crop. Stem rot and root rot from *Erwinia, Fusarium, Pythium, Phytophthora,* and rhizoctonia can be problems. Flowers can become infected with botrytis. *Erwinia* can also cause wet slime or lesions at the base of the stem. *Pythium* root rot can cause pale gray color on foliage when soil is wet. Leaf spots include *Bipolaris* and *Cercospera.*

Troubleshooting

Low fertility or excessive light can cause the foliage to be light green or light yellow foliage. Yellow lower leaves can be due to root loss from soil disease or lack of soil aeration. Ethylene exposure, rapid climate change, or extremes in soil moisture can lead to flower or bud drop. Excessive iron may cause marginal chlorosis. Be sure to maintain pH at 5.5 or higher.

Varieties & Species

Schlumbergera truncata is the Thanksgiving cactus. You can tell it apart from Christmas cactus because it has sharply toothed leaf segments. It also flowers earlier, generally peaking in November. *S.* × *buckleyi* is the Christmas cactus; it flowers close to the Christmas holidays. Leaf segments have rounded margins. The Christmas cactus is an interspecific cross between *S. truncata* and *S. russeliana.*

Today's hybrids are either pure Thanksgiving cacti or complex hybrids developed by crossing Thanksgiving cacti with Christmas cacti. They are sold based on a number of attributes: how well they branch, flower color, plant habit, flowering time, etc. Your supplier can advise you on the proper assortment for your region and climate.

Postharvest

Schlumbergera are highly sensitive to ethylene. When it was legal, growers treated with STS to reduce flower bud drop. Today, some growers are effectively using EthylBloc. Cactus flowers are easily damaged during packing and shipping, so be careful if you are shipping. Growers who ship usually sleeve and box plants that are in the large bud stage for display in typical supermarket or mass-market conditions. Ship at 50–60°F (10–16°C). Avoid low shipping temperatures, as they may cause bud drop.

Retailers should display plants under 300 f.c. (3.2 klux) of light and cool temperatures (60–70°F [16–20°C]). Do not overwater, although plants will drop buds if they dry out. Prevent bud drop by maintaining adequate moisture and not exposing plants to ethylene (ripening fruit and vegetables), high temperatures, or low light. Keep plants groomed by picking off old and dead flowers. Plants will flower over a period of four to six weeks under good light, temperature, and moisture conditions.

Scutellaria

Scutellaria javanica
Common name: Skullcap
Annual

Scutellaria is a member of the mint family, and there are a number of species under this group although this species is best known for the variety 'Veranda'. Plants are known for their heat-loving disposition with glossy-green foliage and blue to purple bicolor flowers. Plants grow from 10–14" (25–35 cm) tall and are best in the shade garden in baskets or containers as opposed to planting in the ground—although that can be done as well.

Propagation

Seed is the most common method of propagation. Use a well-drained, disease-free, soilless medium with a pH of 5.5–6.1 and a medium initial nutrient charge (EC less than 0.75 mS/cm with a 1:2 extraction). Sow seed to a 288-plug tray, leave the seed exposed to light, and germinate at 70–75°F (21–24°C). Light (10 f.c. [100 lux] or more) is required for germination. Keep soil moisture high until radicle emergence, and then reduce moisture levels after the radicle penetrates the medium. Do not allow the seedlings to wilt. At radicle emergence, apply 50–75 ppm nitrogen (N) from 14-0-14 or 13-2-13, alternating with a 20-10-20 fertilizer. The root radicle emerges in two to five days.

Stage 2 (cotyledon expansion) takes from six to ten days at temperatures of 68–72°F (20–22°C). Maintain light levels between 1,000 and 2,500 f.c. (11 and 22 klux). As the seedlings mature, the light levels can be increased up to 5,000 f.c. (54 klux). Increase fertilizer rates to 100–150 ppm N.

For Stage 3 (true leaves development) grow at 68–72°F (20–22°C). To hold plugs (Stage 4) grow at 66–70°F (19–21°C).

The crop time for a 406-plug tray is five to six weeks, while a 288-plug tray takes from seven to eight weeks.

Growing On

Use a well-drained, disease-free, soilless medium with a pH of 5.5–6.2 and a medium initial nutrient charge and temperatures of 66–68°F (19–20°C) nights and 72–78°F (22–26° C) days. *Scutellaria* is a heat-loving crop. The warmer the temperature is, the shorter the crop time. The crop time will be significantly longer when grown in temperatures below 66°F (19°C). Provide shade when light levels are above 5,000 f.c. (54 klux). Avoid both excessive watering and drought. Do not allow plants to wilt. Apply 200 ppm N once a week, alternating between 15-5-15 and 20-10-20 fertilizers.

If plants require a growth regulator, a tank mix spray of B-Nine 2,500 ppm and Cycocel 1,000 ppm every other week starting at two to three weeks after transplanting has been shown to be effective. However, if growing in larger containers and hanging baskets, PGRs may not be needed.

Pests & Diseases

Aphids, thrips, and spider mites are the most common issues. No diseases have been observed.

Table 36. Average Crop Times for *Scutellaria**

CONTAINER SIZE	PLANTS PER POT/BASKET	WEEKS FROM TRANSPLANT	TOTAL WEEKS
306/premium pack	1	7–10	13–16
4–4.5" (10–11 cm) pots	1	8–11	14–17
6–6.5" (15–16 cm) or 1-gal. (4 l) pots	3	8–11	14–17
Hanging baskets	3–4	9–12	15–18

* *Note: Scutellaria* 'Veranda' is a heat-loving crop. Its crop time is very dependent on temperature. Warmer temperatures can speed up the crop time dramatically. It has been finished in as few as thirteen weeks when sown in late May in research trials in Elburn, Illinois, research trials.

Sedum

Sedum spp.
Common name: Stonecrop
Perennial (Hardy to USDA Zones 3–8, depending on species)

The myriad of species and cultivars of sedum thrive in well-drained, sunny locations. They range in height from groundcover selections that are only 1–3" (3–8 cm) tall to selections that are 24–30" (61–76 cm). A wide selection of flower colors is available, from pink and dark pink to burgundy rose, yellow, and silver/white, plus shades in between.

Propagation

The majority of sedum selections are propagated from cuttings, although there are a few species, like *S. acre* and *S. spurium,* which can be propagated by both seed and cuttings.

Seed *(S. acre* and *S. spurium)*

Seed can be sown to any peat-based media, using two or three seeds per cell of a plug tray (288, 128, etc.). Conversely, three or four seeds can be sown to a 32-cell (or similar) flat. Seed does not need to be covered. Germinate at 70–74°F (21–23°C). The root radicle emerges in three to five days for most species.

A 288-plug tray takes from five to seven weeks to be ready from sowing to transplant, while a 32-cell flat takes from eleven to fourteen weeks from sowing.

Cuttings

Stick tip cuttings into a 50, 72, 84, or 105 tray using one cutting per liner on most varieties. On the trailing varieties, one or two cuttings per cell helps to fill out the finished growing container and make a more rounded-looking container that does not require pinches to fill it out. A rooting hormone is not usually needed, but keep mist to a minimum of five to seven days or so. Reduce, and then eliminate, mist as soon as possible. Allow media to dry between watering as rooted cuttings develop.

A 105 tray takes from three to five weeks from sticking the cuttings until ready for transplant to larger containers.

Growing On

Seed

Transplant one plug or liner to a 1-qt. or 1-gal. (1.1 or 4 l) container by late summer or early autumn. Allow plants to fully root in the container and then overwinter cold for spring sales the following year. Plants may be in flower when sold. If not, they will flower later, although flowering is not usually a preference for home gardeners.

While December and early January sowings produce garden plants for the upcoming summer, the plants seldom flower profusely (if at all). These plants may need to be sold from 4" (10 cm) pots, since 1-qt. (1.1 l) or larger containers may not be well rooted in time for spring sales.

Cuttings

Transplant a liner by late summer or early autumn to a 1-qt. (1.1 l) or larger container (tall or robust varieties should be potted to 1-gal. [4 l] containers, while trailing or other dwarf-growing varieties should be potted to 1-qt. [1.1 l] pots). Use one plant per pot. Select a very well-drained, disease-free medium with a moderate initial starter charge and a pH of 5.8–6.5. Sedum prefers dry conditions, so always allow plants to dry down thoroughly between irrigations. Overwinter plants in cold frames for sale the following year.

Plants may also be brought in and potted up during winter with equally good results, as plants do not require cold treatment for flowering. For example, 'Autumn Joy' flowering appears to be triggered by long days (sixteen hours) in combination with warm temperatures. If you are growing plants that will be sold in the spring and your goal is to bulk up the crop as fast as possible, provide fourteen-hour days during the winter months with night interruption lighting.

The crop time for well-rooted 1-gal. (4 l) pots ranges from twelve to fifteen weeks after transplanting one liner per pot.

Varieties

The most famous perennial sedums are the tall-growing hybrids that have become a routine perennial staple for late-summer flowering. 'Autumn Joy' is the best known and puts on a spectacular flower show in a blaze of bronze each fall. Before reaching full flower, 'Autumn Joy' begins with cream-colored flowers that become pink and later bronze red. Both 'Autumn Fire' and 'Matrona' improve on 'Autumn Joy'. 'Autumn Fire' flowers don't fall apart at the height of the flowering season, while 'Matrona' starts flowering a bit earlier. 'Vera Jameson' grows to 12" (30 cm) and has medium-pink flowers, which appear in clusters from summer to fall.

Sedum spectabile 'Brilliant' grows to 24" (61 cm) and produces flat, rose-colored flower heads atop greenish-gray foliage. *Sedum spurium* is a low-growing species with broad leaves. The variety 'Dragon's Blood', long known as a great perennial in rock gardens, has been improved in the variety 'Fulda Glow', which has rose-red flowers in the summer atop bronze-red foliage. *S. acre* has needlelike green leaves with yellow flowers in the spring.

Senecio

(see also *Pericallis*)

Senecio cineraria (also *Cineraria maritime)*
Common name: Dusty miller
Annual

Dusty miller—it's the unassuming plant you see everywhere in bedding plant borders, perennial beds, and combination pots for spring and fall. Consumers enjoy dusty miller's silvery foliage, which acts as the perfect neutral backdrop or complement to any flowering plant. Plants will overwinter easily in milder southern and western climates, taking freezes once they become established and acting as a canvas against which the bright, bold colors of pansies shine. In the Midwest, they will often die in winter.

For growers, dusty miller is a must-have crop. Be sure to have plants in 36-cell packs as well as 4" (10 cm) pots.

Propagation

Sow pelleted seed onto a well-drained, germination medium with a pH of 5.5–5.8 and soluble salts levels below 0.75 mS/cm. Do not cover seed. Germinate at 72–75°F (22–24°C). Keep the medium uniformly moist but not saturated. Radicles will emerge in three to six days.

Move trays to Stage 2 for stem and cotyledon emergence, which will take about seven to fourteen days. Reduce soil temperatures to 70–75°F (21–24°C). Once radicles emerge, begin reducing substrate moisture by allowing trays to dry slightly between irrigations. Provide 400–700 f.c. (4.3–7.5 klux) of light. Once cotyledons are fully expanded, fertilize with 50–75 ppm from 14-0-14 at every other irrigation. Once the tray has a full stand (germination is about 85%), apply a fungicide drench to prevent disease.

Move trays to Stage 3 to develop true leaves, which will take about fourteen to twenty-one days for 288s. Lower soil temperatures to 65–70°F (18–21°C) and allow the medium to dry thoroughly between irrigations. Keep seedling foliage dry to reduce disease. Increase fertilizer to 100–150 ppm at every other irrigation from 20-10-20, alternating with 14-0-14. If height control is needed, B-Nine can be used at 2,500–5,000 ppm as true leaves expand. Apply a preventative fungicide for *Alternaria* control.

Harden plugs for seven days prior to sale or transplanting by lowering the medium temperature to 60–62°F (16–17°C). Allow the medium to dry thoroughly between irrigations. Fertilize with 100–150 ppm from 14-0-14 as needed. Keep foliage as dry as possible to reduce disease.

A 288-plug tray takes from four to five weeks from sowing to be ready to transplant to larger containers.

Growing On

Transplant one plug per cell or 4" (10 cm) pot or three plugs per 6" (15 cm) or 1-gal. (4 l) pot. If adding dusty miller to combination pots, grow it on first in cell packs or pots before placing in large containers. Plant plugs at the same levels as they are currently growing.

Select a disease-free, well-drained medium with a moderate starter charge. Dusty miller responds well to liberal irrigation: Plants grown too dry become stunted and hard. However, be careful to keep foliage dry to prevent disease. Fertilize with 150–200 ppm from 14-0-14, which will help add "fuzz" to foliage.

Do not pinch. Apply B-Nine at 5,000 ppm once stems elongate and every seven to ten days later, as required.

A 288 plug transplanted to a 4" (10 cm) pot requires from eight to nine weeks to be ready to be fully rooted and salable.

Pests & Diseases

Take a preventative approach to disease control. *Alternaria* leaf spot can be a significant problem. Insects rarely bother dusty miller, but aphids or spider mites may be occasional problems.

Varieties

'Silverdust' has deeply notched leaves that are covered with a wooly mat of gray. Plants grow 8–10" (20–25 cm) tall. 'Cirrus' has leaves that are shallowly lobed. Its foliage is greener, and its fuzz is bright white. Plants grow to 10" (25 cm) tall.

Related Species

Related plants include *Chrysanthemum ptarmiciflorum* 'Silver Lace', which has fine, lacy foliage on plants that grow 6–8" (15–20 cm) tall. Foliage is a lighter shade of gray.

Postharvest

Do not ship dusty miller with wet foliage: This is an invitation to disease. Instruct retailers to keep foliage dry when they are irrigating plants.

While grown and sold as an annual, dusty miller is actually a biennial and will produce yellow flowers on upright stems in its second year. Foliage may be dried and used in arrangements.

Setcreasea

Setcreasea purpurea (also known as *Tradescantia pallida*)
Common names: Purple heart; purple wandering Jew
Annual

Talk about revivals—*Setcreasea* has been reborn. Once grown during the heyday of the foliage plant craze as a hanging basket from coast-to-coast, *Setcreasea* has resurfaced in the recent years, this time as a fabulous complement to a variety of flowering plants in combination pots. It also makes a dandy groundcover.

The best part is that plants tolerate just about everything: very hot conditions, sun or shade, and even poor soils. Once established, they will even perform in cool conditions. Their purple color makes a great background for many of today's lighter-colored flowers and foliage.

Propagation

Most growers start by rooting cuttings brought in as liners from January to March. Crops can even be started later for landscape sales through early summer. Plants from which cuttings are taken must be actively growing and lush, because cuttings taken from hardened or flowering plants may be slow to root. *Setcreasea* requires moderate amounts of water while in propagation in order to avoid rotting off due to botrytis or water molds.

A 72 liner requires six to seven weeks from sticking to be ready to transplant to larger containers.

Growing On

Use 1 liner/4" (10 cm) pot; 2–3 liners/6" (15 cm) or 1-gal. (4 l); and 5–7 liners/10" (25 cm) basket. Once plants have begun to root into baskets, you can immediately move them up to the attic, where they will thrive in the spring. Many growers prefer to direct-stick smaller pots of *Setcreasea* using 2–3 cuttings/4" (10 cm) and 3–4 cuttings/6" (15 cm) or 1-gal (4 l) pot. This method works well due to the rugged nature of the crop and its quick, uniform rooting.

Begin fertilizing once plants have begun to root in with 200 ppm from 20-10-20, alternating with 14-0-14 at every other irrigation. Controlled-release fertilizer can be used very successfully to supply all the necessary nutrients to supplement a liquid fertilizer regime.

Setcreasea is perennial in many parts of the South and should be pruned in the spring to shape and remove old growth.

Transplanting a 72 liner into a 4" (10 cm) pot requires from six to eight weeks to be ready for spring sales.

Shasta Daisy (see *Leucanthemum*)

Sinningia

Sinningia speciosa
Common name: Gloxinia
Flowering potted plant

If ever one could imagine a plant stately enough to grace the Victorian parlor, gloxinia would be it. Bold, soft leaves topped by large, velvety, tubular flowers in rich blue, purple, red, pink, and white characterize this old-time favorite.

Gloxinia used to be a pot plant staple, being grown by just about every local retail florist grower in small, medium, and large towns across America for sales during winter and spring. As pot plant production has centralized and consolidated, gloxinia production has fallen off. Today, many growers use it mainly as a summer filler crop, with only a few growers looking at it as the bold, colorful year-round statement that it once was. If you are looking for a crop that is colorful, brings high prices, and comes in colors to match today's upscale interiors, consider gloxinia.

Propagation

Gloxinia is a long plug crop, taking sixteen to seventeen weeks from seed to flower. Most growers leave the job to specialists rather than tie up their bench space for so long. Single-sow pelleted seed onto a disease-free, germination medium with a pH of 6.0–6.5 and a low nutrient charge. Be sure the medium has about 20% perlite for good aeration. Cover very, very lightly with coarse vermiculite to help maintain moisture at the seed coat if humidity levels are below 70–75%. Provide 100–400 f.c. (11–4.3 klux) of light. Cover trays with clear plastic to maintain humidity levels and to keep trays at saturation, or place in germination chambers equipped with fog. Germinate at 72–75°F (22–24°C). Germination will begin in eight days but may take up to twenty-one days for a full stand.

As seedlings develop and cotyledons expand, begin a light fertilizer program of 50–75 ppm from 14-0-14 at every other irrigation, increasing to 100–150 ppm as true leaves emerge. Once cotyledons have expanded, begin to allow trays to dry somewhat between irrigations; allowing them to dry thoroughly before rewatering once true leaves have formed. As leaves expand, gradually increase light levels to 1,500–2,000 f.c. (17–22 klux). Plugs will benefit from supplemental light (400 f.c. [4.3 klux]) in the winter months.

A 288-plug tray will take from six to eight weeks from sowing to be ready to transplant to larger containers.

Growing On

Use one plug per 4, 5, or 6" (10, 13, or 15 cm) pot. Select a well-drained, disease-free, peatlite medium with a pH of 5.5–6.0 and a moderate initial starter charge. Gloxinia plugs sometimes stretch in transit. You will want to add stability to plants by planting plugs deeply: Place the first set of large leaves level with the potting mix line. Do not pack the medium around plants; gloxinia likes loose mixes. Handle seedlings gently, as broken leaves can serve as an entry point for disease. Many growers apply a fungicide drench after transplanting to prevent disease.

Keep the medium moist, but not saturated: Do not allow plants to wilt. Water-stressed plants may set buds prematurely before sizing up. Overhead watering with water below 50°F (10°C) will cause spots on leaves, so avoid overhead watering or use warmed water. Gloxinia performs superbly on ebb-and-flood benches or capillary mats. Use 20-10-20 at 200 ppm for the first three to four weeks, changing to 200 ppm from 15-0-15 for finishing. Some growers avoid all fertilizer during the first couple of weeks after potting, waiting until plants have put out roots before applying fertilizer. Gloxinia that receives too much nitrogen can show twisted, cupped, or curled leaves and off-color, blue-green foliage. During the winter, use calcium fertilizers to avoid problems. Excessive ammonium fertilizer will harden plants and cause them to produce large leaves.

Provide moderate light levels of 2,000–3,000 f.c. (22–32 klux). Higher light levels may cause leaf spotting or hard growth or may prevent buds from opening. Light levels that are too low produce soft plant growth and reduce bud count. Under summer conditions, provide 50–60% shade.

Gloxinia tolerates a wide range of temperatures; however, do not drop below 60°F (16°C) or go above 85°F (29°C). Ideally, maintain 75–80°F (24–26°C) days and 65–70°F (18–21°C) nights. At growing temperatures below 65°F (18°C), add time to the growing schedule. Once buds appear, you can drop temperatures to 70–75°F (21–24°C) days and 60–65°F (16–18°C) nights.

Grow plants on benches that facilitate air movement, such as wire mesh or expanded metal. Maintain good humidity levels for gloxinia: 50–60% relative humidity is ideal. Prevent dry, warm air from entering the growing area. Pad-and-fan cooling should be used during warm-season production.

On compact-growing varieties, growth regulators aren't required. On more vigorous plants, B-Nine at 1,000–1,500 ppm can be used once leaves have expanded to the pot rim. A second application can be made seven to ten days later if needed.

Growers in the past would remove the first one or two crown buds that formed and any surrounding leaves. This would increase the amount of light reaching the interior of the plant and cause buds to develop more uniformly, making a stronger statement of color once flowering began. Only a few growers now go to this extra trouble to make superior plants, but the end result warrants it.

The total crop time from sowing seed to a flowering plant is twenty to twenty-four weeks for winter- or spring-flowering plants. From transplanting a 288 plug to a 6" (15 cm) pot, allow fifteen to eighteen weeks.

Pests & Diseases

Cyclamen mites, spider mites, and thrips can be difficult problems. Diseases that can attack plants include botrytis, *Pythium*, and *Phytophthora*. Gloxinia is very susceptible to INSV (TSWV). If you don't have good control of thrips, do not attempt to produce gloxinia. Rogue and remove infected plants and implement a thrips prevention program.

Troubleshooting

Premature budding of plants can be caused by a number of factors, including holding liners too long before potting, high light combined with high temperatures, and low humidity and too high light. Excessive ammonia or high soluble salts can lead to dark green foliage on stalled plants. Be sure to thoroughly wet the medium and leach pots on a regular basis. High light intensity, low nitrogen levels, fertilizers high in ammonium, and/or high pH and high alkalinity can all result in yellow leaves.

Postharvest

Plants destined for supermarkets are ready to sell when one to three flowers are open; growers selling in local markets should wait until three to five flowers open. Choose compact-growing varieties if you are sleeving, boxing, and shipping plants. Do not keep plants in the dark for longer than three days, as buds can abort. Older varieties of large florist's gloxinias have brittle leaves: Handle plants with care during marketing. Ship gloxinia warm (60°F [16°C]).

Retailers should display plants in at least 100 f.c. (1.1 klux) of light and at temperatures of 65–75°F (18–24°C). Instruct retailers to keep plants moist, watering them from the bottom. Avoid wetting leaves, as cool water on leaves can cause spotting.

Gloxinia is related to African violets, so consumers will find they like similar conditions: A bright window with plants positioned out of direct sun is perfect. Remove dead flowers and leaves from plants to encourage flowering to continue for weeks.

Snapdragon (see *Antirrhinum*)

Solanum

Solanum melongena
Common name: Eggplant
Annual vegetable

Eggplants are upright growing plants from 20–28" (51–71 cm) tall with varieties having from small, egg-shaped fruit to teardrop types to very large fruit. Their skin color is mostly purple, but there are some that are violet striped, others are ivory white, and some have colors in between.

Propagation & Growing On

Sow seed into 288-plug trays. Select a disease-free, germination medium with a pH of 5.5–6.0. Cover seed with vermiculite for germination. Maintain substrate temperatures of 70–75°F (21–24°C) and ensure plug trays do not dry out. The substrate pH should be 5.5–5.8. Radicles will emerge in two to three days.

Move plug trays into Stage 2 and reduce irrigation once the radicle has emerged, allowing the medium to dry slightly before watering. Drop substrate temperatures to 70–72°F (21–22°C) and begin fertilizing with 50–75 ppm from 14-0-14 or 15-5-15, once cotyledons fully expand. Stage 2 will take from seven to ten days.

Stage 3, growth and development of true leaves, will take ten to fourteen days. Drop substrate temperatures to 65–70°F (18–21°C) and allow the substrate to dry out, but avoid permanent wilting between irrigations.

Prepare plugs for sale or transplanting in Stage 4 over seven days by lowering soil temperatures again to 60–62°F (16–17°C) and increasing feed to 100–150 ppm from 14-0-14 once a week. In general, cotyledons emerge in seven to fourteen days.

Allow five to six weeks from sowing to a transplantable 288 plug. Transplant plugs into a well-drained, disease-free medium with a moderate initial starter charge and pH of 5.5–6.2. Grow on at 55–65°F (13–18°C) nights and 60–70°F (16–21°C) nights. Transplanted to a 32-cell pack, plants are ready in five to seven weeks for spring sales.

Solenostemon

Solenostemon scutellarioides (formerly *Coleus hybridus*)
Common name: Coleus
Annual

There are few plant genera in which some varieties excel in sun while others prefer shade; and there are even fewer that one variety can be used in both locations. Coleus is one of these few—and cultivar selection is key to selecting for the right location in the garden or landscape. Another coleus anomaly is that its flowers are rarely appreciated, while its foliage is what customers prefer.

The foliage of coleus is festively patterned and is one of the easiest to grow—both for the grower as well as for the home gardener. Plants range in height from short selections (only 10–12" [25–30 cm] tall) to those that grow as tall as 3' (91 cm). Many of the vegetative varieties do well in full sun, while the majority of seed-propagated selections do better in partial shade—especially during the late morning or early afternoon. However, there are exceptions.

Propagation

Coleus is easily propagated by either cuttings or seed.

Seed

Sow seed onto a well-drained, disease-free, germination media. Cover lightly with coarse vermiculite and keep trays moist, but not fully saturated. Germinate at substrate temperatures of 72–75°F (22–24°C). Radicles will emerge in four to five days.

Move trays to Stage 2 for stem and cotyledon emergence, which occurs over ten days. Maintain substrate temperatures of 72–75°F (22–24°C). Allow trays to begin to dry slightly between irrigations for best germination and rooting. As soon as cotyledons expand, begin fertilizing with 50–75 ppm from 14-0-14 at every other irrigation. Provide 500–1,000 f.c. (5.4–11 klux) of light.

As soon as you have achieved a stand, move trays to Stage 3 to grow and develop true leaves, which will take from fourteen to twenty-one days for 512s. Allow the medium to dry thoroughly between irrigations, but avoid wilting. Fertilize once a week with 100–150 ppm from 20-10-20, alternating with 14-0-14. Slower-growing colors may need more frequent feeding to reach the desired height. If leaves are faded, plants may be receiving too much fertilizer. Increase light levels to 1,500–2,500 f.c. (16–27 klux). If growth regulators are needed, A-Rest, B-Nine, and Bonzi are effective.

Harden plugs for seven days prior to transplanting or sale by lowering the substrate temperatures to 60–62°F (16–17°C). Allow the medium to dry thoroughly between irrigations. Fertilize once with 100–150 ppm from 14-0-14.

A 288-plug tray will take from five to six weeks from sowing to be ready to transplant to larger containers.

Cuttings

Plants root readily in medium or Ellepots. If you bring in cuttings or liners, know your source. Most of the vegetatively propagated varieties on the market are infected with latent viruses, diseases, and sometimes nematodes. While these may not directly affect the performance of the plants in production or the garden, you will be bringing disease-carrying plants into your greenhouse that can serve as a source for subsequent infection to your other crops.

If you cannot stick cuttings right away, you can store them at 50–60°F (10–16°C) for up to twenty-four hours. Stick cuttings into a disease-free medium with a pH of 5.8–6.0 or into Oasis wedges. Provide substrate temperatures of 68–72°F (20–22°C) and ambient air temperatures of 75–80°F (24–27°C) days and 68–70°F (20–21°C) nights. Provide mist for four to six days until rooting begins. Provide 500–1,000 f.c. (5.4–11 klux) of light. As soon as there is any loss of foliage color, apply 50–75 ppm from 20-10-20. Root initials will form in five to seven days. A rooting hormone is not necessary.

When 50% of the cuttings have developed root initials, move cuttings to the next stage of rooting by maintaining soil and air temperatures. Increase light levels to 1,000–2,000 f.c. (11–22 klux). Apply a weekly foliar feed at 100 ppm from 15-0-15, alternating with 20-10-20. Increase the rate rapidly to 200 ppm as roots form. Cuttings will develop roots over seven to fourteen days.

Harden cuttings prior to sale or transplant for seven days by moving the liners from the mist area into an area with lower relative humidity. Increase light levels to 2,000–3,000 f.c. (22–32 klux) and feed once with 150–200 ppm from 15-0-15. Provide shade during midday to reduce temperature stress on the crop.

Some growers prefer to stick unrooted cuttings directly into their final growing container. If you choose this route, provide bottom heat and be dedicated to maintaining substrate moisture levels so the crop is not delayed.

If using a 105- or 84-liner tray, allow three to four weeks for rooting for the majority of varieties.

Growing On

Pot plants one seed-propagated plug per 36 tray, 1801, or 4" (10 cm) pot and one vegetatively propagated liner per 4 or 6" (10 or 15 cm) pot. Some growers also grow 10" (25 cm) hanging baskets using five to seven plugs or three to five liners.

Use a well-drained, disease-free medium with a pH of 5.6–6.0. Select a mix with a moderate initial starter charge for seed coleus and a high initial starter charge for vegetatively propagated coleus.

Fertilize at every other irrigation with 150–200 ppm from 15-0-15, alternating with 20-10-20.

Increase rates for vegetatively propagated coleus to 200–300 ppm as plants mature. Irrigate with clear water to leach pots if soluble salts climb above 1.0 mS/cm. Severe water stress will cause leaf scorch.

Grow on at 70–75°F (21–24°C) days and 62–65°F (17–18°C) nights. Low temperatures cause excessively compact, hard growth. Temperatures above 85°F (29°C) are not recommended.

While seed-grown coleus requires lower light levels (less than 5,000 f.c. [54 klux]), the majority of vegetatively propagated coleus can take full sun (4,000–6,000 f.c. [43–65 klux]), even in the Deep South. Leaf coloring is different under full sun, with plants showing more intense foliage variations under shadier conditions. In addition, on the same plant one leaf will shade another leaf, producing even more color variation and adding to the excitement. Note that high light intensity promotes more rapid flowering.

Pinch vegetatively propagated plants upon planting to larger pots. Seed varieties do not require pinching while growing in the cell pack or pot.

If growth regulators are needed, A-Rest, B-Nine, and Bonzi are effective. A test range for B-Nine on vegetatively propagated coleus is 2,500–5,000 ppm. When excessive B-Nine is applied, laterals fail to elongate. To improve lateral bud growth, increase ammonia fertilizer.

Upon transplanting a 288 plug to a 4" (10 cm) pot, allow six to eight weeks to finish for spring sales. If using a 105-liner tray, the crop time to a 4" (10 cm) pot is five to seven weeks for spring sales.

Pests & Diseases

Aphids, mealy bugs, thrips, and whiteflies can attack plants. Disease problems can include *Alternaria,* botrytis, rhizoctonia, *Pythium,* and *Verticillium.*

Troubleshooting

Excessively low light will cause severe internode elongation. Poorly branching plants may be underfertilized (lacking ammonia), may be growing under too low light or cold temperatures, or may be flowering. Foliage necrosis can be due to plants drying out to wilting between irrigations and/or high soluble salts levels in the soil.

Postharvest

Keep flowers picked off of plants or plants will stop growing.

Solidago

Solidago × *hybrida*
Common name: Goldenrod
Perennial, cut flower (Hardy to USDA Zones 4–8)

Solidago is a perennial whose day has yet to come. Wrongly accused of causing hayfever—ragweed is the true cause—this plant is infrequently used in the American perennial garden. Plants are commonly seen across the eastern and central United States, flowering along roadsides and open areas during mid- to late summer. Their golden-yellow color, appearing in Midwestern and Eastern fields in late summer, represents the "dog days" of summer to many.

Goldenrods form clumps with a small base and a wide crown. The stems are strong, erect to slightly arching, and void of secondary branches until they set flower buds. The blooms are small and scentless and appear in multitudes on the many-branched, flowering stalks. The plants grow 18"–3' (46–90 cm) tall, depending on species, with a spread of 15–24" (38–61 cm) on the broader varieties.

Plants flower in late July or August through to September.

Propagation

Solidago can be propagated by seed or terminal cuttings. Seed varieties are not true to type, are up and down in habit and height, and are not as preferred as those propagated by cuttings.

Seed

Sow seed onto a disease-free, germination medium with a low starter charge and a pH of 5.5–6.2.

Raw seed, without any special treatments, can be variable to germinate. Some varieties will germinate in five to eight days, others may take fifteen to twenty-one days. It is best to use enhanced seed (available from seed companies) for the highest germination.

If using enhanced seed, one seed per 288-plug tray is suggested. If using raw seed, sow two to three seeds per cell. Cover trays lightly with coarse vermiculite to help maintain moisture at the seed-coat level. Germinate at 70°F (21°C).

As cotyledons expand, lower substrate temperatures a few degrees and begin fertilizing with 50–75 ppm from 14-0-14. Slowly increase light levels from about 400 f.c. (4.3 klux) up to 1,000 f.c. (11 klux). Provide long days (more than fourteen hours) throughout seedling production. As seedlings develop true leaves, increase fertilizer to 100–150 ppm once a week from 14-0-14, alternating with 20-10-20, and increase light levels to 1,500–2,000 f.c. (16–22 klux). Drop night temperatures to 50–60°F (10–16°C) nights. Be sure to harden plugs for at least a week prior to sale or transplanting by maintaining lower day temperatures, acclimatizing plants to full sun, and allowing trays to thoroughly dry down between irrigations. Allow about six to eight weeks for plug crop time.

Sow seed to plug trays from June to September. Sowings done in June or July can be transplanted to either cell packs or 1-qt. (1.1 l) pots, or use three 288 plugs per 1-gal. (4 l) pot. These are overwintered dormant for spring sales the following year. For sowings done later, plugs can be transplanted to cell packs that are grown on through the summer and early autumn in cold frames or outside. Moved to a cold frame during autumn, cell packs are grown on cold or dormant until potted up in winter. Conversely, sowings done in October can be potted up to 1-qt. or 1-gal. (1.1 or 4 l) containers and grown cool through the winter. They can be grown colder, but not until the roots have fully massed the containers.

Cuttings

Terminal shoot cuttings are taken from vegetative (non-flowering) plants can be rooted in a 32, 50, 72, or 84 tray. Cuttings dipped in a rooting hormone, transplanted into a medium at 70°F (21°C), and given mist for seven to ten days will root rather quickly. Liner trays will be ready to transplant to larger containers in three to five weeks for an 84 tray. Order cuttings in from commercial perennial companies or take your own. The key is to keep the plants growing vegetatively and not producing flowers. Cuttings taken from plants in midsummer can bud up due to the oncoming short days (day lengths get shorter after June 21). For this reason, plants need to be kept lit to keep them from flowering.

Commercially, growers buy in either liners or bare-root transplants to begin their crops rather than propagating themselves.

Growing On

Regardless if propagated by seed or cuttings, select a well-drained, disease-free medium with a low initial starter charge and a pH of 5.5–6.2.

Transplant a plug or liner in summer to its final container and fertilize with 150–200 ppm from 14-0-14 at every other irrigation. Eliminate feed altogether by late August and grow on until cold weather approaches. Grow dormant, under protection, during winter. Sell plants green the following spring, and they will flower during the summer.

If plugs or liners are potted up during the fall or winter, grow on cool until roots establish in the final container (1 qt. or 1 gal. [1.1–4 l]). Move to a cold frame and grow on with cool or cold nights, but not dormant. Do not allow the pots to freeze. Plants can be sold in the spring and will flower in the summer.

If plugs or liners are potted to 4.5" (11 cm) pots in late winter or early spring, pots will be ready to sell in four to six weeks in the spring, and 1-gal. (4 l) pots in seven to ten weeks, although the roots will be to the side of the pot but not necessarily to the bottom of the container.

Do not pinch spring-produced crops; however, crops grown in the summer can be pinched to increase fullness. Space plants adequately to ensure light is reaching all sides: 1-gal. (4 l) pots need about 1.5 $ft.^2$/pot (0.14 m^2/pot). Pots placed too closely together shade one another and cause weak, stretched growth.

Cut Flowers

Goldenrod tolerates a wide range of soil conditions, but drainage must be good. Adjust the pH to 5.0–7.0 and make sure there is little or no residual fertilizer in beds prior to planting. High salts levels can negatively affect production and stem quality.

Plant at a density of 1.5 plants/$ft.^2$ (16/m^2) in low-light areas; 2 plants/$ft.^2$ (22/m^2) in high-light areas. Space plants at 8" × 12" (20 × 30 cm) or 24" × 24" (61 × 61 cm). Provide a layer of netting for extra support.

Water plants immediately after planting. During the vegetative growth phase, keep plants moist. Once short days begin, gradually allow soil to dry down to help bud set and toning. Once plants become established, they are somewhat drought tolerant. During finishing, it's common practice to allow slight wilting between irrigations. Since airborne diseases can become a serious problem, drip irrigation is ideal to keep foliage dry.

Most goldenrods are short-day plants: They flower naturally in the summer and early fall as days begin to shorten and then become dormant during the shorter days of late fall.

Immediately after planting, maintain long days (more than sixteen hours) to keep growth vegetative. Establish liners for two to three weeks: A good root system is critical for good-quality production. Some growers pinch plants, others do not. If you choose to pinch, do so after plants have been growing for three weeks, leaving two to three sets of leaves on the plant. The pinch will generate two to three top shoots, and additional ground shoots will begin emerging from around the roots.

Grow at night temperatures of 50–60°F (10–16°C) during long days, which will help keep plants growing vegetatively. In total, this vegetative period will take from five to eight weeks, depending on the growing temperatures and variety.

When shoots lengthen to 18" (46 cm), begin short days (twelve hours) to induce flowering. Growers forcing plants in the greenhouse in the winter in some parts of the country may need to use night interruption lighting for one or two hours per night to provide the necessary twelve hours to induce flowering. Winter production in the North will also benefit from supplemental lighting, which will help alleviate bud abortion. During bud initiation, lower nitrogen levels in your feed. Provide short days for four to five weeks at night temperatures of 62–65°F (17–18°C) to speed flower initiation. Higher temperatures give a more complete bud set. When flowering stems have swollen (buds begin to show color), the short-day period is completed.

Discontinue fertilizer during the week prior to harvest. During harvest, provide long days again to ensure that ground shoots that begin developing during harvest will maintain their vegetative-growing state. Harvest generally takes from seven to ten days to clean out a planting.

Total crop time from a liner takes ten to sixteen weeks, depending on temperatures, variety, and the flush from which the plants are being forced. Growers generally get three to five flushes per plant before the production stock is discarded. (For a year of production, you can generally expect three and a half to four and a half total flushes.)

When all flowering stems are harvested from a planted area, cut the plants back all the way to the ground, removing partially cut stems, stubble, and old wood. If old wood is left aboveground, it will produce prematurely budded stems of no value. Total removal of all old stems stimulates the root system to produce underground shoots. These underground shoots develop into the stems and flowers for the next flush or growing period.

In field production, plants will flower naturally from August to September (once a year), depending on the growing conditions and variety. *Solidago* will overwinter in most areas. After overwintering, the field crop should be cut back totally to the ground sometime in May or early June. The ground shoots will develop as described above and bloom again in the fall. In some regions, such as Southern California and Florida, a second flowering crop harvested in the spring may be possible. Natural, warm night temperatures (62–65°F [17–18°C]) will determine this.

Pests & Diseases

Aphids, spider mites, thrips, and whiteflies can become problems. Bees may be a problem in field cut flower crops during harvest. Disease problems that can strike include *Pythium*, *Phytophthora*, rhizoctonia, root rot, downy mildew, powdery mildew, and rust. Maintain good air movement around plants during production and keep foliage dry. Pine trees are hosts for rust that attacks goldenrod, so do not plant *Solidago* near pine trees.

Postharvest

Flowers can be used fresh or dried. Harvest fresh cuts when half the flowers are open. Cut stems about 12–15" (30–38 cm) below the flowers. Place stems in preservative solution immediately after harvest for local sales. Stems may be stored and shipped dry at 36–41°F (2–5°C) for long-distance sales.

Harvest flowers for drying when they are fully open. Dry stems standing up.

Consumers should be sure to select a very well-drained location in full sun and cut plants back to the ground in the spring. Instruct them to divide plants every three to four years for best performance. Goldenrod does *not* cause hay fever.

Spathiphyllum

Spathiphyllum spp.
Common name: Peace lily
Foliage plant

Spathiphyllum is a foliage plant staple. Plants perform in just about any interior environment and have great foliage and attractive flowers. Provided plants get at least good diffused light, consumers will be very successful. Peace lilies turn up everywhere: from mass markets to upscale florists, in dish gardens, and in mall and office interiors nationwide.

Because *Spathiphyllum* is an ultra-long crop that requires a lot of heat, it is mainly grown in Florida and other tropical locations and shipped north to consumption markets.

Propagation

Most foliage growers buy in established liners grown from tissue culture ranging in size from 200- to 32-cell trays. Propagating *Spathiphyllum* by tissue culture offers the grower the advantage of certain named varieties, improved crop uniformity, and year-round availability.

The most common tray size in Florida is 72-cell tray. Larger liners that are ten to fourteen weeks old are used for larger pots sizes. Propagators refer to how plants have been planted, using the terms "produced from clumps" or "plants per cell." The term "clump" refers to a tissue-culture-produced cluster of plantlets held together by a callus tissue base, while "plants per cell" refers to the number of individual microcuttings or seedlings planted per cell. Generally, tissue-cultured clumps produce very full plants but can lack uniformity as a finished product. They are useful for small pots such as 4 or 6" (10 or 15 cm) production, where growing times and chemical flower induction do not allow time for natural branching or flowering.

Young plants produced from individual microcuttings tend to have better uniformity, and, given adequate time or growth regulator treatments, most varieties will produce full plants.

Until recently, seed propagation had lost volume to tissue-culture production. Previously, seed production, although economical, lacked the quality and uniformity demanded by today's growers. Recently, large-scale production of seed crops has gained popularity due to controlled seed production that's backed by excellent production and marketing systems.

Growing On

Spathiphyllum requires a well-drained medium with good water-holding capacity. Generally, a 1:1:1 ratio of peat, perlite, and bark is a common potting mix for the southern United States, while coarse peat has been common in Europe. Maintain the pH at 5.8–6.5.

Use a slow-release or liquid feed in a 3:1:2 nitrogen-phosphorus-potassium (N-P-K) ratio. Slow-release, dry fertilizers, constant liquid feed, or a combination of the two are equally effective methods. Many growers incorporate slow-release fertilizer in the potting mix and supplement later with liquid or additional dry applications. Additionally, many growers use a weekly foliar nutrient spray of 1 lb. (454 g) each of urea, potassium nitrate, and magnesium nitrate per 100 gal. (379 l). A soluble source of trace elements can also be added to this mix. *Spathiphyllum* is a relatively high feeder.

Irrigation frequency should be designed to keep the mix evenly moist during all phases of the crop cycle. *Spathiphyllum* easily tolerates overhead irrigation and does exceptionally well with drip. Plants do not tolerate saturated potting mix conditions for extended periods of time. Various diseases can easily infect overwatered plants, causing wilting or collapsed leaves, necrosis along leaf margins, and extensive root damage.

Production light intensities are somewhat variety dependent, although a range of 800–2,500 f.c. (8.6–27 klux) is commonly used. Plants grown in the lower range tend to have longer petioles, reduced branching, a softer appearance, and darker green color (pending nutrition). Under higher light intensities, the plants tend to be more compact, exhibit more branching, and be lighter in color. Plants grown under excessive light intensities exhibit curled, pale, or chlorotic leaves. Plants grown at the extremes of the light range may produce fewer flowers than those grown in the central range.

The optimum temperature range is 68°F (20°C) nights and up to 90°F (32°C) days, but plants will tolerate lows of 45°F (7°C) and highs of 95°F (35°C). *Spathiphyllum* will not tolerate frost or event short-term freezing temperatures without foliage damage and possible crop loss.

Two growth regulators are commonly used on *Spathiphyllum,* benzyl adenine (BA) and gibberellic acid (GA). BA is very effective at enhancing the branching and fullness of the plant and is generally used at the liner stage. In addition to this application, some growers also apply BA shortly after liners have been planted into a larger pot. Enhancing branching qualities and fullness are especially important to the small pot grower, since the shorter production time limits the impact of natural branching. BA can be applied as a spray or drench at 250–1,000 ppm. BA can inhibit root development if applied before roots are well established. The overall effect of a BA application depends on the variety, BA concentration, stage of growth, application method, and season.

GA is used extensively to force early or year-round flowering. With maturity, *Spathiphyllum* will naturally flower consistently in the spring and sporadically during the rest of the year. Since the market demands that plants be sold with flowers, growers use GA to gain a year-round sales advantage while also allowing the programming of crops for holidays, promotions, or weekly orders. With GA, growers can also force early flowering to allow the production of smaller pot sizes. A standard treatment is a single foliage spray of 150–250 ppm eight to fifteen weeks prior to sale. The spray concentration and time between treatment and flowering depend on varieties and season of the year. Some varieties produce good-quality flowers after treatment, while others do not. Treated plants may exhibit a narrowing of new leaves, a stretching of the petioles, and distorted flowers. Growers need to test GA in their facilities and with their cultivars. Growers also need to determine their market's tolerance to some of these negative quality issues.

Production times are directly related to the variety, pot size, liner size, and cultural environment. Generally, a 3–4" (8–10 cm) pot requires three to five months; a 6" (15 cm) pot, four to nine months; an 8" (20 cm) pot, seven to eleven months; a 10" (25 cm) pot, eight to twelve months; and a 14" (36 cm) pot, twelve to twenty months.

Pests & Diseases

Snails are significant *Spathiphyllum* pests. Aphids, caterpillars, mealybugs, scale, thrips, and whiteflies can also attack plants. Thrips have a voracious affinity for some of the larger-leafed varieties. If left unchecked, their damage can become severe. When treating for thrips, it's important to spray to runoff, because thrips accumulate in the leaf whorl and do their damage on emerging leaves.

Disease problems are directly related to variety, climate, sanitation practices, and cultural conditions. Plants grown in shade-house structures with overhead irrigation, frequent rains, and warm temperatures will be challenged by diseases more frequently than greenhouse-grown plants. *Phytophthora, Myrothecium* leaf spot, and *Pythium* root rot can attack *Spathiphyllum.* However, the most important disease problem is root rot caused by *Cylindrocladium spathiphylli.* This fungus is spread in soil and water and can infect and kill very rapidly. The first symptom is a yellowing of lower leaves, sometimes accompanied by a slight wilting progressing to severe wilting. Splashing water can carry spores onto foliage, resulting in elliptical brown spots on leaves and petioles. Lower portions of the petioles frequently rot, and, at the final stage, roots are severely rotted and the foliage totally collapses.

To combat *Cylindrocladium,* always use pathogen-free plants from tissue culture or seed sources. Use only clean media and new pots. Frequently rogue crops and promptly remove infected or suspect plants from production areas. Always discourage bringing finished plants from other growers into your facility, as these can be sources for the disease. Growing plants on raised benches or concrete blocks is the most effective preventive control in areas where the disease is established.

Troubleshooting

Magnesium deficiency appears as golden-yellow margins on lower leaves. It's easier to prevent the problem with magnesium supplements rather than to reverse a deficiency. Iron and manganese deficiencies are shown as reduced growth rates and chlorotic leaves, both of which can occur during winter months when the substrate temperature is below 65°F (18°C). Sulfur deficiency shows up as an overall chlorosis of the foliage and is sometimes seen when growers use highly refined, low-sulfur fertilizers. Boron deficiency may be a cause of longitudinal ribbing of the leaves, often seen in the new growth.

Plants grown at temperatures above 95°F (35°C) for extended periods can exhibit narrow leaves (strap leaf), loss of color, inhibited root development, and reduced flower quantity and quality.

Varieties

'Mauna Loa' is propagated by seed. 'Sensation' is a tall grower, reaching to about 5' (1.5 m) tall. 'Lynise' and 'Deneve' are also larger growing. 'Tasson' and 'Viscount' are both medium-sized varieties, while 'Petite' and 'Wallisii' are smaller-growing varieties. Additional varieties are also available and can be found from companies specializing in tropical foliage plants.

Postharvest

Plants can be sleeved, boxed, and shipped; however, minimize time in the dark to maximize postharvest quality. Ship at 55–60°F (13–16°C), ideally for no longer than seven to fourteen days. If you have to store plants, hold them under a minimum of 150 f.c. (1.6 klux).

Retailers should unpack and water plants immediately. Display plants at 65–75°F (18–24°C) under a minimum of 150 f.c. (1.6 klux). Maintain uniform moisture in the pots and avoid water with fluoride, as it can cause leaf damage.

Spilanthes (see *Achmella*)

Stachys

Stachys byzantina
Common names: Lamb's ear; woolly betony
Perennial (Hardy to USDA Zones 4–9)

A long-lived perennial, *Stachys byzantina* has woolly, gray foliage similar to dusty miller or other gray-leaved foliage plants. *Stachys* grows 20–28" (51–71 cm) tall when flowering, but the carpet of foliage is only about 8–10" (20–25 cm) tall without blooms. The erect, floral spikes measure 18–24" (46–61 cm) long with the top half dotted with 0.5" (1 cm) rose or rose-purple flowers. The foliage is more sought after than the flowers, although both provide an excellent show.

Propagation

Both seed and cuttings are the common methods of propagation of *Stachys.*

Seed

The majority of perennials propagated by seed should be sown the previous year to produce the best flowering plants the following year.

On *Stachys,* the foliage takes center stage for most gardeners, while the flowers are secondary. For this reason, seed can be sown either the year before for flowering plants during the summer of the following year; or sown during winter for cell pack, 4 or 6" (10 or 15 cm) pot sales in spring. However, these plants will not flower until the following year.

Sow seed to a 288-plug tray using one or two seeds per cell. Do not cover seed, and germinate at a soil temperature of 70–72°F (21–22°C). Plugs are ready to transplant to larger containers in five to seven weeks.

Cuttings

Stick stem tip cuttings into 32, 40, or 72 liners and provide root zone temperatures of 70–72°F (21–22°C). A rooting hormone is not necessary. Provide mist sparingly to allow for rooting, but keep in mind that excessive misting or moisture causes rotting of the pubescent (hairy) foliage. An average of four to six days is usually necessary to get rooting to begin. During high-humidity times of the year a preventative fungicide is recommended to reduce foliage disease. *Stachys* requires four to five weeks from sticking cuttings until it is ready to transplant to larger containers.

Growing On

Either plugs or liners can be transplanted to larger containers. *Stachys* is a perennial that can be potted up in late summer and overwintered dormant in a 1-gal. (4 l) pot to sell the following spring. However, if using a rooted liner (32, 40, 72, etc.) and potting up one liner per a 1-gal. (4 l) pot in winter, plants will be salable in ten to thirteen weeks for spring sales, while plants in 1-qt. (1.1 l) pots will be ready in six to eight weeks.

If sowing from seed and using a 288 plug, transplanting three plugs per 1-gal. (4 l) pot is recommended if potted up in winter for spring sales. Even so, the container may not be fully rooted by spring for planting to the garden. Instead, sow seed no later than October or November, transplant one 288 plug to a 32- or similar cell/liner tray and allow to root for five to seven weeks and then transplant to a larger container. These containers will be better rooted.

Pests & Diseases

Whiteflies and root and stem rots are common with *Stachys.* When growing plants in containers during the summer, high heat and humidity, when combined with overhead watering, often lead to foliar diseases. One or two applications of fungicide may be warranted.

Statice (see *Limonium*)

Stevia

Stevia rebaudiana
Common names: Sweet leaf; sugar leaf
Annual herb (perennial in the Deep South)

The leaves of this plant are incredibly sweet and are used fresh, dried, or powdered to sweeten a number of foods and beverages. Plants grow upright ranging from 15–28" (38–71 cm) in height and prefer full-sun locations.

Propagation & Growing On

For a 288-plug tray, sow one to two seeds per cell, cover lightly with vermiculite, and germinate at 68–70°F (20–21°C). Seedlings will emerge in eight to twelve days. Plug trays will be ready to transplant to larger containers in five to seven weeks.

Transplant plugs to 4" (10 cm) pots and grow on at 58–60°F (14–16°C) nights with days 10°F (6°C) warmer, once roots have established in the pot. Plants will be salable green in six to eight weeks and grow mostly upright, not branching along the stem.

Stock (see *Matthiola*)

Strawberry (see *Fragaria*)

Sunflower (see *Helianthus*)

Sutera

Sutera cordata
Common name: Bacopa
Annual

Bacopa's delicate flowers and trailing plant habit make it an ideal addition to combination pots and baskets. Flowers come in shades of blue, pink, and white; although it is the white-flowering selections that are the most common. Plants grow from 8–12" (20–30 cm) tall.

Avoid overwatering bacopa throughout production. While day neutral, flowering is intensified at higher light levels and temperatures below 65°F (18°C).

Propagation

Bacopa can be propagated by cuttings or by seed. However, cutting propagation is more prevalent.

Cuttings

Maintain sufficient moisture in the rooting medium so that water can be squeezed out during callus formation (the first three to four days). Provide root temperatures of 65–68°F (18–20°C) and air temperatures of 68–72°F (20–22°C) during the day and 65–70°F (18–21°C) at night. Maintain the relative humidity at 75–90% at the base of cutting. Adjust mist frequency as light and ambient air temperatures change. During callus formation, night misting may be required. Maintain 500–1,000 f.c. (5.4–11 klux) of light. Provide retractable shade so light levels can be increased as cuttings mature. Use 20-10-20 at 50–75 ppm applied to foliage as soon as leaves lose color.

When 50% of the cuttings have initiated roots, encourage root development by reducing substrate moisture. Reduce the mist application at night, increase the intervals between mist applications, and shorten the duration. As roots form, increase light levels to 1,500–2000 f.c. (16–22 klux). Apply a foliar feed, alternating between 15-0-15 and 20-10-20 at 100 ppm. Quickly increase feeding to 200 ppm as roots form. To harden cuttings, move liners to an area with lower humidity, lower temperatures, and higher light (3,000–5,000 f.c. [32–54 klux]). Provide shade in the middle of the day to reduce temperature stress. Apply fertilizer twice a week, alternating 15–10–15 and 20-10-20 at 150–200 ppm.

An 84-liner tray with two cuttings per cell requires from four to five weeks to be ready to transplant to larger containers.

Seed

Use one multiseeded pellet per 288-plug tray and do not cover seed. Germinate at 68–73°F (20–23°C). Trays will be ready to transplant to larger containers in four to five weeks.

Growing On

Cuttings

At potting, avoid planting liners too deeply, as *Pythium* or rhizoctonia could become a problem. Use 1 plant/4" (10 cm) pot, 2 plants/6" (15 cm) pot, and 3–4 plants/10" (25 cm) hanging basket. Temperatures determine growth rate: At temperatures above 65°F (18°C), growth is quick. However, the highest quality occurs at temperatures less than 60°F (16°C). Temperatures of less than 65°F (18°C) promote flowering. Optimum light levels are 3,500–5,000 f.c. (38–54 klux). When light levels are too low, plants stretch. Use a well-drained medium with good water-holding capacity and a starting pH of 5.5–6.5. Overwatering can result in chlorosis. If this happens, incorporate additional iron into the mix.

Use a constant liquid feed of 20-10-20 or 15-0-15 at 200–250 ppm. To encourage continued flowering, incorporate a slow-release fertilizer into the medium. Apply iron if plants show chlorosis.

Plants may be pinched once they are established in the final growing container. Pinch *Sutera* back to the edge of the growing container. This helps to develop full plants and avoids open centers. Do not pinch 4" (10 cm) pots. Growth regulators are not necessary. If plants seem to be producing excessive growth, fertilizer rates may be too high and/or light levels may be too low.

Seed

For seed, follow the guidelines for vegetative propagation. As for crop time, transplant one 288 plug to a 4" (10 cm) pot and allow five to seven weeks for a flowering plant.

Pests & Diseases

Sutera can become infested with insects such as whiteflies, aphids, fungus gnats, leaf miners, and thrips. Diseases that may affect plants include botrytis, *Pythium*, rhizoctonia, and powdery mildew.

Varieties

'Snowstorm', 'Abunda', 'Calypso', and 'Bahia' are just a few of the vegetative selections that are available. Contact your plant broker for a list of offerings.

The only seed variety currently available is 'Utopia'. It is offered in either white or lavender-blue flower selections.

Postharvest

Provide temperatures of 65–72°F (18–22°C) days and 55–60°F (13–15°C) nights. Flowering is enhanced by night temperatures below 65°F (18°C). *Sutera* grows best in morning sun to part shade. Maintain medium to dry moisture levels: Overwatering leads to chlorosis and/or botrytis. However, avoid wilting plants. Advise consumers that *Sutera* is best grown in containers where they can enjoy its trailing habit and can better control irrigation.

Sweet Potato Vine (see *Ipomea*)

Syngonium

Syngonium podophyllum
Common names: Syngonium; nephthytis
Foliage plant

Syngonium is a great plant for just about any grower. Its leaves are a great filler in dish gardens, and bedding plant growers will find that it's a reliable filler for combination pots destined for the shade. In production, *Syngonium* is versatile as well: It can be grown in the North as well as Florida and Texas, just as long as temperatures are warm enough.

Most growers start their crops from liners grown from tissue-cultured clumps bought in from specialists. Starting plants only from clean stock is vital, as the most serious production problems can be bacterial diseases.

Growing On

Pot 1 liner/4" (10 cm) pot, 2 liners/6" (15 cm), and 3 liners/8" (20 cm). Select a disease-free, sterile medium with a moderate initial starter charge and a pH of 5.5–6.2. *Syngonium* requires uniformly moist conditions, so make sure the mix has moisture-holding capacity. Irrigate with tempered water to avoid damage to foliage caused by cold water.

Plants will tolerate a wide range of growing temperatures, from 65–95°F (18–35°C). For fastest growth, maintain warmer conditions. While *Syngonium* will tolerate ultra-low light levels, plant growth and habit are best at 1,500–3,500 f.c. (16–38 klux). Fertilize with 200 ppm 20-10-20 at every irrigation. Maintaining good nutrition levels can decrease the severity of *Xanthomonas.*

After potting a liner to a 4" (10 cm) pot, allow eight to ten weeks for salable plants.

Pests & Diseases

The most serious diseases that can strike *Syngonium* are bacterial: *Erwinia* and *Xanthomonas* × *campestris* pv. *syngonii.* Pseudomonas can also strike. Other diseases that can attack include *Cephalosporium, Ceratocystis, Myrothecium,* and rhizoctonia. Insect pests include mealybugs, scale, and spider mites.

Troubleshooting

Phosphorous deficiency can cause brown spots on lower foliage. Water-soaked lesions on leaves may be signs of a bacterial disease or damage caused by cold water. When older leaves yellow and drop suddenly, plants may have been water stressed: *Syngonium* needs consistent moisture. Under ultra-low light levels, *Syngonium* will stretch severely. Maintain 150 f.c. (1.6 klux) as a minimum.

Varieties

'White Butterfly' is one of the most popular *Syngonium* varieties.

Postharvest

Syngonium may be shipped at 55–60°F (13–16°C). Colder temperatures can cause injury. Do not ship plants for more than two weeks; and when shipping for such long time periods, maintain high humidity during storage so plants don't dry out.

Syngonium is sensitive to ethylene and will show epinasty, or severe drooping, when plants are exposed for long periods of time. Retailers should unpack plants as soon as possible, remove sleeves, and place them under a minimum of 150 f.c. (1.6 klux) of light. Plants should recover in a few days. To keep plants from stretching and gaping open, be sure to maintain light levels.

T

Tagetes

Tagetes erecta, American marigold

Tagetes erecta
Common name: African or American marigolds
Tagetes patula
Common name: French marigolds
Tagetes erecta × *patula*
Common names: Triploid marigolds; mule marigolds
Annual

While marigolds are traditionally organized in two main classes, French and American/African, breeders have also crossed these two to form what is called "triploid marigolds." Due to the chromosome count, triploid marigolds cannot reproduce—just like when a horse and a donkey produce a mule offspring; hence the common name mule marigolds.

Marigolds are a mainstay bedding plant. Their yellow, gold, and orange flowers grace yards, public parks, and city streets the world over all season long. They are one of the most versatile of all bedding plants, thriving in both wet and dry conditions. They are also edible, making a fine summertime addition to salads. In southern parts of the United States, marigolds put on their best flower show during cool, fall nights. For growers, marigolds are very easy and one of the most trouble-free bedding plants to grow.

Propagation

Marigolds are propagated from seed. Sow detailed, coated seed onto a disease-free, well-drained, germination medium with a pH of 6.0–6.2. Cover seed lightly with coarse vermiculite after sowing. Germinate seed at substrate temperatures of 72–80°F (22–27°C) and 70–72°F (21–22°C) days and 65–68°F (18–20°C) nights. Maintain moist but not saturated conditions. Keep ammonium levels below 10 ppm. Radicles will emerge in two to five days.

After radicle emergence, move trays to Stage 2 to develop stems and cotyledons. Reduce substrate temperatures to 68–70°F (20–21°C) and continue to maintain day/night air temperatures. Begin to allow trays to dry slightly between irrigations. Begin fertilizing once a week with 50–75 ppm from 14-0-14 once cotyledons are expanded. Be sure foliage is dry by nightfall to prevent disease. Stage 2 takes about seven days.

In Stage 3, develop true leaves by reducing substrate temperatures to 65–68°F (18–20°C) and maintaining day/night air temperatures. Allow the medium to dry thoroughly between irrigations. Fertilize with 100–150 ppm once a week from 14-0-14, alternating with 20-10-20. If growth regulators are required, A-Rest, B-Nine, Bonzi, and Cycocel are effective. Stage 3 will take about fourteen days for 512s and twenty-one days for 384s and 288s.

Harden plants for one week prior to sale or transplanting by lowering substrate temperatures to

60–65°F (16–18°C) and allowing trays to dry thoroughly between irrigations. Fertilize once a week with 100–150 ppm from 14-0-14.

Most growers buy in marigold plugs along with the rest of their bedding plant plugs. However, marigolds are such an easy, trouble-free plug crop that some growers prefer to sow them into 384s or 288s and grow their own.

Keep plants actively growing by planting plugs when they are ready. Holding plugs dry and hungry after they have arrived hardens the plants, sets plugs for premature bud formation, and increases crop time.

Regardless of species, sowings made to 288-plug trays require three to four weeks to be ready to transplant to larger containers.

Growing On

French & triploid marigolds

Marigolds prefer deep cell packs and pots, which will also increase shelf life. Plant one plug per cell pack or 4" (10 cm) pot or use three 384 or 288 plugs per 1-gal (4 l) container. Transplant into a well-drained, disease-free medium with a moderate initial starter charge and a pH of 6.5–6.8.

Fertilize at every other irrigation with 150–200 ppm from 15-0-15, alternating with 20-10-20. Marigolds are susceptible to micronutrient toxicity, especially to iron and manganese, at low pH. Be sure to keep the pH of the growing medium between 6.2–6.5 throughout production. This can be accomplished in two ways, by managing your pH through nitrate fertilizers (14-0-14 or 15-0-15) or by starting with a medium that has a high pH. As pH declines, iron uptake increases, which causes yellow spots on leaves—iron toxicity. Different varieties show more sensitivity. For example, the French marigold 'Boy' series does not show symptoms, while American marigolds are highly sensitive. Reduce the fertilizer rate or frequency at visible bud to increase shelf life.

Grow the crop on at 60–78°F (16–26°C) days and 55–60°F (13–16°C) nights. Reduce temperatures in the final week of production to 50°F (10°C) nights to increase shelf life.

If you need to use growth regulators, A-Rest, B-Nine, and Bonzi are effective. B-Nine or Bonzi can be applied at the second set of true leaves. If you use A-Rest, several applications may be required, spaced fourteen to twenty-one days apart. Marigolds respond very well to negative DIF (days cooler than nights).

For French and triploid varieties, the crop time from sticking a 288 plug to a 32-cell pack is four to six weeks, depending on variety. For 4" (10 cm) pots, it is five to seven weeks.

American/African marigolds

You can follow the culture noted above for American/African marigolds, but there is one additional note.

Provided you can grow your crop under moderate production temperatures, these marigolds like high light: 4,000–6,000 f.c. (43–65 klux). American marigolds are daylength responsive, flowering faster under short days. If you are sowing seed after February 15, grow your crop under short days (nine-hour day lengths) beginning at germination and continuing for two weeks. You can achieve short days by pulling black cloth from 5 P.M.–8 A.M., or using inverted black plastic standard flats to cover germination trays. Providing American marigolds with a short-day treatment will help to produce a smaller overall plant habit and earlier, more uniform flowering packs. If black cloth is unavailable or unrealistic, add two to three weeks of crop time for flowering larger plants, or sell the plants green with color picture tags.

For American/African varieties, the crop time is seven to nine weeks for flowering 4" (10 cm) pots transplanted from a 288-plug tray.

Pests & Diseases

Spider mites have a love affair with marigolds. Other insect problems can include aphids, thrips, and whiteflies. Diseases that attack marigolds include aster yellows, botrytis, and various leaf-spot diseases.

Troubleshooting

Red foliage may be a sign that temperatures are too low (below 50°F [10°C]). Temperatures above 85°F (29°C) can cause plants to stretch. Iron toxicity will show itself as yellow spots on leaves. Maintain pH levels above 6.3 throughout production to avoid problems.

Varieties

T. erecta is commonly called American marigold (also formerly known as African marigold). Plants are usually tall growing, to 24–30" (61–76 cm) in the garden, although dwarf varieties are available as well. Most growers prefer to plant them in 4" (10 cm) pots. If you want to offer them in flats, too, select 1801s, so the crop will have some proportion. American marigolds are best sold green or just as buds begin to open. Plants are long-day sensitive, although newer varieties are less so. When these plants are crowded during production, botrytis and rhizoctonia can become problems.

The most famous variety in this marigold class is 'Inca'. 'Inca II', marketed as an improved 'Inca', has a more compact plant habit and is one week earlier. Stems are stronger, so they will stand up better in shipping too. The 'Marvel' series offers sturdy, very full flowers on 10–12" (25–30 cm) plants. It is great for 1801 or 4" (10 cm) production. Select from 'Gold', 'Orange', 'Yellow', or 'Mix'. Also take a look at 'Taishan' and 'Antigua', in 'Gold', 'Orange', 'Yellow', and 'Mix' as main American marigold series, both are 12–14" (30–36 cm) in the garden and reliable performers. 'Discovery', in 'Yellow' and 'Orange', created the 1801-pack flowering market for this class with its cookie-cutter uniformity and ease of flowering. For something fun, try 'Sweet Cream' with its cream-colored flowers. If you are trying marigolds as a cut flower, try the 'Galore', 'Gold Coin', 'Jubilee', or 'Lady' series. 'Gold Coin' and 'Jubilee' also make excellent hedge marigolds for landscapers.

T. patula, better known as French marigold, is what most gardeners envision when they think about marigolds. These lower-growing plants make a great border or edging flower. French marigolds are classified by their flower types: single, anemone, crested, and double. They are the fastest marigolds to grow—some flower in just seven weeks from sowing the seed.

For a double-flowering series, rely on the 'Bonanza' series in 'Bee', 'Bolero', 'Deep Orange', 'Flame', 'Gold', 'Orange', 'Yellow', and 'Mix'. 'Bonanza' is a great outdoor performer with large flowers. The 'Boy' series, an old, double-flowering standby, is still one of the most important marigolds. It is really fast, but the 'Bonanza' and 'Hero' series offer larger flowers. The 'Hero' series, also a double, is as large-flowered as 'Bonanza' and comes in 'Bee', 'Flame', 'Gold', 'Harmony', 'Orange', 'Spry', 'Yellow', and a 'Mix'. If you're a southern grower, the 'Janie' series will be your mainstay, as it has excellent heat tolerance. However, its flowers are much smaller than 'Bonanza' and 'Hero'.

For anemone-flowered French marigolds choose 'Durango' or 'Safari': Their flowers are supersized and come in wide color ranges. Be sure to offer them in 4" (10 cm) pots: They are spectacular.

In single French marigolds, go with 'Disco', available in separate colors, but most growers buy in a few plug trays of 'Mix'. 'Disco' is a great addition for combination pots, giving an air of country charm.

Triploid marigolds, *T. erecta* × *patula*, are outstanding performers in the garden, outflowering other marigolds with ease. This interspecific cross between American and French marigolds has larger flowers and shows excellent heat tolerance outdoors. In field trials, they are the star marigolds year after year. Germination is lower than their related species and because seed is very difficult to produce, it tends to be expensive. Both of these factors limit their impact in the market. Nevertheless, if you are working to set yourself apart from the crowd with outstanding varieties, be sure to offer triploids and sign them well at retail. 'Zenith' and 'Sunburst' are the series of choice and come in a number of flower colors. Put your triploid seed order in early to be assured of selection.

Related Species

For a real treat in the landscape, try *T. tenuifolia* 'Gem' in 'Golden', 'Lemon', or 'Tangerine'. The foliage of *T. tenuifolia* has a citrus scent. This single marigold has fine, lacy foliage and grows to 14" (36 cm) in the garden. Multiseed (five to seven seeds) directly into its final growing container. Total crop time is eight to ten weeks.

Postharvest

Marigolds are ready to ship as soon as plants begin to flower. The first flower of most French marigolds will be the smallest, so many growers like to wait until the second flower for more show at retail and faster impulse sales. Marigolds can be shipped in the dark for four to seven days at 50°F (10°C). Avoid high temperatures, as they will cause stretching. If you have to hold flats, marigolds are easy to hold back cool (50°F [10°C]) and dry. They can even be cut back and reflowered or bumped into pots if market timing is off.

Retailers can display marigolds in full sun, provided temperatures are moderate: 50°F (10°C) nights are perfect. In high-light, southern locations, providing light shade is beneficial to reduce water requirements.

Talinum

Talinum paniculatum
Common name: Jewels of Opar
Annual

Talinum produces a cloud-like waft of small, rose-pink flowers above its crown of either green or chartreuse foliage. Plants grow from 28–32" (70–80 cm) when in full bloom. Plants prefer full sun, excel in high heat, and can tolerate drought once established.

Propagation

Easily produced from seed, *Talinum* should be sown in 288-cell trays. Use a well-drained, disease-free, soilless medium with a pH of 5.5–6.1 and a medium initial nutrient charge (EC less than 0.75 mS/cm with a 1:2 extraction). A light to medium vermiculite cover is recommended. Germination takes approximately four to six days at 68–74°F (20–24°C). Keep soil wet (level 4) during Stage 1 and maintain 95% or more relative humidity (RH) until radicle emergence.

For Stage 2 (cotyledon development), drop the soil temperature to 68–72°F (20–22°C) and provide light levels of up to 2,500 f.c. (27 klux). Reduce soil moisture slightly (level 3) to allow the roots to penetrate into the medium. Apply fertilizer at rate 1 (less than 100 ppm nitrogen and less than 0.7 mS/cm EC) from nitrate-form fertilizers with low phosphorous.

For Stage 3 (true leaf development), drop soil temperatures to 66–70°F (19–21°C) and maintain light levels at 2,500 f.c. (27 klux). Allow the medium to dry until the surface becomes light brown (level 2) before watering. Increase fertilizer to rate 2 (100–175 ppm nitrogen and 0.7–1.2 mS/cm EC).

For Stage 4 (holding), provide soil temperatures of 66–68°F (19–20°C) and increase light levels to 5,000 f.c. (54 klux), if temperatures can be controlled. Moisture and fertilizer ratios can be followed as noted for Stage 3.

A 288-plug tray requires from five to six weeks from sowing to be ready for transplanting to larger containers.

Growing On

Use a well-drained, disease-free, soilless medium with a pH of 5.5–6.2 and a medium initial nutrient charge. Provide temperatures of 62–66°F (17–19°C) at night and 66–74°F (19–23°C) during the day.

Growing 'Limón' under slightly shady conditions will result in lighter lime-colored foliage. However, plants will tend to stretch under very shady conditions.

Table 37. Crop Scheduling for *Talinum* from Transplant to Finish

CONTAINER SIZE	PLANTS PER POT	WEEKS FROM TRANSPLANT	TOTAL WEEKS
306 premium pack	1	4–5	9–10
4–4.5" (10–11 cm) pot	1	4–5	9–10
6–6.5" (15–16 cm) pot	3	4–5	9–10

Note: Crop times are given for foliage only. Add two more weeks to bring plants into flower. However, it's best to sell the plants before the flower stems elongate, since the flower spikes are tall and will increase the overall crop height. Crop times are longer under cooler conditions.

When grown under high-light conditions, 'Verde' will have slightly darker green foliage.

Apply fertilizer at rate 3 (175–225 ppm nitrogen and 1.2–1.5 mS/cm EC) once a week from a nitrate-form fertilizer with low phosphorus. A balanced ammonium-and-nitrate-form fertilizer may be applied as needed.

Maintain the medium EC at 1.5–2.0 mS/cm and the pH at 5.8–6.2. For a constant fertilizer program, apply fertilizer at rate 2 (100–175 ppm nitrogen and 0.7–1.2 mS/cm EC.

Because 'Verde' is slightly more vigorous in growth than 'Limón', B-Nine/Alar (daminozide) can be applied, if needed, at 2,500–3,500 ppm (3–4.1 g/l of 85% formulation or 3.9–5.5 g/l of 64% formulation) as a foliar spray once after transplant to tone the crop.

Pests & Diseases

Aphids can become problems for *Talinum.* There are no significant disease problems for this crop.

Tanacetum

Tanacetum parthenium (formerly *Chrysanthemum parthenium*)
Common name: Matricaria
Cut flower

The bright, golden-yellow or white, button flowers of matricaria are a staple in mixed bouquets. Matricaria may also be grown in pots and sold as a tender perennial. In the garden, matricaria performs best in full sun and when provided with adequate moisture. Plants are short lived but tend to reseed themselves plentifully.

Propagation

Matricarias are grown from seed. The crop takes about four weeks in a 200-cell tray. After sowing, do not cover seed. Germinate at 70°F (21°C). Seed will emerge in seven to ten days. However, most growers buy in plugs rather than grow their own.

Growing On

Cut flowers

Plant plugs into well-worked, disinfected soil. Matricaria is a moderate to heavy feeder. Incorporate fertilizer into growing beds based on soil tests prior to planting. Fertilize with potassium nitrate, magnesium sulfate, and calcium nitrate until buds develop. After buds develop, switch to potassium nitrate.

Grow the crop at 60–65°F (16–18°C) days and 55–60°F (13–16°C) nights. Be sure to distribute water evenly in the beds, as dry spots will result in short plants that flower before the rest of the crop. Overhead irrigation is fine until visible bud. Once buds appear, additional water is needed, ideally using drip tubes.

Matricaria is a long-day plant: Flowering occurs when days are sixteen hours or longer. Crops planted from September until the end of March must receive additional lighting.

ProGibb can be applied during high-light periods in the late spring and summer to obtain longer stems. Irrigate plants one to two days prior to treating. When applying the spray, be sure to get spray into the interior heart of the plant so it hits the growing tips.

During the summer months, growers producing their crop indoors should apply whitewash or screening.

Count on a crop time of eight to ten weeks from plantings made from May to July, and fourteen to fifteen weeks from plantings made in January and February and from September to December.

Pot plants

For production in pots, plants respond very well to growth regulators such as B-Nine. The crop time from transplanting a 288 plug to a salable 4" (10 cm) pot is seven to nine weeks. Plants will be green or with first flower.

Pests & Diseases

Matricaria is susceptible to root rots during production. Also watch for botrytis as crops finish.

Varieties

For cut flower production, try the 'Vegmo' series, which is available in 'Lime Green', 'Single', 'Snowball', 'Snowball Extra', 'Sunny Ball Gold', and 'Yellow'. The 'Magic' series is available in 'Lime Green', 'Single', 'White', and 'Yellow'.

Postharvest

Stems are ready to be harvested when 75% of the flower buds are fully opened. Most growers wait and harvest an entire bed at once. Matricaria is not sensitive to ethylene.

Tecoma

Tecoma stans
Common names: Esperanza (Spanish for "hope"); yellow trumpet bush
Perennial (Hardy to USDA Zones 7–11)

Tecoma prefers full-sun and high-heat environments. These are upright plants with yellow, spiked flowers that reach 24–36" (60–90 cm) tall.

Propagation

While propagated by both seed and cuttings, seed propagation is covered here. The following culture is based on the only currently available variety: 'Mayan Gold'.

Use a well-drained, disease-free, soilless medium with a pH of 5.8–6.2 and a medium initial nutrient charge (EC of 0.75 mS/cm). Sow coated seed in 288, 200, or larger liners. Cover seed moderately to heavily with vermiculite to maintain medium moisture. While the seed coat sometimes remains on the young seedlings, it will eventually fall off and will not affect the growth rate of the seedlings. Stage 1 takes form three to five days at 68–74°F (20–23°C), and light is optional for radicle emergence. Keep soil wet (level 4) during Stage 1.

For Stage 2, increase the soil temperatures to 72–75°F (21–24°C) and the light up to 2,500 f.c. (27 klux). Reduce soil moisture slightly (levels 3 to 4) to allow the roots to penetrate into the medium. Do not allow the medium to dry out. Apply fertilizer at rate 1 (less than 100 ppm nitrogen and less than 0.7 mS/cm EC) from nitrate-form fertilizers with low phosphorous.

For Stage 3 reduce the soil temperatures to 68–72°F (20–22°C) and maintain the same light levels as Stage 2. Allow medium to dry further until the surface becomes light brown (level 2) before watering. Keep the moisture at a wet-dry cycle (moisture levels 4 to 2). Increase fertilizer to rate 2 (100–175 ppm nitrogen and 0.7–1.2 mS/cm EC). If growth is slow, apply a balanced ammonium-and-nitrate-form fertilizer with every other fertilization. Maintain the medium pH of 5.8–6.2 and the EC between 1.0 and 1.5 mS/cm (1:2 extraction).

For Stage 4, drop soil temperatures to 65–68°F (18–20°C) and increase light levels up to 5,000 f.c. (54 klux), if temperatures can be controlled. As for moisture and fertilizer, follow the same guidelines as noted for Stage 3.

A 288-plug tray takes from five to seven weeks to be ready to transplant to larger containers.

Growing On

Use a well-drained, disease-free, soilless medium with a pH of 5.5–6.2 and a medium initial nutrient charge (EC 0.75 mS/cm) and growing on temperatures of 68–75°F (20–24°C) days and 65–68°F (18–20°C) nights.

Maintain light levels as high as possible. Under low light and short-day conditions, the flower cluster may develop slowly or abort. Tecoma is a facultative, long-day plant that flowers faster and more uniformly at day lengths of fourteen hours or longer. Flowering will be delayed about three weeks when grown at day lengths of twelve hours or less. Light levels can interactively affect the plant's daylength sensitivity. Under high light conditions, plant flowering may require shorter day lengths.

Starting one week after transplant, apply fertilizer at rate 3 (175–225 ppm nitrogen and 1.2–1.5 mS/cm), using predominately nitrate-form fertilizer with low phosphorus and high potassium. Maintain the medium EC at 1.5–2.0 mS/cm and pH at 6.0–6.5. For a constant fertilizer program, apply fertilizer at rate 2 (100–175 ppm nitrogen and 0.7–1.2 mS/cm), while maintaining the same recommended EC and pH ranges.

Plants respond well to a tank mix of B-Nine and Cycocel. Apply B-Nine/Alar (daminozide) 2,500 ppm (3.0 g/l of 85% formulation or 4.0 g/l of 64% formulation) mixed with Cycocel (chlormequat) 1,000 ppm (8.5 ml/l of 11.8% formulation or 1.3 g/l of 75% formulation). Spray every other week starting two to three weeks after transplanting. Do not use Bonzi as a drench or spray, because it will make plants softer and floppy.

Table 38. Total Crop Times for *Tecoma*

CONTAINER SIZE	NUMBER OF PLANTS PER POT	SPRING	SUMMER
4.5–5" (11–12 cm) pot	1	14–15 weeks	12–14 weeks
6" (15 cm) pot	3	14–15 weeks	12–14 weeks
1-gal. (4 l) pot	3	14–15 weeks	12–14 weeks

Tecoma does not branch until after flower bud emergence. Pinching will promote branching but delay flowering. Due to strong apical dominance, if pinched, a hard pinch is recommended, and pinching typically results in only two branches. Be sure to leave two or more nodes below the pinch. The earlier the pinch is done, the less delay in flower time.

Crop time is seven to ten weeks from transplant to flower.

Pests & Diseases

Watch for aphids and spider mites, especially during summer production. There are no diseases that particularly trouble this crop.

Thunbergia

Thunbergia alata
Common name: Black-eyed Susan vine
Annual vine

Thunbergia is a vining plant that thrives in warm, sunny locations and has maintained a niche position in the marketplace. When merchandised on a trellis and in color, *Thunbergia* generates an impulse buy. Achieving this full impact at retail requires an investment in crop time and an attention to detail.

Individual plants can grow 5–6' (1.5–1.8 m) tall over the summer when they're supported on a trellis. With such a colorful performance, your retail customers will return year after year to buy their black-eyed Susan vines. Culture is a snap—you can even direct-sow seed to final growing containers.

Propagation

Thunbergia can be propagated by seed or by cuttings.

Seed

To speed germination, presoak seed in warm (not hot) water overnight before sowing. Sow to a 288 or larger plug and allow four to five weeks from sowing to be ready to transplant to larger containers. Seed can also be sown directly to final growing containers. Sow two to four seeds per 4" (10 cm) pot.

Cuttings

Only a few suppliers offer unrooted cuttings of *Thunbergia.* Handling unrooted cuttings is challenging due to their sensitivity to ethylene in transit and excess moisture in propagation. Specialty propagators have the infrastructure and scale to manage this crop in propagation, and many offer rooted liners.

For those handling unrooted cuttings, here are a few critical factors. Stick cuttings immediately upon arrival. *Thunbergia* cuttings are prone to breakdown and dehydration if stored even an additional twelve hours. Choose a well-drained medium with an EC of 0.75–0.80 mS/cm and a pH of 5.5–5.8. A rooting hormone is not recommended. The soil temperature must be maintained at 68–74°F (20–23°C) until roots are visible. Apply mist at intervals throughout the first three to five days, depending on local conditions. Frequency and run time should be reduced during the dark period, but unrooted cuttings must not become dehydrated. Spray cuttings with a surfactant after sticking, using labeled rates, to help rehydration and reduce the tendency to wilt. This will reduce stress on the cuttings and speed rooting. Begin fertilization with 75–100 ppm nitrogen when roots become visible. Increase to 150–200 ppm nitrogen as roots develop. Avoid phosphorous and ammoniacal nitrogen during the rooting process to reduce stretch and unwanted vegetative growth. *Thunbergia* does not require pinching during propagation. However, to improve branching and habit, pinch seven to ten days before transplanting.

Thunbergia rooted cuttings should be ready for transplanting four to five weeks after sticking and should be transplanted as soon as possible. Rooted cuttings should not be held, as *Thunbergia* will be actively growing, and plants will begin to crowd and stretch quickly.

Growing On

Seed

Transplant one 288 plug per 4" (10 cm) pot. *Thunbergia* begins to vine early, so pack production is not recommended.

Feed with 150–200 ppm nitrogen from 20-10-20, alternated with 15-0-15. Maintain a soil pH of 5.5–6.2 and an EC below 1.0 mS/cm. Provide high light (3,000–5,000 f.c. [32–54 klux]). Ideal growing temperatures are 62–68°F (17–20°C) days and 60–62°F (16–17°C) nights.

Do not pinch, as plants branch along the stem. High temperatures can cause plants to stop flowering.

The crop time from transplanting a plug to a salable 4" (10 cm) pot is eight to ten weeks.

Cuttings

Thunbergia requires relatively high light levels and warm night temperatures. In areas with low light levels in early spring, *Thunbergia* is best planted as a mid- to late-spring and summer crop. Use a light, well-drained, soilless medium with a pH of 5.8–6.2. Maintain night temperatures of 60–65°F (15–18°C) and day temperatures of 70–80°F (21–26°C). Temperatures lower than these will slow plant growth significantly and reduce flowering. The ideal light range is 5,000–8,000 f.c. (54–86 klux). Low light levels promote stem stretch and limit flowering.

Allow the medium to dry moderately between watering. To prevent leaf burn, do not allow the medium to dry completely or the plant to wilt repeatedly. Maintain constant fertilization at 200–250 ppm nitrogen. Excessive phosphorous and ammoniacal nitrogen will cause unwanted vegetative growth. If new growth is chlorotic, add chelated iron to the feed and check the soil moisture, as it may be too wet. Slow-release fertilizer can be incorporated at a moderate rate to supplement a liquid program.

Depending on pot size, *Thunbergia* should be pinched seven to ten days after transplanting. Larger pot sizes will require an additional one to two pinches to produce a well-branched, full plant. Control height by maintaining the recommended soil fertility, allowing the medium to dry slightly between watering, providing maximum light levels, and spacing plants in advance of crowding and stretch. Chemical growth regulators are challenging on *Thunbergia*. B-Nine and Cycocel tank mixes in propagation applied on soft tissue has shown some effect in reducing stretch. Stronger active ingredients in growth regulators easily stunt *Thunbergia*.

Plan for eight to ten weeks to finish a 10–12" (25–30 cm) hanging basket or patio pot in the mid- to late spring. Crop times will be longer in short days.

Pests & Diseases

Spider mites are one of the most common pests on *Thunbergia*. Aphids, whiteflies, and leafminers also favor this crop. Rhizoctonia and *Pythium* will attack roots if excessive moisture levels are maintained in the growing medium.

Troubleshooting

Wet medium for an extended period will lead to *Pythium*, rhizoctonia, and botrytis. Excessive vegetative growth can result from high ammonia in medium, overfertilization under low light, or overwatering under low-light conditions. Low greenhouse temperatures, low-light conditions, or short days can lead to poor flowering.

Varieties

Seed-propagated varieties include 'Alta Mix' and 'Susie Mix'. 'Alta Mix' has a black eye and includes yellow, buff, and orange colors. 'Susie Mix' has dark-eyed, orange and yellow blooms and is excellent for hanging baskets.

The cutting-propagated 'Sunny' series has an orange and a yellow that have been popular with many *Thunbergia* growers. Its large flowers and dark-green foliage make this series a favorite. Hishtil also offers cuttings of *Thunbergia* that are proven to perform and make excellent archways and hanging plants.

Thymus

Thymus vulgaris
Common names: Thyme; garden thyme; common thyme
Perennial herb (Hardy to USDA Zones 5–8)

Thymus is a genus with about 350 species of mostly low-growing perennials. It is one of the most popular herbs and should be part of every herb program. Thyme has been used for medicinal purposes for thousands of years. The most widely used herb in this species is common thyme, *Thymus vulgaris.* Thyme is popular not only for the herb garden, but also for rock gardens, along walls, and in mixed container gardens. Plants grow from 6–15" (15–38 cm) tall and prefer full sun or morning sun and afternoon shade. Although thyme is a perennial, it is short lived in cold and wet winter environments.

Propagation & Growing On

Seed

While seed is readily available, most of the culinary varieties are propagated by cuttings. Thyme seed can be erratic to germinate and germination rates vary from 60–80%. However, those points made, some growers experience few problems producing thyme from seed.

Sow three to four seeds to a 288-plug tray, cover with coarse vermiculite, and germinate at 70–72°F (21–22°C). Seedlings will emerge in six to ten days, and plugs are ready to transplant to larger containers five to seven weeks after sowing.

Transplant to a 4" (10 cm) pot and grow on at 65–70°F (19–21°C) days and 55–60°F (13–16°C) nights. Plants will be salable green in seven to ten weeks.

Cuttings

Thyme can easily be propagated from cuttings. Stock plants should be kept vegetative and in active growth so that cuttings are not too hard. Avoid using woody cuttings, because they take longer to root and result in higher losses. The best time to propagate is in spring and summer, when the plants are in active growth. However, propagation can occur year-round as long as active growth is maintained.

There are also a number of both unrooted and rooted liner suppliers who readily supply this crop throughout autumn and winter to purchase.

Most standard tray sizes for propagation are between 84 and 128 cells per tray. Multiple cuttings per cell (usually two) are recommended to achieve a fuller liner. Use a standard propagation medium with a pH of 6.0–6.5. Temperatures of at least 70–75°F (21–24°C) for the rooting process are best, ideally with bottom heat. Keep the medium moist at all times, but not saturated. Until the plants have developed roots of their own, mist several times a day to avoid wilting. The amount and frequency of misting depend on the temperature and air circulation in the greenhouse. After cuttings have rooted, move plants to a hardening area with increased light levels and lower temperatures.

Stick supple, non-woody cuttings into a 128-liner tray using one or two cuttings per tray. Use a rooting hormone on the cuttings prior to sticking into the tray. Place under mist and allow five to six weeks for trays to be ready to transplant.

After transplanting one liner per 4" (10 cm) pot or three per 6" (15 cm) pot, plants will be salable green in seven to nine weeks when grown on as noted under seed propagation. Use a well-drained, soilless medium such as a peat/perlite mix with a pH of 6.0–7.0.

Provide high light levels. Thyme can be grown outdoors once temperatures are frost-free. High light levels enhance color and fragrance.

Keep the medium on the dry side. Thyme is sensitive to overwatering, especially silver-leafed varieties. Thyme requires little fertilizer. Apply a constant feed of 100–125 from 20-10-20, alternating with 14-0-14 and average amounts of micronutrients.

To encourage branching, give one soft pinch within one to two weeks of planting. Thyme responds well to cutting back, but trimming is not required for plant performance.

Pests & Diseases

Thyme is prone to rot in very hot, humid conditions, but it has few insect problems. Do not use pesticides on herbs.

Varieties

Common thyme varieties include 'Argenteus', an upright variety with gray-green leaves and silver edges, and 'Silver Posie', with mauve-pink flowers atop silver, variegated foliage.

Related Species

Thymus × *citriodorus*, also called lemon thyme, is frequently used as an herb or a scented ornamental. Varieties include 'Archer's Gold', an upright variety with a compact growth habit, golden leaves, and pale purple flowers, and 'Silver Queen', which has silver-green leaves with cream-colored variegation. Plants take on a slight pink tinge in cool temperatures. 'Aureus', another selection, has green leaves splashed with golden yellow, which are most prominent in cool temperatures. 'Aureus' can develop into a small shrub of about 6–8" (15–20 cm) in height.

There are several other noteworthy thyme species. *T. praecox*, or creeping or wild thyme, is a creeping species used for culinary purposes. Varieties are available with flowers ranging from white to pink and almost red. *T. pseudolanuginosus*, or woolly thyme (also known as *T. praecox* var. *pseudolanuginosus*), is mainly used as an ornamental in rock gardens. Plants have hairy stems and sparse, pale-pink flowers. *T. serpyllum*, or mother of thyme (also called wild or creeping thyme), is a species with a prostrate growth habit and is used mainly for medicinal applications. Quite numerous varieties are available, including some with variegated foliage. Flowers range from pale pink to purple, depending on the variety.

Postharvest

Display thyme by type or species and give consumers ideas on how to use the different types. For example, display lemon thyme, which is sold mainly for its scented foliage, separately from common thyme. Provide recipe cards with suggestions on how to use thyme in cooking. Be sure to include some pots in your "Plants for Combinations" display, as thyme makes a great combination plant.

Tomato (see *Lycopersicon*)

Torenia

Torenia fournieri

Common names: Wishbone flower; Florida pansy

Annual

This old-time garden plant got a major facelift in the 1990s, when the seed-propagated variety 'Clown' was introduced, first as a mix and then as separate colors. That's when landscapers discovered how reliably well *Torenia* performs in well-drained, partially shady spots. Add to that the number of vegetatively propagated lines in a wide range of colors now offered, and you can see how this crop has accelerated into a mainstay as a mixed combination component item.

Torenia is an easy crop to grow. And once your customers try it, they will return year after year. Since plants never really become covered with flowers at the point of sale, the trick to convincing consumers of *Torenia*'s charm is color tags and informational signs that tell *Torenia*'s story: Plants love the heat, can take the shade, and will flower continuously all summer long until frost.

Propagation

Both seed and cuttings are commonly used for *Torenia* propagation.

Seed

Sow seed onto a well-drained, disease-free medium with a pH of 5.5–5.8. Do not cover. Provide 100–400 f.c. (1.1–4.3 klux) of light. Germinate at substrate temperatures of 75–80°F (24–27°C). Keep the medium evenly moist but not saturated. Maintain ammonium levels at less than 10 ppm. Radicles will emerge in four to six days.

Move trays to Stage 2 for stem and cotyledon emergence, which will take place over four to seven days. Reduce substrate temperatures to 72–75°F (22–24°C). Once radicles have emerged, begin to allow the medium to dry out slightly between irrigations. Increase light levels to 750–1,000 f.c. (8.1–11 klux). Once cotyledons have expanded, fertilize with 50–75 ppm from 14-0-14, alternating with clear water. To prevent disease problems, be sure foliage is completely dry by nightfall.

Once you have a stand, move trays to Stage 3 for growth and development of true leaves, which will take twenty-one to twenty-eight days. Reduce substrate temperatures to 68–72°F (20–22°C). Allow the medium to dry thoroughly between irrigations, but avoid permanent wilting. Increase fertilizer to 100–150 ppm at every second or third irrigation from 14-0-14. Do not use ammonium fertilizers. If growth regulators are required, B-Nine, Bonzi, and Cycocel are effective.

Harden plugs for seven days prior to sale or transplant by lowering substrate temperatures to 62–65°F (17–18°C). Allow soil to dry thoroughly between irrigations and fertilize with 100–150 ppm 14-0-14 as needed.

Most growers buy in plugs in March and April for spring crops and in May or early June for summer color programs.

The crop time from sowing until a 288-plug tray is ready to transplant to larger containers is four to six weeks.

Cuttings

Choose a well-drained medium with an EC of 0.75–0.80 mS/cm and a pH of 5.5–5.8. Stick cuttings within twelve to twenty-four hours of arrival. Cuttings can be stored overnight, if necessary, at 45–50°F (7–10°C). Soil temperatures should be maintained at 68–74°F (20–23°C) until roots are visible. Begin fertilization with 75–100 ppm nitrogen (N) when roots become visible. Increase to 150–200 ppm N as roots develop. Once roots are visible, the medium should be kept moderately wet and never saturated. This will prevent iron deficiency and the associated chlorotic foliage, which can develop. As the rooted cuttings develop, appropriate water stress and moderate air temperatures should eliminate the need for chemical plant growth regulators (PGRs). *Torenia* does respond to Topflor, Bonzi, or Sumagic sprays to control growth early in the crop cycle. *Torenia* rooted cuttings should be ready for transplanting twenty-four to twenty-eight days after sticking.

Stick cuttings to a 105-liner tray; cuttings do not need a rooting hormone. Allow to root at 68–75°F (20–24°C) soil temperatures. Cuttings root within four to six weeks

Growing On

Seed

Transplant *Torenia* into a well-drained, disease-free, soilless medium with a moderate initial nutrient charge and a pH of 6.0–6.2. Use 1 plant/cell in jumbo packs or 4" (10 cm) pot or 2–3 plants/6" (15 cm) or 1-gal. (4 l) pot. Plants are ready to sell from a 4" (10 cm) pot in seven to eight weeks after transplanting.

Grow on at 65–70°F (18–21°C) days and 60–62°F (16–17°C) nights. Plants can be grown at higher temperatures in high-light areas, but growing plants warm in low-light conditions will cause stretching.

Fertilize at every other irrigation with 150–200 ppm from 15-0-15. Do not use fertilizers containing ammonium. For the best-shaped plants, provide a soft pinch two weeks after planting and maintain light levels at 3,000–5,000 f.c. (32–54 klux). Under low-light production conditions, a second pinch may be needed two weeks later. B-Nine, Bonzi, and Cycocel are effective in controlling height.

Cuttings

Use a well-drained, disease-free, soilless medium with a pH of 5.5–6.0. Grow on with temperatures of 65–72°F (18–22°C) and maintain slightly warmer soils temperatures (70°F [21°C]) after transplanting as roots establish. Maintain temperatures above 55°F (13°C) throughout the crop to avoid bronzing of foliage.

Torenia may require shading in southern markets as light intensities increase throughout the spring.

A pinch is recommended at the time of transplant for 1-qt. (1.1 l) or smaller pots in finishing. For larger containers or hanging baskets, an additional pinch may be needed to increase branching and shape the finished product.

Do not use Florel as a PGR. Sumagic drenches (1–5 ppm) and Bonzi drenches (1–7 ppm) have demonstrated effective growth control in the finish plant stage. Initiate a growth regulator plan prior to stem elongation for the best effect.

A broad-spectrum fungicide regime will prevent root and stem diseases.

After transplanting one 105 liner per 4" (10 cm) pot, plants will start to flower in seven to eight weeks.

Pests & Diseases

Aphids and whiteflies can become problems during production. Powdery mildew can also strike plants. To control powdery mildew, make sure that plants are well ventilated and foliage stays as dry as possible.

Varieties

The 'Clown' series set the standard for seed-propagated *Torenia.* The series is available in 'Blue', 'Blue and White' (a Florastar Award winner), 'Blush', 'Burgundy' (a Florastar Award winner), 'Plum', 'Rose', 'Violet', and 'Mix'. 'Clown' grows to 8–10" (20–25 cm) tall in the garden. 'Duchess', available in 'Blue and White', 'Burgundy', 'Deep Blue', 'Light Blue', 'Pink', and 'Mix' is seed propagated and dwarfer, growing to 6–8" (15–20 cm) in the garden. The newest seed-propagated series is 'Kauai'. It is the most compact variety and is offered in 'Burgundy', 'Lemon Drop', 'Deep Blue', 'Magenta', 'White', and 'Rose' along with a 'Mix'. Unlike the other *Torenia* series available, the 'Kauai' series has full ring color on its flowers. This makes it brighter and more visible as color for the shade.

From cuttings there is a wonderful list of varieties to try in a broader range of flower colors than seed-propagated strains. The 'Summer Wave', 'Catalina', and 'Moon' series are only several inches tall in height but trail or spread, making them excellent additions to hanging baskets and mixed combination containers.

Postharvest

Retailers should display *Torenia* under 50% shade and keep plants moist but not wet. Plants are easily killed by frost, so ensure that the danger of frost has passed before allowing home gardeners to plant them outdoors. When nights are cool, the leaves will bronze. *Torenia* prefers partial shade but will tolerate full sun if moisture is adequate.

Trachelium

Trachelium caeruleum
Common names: Throatwart; trachelium
Cut flower, annual

This lacy blue or white cut flower is a great bouquet filler. Its large umbels make the ideal backdrop to gerberas, roses, and other high-end flowers. New breeding is also introducing dwarfer varieties that are great as flowering pot plants and bedding plants.

Propagation

Sow pelleted seed onto a well-drained, disease-free, germination medium with a low starter charge and a pH of 5.5–6.0. Some growers double-sow standard seed, two per cell. Choose large plug cells (200s) as opposed to small ones. Do not cover seed with medium. Germinate at 65–68°F (18–20°C) with 100–400 f.c. (1.1–4.3 klux) of light. Keep trays moist but not saturated. Crop time for plugs is twelve to sixteen weeks.

Transplant plugs on time: Do not allow them to become root-bound, as this will result in longer crop times.

A 200-plug tray takes from seven to nine weeks from sowing to be ready to transplant.

Growing On

Plant plugs for cut flower crops in steamed ground beds. Steam at 122°F (50°C) for two hours. Do not use higher temperatures, as they can cause the release of manganese, which is toxic to *Trachelium.* Since many weeds and diseases may not be killed at this temperature, treat with fungicides immediately after planting as a precaution against *Pythium* and rhizoctonia root rots. Unless you are growing the crop outdoors in windy conditions, support netting is not required.

Plant out at a density of 64 plants/net $yd.^2$ (80/net m^2). Plants can be irrigated from overhead in the early stages (until they are 6" [15 cm] tall); after that, switch to drip irrigation. Maintain uniformly moist soil conditions without wide swings from wet to dry conditions. As flower buds develop, reduce the irrigation frequency to prevent side shoots from overgrowing the primary flower.

Grow plants on at 52–55°F (11–13°C) for the first few weeks after transplanting to minimize stress on transplants. Shading the crop with whitewash or retractable shade in the summer months will help to reduce plant stress. During the first few weeks, light levels should be less than 3,500 f.c. (38 klux). Once plants begin to root in, increase temperatures to 60°F (15°C) days and 55°F (13°C) nights. As light intensity increases, temperatures can gradually increase to 64°F (18°C) days and 59°F (15°C) nights. During the vegetative growth phases of *Trachelium,* do not allow temperatures to rise above 77°F (25°C).

Immediately after planting, irrigate with clear water for the first three weeks. Then begin feeding with 150 ppm from potassium or calcium nitrate for weeks 3–6. Increase fertilizer in week 7 to 200 ppm. After week 10, irrigate with clear water only.

Trachelium flowers faster under long days (sixteen hours). Begin long days three weeks after planting. Light crops for sixteen continuous hours using supplemental lights or night interruption lighting from 10 P.M.–2 A.M. Continue lighting through harvest, increasing day lengths to eighteen hours during the final three weeks of crop growth.

During the summer months, when temperatures and light levels are high, plugs can go into shock after transplanting. In the Netherlands, growers have found that crops planted during this time require short days during the first two weeks after planting to prevent short flowering stems. Dutch growers use black cloth to apply short days on crops planted from May 15 until July 31 to encourage plants to remain vegetative. Black cloth can be pulled early in the evening as the temperature falls, then opened when it is fully dark outside to allow air circulation, and then pulled again from 5 A.M.–9 A.M. Do not use black cloth for more than three weeks, or flowering can be delayed and yields lowered.

For cut flowers, plants will flower ten to sixteen weeks from transplanting from plugs. For bedding plant varieties, allow nine to eleven weeks from transplanting a plug until flowering in a 4" (10 cm) pot.

Pests & Diseases

Aphids, caterpillars, and thrips can occasionally be problems. Rhizoctonia and *Pythium* can attack plants during production.

Troubleshooting

Spots on foliage that appear just before you are ready to harvest stems can be due to manganese toxicity.

Varieties

Try 'Lake' for a wide range of colors on good stem lengths for cut flowers. It is divided up into winter-, summer-, and spring/autumn-flowering selections.

For bedding plant varieties, look to the 'Devotion' series for production in 4 or 6" (10 or 15 cm) pots. Plants grow to a height of 12–24" (30–60 cm).

Postharvest

Harvest stems when 75% of the umbel shows color. Place stems in clean water immediately after harvesting. Bunch and sleeve in groups of ten. Precool bunches to 34°F (1°C) prior to shipping dry. However, the best-quality stems are wet packed; a Procona system is great for this packaging. *Trachelium* is not sensitive to ethylene.

Tulipa

Tulipa hybrida
Common name: Tulip
Cut flower, flowering potted plant

Tulips are one of the most recognized flowers among consumers. Americans love tulips as a casual bouquet of ten to twenty stems in a vase on their dining table, in mixed bouquets, or planted by the hundreds in gardens, parks, and landscapes nationwide. For growers, tulips are a fabulous opportunity, but one that requires a significant capital investment.

Controlled-temperature rooting rooms are essential to properly program tulips forced as fresh cut flowers and growing and flowering potted plants. Normally, the flowering season is from late December to early May. If desired, tulips can be forced earlier in the fall by utilizing bulbs produced in the southern hemisphere. Production of these bulbs in New Zealand and Chile is increasing, and, thus, bulbs are readily available from reputable suppliers.

The keys to success are: (1) selecting the proper variety for each market product and bloom time; (2) using controlled-temperature reefers and trucks to transport the bulbs; (3) upon arrival, determining the stage of flower development; (4) upon arrival and at planting, inspecting the bulbs for the presence of *Fusarium;* (5) using a well-drained, pathogen-free planting medium with a pH of 6.5–7.0; (6) using the proper temperatures and number of cold weeks in the rooting room; (7) using the optimum greenhouse temperatures for the specific market product; and (8) marketing at the proper stage of floral development.

In addition, in the Netherlands, there are very large operations that are forcing tulips using hydroponics (see sidebar), which requires highly specific equipment and knowledge. If hydroponic production is of interest to you, contact specialists in the Netherlands.

Growing On

For early forcing, 12/up cm (circumference) bulbs must be used. For later forcing, 11/12 cm bulbs of some varieties can be used. The exception is the variety 'Yellow Baby', for which 9/up cm bulbs are primarily used for 4" (10 cm) potted plants. Regardless of their production origin, tulips must be transported in controlled-temperature reefers and trucks. In addition, the relative humidity should also be controlled whenever possible.

When tulip bulbs are harvested, the apical meristem is vegetative and is completing the formation of the complement of leaves (three to five). Subsequently, when bulbs are placed at 63–70°F (17–21°C), flowers form. Normally, this process is completed by mid-August with Dutch-grown bulbs and is complete well in advance of US forcers receiving the bulbs. For early forcing, it is essential that the variety has reached "Stage G" (G = gynoecium) and preferably passed this stage. Once bulbs reach or pass Stage G, they can be precooled at 45–48°F (7–9°C) for up to six weeks before being planted. Non-precooled bulbs should be stored at 55–63°F (13–17°C) until planted.

Upon arrival, it is also critical to examine the bulbs for the presence of *Fusarium.* This disease not only consumes the infected bulb but also causes the bulb to produce ethylene. This gaseous plant growth regulator can cause floral disorders such as abortion and green tips in other healthy bulbs during shipment or in the storage room. Identification symptoms for *Fusarium* include the presence of a very sour smell, white mold growing on the tunic, soft basal plate, and very lightweight bulbs.

Tulips require a well-drained, pathogen-free, planting medium with a pH of 6.5–7.0. Regardless of the container used, tulips can be planted bulb to bulb with the bulb noses protruding slightly out of the medium. After being planted, the containers should remain at 48°F (9°C) until all the bulbs are fully rooted (i.e., the roots are growing through the holes at the bottom of

Hydroponic Tulip Forcing

Basics

The basic procedure with hydroponic tulip forcing is to give approximately 75–80% of the cold requirement to dry, unplanted bulbs. Depending on the cultivar and time of year, this might be twelve to fourteen weeks. Then bulbs are "planted" into the system and a dilute calcium-nitrate solution is added for rooting (about 1.0–1.2 mS/cm). Rooting proceeds at 40ºF (4ºC) for three to four weeks (early crops) or two to three weeks (later crops). The entire cold requirement thus given, the bulbs are moved into the greenhouse for forcing. In the greenhouse, plants are fed with calcium nitrate, with the goal of maintaining an EC of 1.2–1.5 mS/cm.

Roots

It is important to realize that the longer the rooting period (above two to four weeks), the lower the eventual quality of the flower. This is because longer roots cause more rapid oxygen depletion of the solution and become more susceptible to disease. Also, the longer and more entangled the roots are, the more difficult harvest is (harvesting one stem pulls up many more with entangled roots). A key realization of the success of hydroponic forcing is the relatively small root system that is needed to produce a good-quality plant, probably much less than is necessary for cut tulips in soil or peat-based forcing.

Advantages of hydro forcing

Compared to traditional "soil" culture in "boxes" (where bulbs are planted in crates, cooled, and then forced), hydroponic forcing has the following advantages: (1) It is three to five days faster than soil culture; (2) a much less cooler volume is required for chilling bulbs (because most of the cold period is given to densely packed, unplanted tulips in their shipping crates); (3) harvesting is easier and cleaner; and 4) there is no waste soil at the end, greatly reducing materials handling problems.

Why do hydroponic plants force faster than plants grown in traditional soil culture? It isn't due to any inherent superiority of hydroponics, it is simply due to the prevailing temperature (approximately 40°F [4°C]) of the plants during the two- to four-week rooting period. This is 6–8°F (3–4°C) warmer than is typical during the last few weeks of cooling; where, normally, temperatures of 32–33°F (0–0.5°C) might prevail to reduce stem growth in the cooler. These few degrees over a two- to four-week period can easily account reduce the crop time in the greenhouse. Plus, in traditional cut flower forcing in boxes, the mass of the bulbs and soil are substantial, and probably take one or two days to warm to prevailing greenhouse temperatures.

Disadvantages of hydro forcing

The disadvantages of hydroponic forcing are: (1) When grown at the same temperature, the ultimate quality of the stem is not quite as good as when the same cultivar is grown in soil (i.e., hydro stems tend to be 1–2" (2.5–5 cm) shorter and 6–8% lighter compared to substrate-grown stems); (2) not all cultivars are suited to this system; (3) very high-quality and disease-free bulbs are required, especially for later plantings (i.e., careful attention must be placed on proper bulb storage, including temperature, humidity, and ventilation); (4) especially for individual trays, a level bench or tray support system is critical to maintain a level nutrient solution (old, uneven benches won't cut it); and (5) exceptional cleanliness is required, which really means specialized machines for tray washing and sanitizing.

These hydroponic tulips have just been placed into the forcing greenhouse. Note the relatively sparse root development (compared to, for example, a pot tulip).

An excellent crop of hydroponic tulips in Holland

the container). Subsequently, the temperature should be lowered to 41°F (5°C). When shoots of the first variety reach 1" (3 cm), the temperature must be lowered to 33–35°F (0.5–2°C). This is required to keep the shoots short. In the rooting room, always keep the planting medium moist and, if possible, maintain the relative humidity at 90%. This assists in controlling foliar diseases.

The Cut Flower Crop

In general, most varieties used as fresh cut tulips require sixteen to eighteen continuous cold weeks. In the greenhouse, use 50–55°F (10–13°C) night temperatures and low light intensities. These conditions promote longer cut flowers. Always avoid temperatures above 68°F (20°C), which can cause flower abortion and/or stem topple. To assist in preventing these physiological disorders, tulips should be fertilized with calcium nitrate at 2.2 lbs./100 gal. (1 kg/380 l) of water, twice weekly.

Pests & Diseases

Aphids can become a problem during forcing. However, the most important pest problems you may be dealing with are diseases. Among the diseases that can strike are botrytis, *Pythium,* penicillium, rhizoctonia, and *Trichcoderma.*

Troubleshooting

Overwatering can cause veinal streaking. Leaf spots may be the result of botrytis, poor ventilation, or wet foliage (make sure foliage is dry by nightfall). Flower desiccation may occur from using small bulbs, planting too soon (flower parts not developed), a shortened cooling period, *Fusarium* or *Pythium,* underwatering or high soluble salts, and/or compacted soil.

Varieties

There are many cultivars to choose from, and a list is best developed in consultation with a reputable supplier. A significant amount of information is available on the website of the Flower Bulb Research Program at Cornell (www.flowerbulbs.cornell.edu) and from *The Holland Bulb Forcer's Guide* (see Bibliography).

Postharvest

Fresh cut tulips should be pulled from the forcing trays with the bulb attached when the floral bud is 50% colored. Because tulips at this stage of development grow very rapidly in the greenhouse, the forcer should examine the crop at least twice daily. Cut tulips are not sensitive to ethylene from the harvest stage on.

There are three storage methods for cut tulips. They are: (1) store dry and upright with the bulb attached; (2) remove the bulb, wrap the cut flowers, and store them horizontally; and (3) remove the bulb, wrap the flowers, and store them in water. The storage temperature should be 32–35°F (0–2°C), and the relative humidity should be higher than 90%. Special tulip cut flower preservatives are available and should be used by the forcer.

Warning: Do not place freshly cut tulip flowers in the same buckets with freshly cut daffodils (*Narcissus*). The sap that exudes from freshly cut daffodil flowers is toxic to freshly cut tulips.

Retailers should recut stems and place tightly wrapped flowers in a germicide hydration solution or deionized water. Stems elongate after they are cut and can grow out of an arrangement. Keep stems upright to prevent bending. Specific postharvest solutions are available that reduce leaf yellowing and also help reduce growth of the top internode during vase display.

Advise consumers to place cut tulips in the coolest location in the home in order to maximize vase life.

The Pot Plant Crop

For tulips used as potted plants, there are two critical planting factors. First, always place the flat side of the tulip bulb facing the outside of the pot. By doing this, the lower big leaf will face the outside of the pot. Second, do not completely fill the pot with the planting medium. It should be about 0.25" (6 mm) below the rim of the pot. This aids not only in watering but also when applying a plant growth regulator as a drench. This being said, bulbs need to be planted so that at least 1" (2.5 cm) of medium covers the bulbs to help minimize lifting from the pot as rooting proceeds.

In general, most varieties used as flowering potted plants require fourteen to fifteen continuous cold weeks, but forcers often give many more than this due to bulb shipping and planting schedules. Because of this excess cooling, plant growth regulators (PGRs) are a critical production tool.

In the greenhouse, use 60–63°F (16–17°C) night temperatures and medium to high light intensities. These conditions assist in producing short potted plants. Always avoid temperatures above 68°F (20°C), which can cause flower abortion and/or stem topple. To reduce these physiological disorders, tulips should be fertilized with calcium nitrate at 2.2 lbs./100 gal. (1 kg/380 l) of water twice a week. Otherwise use clear water.

For flowering potted plants, PGRs are often needed for height control. For many years, A-Rest drenches of approximately 0.25–0.5 mg/pot were used. From the late 1990s, Bonzi has been used more often, as it is much less expensive per pot than A-Rest. For most tulip cultivars, Bonzi (or equivalent) at 1–2 mg/pot (8 or 16 ppm assuming 4 oz./6" [15 cm] pot) is very useful and economical. In the last few years, Topflor has emerged as an even more effective product and can be used on the range of 0.5–1 mg/pot (4 or 8 ppm, assuming 4 oz./6" [15 cm] pot). In many cases, Topflor provides excellent control of postharvest top internode stretch, a major complaint for pot tulip retailers.

Bonzi or Topflor can be applied within a few days of housing the plants. Previous recommendations state that these growth regulators must be applied on the day after the pots are removed from the rooting room. Research at Cornell University has shown this not to be the case; there is more leeway in terms of application dates, and even for late crops, forcers have at least three or four days to get the PGR on. To be effective, the planting medium must be moist at the time of application. Thus, the planting medium should be thoroughly watered the night before the PGR is applied. In addition, the PGR must be applied in a sufficient volume of water to ensure uniform distribution throughout the planting medium (e.g., 4 oz./6" [15 cm] pot or 2 oz./4" [10 cm] pot). The rates of PGRs are optimally specific to variety and flowering period. Consult www.flowerbulbs.cornell.edu for more details on pot cultivars and Topflor or Bonzi rates, and *The Holland Bulb Forcer's Guide* (see Bibliography) for additional details on A-Rest.

Pests & Diseases

See Pests & Diseases in The Cut Flower Crop section.

Troubleshooting

See Troubleshooting in The Cut Flower Crop section.

Varieties

There are many cultivars to choose from, and a list is best developed in consultation with a reputable supplier. A significant amount of information is available on the website of the Flower Bulb Research Program at Cornell (www.flowerbulbs.cornell.edu) and from *The Holland Bulb Forcer's Guide* (see Bibliography).

Postharvest

Potted tulips should be marketed immediately after they are removed from the rooting room and treated with a PGR. Flowering potted tulips should be marketed when the first flower reaches the "green bud" stage of development. Potted tulips should be stored at 32–35°F (0–2°C) to maintain their flower life.

Advise consumers to place potted tulips in the coolest location in the home in order to maximize their flower life. Forced potted tulips should be discarded once they are spent. They have little or no value for the garden because they are annual replacement bulbs.

V

Verbascum

Verbascum × *hybrida*
Common name: Mullein
Perennial (Hardy to USDA Zones 5–8)

There are a number of species still widely available for *Verbascum*. However, the majority of those offered are the results of crosses between species. The scentless flowers are single, measure 1" (2.5 cm) across, and open in upright racemes that range in height from 16–26" (40–66 cm) tall when blooming for most varieties; some can be as tall as 3–4' (91–122 cm). They come in an array of flower colors including purple, white, apricot, and soft yellow. When not in bloom, the mostly green, rosetting, basal foliage is about 3–4" (7–10 cm) tall. Plants spread from 10–20" (25–51 cm) across. Plants flower in spring or early to midsummer. If the flower stalks are cut off after blooming, many selections will reflower again in three to four weeks.

Propagation

'Southern Charm' is currently the only seed-propagated *Verbascum* × *hybrida* selection. Most of the other selections are propagated by either tissue culture or as bare-root divisions/transplants. A number of other species are also seed propagated (see the Related Species section).

Seed

Due to the size of the leaves, sow 'Southern Charm' to a 200-, 128-, or larger plug tray, using one seed per cell. Lightly cover with vermiculite and germinate at 65–68°F (18–20°C). Seedlings will emerge in three to seven days. Plug trays are ready to transplant to larger containers in four to five weeks.

Tissue culture

Plants from tissue culture are professionally done by commercial labs and/or offered by perennial propagation companies and sold in rooted liner trays such as 20, 32, or 50 plants per standard 11" × 22" flat. Since many tissue-culture varieties are protected, it is best for growers to purchase in liners and transplant to larger containers.

Bare-root transplants

Bare roots are available in late summer to early winter from a number of commercial propagation companies.

Growing On

Seed

While 'Southern Charm' is a first-year flowering perennial from seed and will flower eighteen to twenty-two weeks after sowing in January, the best plants are those sown from seed the previous year.

Seed sown from July to August in plug or liner trays and then transplanted to 1-gal. (4 l) pots can be overwintered dormant for sales the following spring. Plants will be sold green but will flower in the summer.

Seed can also be sown in October or November in 1-gal. (4 l) containers to overwinter cold, but not dormant. Once transplanted to a 1-gal. (4 l) pot, allow roots to develop to the side and down to the bottom of the final container. Grow on cold (45–50°F [7–10°C]) for the remainder of the winter. Plants will develop a rosette of foliage atop the container and can be sold green in late spring. Plants will flower in the summer.

Tissue culture (liners)

Liners are available from commercial propagators in late summer for potting to 1-gal. (4 l) containers to overwinter dormant for green plant sales in the spring of the following year. Liners are also offered in winter as dormant or overwintered-cold plants that can be transplanted to 1-gal. (4 l) containers in January or February. These grow quickly and are ready for green plant sales in the spring. Plants will flower in the garden during the summer.

Bare-root transplants

Roots potted up anytime up until mid-February will grow quickly and be salable as green plants and/or flowering plants in May and June

Related Species

Verbascum bombyciferum is a biennial form with stately, pubescent (covered with fine hairs), gray-green stems and leaves. The plants look gray in overall appearance. The flowers are single and clustered in upright spikes growing 12–20" (30–51 cm) long. The individual blooms are yellow and about 1" (2.5 cm) across. Plants reach 6' (180 cm) tall in bloom when grown under ideal conditions, although they seldom grow over 3' (90 cm) tall in Chicago and are short lived. *V. bombyciferum* is not fond of the Midwest's hot, humid summers and wet winters and often dies out before flowering the second year. Plants are susceptible to crown rot and should be planted in raised beds.

Plants are very dramatic looking when in full bloom, but they won't tolerate frequent overhead watering because it often leads to fatal crown rot and foliar diseases. Plants flower in June and July.

Plants have a similar culture to *V.* × *hybrida.* Allow nine to eleven weeks for salable plants in a 4" (10 cm) pot from winter-sown seed. For larger container sales, sow seed earlier in the year for plants that will flower the following year. *V. bombyciferum* is a true biennial, requiring a cold period prior to flowering.

Verbascum chaixii is similar in its growth and performance to *V.* × *hybrida*, but it has woolly, gray-green leaves with 1" (2.5 cm), light yellow flowers. Plants grow to 3' (90 cm) tall. *V. chaixii* var. *album* is a white-flowering version of the species similar in height and form.

Both of these species can be propagated and grown on as indicated for *V.* × *hybrida.*

Verbena

Verbena × *hybrida*
Common name: Verbena
Annual

Verbenas are one of the brightest parts of most growers' spring plant crops. They excel as garden bedding plants, as combination components in mixed containers, and as landscape plants. They are available in a number of flower colors.

Propagation

Verbenas are commonly propagated by cuttings as well as from seed.

Seed

Germinate standard or enhanced seed onto a well-drained, disease-free, germination medium with a pH of 5.5–5.8. Cover seed lightly with coarse vermiculite to help maintain humidity at the seed coat. Maintain substrate temperatures of 75–80°F (24–27°C).

Run trays dry: Use about half the amount of moisture you would use for another crop. Do not saturate the medium at any time during germination. One old trick to get the right amount of water in your trays is to water the medium in the afternoon before you sow seed. Sow seed the following morning, and do not water it in. Provide 100–400 f.c. (1.1–4.3 klux) of light so that plants will be exposed to light just as soon as radicles emerge. Make sure to keep ammonium levels below 10 ppm. Radicle emergence takes four to seven days.

Once radicles have emerged, move trays to Stage 2 for stem and cotyledon emergence, which will take from ten to fourteen days. Maintain substrate tem-

peratures of 72–75°F (22–24°C). Continue to keep the medium on the dry side by allowing the medium to dry slightly before watering. Provide 1,000–2,500 f.c. (11–27 klux) of light. Begin fertilizing with 50–75 ppm from 14-0-14 as soon as cotyledons have expanded. Alternate feed with clear water. Be sure that foliage is dry by nightfall to prevent disease.

Move trays to the next step, developing true leaves in Stage 3, by lowering substrate temperatures to 68–72°F (20–22°C). Continue to run trays dry. Increase light to 2,000–2,500 f.c. (22–27 klux) and increase feed to 100–150 ppm once a week from 20-10-20, alternating with 14-0-14. If growth regulators are needed, A-Rest, B-Nine, Bonzi, and Cycocel are useful. Stage 3 will last about twenty-one to twenty-eight days for 288s.

Harden plugs for seven days prior to sale or transplanting by lowering substrate temperatures to 65–68°F (18–20°C) and increasing light to 3,000–4,500 f.c. (32–48 klux). Fertilize once with 100–150 ppm from 14-0-14.

A 288-plug tray takes from four to six weeks from sowing to be ready to transplant to larger containers. However, most growers choose to buy in plugs rather than grow their own.

Cuttings

Many growers are buying in unrooted cuttings. However, it is important to verify that you are buying from a producer who is operating a clean-stock program. There are great differences in the performance of finished verbena crops grown from clean cuttings that have been maintained in a clean-stock program and cuttings from any old stock plants. The performance of 'Homestead Purple', which is not patented, can vary widely from one propagator to another depending on how well stock plants are maintained and how regularly mother plants are renewed with indexed material. Clean cuttings will finish much faster. Ask questions before you order and know your sources.

Stick 2–3" (5–8 cm) long, terminal cuttings in Ellepots or medium. Irrigate the medium so that it is uniformly moist. If cuttings are uneven, grade by size and stem caliper: This will make your finished crop more uniform. If you are not able to stick cuttings immediately, place bags of cuttings at 48–50°F (8–10°C) for a maximum of twenty-four hours. Maintain 75–90% relative humidity in the cooler and mist bags with clean water.

Callus will form in five to seven days at a substrate temperature of 68–72°F (20–22°C). Maintain the relative humidity at 75–90% at the base of the cuttings. Use mist only to prevent wilting; increase and decrease frequency as light and air temperatures change during the day. Provide air temperatures of 70–75°F (21–24°C) days and 68–70°F (20–21°C) nights. Give 500–1,000 f.c. (5.4–11 klux) of light during this stage. Higher light intensities will stress plants due to warming. Retractable shade is ideal to allow light intensity to increase as cuttings mature. Use 50–75 ppm 20-10-20 as a foliar feed as soon as a callus forms or there is a loss of foliage color. Maintain a substrate pH of 6.2–6.5 with an EC less than 0.5 mS/cm.

When 50% of the cuttings begin differentiating root initials, move them to the next stage for root development, which will take place over seven to fourteen days. Maintain substrate temperatures of 68–72°F (20–22°C) and air temperatures of 70–75°F (21–24°C) days and 68–70°F (20–21°C) nights. As soon as cuttings begin to form root initials, begin to dry down the medium by reducing the duration and frequency of mist and moving the mist start and end times. Avoid drying out the air, since this will increase evapotranspiration and stress. Begin increasing the light intensity to 1,000–2,000 f.c. (11–22 klux) as cuttings root. Use a foliar feed at 100–150 ppm 15-0-15, alternating with 20-10-20. Increase the frequency and rate at each application to prevent salt problems. Keep the soil EC below 0.5 mS/cm.

In the final stage of rooting and getting plants ready to ship or transplant, maintain 70–75°F (21–24°C) days and 62–68°F (16–20°C) nights. At the beginning of this stage (one week before planting or shipping), pinch or shear liners to four to six nodes. It will have been important to "stack" the nodes in previous stages with high light, harder feed, and, as a last resort, plant growth regulators. There should be several low breaks developing at the end of this final stage. Move cuttings from the mist area to a bench with lower humidity, lower temperatures, and higher light levels (2,000–4,000 f.c. [22–43 klux]). Provide shade during midday to reduce temperature stress. Maintain the soil pH at 6.2–6.5 and EC less than 1.0 mS/cm. Fertilize twice per week using 150–200 ppm 15-0-15, alternating with 20-10-20. This final stage will last about seven days.

A 105 tray requires from three to five weeks from sticking cuttings to be ready to transplant to larger containers.

Growing On

Seed

Pot 1 plug/cell or 4" (10 cm) pot and 5–7 plugs/hanging basket. Select a well-drained, disease-free medium with a moderate initial starter charge and a pH of 6.2–6.5. Water in well after transplanting. Immediately following transplanting, be careful not to overwater or underwater plants, which will damage the delicate root system and can cause leaf-edge damage.

Fertilize with 150–200 ppm at every other irrigation, alternating between 15-0-15 and 20-10-20. Keep fertilizer going to plants to maintain active growth, but avoid high soluble salts, as verbena is very sensitive.

Provide high light, yet cooler temperatures. Provide 4,000–5,000 f.c. (43–54 klux) of light. If you hang your baskets, it is best not to put verbena at the top, in the attic: Hang them a tier or two down so that air temperatures will be a bit cooler. This will help you better manage your water as well.

The crop time from transplanting a plug to a flowering cell pack is five to eight weeks, depending on the variety. For 4" (10 cm) pots, the crop time is six to nine weeks.

Cuttings

Pot liners into a well-drained, disease-free medium with a pH of 6.2–6.5. Make sure the medium has a high initial nutrient charge. Pot 1 plant/4" (10 cm); 1–2 plants/6" (15 cm); and 5 plants/10" (25 cm) hanging basket.

Verbena thrives in cool night temperatures and high light. Maintain 4,500–6,000 f.c. (48–65 klux) of light; however, excessive light and high temperatures will cause leaf necrosis. Low light levels cause stem stretch. Grow on at 68–75°F (20–24°C) days and 62–65°F (16–18°C) nights. Low temperatures will cause temporary leaf yellowing.

Verbenas are very sensitive to over- and underwatering: Water stress either way causes damage to the leaves and root system. Keep plants actively growing with the proper nutrition. Start plants off with a constant liquid feed of 150–200 ppm from 15-0-15, alternating with 20-10-20. As plants mature, increase fertilizer to 200–300 ppm. If you encounter problems with soluble salts, irrigate with clear water at every third irrigation.

Pinch or cutback plants hard once they are rooted to the edge of the containers and the branches from the liner pinch have grown enough. Pinch above the fourth or fifth set of leaves. Continue providing conditions or sprays to the plants so that the internodes are short and the plants are tightly branched. Hanging baskets will require multiple pinches to develop full containers. Shear vegetation two to three times to increase branching and improve plant shape. Four-inch (10 cm), 1-qt. (1.1 l), and smaller containers can't be trimmed multiple times due to a shorter crop time. Thus, it is even more important for you to control growth early on so that the plants stay compact. High light, spacing, cooler production temperatures, and regular growth regulators are all ways to do this. Flowering is not triggered by day length but by total light accumulation. Flowers develop after three to nine leaf pairs are initiated.

Cycocel at 1,500–2,000 ppm sprayed five to six days after pinching and reapplied every two weeks as needed can be used to control height. Florel will promote branching, but discontinue use within eight weeks of sale.

Regularly scout for powdery mildew in your verbena. Though most new varieties are resistant, all verbena can be infected, and once infected it is hard and sometimes impossible to eradicate. Be sure to also check crops that are out of sight, such as baskets hanging up high. The best way to prevent powdery mildew is to control your greenhouse environment. Try to provide as much light and ventilation as possible and space on time. Water only in the mornings and allow foliage and surfaces to dry before the end of the day. Copper-based fungicides such as Phyton 27 have shown good preventative control. Be sure to rotate sprays if you are trying to control an outbreak. Sulfur pots are also still a very effective control measure. Be sure to set them on a timer so they run at night.

Transplanting a 105 liner to a 4" (10 cm) pot requires five to seven weeks to flower when using one liner per pot, while a 10–12" (25–30 cm) basket, with four to five liners, requires ten to thirteen weeks to flower.

Troubleshooting

Lack of flowering and too much vegetative growth can be due to excessive ammonia in the fertilizer, overfertilization combined with low light, and/or low light and overwatering. Drought stress between irrigations, high soluble salts, or powdery mildew can all cause foliage necrosis. Production temperatures that are too low can lead to foliage chlorosis.

Pests & Diseases

Aphids, fungus gnats, spider mites, thrips, and whiteflies can attack verbena. Powdery mildew is an especially bad problem in high humidity. Maintain good ventilation around crops to prevent it. Other diseases that can be a problem include botrytis, *Pythium,* and rhizoctonia.

Varieties

In seed varieties, 'Romance' has been superceded by 'Obsession', which features all of the positive qualities of 'Romance' and even better germination and seed quality. 'Quartz', also with excellent germination and seed quality, has colors that complement 'Obsession'. Between the two series you can offer solid and eyed types, as well as a full range of verbena colors from scarlet to blue, rose, burgundy, and lilac.

'Aztec', 'Superbena', 'Empress', 'Lanai', 'Magalena' and 'Leni' are all major series on the market that are propagated through cuttings. Most colors can be found in these series, including some novelty, speckled colors. Many series can be grown in 4" (10 cm) pots as well as baskets. Ask your broker or breeder representative for any special series adaptations.

Related Species

V. canadensis is highlighted by several varieties, both seed and cutting propagated. Follow the culture for *Verbena* × *hybrida* for similar results.

V. speciosa 'Imagination' has a wild-species look with purple-blue flowers. Plants grow to 18" (46 cm) in the garden and make excellent 4" (10 cm) pots or hanging baskets for growers. 'Imagination' is also a great combination pot addition. The variety will flower a bit later than your bedding plant crop, but you will find that it's much more tolerant to powdery mildew and won't flush into and out of flower, as bedding plant verbenas do. It is also drought tolerant. For the best plants, follow the seed culture for *Verbena* × *hybrida* but sell green. Plants will flower in June or July.

If you are looking for something different that's a tried-and-true performer in most parts of the country, try the perennial verbenas from seed, *V. bonariensis* and *V. rigida.* Plants are more upright growing: *V. bonariensis* can grow to 3' (91 cm) in the garden, and *V. rigida* to 14" (36 cm). Both are best sold green in 4" (10 cm) pots or flowering in 1-gal. (4 l) containers in the summer (twelve to fourteen weeks after sowing).

Veronica

Veronica longifolia
Common name: Speedwell
Perennial, cut flower (Hardy to USDA Zones 4–6)

Veronica 'Sunny Border Blue' is a mainstay perennial and is well known among growers, landscapers, and consumers alike. Deep, dark, violet-blue flowers are the hallmark of this plant. Perennial customers will request plants by name, so be sure to include it in your regular order.

In production, veronica is easy for growers. After a cool treatment, flowering is fast, just eight to ten weeks. Grown at cooler spring temperatures, 'Sunny Border Blue' will produce large flower spikes on well-proportioned plants that will walk off retail shelves.

Propagation

'Sunny Border Blue' is vegetatively propagated. Harvest stem cuttings with two to three nodes from actively growing plants in the spring or summer. Stick into a sterilized, well-drained medium or Oasis, and root under mist. Cuttings stuck to a 105-liner tray will be rooted in fourteen to twenty-one days at substrate temperatures of 72–77°F (22–25°C). Most growers add veronica onto their existing perennial liner orders rather than go to the bother of rooting their own cuttings.

Growing On

Pot one liner per 4" (10 cm), 6" (15 cm), or 1-gal. (4 l) pot. Select a long-lasting, well-drained medium with a pH of 5.8–6.2 and a moderate initial starter charge. Veronica requires a cold treatment in order to flower. For flowering plants in the spring, you can buy in liners or bare-root transplants in the fall and overwinter containers or purchase cool-treated liners in January or February.

Establish plants received in the fall in their final growing containers prior to exposing them to cool treatment. Root plants in and bulk them at 65–70°F (18–21°C) days and 60–65°F (16–18°C) nights.

Fertilize with 100–150 ppm from 20-10-20 at every irrigation. If you are growing your crop in small pots such as 4" (10 cm), be sure to keep plants well watered. As they root in, veronica's large leaf mass and thick leaves have high water requirements. Plants need constant moisture, but do not like wet conditions. Do not allow plants to dry out to wilting once buds are visible in the spring.

Growers potting plants in the fall should provide ten weeks at 41°F (5°C) in a cold frame or cooler to meet the cooling requirements. Provide nine hours of light during cooling using incandescent lights in a cooler or natural light in the cold frame. Keep pots irrigated, but allow them to dry somewhat between irrigations. Michigan State University researchers suggest treating for botrytis control at the beginning of cold treatment.

After cold treatment, to force plants for flowering, move pots into the greenhouse from the cooler or cold frame and force at constant day/night temperatures of 65–68°F (18–20°C). Grow under full-sun conditions (3,000–5,000 f.c. [32–54 klux]). Provide a soft pinch when plants come out of cooling and treat again with a fungicide after the pinch for botrytis prevention. During the winter months, supplemental lighting at 400–500 f.c. [4.3–5.4 klux] is beneficial to improve plant quality. Fertilize with 15-0-15 at 150–200 ppm at every other irrigation and maintain a pH of 5.8–6.2 and an EC of 1.0 mS/cm. A-Rest, B-Nine, Bonzi, and Sumagic can be used for height control two to three weeks after forcing begins. Plants should be ready for sale in eight to ten weeks.

When potted to a 1-gal. (4 l) container, allow nine to thirteen weeks to be fully rooted to sell.

Cut Flowers

V. longifolia can also be grown as a cut flower from bought-in root divisions. Plant at 2.5 divisions/ft.^{2} (27/m^2) into well-drained, friable soils. Heavy soils are acceptable if drainage is good. Adjust the pH to 6.0–7.0. Use one layer of support. Veronica likes moist soil; avoid long periods of drying between irrigations. When plants are actively producing stems and shoots, they are heavier feeders: Use 200 ppm from 20-10-20 as a constant liquid feed at this time. Provide 4,000–7,000 f.c. (43–75 klux) of light; plants will tolerate higher levels at lower temperatures and in moist soil conditions. In warm regions, plant divisions outdoors in January and February and after the danger of frost in the Midwest and northern areas. You can expect from one and a half to two flushes per year. Yields the second year will be greater than the firsts year's harvest. Indoor plantings can be made from January to March with production starting about 100–120 days later. These plantings will produce two or more flushes between spring and late fall. Harvest stems when 10–20% of flowers are open. Place in clean water treated with a bactericide and store cool at 36–40°F (2–4°C). Veronica benefits from an STS treatment, as stems are ethylene sensitive.

Pests & Diseases

Aphids and thrips can attack plants. Botrytis and powdery mildew can also attack plants. *Pythium* and *Verticillium* can cause problems as well.

Varieties

'Sunny Border Blue' is a must-have plant. Customers will request it, so be sure to make this your main veronica. In the garden, plants grow to 18–24" (46–61 cm) tall and flower from midsummer to fall.

For cut flower production, try 'Blauriesin', 'Caya', and 'Dark Martje', the three best-selling cut flower veronicas on the Dutch auctions during the mid-2000s.

V. × 'Goodness Grows' is a low-growing, related, hybrid variety. Plants grow 10–12" (25–30 cm) in the garden and produce typical veronica-blue flowers.

Postharvest

Plants are ready for sale once the very first flowers have opened. Instruct consumers to cut plants back after planting to encourage subsequent reblooming.

Veronica spicata
Perennial (Hardy to USDA Zones 4–6)

To expand your perennial veronica offerings, take a look at *V. spicata*, but stick to named varieties rather than the species, which tend to be gangly and highly irregular in habit and plant form.

Propagation

V. spicata is primarily vegetatively propagated. Harvest cuttings with two to three nodes from actively growing plants and root as outlined for *V. longifolia.*

The popular variety 'Blue Bouquet' is seed grown. Sow pelleted seed onto a well-drained, germination medium with a pH of 5.8–6.2. Cover with coarse vermiculite after sowing. Germinate at 65–75°F (18–24°C) substrate temperatures. Maintain uniform moisture levels; the medium should not be saturated. Keep ammonium levels below 10 ppm and sodium levels below 40 ppm. Seed will germinate in six to nine days, and a 288 plug takes from five to seven weeks from sowing to be ready to transplant to larger containers.

Growing On

Use 1 cutting or plug/4" (10 cm) pot and 3/6" (15 cm) or 1-gal. (4 l) pot. Select a well-drained medium with a pH of 5.8–6.2 and a moderate initial starter charge.

Grow plants on at 60–65°F (15–18°C) days and 50–55°F (10–13°C) nights under full-sun conditions. If growth regulators are required, A-Rest and Sumagic are effective.

Cold treatment is not necessary for flowering, but flowering is more uniform and plants are more compact after ten weeks at 41°F (5°C).

The crop time from a rooted liner to a flowering 1-gal. (4 l) is from eleven to fourteen weeks.

Pests & Diseases

See the Pests & Diseases section for *Veronica longifolia.*

Varieties

'Blue Bouquet', a seed-grown variety, makes an excellent 4 or 6" (10 or 15 cm) pot and flowers in its first year sown from seed. Plants grow 12–15" (30–38 cm) in the garden and produce medium-blue flowers in 6–8" (15–20 cm) long spikes. 'Blue Bouquet' is also suitable as a local market cut flower.

Vegetatively propagated varieties include 'Red Fox', which grows to 12–15" (30–38 cm) in the garden and produces elegant red stems over a long bloom window. Also try some 'Blue Charm', which grows to 24" (61 cm) in the garden and produces mounded plants with lavender-blue flowers.

Vinca

(see also *Catharanthus*)

Vinca major
Common names: Vinca vine; periwinkle; myrtle
Annual vine

Vinca vine is a must-have for bedding plant growers. Most growers make sure they have a few pots of 4" (10 cm) as well as some hanging baskets. Vinca is also a great addition to just about any mixed combination pot, as it thrives in both low and high light.

Propagation

Vinca vine is easily rooted from tip cuttings. However, because cuttings tie up valuable bench space and must be heated during the winter, most growers prefer to buy in liners. Besides, since just about every liner supplier offers it, it is easy to round out boxes to meet ship-week minimums with a tray or two of vinca vine.

Root 2.5" (6 cm) tip cuttings into a sterile medium or Oasis wedges at 75–80°F (24–26°C) days and 68–70°F (20–21 C) nights. Maintain substrate temperatures of 68–72°F (20–22°C) and provide from 500–1,000 f.c. (5.4–11 klux) of light. Provide mist to maintain cutting turgidity. Fertilize with 50–75 ppm nitrogen (N) of 20-10-20 if there is a loss of foliage color. Callus will form in five to seven days.

To develop roots, continue the temperature regime, increase light to 1,000–1,500 f.c. (11–16 klux), and fertilize weekly with 100–150 ppm N from 20-10-20, alternating with 15-0-15. Wean liners from mist by reducing the application at night and reducing water volume, duration, and frequency during the day. This stage will take from fourteen to twenty-one days. To harden plants for shipping or transplanting, drop air temperatures to 75–80°F (24–26°C) days and 62–68°F (16–20°C) nights. Move liners from the mist area and increase light to 1,500–2,000 f.c. (16–22 klux). Maintain fertilizer applications once a week.

A 72 tray will take from five to seven weeks with one or two cuttings per cell to be ready to transplant to larger containers.

Growing On

Transplant into a well-drained, peatlite or bark-peat medium with a pH of 5.6–6.2. Grow on at 70–75°F (21–24°C) days and 62–65°F (16–18°C) nights. Do not grow plants at temperatures above 95°F (35°C).

Provide 1,000–2,500 f.c. (11–27 klux) of light. Vinca produces small flowers in the axils of older leaves. Flowering will not affect branching or overall productivity.

Feed at every other irrigation with 150 ppm of 15-0-15, alternating with 20-10-20. As plants mature, increase feed to 200–300 ppm. If soluble salts become a problem, leach with clear water. Maintain the EC at 1.0 mS/cm. Allow pots to dry thoroughly between irrigations.

Pinch plants to promote branching. Vinca has strong apical dominance. Once the plant is pinched, one or two lateral buds will break. The majority of new shoots originate below the substrate line. Frequently cutting back plants promotes adventitious roots.

Florel decreases internode length and increases laterals. Time the application to occur after pinching. Monitor the crop's response to Florel closely, as it's easy to overdose vinca.

Upon potting a 72 liner to a 4" (10 cm) pot, allow six to eight weeks for sales.

Pests & Diseases

Aphids and fungus gnats can be a problem. INSV/TSWV, rhizoctonia, and *Phytophthora* can also attack vinca.

Troubleshooting

Poor shoot elongation is a sign of excessive Florel. Increase ammonia fertilizer and temperatures to overcome the problem. Too-low light levels can cause excessive internode elongation. Insufficient pinching or lack of Florel causes poor branching.

Varieties

Four types of vinca vine are typically grown by American bedding plant growers: *Vinca major,* with solid green leaves; *V. major* 'Maculata', with variegated green on green leaves; *V. major* 'Variegata', with medium green-and-cream, variegated leaves; and *V. major* 'Elegantissima'. Also try more recently introduced varieties, such as 'Wojo's Jem' (sometimes 'Gem'). 'Wojo's Jem' has lime-green leaves edged in medium green.

Related Species

Vinca minor, commonly called lesser periwinkle or creeping myrtle, is a smaller-leafed version of *Vinca major* and is hardy to USDA Zone 3. Plants grow to 6" (15 cm) tall but spread up to 3' (90 cm) across. There are a number of selections available, but 'Illumination' is the most common one and features bright, golden-yellow leaves that are edged in medium green. Follow the culture noted for *Vinca major* for propagation and growing on.

Postharvest

Allow pots to dry thoroughly between irrigations. Retailers should display under some shade: 1,000–2,500 f.c. (11–27 klux) of light is ideal. Vinca vine is hardy in southern regions.

Viola

Viola × *wittrockiana*
Common name: Pansy
Annual or hardy annual
(*Note:* Plants are often perennial in cool-winter and cool-summer areas. If winters are cold and summers are hot, plants are seldom perennial.)

Pansies, the common name for *Viola* × *wittrockiana,* are creating a solid second season for bedding plants: fall sales. The fall pansy phenomenon started in southern regions such as Atlanta, Georgia, and the Carolinas when landscapers planted commercial buildings with perky, yellow pansies in the fall to attract tenants. Their bright flowers dazzled consumers day after day for months on end—even in the dead of winter. Soon growers began getting requests for pansies from garden centers and mass marketers. Today, the fall pansy craze is heading north up the East Coast into the mid-Atlantic and Northeast, west into the Midwest, and even into Canada. Folks in the West must wonder what all the ruckus is about since California and the Pacific Northwest have enjoyed pansies as a cool-season staple for decades.

While many books and seminar lecturers can group a lot of crops together using similar cultural guidelines under the banner of bedding plants, do not for a moment think that pansies can be so easily categorized. Media pH/nutrient relations, production temperatures, and disease control/avoidance are all critical in producing a successful pansy crop.

In the seedling stage, pansies are so demanding that most growers have given up growing their own and buy in plugs from specialists who have access to germination chambers and cooling equipment to ensure environmental conditions are ideal. From a well-grown plug, most growers producing their crops under cover will experience few disease problems as long as the proper environment is maintained. However, ultrahigh summer temperatures can quickly stretch plants into an unsalvageable mess. Height control is easier when the crop is grown in open fields, but exposing the plants to the elements will also present many challenges to the grower, especially insect and disease control.

Begin planning your fall pansy crop in May, ordering plugs no later than June 1 to ensure availability. Growing 288-size or larger plugs for early plantings takes a minimum of six weeks. Because pansy seed is expensive and growing good plugs is demanding, most plug growers produce to order. Spring crops can be grown from plugs ordered in October or November and planted from December to March, depending on your geographical location.

Propagation

Pansy plugs are produced in four stages. First, sow seed onto a germination medium with a pH of 5.2–5.6, depending on your water alkalinity and fertilizer program. Soil EC should be less than 0.75 mS/cm using the saturated media extract method. Pansies are very sensitive to high salts, particularly ammonium, during germination. Maintain a moist, but not saturated, medium for optimum germination, reducing substrate moisture once the radicle emerges. Excess moisture will have an impact on the germination percentage, especially when using primed seed. After sowing, cover seed lightly with coarse vermiculite (do not use fine vermiculite), to help maintain 90–95% humidity around the seed during the germination process.

To ensure the highest germination possible, many growers prefer to purchase pregerminated or primed seed, which is readily available from your seed supplier.

Stage 1, radicle emergence, takes from three to seven days at substrate temperatures of 65–68°F (18–20°C), depending on the seed form you use. Temperatures above 68°F (20°C) will promote an increased germination rate but a lower germination percentage. Temperatures below 65°F (18°C) will slow down the germination rate, but may not increase the germination percentage. Some growers germinate pansies in a germination chamber using fine mist or fog to maintain high humidity levels. The plug trays need to be removed from the chamber when the radicles begin to point downward on the majority of seed, but before seedlings begin to gooseneck. It is critical to leave the trays in the cooler as long as possible, but the seedlings will stretch very easily when kept in for too long. If temperature stratification occurs in the cooler, the trays on the upper part of the cart may have to be pulled before the trays on the lower part of the cart. Check each variety at least twice a day and pull the trays at the proper development stage. If germinated in the greenhouse, keep the vermiculite cover moist without saturating the growing medium by frequent, light misting. Keep ammonium levels at less than 10 ppm.

Stage 2, cotyledon emergence, takes five to seven days at soil temperatures of 62–65°F (16–18°C). Maintaining cooler temperatures slows and tones plant growth. Reduce moisture levels once the radicle penetrates the growing medium, allowing the medium to dry out slightly before watering for optimum root development and algae control, but never allow the medium to turn tan in color. This is extremely important, as the root cap will die if the medium is too moist due to oxygen depletion or if the medium is too dry from drought. The vermiculite covering will provide the optimum environment if kept slightly moist, but will suffocate the seed if kept too wet. Apply moisture as a fine mist to prevent puddling, as this will saturate the medium and possibly bury the seed or radicle. Irrigate in the morning, under rapid drying conditions, to ensure the foliage is dry by nightfall to prevent disease infection.

Maintain the medium pH at 5.6–6.0 and EC below 0.75 mS/cm. To prevent soft growth, make sure ammonium levels are less than 10 ppm. Begin a fertilization regime using 13-2-13 or calcium/potassium nitrate at 50–75 ppm nitrogen (N) once cotyledons are fully expanded. Alternate liquid feed with clear water when EC is higher than 0.75 mS/cm. Maintain low phosphorous levels to prevent stem and leaf petiole elongation. Some growers supplement their fertilizer program with Solubor or boric acid, depending on the boron level in the medium, fertilizer, and water.

Allow for early morning and late afternoon sun, but maintain light intensity at 2,000 f.c. (22 klux) during the majority of the day until cotyledons have fully expanded. If seedlings appear to be stretching, increase the light intensity to 3,000–3,500 f.c. (32–38 klux). Close the shade curtain from midmorning to early afternoon, closing earlier under warm growing conditions. Excessive light will increase the substrate temperature and potentially kill the weaker seedlings. Low light will stretch seedlings, so work for a balance. Supplemental lighting of 450 f.c. (4.8 klux) for sixteen to eighteen hours per day for at least two weeks will help produce strong-stemmed seedlings when the natural light intensity is low and days are short.

Stage 3, growth and development of true leaves, takes from fourteen to twenty-one days. Maintain substrate temperatures of 60–75°F (16–24°C). If temperatures can be precisely controlled, provide as much light as possible—up to 5,000 f.c. (54 klux). Maintain 3,000–3,500 f.c. (32–38 klux) under warm temperatures to prevent the occurrence of variegated, distorted foliage, commonly called "mottled pansy syndrome." Allow the medium to thoroughly dry between irrigations to encourage root growth and to control shoot growth, but avoid permanent wilting. If needed, acidify water to obtain 60–80 ppm bicarbonates, as higher water alkalinity will slowly raise the soil pH and tie up micronutrients. Maintain the soil pH at 5.6–6.0 and EC at less than 1.0 mS/cm. Increase the fertilizer to 100–125 ppm N from 20-10-20, alternating with 14-0-14 or another calcium/potassium-nitrate fertilizer, or use a complete fertilizer such as 13-2-13. Fertilize every four to six days to prevent nutritional stress, which will quickly predispose the plants to foliar disease. If you choose to use a 15-0-15 fertilizer, supplement with magnesium (Mg) by adding magnesium nitrate to the 15-0-15 to obtain a N-Mg ratio of 5:1, or drench with magnesium sulfate once or twice during the plug production phase at a rate of 1 lb./100 gal. (1.2 g/l). (Do not mix magnesium sulfate with a calcium-based fertilizer, as calcium sulfate with precipitate out of the solution.)

Using a negative DIF during the two hours before and after sunrise can help to control petiole stretch and stem elongation. Growth regulators that are effective on pansy plugs include A-Rest (4–12 ppm), B-Nine (1,000–2,500 ppm), and Bonzi. A B-Nine (1,000–2,500 ppm)/Cycocel (750–1,250 ppm) tank mix *or* a tank mix of B-Nine (1,000–2,500 ppm) and A-Rest (3–6 ppm) can also be used. Exact rates depend on your environmental conditions and fertilizer selection. Use high-nitrate/low-ammonium fertilizers for optimum growth control. Alternate foliar fungicides for leaf spot disease control, spraying every five to fourteen days depending on disease pressure and the environment. Monitor roots for *Pythium,* rhizoctonia, and *Thielaviopsis,* drenching when a suspicious infection is found. Maintain an optimum substrate pH to prevent *Thielaviopsis* infection.

Hardening pansy plugs to get them ready to transplant or ship in Stage 4 takes approximately seven days. Maintain cool soil temperatures of 55–65°F (13–18°C). Allow the medium to dry thoroughly between irrigations, avoiding severe wilting. Maintain a substrate pH of 5.6–6.0 with an EC of 0.60–0.75 mS/cm. Fertilize as needed with 14-0-14 or 13-2-13 at 100–150 ppm. Pansy plugs can be stored in the dark at 38°F (3°C) for at least two weeks if plugs cannot be transplanted or shipped on schedule.

A 288 plug requires from four to six weeks from sowing to be ready to transplant to larger containers, depending on the variety.

Growing On

Most growers in warm regions start their early fall pansy crops from 288 or larger plugs. Using plants with a larger root mass allows for a greater margin of error during production. Once temperatures drop somewhat, later crops can be produced from smaller plugs. Buy plugs that have been single sown: Multiple plants per cell are not advisable, as plants will compete with one another and flowering may be delayed.

Use a disease-free medium with a moderate initial nutrient charge and a pH of 5.6–6.0. To prevent stem rot and slow growth, do not plant plugs too deeply, using only short, toned plugs to prevent finishing a wobbly, unstable plant. Make sure the medium is compressed around the plug at planting to prevent plant wobbling. Use 1 plant/cell in 606s, 1–2 plants/4" (10 cm) pot, and 3–4 plants/6" (15 cm) pot.

Optimum plant quality is obtained when plants are grown outdoors, but avoid moving them outdoors until roots are established. Placing young plants directly outdoors, followed by a heavy rain or excessive irrigation, is a recipe for disaster. This indoor staging period of one to two weeks also helps bulk up plants before they are exposed to the elements.

Plants will finish in approximately six weeks from 288 plugs in 606s or 4" (10 cm) pots. Temperature is the most critical production factor: Pansies must receive cool temperatures. Maintain growing temperatures of 55–65°F (13–18°C) days and 50–55°F (10–13°C) nights. Obtaining cool temperatures for southern growers starting pansy crops in the summer for early fall sales is challenging at best. Many growers use light shade to help reduce temperatures, but excessive shade and warm temperatures will cause stretch. When temperatures are high and plants become stressed, they stretch and flower very quickly. However, it's important to build the plants before allowing them to flower. Variety selection plays a role, as does holding off on planting the first crop turn until temperatures cool down.

Do not overwater pansies once they are planted, as it encourages both disease growth and plant stretch. Maintain a semi-moist soil substrate until roots are seen on the outside of the soil ball, then allow the medium to dry down thoroughly between irrigations for growth control. Apply fungicide drenches to control disease, but wait until new roots are at least 0.5" (1.3 cm) long until doing so.

As with plug production, fertilize with a complete fertilizer such as 13-2-12 or 17-5-17, or alternate a calcium-based fertilizer with 20-10-20 at 150–200 ppm as needed to maintain an EC between 1.0–1.5 mS/cm. Prevent nutritional deficiency to help maintain leaf spot disease control, but avoid soft, lush foliage.

Pansies respond wonderfully to DIF to control plant height. If DIF is not practical, control plant growth through low substrate moisture, high light intensity, fertilizer selection, and growth regulators. A-Rest (10–33 ppm spray or 0.25–1.00 ppm drench), B-Nine (2,500–5,000 ppm), Bonzi (1–5 ppm), and Sumagic (3–6 ppm) are all effective in helping to control height. Additionally, tank mixes of B-Nine and Cycocel or B-Nine and A-Rest as outlined for plugs are effective. Excessive ammonia fertilizer and wet soils promote stem stretch, which requires additional growth regulators. Allow the crop to dry down moderately between irrigations for better root growth, disease management, and height control. When using ammonia fertilizers, apply the growth regulator before fertilizing for better height control. Since growth regulators require two days or more to become effective, apply the growth regulator as the plant starts reaching the allowable size, not after it is too tall. Apply growth regulators during the crop cycle to maintain foliage tone so that high rates are not required at the end to stop plant growth, which affects garden performance. Apply only as much growth regulator as is needed to control growth for one to two weeks, adjusting the rate based on your growing conditions.

Maintain light levels as high as possible (3,000–5,000 f.c. [32–54 klux]) while maintaining moderate temperatures. Pansies are not daylength sensitive in their flowering, although flowering is more profuse under longer days. However, stem length is shorter under short days. Breeders have selected pansy strains that perform better in production under the high temperatures and longer days of late summer/early fall and other strains that flower quickly and perform better under the short days of winter. A reputable salesperson can guide you in selecting varieties to plant for best flowering at different times of the year in your region.

Pests & Diseases

Pansies are susceptible to a wide range of diseases. Control disease before it starts by understanding the environmental conditions that cause outbreaks and managing the environment to prevent it. Growers producing pansies outdoors frequently experience root and crown rot pathogens when excess rains occur; delay moving plants outside until the root system is developed enough to handle the excess moisture.

Thielaviopsis, commonly called black root rot, develops black chlamydospores inside the roots, which causes sections of the roots to turn black. However, don't be alarmed if you find black roots while checking your pansies. The pansy roots will be transparent, almost black in color, when the medium is wet. The lower foliage of infected plants will turn chlorotic, eventually progressing to the young foliage, due to nutritional deficiency resulting from the inability of the roots to take up adequate fertilizer. Do not reuse flats or pots on *Thielaviopsis*-sensitive plants, as most greenhouse disinfectants will not kill the chlamydospores. Maintaining wet substrate conditions along with a pH of 6.2 or higher promotes disease infection, so allow the medium to dry out between irrigations and maintain a medium pH between 5.6 and 6.0. Iron-sulfate drenches or increased water acidification will help keep the pH below 6.0. Only a few fungicides are labeled for *Thielaviopsis* control. It is very important to maintain the proper application interval stated on the label. When *Thielaviopsis* is confirmed, follow the minimum application interval for best results.

Myrothecium crown rot can be a very serious disease on recently transplanted plugs, causing a general yellowing of the plant that's followed by wilting and death. The plant can be easily broken off at the substrate line, and it is commonly seen when heat-stressed young plants are overwatered. Black spores with a white fringe of mycelium will be found on the diseased tissue when closely examined. The spores are long lived and survive in the medium and plant debris. Sanitation plays a key role in controlling the disease. Remove all infected plants on a timely basis, including the medium and plant debris in the growing area to control the spread of this disease. Avoid touching other plants and bag the infected plants onsite to prevent disseminating the disease. Rotate crops if at all possible, so pansies are not grown on the same bench or greenhouse every year. Eradicate fungus gnats and shore flies, known vectors of fungal diseases. If *Myrothecium* has been a problem on past crops, weekly preventive fungicide applications are required.

Plants that are underfertilized or grown under less-than-optimal conditions, such as high relative humidity, will be predisposed to a multitude of leaf spot pathogens. Only water the plants under rapid drying conditions to ensure that the foliage has dried off prior to sunset. Maintain a regular fertilization program, which includes calcium, to encourage strong, toned growth to reduce the incidence of leaf disease.

Pansies are especially attractive to aphids, which can cause leaf distortion when present in high numbers. Spider mites and thrips are other insect pests to watch out for as well as fungus gnats and shore flies when the soil is kept too wet.

Troubleshooting

Yellow growing points are not a symptom of boron deficiency, as some might imagine: A pansy's growing point is cream colored. Boron-deficient plants have green growing points. Yellow growing points appear when plants are placed in high temperatures. High temperatures can also cause plants to stall.

Boron deficiency causes young foliage to become thick and puckered, yet it remains green. Boron deficiency is common with summer-grown pansies, especially in plug trays. The tip may abort with numerous small shoots developing below the meristem. Immature foliage will be hard, leathery, and distorted and will not expand. Maintain a pH below 6.0 to ensure boron availability. Boron is readily leached, so practice good water-management techniques, especially in the plug tray. Supplement fertilization with one or two applications of a boron source during production, especially if temperatures are above 85°F (29°C). Solubor at 0.25 oz./100 gal. (19 mg/l) or borax dissolved in hot water at 0.5 oz./100 gal. (38 mg/l) will prevent deficiency. The exact rate you apply will depend on your irrigation water—the more boron you have in your water, the less additional boron will be required. Correcting boron deficiency is difficult, so it's better to prevent the problem from occurring. If you see a problem, treat as above and wait one week to determine if an additional application is needed.

Pansies are also sensitive to boron toxicity, which is common when too much boron is applied. Symptoms include stunted growth, reduced leaf size, and lower chlorotic foliage with marginal necrosis. Do not apply more than 1.5 ppm total boron on a continual basis. Do not supplement fertilizer with boron without checking levels in your irrigation water and fertilizer. Poor grades of fertilizer may contain excessive amounts of boron as well as other horticultural products. Correct boron toxicity by increasing the substrate pH to 6.0–6.2, leaching excess boron from the medium, and fertilizing with calcium-based fertilizers containing little boron.

Iron and/or manganese deficiency is another common problem, especially when the substrate pH is higher than 6.0 and the substrate is being kept too wet. Iron deficiency symptoms develop on the young, immature foliage as interveinal chlorosis, followed by complete yellowing of the foliage. Manganese deficiency symptoms look similar to iron deficiency symptoms but are less likely to cause the leaf to turn completely yellow. Marginal necrosis or spotting may eventually develop with severe manganese deficiency. Maintain the medium pH between 5.6 and 6.0, and allow the medium to dry down moderately to prevent these deficiency symptoms. Lower soil pH by drenching with iron sulfate at 8–24 oz./100 gal. (0.6–1.8 g/l), depending on the container size and how much the substrate pH needs to be lowered. *Always* wash the foliage off immediately to prevent foliar phytotoxicity problems. Using a fertilizer with a high potential acidity value, such as 21-7-7 Acid Special, will help lower the substrate pH as well, but will cause soft growth when used at rates greater than 150 ppm nitrogen (N). Many growers are now acidifying 20-10-20 or 17–5–17 with citric acid to prevent having to purchase an acid fertilizer and lowering the pH of the water coming out the end of the hose to 4.0–4.5. Supplement fertilizer with iron-chelate EDDHA to increase the availability of the iron (Fe) to the plants, typically starting at the 1 ppm Fe rate if iron deficiency is a common problem. Check for root rot diseases, as both *Pythium* and *Thielaviopsis* infections will restrict nutrient absorption and cause deficiency problems.

Magnesium deficiency will also cause interveinal chlorosis, but it will begin to develop on the recently expanded middle leaves instead of the very young leaves. The chlorosis is more pronounced along the leaf margins, spreading toward the mid-vein as the deficiency becomes more severe. Magnesium deficiency is more prevalent with high calcium and/or potassium levels in the medium. Supplement fertilizer with magnesium to maintain a N-Mg ratio of 5:1 via magnesium nitrate or magnesium sulfate, if magnesium deficiency is common. Do not add magnesium sulfate to a calcium-based fertilizer to prevent precipitation. Drench with Epsom salts at 1–2 lbs./100 gal. (1.2–2.4 g/l) to correct magnesium deficiency.

Desiccation can cause dry, shriveled up flowers. Plants being produced in an unheated cold frame or outdoors can become desiccated when the substrate is frozen yet air temperatures are high enough for plants to respire.

Lower leaf edge necrosis may be due to potassium deficiency or micronutrient toxicity. Submit plant tissue samples to a reliable diagnostic lab for a complete nutrient analysis to help determine corrective actions.

Mottled pansy syndrome can be a problem when young plants are grown under high light levels and temperatures above 85°F (29°C). Symptoms include mottled white/green, distinctly variegated leaves with an elongated, twisted appearance. Do not confuse this with genetic variegation, out of which the plants will eventually grow. Rogue out all suspect variegated plants from the plug trays before they are transplanted. Prevent the occurrence of this problem by maintaining optimum light levels and substrate temperatures until the plants have at least eight leaves.

Varieties

The pansy market has moved to almost exclusively the domain of F_1 varieties. Compared to open-pollinated varieties, they are faster to flower, produce larger flowers with more consistent coloration, and are more uniform in plant habit and flowering response. Refer to the catalogs of the various seed and plug companies for details of the varieties available.

Related Species

Viola cornuta are the smaller-flowered selections of violas and are highlighted by variety names like 'Penny', 'Rocky', and 'Sorbet'. Following the culture detailed for pansies, *Viola cornuta* plants will be both earlier to flower as well as more hardy in the garden. A 288-plug tray takes from four to six weeks to be ready to transplant to larger containers. Transplanted to a 4" (10 cm) pot, plants will flower in four to six weeks.

Postharvest

Maintain cool temperatures for retail displays: 65–75°F (18–24°C) days and 60–65°F (16–18°C) nights. For the highest-quality plants that can be held the longest, maintain temperatures at 50–60°F (10–16°C). Provided cool temperatures can be maintained, display plants in full sun to partial shade. If temperatures are high, supply partial shading to help control temperatures.

Do not overwater pansy displays: Plants do not transpire quickly and therefore do not dry out as quickly as other plants do.

Pansies are somewhat sensitive to ethylene, so avoid exposure. Ethylene exposure will result in droopy flowers, bud abortion, and cupped foliage.

Encourage consumer and landscape customers to wait to purchase pansies until outdoor temperatures have cooled down for best performance.

Z

Zantedeschia

Zantedeschia rehmannii
Zantedeschia elliotiana
Zantedeschia hybrids
Common name: Calla lily
Flowering potted plant, cut flower

Colored calla lilies have become an increasingly popular potted plant and cut flower for their classic chalice shape and variety of new colors and plant sizes. They add a tremendous new dimension to the classic white *Z. aethiopica* wedding/funeral flower. Colored callas have brightened their way as perennial tubers into gardens year-round in warmer climates and as lifted and stored perennial bulbs in areas of hard-soil frosts.

The highest-quality and most-uniform calla culture requires special attention to media selection, water management, and forcing temperatures, as well as a strong preventative fungicide program.

If you are new to callas, try them on a limited scale first.

Propagation

Most callas sold today in the United States are two-year-old, true seed hybrids. Treated seed virtually eliminates viral, fungal, and bacterial diseases at the time of planting, greatly reducing disease in professional grower products. Seed-grown tubers generally produce more eyes, which means plants will have a higher number of flowers and leaf sprouts.

Callas produced in New Zealand and the Netherlands are started from tissue-cultured plantlets. Again, it is a long process that takes about two years.

Propagators must maintain a rigorous, clean-stock program ensuring that latent diseases and viruses are not spread.

Since disease is the most important aspect of growing callas, starting from disease-free stock is critical. From tissue culture to salable tuber takes a minimum of one year, longer for larger tubers. Once tubers are lifted from the field, they also require a minimum of six weeks storage at the correct temperature regimes. That's why growers start their crops from tubers produced by specialists in the United States, New Zealand, or the Netherlands. Do not purchase old tubers from cut flower growers—these tubers are very likely to be infested with bacteria and viruses that cannot be cleaned up.

Growing On

The best success will result from buying in grower-pretreated calla tubers. Many suppliers treat tubers with gibberellic acid (GA) for increased flowering and a multipart fungicide-bactericide solution for strong disease prevention. If grower-pretreated tubers are not available in your area, spray tubers with GA (either GA_3-

ProGibb or GA_{4+7}-Promalin) at 100–150 ppm. Callas will respond to GA concentrations up to 500 ppm, but flower abnormalities increase with greater than recommended rates. Combine the GA solution with a fixed copper fungicide such as Kocide at 1 lb./50 gal. (2.4 g/l). Spraying tubers in trays, in lieu of tank dipping, reduces the potential for disease spread and eliminates the disposal of used dip-tank pesticide.

Unpack tubers immediately and ventilate for a day or two at 60–65°F (16–18° C) to dry off any abrasions. Inspect tubers and discard any soft, rotted, or chalked tubers. If storing for longer than a month, hold at 45°F (7°C) and 80% relative humidity with good air circulation. Order tubers as early as possible to ensure the best selection and the desired quantities.

Available tuber sizes are in five principal grades of 1.5–1.75" diameter (12–14 cm circumference), 1.75–2" (14–16 cm circumference), 2–2.25" (16–18 cm circumference), 2.25–2.5" (18–20 cm circumference), and 2.5" and up (20+ cm circumference). Other specialty sizes, larger or smaller, can be special ordered. Potted plant production can utilize most tuber sizes, singly or in multiples of smaller grades. The smallest recommended pot size is 4.5" (11 cm) using one 1.75–2" (14–16 cm) tuber for high-quality production. Use a single 2.25–2.5" or 2.5" (18–20 or 20+ cm) tuber; two 1.75–2" (14–16 cm) tubers, or three 1.5–1.75" (12–14 cm) tubers for 6" (15 cm) pots. Eight-inch or larger pots need multiples of two to three tubers, or extremely large grades to balance and fill out pots. Ask your supplier for recommendations by tuber size and variety.

Pot crops can be grown pot tight until the final two to three weeks of production, when they need spacing.

Many cut flower calla growers prefer to grow the crop in containers so plants can be moved around. If you choose this route, plant tubers in the container in which they will grow for several years; do not plan to repot. Other growers prefer to use mulched ground beds in the greenhouse, outdoors in the field, or under lath or shade. A semi-hydroponic system for production involves growing the crop in a thick layer of pine sawdust that is placed in a container or on top of soil. Space tubers at a minimum of 4–8" (10–20 cm) for cut flower production in beds to enhance root development and tuber growth. If you choose to grow in ground beds, change out beds every three years, allowing beds to lie fallow before replanting. Treat for pests and diseases before replanting.

Cut flower growers should change out stock every two to three years, rotating in fresh material to maintain high yields. As tubers age, they become more dominant-eyed and tuber losses increase. Growers may get a higher percentage of No. 1–grade flowers, but greatly lower yields. Lift and destroy tubers away from production areas, as they will be infected with disease.

Scheduling callas is challenging. Differences of growth time on the bench or container may vary by variety by up to two weeks. Ask your supplier for bench times for individual variety "days to first flower" and "peak bloom," which include differences for time of year, temperatures, and tuber sizes. Generally speaking, earlier types that include, for instance, *Z. rehmannii* 'Gem' hybrids require sixty-five to seventy days to peak (GA-induced) bloom for plantings from November to February. Most other varieties require seventy to eighty days to bloom during the same period when greenhouses are kept in the 60°F (16°C) or higher range at night. Plantings after March, or using only large-sized tubers, can reduce bench time by one week. Warmer temperatures and longer days accelerate growth in late-season plantings.

Select a well-drained medium with a moderate to high nutrient charge using coarse peat with good air porosity and a pH of 6.0–6.5. Use gypsum and lime to adjust pH, as this also promotes good root health. Keep peat to 50% or less, especially in fall and winter plantings. For the remainder of the mix, use other coarse drainage materials such as graded 0.125–0.25" (3–6 mm) fir bark, No. 2 sand, or perlite. Incorporating a biological such RootShield (*Trichoderma* spp.) or Soil Guard (*Gliocladium* spp.) at label rates can help manage root disorders. Cover tubers with 1" (3 cm) of medium at planting.

The most critical factor in producing a great calla crop is disease control and maintaining plant health. Focus on preventing diseases: Once established, diseases are difficult to control. Immediately remove any diseased tubers or pots from healthy areas.

Fungicide drenches must start within two days of planting and are repeated at seventeen- to twenty-eight-day intervals. (See the Pests & Diseases section for a program outline.)

The best cultural conditions for callas involve good air circulation, moderately high relative humidity, consistent temperature regimes, and the highest possible light (in hot, bright summer conditions, provide 30–50% shade, reducing light to 2,500–3,500 f.c. [27–38 klux]). Recommended temperatures are growth-stage dependent. Avoid temperature extremes to enhance plant performance, health, and growth regulation.

Stage 1 of your calla crop, from planting until sprouting vegetation is 1–3" (3–8 cm) tall, will take twelve to twenty-five days at 75°F (24°C) days and 65°F (18°C) nights or a constant 68°F (20°C). Note that plant growth regulator (PGR) drenches of either Bonzi or Topflor are applied during this 1–3" (3–8 cm) sprout stage; see the section on PGR usage.

Stage 2 of the crop, from 3" (8 cm) to flower bud (about one month after Bonzi treatment) will last from about days 28–50 at 70–75°F (21–24°C) days and 60°F (16°C) nights.

Stage 3 is marked by flower buds beginning to push and color, approximately days 50–75 at cool temperatures of 70°F (21°C) days and 50–55°F (10–13°C) nights.

Run temperatures 2–5°F (1–3°C) cooler during Stage 2 and under poor light conditions or if you do not use Bonzi for potted callas. Cooler temperatures at Stage 2 can improve plant habit, often reducing Bonzi requirements, but this will result in longer bench times and water retention in the pots. Warmer temperatures can reduce bench times but will also produce taller, softer plants.

If you are producing a cut flower crop, strive for 65–77°F (18–25°C) days and 54–65°F (12–18°C) nights. Maintain soil temperatures at 65–68°F (18–20°C). Excessively higher summer temperatures stress plants and can spur disease.

Fertilizing early during rooting improves crop uniformity and vigor. Use a balanced liquid feed of 300 ppm or a precharge granular in the substrate for two weeks, and then go to 200 ppm for the duration. Water management is critical. Keep pots moist, but avoid excessively wet or dry conditions. Apply clear water, leaching weekly to reduce salts. Schedule drenches and Bonzi around irrigation timing to minimize duplicate wetting applications. Avoid pooling and splashing to reduce disease spread. Ebb-and-flood systems can be risky because of the potential to spread disease.

A crop time of nine to thirteen weeks is required for flowering plants.

PGR Use

Bonzi or Topflor are both effective growth regulators for potted callas. Using them is both an art and science. Be sure to take notes on your environmental, plant, and medium conditions, as well as ppm rates and volumes you are applying because various factors greatly influence PGR efficacy. Apply Bonzi of Topflor when sprouts are 1–3" (3–8 cm) tall. Pots must be uniformly moist, so apply them within one day of a previous irrigation. Segregate pots by sprout size, treating the tallest plants first. Second or subsequent PGR applications are usually made six to ten days later. Do not treat beyond forty days after emergence. Too much Bonzi or Topflor reduces flower count. Change temperatures incrementally at the time of PGR application since radical temperature changes can affect plant response reaction.

PGR rates vary, depending on the variety's height or vigor and how conducive your environment is to tall, soft growth. Research at Cornell University indicates Bonzi and Topflor rates are very similar across a range of cultivars. Generally, use 8–10 ppm (of either PGR) on compact- and moderate-sized varieties and 10–15 ppm for intermediate-height varieties or for softer growing conditions, assuming 4 oz. (113 g) applied per 6" (15 cm) pot or 2 oz. (57 g) per 4" (10 cm) pot. Note that certain media components such as pine bark can tie up PGRs and reduce effectiveness. Cold, dark, or poor growing conditions after application will affect uptake and assimilation of Bonzi or Topflor, thereby greatly affecting its efficacy; optimal conditions for plant growth are critical postapplication.

Pests & Diseases

Your success in managing diseases depends on simultaneously controlling three primary pathogens that independently or in concert can cause root rot and soft rot bulb syndrome. They are *Erwinia* bacteria, water mold fungi (primarily *Pythium*), and rhizoctonia root fungus. Not one or even two products can successfully control all three pathogens; and, controlling only a portion of the complex will sometimes lead to a worsening of the other untreated pathogens. Product labeling and registrations vary by locale. If the following recommended products are not available, then other drenches need to be tested to manage the three pathogens. Callas are generally tolerant to most applications, but smaller trials are encouraged before using any new chemical or growth regulator on a large scale.

The trichoderma-type biologicals, RootShield at 1 lb./yd.^{3} (460 g/m^3) or Soil Guard at 12 oz./yd.^{3} (340 g/m^3), are compatible with most drenches. Fungicide drenches must start within two days of planting and are repeated at seventeen- to twenty-eight-day intervals. A three-part, tank-mix solution must address the three major pathogens. *Erwinia* is best controlled by streptomycin sulfate (i.e., Agrimycin-17) at 100–200 ppm. Few bacteria control alternatives exist; however, other antibiotics or Phyton-27 (copper sulfate pentahydrate) at 13–20 oz./100 gal. (100–150 ml/100 l) can help. Water molds can best be drenched with metalaxyl (Subdue) at 1–2 oz./100 gal. (8–16 ml/100 l) or, alternatively, with Aliette (fosetyl-aluminum) at 13 oz./100 gal. (1 ml/l). Rhizoctonia can best be addressed with Heritage (azoxystrobin) at 1.0 oz./100 gal. (7.4 g/100 l) or poorer alternatives include thiophanatemethyl (i.e., Cleary's 3336) at 1 lb./100 gal. (120 g/100 l) or iprodione (Chipco 26019) at 6.5 oz./100 gal. (49 g/100 l). Any other alternatives need to be trialed on a small scale for compatibility or toxicity.

Inspect pots weekly for cleared or browning roots and drench again at seventeen to twenty-eight days, including biologicals, as these products do lose their effectiveness with time.

Insects may also attack callas. Control shore flies, fungus gnats, or *Lepidoptera* larvae, if present, using labeled insecticides or biologicals. You must control thrips, or INSV can spread rapidly. This is not bulb-borne, so remove leaves or plants immediately.

Troubleshooting

Narrow leaves and stretched leaves and excessive flower stems may indicate excessive GA. Very dark, rosetted, and squat plants can be due to excessive Bonzi.

Varieties

Z. elliotiana is the yellow calla, and *Z. rehmannii* is the pink calla. Today, many of the varieties on the market are hybrids of these two, sometimes including other species as well. The 'Gem' callas, *Z. rehmannii violacea,* are available in lavender, pink, and rose. 'Rubylite', also *Z. rehmannii violacea,* is available in 'Pink Ice' and 'Rose'.

Z. aethiopica is the traditional white calla. Plants flower naturally in mild, frost-free climates from fall to early spring when temperatures are cool; they have even naturalized in some mild climates. *Z. aethiopica* will flower continuously under mild temperatures; however, colored callas only flower once a year.

Postharvest

Harvest cut flowers when the spathe is three-quarters to fully open and prior to pollen shed. Generally, flowering begins ninety days after planting. Flowers can be harvested by pulling or cutting stems. Pulling stems carefully, under adequate soil moisture, is more accepted, and produces a longer stem without leaving a stem stub that could later rot. Do not pull the stems of poorly rooted plants. If you must or simply choose to cut stems, sterilize knives between each plant to avoid virus or other disease spread.

Flowers should be recut, graded, and placed into a preservative solution with a disinfectant to reduce stem splitting. Store colored cut callas at 43–47°F (6–8°C). Every two days, recut stems and change vase water. Store *Z. aethiopica* at 38°F (3°C).

Pot plant callas are ready to sell when approximately half of the total number of expected blooms show color or are in spike. Ship long distances in refrigerated conditions (40°F [4°C]). Do not store plants in the dark for more than one or two days. Retail under cool, bright conditions. Consumers will have the most success placing their pot callas in cool, bright windows.

Zebrina

Zebrina pendula
Common names: Wandering Jew; inch plant
Annual, foliage plant

In just about every bedding plant grower's greenhouse you will find a line of wandering Jew plants hanging in the back of one of the Quonsets. They are the stock plants from which cuttings are pulled each winter for 4" (10 cm) pots and hanging baskets sold in the spring and summer.

Since wandering Jews can tolerate really low light levels (down to 300 f.c. [3.2 klux] or so), they are great to sell to consumers looking for plants that will thrive on shady decks, on terraces, or under thick oak or maple shade. Wandering Jew also makes a great trailing plant to add to combination pots and baskets marketed for either sun or shade. Plants perform in the interior environment as well, having the reputation among consumers, young and old alike, as one of those plants that's hard to kill.

Zebrina is so easy to propagate that some growers direct-stick cuttings into hanging baskets. Take tip cuttings measuring 3–4" (8–10 cm) long, snaking them off with your hands or using a knife. Pull leaves off the lower end that will be stuck into the medium. Mist is not necessary but will help cuttings root faster. Keep the basket medium moist, not wet, and keep the medium warm at 70–74°F (21–23°C) and air temperatures at 70–85°F (21–29°C). Once cuttings have rooted in (five to six weeks after sticking cuttings), you can move baskets off the bench and onto hangers.

Begin fertilizing with constant liquid feed at every irrigation. Start with 100–150 ppm from 20-10-20 and increase the rate to 200–250 ppm as plants grow. About every fifth or sixth irrigation, leach with clear water to avoid salts buildup. Plants are sensitive to fluoride; ideally, use fluoride-free or low-fluoride water sources. Do not allow plants to dry out, or older leaves will begin to die. For fastest production, grow plants at 3,500–4,500 f.c. (37–48 klux). During the spring, they can take hanging high in the attic. Later in the summer, provide shade and/or move plants to lower tiers.

If your plants are showing tip burn, the cause may be high soluble salts, low humidity, fluoride, or plants being maintained too dry.

There is a range of wandering Jew varieties available, and your customers can use them to color coordinate in combinations.

From sticking rooted liners to 4" (10 cm) pots, allow six to eight weeks for well-rooted pots.

Zinnia

Zinnia spp.
Common name: Zinnia
Annual

Zinnias are highlighted by three groups: *Zinnia angustifolia, Zinnia elegans,* and *Zinnia marylandica.* This culture is based on *Z. elegans* (but the other two are addressed in the Related Species section), the tall growing (3' [90 cm]) plant with large, 3–4" (8–10 cm) flowers in a number of colors. It is personified by every farmer's wife sowing rows of zinnias beside fields for summer color. These old-fashioned zinnias have been modernized and updated with dwarfer plants that produce equally electric flower colors.

For growers, zinnias are easy, provided they are careful about watering and don't try to produce the crop when temperatures are too cold.

Propagation

Germinate coated seed onto a well-drained, disease-free, germination medium with a pH of 5.5–5.8. Cover seed lightly with coarse vermiculite to help maintain humidity at the seed coat. Maintain substrate temperatures of 68–70°F (20–21°C). Run trays dry; use about half the amount of moisture you would use for another crop. Do not saturate the medium at any time during germination. One old trick to use to get the right amount of water in your trays is to water the medium in the afternoon before you sow seed. Sow seed the following morning and do not water it in. Provide 100–400 f.c. (1.1–4.3 klux) of light in the germination area. Make sure to keep ammonium levels below 10 ppm. Radicle emergence takes one to two days.

Once radicles have emerged, move trays to Stage 2 for stem and cotyledon emergence, which will take from three to five days. Maintain substrate temperatures of 68–70°F (20–21°C). Continue to keep the medium on the dry side by allowing the medium to dry slightly before watering. Provide 500–1,000 f.c. (5.4–11 klux) of light. Begin fertilizing with 50–75 ppm from 14-0-14 as soon as cotyledons have expanded. Alternate feed with clear water. Be sure that foliage is dry by nightfall to prevent disease. Zinnias are susceptible to a number of foliage diseases that require free water on the foliage to become established.

Move trays to the next step, developing true leaves in Stage 3, by lowering substrate temperatures to 65–68°F (18–20°C). Continue to run trays dry. Increase light to 1,500–2,000 f.c. (16–22 klux) and increase feed to 100–150 ppm once a week from 20-10-20, alternating with 14-0-14. If growth regulators are needed, A-Rest, B-Nine (2,500 ppm), and Bonzi are useful. Stage 3 will last about fourteen days.

Harden plugs for seven days prior to sale or transplanting by lowering substrate temperatures to 62–65°F (16–18°C) and increasing light to 2,000–3,000 f.c. (22–32 klux). Fertilize once with 100–150 ppm from 14-0-14. Continue to run trays dry.

A 288-plug tray takes from four to five weeks from sowing to be ready to transplant to larger containers.

Growing On

Do not delay planting zinnia plugs, especially under short-day conditions, as delays can cause plants to prematurely bud up. Pot 1 plug/cell or 4" (10 cm) pot and 2–3 plugs/6" (15 cm) or 1-gal. (4 l) pot. Select a well-drained, disease-free medium with a moderate initial starter charge and a pH of 6.2–6.5. Water in well after transplanting; however, be careful to keep foliage dry throughout production.

Fertilize with 100–150 ppm at every irrigation, alternating between 14-0-14 and 20-10-20. Maintain pH levels at or near 6.0 to avoid nutrient deficiencies or toxicities.

Provide high light, yet cooler temperatures: 60–65°F (16–18°C) days and 55–60°F (13–16°C) nights. Grow under full-sun conditions.

If growth regulators are needed, A-Rest (33–66 ppm), B-Nine (2,500–5,000 ppm), and Bonzi can be used as needed.

Transplanting a plug to a 4" (10 cm) pot takes from six to eight weeks to be ready for spring sales. Plants will not be in flower but sold green. Plants will flower three to four weeks later.

Cut Flowers

Transplant once the danger of frost has passed. Plants will take off the fastest once soil temperatures have reached 70°F (21°C). Amend soils to a pH of 6.3–6.8. Fertilize with 100 ppm from 20-10-20 at every irrigation. Apply ground cloth or plastic mulch to control weeds and prevent dirt from splashing back up on foliage.

While zinnias flower faster under short days, most commercial growers do not pull black cloth over plants. Plants will flower under long days, but flowering takes longer.

Irrigate zinnias with drip tubes or tape, because wetting foliage is an invitation to foliar disease.

Be sure to schedule successive crops, planting seedlings every two weeks, so you will have a continuous crop to harvest once flowering begins. When you cut the first flower off the plants, it will take about four weeks for them to branch and flower again. Plant from four to eight plantings to ensure continuous harvest.

Pests & Diseases

Aphids, leaf miners, thrips, and whiteflies can attack plants. Budworms and grasshoppers can attack outdoor cut flower crops as well. Foliar diseases can be serious and include *Alternaria,* bacteria, botrytis, and a range of leaf spots. Control diseases by maintaining excellent air circulation, keeping foliage dry, and allowing plants to dry between irrigations. *Pythium* root rot can also be a problem.

Troubleshooting

Boron deficiency may cause bud blasting, while boron toxicity can delay flowering by up to two weeks. Symptoms may show up as distorted foliage, flower bud death, and/or a proliferation of side shoots below the growing point. Maintain the pH at 5.5–6.3. Cut flower growers should also check well water for boron levels and use boron-free water throughout postharvest to avoid problems. Iron deficiency can appear as interveinal chlorosis in upper foliage. If the pH is above 6.5, lower it with iron sulfate; but if it is below 6.2, apply chelated iron to raise it. Low temperatures (below 60°F [15°C]) can cause chlorotic leaves. High ammonium levels can lead to soft, weak growth.

Varieties

For bedding plant pack or pot performance, you can't beat the 'Dreamland' or 'Magellan' series, available in a number of separate colors. Plants grow 10–14" (25–35 cm) tall and have fully double flowers.

Varieties for cut flowers include Benary's 'Giant Mix', with 30–36" (76–91 cm) and fully double flowers. The 'Oklahoma' series is available in five separate colors and a mix. Plants grow to 30–40" (76–102 cm) and bear double and semi-double blooms. Plants are tolerant of powdery mildew. 'Ruffles Mix' is an F_1 hybrid, with 24–30" (61–76 cm) stems. 'State Fair Mix', a very well-known variety, grows to 30–36" (76–91 cm) tall, yet yields less than half the number of stems per plant as newer varieties. However, none of the newer breeding has the phenomenal flowers of 'State Fair'.

Related Species

Z. angustifolia is more disease resistant than *Z. elegans* is and has become a favorite among landscapers, parks departments, municipalities, and anyone looking for reliable color in the heat. Sell plants green with color tags in packs, or in color in 4" (10 cm), 6" (15 cm), or 1-gal. (4 l) pots. The 'Star' series is available in 'Gold', 'Orange', 'White', and 'Starbright Mix'. The well-branching plants grow 10–12" (25–30 cm) tall, and a 288-plug tray takes from four to six weeks to be ready to transplant to larger containers. Once transplanted to a 4" (10 cm) pot, plants will be salable in seven to eight weeks. If plant growth regulators are needed, use 1,250–1,800 ppm of B-Nine.

The All-America Selections Award-winning 'Profusion' series is a *Z. marylandica* and is available in a number of colors. These zinnias are early and continuously bloom all summer long, no matter what the temperature or humidity conditions. Plants are very tolerant of powdery mildew and bacterial leaf spot. The newest notable series in this class is 'Zahara'. It includes a few All-American Selections Award winners and is available in both single- and double-flowering lines. For *Z. marylandica*, follow the culture outlined for *Zinnia elegans*: Plants flower seven to nine weeks after transplanting each 288 plug to a 4" (10 cm) pot.

Zinnia angustifolia

Zinnia 'Profusion'

Postharvest

Cut flowers are ready for harvest once pollen begins to form. Leave two to three leave pairs on the stem. Strip excess foliage from the lower portion of stems in the field. Harvest early in the morning or late in the evening, when air temperatures are cooler. Store stems in water with a floral preservative at 42°F (6°C). Vase life is seven to eighteen days using floral preservatives such as Chrysal OVB or Floralife.

Zinnia flowers may also be dried. Harvest flowers fully open and remove foliage and stems. Place flowers in a shallow container (3–4" [8–10 cm] deep) on top of a 0.50–0.75" (13–19 mm) layer of a drying substance, such as white cornmeal, sand, borax, cat litter, silica gel, or a commercial drying formula. Carefully pour drying material over the flowers, filling in the container 3–4" (8–10 cm) deep.

Appendix

USDA Hardiness Zone Map

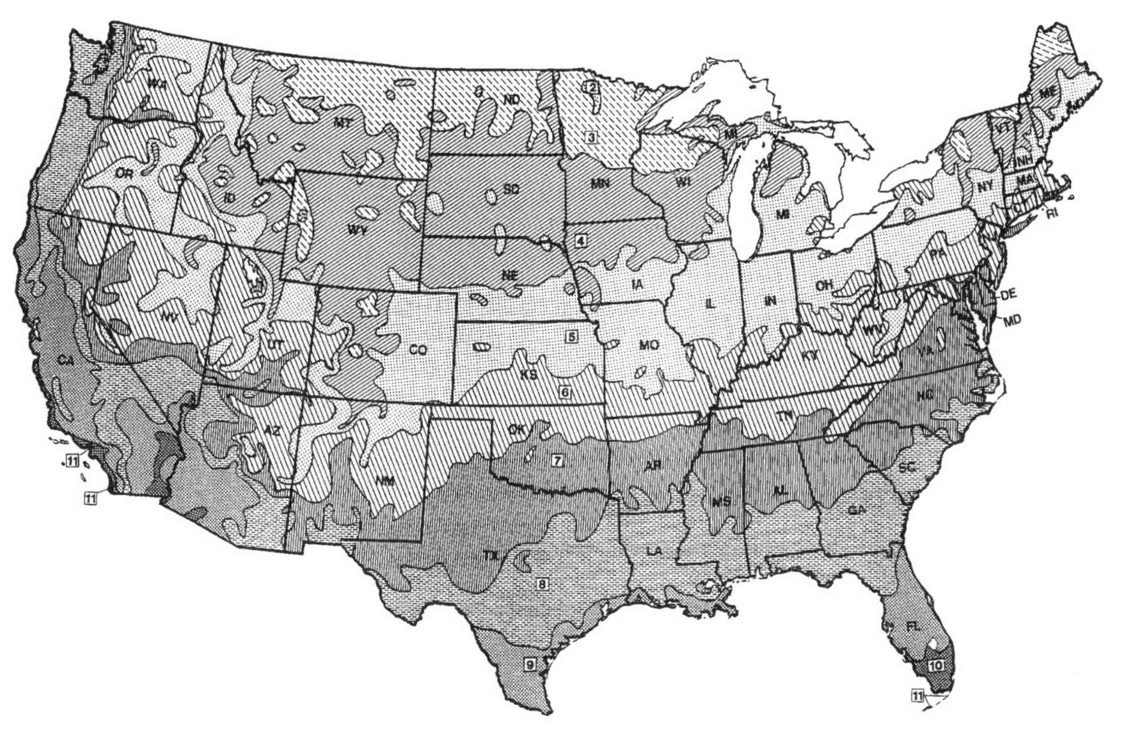

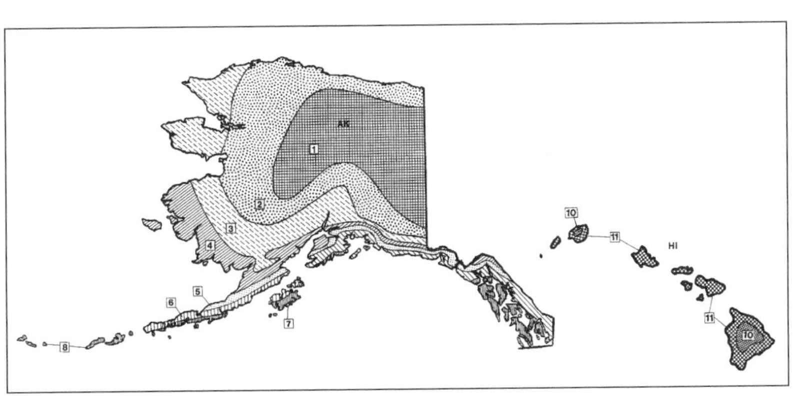

RANGE OF AVERAGE ANNUAL MINIMUM TEMPERATURES FOR EACH ZONE	
ZONE 1	BELOW -50°F
ZONE 2	-50° TO -40°
ZONE 3	-40° TO -30°
ZONE 4	-30° TO -20°
ZONE 5	-20° TO -10°
ZONE 6	-10° TO 0°
ZONE 7	0° TO 10°
ZONE 8	10° TO 20°
ZONE 9	20° TO 30°
ZONE 10	30° TO 40°
ZONE 11	ABOVE 40°

Bibliography

Books

Armitage, Allan M. *Bedding Plants: Prolonging Shelf Performance.* Batavia, Illinois: Ball Publishing, 1993.
———. *Herbaceous Perennial Plants.* 2nd ed. Champaign, Illinois: Stipes Publishing, 1997.
———. *Manual of Annuals, Biennials and Half-Hardy Perennials.* Portland: Timber Press, 2001.
———. *Specialty Cut Flowers.* Portland: Varsity Press/Timber Press, 1993.
Bailey, Douglas A. *Hydrangea Production.* Portland: Timber Press, 1989.
Banner, Warren, and Mike Klopmeyer. *New Guinea Impatiens: A Ball Guide.* Batavia, Illinois: Ball Publishing, 1995.
Blanchette, Rick, and Jayne N. La Scola. *GrowerTalks on Crop Culture 2.* Batavia, Illinois: Ball Publishing, 1999.
Blessington, Thomas M., and Pamela C. Collins. *Foliage Plants: Prolonging Quality.* Batavia, Illinois: Ball Publishing, 1993.
Boodley, James W. *The Commercial Greenhouse.* Clifton Park, New York: Delmar Publishers Inc., 1981.
Bown, Deni. *The Herb Society of America Encyclopedia of Herbs and Their Uses.* New York: DK Publishing, 1995.
Darden, Jim. *Great American Azaleas: A Guide to the Finest Azalea Varieties.* Clinton, North Carolina: The Greenhouse Press, 1985.
De Hertogh, August A. *Holland Bulb Forcer's Guide.* 5th ed. Batavia, Illinois: Ball Publishing, 1996.
De Hertogh, August A., and M. Le Nord. *The Physiology of Flower Bulbs.* Amsterdam: Elsevier Science Publishers, 1993.
Dole, John M., and Harold F. Wilkins. *Floriculture: Principles and Species.* Upper Saddle River, New Jersey: Prentice Hall, 1999.
Forcing Flower Bulbs. Hillegom, the Netherlands: International Flower Bulb Centre, n.d.
Griffith, Lynn P. *Tropical Foliage Plants.* Batavia, Illinois: Ball Publishing, 1998.
Hanan, Joe J. *Greenhouses: Advanced Technology for Protected Horticulture.* New York: CRC Press, 1998.
Herwig, Rob. *How to Grow Healthy House Plants.* Tucson: H.P. Books, 1979.
Information on Special Bulbs. Hillegom, the Netherlands: International Flower Bulb Centre, n.d.
Joiner, Jasper N. *Foliage Plant Production.* Upper Saddle River, New Jersey: Prentice Hall, 1981.
van de Laar, H.J., G. Fortgens, M.H.A. Hoffman, and P.C. de Jong. *List of Names of Perennials.* Boskoop, the Netherlands: Proefstation voor de Boomkwekerij, 1995.
Larson, Roy A. *Introduction to Floriculture.* New York: Academic Press, 1980.
———. *Introduction to Floriculture.* 2nd ed. New York: Academic Press, 1992.
———. *Production of Florist Azaleas.* Portland: Timber Press, 1993.
Nau, Jim. *Ball Culture Guide: The Encyclopedia of Seed Germination.* Batavia, Illinois: Ball Publishing, 1999.
———. *Ball Perennial Manual: Propagation and Production.* Batavia, Illinois: Ball Publishing, 1996.
Nell, Terril A. *Flowering Potted Plants: Prolonging Shelf Performance.* Batavia, Illinois: Ball Publishing, 1993.
Nelson, Paul V. *Greenhouse Operation and Management.* 6th ed. Upper Saddle River, New Jersey: Prentice Hall, 2003.
Perry, Leonard P. *Herbaceous Perennials Production: A Guide from Propagation to Marketing.* Ithaca, New York: Northeast Regional Agricultural Engineering Service, 1998.
Post, Kenneth. *Florist Crop Production and Marketing.* New York: Orange Judd Publishing Company, 1950.

Scalis, John N. *Cut Flowers: Prolonging Freshness.* Batavia, Illinois: Ball Publishing, 1993.
Styer, Roger C., and David S. Koranski. *Plug and Transplant Production: A Grower's Guide.* Batavia, Illinois: Ball Publishing, 1997.
Sunset New Western Garden Book. Menlo Park, California: Lane Publishing Co., 1981.
Swiader, John M., George W. Ware, and J.P. McCollum. *Producing Vegetable Crops.* Danville, Illinois: Interstate Publishers Inc., 1992.
White, John W. *Geraniums IV.* Batavia, Illinois: Ball Publishing Co., 1993.

Databases

Ball Seed Co. "Ball Seed Culture Advisors." *Ball Seed Cultural Database.* West Chicago, Illinois: Ball Seed Co., n.d.

Electronic versions of "Ball Seed Culture Advisors" were used as the base of the following topics and crops: *Achillea,* African violet, *Ageratum, Aglaonema, Allium, Alstroemeria, Alternanthera, Amaryllis, Anemone, Angelonia, Aquilegia, Asparagus, Aster, Astilbe,* bacopa, basil, begonia (fibrous), begonia 'Charisma', Boston ferns, *Bougainvillea, Brassica,* bromeliads, *Browallia, Caladium, Calathea, Calceolaria, Calendula, Calibrachoa, Campanula,* Cape primrose, *Capsicum,* 'Celebrette' and 'Celebration' New Guinea impatiens, *Celosia, Centaurea, Cineraria, Clematis, Cleome, Coleus, Consolida, Cordyline australis, Coreopsis, Crossandra, Cuphea, Cyclamen, Dahlia,* daylily, *Delphinium, Dianthus, Diascia, Dicentra, Dieffenbachia, Digitalis,* dusty miller, *Echinacea, Eustoma, Evolvulus, Exacum, Ficus,* 'Fiesta' double impatiens, *Freesia, Fuchsia, Gaillardia,* 'Galleria' ivy geranium, *Gaura,* garden mums (spring), *Gardenia, Gazania, Geranium, Geranium dalmaticum,* ten-step fast-crop geraniums, geranium trees, *Gerbera, Gerbera* (cut), *Gladiola* "Parigo hybrid" cut flowers, Gloriosa daisy 'Goldilocks', *Gloxinia, Godetia, Gomphrena,* 'Good Morning' morning glory, *Gypsophila,* greenhouse sanitation for vegetative geraniums, handling unrooted geranium cuttings, direct-stick program, 'Heimalis' begonia, *Helichrysum,* heliotrope, *Hemerocallis, Heuchera, Hibiscus,* holiday cactus, *Hosta, Hydrangea, Hypericum, Hypoestes, Iberis sempervirens, Impatiens walleriana, Iris hollandica,* ivy geranium, 'Java' New Guinea impatiens, *Kalanchoe, Lantana, Lathyrus,* lavender, *Leucanthemum, Liatris,* lily (Asiatic), lily (Easter), lily (Oriental), *Lobelia erinus,* lupine, *Lysimachia,* Marguerite daisy, matricaria, miniature pot roses, mint, *Nicotiana,* Oriental poppy, ornamental pepper, *Osteospermum, Oxalis, Paeonia* cut flowers, pansy, pansy plug culture, *Pennisetum setaceum, Pentas, Petunia, Phlox paniculata,* phlox for cutting, *Platycodon,* poinsettias, *Portulaca grandiflora,* pot mums, *Primula,* purslane, *Ranunculus,* regal geranium, rex begonias, *Rhododendron,* rose, *Rudbeckia, Salvia farinacea, Salvia nemorosa* 'May Night', *Salvia splendens, Salvia superba* 'Blue Queen', satin flower, *Scabiosa, Scaevola, Schizanthus, Sedum, Sinningia,* snapdragon (cut), snapdragon (trailing), *Solanum, Solidago, Spathiphyllum,* spray mum cultural information, standard mum cultural information, statice (perennial), statice (*Limonium sinuatum*), stock, strawberries, sunflower, sweet alyssum, sweet potato 'Blackie', *Thunbergia,* tomato, *Torenia,* trailing petunias, tuberous begonia, *Tulipa,* vegetative geranium culture, vegetative petunias, *Verbena, Veronica, Vinca,* vinca vine, vinca plug culture, *Viola,* 'Wave' petunia, *Zantedeschia* 'Kiwi Callas', and *Zinnia.*

Internet

"Florel on Double Impatiens." *The Ohio State University.* Florinet dialogue (Florinet@agvax2.ag.ohio-state.edu) (April 1998).
Hoff, Peter. "The Growing of Lilies: Cut Lilies." *Hoff Quality First.* January 2002. www.hoffqualityfirst.nl (February 2002).
"Hosta Popularity Lists." *American Hosta Society.* www.hosta.org (March 2002).
"Introductory Information." *American Peony Society.* www.americanpeonysociety.org (March 2002).

"Kale." *The Ohio State University.* Florinet dialogue (Florinet@agvax2.ag.ohio-state.edu) (September 1998).
"Pot Chrysanthemum Cultural Notes." *Fides Goldstock Breeding.* www.fgb.nl (January 2002).
Richter, Conrad. "Getting Oregano Right." *Richter's Herb Specialists.* www.richters.com (August 2002).
"Schlumbergera." *J. De Vries Potplantencultures.* www.jdevriespotplantencultures.nl (August 2002).
"Spray Chrysanthemum Cultural Notes." *Fides Goldstock Breeding.* www.fgb.nl (January 2002).
"Successful Culture of Pot Anthuriums, A." *RijnPlant Anthurium.* www.rijnplant.com (January 2002).
Wood, Betty. "Growing Information: Planting and Growing Iris." *American Iris Society.* www.irises.org (March 2002)

Periodicals

Adam, Sinclair A., Jr. "*Phlox paniculata* 'David': 2002 Perennial Plant of the Year." *Perennial Plants* (Autumn 2001): 53–62.
Aimone, Teresa. "This month . . . *Primula obconica.*" *GrowerTalks* (June 1998): 82.
Anderson, Robert G. "Cut Flower Production of Bachelor Button or Corn Flower." *International Cut Flower Growers Association (ICFGA) Bulletin* (January 2002): 21–23.
———. "Performance of Angelonia Cultivars as a Summer Greenhouse Cut Flower." *The Cut Flower Quarterly* 14, no. 1 (2002): 26–28.
Apps, Darrel. "The Quest for Reblooming Daylilies." *American Nurseryman* (June 1, 2001): 44–48.
Armitage, Allan. "Hostas: Growing Them Faster." *Greenhouse Grower* (May 1998): 130–132.
"Avoiding Iron-Manganese Syndrome on Geraniums." *GrowerTalks* (July 1999): 88.
Barnes, Larry W., Mary K. Hausbeck, and Margery Daughtrey. "Managing Powdery Mildew on Gerberas." *GrowerTalks* (July 1999): 106–114.
Batschke, Karl. "Culture Tips for Regal Geranium." *GPN* (January 2002): 60–62.
———. "Doing Poinsettias Right." *Greenhouse Grower* (July 2001): 64–68.
———. "Kalanchoes: The 'New' Fall Crop." *GrowerTalks* (Fall 2000): 13–14.
Behe, Bridget, Art Cameron, Rachel Walden, Beth Fausey, Kathy Kelley, Liz Moore, Erin Nausieda, Kevin Kern, and Will Carlson. "Turning Perennials Inside Out: Echinacea." *GM PRO* (May 2002): 52–54.
Beytes, Chris, and Andrew Britten. "Getting Your Petunias into Flower." *GrowerTalks* (December 2000): 70.
———. "Obsessed with Quality." *GrowerTalks* (August 1999): 36–40.
———. "Quiet Excellence." *GrowerTalks* (February 2000): 30–37.
Burnette, Stephanie E., Gary J. Keever, Charles H. Gilliam, Raymond Kessler, and Charles Hesselein. "Slowing growth of *Achillea* × 'Coronation Gold' and *Guara lindheimeri* 'Corrie's Gold.'" *American Nurseryman* (June 15, 2001): 61.
Cameron, Art, Beth Fausey, Bridget Behe, Kevin Kern, Liz Moore, and Will Carlson. "Turning Perennials Inside Out: Pennisetum." *GM PRO* (July 2002): 68–70.
Chase, A.R. "Geranium Rust in the New Century." *GrowerTalks* (August 1999): 150.
Corr, Brian, and Philip Katz. "A grower's guide to lisianthus production." *FloraCulture International* (May 1997): 16–20.
Croft, Bob. "Maximize Your Mimulus." *GrowerTalks* (January 2000): 101.
Croft, Bob, and Corinne Marshall. "MGII: The Next Generation." *GrowerTalks* (October 2001): 78–79.
Cunliffe, Bruce A., Mary Hockenberry Meyer, and Peter D. Ascher. "Propagation of *Pennisetum setaceum* 'Rubrum' by cuttings." *American Nurseryman* (August 15, 2001): 62.
Daughtrey, Margery. "Desperately Seeking Xanthomonas." *GrowerTalks* (October 2000): 52–55
De Hertogh, Gus. "Growing Hybrid Lilies." *GrowerTalks* (January 2000): 72–80.

———. “The Y2K Easter Lily Challenge.” *GrowerTalks* (August 1999): 120–132.
Derrig, Ron. “Combating Pansy Diseases Culturally, Chemically.” *GrowerTalks* (July 2000): 118–119.
———. “Feed for the Right Height.” *GrowerTalks* (August 2000): 134–136.
———. “More on Mum Height.” *GrowerTalks* (August 2000): 124.
Dole, John M., Brian E. Whipker, and Paul V. Nelson. “Producing Vegetative Petunias and Calibrachoa.” *GPN* (February 2002): 30–34.
———. “Success with Vegetative Petunias and Calibrachoa.” *North Carolina Flower Growers’ Bulletin* (April 2002): 8–11.
Dole, John M., Todd Cavins, and Theresa Bosma. “Success with Campanulas.” *GPN* (December 2001): 28–35.
Donahue, Tim. “‘Vine’-tune Your Clematis.” *GrowerTalks* (March 2001): 77.
Dummen and Plantpeddler. “How We Grow: Hiemalis Begonias.” *GrowerTalks* (November 2001): 68–70.
Eddy, Robert T., and Brian Whipker. “Guide to Successful Outdoor Garden Aster Production.” *North Carolina Flower Growers’ Bulletin* (June 1998): 6–9.
“Fantastic Phlox Mildew Resistance.” *GrowerTalks* (June 1999): 107.
Fausey, Beth, Erik Runkle, Art Cameron, Royal Heins, and Will Carlson. “Herbaceous Perennials: Heuchera.” *Greenhouse Grower* (June 2001): 50–61.
Faust, James E., and Kelly P. Lewis. “Cool Temps and Bright Light Treat Pansies Right.” *GPN* (December 2001): 10–14.
Ferry, Shannen. “Classy Coleus.” *GrowerTalks* (November 2001): 65, 70.
Fisher, Paul. “Timing Is Everything.” *GrowerTalks* (October 2002): 28.
Foss, Janet. “Lemons and Lemonade Iceland Poppies.” *The Cut Flower Quarterly* 13, no. 3 (2001): 14.
Frye, Bob. “Florel: A Different Perspective.” *GrowerTalks* (March 2001): 80.
———. “How We Grow: Premium Seven-Inch Geraniums, Part I.” *GrowerTalks* (September 1999): 51, 56.
———. “How We Grow: Premium Seven-Inch Geraniums, Part II.” *GrowerTalks* (Fall 1999): 22–26.
Garner, Jim, and Pam Lewis. “Greenhouse Production of Herbaceous Perennials for Cut Flowers.” *The Cut Flower Quarterly* 11, no. 4 (1999): 16–17.
Gaydos, John P. “Cool Tips for Argyranthemum.” *GrowerTalks* (October 2001): 95.
Gentsch, Bryan. “Hot Pots—Mini Roses.” *GrowerTalks* (April 2000): 67, 72.
“Getting Consistent, Clear Hydrangea Color.” *GrowerTalks* (March 2000): 77.
Gibson, James L., and Brian E. Whipker. “Producing Top-Quality Ornamental Vegetables.” *Ohio Florists Association Bulletin* (June/July 2001): 12–16.
———. “Research Progress Report: Ornamental Cabbage and Kale Fertilization Study.” *North Carolina Flower Growers’ Bulletin* (December 1999): 6–7.
———. “Research Progress Report: The Effect of B-Nine, Bonzi, and Sumagic on the Growth of Ornamental Cabbage and Kale.” *North Carolina Flower Growers’ Bulletin* (December 1999): 8–9.
Gillum, Russ. “Champion Campanula.” *GrowerTalks* (Fall 2000): 15–16.
Gooder, Mike. “How We Grow: Hiemalis Begonias.” *GrowerTalks* (September 2000): 83, 88.
Hamaker, Cheryl K., Royal D. Heins, Arthur Cameron, and Will Carlson. “Forcing Perennials Crop by Crop: *Coreopsis verticillata*.” *Greenhouse Grower* (July 1996): 43–46.
Hammer, P. Allen, and Karen Rane. “Southern Bacterial Wilt Found in Geraniums.” *GrowerTalks* (July 1999): 80–82.
Harada, Daijiro. “Limonium.” *GrowerTalks* (April 1998): 60.
Hawke, Richard G. “An Evaluation of Goldenrods for the Garden.” *Perennial Plants* (Spring 2001): 11–24.
Heaton, Thomas C., and Peter Denney. “Growing the New Sunflowers.” *GrowerTalks* (May 1999): 95, 100.
Henne, Laura. “Toughening Up Fall Pansies.” *Greenhouse Grower* (July 2002): 52–54.

Hopper, Douglas A. "Scheduling Lupine Flowering in Response to Long Day Photoperiod Lighting." *International Cut Flower Growers Association (ICFGA) Bulletin* (September 1999): 44. First published in *Rocky Mountain and High Plains Horticulture Research Quarterly* 42, no. 2 (June 1999).
Hunter, Don, Larissa Clark, and Michael Reid. "Post Harvest Handling of Campanulas—Watch out for Ethylene!" *The Cut Flower Quarterly* 13, no. 3 (2001): 25.
Jelitto Seeds. "GM PRO Grower's Notebook: *Iberis sempervirens.*" *GM PRO* (June 2002): 4.
Jepsen, Knud. "How We Grow: Kalanchoes." *GrowerTalks* (April 2000): 68–69.
Karlsson, Meriam. "Black-Eyed Susan: A Novelty Potted Plant." *GPN* (November 2001): 30–36.
———. "Cyclamen—A Versatile Cool-Season Crop." *Ohio Florists Association Bulletin* (October 2001): 13–15.
———. "Primula Is Still a Cool Crop." *Ohio Florists Association Bulletin* (November 2001): 8–9.
Kessler, Ray. "Greenhouse Production of Boston Ferns." *Ohio Floriculture Update* (August 1999).
Kim, Seung-Hyun, A.A. De Hertogh, and P.V. Nelson. "Effects of Plant Growth Regulators Applied as Sprays or Media Drenches on Forcing of Dutch-grown Bleeding Heart as a Flowering Pot Plant." *HortTechnology* (October–December 1999): 629–633.
Konjoian, Peter. "The Care and Feeding of Patriot Lantana." *GrowerTalks* (April 1998): 64–73.
———. "Geranium Production." *Ohio Florists Association Bulletin* (December 2001): 9, 16.
———. "Production Pointers for Petunias: Seed and Vegetative." *Ohio Florists Association Bulletin* (February 2001): 1, 8–10.
———. "Surfactants to the Rescue." *GrowerTalks* (June 2000): 71, 76.
Kristl, Ginny. "Culture Profile: From 'Sunrise' to 'Sunset,' Growing the New Cut Flower Kales." *The Cut Flower Quarterly* 13, no. 3 (2001): 5.
Lang, Harvey. "Ivy Geranium Take-Home Tips." *GrowerTalks* (December 2000): 69, 73.
Legnani, Garry, and William B. Miller. "Manipulating Dahlias." *GPN* (December 2001): 36–43.
Lieth, J. Heinrich, Loren Oki, and Soo Hyung Kim. "Development of a Model for Rose Productivity." *International Cut Flower Growers Association (ICFGA) Bulletin* (February 2002): 28–39.
Ling, Peter. "Humidity Management." *Ohio Florists Association Bulletin* (November 2002): 8–9.
Linwick, Tom. "Roll out the Hypoestes." *GrowerTalks* (December 1999): 79, 84.
Mackay, Wayne A., and Tim D. Davis. "Big Bend Bluebonnets as a Cut Crop." *GM PRO* (May 2002): 34–37.
Marquardt, Bonnie. "Get a Line on Linaria." *GrowerTalks* (February 2000): 77.
McGrew, Jeff. "Growing Cut Hypericum." *GrowerTalks* (April 1998): 57.
McKeegan, Tom. "A Grower's Guide to Ferns." *GrowerTalks* (August 2000): 121–122.
Michigan State University. "Forcing Perennials Crop by Crop: Hosta." *Greenhouse Grower* (November 1999): 84–90.
Miller, Bob. "How We Grow: Summer Dahlias." *GrowerTalks* (March 2000): 75, 80.
Miller, William B. "Easter Lily Odyssey." *GrowerTalks* (November 2000): 62–72.
Morey, Lesa. "Garden Mums by the Experts." *GrowerTalks* (July 2000): 117, 121.
Muraoka, Kieth. "Prime Primula." *GrowerTalks* (Fall 2000): 8.
Nameth, Stephen G.P. "The Grower's Role: How to Keep Xanthomonas out of the Greenhouse." *GrowerTalks* (October 2000): 56–58.
Nau, Jim. "Succeeding with Daylilies." *GrowerTalks* (June 1998): 79, 84.
Niklas, David W. "Cineraria Made Simple." *GrowerTalks* (September 2001): 93, 97.
Niu, Genhua, Thomas Griffing, Erik Runkle, Royal Heins, Art Cameron, and Will Carlson. "Herbaceous Perennials: *Ceratostigma plumbaginoides.*" *Greenhouse Grower* (October 2001): 96–100.
Ohkawa, K., and M. Suematsu. "Arching Cultivation Technique for Growing Cut Roses." *International Cut Flower Growers Association (ICFGA) Bulletin* (October 1999): 25–28.
Ouellet, Kerstin. "Diascia." *GrowerTalks* (Fall 2000): 18.

Phillips, Martyn. "Cleome in 4½-in. Pots." *GrowerTalks* (October 2001): 81.
Preston, Bill. "Culture Profile: Dahlia." *The Cut Flower Quarterly* 14, no. 2 (2002): 6–7.
Ranwala, Anil P., and William B. Miller. "Using Gibberellins to Prevent Leaf Yellowing in Cut Lilies." *GPN* (January 2002): 30–36.
Reiner, Stefan. "Blooming Pot Carnations." *GrowerTalks* (October 2001): 83.
Reynolds, Dennis. "Rules to Begonia Production." *GrowerTalks* (Fall 2000): 18–19.
Richter, Conrad. "Success with Mints." *GrowerTalks* (June 1999): 105–110.
Roh, Mark S. "Lisianthus (Eustoma)." *Ohio Florists Association Bulletin* (September 1999): 3–6.
Runkle, Erik, Royal Heins, and Hiroshi Shimizu. "How Low Can You Go?" *GrowerTalks* (February 2002): 63, 68.
Runkle, Erik, Royal Heins, Art Cameron, and Will Carlson. "Herbaceous Perennials: *Scabiosa columbaria*." *Greenhouse Grower* (September 2001): 70–76.
———. "Herbaceous Perennials: *Phlox subulata*." *Greenhouse Grower* (August 2001): 80–88.
———. "Forcing Perennials Crop by Crop: *Veronica longifolia* 'Sunny Border Blue.'" *Greenhouse Grower* (September 1996): 41–42.
Sari, Ali O., Mario R. Morales, and James E. Simon. "Overcoming dormancy and improving germination of Echinacea seed." *American Nurseryman* (December 15, 2001): 73.
Shoellhorn, Rick. "Plectranthus—Coleus' Cousin" *GPN* (February 2002): 78–82.
Stamback, Vicki. "Zinnia: Perfect Cut for Local Markets." *GrowerTalks* (May 1999): 56–60.
Trellinger, Karl. "New Guinea Impatiens: 25 Growing Tips." *Greenhouse Grower* (February 2002): 50–56.
"Triple Your Fall Color." *GrowerTalks* (October 2000): 89.
Venzke, Jon. "How We Grow: Pot Cyclamen." *GrowerTalks* (August 1999): 97, 102.
Walters Gardens. "GM PRO Grower's Notebook: *Leucanthemum*." *GM PRO* (August 2002): 4
Walters, Mary. "Summer Potting Gives Better Daylilies." *GrowerTalks* (September 2001): 95.
Warner, Ryan M., John E. Erwin, Michael J. McDonough, and Neil Mattson. "The Pros and Cons of Growing Plants Cooler." *Ohio Florists Association Bulletin* (January 2002): 5–10.
"What's Wrong with Your New Guineas?" *GrowerTalks* (August 1999): 100.
Whipker, Brian E., James L. Gibson, Ingram McCall, Todd J. Cavins, Colleen Warfield, and Raymond Cloyd. "Success with Ornamental Peppers." *North Carolina Flower Growers' Bulletin* (June 2002): 10–12.
Whipker, Brian E., Shravan Dasoju, and Ingram McCall. "Success with Pot Sunflowers." *North Carolina Flower Growers' Bulletin* (December 1998): 7–10.
Whipker, Brian E., William C. Fonteno, Todd J. Cavins, and James L. Gibson. "Establishing a PourThru Sampling Program for Geraniums." *North Carolina Flower Growers' Bulletin* (February 2000): 10–21.
White, Jennifer Duffield. "Get Your BA in Hosta Propagation." *GrowerTalks* (March 2001): 79.
Whitman, Catherine, Royal Heins, Art Cameron, and Will Carlson. "Herbaceous Perennials: *Oxalis crassipes* 'Rosea'." *Greenhouse Grower* (December 2001): 77–84.
———. "Forcing Perennials Crop by Crop: *Campanula carpatica* 'Blue Clips'." *Greenhouse Grower* (August 1996): 17–20.
———. "Forcing Perennials Crop by Crop: *Lavandula angustifolia*." (May 1996): 37–40.
———. "Forcing Perennials Crop by Crop: *Platycodon grandiflorus* 'Sentimental Blue'." *Greenhouse Grower* (September 1996): 39–42.
Williams, Jack, and Shannen Ferry. "Planning for Success." *Greenhouse Grower* (July 2001): 74–82.
Williams, Jack. "Maximum Mimulus." *GrowerTalks* (December 2000): 74.
Williams, Kim. "Calceolaria: A Cool Temperature Crop to Generate Cold Cash." *Ohio Florists Association Bulletin* (December 2001): 4–5.
Willmott, Jim, Richard Merritt, and Tom Gianfagna. "Aquilegias in Four Months." *GrowerTalks* (July 2000): 128, 130.
Wilson, Bill. "Glox that Rock." *GrowerTalks* (October 2001): 98–100.

———. "Growing Perfect Gerbera." *GrowerTalks* (Fall 2000): 6.
Yoder Green Leaf Perennials. "GM PRO Grower's Notebook: Phlox." *GM PRO* (August 2001): 4.
Yuan, Mei, Erik Runkle, Royal Heins, Art Cameron, and Will Carlson. "Forcing Perennials Crop by Crop: *Rudbeckia fulgida* 'Goldsturm.'" *Greenhouse Grower* (December 1996): 57–61.
———. "Forcing Perennials Crop by Crop: *Gaillardia* × *Grandiflora* 'Goblin'." *Greenhouse Grower* (December 1996): 57–60.
Yuan, Mei, Royal Heins, Art Cameron, and Will Carlson. "Forcing Perennials Crop by Crop: *Coreopsis grandiflora*." *Greenhouse Grower* (June 1996): 57–60.

Other Publications

British Columbia Ministry of Agriculture and Food. *Floriculture Production Guide for Commercial Growers*. 1999.
Buschman, J.C.M. *Gladiolus as a Cut Flower in Subtropical and Tropical Regions*. Hillegom, the Netherlands: International Flowerbulb Center, n.d.
"Cut Flower Production of Iris." *Bulb Flower Production: Service Bulletin Cut Flowers*. Hillegom, the Netherlands: International Flower Bulb Center, September 1999.
"Cut Flower Production of Iris." *Bulb Flower Production: Service Bulletin Cut Flowers*. Hillegom, the Netherlands: International Flower Bulb Center, September 2000.
Dole, John. "Growing Sunflowers as Potted Plants." Handout at the 2000 International Ohio Florists Association Short Course, Columbus, Ohio, July 2000.
Dole, John. "Gypsophila." *Floriculture—Principles and Species*. 2nd ed. Upper Saddle River, New Jersey: Prentice Hall, 2002.
Erwin, John E. "Temperature." A handout at the 2002 International Plug and Cutting Conference, Orlando, Florida, October 2002.
Erwin, John E., and Ryan Warner. "What Causes a Plant to Flower?" A handout at the 2002 International Plug and Cutting Conference, Orlando, Florida, July 2002.
Eustoma (Lisianthus). Frinton on Sea, Essex, UK: Pathfast Publishing, July 1996.
Forcing Flower Bulbs: Bulb Flowers—Bulbs in Pots. Hillegom, the Netherlands: International Flower Bulb Centre, November/December 1999.
Frye, Bob. *Plantation Geraniums*. Lincoln, Nebraska: Plantation Perfect Plants, 2000.
Handbook for Modern Greenhouse Rose Cultivation (English version). Applied Plant Research Praktijkonderzoek Plant and Omgeving. 2001.
Hosta Adventure: A Grower's Guide. Snellville, Georgia: American Hosta Society, 2001.
Jepsen, Knud. "Kalanchoe production." Handout at the 2002 International Ohio Florists Association Short Course, Columbus, Ohio, 2002.
Kansas State University Cooperative Extension Service. *Aster Yellows*. Manhattan, Kansas:: Kansas State University, n.d.
Kansas State University Cooperative Extension Service. *Commercial Specialty Cut Flower Production: Sunflowers*. Manhattan, Kansas: Kansas State University, n.d.
Kansas State University Cooperative Extension Service. *Zinnias*. Manhattan, Kansas: Kansas State University, August 1993.
Karlsson, Meriam. "Temperature and Light Requirements for Flowering and Development of Ranunculus." *Bedding Plants Foundation Inc. Research Report*. Lansing, Michigan: Bedding Plants Foundation, November 1998.
Nameth, Steven G.P., and Margery Daughtrey. "I See Leaf Spots or Wilting in My Geraniums: A Flow Chart of What to Do When You Think You May Have Xcp in Your Greenhouse." Handout at the 2001 International Ohio Florists Association Short Course, Columbus, Ohio, July 2001.

National Garden Bureau. "The National Garden Bureau Celebrates 1998 as the Year of the Geranium." Downers Grove, Illinois: National Garden Bureau, 1998.
National Garden Bureau. "The National Garden Bureau Celebrates 1998 as the Year of the Tomato." Downers Grove, Illinois: National Garden Bureau, 1998.
National Garden Bureau. "National Garden Bureau Celebrates 2001 as the Year of the Basil." Downers Grove, Illinois: National Garden Bureau, 2001.
National Garden Bureau. "The National Garden Bureau Celebrates 2001 as the Year of the Centaurea." Downers Grove, Illinois: National Garden Bureau, 2001.
National Garden Bureau. "The National Garden Bureau Celebrates 2002 as the Year of the Vinca." Downers Grove, Illinois: National Garden Bureau, 2002.
National Garden Bureau. "Parsley." *Today's Garden.* Downers Grove, Illinois: National Garden Bureau, March 1999.
North Carolina Commercial Flower Growers' Association and North Carolina State University. *Commercial Poinsettia Production Guide.* Raleigh: North Carolina Flower Growers' Association, 1999.
Ohio Florists' Association. *Tips on Growing and Marketing Garden Mums.* Columbus: Ohio Florists' Association, 1996.
Ohio State University, The. *Tips on Growing Potted Chrysanthemums.* Columbus: The Ohio State University, 1989.
Ohio State University, The. *Tips on Growing Zonal Geraniums.* 2nd ed. Columbus: The Ohio State University, 1991.
"Pay Attention to the Following Points." *Bulb Flower Production: Service Bulletin Cut Flowers.* Hillegom, the Netherlands: International Flower Bulb Center, October 1999.
Proceedings Rose Seminar. Proefstation voor Bloemisterij en Glasgroente, Aalsmeer, the Netherlands, November 1999.
Scullin, Laurie. "Poinsettia Color Trends in the US Market." Personal correspondence, January 2002.
Song Sparrow Perennial Farm Mail Order Peony Catalog. Avalon, Wisconsin: Roy G. Klemm, 2000.
University of Minnesota. *Geranium Production.* Handout at the 2000 North Dakota Greenhouse Conference, November 2000.
Whealy, C. Anne, and Ike Vlielander. "Kalanchoes." Roanoke, Texas: Proprietary Rights International, December 1998.
Whipker, Brian E., Paul A. Thomas, and Todd J. Cavins. *Pansy Production Handbook.* Raleigh: North Carolina Commercial Flower Growers Association, 2000.
Whitman, Catherine, Royal Heins, Art Cameron, and Will Carlson. "Production Guide for *Campanula carpatica* as a Flowering Potted Plant." *PPGA News* (April 1995): 2–4.

Supplier Publications

Bay City Flower Company, Half Moon Bay, California.
"Achieving Superior Hydrangea Root Development from 2 or 4 Dormant Plants." 1993.
Ball FloraPlant, West Chicago, Illinois.
"Ivy Geranium Finishing Trial Response to Florel and Cycocel." May 1998.
"Royal Dahlietta and Jackpot Series Dahlia Culture." August 1997.
Ball Seed Co., West Chicago, Illinois.
Danielson, Bob. "Garden Mum Culture Guide." October 1994.
"Your Source New 21st Century Phlox." 1997.
"Your Source Berries Galore and Berri Basket." 1999.
Bartelsstek, Aalsmeer, the Netherlands.
"Hypericum Cultural Description." 1997.

Benary Samenzucht. Hann, Munden, Germany.
"Benary's Rudbeckia Assortment." May 2002.
"*Echinacea purpurea* Primadonna Deep Rose." May 2002.
"*Limonium sinuatum* Compindi White." N.d.
"Sunflower Florenza." May 2002.
C. Raker & Sons. Inc., Litchfield, Michigan.
"Gerbera Culture Guide. " August 2001.
Casa Flora, Dallas, Texas.
"Garden Ferns/Tropical Ferns." N.d.
Clearview Horticultural Products, Inc., Aldergrove, British Columbia, Canada.
Concise Guide to Clematis in North America. 2001.
Donahue's Clematis Specialists, Faribault, Minnesota.
Clematis. 1999.
Earl J. Small Growers Inc., Pinellas Park, Florida.
Cummiskey, Paul, and C.C. Powell. "*Control of Thrips on Gloxinias and Other Flowering Crops in the Greenhouse.*" N.d.
Daehnfeldt Cultural Guidelines. Daehnfeldt, Odense, Denmark.
"*Campanula carpatica* Star." N.d.
"*Campanula isophylla* Topstar." N.d.
"Confetti Hypoestes." N.d.
"Elatior Begonia." N.d.
"Festival, Mardi Gras, and Masquerade Gerbera." N.d.
"Fortune and Galaxy *Begonia tuberosa.*" N.d.
"Gypsy—Garden Bride." N.d.
"Kalanchoe." N.d.
"Passion in Violet: *Trachelium caeruleum.*" N.d.
"Platycodon Astra." N.d.
"*Primula veris* Concorde." N.d.
"*Senecio cruentus* (Cineraria) Star Wars-Satellite." N.d.
"Sinningia Avanti." N.d.
Dahlstrom & Watt Bulb Farms, Inc., Smith River, California.
"1994 Hydrangea Notes." 1994.
Danziger "Dan" Flower Farm, Beit Dagen, Israel.
"Gypsophila: Cultivation Practices in Israel." *Danzinger Newsletter.* 1998.
"New Love is in the Air." *Danzinger Newsletter.* October 2001.
Golden State Bulb Growers, Watsonville, California.
"California Callas: Guidelines for Pot Growers." September 1997.
The Paul Ecke Ranch. Encinitas, California.
"Bacopa." *Fastfax Growing Guidelines.* N.d.
"Duet Portulaca." *Fastfax Growing Guidelines.* N.d.
"Helichrysum." *Fastfax Growing Guidelines.* N.d.
"Summer Daisy Argyranthemum." *Fastfax Growing Guidelines.* N.d.
"Sunscape Daisy." *Fastfax Growing Guidelines.* N.d.
"Technical Information Bulletin: Mini Poinsettia Trees." June 2002.
"Technical Information Bulletins: Poinsettia Wreaths." June 2002.
"Technical Information Bulletins: Single Stem Poinsettia Production." N.d.
"Tioga Double Impatiens." *Fastfax Growing Guidelines.* N.d.
"Vegetative Petunia." *Fastfax Growing Guidelines.* N.d.

Emerald Coast Growers, Pensacola, Florida.
"Hosta." N.d.
Fischer USA, Boulder, Colorado.
"Geranium Culture." 1997.
Fides B.V., De Lier, the Netherlands.
Fides Kalanchoe Culture Guide. 1998.
Florist de Kwakel B.V., de Kwakel, the Netherlands.
Mieremet, Wim. "Growing Floripot Pot Gerbera." 1998.
Goldsmith Seeds, Gilroy, California.
"Colorcade Ivy Geraniums." 1998.
"Ramblin' Trailing Petunia from Seed." 2002.
Gartneriet Naeldebakken, ApS, Denmark.
"Growing Programme: Gerbera Hummingbird." 1999.
Gilberg Farms, Pacific, Missouri.
"Hardy Hibiscus." N.d.
"Hibiscus Growing and Cultural Instructions." N.d.
Holtkamp Greenhouses, Nashville, Tennessee.
"Optimara Growing Tips from Start to Finish." N.d.
Kieft Seed Co., Venhuizen, the Netherlands.
"Celosia Bombay Series." January 2001.
Marflor Farms, Encinitas, California.
"*Hypericum androsaemum* Flair." N.d.
PanAmerican Seed, West Chicago, Illinois.
2002–2003 Product Information Guide. 2002.
"Aster Culture Guide." February 2001.
"Champagne Bubbles Poppy." Cultural guide. Waller Genetics, n.d.
"Delphinium Culture Guide." February 2001.
"Grower Facts Bella Mix Abutilon." 1998.
"Grower Facts Dragon Wing Begonia." 1998.
"Grower Facts Impatiens Disease Identification and Control." N.d.
"Grower Facts Java New Guinea Impatiens." N.d.
"Grower Facts Medusa Ornamental Pepper." N.d.
"Grower Facts Pentas Butterfly Series." N.d.
"Grower Facts Schizanthus Royal Pierrot." N.d.
"Grower Facts Wave Spreading Petunias." N.d.
"Helianthus Culture." February 2001.
"Limonium Culture Guide." February 2001.
"Lisianthus Culture Guide." February 2001.
"Matthiola Culture." February 2001.
"Mona Lisa Anemone." N.d.
"Snapdragon Culture Guide." September 2001.
"Trachelium Culture Guide." February 2001.
S&G Seeds, Downers Grove, Illinois.
"Jaguar Hybrid Gerbera." N.d.

Sakata Seed America, Morgan Hill, California.
"Campanula Champion Series Pot Culture." 2002.
"Cineraria Series." N.d.
"Culture Details: Lisianthus." N.d.
"Mimulus Mystic Series." N.d.
"Ranunculus Bloomingdale Series." N.d.
"Zinnia Profusion." February 1998.
Santa Rosa Tropicals, Santa Rosa, California.
"Culture Details: Petunia Explorer Series." N.d.
"Fern Growing Tips." N.d.
Skagit Gardens, Mount Vernon, Washington.
"Double Impatiens." 1998.
"Fuchsias." 1998.
"Hardy Fuchsias." 1998.
"Marguerite Daisies." 1998.
"Osteospermum." 1998.
"Petunias." 1998.
"Primula." 1998.
"*Primula malacoides.*" 1998.
"*Primula obconica.*" 1998.
"Ranunculus." 1997.
Stokman Rozen, Aalsmeer, the Netherlands.
"Your World in Roses." N.d.
Takii, Salinas, California.
"Cultural Information on Flowering Kale." N.d.
Yoder Bros. Inc./Yoder-Green Leaf, Barberton, Ohio.
"Garden Mums." 2000/2001.
"Miniature Potted Roses." *The Cutting Edge* (January 1998): 3–6
"New Ovation Roses from Yoder." *The Cutting Edge* (July 1998): 3.
"Tips: Chrysanthemum Pest Control." 1997.
"Tips: Daylilies." 2000.
"Tips: Dwarf Dahlias." 1998.
"Tips: Improving Pot Mum Winter Quality." 1997
"Tips: Miniature Roses." 1998.
"Tips: *Phlox subulata.*" 2001.
"Tips: Trade Winds Everblooming Hibiscus Liners and Quick Starts." 1997.
Walters Gardens, Zeeland, Michigan.
"Valuable Advice for Better Success with Hosta." N.d.
Walter's Gardens Inc. Wholesale Catalog. 2002.

Subject Index

Page numbers in *italics* indicate tables; page numbers in **bold** indicate figures.

A

abscisic acid (ABA), 98
acaricides (miticides), 139
acclimatization of tissue-cultured plantlets, 162
acid. *See also* pH
  in neutralizing alkalinity, 12–13
acid drenches, 48
acidic fertilizers, 45
acidification, 15
acid injection
  in eliminating carbonates and bicarbonates, 14
  in neutralizing alkalinity, 12
Actinovate, 123, 153
Adept, 154
ADT. *See* average daily temperature (ADT)
agrochemicals, 131
air porosity, container size effect on, *25*
air temperature, 67
  sensors for, 67–68
algae, 30
  control of, 14–16, 154
Aliette, 123
alkalinity. *See also* pH
  acid injection in neutralizing, 12–13
  in high-quality water, 9
  in irrigation water, 11–12, *12*
  in media, 28
  pH and, 49–50
  water quality and, 9, 148
Alternaria leaf spot, 126–27
  environmental influences on, 126
  management of, 127
  symptoms of, 126
alternative technologies, search for, 193–94
Altierri, Miguel, 136
Alude, 123
aluminum sulfate in dropping pH, 48
ammonium, 45
  toxicity of, 39
ancymidol, 97, 101, 102
annual renewal, 175
aphids, 113–14
applications
  spray, 101–2
  timing
    to plant growth regulators, 99, 101
    in targeting most vulnerable life stage, 142
AquaGrow L, 189
A-Rest, 152
artificial lighting, 90–91, 94
Atheta beetles, 154
automated irrigation systems, 9
average daily temperature (ADT), 69–70
  crop maturation and, *73*
  flower development rates and, **71**
Azatin, 154
azoxystrobin (Heritage), 125, 126, 129

B

bacteria, 121
bacterial diseases
  caused by *Xanthomonas,* 127–28
    environmental influences on, 127
    management of, 127–28
    symptoms of, 127
  suspicion of, 586
bacterial leaf spots, 153
bacterial pathogens, indexing for, 173
bafter shipping, 190
bagged mix, adding moisture to, 26
Ball Field Guide to Diseases of Greenhouse Ornamentals, 122
Banrot, 153
bark
  pine, 23
  plant growth regulators and, 103
bark mixes, disease suppression and, 29
base temperature, 70–72, *71*
battery acid in alkalinity control, 13
Behe, Bridget, 194
benching system, vegetative propagation and, 161
benzimidazoles, 129, 137
benzyladenine (BA), 98–99, **99**
benzyladenine + gibberellin combinations, 99
bicarbonates, acid injection in eliminating, 14
bindweed, 107
biofilm, control of, 14–16
biofumigation, 135, **135**

biological control, 123
in pest management, 111–12, *113*
in plant growth regulation, 95
biological indicator plants in virus detection, 174
blending in neutralizing alkalinity, 12
B-Nine, 152
Bonzi, 152, *153*
boron, 10, 14
deficiency of, 37, 40–41
toxicity of, 41
botrytis
bracts susceptibility to, 180
disease management and, 128–29
environmental influences on, 129
management of, 129
in plug production, 153
symptoms of, 129
Botrytis cinerea, 128–29, 137
Botrytis elliptica, 129
bract drop, 178
brown rust *(Puccinia chrysanthemi),* 129
Brundtland Commission, 193
buffering capacity, 28–29
Burbank, Luther, 169
Burnett, Stephanie, 24

C

calcium
as antagonistic to other nutrients, 10
deficiency of, 39–40
effect on postproduction life, 179
in hard water, 14
calcium carbonate ($CaCO_3$), 46
calcium carbonate equivalency (CCE), 45
calcium hypochlorite, 15
calcium sprays in quality, 180
calibrachoa mottle virus (CbMV), 174
callusing, 165–66
Camelot, 153
Campbell, Benjamin, 194
Canadian sphagnum peat moss, 24
cankers, 122
canopy closure, 90
carbonates, acid injection in eliminating, 14
Cease, 123
certifications in sustainability, 195
charts, 56
chemical growth regulators, 96, 154
in plug production, 152–53
chickweed, 107
chilling injury, 75
Chipco 26019, 153
chloride, water electrical conductivity and, 10–11
chlorine
checking concentration of, 19
as hazardous to worker safety, 19
chlorine dioxide
injecting, 15
in removing biofilm, 15
chlormequat chloride, 96–97, 99, 100
chloropicrin, 134
chlorosis, 33, 39, 40
chlorothalonil (Daconil and PathGuard), 126, 127, 130
Citation, 154
citric acid in controlling alkalinity, 13
clean-stock production, 175–76
coir, 23, 34
cold growing, 378
commercial labs in testing water, 11, *11*
community in sustainability, 198
compact fluorescent lamps, 94
Companion, 123
Compass, 153
Compendium of Flowering Potted Plant Diseases, 122
condensation, 185
conductivity meter, 33
Conserve, 154
constant liquid feed, 28
consumer care, 191
container media, managing pH for, **43,** 43–48
correcting problems, 46–48
prevention of, 48
recognizing problem, 43–44
regular testing, 46
containers, prefilling, 30
container size
effect on air porosity, *25*
as growth regulator, 95
convection, 68
conventional and alternative pesticides, 109–11, *110*
copper

checking concentration of, 19
deficiency of, 40
copper (Camelot and Phyton 27), 127
copper ionization, used by plug and liner growers, 15
Cornell mix, 24, **24**
Cornell Mix A, 28
crop culture in sustainability, 197
crop development stage, effect on electrical conductivity (EC), 34
crop factors, effect on electrical conductivity (EC), 34
crop maturation, effects of average daily temperature (ADT) on, *73*
cropping time, temperature and, 73–74
crop timing, 69
crown rot, 153
cucumber mosaic virus (CMV), 174
culture indexing, 173
cuttings
advantages of buying unrooted, 155–56
rooting, 158–67
toning rooted, 166–67
cyazofamid (Segway), 123
Cycocel, 152

D

Daconil, 153
daily light integral (DLI), 69, 84, **84, 85**
effect of, zinnias, **86**
plant growth responses to, 84, 86–87
for various greenhouse crops, *87–89*
daminozide, 96, 99, 101
damping-off, 153
day-neutral plants, 89, 90
Decathlon, 154
Decree, 153
dehydration, shipping and, 187
deionization in treating water with high ion control, 13
Dennis, Jennifer, 194
dibblers, 144
dibromochloropropane, 131
dicarboximides (Chipco 26019), 129, 137
1,3-dichloropropene, 134
differential temperatures (DIF), 77–80, 384
in controlling height in production, 179
effect of, on stem elongation, *78–79*
influence on plant growth and quality, 96, 180–81
dikegulac sodium, 98
dilution levels, electrical conductivity (EC) values for, for manufacturers' fertilizers, *60–64*
dimethomorph (Stature SC), 124
DIP. *See* temperature drop (DIP)
disease(s). *See also specific by name*
bacterial, caused by *Xanthomonas,* 127–28
environmental influences on, 127
management of, 127–28
symptoms of, 127
foliar, 39, 121, 125–27
water management in, 121
fungal, 39, 153
isolating and separating crops sensitive to, 175
phytoplasma, 122
in plug production, 153–54
in roots, 121
in seed crop propagation, 153–54
for stock plants, 157
vascular, 122
disease indexing, 173–76
disease management, 121–30
bacterial diseases caused by *Xanthomonas* in, 127–28
bark mixes and, 29
botrytis in, 128–29
foliar diseases in, 121, 125–27
pesticides in, 121
root diseases in, 121, 122–25
rust in, 129–30
scouting and symptoms in, 121–22
tospoviruses in, 128
vegetative propagation in, 161
water management in, 121
disinfectants, vegetative propagation and, 161
Distance, 154
distillation in treating water with high ion control, 13
DLI. *See* daily light integral (DLI)
DNA analysis techniques, 19
dormancy, breaking, in germination, 146
downy mildews, 122
drainage, effect of, on texture, 24
drenches, applying, 102
drought stress, 9, 87

E

Ecke. *See Paul Ecke Ranch*
Ecke Poinsettia Bract Meter, 378
Edison, Thomas A., 169
electrical conductivity (EC), 33
  calibrating meters, 51, 53
  collecting leachate for, 51
  conversion factors among units, 59
  effects of, 9
  electrode calibration and use, 58
  measuring injector accuracy via, 58–59
  steps to increase or lower, 55, *55*
  in testing, interpreting, and managing medium, 54–56
  testing samples for, 51
  units for expressing, *57*
  values for dilution levels
    for fertilizers, *65*
    for manufacturers' fertilizers, *60–64*
  water quality and, 9
electrodialysis in treating water with high ion control, 13
ELISA technique in virus indexing, 173–74
employees and sustainability, 198
energy curtains, 196
energy-efficient production, 80–81
energy in sustainability, 196
environmental conditions
  effect on plant responses to light, 87
  impact on efficacy of plant growth regulators (PGRs), 103–4
environment in plug production, *151,* 151–52, *152*
environment monitoring systems, vegetative propagation and, 161
EPA-registered products, 15–16
epinasty, 178, 180
equipment failure as cause of nutritional problems, 38
Escherichia coli, 15
ethephon phosphonic acid, 98, 101, 167
EthylBloc, 188–89
ethylene
  reducing plant exposure to, 187–88
  sensitivity to, *183–84*
ethylene dibromide, 131
ethylene protection, 188–89
etridiazole (Truban, Terrazole, and Banrot), 123
evaporative cooling pads, 92

F

Federal Insecticide Fungicide and Rodenticide Act (FIFRA), 135
fenamidone (Fenstop), 123
fenhexamid (Decree), 129
fertility in pest management, 107
fertilizers, 179
  acidic, 45
  acidic versus basic, 34–35
  common commercial, *150*
  in controlling growth, *150,* 150–51
  effect on electrical conductivity (EC), 33
  electrical conductivity values for dilution levels for manufacturers, *60–65*
  formulations for, *66*
  high-ammonium, 47
  modifications of, in neutralizing alkalinity, 12
  neutral, 45
  nitrate-based, 157
  selecting, 34–35
  for stock plants, 157
  in sustainability, 196
  terminating, 179–80
  for toning rooted cuttings, 166–67
  type of, 45, *46*
  urea, 45
fertilizer salts, electrical conductivity (EC) values for dilution levels of, *65*
fertilizer solution testing, 50
filtration, 16, *18,* 19
flat filling, 144
Florel, 157, 167, 581
floriculture crops, 72
flower automation, future of, 421
flowering, effect of photoperiod on, 89–90
flower longevity, 178
fludioxonil (Medallion and Hurricane), 124, 125, 129
fluopicolide (Adorn), 123
fluoride, 14
flurprimidol, 97, 101, 102
foliar analysis, 49
foliar diseases, 39, 121, 125–27
  water management in, 121
foliar sprays, 48
  mixing rates for plant growth regulators (PGRs) used as, *105*

Fonteno, William, 21
Food Quality Protection Act (FQPA), 135
Fosphite, 123
FRAC Codes, 139
Frantz, Jonathan, 80
freezing injury, 75
fumigant pesticides, 134–35
fungal diseases, 39, 153
fungal pathogens, indexing for, 173
Fungicide Resistance Action Committee, 139
fungicides, 130, 135, 139
  chemical class and mode of action of, for greenhouse use, *140*
fungus, 122
fungus gnats, 30, 117–18, 154
Fusarium wilt, 124–25
  environmental influences on, 125
  management of, 125
  positive presence of, 20
  symptoms of, 125
  testing for, 19

G

General Agreement of Tariffs and Trade (GATT) Uruguay Round, 170
germination
  breaking dormancy in, 146
  facilities for, 148
  light in, 146, *148*
  moisture in, 146, *147*
  optimal temperature for, 145–46
  of seed, 144, 148
gibberellins (GA), 99, **99**
  influence on plant growth, 96
Gnatrol, 154
graphical tracking, 78
greenhouses
  daily light integral for various crops, *87–89*
  filtration options for, *18*
  insects in, 112–19
  location of, for stock program, 156
  managing light in, 91–94
  media testing in, 49, *50*
    by saturated media extraction (SME) method, *50*
  retractable-roof, 92
  tissue nutrient levels of high-quality plants in, *52–53*
green marketing, 193
greenwashing, 194
ground wood, 23–24
growth regulators, 95–105
  alternative methods of applying plant growth regulators of, 102–4
    liner dips, 103
    media surface application, 103
    sprenches, 103
  application timing, 99, 101
  applying drenches, 102
  biological control, 95
  chemical growth control, 96
  commonly used, 96–98
    abscisic acid (ABA), 98
    ancymidol, 97
    benzyladenine, 98–99, **99**
    benzyladenine + gibberellin combinations, 99
    chlormequat chloride, 96–97
    daminozide, 96
    dikegulac-sodium, 98
    ethephon phosphonic acid, 98
    flurprimidol, 97
    gibberellins, 99, **99**
    paclobutrazol, 97–98
    uniconazole, 98
  physical control, 95–96
  spray applications/equipment, 101–2
  for stock plants, 157

H

Hall, Charles, 194
hand washing, 176
hand watering, 9
hanging baskets, **93,** 93–94
hard water, 10
  calcium in, 14
  magnesium in, 14
harvesting of stock plants, 157–58
haustoria, 126
Hawley-Smoot Tariff (1920), 169
heat
  in controlling soilborne pests, **131,** 131–34
  disadvantages of, 134
  natural, **133,** 133–34
heat delay, 74

heating system, 196
  root-zone, 68
heptachlor, 131
herbaceous perennials requiring vernalization, *76*
herbicides, 135
Heritage, 153
HID. *See* high intensity discharge (HID) lighting
high-ammonium fertilizers in correcting high pH, 47
high intensity discharge (HID) lighting, 181
  for stock plants, 156
high-quality irrigation water, characteristics of, *10*
high soluble salts, 13–14
high temperature, 74
Holland Bulb Forcers Guide (A.A. De Hertogh), 182
home water softening, 14
horizontal airflow (HAF) fans, 68, 111
Hurricane, 123
hydrogen dioxide, 15
  checking concentration of, 19
hydrogen dioxide/activated peroxygen chemistry, 15
hydrogen peroxide, 15
hydroponic production of *Alstromeria,* 223
Hypoaspis mites, 154
hypochlorite, 15
hypochlorous acid, 16

I

impatiens necrotic spot virus (INSV)
  detection of, 174
  disease management and, 128
  eliminating insect populations to protect stock plants from, 161
  sanitation and, 107
  western flower thrips causing damage by vectoring, 112
incandescent lamps, 94
infrared thermometer, 69
in-house testing, 51
  calibrating meters, 56–58
    calibration, 57–58
    calibration solutions, 56–57
    electrode care, 57
    measuring injector accuracy via electrical conductivity (EC), 58–59
    pH and electrical conductivity (EC) meter principles, 56
    power source, 57
    temperature, 57
  crop factors, 51
  frequency, 51
  in managing medium electrical conductivity, 54–56
  PourThru method, 51, 53–54
  sampling for, 51
  of water, 11
injector accuracy, measuring via electrical conductivity (EC), 58–59
Insecticide Resistance Action Committee (IRAC), 139
insecticides, 139
  chemical class and mode of action of, for greenhouse use, *141–42*
  using minimum label rates of, 139
insects. *See* pest management
integrated pest management (IPM), 197
  for soilborne pests, 134–35
interveinal chlorosis, 33, 40
ions, water treatments for, 13–14
IPM. *See* integrated pest management (IPM)
iprodione (Chipco 26019, 26 GT, and Sextant), 124, 127
iron
  as antagonistic to other nutrients, 10
  deficiency of, 39, 40
  in subsurface water, 14
  toxicity of, 40
iron sulfate, applying, 48
irrigation
  automated, 9
  effect on electrical conductivity, 34
irrigation expert, obtaining advice from, before installing filtration system, 16
irrigation water
  alkalinity of, 11–12
  classification of quality, based on alkalinity, *12*
  effect on electrical conductivity, 33–34
  recycling, 14
  testing of, 49–50
isolation, 175

K

kalanchoe, 40
Kieft Seeds, 306
K-Phite, 123

L

laboratory testing
- fertilizer solution testing, 50
- greenhouse media testing, 49, *50*
- irrigation water testing, 49–50
- plant tissue testing, 50–51

lambsquarters, 107
Lansbergen, Hein, 421
Lansbergen Gerberas Moving Flowers system (the Netherlands), 421
leaching, 181
- effect on electrical conductivity (EC), 34

leaf drop, 178, 180
leafhoppers, 122
leaf interveinal chlorosis, 33
leaf spots, 122
leaf yellowing, 178
LED lighting, 94
Len Busch Roses (Plymouth, Minnesota), hydroponic cut *Alstroemeria* at, 223
license agreements, 156
life cycle, 197
life stage, timing applications to target most vulnerable, 142
light intensity, **83,** 83–84
light/lighting, 83–94
- artificial, 90–91, 94
- canopy closure, 90
- crops that benefit from supplemental, in plug trays, *151*
- distribution patterns of, 92
- in germination, 146, *148*
- hanging baskets, **93,** 93–94
- influence on plant growth and quality, 180–81
- LED, 94
- managing, in greenhouses, 91–94
- in media, 28
- natural, 91
- night interruption, 94
- outdoor levels, 91
- photoperiod, 89–90, 94
- plant growth responses to daily integrals, 84, 86–87
- plant responses to, **83,** 83–84, **84**
- shading, 92–93, **93**
- spectral quality, 90
- for stock plants, 156
- supplemental, 94
- supplemental high intensity discharge (HID), 152
- transmission, 91–92

light quality
- and quantity as growth regulator, 96
- spectral, 90

light transmission, 91–92
lime in raising pH, 44
liner dips, 103
living mulches, 135–36
long-day plants, 89
Lopez, Roberto, 194
low temperature, 75–77

M

magnesium
- deficiency of, 33, 40
- in hard water, 14

maintenance of stock plants, 157
mancozeb (Protect T/O and Dithane), 127, 130
manganese, 10
- deficiency of, 40
- in subsurface water, 14
- toxicity of, 40

marginal leaf burn, 9
marginal necrosis, 33
market identity, 170–71
marketing
- green, 193
- in sustainability, 194–95

market trends in stock plants, 155
Mattson, Neil, 24
mechanical conditioning as growth regulator, 96
mechanical heat, 132–33
Medallion, 153
media, 21–31
- alkalinity in, 28
- biology, 21, 29
- chemical properties, 21, 28–29
  - alkalinity, 28
  - buffering capacity, 28–29
  - light level, 28
  - mix temperature, 28
  - pH, 28
  - starter charge, 28
- compacting, 27, *27*

components in, 23, 25
composition of growing, 22, *25*
general, 21–24
ground wood in, 23–24
handling issues, 21, 29–30
light level in, 28
managing pH for container, **43,** 43–48
correcting problems, 46–48
prevention of, 48
recognizing problem, 43–44
regular testing, 46
mineral soil in growing, 22–23
mix temperature in, 28
pH in, 28
physical properties, 21
moisture level, 26
texture, 24
porosity and, 27–28
postproduction, 21, 31
quality of, in plug production, *149,* 149–50
for stock plants, 156–57
total pore space and solid percentages for various, 22, **22**
vegetative propagation and, 161–62
media biology, 29
media mix
adding wetting agents to, 26–27
choosing, 21–22
storage of, 29–30
media surface application, 103
media testing, greenhouse, 49, *50*
medium temperature, 68
medium type, effect on electrical conductivity (EC), 34
metalaxyl/mefenoxam, 123
metamsodium, 134
methyl bromide, 131
1-methylcyclopropene (1-MCP), 188
microbes, testing for, 19–20
microbial analysis, sampling water for, 20
micronutrients, 38
correcting deficiencies, 47–48
microorganisms, water treatment for, 14
mineral antagonisms as nutritional problems, 39
mineral soil in growing media, 22–23
misting during propagation, 68
mites, 116–17. *See also* pest management
spot spray for, 142
miticides
chemical class and mode of action of, for greenhouse use, *141–42*
using minimum label rates of, 139
mix temperature in media, 28
moisture
adding, to bagged mix, 26
effect of, on porosity, 26
effect on shrinkage, 27
in germination, 146, *147*
testing for, 26
mold growth, 29
molydenum deficiency, 41
MPS certification, 195
mulches, living, 135–36
mutations, ownership of, 170
myclobutanil (Hoist and Eagle), 130

N

National Greenhouse Manufacturers Association (NGMA), 108
natural heat, **133,** 133–34
natural light, 91
necrotic leaf tips, 33
necrotic spots, 180
Nemashield, 154
Nemasys, 154
neutral fertilizer, 45
night interruption lighting, 94
nightshade, 107
nitrate, 45
nitrate-based fertilizers, 157
nitric acid in controlling alkalinity, 13
nitrogen deficiency, 39
in promoting plant growth, 95
nitrogen-potassium ratio, 179
nitrogen tie up, 23
non-fumigant pesticides, 135
nurseries, filtration options for, *18*
nutrient demands, effect on electrical conductivity (EC), 34
nutrient disorders, typical symptoms of, 39–41
nutrient ratios in fertilizer, 179
nutrient stress as growth regulator, 95
nutritional problems, reasons for, 38

O

Oasis foam, 28
Oasis Soax, 189
octanoic acid, 15–16
offices in sustainability, 197
OHP 6672, 153
On Target (web-based program), 384
oomycete, 122
optimum temperature, 72–73
Organic Materials Review Institute (OMRI), 195
organic mix, 24, *24*
Orzanin, 154
osmocote, 180
Ostracoderma slime mold, 29
outdoor light levels, 91
overwatering. *See* water stress (overwatering)
oxalis, 107
Oxcide, 16
oxidation-reduction potential (ORP), 15, 16
  checking concentration of, 19
ozone systems, 16

P

paclobutrazol, 97–98, 101, 102
Pageant, 153
parasitic wasps (parasitoids), 111
patents
  infringement of, 171
  length of, 169–70
  rights granted by, 170
pathogens, 121
Paul Ecke Ranch (Encinitas, CA), 372, 378
PCNB (Terraclor), 124
peat-based medium, nutrient availability changes with pH in, *149*
peat moss, 23
Pelargonium, 127
perlite, 23, 24
peroxyacetic acid (PAA), 15
pesticide break, 142
pesticides, 134–35
  in disease management, 121
  fumigant, 134–35
  minimizing use of mixtures, 139
  non-fumigant, 135
  rotating, 138–39
  in sustainability, 196–97
  using, with nonspecific modes of action, 139
pest management, 107–19
  biological control, 111–12, *113*
  conventional and alternative pesticides, 109–11, *110*
  cultural, 107
    fertility, 107
    sanitation, 107
    watering, 107
  major greenhouse, 112–19
    aphids, 113–14
    fungus gnats and shore flies, 117–18
    mites, 116–17
    snails and slugs, 118–19
    thrips, 112–13
    whiteflies, 114–15
  physical/mechanical, 107–11
    scouting, 108–9
    screening, 107–8
  in plug production, 153–54
  rotating materials for, 138–39
  spot spray for, 142
  varying strategies, 138
  vegetative propagation and, 161
pests for stock plants, 157
Pests of the Garden and Small Farm: A Growers Guide to Using Less Pesticide (Flint), 136
PGRs. *See* plant growth regulators (PGRs)
pH, 9
  collecting leachate for, 51
  correcting problems with, 46–48
  ideal water, 9
  lime in raising, 44
  managing, for container media, **43,** 43–48
    correcting problems, 46–48
    prevention of, 48
    recognizing problem, 43–44
    regular testing, 46
  in media, 28
  as nutritional problems, 38
  reasons for problems with, 44–46
  testing samples for, 51
pH electrode calibration and use, 58
pH meters, calibrating, 51, 53
phosphonates, 123
phosphoric acid in alkalinity control, 13
phosphorus (P)

amount of, to apply, 35, 37
deficiency of, 33, 39
in promoting plant growth, 95
photoperiod, 89–90
photoperiodic lighting, 94
photosynthesis
impact of water stress on, 9
light as driving force for, 83
physical control in regulating plant growth, 95–96
Phyton 27, 153
Phytophthora, 14–15, 124, 174
environmental influences on, 123–24
management of, 124
positive presence of, 20
symptoms of, 123
testing for, 19
phytoplasmas, 122
phytotoxicity, 48, 97
minimizing, 46, 48
pigweed, 107
pinching as growth regulator, 96
pine bark, 23
Plantation, The, 582–84
plant breeding, 170
plant growth, responses to daily light integrals, 84, 86–87
plant growth regulators (PGRs), 95. *See also* growth regulators
alternative methods of applying, 102–4
bark and, 103
commonly used, 96–98
comparing attributes of, *100*
environmental conditions impact on efficacy of, 103–4
in greenhouse production, 98–99
mixing rates for, used as foliar sprays, *105*
preparing solutions, 104
recordkeeping on treatment, 104
for vegetatively propagated crops, *168*
plant growth retardants, 96
plant nutrition, 33–41
amount of phosphorus to apply, 35, 37
electrical conductivity (EC), 33
factors affecting, 33–34
nutritional problems, 38–40
selecting fertilizer, 34–35
Plant Patent Act (1930), 169
plant pathogens, testing for, 19–20
plant pathology laboratories, 19
Plant Protection Act (1930), 170
plants
day-neutral, 89, 90
improving postharvest handling of, 31
long-day, 89
responses to light, **83,** 83–84, **84,** *85*
short-day, 89, 90
plant temperature, 68–69
plant tissue analysis, 49, 50–51
Plant Variety Protection Act (1970), 171
plug crops, classification of, by optimum rate range for Bonzi and Sumagic, *153*
plug growth, stages of, 143, *143*
plug production, 143–54
chemical growth regulators in, 152–53
environment in, *151,* 151–52, *152*
fertilization in, *150,* 150–51
germination substrate temperatures for Stage 1 of, *145*
media quality in, *149,* 149–50
pests and diseases in, 153–54
selecting fertilizer to control growth, 150–51
Stage 0, equipment and techniques, 143–44
Stages 1 and 2, germination, 144–48
stages of growth, 143, *143*
transplanting in, 154
water quality in, 148–49
plug trays, 143–44
crops that benefit most from supplemental lighting in, *151*
polyacrylamide gels (PAM), 189
polymerase chain reaction (PCR), in virus indexing, 174
polystyrene, 23
polystyrene beads, 23
porosity, 27–28
effect of moisture level on, 26
postharvest care and handling of potted plants, 177–92
consumer care, 191
factors that affect quality, 177–81
importance of quality, 177
nutrient ratios in fertilizer, 179
retail handling and display, 189–90, *190*
shipping and transportation, 182–88

specialized treatments, 188–89
postproduction, 31
potassium
deficiency of, 33, 39
in water softening, 14
potassium-calcium-magnesium ratio, 38
potato sticks, 109
potato virus Y (PVY), 174
potato wedges, 109
PourThru method, 51, 53–54
powdery mildews
disease management and, 125–26
environmental influences on, 126
showing up after shipping, 190
symptoms of, 125–26
visibility of, 122
processed bark fines, 24
propagation, misting during, 68
propamocarb hydrochloride (Banol), 123
PsiMatric Technology, 189
Puccinia pelargonii-zonalis, 129
pyraclostrobin + boscalid (Pageant), 129
pyrethrum, 154
Pythium root rot, 14–15, 27, 33, 39, 121, 122–23, 127
environmental influences on, 123
management of, 123
positive presence of, 20
symptoms of, 123
testing for, 19

Q

quality. *See also* water quality
factors that affect postharvest, 177–81
quaternary ammonium compounds (QAC), 161

R

radiation, shortwave, 68
radio-frequency identification (RFID) technology in tracking, 186
Ralstonia solanacearum, 15, 127, 173, 174
recordkeeping on plant growth regulators (PGRs) treatment, 104
residual control, 15
resistance mitigation, 137–42
avoiding problems associated with, 138–39
basics in, 137–38
resistant temperature detectors (RTD), 67
restricted entry interval (REI), 103
retail handling and display, 189–90, *190*
retractable-roof greenhouses, light provided by, 92
retractable shade curtains, 92, 93
reverse osmosis, 9, 13, 15, 16
Rhizoctonia solani, 123, 124
environmental influences on, 124
management of, 124
positive presence of, 20
symptoms of, 124
testing for, 19
ribgrass mosaic virus (RMV), 174
rooted cuttings, toning, 166–67
rooting hormone, 164
root rots, 122, 153
roots
development of, 166
watering and, 181
environmental and cultural factors promoting growth of, *152*
management of disease in, 121
RootShield, 123, 153
root-tip dieback, 33
root-zone heating system, 68
rust, 122, 129–30
environmental influences on, 129
management of, 129–30
symptoms of, 129

S

salts, high soluble, 13–14
sanitation
in pest management, 107
water technology options in, 15–16
sanitizing agents, testing for, 19–20
saturated media extraction (SME) method, greenhouse media testing by, *50*
scouting, 157
in disease management, 121–22
in pest management, 108–9
screening in pest management, 107–8
SCS Certified, 195
seed, 144
optimal temperature for germination of, 145–46
seed covering, 145

seed crop propagation, 143–54. *See also* plug production
chemical growth regulators in, 152–53
flat filling, 144
media quality in, *149,* 149–50
pests and diseases in, 153–54
plug trays, 143–44
seeders, 144
selecting fertilizer to control growth, 150–51
stage 0, equipment and techniques, 143–44
stages 1 and 2, germination, 144–48
stages of growth, 143, *143*
transplanting in, 154
water quality in, 148–49
seeders, 144
shadecloth, 92
shade curtains, retractable, 93
shade systems, retractable, 92
shading, 92–93, **93**
shepherds purse, 107
shipping and transportation, 182–88
shipping duration, 187
shoot growth, environmental and cultural factors promoting, *152*
shore flies, 30, 117–18, 154
short-day plants, 89, 90
shortwave radiation, 68
shrinkage, effect of moisture level on, 27
slugs, 118–19
SME. *See* saturated media extraction (SME) method
snails, 118–19
sodium, water electrical conductivity and, 10–11
sodium adsorption ratio (SAR), 148–49
soft rots, 122
soilborne pest control, 131–36
biofumigation, 135, **135**
heating in, **131,** 131–34
integrated approach, 134–35
living mulches, 135–36
SoilGard, 123
soilless media
poor buffering of, 44
suggested media PourThru electrical conductivity (EC) ranges for floricultural crops grown in, *36*
solarization, **133,** 133–34
comparison of steam pasteurization and, *133*
pests managed by, *134*
soluble salts levels as nutritional problems, 38–39
specialized treatments, 188–89
spectral light quality, 90
sphagnum peat moss, 29
sports, ownership of, 170
spray applications, 101–2
sprenches, 103
starter charge, 28
steam pasteurization, 132
comparison of solarization and, *133*
stem elongation, effect of DIF and DIP on, *78–79*
stick order, 159–60
stock plants
dealing with, 155–58
harvesting of, 157–58
light and temperature requirements for, 156
location of greenhouse and, 156
maintenance of, 157
media for, 156–57
source of, 156
stomatal closure, 98
storage cooler, 160
strobilurin fungicides (Compass O, Cygnus, Disarm, Heritage, Insignia, and Pageant), 124, 130
Subdue, 123, 124
Subdue MAXX, 123, 124, 153
substrates
drenches in controlling growth, 96–97
wood in, 23
sudden oak death, 15
sulfur deficiency, 40
sulfuric acid in controlling alkalinity, 13
Sumagic, 152, *153*
supplemental high intensity discharge (HID) lighting, 152
supplemental lighting, 94
sustainability, 193–98
achieving, 195–96
certifications in, 195
crop culture in, 197
employees and community in, 198
energy in, 196
marketing in, 194–95
offices and transportation in, 197

pesticides in, 196–97
water and fertilizer in, 196
Sustainable Agriculture Service, 193
sustainable floriculture, 193

T

Talstar, 154
Tame, 154
temperature, 67–81
air, 67–68
average daily, 69–70
base, 70–72, *71*
chilling and freezing injury, 75
cropping time and, 73–74
differential temperatures, 77–80
energy-efficient production, 80–81
as growth regulator, 96
high, 74
influence on plant growth and quality, 180–81
low, 75–77
as nutritional problems, 39
medium, 68
optimal, 72–73
for germination, 145
plant, 68–69
for stock plants, 156
temperature drop (DIP), 77–80
during transit, 186–87
vernalization, 75–77
temperature drop (DIP), 79–80
effect of, on stem elongation, *78–79*
Terrazole + Clearys 3336, 153
testing. *See* in-house testing; laboratory testing
texture, effect of drainage on, 24
thermistors, 67
thermocouples, 67
thermometers, 67
infrared, 69
Thielaviopsis basicola, 122
thiophanate-methyl (3336, 6672, Fungo, Alban, 26/36, and Banrot), 124, 125
thrips, 112–13
timing as growth regulator, 95
tissue-cultured plantlets, acclimatizing, 162
tissue nutrient levels of high-quality greenhouse plants, *52–53*
tobacco mosaic virus (TMV), 174
tomato spotted wilt virus (TSWV), 107, 112, 128, 174
top coaters, 144
tospoviruses, 107
environmental influences on, 128
management of, 128
symptoms of, 128
transit, temperature during, 186
transpiration, 69
transplanting in plug production, 154
transportation in sustainability, 197
transport systems, 185–86
triadimefon (Strike), 130
triazoles, 101
as plant growth regulator, 97
trifloxystrobin (CompassO), 129
triflumizole (Terraguard), 124, 125
Truban, 153

U

ultraviolet (UV) light, 15
as water treatment, 16
underwatering, 181
uniconazole, 98, 101, 102
as plant growth regulator, 97
unidirectional flow, 175
unrooted cuttings, advantages of buying, 155–56
urea fertilizers, 45
United States Department of Agriculture (USDA), 3
Floriculture Crops Summary, 3–7, *7*
organic, 195
US Plant Patents, 171
application for, 169
legal issues and, 167–69

V

variety selection in enhancing postharvest performance, 177–79
vascular wilt disease, 122
vegetative crops propagation, 155–71
dealing with stock plants, 155–58
market trends in, 155
plant growth regulators (PGRs) for, *168*
stages of, 158–67
VeriFlora certification, 195
vermiculite, 23, 24
vernalization, 75–77

herbaceous perennials that require, *76*
Verticillium wilt, 125
viroids, 122
Virtual Grower software program, 80, 81
viruses, 122
eliminating, 174
virus indexing, 173–74

W

wash stations, 176
water
hard, 10, 14
irrigation, 11–12, *12,* 14, 33–34, 49–50
sampling, for microbial analysis, 20
in sustainability, 196
water alkalinity, 45–46, *47*
waterborne pathogens
control of, 14–16
water quality and, 9
watering
hand, 9
in pest management, 107
practice of, 9
root development and, 181
watering tunnels, 144
water management
in foliar diseases, 121
for toning rooted cuttings, 166
water quality, 9–20
acid, 12–13
alkalinity, 11–12, *12*
control of waterborne pathogens, algae, and biofilm, 14–16
defining, 9–11, *10*
filtration, 16, *18,* 19
in plug production, 148–49
testing
in commercial labs, 11, *11*
in-house, 11
for plant pathogens and sanitizing agents in, 19
microbes and pathogens, 19–20
onsite tests of active ingredients, 19
sampling water for microbial analysis, 20
treatments for ions, 13–14
water softening, 14
home, 14
water-soluble fertilizers, acidity or basicity of, *46*
water stress (overwatering), 9, 181
as growth regulator, 95
as nutritional problems, 38
water technology, sanitizing options in, 15–16
water testing, irrigation, 49–50
water-testing labs, 11, *11*
water treatments
comparison of installation and operating cost and residual activity of several technologies, *17*
for ions, 13–14
western flower thrips (WFT), 128, 137
wetting agents
adding to media mix, 26–27
drenching media with, 31
whiteflies, 114–15
white rust *(Puccinia horiana),* 129
whole trees (WT), 23
Williams, Jack, 372
wilting, delaying, 189
wood in substrates, 23
wood sawdust, 23

X

X3, 154
Xanthomonas, bacterial diseases caused by, 127–28
environmental influences on, 127
management of, 127–28
symptoms of, 127
Xanthomonas campestris pv. *pelargonii,* 127, 173, 174

Y

Yue, Chengyan, 194

Z

ZeroTol, 29, 154
zinc deficiency, 40

Plant Index

Page numbers in *italics* indicate tables; page numbers in **bold** indicate figures.

A

Abelmoschus esculentus (okra; gumbo; lady's fingers), 201
Abutilon × *hybridum* (flowering maple; Chinese lantern), 201–203, **202**
  solarization and, *134*
Achillea sp. (yarrow), **203,** 203–205
  A. clypeoplata, 205
  A. filipendulina, 205
  A. hybrid, 205
  A. millefolium, 205
  A. ptarmica, 205
  A. taygetea, 205
  A. × 'Coronation Gold,' 205
Achimenes sp. (Star of India; monkey-faced pansy; orchid pansy; hot-water plant), 206–207
  ethylene sensitivity and, 183
Achmella oleracea (eyeball plant; toothache plant), **207,** 207–208, **208**
Aconitum spp., *35*
Adiantum spp., *52*
  daily light integral (DLI) for, *87*
Aechmea spp., *52*
African daisy. *See Osteospermum* hybrids
African marigold. *See Tagetes erecta*
African violet. *See Saintpaulia ionantha*
Agastache spp.
  A. foeniculum (anise hyssop; giant hyssop; blue giant hyssop), 209, **209**
  A. rugosa, 209
Ageratum houstonianum (floss flower), 80, **210,** 210–211
  base temperature and, *72*
  clinical growth regulators for, *153*
  daily light integral (DLI) for, *88*
  ethylene sensitivity and, *183*
  germination of, *145*
  media moisture level for, *147*
  need for supplemental lighting, *151*
Aglaonema spp. (Chinese evergreen), *52,* **212,** 212–213
  daily light integral (DLI) for, *87*
  ethylene sensitivity and, *184*
  low temperature and, 75
Ajuga reptans (bugle weed), 213, **213**
  vernalization and, *76*
Alcea rosea (hollyhock), **214,** 214–215
Alchemilla mollis (lady's mantle), **215,** 215–216
  A. erythropoda, 216
  vernalization and, *76*
Allium spp., *35,* 216–218, **217,** *218*
  A. ampeloprasum var. *porrum* (leeks), 216–217
  A. cepa (onion), 216
  A. giganteum, 217–218
  A. schoenoprasum (chive), 217
  ethylene sensitivity and, *184*
alpine wallflower. *See Erysimum linifolium*
Alstroemeria hybrids (Inca lily; princess lily), **219,** 219–223
  relative nutrient requirements of, *35*
  tissue nutrient levels of, *52*
  watering solution developed by Aalsmeer Research Station, *223*
Alternanthera spp. (Joseph's coat), **224,** 224–225
  A. dentata, 224–225
  A. ficoidea, 224–225
alumroot. *See Heuchera* hybrids (alumroot; coral bells)
alyssum. *See Lobularia maritime*
Alyssum montanum (madwort; yellow tuft; mountain gold), 225
Alyssum saxatilis. See Aurinia saxatilis
amaryllis. *See Hippeastrum* spp.
American marigold. *See Tagetes erecta*
Ammi majus, 617
Amsonia (bluestar), vernalization and, *76*
Anemone coronaria (windflower), *35,* **226,** 226–228
  ethylene sensitivity and, *183*
Anethum graveolens (dill), 229, **229**
Angelonia angustifolia (summer snapdragon, summer orchid), **70, 230,** 230–233, *232*
  average daily temperature (ADT) and, *73*
  base temperature and, 70, 72
  chilling and freezing injury and, 75
  daily light integral (DLI) for, *89*
  plant growth regulators (PGRs) for, *168*
  sticking priorities for, *160*
  vegetative propagation of, *158*

Anigozanthos spp., *35*
anise. *See Pimpinella anisum*
anise hyssop. *See Agastache foeniculum*
annual phlox. *See Phlox drummondii*; *Phlox maculata*
Anthurium spp., 233–236, **234**
  A. andreanum, 234–235
  A. scherzerianum, 234
    ethylene sensitivity and, *184*
Antirrhinum majus (snapdragon), **236,** 236–244
  average daily temperature (ADT) and, *73*
  base temperature and, 71, *71*
  stem elongation and, *78*
  clinical growth regulators for, *153*
  daily light integral (DLI) for, *88*
  ethylene sensitivity and, *184*
  flowering response for varieties of, *243–244*
  light for germination for, *148*
  media moisture level for, *147*
  need for supplemental lighting, *151*
  nutritional needs of, 44
  plant growth regulators (PGRs) for, *168*
  relative nutrient requirements of, *35*
  rust disease in, 129
  scheduling of sowing and harvest dates of, *242*
  sticking priorities for, *160*
  tissue nutrient levels of, *52*
  variety groupings of, *241*
  vegetative propagation of, *158*
Aphelandra squarrosa
  ethylene exposure in, **188**
  ethylene sensitivity and, *184*
Aquilegia × *hybrida* (columbine), **245,** 245–247, **246,** 247
  A. canadensis (red)
    base temperature and, *71*
  A. flabellata (fan), 247
    base temperature and, *72*
  A. vulgaris, 247
  base temperature and, *71*
  chilling and freezing injury and, 75
  forcing crops, 246
  vernalization and, *76*
Arabis caucasica (rock cress; wall cress), **247,** 247–248, **248**
  A. blepharophylla, 248
Argyranthemum spp. (also *A. frutescens*) (marguerite daisy), **249,** 249–250
  plant growth regulators (PGRs) for, *168*
  sticking priorities for, *160*
  vegetative propagation of, *158*
Armeria maritima (thrift; sea pinks), **250,** 250–252, **251**
  A. latifolia, 251
  A. pseudoarmeria, 251
  vernalization and, *76*
Asclepias tuberosa (butterfly weed; butterfly flower), *35,* **252,** 252–253
Asiatic lily. *See Lilium* × *hybrida*
asparagus fern. *See Asparagus* spp.
Asparagus densiflorus var. 'Sprengeri' (asparagus fern; Sprengeri fern), *52,* **253,** 253–254
  A. falcatus, 254
  A. meyeri, 254
  A. pseudoscaber, 254
  A. pyramidalis, 254
  A. setaceus, 254
Asplenium nidas (Japanese bird's nest fern), 559
  ethylene sensitivity and, *184*
aster, China. *See Callistephus chinesis*
Aster spp. (aster), *35, 52,* **255,** *256*
  A. novae-angliae (New England aster), 254–258
  A. novi-belgii (New York aster), 254–258
    stem elongation and, *78*
  daily light integral (DLI) for, *89*
  ethylene sensitivity and, *183*
  media moisture level for, *147*
  rust disease in, 129
  seed for, 145
Astilbe spp. (false spirea), *35,* **259,** 259–260
  A. chinensis, 260
    base temperature and, *71*
  A. japonica, 260
  A. simplicifolia, 260
  A. × *arendsii,* 260
  A. × *hybrida,* vernalization and, *76*
  ethylene sensitivity and, *184*
Athyrium spp.
  A. filix-femina (Miss Vernon's crested lady fern), 560
  A. nipponicum (Japanese painted fern), 560
Aurinia saxatilis (basket of gold; gold dust), 227, 261, **261**

autumn fern. *See Dryopteris erythrosora*
avens. *See Geum* spp.
azalea. *See Rhododendron* spp.

B

baby's breath. *See Gypsophila paniculata*
baby snapdragon. *See Linaria hybrida*
bachelor's button. *See Centaurea cyanus*
bacopa. *See Sutera* spp.
balcony geranium. *See Pelargonium peltatum*
balloon flower. *See Platycodon grandiflorus*
Baptisia australis (false indigo), vernalization and, *76*
basil. *See Ocimum* spp.
basket of gold. *See Aurinia saxatilis*
bearded iris. *See Iris* hybrids
Begonia spp., 263–272
  B. rex hybrids (Rex begonia; painted begonia), 267–268
  B. semperflorens, need for supplemental lighting, *151*
  B. × *hiemalis* (Rieger begonia; Elatior begonia; hiemalis begonia), *52,* **263,** 263–267, *266*
    base temperature and, *72*
    fertilizer termination in, 180
    daily light integral (DLI) for, *88*
    shipping, 182, 185
    stem elongation and, *78*
  B. × *hybrida,* 269–270
  B. × *semperflorens-cultorum* (wax leaf begonia; fibrous begonia), *52,* 268–269
    base temperature and, *72*
    daily light integral (DLI) for, *88*
  B. × *tuberhybrida* ('Non-Stop' begonia; tuberous begonia), 264, 270–272
    stem elongation and, *78*
  ethylene sensitivity and, *183*
  germination of, *145*
  light for germination of, *148*
  media moisture level for, *147*
  need for supplemental lighting, *151*
  nitrogen deficiency in, 39
  plant management in, 119
  powdery mildew in, 125
  relative nutrient requirements of, *35*
  sales, 4, **7**
  shipping, 182, 185, 186
  tissue nutrient levels of, *52*
Bellis perennis (English daisy), 272
Beta vulgaris (Cicla group) (Swiss chard), 273
Bidens hybrids, 273–274, 587
  plant growth regulators (PGRs) for, *168*
  sticking priorities for, *160*
  vegetative propagation of, *158*
bigleaf goldenray. *See Ligularia dentata*
bigleaf ligularia. *See Ligularia dentata*
black-eyed Susan. *See Rudbeckia* spp.
black-eyed Susan vine. *See Thunbergia alata*
blanket flower. *See Gaillardia* spp.
blazing star. *See Liatris spicata*
bleeding heart. *See Dicentra spectabilis*
blue arrows. *See Juncus inflexus*
blue festuca. *See Festuca glauca*
blue giant hyssop. *See Agastache foeniculum*
blue hair grass. *See Koeleria*
bluestar. *See Amsonia*
Boltonia asteroides (boltonia; false chamomile), 274–275
  var. *latisquama,* 275
borage. *See Borago officinalis*
Borago officinalis (borage), 275
Boston fern. *See Nephrolepsis exaltata*
Bougainvillea spp., *35*
  B. glabra, 275–277, **276**
  ethylene sensitivity and, *183*
  shipping, 186
Bracteantha bracteata (strawflower), **277,** 277–279
  iron toxicity in, 40
  manganese deficiency in, 40
  plant growth regulators (PGRs) for, *168*
  sticking priorities for, *160*
  vegetative propagation of, *158*
Brassica spp.
  B. oleracea var. *acephala* (kale; collards; flowering or ornamental kale and cabbage), **279,** 279–281
  B. oleracea var. *botrytis* (cauliflower), **279,** 279–281
  B. oleracea var. *capitata* (cabbage), **279,** 279–281
  B. oleracea var. *gemmifera* (Brussels sprouts), **279,** 279–281
  B. oleracea var. *gongylodes* (kohlrabi), **279,** 279–281

B. oleracea var. *italica* (broccoli), *35,* **279,** 279–281
media moisture level for, *147*
seed for, 145
solarization and, *134*
broccoli. *See Brassica oleracea* var. *italica*
Bromeliaceae (bromeliads), 281–283
daily light integral (DLI) for, *87*
bromeliads. *See* Bromeliaceae (bromeliads)
browallia. *See Browallia speciosa* (browallia)
Browallia speciosa, **283,** 283–284
base temperature and, 70, *72*
germination of, *145*
media moisture level for, *147*
Brussels sprouts. See *Brassica oleracea* var. *gemmifera*
bugbane. *See Cimicifuga racemosa*
bugleweed. *See Ajuga reptans*
bulb fennel. *See Foeniculum vulgare dulce*
buttercup. *See Ranunculus asiaticus*
butterfly flower. *See Asclepias tuberosa*; *Schizanthus* × *wisetonensis*
butterfly weed. *See Asclepias tuberosa*
button fern. *See Pellaea rotundifolia*

C

cabbage. *See Brassica oleracea* var. *capitata*
cactus
Christmas; holiday; Thanksgiving. *See Schlumbergera* spp.
ethylene sensitivity and, *183*
Caladium spp. (caladium), *35, 52*
C. bicolor, **285,** 285–287, *286*
base temperature and, *72*
daily light integral (DLI) for, *87*
ethylene sensitivity and, *184*
Calceolaria spp. (pocketbook plant), *35*
C. × *herbeohybrida,* **287,** 287–288
stem elongation and, *78*
ethylene sensitivity and, *183*
calendiva, ethylene sensitivity and, *183*
Calendula spp., *35*
C. officinalis (pot marigold), **289,** 289–290
base temperature and, *71*
ethylene sensitivity and, *183*
media moisture level for, *147*
seed for, 145
Calibrachoa (million bells), 44, 608
cropping time and, 73, 74
C. × *hybrida,* **290,** 290–292
base temperature and, *72*
plant growth regulators (PGRs) for, *168*
sticking priorities for, *160*
vegetative propagation of, *158*
calla lily. *See Zantedeschia* spp.
Callistephus chinesis (aster; China aster), **292,** 292–293
Campanula spp., *35,* 294–298
C. carpatica (Carpathian harebell; Carpathian bellflower), 294–295
base temperature and, *71*
stem elongation and, *78*
C. isophylla (Italian bellflower; Star of Bethlehem), 295–296
stem elongation and, *78*
C. longistyla, 296
C. medium (Canterbury bells), 297–298
ethylene sensitivity and, *183*
high temperature and, 74
candytuft. *See Iberis sempervirens* (candytuft; hardy candytuft)
Canna × *generalis* (canna; canna lily; Indian shot), **298,** 298–299
Canterbury bells. *See Campanula longistyla*
cape jasmine. *See Gardenia* spp. (cape jasmine, gardenia)
Capsicum (pepper), *35*
base temperature and, *72*
C. annum (pepper; ornamental pepper), **299,** 299–302, *301*
C. frutescens (tabasco pepper), 301
ethylene sensitivity and, *184*
germination of, *145*
iron toxicity in, 40
media moisture level for, *147*
seed for, 145
stem elongation and, *78*
cardinal flower. *See Lobelia cardinalis*
Carex spp., 302–303
C. comans (leatherleaf sedge), 302
C. flagellifera (carex), 303
C. morrowii, 302
carnation. *See Dianthus caryophyllus*
Carpathian bellflower. *See Campanula carpatica*

Carpathian harebell. *See Campanula carpatica*
Catharanthus roseus (vinca; Madagascar periwinkle), **303,** 303–305, 722–723
  base temperature and, *72*
  clinical growth regulators for, *153*
  daily light integral (DLI) for, *89*
  effect on temperature on, **70**
  ethylene exposure in, **188**
  ethylene sensitivity and, *183*
  germination of, *145,* 146
  light for germination of, *148*
  media moisture level for, *147*
  need for supplemental lighting, *151*
  nutritional needs of, 44
  pests and diseases in, 153
  pH of medium, 150
  plant temperature and, 68
  seed for, 145
  stem elongation and, *78*
  tissue nutrient levels of, *52*
catnip. *See Nepeta cataria*
cauliflower. *See Brassica oleracea* var. *botrytis*
Celosia spp., *35*
  C. argentea (cockscomb; plumed cockscomb; feathery amaranth), **305,** 305–307, **306**
    base temperature and, *72*
    stem elongation and, *78*
  clinical growth regulators for, *153*
  C. spicata (wheat celosia), 307
  daily light integral (DLI) for, *89*
  germination of, *145*
  media moisture level for, *147*
  need for supplemental lighting, *151*
  pests and diseases in, 153
Centaurea spp. (bachelor's button; cornflower), *35*
  C. cyanus, **307,** 307–309
Chamaedorea elegans, ethylene sensitivity and, *184*
China aster. *See Callistephus* spp.
Chinese evergreen. *See Aglaonema* spp.
Chinese lantern. *See Abutilon* × *hybridum*
chive. *See Allium schoenoprasum*
Chloris gayana (Rhodes grass) as groundcover, 136
Chlorophytum comosum, ethylene sensitivity and, *184*
Christmas cactus. *See Schlumbergera* spp.
Christmas cherry. *See Solanum pseudocapsicum*
Chrysanthemum spp., *35, 52,* 174
  C. frutescens. See Argyranthemem spp.
  C. parthenium. See Tanacetum parthenium
  C. ptarmiciflorum, 682
  C. superbum. See Leucanthemum × *superbum*
  C. × *grandiflorum,* stem elongation and, *78*
  C. × *morifolium* (chrysanthemum; mum; dendranthema; daisy), **309,** 309–319, *313*
  daily light integral (DLI) for, *89*
  ethylene sensitivity and, *183*
  fertilizer termination in, 179–180
  pest management in, 114
  plant management in, 119
  potted, 178
    daily light integral (DLI) for, *88*
    starter charge for, 28
  premature leaf yellowing in, 178
  rust disease in, 129
  shipping, 182, 185, 186
  symptoms of disease in, 129
cigar plant. *See Cuphea ignea*
cilantro. *See Coriandrum sativum*
Cimicifuga racemosa (bugbane), vernalization and, *76*
Cineraria maritime. See Senecio cineraria
cinnamon fern. *See Osmunda cinnamonea* (cinnamon fern)
Citrullus lanatus (watermelon), 320
  stem elongation and, *78*
clarkia. *See Godetia whitney*
Clematis spp. (clematis; virgin's bower), **320,** 320–322
Cleome hassleriana (spider flower), *35,* **322,** 322–323
  base temperature and, *72*
Clerodendrum spp., *35*
clubawn grass. *See Corynephorus canescens*
Codiaeum variegatum pictum (garden croton)
  ethylene sensitivity and, *184*
coleus. *See Solenostemon* spp.
Coleus hybridus, 35, 155
  daily light integral (DLI) for, *87, 89*
  germination of, *145*
  media moisture level for, *147*
  plant growth regulators (PGRs) for, *153, 168*
  sticking priorities for, *160*
  vegetative propagation of, *158*

collards. *See Brassica oleracea* var. *acephala*
Colocasia spp. (elephant ears), base temperature and, *72*
columbine. *See Aquilegia* spp.
coneflower. *See Echinacea purpurea*
Consolida ambigua (larkspur), *35,* **324,** 324–325
coral bells. *See Heuchera* hybrids
Cordyline indivisa (dracaena spike; spikes), 325, **325**
Coreopsis spp., 326–329
C. grandiflora (tickseed), **326,** 326–328
base temperature and, *71*
C. rosea (pink coreopsis; pink threadleaf coreopsis), 329
C. verticillata (threadleaf coreopsis; tickseed), 328–329
base temperature and, *71*
ethylene sensitivity and, *183*
liner dips for, 103, **103**
media moisture level for, *147*
coriander. *See Coriandrum sativum*
Coriandrum sativum (cilantro; coriander), 329
cornflower. *See Centaurea cyanus*
Corsican mint. *See Mentha requienii*
Cortaderia selloana (pampas grass), 330
Corynephorus canescens (clubawn grass; grey hair grass), 330
Cosmos spp.
base temperature and, *72*
C. bipinnatus (cosmos), 330–331
C. sulphureus, 35
daily light integral (DLI) for, *89*
iron toxicity in, 40
media moisture level for, *147*
need for supplemental lighting, *151*
seed for, 145
stem elongation and, *78*
cranesbill. *See Geranium sanguineum*
creeping phlox. *See Phlox subulata*
creeping myrtle. *See Vinca minor*
creeping thyme. *See Thymus praecox*
creeping wirevine. *See Muehlenbeckia axillaris*
Crocus spp., *35*
ethylene sensitivity and, *184*
Crossandra spp., *35*
C. infundibuliformis, 331–332
ethylene sensitivity and, *183*
croton, daily light integral (DLI) for, *89*
cucumber. *See Cucumis sativas*
Cucumis spp., 332–333
Cucumis melo (melon; cantaloupe; rockmelon), 332–333
Cucumis sativas (cucumber), 333
Cucurbita spp. (cucumber; melon; squash), stem elongation and, *78*
Cuphea spp.
C. hyssopifola (Mexican heather), 333–334, **334**
C. ignea (cigar plant), 334
sticking priorities for, *160*
vegetative propagation of, *158*
curcubits, DIF and, 78
cuttings (during rooting), *35*
cyathia, shipping, 185
Cyclamen spp., *35, 52*
C. persicum, **334,** 334–337
base temperature and, *72*
stem elongation and, *78*
daily light integral (DLI) for, *88*
ethylene sensitivity and, *183*
fertilizer termination in, 179–180
germination of, 146
light for germination of, *148*
shipping, 185
Cylindrocladium spathiphylli, 692
Cymbidium
ethylene sensitivity and, *183*
shipping, 186
Cyperus spp.
C. alternifolia (umbrella plant), 337–338
C. papyrus, 338
Cyrtomium falcatum (Japanese holly fern), 560

D

Dahlia × *hybrida, 35,* **339,** 339–342
base temperature and, *72*
clinical growth regulators for, *153*
daily light integral (DLI) for, *89*
ethylene sensitivity and, *184*
iron toxicity in, 40
media moisture level for, *147*
need for supplemental lighting, *151*
plant growth regulators (PGRs) for, *168*
seed for, 144, 145
stem elongation and, *78*

sticking priorities for, *160*
vegetative propagation of, *158*
Daucus carota, 617
Davallia trichomanoides (rabbit's foot fern), 559
daylily. *See Hemerocallis* hybrids
dead nettle. *See Lamium maculatum*
Delphinium spp., **343,** 343–345
D. elatum, 344
D. grandiflorum (Siberian larkspur), 344
base temperature and, *71*
Dendranthema × *grandiflora. See Chrysanthemum* × *morifolium*
Dendrobium hybrids, **345,** 345–347
D. bigibbum, 346
D. phalaenopsis, 346
Dianthus spp., *35, 52,* 174
daily light integral (DLI) for, *88*
D. barbatus (pinks, sweet william), 347–348
D. barbatus interspecific, 347
D. carthusianorum (Carthusian pink)
stem elongation and, *78*
D. caryophyllus (carnation), **350,** 350–354
D. chinensis (garden dianthus; dianthus), **349,** 349–350
average daily temperature (ADT) and, *73*
base temperature and, *71*
stem elongation and, *78*
D. deltoides (garden pinks; maiden pinks), 354–355
vernalization and, *76*
D. hybrida, 349, **349**
D. gratianopolitanus (cheddar pink), vernalization and, *76*
media moisture level for, *147*
need for supplemental lighting, *151*
potted, fertilizer terminated in, 180
seed for, 145
values for nutrients for, *351*
Diascia spp. (twinspur), 44
base temperature and, *71*
D. barberae, **356,** 356–357
plant growth regulators (PGRs) for, *168*
sticking priorities for, *160*
vegetative propagation of, *158*
Dicentra spp.
D. eximia (fringed bleeding heart), 358
vernalization and, *76*
D. spectabilis (bleeding heart), 357–358
Dichondra argentea, **358,** 358–359, **359**
Dieffenbachia spp. (dumb cane), *52,* **360,** 360–361
daily light integral (DLI) for, *87*
D. amoena, 361
D. maculata, 361
ethylene sensitivity and, *184*
low temperature and, 75
Digitalis purpurea (foxglove), **361,** 361–363
dill. *See Anethum graveolens*
Doronicum orientale (leopard's bane), 363
dracaena spike. *See Cordyline indivisa*
Dracaena spp., **364,** 364–365
daily light integral (DLI) for, *87*
D. deremensis, 52, 364–365
D. fragrans, 52, 364–365
D. marginata, 364–365
ethylene sensitivity and, *184*
Dryopteris spp.
D. erythrosora (autumn fern), 560
D. × *complexa* (robust male fern), 560
dumb cane. *See Dieffenbachia* spp.
Dusty miller. *See Senecio*
Dutch iris. *See Iris* spp.

E

Easter cactus, ethylene sensitivity and, *183*
Easter lily. *See Lilium longiflorum*
Echinacea purpurea (coneflower; purple coneflower), **367,** 367–369
eggplant. *See Solanum melongena*
Egyptian starflower. *See Pentas lanceolata*
Elatior begonia. *See Begonia* × *hiemalis*
elephant ears. *See Colocasia* spp.
Eleusine spp., solarization and, *134*
English daisy. *See Bellis perennis*
English ivy. *See Hedera helix*
English lavender. *See Lavandula angustifolia*
Epipremnum aureum (pothos), 611
calcium deficiency in, 40
ethylene sensitivity and, *184*
Eragrostis elliotii (love grass), 369, **369**
Erysimum linifolium (wallflower; alpine wallflower), **369,** 369–370
esperanza. *See Tecoma stans*
Euphorbia spp., 370–391

E. amygdaloides (wood spurge), vernalization and, 76
E. hypericifolia, 370–372
E. polychroma (cushion spurge), vernalization and, 76
E. pulcherrima (poinsettia), 372, 372–391
base temperature and, 70, 72
color trends of, 389
daily light integral (DLI) for, 88
DIF in, 180
disease in, 123
ethylene sensitivity and, 183
fertilizer termination in, 179–180
graphical tracking of, 385
growth regulator for, 96
leaf drop in, 178
nutritional needs of, 44
pest management in, 115
plant management in, 119
plant temperature and, 68
postproduction life of, 179
powdery mildew in, 125
prolonged sleeving of, 190
relative nutrient requirements of, 35
sales, 5
schedule samples, 379–382
shipping, 185, 186
starter charge for, 28
stem elongation and, 78
sticking priorities for, 160
stock, in, 156
tissue nutrient levels of, 52
varieties of, 178–179
by breeder, 390
E. splendens, ethylene sensitivity and, *184*
Eustoma grandiflorum (lisianthus), 44, **391,** 391–395
ethylene sensitivity and, *183*
light for germination of, *148*
media moisture level for, *147*
need for supplemental lighting, *151*
pH of medium, 149
evergreen candytuft. *See Iberis sempervirens*
Evolvulus nuttallianus, 395–396
sticking priorities for, *160*
vegetative propagation of, *158*
Exacum affine, 35, 52, **396,** 396–397
daily light integral (DLI) for, *88*
ethylene sensitivity and, *183*
shipping, 186
exotic impatiens. *See Impatiens hybrida*
eyeball plant. *See Achmella oleracea*

F

false chamomile. *See Boltonia asteroides*
false indigo. *See Baptisia australis*
false spirea. *See Astilbe* spp.
fan flower. *See Scaevola aemula*
fennel. *See Foeniculum* spp.
fern leaf peony. *See Paeonia tenuifolia*
ferns. *See also* specific species
marginal leaf burn in, 9
Festuca spp.
F. cinerea, 399
F. glauca (blue fescue; glaucus fescue), 399, **399**
F. ovina var. *glauca,* 399
fiber-optic grass. *See Isolepsis cernua*
Ficus spp. (ficus)
ethylene sensitivity and, *184*
F. benjamina (weeping fig), *52,* **400,** 400–401
daily light integral (DLI) for, *89*
F. elastica, 401
F. lyrata, 401
F. maclellandii, 401
F. pumila, 401
F. retusa nitida, 401
high temperature and, 74
leaf drop in, 181
Filipendula purpurea (meadowsweet), vernalization and, *76*
finocchio. *See F. vulgare dulce*
fishtail fern. *See Nephrolepsis biserrata*
fleur-de-lis. *See Iris* hybrids
Florida pansy. *See Torenia fournieri*
floss flower. *See Ageratum houstonianum*
flowering kale and cabbage. *See Brassica oleracea* var. *acephala*
flowering maple. *See Abutilon* × *hybridum*
flowering tobacco. *See Nicotiana alata*
Foeniculum spp.
F. vulgare (fennel; leaf fennel), 401
F. vulgare dulce (sweet fennel; bulb fennel; finocchio) 401–402

foliage plants, *52*
sales, 5
forget-me-not. *See Myositis sylvatica*
fountain grass. *See Pennisetum setaceum*
foxglove. *See Digitalis purpurea*
Fragaria × *hybrida* (strawberry), **402,** 402–403
Freesia hybrids, *35, 52,* **403,** 403–405
ethylene sensitivity and, *184*
French lavender. *See Lavandula dentata*
French marigold. *See Tagetes patula*
fringed bleeding heart. *See Dicentra eximia*
Fuchsia × *hybrida* (fuchsia), 155, **405,** 405–408
daily light integral (DLI) for, 87, *88*
disease in, 123
ethylene sensitivity and, *183*
rust disease in, 129
stem elongation and, *78*
tissue nutirent levels of, *52*

G

Gaillardia spp.
G. aristata, **98**
G. pulchella, 411
G. × *grandiflora* (blanket flower), **409,** 409–411
base temperature and, *71*
Gardenia jasminoides (cape jasmine; gardenia), 412–413
ethylene sensitivity and, *183*
garden phlox. *See Phlox paniculata*
garden pinks. *See Dianthus deltoides*
Gaura lindheimeri (whirling butterflies; wandflower), **99, 413,** 413–414
base temperature and, *71*
sticking priorities for, *160*
gay feather. *See Liatris spicata*
Gazania rigens (gazania), **415,** 415–416
base temperature and, *72*
daily light integral (DLI) for, *88*
media moisture level for, *147*
Seed for, 145
geranium. *See Pelargonium* (geranium)
Geranium spp.
G. dalmaticum (Damaltian cranesbill), vernalization and, *76*
G. psilostemon, 416
G. sanguineum (hardy geranium; cranesbill), 416–417
vernalization and, *76*
G. sanguineum striatum, 416
G. × *cantabrigiense,* 416
G. × 'Johnson's Blue,' 416–417
G. × *oxonianum,* 416
Gerbera jamesonii (gerbera), **417,** 417–422
base temperature and, *71*
daily light integral (DLI) for, *88*
ethylene sensitivity and, *183*
fertilizer termination in, 180
flower automation and, *421*
light for germination for, *148*
need for supplemental lighting, *151*
plant management in, 119
pot life in, 178
powdery mildew in, 125
relative nutrient requirements of, *35*
shipping, 185
stem elongation and, *78*
tissue nutrient levels of, *52*
German iris. *See Iris* hybrids
Geum spp. (avens), 423–424
G. chiloense (Grecian rose), vernalization and, *76*
Giant hyssop. *See Agastache foeniculum*
ginger mint. *See Mentha* × *gracilis*
Gladiolus spp. (gladiolus; glads), **424,** 424–425
cut flower sales, 5
glads. *See Gladiolus* spp.
glaucus fescue. *See Festuca glauca*
globe amaranth. *See Gomphrena globosa*
gloriosa daisy. *See Rudbeckia hirta*
gloxinia. *See Sinningia speciosa*
Godetia whitney (Clarkia amoena) (satin flower), **425,** 425–428
gold dust. *See Aurinia saxatilis*
goldenrod. *See Solidago* × *hybrida*
Gomphrena spp.
base temperature and, *72*
daily light integral (DLI) for, *89*
germination of, *145*
G. globosa (globe amaranth), **428,** 428–429
G. haageana, 429
high temperature and, 74
media moisture level for, *147*
Goniolimon tataricum. See Limonium tataricum
Greek oregano. *See Origanum vulgare hirtum*
grey hair grass. *See Corynephorus canescens*

ground phlox. *See Phlox subulata*
Gypsophila paniculata (baby's breath; gyp), **429,** 429–432, **430**
G. muralis, 432
G. repens, 432

H

hardy candytuft. *See Iberis sempervirens*
hardy geranium. *See Geranium sanguineum*
hardy hibiscus. *See Hibiscus moscheutos*
heart-leafed philodendron. *See Philodendron scandens oxycardium*
Hedera helix (English ivy), 433, **433**
daily light integral (DLI) for, *88*
ethylene sensitivity and, *184*
Helenium spp.
H. amarum (tick weed), **434,** 434–435, *435*
H. autumnale (sneezeweed), 432
Helianthus annuus (sunflower), *35, 52,* **436,** 436–437
Helichrysum spp.
H. bracteantha. See Bracteantha bracteata
H. petiolare, **438,** 438–439
H. thianschanicum, 439
sticking priorities for, *160*
vegetative propagation of, *158*
Heliopsis helianthoides (false sunflower; hardy sunflower; sunflower heliopsis), **439,** 439–440
Heliotropium arborescens (heliotrope), **440,** 440–442
base temperature and, *72*
sticking priorities for, *160*
vegetative propagation of, *158*
Hemerocallis hybrids (daylily), **442,** 442–444
base temperature and, *71*
herbs, daily light integral (DLI) for, *88*
Heuchera hybrids (coral bells; alumroot), **444,** 444–447
daily light integral (DLI) for, *88*
H. micrantha, 446
H. sanguinea, 444, 446
H. villosa, 444
vernalization and, *76*
Hibiscus spp. (hibiscus), *35, 52,* 96
base temperature and, *72*
ethylene sensitivity and, *183*
H. moscheutos (swamp rose; rose mallow; hardy hibiscus), 448–450
H. rosa-sinensis (tropical hibiscus), **447,** 447–450, **449**
daily light integral (DLI) for, *88*
optimum temperature and, 73
shipping, 182, 186
hiemalis begonia. *See Begonia × hiemalis*
Hippeastrum hybrids (amaryllis), *35,* **450,** 450–452
ethylene sensitivity and, *184*
Hosta spp. (hosta; plantain lily; funkia), **452,** 452–454
base temperature and, *71*
daily light integral (DLI) for, *87*
H. fortunei, 454
H. montana, 454
H. undulata, 454
H. ventricosa, 453
pest management in, 119
plant management in, 119
Hyacinth orientalis (hyacinth), *35, 52,* 454–456, **455**
daily light integral (DLI) for, *88*
ethylene sensitivity and, *183, 184*
fertilizer termination in, 180
pot life in, 178
powdery mildew in, 125
shipping, 185
Hydrangea macrophylla, **456,** 456–462
blooms and pot sizes, *458*
characteristics of, *461*
picking hydrangeas, *459*
Hypericum spp., **463,** 463–464
Hypoestes phyllostachya (polka dot plant), **464,** 464–465
germination of, *145*
media moisture level for, *147*

I

Iberis sempervirens (candytuft; hardy candytuft; evergreen candytuft), **467,** 467–468
vernalization and, *76*
Iceland poppy. *See Papaver nudicaule*
Impatiens spp., **468,** 468–477, **472, 475**
clinical growth regulators for, *153*
daily light integral (DLI) for, *88*
ethylene sensitivity and, *183*

germination of, *145*
I. hawkeri (New Guinea impatiens), 44, *52,* 155, 156, **468,** 468–472
base temperature and, 70, *72*
daily light integral (DLI) for, *88*
delaying wilting in, 189
disease in, 123
marginal leaf burn in, 9
plant growth regulators (PGRs) for, *168*
plant temperature and, 69
powdery mildew in, 125
sales, 4, 5, **7**
shipping and, 187
starter charge for, 28
sticking priorities for, *160*
vegetative propagation of, *158*
I. hybrida (exotic impatiens), 477
I. walleriana (seed impatiens; bedding plant impatiens; double-flowering impatiens), **472,** 472–475, **475,** 475–477
base temperature and, *72*
medium temperature for, 68
plant growth regulators (PGRs) for, *168*
sales, 4, 5, **7**
stem elongation and, *78*
sticking priorities for, *160*
vegetative propagation of, *159*
light for germination of, *148*
media moisture level for, *147*
mini
sticking priorities for, *160*
vegetative propagation of, *158*
need for supplemental lighting, *151*
pH of medium, 149
plant management in, 119
relative nutrient requirements of, *35*
storage of, 160
trailing
plant growth regulators (PGRs) for, *168*
sticking priorities for, *160*
vegetative propagation of, *159*
Inca lily. *See Alstroemeria* hybrids
inch plant. *See Zebrina pendula*
Indian shot. *See Canna* × *generalis*
India pink. *See Lobelia cardinalis*
Ipomoea spp., *35,* **477,** 477–479, **478**
I. batatas (sweet potato vine), **477,** 477–478
I. tricolor (morning glory), **478,** 478–479
pest management in, 114
storage of, 160
vegetative propagation of, *159*
Iresine herbstii (iresine), **479,** 479–481
Iris spp., **479,** 481–484
dwarf, ethylene sensitivity and, *184*
I. ensata, 484
Iris hybrids (German iris; bearded iris; fleur-de-lis), 482–484
I. pallida, 484
I. sibirica, 484
I. × *hollandica* (Dutch iris), *88,* 481–482
Isolepsis cernua (fiber-optic grass), **484,** 484–485, *485*
Italian bellflower. *See Carpanula isophylla*
ivy geranium. *See Pelargonium peltatum*

J

jade plant, disease in, 126
Jamesbrittenia, sticking priorities for, *160*
Japanese bird's nest fern. *See Asplenium nidas*
Japanese holly fern. *See Cyrtomium falcatum*
Japanese painted fern. *See Athyrium nipponicum*
Jerusalem cherry. *See Solanum pseudocapsicum*
Jewels of Opar. *See Talinum paniculatum*
Joseph's coat. *See Alternanthera* spp.
Juncus spp., **487,** 487–488
J. effuses (twisted juncus; spiraled juncus), 488
J. inflexus (juncus; blue arrows), 487, **487,** 488
J. pallidus (juncus), 487–488

K

Kalanachoe blossfeldiana (kalanchoe), **489,** 489–493
crop schedule of, *492*
daily light integral (DLI) for, *88*
DIF and, 78
disease in, 126
ethylene sensitivity and, *183*
fertilizer termination in, 180
pot life in, 178
relative nutrient requirements of, *35*
shipping, 182, 185
spacing of, *491*
stem elongation and, *79*
tissue nutrient levels of, *52*

kale. *See Brassica oleracea* var. *acephala*
Kniphofia uvaria (red hot poker; torchlily), 494–495
Koeleria (blue hair grass), 495, **495**
Kohlrabi. *See Brassica oleracea* var. *gongylodes*

L

lady's mantle. *See Alchemilla mollis*
lamb's ear. *See Stachys byzantina*
Lamium spp.
  L. galeobdolon, 498
  L. maculatum (dead nettle; spotted dead nettle), **497,** 497–498
  sticking priorities for, *160*
  vegetative propagation of, *159*
Lantana spp.
  daily light integral (DLI) for, *89*
  ethylene sensitivity and, *183*
  L. camara, **498,** 498–500
  L. montevidensis, 500
  sticking priorities for, *160*
  storage of, 160
  vegetative propagation of, *159*
larkspur. *See Consolida ambigua*
Lathyrus odoratus (sweet pea), 500–502, **501**
Lavandula spp. (lavender), **502,** 502–504
  L. angustifolia (English lavender), 502, 503, 504
  L. dentata (French lavender), 504
  L. intermedia, 504
  L. latifolia, 504
  L. pinnata, 504
  L. stoechas, 504
lavender. *See Lavandula* spp.
lavender cotton. *See Santolina chamaecyparissus*
leadwort. *See Plumbago auriculata*
leaf fennel. *See Foeniculum vulgare*
leatherleaf fern, cut cultivated greens sales, 5
leatherleaf sedge. *See Carex comans*
leeks. *See Allium ampeloprasum* var. *porrum*
lemon mint. *See Mentha* × *piperita*
lemon thyme. *See Thymus* × *citriodorus*
leopard's bane. *See Doronicum orientale*
lesser periwinkle. *See Vinca minor*
Leucanthemum × *superbum* (Shasta daisy), **504,** 504–506
  base temperature and, *71*
  plant management in, 119
Liatris spicata (gay feather; blazing star), *53,* **506,** 506–508
Ligularia spp.
  L. dentata (bigleaf goldenray; bigleaf ligularia), 508–509
  L. przewalskii, 509
  L. stenocephala, 509
Lilium × *hybrida* (Asiatic and Oriental lily), **509,** 509–513, 509–525
  base temperature and, *71*
  bulb size and flower number, *517*
  controlled pot-cooling schedule, *516*
  controlled temperature forcing schedule, *523*
  cooling and flowering of, *515*
  daily light integral (DLI) for, *88*
  development schedule, *519*
  ethylene sensitivity and, *184*
  L. candidum (Madonna lily), 513
  L. longiflorum (Easter lily), **513,** 513–525
    base temperature and, *71*
    ethylene sensitivity and, *184*
    fertilizer termination in, 180
    medium temperature for, 68
    pest management in, 109
    relative nutrient requirements of, *35*
    shipping, 182
  L. maroccanna, 529
  packing standards of, *514*
  plant management in, 119
  relative nutrient requirements of, *35*
  shipping, 182, 186
  stem elongation and, *79*
  tissue nutrient levels of, *53*
Limonium spp., 525–528
  L. altaica, 528
  L. gmelinii (Siberian statice), 528
  L. latifolium (sea lavender; wideleaf sea lavender), 527–528
  L. peregrinum, 526
  L. perezii, 526
  L. sinuatum (statice; common statice; garden statice), 525–527
  L. suworowii (Russian statice; rat-tail statice), 526
  L. tataricum (*Goniolimon tataricum*), 528
Linaria hybrida (linaria; baby snapdragon; spurred snapdragon), 528–529, *529*

Lippia graveolens (Mexican oregano), 566
Lippia palmeri, 566
Lisianthus. *See Eustoma grandiflorum*
Lobelia spp., **530,** 530–533, 532
daily light integral (DLI) for, *88*
full sunlight and, 87
germination of, *145*
L. amoena, 532
L. cardinalis (cardinal flower; red lobelia; scarlet lobelia; India pink), 532
L. cardinalis atrosanguinea, 532
L. erinus, **530,** 530–531
base temperature and, *72*
L. siphilitica (blue cardinal flower; great lobelia; great blue lobelia), 532, 533
L. splendens (*L. fulgens*) (Mexican lobelia), 532
media moisture level for, *147*
need for supplemental lighting, *151*
plant growth regulators (PGRs) for, *168*
plant management in, 119
relative nutrient requirements of, *35*
result of cross species, 531
sticking priorities for, *160*
vegetative propagation of, *159*
loblolly pine. *See Pinus taeda*
Lobularia maritima (alyssum; sweet alyssum), *35,* 534, **534**
base temperature and, *71*
daily light integral (DLI) for, *88*
germination of, *145*
media moisture level for, *147*
optimum temperature and, 73
love grass. *See Eragrostis elliotii*
lucky shamrock. *See Oxalis tetraphylla*
Lupinus polyphyllus (lupine), **535,** 535–536
ethylene sensitivity and, *183*
Lychnis coronaria (rose campion), vernalization and, *76*
Lycopersicon esculentum (tomato), *35,* 536–538
base temperature and, *72*
chilling and freezing injury and, 75
effect of low temperature, 39
iron toxicity in, 40
low temperature and, 75
media moisture level for, *147*
powdery mildew in, 125
seed for, 144, 145
stem elongation and, *79*
Lysimachia spp.
L. clethroides, 539
L. congestiflora, 538–539
L. nummularia, 539, **539**
L. procubens, 538–539
L. punctata, 539
sticking priorities for, *160*

M

Madagascar periwinkle. *See Catharanthus roseus*
Madonna lily. *See Lilium candidum*
madwort. *See Alyssum montanum*
Maranta, daily light integral (DLI) for, *87*
marguerite daisy. *See Argyranthemum* spp.
marigold. *See Tagetes* spp.
marjoram. *See Origanum* spp.
Martha Washington geranium. *See Pelargonium × domesticum*
matricaria. *See Tanacetum parthenium*
Matthiola incana (stock), **541,** 541–544
base temperature and, *71*
media moisture level for, *147*
mealycup sage. *See Salvia farinacea*
Mentha spp. (mint), 544–545
M. aquatica (orange mint), 545
M. requienii (Corsican mint), 545
M. spicata (spearmint), 545
M. suaveolens (pineapple mint), 545
M. × gracilis (ginger mint), 545
M. × piperita (lemon mint), 545
Mexican heather. *See Cuphea hyssopifola*
Mexican lobelia. *See Lobelia splendens*
Mexican oregano. *See Lippia graveolens*
Mimulus × hybridus, **546,** 546–547
vegetative propagation of, *159*
miniature rose, daily light integral (DLI) for, *88*
mint. *See Mentha* spp.
Missouri evening primrose. *See Oenothera macrocarpa*
Monopsis
sticking priorities for, *160*
vegetative propagation of, *159*
morning glory. *See Ipomoea tricolor*
moss pink phlox. *See Phlox subulata*
moss rose. *See Portulaca grandiflora*
mother of thyme. *See Thymus serphyllum*

moth orchid. *See Phalaenopsis* hybrids
mountain gold. *See Alyssum montanum*
Muehlenbeckia axillaris (creeping wirevine), 547–548
mule marigold. *See Tagetes erecta × patula*
Musa ornata (banana), base temperature and, *72*
Myositis sylvatica (forget-me-not), **548,** 548–549
myrtle. *See Vinca major*

N

Narcissus spp., *35,* 551–554, **552, 553**
  ethylene sensitivity and, *184*
  N. hybrids (daffodils), 551–552
  N. tazetta (paperwhites), 552–554, **553**
  plant management in, 119
  shipping, 182
  stem elongation and, *79*
nasturtium, iron toxicity in, 40
Nemesia spp., 44
  N. foetans, 554–556, *555*
  N. fruticans, 555–556
  N. strumosa
    base temperature and, *71*
    medium tem, 68
  plant growth regulators (PGRs) for, *168*
  sticking priorities for, *160*
  vegetative propagation of, *159*
Nepeta cataria (catnip), 556–557
Nephrolepsis spp. (fern)
  daily light integral (DLI) for, *87*
  N. biserrata (fishtail fern), 559
  N. exaltata (Boston fern; sword fern), **557,** 557–560, *558*
    ethylene sensitivity and, *184*
nephthytis. *See Syngonium podophyllum*
New England aster. *See Aster novae-angliae*
New Guinea impatiens. *See Impatiens hawkeri*
Nicotiana spp.
  germination of, *145*
  light for germination of, *148*
  media moisture level for, *147*
  N. alata (flowering tobacco), **560,** 560–561
  N. × sanderae, base temperature and, *72*
Nierembergia, need for supplemental lighting, *151*
Nolana, vegetative propagation of, *159*

O

Ocimum spp. (basil), 40, **563,** 563–565
  iron toxicity in, 40
  O. basilicum (sweet basil), 564
  O. basilicum citriodorum, 564
  O. basilicum purpurescens, 564
Oenothera spp.
  O. fruticosa (sundrops), base temperature and, *72*
  O. macrocarpa (formerly *Oenothera missouriensis*) (Ozark sundrop; Missouri evening primrose), 565
onion. *See Allium cepa*
orchids, *35*
  sales, 5
  shipping, 186
oregano. *See Origanum vulgare*
Oriental lilies. *See Lilium × hybrida*
Oriental poppy. *See Papaver orientale*
Origanum spp. (marjoram; sweet marjoram), 566
  O. syriacum, 566
  O. vulgare (oregano), 566
  O. vulgare hirtum (Greek oregano), 566
ornamental kale and cabbage. *See Brassica oleracea* var. *acephala*
ornamental millet. *See Pennisetum glaucum*
ornamental pepper. *See Capsicum annum*
Osmunda cinnamonea (cinnamon fern), 560
Osteospermum hybrids, **97**
  base temperature and, *71*
  ethylene sensitivity and, *183*
  growth regulator for, 96
  O. ecklonis (African daisy), **567,** 567–570
  plant growth regulators (PGRs) for, *168*
  sticking priorities for, *160*
  vegetative propagation of, *159*
Oxalis sp. (oxalis; shamrock; wood sorrel), **570,** 570–572, **571**
  O. adenophylla (silver shamrock), 570, *571*
  O. regnellii (shamrock plant), 570, 571, *571,* 572
  O. tetraphylla (lucky shamrock), 570, *571,* 572
  O. triangularis (purple shamrock), 570, 571, *571,* 572
  O. versicolor, 571, 572
  O. vulcanicola, 571, 572
  relative nutrient requirements of, *35*
  storage, forcing temperatures, weeks to force, and flower color, *571*
Ozark sundrop. *See Oenothera macrocarpa*

P

Paeonia spp. (peony), **573,** 573–574
P. lactiflora, 574
P. officinalis, 574
P. suffruticosa (tree peony), 574
P. tenuifolia (fern leaf peony), 574
pampas grass. *See Cortaderia selloanan*
pansy. *See Viola* spp.
Papaver spp. (poppy), 574–576
media moisture level for, *147*
P. nudicaule (Iceland poppy), 574–575
P. orientale (Oriental poppy), 575–576
paperwhites. *See Narcissus tazetta*
parsley. *See Petroselinum crispum*
peace lily. *See Spathiphyllum* spp.
Pelargonium spp. (geranium), 576–590, **578, 587**
clinical growth regulators for, *153*
DIF and, 78
effect of low temperature, 39
eliminating viruses in, 174
ethylene sensitivity and, *183*
fertilizer for, 157
growing mix for, 22
growth regulator for, 96
iron toxicity in, 40
leaf spot disease in, 127
low temperature and, 75
media moisture level for, *147*
need for supplemental lighting, *151*
nutritional needs of, 44
pest management in, 112
pH of medium, 149, 150
P. peltatum (ivy geranium; balcony geranium), *53,* 578, **587,** 587–590
daily light integral (DLI) for, *88*
growing mix for, 22
nutritional needs of, 44
plant growth regulators (PGRs) for, *168*
sticking priorities for, *160*
vegetative propagation of, *158*
P. × *domesticum* (Martha Washington geranium; regal geranium), *53,* 182, 576–578
P. × *hortorum* (geranium; zonal geranium), *53,* **578,** 578–587
average daily temperature (ADT) and, *73*
base temperature and, *72*
daily light integral (DLI) for, *88*
plant growth regulators (PGRs) for, *168*
sticking priorities for, *160*
vegetative propagation of, *158*
relative nutrient requirements of, *35*
response to light and temperature, 156
rust disease in, 129
sales, 4, 5, **7**
seed for, *53,* 145
self-propagation of, 155
stem elongation and, *79*
in stock program, 156
symptoms of disease in, 129
Pellaea rotundifolia (button fern), 559
Pennisetum spp., **590,** 590–593, *591,* **593**
base temperature and, *72*
P. glaucum (ornamental millet), **590,** 590–592, *591*
P. setaceum (fountain grass; pink fountain grass; purple fountain grass), 592–593, **593**
Pentas lanceolata (Egyptian starflower), 155, **593,** 593–595
average daily temperature (ADT) and, *73*
base temperature and, *72*
clinical growth regulators for, *153*
ethylene sensitivity and, *183*
marginal leaf burn in, 9
need for supplemental lighting, *151*
peony. *See Paeonia* spp.
pepper. *See Capsicum*
perennial phlox. *See Phlox paniculata*
perennials (full sun), daily light integral (DLI) for, *89*
perennial sage. *See Salvia nemorosa*
perennial salvia. *See Salvia nemorosa*
Pericallis cruenta (cineraria), *35,* 596–597. *See also Senecio cruentus*
ethylene sensitivity and, *183*
P. × *hybrida,* base temperature and, *71*
Perilla frutescens (perilla), **597,** 597–598, *598*
periwinkle. *See Vinca major*
Perovskia atriplicifolia (Russian sage), **599,** 599–600
Petroselinum crispum (parsley), 600, **600**
Petunia × *hybrida* (petunia), **70, 601,** 601–608, **605**
average daily temperature (ADT) and, *73*

base temperature and, *71, 72*
clinical growth regulators for, *153*
daily light integral (DLI) for, *89*
ethylene sensitivity and, *183*
germination of, *145*
light for germination for, *148*
media moisture level for, *147*
need for supplemental lighting, *151*
nutritional needs of, 44
plant growth regulators (PGRs) for, *168*
plant management in, 119
plant size control of, **604**
powdery mildew in, 125
relative nutrient requirements of, *35*
sales, 4, 5, **7**
stem elongation and, *79*
sticking priorities for, *160*
tissue nutrient levels of, *53*
vegetative propagation of, *35, 159*
Phalaenopsis hybrids (moth orchid), **609,** 609–610
base temperature and, *72*
daily light integral (DLI) for, *87*
ethylene sensitivity and, *183*
Phaseolus sp. (bean), stem elongation and, *79*
Philodendron spp. (philodendron), *53,* **610,** 610–611
high temperature and, 74
low temperature and, 75
P. scandens, ethylene sensitivity and, *184*
P. scandens oxycardium (heart-leafed philodendron), 610, 611
calcium deficiency in, 40
P. selloum, 610, **610**
Phlox spp., *35,* 611–616, **614**
light for germination of, *148*
media moisture level for, *147*
P. divaricata (woodland phlox), vernalization and, *76*
P. drummondii (annual phlox), 611–612
P. maculate (annual phlox), 612–613
P. paniculata (perennial phlox; garden phlox), **614,** 614–615
base temperature and, *72*
P. subulata (moss pink phlox; creeping phlox; ground phlox), 616
base temperature and, *71*
vernalization and, *76*
seed for, 145
Piectranthus, sticking priorities for, *160*
Pimpinella anisum (anise), 617
pincushion flower. *See Scabiosa caucasia*
pineapple mint. *See Mentha suaveolens*
pink calla lily. *See Zantedeschia rehmannii*
pink coreopsis. *See Coreopsis rosea*
pink fountain grass. *See Pennisetum setaceum*
pinks. *See Dianthus barbatus*
pink threadleaf coreopsis. *See Coreopsis rosea*
Pinus taeda (loblolly pine), 23
Pisum sativum (pea)
base temperature and, *71*
stem elongation and, *79*
Platycerium bifurcatum (staghorn fern), 559
Platycodon grandiflorus (balloon flower), *35,* **617,** 617–618
base temperature and, *71*
Plectranthus spp. (Swedish ivy), **619,** 619–620, **620**
P. amboinicus, 619–620
P. coleoides, 619, 620
vegetative propagation of, *159*
Plumbago auriculata (plumbago; leadwort), 620–621
pocketbook plant. *See Calceolaria* spp.
poinsettia. *See Euphorbia pulcherrima*
polka dot plant. *See Hypoestes phyllostachya*
Polyscias fruticosa, ethylene sensitivity and, *184*
Polystichum polyblepharum (tassel fern), 560
poor man's orchid. *See Schizanthus × wisetonensis*
Porphyrocoma pohliana (Brazilian fireworks), 621–622, *622*
Portulaca spp., **623,** 623–626, **624**
crop times for, *625*
ethylene sensitivity and, *183*
germination of, *145*
media moisture level for, *147*
pests and diseases in, 153
P. grandiflora (moss rose), **623,** 623–624
base temperature and, *72*
P. oleracea (purslane), **624,** 624–626, *625*
relative nutrient requirements of, *35*
sticking priorities for, *160*
Potentilla atrosanguinea (cinquefoil), vernalization and, *76*
pothos. *See Epipremnum aureum*
pot marigold. *See Calendula officinalis*

primrose. *See Primula acaulis*
Primula spp., **626,** 626–630, **629**
daily light integral (DLI) for, *88*
light for germination for, *148*
media moisture level for, *147*
P. acaulis (primrose), **626,** 626–629
plant management in, 119
P. auricula, 628
P. denticulata, 628
P. japonica, 628
P. malacoides, 629
P. obconica, **629,** 629–630
P. vialii, 628
P. vulgaris, 629
P. × *polyantha,* 628
relative nutrient requirements of, *35*
seed for, 145
tissue nutrient levels of, *53*
princess lily. *See Alstroemeria* hybrids
Psylliostachys suworowii. See Limonium suworowii
Pteris (fern), daily light integral (DLI) for, *87*
Pulmonaria saccharata (lungwort), vernalization and, *76*
purple coneflower. *See Echinacea purpurea*
purple fountain grass. *See Pennisetum setaceum*
purple heart. *See Setcreasea purpurea*
purple shamrock. *See Oxalis triangularis*
purple wandering Jew. *See Setcreasea purpurea*
purslane. *See Portulaca oleracea*

R

rabbit's foot fern. *See Davallia trichomanoides*
Radermachera sinica, ethylene sensitivity and, *184*
Ranunculus asiaticus (buttercup), *35,* **631,** 631–633
media moisture level for, *147*
seed for, 145
rat-tail statice. *See Limonium suworowii*
red hot poker. *See Kniphofia uvaria*
red lobelia. *See Lobelia cardinalis*
regal geranium. *See Pelargonium* × *domesticum*
Rhododendron spp. (azalea), *35, 53*
dikegulac sodium as pinching agent for, 98
disease in, 123
ethylene sensitivity and, *183*
R. simsii, 633–636
R. × *obtusum,* 633–636
shipping, 186
robust male fern. *See Dryopteris* × *complexa*
rock cress. *See Arabis caucasica*
Rosa hybrids (rose), **636,** 636–656, **640**
base temperature and, *72*
ethylene sensitivity and, *184*
guide to finishing pot roses, *639*
pot rose
color mixes, *640*
pot life in, 178
production greenhouse in Denmark, **638**
powdery mildew in, 125
production of, at Zuurbier, Heerhugowaard, the Netherlands, **642**
R. canina, 640, 648
relative nutrient requirements of, *35*
R. indica, 640, 641
R. laxa, 648, 649
R. multiflora, 648, 649
shipping, 186
stem elongation and, *79*
tissue nutrient levels of, *53*
water and fertilizers are recycled in new system in Europe, **654**
rose. *See Rosa* hybrids
rose mallow. *See Hibiscus moscheutos*
Rosmarinus officinalis (rosemary), **656,** 656–657
powdery mildew in, 125
rudbeckia. *See Rudbeckia fulgida*
Rudbeckia spp. (coneflower)
R. fulgida (black-eyed Susan; rudbeckia), **657,** 657–659
base temperature and, *71*
R. hirta (black-eyed Susan; gloriosa daisy), **659,** 659–660
base temperature and, *72*
high temperature and, 74
Russian sage. *See Perovskia atriplicifolia*
Russian statice. *See Limonium suworowii*

S

sage. *See Salvia* spp.
Saintpaulia ionantha (African violet), *35, 53,* **661,** 661–663
base temperature and, *72*
daily light integral (DLI) for, *87*
ethylene sensitivity and, *183*
home care for, 191

low temperature and, 75
powdery mildew in, 125
shipping, 186
stem elongation and, *79*
Salvia spp. (sage), *35, 53,* **90, 97, 663,** 663–669, **666**
clinical growth regulators for, *153*
ethylene sensitivity and, *184*
germination of, *145*
media moisture level for, *147*
seed for, 145
S. farinacea (mealycup sage), **663,** 663–664
base temperature and, *72*
daily light integral (DLI) for, *89*
need for supplemental lighting, *151*
S. nemorosa (formerly *S. superba*) (perennial salvia; perennial sage), 664–666
S. officinalis (sage; garden sage; common sage), **666,** 666–667
S. splendens (scarlet sage), 668–669
base temperature and, *72*
daily light inegral (DLI) for, *88*
need for supplemental lighting, *151*
stem elongation and, *79*
sticking priorities for, *160*
vegetative propagation of, *159*
Santolina spp.
S. chamaecyparissus (lavender cotton), 669–670
S. rosmarinifolia, 670
S. virens, 670
Sanvitalia speciosa (sanvitalia), 670–671
plant growth regulators (PGRs) for, *168*
sticking priorities for, *160*
vegetative propagation of, *159*
satin flower. *See Godetia whitney*
SaxifragaI (London pride), vernalization and, *76*
Scabiosa spp.
S. caucasia (pincushion flower), 671–672
base temperature and, *71*
S. columbaria (pincushion flower), 672
Scaevola aemula (fan flower), 587, 608, **673,** 673–674
daily light integral (DLI) for, *89*
sticking priorities for, *160*
vegetative propagation of, *159*
scarlet lobelia. *See Lobelia cardinalis*
scarlet sage. *See Salvia splendens*
Schefflera spp.
Brassaia, 53
daily light integral (DLI) for, *88*
Dizygotheca, 53
S. arboricola
ethylene sensitivity and, *184*
S. elegantissima
ethylene sensitivity and, *184*
shipping, 186
Schizanthus × *wisetonensis* (poor man's orchid; butterfly flower), 675–676
Schlumbergera spp. (Christmas cactus; holiday cactus; Thanksgiving cactus), **676,** 676–678
daily light integral (DLI) for, *87*
ethylene sensitivity and, *183*
promoting flower set in, 99, **99**
relative nutrient requirements of, *35*
shipping, 182
S. russeliana, 678
S. truncata (Thanksgiving cactus), 676–678
base temperature and, *71*
S. × *buckleyi,* **676,** 676–678
tissue nutrient levels of, *53*
Scutellaria javanica (skullcap), 679, *679*
sea lavender. *See Limonium latifolium*
sea pinks. *See Armeria maritima*
Sedum spp. (stonecrop), 680–681
base temperature and, *71*
disease in, 126
S. acre, 680–681
S. spectabile, 681
S. spurium, 680–681
Senecio cineraria (dusty miller), *35,* **681,** 681–682
daily light integral (DLI) for, *88*
stem elongation and, *79*
Senecio cruentus. See Pericallis cruenta
Setcreasea purpurea (purple heart; purple wandering Jew), 682
shamrock plant. *See Oxalis regnellii*
Shasta daisy. *See Leucanthemum* × *superbum*
Siberian statice. *See Limonium gmelinii*
silver shamrock. *See Oxalis adenophylla*
Sinningia speciosa (gloxinia), *35, 53,* **683,** 683–684
chilling and freezing injury and, 75
daily light integral (DLI) for, *88*

disease in, 123
ethylene sensitivity and, *183*
shipping, 186
skullcap. *See Scutellaria javanica*
snapdragon. *See Antirrhinum* spp.
sneezeweed. *See Helenium*
Solanum lycopersicum. See Lycopersicon esculentum
Solanum spp.
S. (potato), 174
S. melongena (eggplant), 685
S. pseudocapsicum (Jerusalem cherry; Christmas cherry), 302
S. tuberosum, stem elongation and, *79*
Solenostemon scutellarioides (coleus), *35,* 155, **685,** 685–687
chilling and freezing injury and, 75
Solidago × *hybrida* (goldenrod), **687,** 687–690
southern wood fern. *See Thelypteris kunthii*
Spathiphyllum spp. (peace lily), **690,** 690–692
daily light integral (DLI) for, *87*
ethylene sensitivity and, *184*
spearmint. *See Mentha spicata*
speedwell. *See Veronica longifolia*
spider flower. *See Cleome hassleriana*
spikes. *See Cordyline indivisa*
Spilanthes. See Achmella oleracea
spotted dead nettle. *See Lamium maculatum*
Sprengeri fern. *See Asparagus densiflorus* var. 'Sprengeri'
spurred snapdragon. *See Linaria hybrida*
squash. *See Cucurbita* spp.
Stachys byzantina (lamb's ear; woolly betony), 692–693
staghorn fern. *See Platycerium bifurcatum*
statice. See Limonium spp.
Stevia rebaudiana (sweet leaf; sugar leaf), 693
stock. *See Matthiola incana*
stonecrop. *See Sedum* spp.
strawberry. *See Fragaria* × *hybrida*
strawflower. *See Bracteantha bracteata*
Streptocarpus nobilis (streptocarpus), *35, 53*
daily light integral (DLI) for, *87*
ethylene sensitivity and, *184*
stem elongation and, *79*
Strobilanthes, sticking priorities for, *160*
sugar leaf. *See Stevia rebaudiana*
summer orchid. *See Angelonia* spp.
summer snapdragon. *See Angelonia* spp.
sunflower. *See Helianthus*
Sutera cordata (bacopa), 608, **694,** 694–695
sticking priorities for, *160*
vegetative propagation of, *158*
swamp rose mallow. *See Hibiscus moscheutos*
Swedish ivy. *See Plectranthus* spp.
sweet alyssum. *See Lobularia maritima*
sweet basil. *See Ocimum basilicum*
sweet fennel. *See Foeniculum vulgare dulce*
sweet leaf. *See Stevia rebaudiana*
sweet marjoram. *See Origanum* spp.
sweet pea. *See Lathyrus odoratus*
sweet potato vine. *See Ipomea* spp.
sweet william. *See Dianthus barbatus*
sword fern. *See Nephrolepsis exaltata*
Syngonium podophyllum (syngonium; nephthytis), *53,* 696, **696**
ethylene sensitivity and, *184*

T

tabasco pepper. *See Capsicum frutescens*
Tagetes spp. (marigold), *35,* 44, **697,** 697–699
clinical growth regulators for, *153*
daily light integral (DLI) for, *89*
DIF and, 78
ethylene sensitivity and, *183*
iron toxicity in, 40
media moisture level for, *147*
nitrogen deficiency in, 39
pH of medium, 149
plant management in, 119
sales, 4, **7**
seed for, 144, 145
T. erecta (African marigold; American marigold), **697,** 697–699
base temperature and, *71, 72*
clinical growth regulators for, *153*
iron toxicity in, 40
need for supplemental lighting, *151*
T. erecta × *patula* (triploid marigold; mule marigold), **697,** 697–699
T. patula (French marigold), **70, 697,** 697–699
average daily temperature (ADT) and, *73*
base temperature and, 71, *71*
clinical growth regulators for, *153*
stem elongation and, *79*

T. tenuifolia, 699
Talinum paniculatum (jewels of Opar), **700,** *700,* 700–701
Tanacetum parthenium (matricaria), 701
tassel fern. *See Polystichum polyblepharum*
Tecoma stans (esperanza; yellow trumpet bush), *702,* 702–703
Thalictrum aquilegifolium (meadow rue), vernalization and, *76*
Thelypteris kunthii (southern wood fern), 560
threadleaf coreopsis. *See Coreopsis verticillata*
thrift. *See Armeria maritima*
throatwart. *See Trachelium caeruleum*
Thunbergia alata (black-eyed Susan vine), **703,** 703–704
sticking priorities for, *160*
vegetative propagation of, *159*
Thymus spp. (thyme), 705–706
T. praecox (creeping or wild thyme), 706
T. praecox var. *pseudolanuginosus,* 706
T. pseudolanuginosus (woolly thyme), 706
T. serpyllum (mother of thyme), 706
T. vulgaris (garden thyme; common thyme), **705,** 705–706
T. × *citriodorus* (lemon thyme), 706
tickseed. *See Coreopsis grandiflora*
tick weed. *See Helenium amarum*
tomato. *See Lycopersicon*
toothache plant. *See Achmella oleracea*
torchlily. *See Kniphofia uvaria*
Torenia fournieri (wishbone flower), **706,** 706–708
base temperature and, *72*
Trachelium caeruleum (throatwart; trachelium), 708–709
Tradescantia pallida. See Setcreasa purpurea
tree peony. *See Paeonia suffruticosa*
triploid marigold. *See Tagetes erecta* × *patula*
tropical hibiscus. *See Hibiscus rosa-sinensis*
tulip. *See Tulipa hybrida*
Tulipa hybrida (tulip), **710,** 710–713, **711**
ethylene sensitivity and, *184*
hydroponic forcing, 711
shipping, 182
stem elongation and, *79*
twinspur. *See Diascia* spp.

U

umbrella plant. *See Cyperus alternifolia*

V

Verbascum × *hybrida* (mullein), **715,** 715–716
V. bombyciferum, 716
V. chaixii, 716
Verbena × *hybrida* (verbena), *35,* 155, 716–719, **717**
average daily temperature (ADT) and, *73*
clinical growth regulators for, *153*
daily light integral (DLI) for, *89*
germination of, *145*
media moisture level for, *147*
need for supplemental lighting, *151*
plant growth regulators (PGRs) for, *168*
powdery mildew in, 125
seed for, 145
stem elongation and, *79*
sticking priorities for, *160*
vegetative propagation of, *159*
V. bonariensis (tall), 719
stem elongation and, *79*
V. canadensis, 719
V. rigida, 719
V. speciosa, 719
Veronica spp., **720,** 720–721
V. longifolia (speedwell), **720,** 720–721
V. spicata, 721
vinca. *See Catharanthus roseus*
Vinca spp.
V. major (vinca vine; periwinkle; myrtle), 722–723
V. minor (lesser periwinkle; creeping myrtle), 723
vinca vine. *See Vinca major*
Viola spp. (pansy), *35,* 37, 44
clinical growth regulators for, *153*
daily light integral (DLI) for, *89*
disease in, 123
ethylene sensitivity and, *183*
media moisture level for, *147*
nitrogen deficiency in, 39
plant growth regulators (PGRs) for, 101
sales, 4, 5, **7**
seed for, 144, 145
V. cornuta, 728

V. × *wittrockiana, 53,* **723,** 723–728, **726**
base temperature and, 71, *71*
stem elongation and, *79*
virgin's bower. *See Clematis* spp.

W

wall cress. *See Arabis caucasica*
wallflower. *See Erysimum linifolium*
Wandering Jew. *See Zebrina pendula*
wandflower. *See Gaura lindheimeri*
watermelon. *See Citrullus lanatus*
weeping fig. *See Ficus benjamina*
wheat celosia. *See Celosia spicata*
whirling butterflies. *See Gaura lindheimeri*
wideleaf sea lavender. *See Limonium latifolium*
wild thyme. *See Thymus praecox*
windflower. *See Anemone coronaria*
wishbone flower. *See Torenia fournieri*
woolly betony. *See Stachys byzantina*
woolly thyme. *See Thymus pseudolanuginosus*

Y

yarrow. *See Achillea* spp.
yellow trumpet bush. *See Tecoma stans*
yellow tuft. *See Alyssum montanum*

Z

Zantedeschia spp. (calla lily), *35,* **98, 729,** 729–732
Z. aethiopica (white), 729, 732
Z. elliotiana (yellow), 729, 732
Z. hybrids, **729,** 729–732
Z. rehmannii (pink), 729, 730, 732
Z. rehmannii violacea, 732
Zebrina pendula (wandering Jew; inch plant), 732–733
Zinnia spp., (zinnia) *35,* **733,** 733–735, **735**
daily light integral (DLI) for, *89*
effect of, **86**
ethylene sensitivity and, *184*
iron toxicity in, 40
media moisture level for, *147*
seed for, 144, 145
Z. angustifolia, 733, 735
Z. elegans, 733, 735
average daily temperature (ADT) and, *73*
base temperature and, 70, *72*
high temperature and, 74
stem elongation and, *79*
Z. marylandica, 733, 735
zonal geranium. *See Pelargonium* × *hortorum*